HUMAN REPRODUCTIVE
AND PRENATAL GENETICS

HUMAN REPRODUCTIVE AND PRENATAL GENETICS

Edited by

PETER C. K. LEUNG
Department of Obstetrics and Gynecology,
University of British Columbia,
Vancouver, BC, Canada

JIE QIAO
Center for Reproductive Medicine,
Department of Obstetrics and Gynecology,
Peking University Third Hospital, Beijing, China

Fellow of the Chinese Academy of Engineering, Beijing, China

ACADEMIC PRESS
An imprint of Elsevier

Academic Press is an imprint of Elsevier
125 London Wall, London EC2Y 5AS, United Kingdom
525 B Street, Suite 1650, San Diego, CA 92101, United States
50 Hampshire Street, 5th Floor, Cambridge, MA 02139, United States
The Boulevard, Langford Lane, Kidlington, Oxford OX5 1GB, United Kingdom

Notices

Knowledge and best practice in this field are constantly changing. As new research and experience broaden our understanding, changes in research methods, professional practices, or medical treatment may become necessary.

Practitioners and researchers must always rely on their own experience and knowledge in evaluating and using any information, methods, compounds, or experiments described herein. In using such information or methods they should be mindful of their own safety and the safety of others, including parties for whom they have a professional responsibility.

To the fullest extent of the law, neither the Publisher nor the authors, contributors, or editors, assume any liability for any injury and/or damage to persons or property as a matter of products liability, negligence or otherwise, or from any use or operation of any methods, products, instructions, or ideas contained in the material herein.

Library of Congress Cataloging-in-Publication Data
A catalog record for this book is available from the Library of Congress

British Library Cataloguing-in-Publication Data
A catalogue record for this book is available from the British Library

ISBN: 978-0-12-813570-9

For information on all Academic Press publications visit our
website at https://www.elsevier.com/books-and-journals

Working together
to grow libraries in
developing countries

www.elsevier.com • www.bookaid.org

Publisher: Andre G. Wolff
Acquisition Editor: Peter B. Linsley
Editorial Project Manager: Kristi Anderson
Production Project Manager: Punithavathy Govindaradjane
Cover Designer: Christian J. Bilbow

Typeset by SPi Global, India

Contents

II

EMBRYO IMPLANTATION, PLACENTA DEVELOPMENT AND PREGNANCY

IV

PREIMPLANTATION/ PRENATAL GENETIC DIAGNOSIS AND SCREENING

Contributors

Svetlana Arbuzova Center of Medical Genetics and Prenatal Diagnosis, Mariupol, Ukraine

Komal Bajaj New York City Health + Hospitals; Albert Einstein College of Medicine, New York, NY, United States

Christian Becker Nuffield Department of Obstetrics and Gynaecology, University of Oxford, Women's Centre, John Radcliffe Hospital, Oxford, United Kingdom

Tanmoy Bhattacharyya The Jackson Laboratory, Bar Harbor, ME, United States

Xiaotao Bian State Key Laboratory of Stem Cell and Reproductive Biology, Institute of Zoology, Chinese Academy of Sciences; University of Chinese Academy of Sciences, Beijing, China

Mats Brännström Department of Obstetrics and Gynecology, Sahlgrenska Academy, University of Gothenburg, Gothenburg, Sweden; Stockholm IVF, Stockholm, Sweden

Richard O. Burney Department of Clinical Investigation, Madigan Army Medical Center, Tacoma, WA, United States

Dan D. Cao School of Biomedical Sciences, Faculty of Medicine, The Chinese University of Hong Kong, Shatin, Hong Kong

Wai Y. Chan School of Biomedical Sciences, Faculty of Medicine, The Chinese University of Hong Kong, Shatin, Hong Kong

Chien-Wen Chen Department of Obstetrics and Gynecology, Shuang Ho Hospital, Taipei Medical University, New Taipei, Taiwan

Ya-Ching Chou Department of Obstetrics and Gynecology, College of Medicine, Taipei Medical University; Center for Reproductive Medicine, Taipei Medical University Hospital, Taipei, Taiwan

Marco Conti Department of OBGYN-RS, University of California, San Francisco, San Francisco, CA, United States

Zebulun S. Cope Michigan State University, College of Human Medicine, Department of Obstetrics, Gynecology and Reproductive Biology; Spectrum Health, Butterworth Hospital, Women's Center, Department of Obstetrics and Gynecology, Grand Rapids, MI, United States

Howard Cuckle Faculty of Medicine, Tel Aviv University, Israel

Mo-Yu Dai State Key Laboratory of Stem Cell and Reproductive Biology, Institute of Zoology, Chinese Academy of Sciences; University of Chinese Academy of Sciences, Beijing, China

Rabindranath De La Fuente Department of Physiology and Pharmacology, College of Veterinary Medicine; Regenerative Biosciences Center (RBC), University of Georgia, Athens, GA, United States

Guo-Lian Ding Institute of Embryo-Fetal Original Adult Disease and Shanghai Key Laboratory for Reproductive Medicine; The International Peace Maternity and Child Health Hospital, School of Medicine, Shanghai Jiao Tong University, Shanghai, People's Republic of China

Savina Dipresa Department of Medicine, Unit of Andrology and Reproductive Medicine, University of Padova, Padova, Italy

Jin Du State Key Laboratory of Stem Cell and Reproductive Biology, Institute of Zoology, Chinese Academy of Sciences; University of Chinese Academy of Sciences, Beijing, China

Cristina Eguizabal Cell Therapy and Stem Cell Group, Basque Center for Blood Transfusion and Human Tissues, Galdakao, Spain

Ecem Esencan Department of Obstetrics, Gynecology, and Reproductive Sciences, Yale School of Medicine, New Haven, CT, United States

Alberto Ferlin Department of Medicine, Unit of Andrology and Reproductive Medicine, University of Padova, Padova, Italy

Jose Carlos Pinto B. Ferreira Genomed S.A., Warsaw, Poland; Central Hospital of Maputo; ICOR—Heart Institute; Faculty of Medicine, Eduardo Mondlane University, Maputo, Mozambique

Heather Fice Departments of Pharmacology and Therapeutics and of Obstetrics and Gynecology, McGill University, Montréal, QC, Canada

Carlo Foresta Department of Medicine, Unit of Andrology and Reproductive Medicine, University of Padova, Padova, Italy

Qing-Qin Gao State Key Laboratory of Stem Cell and Reproductive Biology, Institute of Zoology, Chinese Academy of Sciences; University of Chinese Academy of Sciences, Beijing, China

Jessica Giordano Department of Obstetrics and Gynecology, Columbia University Medical Center, New York, NY, United States

Linda C. Giudice Center for Reproductive Sciences, Center for Reproductive Health, University of California, San Francisco, San Francisco, CA, United States

Francesca Romana Grati TOMA Advanced Biomedical Assays S.p.A., Busto Arsizio, Italy

Susan J Gross Department of Genetics and Genomic Sciences, Icahn School of Medicine at Mount Sinai, New York, NY, United States

Mary Ann Handel The Jackson Laboratory, Bar Harbor, ME, United States

Tristan Hardy Genetics and Molecular Pathology, SA Pathology, Adelaide; Repromed, Dulwich, SA, Australia

Cheng Huang State Key Laboratory of Stem Cell and Reproductive Biology, Institute of Zoology, Chinese Academy of Sciences; University of Chinese Academy of Sciences, Beijing, China

He-Feng Huang Institute of Embryo-Fetal Original Adult Disease and Shanghai Key Laboratory for Reproductive Medicine; The International Peace Maternity and Child Health Hospital, School of Medicine, Shanghai Jiao Tong University, Shanghai, People's Republic of China

Juan Carlos Izpisua Belmonte Gene Expression Laboratory, The Salk Institute for Biological Studies, La Jolla, CA, United States

Sylvie Jaillard Murdoch Children's Research Institute, Royal Children's Hospital, Melbourne, VIC, Australia; Rennes University Hospital, Cytogenetics and Cell Biology Department; INSERM U1085-IRSET, Rennes 1 University, Rennes, France

Hai-Ping Jiang State Key Laboratory of Stem Cell and Reproductive Biology, Institute of Zoology, Chinese Academy of Sciences; University of Chinese Academy of Sciences, Beijing, China

Zi-Ru Jiang Institute of Embryo-Fetal Original Adult Disease and Shanghai Key Laboratory for Reproductive Medicine; The International Peace Maternity and Child Health Hospital, School of Medicine, Shanghai Jiao Tong University, Shanghai, People's Republic of China

Laura Kasak Institute of Biomedicine and Translational Medicine, University of Tartu, Tartu, Estonia

Travis Kent The Jackson Laboratory, Bar Harbor, ME, United States

Ahmed Khattab Icahn School of Medicine at Mount Sinai Hospital, New York, NY, United States

Chaini Konwar Department of Medical Genetics, University of British Columbia; BC Children's Hospital Research, Vancouver, BC, Canada

Maris Laan Institute of Biomedicine and Translational Medicine, University of Tartu, Tartu, Estonia

Guan-Lin Lai School of Public Health, College of Public health and Nutrition, Taipei Medical University, Taipei, Taiwan

Jonathan LaMarre Department of Biomedical Sciences, University of Guelph, Guelph, ON, Canada

Dolores J. Lamb Center for Reproductive Medicine; Department of Molecular and Cellular Biology; Scott Department of Urology, Baylor College of Medicine, Houston, TX; Department of Urology, Weill Cornell Medical College, New York, NY, United States

Yi-Xuan Lee Center for Reproductive Medicine, Taipei Medical University Hospital, Taipei, Taiwan

Brynn Levy Department of Pathology and Cell Biology, Columbia University Medical Center, New York, NY, United States

Yu-Fei Li State Key Laboratory of Stem Cell and Reproductive Biology, Institute of Zoology, Chinese Academy of Sciences; University of Chinese Academy of Sciences, Beijing, China

Ming Liu State Key Laboratory of Stem Cell and Reproductive Biology, Institute of Zoology, Chinese Academy of Sciences, Beijing, China

Xin-Mei Liu Institute of Embryo-Fetal Original Adult Disease and Shanghai Key Laboratory for Reproductive Medicine; The International Peace Maternity and Child Health Hospital, School of Medicine, Shanghai Jiao Tong University, Shanghai, People's Republic of China

Y.M. Dennis Lo Li Ka Shing Institute of Health Sciences; Department of Chemical Pathology, The Chinese University of Hong Kong, Shatin, Hong Kong, China

Gang Lu School of Biomedical Sciences, Faculty of Medicine, The Chinese University of Hong Kong, Shatin, Hong Kong

Xuan G. Luong Department of OBGYN-RS, University of California, San Francisco, San Francisco, CA, United States

Stephen J. Lye Lunenfeld-Tanenbaum Research Institute at Sinai Health System; Department of Physiology; Department of Obstetrics and Gynecology, University of Toronto, Toronto, ON, Canada

Xinyi Ma Center for Reproductive Medicine, Department of Obstetrics and Gynecology, Peking University Third Hospital, Beijing, China

Yun-Yi Ma Department of Obstetrics and Gynecology, College of Medicine, Taipei Medical University; Center for Reproductive Medicine, Taipei Medical University Hospital, Taipei, Taiwan

Federico Maggi TOMA Advanced Biomedical Assays S.p.A., Busto Arsizio, Italy

Jose Miravet-Valenciano IGENOMIX, Valencia, Spain

Stacey Missmer Department of Epidemiology, Harvard T.H. Chan School of Public Health; Boston Center for Endometriosis, Boston Children's and Brigham and Women's Hospitals, Boston, MA; Department of Obstetrics, Gynecology and Reproductive Biology, Michigan State University, East Lansing, MI, United States

Kai K. Miu School of Biomedical Sciences, Faculty of Medicine, The Chinese University of Hong Kong, Shatin, Hong Kong

Grant Montgomery Institute for Molecular Bioscience, The University of Queensland, Brisbane, QLD, Australia

Nuria Montserrat Pluripotent Stem Cells and Activation of Endogenous Tissue Programs for Organ Regeneration (PR Lab), Institute for Bioengineering of Catalonia (IBEC), Barcelona, Spain

Lubna Nadeem Lunenfeld-Tanenbaum Research Institute at Sinai Health System, Toronto, ON, Canada

Kavita Narang Michigan State University, College of Human Medicine, Department of Obstetrics, Gynecology and Reproductive Biology; Spectrum Health, Butterworth Hospital, Women's Center, Department of Obstetrics and Gynecology, Grand Rapids, MI, United States

Maria New Icahn School of Medicine at Mount Sinai Hospital, New York, NY, United States

Anaïs Noblanc Departments of Pharmacology and Therapeutics and of Obstetrics and Gynecology, McGill University, Montréal, QC, Canada

Robert J. Norman University of Adelaide; Fertility SA; Royal Adelaide Hospital; NHMRC Centre for Research Excellence in Polycystic Ovary Syndrome, Adelaide, SA, Australia

Elizabeth A. Normand Department of Molecular and Human Genetics, Baylor College of Medicine, Houston, TX, United States

Marisol O'Neill Center for Reproductive Medicine; Department of Molecular and Cellular Biology, Baylor College of Medicine, Houston, TX, United States

Maria S. Peñaherrera Department of Medical Genetics, University of British Columbia; BC Children's Hospital Research, Vancouver, BC, Canada

Jie Qiao Center for Reproductive Medicine, Department of Obstetrics and Gynecology, Peking University Third Hospital; Fellow of the Chinese Academy of Engineering, Beijing, China

Endah Rahmawati Graduate Institute of Clinical Medicine, College of Medicine, Taipei Medical University, Taipei, Taiwan

Nilufer Rahmioglu Wellcome Trust Center for Human Genetics, University of Oxford, Oxford, United Kingdom

Svetlana Rechitsky Reproductive Genetic Innovations (RGI), Northbrook, IL, United States

Bernard Robaire Departments of Pharmacology and Therapeutics and of Obstetrics and Gynecology, McGill University, Montréal, QC, Canada

Wendy P. Robinson Department of Medical Genetics, University of British Columbia; BC Children's Hospital Research, Vancouver, BC, Canada

Peter A.W. Rogers Department of Obstetrics and Gynaecology, University of Melbourne, Royal Women's Hospital, Parkville, VIC, Australia

María Ruiz-Alonso IGENOMIX, Valencia, Spain

Kristiina Rull Institute of Biomedicine and Translational Medicine, University of Tartu; Department of Obstetrics and Gynecology, Women's Clinic of Tartu University Hospital, Tartu, Estonia

Emre Seli Department of Obstetrics, Gynecology, and Reproductive Sciences, Yale School of Medicine, New Haven, CT, United States

Johanna Selvaratnam Departments of Pharmacology and Therapeutics and of Obstetrics and Gynecology, McGill University, Montréal, QC, Canada

Oksana Shynlova Lunenfeld-Tanenbaum Research Institute at Sinai Health System; Department of Obstetrics and Gynecology, University of Toronto, Toronto, ON, Canada

Carlos Simón University of Valencia/ INCLIVA; Igenomix S.L, Valencia, Spain; Department of Ob/Gyn, Stanford University, Stanford, CA; Department of Ob/Gyn, Baylor College of Medicine, Houston, TX, United States

Giuseppe Simoni TOMA Advanced Biomedical Assays S.p.A., Busto Arsizio, Italy

Joe Leigh Simpson Herbert Wertheim College of Medicine, Florida International University, Miami, FL, United States

Andrew H. Sinclair Murdoch Children's Research Institute, Royal Children's Hospital; Department of Paediatrics, University of Melbourne, Melbourne, VIC, Australia

Leanne Stalker Department of Biomedical Sciences, University of Guelph, Guelph, ON, Canada

Melissa Stosic Department of Obstetrics and Gynecology, Columbia University Medical Center, New York, NY, United States

Jose M. Teixeira Michigan State University, College of Human Medicine, Department of Obstetrics, Gynecology and Reproductive Biology, Grand Rapids, MI, United States

Nannan Thirumavalavan   Center for Reproductive Medicine; Scott Department of Urology, Baylor College of Medicine, Houston, TX, United States

Jason C.H. Tsang   Li Ka Shing Institute of Health Sciences; Department of Chemical Pathology, The Chinese University of Hong Kong, Shatin, Hong Kong, China

Allison Tscherner   Department of Biomedical Sciences, University of Guelph, Guelph, ON, Canada

Elena J. Tucker   Murdoch Children's Research Institute, Royal Children's Hospital; Department of Paediatrics, University of Melbourne, Melbourne, VIC, Australia

Chii-Ruey Tzeng   Department of Obstetrics and Gynecology, College of Medicine, Taipei Medical University; Center for Reproductive Medicine, Taipei Medical University Hospital, Taipei, Taiwan

Ignatia B. Van den Veyver   Department of Molecular and Human Genetics; Department of Obstetrics and Gynecology, Baylor College of Medicine, Houston, TX, United States

Maria M. Viveiros   Department of Physiology and Pharmacology, College of Veterinary Medicine; Regenerative Biosciences Center (RBC), University of Georgia, Athens, GA, United States

Hao Wang   Medical Research Center, Peking University Third Hospital, Beijing, China

Yan-Ling Wang   State Key Laboratory of Stem Cell and Reproductive Biology, Institute of Zoology, Chinese Academy of Sciences; University of Chinese Academy of Sciences, Beijing, China

Ronald Wapner   Department of Obstetrics and Gynecology, Columbia University Medical Center, New York, NY, United States

Jeffrey T. White   Center for Reproductive Medicine; Scott Department of Urology, Baylor College of Medicine; Texas Children's Hospital, Houston, TX, United States

Samantha L. Wilson   Department of Medical Genetics, University of British Columbia; BC Children's Hospital Research, Vancouver, BC, Canada

Liying Yan   Center for Reproductive Medicine, Department of Obstetrics and Gynecology, Peking University Third Hospital, Beijing, China

Victor Yuan   Department of Medical Genetics, University of British Columbia; BC Children's Hospital Research, Vancouver, BC, Canada

Fan Zhai   Center for Reproductive Medicine, Department of Obstetrics and Gynecology, Peking University Third Hospital, Beijing, China

Boryana Zhelyazkova   Center for Reproductive Medicine; Department of Molecular and Cellular Biology, Baylor College of Medicine, Houston, TX, United States

Qi Zhou   State Key Laboratory of Stem Cell and Reproductive Biology, Institute of Zoology; Institute for Stem Cell and Regeneration, Chinese Academy of Sciences; University of Chinese Academy of Sciences, Beijing, China

Krina Zondervan   Wellcome Trust Center for Human Genetics; Nuffield Department of Obstetrics and Gynaecology, John Radcliffe Hospital, University of Oxford, Oxford, United Kingdom

PART I

REPRODUCTIVE TRACT DEVELOPMENT AND GAMETOGENESIS

Developmental Genetics of the Male Reproductive System

Marisol O'Neill*,†, Boryana Zhelyazkova*,†,
Jeffrey T. White*,‡,§, Nannan Thirumavalavan*,‡,
Dolores J. Lamb*,†,‡,¶

*Center for Reproductive Medicine, Baylor College of Medicine, Houston, TX, United States
†Department of Molecular and Cellular Biology, Baylor College of Medicine, Houston, TX,
United States ‡Scott Department of Urology, Baylor College of Medicine, Houston, TX,
United States §Texas Children's Hospital, Houston, TX, United States
¶Department of Urology, Weill Cornell Medical College, New York, NY, United States

O U T L I N E

Abbreviations

aCGH	array comparative genomic hybridization
AER	apical ectodermal ridge
AHR	aryl hydrocarbon receptor
AMH/Amh	anti-Mullerian hormone
AR	androgen receptor
ARNT2	aryl hydrocarbon receptor nuclear translocator 2
Arx	aristaless related homeobox
ATF3	activating transcription factor 3
BMP	bone morphogenic protein
cAMP	cyclic adenosine monophosphate
Cbx2	chromatin modification and remodeling factor also known as M33
CGRP	calcitonin gene related peptide
CNV	copy number variant
CREBP	cyclic AMP response element binding protein
CRKL	CRK like proto oncogene
CYP17A1	cytochrome P450 family 17 subfamily A1
CYP1A2	cytochrome P450 family 1 subfamily A2
Dax1	dosage sensitive sex reversal region on the X chromosome
DGKK	diacylglycerol kinase kappa
Dhh	desert hedgehog
Dlx	distal less
E	embryonic day
EphB2	ephrin
ESR1	estrogen receptor 1
FGD1	faciogenital dysplasia protein 1
FGF	fibroblast growth factor
FGFR	fibroblast growth factor receptor
Gata4	GATA binding protein
GFN	genitofemoral nerve
Gli	glioma associated oncogene
hCG	human chorionic gonadotropin
Hox	homeobox
IGF	insulin-like growth factor
Igfr1	insulin-like growth factor receptor
INSL3	insulin-like 3
Insr	insulin receptor
JNK	Jun N terminal kinase
LHR	luteinizing hormone receptor
Lhx9	Lim homeobox protein
M33	chromatin modification and remodeling factor also known as Cbx2
MAMLD1	mastermind like domain containing 1
MAPK	mitogen activated protein kinase
Mis	Mullerian inhibiting substance
NR5A1/Nr5a1	nuclear receptor 5 A1 also known as steroidogenic factor 1
PDE4B	phosphodiesterase 4B
Pdgfra	platelet derived growth factor receptor
PGCs	primordial germ cells
PTPN11	protein tyrosine phosphatase receptor
Ras	rat sarcoma
RXFP2	relaxin family of peptide receptor
Sf1	nuclear receptor 5 A1 also known as steroidogenic factor 1

Shh	sonic hedgehog
SNP	single nucleotide polymorphism
Sox9	SRY box 9
SPAG5	sperm associated antigen 5
SRY/Sry	sex determining region
STRBP	spermatid perinuclear RNA binding protein
Tbx	T box transcription factor
Tfm	testicular feminization mutation
VAMP7	vesicle associated membrane protein
WNT	wingless
WT1/Wt1	Wilm's tumor 1

INTRODUCTION

Early embryonic development occurs in a sex-independent manner. This gender-neutral phase is brief. As the gonads begin to differentiate, male and female structures differentiate. In humans as well as many other mammals, sex is determined by the presence of X and Y chromosomes. The initial driver of male sexual differentiation is the presence of a Y chromosome that contains genes necessary for masculinization. However, the genes that drive sexual differentiation are not restricted to the Y chromosome. Genes across all chromosomes regulate sexual differentiation and are expressed in a sex-specific manner. This type of genetic regulation often occurs in a hormone-dependent manner. In this chapter, we will explore the genetic drivers of sexual differentiation throughout embryonic development.

EARLY TESTICULAR DEVELOPMENT

Sexual reproduction, by definition, is the production of a new living organism by combining the genetic information of two individuals from different sexes. The genetic sex of the embryo (the presence or absence of the Y chromosome) determines the gonadal sex (testis or ovary), which in turn leads to the development of the phenotypic sex (secondary sexual characteristics, such as external genitalia).

The reproductive system ontogeny begins with the formation of the genital ridge on the ventral surface of the mesonephros as paired thickenings of the intermediate mesoderm. The genital ridge is composed of somatic and germ cells. In mice, primordial germ cells (PGCs) are specified in the proximal epiblast and migrate from the primitive streak to the endoderm, which will form the future hindgut. This happens at embryonic day 7.5 (E7.5). Then PGCs migrate along the endoderm to reach the genital ridge at E10.5 to consequently form the embryonic gonad [1–4]. The decision to develop either a testis or an ovary comes from the differentiation of the supporting cells (Sertoli cells in males or granulosa cells in females) according to the genetic sex of the embryo.

The initial stages of genital development are the same for both sexes with sex determination occurring based on the presence of the X or Y chromosomes at E32 for human and E10.5 in mice. The proper development of the gonads is a tightly regulated process and, if disrupted, can lead to the development of disorders of sex development such as gonadal dysfunction, infertility, ambiguous genitalia, etc.

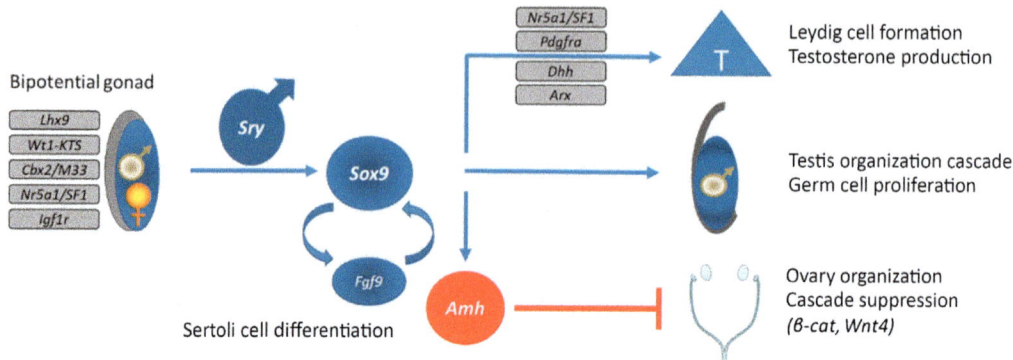

FIG. 1.1   Genes and pathways in the developing fetal testis. Multiple genes are involved in the early development of the bipotential gonad. The *Sry* gene expression induces the development of Sertoli cells, which trigger the signaling both for the rest of the testicular organization and the regression of the Mullerian duct.

During the early stages of embryonic development, the gonads are bipotential (Fig. 1.1). There are key genes involved in the early growth and survival of the gonads, some of which are *Nr5α1, Lhx9, Wt1, Cbx2,* and *Igf1r*. The orphan nuclear receptor *Nr5α1* [5] (also known as *Ad4BP/Sf1*) is a transcription factor that is expressed in the gonads and all primary steroidogenic tissues. Deletion in mice results in complete gonadal and adrenal gland agenesis. One of the genes regulating *Nr5α1* is *Lhx9* [6] (LIM homeobox protein), which is necessary for gonad formation. The zinc finger transcription factor Wilm's tumor 1 (*Wt1*) [7], specifically its *Wt1-KTS* isoform, is also essential for development of the bipotential gonad and kidneys. *Wt1-KTS* binds to the *Nr5α1* promoter in cooperation with *Lhx9*. The chromatin modification and remodeling factor *Cbx2* [8] (also known as *M33*) also plays a key role. Embryos lacking this gene have reduced expression of *Nr5α1* and *Lhx9*, resulting in gonadal growth defects. Impaired gonadal development is also shown in mice lacking insulin/insulin-like growth factor (IGF) pathways, more specifically an insulin receptor (*Insr*) [9] and an IGF receptor (*Igf1r*). Mice lacking these two genes have reduced proliferation rates of the somatic progenitor, agenesis of the adrenal gland, and failure of testis development.

The cells of the early gonad are bipotential. The Y chromosome has a specific region, the *Sry* or sex-determining region, that shifts the bipotential gonad to differentiate into the testis by inducing the differentiation of the Sertoli cells [10, 11]. *Sry* upregulates *Sox9* to stimulate the expression of fibroblast growth factor 9 (*Fgf9*) [12]. Together, *Sox9* and *Fgf9* act in a positive feedback loop to drive expression of *Sox9*, promote PGC survival, prevent entry of PGC into meiosis, and suppress development of ovaries. Additional factors upregulating the expression of *Sox9* are *Dax1* and *Nr5α1*. Sox9 and Nr5a1 activate expression of the *Anti-Mullerian hormone/Mullerian inhibiting substance* (*Amh/Mis*) [13]. *Wt1-KTS* and *Gata4* also regulate *Amh* expression. Sertoli cells secrete *Amh* into the embryonic genital ridge, causing regression of the Mullerian ducts in males. In mouse models, ectopic expression of *Sox9* introduced in female mice leads to female-to-male sex reversal, suggesting that *Sox9* can induce male development in female mice [14]. Conversely, homozygous deletion of *Sox9* in XY gonads leads to complete male-to-female sex reversal. Overall, *Sox9* is necessary for Sertoli cell proliferation, somatic cell expansion, migration and differentiation of the developing testis, regression of the Mullerian duct, and somatic differentiation.

Once the Sertoli cell fate has been determined, four key genes lead Leydig cell differentiation: *Dhh*, *Pdgfra*, *Nr5α1/Sf1*, and *Arx*. Desert hedgehog (*Dhh*) is expressed by the Sertoli cells and is responsible for the differentiation of the peritubular myeloid cells and the consequent formation of testis cords [15]. DHH signals from Sertoli cells to the Patched1 receptor on the fetal Leydig cell precursors in the interstitium of the XY gonad to trigger Leydig cell differentiation by upregulating *Nr5α1/Sf1* and P450 side chain cleavage [16].

TESTICULAR DESCENT

During embryonic development, the gonads are initially anchored high in the abdomen, near the kidneys by the cranial suspensory neuron and attached to the inguinal canal via a cylindrical ligament called the gubernaculum. While the gonads are still undifferentiated, there is a passive, hormone-independent descent that occurs in both sexes. This initial descent is closely related to the descent of the diaphragm and marks the only descent that will occur in female gonadal development [17]. In males, maturation of the gonads requires further descent into the scrotum. Subsequent descent is hormonally regulated and is biphasic, consisting of a transabdominal phase and an inguinoscrotal phase.

Transabdominal Phase

The transabdominal stage of testicular descent occurs between 10 and 15 weeks gestational age in humans and 13–17 days in rodents. The transabdominal phase of testicular descent relies on hormone-dependent gubernacular development and reorganization. Development of the testis occurs on the anteromedial surface of the mesonephros in the urogenital ridge.

The gubernaculum develops as the mesonephros is regressing at the onset of sexual differentiation. The gubernaculum then attaches to the testis and the Wolffian duct. The cranial Wolffian duct will become the epididymis and the vas deferens while the distal Wolffian duct persists as the ureter, trigone, and proximal urethra. Descent of the testis is initiated by gubernacular swelling, which occurs in response to hyaluronic acid and glycosaminoglycans [18]. The gubernaculum first elongates cranially, pushing the testis and epididymis to the inguinal ring. During this time, the caudal portion of the gubernaculum also swells, widening the inguinal channel to prepare for transinguinal passage of the testis [19]. Growth of the gubernaculum requires cell proliferation, uptake of water, and increase in glycosaminoglycans and hyaluronic acid.

Inguinoscrotal Phase

By gestational weeks 22–25, the testis and epididymides are positioned at the internal rings of the inguinal canal. The second phase of descent, the inguinal-scrotal phase, occurs after a brief pause until the 25th gestation week in humans; it will occur between weeks 25 and 35. In rodents, inguinal-scrotal descent occurs postnatally in weeks 3–4. The testis, epididymis, and gubernaculum will begin to move through the inguinal canal as a single entity. The gubernaculum will then shrink, allowing for a pressure change that permits transinguinal passage. After passage, the gubernaculum will become fibrous and involute.

Genetic Control of Testicular Descent

Insulin Like 3 (INSL3)

INSL3, the key hormone in transabdominal migration [20, 21], is a member of the insulin superfamily of structurally related hormones produced by Leydig cells (Fig. 1.2). Insl3 experiments in mice have led to the understanding of its importance in testicular descent [22]. *Insl3* knockout mice are born with abdominal testes due to a lack of gubernacular expansion [22, 23]. However, overexpression of Insl3 in female mice leads to gubernacular swelling and ovarian descent, causing localization of the ovaries near the bladder neck. INSL3 is produced in the testis by the Leydig cells and signals through the relaxin family of peptide receptor 2 (*RXFP2*) [24]. RXFP2 is a G protein coupled receptor expressed on the gubernaculum that increases intracellular cAMP, initiating signaling through β-catenin, WNT, and NOTCH pathways [23, 25].

INSL3 is also believed to interact with testosterone. *Rxfp2* knockout mice have apoptosis of androgen receptor (AR) positive cells in the cranial gubernaculum and a decrease in gubernacular swelling [23]. Mice lacking the luteinizing hormone receptor (*Lhr*) show that testosterone treatment upregulates expression of RXFP2 in the gubernaculum and cremaster muscle in an AR-dependent fashion. Testosterone has also been shown to stimulate INSL3 production in Leydig cell lines [26]. These in vitro experiments provide further evidence that

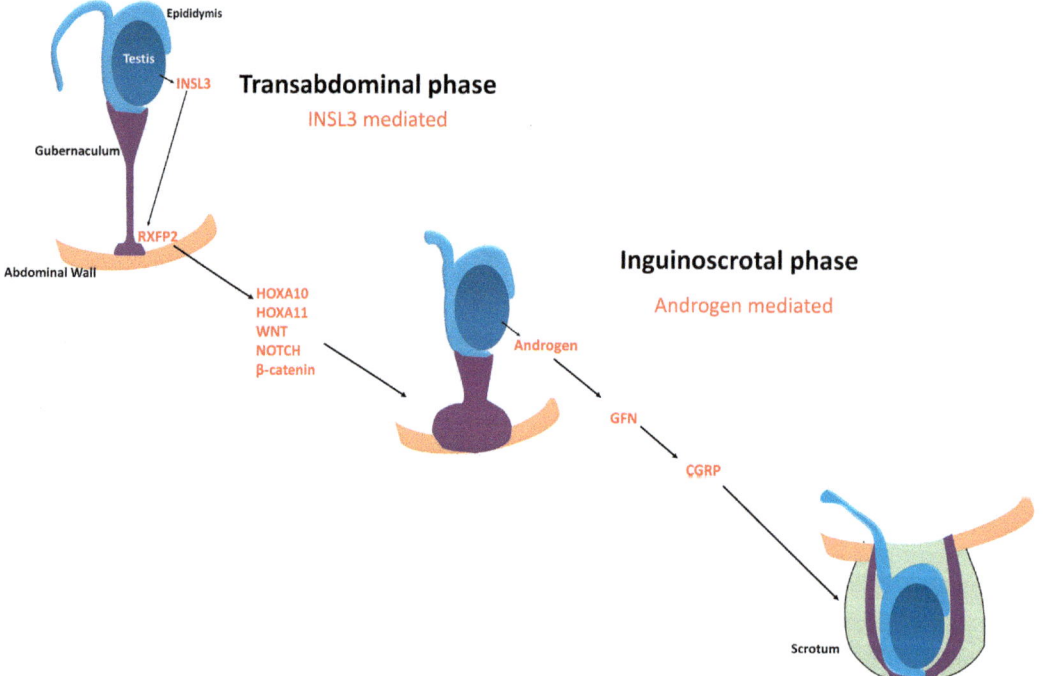

FIG. 1.2 Biphasic testicular descent consists of a transabdominal phase and an inguinoscrotal phase. During the transabdominal phase, INSLS3 binds its receptor RXFP2 in the gubernaculum, leading to expression of various genes that induce expansion of the gubernaculum. The second phase, the inguinoscrotal phase, is androgen-mediated. Androgen signals to the GFN, a nerve, which will in turn produce CGRP. Production of CGRP induces contraction of the cremaster muscle and acts as a chemotactic factor leading to descent into the scrotum.

there is an interaction between INSL3 and testosterone necessary for reorganization of the gubernaculum.

Fibroblast Growth Factors (FGF), Homeobox (Hox), Wingless (Wnt), Notch

The outgrowth of the gubernaculum from the inguinal wall is analogous to the growth of the embryonic limb bud. Limb bud formation requires growth in three dimensions: the anterior-posterior axis, the dorsal-ventral axis, and the proximal-distal axis. This type of growth requires regulation of highly specific expression patterns of different genes. The limb bud arises from a mesoderm core surrounded by a layer of ectoderm. The mesoderm is made from undifferentiated mesenchymal cells [27]. The initial proliferation in the formation of the limb bud is initiated by fibroblast growth factors (FGFs) and homeobox transcription factor (Hox) genes [28]. FGF 4, 8, and 10 are expressed by the apical ectodermal ridge (AER) promoting mesenchymal growth in the proximal distal axis toward the progress zone [29].

The gubernaculum is an elongated conical structure composed of loose extracellular matrix and mesenchymal cells comprising mainly fibroblasts and muscle cells. The rodent growth of the gubernaculum occurs in an oval-shaped distal region of the gubernacular bulb, which acts similarly to the progress zone in the limb bud. Gubernacular proliferation is controlled by the calcitonin gene-related peptide (CGRP), which functions similarly to the AER. Fgf10 and Hoxa10 are distinctively expressed in the caudal end of the gubernaculum at E15 in rodents, suggesting proximodistal elongation of the gubernaculum prior to inguinoscrotal descent [30]. Knockout studies of *Hox11* in mice lead to cryptorchidism, suggesting a role for Hox11 in testicular descent. β-catenin, WNT, and NOTCH are also important for gubernacular mesenchymal proliferation downstream of INSL3 [25]. Notch signaling is also necessary for growth of the processus vaginalis, a specialized peritoneal diverticulum in the gubernaculum that provides a route of passage for the descending testis. The processus vaginalis may have a role in the second phase of descent, especially in rodents. Evidence in rodents suggests that it forms from a specialized peritoneal epithelium that covers the urogenital ridge early in the embryo. Its role in testicular descent is being further investigated. In humans, the processus is believed to form passively due to weakness in the inguinal triangle. Clinically, if the processus vaginalis does not close, there is an open communication from the abdominal cavity to the scrotum, leading to formation of a hydrocele/hernia.

Anti-Mullerian Hormone (AMH)

Anti-Mullerian hormone (AMH) is produced by the Sertoli cells in the testis and was previously believed to regulate transabdominal testicular descent. *AMH* mutations can lead to persistent Mullerian duct syndrome in males, which consists of different genital anomalies and failure of testicular descent. The failure to descend is, however, believed to be caused by an anatomical obstruction. Evidence comes from Anti-Mullerian hormone receptor (*Amhr*) knockout mice that have normal gubernaculum and complete testicular descent [31], suggesting AMH itself is not required for testicular descent.

Androgen Receptor (AR)

Mice with complete androgen resistance and absence of external virilization have complete failure of inguinoscrotal testicular descent. Treatment of E16-E17 rats with an antiandrogen (flutamide) causes deranged gubernacular migration during the inguinoscrotal period, which leads to cryptorchidism [32]. Gonadotropins from the placenta and fetal pituitary are important in regulating testicular production of androgens as well as INSL3 and AMH, respectively. INSL3 is dependent on Leydig cell development during the transabdominal phase. During weeks 10–25, human chorionic gonadotropin (hCG) produced by the placenta induces the initial production of INSL3 via the LHR. INSL3 production then becomes constitutive. Pituitary gonadotropes are necessary for the androgen-mediated inguinoscrotal phase at weeks 25–35. The hypothalamic-pituitary-gonadal axis is important for testosterone production as axis anomalies lead to androgen deficiency manifested as micropenis and cryptorchidism. Interestingly, bilateral undescended testicles are seen in subgroups with high gonadotropes, suggesting a lack of negative feedback from androgens. Males with androgen insensitivity syndrome have intraabdominal testes positioned near the internal inguinal ring [33]. This provides evidence for androgen signaling being involved in the second phase of testicular descent. Rodent studies have suggested indirect action of testosterone via the androgen receptor-containing genitofemoral nerve (GFN), which releases the calcitonin gene-related peptide (CGRP). CGRP acts as a chemotactic signal, guiding the gubernaculum to grow into the scrotum.

Calcitonin Gene Related Peptide (CGRP)

CGRP is a proposed second messenger involved in inguinal scrotal descent. Transection of GFN neurons in neonatal rats leads to failure of the gubernaculum to migrate into the scrotum. Exogenous CGRP stimulates migration of undescended inguinal testes (Fig. 1.2). In ectopic and intrabdominal testes, however, CGRP does not have an effect. The genitofemoral nerve (GFN) is a sexually dimorphic nerve in the L1–L2 dorsal root ganglia and more CGRP positive neurons are found in males [29]. CGRP released from the sensory nerve terminal of the GFN induces rhythmic contractility of the developing cremaster muscle of the gubernaculum [34]. Contraction may be important to orient the gubernacular tip toward the scrotum and assist in migration [35]. In E17 mice, an increase in contraction is observed and lasts through the first postnatal week.

DEVELOPMENT OF THE EXTERNAL GENITALIA

Building the external genitalia requires specification of a competent tissue field, induction of budding (swellings), reciprocal maintenance of outgrowth (tubercle), and coordination with dorsoventral and proximodistal patterning. This developmental program is similar to that of limb ontogeny [36].

Formation of the Genital Tubercle

Interestingly, the initial fating and source of the genital mesenchyme is not well understood. Only *Wnt5a* mouse mutants that fail to form genital swellings have been used to study early genital tubercle development [37]. In these mice, the endodermal urorectal septum fails to contact the cloacal ectoderm and therefore the cloacal membrane fails to form [37]. Thus, the cloacal membrane is required for development of the genital swellings and therefore the genital tubercle.

The competent field of the external genitalia forms as a pair of genital swellings on either side of the cloacal membrane; they are visible at E10.5 in the mouse [38]. Experts postulate that a diffusible signal likely emanates from the cloacal membrane to spatiotemporally specify the early genital mesenchyme. However, no such signal has been discovered [36]. These swellings then fuse to form the genital tubercle by E11.5 in the mouse [38].

The genital tubercle is composed of all three germ layers. Similar to a limb, the majority of the genital tubercle is composed of mesoderm wrapped with an ectodermal covering. This ectodermal covering is required for proximodistal patterning and maintenance of outgrowth of the genital tubercle. Removal of this epithelium results in stage-dependent truncation of the baculum and erectile tissues [39].

Patterning of the other genital tubercle axes is not well understood. Medial-lateral patterning is a mirror image centered on the urethral plate. Furthermore, the only structural distinction of the dorsoventral axis is the urethral plate. Therefore, the urethral plate has been suspected to be the organizing center of the genital tubercle. Based on experiments where the urethral plate, which was transplanted to the anterior surface of a developing limb, was competent to induce mirror-imaged digits and muscle tissue from the mesenchyme, the urethral plate was identified as an organizing region with the potential to induce tissue polarity and tubulognesis [38]. Importantly, this organizing function has been demonstrated only in the limb and not in the genital tubercle itself [36].

Formation of the Urethra

The urethral plate forms from an extension of the cloacal endoderm. Past theories of dual origin urethral ontogeny have been disproven; lineage studies have demonstrated that the entire proximal and distal urethra is formed by an endodermal urethral plate, not an ingrowth of the ectoderm [39a]. The urethra and urethral plate have been investigated extensively regarding tubulogenesis, that is, urethral closure. Several genetic pathways have been implicated including Shh, Fgf10-Fgfr2, EphB2-EphrinB2, Hoxa13, and Dlx5/6. Each will be reviewed below in the genetics component of the chapter. Urethral closure occurs at E16 when the proximal urethral plate edges fuse, forming a lumen; this process progresses to the glans at E18 when the lumen opens on the glans [40].

Two processes are required for urethral closure: ventrolateral growth of the preputial swellings and remodeling of the bilaminar urethral plate into a tube. The phenotypic result of failure of either of these two processes is hypospadias (described in detail later in the chapter). These processes are tightly controlled with regards to a fine balance between proliferation and apoptosis. Additionally, these processes are under hormonal regulation.

Differentiation of the Male and Female Genitalia

The external genitalia remains gender-neutral until approximately E15 in the mouse. Female versus male phenotype is governed by the balance of androgens and estrogens. It was previously thought that a default feminine genital phenotype existed, resulting in development of a clitoris, labia majora and minora, a urethral orifice posterior to the clitoris in the midline, and a separate, distinct vaginal introitus posterior to the urethra in the midline.

Recent reports have refuted the default programming theory: loss of estrogen signaling decreases the estrogen:androgen ratio, resulting in masculinization of female genitalia [41]. Alternatively, increased androgen signaling in a female increases the estrogen:androgen ratio, resulting in genital tubercle elongation, complete tubularization of the urethra with some degree of labioscrotal fold fusion resulting in clitoromegaly or a pseudophallus, an orifice on the enlarged clitoris or pseudophallus, and a single perineal opening that branches to the urethra and vagina (i.e., common genitourinary sinus). Decreased androgenization of the male increases the estrogen:androgen ratio, resulting in feminization of the male genitalia: a smaller genital tubercle, incomplete formation of the urethra, and incomplete fusion of the preputial folds, resulting in micropenis, hypospadias, and incomplete foreskin. These states of abnormal androgenization result in a phenotype of ambiguous genitalia.

Genetic Control of External Genitalia Ontogeny

Several genes have been described as important in the development of the external genitalia. Each is reviewed with a focus on developmental mechanisms.

Homeobox 13 (Hox13)

Hox13 isoforms have dual roles in the development of the genital tubercle: an early role in the genital mesenchyme and a later role in urethral tube closure. Hoxa13 and Hoxd13 are expressed during outgrowth of the genital tubercle as well as the urethral plate and surrounding mesenchyme [42–44]. Homozygous deletion of either Hox13 isoform creates genital tubercle patterning defects; double knockout for both Hoxd13 and Hoxa13 develop aphallia [45, 46].

Fibroblast Growth Factors (Fgfs)

In limb outgrowth, T-box transcription factors control *Fgf* expression during limb outgrowth [47]. Similarly, expression of *Tbx2*, *Tbx3*, *Tbx4*, and *Tbx5* have been described in the genitalia but have not been found as required during genital development [36, 48].

No clear role for Fgf ligands has been shown in initiation or maintenance of genital tubercle outgrowth. While Fgf8 and Shh function in a positive feedback loop to complete limb outgrowth and patterning, no such role has been found in the genital tubercle. *Fgf8* knockout mice demonstrate no genital tubercle phenotype.

A role for only Fgf10, which binds the Fgfr2 receptor, is described as necessary for urethral tube closure [49, 50]. *Fgfr2* is expressed in both the preputial folds and the urethral epithelium. *Fgf10* is expressed in the genital tubercle mesenchyme adjacent to the urethral cells. The interaction of the ligand with the Fgfr2 receptor is required for ventral growth of the preputial folds and formation of the complete foreskin [49]. Additionally, conditional loss of the *Fgfr2* receptor from the ectoderm results in severe hypospadias and loss of the ventral prepuce while conditional loss of *Fgfr2* from the endoderm causes mild hypospadias [49, 51, 52].

Further studies where knockouts can be controlled both spatially and temporally will be required to dissect the role of the Fgf in development of the external genitalia. Currently, Fgf signaling is required for mesenchymal outgrowth on the ventral surface and urethral maturation and closure.

Wingless (Wnt) Signaling

Canonical Wnt signaling is active during budding of the external genitalia and *Wnt2*, *Wnt3*, *Wnt4*, *Wnt5a*, *Wnt9b*, and *Wnt11* are expressed by E11.5 in the budding genital tubercle [53]. The importance of Wnt signaling is further supported by the fact that conditional *β-Catenin* knockouts do not have external genitalia [53]. Initiation of the paired genital swellings does occur in these mutants, suggesting that canonical Wnt signaling is required for initiation and/ or maintenance of genital outgrowth. Interestingly, the loss of *β-Catenin* expression reduces the expression of cellular pathway gene expression, including *Shh*. Further evidence for the importance of canonical Wnt signaling abounds: (1) loss of *Lrp6*, a coreceptor of canonical Wnt signaling, mimics this phenotype, and (2) target genes of canonical Wnt signaling, such as *Dkk1*, are expressed in the distal urethral epithelium [54, 55].

In contrast, there is no clear role for noncanonical Wnt signaling via Wnt5a and Wnt11 in the genital tubercle. Loss of *Wnt5a* demonstrates a variable phenotype ranging from shortened to normal-sized genital tubercles [37, 56]. Similarly, loss of a noncanonical Wnt signaling receptor *Ror2* results in an underdeveloped genital tubercle [57]. Finally, despite proper spatiotemporal expression patterns of another noncanonical Wnt, *Wnt11*, no clear role could be distinguished [58]. Noncanonical Wnt signaling may play a role in maintenance of genital tubercle outgrowth.

Sonic Hedgehog (Shh)

Sonic hedgehog (Shh) is expressed in the cloacal endoderm prior to genital budding and persists in the urethral epithelium during outgrowth [37, 38, 50]. Though Shh knockouts do not develop a genital tubercle, the paired genital swellings do form, suggesting that Shh signaling is important in initiation and/or maintenance of genital outgrowth [38, 50]. Interestingly, the loss-of-*Shh* phenotype can be partially rescued by *β-Catenin* overexpression in the urethral epithelium [53].

Time-dependent conditional knockout models have demonstrated the intricate role of *Shh* [37, 53]. It is both a proliferative cue for the elongating mesenchyme as well as an inhibitor of apoptosis of the ectoderm. Early deletion results in apoptosis of condensing mesenchyme. If prior to E13.5, the embryo develops a persistent cloaca, *Shh* is also required for urethral closure. Late deletion up to E16.5 results in hypospadias and a feminized genital appearance [37]. To understand where Shh acts, conditional deletions of *Smoothened* (*Smo*), a coreceptor required for Shh intracellular transduction pathways, have been performed [37, 53]. Deletion from the ectoderm epithelial cap or the mesenchyme, but not the urethral epithelium, disrupts genital tubercle formation. When removed from the mesenchyme, the genital tubercle is shortened and has a hypospadic phenotype. If removed from the ectoderm, the genital tubercle develops a hypospadic urethra. Therefore, *Shh* is required for integration of the mesoderm and ectoderm to form a normal phallus: it is a proliferative and antiapoptotic signal for both the mesenchyme and the ectoderm.

Androgen Receptor (AR)

AR is expressed in the genital mesenchyme and the ventral surface of the growing genital tubercle, and the ventral surface of the urethral folds [59, 60]. Expression is identical in male and female embryos until E12 [60]; thereafter, expression is localized to the ventral surface to mediate urethral plate growth and closure [59].

Mutation of the androgen receptors has been characterized with the *Tfm* (*testicular feminization mutation*) mouse [41]. This mutation results in external genitalia that are indistinguishable from female mice. Additional evidence from deletion of *5-alpha reductase*, an enzyme converting testosterone to the more active androgen dihydrotestosterone, results in a range of phenotypes from hypospadias to micropenis to complete feminization [61].

There is also evidence of AR-mediated control of the genetic mediators of genitalia ontogeny. For example, *Fgfr2* contains an androgen-response element in its promoter; deletion of *AR* also results in a decrease in *Fgfr2* expression, resulting in hypospadias [49]. Similar changes and phenotypes occur with AR-mediated modulation of *EphrinB2* [62].

BIRTH DEFECTS

Disorders of sexual development are among the most common congenital anomalies. The majority of cases can be corrected through surgical intervention. Multiple surgeries are, however, often required, increasing the risk for surgical complications as well as sexual dysfunction and psychological problems. We will explore the genetics of the two most common disorders of sexual development, hypospadias and cryptorchidism, in further detail.

Hypospadias

Hypospadias is a condition in which the urethral meatus opens more proximally on the penile shaft. Hypospadias is one of the most common genitourinary birth defects, occurring in ~1 in 300 male births. Of note, hypospadias occurs with increased frequency, at an odds ratio of 1.77, in children born using assisted reproductive technologies [63]. Interestingly its incidence is increasing, suggesting that environmental cues may have an effect. Hypospadias is commonly associated with incomplete foreskin and chordee, a ventral curvature of the penis. The urethral meatus can be located at the glans in the mildest of phenotypes, or as proximal as the scrotum in more severe cases. Correction requires surgical repair, sometimes in multiple stages, and is associated with significant morbidity. It can be associated with long-term urinary difficulties, difficulty with sexual activity and fertility in the future, and decreased quality of life [64, 65].

Though not fully elucidated, the etiology of hypospadias is thought to be multifactorial, with hormonal, environmental, and genetic factors. Up to 25% of hypospadias cases can have some familial clustering. One study found that the incidence of hypospadias in monochorionic twins is 4% while only 1% in dichorionic twins [66]. Prematurity and low birth weight are also strongly associated with hypospadias [67]. Many genes have been implicated in being associated with hypospadias due to changes in both steroidogenic and nonsteroidogenic processes involved in genital tubercle development. Mutations in genes involved in the development of the urogenital system can potentially lead to hypospadias. Mutations in both Wilm's Tumor 1 (*WT1*) and Steroidogenic factor 1 (*SF1/NR5A1*), both of which have roles in early embryonic development of the urogenital system, can lead to hypospadias as well as more severe birth defects. *WT1* and *NR5A1* mutations are both associated with penoscrotal hypospadias while WT1 mutations are also associated with micropenis and *NR5A1* mutations are also associated with cryptorchidism [68, 69].

Genes involved in genital tubercle patterning, including BMP, HOX, SHH, and FGF genes, have been identified as hypospadias genes through rodent [28, 70, 71] studies (Table 1.1). Specifically, *Hox13* knockout models in mice develop hypospadias as a result of FGF8 and BMP7 signaling in the urethral plate epithelium [70]. *Fgf10* and *Fgfr2* deficient mice have also been shown to develop hypospadias due to cell proliferation arrest and premature maturation of the urethral epithelium [28, 70, 71]. These FGF genes signal downstream of AR. Loss of *Gli2* and *Gli3*, which function downstream of SHH, has also been shown to cause hypospadias in mice [72]. SNPs in *BMP7, FGF10, Gli1,2,3, SHH,* and *WT1* are more prevalent in patients with hypospadias, validating the link between human hypospadias and rodent studies. The Wnt/β-catenin pathway has also been implicated as a sexually dimorphic gene that may be involved in the development of hypospadias. *β-catenin* is expressed in the bilateral mesenchymal region adjacent to urethral epithelium and loss and gain of function mutations in *β-catenin* have been associated with altered development of the external genitalia [28].

Several candidate genes for hypospadias have been identified through genome-wide association studies. One such gene is Diacylglycerol Kinase Kappa (*DGKK*), a gene expressed in the preputial skin, for which an intronic SNP is associated with a 2.5 times increased risk for hypospadias [73]. Defects in the androgen receptor have also been implicated as well as defects in 5 alpha reductase [74, 75]. Given the interplay between estrogen and androgens in the developing genitourinary tract, defects in the estrogen receptor have been investigated, and also found in a metaanalysis to be associated with an increased risk of hypospadias [76]. Activating transcription factor 3 (*ATF3*), an estrogen-responsive gene, has been shown to be upregulated in several patients with hypospadias [77]. Another gene with strong evidence for its involvement in hypospadias is the mastermind-like domain containing 1 (*MAMLD1*). MAMLD1 contains the SF1 target sequence mutations in *MAMLD1*, and is highly associated with hypospadias [78].

Sex chromosome abnormalities are highly associated with hypospadias development, but further studies are needed to fully explore the regions of the sex chromosomes that are important in genital development [79]. *SRY*, which is located on the Y chromosome and is necessary for testicular development, has been examined as a potential candidate, but a strong association between SRY SNPs and hypospadias has not been verified [28]. *Vamp7*, a vesicle trafficking protein located on Xq28, has been verified as having a role in hypospadias. Array comparative genomic hybridization (aCGH) studies in children with congenital GU abnormalities revealed an enrichment for duplications on the long arm of the X chromosome encompassing *VAMP7* when compared to patients with normal GU as well as the general population [80]. Mice with genetically elevated VAMP7 levels revealed penile defects, hypospadias, abnormalities of the epithelial-lined prepuce housing the penis, and a reduction in the thickness of the tunica albuginea capsule surrounding the corpora cavernosa. Penile length and anogenital distance were both decreased. Anogenital distance has been evaluated elsewhere as a marker for androgen exposure in humans [81]. Shorter anogenital distances have been linked with decreased testosterone levels, decreased semen quality, and lower fertility. It was also found that increased gene dosage of *VAMP7* increases ESR1 transcriptional activity, thus increasing estrogenic activity. This increase in estrogenic and decrease in androgenic activity contributes to the phenotypes associated with *Vamp7*. Beyond the initial birth defects, elevated levels of VAMP7 lead to more chronic issues. For example, transgenic mice had decreased testis size and decreases in spermatogenesis. The *Vamp7* mice were noted to have smaller litter sizes as well.

Cryptorchidism

Cryptorchidism is the failure of one or both testicles to descend. It is one of the most common congenital anomalies, occurring in 1%–9% of boys worldwide [82]. While most of these cases will spontaneously descend within the first 3 months, 1% of boys are still cryptorchid by the first year of age. The primary treatment for cryptorchidism is surgical intervention called an orchidopexy. Descent of the testicles is important for proper testicular function due to the temperature difference between the body and the scrotum. Cryptorchidism increases one's risk of subfertility as men who have a history of cryptorchidism are two times more likely to be subfertile. The risk of developing testicular cancer is also linked to cryptorchidism. Any history of cryptorchidism makes one 3–4 times more likely to develop testicular cancer. This risk is significantly higher if the testicles have not descended prior to the onset of puberty; males who still have undescended testicles by age 12 are 2–6 times more likely to develop testicular cancer compared to cryptorchid males whose testicles have descended.

There is evidence for a genetic basis of cryptorchidism because brothers and sons of men with cryptorchidism have a higher risk of developing cryptorchidism themselves. The most prevalent genetic mutations associated with cryptorchidism are of either *INSL3* or its receptor *RXFP2* [83]. *INSL3-RXFP2* mutations account for 1%–4% of cryptorchidism cases [82] and result in bilateral abdominal cryptorchidism due to the role of INSL3 in gubernacular expansion during the transabdominal phase of testicular descent [21, 22, 24]. Other genes include *AHR* [84], *AR* [85], *ARNT2* [84], *AXIN1* [86], *CRKL* [87], *CYP17A1* [84], *ESR1* [88], *HOXA10* [89], *NR5A1* [90], *PDE4B* [91], *SPAG5* [92], *STRBP* [93], and *VAMP7* [80] (Table 1.1). Many of these genes were identified through SNP studies and have differential risk factors in men, depending on their country of origin [84]. It is unknown if this risk is due to mutations that have been enriched in specific populations or if these mutations are interacting with environmental factors in different countries, leading to the changes in different populations. *ARNT2* SNPS were specifically found to have an association with cryptorchidism in Japanese and Italian men. However, *CYP1A2* and *CYP17A1* SNPs were only found to be significantly higher in Japanese men and AHR SNPs were only found to be enriched in Italian men with cryptorchidism [84]. ARNT2 is a transcription factor that regulates several physiological pathways, including the response to environmental contaminants, and provides a possible gene candidate for a genetic-environmental interaction that is associated with cryptorchidism. Many of the other genes identified are important in the regulation and maintenance of normal steroid homeostasis, including *AR, CYP17A1, ESR1, NR5A1, PDE4B, STRBP,* and *VAMP7*.

Cryptorchidism is also associated with several other congenital anomalies, including hypospadias, and is often syndromic [19]. One syndrome in which cryptorchidism is common is Robinow syndrome, which arises from disruption of WNT5a, a noncanonical WNT that signals through the JNK pathway [94]. JNK modulates the migration of cells involved in spatial organization. Expression of WNT5a occurs downstream of INSL3 signaling in the gubernaculum [95]. Knockout mouse studies of Wnt5a show that loss of Wnt5a results in abdominal cryptorchidism due to a hypoplastic gubernaculum. Another syndrome that involves alteration of the JNK pathway and is associated with cryptorchidism is Aarskog Syndrome [96], an X-linked syndrome caused by faciogenital dysplasia protein 1 (*FGD1*) mutations. FGD1 is a guanine exchange factor that activates cell division control protein 42 (CDC42). CDC42 is involved in activation of JNK. Rubenstein Taybi syndrome, which

TABLE 1.1 Genes Associated With Anomalies of the External Genitalia

Gene	Locus	Congenital Anomaly	Species	Polymorphism type	Population	Reference
AHR	7p21.1	Cryptorchidism	Human	SNP	334 Japanese men (141 controls, 95 cryptorchidism, 98 hypospadias) 187 Italian men (129 controls, 58 cryptorchidism)	[84]
AR	Xq12	Inguinal cryptorchidism	Human	CAG and GGN repeat lengths		[85]
ARNT2	15q25.1	Cryptorchidism	Human	SNP	334 Japanese men (141 controls, 95 cryptorchidism, 98 hypospadias) 187 Italian men (129 controls, 58 cryptorchidism)	[84]
ATF3	1q32.3	Penoscrotal hypospadias	Mice		C57BL/6 mice treated with corn oil (control) or DEHP	[93a]
			Human	SNP	41 boys with hypospadias 30 controls	[93b]
			Human	SNP	330 boys with hypospadias 380 controls	[77]
Axin1	16p13.3	Cryptorchidism	Human	SNP	113 cryptorchidism cases (92 unilateral, 21 bilateral) 179 controls	[86]
CRKL	22q11.21	Cryptorchidism	Human	CNV	277 GU abnormal patients 6813 general population controls	[87]
			Mice	Gene deletion	$CRKL^{-/-}$ embryos $CRKL^{+/-}$ adults C56BL/6 (controls)	
CYP17A1	10q24.32	Cryptorchidism	Human	SNP	334 Japanese men (141 controls, 95 cryptorchidism, 98 hypospadias) 187 Italian men (129 controls, 58 cryptorchidism)	[84]
CYP1A2	15q24.1	Cryptorchidism	Human	SNP	334 Japanese men (141 controls, 95 cryptorchidism, 98 hypospadias) 187 Italian men (129 controls, 58 cryptorchidism)	[84]
DGKK	Xq11.22	Hypospadias	Human	SNP	166 boys with hypospadias 285 controls	[73]
ESR1	6q25.1	Cryptorchidism	Human	SNP and CA repeats	60 boys with hypospadias 90 controls	[93c]

Continued

I. REPRODUCTIVE TRACT DEVELOPMENT AND GAMETOGENESIS

TABLE 1.1 Genes Associated With Anomalies of the External Genitalia—cont'd

Gene	Locus	Congenital Anomaly	Species	Polymorphism type	Population	Reference
ESR2	14q23.2	Hypospadias	Human	SNP and CA repeats	60 boys with hypospadias 90 controls	[93c]
Gli2	2q14.2	Hypospadias	Mice	Gene deletion	Gli2$^{+/-}$: Gli3$^{\Delta699/+}$ mice on CD1 black background	[72]
Gli3	7p14.1	Hypospadias	Mice	Gene deletion	Gli2$^{+/-}$: Gli3$^{\Delta699/+}$ mice on CD1 black background	[72]
HOXA13	7p15.2	Hypospadias	Mice	Gene deletion	Hoxa13$^{-/-}$ mouse embryos	[43]
HOXA10	7p15.2	Cryptorchidism	Mice	Gene deletion	Hoxa10$^{-/-}$ mice	[89]
INSL3	19p13.11	Abdominal cryptorchidism	Mice	Gene deletion	Insl3$^{-/-}$ mice	[21, 22]
MAMLD1	Xq28	Hypospadias	Human	SNP	3 different SNPs with abnormal genitalia 200 controls	[78]
NR5A1	9q33.3	Penoscrotal hypospadias Cryptorchidism	Human	SNP	60 individuals with hypospadias 20 also had at least one undescended testes 100 controls	[69]
PDE4B	1p31.3	Cryptorchidism	Human	mRNA expression levels	5 patients with cryptorchidism and Down's syndrome 6 patients with Down's syndrome without cryptorchidism 5 controls	[91]
RXFP2	13q13.1	Abdominal cryptorchidism	Mice	Transgene mutation	RXFP2 (GREAT) transgenic mice	[24]
SPAG5	17q11.2	Cryptorchidism	Human	mRNA expression levels	5 men with Down's syndrome and cryptorchidism 5 controls	[91]
SHH	7q36.3	Agenesis of external genitalia	Mice	Deletion	Shh$^{-/-}$ mouse embryos	[38]
SRD5A2	2p23.1	Hypospadias	Human	SNP	109 patients with hypospadias	[74]
STRBP	9q33.3	Cryptorchidism	Human	mRNA expression levels	5 men with Down's syndrome and cryptorchidism 5 controls	[92]
VAMP7	Xq28	Hypospadias Cryptorchidism	Human	CNV	116 GU abnormal patients 8951 control patients	[80]
			Mice	Transgenic overexpression	Mice overexpressing VAMP7	

TABLE 1.1 Genes Associated With Anomalies of the External Genitalia—cont'd

Gene	Locus	Congenital Anomaly	Species	Polymorphism type	Population	Reference
WNT5A	3p14.3	Cryptorchidism	Human	Missense mutation	Family with Robinow syndrome	[94]
		Genital hypoplasia	Mice	Gene deletion	Wnt5a$^{-/-}$ mice Ror2$^{-/-}$ mice (Wnt5a receptor)	
WT1	11p13	Penoscrotal hypospadias Micropenis	Human	SNPs	90 Chinese hypospadias patients 276 Chinese controls	[28]

is caused by mutations of the cyclic AMP response element binding protein (*CREBP*) on chromosome 9, is also associated with cryptorchidism [97]. CREBP is a transcription factor that is activated downstream on G protein coupled receptors in response to an increase in intracellular cAMP, and is also a coactivator of AR. Noonan Syndrome is also highly associated with cryptorchidism. Noonan syndrome is caused by mutations in the Ras/MAPK pathway [98]. The most common mutations that lead to Noonan syndrome are in *PTPN11* [98], a known activator of HOXA10 that is expressed in the caudal end of the gubernaculum. Loss of *Hoxa10* in mice leads to cryptorchidism [89].

CONCLUSION

In this chapter, we explored the genetic drivers of male reproductive development. We have learned the key factors in reproductive development by studying genetic anomalies that lead to genital birth defects in humans as well as rodents. There are a vast amount of genes that regulate reproductive development. Many cases of cryptorchidism and hypospadias occur due to genetic changes in genes that have no prior known role in genital development. Furthermore, many of the genes can interact with endocrine and environmental signals in mechanisms that remain heretofore elusive. The interactions between environment and genes involved in sexual development is an area that is quickly gaining interest but still requires further study.

References

[1] Anderson R, Copeland TK, Scholer H, Heasman J, Wylie C. The onset of germ cell migration in the mouse embryo. Mech Dev 2000;91:61–8.
[2] Ginsburg M, Snow MHL, Mclaren A. Primordial germ cells in the mouse embryo during gastrulation. Development 1990;110:521–8.
[3] Hara K, Kanai-azuma M, Uemura M, Shitara H, Taya C, Yonekawa H, et al. Evidence for crucial role of hindgut expansion in directing proper migration of primordial germ cells in mouse early embryogenesis. Dev Biol 2009;330:427–39. https://dx.doi.org/10.1016/j.ydbio.2009.04.012.
[4] Molyneaux KA, Stallock J, Schaible K, Wylie C. Time-lapse analysis of living mouse germ cell migration. Dev Biol 2001;240:488–98. https://dx.doi.org/10.1006/dbio.2001.0436.

[5] Luo X, Ikeda Y, Parker KL. A cell-specific nuclear receptor is essential for adrenal and gonadal development and sexual differentiation. Cell 1994;77:481–90.

[6] Birk OS, Casiano DE, Wassif CA, Huang S, Kreidberg JA, Parker KL, et al. The LIM homeobox gene Lhx9 is essential for mouse gonad formation. Nature 2000;403.

[7] Kreidberg JA, Sarioia H, Loring JM, Maeda Y, Peiietier J, Housman D, et al. WT-1 is required for early kidney development. Cell 1993;74:679–91.

[8] Katoh-fukui Y, Tsuchiya R, Shiroishi T, Nakahara Y, Hashimoto N, Noguchi K, et al. Male-to-female sex reversal in M33 mutant mice. Nature 1998;393:688–92.

[9] Nef S, Verma-kurvari S, Merenmies J, Vassalli J-D, Efstratiadis A, Accili D, et al. Testis determination requires insulin receptor family function in mice. Nature 2003;426:291–5.

[10] Gubbay J, Collington J, Koopman P, Capel B, Economou A, Munsterberg A, et al. A gene mapping to the sex-determining region of the mouse Y chromosome is a member of a novel family of embryonically expressed genes. Nature 1990;346:245–50.

[11] Sinclair A, Berta P, Palmer M, Hawkins JR, Griffiths B, Smith M, et al. A gene from the human sex-determining region encodes a protein with homology to a conserved DNA-binding motif. Nature 1990;346:240–4.

[12] Hacker A, Capel B, Goodfellow P, Lovell-Badge R. Expression of Sry, the mouse sex determining gene. Development 1995;121:1603–14.

[13] Barbara PDES, Bonneaud N, Boizet B, Desclozeaux M, Moniot B, Sudbeck P, et al. Direct interaction of SRY-related protein SOX9 and steroidogenic factor 1 regulates transcription of the human anti-mullerian hormone gene. Mol Cell Biol 1998;18:6653–65.

[14] Vidal VP, Chaboissier M-C, de Rooij DG, Schedl A. Sox9 induces testis development in XX transgenic mice. Nat Genet 2001;28:216–7.

[15] Bitgood MJ, Shen L, Mcmahon AP. Sertoli cell signaling by desert hedgehog regulates the male germline. Curr Biol 1996;6:298–304.

[16] Yao HH, Whoriskey W, Capel B. Desert hedgehog/patched 1 signaling specifies fetal Leydig cell fate in testis organogenesis. Genes Dev 2002;16:1433–40. https://dx.doi.org/10.1101/gad.981202.interstitium.

[17] Bay K, Main KM, Toppari J, Skakkebæk NE. Testicular descent: INSL3, testosterone, genes and the intrauterine milieu. Nat Rev Urol 2011;8:187–96. https://dx.doi.org/10.1038/nrurol.2011.23.

[18] Heyns CF, de Klerk DP. The gubernaculum during testicular descent in the pig fetus. J Urol 1985;133:694–9.

[19] Hutson JM, Southwell BR, Li R, Lie G, Ismail K, Harisis G, et al. The regulation of testicular descent and the effects of cryptorchidism. Endocr Rev 2013;34:725–52. https://dx.doi.org/10.1210/er.2012-1089.

[20] Ivell R, Balvers M, Pohnke Y, Telgmann R, Bartsch O, Milde-Langosch K, et al. Immunoexpression of the relaxin receptor LGR7 in breast and uterine tissues of humans and primates. Reprod Biol Endocrinol 2003;1:114. https://dx.doi.org/10.1186/1477-7827-1-114.

[21] Zimmermann S, Steding G, Emmen JMA, Brinkmann AO, Nayernia K, Holstein AF, et al. Targeted disruption of the Insl3 gene causes bilateral cryptorchidism. Mol Endocrinol 1999;13:681–95.

[22] Nef S, Parada LF. Cryptorchidism in mice mutant for Insl3. Nat Genet 1999;22:295–9.

[23] Kaftanovskaya EM, Feng S, Huang Z, Tan Y, Barbara AM, Kaur S, Truong A, Gorlov IP, Agoulnik AI. Suppression of insulin-like3 receptor reveals the role of β-catenin and Notch signaling in gubernaculum development, Mol Endocrinol 2011;25(1):170–83. http://doi.org/10.1210/me.2010-0330.

[24] Gorlov IP, Kamat A, Bogatcheva NV, Jones E, Lamb DJ, Truong A, et al. Mutations of the GREAT gene cause cryptorchidism. Hum Mol Genet 2002;11:2309–18.

[25] Chen N, Harisis GN, Farmer P, Buraundi S, Sourial M, Southwell DR, et al. Gone with the Wnt: the canonical Wnt signaling axis is present and androgen dependent in the rodent gubernaculum. J Pediatr Surg 2011;46:2363–9. https://dx.doi.org/10.1016/J.JPEDSURG.2011.09.032.

[26] Laguë É, Tremblay JJ. Antagonistic effects of testosterone and the endocrine disruptor mono-(2-Ethylhexyl) phthalate on INSL3 transcription in Leydig cells. Endocrinology 2008;149:4688–94. https://dx.doi.org/10.1210/en.2008-0310.

[27] Mariani FV, Martin GR. Deciphering skeletal patterning: clues from the limb. Nature 2003;423:319–25. https://dx.doi.org/10.1038/nature01655.

[28] Carmichael SL, Ma C, Choudhry S, Lammer EJ, Witte JS, Shaw GM. Hypospadias and genes related to genital tubercle and early urethral development. J Urol 2013;190:1884–92. https://dx.doi.org/10.1016/j.juro.2013.05.061.

[29] Na AF, Harnaen EJ, Farmer PJ, Sourial M, Southwell BR, Hutson JM. Cell membrane and mitotic markers show that the neonatal rat gubernaculum grows in a similar way to an embryonic limb bud. J Pediatr Surg 2007;42:1566–73. https://dx.doi.org/10.1016/J.JPEDSURG.2007.04.035.

[30] Ivell R, Hartung S. The molecular basis of cryptorchidism. Mol Hum Reprod 2003;9:175–81.

[31] Tomiyama H, Hutson JM, Truong A, Agoulnik AI. Transabdominal testicular descent is disrupted in mice with deletion of insulinlike factor 3 receptor. J Pediatr Surg 2003;38:1793–8. https://dx.doi.org/10.1016/J.JPEDSURG.2003.08.047.

[32] Welsh M, Saunders PTK, Fisken M, Scott HM, Hutchison GR, Smith LB, et al. Identification in rats of a programming window for reproductive tract masculinization, disruption of which leads to hypospadias and cryptorchidism. J Clin Invest 2008;118:1479–90. https://dx.doi.org/10.1172/JCI34241.

[33] Quigley CA, De Bellis A, Marschke KB, El-Awady MK, Wilson EM, French FS. Androgen receptor defects: Historical, clinical, and molecular perspectives. Endocr Rev 1995;16:271–321. https://dx.doi.org/10.1210/edrv-16-3-271.

[34] Tomiyama H, Hutson JM. Contractility of rat gubernacula affected by calcitonin gene-related peptide and β-agonist. J Pediatr Surg 2005;40:683–7. https://dx.doi.org/10.1016/J.JPEDSURG.2004.12.006.

[35] Yong EXZ, Huynh J, Farmer P, Ong SY, Sourial M, Donath S, et al. Calcitonin gene–related peptide stimulates mitosis in the tip of the rat gubernaculum in vitro and provides the chemotactic signals to control gubernacular migration during testicular descent. J Pediatr Surg 2008;43:1533–9. https://dx.doi.org/10.1016/J.JPEDSURG.2007.11.037.

[36] Cohn MJ. Development of the external genitalia: Conserved and divergent mechanisms of appendage patterning. Dev Dyn 2011;240:1108–15. https://dx.doi.org/10.1002/dvdy.22631.

[37] Seifert AW, Bouldin CM, Choi K-S, Harfe BD, Cohn MJ. Multiphasic and tissue-specific roles of sonic hedgehog in cloacal septation and external genitalia development. Development 2009;136:3949–57. https://dx.doi.org/10.1242/dev.042291.

[38] Perriton CL, Powles N, Chiang C, Maconochie MK, Cohn MJ. Sonic hedgehog signaling from the urethral epithelium controls external genital development. Dev Biol 2002;247:26–46. https://dx.doi.org/10.1006/dbio.2002.0668.

[39] Murakami R, Mizuno T. Proximal-distal sequence of development of the skeletal tissues in the penis of rat and the inductive effect of epithelium. J Embryol Exp Morpholog 1986;92:133–43.

[39a] Seifert AW, Harfe BD, Cohn MJ. Cell lineage analysis demonstrates an endodermal origin of the distal urethra and perineum. Dev Biol 2008;318:143–52. https://dx.doi.org/10.1016/j.ydbio.2008.03.017.

[40] Hynes PJ, Fraher JP. The development of the male genitourinary systems: III. The formation of the spongiose and glandar urethra. Br J Plast Surg 2004;57:203–14. https://dx.doi.org/10.1016/j.bjps.2003.08.017.

[41] Yang JH, Menshenina J, Cunha GR, Place N, Baskin LS. Morphology of mouse external genitalia: implications for a role of estrogen in sexual dimorphism of the mouse genital tubercle. J Urol 2010;184:1604–9. https://dx.doi.org/10.1016/j.juro.2010.03.079.

[42] Spitz F, Gonzalez F, Duboule D. A global control region defines a chromosomal regulatory landscape containing the HoxD cluster. Cell 2003;113:405–17. https://dx.doi.org/10.1016/S0092-8674(03)00310-6.

[43] Morgan EA, Nguyen SB, Scott V, Stadler HS. Loss of Bmp7 and Fgf8 signaling in Hoxa13-mutant mice causes hypospadia. Development 2003;130:3095–109. https://dx.doi.org/10.1242/dev.00530.

[44] Dollé P, Izpisúa-Belmonte JC, Brown JM, Tickle C, Duboule D. Hox-4 genes and the morphogenesis of mammalian genitalia. Genes Dev 1991;5:1767–76. https://dx.doi.org/10.1101/gad.5.10.1767.

[45] Warot X, Fromental-Ramain C, Fraulob V, Chambon P, Dollé P. Gene dosage-dependent effects of the Hoxa-13 and Hoxd-13 mutations on morphogenesis of the terminal parts of the digestive and urogenital tracts. Development 1997;124:4781–91.

[46] Kondo T, Zakany J, Innis JW, Duboule D. Of fingers, toes and penises. Nature 1997;390:29. https://dx.doi.org/10.1038/36234.

[47] Agarwal P. Tbx5 is essential for forelimb bud initiation following patterning of the limb field in the mouse embryo. Development 2003;130:623–33. https://dx.doi.org/10.1242/dev.00191.

[48] Chapman DL, Garvey N, Hancock S, Alexiou M, Agulnik SI, Gibson-Brown JJ, et al. Expression of the T-box family genes, Tbx1-Tbx5, during early mouse development. Dev Dyn 1996;206:379–90. https://dx.doi.org/10.1002/(SICI)1097-0177(199608)206:4<379::AID-AJA4>3.0.CO;2-F.

[49] Petiot A. Development of the mammalian urethra is controlled by Fgfr2-IIIb. Development 2005;132:2441–50. https://dx.doi.org/10.1242/dev.01778.

[50] Haraguchi R, Mo R, Hui C, Motoyama J, Makino S, Shiroishi T, et al. Unique functions of sonic hedgehog signaling during external genitalia development. Development 2001;128:4241–50.

[51] Harada M, Omori A, Nakahara C, Nakagata N, Akita K, Yamada G. Tissue-specific roles of FGF signaling in external genitalia development. Dev Dyn 2015;https://dx.doi.org/10.1002/dvdy.24277.

[52] Gredler ML, Seifert AW, Cohn MJ. Tissue-specific roles of Fgfr2 in development of the external genitalia. Development 2015;142:2203–12. https://dx.doi.org/10.1242/dev.119891.

[53] Lin C, Yin Y, Veith GM, Fisher AV, Long F, Ma L. Temporal and spatial dissection of Shh signaling in genital tubercle development. Development 2009;136:3959–67. https://dx.doi.org/10.1242/dev.039768.

[54] Zhou CJ, Wang YZ, Yamagami T, Zhao T, Song L, Wang K. Generation of Lrp6 conditional gene-targeting mouse line for modeling and dissecting multiple birth defects/congenital anomalies. Dev Dyn 2010;239:318–26. https://dx.doi.org/10.1002/dvdy.22054.

[55] Lieven O, Knobloch J, Rüther U. The regulation of Dkk1 expression during embryonic development. Dev Biol 2010;340:256–68. https://dx.doi.org/10.1016/j.ydbio.2010.01.037.

[56] Yamaguchi TP, Bradley A, McMahon AP, Jones S. A Wnt5a pathway underlies outgrowth of multiple structures in the vertebrate embryo. Development 1999;126:1211–23. https://dx.doi.org/10.1242/dev.00463.

[57] Schwabe GC, Trepczik B, Süring K, Brieske N, Tucker AS, Sharpe PT, et al. Ror2 knockout mouse as a model for the developmental pathology of autosomal recessive Robinow syndrome. Dev Dyn 2004;229:400–10. https://dx.doi.org/10.1002/dvdy.10466.

[58] Lin C, Yin Y, Long F, Ma L. Tissue-specific requirements of beta-catenin in external genitalia development. Development 2008;135:2815–25. https://dx.doi.org/10.1242/dev.020586.

[59] Kim KS, Liu W, Cunha GR, Russell DW, Huang H, Shapiro E, et al. Expression of the androgen receptor and 5 alpha-reductase type 2 in the developing human fetal penis and urethra. Cell Tissue Res 2002;307:145–53. https://dx.doi.org/10.1007/s004410100464.

[60] Sajjad Y, Quenby S, Nickson P, Lewis-Jones DI, Vince G. Immunohistochemical localization of androgen receptors in the urogenital tracts of human embryos. Reproduction 2004;128:331–9. https://dx.doi.org/10.1530/rep.1.00227.

[61] Sinnecker GH, Hiort O, Dibbelt L, Albers N, Dörr HG, Hauss H, et al. Phenotypic classification of male pseudohermaphroditism due to steroid 5 alpha-reductase 2 deficiency. Am J Med Genet 1996;63:223. https://dx.doi.org/10.1002/(SICI)1096-8628(19960503)63:1<223::AID-AJMG39>3.0.CO;2-O.

[62] Lorenzo AJ, Nguyen MT, Sozubir S, Henkemeyer M, Kropp B, Baker LA. Dihydrotestesterone induction of EphB2 expression in the female genital tubercle mimics male pattern of expression during embryogenesis. J Urol 2003;170:1618–23. https://dx.doi.org/10.1097/01.ju.0000087423.89813.64.

[63] Shechter-Maor G, Czuzoj-Shulman N, Spence AR, Abenhaim HA. The effect of assisted reproductive technology on the incidence of birth defects among livebirths. Arch Gynecol Obstet 2018. https://dx.doi.org/10.1007/s00404-018-4694-8.

[64] Stein DM, Gonzalez CM, Barbagli G, Cimino S, Madonia M, Sansalone S. Erectile function in men with failed hypospadias repair. Arch Esp Urol 2014;67:152–6.

[65] Aulagne MB, Harper L, de Napoli-Cocci S, Bondonny JM, Dobremez E. Long-term outcome of severe hypospadias. J Pediatr Urol 2010;6:469–72. https://dx.doi.org/10.1016/j.jpurol.2009.12.005.

[66] Visser R, Burger NCM, Van Zwet EW, Hilhorst-Hofstee Y, Haak MC, Van Den Hoek J, et al. Higher incidence of hypospadias in monochorionic twins. Twin Res Hum Genet 2015;18:591–4. https://dx.doi.org/10.1017/thg.2015.55.

[67] Jensen M Søndergaard, Wilcox AJ, Olsen J, Bonde JP, Thulstrup AM, Høst Ramlau-Hansen C, et al. Cryptorchidism and hypospadias in a cohort of 934,538 Danish boys: the role of birth weight, gestational age, body dimensions, and fetal growth. Am J Epidemiol 2012;175. https://dx.doi.org/10.1093/aje/kwr421.

[68] Peycelon M, Mansour-Hendili L, Hyon C, Collot N, Houang M, Legendre M, et al. Recurrent intragenic duplication within the NR5A1 gene and severe proximal hypospadias. Sex Dev 2018;11:293–7. https://dx.doi.org/10.1159/000485909.

[69] Köhler B, Lin L, Mazen I, Cetindag C, Biebermann H, Akkurt I, et al. The spectrum of phenotypes associated with mutations in steroidogenic factor 1 (SF-1, NR5A1, Ad4BP) includes severe penoscrotal hypospadias in 46,XY males without adrenal insufficiency. Eur J Endocrinol 2009;161:237–42. https://dx.doi.org/10.1530/EJE-09-0067.

[70] Beleza-Meireles A, Lundberg F, Lagerstedt K, Zhou X, Omrani D, Frisén L, et al. FGFR2, FGF8, FGF10 and BMP7 as candidate genes for hypospadias. Eur J Hum Genet 2007;15:405–10. https://dx.doi.org/10.1038/sj.ejhg.5201777.

[71] Petiot A, Perriton CL, Dickson C, Cohn MJ. Development of the mammalian urethra is controlled by Fgfr2-IIIb. Development 2005;132:2441–50. https://dx.doi.org/10.1242/dev.01778.

[72] He F, Akbari P, Mo R, Zhang JJ, Hui C-C, Kim PC, et al. Adult Gli2+/−;Gli3Δ699/+ male and female mice display a spectrum of genital malformation. PLoS One 2016;11:e0165958. https://dx.doi.org/10.1371/journal.pone.0165958.

[73] Hozyasz KK, Mostowska A, Kowal A, Mydlak D, Tsibulski A, Jagodzi?ski PP. Further evidence of the Association of the Diacylglycerol Kinase Kappa (DGKK) gene with hypospadias. Urol J 2018. https://dx.doi.org/10.22037/UJ.V0I0.4061.

[74] Rahimi M, Ghanbari M, Fazeli Z, Rouzrokh M, Omrani S, Mirfakhraie R, et al. Association of SRD5A2 gene mutations with risk of hypospadias in the Iranian population. J Endocrinol Investig 2017;40:391–6. https://dx.doi.org/10.1007/s40618-016-0573-y.

[75] Acar S, Tuhan H, Bora E, Demir K, Onay H, Erçal D, et al. Identification of an AR mutation in klinefelter syndrome during evaluation for penoscrotal hypospadias. Hormones 2017;16:313–7. https://dx.doi.org/10.14310/horm.2002.1741.

[76] Deng C, Dai R, Li X, Liu F. Association between SNP12 in estrogen receptor α gene and hypospadias: a systematic review and meta-analysis. Springerplus 2016;5:587. https://dx.doi.org/10.1186/s40064-016-2288-0.

[77] Beleza-Meireles A, Töhönen V, Söderhäll C, Schwentner C, Radmayr C, Kockum I, et al. Activating transcription factor 3: a hormone responsive gene in the etiology of hypospadias. Eur J Endocrinol 2008;158:729–39. https://dx.doi.org/10.1530/EJE-07-0793.

[78] Fukami M, Wada Y, Miyabayashi K, Nishino I, Hasegawa T, Nordenskjöld A, et al. CXorf6 is a causative gene for hypospadias. Nat Genet 2006;38:1369–71. https://dx.doi.org/10.1038/ng1900.

[79] Van der Zanden LFM, Van Rooij IALM, Feitz WFJ, Franke B, Knoers NV, Roeleveld N. Aetiology of hypospadias: a systematic review of genes and environment. Hum Reprod Update 2012;18:260–83. https://dx.doi.org/10.1093/humupd/dms002.

[80] Tannour-Louet M, Han S, Louet J-F, Zhang B, Romero K, Addai J, et al. Increased gene copy number of VAMP7 disrupts human male urogenital development through altered estrogen action. Nat Med 2014;20:715–24. https://dx.doi.org/10.1038/nm.3580.

[81] Thankamony A, Pasterski V, Ong KK, Acerini CL, Hughes IA. Anogenital distance as a marker of androgen exposure in humans. Andrology 2016;4:616–25. https://dx.doi.org/10.1111/andr.12156.

[82] Gurney JK, McGlynn KA, Stanley J, Merriman T, Signal V, Shaw C, et al. Risk factors for cryptorchidism. Nat Rev Urol 2017;14:534–48. https://dx.doi.org/10.1038/nrurol.2017.90.

[83] Bogatcheva NV, Agoulnik AI. INSL3/LGR8 role in testicular descent and cryptorchidism. Reprod BioMed Online 2005;10:49–54.

[84] Qin X-Y, Kojima Y, Mizuno K, Ueoka K, Massart F, Spinelli C, et al. Association of variants in genes involved in environmental chemical metabolism and risk of cryptorchidism and hypospadias. J Hum Genet 2012;57:434–41. https://dx.doi.org/10.1038/jhg.2012.48.

[85] Aschim EL, Nordenskjöld A, Giwercman A, Lundin KB, Ruhayel Y, Haugen TB, et al. Linkage between cryptorchidism, hypospadias, and GGN repeat length in the androgen receptor gene. J Clin Endocrinol Metab 2004;89:5105–9. https://dx.doi.org/10.1210/jc.2004-0293.

[86] Zhou B, Tang T, Chen P, Pu Y, Ma M, Zhang D, et al. The variations in the AXIN1 gene and susceptibility to cryptorchidism, J Pediatr Urol 2015;11: 132.e1–5. https://doi.org/10.1016/j.jpurol.2015.02.007.

[87] Haller M, Mo Q, Imamoto A, Lamb DJ. Murine model indicates 22q11.2 signaling adaptorCRKLis a dosage-sensitive regulator of genitourinary development. Proc Natl Acad Sci U S A 2017;114:4981–6. https://dx.doi.org/10.1073/pnas.1619523114.

[88] Yoshida R, Fukami M, Sasagawa I, Hasegawa T, Kamatani N, Ogata T. Association of cryptorchidism with a specific haplotype of the estrogen receptor α gene: implication for the susceptibility to estrogenic environmental endocrine disruptors. J Clin Endocrinol Metab 2005;90:4716–21. https://dx.doi.org/10.1210/jc.2005-0211.

[89] Kolon TF, Wiener JS, Lewitton M, Roth DR, Gonzales ET, Lamb DJ. Analysis of homeobox gene HOXA10 mutations in cryptorchidism. J Urol 1999;161:275–80. https://dx.doi.org/10.1016/S0022-5347(01)62132-3.

[90] Wada Y, Okada M, Fukami M, Sasagawa I, Ogata T. Association of cryptorchidism with Gly146Ala polymorphism in the gene for steroidogenic factor-1. Fertil Steril 2006;85:787–90. https://dx.doi.org/10.1016/J.FERTNSTERT.2005.09.016.

[91] Salemi M, Condorelli RA, La Vignera S, Castiglione R, Salluzzo MG, Bonaccorso CM, et al. Expression of phosphodiesterase 4B cAMP-specific gene in subjects with cryptorchidism and Down's syndrome. J Clin Lab Anal 2016;30:196–9. https://dx.doi.org/10.1002/jcla.21835.

[92] Salemi M, Longo GA, La Vignera S, Romano C, Condorelli RA, Romano C, et al. SPAG5 mRNA is overexpressed in peripheral blood leukocytes of patients with Down's syndrome and cryptorchidism. Neurol Sci 2013;34:549–51. https://dx.doi.org/10.1007/s10072-012-1152-4.

[93] Salemi M, La Vignera S, Castiglione R, Condorelli RA, Cimino L, Bosco P, et al. Expression of STRBP mRNA in patients with cryptorchidism and Down's syndrome. J Endocrinol Investig 2012;35:5–7. https://dx.doi.org/10.1007/BF03345414.

[93a] Liu X, Zhang D-Y, Li Y-S, Xiong J, He D-W, Lin T, et al. Di-(2-ethylhexyl) phthalate upregulates ATF3 expression and suppresses apoptosis in mouse genital tubercle. J Occup Health 2009;51:57–63. https://dx.doi.org/10.1539/joh.L8091.

[93b] Kalfa N, Liu B, Klein O, Wang M-H, Cao M, Baskin LS. Genomic variants of ATF3 in patients with hypospadias. J Urol 2008;180:2183–8. https://dx.doi.org/10.1016/J.JURO.2008.07.066.

[93c] Beleza-Meireles A, Omrani D, Kockum I, Frisen L, Lagerstedt K, Nordenskjold A. Polymorphisms of estrogen receptor beta gene are associated with hypospadias. J Endocrinol Invest 2006;29(1):5–10.

[94] Person AD, Beiraghi S, Sieben CM, Hermanson S, Neumann AN, Robu ME, et al. WNT5A mutations in patients with autosomal dominant Robinow syndrome. Dev Dyn 2009;239. https://dx.doi.org/10.1002/dvdy.22156.

[95] Johnson KJ, Robbins AK, Wang Y, McCahan SM, Chacko JK, Barthold JS. Insulin-like 3 exposure of the fetal rat gubernaculum modulates expression of genes involved in neural Pathways1. Biol Reprod 2010;83:774–82. https://dx.doi.org/10.1095/biolreprod.110.085175.

[96] Porteous ME, Goudie DR. Aarskog syndrome. J Med Genet 1991;28:44–7. https://dx.doi.org/10.1136/JMG.28.1.44.

[97] Petrif F, Giles RH, Dauwerse HG, Saris JJ, Hennekam RCM, Masuno M, et al. Rubinstein-Taybi syndrome caused by mutations in the transcriptional co-activator CBP. Nature 1995;376:348–51. https://dx.doi.org/10.1038/376348a0.

[98] Tartaglia M, Gelb BD. Noonan syndrome and related disorders: genetics and pathogenesis. Annu Rev Genomics Hum Genet 2005;6:45–68. https://dx.doi.org/10.1146/annurev.genom.6.080604.162305.

Glossary

Ambiguous genitalia	A rare condition in which one's external genitalia cannot be clearly distinguished as either male or female.
Apical ectodermal ridge (AER)	A structure that forms from ectodermal cells at the distal end of the limb bud, acting as a signaling center during limb development.
Calcitonin gene related peptide (CGRP)	A member of the calcitonin family of peptides that is produced by both peripheral and central neurons. CGRP is a vasodilator that is involved in various physiological processes, including androgen-mediated testicular descent.
Cranial suspensory ligament (CSL)	A fibromuscular structure that anchors the gonads to the posterior abdominal wall in mammals. The CSL is present in both males and females.
Cryptorchidism	A congenital anomaly in which the testis fail to descend into the scrotum prior to birth.
Epididymis	A convoluted duct through which sperm exits the testis.
Genital tubercle	A conical protuberance on the lower abdominal wall of an embryo. The genital tubercle develops into the phallus.
Gubernaculum	A structure that serves as a guide during testicular descent. The gubernaculum connects the testis to the scrotum.
Human chorionic gonadotropin (hCG)	A hormone produced by placenta that functions as a gonadotrope during early embryonic development.
Hypospadias	A congenital condition in which the urethral opening is misplaced on the ventral side of the penis.
Inguinal ring	An opening at either end of the abdominal muscle that provides passage to the inguinal canal.

INSL3	Insulin like 3 is a protein in humans that mediates the transabdominal phase of testicular descent.
Leydig cells	A type of cell found in the interstitium of the testes, adjacent to seminiferous tubules. A distinguishing characteristic of Leydig cells is their ability to produce testosterone in the presence of luteinizing hormone.
Luteinizing hormone	A hormone secreted by the pituitary gland that stimulates synthesis of androgen in males and ovulation in females.
Mullerian duct	An embryonic duct that gives rise to the fallopian tubes, uterus, cervix, and upper portion of the vagina in females. The Mullerian duct parallels the Wolffian duct present in males.
Primordial germ cell	Large spherical diploid cells that are formed in the early stages of embryonic development and are the precursors of oogonia and spermatogonia.
Processus vaginalis	A pouch of peritoneum that is carried into the scrotum during testicular descent.
Progress zone	A layer of mesoderm beneath the AER in the developing limb that is involved in limb patterning.
Sertoli cell	A type of somatic cell that is present in the tubules of the testis. Sertoli cells are the nurse cells of the testes. During spermatogenesis, spermatids become attached to the Sertoli cells for nourishment.
Sox9	A gene on chromosome 17q23 that encodes a member of the SRY-related HMG box transcription factors. Sox9 is an important factor for male development. Sox9 regulates transcription of the anti-Mullerian hormone gene.
SRY	Sex determining region Y is a gene located on the Y chromosome that encodes a DNA binding protein necessary for the initiation of male sex determination.
Urogenital ridge	The precursor to the gonads. It consists mainly of cells of mesonephric origin.
Wolffian duct	An embryonic duct that gives rise to the testis, epididymis, vas deferens, seminal vesicle, and ejaculatory duct. The Wolffian duct parallels the Mullerian duct present in females.

Genetics and Genomics of Early Gonad Development

Kai K. Miu, Dan D. Cao, Gang Lu, Wai Y. Chan

School of Biomedical Sciences, Faculty of Medicine, The Chinese University of Hong Kong, Shatin, Hong Kong

INTRODUCTION

The primary function of the gonad is to provide an anatomical niche to nourish and promote the growth of the germ cells. During the initial stages of development, male and female gonads remain indifferent and the presence of hormone signals later in development dictate the sexual identity. Gonad development mainly involves both residency of primordial germ cells and defining the sexual identity for reproductive tract development. On the course of gonad development, primordial germ cells (PGCs) migrate via the hindgut and mesentery to colonize their anatomical niches with active mitotic proliferation in the male while entering the meiotic cell cycle in the female to produce corresponding gametes. It is now clear that both the sexual identity and site of migration for PGCs were not predefined by the indifferent

Human Reproductive and Prenatal Genetics
https://doi.org/10.1016/B978-0-12-813570-9.00002-4

gonad. Indeed, PGCs respond to chemical cues and migrate to nearby regions irrespective of whether the developing gonad is present [1].

Unsurprisingly, the possession of the Y chromosome, most importantly the presence of the sex-determinant SRY gene on the chromosome in the embryo, will carry sufficient information to govern the male-specific differentiation program in the early bipotent gonad [2, 3]. In fact, gonad formation staged the irreversible path to male- or female-specific differentiation of the reproductive tract. Hormonal instructions such as steroid hormones and anti-Müllerian hormones (AMH) present in the male lead to the forming of the male reproductive tract that signals for Müllerian duct regression, a structure that the female reproductive tract will differentiate from without the hormones [4]. PGCs await their cues from the formed organ to be driven into either meiotic or mitotic processes. In fact, the default female anatomy will allow PGCs to become oogonia in the cortex upon meiotic arrest [5]. In the male, PGCs colonize the newly formed seminiferous cords of the medulla and repeatedly undergo mitotic divisions in becoming resident spermatogonia.

Past research indicates that the mammalian adrenal cortex and gonad are derived from the same primordium present during early urogenital development [6–8]. They were similar in a sense that they both represent major sites for steroidogenesis in the organism. Our chapter focuses on the genetics governing early development of the indifferent gonads by referencing steroidogenesis in both organs and the preparations required in the structure before staging in sexual divergence.

Gonad development relies on a series of transcriptional machinery while inadequacies will often result in infertility or sex reversal. The key transcription factors underlying gonad development mainly fall into one of these functions: (1) establishment of the steroidogenic adrenogonadal primordium (AGP); (2) partition/differentiation of the genital ridge from AGP; (3) male-specific differentiation to diverge from the default female program; and (4) production of cues to nurture PGCs. It is not surprising that transcription factors controlling development of the indifferent gonad are evolutionarily conserved while many of these genes shared unified function in the sexually defined gonads of both sexes, despite the fact that their expression changes are differential [9–11]. Indeed, such an arrangement can also explain why the expressions of these transcription factors are with high spatiotemporal precision in the gonads. Nevertheless, appropriate gene dosage is the key to prevent sexual reversal. With the longstanding argument for SRY as the unique sex determinant, knowing how these transcription factors affect SRY expression during gonad development may provide clues to the spatiotemporal requirements that dictate the male fate.

Our group produced a comprehensive serial analysis of gene expression (SAGE) profiling of gonad development, that is, GonadSAGE, back in 2009 to demonstrate the temporal patterns of the male embryonic gonad gene expression at different developmental time points [12]. It was the first public interactive database that contains transcriptome information on male embryonic gonad development. The output of SAGE is an unbiased, count-based list of short sequence tags ideal for statistical comparison. It was a powerful database with the potential to identify novel genes because the transcript sequences do not need to be known a priori. Also, it is handy to perform comparisons in the relative expression level of gene of interest at specific developmental time points. In fact, the database aided in identifying several novel genes related to sex determination, meiosis, and steroidogenesis. Furthermore, some of these observations had been correlated with developmental defects in animal models, as reported by previous publications. A total of six male mouse embryonic gonad stages were included (E10.5, E11.5, E12.5, E13.5, E15.5, and E17.5) in the database. The total tag counts were comparable among the sequencing samples while gene coverage was also balanced along the serial analysis. The expression profile from E10.5 to E12.5 was selected in this chapter as representative time references for early gonad development.

Within this time window, *Sry* is transiently overexpressed in the gonads for the preparation of male sex determination [2, 13]. In male gonads, *Sry* expression promotes the differentiation of the gonad precursors to Sertoli cells, mainly via fibroblast growth factor 9 (FGF9) signaling contributing to the down-regulation of Wnt signaling. At this stage, bipotency in the gonad ridge appears to cease in E11.5, coinciding with the time of appearance in the transcription factor SOX9, the male-specific differentiation factor. Hence, phenotypic gender can be reversible upon defects occurring before this time point. We had identified the essential network governing the development of the reproductive system in each of the three time-points and studied the transcription factors that are related to gonad development. In an attempt to identify the network related to the development of the reproduction system, we filtered off the expressed gene tags from the GonadSAGE database with a cut-off of tag counts lower than 30, followed by running the core analysis function via the Ingenuity Pathway Analysis (IPA) database.

Also, we will revisit the key transcription factors and classify them according to their major functions essential for steroidogenesis, maintenance of gonad identity, and sex determination (reversal) across these three time-points, as analyzed through IPA. Gene ontology and signaling pathway enrichment were performed by clustering retrieved differential expressed genes that are either directly or indirectly downstream of the studied transcription factors annotated by the IPA database. The downstream genes were then adopted for pathway enrichment analysis if they presented differential tag counts among sequenced samples of adjacent chronological order.

MOLECULAR PHYSIOLOGY IN GONAD DEVELOPMENT

The primordial gonad formed as a pair of coelomic epithelial layers on the ventromedial surface of the mesonephroi running along both sides of the dorsal mesentery [14, 15]. The onset of gonadal development occurs after the differentiation of pronephros and mesonephros from the intermediate mesoderm by E9.5. Despite past efforts in experimentation on genetic knock-out models, the molecular pathway or chemical cues leading to the onset of genital ridge development remains largely unknown.

A pioneer structure called the adrenogonadal primordium (AGP) was formed soon after the mesonephros was formed. In AGP, the transcriptional network was marked by *Gata4* and *Wt1* [16]. Shortly afterward, *Sf1* expression begins in the AGP, which will signal for steroidogenesis in the primordium of both the future adrenal glands and gonads [6]. Because the expression of this orphan nuclear receptor is restricted to the AGP, undoubtedly SF1 protein marks the identity of true gonadal somatic precursor cells. Indeed, AGP constitutes all cells that form the steroidogenic structures in the body while its split lies in the discrepancy in the local expression of *Gata4* and *Wt1*. At the anterior part of the mesonephros, thickening of the local coelomic epithelium that will become the future genital ridge proceeds to the posterior half of the mesonephros. The expression of *Gata4* and *Wt1* was restricted later to the genital ridge while they were silenced in the other part of the AGP, which will form the adrenal cortex primordium at E10.5.

From IPA (Diagram 1 below), the reproductive system development and function network for E10.5 is related to "Cellular assembly and organization" and "Cell-to-cell signaling and interaction." Indeed, most gene nodes in this network are related to chaperonin containing T-complex protein (CCT/TCP) complex. For example, *Cct3* in the node with most edges encodes T-complex protein 1 (*Tcp1*) subunit γ, which is a molecular chaperonin belonging to the TCP1 ring complex (TRiC) [17]. Unfolded cytoskeleton subunits including actin and tubulin polypeptides will enter the central cavity of the complex to be folded in an ATP-dependent manner. Other gene nodes in the complex include the prefoldin (*Pfdn*) family members [18]. These molecular cofactors assist the chaperonin in folding other nascent proteins, especially common in actin folding. Besides, *Cct* family members can interact with Dkc1 [19]. This gene

Network 9 : E10.5 - 2017-10-06 03:13 PM : E10.5 : E10.5 - 2017-10-06 03:13 PM

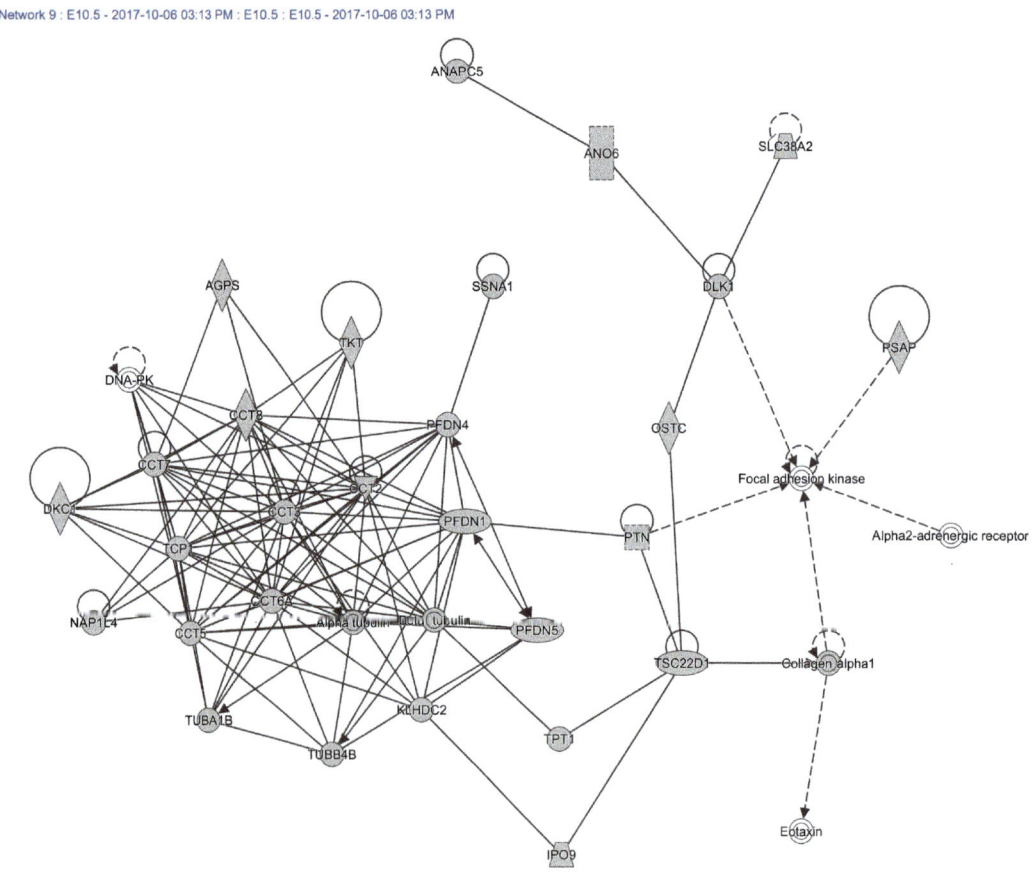

DIAGRAM 1   The reproductive system development and function network (E10.5).

has been implicated in maintaining telomerase stabilization and maintenance as well as providing stability during biogenesis and assembly of small nucleolar RNA ribonucleoproteins (snoRNPs). It is also involved in nucleocytoplasmic shuttling, DNA damage response, and cell adhesion.

Furthermore, the CCT/TCP complex is linked to focal adhesion kinase (*Fak*) through nodes branched from pleiotrophin (*Ptn*) and *Tsc22d1*. Ptn encoded a secreted heparin-binding growth factor stimulated by phospholipase C and the GPCR pathway [20]. It plays significant roles in cell proliferation, cell migration, and angiogenesis. *Tsc22d1* encodes a leucine zipper transcription factor belonging to the TSC22 domain family regulated by transforming growth factor β (TGFβ), signaling for the expression of other genes including C-type natriuretic peptide [21]. *Fak* encodes a tyrosine kinase found in the focal adhesions between cells that interacted with the extracellular matrix (ECM). This kinase is important in relaying the ECM to intracellular signaling to regulate cell growth. In addition, another node linked to *Fak* is *Dlk1*, which encodes a transmembrane protein with multiple epidermal growth factor repeats that also regulates cell growth and is involved in cell differentiation [22].

Summarizing from here, formation of the gonadal ridge at E10.5 requires cell-to-cell signaling/focal adhesion-to-ECM interactions for the decision of active proliferation. This process is potentially regulated by TGFβ signaling, upon which the process of epithelial-mesenchymal (EMT) transition relies [23].

Indeed, some coelomic cells thereafter lose their epithelial features by the EMT process and transform into SF1-positive gonadal precursor cells. These cells then ingress through the disintegrating basement membrane [24]. After the coelomic epithelium thickening, CBX2/M33 together with a mesenchyme-expressed transcription factor POD1 modulate *Sf1* expression to regulate cell proliferation and differentiation of the genital ridges [25, 26]. The early genital ridge thus forms as a bulge of proliferating gonadal precursors covered partially by the original coelomic epithelium. Soon, the gonad expanded, following cell migration from the nearby mesonephric mesenchyme at E11.5.

Similar to that of E10.5, the CCT/TCP complex, is important in the reproductive system development and function network also related to cell assembly and cell-to-cell interactions as analyzed by IPA (Diagram 2). The interacting partners of the complex include stathmin 1 (Stmn1), which encodes a ubiquitous phosphoprotein involved in the destabilization of microtubules by preventing their assembly [27]. Other partners include key players of the Rho GTPase signaling pathway in regulating the decision for contact inhibition or anoikis in anchorage dependency [28]. In particularly, p21 activated kinase (*Pak*) serves as the bridge to link Rho GTPases to cytoskeleton reorganization and nuclear signaling. In fact, the network demonstrated a relay for cytoskeleton rearrangement to cell proliferation via the mitogen-activated protein kinase (MAPK) pathway, with nodes represented by stress-activated protein kinase (*Sapk/Jnk*), *Sos*, *Raf*, etc. In fact, SAPK activity can also be activated by other cell stimuli, but the major function is for the expression of immediate early genes to regulate the decision for either cell proliferation or intrinsic apotosis–the cytochrome *c*-mediated programmed cell death pathway [29].

The network demonstrated the reliance of anchorage dependency to make decisions for either cell proliferation or programmed cell death channeled through SAPK activity. Some gonad cells maintain survival through their mesenchymal transition to avoid contact inhibition, while remaining cells that did not undergo EMT was signaled for apoptosis under Rho

Network 20 : E11.5 - 2017-10-06 03:12 PM : E11.5 : E11.5 - 2017-10-06 03:12 PM

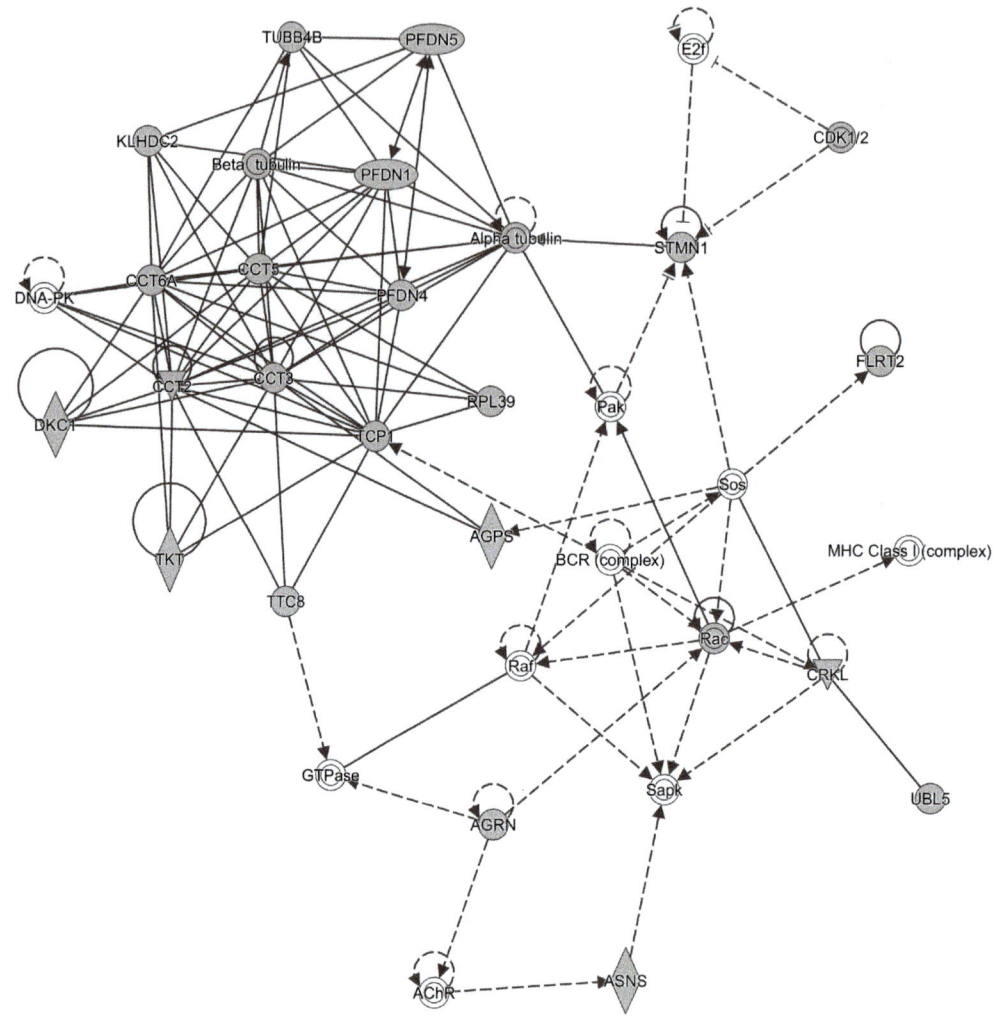

DIAGRAM 2   The reproductive system development and function network (E11.5).

GTPase signaling. This reliance on cell polarity reinforced the idea that both EMT and cell proliferation are the active processes occurring in the gonad at E11.5.

The migrated stromal cells penetrate the gonads toward their surface, separating the sex cords, where they induce a true epithelium to cover the gonads [30]. Similarly, mesonephros-derived cells formed the gonad vasculature. From this time, PGCs migrated in and settled among the SF1-positive gonadal precursor cells. Shortly afterward, *Sf1* expression ceased in the coelomic epithelial cells and these cells gave rise to interstitial cells [31]. Finally, the gonadal precursor cells enclosed the PGCs within the sex cords and slowly differentiated either into Sertoli cells in the developing testes or follicular (granulosa) cells in the developing ovaries at around E12.5 [13].

The reproductive system development and function network for E12.5 is related to "Cell Morphology" and "Organ Morphology," and it involved mainly cytochrome C and the process of mitochondrial permeability transition (Diagram 3 below). PKA signaling regulates ITPR essential for calcium signaling in the second messenger inositol trisphosphate (IP3) [32]. Calcium signaling is also essential to govern cytochrome C release from mitochondria to regulate cell apoptosis. Besides, cytochrome C can interact with phosphatases in the network, particularly with phosphatase *Ppp1C* as the connecting node. PP1 family members are essential phosphatases required for cell division [33]. Other actions of phosphatase include

Network 21 : E12.5 - 2017-10-06 03:14 PM : E12.5 : E12.5 - 2017-10-06 03:14 PM

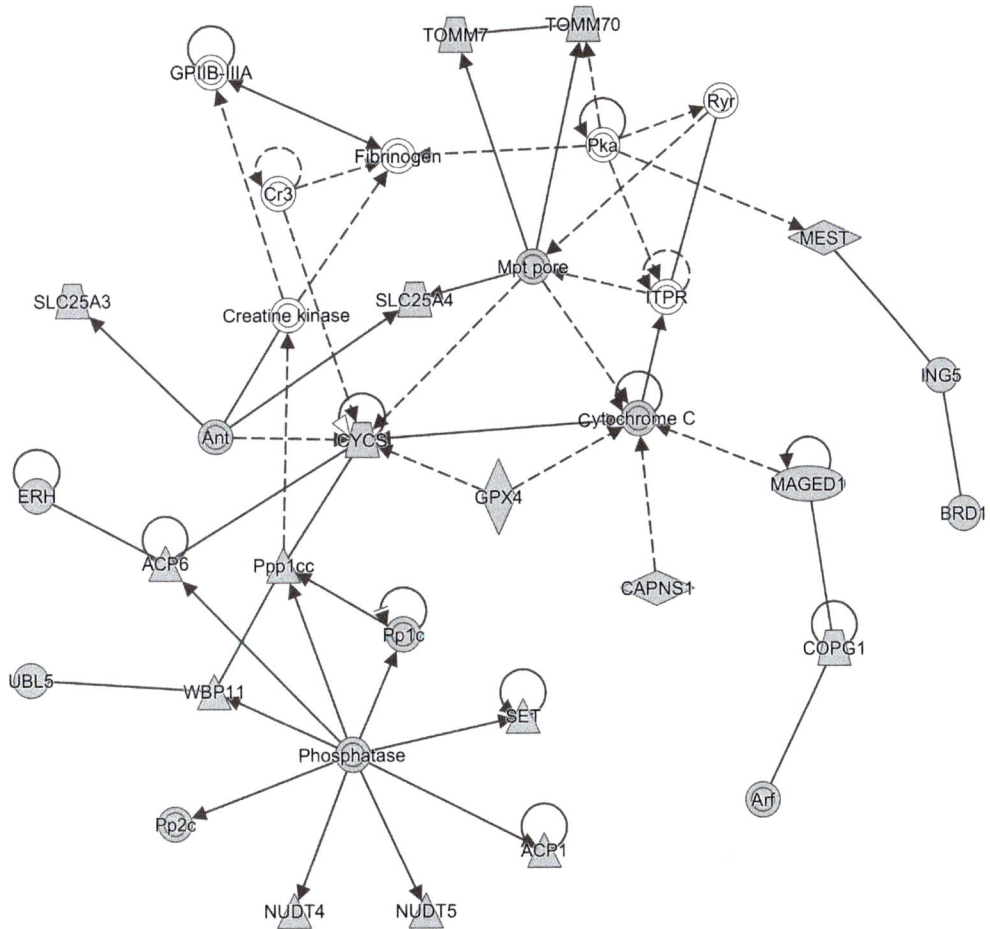

DIAGRAM 3 The reproductive system development and function network (E12.5).

regulating *Wbp11*, *Nudt4*, and *Nudt5* in the network. *Wbp11* encodes a nuclear protein that colocalizes with mRNA splicing factors and intermediate filament-containing perinuclear networks [34]. This protein can bind to the WW domain of another nuclear protein, NPW38, to function as a component of an mRNA factory. NUDT family members regulate the turnover of high-energy diphosphoinositol polyphosphates that are important for intracellular trafficking [35].

The network illustrated that organ morphology changes are taking place in gonads at E12.5. For the male gonad, Sertoli cells are formed from the gonad precursors and they enclose the PGCs to form the sex cords while migrated mesenchymal cells differentiate and fuse to form the vasculature. PKA and its crosstalk for calcium signaling is required for modulating organ morphology while phosphatases regulate various intracellular processes, including transcript processing and intracellular material transport.

Sexual dimorphism in the gonad in terms of both cellular processes and morphology becomes apparent starting from E11.5, due to the onset of differentiation in the gonad precursors. The divergence in morphology is characterized by the appearance of sex cords. Though the precise origin of Sertoli cells in the testis is still unclear, lineage tracing suggested that the majority of them originated from these SF1-positive gonadal precursor cells that were derived from ingressed coelomic epithelial cells [36, 37]. These cells express the *Sry* gene transiently, followed by the expression of *Sox9* starting to obtain their lineage determination at around E11.5.

In the male gonad, the coelomic epithelium continued to undergo active proliferating of the genital ridges by E11.5. This phenomenon is to propagate the formation of more SF1-positive gonadal precursors to differentiate into future Sertoli cells [36]. The newly formed Sertoli cells signal for the organization of two main compartments in the testis, which includes the testis cords with PGCs aggregated in a layer of Sertoli cells surrounded by peritubular myoid cells and the testis interstitium containing the vasculature and fetal Leydig cells that produce steroid hormones [38]. The sex cord will become the seminiferous tubules in later development. After the testis is developed, the morphology of external genitalia is directed by its secretion of anti-Müllerian hormone that leads to the regression of the Müllerian duct [4].

It is known that fetal Leydig cell are recruited and transdifferentiated from other cells rather than formed from repeated mitotic divisions from prior differentiated Leydig cells [39]. It was, however, just recently discovered that both Sertoli cells and fetal Leydig cells were derived from a single population of steroidogenic gonadal precursors [40]. To date, the absence of more defined markers to mark precursor population hampered the characterization of cell heterogeneity in the gonad prior to sex determination. Therefore, single-cell RNA sequencing was performed followed by pseudotime ordering to trace the lineage relationship and expression changes of individual cells in the pool [40]. A single uncommitted cell population found at E10.5 was found to first give rise to Sertoli cells rapidly around E11.5 while some of the remaining progenitors evolved to express steroidogenic precursor marker genes that soon become fetal Leydig cells. The remaining precursors would not revert to form extra Sertoli cells because this requires the expression of several genes, including *Wt1* and *Cbx2*, to promote *Sry* and *Sox9* expression for differentiating to the Sertoli lineage. However, many of these genes were silenced after E11.5, so there will not be transdifferentiation of these steroidogenic progenitors back to Sertoli cells [40].

Importantly, retinoic acid (RA) is the chemical cue to trigger meiosis in PGCs, which is prevented in male gonad development [41]. Indeed, prominent levels of FGF9 in the gonad signaled for inhibition of RA secretion while active transcription in *Cyp26b1* produced metabolic enzymes that actively degrade RA to maintain its low level. Therefore, PGCs remained proliferate and soon become the spermatogonia in the sex cords.

In contrast to male gonad development, proliferation ceased quickly in developing female gonads, with the ingressed coelomic epithelium-derived cells differentiated into follicular granulosa cells while the mixed pool of either coelomic or mesonephric multipotent precursors formed the theca cells [42]. After the sex divergence in the absence of *Sox9* expression, *Rspo1* and *Wnt4* expression are upregulated while SF1-positive gonadal precursor cells gather into ovarian follicles, each enclosing a single oocyte derived from the PGCs that underwent meiosis induced by RA [43]. Without active proliferation, it is thus not surprising that SF1-positive gonadal precursors disappeared in the female gonad immediately at around E13.0. The default state without steroid hormone masculinization allowed the differentiation of the Müllerian tube to differentiate into organs of the reproductive tract, including the uterus and upper vagina.

ESSENTIAL TRANSCRIPTION FACTORS IN GONAD DEVELOPMENT

Transcription Factors Required for Steroidogenesis

SF1—The Transcription Factor for Steroidogenic Enzymes

The *Sf1* gene encodes the protein Steroidogenic Factor 1. *Sf1* expression appeared immediately after *Gata4* in the AGP while its transcripts remained detectable in all cells that give rise to either the adrenal cortex or the genital ridge [6]. Biallelic loss of *Sf1* resulted in failure of adrenal and gonadal development beyond the indifferent stage [44]. The *Sf1* gene encodes an orphan nuclear receptor that regulates the expression of several genes involved in steroidogenesis as well as genital development, such as members of the steroid precursor synthetic enzymes, including cytochrome P-450 hydroxylases and the negative regulator for steroidogenesis, *Dax1* [45].

The gene function is not limited to steroidogenesis. Mutant *Sf1* in XY (male) embryos initiates partial default female development with remnants derived from the supposed Müllerian duct, including the upper vagina and uterus [6]. It was found to be one of the genes in sensitizing a low, sex-independent *Sox9* expression otherwise abolished in the SF1-null mutant gonads of both sexes at E10.5 [46, 47]. SRY binds SF1 to form a complex, which triggers the expression of *Sox9* for the specification of male differentiation through Sertoli cell formation [48].

Revisiting the downstream targets of SF1 resulted in key genes regulating lipid metabolism and adipogenesis at E10.5, including *Hdac4, Hdac5, Kat2a,* and *Sox9*. During the same period, Wnt signaling is active with SF1-induced β-catenin (*Ctnnb1*) expression and increasing expression of *Pou5f1* (*Oct4*), *Sox2, Sox8,* and *Sox9*. This signaling pathway is essential for the proliferation of gonad cells [49]. The developmental process thereafter relied on SF1-directed retinoic acid (RA) signaling, which involved a high expression of *Nr2f1* and *Cdk7* at E11.5. RA as a local paracrine may stimulate neighboring cell types to respond during the gonad developmental processes [50]. Activated *Cyp26b1* along this pathway may serve partly as the key determinant for gamete meiotic events. In the developing testis, the active degradation of RA in the presence of this enzyme prevented meiosis during gonad development [51, 52].

Following through E12.5, SF1 upregulated *Amh, Amhr2, Cyp11a1, Gata4, Lhcgr,* and *Sox9* expression. This, together with high levels of Esr1, Gal, and Cyp26b1, will lead to differentiation and maturation of the male reproductive phenotype. The most important function of SF1 then is to activate key enzymes responsible for steroid hormone synthesis via catalyzing the production of the steroid hormone precursor pregnenolone. Indeed, SF1-mediated transcriptional activation of steroidogenic gene-encoding enzymes, such as *Cyp11a, Cyp17,* and *Cyp19* is essential for the synthesis of androgens and estrogens [53]. SF1 is also the upstream of StAR, the transport protein regulating cholesterol transfer at the mitochondria [8].

To conclude, SF1 participates in regulating gonad proliferation under the action of the Wnt signaling pathway at the beginning of gonad development. It then turned toward the transcription of metabolic enzymes essential for steroid hormone synthesis and upheld the maturation of the male phenotype after the formation of Sertoli cells.

GATA4—Its Presence is Crucial for the Development of Genital Ridge After AGP Split

The Gata4 gene encodes the protein GATA binding protein 4. GATA4 marked the site for future gonad development as early as the formation of AGP while its expression ceased in the adrenal cortex primordium after the split [16]. The cells that contribute to the anteroposterior thickening of the coelomic epithelium bulged the anterior part of AGP in formation of genital ridges are GATA4-positive. This clearly demonstrated an essential action unique in the genital ridge. Previous reports suggested that Gata4 expression is essential for the disintegration of the basement membrane underlying the coelomic epithelium and the proliferation of cells in this epithelium [54, 55].

With cooperative action with its cofactor FOG2, GATA4 forms a multitude of protein complexes with key developmental determinants such as SF1 to regulate the expression of sex-determining genes, including *Sry, Sox9,* and *Amh,* and key steroidogenic genes such as *Star* and *Cyp19a1* [8, 56]. It is crucial to understand that Gata4 is the only member within its family to express only in the somatic cell type within the bipotential gonad while never existing in the PGCs [57]. Its expression remained until E11.5 in both sexes while it is downregulated thereafter in female embryonic gonads. In male embryos, they continued to be expressed in the Sertoli cells up to E13.5 [54]. This marker expression pattern is similar to SF1, indicating GATA4 as a reliable marker to the same gonadal precursor cells.

There were reports on a reduction in the basal Sry transcript level in *Fog2*-deficient male gonads as of E11.5, which is followed by the absence of *Sox9* expression [56]. Such a phenomenon likely suggested the involvement of *Gata4* in sex differentiation. In effect, GATA4 was confirmed to demonstrate enhanced cooperativity with WT1 to bind the *Sry* promoter [56]. Nevertheless, similar cooperativity was observed in *Amh* promoters [56, 58]. It is apparent from the fact that a lack of expression from *Gata4* or its cofactor *Fog2* prevented male-specific differentiation [59]. However, in mutants where there is abrogated interaction between the two factors, they had surprisingly elevated levels of WNT4, which is essential for commitment to ovarian development. It demonstrates that the interaction between the two factors is required for sex determination while GATA4 alone is required for female differentiation after sex determination [60].

In our serial profile, GATA4 was found to regulate both calcium signaling and the integrin-linked kinase (ILK) signaling pathway in the indifferent gonad after staging the formation of the genital ridge at E10.5. These genes, including *Actc1*, *Ep300*, *Gsk3b*, *Hdac2*, *Med2d*, *Nfatc4*, and *Slc8a1*, generally demonstrated enhanced activity until E11.5. ILK signaling was previously demonstrated to be essential for nematode gonad formation [61]. The cytoskeleton, particularly actin networks, could be disorganized if ILK is deregulated to affect the gonad structure. GATA4 continued to modulate the actin assembly by E12.5, together with inducing the expression of *Amh*, *Inhb*, *Lhcgr*, and *Ntf4* along the Activin/FSH signaling pathway essential for male reproductive tract development.

To conclude, GATA4 participates in governing the actin cytoskeleton network through both calcium signaling and the ILK signaling pathway throughout early gonad development. This reinstated the importance of cell-to-cell contact signaling for the remodeling of the cytoskeleton architecture. The continuous *Gata4* expression modulated Sox9 expression in the indifferent gonad and remained present in low levels in male gonads to reinforce the subsequent male-specific differentiation program.

DAX1—A Negative Regulator for Steroidogenesis

The *Dax1* gene encodes the protein DSS-AHC critical region on the X chromosome protein 1, which indicates that its mutation will give rise to development disorders, including X-linked congenital adrenal hypoplasia (AHC) and hypogonadotropic hypogonadism [62]. Dax1 is a negative regulator to steroidogenesis by inhibiting the expression of *Cyp11A*, *Cyp17*, and *Cyp19* [63]. Dax1 is prominent in the repression of *Star* expression that could result in a pathologic accumulation of lipids in cells, despite the fact that it was less severe in early female gonad development because ovaries do not express the *Star* gene until puberty [64].

Such a mode of action is conserved in both the development of the adrenal cortex and the gonads, where *Dax1* expression is thought to be down-regulated as these tissues develop. However, a host of testis defects such as abnormal organization of Sertoli cells, lack of basal laminae in the testis, and disrupted peritubular myoid cell development may result from the absence of Dax1 [65]. This is partly because Sertoli cell differentiation depends on the coordinated expression of *Dax1*, *Sry*, and *Tda1* [66]. Initially, testis development appeared normal until the testis cord was disorganized or incomplete by E13.5.

Dax1 expression strikes a balance in the antitestis determination to the default female state in mammals [65]. The gene is responsible for the dosage sensitive male-to-female sex reversal observed in patients with a duplicated region of Xp21, of which it is embedded even with the presence of an intact SRY gene [11]. Such activity clearly demonstrated that its high expression prevented testis development. Overexpression of the gene in mice resulted in ovotesticular disorder, that is, coexistence of ovarian and testicular tissues within the same gonad in mice by inhibiting SF1 activation of the Sox9 gene in the locus named testis-specific enhancer of the Sox9 core element (TESCO). Dax1 inhibits the synergic activation of TESCO either by SF1 and SRY or by SF1 and SOX9 [11]. However, its activity is absolutely required for testis formation, probably by its action in sensitizing basal levels of *Sox9* expression [67]. In fact, a nuclear receptor corepressor alien expressed in the testis was found to be able to interact with Dax1 despite the fact that the action of such interaction is unknown [68].

Dax1 is found to be able to mediate estrogen signaling via modulating the expression of *Sra1* and *Nr3c1* at E10.5. It also participates in retinoic acid (RA) signaling via altering the expression of *Prmt1* and *Snw1* at E11.5. Finally, Dax1 regulates the expression of key enzymatic reactions governing steroid hormone synthesis at E12.5. These serve as the prerequisite enzymes in steroid hormone synthesis, including *Cyp17a1* and *Hsd3b2*. At this time point, Dax1 also modulated the expression of nuclear receptors involved in reproductive functions, including *Nr3c1* and *Nr4a1*. Finally, *Amh* expression driven by the male determinant Sox9 could be counteracted by Dax1.

To conclude, Dax1 appeared in early gonad development to modulate steroidogenesis by antagonizing the effects of SF1. It also counteracted the male-specific differentiation program by inhibiting the binding of SF1 to the TESCO enhancer for *Sox9* expression and subsequent autofeedback loop during sex determination. Furthermore, it modulated RA signaling potentially to interfere with the decision of PGCs for meiotic events. However, it is not just an antitestis gene because there was evidence that its absence abrogated male differentiation without its sensitizing action to the basal levels of Sox9 expression.

Transcription Factors Required for the Maintenance of Gonad Identity

WT1—The Regulator for Gonad Precursor Cell Differentiation

Wt1 encodes the protein Wilm's Tumor 1. It was initially identified as a gene inactivated in a subset of Wilm's tumors, which is a form of pediatric kidney cancer [69, 70]. It is initially expressed ubiquitously in the AGP, then restricted to the genital ridge afterwards. WT1 was generally considered a potent transcriptional repressor. However, accumulating reports suggested that it can also activate downstream targets, including the cyclin-dependent kinase inhibitor p21, BCL2, Dax1, etc. [71–73].

Inactivation of both *Wt1* alleles was confirmed to result in embryonic lethality [74]. The *Wt1* gene locus encodes for alternative spliced isoforms with or without the insertion of three amino acids lysine, threonine, and serine, i.e. collectively known as KTS, between its zinc finger domains to disrupt DNA binding activity. Such +KTS isoform was found likely to participate in mRNA processing [75]. The short isoform of WT1 that lacks the KTS, that is, WT1 – KTS, retained its DNA binding activity and was found to be essential for gonad development [9]. Indeed, mice devoid of functional WT1 – KTS have increased cell death in the genital ridge, eventually resulting in impaired development.

WT1 cooperated with its transcription cofactor CITED2 to stimulate the expression of *Sf1* in the AGP to ensure cortex adrenal development [7]. Likewise, the –KTS isoform of WT1 was shown to bind sequences within the *Sf1* promoter in the gonads [9]. Moreover, WT1, CITED2, and SF1 coerced to increase the expression of *Sry* levels required for testis development [76,77]. A lack of *Wt1* prevented gonad development further downstream of coelomic thickening. It was demonstrated that overexpression of *Wt1* in Leydig cells led to the increased Sertoli cell-like marker changes and reduced steroidogenic gene expression, indicating that WT1 governed the lineage relationship during gonad development of Sertoli cells and Leydig cells [78]. Similarly, deletion of *Wt1* in the genital ridge before sex determination completely blocked the differentiation of either the Sertoli cell in the male or granulosa cells in the female [79].

The most obvious action by WT1 at E10.5 is the regulation of genes along the mitogen-activated kinase (MAPK) pathway to control cell proliferation in the gonads. In fact, when genes involved in this pathway, i.e. *Ccne1*, *Cdk2*, *Cdk4*, *Mdm2*, *Smad3*, and *Tgfb3* are regulated, they promote the entry into cell cycle at the G_1/S transition. Concurrently, cell proliferation can be finetuned by aryl hydrocarbon receptor (AhR) signaling. AhR is a ligand-activated transcription factor initially involved in the regulation of biological responses to xenobiotic aromatic hydrocarbons [80]. In developing vertebrates, *Ahr* seemingly plays a different role in governing cellular proliferation and differentiation, including immune, neural, and liver development. WT1 can modify key genes along the AhR signaling pathway with shared signal transducers to the cell cycle transition, including *Ccnd1, Cdk2, Cdk4, Hsp90b1, Mdm2, Myc,* and *Nqo2a*.

From E11.5 onward, WT1 is involved in TGF-β signaling, which is crucial to the activation of EMT, as that seen in Emx2 [81]. Because epithelial cells are closely connected to each other by junction machineries to elicit cell polarity, EMT activation changes the cell-cell contact and the cytoskeleton in the cells to increase mobility. In fact, the advantage of cells undergoing EMT was not limited to their facilitated dissemination to various parts of the embryo but also to the promotion of their plasticity during development [82].

Changes along the TGF-β pathway included the upregulation of genes such as *Amh, Amhr2, Mapk8,* and *Vdr*. Besides, genes related to junction machineries such as *Cdh1* and *Dvl3* while migratory signaling such as *Mmp9* and *Stat3* were found upregulated in the gonads by E11.5. Wt1 also promoted changes in the expression of *Dvl2, Egr1, Ets1, Rpbj, Mmp2, Smad3, Tgfb2,* and *Wnt4* by E12.5.

To conclude, because WT1 can regulate the expression of the signal transducers along the G_1/S cell cycle transition mediated by AhR signaling, it initiates cell proliferation in early developing gonads. After proliferation, WT1 also signals for EMT to promote the ingression of gonadal precursors via the TGF-β pathway.

CBX2—Amplifier for Gonad Development and Male-Specific Differentiation

CBX2 (chromobox homolog 2, M33) belongs to the polycomb group (PcG) family of epigenetic modifiers usually regulating the chromatin structure to allow local silencing of gene expression that is usually involved in development and cell proliferation. It was clearly noted to be expressed in the indifferent gonad prior to sex determination. Studies in the mouse model suggested the involvement of this gene in the specification of the anterior-posterior axis as well as in cell proliferation in early development. Mice lacking functional *Cbx2* demonstrated reduced *Lhx9, Sf1,* and *Gata4* expression, the marker proteins expressed in the gonadal precursors [83]. Therefore, it is generally accepted that its presence helps maintain the active proliferation of SF1-positive gonadal precursors. Because these cells eventually contribute to somatic gonad structures, the failure of their active proliferation led to subsequent defects such as male-to-female sex reversal and hypoplastic gonads of both sexes [83]. Its activity, however, is not essential for the formation of gonads.

Nonetheless, *Cbx2* knockout in SRY-positive mice could be sexually reversed to females with normal reproductive tract development. Therefore, the gene was considered a transactivator upstream of *Sry* in the development cascade [84]. *Cbx2* also facilitates male sex determination with its reported upregulation of male-related genes, for example, *Sox9, Sox3, Fgf2, Insl3, Sf1,* and *Sry*. Moreover, Cbx2 negatively regulates female-related genes, including *Fzd1, Pbx1,* and *Foxl2* [84].

This gene was found to regulate the β-catenin level in the canonical Wnt signaling at E10.5 to regulate proliferation of the gonadal precursors. Starting at E11.5, it regulated ubiquitination via *Rnf2* expression. RNF2 itself also belongs to the PcG family and it suppresses the activity of transcription factor CP2 (TFCP2/CP2) [85]. Interestingly, it was found to be essential for the differentiation of PGCs. CBX2 also regulated downstream genes related to the development of the reproductive system, including *Lhx9*, *Sf1*, *Csnk2a* and *Csnk2b*, *Cdkn2a*, *Gtf2i*, and *Cbx1* at this time point. Cbx2 also regulated *Nanog*, a homeodomain-containing transcription factor commonly expressed in pluripotent cells such as preimplantation embryos [86]. Importantly, *Nanog* expression is highly expressed in primordial germ cells (PGCs) of E11.5 and E12.5 mouse embryos of both sexes while its expression pattern differed between the sexes in the later developmental stage. It was understood that NANOG is only present in the proliferating PGCs and partially mediates its migration through signaling via *Cxcr4b* [87].

To conclude, CBX2 does not affect the formation of the indifferent gonad. However, it regulates the epigenetic landscape via local silencing of gene expression to favor gonad development. It also amplifies proliferative signals via the Wnt signaling and the male-specific differentiation program. Moreover, because mitotic division of PGCs is essential to form spermatogonia in male gonads, CBX2 reinforced the pluripotency of these cells to differentiate to the male-specific gamete.

Transcription Factors Essential for Sex Determination

SRY—The Testis-Determining Factor, a Key to Unlocking Male Sex Determination

SRY is the acronym gene name for "the **S**ex-determining **R**egion of the **Y** chromosome," which is the only gene identified in the Y chromosome to trigger the onset of male development in mammalian embryos act transiently by inducing a bend in the chromatin it binds [88]. It is believed that the gonad precursors autonomously differentiate into Sertoli cells under the influence of SRY and thereafter SOX9 [89]. SRY lacks protein sequence conservation across mammalian species where most structural domains are poorly conserved [3]. The only conserved domain between human, mouse, and other eutherian organisms was found to be the high mobility group (HMG) domain [90]. In fact, this domain is the active DNA binding domain found on all SRY-related HMG-box (SOX) family members to bend the DNA. Chromatin rearrangements directed by the bend likely facilitate recognition for transcription [91].

SP1 is a ubiquitously expressed transcription factor acting in combination with tissue-specific transcription factors [92]. Such expression is extinguished by E12.5, which is followed by induced levels of *Sox9* expression in an autofeedforward manner. *Sry* expression is not rigid spatially but begins around the center of the gonad and expands outward.

It was found that WT1 and SF1 likely contributed to how SP1 regulates *Sry* expression as early as E10.5. Their gonad-specific expression provided a high local expression specificity of *Sry* in the gonads. In fact, similar modes of coregulation between these genes were also found in the *Amh* promoter later at E12.5. It was previously demonstrated that WT1 + KTS indirectly regulated the transient expression of *Sry*, which peaked at E11.5 after its initiation at E10.5 [77]. Because DAX1 appears to demonstrate an overlapping expression profile with the absent SRY spatiotemporally in female gonads to inhibit SF1-mediated transcription, likely it served as a negative regulator to *Sry* expression [93].

Until now, there have been no reported clear downstream targets of SRY. The only confirmed gonad-specific target is the TESCO enhancer, which favors the expression of *Sox9* described below.

SOX9—The Initiator for the Male-Specific Differentiation Program

Because SRY was found to bind the TESCO enhancer, therefore this provided evidence for its activation of *Sox9*. Interestingly, it was previously demonstrated that SOX9 could replace SRY in triggering Sertoli cell differentiation [94]. Its expression is up-regulated in the male gonads and suppressed in the female gonad with transcription activity associated with its nuclear localization [95]. In fact, SOX9 could potentially feedback amplify the *Sry* gene [96].

SOX9 plays a pivotal role in male sex-determination by acting as a potent transcriptional activator during sex determination. It was known that SOX9 could lead to the transcription of two genes in the gonads that activate the male-specific program, namely *Sf1* and *Amh* [97]. Indeed, SOX9 sustained *Sf1* expression in developing the male gonad for steroidogenesis, which was otherwise down-regulated in the female counterparts. *Amh* was induced by the interplay of SOX9 with SF1, WT1, and GATA4 while its gene product anti-Müllerian hormone is crucial in male sexual differentiation that signals the regression of the Müllerian duct secreted from the Sertoli cell lineage [97, 98].

Positive regulatory loops exist to reinforce initial decisions while the maintenance of the gonadal phenotype is governed by the active repression of the opposite pathway. Indeed, SOX9 is maintained at high levels by FGF9 signaling [99]. In parallel, the gene product of *Ptgds* prostaglandin D2 synthase promotes the accumulation of SOX9 protein within the nucleus, after its active transcription by SOX9 [100]. Furthermore, SOX8 seemed to have redundant roles to its own upstream transcription factor SOX9 and served to maintain an active male-specific differentiation program [101].

In contrast, *Sox9* expression is downregulated in female gonads; the repression is maintained throughout fetal development and even up to adult life. Such suppression was realized by high expression of *Dax1* that repressed *Sf1* expression and its subsequent binding to TESCO while *Foxl2* and the estrogen receptor α (ERα) were physically interacting to bind TESCO in adult ovaries [102]. Finally, Wnt signaling-induced β-catenin expression could antagonize the feedback induction of *Sox9* by FGF9 signaling [103].

SOX9 is involved in pigmentation signaling through its regulation of downstream genes along the protein kinase A (PKA) pathway, including *Creb1*, *Dct*, *Kit*, *Mapk3*, *Prkaca*, *Sox10*, and *Smad3* at E10.5. By E11.5, it regulates the EMT via the canonical Wnt signaling pathway by altering *Cdh2*, *Ctnnb1*, *Smad2*, *Smad3*, *Gsk3b*, *Mapk3*, and *Tcf7l2*. The expression of Sox9 thereafter is reinforced by the autofeedback activation and maintained at high levels. Subsequently, the regulation of Wnt signaling continued by E12.5 through acting on *Btrc*, *Lrp6*, and *Sox5*. Besides the aforementioned pathways, this gene also regulates TGF-β signaling through positive regulation of *Amh*, *Mapk3*, and *Smad2*.

To conclude, SOX9 regulated the PKA pathway to modulate cell morphology and the Wnt signaling pathway for EMT. Its expression is governed by the recruitment of gonad-specific transcription factors, including SRY and SF1, to the TESCO enhancer. As the male sex determinant, SOX9 expression is subjected to a cooperative autofeedback loop by FGF9 and prostaglandin D2 signaling to promote *Sox8* or *Sox9* expression itself for the reinforcement of the male-specific Sertoli lineage to form the testes, which is otherwise repressed in female gonad development.

Homeodomain Proteins in Early Gonad Development

Homeodomain proteins are a class of evolutionary conserved proteins that functions essentially in developing adult organisms [104]. Although homeodomain proteins have similar DNA binding specificity, their functions are highly diverse and context-dependent. In gonad development, the involvement of some of these homeodomain proteins has been confirmed [105]. The most well-studied homeodomain proteins in gonad development include Lhx9 and Emx2, which have been associated with cell proliferation and cell death, respectively. Despite their importance in gonad development, there are limited reports on the downstream effectors of these transcription factors and their likely functions in gonad development. We hereby summarize past studies that aimed to study their involvement in regulating reproductive functions.

LHX9—A Likely Regulator of Cell Stemness

Lhx9 is expressed only transiently in the gonad as early as E9.5 and will disappear along gonad differentiation. The major function for LHX9 in the indifferent gonad is to enhance proliferation of SF1-positive gonad precursors. Interestingly, its presence is required for genital ridge formation through cooperative binding to the *Sf1* promoter with WT1 − KTS while, when there is a deficiency in gene expression, *Sf1* expression is reduced to minimal levels [31]. The absence of LHX9 protein expression in somatic cell types such as Sertoli and interstitial cells in the male and granulosa cells in the female gonads indicated that its presence likely regulated cell plasticity or maintaining the stemness of gonadal precursors [106]. In other words, low *Lhx9* expression promotes gonadal precursors to differentiate.

EMX2—Its Presence Ensures That the Genital Ridge Could Bulge in for Cell Ingression

Emx2 expression is ubiquitously expressed in epithelial components in tissues originating from mesonephros. Because its absence only affected the thickening of the coelomic epithelium, it clearly points out that its main action is to maintain the genital ridge by preventing its regression [81]. During the anteroposterior thickening, EMX2 likely signals for epithelial-mesenchymal transition (EMT) in the coelomic epithelium, therefore promoting the differentiation of ingressed cells to form gonadal precursor cells [107]. At the same time, the epithelium is maintained in the presence of EMX2 to form the primitive gonad structure. It also participated in regulating tight junction assemblies. Its presence also correlated to reduced *Egfr* gene expression and therefore prevented active proliferation of gonadal cells [24]. Nevertheless, CDH6 and CDH8 were identified as downstream targets of EMX2 that favor EMT [108].

PBX1—Essential Early Expression of This Gene for the Induction of SF1

Pre-B-Cell Leukemia Homeobox 1 (*Pbx1*) gene encodes a TALE (three amino acid loop extension) class homeodomain protein that participates in multimeric transcriptional complexes to regulate developmental gene expression [109]. The embryonic expression of *Pbx1* has been reported in urogenital organs [110]. Moreover, the expression of *Pbx1* is dynamic and tissue-restricted throughout urogenital development, including at AGP by E10.0, the bipotential gonad by E11.0, the epithelium of the Müllerian duct by E13.5, and the interstitium of testis and ovary by E14.5 [109]. A *Pbx1* knockout mice study showed a range of anomalies during urogenital development: lack of adrenal glands and gonads, impaired differentiation of the mesonephros and kidneys, and the absence of the Müllerian duct [109]. Molecular examination showed that expression of *Sf1* was dramatically reduced in *Pbx1* mutants [109]. Because SF1 is essential for adrenal formation and gonadal development, expression of *Pbx1* is probably an early prerequisite to up-regulate *Sf1* expression and SF1-postive cell proliferation and expansion to secure early genital ridge formation and differentiation.

SIX1/SIX4—Redundant Gene Pair to Finetune Cell Proliferation and Differentiation

The sine oculis homeobox homologs 1 (*Six1*) and 4 (*Six4*) genes belong to the mammalian homolog of the Drosophila sine oculis homeobox (Six) family with six member genes (*Six1–Six6*) in the mouse genome [111]. All the member genes of the Six family encode for transcription factors that carry the characteristic Six domains and homeodomains. In fact, these family members are functionally redundant to each other during mouse embryonic development with their overlapping tissue expression and equivalent binding to downstream target gene, for example, *Mef3* for transactivation [111–113]. Moreover, Six1 single-knockout (KO) embryos showed kidney malformation that is less severe than the kidney agenesis phenotype observed in *Six1/Six4* double-KO embryos [114]. In contrast, *Six4* single-knockout embryos demonstrated normal kidney formation [115].

Indeed, such functional redundancy is also found in gonad development. Fujimoto et al. reported that in *Six1−/−* or *Six4−/−* single-mutant mice, no gonad-specific defective phenotype was manifested, whereas in *Six1* and *Six4* double-mutants (*Six1−/−*; *Six4−/−*), smaller gonads and adrenal glands were observed compared to controls [116]. Specifically, *Six1* and *Six4* showed an overlapping expression pattern in the coelomic epithelium of the forming genital ridges. Double KOs had delayed/reduced endothelial-to-mesenchymal transition (EMT) and subsequent ingression of gonadal progenitor cells in the genital ridges, which ended up with reduced gonadal size and disrupted testicular differentiation. Further examination showed a reduction of *Sry* expression in *Six1*/Six4 double-KO gonads. Overexpression of *Sry* in these double-KO gonads could rescue testicular development but not gonadal size. These results suggested that *Six1/Six4* simultaneously regulated genital ridge development and testicular development. A mechanistic study identified *Fog2* and *Sf1* as two downstream targets of *Six1/Six4*. FOG2 is a known regulator of *Sry* while SF1 is indispensible for gonad determination and differentiation. Taken together, the *Six1-Six4-Fog2* trio is required for precise spatiotemporal regulation of *Sry* expression to ensure normal sex differentiation. Also, the *Six1-Six4-Sf1* trio is critical to controlling the initial growth of gonad precursor cells and subsequent determination of gonad size.

HOX Genes—A Family of Genes Governing Spatiotemporal Reproductive Development

HOX genes are conserved transcription factors that regulate embryonic morphogenesis and differentiation [105, 117, 118]. In Drosophila, the *Hox* gene domains define the position of the gonad [119]. In mammals, no HOX genes are currently known to specify the position of the gonad. However, we cannot rule out the possibility that these genes might also be important regulators for spatiotemporal patterning of gonadal development.

Several Hox genes (e.g., *Hox-5*, *Hoxd1*, *Hox-1.4*, *Hoxa11*) are expressed in both fetal and adult reproductive organs [120–123]. While the evidence for Hox genes in gonadogenesis is limited, a bunch of studies clearly identified their roles in sex differentiation. Detection of *Hoxa9, 10, 11 and 13* in the paramesonephric duct of E15.5 mice implicated their function in Müllerian duct formation and differentiation [124]. Specifically, *Hoxa9* is expressed in the oviduct. *Hoxa10* and *Hoxa11* are both expressed in the uterus. *Hoxa11* can also be found in the cervix and anterior vagina. *Hoxa13* is expressed in the cervix and the Müllerian vagina but not in the uterus.

Hoxa10 homozygous mutant male and female mice displayed sterility [125]. Morphologically, *Hoxa10*-deficient male mice showed bilateral cryptorchidism with decreased seminiferous tubule formation and disrupted spermatogenesis. No histological difference was observed in *Hoxa10*-deficient females. Normal ovulatory cycles retained in mutant females suggest that *Hoxa10* is not required for gonadal development in females. Further studies on *Hoxa10* mutants have demonstrated that the reduced fertility in females should be caused by homeotic transformation of the anterior part of the oviduct into uterus [126].

Similarly, mutation of *Hoxa11* resulted in sterility in the male and female [122]. The cause for female sterility is also a defective uterine environment without any defects in ovulation. As for male sterility, homeotic transformation from the vas deferens to an epididymis with an abnormal testis development is regarded as the cause. All similar phenotypes further identified the consistent roles of *Hoxa10* and *Hoxa11* in sex differentiation. Finally, *Hoxa13*-null mice showed agenesis of the distal portion of the Müllerian ducts, indicating a role for *Hoxa13* not only in differentiation but also in the formation of Müllerian ducts [127].

In addition to murine models, mutation screening in human *HOX* genes in independent case studies provided further evidence to their function in secondary sexual organogenesis. Mutations in *HOXA10* and *HOXA13* were identified in human female genital malformations with congenital absence of the uterus and vagina [128]. *HOXA10* mutations are also found in cases with Müllerian anomalies [129].

To date, transcriptional targets of these mammalian HOX proteins remain largely unknown. There are some suggested regulatory roles of *Hoxa11*, *Hoxa10*, and *Hoxa13* in sex differentiation [124, 127, 130–132]. *Hoxa13* appears to upregulate *Bmp4* in the uterus [62]. Loss of *Hoxa13* in mice induces the loss of *Bmp7* and *Fgf8* signaling in the developing genital tubercle, which causes hypospadias [127]. Reciprocally, *Hoxa10* and *Hoxa11* was possibly regulated by WNT7A because knockout of *Wnt7a* decreased their expression [132].

In summary, the roles of HOX genes in sex differentiation are verified by several lines of evidence, described above. However, evidence for their role in early gonadal development is rare. Likely, this implies that they are not playing essential roles in gonadogenesis. However, it is also possible that their functional role could be masked by functional redundancy among the wide spectrum of HOX members. Conditional overexpression remains the only research strategy to justify the statement. In fact, the functions of these homeodomain proteins are not independent. For example, PBX1, as a TALE heterodimer, has the capacity to cooperate with a subset of HOX proteins to bind DNA [109]. The crystal structure of HOXA9 and PBX1 interacting together with DNA has already been revealed [133].

Gonad development and subsequent sex differentiation are well-coordinated complex processes that involve the participation of homeodomain proteins. Unsurprisingly, many more such genes are seemingly crucial yet to be described. Thus, further study on additional homeodomain proteins may reveal new insights that finally can contribute to human urogenital diseases.

CONCLUSIONS

Most genes we introduced were transcription factors expressed with spatiotemporal precision prior and during embryonic gonad development. The restricted expression of *Gata4*, *Sf1*, and *Dax1* in the AGP with tightly regulated expression level ensures proper differentiation of the tissues that will become the future adrenal cortex and gonads. The subsequent formation of these steroidogenic tissues will promote appropriate hormonal signals for the development of the whole reproductive system and maintain the subsequent gender identity. *Wt1* and *Cbx2* are important in maintaining the identity of the subsequent genital ridge after its formation by reinforcing the migration and proliferation of newly formed gonadal precursors from the coelomic epithelium by EMT. *Sry* was expressed early in the indifferent gonad, spreading outward from the center to unlock the expression of *Sox9*, which will define the Sertoli lineage essential for male-specific differentiation, including formation of the sex cord and testis vasculature. PGCs were nurtured in the sex cords and divided repeatedly to form the spermatogonia. In contrast, the absence in *Sry* expression as in female gonads bypassed the formation of these structures wherein the germ cells enter meiotic division to form the oogonia. Finally, various homeodomain proteins are expressed early during gonad formation, either to regulate the above processes or be involved in defining the structural integrity of developing organs in the reproductive system.

References

[1] Doitsidou M, Reichman-Fried M, Stebler J, Köprunner M, Dörries J, Meyer D, et al. Guidance of primordial germ cell migration by the chemokine SDF-1. Cell 2002;111(5):647–59.

[2] Sinclair AH, Berta P, Palmer MS, Hawkins JR, Griffiths BL, Smith MJ, et al. A gene from the human sex-determining region encodes a protein with homology to a conserved DNA-binding motif. Nature 1990;346 (6281):240–4.

[3] Tiersch TR, Mitchell MJ, Wachtel SS. Studies on the phylogenetic conservation of the SRY gene. Hum Genet 1991;87(5):571–3.

[4] Wilson JD, Griffin JE, George FW. Sexual differentiation: early hormone synthesis and action. Biol Reprod 1980;22(1):9–17.

[5] McLaren A. Studies on mouse germ cells inside and outside the gonad. J Exp Zool 1983;228(2):167–71.

[6] Ikeda Y, Shen WH, Ingraham HA, Parker KL. Developmental expression of mouse steroidogenic factor-1, an essential regulator of the steroid hydroxylases. Mol Endocrinol 1994;8(5):654–62.

[7] Val P, Martinez-Barbera J-P, Swain A. Adrenal development is initiated by Cited2 and Wt1 through modulation of Sf-1 dosage. Development 2007;134(12):2349–58.

[8] Nishida H, Miyagawa S, Vieux-Rochas M, Morini M, Ogino Y, Suzuki K, et al. Positive regulation of steroidogenic acute regulatory protein gene expression through the interaction between Dlx and GATA-4 for testicular steroidogenesis. Endocrinology 2008;149(5):2090–7.

[9] Wilhelm D, Englert C. The Wilms tumor suppressor WT1 regulates early gonad development by activation of Sf1. Genes Dev 2002;16(14):1839–51.

[10] Combes AN, Spiller CM, Harley VR, Sinclair AH, Dunwoodie SL, Wilhelm D, et al. Gonadal defects in Cited2-mutant mice indicate a role for SF1 in both testis and ovary differentiation. Int J Dev Biol 2010;54(4):683–9.

[11] Ludbrook LM, Bernard P, Bagheri-Fam S, Ryan J, Sekido R, Wilhelm D, et al. Excess DAX1 leads to XY ovotesticular disorder of sex development (DSD) in mice by inhibiting steroidogenic factor-1 (SF1) activation of the testis enhancer of SRY-box-9 (Sox9). Endocrinology 2012;153(4):1948–58.

[12] Lee T-L, Li Y, Cheung H-H, Claus J, Singh S, Sastry C, et al. GonadSAGE: a comprehensive SAGE database for transcript discovery on male embryonic gonad development. Bioinformatics 2010;26(4):585–6.

[13] Hanley NA, Hagan DM, Clement-Jones M, Ball SG, Strachan T, Salas-Cortés L, et al. SRY, SOX9, and DAX1 expression patterns during human sex determination and gonadal development. Mech Dev 2000;91(1–2):403–7.

[14] Gropp A, Ohno S. The presence of a common embryonic blastema for ovarian and testicular parenchymal (follicular, interstitial and tubular) cells in cattle Bos taurus. Z Zellforsch Mikrosk Anat 1966;74(4):505–28.

[15] Lucas-Herald AK, Bashamboo A. Gonadal Development; 2014 p.1–16.

[16] Bandiera R, Vidal VPI, Motamedi FJ, Clarkson M, Sahut-Barnola I, von Gise A, et al. WT1 maintains adrenal-gonadal primordium identity and marks a population of AGP-like progenitors within the adrenal gland. Dev Cell 2013;27(1):5–18.

[17] Spiess C, Meyer AS, Reissmann S, Frydman J. Mechanism of the eukaryotic chaperonin: protein folding in the chamber of secrets. Trends Cell Biol 2004;14(11):598–604.

[18] Millan-Zambrano G, Chavez S. Nuclear functions of prefoldin. Open Biol 2014;4(7):140085.

[19] Freund A, Zhong FL, Venteicher AS, Meng Z, Veenstra TD, Frydman J, et al. Proteostatic control of telomerase function through TRiC-mediated folding of TCAB1. Cell 2014;159(6):1389–403.

[20] Zhang N, Deuel TF. Pleiotrophin and midkine, a family of mitogenic and angiogenic heparin-binding growth and differentiation factors. Curr Opin Hematol 1999;6(1):44–50.

[21] Ohta S, Shimekake Y, Nagata K. Molecular cloning and characterization of a transcription factor for the C-type natriuretic peptide gene promoter. Eur J Biochem 1996;242(3):460–6.

[22] Miller AJ, Cole SE. Multiple Dlk1 splice variants are expressed during early mouse embryogenesis. Int J Dev Biol 2014;58(1):65–70.

[23] Xu J, Lamouille S, Derynck R. TGF-β-induced epithelial to mesenchymal transition. Cell Res 2009;19(2):156–72.

[24] Kusaka M, Katoh-Fukui Y, Ogawa H, Miyabayashi K, Baba T, Shima Y, et al. Abnormal epithelial cell polarity and ectopic epidermal growth factor receptor (EGFR) expression induced in Emx2 KO embryonic gonads. Endocrinology 2010;151(12):5893–904.

[25] Cui S, Ross A, Stallings N, Parker KL, Capel B, Quaggin SE. Disrupted gonadogenesis and male-to-female sex reversal in Pod1 knockout mice. Development 2004;131(16):4095–105.

[26] França M, Lerario A, Fragoso M, Lotfi C. New evidences on the regulation of SF-1 expression by POD1/TCF21 in adrenocortical tumor cells. Clinics 2017;72(6):391–4.

[27] Amayed P, Pantaloni D, Carlier M-F. The effect of stathmin phosphorylation on microtubule assembly depends on tubulin critical concentration. J Biol Chem 2002;277(25):22718–24.

[28] Ishikawa F, Ushida K, Mori K, Shibanuma M. Loss of anchorage primarily induces nonapoptotic cell death in a human mammary epithelial cell line under atypical focal adhesion kinase signaling. Cell Death Dis 2015;6(1):e1619.

[29] Tournier C, Hess P, Yang DD, Xu J, Turner TK, Nimnual A, et al. Requirement of JNK for stress-induced activation of the cytochrome c-mediated death pathway. Science 2000;288(5467):870–4.

[30] Wilhelm D, Palmer S, Koopman P. Sex determination and gonadal development in mammals. Physiol Rev 2007;87(1):1–28.

[31] Birk OS, Casiano DE, Wassif CA, Cogliati T, Zhao L, Zhao Y, et al. The LIM homeobox gene Lhx9 is essential for mouse gonad formation. Nature 2000;403(6772):909–13.

[32] Ermakov A, Pells S, Freile P, Ganeva VV, Wildenhain J, Bradley M, et al. A role for intracellular calcium downstream of G-protein signaling in undifferentiated human embryonic stem cell culture. Stem Cell Res 2012;9(3):171–84.

[33] Mochida S, Hunt T. Protein phosphatases and their regulation in the control of mitosis. EMBO Rep 2012;13(3):197–203.

[34] Craggs G, Finan PM, Lawson D, Wingfield J, Perera T, Gadher S, et al. A nuclear SH3 domain-binding protein that colocalizes with mRNA splicing factors and intermediate filament-containing perinuclear networks. J Biol Chem 2001;276(32):30552–60.

[35] Caffrey JJ, Safrany ST, Yang X, Shears SB. Discovery of molecular and catalytic diversity among human diphosphoinositol-polyphosphate phosphohydrolases. An expanding Nudt family. J Biol Chem 2000;275(17):12730–6.

[36] Karl J, Capel B. Sertoli cells of the mouse testis originate from the Coelomic epithelium. Dev Biol 1998;203(2):323–33.

[37] Schmahl J, Eicher EM, Washburn LL, Capel B. Sry induces cell proliferation in the mouse gonad. Development 2000;127(1):65–73.

[38] Tilmann C, Capel B. Mesonephric cell migration induces testis cord formation and Sertoli cell differentiation in the mammalian gonad. Development 1999;126(13):2883–90.

[39] Barsoum IB, Kaur J, Ge RS, Cooke PS, HH-C Y. Dynamic changes in fetal Leydig cell populations influence adult Leydig cell populations in mice. FASEB J 2013;27(7):2657–66.

[40] Stevant I, Neirijnck Y, Borel C, Escoffier J, Smith LB, Antonarakis SE, et al. Deciphering cell lineage specification during male sex determination with single-cell RNA sequencing. Cell Rep 2018;22(6):1589–99.

[41] Bowles J, Koopman P. Retinoic acid, meiosis and germ cell fate in mammals. Development 2007;134(19):3401–11.

[42] Liu C, Peng J, Matzuk MM, HH-C Y. Lineage specification of ovarian theca cells requires multicellular interactions via oocyte and granulosa cells. Nat Commun 2015;6:6934.

[43] Piprek RP. Molecular mechanisms underlying female sex determination—antagonism between female and male pathway. Folia Biol (Praha) 2009;57(3–4):105–13.

[44] Luo X, Ikeda Y, Parker KL. A cell-specific nuclear receptor is essential for adrenal and gonadal development and sexual differentiation. Cell 1994;77(4):481–90.

[45] Yu RN, Ito M, Jameson JL. The murine *Dax-1* promoter is stimulated by SF-1 (steroidogenic factor-1) and inhibited by COUP-TF (chicken ovalbumin upstream promoter-transcription factor) via a composite nuclear receptor-regulatory element. Mol Endocrinol 1998;12(7):1010–22.

[46] Lovell-Badge R, Canning C, Sekido R. Sex-determining genes in mice: building pathways. Novartis Found Symp 2002;244:4–18. 18–22, 35–42, 253–7.

[47] Sekido R, Bar I, Narváez V, Penny G, Lovell-Badge R. SOX9 is up-regulated by the transient expression of SRY specifically in Sertoli cell precursors. Dev Biol 2004;274(2):271–9.

[48] Sekido R, Lovell-Badge R. Sex determination involves synergistic action of SRY and SF1 on a specific Sox9 enhancer. Nature 2008;453(7197):930–4.

[49] Gummow BM, Winnay JN, Hammer GD. Convergence of Wnt signaling and steroidogenic factor-1 (SF-1) on transcription of the rat inhibin α gene. J Biol Chem 2003;278(29):26572–9.

[50] Raverdeau M, Gely-Pernot A, Féret B, Dennefeld C, Benoit G, Davidson I, et al. Retinoic acid induces sertoli cell paracrine signals for spermatogonia differentiation but cell autonomously drives spermatocyte meiosis. Proc Natl Acad Sci U S A 2012;109(41):16582–7.

[51] Menke DB, Page DC. Sexually dimorphic gene expression in the developing mouse gonad. Gene Expr Patterns 2002;(3–4):359–67.

[52] Saba R, Wu Q, Saga Y. CYP26B1 promotes male germ cell differentiation by suppressing STRA8-dependent meiotic and STRA8-independent mitotic pathways. Dev Biol 2014;389(2):173–81.

[53] Li L-A, Chang Y-C, Wang C-J, Tsai F-Y, Jong S-B, Chung B-C. Steroidogenic factor 1 differentially regulates basal and inducible steroidogenic gene expression and steroid synthesis in human adrenocortical H295R cells. J Steroid Biochem Mol Biol 2004;91(1–2):11–20.

[54] Tremblay JJ, Robert NM, Viger RS. Modulation of endogenous GATA-4 activity reveals its dual contribution to Müllerian inhibiting substance gene transcription in sertoli cells. Mol Endocrinol 2001;15(9):1636–50.

[55] Hu Y-C, Okumura LM, Page DC. Gata4 is required for formation of the genital ridge in mice. PLoS Genet 2013;9(7):e1003629.

[56] Miyamoto Y, Taniguchi H, Hamel F, Silversides DW, Viger RS. A GATA4/WT1 cooperation regulates transcription of genes required for mammalian sex determination and differentiation. BMC Mol Biol 2008;9:44.

[57] Hu Y-C, Nicholls PK, Soh YQS, Daniele JR, Junker JP, van Oudenaarden A, et al. Licensing of primordial germ cells for gametogenesis depends on genital ridge signaling. PLoS Genet 2015;11(3):e1005019.

[58] Lourenço D, Brauner R, Rybczynska M, Nihoul-Fékété C, McElreavey K, Bashamboo A. Loss-of-function mutation in GATA4 causes anomalies of human testicular development. Proc Natl Acad Sci U S A 2011;108(4):1597–602.

[59] Manuylov NL, Zhou B, Ma Q, Fox SC, Pu WT, Tevosian SG. Conditional ablation of Gata4 and Fog2 genes in mice reveals their distinct roles in mammalian sexual differentiation. Dev Biol 2011;353(2):229–41.

[60] Efimenko E, Padua MB, Manuylov NL, Fox SC, Morse DA, Tevosian SG. The transcription factor GATA4 is required for follicular development and normal ovarian function. Dev Biol 2013;381(1):144–58.

[61] Xu X, Rongali SC, Miles JP, Lee KD, Lee M. pat-4/ILK and unc-112/Mig-2 are required for gonad function in Caenorhabditis elegans. Exp Cell Res 2006;312(9):1475–83.

[62] Muscatelli F, Strom TM, Walker AP, Zanaria E, Récan D, Meindl A, et al. Mutations in the DAX-1 gene give rise to both X-linked adrenal hypoplasia congenita and hypogonadotropic hypogonadism. Nature 1994;372(6507):672–6.

[63] Iyer AK, McCabe ERB. Molecular mechanisms of DAX1 action. Mol Genet Metab 2004;83(1–2):60–73.

[64] Kim CJ. Congenital lipoid adrenal hyperplasia. Ann Pediatr Endocrinol Metab 2014;19(4):179.

[65] Meeks JJ, Weiss J, Jameson JL. Dax1 is required for testis determination. Nat Genet 2003;34(1):32–3.

[66] Bouma GJ, Albrecht KH, Washburn LL, Recknagel AK, Churchill GA, Eicher EM. Gonadal sex reversal in mutant Dax1 XY mice: a failure to upregulate Sox9 in pre-Sertoli cells. Development 2005;132(13):3045–54.

[67] Park SY, Lee E-J, Emge D, Jahn CL, Jameson JL. A phenotypic spectrum of sexual development in Dax1 (Nr0b1)-deficient mice: consequence of the C57BL/6J strain on sex determination. Biol Reprod 2008;79(6):1038–45.

[68] Ritchie HH, Wang LH, Tsai S, O'Malley BW, Tsai MJ. COUP-TF gene: a structure unique for the steroid/thyroid receptor superfamily. Nucleic Acids Res 1990;18(23):6857–62.

[69] Pelletier J, Bruening W, Li FP, Haber DA, Glaser T, Housman DE. WT1 mutations contribute to abnormal genital system development and hereditary Wilms' tumour. Nature 1991;353(6343):431–4.

[70] Armstrong JF, Pritchard-Jones K, Bickmore WA, Hastie ND, Bard JB. The expression of the Wilms' tumour gene, WT1, in the developing mammalian embryo. Mech Dev 1993;40(1–2):85–97.

[71] Englert C, Maheswaran S, Garvin AJ, Kreidberg J, Haber DA. Induction of p21 by the Wilms' tumor suppressor gene WT1. Cancer Res 1997;57(8):1429–34.

[72] Loeb DM. WT1 influences apoptosis through transcriptional regulation of Bcl-2 family members. Cell Cycle 2006;5(12):1249–53.

[73] Kim J, Prawitt D, Bardeesy N, Torban E, Vicaner C, Goodyer P, et al. The Wilms' tumor suppressor gene (wt1) product regulates Dax-1 gene expression during gonadal differentiation. Mol Cell Biol 1999;19(3):2289–99.

[74] Kreidberg JA, Sariola H, Loring JM, Maeda M, Pelletier J, Housman D, et al. WT-1 is required for early kidney development. Cell 1993;74(4):679–91.

[75] Morrison AA, Venables JP, Dellaire G, Ladomery MR. The Wilms tumour suppressor protein WT1 (+KTS isoform) binds alpha-actinin 1 mRNA via its zinc-finger domain. Biochem Cell Biol 2006;84(5):789–98.

[76] Bullejos M, Koopman P. Spatially dynamic expression of Sry in mouse genital ridges. Dev Dyn 2001; 221(2):201–5.

[77] Bradford ST, Wilhelm D, Bandiera R, Vidal V, Schedl A, Koopman P. A cell-autonomous role for WT1 in regulating Sry in vivo. Hum Mol Genet 2009;18(18):3429–38.

[78] Zhang L, Chen M, Wen Q, Li Y, Wang Y, Wang Y, et al. Reprogramming of Sertoli cells to fetal-like Leydig cells by Wt1 ablation. Proc Natl Acad Sci U S A 2015;112(13):4003–8.

[79] Chen M, Zhang L, Cui X, Lin X, Li Y, Wang Y, et al. Wt1 directs the lineage specification of sertoli and granulosa cells by repressing Sf1 expression. Development 2017;144(1):44–53.

[80] Beischlag TV, Luis Morales J, Hollingshead BD, Perdew GH. The aryl hydrocarbon receptor complex and the control of gene expression. Crit Rev Eukaryot Gene Expr 2008;18(3):207–50.

[81] Miyamoto N, Yoshida M, Kuratani S, Matsuo I, Aizawa S. Defects of urogenital development in mice lacking Emx2. Development 1997;124(9):1653–64.

[82] Voon DC-C, Wang H, Koo JKW, Chai JH, Hor YT, Tan TZ, et al. EMT-induced stemness and tumorigenicity are fueled by the EGFR/Ras pathway. PLoS One 2013;8(8):e70427.

[83] Katoh-Fukui Y, Miyabayashi K, Komatsu T, Owaki A, Baba T, Shima Y, et al. Cbx2, a polycomb group gene, is required for Sry gene expression in mice. Endocrinology 2012;153(2):913–24.

[84] Eid W, Opitz L, Biason-Lauber A. Genome-wide identification of CBX2 targets: insights in the human sex development network. Mol Endocrinol 2015;29(2):247–57.

[85] Lee S-J, Choi D, Rhim H, Choo H-J, Ko Y-G, Kim CG, et al. PHB2 interacts with RNF2 and represses CP2c-stimulated transcription. Mol Cell Biochem 2008;319(1–2):69–77.

[86] Hambiliki F, Ström S, Zhang P, Stavreus-Evers A. Colocalization of NANOG and OCT4 in human preimplantation embryos and in human embryonic stem cells. J Assist Reprod Genet 2012;29(10):1021–8.

[87] Sánchez-Sánchez AV, Camp E, Leal-Tassias A, Atkinson SP, Armstrong L, Díaz-Llopis M, et al. Nanog regulates primordial germ cell migration through Cxcr4b. Stem Cells 2010;28(9):1457–64.

[88] Giese K, Pagel J, Grosschedl R. Distinct DNA-binding properties of the high mobility group domain of murine and human SRY sex-determining factors. Proc Natl Acad Sci U S A 1994;91(8):3368–72.

[89] Svingen T, Koopman P. Building the mammalian testis: origins, differentiation, and assembly of the component cell populations. Genes Dev 2013;27(22):2409–26.

[90] Whitfield LS, Lovell-Badge R, Goodfellow PN. Rapid sequence evolution of the mammalian sex-determining gene SRY. Nature 1993;364(6439):713–5.

[91] Hou L, Srivastava Y, Jauch R. Molecular basis for the genome engagement by Sox proteins. Semin Cell Dev Biol 2017;63:2–12.

[92] Perkins ND, Edwards NL, Duckett CS, Agranoff AB, Schmid RM, Nabel GJ. A cooperative interaction between NF-kappa B and Sp1 is required for HIV-1 enhancer activation. EMBO J 1993;12(9):3551–8.

[93] Swain A, Narvaez V, Burgoyne P, Camerino G, Lovell-Badge R. Dax1 antagonizes Sry action in mammalian sex determination. Nature 1998;391(6669):761–7.

[94] Vidal VPI, Chaboissier M-C, de Rooij DG, Schedl A. Sox9 induces testis development in XX transgenic mice. Nat Genet 2001;28(3):216–7.

[95] da Silva SM, Hacker A, Harley V, Goodfellow P, Swain A, Lovell-Badge R. Sox9 expression during gonadal development implies a conserved role for the gene in testis differentiation in mammals and birds. Nat Genet 1996;14(1):62–8.

[96] Daneau I, Pilon N, Boyer A, Behdjani R, Overbeek PA, Viger R, et al. The porcine SRY promoter is transactivated within a male genital ridge environment. Genesis 2002;33(4):170–80.

[97] De Santa Barbara P, Bonneaud N, Boizet B, Desclozeaux M, Moniot B, Sudbeck P, et al. Direct interaction of SRY-related protein SOX9 and steroidogenic factor 1 regulates transcription of the human anti-Müllerian hormone gene. Mol Cell Biol 1998 Nov;18(11):6653–65.

[98] Shen WH, Moore CC, Ikeda Y, Parker KL, Ingraham HA. Nuclear receptor steroidogenic factor 1 regulates the müllerian inhibiting substance gene: a link to the sex determination cascade. Cell 1994;77(5):651–61.

[99] Kim Y, Kobayashi A, Sekido R, DiNapoli L, Brennan J, Chaboissier M-C, et al. Fgf9 and Wnt4 act as antagonistic signals to regulate mammalian sex determination. Hamada H, editor, PLoS Biol 2006;4(6):e187.

[100] Moniot B, Declosmenil F, Barrionuevo F, Scherer G, Aritake K, Malki S, et al. The PGD2 pathway, independently of FGF9, amplifies SOX9 activity in Sertoli cells during male sexual differentiation. Development 2009;136(11):1813–21.

[101] Schepers G, Wilson M, Wilhelm D, Koopman P. SOX8 is expressed during testis differentiation in mice and synergizes with SF1 to activate the Amh promoter in vitro. J Biol Chem 2003;278(30):28101–8.

[102] Uhlenhaut NH, Jakob S, Anlag K, Eisenberger T, Sekido R, Kress J, et al. Somatic sex reprogramming of adult ovaries to testes by FOXL2 ablation. Cell 2009;139(6):1130–42.

[103] Bernard P, Ryan J, Sim H, Czech DP, Sinclair AH, Koopman P, et al. Wnt signaling in ovarian development inhibits Sf1 activation of Sox9 via the Tesco enhancer. Endocrinology 2012;153(2):901–12.

[104] Duverger O, Morasso MI. Role of homeobox genes in the patterning, specification, and differentiation of ectodermal appendages in mammals. J Cell Physiol 2008;216(2):337–46.

[105] Svingen T, Koopman P. Involvement of homeobox genes in mammalian sexual development. Sex Dev 2007;1(1):12–23.

[106] Mazaud S, Oréal E, Guigon CJ, Carré-Eusèbe D, Magre S. Lhx9 expression during gonadal morphogenesis as related to the state of cell differentiation. Gene Expr Patterns 2002;2(3–4):373–7.

[107] Yue D, Li H, Che J, Zhang Y, Tolani B, Mo M, et al. EMX2 is a predictive marker for adjuvant chemotherapy in lung squamous cell carcinomas. de Mello RA, editor, PLoS One 2015;10(7):e0132134.

[108] Fukuchi-Shimogori T, Grove EA. Emx2 patterns the neocortex by regulating FGF positional signaling. Nat Neurosci 2003;6(8):825–31.

[109] Schnabel CA, Selleri L, Cleary ML. Pbx1 is essential for adrenal development and urogenital differentiation. Genesis 2003;37(3):123–30.

[110] Schnabel CA, Selleri L, Jacobs Y, Warnke R, Cleary ML. Expression of Pbx1b during mammalian organogenesis. Mech Dev 2001;100(1):131–5.

[111] Tanaka SS, Nishinakamura R. Regulation of male sex determination: genital ridge formation and Sry activation in mice. Cell Mol Life Sci 2014;71(24):4781–802.

[112] Eggers S, Ohnesorg T, Sinclair A. Genetic regulation of mammalian gonad development. Nat Rev Endocrinol 2014;10(11):673–83.

[113] Bashamboo A, Eozenou C, Rojo S, McElreavey K. Anomalies in human sex determination provide unique insights into the complex genetic interactions of early gonad development. Clin Genet 2017;91(2):143–56.

[114] Kobayashi H, Kawakami K, Asashima M, Nishinakamura R. Six1 and Six4 are essential for Gdnf expression in the metanephric mesenchyme and ureteric bud formation, while Six1 deficiency alone causes mesonephric-tubule defects. Mech Dev 2007;124(4):290–303.

[115] Ozaki H, Watanabe Y, Takahashi K, Kitamura K, Tanaka A, Urase K, et al. Six4, a putative myogenin gene regulator, is not essential for mouse embryonal development. Mol Cell Biol 2001;21(10):3343–50.

[116] Fujimoto Y, Tanaka SS, Yamaguchi YL, Kobayashi H, Kuroki S, Tachibana M, et al. Homeoproteins Six1 and Six4 regulate male sex determination and mouse gonadal development. Dev Cell 2013;26(4):416–30.

[117] Duboule D, Boncinelli E, DeRobertis E, Featherstone M, Lonai P, Oliver G, et al. An update of mouse and human HOX gene nomenclature. Genomics 1990;7(3):458–9.

[118] Scott MP. Vertebrate homeobox gene nomenclature. Cell 1992;71(4):551–3.

[119] Greig S, Akam M. The role of homeotic genes in the specification of the Drosophila gonad. Curr Biol 1995;5(9):1057–62.

[120] Wolgemuth DJ, Viviano CM, Gizang-Ginsberg E, Frohman MA, Joyner AL, Martin GR. Differential expression of the mouse homeobox-containing gene Hox-1.4 during male germ cell differentiation and embryonic development. Proc Natl Acad Sci U S A 1987;84(16):5813–7.

[121] Dollé P, Duboule D. Two gene members of the murine HOX-5 complex show regional and cell-type specific expression in developing limbs and gonads. EMBO J 1989;8(5):1507–15.

[122] Hsieh-Li HM, Witte DP, Weinstein M, Branford W, Li H, Small K, et al. Hoxa 11 structure, extensive antisense transcription, and function in male and female fertility. Development 1995;121(5):1373–85.

[123] Pitera JE, Milla PJ, Scambler P, Adjaye J. Cloning of HOXD1 from unfertilised human oocytes and expression analyses during murine oogenesis and embryogenesis. Mech Dev 2001;109(2):377–81.

[124] Taylor HS, Vanden Heuvel GB, Igarashi P. A conserved Hox axis in the mouse and human female reproductive system: Late establishment and persistent adult expression of the Hoxa cluster genes. Biol Reprod 1997;57(6):1338–45.

[125] Satokata I, Benson G, Maas R. Sexually dimorphic sterility phenotypes in Hoxa10-deficient mice. Nature 1995;374(6521):460–3.

[126] Benson GV, Lim H, Paria BC, Satokata I, Dey SK, Maas RL. Mechanisms of reduced fertility in Hoxa-10 mutant mice: uterine homeosis and loss of maternal Hoxa-10 expression. Development 1996;122(9):2687–96.

[127] Morgan EA, Nguyen SB, Scott V, Stadler HS. Loss of Bmp7 and Fgf8 signaling in Hoxa13-mutant mice causes hypospadia. Development 2003;130(14):3095–109.

[128] Ekici AB, Strissel PL, Oppelt PG, Renner SP, Brucker S, Beckmann MW, et al. HOXA10 and HOXA13 sequence variations in human female genital malformations including congenital absence of the uterus and vagina. Gene 2013;518(2):267–72.

[129] Cheng Z, Zhu Y, Su D, Wang J, Cheng L, Chen B, et al. A novel mutation of HOXA10 in a Chinese woman with a Mullerian duct anomaly. Hum Reprod 2011;26(11):3197–201.

[130] Suzuki M, Ueno N, Kuroiwa A. Hox proteins functionally cooperate with the GC box-binding protein system through distinct domains. J Biol Chem 2003;278(32):30148–56.

[131] Connell M, Owen C, Segars J. Genetic syndromes and genes involved in the development of the female reproductive tract: a possible role for gene therapy. J Genet Syndr Gene Ther 2013;4.

[132] Miller C, Sassoon DA. Wnt-7a maintains appropriate uterine patterning during the development of the mouse female reproductive tract. Development 1998;125(16):3201–11.

[133] LaRonde-LeBlanc NA, Wolberger C. Structure of HoxA9 and Pbx1 bound to DNA: Hox hexapeptide and DNA recognition anterior to posterior. Genes Dev 2003;17(16):2060–72.

Genetics of Meiotic Chromosome Dynamics and Fertility

Travis Kent, Tanmoy Bhattacharyya, Mary Ann Handel

The Jackson Laboratory, Bar Harbor, ME, United States

OUTLINE

Abbreviations

AE	axial elements
BPA	bis-phenol A
C	DNA content
CE	central element of the SC

CO	crossover
DSB	DNA double-strand break
GWAS	genome-wide association studies
HR	homologous recombination
HS	recombination hot spot
LE	lateral elements
MI	first meiotic division
MII	second meiotic division
MPF	metaphase promoting factor
N	chromosome number (ploidy)
NCO	noncrossover
RA	retinoic acid
RN	recombination nodule
SC	synaptonemal complex
SNP	single nucleotide polymorphism
TFs	transverse filaments

OVERVIEW AND SCOPE

Meiosis is not only a defining event of gametogenesis and central to reproductive success and the production of offspring, but also a basic feature of eukaryotic genetics. It provides for the "gene shuffling" that leads to genetic diversity in populations, and also ensures the genomic integrity and correct number of chromosomes that produce healthy offspring. Meiosis is an area of intense investigation in the fields of developmental and reproductive medicine. Given the space limitations of this chapter, readers are frequently referred to many excellent reviews rather than primary literature. The goal of this chapter is to provide genetic perspectives on meiosis that are useful in the context of human health, fertility, and disease. The emphasis is on how the complex chromosomal dynamics of meiosis are revealed through the lens of insights derived from the analysis of mutations in the controlling genes. The purview extends beyond human biology to studies of meiosis in model animal organisms (particularly the laboratory mouse), which have identified many of the genes and proteins mentioned in this review (Table 3.1) and thereby informed our understanding of mammalian meiosis. Finally, human health concerns relevant to meiotic processes are discussed with the perspective that knowledge of meiotic mechanisms will enhance the understanding of human fertility, infertility, health, and disease.

MEIOSIS AND ITS REGULATION IN MALES AND FEMALES

Meiosis is an essential prerequisite to mammalian fertility. In order to preserve the correct chromosome number through generations, the number of chromosomes must be precisely reduced by half in the production of mammalian gametes (Fig. 3.1A); this allows diploidy to be restored upon fertilization. In this section, the basic stages of mammalian meiosis that lead to this outcome are reviewed, with attention to both sexual dimorphism in events and tempo (Fig. 3.1B) and the gonadal context in which meiosis unfolds and how that provides insight into regulatory mechanisms.

TABLE 3.1 Proteins Involved in Meiosis

Protein	Human Gene	Mouse Gene	Mutant Phenotypes	Infertility	References
AGO4	AGO4	Ago4	Premature entry into meiosis and abnormal of sex body; males and females are fertile	Both fertile	[1]
ATM	ATM	Atm	Pachynema arrest	M, F	[2]
ATR	ATR	Atr	Pachytene spermatocyte losses		[3]
BRCA1	BRCA1	Brca1	Persistent DSBs, no COs, and sterility in males	M	[4]
BRCA2	BRCA2	Brca2	Pachynema arrest, persistent DSBs, and sterility in males. Reduced numbers of oocytes in females	M	[5]
CCNA1	CCNA1	Ccna1	Male sterility due to abnormal spermiogenesis	M	[6]
CDK2	CDK2	Cdk2	Germ cell death during meiotic prophase	M, F	[7, 8]
CYP26B1	CYP26B1	Cyp26b1	Precocious meiotic entry in fetal testes	Embryonic lethal	[9, 10]
DMC1	DMC1	Dmc1	Zygonema arrest, no synapsis, unrepaired DSBs, no crossover	M, F	[11, 12]
DMRT1	DMRT1	Dmrt1	Precocious entry into meiosis in testis	M	[13]
EXO1	EXO1	Exo1	Arrest at metaphase I in males, nondisjunction at first meiotic division in females	M, F	[14]
EIF4G3	EIF4G3	Eif4g3	Spermatocytes fail to exit meiotic prophase	M	[15]
FKBP9	FKBP9	Fkbp6	Pachytene arrest in males	M	[16]
H2AFX	H2AFX	H2afx	Pachytene arrest in males	M	[17, 18]
HEI10	HEI10	Hei10	Failure of crossover formation; meiotic arrest	M, F	[19]
HOP2	HOP2	Hop2	Zygonema/pachynema arrest, defective synapsis and recombination	M, F	[20]
HORMAD1	HORMAD1	Hormad1	Zygonema arrest, no synapsis, unrepaired DSBs, no crossover	M, F	[21, 22]
HORMAD2	HORMAD2	Hormad2	Pachynema arrest in males; failure of MSCI	M	[23, 24]
HSPA2	HSPA2	Hspa2	Pachynema/diplonema arrest in males	M	[25]
IHO1	IHO1	Iho1	Failure of meiotic recombination initiation, no chromosomal asynapsis	M, F	[26]
KASH5	KASH5	Kash5	Meiotic arrest, no synapsis, unrepaired DSBs, no crossover	M, F	[27]

Continued

TABLE 3.1   Proteins Involved in Meiosis—cont'd

Protein	Human Gene	Mouse Gene	Mutant Phenotypes	Infertility	References
MAJIN	MAJIN	Majin	Zygonema arrest, no chromosomal pairing or synapsis	M, F	[28]
MCM9	MCM9	Mcm9	Progressive germ cell loss, meiotic block in some testes tubules	Both fertile	[29]
MDC1	MDC1	Mdc1	Pachynema arrest in males	M	[30]
MEI1	MEI1	Mei1	Zygonema arrest, failure of meiotic recombination initiation, no chromosomal asynapsis	M, F	[31–33]
MEI4	MEI4	Mei4	Zygonema arrest, failure of meiotic recombination initiation, no chromosomal asynapsis	M, F	[34]
MEIOB	MEIOB	Meiob	Zygonema arrest, no synapsis, unrepaired DSBs; no crossover	M, F	[35, 36]
MEIOC	MEIOC	Meioc	Abnormal and short meiotic prophase, premature condensation, recombination failure, no crossover	M, F	[37, 38]
MLH1	MLH1	Mlh1	Normal synapsis, crossover failure	M, F	[39–41]
MLH3	MLH3	Mlh3	Metaphase arrest, no crossover; metaphase arrest	M, F	[42]
MRE11	MRE11	Mre11	Severely reduced female fertility	F	[43]
MSH4	MSH4	Msh4	Zygonema arrest, no synapsis, unrepaired DSBs, no crossover	M, F	[44]
MSH5	MSH5	Msh5	Zygonema arrest, no synapsis, unrepaired DSBs, no crossover	M, F	[45]
MTAP2	MTAP2	Mtap2	Metaphase arrest, normal synapsis; females fertile	M	[46]
MYBL1	MYBL1	Mybl1	Pachynema arrest in males, lack of CO, and disruption of cell cycle. Females fertile	M	[47]
PLK1	PLK1	Plk1	Failure to desynapse SCs, unable to properly exit prophase I	Not tested	[48]
PRDM9	PRDM9	Prdm9	Pachynema arrest, chromosomal asynapsis and recombination defects	M, F	[49]
RAD21L	RAD21L	Rad21l	Zygonema arrest, abnormal chromosomal pairing, synaptic and DSB repair failure	M	[50, 51]
RAD51	RAD51	Rad51	Knockdown leads to apoptosis of spermatocytes at late prophase. Fertility parameters unknown	Not tested	[52]

TABLE 3.1 Proteins Involved in Meiosis—cont'd

Protein	Human Gene	Mouse Gene	Mutant Phenotypes	Infertility	References
REC8	REC8	Rec8	Zygonema arrest, abnormal SC, synapsis between sister chromatids, failure of recombination, no crossover	M, F	[53, 54]
RNF212	RNF212	Rnf212	Metaphase arrest in males	M	[55, 56]
SIX6OS1	SIX6OS1	Six6os1	Pachynema arrest, no synapsis, unrepaired DSBs	M, F	[57]
SMC1B	SMC1B	Smc1β	Pachynema arrest, no crossover, unrepaired DSBs in males. Aneuploid oocytes in females	M, F	[58]
SPATA22	SPATA22	Spata22	Zygonema arrest in males, pachynema arrest in females, impaired synapsis and DNA repair	M, F	[59]
SPO11	SPO11	Spo11	Failure of meiotic recombination initiation, no chromosomal asynapsis	M, F	[60, 61]
STAG3	STAG3	Stag3	Zygonema arrest, abnormal SC, failure of meiotic recombination; no crossover	M, F	[62–66]
STRA8	STRA8	Stra8	Failure to properly initiate meiotic prophase	M, F	[67–69]
SUN1	SUN1	Sun1	Zygonema arrest, no chromosomal pairing or synapsis	M, F	[70]
SYCE1	SYCE1	Syce1	Pachynema arrest, synapsis and recombination failure	M, F	[71]
SYCE2	SYCE2	Syce2	Pachynema arrest, synapsis and recombination failure	M, F	[72]
SYCE3	SYCE3	Syce3	Pachynema arrest, synapsis and recombination failure	M, F	[73]
SYCP1	SYCP1	Sycp1	Pachynema arrest, synapsis and recombination failure	M, F	[74]
SYCP2	SYCP2	Sycp2	Zygonema arrest in males, abnormal SC, failure of meiotic recombination; no crossover; females are subfertile due to enrichment of aneuploid oocytes in females	M	[75]
SYCP3	SYCP3	Sycp3	Zygonema arrest in males, abnormal SC, failure of meiotic recombination; no crossover; females are subfertile due to enrichment of aneuploid oocytes in females	M	[76]

Continued

TABLE 3.1 Proteins Involved in Meiosis—cont'd

Protein	Human Gene	Mouse Gene	Mutant Phenotypes	Infertility	References
TERB1	TERB1	Terb1	Zygonema arrest, no chromosomal pairing or synapsis	M, F	[77, 78]
TEX12	TEX12	Tex12	Pachynema arrest, synapsis and recombination failure	M, F	[79, 80]
TEX15	TEX15	Tex15	Pachynema arrest in males, synapsis and recombination failure	M	[81]
TOPOVIBL	TOPOVIBL	TopoVIBL	Failure of meiotic recombination initiation, no chromosomal asynapsis	M, F	[82]
TRIP13	TRIP13	Trip13	Unrepaired DSBs	M, F	[83, 84]

Meiotic Stages

In eutherian mammals, each cell has two copies of each autosomal chromosome, one paternally derived and the other maternally derived, as well as two sex chromosomes (XX in females, or XY in males). While each S phase of DNA replication in mitotically proliferating cells is followed by a single mitotic division, meiosis is unique in that the S phase is followed by two meiotic divisions (Fig. 3.1A). Additionally, mitosis yields genetically identical cells whereas meiosis produces genetically distinct cells that become the gametes. These defining features of meiosis are enabled by pairing and recombination between homologous chromosomes, which sets the stage for the two meiotic divisions. The first meiotic division (MI) is a reductive division that separates the homologous chromosomes while the second meiotic division (MII) is an equational division that results in the segregation of sister chromatids (Fig. 3.1A). The end products form the haploid gametes, with one copy of each homologous chromosome.

Meiosis is initiated with an extended S phase, replicating DNA and generating a cell with 4C DNA content where "C" represents the DNA of a single chromatid from each chromosome. After the premeiotic DNA replication, germ cells are known as primary gonocytes, either spermatocytes (in the male) or oocytes (in the female). Gonocytes are 2N, or diploid, with respect to chromosome number (Fig. 3.1A); however, each chromosome is composed of two chromatids (DNA molecules with associated proteins).

Events of the first meiotic prophase (Fig. 3.2) set the stage for the subsequent segregation of chromosomes to form haploid gametes. In this extended prophase, during the leptotene and zygotene substages (Fig. 3.2), chromosomes condense and then the maternal and paternal homologs align in homologous pairs (Figs. 3.1A and 3.2). Exactly how homology is recognized is one of the lingering mysteries of meiosis. The final intimate juxtaposition of homologs, called synapsis, marks the beginning of the pachytene substage of meiotic prophase I (Fig. 3.2). It is achieved with the aid of a proteinaceous zipper-like structure called the synaptonemal complex (SC). As the homolog pairing and synapsis processes progress, molecular recombination occurs between nonsister chromatids, a process beginning during the leptotene substage and lasting through the pachytene substage. At least one recombination event or crossover (CO),

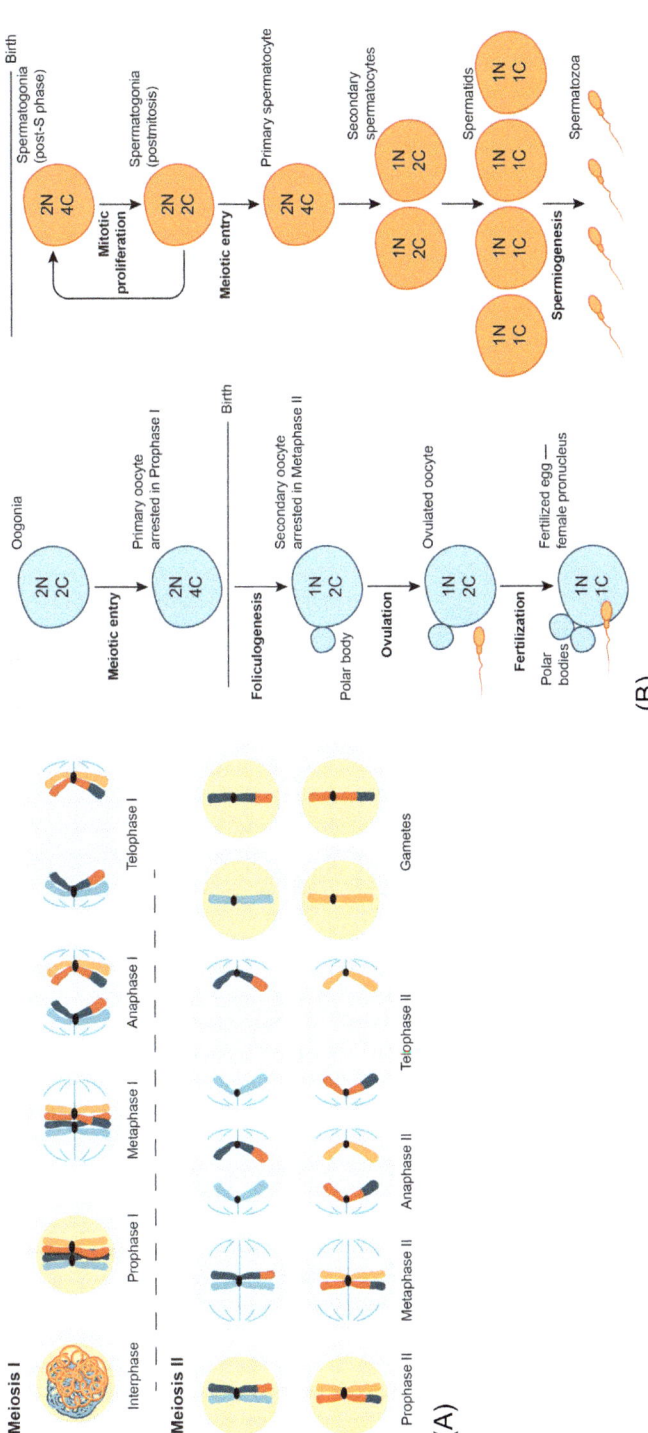

FIG 3.1 Mammalian meiosis and sexual dimorphism. (A) Schematic overview of meiosis. This schematic illustrates the key events of mammalian meiosis. During the interphase before meiosis, diploid (2N) germ cells undergo DNA replication, and are seen here with two identical copies of both the maternal (*red and orange*) and paternal (*dark and light blue*) chromosomes, thus with 4C DNA content. During the first meiotic prophase, chromatin condenses and the homologs pair and undergo DNA breakage and exchange. After nuclear envelope breakdown, microtubules attach to kinetochores in metaphase I, followed by separation of homologs during anaphase I and cytokinesis in telophase at the end of the first meiotic division. This produces two haploid (1N) cells, each with 2C DNA content. In the ensuing second meiotic division, sister chromatids align at metaphase and segregate during anaphase and telophase, producing haploid germ cells (1N, 1C). (B) Sexual dimorphism in mammalian meiosis. There is considerable sexual dimorphism in the timing of meiotic events in mammals. In females (*left panel*), meiotic prophase is initiated during fetal development and then is arrested at the end of the first meiotic prophase, around the time of birth in mice. Meiotic progress is resumed when the oocyte is in large, almost fully developed follicles, but arrests at MII just prior to ovulation. After ovulation and fertilization of the oocyte, meiosis is completed, resulting in a haploid female pronucleus. In males (*right panel*), from puberty onward continuous waves of germ cells enter meiosis. There is no arrest of meiosis in males, and it is completed over the course of a few weeks. Following the second meiotic division, haploid germ cells undergo extensive differentiation to form spermatozoa.

Prophase I

 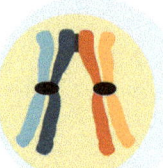 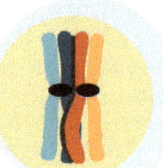 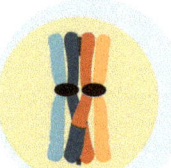 

Leptonema
- Replicated homologous chromosomes condense and begin to align
- DSBs formed

Zygonema
- Homologous pairing progresses; synapsis initiated with SC formation
- Processing of DSBs begins

Pachynema
- Homologous chromosomes fully synapsed and SC visible
- DSBs repaired with formation of COs

Diplonema
- Disassembly of SC
- Chiasmata between homologous chromosomes visible

Diakinesis
- Chromosome bivalents fully condensed
- Nuclear envelope breaks down

Metaphase I

Metaphase
- Spindle microtubules attach to nonsister kinetochores
- Tension aligns chromosome bivalents in bipolar orientation

FIG. 3.2 Events of the first meiotic prophase. Many of the key events of meiosis occur during the first lengthy meiotic prophase in five distinct substages, diagrammatically illustrated here. During leptonema, chromosomes condense and begin to pair homologously with initiation of DSBs occurring. In zygonema, homologous pairing progresses, DSB repair begins, and synapsis is initiated with the formation of short lengths of the SC. Pachynema is marked by full and homologous synapsis with SC formed along the entire length of each homologous chromosome pair. Subsets of DSBs are repaired as reciprocal COs. During diplonema, the SC is disassembled and COs become visible as chiasmata holding the homologs together. By diakinesis, the chromosome bivalents have fully condensed and the nuclear envelope breaks down, marking metaphase when the spindle microtubules attach to nonsister kinetochores, aligning the homologous pairs on the spindle, poised for separation.

visualized as a chiasma, must occur on each homologous chromosome pair for meiosis to be resolved successfully. This is because the chiasmata holds the homologous chromosomes poised for segregation from each other at the onset of the first meiotic division. At the end of meiotic prophase I (Fig. 3.2), during the diplotene substage, the SC is disassembled and the homologous chromosomes are held together by chiasmata. At the onset of MI, the microtubules of the forming spindle apparatus attach to kinetochores located at chromosomal centromeres, aligning the chromosomes. This, after the disassembly of chiasmata, allows the homologs to be pulled to opposite ends of the spindle at anaphase. After completion of cytokinesis, the "reductional" division is complete and germ cells are haploid (1N), but each chromosome is composed of two sister chromatids (2C) (Fig. 3.1A). Sister chromatids are separated from each other in MII, an equational division that is very similar to mitosis (Fig. 3.1) and results in haploid (1N) gametes, each containing one sister chromatid from each chromosome (1C) [85].

Sexual Dimorphism in Meiosis

The timing of meiotic stages and events between males and females is very different (Fig. 3.1B) [86]. In male mammals, meiosis is initiated throughout postpubertal life and proceeds without arrest from the premeiotic S-phase to the completion of MII. In contrast, in female mammals, progress through meiosis is punctuated. Oocytes enter the first meiotic prophase embryonically and become arrested at the end of prophase in an extended diplotene configuration known as the dictyate stage. Massive numbers of oocytes undergo atresia at this point (and later during the ovarian cycles), a phenomenon that is not found in males. The dictyate stage in surviving oocytes can last for weeks and months (in the laboratory mouse) or decades (in humans). Following puberty, cohorts of follicles are recruited in each ovarian (estrus or menstrual) cycle to develop through the first division ("meiotic maturation") and arrest again at metaphase II, an arrest that is broken by fertilization. In addition to sexual dimorphism in the timing of meiotic events, the products of meiosis differ between males and females. In males, meiosis results in four haploid spermatids that differentiate to form sperm (Fig. 3.1B). However, in females, oocyte cytokinesis is asymmetric, producing only one functional oocyte and two small polar bodies containing complete nuclei, but very little cytoplasm (Fig. 3.1B). In human females, gametogenesis ceases after roughly 50 years of age, but in males, sperm production can continue well into old age. Other aspects of sexual dimorphism in meiosis will be noted throughout this chapter.

Gonadal Context of Meiosis

Mammalian germ cells undergo meiosis only in the context of the gonad—the testis or the ovary—where they are in intimate contact with surrounding somatic cells, facilitating bidirectional signaling regulating the tempo of meiosis (The requirement for gonadal signaling is one reason underlying the difficulties in recapitulating germ-cell development ex vivo [87]). In the ovary, the oocyte's somatic "nurse" cells are the granulosa cells of the follicle that surround the oocyte; in the testis, it is the Sertoli cells of the seminiferous epithelium that surround and nurture the spermatocytes. There are many excellent reviews on the contact and communication between germ cells and their surrounding somatic cells that are essential for their development [88–95]. It is not only the somatic cells immediately surrounding the germ cells that are crucial for meiosis, but also more distant cells are important because they are responsible for producing gonadal hormones essential for normal meiosis. The theca and Leydig cells, in the ovary and the testis, respectively, are the primary gonadal hormone-producing cells. Theca cells comprise the outer layer of developing follicles, where they produce and transfer androstenedione to the granulosa cells for conversion by aromatase to 17β-estradiol [96]. In the testis, the primary gonadal hormone produced by Leydig cells is testosterone, which is transported throughout the body. In Sertoli cells, testosterone signaling via the androgen receptor (AR) is required for completion of meiosis [97, 98].

In the testis, germ cells develop in seminiferous tubules, where they are intimately associated with Sertoli cells, any one of which supports multiple germ cells at different developmental stages. Communication between Sertoli cells and germ cells is important for meiosis and development. For example, ablation of AR in Sertoli cells causes meiotic arrest [97, 98]. These interactions occur in the complex anatomical environment of the seminiferous epithelium, which regularly cycles through stages morphologically defined by the associations of different developmental states of the germ cells [89]. This cycle is initiated and set by iterative recruitment of cohorts of immature spermatogonia, which become committed to complete spermatogenesis and eventually develop into spermatozoa before their release into the lumen of the seminiferous epithelium. In the mouse, new cohorts of cells undergo spermatogonial differentiation every 8.6 days (d) [99], thus forming a new layer of germ cells along the basement membrane, basal to the progressing germ cells of the previous cohort. These layers form predictable, repeatable germ-cell associations, set by the kinetics of both differentiation and the repeating 8.6 d cycle. In the mouse, there are 12 such germ-cell associations, or "stages" of the seminiferous epithelium. Over an 8.6 d cycle, one tubule cross-section progresses through all 12 stages before returning to its initial stage. During this 8.6 d window, each germ cell in the tubule cross-section progresses further in its differentiation and is consequently one layer closer to the lumen of the tubule at the end of the 8.6 d period. The process is quite similar in humans, but with a ~16 d cycle between spermatogonial differentiation events and with orientation of the stages of the seminiferous epithelium in a spiral pattern around the tubule, resulting in epithelial cross-sections displaying 3–4 stages rather than one [100–102]. Regardless of the species, the events of meiosis entail ~1.5 cycles (~12 days in mouse and ~24 days in humans). This temporal cycling, with the existence of specific stages of cell associations, is an important component in regulating meiosis. For example, as discussed in detail below, signaling from retinoic acid (RA) initiates meiosis and the availability of RA is stage-dependent [103, 104], thus helping to explain how the cycle and its component stages are set. Additionally, *Ar* expression in the Sertoli cell varies across the spermatogenic cycle, suggesting a stage-specific role for AR signaling [105].

In the ovary, meiosis occurs in the context of follicles, which are assembled following birth as granulosa cells surround the differentiating oocytes that are at the stage of dictyate arrest. Prior to follicle formation, oocytes are conjoined by cytoplasmic bridges in cysts, and thus share cytoplasmic contents. As newly formed follicles develop, the granulosa cells proliferate with the morphology progressing from the primordial follicle to primary, secondary, and eventually the antral follicle, which has a lumen, suggesting secretory activity of the granulosa cells. Throughout follicular growth to the stage of ovulation, the germ cells and granulosa cells are engaged in constant bidirectional communication, each instructing the other to provide molecules required for the next stage of development [91–94]. As discussed above, oocyte meiosis is not continuously progressive but is marked by arrests and restarts ultimately tied to ovarian cyclicity. The establishment of a hormonal cycle at puberty is the catalyst for the resumption of meiosis. Gonadotropin activity in the hypothalamus, pituitary, and ovary results in establishment of a dominant follicle(s) with an oocyte that undergoes meiotic maturation through the first division and is ovulated.

GENETIC APPROACHES TO STUDY MEIOSIS

For many years, the molecular mechanisms underlying the fundamental processes of meiosis were not understood, but both biased and unbiased genetic screens in model organisms have led to important and surprising insights derived from the many genes thus discovered (Table 3.1).

Strategies to Identify Meiotic Genes in Model Organisms

Until the past few decades, a primary obstacle to progress in elucidating meiotic mechanisms was the absence of resources and technologies for viewing and tracking meiosis, even in genetically well-studied organisms such as fruit flies (*Drosophila melanogaster*) and nematode worms (*Caenorhabditis elegans*) [106]. Arguably, one of the most important enabling technologies for tracking the progress of meiosis was the discovery of the SC. This proteinaceous structure, which holds homologous chromosomes together, was first visualized by electron microscopy in the middle of the last century [107–110]. By the turn of this century, there were antibodies to many SC proteins, allowing tracking and quantification of meiotic progress by immunofluorescence microscopy [111]. This greatly propelled the field forward and enabled the genetic screens that have identified the regulatory, structural, and catalytic proteins that mediate meiosis.

In conjunction with antibody and protein interaction technologies, both forward and reverse genetics have been successfully used to identify proteins in model organisms that are essential for meiosis. Biochemical dissection of the SC led to the identification of component proteins, the functions of which were subsequently validated by reverse genetic approaches, including transgenesis and targeted gene interruption. Model organisms such as yeast, flies, worms, and laboratory mice contributed greatly to these approaches, which have informed our knowledge about mammalian, particularly human, orthologs [106]. For the discovery of mammalian genes essential for meiosis and fertility, massively high-throughput international efforts (such as KOMP, the Knockout Mouse Project) to ablate all mouse genes are reaping unexpected discoveries in many realms of biomedicine [112]. Resources such as advanced mouse genetic diversity populations [113] are illustrating the importance of genetic backgrounds and modifier genes in phenotypes. The advent of CRISPR/*Cas9* technology is making such gene targeting quicker, easier, and more precise [114]. Finally, RNAi screens targeting known genes are proving to be an important and informative ex vivo approach to identifying meiotically functioning genes [115].

While useful for testing known candidates, reverse genetic approaches can be limiting as the researcher must have a candidate gene in mind. Unbiased, forward genetic screens are founded on a principle of genome-wide mutagenesis with subsequent screening for phenotypes of interest. As one example of this approach, the Reproductive Genomics program at the Jackson Laboratory used chemical mutagenesis coupled with a screen for fertility in potential homozygous mutant mice [116]. Infertile phenotypes were mapped by traditional genetic approaches to identify the causative genes. This forward-genetics screen was validated by the identification of phenotypes caused by genes previously known to be essential for meiosis, but proved its worth by the identification of many novel genes not previously known to be associated with meiosis or fertility [47, 59, 116–118]. In spite of their tremendous value, such forward genetics screens using mammalian models entail investments of time and resources beyond what would be required for a similar experiment in "lower" organisms, and are limited to institutions and consortia enabled for such programs [116, 119–121]. For this reason, there has been considerable and productive attention devoted to validating mammalian function for genes determined to be essential for meiosis in yeast, *C. elegans*, and *Drosophila*.

Genetic Analyses of Human Meiosis

Unbiased genetic screens are, of course, impossible in humans, where researchers are restricted to less invasive and more limited approaches to understanding the genetic regulation of meiosis and fertility. Even given these impediments, cytological studies have been of tremendous value. For example, 50 years ago, karyotype analyses for the first time linked a clinical syndrome to a chromosomal disorder with the discovery of trisomy 21 in individuals with Down syndrome [122, 123]. By extension, this finding implicated meiotic error as the underlying cause. Because of this discovery, chromosomal and karyotype analyses have been a useful readout to detect abnormal meiosis. Trisomies associated with chromosomes 13, 15, 16, 17, 18, 20, 21, and 22 as well as both sex chromosomes have been identified using karyotyping at various points during gametogenesis and embryogenesis [124]. Moreover, the advent and increased prevalence of assisted reproductive technologies has provided material that is useful in understanding and tracking the incidence of aneuploidy from gametogenesis through parturition. Importantly, studies of chromosomal material have progressed far beyond tracking the origin and etiology of aneuploidy, and now decipher basic mechanisms of meiosis in human germ cells. For example, it is possible to analyze correlations between meiotic recombination in human meiocytes and initiation of synapsis to provide information on sex- and species-specific mechanisms [125, 126].

In spite of these advances founded in modern cytogenetics, identifying the human genes important for meiosis is more problematic. Because access to clinical material for staging and marking progress of meiosis is limited, other tools must be exploited. One such tool now being applied to study human infertility is genome-wide association studies (GWAS). These observational analyses of phenotypic and genetic variation require large and genetically diverse panels of individuals for relating a phenotype of interest to specific genomic regions. The ultimate goal of GWAS is discovery of single nucleotide polymorphisms (SNPs) that are more likely to occur, or be associated, with a phenotype of interest. While such genetic association does not demonstrate that any specific SNP is causative for the noted phenotype, it can reveal a genomic region that is a good target for future interrogation. For example, by analyzing data from more than 1000 two-generation families, one study identified genes associated with five parameters of recombination such as number and placement of recombination events [127]. The candidates thus identified included not only genes previously known to play roles in recombination (including *RNF212* and *PRDM9*; see below), but also novel genes with no previously known roles in meiotic recombination. However, results from GWAS must be approached with caution. Stringent testing of candidate genes identified via GWAS is required to validate their significance because SNPs identified as associated with a phenotype of interest may not play a biological role. One such validation analysis exploited rapid-throughput and precise CRISPR/*Cas9* gene editing in mice to test the functions of nonsynonymous mutations that had been associated with meiosis and infertility by human GWAS. Instructively, only one of the four mutations so tested resulted in infertility [128]. In a slightly different strategy, 54 evolutionarily conserved and testis-expressed genes were tested by CRISPR/*Cas9* targeting and found to be not required for male fertility [129]. Thus, cross-species strategies to identify human fertility genes putatively required for meiosis must be carefully validated.

In spite of the caveats, the genetic strategies reviewed here have resulted in insights that greatly contribute to our current understanding of mammalian meiosis and its regulation. This is particularly the case for insights into the chromosomal dynamics of meiotic prophase and the meiotic division phases, the two most essential and defining processes of meiosis, as reviewed in the following sections.

MEIOTIC PROPHASE CHROMOSOME DYNAMICS: THE FOUNDATION OF MEIOSIS

At the heart of meiosis are the unique events in the first prophase: initiation of a meiotic (rather than mitotic) division, homologous chromosome pairing, and molecular recombination between homologous (but nonsister) chromatids. These processes are initiated by gonadal signals and orchestrated by unique aspects of chromosomal axes (Fig. 3.3). Their mechanisms and regulation have been revealed by genetic approaches (Table 3.1).

Initiation of Meiosis in Germ Cells

The nature of the instructive signals leading to the onset of meiotic chromosome dynamics in the germ cells of laboratory mice is one of the mysteries that has been resolved, at least in part, by genetic strategies. The morphogen retinoic acid (RA), the active metabolite of vitamin A (retinol), initiates the meiotic program of germ cells, albeit in a sexually dimorphic manner.

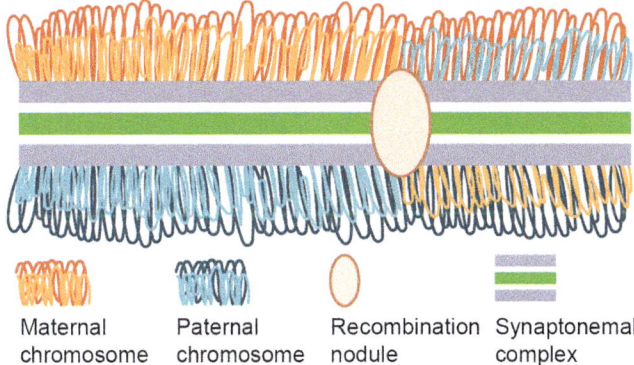

Maternal chromosome   Paternal chromosome   Recombination nodule   Synaptonemal complex

FIG. 3.3 The synaptonemal complex: This figure is a diagram of the major components of the SC, conceptualized in the pachytene substage of the first meiotic prophase. Here the two genetically identical sister chromatids of the maternal homolog are depicted in *red* and *orange* loops while the sister chromatids of the paternal homolog are depicted in *dark* and *light blue*. Each of the two homologously paired chromosomes are attached to the lateral elements (LE) of the SC, depicted in *lavender*. The LEs are maintained in synapsis configuration by SC central elements (CEs), depicted in *green*. This configuration allows precisely aligned segments of DNA from the two homologs to undergo recombination, catalyzed in protein complexes known as recombination nodules (RN), depicted as a *tan oval*. The outcome of recombination is diagrammatically illustrated here as a *light blue* paternal chromatid joined into the maternal homolog, and the reciprocal, *light orange* maternal chromatid in the paternal homolog.

Retinol, which is of broad developmental significance, is widely available throughout the body via the circulatory system and is transported into target cells, where it undergoes a two-step enzymatic conversion to retinaldehyde and RA [130]. RA is either degraded by cytochrome P450 family 26 (CYP26) enzymes, or binds to retinoic acid receptors (RARs) or retinoid X receptors (RXRs). Upon ligand binding, RAR and RXR heterodimerize and promote gene expression by binding to RA-response elements in DNA upstream of RA-regulated genes. An important target for RA action on meiosis in the mouse is the gene *Stra8* (stimulated by retinoic acid, gene 8) [131]; indeed, the onset of meiosis in germ cells of both sexes is dependent on expression of the STRA8 protein [67–69].

RA availability in the fetal gonad is responsible for the decision of sexually undifferentiated primordial germ cells to follow either the female path of entry into meiosis, or the male path of mitotic quiescence without meiotic initiation. By 13.5 days postcoitus in mouse embryos, RA-promoted sexual dimorphism is apparent, with ovarian germ cells entering meiosis and testicular germ cells ceasing their mitotic proliferation. Although RARs are present in both testes and ovaries in mice, there is differential expression—in fetal testes but not ovaries—of *Cyp26b1*, a gene that encodes a protein responsible for RA degradation [9, 132–134]. Evidence suggests that this may account for the resistance of fetal male germ cells to the meiosis-promoting activity of RA. Studies utilizing gene knockouts of *Cyp26b1* or CYP26 inhibitors have shown that failure to degrade RA in embryonic testes results in precocious expression of the *Stra8* target of RA and entry of the male germ cells into meiosis [9, 132, 134]. Numerous studies with widely differing approaches of inhibitors, gene knockouts, and dietary restrictions have shown that RA in postnatal testes is critical for male germ cells to become differentiated spermatogonia, committed to ultimately initiate meiosis [135–139].

As stated above, activation of the *Stra8* gene is essential for meiosis in both sexes but, interestingly, *Stra8* mutant phenotypes are susceptible to the action of modifier genes. This is known because *Stra8* knockouts on different genetic backgrounds produce variable phenotypes with respect to initial stages of meiosis [67, 69]. Although STRA8 is essential for normal meiosis, its precise function and downstream targets are not yet fully understood. Genetic analyses suggest that it is a key activator of downstream genes required for multiple aspects of meiosis, including chromosome dynamics and recombination [67, 68]. STRA8 may also function as a switch between the mitotic and meiotic cell cycle programs, suggested by the finding that one gene it activates is *Meioc* (Meiosis-specific with coiled-coil domain), which is critical for the extended meiotic prophase required to complete chromosome dynamics and meiotic recombination processes [37]. MEIOC may act by preventing degradation of transcripts of genes essential for meiotic prophase [38]. Finally, RA is not the only regulator of the *Stra8* gene. The DMRT1 (doublesex related transcription factor 1) protein regulates *Stra8* in complex and sexually dimorphic manners [13]. Overall, these genetic analyses have identified key regulators of meiotic initiation, demonstrating that the regulation of meiotic entry is complex and finely tuned.

Meiotic Chromosome Axes and Homologous Interaction

Homology pairing and synapsis of chromosomes in prophase are mediated by organizational and structural features of meiotic chromosomes that differ from those of mitotic chromosomes. Most notable of these are the chromosomal axes, which form the proteinaceous scaffold to which DNA loops are attached [140–142] (Fig. 3.3).

Initial Formation of the Meiotic Chromosome Axis

The meiotic chromosome axis, initially similar to the mitotic chromosomes axis, evolves through meiotic prophase into the unique protein superstructure that is the SC [62, 140–147]. The early axis, or axial elements (AEs) of meiotic chromosomes, contains cohesin complexes. Cohesin complexes are ring-like structures composed of four protein subunits: two structural maintenance of chromosome proteins (SMC1A or SMC1B and SMC3), a kleisin protein (RAD21, RAD21L, or REC8), and a stromal antigen protein (STAG1, 2, or 3) [141, 143]. In mitotic cells, cohesin complexes function in sister-chromatid cohesion, contributing to precise chromosome segregation [148]. While SMC1A, SMC3, RAD21, STAG1, and STAG2 are expressed in mitotic cells, SMC1B, RAD21L, REC8, and STAG3 are cohesin subunits required specifically for meiosis. Genetic targeting of each meiotic cohesin subunit in mouse models has established their distinct and individual roles and phenotypes. All meiotic cohesin subunits are required for normal axis length, completion of meiosis, and fertility, but the phenotypes can be sexually dimorphic and subject to modification by background genes. For example, deficiency of SMC1B causes infertility with meiotic arrest at the pachytene stage in males, but arrest as late as MII in females [58]. Also, deletion of RAD21L causes reduced axis length and meiotic arrest in the midzygotene stage in males, but mutant females are fertile [50, 51]. However, overall there are major abnormalities of axis structure and subsequent SC formation in germ cells with mutations of one or more meiotic cohesin proteins [141, 143, 149], revealing the essential nature of these proteins for development and maintenance of meiotic chromatin organization. Moreover, these proteins are of major clinical significance for understanding the etiology of human age-related gametic aneuploidy, as revealed by findings that oocytes are sensitive to cohesin protein dosage [150] and also exhibit age-related deterioration of cohesin complexes related to errors of meiotic chromosome segregation [151].

Homologous Interactions of Meiotic Chromosomes

The cohesin axes are assembled concurrently with, and facilitate, the processes by which meiotic chromosomes find each other and undergo homologous pairing. There is considerable interest in the concept of a chromosome "bar code" as a mechanism mediating the homology search and recognition process, and this might be based on the proteins composing the axis. For example, the meiosis-specific cohesin complex proteins REC8 and RAD21L have mutually exclusive localization along the chromosomal axis, potentially providing such a bar code for homology recognition [152]. Deletion of RAD21L impedes the pairing of homologous chromosomes and homolog recognition during early meiotic prophase [51, 152].

Although the axes are important, chromosome movement also appears to contribute to homologous interactions. In germ cells of laboratory mice, chromosomal telomeres exhibit dynamic movement and tethering to the nuclear envelope at the preleptotene stage, forming a cluster termed the "bouquet" [153], which facilitates homologous interactions between chromosomes. Clustering of chromosome ends has the effect of reducing the size of nuclear space that must be searched for homology recognition. Deletion of the gene encoding the nuclear envelope protein SUN1 (Sad1 and UNC84 domain containing 1) causes infertility in both sexes. The chromosomal telomeres of SUN1-deficient germ cells fail to attach to the nuclear envelope and lose their mobility, with subsequent failure of homologous chromosome pairing [70, 154]. The essential nature of this relationship between telomeres and the nuclear envelope is reinforced by the meiotic requirement for many other proteins that interact with SUN1 and/or telomeres [78].

In summary, genetic approaches suggest that both axis architecture and chromosomal movements play essential roles in pairing and homolog recognition during mammalian meiosis.

Homologous Synapsis and the Role of the SC

The processes leading to homology-based pairing are reinforced by maturation of the chromosome axis to the mature SC. Because the SC is the morphological bridge between paired chromosomes, investigations of it have been central to efforts to decipher the mechanisms by which homologous chromosomes interact. The SC is a player in the final intimate synapsis or "gluing" of homologs to each other (Fig. 3.3), which is the culmination of homology recognition and pairing, completed by the onset of the pachytene stage. The past few decades have seen an explosion of work to define both the structure and function of the SC, and although it is beyond the scope of this chapter to cover its structure in detail, the reader is referred to several excellent and comprehensive reviews [140, 147, 155–157]. The multiprotein SC axis, or scaffold, is comprised of lateral elements (LEs) interacting with and derived from the cohesin-rich AEs described above, and a central element (CE) with ladder-like transverse filaments (TFs). In general, SC proteins are nearly all specific to meiotic prophase, and their genetic disruption results in abnormal meiosis and, frequently, infertility. Major proteins of the LEs are SYCP2 and SYCP3 (synaptonemal complex proteins 2 and 3). Deletion of either SYCP3 or SYCP2 during meiotic prophase perturbs the formation of the SC [75, 76, 158, 159]. Deletion of SYCP2 impedes recruitment of SYCP3 to the LEs [75], although SYCP2 interacts with TF and CE proteins but not with SYCP3 [75, 159], giving a picture of the complexity of the assembly of the mature SC. Although deletion of SYCP3 or SYCP2 in males leads to meiotic arrest and sterility, deletion in females causes only subfertility but a high rate of oocyte aneuploidy, a fact that may provide insight into the etiology of some human gametic aneuploidy [75, 76, 158, 159].

Proteins that are not actually part of the SC structure may monitor homologous synapsis, and two SC-associated proteins that are especially intriguing in this regard are the HORMA-domain proteins, HORMAD1 and HORMAD2. These proteins are associated with regions that fail to synapse or do not maintain synapsis along chromosomes. When synapsis is complete, HORMADs are depleted from the SC by an ATPase designated TRIP13 (thyroid hormone receptor interactor 13) [22, 160]. Mouse mutants for *Hormad1* are sterile in both sexes with meiotic arrest reflected by disrupted homologous pairing, chromosomal synapsis, and meiotic recombination [22]. Therefore, the HORMADs are thought to be key in assembly and/or signaling from the SC, although their precise function is not understood.

Arguably, the CE is the most interesting component of the SC because it is at the heart of homologous synapsis and is where recombination takes place (as reviewed in the next section). An essential protein of the CE is SYCP1 (synaptonemal complex protein 1). In fact, when SYCP1 is deleted, there is failure in both sexes of homologous pairing to culminate in synapsis and SC formation at the pachytene stage [74]. The C-terminal of SYCP1 interacts with proteins associated with the LEs while the N-terminal is in association with CE proteins [147].

Other CE proteins identified and validated by genetic analyses include SYCE1, SYCE2, SYCE3, TEX12, and SIX60S1. Mutations of genes encoding these proteins lead to failure to form the CE, synapsis failure, and sterility in both sexes [57, 71–73, 79, 80]. Of all the axis and SC proteins studied by genetic targeting, the CE genes are the least sexually dimorphic in phenotype, demonstrating conserved function of this region and suggesting that mutations in the human CE genes might cause infertility. One interesting exception to this generalization is FKBP6 (FK506 binding protein), which is also present in regions of homologous synapsis but has sexually dimorphic function. Targeted deletion of *Fkbp6* affects meiotic synapsis in males but not females [16]. Therefore, further investigation of this protein might clarify some sex differences in mammalian meiosis. Moreover, it can be expected that the many ongoing studies of interactions and dependence relationships among CE and LE proteins will lead to greater understanding of the assembly of the SC and the nature of homology-based interactions.

In spite of this wealth of information about the constituents of the SC, its precise function is not yet understood, in part because biophysical understanding of its properties is lacking. Originally, from both light and electron microscopy evidence, the SC was thought to be a rigid ladder-like structure. However, recent studies of the nematode SC demonstrated liquid-crystal properties (including the ability to flow, but with intrinsic ordering of components), which suggests phase transitions in its assembly and function [161]. As homologous chromosomes pair, they appear to "zip up" by the dynamic extension of the SC [140, 147, 162], a process that could be enabled by a less rigid structure facilitated by liquid-crystal properties.

In summary, these studies show how dramatically the dynamic and exquisitely intricate behavior of meiotic chromosomes is firmly rooted in the chromosomes themselves, especially the associated axis elements that continually evolve throughout meiotic prophase. The axis elements not only facilitate homologous synapsis but also are the scaffold upon which much of recombination occurs.

Meiotic Recombination

Recombination, the other major and defining event of the first meiotic prophase, plays out in synchrony with axis formation, homology recognition, and synapsis. The extent to which all these events are interdependent is not yet fully appreciated, but genetic disruption of any one of the pathways frequently affects all. Aside from its genetic function to create diversity with new and shuffled gene allele combinations, meiotic recombination serves the important cell biological function of creating the linkages between homologous chromosomes that are required for their accurate separation (or "segregation") from each other, thereby ensuring accuracy of MI. Aneuploidy and loss of genomic integrity are frequently the penalties of meiotic recombination gone wrong. The emphasis in this section is on the genetic phenotypes that have led to our appreciation of the major players in mammalian recombination; there are many excellent reviews covering the molecular biology of recombination in greater depth [163–168]. Here, an overview is provided of the three major phases of recombination in mammals: (1) activation of recombination sites by PRDM9 (PR domain containing 9), (2) formation of DNA DSBs by SPO11 (SPO11 meiotic protein covalently bound to DSB), and (3) DSB repair and CO by the many proteins that *en masse* contribute to subsequent steps of recombination in recombination nodules (RNs), the physical protein aggregates associated with the axis (Fig. 3.3).

Activation of Meiotic Recombination

Meiotic recombination is a chromatin-associated process that generally is limited to specific genomic locations called hotspots (HS) [169]. In humans and mice, the HS are determined and activated by the DNA-binding, zinc finger PRDM9 protein [165, 170–173]. The genetically variable zinc-finger domain of PRDM9 determines the genomic sites for recombination and activates them via its SET domain, which trimethylates histone 3 on lysine 4 (H3K4me3) and lysine 36 (H3K36me3) [49, 165, 174, 175]. In mice, targeted deletion of PRDM9 causes meiotic arrest with consequent sterility in both sexes [49]. The requirement is not absolute; canids lack PRDM9 and there is a known case of a fertile woman lacking functional PRDM9 [176]. Interestingly, mouse germ cells lacking PRDM9 exhibit ectopic DSBs (for instance, at transcription start sites) that fail to be repaired, and also some limited chromosomal synapsis in spite of failed recombination [49, 177].

Formation of DSBs

The second crucial step of recombination is creation of DSBs at the activated HS. This is accomplished by SPO11, a conserved topoisomerase, in association with TOPOVIBL (TopoVI DNA topoisomerase B like) [60, 61, 82]. Deletion of SPO11 or TOPOVIBL affects meiotic DSB formation and homologous synapsis, leading to sterility in both sexes [60, 61, 82]. There are two isoforms of SPO11: SPO11A and SPO11B, each with a distinct function [178]. SPO11B is involved in global DSB formation while SPO11A is uniquely involved in catalyzing DSBs in the pairing region of the X and Y sex chromosomes, thereby promoting their pairing and synapsis [178]. Although SPO11 is the primary enzyme to catalyze DSBs, its activity is regulated by negative feedback from ATM (ataxia telangiectasia mutated) kinase via phosphorylation [179]. Although it is not known how the axis forms adjacent to activated HS, it is apparent that a multiprotein complex, known as the recombinosome, is required for SPO11-mediated DSB formation. The recombinosome, a structure that assembles and evolves in its protein components during early meiotic prophase, is formed by MEI4 (meiotic double-stranded break formation protein 4) and REC114 (meiotic recombination protein) [26, 180]. Reflecting the importance of the recombinosome for tethering recombination and homology recognition, both male and female mice with mutation of the *Mei4* gene are sterile, with failure of germ cells to undergo homologous synapsis and recombination [34, 180]. Recently, another axis-associated protein, IHO1 (interactor of HORMAD1), has also been shown to play a critical role in meiotic DSB formation [26]. Deletion of IHO1 leads to complete failure of meiotic recombination initiation, similar to that observed in *Spo11* mutant spermatocytes [26]. Mice with mutation of the *Iho1* gene exhibit meiotic arrest at the midpachytene stage, with failure of synapsis [26]. These observations, and the finding that HORMAD1 itself regulates efficient formation of meiotic DSBs [21, 26], demonstrate the essential nature of the chromosomal axis for initiation of both recombination and homologous synapsis.

Repair of DSBs and CO

The third important step of meiotic recombination is repair of the DSBs and their resolution as either reciprocal recombination crossovers (COs) or noncrossover events (NCOs). COs are the outcome of homologous recombination whereby single-stranded DNA ends from a DSB invade a homologous (but nonsister) chromatid, using it as a repair template. This repair event forms the interhomolog connections visualized as chiasmata during MI [140, 147]. NCO-associated repair mechanisms involve synthesis-dependent strand annealing with a sister chromatid as the repair template, leading to gene conversion events [140, 147]. Of the large number of DSBs created in early prophase (250–300 in mice), only about 10% are resolved as COs while the remainder are repaired as NCOs. The molecular biology of these repair processes is under active investigation and is viewed as being of broad medical importance because of the similarity of many steps to the processes of mitotic DSB repair that are so important in the etiology of cancer [181, 182]. Here, a very brief overview of the players in HR is presented because these are genetically required for the COs that are essential for normal meiotic chromosome segregation. During meiotic recombination, chromatin surrounding DSBs is modified by histone H2AFX phosphorylated on serine 139 (pH2AFX) by ATM and ATR (ataxia telangiectasia and *Rad3* related) [179, 183]. Deletion of these proteins impacts meiotic processes, albeit with sexually dimorphic outcomes [3, 17]. Subsequent to this chromatin modification, other proteins move in and out of the RNs. The early and transitional RNs promote processing of DSBs, and member proteins include RPA (replication protein A) [184], MEIOB (meiosis-specific with OB domains) [35], SPATA22 (spermatogenesis associated 22) [59, 184], RAD51 (RAD51 recombinase) [52], DMC1 (DNA meiotic recombinase 1) [11, 140, 147, 184], and TEX15 (testis expressed gene 15) [81]. Because the activity of these proteins parallels the early steps of homologous chromosome pairing, it is assumed that pairing is to some degree dependent on recombination events. By the late zygotene and early pachytene stages, when homologous synapsis is established, the composition of the RNs associated with the SC changes to proteins involved in the final stages of meiotic recombination. These include the meiosis-specific heterodimer of MSH4 and MSH5 (MutS homologs 4 and 5) at the DSB sites [182]. The number of MSH4 and MSH5 foci declines about threefold from the zygotene to pachytene stages, and it has been suggested that MSH4/MSH5 determine the final repertoire of sites to be resolved as COs in cooperation with RNF212 (ring finger protein 212) [182]. These repair events are intimately associated with the ongoing events of homologous synapsis because deletion of MSH4 or MSH5 during meiotic prophase impacts both homologous synapsis and meiotic CO formation, causing meiotic arrest and sterility [44, 45, 185, 186]. The final sites designated for CO are marked at the midpachytene stage by a second set of MutL protein heterodimers, composed of mismatch repair proteins MLH1 and MLH3 (MutL homologs 1 and 3) [181, 182]. Mice with germ cells lacking either MLH1 or MLH3 are sterile with a lack of chiasmata in germ cells, which arrest in late prophase or metaphase [39, 42, 187]. Interestingly, recruitment of MLH3 to CO sites is not limited by deficiency of MLH1 [39, 182, 187, 188]. The relationship between foci of MLH1/3 and COs is so tight that MLH1/3 foci have become the preferred method for determining recombination frequency in both mouse and human germ cells [111].

MEIOTIC DIVISION PHASES: THE CULMINATION OF MEIOSIS

The complex chromosome dynamics of meiotic prophase poise the germ cell for the meiotic division phases. Recent studies are shedding light on the proteins (Table 3.1) and mechanisms by which completion of prophase and initiation of meiotic metaphase are signaled, and by which a reductive division is followed by an equational division, ensuring haploid gametes.

Regulating the Onset of Meiotic Divisions

The first step on the way to the division phase is starting the cell-cycle machinery that promotes metaphase. Throughout prophase, the progress of homologous synapsis and recombination are monitored by checkpoint mechanisms ensuring quality control [85]. Although is it not known what finally signals the completion of recombination in either male or female germ cells, the mechanisms activating the division phase have been intensely investigated and much of our knowledge derives from genetic approaches.

A key initiating event in the transition from prophase I to metaphase I is the disassembly of the SC, a process much better understood as it plays out in male germ cells. The first step, marking the onset of diplonema, is disassembly of the CE of the SC, visualized as loss of SYCP1 [189, 190]. Rapidly, SYCP3 redistributes to the centromeres and the HORMAD proteins re-associate with the chromosomal axes [160, 191, 192]. The disassembly of the SC is thought to begin with the phosphorylation of SYCP1 and TEX12 (testis expressed gene 12) via polo-like kinase (PLK1), thereby facilitating SC removal [48]. SYCP3 removal has also been shown to be under the control of the cyclin-dependent kinases (CDKs) and aurora kinases (AURKs) [190, 193].

The onset of metaphase in germ cells is prompted by many of the cyclin-dependent kinases (CDKs) and phosphatases that function in mitosis. In particular, in both mammalian oocytes and spermatocytes, this is MPF, or metaphase-promoting factor. Active MPF is a heterodimer composed of a CDK and a cyclin (CCN). Although functional in both oocytes and spermatocytes in prompting the first meiotic metaphase, the mechanisms regulating assembly and activation of MPF are sexually dimorphic. Additionally, other kinases, such as the AURKs and PLKs, are responsible for some events of progress from meiotic prophase to MI [48, 190, 194, 195]. Moreover, in male germ cells, the progress from late prophase I (the pachytene substage) to metaphase is continuous (Fig. 3.1B) while in mammalian oocytes, there is an arrest, frequently prolonged, at the end of meiotic prophase (Fig. 3.1B). And, interestingly, in both oocytes and spermatocytes, progress to metaphase and activation of MPF is under acute translational regulation.

In male germ cells, a spermatocyte-specific cyclin A1 (CCNA1) constituent of MPF promotes MI. *Ccna1*-mutant spermatocytes undergo arrest after desynapsis, but before division, failing to activate CDK1 (cyclin-dependent kinase 1) [6], although females lacking CCNA1 are fertile. CDK1 is required in spermatocytes for the final stages of chromosome condensation before metaphase [196], although not for desynapsis. One of the more interesting spermatocyte regulators of MPF is a heat shock protein, HSPA2. HSPs act both as transcription factors and as molecular chaperones, promoting protein folding. Spermatocytes with null mutation of *Hspa2* are arrested in late prophase before homolog desynapsis [25], with failure to activate

MPF due to lack of chaperone activity that juxtaposes the CDK and CCN [25]. In oocytes, MPF is also a main factor in the resumption of meiosis following dictyate arrest, a phenomenon known as meiotic maturation (as a historical note, the metaphase promoting factor MPF was originally known as "maturation promoting factor" due to its initial discovery in oocytes [197]). Oocyte meiotic arrest is maintained by a cAMP-PKA (protein kinase A) pathway, in part by the production of cGMP by the granulosa cells surrounding the oocyte [92, 93, 95, 198]. A surge of luteinizing hormone (LH) breaks this cAMP-mediated arrest, allowing activation of MPF, the oocytes progress through the remainder of MI.

In both oocytes and spermatocytes, translational control is a primary mechanism for activating MPF and other metaphase-promoting pathway proteins. Perhaps this reflects the fact that the meiotic chromatin, packaged for pairing and recombination as described above, may not be appropriately active transcriptionally. The discovery of this mode of regulation in spermatocytes derived from an unbiased mutagenesis screen [116] revealing a phenotype of meiotic arrest in spermatocytes (but not oocyte) due to a lack of EIF4G3 (eukaryotic translation initiation factor 4 gamma 3) [15]. The *Hspa2* mRNA is a specific substrate for EIF4G3; *Eif4g3* mutant spermatocytes have normal levels of *Hspa2* mRNA expression but fail to synthesize the HSPA2 protein that is required for activation of MPF [15]. Because the EIF4G3 translation initiation factor is localized in the nuclei of spermatocytes, suggesting it may be poised to act when recombination is completed [199], it will be fascinating to learn of its other substrate mRNAs and their potential role in regulating the meiotic division phase. And, as reviewed elsewhere [200], there is also translational control of the resumption of meiosis in oocytes after dictyate arrest, as shown by genetic mutation approaches and genetically engineered models such as the RiboTag mouse to enrich for cell- and stage-specific populations of actively translating polysomes [201]. For example, RiboTag immunoprecipitation of specific polysome populations has revealed differential translation of cyclins in oocytes that contributes to the tempo of entry into meiotic metaphase [202].

Reductional and Equational Segregation in Meiosis I and II

The meiotic divisions bring about "reductional" segregation of homologs from each other in the first division, to produce 1N cells that are still 2C with respect to DNA content, followed by "equational" segregation of the chromatids of each chromosome to produce gametes that are 1N (haploid) and 1C (Fig. 3.1A). This key feature of meiosis ensures euploidy of gamete cells so that fertilization can restore diploidy. Although genetic approaches have been very productive in identifying proteins and mechanisms controlling meiotic prophase chromosome behavior (Table 3.1), unbiased screens (or spontaneous mutations) have been less informative about genetic control of the meiotic division phase in mammals. It certainly appears that the mechanisms of the meiotic division phases "repurpose" highly conserved mitotically acting proteins, thus phenotypes of mutations in such genes would be expected to be manifest as early developmental lethality. For this reason, screens not requiring organismal viability are valuable. For instance, a large-scale, high-content RNAi screen coupled with live imaging of meiotic chromosome dynamics in mouse oocytes identified genes essential for normal first and second meiotic divisions [115]. Thus, in spite of difficulties, clever screens are yielding encouraging progress in understanding the genetics of the meiotic division phases.

Not surprisingly, the proteinaceous chromosomal axes are important guardians of accurate segregation. As revealed by genetic analyses of proteins required for maturation of DSBs into COs, we know that the presence of chiasmata is essential for holding homologs together until they are properly oriented on the spindle (Fig. 3.2). Prior to the division phase, cohesin is removed from the chromosome, leaving homologs held together by the chiasmata and facilitating their separation as the chiasmata are resolved in the first meiotic division (Fig. 3.1A). Cohesin is subsequently removed from between the sister centromeres, allowing sister chromatids to separate in the second meiotic division (Fig. 3.1A) [151]. Because of their long meiotic arrest (Fig. 3.1B), accurate chromosome segregation in oocytes is susceptible to age, and one contributing factor is thought to be age-related degradation of cohesins. Elegant studies using oocytes with genetically engineered cleavable REC8 revealed that cohesion complexes established at the time of premeiotic DNA replication are not replaced in growing oocytes [151, 203–205]. Therefore, inability to regenerate cohesion may be a factor in age-related aneuploidy due to missegregation in the MI division. This lack of turnover may also contribute to the sensitivity of oocytes to reduced dosage of cohesin components [205, 206]. Although their role is less well understood, other division-phase proteins also contribute to accurate reductional segregation in the first meiotic division. This may be particularly the case in oocytes, where the spindle is highly asymmetric, leading to retention of most of the cytoplasm in the future egg cell with not much more than a nucleus segregated to the polar body (Fig. 3.1B). A fascinating study used genetic and high-resolution optical approaches to demonstrate cooperation between chromosomes and cytoplasmic elements to set up spindle microtubule asymmetry [207]. This spindle asymmetry is a major contributing factor to biased (nonrandom) segregation of homologs in either the first meiotic division [207] or in the second division [208], either of which can have significant genetic consequences [209]. Indeed, recent cell-biological analyses indicate that the spindle and its component microtubules, in addition to cohesin dynamics, contribute to the success of the meiotic divisions [210–212].

MEIOSIS, FERTILITY, AND HUMAN HEALTH

Clearly, meiosis is essential for fertility, as without it, there are no gametes. Therefore, it is expected that as more precise genetic and genomic information is acquired across large cohorts of humans, mutations in genes known to be required for meiosis (Table 3.1, and discussed above) will continue to be associated with cases of human infertility. In addition to the direct relationship between meiosis genes and fertility, meiosis is also essential for genomic integrity across generations, thereby affecting both fertility and offspring well-being. Moreover, intriguing data relate abnormalities in meiosis that might be reflected in infertility to downstream health consequences in the infertile individuals.

Human Meiotic Error and Aneuploidy

Because of the essential nature of meiosis for gamete formation, the penalties of meiotic error include infertility and aneuploidy. Arrested meiosis results in the failure to produce gametes, as discussed earlier in this chapter, but errors in the meiotic divisions can lead to aneuploid gametes. A leading cause of human genetic birth defects, still births, and sponta-neous pregnancy loss is fetal aneuploidy due to meiotic chromosome nondisjunction at MI or MII. While the frequency of aneuploidy in live births is ~0.3%, the true impact of aneuploidy on human pregnancy loss is considerably greater [124]. Some aneuploidies are compatible with postgestational survival, including well-known conditions such as Down syndrome, Turner syndrome, and XXY syndrome; these and other aneuploidies more frequently asso-ciated with premature death impose considerable cost on the healthcare system.

Like many aspects of meiosis, there is significant sexual dimorphism in the incidence of gametic aneuploidy, and the maternal parent contributes to the majority of conceptus aneu-ploidy [205]. The aneuploidy rate in human sperm samples is between 1% and 4% [213–215]; the incidence of aneuploidy in human oocytes is between 20% and 70% [216–218].

Cellular and genetic mechanisms behind this sexual dimorphism in the incidence of ga-metic aneuploidy are not fully elucidated, but a number of testable hypotheses are being ex-plored experimentally. Much attention is devoted to the role of aging in the incidence of oocyte aneuploidy because the fetal initiation of meiosis followed by long prophase arrest leads to meiotically maturing oocytes that are quite "old" compared to sperm cells that are weeks from their "birth" from adult proliferating spermatogonia (Fig. 3.1B). This is an impor-tant consideration because cohesin complexes, which are essential for the step-wise reduc-tional segregation of homologs before the equational separation of chromatids, are not regenerated over time [151]. Thus, cohesin complexes that have been damaged or degraded over decades of arrest may not be repaired or replaced, leading to increased incidence of non-disjunction in the oocyte. This may, at least in part, account for the well-known maternal age effect in incidence of aneuploid pregnancies. It has also been hypothesized that the number and distribution of COs could play a role in age-related increase in aneuploidy. High-resolution mapping demonstrated a decrease in recombination frequency and abnormal lo-calization of chiasmata in oocytes from older women [219, 220].

Not surprisingly, environmental agents impact meiotic progression and gametic aneu-ploidy. Although diet is a factor determining overall sperm quality, its effect on meiosis in males is not known. In females, diet can affect follicular development; high lipid content and maternal obesity are negatively correlated with oocyte health and release from meiotic arrest for maturation [221–225]. The best characterized environmental toxin impacting mei-osis is bisphenol A (BPA). BPA is an endocrine disruptor commonly used as a plasticizing agent in many commercial products. Fetal and neonatal exposure of mice to BPA alters re-combination frequency and increases synaptonemal defects in both male and female germ cells [226–228]. While BPA usage has declined in the United States, studies show that replace-ments for BPA can have similar impacts on health [229, 230].

Meiosis and Human Disease

One developing concept that impacts our understanding of the significance of meiosis is the idea that fertility may be a biomarker for overall health [231]. A growing body of epide-miological data suggests that both male and female infertility can be a harbinger of, and as-sociated with, somatic diseases presenting later in life. Although there are many diseases, such as diabetes and cystic fibrosis, known to have a negative effect on fertility, a current goal is to determine the reverse: whether fertility status is a biomarker of the overall health and/or later-in-life health concerns [232]. For example, epidemiological data correlate infertility in males with other chronic medical conditions [231, 233–235]. Single gene mutations causing infertility in both mice and humans have also been linked to other diseases, and these include genes important for meiosis. As one example, an *Mybl1* mutation results in abnormal serum antibody response as well as aberrant breast development, but also results in male infertility due to meiotic defects [47]. Both *Brca2* and *Mcm9* are associated with meiotic defects and infertility, but also increased likelihood for cancer later in life [5, 29]. These lines of research and disease association are still in a rather nascent stage, one reason being that animals being studied for meiosis and infertility phenotypes are rarely aged to the point where aging-related diseases might be observed. Nonetheless, determination of how and to what extent meiotic impairment and infertility are associated with other pathologies will be of significant importance to clinicians, allowing for predictive, proactive, and prophylactic approaches to preventing or mitigating disease.

SUMMARY AND PERSPECTIVES

Meiosis defines and dictates much of the success of gametogenesis and is pivotal to our understanding of causes of infertility. Aside from the obvious importance for the field of human reproduction, studies of the basic chromosomal mechanisms of meiosis have greatly informed our concepts of chromosome structure and behavior in mitotically proliferating somatic cells as well as in meiotic germ cells. In this chapter, the importance of the chromosomal axis as a determinant of meiotic chromosome dynamics has been highlighted. Genetic analyses have been key to our unfolding knowledge and we look forward to continued and new insights deriving from precision mutagenesis, ex vivo germ-cell screens, and ultimately the experimental analyses that will be newly possible with the advent of systems of in vivo gametogenesis.

Acknowledgments

Work in the laboratory of the authors is supported by grants from the NIH (HD33816, HD73077, GM99640). We thank our laboratory members for input and thoughtful discussion. We are especially grateful for the contributions (those cited as well as those not cited due to space limitations) of our many colleagues in the fields of meiosis and mammalian reproduction.

References

[1] Modzelewski AJ, Holmes RJ, Hilz S, Grimson A, Cohen PE. AGO4 regulates entry into meiosis and influences silencing of sex chromosomes in the male mouse germline. Dev Cell 2012;23(2):251–64.
[2] Xu Y, Ashley T, Brainerd EE, Bronson RT, Meyn MS, Baltimore D. Targeted disruption of ATM leads to growth retardation, chromosomal fragmentation during meiosis, immune defects, and thymic lymphoma. Genes Dev 1996;10(19):2411–22.
[3] Royo H, Prosser H, Ruzankina Y, Mahadevaiah SK, Cloutier JM, Baumann M, et al. ATR acts stage specifically to regulate multiple aspects of mammalian meiotic silencing. Genes Dev 2013;27(13):1484–94.

[4] Hakem R, de la Pompa JL, Elia A, Potter J, Mak TW. Partial rescue of Brca1 (5-6) early embryonic lethality by p53 or p21 null mutation. Nat Genet 1997;16(3):298–302.

[5] Sharan SK, Pyle A, Coppola V, Babus J, Swaminathan S, Benedict J, et al. BRCA2 deficiency in mice leads to meiotic impairment and infertility. Development 2004;131(1):131–42.

[6] Liu D, Matzuk MM, Sung WK, Guo Q, Wang P, Wolgemuth DJ. Cyclin A1 is required for meiosis in the male mouse. Nat Genet 1998;20(4):377–80.

[7] Ortega S, Prieto I, Odajima J, Martín A, Dubus P, Sotillo R, et al. Cyclin-dependent kinase 2 is essential for meiosis but not for mitotic cell division in mice. Nat Genet 2003;35(1):25–31.

[8] Viera A, Rufas JS, Martínez I, Barbero JL, Ortega S, Suja JA. CDK2 is required for proper homologous pairing, recombination and sex-body formation during male mouse meiosis. J Cell Sci 2009;122(Pt 12):2149–59.

[9] MacLean G, Li H, Metzger D, Chambon P, Petkovich M. Apoptotic extinction of germ cells in testes of Cyp26b1 knockout mice. Endocrinology 2007;148(10):4560–7.

[10] Yashiro K, Zhao X, Uehara M, Yamashita K, Nishijima M, Nishino J, et al. Regulation of retinoic acid distribution is required for proximodistal patterning and outgrowth of the developing mouse limb. Dev Cell 2004; 6(3):411–22.

[11] Pittman DL, Cobb J, Schimenti KJ, Wilson LA, Cooper DM, Brignull E, et al. Meiotic prophase arrest with failure of chromosome synapsis in mice deficient for Dmc1, a germline-specific RecA homolog. Mol Cell 1998; 1(5):697–705.

[12] Yoshida K, Kondoh G, Matsuda Y, Habu T, Nishimune Y, Morita T. The mouse RecA-like gene Dmc1 is required for homologous chromosome synapsis during meiosis. Mol Cell 1998;1(5):707–18.

[13] Matson CK, Murphy MW, Griswold MD, Yoshida S, Bardwell VJ, Zarkower D. The mammalian doublesex homolog DMRT1 is a transcriptional gatekeeper that controls the mitosis versus meiosis decision in male germ cells. Dev Cell 2010;19(4):612–24.

[14] Wei K, Clark AB, Wong E, Kane MF, Mazur DJ, Parris T, et al. Inactivation of Exonuclease 1 in mice results in DNA mismatch repair defects, increased cancer susceptibility, and male and female sterility. Genes Dev 2003; 17(5):603–14.

[15] Sun F, Palmer K, Handel MA. Mutation of *Eif4g3*, encoding a eukaryotic translation initiation factor, causes male infertility and meiotic arrest of mouse spermatocytes. Development 2010;137(10):1699–707.

[16] Crackower MA, Kolas NK, Noguchi J, Sarao R, Kikuchi K, Kaneko H, et al. Essential role of *Fkbp6* in male fertility and homologous chromosome pairing in meiosis. Science 2003;300(5623):1291–5.

[17] Celeste A, Petersen S, Romanienko PJ, Fernandez-Capetillo O, Chen HT, Sedelnikova OA, et al. Genomic instability in mice lacking histone H2AX. Science 2002;296(5569):922–7.

[18] Fernandez-Capetillo O, Liebe B, Scherthan H, Nussenzweig A. H2AX regulates meiotic telomere clustering. J Cell Biol 2003;163(1):15–20.

[19] Ward JO, Reinholdt LG, Motley WW, Niswander LM, Deacon DC, Griffin LB, et al. Mutation in mouse *Hei10*, an e3 ubiquitin ligase, disrupts meiotic crossing over. PLoS Genet 2007;3(8):e139.

[20] Petukhova GV, Romanienko PJ, Camerini-Otero RD. The Hop2 protein has a direct role in promoting interhomolog interactions during mouse meiosis. Dev Cell 2003;5(6):927–36.

[21] Daniel K, Lange J, Hached K, Fu J, Anastassiadis K, Roig I, et al. Meiotic homologue alignment and its quality surveillance are controlled by mouse HORMAD1. Nat Cell Biol 2011;13(5):599–610.

[22] Shin Y-H, Choi Y, Erdin SU, Yatsenko SA, Kloc M, Yang F, et al. *Hormad1* mutation disrupts synaptonemal complex formation, recombination, and chromosome segregation in mammalian meiosis. PLoS Genet 2010; 6(11):e1001190.

[23] Kogo H, Tsutsumi M, Inagaki H, Ohye T, Kiyonari H, Kurahashi H. HORMAD2 is essential for synapsis surveillance during meiotic prophase via the recruitment of ATR activity. Genes Cells 2012;17(11):897–912.

[24] Wojtasz L, Cloutier JM, Baumann M, Daniel K, Varga J, Fu J, et al. Meiotic DNA double-strand breaks and chromosome asynapsis in mice are monitored by distinct HORMAD2-independent and -dependent mechanisms. Genes Dev 2012;26(9):958–73.

[25] Dix DJ, Allen JW, Collins BW, Poorman-Allen P, Mori C, Blizard DR, et al. HSP70-2 is required for desynapsis of synaptonemal complexes during meiotic prophase in juvenile and adult mouse spermatocytes. Development 1997;124(22):4595–603.

[26] Stanzione M, Baumann M, Papanikos F, Dereli I, Lange J, Ramlal A, et al. Meiotic DNA break formation requires the unsynapsed chromosome axis-binding protein IHO1 (CCDC36) in mice. Nat Cell Biol 2016;18(11): 1208–20.

[27] Horn HF, Kim DI, Wright GD, Wong ESM, Stewart CL, Burke B, et al. A mammalian KASH domain protein coupling meiotic chromosomes to the cytoskeleton. J Cell Biol 2013;202(7):1023–39.

[28] Shibuya H, Hernández-Hernández A, Morimoto A, Negishi L, Höög C, Watanabe Y. MAJIN links telomeric DNA to the nuclear membrane by exchanging telomere cap. Cell 2015;163(5):1252–66.

[29] Hartford SA, Luo Y, Southard TL, Min IM, Lis JT, Schimenti JC. Minichromosome maintenance helicase paralog MCM9 is dispensable for DNA replication but functions in germ-line stem cells and tumor suppression. Proc Natl Acad Sci U S A 2011;108(43):17702–7.

[30] Ichijima Y, Ichijima M, Lou Z, Nussenzweig A, Camerini-Otero RD, Chen J, et al. MDC1 directs chromosome-wide silencing of the sex chromosomes in male germ cells. Genes Dev 2011;25(9):959–71.

[31] Libby BJ, De La Fuente R, O'Brien MJ, Wigglesworth K, Cobb J, Inselman A, et al. The mouse meiotic mutation mei1 disrupts chromosome synapsis with sexually dimorphic consequences for meiotic progression. Dev Biol 2002;242(2):174–87.

[32] Libby BJ, Reinholdt LG, Schimenti JC. Positional cloning and characterization of Mei1, a vertebrate-specific gene required for normal meiotic chromosome synapsis in mice. Proc Natl Acad Sci U S A 2003;100(26): 15706–11.

[33] Reinholdt LG, Schimenti JC. *Mei1* is epistatic to *Dmc1* during mouse meiosis. Chromosoma 2005;114(2):127–34.

[34] Kumar R, Bourbon H-M, de Massy B. Functional conservation of *Mei4* for meiotic DNA double-strand break formation from yeasts to mice. Genes Dev 2010;24(12):1266–80.

[35] Luo M, Yang F, Leu NA, Landaiche J, Handel MA, Benavente R, et al. MEIOB exhibits single-stranded DNA-binding and exonuclease activities and is essential for meiotic recombination. Nat Commun 2013;4:2788.

[36] Souquet B, Abby E, Hervé R, Finsterbusch F, Tourpin S, Le Bouffant R, et al. MEIOB targets single-strand DNA and is necessary for meiotic recombination. PLoS Genet 2013;9(9):e1003784.

[37] Soh YQS, Mikedis MM, Kojima M, Godfrey AK, de Rooij DG, Page DC. *Meioc* maintains an extended meiotic prophase I in mice. PLoS Genet 2017;13(4)e1006704.

[38] Abby E, Tourpin S, Ribeiro J, Daniel K, Messiaen S, Moison D, et al. Implementation of meiosis prophase I programme requires a conserved retinoid-independent stabilizer of meiotic transcripts. Nat Commun 2016;7:10324.

[39] Edelmann W, Cohen PE, Kane M, Lau K, Morrow B, Bennett S, et al. Meiotic pachytene arrest in MLH1-deficient mice. Cell 1996;85(7):1125–34.

[40] Baker SM, Plug AW, Prolla TA, Bronner CE, Harris AC, Yao X, et al. Involvement of mouse Mlh1 in DNA mismatch repair and meiotic crossing over. Nat Genet 1996;13(3):336–42.

[41] Eaker S, Cobb J, Pyle A, Handel MA. Meiotic prophase abnormalities and metaphase cell death in MLH1-deficient mouse spermatocytes: insights into regulation of spermatogenic progress. Dev Biol 2002;249(1):85–95.

[42] Lipkin SM, Moens PB, Wang V, Lenzi M, Shanmugarajah D, Gilgeous A, et al. Meiotic arrest and aneuploidy in MLH3-deficient mice. Nat Genet 2002;31(4):385–90.

[43] Theunissen J-WF, Kaplan MI, Hunt PA, Williams BR, Ferguson DO, Alt FW, et al. Checkpoint failure and chromosomal instability without lymphomagenesis in Mre11(ATLD1/ATLD1) mice. Mol Cell 2003;12(6):1511–23.

[44] Kneitz B, Cohen PE, Avdievich E, Zhu L, Kane MF, Hou H, et al. MutS homolog 4 localization to meiotic chromosomes is required for chromosome pairing during meiosis in male and female mice. Genes Dev 2000;14(9): 1085–97.

[45] De Vries SS, Baart EB, Dekker M, Siezen A, de Rooij DG, de Boer P, et al. Mouse MutS-like protein *Msh5* is required for proper chromosome synapsis in male and female meiosis. Genes Dev 1999;13(5):523–31.

[46] Sun F, Handel MA. A mutation in *Mtap2* is associated with arrest of mammalian spermatocytes before the first meiotic division. Genes (Basel) 2011;2(1):21–35.

[47] Bolcun-Filas E, Bannister LA, Barash A, Schimenti KJ, Hartford SA, Eppig JJ, et al. A-MYB (MYBL1) transcription factor is a master regulator of male meiosis. Development 2011;138(15):3319–30.

[48] Jordan PW, Karppinen J, Handel MA. Polo-like kinase is required for synaptonemal complex disassembly and phosphorylation in mouse spermatocytes. J Cell Sci 2012;125(Pt 21):5061–72.

[49] Hayashi K, Yoshida K, Matsui Y. A histone H3 methyltransferase controls epigenetic events required for meiotic prophase. Nature 2005;438(7066):374–8.

[50] Herrán Y, Gutiérrez-Caballero C, Sánchez-Martín M, Hernández T, Viera A, Barbero JL, et al. The cohesin subunit RAD21L functions in meiotic synapsis and exhibits sexual dimorphism in fertility. EMBO J 2011; 30(15):3091–105.

[51] Ishiguro K, Kim J, Fujiyama-Nakamura S, Kato S, Watanabe Y. A new meiosis-specific cohesin complex impli-cated in the cohesin code for homologous pairing. EMBO Rep 2011;12(3):267–75.

[52] Dai J, Voloshin O, Potapova S, Camerini-Otero RD. Meiotic knockdown and complementation reveals essential role of RAD51 in mouse spermatogenesis. Cell Rep 2017;18(6):1383–94.

[53] Bannister LA, Reinholdt LG, Munroe RJ, Schimenti JC. Positional cloning and characterization of mouse *mei8*, a disrupted allelle of the meiotic cohesin *Rec8*. Genesis 2004;40(3):184–94.

[54] Xu H, Beasley MD, Warren WD, van der Horst GTJ, McKay MJ. Absence of mouse REC8 cohesin promotes synapsis of sister chromatids in meiosis. Dev Cell 2005;8(6):949–61.

[55] Reynolds A, Qiao H, Yang Y, Chen JK, Jackson N, Biswas K, et al. RNF212 is a dosage-sensitive regulator of crossing-over during mammalian meiosis. Nat Genet 2013;45(3):269–78.

[56] Fujiwara Y, Matsumoto H, Akiyama K, Srivastava A, Chikushi M, Ann Handel M, et al. An ENU-induced mutation in the mouse *Rnf212* gene is associated with male meiotic failure and infertility. Reproduction 2015;149(1):67–74.

[57] Gómez HL, Felipe-Medina N, Sánchez-Martín M, Davies OR, Ramos I, García-Tuñón I, et al. C14ORF39/ SIX6OS1 is a constituent of the synaptonemal complex and is essential for mouse fertility. Nat Commun 2016;7:13298.

[58] Revenkova E, Eijpe M, Heyting C, Hodges CA, Hunt PA, Liebe B, et al. Cohesin SMC1 beta is required for mei-otic chromosome dynamics, sister chromatid cohesion and DNA recombination. Nat Cell Biol 2004;6(6):555–62.

[59] La Salle S, Palmer K, O'Brien M, Schimenti JC, Eppig J, Handel MA. *Spata22*, a novel vertebrate-specific gene, is required for meiotic progress in mouse germ cells. Biol Reprod 2012;86(2):45.

[60] Baudat F, Manova K, Yuen JP, Jasin M, Keeney S. Chromosome synapsis defects and sexually dimorphic mei-otic progression in mice lacking *Spo11*. Mol Cell 2000;6(5):989–98.

[61] Romanienko PJ, Camerini-Otero RD. The mouse *Spo11* gene is required for meiotic chromosome synapsis. Mol Cell 2000;6(5):975–87.

[62] Hopkins J, Hwang G, Jacob J, Sapp N, Bedigian R, Oka K, et al. Meiosis-specific cohesin component, *Stag3* is essential for maintaining centromere chromatid cohesion, and required for DNA repair and synapsis between homologous chromosomes. PLoS Genet 2014;10(7):e1004413.

[63] Caburet S, Arboleda VA, Llano E, Overbeek PA, Barbero JL, Oka K, et al. Mutant cohesin in premature ovarian failure. N Engl J Med 2014;370(10):943–9.

[64] Winters T, McNicoll F, Jessberger R. Meiotic cohesin STAG3 is required for chromosome axis formation and sister chromatid cohesion. EMBO J 2014;33(11):1256–70.

[65] Llano E, Gomez HL, García-Tuñón I, Sánchez-Martín M, Caburet S, Barbero JL, et al. STAG3 is a strong can-didate gene for male infertility. Hum Mol Genet 2014;23(13):3421–31.

[66] Fukuda T, Fukuda N, Agostinho A, Hernández-Hernández A, Kouznetsova A, Höög C. STAG3-mediated sta-bilization of REC8 cohesin complexes promotes chromosome synapsis during meiosis. EMBO J 2014;33 (11):1243–55.

[67] Anderson EL, Baltus AE, Roepers-Gajadien HL, Hassold TJ, de Rooij DG, van Pelt AMM, et al. *Stra8* and its inducer, retinoic acid, regulate meiotic initiation in both spermatogenesis and oogenesis in mice. Proc Natl Acad Sci U S A 2008;105(39):14976–80.

[68] Baltus AE, Menke DB, Hu Y-C, Goodheart ML, Carpenter AE, de Rooij DG, et al. In germ cells of mouse em-bryonic ovaries, the decision to enter meiosis precedes premeiotic DNA replication. Nat Genet 2006;38(12): 1430–4.

[69] Mark M, Jacobs H, Oulad-Abdelghani M, Dennefeld C, Féret B, Vernet N, et al. STRA8-deficient spermatocytes initiate, but fail to complete, meiosis and undergo premature chromosome condensation. J Cell Sci 2008;121(Pt 19):3233–42.

[70] Ding X, Xu R, Yu J, Xu T, Zhuang Y, Han M. SUN1 is required for telomere attachment to nuclear envelope and gametogenesis in mice. Dev Cell 2007;12(6):863–72.

[71] Bolcun-Filas E, Hall E, Speed R, Taggart M, Grey C, de Massy B, et al. Mutation of the mouse *Syce1* gene disrupts synapsis and suggests a link between synaptonemal complex structural components and DNA repair. PLoS Genet 2009;5(2):e1000393.

[72] Bolcun-Filas E, Costa Y, Speed R, Taggart M, Benavente R, De Rooij DG, et al. SYCE2 is required for synaptonemal complex assembly, double strand break repair, and homologous recombination. J Cell Biol 2007;176(6):741–7.

[73] Schramm S, Fraune J, Naumann R, Hernandez-Hernandez A, Höög C, Cooke HJ, et al. A novel mouse synaptonemal complex protein is essential for loading of central element proteins, recombination, and fertility. PLoS Genet 2011;7(5):e1002088.

[74] De Vries FAT, de Boer E, van den Bosch M, Baarends WM, Ooms M, Yuan L, et al. Mouse *Sycp1* functions in synaptonemal complex assembly, meiotic recombination, and XY body formation. Genes Dev 2005;19(11): 1376–89.

[75] Yang F, De La Fuente R, Leu NA, Baumann C, McLaughlin KJ, Wang PJ. Mouse SYCP2 is required for synaptonemal complex assembly and chromosomal synapsis during male meiosis. J Cell Biol 2006;173(4): 497–507.

[76] Yuan L, Liu JG, Zhao J, Brundell E, Daneholt B, Höög C. The murine *Scp3* gene is required for synaptonemal complex assembly, chromosome synapsis, and male fertility. Mol Cell 2000;5(1):73–83.

[77] Daniel K, Tränkner D, Wojtasz L, Shibuya H, Watanabe Y, Alsheimer M, et al. Mouse CCDC79 (TERB1) is a meiosis-specific telomere associated protein. BMC Cell Biol 2014;15:17.

[78] Shibuya H, Watanabe Y. The meiosis-specific modification of mammalian telomeres. Cell Cycle 2014;13(13): 2024–8.

[79] Hamer G, Wang H, Bolcun-Filas E, Cooke HJ, Benavente R, Höög C. Progression of meiotic recombination requires structural maturation of the central element of the synaptonemal complex. J Cell Sci 2008;121(Pt 15): 2445–51.

[80] Hamer G, Gell K, Kouznetsova A, Novak I, Benavente R, Höög C. Characterization of a novel meiosis-specific protein within the central element of the synaptonemal complex. J Cell Sci 2006;119(Pt 19):4025–32.

[81] Yang F, Eckardt S, Leu NA, McLaughlin KJ, Wang PJ. Mouse TEX15 is essential for DNA double-strand break repair and chromosomal synapsis during male meiosis. J Cell Biol 2008;180(4):673–9.

[82] Robert T, Nore A, Brun C, Maffre C, Crimi B, Bourbon HM, et al. The TopoVIB-Like protein family is required for meiotic DNA double-strand break formation. Science 2016;351(6276):943–9.

[83] Li XC, Schimenti JC. Mouse pachytene checkpoint 2 (*Trip13*) is required for completing meiotic recombination but not synapsis. PLoS Genet 2007;3(8):e130.

[84] Roig I, Dowdle JA, Toth A, de Rooij DG, Jasin M, Keeney S. Mouse TRIP13/PCH2 is required for recombination and normal higher-order chromosome structure during meiosis. PLoS Genet 2010;6(8).

[85] Cohen PE, Holloway JK. Mammalian meiosis. In: Plant TM, Zeleznik AJ, editors. Knobil and Neill's physiology of reproduction. 4th ed: 2015. p. 5–57.

[86] Lesch BJ, Page DC. Genetics of germ cell development. Nat Rev Genet 2012;13(11):781–94.

[87] Zhou Q, Wang M, Yuan Y, Wang X, Fu R, Wan H, et al. Complete meiosis from embryonic stem cell-derived germ cells in vitro. Cell Stem Cell 2016;18(3):330–40.

[88] Suzuki H, Kanai-Azuma M, Kanai Y. From sex determination to initial folliculogenesis in mammalian ovaries: morphogenetic waves along the anteroposterior and dorsoventral axes. Sex Dev 2015;9(4):190–204.

[89] Griswold MD. Spermatogenesis: the commitment to meiosis. Physiol Rev 2016;96(1):1–17.

[90] Chen S-R, Liu Y-X. Regulation of spermatogonial stem cell self-renewal and spermatocyte meiosis by Sertoli cell signaling. Reproduction 2015;149(4):R159–67.

[91] Jaffe LA, Egbert JR. Regulation of mammalian oocyte meiosis by intercellular communication within the ovarian follicle. Annu Rev Physiol 2017;79:237–60.

[92] Matzuk MM, Burns KH, Viveiros MM, Eppig JJ. Intercellular communication in the mammalian ovary: oocytes carry the conversation. Science 2002;296(5576):2178–80.

[93] Su Y-Q, Sugiura K, Eppig JJ. Mouse oocyte control of granulosa cell development and function: paracrine regulation of cumulus cell metabolism. Semin Reprod Med 2009;27(1):32–42.

[94] Monniaux D. Driving folliculogenesis by the oocyte-somatic cell dialog: lessons from genetic models. Theriogenology 2016;86(1):41–53.

[95] Wigglesworth K, Lee K-B, O'Brien MJ, Peng J, Matzuk MM, Eppig JJ. Bidirectional communication between oocytes and ovarian follicular somatic cells is required for meiotic arrest of mammalian oocytes. Proc Natl Acad Sci U S A 2013;110(39):E3723–9.

[96] Young JM, McNeilly AS. Theca: the forgotten cell of the ovarian follicle. Reproduction 2010;140(4):489–504.

[97] Chang C, Chen Y-T, Yeh S-D, Xu Q, Wang R-S, Guillou F, et al. Infertility with defective spermatogenesis and hypotestosteronemia in male mice lacking the androgen receptor in Sertoli cells. Proc Natl Acad Sci U S A 2004;101(18):6876–81.

[98] Tan KAL, De Gendt K, Atanassova N, Walker M, Sharpe RM, Saunders PTK, et al. The role of androgens in Sertoli cell proliferation and functional maturation: studies in mice with total or Sertoli cell-selective ablation of the androgen receptor. Endocrinology 2005;146(6):2674–83.

[99] Clermont Y. Kinetics of spermatogenesis in mammals: seminiferous epithelium cycle and spermatogonial renewal. Physiol Rev 1972;52(1):198–236.

[100] Schulze W, Rehder U. Organization and morphogenesis of the human seminiferous epithelium. Cell Tissue Res 1984;237(3):395–407.

[101] Wistuba J, Schrod A, Greve B, Hodges JK, Aslam H, Weinbauer GF, et al. Organization of seminiferous epithelium in primates: relationship to spermatogenic efficiency, phylogeny, and mating system. Biol Reprod 2003; 69(2):582–91.

[102] Luetjens CM, Weinbauer GF, Wistuba J. Primate spermatogenesis: new insights into comparative testicular organisation, spermatogenic efficiency and endocrine control. Biol Rev Camb Philos Soc 2005;80(3):475–88.

[103] Endo T, Freinkman E, de Rooij DG, Page DC. Periodic production of retinoic acid by meiotic and somatic cells coordinates four transitions in mouse spermatogenesis. Proc Natl Acad Sci U S A 2017;114(47):E10132–41.

[104] Hogarth CA, Arnold S, Kent T, Mitchell D, Isoherranen N, Griswold MD. Processive pulses of retinoic acid propel asynchronous and continuous murine sperm production. Biol Reprod 2015;92(2):37.

[105] Zhou Q, Nie R, Prins GS, Saunders PTK, Katzenellenbogen BS, Hess RA. Localization of androgen and estrogen receptors in adult male mouse reproductive tract. J Androl 2002;23(6):870–81.

[106] Matzuk MM, Lamb DJ. The biology of infertility: research advances and clinical challenges. Nat Med 2008; 14(11):1197–213.

[107] Fawcett DW. The fine structure of chromosomes in the meiotic prophase of vertebrate spermatocytes. J Biophys Biochem Cytol 1956;2(4):403–6.

[108] Moses MJ, Solari AJ. Positive contrast staining and protected drying of surface spreads: electron microscopy of the synaptonemal complex by a new method. J Ultrastruct Res 1976;54(1):109–14.

[109] Moses MJ. Chromosomal structures in crayfish spermatocytes. J Biophys Biochem Cytol 1956;2(2):215–8.

[110] Moses MJ. Structure and function of the synaptonemal complex. Genetics 1969;61(1):41–51. suppl.

[111] Lynn A, Ashley T, Hassold T. Variation in human meiotic recombination. Annu Rev Genomics Hum Genet 2004;5:317–49.

[112] Dickinson ME, Flenniken AM, Ji X, Teboul L, Wong MD, White JK, et al. High-throughput discovery of novel developmental phenotypes. Nature 2016;537(7621):508–14.

[113] Simecek P, Forejt J, Williams RW, Shiroishi T, Takada T, Lu L, et al. High-resolution maps of mouse reference populations. G3 (Bethesda) 2017;7(10):3427–34.

[114] Singh P, Schimenti JC, Bolcun-Filas E. A mouse geneticist's practical guide to CRISPR applications. Genetics 2015;199(1):1–15.

[115] Pfender S, Kuznetsov V, Pasternak M, Tischer T, Santhanam B, Schuh M. Live imaging RNAi screen reveals genes essential for meiosis in mammalian oocytes. Nature 2015;524(7564):239–42.

[116] Handel MA, Lessard C, Reinholdt L, Schimenti J, Eppig JJ. Mutagenesis as an unbiased approach to identify novel contraceptive targets. Mol Cell Endocrinol 2006;250(1-2):201–5.

[117] Su Y-Q, Sugiura K, Sun F, Pendola JK, Cox GA, Handel MA, et al. MARF1 regulates essential oogenic processes in mice. Science 2012;335(6075):1496–9.

[118] Su Y-Q, Sun F, Handel MA, Schimenti JC, Eppig JJ. Meiosis arrest female 1 (MARF1) has nuage-like function in mammalian oocytes. Proc Natl Acad Sci U S A 2012;109(46):18653–60.

[119] Jamsai D, O'Bryan MK. Genome-wide ENU mutagenesis for the discovery of novel male fertility regulators. Syst Biol Reprod Med 2010;56(3):246–59.

[120] Jamsai D, O'Bryan MK. Mouse models as tools in fertility research and male-based contraceptive development. Handb Exp Pharmacol 2010;198:179–94.

[121] Jamsai D, O'Bryan MK. Mouse models in male fertility research. Asian J Androl 2011;13(1):139–51.

[122] Lejeune J, Turpin R, Gautier M. Mongolism; a chromosomal disease (trisomy). Bull Acad Natl Med 1959;143 (11–12):256–65.

[123] Jacobs PA, Baikie AG, Court Brown WM, Strong JA. The somatic chromosomes in mongolism. Lancet 1959; 1(7075):710.

[124] Hassold T, Abruzzo M, Adkins K, Griffin D, Merrill M, Millie E, et al. Human aneuploidy: incidence, origin, and etiology. Environ Mol Mutagen 1996;28(3):167–75.

[125] Gruhn JR, Rubio C, Broman KW, Hunt PA, Hassold T. Cytological studies of human meiosis: sex-specific differences in recombination originate at, or prior to, establishment of double-strand breaks. PLoS One 2013;8(12): e85075.

[126] Gruhn JR, Al-Asmar N, Fasnacht R, Maylor-Hagen H, Peinado V, Rubio C, et al. Correlations between synaptic initiation and meiotic recombination: a study of humans and mice. Am J Hum Genet 2016;98(1):102–15.

[127] Begum F, Chowdhury R, Cheung VG, Sherman SL, Feingold E. Genome-wide association study of meiotic recombination phenotypes. G3 (Bethesda) 2016;6(12):3995–4007.

[128] Singh P, Schimenti JC. The genetics of human infertility by functional interrogation of SNPs in mice. Proc Natl Acad Sci U S A 2015;112(33):10431–6.

[129] Miyata H, Castaneda JM, Fujihara Y, Yu Z, Archambeault DR, Isotani A, et al. Genome engineering uncovers 54 evolutionarily conserved and testis-enriched genes that are not required for male fertility in mice. Proc Natl Acad Sci U S A 2016;113(28):7704–10.

[130] Kumar S, Sandell LL, Trainor PA, Koentgen F, Duester G. Alcohol and aldehyde dehydrogenases: retinoid metabolic effects in mouse knockout models. Biochim Biophys Acta 2012;1821(1):198–205.

[131] Oulad-Abdelghani M, Bouillet P, Décimo D, Gansmuller A, Heyberger S, Dollé P, et al. Characterization of a premeiotic germ cell-specific cytoplasmic protein encoded by Stra8, a novel retinoic acid-responsive gene. J Cell Biol 1996;135(2):469–77.

[132] Koubova J, Menke DB, Zhou Q, Capel B, Griswold MD, Page DC. Retinoic acid regulates sex-specific timing of meiotic initiation in mice. Proc Natl Acad Sci U S A 2006;103(8):2474–9.

[133] Menke DB, Page DC. Sexually dimorphic gene expression in the developing mouse gonad. Gene Expr Patterns 2002;2(3–4):359–67.

[134] Bowles J, Knight D, Smith C, Wilhelm D, Richman J, Mamiya S, et al. Retinoid signaling determines germ cell fate in mice. Science 2006;312(5773):596–600.

[135] Griswold MD, Bishop PD, Kim KH, Ping R, Siiteri JE, Morales C. Function of vitamin A in normal and synchronized seminiferous tubules. Ann N Y Acad Sci 1989;564:154–72.

[136] Huang HF, Hembree WC. Spermatogenic response to vitamin A in vitamin A deficient rats. Biol Reprod 1979; 21(4):891–904.

[137] Morales C, Griswold MD. Retinol-induced stage synchronization in seminiferous tubules of the rat. Endocrinology 1987;121(1):432–4.

[138] Thompson JN, Howell JM, Pitt GA. Vitamin A and reproduction in rats. Proc R Soc Lond B Biol Sci 1964;159:510–35.

[139] Lin Y, Gill ME, Koubova J, Page DC. Germ cell-intrinsic and -extrinsic factors govern meiotic initiation in mouse embryos. Science 2008;322(5908):1685–7.

[140] Handel MA, Schimenti JC. Genetics of mammalian meiosis: regulation, dynamics and impact on fertility. Nat Rev Genet 2010;11(2):124–36.

[141] Llano E, Herrán Y, García-Tuñón I, Gutiérrez-Caballero C, de Álava E, Barbero JL, et al. Meiotic cohesin complexes are essential for the formation of the axial element in mice. J Cell Biol 2012;197(7):877–85.

[142] Pelttari J, Hoja MR, Yuan L, Liu JG, Brundell E, Moens P, et al. A meiotic chromosomal core consisting of cohesin complex proteins recruits DNA recombination proteins and promotes synapsis in the absence of an axial element in mammalian meiotic cells. Mol Cell Biol 2001;21(16):5667–77.

[143] Biswas U, Hempel K, Llano E, Pendas A, Jessberger R. Distinct roles of meiosis-specific cohesin complexes in mammalian spermatogenesis. PLoS Genet 2016;12(10)e1006389.

[144] Lee J, Hirano T. RAD21L, a novel cohesin subunit implicated in linking homologous chromosomes in mammalian meiosis. J Cell Biol 2011;192(2):263–76.

[145] Phadnis N, Cipak L, Polakova S, Hyppa RW, Cipakova I, Anrather D, et al. Casein Kinase 1 and Phosphorylation of Cohesin Subunit Rec11 (SA3) Promote Meiotic Recombination through Linear Element Formation. PLoS Genet 2015;11(5):e1005225.

[146] Prieto I, Suja JA, Pezzi N, Kremer L, Martínez AC, Rufas JS, et al. Mammalian STAG3 is a cohesin specific to sister chromatid arms in meiosis I. Nat Cell Biol 2001;3(8):761–6.

[147] Bolcun-Filas E, Schimenti JC. Genetics of meiosis and recombination in mice. Int Rev Cell Mol Biol 2012;298:179–227.

[148] Losada A. Cohesin in cancer: chromosome segregation and beyond. Nat Rev Cancer 2014;14(6):389–93.

[149] Ward A, Hopkins J, Mckay M, Murray S, Jordan PW. Genetic interactions between the meiosis-specific cohesin components, STAG3, REC8, and RAD21L. G3 (Bethesda) 2016;6(6):1713–24.

[150] Murdoch B, Owen N, Stevense M, Smith H, Nagaoka S, Hassold T, et al. Altered cohesin gene dosage affects Mammalian meiotic chromosome structure and behavior. PLoS Genet 2013;9(2):e1003241.

[151] Tachibana-Konwalski K, Godwin J, van der Weyden L, Champion L, Kudo NR, Adams DJ, et al. Rec8-containing cohesin maintains bivalents without turnover during the growing phase of mouse oocytes. Genes Dev 2010;24(22):2505–16.

[152] Ishiguro K-I, Kim J, Shibuya H, Hernández-Hernández A, Suzuki A, Fukagawa T, et al. Meiosis-specific cohesin mediates homolog recognition in mouse spermatocytes. Genes Dev 2014;28(6):594–607.

[153] Scherthan H, Sfeir A, de Lange T. Rap1-independent telomere attachment and bouquet formation in mammalian meiosis. Chromosoma 2011;120(2):151–7.

[154] Boateng KA, Bellani MA, Gregoretti IV, Pratto F, Camerini-Otero RD. Homologous pairing preceding SPO11-mediated double-strand breaks in mice. Dev Cell 2013;24(2):196–205.

[155] Yang F, Wang PJ. The mammalian synaptonemal complex: a scaffold and beyond. Genome Dyn 2009;5:69–80.

[156] Fraune J, Schramm S, Alsheimer M, Benavente R. The mammalian synaptonemal complex: protein components, assembly and role in meiotic recombination. Exp Cell Res 2012;318(12):1340–6.

[157] Gao J, Colaiácovo MP. Zipping and unzipping: protein modifications regulating synaptonemal complex dynamics. Trends Genet 2018;34(3):232–45.

[158] Yuan L, Liu J-G, Hoja M-R, Wilbertz J, Nordqvist K, Höög C. Female germ cell aneuploidy and embryo death in mice lacking the meiosis-specific protein SCP3. Science 2002;296(5570):1115–8.

[159] Kolas NK, Yuan L, Hoog C, Heng HHQ, Marcon E, Moens PB. Male mouse meiotic chromosome cores deficient in structural proteins SYCP3 and SYCP2 align by homology but fail to synapse and have possible impaired specificity of chromatin loop attachment. Cytogenet Genome Res 2004;105(2–4):182–8.

[160] Wojtasz L, Daniel K, Roig I, Bolcun-Filas E, Xu H, Boonsanay V, et al. Mouse HORMAD1 and HORMAD2, two conserved meiotic chromosomal proteins, are depleted from synapsed chromosome axes with the help of TRIP13 AAA-ATPase. PLoS Genet 2009;5(10):e1000702.

[161] Rog O, Köhler S, Dernburg AF. The synaptonemal complex has liquid crystalline properties and spatially regulates meiotic recombination factors. elife 2017;6.

[162] Huang F, Sirinakis G, Allgeyer ES, Schroeder LK, Duim WC, Kromann EB, et al. Ultra-high resolution 3D imaging of whole cells. Cell 2016;166(4):1028–40.

[163] Zickler D, Kleckner N. Recombination, pairing, and synapsis of homologs during meiosis. Cold Spring Harb Perspect Biol 2015;7(6).

[164] Paigen K, Petkov P. Mammalian recombination hot spots: properties, control and evolution. Nat Rev Genet 2010;11(3):221–33.

[165] Paigen K, Petkov PM. PRDM9 and its role in genetic recombination. Trends Genet 2018;34(4):291–300.

[166] Ranjha L, Howard SM, Cejka P. Main steps in DNA double-strand break repair: an introduction to homologous recombination and related processes. Chromosoma 2018;127(2):187–214.

[167] Reichman R, Alleva B, Smolikove S. Prophase I: preparing chromosomes for segregation in the developing oocyte. Results Probl Cell Differ 2017;59:125–73.

[168] Hunter N. Meiotic recombination: the essence of heredity. Cold Spring Harb Perspect Biol 2015;7(12).

[169] Baudat F, Imai Y, de Massy B. Meiotic recombination in mammals: localization and regulation. Nat Rev Genet 2013;14(11):794–806.

[170] Baudat F, Buard J, Grey C, Fledel-Alon A, Ober C, Przeworski M, et al. PRDM9 is a major determinant of meiotic recombination hotspots in humans and mice. Science 2010;327(5967):836–40.

[171] Berg IL, Neumann R, Lam K-WG, Sarbajna S, Odenthal-Hesse L, May CA, et al. PRDM9 variation strongly influences recombination hot-spot activity and meiotic instability in humans. Nat Genet 2010;42(10):859–63.

[172] Myers S, Bowden R, Tumian A, Bontrop RE, Freeman C, MacFie TS, et al. Drive against hotspot motifs in primates implicates the PRDM9 gene in meiotic recombination. Science 2010;327(5967):876–9.

[173] Parvanov ED, Petkov PM, Paigen K. *Prdm9* controls activation of mammalian recombination hotspots. Science 2010;327(5967):835.

[174] Powers NR, Parvanov ED, Baker CL, Walker M, Petkov PM, Paigen K. The meiotic recombination activator PRDM9 trimethylates both H3K36 and H3K4 at recombination hotspots in vivo. PLoS Genet 2016;12(6):e1006146.

[175] Baker CL, Petkova P, Walker M, Flachs P, Mihola O, Trachtulec Z, et al. Multimer formation explains allelic suppression of PRDM9 recombination hotspots. PLoS Genet 2015;11(9):e1005512.

[176] Narasimhan VM, Hunt KA, Mason D, Baker CL, Karczewski KJ, Barnes MR, et al. Health and population effects of rare gene knockouts in adult humans with related parents. Science 2016;352(6284):474–7.

[177] Brick K, Smagulova F, Khil P, Camerini-Otero RD, Petukhova GV. Genetic recombination is directed away from functional genomic elements in mice. Nature 2012;485(7400):642–5.

[178] Kauppi L, Barchi M, Baudat F, Romanienko PJ, Keeney S, Jasin M. Distinct properties of the XY pseudoautosomal region crucial for male meiosis. Science 2011;331(6019):916–20.

[179] Bellani MA, Romanienko PJ, Cairatti DA, Camerini-Otero RD. SPO11 is required for sex-body formation, and *Spo11* heterozygosity rescues the prophase arrest of *Atm* −/− spermatocytes. J Cell Sci 2005;118(Pt 15):3233–45.

[180] Kumar R, Ghyselinck N, Ishiguro K, Watanabe Y, Kouznetsova A, Höög C, et al. MEI4–a central player in the regulation of meiotic DNA double-strand break formation in the mouse. J Cell Sci 2015;128(9):1800–11.

[181] Cohen PE, Pollack SE, Pollard JW. Genetic analysis of chromosome pairing, recombination, and cell cycle control during first meiotic prophase in mammals. Endocr Rev 2006;27(4):398–426.

[182] Gray S, Cohen PE. Control of meiotic crossovers: from double-strand break formation to designation. Annu Rev Genet 2016;50:175–210.

[183] Turner JMA, Aprelikova O, Xu X, Wang R, Kim S, Chandramouli GVR, et al. BRCA1, histone H2AX phosphorylation, and male meiotic sex chromosome inactivation. Curr Biol 2004;14(23):2135–42.

[184] Ribeiro J, Abby E, Livera G, Martini E. RPA homologs and ssDNA processing during meiotic recombination. Chromosoma 2016;125(2):265–76.

[185] Edelmann W, Cohen PE, Kneitz B, Winand N, Lia M, Heyer J, et al. Mammalian MutS homologue 5 is required for chromosome pairing in meiosis. Nat Genet 1999;21(1):123–7.

[186] Her C, Wu X, Wan W, Doggett NA. Identification and characterization of the mouse MutS homolog 5: *Msh5*. Mamm Genome 1999;10(11):1054–61.

[187] Avdievich E, Reiss C, Scherer SJ, Zhang Y, Maier SM, Jin B, et al. Distinct effects of the recurrent *Mlh1*^{G67R} mutation on MMR functions, cancer, and meiosis. Proc Natl Acad Sci U S A 2008;105(11):4247–52.

[188] Anderson LK, Reeves A, Webb LM, Ashley T. Distribution of crossing over on mouse synaptonemal complexes using immunofluorescent localization of MLH1 protein. Genetics 1999;151(4):1569–79.

[189] Moens PB, Spyropoulos B. Immunocytology of chiasmata and chromosomal disjunction at mouse meiosis. Chromosoma 1995;104(3):175–82.

[190] Sun F, Handel MA. Regulation of the meiotic prophase I to metaphase I transition in mouse spermatocytes. Chromosoma 2008;117(5):471–85.

[191] Bisig CG, Guiraldelli MF, Kouznetsova A, Scherthan H, Höög C, Dawson DS, et al. Synaptonemal complex components persist at centromeres and are required for homologous centromere pairing in mouse spermatocytes. PLoS Genet 2012;8(6):e1002701.

[192] Dobson MJ, Pearlman RE, Karaiskakis A, Spyropoulos B, Moens PB. Synaptonemal complex proteins: occurrence, epitope mapping and chromosome disjunction. J Cell Sci 1994;107(Pt 10):2749–60.

[193] Cobb J, Cargile B, Handel MA. Acquisition of competence to condense metaphase I chromosomes during spermatogenesis. Dev Biol 1999;205(1):49–64.

[194] Schindler K. Protein kinases and protein phosphatases that regulate meiotic maturation in mouse oocytes. Results Probl Cell Differ 2011;53:309–41.

[195] Nguyen AL, Schindler K. Specialize and divide (twice): functions of three aurora kinase homologs in mammalian oocyte meiotic maturation. Trends Genet 2017;33(5):349–63.

[196] Clement TM, Inselman AL, Goulding EH, Willis WD, Eddy EM. Disrupting cyclin dependent kinase 1 in spermatocytes causes late meiotic arrest and infertility in mice. Biol Reprod 2015;93(6):137.

[197] Masui Y, Markert CL. Cytoplasmic control of nuclear behavior during meiotic maturation of frog oocytes. J Exp Zool 1971;177(2):129–45.

[198] Sun Q-Y, Miao Y-L, Schatten H. Towards a new understanding on the regulation of mammalian oocyte meiosis resumption. Cell Cycle 2009;8(17):2741–7.

[199] Hu J, Sun F, Handel MA. Nuclear localization of EIF4G3 suggests a role for the XY body in translational regulation during spermatogenesis in mice. Biol Reprod 2017;98(1):102–14.

[200] Conti M, Hsieh M, Zamah AM, Oh JS. Novel signaling mechanisms in the ovary during oocyte maturation and ovulation. Mol Cell Endocrinol 2012;356(1–2):65–73.

[201] Sanz E, Yang L, Su T, Morris DR, McKnight GS, Amieux PS. Cell-type-specific isolation of ribosome-associated mRNA from complex tissues. Proc Natl Acad Sci U S A 2009;106(33):13939–44.

·

[202] Han SJ, Martins JPS, Yang Y, Kang MK, Daldello EM, Conti M. The translation of cyclin B1 and B2 is differentially regulated during mouse oocyte reentry into the meiotic cell cycle. Sci Rep 2017;7(1):14077.

[203] Burkhardt S, Borsos M, Szydlowska A, Godwin J, Williams SA, Cohen PE, et al. Chromosome cohesion established by REC8-cohesin in fetal oocytes is maintained without detectable turnover in oocytes arrested for months in mice. Curr Biol 2016;26(5):678–85.

[204] Revenkova E, Herrmann K, Adelfalk C, Jessberger R. Oocyte cohesin expression restricted to predictyate stages provides full fertility and prevents aneuploidy. Curr Biol 2010;20(17):1529–33.

[205] Nagaoka SI, Hassold TJ, Hunt PA. Human aneuploidy: mechanisms and new insights into an age-old problem. Nat Rev Genet 2012;13(7):493–504.

[206] Hodges CA, Revenkova E, Jessberger R, Hassold TJ, Hunt PA. SMC1beta-deficient female mice provide evidence that cohesins are a missing link in age-related nondisjunction. Nat Genet 2005;37(12):1351–5.

[207] Akera T, Chmátal L, Trimm E, Yang K, Aonbangkhen C, Chenoweth DM, et al. Spindle asymmetry drives non-Mendelian chromosome segregation. Science 2017;358(6363):668–72.

[208] Ottolini CS, Newnham L, Capalbo A, Natesan SA, Joshi HA, Cimadomo D, et al. Genome-wide maps of recombination and chromosome segregation in human oocytes and embryos show selection for maternal recombination rates. Nat Genet 2015;47(7):727–35.

[209] Bolcun-Filas E, Handel MA. Meiosis: the chromosomal foundation of reproduction. Biol Reprod 2018; https://dx.doi.org/10.1093/biolre/ioy021.

[210] Howe K, FitzHarris G. Recent insights into spindle function in mammalian oocytes and early embryos. Biol Reprod 2013;89(3):71.

[211] Nakagawa S, FitzHarris G. Intrinsically defective microtubule dynamics contribute to age-related chromosome segregation errors in mouse oocyte meiosis-I. Curr Biol 2017;27(7):1040–7.

[212] Webster A, Schuh M. Mechanisms of aneuploidy in human eggs. Trends Cell Biol 2017;27(1):55–68.

[213] Templado C, Vidal F, Estop A. Aneuploidy in human spermatozoa. Cytogenet Genome Res 2011; 133(2-4):91–9.

[214] Martin RH, Ko E, Rademaker A. Distribution of aneuploidy in human gametes: comparison between human sperm and oocytes. Am J Med Genet 1991;39(3):321–31.

[215] Martin RH, Rademaker A. The frequency of aneuploidy among individual chromosomes in 6,821 human sperm chromosome complements. Cytogenet Cell Genet 1990;53(2–3):103–7.

[216] Pellestor F, Andréo B, Anahory T, Hamamah S. The occurrence of aneuploidy in human: lessons from the cytogenetic studies of human oocytes. Eur J Med Genet 2006;49(2):103–16.

[217] Pacchierotti F, Adler ID, Eichenlaub-Ritter U, Mailhes JB. Gender effects on the incidence of aneuploidy in mammalian germ cells. Environ Res 2007;104(1):46–69.

[218] Fragouli E, Alfarawati S, Goodall N-N, Sánchez-García JF, Colls P, Wells D. The cytogenetics of polar bodies: insights into female meiosis and the diagnosis of aneuploidy. Mol Hum Reprod 2011;17(5):286–95.

[219] Hussin J, Roy-Gagnon M-H, Gendron R, Andelfinger G, Awadalla P. Age-dependent recombination rates in human pedigrees. PLoS Genet 2011;7(9)e1002251.

[220] Bleazard T, Ju YS, Sung J, Seo J-S. Fine-scale mapping of meiotic recombination in Asians. BMC Genet 2013;14:19.

[221] Bermejo-Alvarez P, Rosenfeld CS, Roberts RM. Effect of maternal obesity on estrous cyclicity, embryo development and blastocyst gene expression in a mouse model. Hum Reprod 2012;27(12):3513–22.

[222] Sutton ML, Gilchrist RB, Thompson JG. Effects of in-vivo and in-vitro environments on the metabolism of the cumulus-oocyte complex and its influence on oocyte developmental capacity. Hum Reprod Update 2003; 9(1):35–48.

[223] Boudoures AL, Chi M, Thompson A, Zhang W, Moley KH. The effects of voluntary exercise on oocyte quality in a diet-induced obese murine model. Reproduction 2016;151(3):261–70.

[224] Broughton DE, Moley KH. Obesity and female infertility: potential mediators of obesity's impact. Fertil Steril 2017;107(4):840–7.

[225] Minge CE, Bennett BD, Norman RJ, Robker RL. Peroxisome proliferator-activated receptor-gamma agonist rosiglitazone reverses the adverse effects of diet-induced obesity on oocyte quality. Endocrinology 2008; 149(5):2646–56.

[226] Hunt PA, Koehler KE, Susiarjo M, Hodges CA, Ilagan A, Voigt RC, et al. Bisphenol a exposure causes meiotic aneuploidy in the female mouse. Curr Biol 2003;13(7):546–53.

[227] Hunt PA, Lawson C, Gieske M, Murdoch B, Smith H, Marre A, et al. Bisphenol A alters early oogenesis and follicle formation in the fetal ovary of the rhesus monkey. Proc Natl Acad Sci U S A 2012;109(43):17525–30.

[228] Vrooman LA, Oatley JM, Griswold JE, Hassold TJ, Hunt PA. Estrogenic exposure alters the spermatogonial stem cells in the developing testis, permanently reducing crossover levels in the adult. PLoS Genet 2015; 11(1):e1004949.

[229] Nakano K, Nishio M, Kobayashi N, Hiradate Y, Hoshino Y, Sato E, et al. Comparison of the effects of BPA and BPAF on oocyte spindle assembly and polar body release in mice. Zygote 2016;24(2):172–80.

[230] Rochester JR, Bolden AL. Bisphenol S and F: a systematic review and comparison of the hormonal activity of bisphenol A substitutes. Environ Health Perspect 2015;123(7):643–50.

[231] Eisenberg ML, Li S, Cullen MR, Baker LC. Increased risk of incident chronic medical conditions in infertile men: analysis of United States claims data. Fertil Steril 2016;105(3):629–36.

[232] Cedars MI, Taymans SE, DePaolo LV, Warner L, Moss SB, Eisenberg ML. The sixth vital sign: what reproduction tells us about overall health. Proceedings from a NICHD/CDC workshop. Hum Reprod Open 2017;2017(2).

[233] Tarín JJ, García-Pérez MA, Hamatani T, Cano A. Infertility etiologies are genetically and clinically linked with other diseases in single meta-diseases. Reprod Biol Endocrinol 2015;13:31.

[234] Eisenberg ML, Li S, Brooks JD, Cullen MR, Baker LC. Increased risk of cancer in infertile men: analysis of U.S. claims data. J Urol 2015;193(5):1596–601.

[235] Glazer CH, Bonde JP, Eisenberg ML, Giwercman A, Hærvig KK, Rimborg S, et al. Male infertility and risk of nonmalignant chronic diseases: a systematic review of the epidemiological evidence. Semin Reprod Med 2017;35(3):282–90.

4

Effects of Aging on Sperm Chromatin

Johanna Selvaratnam, Heather Fice, Anaïs Noblanc,
Bernard Robaire

Departments of Pharmacology and Therapeutics and of Obstetrics and Gynecology, McGill
University, Montréal, QC, Canada

INTRODUCTION

The effects of aging on male reproductive health have received less attention than the well-established effects of aging on the female reproductive system. However, over the last 20 years, advanced paternal age has been demonstrated to contribute to an increasing number of health concerns including reduced fertility [1], increased pregnancy-related complications [2], and a growing number of diseases in offspring [3–7]. The majority of studies have focused on damage in spermatozoa [8] because DNA-damaged sperm are the cause of conditions ranging from spontaneous abortion to congenital malformations and genetic defects. However, more recent studies have examined germ cells as they go through spermatogenesis, a process during which many errors can be introduced into the genome through DNA damage and repair as well as chromatin compaction [9]. Behaviors such as smoking and lifestyle choices as well as medications all increase exposure to genotoxic compounds that, over several years, can contribute to a decrease in the quality of sperm as men age.

The male germ cell nucleus undergoes a remarkable series of changes as germ cells progress through spermatogenesis. After undergoing several mitotic divisions as spermatogonia, germ cells transit to proceed through meiotic divisions (described in the previous chapter) where, as spermatocytes, the complex processes of chromosome meiotic recombination and reduction divisions take place. Once haploid, male germ cells, as spermatids, go through a dramatic cellular and nuclear reorganization. The transformation from a haploid but normally appearing nucleus in early round spermatids to the completed mature spermatozoa found in the tail of the epididymis involves a series of changes that are only beginning to be unraveled.

In order to place in context the impact of aging on male germ cells, we will first discuss the structural and epigenetic changes that take place in germ cell nuclei as they proceed from early germ cells as spermatogonia to their complete maturation in the tail of the epididymis. We will then discuss existing evidence that demonstrates how these processes are altered in germ cells from aging males. While much has been learned about these changes from nonmammalian species, the discussion below focuses on studies done in mammals, primarily in mice, rats, and humans.

TRANSFORMATIONS OF THE NUCLEUS DURING SPERMATOGENESIS

Nuclear Protein Dynamics From Spermatogonia to Spermatozoa

Chromatin packaging throughout spermatogenesis is transitional. The diploid spermatogonial stem cells undergo a series of transitions where most histones are removed and are replaced with protamines. The transition from histones to protamines involves many changes to histone dynamics, primarily through the hyperacetylation of histone variant four (H4). This is controlled by histone acetyl transferases (HAT) and histone deacetylases (HDAC) [10]. Briefly, hyperacetylation of histone H4 (recognized as H4 acetylated on lysines 5 and 8) serves as a signal platform for the recruitment of a testis-specific bromodomain (BRDT) protein that, through an unknown mechanism, directs histone removal [11] and replacement first with arginine and lysine-rich transition proteins (TPs) and then with other packaging proteins called protamines (Fig. 4.1) [12]. When H4 hyperacetylation is impaired, infertility ensues [13], suggesting this is a critical step in the transition from histone to protamine bound DNA in the sperm nucleus. Transition proteins are also critical for functional sperm packaging; however, there may be functional redundancy between the two variants. Mice having a null mutation for either of the transition proteins [14, 15] are subfertile while mice with null mutations for both TPs are infertile [16].

Protamines, first discovered in 1874, are small arginine-rich proteins. There are two different protamines in mammals; protamine 1, which is expressed in all animals, and protamine 2, which is expressed in mice and humans [17]. Protamines increase DNA condensation during

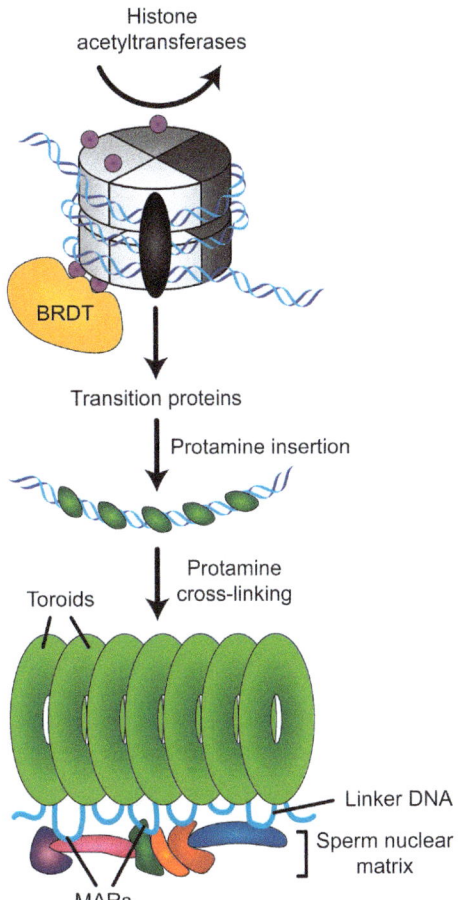

FIG. 4.1 Histone to protamine transition in sperm. The histone-to-protamine transition begins with nucleosome bound DNA, wherein DNA is coiled around two tetramers comprised of four histone variants shown in the four shades of *gray*. Histone 1 is shown in *black*. Histones 3 and 4 are hyperacetylated by histone acetyltransferases, and recognized by BRDT for transition to transition proteins, and insertion of protamines shown in *green*. The protamines then undergo cross-linking and form toroids with linker DNA bound to the sperm nuclear matrix at matrix-associated regions (MARs).

spermiogenesis. As these proteins are arginine-rich, the cysteine groups undergo oxidization during transit through the epididymis to form disulfide bonds. These bonds allow for a much tighter conformation than that found in nucleosome-containing chromatin. Ultimately, these changes in compaction lead to a transcriptionally inactive state and provide a higher degree of protection to the sperm DNA. When protamine knockout mice are bred, they are infertile [18–20]. In humans, ~8%–15% of the genome remains bound by histones, including the ε-globin, γ globin, and telomeric DNA [21] while in rodents the percent of histone remaining in mature sperm is 1%–2%. Additionally, it seems that in humans the ratio of the two protamines matters as sperm from infertile subjects have altered P1:P2 ratios [22, 23].

Spermatogenesis and the Nuclear Matrix

The formidable amount of chromatin in the average somatic cell requires tight compaction for containment within the nucleus; in spermatozoa, the compaction of chromatin is even several-fold greater. In somatic cells, histones bind the DNA to form nucleosomes that allow for the nucleosomal beads-on-a-string conformation of DNA. The packaging of chromatin in sperm differs markedly due to the insertion of protamines, resulting in a unique structure similar to that of a toroid where the protamines are wound in toroid loops. The histones that remain in low abundance are also bound, and there are linker DNA sequences between these loops [24].

This conformation of chromatin lends itself well to the concept of organization around a nuclear matrix, where the linker DNA regions are attached to elements of the matrix. This matrix can be compared to the cytoskeleton; however, it is a transient structure that is in flux in order to accommodate genetic activity [25]. The nuclear matrix is a nuclear component that has most prominently been studied in cancer cells using electron microscopy after high salt extraction. Initially, when discovered, the matrix was found through high salt protein extraction as some proteins remained bound to DNA, forming the core nuclear matrix. Although the process of salting out has been criticized, when using physiologically relevant buffers very similar results can be obtained [25]. The core nuclear matrix proteins are bound to linker DNA at specific sites to form matrix-associated regions (MARs). The linker DNA has been studied to assess the nature of the DNA at these sites. Examining the genes near MARs has led to the observation that these regions are often origins of replication, topoisomerase II sites, and AT-rich sequences involved in DNA replication [25]. The limited literature on this subject suggests that the MARs allow for priming of specific genes that are required shortly after fertilization. This includes developmentally relevant genes and those required to help unpack the paternal genome and allow for blastocyst development.

After first elucidating the structure of the nuclear matrix, it was shown that it is essential for fertilization after intracytoplasmic sperm injection (ICSI). This was shown by doing ICSI with various components of the DNA and nuclear matrix complex. Successful fertilization and pronuclear formation was obtained with control sperm as well as with the nuclear matrix and minimal associated DNA after enzymatic digestion. However, when either isolated DNA or isolated DNA with a reconstituted matrix was injected, fertilization was unsuccessful, and there was no pronuclear formation [26]. Similarly, if the nuclear matrix is damaged, DNA replication in the paternal pronucleus is delayed after fertilization [27]. Most studies are consistent with these findings and support the hypothesis that genome association with the matrix provides the template for protamine replacement after fertilization by maternal histones [28], thus priming sites required for early transcription by anchoring them to MARs. Together, these studies establish the necessity of the nuclear matrix and its DNA association for fertility.

Sperm Chromatin Organization

Initial studies examining how chromatin is spatially organized were based on the use of fluorescence in situ hybridization (FISH). Starting in the early 1990s, studies focused on the examination of the positioning of telomeres and centromeres. Haaf and Ward first described the distinct locations for each chromosome within the sperm nucleus [29]. These chromosome territories have been studied in many cell types and could be of critical importance for the developing embryo. Zalensky et al. [30] found that centromeres cluster toward the center of the nucleus and form a chromocenter while the telomeres extend outward toward the periphery of the nucleus. They suggest that telomere sequences form dimers and tetramers, linking the chromosomes in a hairpin configuration [30]. This finding fits with the hypothesis that a nucleus has gene-rich internal and gene-poor peripheral zones [31]. Additionally, the dispersal of telomeres aligns with the observation that telomeres in somatic cells form a bouquet within the nucleus [32]. Recently, a newly proposed model has been put forward in which there are multiple chromocenters and interspersed telomeres throughout the nucleus [33].

These methods of assessing chromatin organization in sperm are often met with skepticism as the sperm DNA must be manipulated in order to allow access of antibodies or probes. Currently, sperm chromatin conformation research is making major strides with the development of a technique called chromatin conformation capture (3C). This has allowed for a higher resolution assessment of the genome and the interactions that exist within the nucleus, breaking down chromosome territories into topologically associated domains, compartments, and chromatin loops [34]. Different conformation of capture methods exist, each with different goals, but all using a proximity-based approach to better understand the spatial organization of the genome after ligation of fragmented products.

The first of these methods, 3C, examines the genome by assessing the association between individual loci, making it useful when examining the relationship between two genes or a gene and its regulators. 4C allows for more components to be examined, where chromatin conformation capture is performed on a chip. 5C is a high throughput version of 3C, wherein the proximity of many different loci can be examined [35]. Though these methods allow for the contact frequency of loci to be examined, the sequencing of the fragments in the most recent variation of the protocol, Hi-C, allows for the various spatial relationships observed to gain functional relevance [36]. Hi-C methods have now been applied to examining the sperm genome. Battulin et al. [37] used Hi-C to examine the similarity in structure between fibroblasts and mature sperm, and found that the structure is similar; however, sperm show more long-range contacts, consistent with the tight packaging. The structural similarity remains true for mouse embryonic stem cells and sperm as well, as shown by Jung et al. [38]. They also showed that many of the proteins necessary for transcriptional activity are maintained on the DNA in mature sperm. Ke et al. [39] confirmed the results of Battulin et al. [37] by displaying that long-range interactions are more frequent in sperm. They also examined the positioning of chromatin relative to oocytes and early embryos. Although very powerful, research using chromatin conformation capture methods can be met with some skepticism as unpacking the tightly compacted sperm DNA may alter the normal spatial relationships existing in intact sperm nuclei.

Epigenetic Changes Occurring During Spermatogenesis

Epigenetics is defined as the inheritance of changes in gene function without any alterations in the DNA sequence. The three major mechanisms by which epigenetic information can be passed on from one cell generation to the next are DNA methylation, histone modification, and the expression and activities of noncoding RNAs. Because sperm chromatin goes through the removal of most histones, replacement by protamines, and compaction of DNA during spermiogenesis [40, 41], it is not surprising that spermatozoa are vulnerable to aberrant epigenetic alterations. This vulnerability becomes clear in cases of oligozoospermia (abnormally low concentration of spermatozoa in semen) associated with epigenetic factors

[42], such as DNA methylation, histone modifications, and chromatin remodeling [43, 44]. Furthermore, evidence from babies conceived using assisted reproductive technology (ART) links aberrant methylation in spermatozoa to increased incidence of rare imprinting disorders such as Beckwith-Weidemann syndrome (BWS) and Angelman syndrome (AS) [45–48].

DNA Methylation

DNA methylation is a common epigenetic mechanism used in cells to regulate gene expression. It involves the methylation of the 5-carbon cytosine residues (5-mC) at cytosine-phosphate-guanine (CpG) dinucleotides. These epigenetic marks can either activate or inactivate gene expression, with hypermethylation at a promoter usually blocking access to the transcriptional machinery and leading to gene silencing [49]. Gene promoter hypomethylation facilitates access of DNA polymerase to transcribe the gene; this is associated with increased gene expression [49, 50]. There has been growing concern over aberrant DNA methylation patterns in spermatozoa due to the normal process of germ-line reprogramming that occurs during germ cell development and spermatogenesis [51]. During this process, existing DNA methylation patterns are erased and reestablished in developing male germ cells [51], creating a window of susceptibility within which damage can occur due to aging or exposure to xenobiotics [52].

The significance of the DNA methylome in sperm cells has been revealed by the accumulation of studies showing intergenerational (F0 to F1) and transgenerational (F0 to F1, F2, F3 …) epigenetic inheritance mediated by paternal DNA methylation [53–57]. The transmission of metabolic diseases to the offspring is one of the most studied transgenerational phenotype inheritances via the paternal DNA methylome [56]. Radford and colleagues [58] demonstrated the role of the sperm DNA methylome in the intergenerational transmission of metabolic disorders generated by undernourishment of F0-dams during the third week of gestation, when F1-primordial germ cell DNA methylation is reprogrammed. The F1-males, who presented some metabolic disorders, were bred to control females. An increase in hepatic triglyceride abundance and in gene expression involved in lipid oxidation as well as a trend toward a decrease in gene expression involved in lipid synthesis were observed in F2 embryos from F0-undernourished dams in comparison to F2 embryo from F0-control dams. At 8 months, the F2 mice presented an increase in the adiposity index, a decrease in muscle mass, and glucose intolerance compared to F2 control mice. F1-spermatozoa and the F2-liver and brain of E16.5 embryos were analyzed for DNA methylation. Numerous differentially methylated regions (DMR) were identified in F1-sperm DNA, but they were not conserved in the F2-E16.5 embryo DNA. However, the expression of genes in the neighborhood of 8 of the 15 validated F1-hypomethylated DMRs was modified in F2-E16.5 embryo tissues compared to the control F2-E16.5 embryo, including genes involved in insulin secretion. The direct consequence is an altered regulation of insulin secretion by pancreatic islets isolated from 4-month-old F2 mice. Even if the intermediate mechanism is still unknown, this study proves the significance of the sperm DNA methylome in the intergenerational phenotypic inheritance.

Histone Variants and Histone Posttranslational Modifications in the Germline

During spermatogenesis, an extended remodeling of the germ cell chromatin occurs due to the incorporation of numerous histones variants and to posttranslational modifications (PTMs). There are a remarkable number of histone variants and PTMs with at least 18 PTMs (acetylation, β-N-acetyl-glucosaminylation, ADP ribosylation, butyrylation, citrullination/deimination, crotonylation, formylation, glutarylation, hydroxylation, 2-hydroxyisobutyrylation, malonylation, mono-, di-, and tri-methylation, phosphorylation, proline isomerization, propionylation, soccinylation, sumoylation, ubiquitination) of canonical histones and their variants that have thus far been identified [59, 60]. These changes alter the physicochemical properties of the nucleosomes, modify the 3D conformation of the germ cell chromatin, and play a role in progression through the various steps of meiosis. For example, in pachytene spermatocytes, the paired chromosomes must undergo homologous recombination and gene transcription starts again in autosomal chromosomes, which have been highly condensed because of leptonema. The opening of the autosomal chromatin is due to the incorporation of different histones variants, including TH2A, TH2B, and H1t, and to the H2A ubiquitinylation, the H3K9 acetylation, and the replacement of H3 by its variant H3.3 [61]. At the same time, this process is delayed in sex chromosomes that are still highly condensed and inactivated, forming the sex or XY body. This phenomenon is called meiotic sex chromosome inactivation and involves numerous histone repressive marks such as H3K9me2/3 and macroH2A variant [62, 63].

Recently, using different mass spectrometry-based proteomics approaches, a systematic, wide-ranging detection of histones and variant PTMs has been undertaken using the entire testis or using isolated germ cells at several points during differentiation [63–66]. The most conserved histone PTMs between mouse and human germ cells occur on H3 and H4, and the relative abundance of most PTMs in normozoospermic human semen samples is consistent between individuals despite the interindividual variability of semen parameters [65]. Although these studies have provided important new information about the numerous PTMs that occur during spermatogenesis, a limitation of this approach is the lack of information regarding the exact location, interactions, and roles of these histone and variant PTMs. Germline proteomics is still a largely underdeveloped field that deserves greater attention in order to yield a more comprehensive understanding of the mechanisms operating during spermatogenesis.

Noncoding RNAs in the Germline

The number of RNA identified categories, the roles they play, and the mechanisms by which they act have remarkably increased over the past 15 years. Their functions in the regulation of gene expression, chromatin conformation, or genome stability have been demonstrated in numerous cell types, tissues, and biological functions, including the process of male germ cell differentiation. Various studies of the testis and male germ cells have revealed that spermatogenesis would not be possible without many RNAs, both coding and noncoding. New categories of RNAs have been discovered in germ cells, such as PIWI-RNAs, and numerous genes express testis-specific transcripts and the corresponding testis-specific protein isoforms.

Due to the numerous chromatin rearrangements during male germ cell differentiation, the transcriptional activity is not constant and is highly reduced and eventually halted during leptonema/zygonema of meiosis I and during the histone-to-protamine transition. However, RNAs are still essential to the completion of these steps. To remedy this problem, numerous genes are transcribed in advance and transcripts are translationally repressed by RNA-binding proteins. These transcripts are stored in germ granules such as the chromatoid body found in round spermatids [67]. After spermatogenesis, the transcriptionally silent spermatozoa go through a reduction in RNA content following the removal of a large volume of cytoplasm in the residual body prior to spermiation, with only ~50 fg of total RNA remaining in a spermatozoon [68–70] when somatic cells contain ~12 pg of RNA/cell.

The transcriptome of germ cells is highly dynamic and step-dependant [71–75]. In particular, various noncoding RNAs tightly regulate these steps. For example, two microRNAs [76, 77], miR-146 and miR-221, inhibit spermatogonial stem cell differentiation. They both target the c-Kit mRNA that codes for a tyrosine kinase receptor essential to initiate spermatogonial differentiation, by cleaving it and preventing its translation [75, 78]. One characteristic of germ cells before meiosis is their absence of full DNA methylation reprogramming. The consequence is the lack of transposon element inhibition. Some noncoding RNAs target transposable elements to enhance degradation and prevent their insertion into the genome. The most prominent are prepachytene PIWI-RNAs, which are 24–31 nt-long single-stranded noncoding RNAs associated with PIWI proteins [79–82]. These PIWI-RNAs are organized in clusters in the genome and recognize their target by complementarity. They are also able to drive the methylation of their target in the genome for a long-term effect. Recently, another small noncoding RNA class, the transfer RNA-derived small RNAs or tRFs [83, 84], has been identified as an inhibitor of transposable elements in a mouse stem cell line [85]. Further studies will be necessary to evaluate their function in premeiotic germ cells.

Once germ cells enter meiosis, different noncoding RNAs inhibit specific mRNAs to control their translational role in protein synthesis. Among these, miR-18 inhibits the expression of heat shock factor 2 (Hsf2) [86] and miR-34b-5p regulates the presence of cyclin-dependent kinase 6 (Cdk6) [75]. Pachytene spermatocytes also express a large diversity of PIWI-RNAs derived from nontransposon intergenic regions. These pachytene PIWI-RNAs have been demonstrated to be involved in the mRNA decay observed during this step [87, 88] and could potentially explain the dramatic changes in the gene expression profile that occur between late pachytene spermatocytes and round spermatids [73]. Later, during spermiogenesis, two clusters of microRNAs from the same family, miR-449 and miR-34b/c clusters, play a redundant role in the condensation of the sperm chromatin and the formation of the flagellum. The male double knockout mice for these two microRNA clusters are sterile due to the production of abnormal, immotile, and decondensed spermatozoa [89, 90].

In mature epididymal spermatozoa, despite the removal of most RNAs, a wide diversity of noncoding RNAs is still detectable using RNA sequencing [70, 91, 92]: long noncoding RNAs, fragmented ribosomal RNAs, tRFs, microRNAs, PIWI-RNAs, and mitochondrial RNAs. Repetitive elements, quiescent RNAs, intronic retained elements, snoRNAs, and YRNAs have also been identified in lower quantities. The origin and function of these noncoding RNAs are not yet understood. For a long time, they were considered as functionless leftovers from spermatogenesis, but their pattern of consistency among individuals and the fact that some of these RNA species are acquired during the epididymal maturation of spermatozoa [93] indicate that they are likely to be playing functional roles during postfertilization events. Some studies support this hypothesis and point out a possible role for noncoding RNAs in intergenerational epigenetic transmission [94, 95]. Several studies have evaluated the potential value of the sperm noncoding RNA profile as a potential diagnostic marker of male infertility. Their results support this suggestion as they identified significant differentially expressed noncoding RNAs in different groups of infertile patients showing altered seminal parameters [96–101].

DNA Damage/Repair in Germ Cells During Spermatogenesis

The highly replicative nature of the male germ line exposes it to the possibility of numerous types of DNA lesions. These include single- or double-strand breaks, interstrand or intrastrand DNA cross-links, base mismatches, and modifications. In order to respond to the array of damage that can occur in DNA during cell divisions and chromatin compaction, germ cells have evolved a wide range of repair pathways, together referred to as the DNA damage response (DDR). The process of DDR involves sensors of damage, mediators of sensor signaling, transducers, and effector classes of proteins. The major repair pathways found in germ cells during spermatogenesis include base-excision repair (BER), nucleotide-excision repair (NER), mismatch repair (MMR), homologous recombination (HR), and nonhomologous end joining (NHEJ).

Spermatogonial stem cells divide mitotically, thus either NHEJ or HR can be used to repair DSBs that occur in these cells. High levels of mitotic activity at this stage increase the rate of DNA damage sites (potential mutations). During this stage, there are increased DNA replication error derived damages, but also the potential for increased DDR leading to the repair of these errors. Furthermore, MMR-related genes are expressed in spermatogonia as well as primary and secondary spermatocytes [102], suggesting that several repair mechanisms are active together at this early stage of spermatogenesis.

DNA lesions that form in spermatogonia can be repaired or left to replicate and give rise to sperm carrying these mutations. Because germ cells do not replicate during the late spermatogenic stages, mutations that may have been acquired are not removed, and DNA lesion/mutation removal relies on replication-independent DNA synthesis involving DNA recombination or repair.

The main type of repair occurring in leptotene-midpachytene spermatocytes is HR, with necessary components Rad51 and Dmc1 expressed in primary spermatocytes. Studies show that there is less NER in spermatocytes compared to somatic cells, and this is mainly because damaged cells at this stage can be eliminated by apoptosis rather than repaired [103].

The expression of H2AX in elongating spermatids reveals that there is potential for DNA damage response in the spermatids with suggestions of NHEJ occurring during spermiogenesis [104]. However, this repair is slow in round spermatids and usually incomplete [105]. Genes involved in MMR are not present in spermatids, indicating that this is not an important DDR in these cells [102]. Round spermatids contain Ku70 [106] as expected in haploid cells, indicating that NHEJ is the primary mode of DNA DSB repair in these germ cells. Clinical resistance to apoptosis, low levels of DDR, and high chromosome aberrations in round spermatids can reduce the success of in vitro fertilization (IVF) techniques, such as round spermatid injections (ROSI) [107].

Elongating spermatids are known for their down-regulation of DNA repair-associated transcripts, with some studies suggesting that elongated spermatids may have no DNA repair occurring [108, 109]. These findings are mainly attributed to the condensation and replacement of histones with protamines at this stage of spermatogenesis. The compaction by protamines is expected to prevent further DNA damage past this stage, but defects in protamination can lead to all sorts of DNA damage. To add to this, DDR like BER cannot function past the elongating spermatid stage [110]. Furthermore, the apoptotic pathway is not accessible to late spermatogenic cells [111], thus increasing the number of sperm with accumulated DNA damage that are not eliminated [112, 113].

Apurinic endonucleases are absent in spermatozoa, meaning BER repair is not possible until the S phase of mitotic division in the zygote [114]. The spermatozoa are terminally differentiated cells that shed their cytoplasm during the process of spermiogenesis, thereby leaving late spermatids and spermatozoa devoid of any repair machinery. If repair factors remain, it is unlikely that they will be able to make contact with the sites of DNA damage due to protamines that tightly compact DNA in spermatids and spermatozoa.

AGING AND MALE GERM CELLS

Consequences of Advanced Paternal Age

The age of parenthood has been increasing in most of the industrialized world [115], due to societal changes that have led to a delay in the age of childbearing [116]. These societal factors include, but are not limited to, increased life expectancy [115, 117]; development of effective and reversible contraception [118]; advancements in assisted reproductive technologies (ARTs) [119]; and the desire for increased financial security prior to starting a family [120]. Advanced maternal age has been well documented and linked to reduced fertility [121], increased risk of miscarriage, spontaneous abortions [122], and birth defects [121–124]. These negative effects of aging on reproduction are evident as women reach their natural reproductive limits. However, men do not reach clear-cut reproductive limits as they continue to generate spermatozoa throughout their lives and can sire children well into old age [125].

There has been increased scientific interest in the study of paternal age due partly to higher birthrates among older fathers. In addition, growing evidence that aged paternity is associated with increased incidence of diseases in offspring, such as autism [3, 4], schizophrenia [5], achondroplasia [6], and attention deficit hyperactivity disorder (ADHD) [7], has generated interest in the study of spermatozoa from aged fathers.

The geneticist James Crow emphasized that while maternal mutations were more striking, there should be greater concern about the accumulated effects of less striking, but more frequent, mutations in the paternal germ line [126–128]. Recent studies have renewed interest in germ-line mutations, pointing to increased de novo germ-line mutations in aged fathers [129]. These studies suggest that with each additional year in a father's age, there is the addition of two de novo mutations in the child [126–128, 130].

Effects of Aging on DNA Damage/Repair in Male Germ Cells

DNA damage is three-times higher in the spermatozoa of males 36–57 old compared to that in men below 35 [131]. The DNA damage load of a mammalian genome due to replication/repair errors and cellular metabolism is estimated to be 105 DNA lesions per cell per day [132]. This makes the proper repair of DNA damage vital as it prevents the accumulation of damage and thus limits mutations from being passed on to progeny (Fig. 4.2). Comet assays reveal high levels of spermatozoa with DNA fragmentation that have been found in the ejaculates of subfertile males [133].

There are several possible sources for the age-associated DNA damage/anomaly in the germ cells. These include the accumulation of DNA mutations due to continuous cell

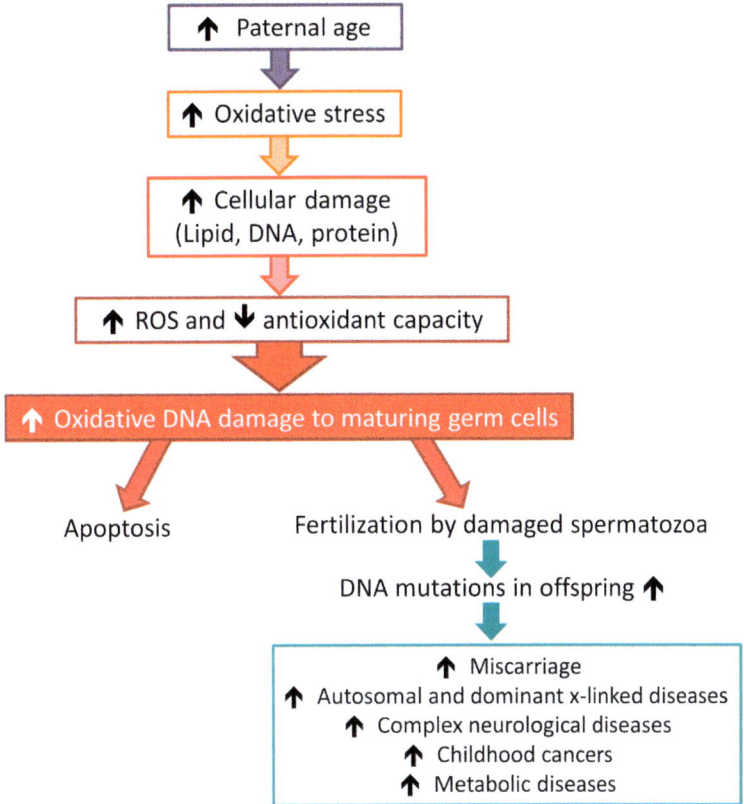

FIG. 4.2 Advanced paternal age affects sperm and progeny outcome. A proposed mechanism linking advanced paternal to age, increased oxidative damage in sperm, reduced fertility, and altered progeny outcome.

divisions, the accumulation of environmental toxin exposure and harmful medical treatments, the decreased ability to check and correct chromosome segregation during mitosis/meiosis, increased oxidative stress, and decreased ability to repair DNA damage (breaks, crosslinks) that occurs during the normal process of spermatogenesis.

Spermatozoa from aged animals have increased DNA damage, which is associated with decreased conception and increased rates of miscarriage (Fig. 4.2). While mutations can be repaired after fertilization in the zygote, the spermatozoa of aged and infertile men are severely damaged, which can lead to preimplantation loss [134]. Aging results in poor repair and removal of DNA adducts and oxidized purines in testicular cells, resulting in the accumulation of DNA damage such as 8-oxodG [135]. Moreover, there is evidence that some DNA adducts in spermatogenic cells accumulate and do not induce apoptosis, leading to spermatozoa that retain these adducts [136].

Aging studies in rats have revealed that genes involved in BER are down-regulated, leading to deficient repair in the spermatocytes of the aged animals [135]. While NHEJ does not show compelling evidence of change with aging in spermatocytes, the presence of Ku70 in mid-late pachytene and secondary spermatocytes indicates that NHEJ can occur in these cells.

Deficient protamination is known to occur with paternal aging [137], and genotoxic 8-oxodG lesions increase in the spermatozoa of aged animals [135].

One of the mechanisms responsible for the increased DNA damage in germ cells from aged males is an increase in reactive oxygen species (ROS) due to induction of oxidative stress.

While the potential damage caused by ROS in the early phases of spermatogenesis remain unclear, it is well established that the later phases of spermatogenesis, and in particular mature spermatozoa, are vulnerable to ROS because most of their antioxidant defenses and repair machinery are no longer there [138]. The expression of transcripts and proteins for several enzymes having antioxidant activities decreases in pachytene spermatocytes from aged rats, including SOD1. Oxidative stress of mature spermatozoa [139, 140] reveals that increased ROS is inversely related to spermatozoa quality [141–143]. Decreased antioxidant capacity has been identified in mature spermatozoa from aged animals [139].

Aging Affects Epigenetic Marks in Male Germ Cells

Aging is also thought to affect the epigenome of germ cells. Currently, only the DNA methylome and the length of telomeres have been reported to be modified with advanced paternal age. Further studies are necessary to evaluate its effect of paternal aging on noncoding RNA profiles, histone modifications, and the nuclear proteome in the germline.

The normal patterns of DNA methylation are critical for genomic imprinting and separation of the genome into transcriptionally active and inactive regions. Aging and aging-associated diseases [144] have been linked to alterations in DNA methylation in somatic cells [145, 146]. Early studies revealed abnormalities in the embryos sired by aged rats [147] and similar results in those sired by rats treated using the DNA demethylating agent, 5-azacytidine [148], thus suggesting a link between aging and altered germ cell DNA methylation. Studies by Oakes and colleagues demonstrated age-related epigenetic defects in male germ cells as aging resulted in hypermethylation of ribosomal DNA in the sperm of male rats [149].

Data indicate that the global human sperm DNA methylation and hydroxymethylation levels significantly increase by 1.76% and 5% per year, respectively [150]. The evolution with age of the sperm DNA methylation is opposed to what is observed in somatic cells, which undergo a global DNA demethylation. Some short regions (~1000 bp) of the human sperm genome are demethylated and these sequences are often close to nucleosome retention regions, preferentially occurring at CpG shores, and are enriched in genes associated with bipolar disorder [151]. In mouse sperm, the global sperm DNA methylation and hydroxymethylation levels are not altered with aging [152, 153]. However, Kobayashi and colleagues reported that the sperm DNA methylation level varies at some promoters. The ones that are hypomethylated with aging are enriched in genes involved in meiosis and spermatogenesis, and half these promoters overlap with CpG islands. The promoters that are hypermethylated with aging are not associated with CpG islands [152]. The CpG shores surrounding transcription start sites and splice junctions are hypomethylated in the sperm genome of older mice [153].

The intergenerational transmission of epigenetic defects due to advanced paternal age has been demonstrated in mice by Milekic and colleagues. The brain DNA of the older fathers' offspring revealed a parallel and smaller DNA hypomethylation at CpG shores surrounding the transcriptional start site than observed in paternal sperm genome, but a slight hypermethylation at splice junctions compared to the young fathers' offspring [153]. The older fathers' offspring also demonstrate a significantly altered behavior compared to the young father's offspring, and a differential expression of 17 genes in their brain tissue [153].

As age increases, telomere dynamics in somatic cells are well established with a consistent decline in telomere length in most cell types due to shortening with each mitotic division. This decrease in telomere length is debated with regard to the male germ cells, as these cells are among the only cell type within the body to have active telomere lengthening by the enzyme telomerase. There is limited literature on the subject, with conflicting and inconclusive results. Evidence in mice suggests that sperm telomeres are getting shorter, consistent with somatic cell telomere dynamics [154]. When a similar study was conducted in humans, the evidence was contrary to what is known in somatic cells, with the telomeres in sperm increasing in length with age [155]. Unfortunately, it is unclear if the trends observed are species-specific or due to different methods used for assessing telomere length.

CONCLUSION

The information supporting reproductive changes with age is limited, with very little data to support the observed changes. Epigenetic marks are beginning to gain momentum in the study of reproductive aging, including changes in DNA methylation, histone marks. and noncoding RNAs. Further robust studies on the various epigenetic marks, telomere length, and chromatin structure are necessary to understand the effects of advanced paternal age on fertility and outcome for the progeny.

Acknowledgments

We thank Elise Boivin-Ford for assistance in the preparation of the manuscript and CIHR-IGH Team Grant—Boys and Men's Health TEI-138298.

References

[1] Ford WC, North K, Taylor H, Farrow A, Hull MG, Golding J. Increasing paternal age is associated with delayed conception in a large population of fertile couples: evidence for declining fecundity in older men. The ALSPAC study team (Avon longitudinal study of pregnancy and childhood). Hum Reprod 2000;15(8):1703–8.

[2] De La Rochebrochard E, McElreavey K, Thonneau P. Paternal age over 40 years: the "amber light" in the reproductive life of men? J Androl 2003;24(4):459–65.

[3] Frans EM, Sandin S, Reichenberg A, Langstrom N, Lichtenstein P, McGrath JJ, et al. Autism risk across generations: a population-based study of advancing grandpaternal and paternal age. JAMA Psychiatry 2013;70(5):516–21.

[4] Grether JK, Anderson MC, Croen LA, Smith D, Windham GC. Risk of autism and increasing maternal and paternal age in a large north American population. Am J Epidemiol 2009;170(9):1118–26.

[5] Hubert A, Szoke A, Leboyer M, Schurhoff F. Influence of paternal age in schizophrenia. L'Encéphale 2011;37 (3):199–206.

[6] Orioli IM, Castilla EE, Scarano G, Mastroiacovo P. Effect of paternal age in achondroplasia, thanatophoric dysplasia, and osteogenesis imperfecta. Am J Med Genet 1995;59(2):209–17.

[7] D'Onofrio BM, Rickert ME, Frans E, Kuja-Halkola R, Almqvist C, Sjolander A, et al. Paternal age at childbearing and offspring psychiatric and academic morbidity. JAMA Psychiatry 2014;71(4):432–8.

[8] Aitken RJ, Koopman P, Lewis SE. Seeds of concern. Nature 2004;432:48–52.

[9] Selvaratnam J, Paul C, Robaire B. Male rat germ cells display age-dependent and cell-specific susceptibility in response to oxidative stress challenges. Biol Reprod 2015;93(3):72.

[10] Davie JR, Chadee DN. Regulation and regulatory parameters of histone modifications. J Cell Biochem Suppl 1998;30-31:203–13.

[11] Pivot-Pajot C, Caron C, Govin J, Vion A, Rousseaux S, Khochbin S. Acetylation-dependent chromatin reorganization by BRDT, a testis-specific bromodomain-containing protein. Mol Cell Biol 2003;23(15):5354–65.

[12] Meistrich ML, Mohapatra B, Shirley CR, Zhao M. Roles of transition nuclear proteins in spermiogenesis. Chromosoma 2003;111(8):483–8.

[13] Sonnack V, Failing K, Bergmann M, Steger K. Expression of hyperacetylated histone H4 during normal and impaired human spermatogenesis. Andrologia 2002;34(6):384–90.

[14] Yu YE, Zhang Y, Unni E, Shirley CR, Deng JM, Russell LD, et al. Abnormal spermatogenesis and reduced fertility in transition nuclear protein 1-deficient mice. Proc Natl Acad Sci U S A 2000;97(9):4683–8.

[15] Zhao M, Shirley CR, Yu YE, Mohapatra B, Zhang Y, Unni E, et al. Targeted disruption of the transition protein 2 gene affects sperm chromatin structure and reduces fertility in mice. Mol Cell Biol 2001;21(21):7243–55.

[16] Zhao M, Shirley CR, Mounsey S, Meistrich ML. Nucleoprotein transitions during spermiogenesis in mice with transition nuclear protein Tnp1 and Tnp2 mutations. Biol Reprod 2004;71(3):1016–25.

[17] Oliva R. Protamines and male infertility. Hum Reprod Update 2006;12(4):417–35.

[18] Braun RE, Peschon JJ, Behringer RR, Brinster RL, Palmiter RD. Protamine 3′-untranslated sequences regulate temporal translational control and subcellular localization of growth hormone in spermatids of transgenic mice. Genes Dev 1989;3(6):793–802.

[19] Oliva R, Dixon GH. Vertebrate protamine genes and the histone-to-protamine replacement reaction. Prog Nucleic Acid Res Mol Biol 1991;40:25–94.

[20] Cho C, Willis WD, Goulding EH, Jung-Ha H, Choi YC, Hecht NB, et al. Haploinsufficiency of protamine-1 or −2 causes infertility in mice. Nat Genet 2001;28(1):82–6.

[21] Balhorn R. The protamine family of sperm nuclear proteins. Genome Biol 2007;8(9):227.

[22] Balhorn R, Reed S, Tanphaichitr N. Aberrant protamine 1/protamine 2 ratios in sperm of infertile human males. Experientia 1988;44(1):52–5.

[23] Belokopytova IA, Kostyleva EI, Tomilin AN, Vorob'ev VI. Human male infertility may be due to a decrease of the protamine P2 content in sperm chromatin. Mol Reprod Dev 1993;34(1):53–7.

[24] Ward WS, Coffey DS. DNA packaging and organization in mammalian spermatozoa: comparison with somatic cells. Biol Reprod 1991;44(4):569–74.

[25] Wilson RH, Coverley D. Relationship between DNA replication and the nuclear matrix. Genes Cells 2013;18 (1):17–31.

[26] Shaman JA, Yamauchi Y, Ward WS. The sperm nuclear matrix is required for paternal DNA replication. J Cell Biochem 2007;102(3):680–8.

[27] Gawecka JE, Marh J, Ortega M, Yamauchi Y, Ward MA, Ward WS. Mouse zygotes respond to severe sperm DNA damage by delaying paternal DNA replication and embryonic development. PLoS One 2013;8(2): e56385.

[28] Kramer JA, Krawetz SA. Nuclear matrix interactions within the sperm genome. J Biol Chem 1996;271(20):11619–22.

[29] Haaf T, Ward DC. Higher order nuclear structure in mammalian sperm revealed by in situ hybridization and extended chromatin fibers. Exp Cell Res 1995;219(2):604–11.

[30] Zalensky AO, Allen MJ, Kobayashi A, Zalenskaya IA, Balhorn R, Bradbury EM. Well-defined genome architecture in the human sperm nucleus. Chromosoma 1995;103(9):577–90.

[31] Zink D, Fischer AH, Nickerson JA. Nuclear structure in cancer cells. Nat Rev Cancer 2004;4(9):677–87.

[32] Scherthan H. A bouquet makes ends meet. Nat Rev Mol Cell Biol 2001;2(8):621–7.

[33] Ioannou D, Millan NM, Jordan E, Tempest HG. A new model of sperm nuclear architecture following assessment of the organization of centromeres and telomeres in three-dimensions. Sci Rep 2017;7:41585.

[34] Fraser J, Williamson I, Bickmore WA, Dostie J. An overview of genome organization and how we got there: from FISH to Hi-C. Microbiol Mol Biol Rev 2015;79(3):347–72.

[35] Dekker J, Marti-Renom MA, Mirny LA. Exploring the three-dimensional organization of genomes: interpreting chromatin interaction data. Nat Rev Genet 2013;14(6):390–403.

[36] Dostie J, Bickmore WA. Chromosome organization in the nucleus - charting new territory across the Hi-Cs. Curr Opin Genet Dev 2012;22(2):125–31.

[37] Battulin N, Fishman VS, Mazur AM, Pomaznoy M, Khabarova AA, Afonnikov DA, et al. Comparison of the three-dimensional organization of sperm and fibroblast genomes using the Hi-C approach. Genome Biol 2015;16:77.

[38] Jung YH, Sauria MEG, Lyu X, Cheema MS, Ausio J, Taylor J, et al. Chromatin states in mouse sperm correlate with embryonic and adult regulatory landscapes. Cell Rep 2017;18(6):1366–82.

[39] Ke Y, Xu Y, Chen X, Feng S, Liu Z, Sun Y, et al. 3D chromatin structures of mature gametes and structural reprogramming during mammalian embryogenesis. Cell 2017;170(2):367–81 e20.

[40] Braun RE. Packaging paternal chromosomes with protamine. Nat Genet 2001;28(1):10–2.

[41] Balhorn R. A model for the structure of chromatin in mammalian sperm. J Cell Biol 1982;93(2):298–305.

[42] Kitamura A, Miyauchi N, Hamada H, Hiura H, Chiba H, Okae H, et al. Epigenetic alterations in sperm associated with male infertility. Congenit Anom (Kyoto) 2015;55(3):133–44.

[43] Wilson VL, Jones PA. DNA methylation decreases in aging but not in immortal cells. Science 1983;220 (4601):1055–7.

[44] Dada R, Kumar M, Jesudasan R, Fernandez JL, Gosalvez J, Agarwal A. Epigenetics and its role in male infertility. J Assist Reprod Genet 2012;29:213–23.

[45] DeBaun MR, Niemitz EL, Feinberg AP. Association of in vitro fertilization with Beckwith-Wiedemann syndrome and epigenetic alterations of LIT1 and H19. Am J Hum Genet 2003;72:156–60.

[46] Gosden R, Trasler J, Lucifero D, Faddy M. Rare congenital disorders, imprinted genes, and assisted reproductive technology. Lancet 2003;361:1975–7.

[47] Maher ER. Imprinting and assisted reproductive technology. Hum Mol Genet 2005;14(1):R133–8.

[48] Lazaraviciute G, Kauser M, Bhattacharya S, Haggarty P, Bhattacharya S. A systematic review and meta-analysis of DNA methylation levels and imprinting disorders in children conceived by IVF/ICSI compared with children conceived spontaneously. Hum Reprod Update 2014;20:840–52.

[49] Jones PA. Functions of DNA methylation: islands, start sites, gene bodies and beyond. Nat Rev Genet 2012;13 (7):484–92.

[50] Portela A, Esteller M. Epigenetic modifications and human disease. Nat Biotechnol 2010;28(10):1057–68.

[51] Carrell DT. Epigenetics of the male gamete. Fertil Steril 2012;97(2):267–74.

[52] Prados J, Stenz L, Somm E, Stouder C, Dayer A, Paoloni-Giacobino A. Prenatal exposure to DEHP affects spermatogenesis and sperm DNA methylation in a strain-dependent manner. PLoS One 2015;10(7): e0132136.

[53] Chastain LG, Sarkar DK. Alcohol effects on the epigenome in the germline: role in the inheritance of alcohol-related pathology. Alcohol 2017;60:53–66.

[54] de Castro Barbosa T, Ingerslev LR, Alm PS, Versteyhe S, Massart J, Rasmussen M, et al. High-fat diet reprograms the epigenome of rat spermatozoa and transgenerationally affects metabolism of the offspring. Mol Metab 2016;5(3):184–97.

[55] Dunn GA, Bale TL. Maternal high-fat diet effects on third-generation female body size via the paternal lineage. Endocrinology 2011;152(6):2228–36.

[56] Illum LRH, Bak ST, Lund S, Nielsen AL. DNA methylation in epigenetic inheritance of metabolic diseases through the male germ line. J Mol Endocrinol 2018;60(2):R39–56.

[57] Lambrot R, Xu C, Saint-Phar S, Chountalos G, Cohen T, Paquet M, et al. Low paternal dietary folate alters the mouse sperm epigenome and is associated with negative pregnancy outcomes. Nat Commun 2013;4:2889.

[58] Radford EJ, Ito M, Shi H, Corish JA, Yamazawa K, Isganaitis E, et al. In utero effects. In utero undernourishment perturbs the adult sperm methylome and intergenerational metabolism. Science 2014;345(6198):1255903.

[59] Huang H, Lin S, Garcia BA, Zhao Y. Quantitative proteomic analysis of histone modifications. Chem Rev 2015;115(6):2376–418.

[60] Zhao Y, Garcia BA. Comprehensive catalog of currently documented histone modifications. Cold Spring Harb Perspect Biol 2015;7(9)a025064.

[61] Page J, de la Fuente R, Manterola M, Parra MT, Viera A, Berrios S, et al. Inactivation or nonreactivation: what accounts better for the silence of sex chromosomes during mammalian male meiosis? Chromosoma 2012;121 (3):307–26.

[62] Fernandez-Capetillo O, Mahadevaiah SK, Celeste A, Romanienko PJ, Camerini-Otero RD, Bonner WM, et al. H2AX is required for chromatin remodeling and inactivation of sex chromosomes in male mouse meiosis. Dev Cell 2003;4(4):497–508.

[63] Turner JM, Aprelikova O, Xu X, Wang R, Kim S, Chandramouli GV, et al. BRCA1, histone H2AX phosphorylation, and male meiotic sex chromosome inactivation. Curr Biol 2004;14(23):2135–42.

[64] Kwak HG, Dohmae N. Proteomic characterization of histone variants in the mouse testis by mass spectrometry-based top-down analysis. Biosci Trends 2016;10(5):357–64.

[65] Luense LJ, Wang X, Schon SB, Weller AH, Lin Shiao E, Bryant JM, et al. Comprehensive analysis of histone post-translational modifications in mouse and human male germ cells. Epigenetics Chromatin 2016;9:24.

[66] Tan M, Luo H, Lee S, Jin F, Yang JS, Montellier E, et al. Identification of 67 histone marks and histone lysine crotonylation as a new type of histone modification. Cell 2011;146(6):1016–28.

[67] Lehtiniemi T, Kotaja N. Germ granule-mediated RNA regulation in male germ cells. Reproduction 2018;155(2): R77–91.

[68] Concha II, Urzua U, Yañez A, Schroeder R, Pessot C, Burzio LO. U1 and U2 snRNA are localized in the sperm nucleus. Exp Cell Res 1993;204:378–81.

[69] Krawetz SA. Paternal contribution: new insights and future challenges. Nat Rev Genet 2005;6(8):633–42.

[70] Schuster A, Tang C, Xie Y, Ortogero N, Yuan S, Yan W. SpermBase: a database for sperm-borne RNA contents. Biol Reprod 2016;95(5):99.

[71] Ball RL, Fujiwara Y, Sun F, Hu J, Hibbs MA, Handel MA, et al. Regulatory complexity revealed by integrated cytological and RNA-seq analyses of meiotic substages in mouse spermatocytes. BMC Genomics 2016;17(1):628.

[72] Chen X, Che D, Zhang P, Li X, Yuan Q, Liu T, et al. Profiling of miRNAs in porcine germ cells during spermatogenesis. Reproduction 2017;154(6):789–98.

[73] Jan SZ, Vormer TL, Jongejan A, Roling MD, Silber SJ, de Rooij DG, et al. Unraveling transcriptome dynamics in human spermatogenesis. Development 2017;144(20):3659–73.

[74] Laiho A, Kotaja N, Gyenesei A, Sironen A. Transcriptome profiling of the murine testis during the first wave of spermatogenesis. PLoS One 2013;8(4): e61558.

[75] Smorag L, Zheng Y, Nolte J, Zechner U, Engel W, Pantakani DV. MicroRNA signature in various cell types of mouse spermatogenesis: Evidence for stage-specifically expressed miRNA-221, −203 and -34b-5p mediated spermatogenesis regulation. Biol Cell 2012;104(11):677–92.

[76] Chen X, Li X, Guo J, Zhang P, Zeng W. The roles of microRNAs in regulation of mammalian spermatogenesis. J Anim Sci Biotechnol 2017;8:35.

[77] Kotaja N. MicroRNAs and spermatogenesis. Fertil Steril 2014;101(6):1552–62.

[78] Huszar JM, Payne CJ. MicroRNA 146 (Mir146) modulates spermatogonial differentiation by retinoic acid in mice. Biol Reprod 2013;88(1):15.

[79] Hirakata S, Siomi MC. piRNA biogenesis in the germline: from transcription of piRNA genomic sources to piRNA maturation. Biochim Biophys Acta 2016;1859(1):82–92.

[80] Iwasaki YW, Siomi MC, Siomi H. PIWI-interacting RNA: its biogenesis and functions. Annu Rev Biochem 2015;84:405–33.

[81] Russell SJ, Stalker L, LaMarre J. PIWIs, piRNAs and retrotransposons: complex battles during reprogramming in gametes and early embryos. Reprod Domest Anim 2017;52(Suppl. 4):28–38.

[82] Sarkar A, Volff JN, Vaury C. piRNAs and their diverse roles: a transposable element-driven tactic for gene regulation? FASEB J 2017;31(2):436–46.

[83] Kumar P, Kuscu C, Dutta A. Biogenesis and function of transfer RNA related fragments (tRFs). Trends Biochem Sci 2016;41(8):679–89.

[84] Sobala A, Hutvagner G. Transfer RNA-derived fragments: Origins, processing, and functions. Wiley Interdiscip Rev RNA 2011;2(6):853–62.

[85] Schorn AJ, Gutbrod MJ, LeBlanc C, Martienssen R. LTR-retrotransposon control by tRNA-derived small RNAs. Cell 2017;170(1):61–71 (e11).

[86] Bjork JK, Sandqvist A, Elsing AN, Kotaja N, Sistonen L. miR-18, a member of Oncomir-1, targets heat shock transcription factor 2 in spermatogenesis. Development 2010;137(19):3177–84.

[87] Gou LT, Dai P, Yang JH, Xue Y, Hu YP, Zhou Y, et al. Pachytene piRNAs instruct massive mRNA elimination during late spermiogenesis. Cell Res 2014;24(6):680–700.

[88] Gou LT, Dai P, Yang JH, Xue Y, Hu YP, Zhou Y, et al. Pachytene piRNAs instruct massive mRNA elimination during late spermiogenesis. Cell Res 2015;25(2):266.

[89] Wu J, Bao J, Kim M, Yuan S, Tang C, Zheng H, et al. Two miRNA clusters, miR-34b/c and miR-449, are essential for normal brain development, motile ciliogenesis, and spermatogenesis. Proc Natl Acad Sci U S A 2014;111(28): E2851–7.

[90] Yuan S, Tang C, Zhang Y, Wu J, Bao J, Zheng H, et al. Mir-34b/c and mir-449a/b/c are required for spermatogenesis, but not for the first cleavage division in mice. Biol Open 2015;4(2):212–23.

[91] Jodar M, Selvaraju S, Sendler E, Diamond MP, Krawetz SA, Reproductive Medicine N. The presence, role and clinical use of spermatozoal RNAs. Hum Reprod Update 2013;19(6):604–24.

[92] Krawetz SA, Kruger A, Lalancette C, Tagett R, Anton E, Draghici S, et al. A survey of small RNAs in human sperm. Hum Reprod 2011;26(12):3401–12.

[93] Sharma U, Conine CC, Shea JM, Boskovic A, Derr AG, Bing XY, et al. Biogenesis and function of tRNA fragments during sperm maturation and fertilization in mammals. Science 2016;351(6271):391–6.

[94] Chen Q, Yan M, Cao Z, Li X, Zhang Y, Shi J, et al. Sperm tsRNAs contribute to intergenerational inheritance of an acquired metabolic disorder. Science 2016;351(6271):397–400.

[95] T1 F, Ohlsson Teague EM, Palmer NO, DeBlasio MJ, Mitchell M, Corbett M, Print CG, Owens JA, Lane M. Paternal obesity initiates metabolic disturbances in two generations of mice with incomplete penetrance to the F2 generation and alters the transcriptional profile of testis and sperm microRNA content. FASEB J 2013;27(10):4226–43.

[96] Abu-Halima M, Hammadeh M, Schmitt J, Leidinger P, Keller A, Meese E, et al. Altered microRNA expression profiles of human spermatozoa in patients with different spermatogenic impairments. Fertil Steril 2013;99 (5):1249–55 e16.

[97] Burl RB, Clough S, Sendler E, Estill M, Krawetz SA. Sperm RNA elements as markers of health. Syst Biol Reprod Med 2018;64(1):25–38.

[98] Jodar M, Sendler E, Moskovtsev SI, Librach CL, Goodrich R, Swanson S, Hauser R, Diamond MP, Krawetz SA. Absence of sperm RNA elements correlates with idiopathic male infertility. Sci Transl Med 2015;7(295). 295re6.

[99] Salas-Huetos A, Blanco J, Vidal F, Godo A, Grossmann M, Pons MC, et al. Spermatozoa from patients with seminal alterations exhibit a differential micro-ribonucleic acid profile. Fertil Steril 2015;104(3):591–601.

[100] Wang C, Yang C, Chen X, Yao B, Yang C, Zhu C, et al. Altered profile of seminal plasma microRNAs in the molecular diagnosis of male infertility. Clin Chem 2011;57(12):1722–31.

[101] Yao C, Yuan Q, Niu M, Fu H, Zhou F, Zhang W, et al. Distinct expression profiles and novel targets of MicroRNAs in human Spermatogonia, Pachytene spermatocytes, and round spermatids between OA patients and NOA patients. Mol Ther Nucleic Acids 2017;9:182–94.

[102] Richardson LL, Pedigo C, Ann Handel M. Expression of deoxyribonucleic acid repair enzymes during spermatogenesis in mice. Biol Reprod 2000;62(3):789–96.

[103] Xu G, Spivak G, Mitchell DL, Mori T, McCarrey JR, McMahan CA, et al. Nucleotide excision repair activity varies among murine spermatogenic cell types. Biol Reprod 2005;73(1):123–30.

[104] Leduc F, Maquennehan V, Nkoma GB, Boissonneault G. DNA damage response during chromatin remodeling in elongating spermatids of mice. Biol Reprod 2008;78(2):324–32.

[105] Ahmed EA, van der Vaart A, Barten A, Kal HB, Chen J, Lou Z, et al. Differences in DNA double strand breaks repair in male germ cell types: lessons learned from a differential expression of Mdc1 and 53BP1. DNA Repair 2007;6(9):1243–54.

[106] Ahmed EA, Sfeir A, Takai H, Scherthan H. Ku70 and nonhomologous end joining protect testicular cells from DNA damage. J Cell Sci 2013;126:3095–104. Pt 14.

[107] Levran D, Nahum H, Farhi J, Weissman A. Poor outcome with round spermatid injection in azoospermic patients with maturation arrest. Fertil Steril 2000;74(3):443–9.

[108] van Loon AA, Sonneveld E, Hoogerbrugge J, van der Schans GP, Grootegoed JA, Lohman PH, et al. Induction and repair of DNA single-strand breaks and DNA base damage at different cellular stages of spermatogenesis of the hamster upon in vitro exposure to ionizing radiation. Mutat Res 1993;294(2):139–48.

[109] Olsen AK, Duale N, Bjoras M, Larsen CT, Wiger R, Holme JA, et al. Limited repair of 8-hydroxy-7,8-dihydroguanine residues in human testicular cells. Nucleic Acids Res 2003;31(4):1351–63.

[110] Aguilar-Mahecha A, Hales BF, Robaire B. Expression of stress response genes in germ cells during spermatogenesis. Biol Reprod 2001;65(1):119–27.

[111] Oldereid NB, Angelis PD, Wiger R, Clausen OP. Expression of Bcl-2 family proteins and spontaneous apoptosis in normal human testis. Mol Hum Reprod 2001;7(5):403–8.

[112] Aitken RJ, Bronson R, Smith TB, De Iuliis GN. The source and significance of DNA damage in human sperma-tozoa; a commentary on diagnostic strategies and straw man fallacies. Mol Hum Reprod 2013;19(8):475–85.

[113] Aitken RJ, Koppers AJ. Apoptosis and DNA damage in human spermatozoa. Asian J Androl 2011;13(1):36–42.

[114] Aitken RJ, Smith TB, Jobling MS, Baker MA, De Iuliis GN. Oxidative stress and male reproductive health. Asian J Androl 2014;16(1):31–8.

[115] Roush W. Live long and prosper? Science 1996;273(5271):42–6.

[116] Francis HH. Delayed childbearing. IPPF Med Bull 1985;19(3):3–4.

[117] Laufer N. Introduction: fertility and longevity. Fertil Steril 2015;103(5):1107–8.

[118] Sharma R, Agarwal A, Rohra VK, Assidi M, Abu-Elmagd M, Turki RF. Effects of increased paternal age on sperm quality, reproductive outcome and associated epigenetic risks to offspring. Reprod Biol Endocrinol 2015;13:35.

[119] Toriello HV, Meck JM, Professional P, Guidelines C. Statement on guidance for genetic counseling in advanced paternal age. Genet Med 2008;10(6):457–60.

[120] Lawson G, Fletcher R. Delayed fatherhood. J Fam Plann Reprod Health Care 2014;40:283–8.

[121] Sauer MV. Reproduction at an advanced maternal age and maternal health. Fertil Steril 2015;103(5):1136–43.

[122] Heffner LJ. Advanced maternal age—how old is too old? N Engl J Med 2004;351(19):1927–9.

[123] Herbert M, Kalleas D, Cooney D, Lamb M, Lister L. Meiosis and maternal aging: insights from aneuploid oo-cytes and trisomy births. Cold Spring Harb Perspect Biol 2015;7:a017970.

[124] Djahanbakhch O, Ezzati M, Zosmer A. Reproductive ageing in women. J Pathol 2007;211(2):219–31.

[125] Seymour FI, Duffy C, Korner A. A case of authentic fertility in a man of 94. JAMA 1935;105:1423–4.

[126] Crow JF. The high spontaneous mutation rate: is it a health risk? Proc Natl Acad Sci U S A 1997;94(16):8380–6.

[127] Callaway E. Fathers bequeath more mutations as they age. Nature 2012;488(7412):439.

[128] Kong A, Frigge ML, Masson G, Besenbacher S, Sulem P, Magnusson G, et al. Rate of de novo mutations and the importance of father's age to disease risk. Nature 2012;488(7412):471–5.

[129] Rahbari R, Wuster A, Lindsay SJ, Hardwick RJ, Alexandrov LB, Al Turki S, et al. Timing, rates and spectra of human germline mutation. Nat Genet 2016;48(2):126–33.

[130] Kong X, Wang S, Jiang H, Nie G, Li X. Responses of acid/alkaline phosphatase, lysozyme, and catalase activities and lipid peroxidation to mercury exposure during the embryonic development of goldfish Carassius auratus. Aquat Toxicol 2012;120-121:119–25.

[131] Singh NP, Muller CH, Berger RE. Effects of age on DNA double-strand breaks and apoptosis in human sperm. Fertil Steril 2003;80(6):1420–30.

[132] Marteijn JA, Lans H, Vermeulen W, Hoeijmakers JH. Understanding nucleotide excision repair and its roles in cancer and ageing. Nat Rev Mol Cell Biol 2014;15(7):465–81.

[133] Enciso M, Sarasa J, Agarwal A, Fernandez JL, Gosalvez J. A two-tailed comet assay for assessing DNA damage in spermatozoa. Reprod BioMed Online 2009;18(5):609–16.

[134] Zenzes MT, Puy LA, Bielecki R, Reed TE. Detection of benzo(a)pyrene diol epoxide-DNA adducts in embryos from smoking couples: evidence for transmission by spermatozoa. Mol Hum Reprod 1999;5(2):125–31.

[135] Paul C, Nagano M, Robaire B. Aging results in differential regulation of DNA repair pathways in pachytene spermatocytes in the Brown Norway rat. Biol Reprod 2011;85(6):1269–78.

[136] Zenzes MT, Bielecki R, Reed TE. Detection of benzo(a)pyrene diol epoxide-DNA adducts in sperm of men exposed to cigarette smoke. Fertil Steril 1999;72(2):330–5.

[137] Zubkova EV, Wade M, Robaire B. Changes in spermatozoal chromatin packaging and susceptibility to oxida-tive challenge during aging. Fertil Steril 2005;84(Suppl. 2):1191–8.

[138] Aitken RJ, Baker MA. Causes and consequences of apoptosis in spermatozoa; contributions to infertility and impacts on development. Int J Dev Biol 2013;57(2–4):265–72.

[139] Weir CP, Robaire B. Spermatozoa have decreased antioxidant enzymatic capacity and increased reactive oxygen species production during aging in the Brown Norway rat. J Androl 2007;28(2):229–40.

[140] Zubkova EV, Robaire B. Effect of glutathione depletion on antioxidant enzymes in the epididymis, seminal vesicles, and liver and on spermatozoa motility in the aging brown Norway rat. Biol Reprod 2004;71(3):1002–8.

[141] Chianese C, Brilli S, Krausz C. Genomic changes in spermatozoa of the aging male. Advances in experimental medicine and biology. vol. 791;2014. p. 13–26.

[142] Haidl G, Jung A, Schill WB. Ageing and sperm function. Hum Reprod 1996;11(3):558–60.

[143] Hudson WC. Sperm banking as a strategy to reduce harms associated with advancing paternal age. Food Drug Law J 2015;70(4):573–91 (ii).

[144] Issa JP. CpG-island methylation in aging and cancer. Curr Top Microbiol Immunol 2000;249:101–18.

[145] Wilson VL, Smith RA, Ma S, Cutler RG. Genomic 5-methyldeoxycytidine decreases with age. J Biol Chem 1987;262(21):9948–51.

[146] Ono T, Takahashi N, Okada S. Age-associated changes in DNA methylation and mRNA level of the c-myc gene in spleen and liver of mice. Mutat Res 1989;219(1):39–50.

[147] Serre V, Robaire B. Paternal age affects fertility and progeny outcome in the Brown Norway rat. Fertil Steril 1998;70(4):625–31.

[148] Doerksen T, Trasler JM. Developmental exposure of male germ cells to 5-azacytidine results in abnormal preimplantation development in rats. Biol Reprod 1996;55(5):1155–62.

[149] Oakes CC, Smiraglia DJ, Plass C, Trasler JM, Robaire B. Aging results in hypermethylation of ribosomal DNA in sperm and liver of male rats. Proc Natl Acad Sci U S A 2003;100(4):1775–80.

[150] Jenkins TG, Aston KI, Cairns BR, Carrell DT. Paternal aging and associated intraindividual alterations of global sperm 5-methylcytosine and 5-hydroxymethylcytosine levels. Fertil Steril 2013;100(4):945–51.

[151] Jenkins TG, Aston KI, Pflueger C, Cairns BR, Carrell DT. Age-associated sperm DNA methylation alterations: possible implications in offspring disease susceptibility. PLoS Genet 2014;10(7): e1004458.

[152] Kobayashi N, Okae H, Hiura H, Chiba H, Shirakata Y, Hara K, et al. Genome-scale assessment of age-related DNA methylation changes in mouse spermatozoa. PLoS One 2016;11(11): e0167127.

[153] Milekic MH, Xin Y, O'Donnell A, Kumar KK, Bradley-Moore M, Malaspina D, et al. Age-related sperm DNA methylation changes are transmitted to offspring and associated with abnormal behavior and dysregulated gene expression. Mol Psychiatry 2015;20(8):995–1001.

[154] de Frutos C, López-Cardona AP, Fonseca Balvís N, Laguna-Barraza R, Rizos D, Gutierrez-Adán A, Bermejo-Álvarez P. Spermatozoa telomeres determine telomere length in early embryos and offspring. Reproduction 2016;151(1):1–7.

[155] Kimura M, Cherkas LF, Kato BS, Demissie S, Hjelmborg JB, Brimacombe M, et al. Offspring's leukocyte telomere length, paternal age, and telomere elongation in sperm. PLoS Genet 2008;4(2):e37.

Further Reading

Kwak HG, Suzuki T, Dohmae N. Global mapping of post-translational modifications on histone H3 variants in mouse testes. Biochem Biophys Rep 2017;11:1–8.

In Vitro Spermatogenesis From Pluripotent Stem Cells

Jin Du*,‡, Qing-Qin Gao*,‡, Cheng Huang*,‡, Hai-Ping Jiang*,‡, Mo-Yu Dai*,‡, Yu-Fei Li*,‡, Qi Zhou*,†,‡

*State Key Laboratory of Stem Cell and Reproductive Biology, Institute of Zoology, Chinese Academy of Sciences, Beijing, China †Institute for Stem Cell and Regeneration, Chinese Academy of Sciences, Beijing, China ‡University of Chinese Academy of Sciences, Beijing, China

O U T L I N E

Abbreviations

5mC	5-methylcytosine
AP	alkaline phosphatase
AVE	anterior visceral endoderm
bFGF	basic fibroblast growth factor
BMP4	bone morphogenetic protein 4
BPE	bovine pituitary extract
BVSC	Blimp1-mVenus and stella-ECFP
CER1	Cerberus 1
Dazl	deleted in azoospermia-like

DMR	differentially methylated region
EGCs	embryonic germ cells
EpiLCs	pregastrulation epiblast-like cells
EpiSCs	epiblast-derived stem cells
ESCs	embryonic stem cells
ExM	extraembryonic mesoderm
FSH	follicle-stimulating hormone
GDNF	glial cell-derived growth factor
GK15	GMEM with 15% KSR
GSCs	germline stem cells
H3K27me3	histone H3 lysine 27 trimethylation
H3K9me2	histone H3 lysine 9 dimethylation
H3K9me2	histone H3 lysine 9 di-methylation
H3K9me3	histone H3 lysine 9 tri-methylation
HUMSCs	human umbilical-cord Wharton's jelly-derived mesenchymal stem cells
ICM	inner cell mass
ICSI	intracytoplasmic sperm injection
iMeLCs	incipient mesoderm-like cells
iPSCs	induced pluripotent stem cells
JNK	c-Jun N-terminal kinase
LH	luteinizing hormone
PGCLCs	primordial germ cell-like cells
PGCs	primordial germ cells
PRDM1	PR domain zincfinger protein 1
PSC	pluripotent stem cells
RA	retinoic acid
SCF	stem cell factor
SLCs	spermatid-like cells
SSCs	spermatogonial stem cells
SSEA	stage specific embryonic antigen
VE	visceral endoderm

INTRODUCTION

Mammalian life originates from one single cell, a zygote, which comes from the fertilization of a sperm and an egg. But where do **sperm** and eggs come from? The process from fertilization to the production of fertile gametes, **sperm**, and eggs, is referred to as gametogenesis, which can roughly be divided into four steps: (i) the origination and specification of primordial germ cells (PGCs); (ii) PGC migration and proliferation; (iii) development into haploid germ cells through meiotic division; and (iv) maturation of gametes.

In mice, before gastrulation, the mouse embryo consists of three distinct cell lineages, which were established in the blastocyst during the preimplantation period, that is, epiblast, extraembryonic endoderm, and trophectoderm. The epiblast, from which the entire fetus as well as the extraembryonic mesoderm (ExM) and amnion ectoderm will form, is a cup-shaped epithelium apposed on its open end to the extraembryonic ectoderm, a trophectoderm derivative. Both the epiblast and the extraembryonic ectoderm are covered by a visceral endoderm, which is part of the extraembryonic endoderm lineage.

At around embryonic day (E) 6.0, due to the induction of bone morphogenetic protein 4 (BMP4) [1], a growth factor secreted by the most proximal posterior epiblasts, ~30–40 mPGCs are generated within the ExM at around E7.0 [2, 3]. This process is called PGC specification, in which **some gene expression as well as some epigenetic reprogramming** are very important for directing cells to specify toward the germ cell lineage. For example, a set of transcription factors (TFs), such as *Blimp1/Prdm1* and *Prdm14* [2–4], are involved in **PGC** specification. Previous studies showed that *Blimp1*-deficient PGCs suffered severe disruption of PGC development at the early stage [5]. *Prdm14*-deficient embryos also showed defective PGC specification with impaired *Sox2* expression and aberrant histone modification at a genome-wide level [4]. Besides, Wnt3, which is secreted from the epiblast cells themselves partly in response to BMP4, is also essential for the induction of *Blimp1* and *Prdm14*.

At around E10.5, the mPGCs start to migrate through the hindgut endoderm and mesentery and colonize the embryonic gonadal primordia, which then forms either the ovaries or testes. At the same time, mPGCs continue to proliferate and reach a number of nearly 25,000 at around E13.5 [6] while at E10.5, there are only ~1000 cells in number [6]. Along with the migration and proliferation, extensive and dynamic changes of epigenetic modifications on the genome are happening. This includes erasure of CpG methylation in the differentially methylated region (DMR) of imprinting gene loci in both male and female by E13.5, and a decrease in histone H3 lysine 9 dimethylation (H3K9me2) and an increase in histone H3 lysine 27 trimethylation (H3K27me3) in the PGC genome [7, 8].

When PGCs colonize in the genital ridge, they show their first sex-based differences. In females, PGCs continue to proliferate and then enter into meiosis at around E13.5. **They arrest at the diplotene stage of the first meiotic prophase until fertilization by sperm, which leads to the meiosis of the primary oocytes** [9]. At around birth, a layer of granulosa cells surrounds the oocytes to form primordial follicles, the majority of which are maintained in a dormant state after birth. The follicles develop into primary, secondary, and antral follicles, and the oocytes grow in size and gradually acquire a gynogenetic epigenome, including maternal imprints [10, 11]. When estrus cycles commence at around 6 weeks after birth in response to hormonal stimulation, fully grown oocytes resume the first meiotic division, extrude the first polar bodies, and form secondary oocytes. The secondary oocytes ovulate, are fertilized with spermatozoa, and complete the second meiotic division to form zygotes, whereas in males, this meiosis process is quite different from the oocytes.

In males, after substantial proliferation, PGCs enter into mitotic arrest to become gonocytes from around E15.5; they locate within the luminal compartment of seminiferous tubules during the fetal period and acquire an androgenic epigenome, including paternal imprints [12, 13]. Male gonadal somatic cells expressing Sry differentiate into a fetal Sertoli cell lineage, which establishes a microenvironment for the maintenance of spermatogonial stem cells by secreting glial cell-derived growth factor (GDNF). After birth, the prospermatogonia translocate to the basal compartment of seminiferous tubules to differentiate into spermatogonia. At around postnatal day (P) 10, the vast majority of spermatogonia proceed into the first wave of spermatogenesis, resulting in the formation of the first population of haploid spermatozoa at around 3 weeks [14], whereas a small population of spermatogonia generate spermatogonial stem cells (SSCs), which sustain spermatogenesis throughout adulthood [15]. These haploid spermatozoas, though having completed the entire meiosis cycle, have no fertile ability and still need to mature from the spermatid cell to a fertile sperm through a process called spermiogenesis, which mainly includes the sharp change of the composition of the nuclei and cell organelles.

In humans, the process of gametogenesis is quite similar to that of mice, although there are several differences.

Human primordial germ cells (hPGCs) migrate from the yolk sac endoderm through the hindgut endoderm and dorsal mesentery to the genital ridges, and the process is analogous to that of mPGCs. In embryonic testes, which begin to differentiate from around weeks 5–6 with the expression of SRY and SOX9 [16], hPGCs continue to express key pluripotency markers such as POU5F1, NANOG, and SSEA1, and to proliferate, at least during the first trimester (weeks 10–12) [17]. Thereafter, during the second trimester, most, but not all, hPGCs appear to enter progressively into mitotic arrest and initiate differentiation into prospermatogonia. Some of the prospermatogonia translocate to the basal compartment of seminiferous tubules and differentiate into spermatogonia during the embryonic period. In embryonic ovaries, which begin to differentiate from around weeks 6–8 with the expression of FOXL2 and RSPO1 [18], hPGCs continue to proliferate at least until weeks 10–11, when they begin to enter into meiosis to differentiate into oocytes. The proliferation of hPGCs appears to continue at least until week 20. Therefore, mitotically active hPGCs and oocytes in the first prophase of meiosis coexist for a relatively long period of time in embryonic ovaries. Thus, the overall developmental dynamics of human germ cells in the embryonic period are similar to those of mouse germ cells, but human germ cells appear to show greater heterogeneity/asynchronicity in their developmental timing and require a much longer time for their developmental transitions. The number of germ cells is around ~100 at week 3, which increases to around ~1000 at week 4 and ~150,000 in males and ~450,000 in females at week 9 [19]. hPGCs exhibit substantial demethylation as early as week 5 around their colonization of embryonic gonads (genome-wide 5-methylcytosine (5mC) level, ~20%) and undergo further demethylation thereafter (genome-wide 5mC level, ~5% at week 9), and, as a consequence, hPGCs exhibit much lower genome-wide 5mC levels than inner cell mass cells of the blastocysts. In parallel, the erasure of parental imprints and X chromosome reactivation in females occurs with some human-specific dynamics. There are also some regions that evade genome-wide DNA demethylation, and these DNA demethylation circumstances may contribute to the transgenerational epigenetic inheritance. hPGCs/gonadal germ cells display low levels of histone H3 lysine 9 di-methylation (H3K9me2) and histone H3 lysine 9 tri-methylation (H3K9me3) as well as of DNMT3A/3B and UHRF1, suggesting that the mechanism for epigenetic reprogramming in human and mouse germ cells is conserved.

However, due to exposure to polluted environments or for other reasons, many people cannot produce healthy and fertile eggs or sperm, and that causes infertility. To conquer the problem, scientists have made various efforts, including reconstituting the entire cycle of germ cells in vitro.

Because PSCs, including ESCs and induced pluripotent stem cells (iPSCs), are capable of differentiating into multiple cell lineages, including germ cells, it seems feasible that an optimal set of culture conditions could direct ESCs/iPSCs toward a germ cell lineage. Therefore, researchers have tried to induce germ cells from ESCs/iPSCs. The significance of reconstitution of spermatogenesis and oogenesis in vitro includes obtaining ESCs/iPSCs; setting up standards to evaluate the function of germ cells, for example, transplanting cells to produce offspring and testing the level of gene expression; and studying epigenetic reprogramming and the mechanism of germ cell specification and maturation.

In order to mimic the generation of germ cells in vivo, which is a process that begins with PSCs then specifies to PGCs and finally becomes haploid germ cells through meiosis, researchers first tried to induce the production of PGCs from PSCs. But during the preparation of ESCs as the starting material, researchers found that ESCs could be divided into two categories, the naïve pluripotent stem cells and the primed pluripotent stem cells, according to their different pluripotent state. Naïve pluripotent stem cells, for example, mESCs under the 2i+LIF conditions, are closest to those of the E4.5 inner cell mass (ICM). Also, they can differentiate into all kinds of cell lineages and form chimeras following blastocyst injection [20, 21]. Primed pluripotent stem cells, such as epiblast-derived stem cells (EpiSCs) [21–23], seldom contribute to chimeras following blastocyst injection, although they are capable of multilineage differentiation. Gene expression and epigenetic profiles of ESCs and EpiSCs are also different. In addition, studies have shown that the pluripotent state of E6 epiblast with PGC competence in mice is in the situation between the naïve and primed pluripotent states. Trying to induce ESCs/iPSCs in a naïve state to an epiblast-like state with PGC-competence, Hayashi et al. [24] developed a culture system in which the PGC specification process was reconstructed in vitro by using mouse ESCs/iPSCs under a defined set of conditions, with the existence of bFGF and activin A (ActA). As a result, mouse ESCs/iPSCs at day 2 of culture differentiate efficiently into PGC-like cells (PGCLCs) in response to BMP4. This was confirmed by the expression of PGC-specific reporter genes, including *Blimp1-mVenus (BV)* and *stella-CFP (SC)*; the expression of PGC-specific surface proteins, including SSEA1 and ITGβ3; and a PGC-like epigenetic modification, including a decrease in H3K9me2 and an increase in H3K27me3 in the PGCLC genome. To test whether the PGCLCs have the potential to contribute to gametogenesis and to generate offspring, PGCLCs derived from male mouse ESCs were transplanted into the seminiferous tubules of neonatal *W/Wv* males that did not have their own germ cells. The results confirmed that the PGCLCs did contribute to spermatogenesis. Also, mature spermatozoa yielded from the testes were functional, as fertilized eggs with the spermatozoa had developed fully to healthy offspring with normal-sized placentas. The offspring, both male and female, grew normally and had the ability to bear the next generation.

Similar to PGCLCs derived from ESCs, those from iPSCs are also capable of differentiating into fully functional spermatozoa that eventually give rise to healthy and fertile offspring [24]. However, it appears that PGCLCs from two out of three iPSC lines, for which the germline transmission through chimera mice has been proven, were aberrant, for they did not contribute to spermatogenesis but rather formed teratomas in the transplanted testes. In contrast, all three ESC lines tested gave rise to fully potent PGCLCs. This suggests that the capacity of iPSCs to differentiate into PGCLCs is different among iPSC lines, consistent with previous reports showing that each iPSC line had different properties. It is still unclear whether the limited capability for PGCLC differentiation is caused by genetic or epigenetic alteration. Nevertheless, it is important to choose iPSCs that have a good propensity to differentiate into PGCLCs.

The obtained mPGCLCs could be maintained for no more than 10 days [25]. A better understanding of the mechanism and the cytokines that stimulate mPGC survival/proliferation would be helpful to advance these efforts. The ability to maintain long-term proliferation of mPGCLCs/mPGCs in culture should provide a key opportunity to identify signals, based on the knowledge of the mechanism for sex determination of germ cells that induce mPGCLCs into either spermatogenic or oogenic pathways without gonadal somatic cells [26]. To reconstitute the male pathway, a critical next step will be inducing a population of cells with SSC activity from mPGCLCs. Currently, SSC activity can only be defined by the transplantation assay, and there has been no single marker or combination of markers that can prospectively define a "pure" SSC population, which makes it a subject with intensive debates.

For a long time, researchers were not able to complete the entire cycle of gametogenesis in vitro due to the difficulty in initiating meiosis properly, which could form haploid sex-specific germ cells. In 2016, Zhou et al. [27] reported complete in vitro meiosis from ESC-derived PGCs by coculturing PGCLCs with neonatal testicular somatic cells and sequentially exposing them to morphogens and sex hormones. It is known that there is a "gold standard" panel to evaluate meiosis occurring in vitro, including correct nuclear DNA content at specific meiotic stages (for male cells, premeiotic, meiotic S phase, first reductional, and second meiotic division stages), normal chromosome number and organization, appropriate nuclear and chromosomal localization of proteins involved in homologous synapsis and recombination, and capacity of the in vitro-produced germ cells to produce viable euploid offspring. In this research, the artificial male spermatid-like cells (SLCs) showed key hallmarks of meiosis, including erasure of genetic imprinting, chromosomal synapsis and recombination, and correct nuclear DNA and chromosomal content. Intracytoplasmic injection of the resulting SLCs into oocytes produced viable and fertile offspring, which indicated that the robust stepwise approach can functionally recapitulate male gametogenesis in vitro.

Also in 2016, Hikabe et al. [28] reported successful reconstitution in vitro of the entire process of oogenesis from mouse pluripotent stem cells, including embryonic stem cells and induced pluripotent stem cells derived from both embryonic fibroblasts and adult tail tip fibroblasts. Moreover, pluripotent stem cell lines were rederived from the eggs that were generated in vitro, thereby reconstituting the full female germline cycle in a dish. Many experiments were carried out to evaluate the function and quality of the MII oocytes produced in vitro, such as RNA sequencing to test the global transcription dynamics and in vitro fertilization to test the capability of developing into offspring. Some sporadic flaws were also observed. For example, the placentas of the pups coming from the in vitro oocytes were heavier than those of wild-type mice, and the pups tended to be slightly heavier than the wild-type mice.

As reconstitution of the entire cycle of mouse gametogenesis in vitro has been realized, studies on the in vitro gametogenesis of human PSCs are on the stage [29, 30].

hPSCs cultured under conventional conditions bear characteristics similar to EpiSCs but not to mESCs, with regard to gene expression, epigenetic properties, and cytokine dependence. Because these represent a primed pluripotency state with biased differentiation potential, they bear little, if any, capacity to contribute to chimeras. In 2015, Irie et al. [31] confirmed a culture condition under which, subsequently, hESCs/hiPSCs could be efficiently induced into hPGCLCs using a procedure identical to that for mPGCLC induction, whereas hPSCs cultured under conventional conditions are unable to form hPGCLCs. The hPGCLCs induced by Monk and McLaren [32], however, are different from mPGCLCs and are negative for genes such as DDX4 and DAZL. Remarkably, the authors found that SOX17 acts upstream of BLIMP1 and is essential for hPGCLC specification. According to their model, SOX17 promotes the expression of genes for hPGC and endoderm development whereas BLIMP1 promotes the expression of genes for hPGC development and represses genes for mesoderm and endoderm development. On the other hand, Sasaki et al. [33] demonstrated that hiPSCs cultured under a feeder-free, defined condition with a primed state of pluripotency could first be differentiated into incipient mesoderm-like cells (iMeLCs) and then into hPGCLCs. The authors also found that hiPSCs bear intermediate properties between EpiLCs and EpiSCs. These hPGCLCs induced by Sasaki et al. [33], similar to the hPGCLCs induced by Monk and McLaren [32], were still negative for DDX4 and DAZL.

Successful in vitro gametogenesis from hPSCs should provide highly valuable information on all aspects of in vivo human germ cell development that are difficult to analyze, including the mechanisms for transcriptional control, epigenetic reprogramming, meiosis, and genome stability. Such knowledge, in turn, is also useful in discerning the causes and consequences of diseases arising from anomalies in germ cell development, including infertility, impaired development/physiology, and a diverse array of genetic/epigenetic disorders within offspring. A system for in vitro gametogenesis should also serve as a platform to identify potential drugs for such diseases as well as to screen chemicals for reproductive toxicity [34].

Another important application should be using the gametes generated from hPSCs to treat infertility. Even after we can attain mature gametes, a critical technological concern would be the genetic mutations that may accumulate in the somatic cells from which hiPSCs are derived. It has been shown that a germ cell lineage such as SSCs exhibits a lower mutation rate than somatic lineages [35]. The meiosis system of male germ cells can minimize genetic mutations whereas female germ cells cease mitotic divisions early, therefore preventing the accumulation of such mutations in their development [36]. In considering the use of such cells, therefore, it will be essential to scrutinize the genome sequences of the hiPSCs that could be used as starting material as well as of the resultant gametes. The epigenetic profiles of the resultant gametes or embryos should also be carefully monitored to prevent the transmission of any deleterious epimutations. These two technological requirements should at least be strictly met when considering the use of artificial gametes. Moreover, even when all the technological concerns are resolved, the use of artificial gametes presents huge ethical and sociological concerns that have been discussed in detail elsewhere and should also be widely discussed at the level of the individual and society [37]. The rapidly evolving genome-editing technologies may also be applied to hiPSCs and gametes derived from them, but the use of such cells should also require careful discussion and guidance.

INDUCTION OF SPERMATOGENESIS IN MICE

Induction of mPGCLCs From mESCs

Process of mPGC Development In Vivo

mPGC specification occurs in the egg cylinder at embryonic day 6.25 (E6.25), just before gastrulation [38–40]. At the beginning of PGC specification, BMP and WNT signals secreted from the extraembryonic ectoderm and the proximal visceral endoderm (VE) induce the expression of PR domain zinc finger protein 1 (PRDM1, also known as BLIMP1), a key regulator of PGC fate [41]. Then, the expressions of two key specification factors, PRDM14 and transcription factor AP2γ [42], are upregulated, and PRDM1, PRDM14, and AP2γ form a core regulatory network that induces germ cell fates [43–45]. At E7.5, approximately 40 mPGCs are formed at the base of allantois [46] and subsequently migrate into the hindgut and colonize the genital ridge by E11.5. In the process of migration, mPGCs undergo genome-wide epigenetic reprogramming, including global DNA demethylation [8], genomic imprint erasure [47, 48], X chromosome reactivation [32], and reorganization of chromatin modification [35]. Finally, XY and XX germ cells undergo mitotic arrest and meiotic entry by E13.5, respectively. This marks the end of the PGC stage of germline development [49].

Regulatory Network of mPGC Specification

BMP-SMAD and WNT3-β-catenin receive signals from extraembryonic tissues and induce the expression of PRDM1 and PRDM14, simultaneously mesoderm genes T in PGC precursors [40]. The precise mechanism of PRDM initiation at E6.25 remains unclear. Mesodermal factor T is required for sustaining PRDM1 expression and triggering PRDM14 expression [40]. The three transcription factors, PRDM1, PRDM14, and AP2γ, form the core regulatory network of PGC specification [43, 45]. In this network, PRDM1 and PRDM14 maintain each expression mutually [4] while also inducing the expression of AP2γ [43]. The three-core-factor network upregulates the expression of germ cell genes and pluripotency genes, inhibits mesoderm genes and represses cell proliferation [50], and initiates epigenetic modification and migration.

In the mouse embryo, the knockout of Prdm1, Prdm14, or Tfap2c impairs PGC specification [3, 4, 41, 42] and results in a reduced number of alkaline phosphatase-positive mutant mPGCs. Depression of mesodermal genes is also observed in such an embryo [5, 41, 42]. These results indicate that the three transcription factors together are essential for the repression of somatic programming. Meanwhile, none of these three factors has been demonstrated to have histone-modifying activities, and this may suggest that they probably regulate transcriptional activities by recruiting epigenetic modifiers. Overexpression of any two of the three transcription factors, or of Prdm14 alone, is sufficient to induce mPGCLCs from competent mEpiLCs [43, 45]. In particular, PRDM1 predominantly binds to promoters [43] whereas PRDM14 is enriched in distal regulatory elements [51]. AP2γ binds to targets of both PRDM1 and PRDM14 to activate or repress target genes [43, 44].

Early Studies of PGC Induction In Vitro

As aforementioned, EpiSCs derive from epiblast, which differentiates ICM after embryo implantation. ESCs have a naïve pluripotency and can develop into all lineages when injected into blastocysts, but EpiSCs exhibit a primed pluripotency and are unable to contribute to chimeras when introduced into blastocysts [52].

There have been several attempts to generate gametes or PGCs in vitro from ESCs, both in mice and humans [53]. These studies were based on the strategy of continuously isolating cells expressing germ cell markers from embryoid bodies that can differentiate spontaneously under undefined conditions.

However, this strategy was inefficient for obtaining cells containing germ cell markers (<1.0%) [53]. What's worse, these isolated cells cannot contribute to healthy offspring. Only one study reported that they obtained a live offspring from gamete-like cells derived from ESCs, but that offspring was abnormal [54].

In in vivo development, mPGCs are specified from E5.75 to E6.25 postimplantation epiblasts, and this fact hints that mEpiSCs might be germline competent. However, mEpiSCs were claimed to have a very limited capacity for PGCLC specification (<1.5%) [23]. In order to achieve robust mouse PGCLC induction, an intermediate stage is required between naïve mESCs and primed mEpiSCs. And such cells have been established in vitro [24], which will be described in detail next.

Strategy for mPGC Induction

ESCs cultured under a serum- and feeder-free condition with a combination of MARK inhibitor, GSK inhibitor, and leukemia inhibitory factor (2i + LIF) exhibit properties similar to the ICM/preimplantation (E4.5) state [55]. Saitou and colleagues derived ESCs from the E3.5 blastocysts bearing *Blimp1-mVenus* and *stella-ECFP* transgenes [56], and stimulated them with ActA and 1% KSR. The cells grew rapidly in the first 2 days. Nevertheless, they exhibited mass mortality. During 3 days of induction, cells with no BVSC expression were observed, and were called epiblast-like cells (EpiLCs).

To test the properties of EpiLCs, Saitou et al. quantified the expression of key genes in EpiLCs, EpiSCs, and E5.75 epiblast by q-PCR and found that EpiLCs showed properties that were consistent with pregastrulating epiblasts whereas EpiSCs bore distinct characteristics.

They induced the ESCs and d1/2/3 EpiLCs (~1000 cells) for 2 days under a floating condition in GMEM with 15% KSR (GK15), and with cytokines including BMP4, which fostered the induction of epiblast cells to the PGC state [38]. They found that d2 EpiLCs were highly competent to express Blimp1 in response to BMP4 and for subsequent healthy growth. Comparing d2 EpiLCs with epiblasts, d2 EpiLCs bore similar properties to the pregastrulating epiblast cells.

Then a gene expression dynamics test by q-PCR was performed on PGCLCs from d2 EpiLCs. The results indicated that they were similar to those associated with PGC specification [5, 57]. To determine the global transcription dynamics for PGCLC induction, the authors isolated total RNAs from ESCs, d1/2/3 EpiLCs, EpiSCs, E5.75 epiblasts, and BVSC-positive PGCLCs at day six of induction and stella-EGFP(+) PGCs at E9.5 [58]. Notably, the pathway of PGCLC induction from d2 EpiLCs was parallel to that of E9.5 PGC formation from E5.75 epiblasts EpiSCs, but upregulated more genes associated with the development of a variety of organs (heart, blood vessels, kidneys, muscles, and bone) than E5.75 epiblasts and d2 EpiLCs. The authors also evaluated the epigenetic profile of PGCLCs and found that the dynamics of histone modification and 5mC changes during PGCLC formation were the recapitulation of those observed during PGC formation [7].

The best way to evaluate whether PGCLCs have similar functions to PGCs in vivo would be to test their ability to contribute to spermatogenesis and normal offspring. Saitou et al. transplanted PGCLCs into the seminiferous tubules of W/Wv neonatal mice lacking endogenous germ cells [59]. Meanwhile, they fertilized the oocytes with the spermatozoa derived from PGCLCs by intracytoplasmic sperm injection (ICSI). The embryos acquired were transferred to foster mothers, and eventually produced healthy offspring with normal placentas and imprinting patterns. These male and female offspring from the PGCLCs developed normally and grew into fertile adults. PGCLCs have normal capacity of spermatogenesis and offspring generation, which demonstrates that they have comparable functions of specializing into male germ cells as PGCs.

Saitou et al. also obtained PGCLCs derived from iPSCs and induced them into germ cells successfully, followed by achieving live offspring. Notably, some of the offspring died prematurely, apparently due to tumors around the neck region. This finding demonstrates that, although iPSCs exhibit different induction properties depending on cell lines, they can nonetheless form PGCLCs with proper function.

Although some problems still exist, there is no doubt that these achievements laid the foundation for further investigations.

Induction of Mature Germ Cells and Gametes From mPGCLCs

After the induction of mPGCLCs from mESCs/miPSCs, a further scientific question of whether we can induce mature germ cells and gametes from mPGCLCs—in other words, whether we can reconstitute the entire process of spermatogenesis in vitro—becomes worth considering and exploring. Hayashi et al. [24] found that the maintenance of induced mPGCLCs could only last for <10 days because the process of mPGCLC induction from mESCs was quite fleeting. This result limits the possibility of induction of mature germ cells or gametes from mPGCLCs. As mentioned above, after inducing the PGCLCs for 6 days, Hayashi et al. [24] transplanted dissociated single cells into the seminiferous tubules of W/Wv neonatal mice [59], and obtained the results indicating the capacity to further develop into haploid spermatozoa in the testis environment of those cells [24]. Under such circumstances, we may call it a solution called "semiartificial" (in vitro at the previous stage of PGCLCs and in vivo at the stage after PGCLCs). Obviously, prolonging the proliferation of mPGCLCs/mPGCs in culture will be a pivotal issue to be resolved. In 1998, Tam and Snow [6] discovered the mechanisms that supported the proliferation of mPGCs in vivo, which stressed that the number of mPGCs would increase 100 times from E9.5 to E12.5 especially. Such specific events happening at a particular time guarantee a sufficient number of PGCs, which can help to increase the possibility of reconstitution of spermatozoa in vitro. Therefore, to study and understand more similar mechanisms may provide further help to promote the development of spermatogenesis in vitro.

Previous studies showed that mPGCLCs at day 6 only expressed little alleged "germline genes" [60], including *Ddx4*, *Dazl*, *Piwil2*, and *Mael*, which had been proved to have functions during migration of mPGCs from embryonic gonads to genital warts. These are important for not only initiation of female meiosis but also male repression of transposon. Also, Kurimoto et al. confirmed that H3K27me3 repressed these germline genes in mPSCLCs. Interestingly, researchers found that the processes of either mPGCs entering meiosis or the access of mPGCs to spermatogonial cells for differentiation can happen normally on the condition that *Dazl* is expressed normally in C57BL/6 backgrounds [61]. Gradually, through studies, scientists determined that, according to the current knowledge and understanding of the mechanisms of sex determination of germ cells [26], maintaining mPGCLC/mPGC proliferation over the long term in culture depends on whether we can find an appropriate occasion to authenticate signals. In the pathways of spermatogenesis, there are two critical markers: one is *Nanos2*, which is significant for the differentiation of prospermatogonias as a translational regulator [62], and the other is *Plzf*, which is essential for SSC maintenance as a transcription factor [63]. And for meiotic recombination, some key genes such as *Spo11* and *Sycp3* [64, 65] serve as markers.

To reconstitute the male pathway, another concern is whether we can achieve the induction from mPGCLCs to cells bearing SSC-like ability, which can only be identified by transplantation [15]. The challenge is that there is not yet one or several combined markers for defining a real and pure population of SSCs. With the technology of live image analyses, researchers found that GFRα1, which is one of the receptors for glial cell-derived neurotrophic factor (GDNF), was positive with some spermatogonia and might have functions in SSCs physiologically. Primary culture cell lines having the activity of SSCs by transplantation are regarded as germline stem cells (GSCs) [66]. Surprisingly, GSCs in culture can not only increase exponentially in connection with GNDF [67], but also can hold their activity of SSC even far more than 2 years [68]. Therefore, whether we can induce mPGCLCs into a state like GSCs may be an option to consider.

It is worth noting that we can complete and maintain a process of spermatogenesis over the long term, even up to 6 months, under the defined culture conditions of ex vivo neonatal mouse testes, which only contain spermatogonia initially. At the same time, the acquired spermatozoa can have the ability to contribute to fertile offspring [69, 70]. What's more, in the transplanted testes of neonatal W/Wv mice, cultured under the same ex vivo conditions, GSCs can undergo spermatogenesis normally [71]. Therefore, these cells can be used for reconstitution of spermatogenesis in vitro upon induction from mPGCLCs to a GSC-like state, which can provide a new view for us to make the completion of spermatogenesis from mPSCs under the condition of in vitro or ex vivo. Also, there is a need to explore whether we can let GSCs or cells of a GSC-like state enter the process of meiosis in a direct way, using key factors such as retinoic acid (RA) [72, 73].

In 2016, Zhou et al. [27] found that day 6 mPGCLCs could enter into the first prophase of meiosis during the first 6 days, through coculture with dissociated cells from testes of neonatal W/Wv mice in the presence of RA, BMP2/4/7, and ActA. When the coculture medium was supplemented with a follicle-stimulating hormone (FSH), bovine pituitary extract (BPE), and testosterone from day 6 of the coculture onward, the day 6 mPGCLC-derived cells completed meiosis over the subsequent 8 days, resulting in the generation of spermatid-like cells. Furthermore, these cells can contribute to fertile offspring upon ICSI. We may call this a solution that is "nearly completely artificial." The most valuable point is that this in vitro meiosis fully complies with the gold standards of meiosis, including erasure of imprints, synapsis, and recombination [74].

Spermatogenesis in vivo has a period of time during which male germ cells undergo epigenetic reprogramming, this is important for a proper androgenetic epigenome [12, 13]. Spermatogenesis in vitro may have skipped this period, which remains the space to be explored later. But this work of complete meiosis in vitro is still one of the epoch-making tasks, which provides significant basis for studying germ cell development and related mechanisms.

Important Genes Functioning in the Pathway

To establish an ideal culture system for gamete-like cells in vitro, several key points need to be considered: (i) how to differentiate the germ cell lineage in vivo; (ii) what the state of pluripotent stem cells is; and (iii) how to produce and validate in vitro-derived germ cells. The effective production of functional gametes in culture would not only provide a system to investigate the genetic, epigenetic, and environmental factors that shape germ cell development, but may also lead to clinical applications addressing infertility resulting from defects in gametogenesis. As described in detail below, germ cell development is highly orchestrated by a unique set of genetic and epigenetic regulations, many of which remain to be investigated [75, 76].

The germ cell lineage ensures the creation of new individuals in most multicellular organisms, perpetuating the genetic and epigenetic information across the generations. Consequently, in vitro reconstitution of germ cell development is one of the most fundamental challenges in biology. In the case of mice, the necessary conditions for successful production for functional PGCLCs from pluripotent stem cells must be considered: (i) establishment of the ground state; (ii) knowledge of early germ cell development in vivo; (iii) growth factors for inducing EpiLCs and PGCs; and (iv) sophisticated transplantation methods to validate the functionality of PGCLCs. Recent advances in understanding the PGC specification mechanisms have laid the groundwork for the induction of PGCLCs from PSCs.

Blimp1 and Prdm14 expression marks cells specified toward the germ cell lineage in the most proximal posterior epiblast cells [3, 4]. Their specification into PGCs involves three key events: repression of the somatic mesodermal program, reacquisition of pluripotent potential, and genome-wide epigenetic reprogramming [7, 57, 77]. Blimp1 is required for all three events to occur whereas Prdm14 is essential for at least the latter two [4, 78].

BMP4, secreted from the extraembryonic ectoderm (ExE), is essential for conferring germ cell fate to the most proximal posterior epiblast cells [1]. A subsequent study revealed a signaling principle for the induction of mPGCs in the epiblasts [38]. In normal development, antagonists of BMP4, such as Cerberus 1 (CER1), which is secreted from the anterior visceral endoderm (AVE), prohibit the mPGC specification Blimp1 of anterior epiblast cells. Usually, in mutants that are defective in AVE formation, such as Smad2 knockouts, Blimp1 expression becomes widespread in epiblast cells. Accordingly, ex vivo culture experiments demonstrated that, when separated from the VE, epiblast cells derived at E5.5–E6.0, but not those after E6.5, were competent for the expression of Blimp1 and, subsequently, Stella, a marker for established mPGCs [57, 79], in response to BMP4. WNT3 is secreted from the epiblast cells themselves partly in response to BMP4. It is also essential for the induction of Blimp1 and Prdm14. When cultured as floating aggregates with the presence of BMP4, stem cell factor (SCF), and other cytokines critical for the growth/survival of mPGCs, isolated E6.0 epiblast cells acquire properties resembling mPGCs with regard to gene expression and epigenetic profiles [38].

Among key regulators for pluripotency, Pou5f1 shows continuous expression whereas Sox2 and Nanog are reactivated during PGC specification. Blimp1 and Prdm14 are activated upon mPGC specification whereas Stella serves as a key marker for mPGCs and oocytes. Dazl and Ddx4 initiate their expression in mPGCs from around E10.5 and continue to be expressed subsequently. In male pathways, Nanos2 is a key determinant for differentiation into prospermatogonia [62, 80], which express genes such as Dnmt3l [81], Plzf, and Gfra1 [67]. Plzf and Gfra1 are markers for spermatogonia.

Moreover, the induction of mPGCLCs from day 2 EpiLCs involves three key events associated with mPGC specification: reacquisition of pluripotency gene expression, such as Sox2 and Nanog; transient upregulation of somatic mesodermal genes, such as T (Brachyury), Hoxa1, and Hoxb1; and downregulation of key epigenetic modifiers such as Dnmt3a, Dnmt3b, Uhrf1, and Ehmt1. Consequently, day 6 mPGCLCs exhibit transcriptome and epigenetic properties highly similar to those of E9.5 mPGCs at a migrating stage. It is important to note that mPGCLCs do not upregulate genes such as Dazl (deleted in azoospermia-like) [82] and Ddx4 [83], which are highly upregulated in mPGCs colonized in embryonic gonads [24].

By comparing the expression profiles of SGPD ESC-derived EpiLCs and day 6 PGCLC aggregates, a study [57] revealed the upregulation of transcripts of pluripotency genes, including *Oct4*, *Sox2*, and *Nanog*, in PGCLCs. Similarly, transcript levels of PGC-related genes, including *Blimp1*, *Prdm14*, *Tcfap2c*, *Nanos3*, *Stella*, *Tdrd5*, *Dnd1*, *Dnmt1*, *Ddx4*, and *Dazl*, increased whereas those of somatic cell-related genes such as *Hoxa1* and *Hoxb1* and other genes, including *Dnmt3a/3b*, *Np95*, and *c-Myc*, became downregulated. This gene expression profile resembles that of in vivo PGCs [5, 57], confirming in vitro PGCLC specification [24].

Based on the knowledge of the mechanisms for sex determination of germ cells [26], the ability to maintain long-term proliferation of mPGCLCs/mPGCs in culture should provide a key opportunity to identify signals that help to induce mPGCLCs into either spermatogenic or oogenic pathways. Crucial markers for spermatogenic pathways include Nanos2, which is a translational regulator essential for prospermatogonia differentiation [62], and Plzf, a TF critical for the maintenance of SSCs [63, 84], whereas key markers for oogenic pathways include Stra8, a critical regulator for the initiation of meiosis [85], and key genes for meiotic recombination such as Spo11 and Sycp3 [64, 65].

What we know now is that simultaneous exposure of PGCLCs to ActA, BMPs, and RA resulted in rapid silencing of Blimp1 and Stella as well as subsequent upregulation of Stra8 expression. This then resulted in initiation of meiosis and changes in gene expression similar with that in in vivo differentiating germ cells [5, 57]. Consistent with previous observations demonstrating that BMPs and ActA were required for the self-renewal [86, 87] and proliferation of neonatal germ cells [88], it is also suggested that BMPs and ActA were essential for the proliferation of meiosis-competent PGCLCs in culture whereas RA was required to induce regulatory networks, leading to meiotic entry and differentiation.

In recent years, there has been rapid progress in gametogenesis from PSCs in vitro, especially through the mouse system. Successful in vitro gametogenesis provides highly valuable information on all aspects of germ cell development that are difficult to analyze, including the mechanisms for transcriptional control, epigenetic reprogramming, meiosis, and genome stability. Understanding more mechanisms involved in the pathway can help to get more information on germ cell development and extend more technologies with various applications. Moreover, these findings could facilitate the generation of haploid human spermatids in vitro with the prospect of treating male infertility.

INDUCTION OF SPERMATOGENESIS IN HUMANS

Induction of hPGCLCs

Although the mouse has been serving as a model animal for a long time, for germ cell development, different species have their own features. Therefore, scientists began to put efforts into investigating the development of human germ cells as early as a century ago. Like the mouse, the process of human germ cell development can also be divided into three phases: PGCs specification, sex determination, and gametogenesis. Analyzing embryos is a basic method to study germ cells. Through this way, hPGCs, which are differentiated from epiblast cells and are identifiable in the hind gut at 4 weeks of gestation, were defined to undergo such a migratory pathway as from the yolk sac endoderm through the hindgut endoderm in vivo, and reside in their final destination, the male or female gonads. Then they will colonize the developing gonads by 7 weeks of gestation. In 2004, Clark discovered the spontaneous differentiation of germ cells in embryoid bodies (cell aggregations that permit random differentiation) from hESCs [89].

As a specific type of cells, hPGCs were able to be isolated and confirmed in vitro according to their migrating activity. Compared to the mouse, it is difficult to establish a certain genetic basis of PGC specification in humans, partly due to the need for E9–E16 embryos, which is not practicable. However, embryonic hPGCs at approximately week 5–10, which correspond to mouse PGCs at E10.5–E13.5, can be examined in principle. These cells retain the characteristics of PGCs while resetting the epigenome and global DNA demethylation, making it possible to do similar studies with human backgrounds.

In 1998, significant steps were achieved in producing pluripotent stem-cell lines. Human ESCs from blastocysts and human embryonic germ cells (EGCs) from primordial germ cells (PGCs) were first established [90]. These advances helped scientists to carry out further studies on germ cell development in vitro.

With the presence of feeder cells, LIF, and bFGF, cultured hPGCs can be turned into hEGCs in vitro. And their features of self-renewal and pluripotency were able to maintain and may begin to express SOX2, FGF4, TRA1-60, and TRA1-81. hEGCs show some differences from hPGCs. hEGCs and hESCs both express tumor rejection antigens TRA-1-60 and TRA-1-81, which are not seen in hPGCs. There are also markers only expressed in hESCs and hPGCs, such as FGF4 and SOX2. These differences enable us to characterize one kind of cells from another.

In 2006, another significant advance was achieved, which created iPSCs in mice [91]. And then in 2007, human iPSCs (hiPSCs) were also established by transducting several transcription factors, the same as in mice (OCT3/4, SOX2, KLF4, and MYC) or a little different from mice (OCT3/4, SOX2, NANOG, and LIN28), into human somatic cells [92].

hiPSCs were found to be similar to hESCs in many aspects, such as morphology, gene expression pattern, and epigenetic status of pluripotent cell-specific genes [92]. Furthermore, they can also differentiate into cell types of all three germ-layers in vitro and form teratoma in vivo. These results imply that it is possible to use hiPSCs to take the place of hESCs.

The development of germ cells requires two main factors: the signaling pathways and the inner environment. Therefore, adding appropriate signaling molecules would be a promising method to help mammalian epiblast cells acquire the germ-cell fate.

Studies have shown that the formation and proliferation of the PGC population in the mouse is dependent on BMP2, BMP4, and BMP8b [1]. In human cells, the addition of recombinant BMP4 is able to increase the number of human PGCs after 1 week of culture in a dose-dependent way [93]. The efficiency of EGC derivation and maintenance in culture was also enhanced by the presence of recombinant BMP4. Furthermore, the importance of BMP4 lies in its ability to promote the differentiation of hESCs/iPSCs into PGC-like cells, which will be further discussed next.

A number of studies have shown that hESCs or hiPSCs can be induced into human germ cells or germ cell-like cells through different methods, including random differentiation in an embryoid body, a two-dimensional culture, and differentiation on feeder cells prepared from embryonic testes [29, 89]. Adding BMPs, as just mentioned, can also induce such differentiation.

All these studies used the same way to identify desired cells-detecting DDX4, a gene expressed in hPGCs colonized in embryonic gonads. But validation of the DDX4 reporter was insufficient, as those cells needed were poorly characterized. Also, some studies could only achieve induction at low efficiency.

Other factors were also investigated. The DAZ family, DAZL and DDX4, were shown to be able to play a role in the development of germ cells. By overexpressing these genes, hESCs/hiPSCs could be promoted to enter meiosis and generate haploid cells within 2 weeks under differentiation culture conditions, with an efficiency of 2% [94]. Other studies also reported induction of meiosis and haploid cells through spontaneous differentiation of hiPSCs for 3 weeks, and then through further differentiation with RA for another 3 weeks, or by culturing hESCs under GSC culture conditions [95]. However, the problems of low efficiency and insufficient characterization of the induced cells still exist.

One research milestone was the demonstration of in vitro reconstitution of specification and development of a mouse germline by PSCs, published by Hayashi et al. [24]. In this study, mESCs/iPSCs, which possess ground state pluripotency, were induced into PGCLCs through an intermediate state called pregastrulation EpiLCs. The method is mainly dependent on adding signaling molecules. The PGCLCs attained were shown to have robust capacity for both spermatogenesis and oogenesis as well as for the generation of offspring. These findings provided inspiration for reconstituting human germ cell development through a similar method.

However, studies have shown that hESCs/iPSCs, different from mESCs/iPSCs, bear many characteristics similar to mouse EpiSCs. These include gene expression, epigenetic properties, and cytokine dependence, resembling postgastrulation epiblasts with limited potential for transferring into the germ-cell fate [24].

Although many obstacles exist in gametogenesis in vitro, scientists did find a feasible way to achieve human germ cell development, based on the research in mice. In 2015, Irie et al. reported that under a culture condition with inhibitors to four kinases (MAPK, GSK3, p38, and c-Jun N-terminal kinase (JNK)), called 4i for short, hESCs/hiPSCs could be induced into hPGCLCs [33]. In this study, the authors defined TNAP/NANOS3(+) cells as hPGCLCs, which showed gene expression very similar to that of gonadal hPGCs at 7 weeks of gestation. The procedure used in this study was the same as that in mPGCLC induction. It is worth noting that culture condition with 4i has been used to culture human pluripotent cells into a naive state [24]. If changed back into conventional conditions, hESCs/hiPSCs were not able to be induced into hPGCLCs.

The efficiency of induction could be up to 50%, which showed a successful aspect of this study. But one point is that, although hPGCLCs did show a similar gene expression pattern to that of hPGCs at 7 weeks of gestation, they lacked the expression of DDX4 and DAZL, meaning that they did not enter a late stage corresponding to hPGCs, and that was different from mPGCLCs.

SOX17 was found to be essential for hPGCLC induction by promoting the expression of genes for hPGC and endoderm development. SOX17 acts at the upstream of BLIMP1, which promotes the expression of genes for hPGC development but represses genes for mesoderm and endoderm development, and other somatic genes during specification of hPGCLCs.

Researchers also found that CD38 expresses in hPGCLCs, gonadal hPGCs, and TCam-2, but not in hESCs or gonadal somatic cells, suggesting that the characteristics of hPGCLCs are consistent with the embryonic hPGCs. As an established cell-surface glycoprotein on leukocytes, CD38 is defined as a prognostic marker of leukemia [33]. The results showed that CD38 was present on all the TNAP-positive embryonic hPGCs and on TCam-2 with some heterogeneity. Although CD38 is absent on hESCs, 50% of the NANOS3-mCherry-positive hPGCLCs were CD38 positive on day 4, and this proportion even increased to 70% by day 5. The NANOS3-mCherry/CD38 cells had a higher expression of NANOS3, BLIMP1, SOX17, OCT4, and NANOG whereas hESCs and embryonic carcinoma cells exhibit CD30 expression, which is also known as TNFRSF8 and SOX2. The results suggest that CD38 and CD30 may potentially be used as additional cell-surface markers to define germ cell tumors in vivo.

In addition to the above findings, the authors also made some discoveries on epigenetic changes during gametogenesis. They found downregulation of UHRF1, DNMT3A, and DNMT3B, and upregulation of TET1 and TET2 in hPGCLCs through RNA-seq. They also found a significant increase in 5-hydroxymethylacytosine (5hmC) in hPGCLCs, which is consistent with an increase in the expression of TET1, an enzyme that converts 5-methylcytosine (5mC) to 5hmC, and notably, a small decline in 5mC at the same time. Compared with the results in mice, it suggests that loss of 5mC might be coupled with the conversion of 5mC to 5hmC.

Also in 2015, Sasaki et al. demonstrated another method to induce hESCs/hiPSCs into hPGCLCs through a midterm situation of incipient mesoderm-like cells (iMeLCs) [34]. They set out to establish hiPSC lines bearing reporters that can be used to mark hPGC specification. Different from the method used by Irie et al. and Sasaki et al. chose to select BLIMP1/PRDM1 and TFAP2C/AP2g as candidates to achieve hPGC specification [41], as Blimp1 and Tfap2c encode transcription factors necessary and sufficient for mouse PGC specification, and BLIMP1 and TFAP2C have been reported to be expressed in human fetal germ cells.

The procedure used was identical to that for mPGCLC induction, under a feeder-free condition and with GK15 adding ActA and a WNT signaling agonist (CHIR99021 (CHIR)). hESCs/hiPSCs were turned into iMeLCs after 2 days, and into BTAG(+) cells under the stimulation of LIF, BMP4, SCF, and EGF after another 2 days. The BTAG(+) cells were defined as hPGCLCs. The authors showed evidence that the hPGCLCs induced here exhibited gene expression similar to early gonadal PGCs of cynomolgus monkeys (cyPGCs) and week seven hPGCs as well as the hPGCLCs induced by Sasaki et al. [33]. However, mPGCLCs showed markedly different gene expression from cyPGCs, which might be due to the species difference. Also, the heat map showed that genes upregulated in hPGCs, except for the main late PGC genes, were upregulated similarly in BTAG(+) cells and TNAP/NANOS3(+) cells, demonstrating again that these two kinds of cells were hPGCLCs. Another similarity to the hPGCLCs induced by Irie et al. was that they were negative for DDX4 and DAZL, meaning hPGCLCs here still corresponded to early stage of hPGCs.

Sasaki et al. provided evidence that hiPSCs cultured under the conditions in their paper had a property intermediate between EpiLCs and EpiSCs. And the hiPSCs/hESCs in 4i by Irie et al. did not show consistent upregulation of genes for naive pluripotency; instead, they exhibited upregulation of mesodermal markers, which suggested that the hPSCs in 4i were not in the postulated naive state but rather in a type of peri-gastrulating epiblast-like state, similar to iMeLCs. The different competence of hiPSCs/hESCs used for hPGCLC induction in these two studies might be due to the different culture conditions used.

Because hiPSCs/hESCs are similar to the situation of primed pluripotency state, which shows little, if any, competence for the germ-cell fate, it is already a great achievement to be able to robustly induce hiPSCs/hESCs into hPGCLCs. It is worth noting that some differences do exist between hiPSCs/hESCs and EpiSCs. Taking one gene, for example, hiPSCs/hESCs express PRDM14 but EpiSCs do not. This may result from the species difference, which may explain current discoveries.

To define clearly which stage and situation hPGCLCs can truly represent in vitro, more experiments need to be carried out, including epigenetic ones, and the competence of hPGCLCs for further meiosis needs to be demonstrated. In addition, investigations into other species would also be needed to acquire comparison. Because further research of spermatogenesis in mice toward functional sperm was based on the results of mPGCLCs, more studies on whether hPGCLCs are able to be induced into mature gametes need to be done.

Important Factors Involved in the Pathway

Based on further investigations, Sasaki et al. [33] discovered that hiPSCs/hESCs under culture condition of 4i expressed many genes associated with mesoderm development. There were also many gene expression profiles related to "pattern specification process," "muscle organ development," and "regulation of cell migration." At the same time, in the study by Hayashi et al. [34], genes upregulated in iMeLCs compared to their levels in hiPSCs showed a similar tendency, which was mostly related to cell migration and pattern specification process. That is why it was speculated that a peri-gastrulating, mesoderm-like state might serve as an efficient precursor for human germ cell lineage.

Although the two-step method used in the study by Hayashi et al. [34] is similar to that in mice, the number of genes up- or down-regulated between the cell state transitions during the induction process of hPGCLCs, compared with the results of gene expression in mPGCLC induction, was substantially smaller. Genes upregulated during hiPSC-to-iMeLC transition were enriched with those bearing GO terms such as "cell migration" and "pattern specification process" whereas genes downregulated were enriched with those for "chemical homeostasis" and "cell adhesion." And the genes upregulated during the iMeLC-to-d2 hPGCLC transition include potential regulators for hPGCLC specification, such as TFAP2C, PRDM1, SOX17, SOX15, KLF4, KIT, TCL1A, and DND1, many of which were related with "stem cell maintainance" and "regulation of cell migration." Meanwhile, those genes downregulated were mostly related with "pattern specification process" and "neuron development." From studies in mice, numerous genes have been identified to be specific and functional in PGCs, such as Fragilis, Stella/PGC7, Blimp1/Prdm1, and Nanos3. However, the expression and function of some of the orthologues in human PGCs remain unclear.

BLIMP1 plays an important role in the specification and maintenance of hPGCLCs. The key functions of BLIMP1 include repression of genes related to "neuron differentiation" in a dose-dependent way and to appropriate levels, no matter whether such genes are up- or down-regulated in hPGCLCs. BLIMP1 can exert differential effects on genes associated with mesoderm development, but the regulation of key pluripotency genes in hPGCLCs is independent of BLIMP1. Although BLIMP1 represses somatic cell differentiation programs in mice, it activates and stabilizes a germline transcriptional circuit in human cells.

SOX17 is another key factor that has previously been considered very important in endoderm development, both in mice and humans. It was identified as one of the most upstream TFs for hPGCLC specification. In mouse, Sox17 exhibits transient upregulation upon mPGC specification, but does not have effects in mPGCs [96]. For hPGC specification, the recruitment of SOX17 also plays an important part.

Expression of BLIMP1 is intimately associated with SOX17 during hPGCLC specification. BLIMP1 represses somatic genes, including mesendodermal genes, which may allow SOX17 to function as the regulator of hPGCLC specification. A mutation in BLIMP1 will abrogate hPGCLC specification, but the expression of SOX17 will not be completely abolished. In TNAP-positive cells, PGC-specific genes were repressed whereas some endodermal and other somatic genes were upregulated. This implies that BLIMP1 may play a role in hPGCLC specification with the existence of SOX17, although the role might be repressing those genes. In mice, BLIMP1 also represses somatic genes in PGCs [41], but it acts with PRDM14 and TFAP2Cit to make effects during PGC specification.

SOX2 is also a core TF for pluripotency in mice and humans. In mouse germ cell development, the expression of SOX2 will regain, and it is considered to be essential. However, SOX2 is downregulated, most likely by BMP4 rather than by BLIMP1 [34], and appears to be dispensable in hPGCs and upon hPGCLC specification [33, 34]. These differences can partly show mechanistic divergences between mice and humans.

DISCUSSION

It has been found that differentiation of mPGCLCs into male postmeiotic germ cells in vitro requires simultaneous exposure to the sex hormones testosterone, FSH, and BPE [27]. This reflects that spermatogenesis in vivo may also need FSH and locally produced testicular androgens, including testosterone.

Studies have shown that FSH can support the proliferation of Sertoli cells and stimulate mitotic division of spermatogonia to maintain adequate cell counts. And BPE may promote meiotic progression and spermatid differentiation, similar to some other factors, such as luteinizing hormone (LH), which was discovered to be able to stimulate the secretion of testosterone from Leydig cells and to help maintain meiotic germ cells [27]. The analysis may further improve protocols for in vitro spermatogenesis.

Although great advances have been achieved in mice, and the translation of similar technologies to humans is underway, the current situation is still not ideal. For example, mechanisms underlying human germ cell development are still not clear enough. Other obstacles include the restricted access to human embryos and the intractability of human tissues in culture. To overcome these difficulties, nonhuman primates may serve as the model to accumulate more information on germ cell development.

According to the research results attained to date, in mPGCLCs, the late PGC genes can be repressed by H3K27me3 and H3K9me2 as well as by DNA methylation [96]. These genes exhibit strong upregulation upon aggregation with embryonic gonads, particularly with female embryonic gonads [27]. In contrast, hPGCLCs are negative in expression of genes such as DDX4, DAZL, and PIWIL2. They are also in a state prior to extensive epigenetic reprogramming, which implies a distance from the late hPGC state. There are similar signaling requirements for the induction and proliferation of hPGCLCs and mPGCLCs, but downstream transcriptional programs show high differences between hPGCLCs and mPGCLCs. Also, transient activation and subsequent repression of the somatic mesodermal program are not eminent in hPGCLC specification, which is different from mPGCLC specification. Therefore, the induction of hPGCLCs to more mature germ cells is still a challenge to be conquered. To further reconstitute human germ cell development in vitro, it is important to establish a method to promote hPGCLCs into the later hPGC state as well as investigate the mechanisms of epigenetic reprogramming in the human germ cell lineage. It might also be necessary to conduct studies into different requirements and embryonic structures between mice or other mammalian models and humans. Referring to relevant periods of embryo development and PGC specification can give some help.

Because mammalian germline development is dependent on signaling molecules and the gonadal microenvironment, to mimic a suitable microenvironment for germ cell development, coculture systems or conditioned medium have often been used to differentiate mouse or human ESCs/iPSCs. As for mPGCLCs, aggregation culture with somatic cells of embryonic gonads or reconstituted testis or ovary could be a method to get further developed germ cells. Through coculturing with testis cells, mESCs/iPSCs can be induced into functional gametes. However, as with humans, it is difficult to acquire appropriate embryonic gonads or some specific somatic cells. To explore the method of induction of gonadal somatic cells using mice and appropriate primates as model organisms might be a feasible alternative, yet such studies will require careful investigation. Research of long-term culture and differentiation of hSSCs may be an option as well.

A disadvantage of such a coculturing system is that it is difficult to learn which trophic factors affect differentiation and how, as some specific factors or conditions in the cocultured cells or organs are still not clear. This has prevented us from reconstituting germ cell development totally in vitro. Another limiting factor would be the number of cells that could be obtained. For disease or infertility treatment, a large amount of germ cells is required.

In the differentiation culture systems of pluripotent stem cells, it is sometimes observed that cells may exhibit different behaviors and patterns of gene expression from that of in vivo development. These suggest the inconsistency between in vivo development and in vitro differentiation of pluripotent stem cells. It was reported that in germ cell development using hESCs/iPSCs, haploid cells can be induced just 14 days after induction by overexpression of DAZL, BOULE, and DAZ [94]. Thus, further and extensive investigation and consideration are needed before transferring the advances of in vitro development into applications.

An alternative way to obtain germ cells might be to convert other types of cells to germ cells instead of differentiating germ cells from pluripotent stem cells. Previous studies reported that a subset of cells in human fetal bone marrow express early germ cell marker [97], OCT4, FRAGILIS, STELLA, and VASA, and germ cell-specific markers, male DAZL and STRA8 [98]. Huang et al. [99] also reported that human umbilical-cord Wharton's jelly-derived mesenchymal stem cells (HUMSCs) could form "tadpole-like" cells, which expressed germ cell-specific markers OCT4, C-KIT, CD49F, STELLA, and VASA in the culture conditions with all-trans RA, testosterone, and testicular cell-conditioned medium prepared from newborn male mouse testes after being inducted with different reagents. These results suggest that methods such as transdifferentiation or other traditional methods combined with new technologies could also be a way to get germ cells.

Ethical issues are another concern in the complete reconstitution of germ cells in vitro, especially human germ cells. Administrative guidelines are necessary to instruct the application of the relevant technologies and methods.

In recent years, research of gametogenesis from PSCs in vitro has seen rapid progress both in mice and humans, which contributed a lot to the deeper understanding of the mechanisms of germ cell development and to the advancement of reproductive technologies and culture protocols. The process of spermatogenesis is complicated, but it is worth putting effort into research of this aspect. Obtaining high quality gametes is still the ultimate goal of reproductive biology in the future. Reconstituting human germ cell development in vitro would give us new perspectives of basic research on germ cells, and provide new possibilities to treat diseases arising from its anomalies. The knowledge and technologies can be used not only to help humans, but also to preserve rare animal species and maintain species diversity.

References

[1] Lawson KA, Dunn NR, Roelen BA, Zeinstra LM, Davis AM, Wright CV, et al. Bmp4 is required for the generation of primordial germ cells in the mouse embryo. Genes Dev 1999;13(4):424–36.

[2] Saitou M, Payer B, O'Carroll D, Ohinata Y, Surani MA. Blimp1 and the emergence of the germ line during development in the mouse. Cell Cycle 2005;4(12):1736–40.

[3] Vincent SD, Dunn NR, Sciammas R, Shapiro-Shalef M, Davis MM, Calame K, et al. The zinc finger transcriptional repressor Blimp1/Prdm1 is dispensable for early axis formation but is required for specification of primordial germ cells in the mouse. Development 2005;132(6):1315–25.

[4] Yamaji M, Seki Y, Kurimoto K, Yabuta Y, Yuasa M, Shigeta M, et al. Critical function of Prdm14 for the establishment of the germ cell lineage in mice. Nat Genet 2008;40(8):1016–22.

[5] Kurimoto K, Yabuta Y, Ohinata Y, Shigeta M, Yamanaka K, Saitou M. Complex genome-wide transcription dynamics orchestrated by Blimp1 for the specification of the germ cell lineage in mice. Genes Dev 2008; 22(12):1617–35.

[6] Tam PP, Snow MH. Proliferation and migration of primordial germ cells during compensatory growth in mouse embryos. J Embryol Exp Morpholog 1981;64:133–47.

[7] Seki Y, Hayashi K, Itoh K, Mizugaki M, Saitou M, Matsui Y. Extensive and orderly reprogramming of genome-wide chromatin modifications associated with specification and early development of germ cells in mice. Dev Biol 2005;278(2):440–58.

[8] Guibert S, Forne T, Weber M. Global profiling of DNA methylation erasure in mouse primordial germ cells. Genome Res 2012;22(4):633–41.

[9] McGee EA, Hsueh AJ. Initial and cyclic recruitment of ovarian follicles. Endocr Rev 2000;21(2):200–14.

[10] Kobayashi H, Sakurai T, Imai M, Takahashi N, Fukuda A, Yayoi O, et al. Contribution of intragenic DNA methylation in mouse gametic DNA methylomes to establish oocyte-specific heritable marks. PLoS Genet 2012;8(1): e1002440.

[11] Lucifero D, Mertineit C, Clarke HJ, Bestor TH, Trasler JM. Methylation dynamics of imprinted genes in mouse germ cells. Genomics 2002;79(4):530–8.

[12] Kato Y, Kaneda M, Hata K, Kumaki K, Hisano M, Kohara Y, et al. Role of the Dnmt3 family in de novo methylation of imprinted and repetitive sequences during male germ cell development in the mouse. Hum Mol Genet 2007;16(19):2272–80.

[13] Kubo N, Toh H, Shirane K, Shirakawa T, Kobayashi H, Sato T, et al. DNA methylation and gene expression dynamics during spermatogonial stem cell differentiation in the early postnatal mouse testis. BMC Genomics 2015;16:624.

[14] Bellve AR, Cavicchia JC, Millette CF, O'Brien DA, Bhatnagar YM, Dym M. Spermatogenic cells of the prepuberal mouse. Isolation and morphological characterization. J Cell Biol 1977;74(1):68–85.

[15] Kanatsu-Shinohara M, Shinohara T. Spermatogonial stem cell self-renewal and development. Annu Rev Cell Dev Biol 2013;29:163–87.

[16] Hanley NA, Hagan DM, Clement-Jones M, Ball SG, Strachan T, Salas-Cortes L, et al. SRY, SOX9, and DAX1 expression patterns during human sex determination and gonadal development. Mech Dev 2000;91(1–2):403–7.

[17] Culty M. Gonocytes, the forgotten cells of the germ cell lineage. Birth Defects Res C Embryo Today 2009;87 (1):1–26.

[18] Kurilo LF. Oogenesis in antenatal development in man. Hum Genet 1981;57(1):86–92.

[19] Mamsen LS, Lutterodt MC, Andersen EW, Byskov AG, Andersen CY. Germ cell numbers in human embryonic and fetal gonads during the first two trimesters of pregnancy: analysis of six published studies. Hum Reprod 2011;26(8):2140–5.

[20] Evans MJ, Kaufman MH. Establishment in culture of pluripotential cells from mouse embryos. Nature 1981;292 (5819):154–6.

[21] Nichols J, Smith A. Naive and primed pluripotent states. Cell Stem Cell 2009;4(6):487–92.

[22] Brons IG, Smithers LE, Trotter MW, Rugg-Gunn P, Sun B, Chuva de Sousa Lopes SM, et al. Derivation of pluripotent epiblast stem cells from mammalian embryos. Nature 2007;448(7150):191–5.

[23] Hayashi K, Surani MA. Self-renewing epiblast stem cells exhibit continual delineation of germ cells with epigenetic reprogramming in vitro. Development 2009;136(21):3549–56.

[24] Hayashi K, Ohta H, Kurimoto K, Aramaki S, Saitou M. Reconstitution of the mouse germ cell specification pathway in culture by pluripotent stem cells. Cell 2011;146(4):519–32.

[25] Hayashi K, Ogushi S, Kurimoto K, Shimamoto S, Ohta H, Saitou M. Offspring from oocytes derived from in vitro primordial germ cell-like cells in mice. Science 2012;338(6109):971–5.

[26] Lin YT, Capel B. Cell fate commitment during mammalian sex determination. Curr Opin Genet Dev 2015;32:144–52.

[27] Zhou Q, Wang M, Yuan Y, Wang X, Fu R, Wan H, et al. Complete meiosis from embryonic stem cell-derived germ cells in vitro. Cell Stem Cell 2016;18(3):330–40.

[28] Hikabe O, Hamazaki N, Nagamatsu G, Obata Y, Hirao Y, Hamada N, et al. Reconstitution in vitro of the entire cycle of the mouse female germ line. Nature 2016;539(7628):299–303.

[29] Park TS, Galic Z, Conway AE, Lindgren A, van Handel BJ, Magnusson M, et al. Derivation of primordial germ cells from human embryonic and induced pluripotent stem cells is significantly improved by coculture with human fetal gonadal cells. Stem Cells 2009;27(4):783–95.

[30] West FD, Roche-Rios MI, Abraham S, Rao RR, Natrajan MS, Bacanamwo M, et al. KIT ligand and bone morphogenetic protein signaling enhances human embryonic stem cell to germ-like cell differentiation. Hum Reprod 2010;25(1):168–78.

[31] Irie N, Weinberger L, Tang WWC, Kobayashi T, Viukov S, Manor YS, et al. SOX17 is a critical specifier of human primordial germ cell fate. Cell 2015;160(1–2):253–68.

[32] Monk M, McLaren A. X-chromosome activity in foetal germ cells of the mouse. J Embryol Exp Morpholog 1981;63:75–84.

[33] Sasaki K, Yokobayashi S, Nakamura T, Okamoto I, Yabuta Y, Kurimoto K, et al. Robust in vitro induction of human germ cell fate from pluripotent stem cells. Cell Stem Cell 2015;17(2):178–94.

[34] Hayashi Y, Saitou M, Yamanaka S. Germline development from human pluripotent stem cells toward disease modeling of infertility. Fertil Steril 2012;97(6):1250–9.

[35] Murphey P, McLean DJ, McMahan CA, Walter CA, McCarrey JR. Enhanced genetic integrity in mouse germ cells. Biol Reprod 2013;88(1).

[36] Rahbari R, Wuster A, Lindsay SJ, Hardwick RJ, Alexandrov LB, Al Turki S, et al. Timing, rates and spectra of human germline mutation. Nat Genet 2016;48(2):126–33.

[37] Pearson H, Solter D, Trounson A, Baruch S, Sutcliffe A, Gelfand S, et al. Special report: making babies: the next 30 years. Nature 2008;454(7202):260–2.

[38] Ohinata Y, Ohta H, Shigeta M, Yamanaka K, Wakayama T, Saitou M. A signaling principle for the specification of the germ cell lineage in mice. Cell 2009;137(3):571–84.

[39] Tam PP, Loebel DA. Gene function in mouse embryogenesis: get set for gastrulation. Nat Rev Genet 2007;8 (5):368–81.

[40] Aramaki S, Hayashi K, Kurimoto K, Ohta H, Yabuta Y, Iwanari H, et al. A mesodermal factor, T, specifies mouse germ cell fate by directly activating germline determinants. Dev Cell 2013;27(5):516–29.

[41] Ohinata Y, Payer B, O'Carroll D, Ancelin K, Ono Y, Sano M, et al. Blimp1 is a critical determinant of the germ cell lineage in mice. Nature 2005;436(7048):207–13.

[42] Weber S, Eckert D, Nettersheim D, Gillis AJM, Schafer S, Kuckenberg P, et al. Critical function of AP-2gamma/TCFAP2C in mouse embryonic germ cell maintenance. Biol Reprod 2010;82(1):214–23.

[43] Magnusdottir E, Dietmann S, Murakami K, Gunesdogan U, Tang FC, Bao SQ, et al. A tripartite transcription factor network regulates primordial germ cell specification in mice. Nat Cell Biol 2013;15(8):905–15.

[44] Magnusdottir E, Surani MA. How to make a primordial germ cell. Development 2014;141(2):245–52.

I. REPRODUCTIVE TRACT DEVELOPMENT AND GAMETOGENESIS

[45] Nakaki F, Hayashi K, Ohta H, Kurimoto K, Yabuta Y, Saitou M. Induction of mouse germ-cell fate by transcription factors in vitro. Nature 2013;501(7466):222–6.

[46] Lawson KA, Hage WJ. Clonal analysis of the origin of primordial germ-cells in the mouse. Germline Dev 1994;182:68–84.

[47] Hajkova P, Erhardt S, Lane N, Haaf T, El-Maarri O, Reik W, et al. Epigenetic reprogramming in mouse primordial germ cells. Mech Dev 2002;117(1–2):15–23.

[48] Lee J, Inoue K, Ono R, Ogonuki N, Kohda T, Kaneko-Ishino T, et al. Erasing genomic imprinting memory in mouse clone embryos produced from day 11.5 primordial germ cells. Development 2002;129(8):1807–17.

[49] McLaren A. Primordial germ cells in the mouse. Dev Biol 2003;262(1):1–15.

[50] Seki Y, Yamaji M, Yabuta Y, Sano M, Shigeta M, Matsui Y, et al. Cellular dynamics associated with the genome-wide epigenetic reprogramming in migrating primordial germ cells in mice. Development 2007;134(14):2627–38.

[51] Ma Z, Swigut T, Valouev A, Rada-Iglesias A, Wysocka J. Sequence-specific regulator Prdm14 safeguards mouse ESCs from entering extraembryonic endoderm fates. Nat Struct Mol Biol 2011;18(2):120–7.

[52] Nichols J, Silva J, Roode M, Smith A. Suppression of Erk signalling promotes ground state pluripotency in the mouse embryo. Development 2009;136(19):3215–22.

[53] Daley GQ. Gametes from embryonic stem cells: a cup half empty or half full? Science 2007;316(5823):409–10.

[54] Nayernia K, Nolte J, Michelmann HW, Lee JH, Rathsack K, Drusenheimer N, et al. In vitro-differentiated embryonic stem cells give rise to male gametes that can generate offspring mice. Dev Cell 2006;11(1):125–32.

[55] Ying QL, Wray J, Nichols J, Batlle-Morera L, Doble B, Woodgett J, et al. The ground state of embryonic stem cell self-renewal. Nature 2008;453(7194):519–23.

[56] Ohinata Y, Sano M, Shigeta M, Yamanaka K, Saitou M. A comprehensive, non-invasive visualization of primordial germ cell development in mice by the Prdm1-mVenus and Dppa3-ECFP double transgenic reporter. Reproduction 2008;136(4):503–14.

[57] Saitou M, Barton SC, Surani MA. A molecular programme for the specification of germ cell fate in mice. Nature 2002;418(6895):293–300.

[58] Payer B, Chuva de Sousa Lopes SM, Barton SC, Lee C, Saitou M, Surani MA. Generation of stella-GFP transgenic mice: a novel tool to study germ cell development. Genesis 2006;44(2):75–83.

[59] Chuma S, Kanatsu-Shinohara M, Inoue K, Ogonuki N, Miki H, Toyokuni S, et al. Spermatogenesis from epiblast and primordial germ cells following transplantation into postnatal mouse testis. Development 2005; 132(1):117–22.

[60] Borgel J, Guibert S, Li YF, Chiba H, Schubeler D, Sasaki H, et al. Targets and dynamics of promoter DNA methylation during early mouse development. Nat Genet 2010;42(12):1093–100.

[61] Gill ME, Hu YC, Lin YF, Page DC. Licensing of gametogenesis, dependent on RNA binding protein DAZL, as a gateway to sexual differentiation of fetal germ cells. Proc Natl Acad Sci U S A 2011;108(18):7443–8.

[62] Suzuki A, Saga Y. Nanos2 suppresses meiosis and promotes male germ cell differentiation. Genes Dev 2008;22 (4):430–5.

[63] Buaas FW, Kirsh AL, Sharma M, McLean DJ, Morris JL, Griswold MD, et al. Plzf is required in adult male germ cells for stem cell self-renewal. Nat Genet 2004;36(6):647–52.

[64] Baudat F, Manova K, Yuen JP, Jasin M, Keeney S. Chromosome synapsis defects and sexually dimorphic meiotic progression in mice lacking Spo11. Mol Cell 2000;6(5):989–98.

[65] Yuan L, Liu JG, Zhao J, Brundell E, Daneholt B, Hoog C. The murine SCP3 gene is required for synaptonemal complex assembly, chromosome synapsis, and male fertility. Mol Cell 2000;5(1):73–83.

[66] Kanatsu-Shinohara M, Ogonuki N, Inoue K, Miki H, Ogura A, Toyokuni S, et al. Long-term proliferation in culture and germline transmission of mouse male germline stem cells. Biol Reprod 2003;69(2):612–6.

[67] Meng X, Lindahl M, Hyvonen ME, Parvinen M, de Rooij DG, Hess MW, et al. Regulation of cell fate decision of undifferentiated spermatogonia by GDNF. Science 2000;287(5457):1489–93.

[68] Kanatsu-Shinohara M, Ogonuki N, Iwano T, Lee J, Kazuki Y, Inoue K, et al. Genetic and epigenetic properties of mouse male germline stem cells during long-term culture. Development 2005;132(18):4155–63.

[69] Komeya M, Kimura H, Nakamura H, Yokonishi T, Sato T, Kojima K, et al. Long-term ex vivo maintenance of testis tissues producing fertile sperm in a microfluidic device. Sci Rep 2016;6.

[70] Sato T, Katagiri K, Inoue K, Ogonuki N, Ogura A, Kubota Y, et al. In vitro production of functional sperm in cultured neonatal mouse testes. Biol Reprod 2011;85.

[71] Sato T, Katagiri K, Yokonishi T, Kubota Y, Inoue K, Ogonuki N, et al. In vitro production of fertile sperm from murine spermatogonial stem cell lines. Nat Commun 2011;2.

[72] Anderson EL, Baltus AE, Roepers-Gajadien HL, Hassold TJ, de Rooij DG, van Pelt AMM, et al. Stra8 and its inducer, retinoic acid, regulate meiotic initiation in both spermatogenesis and oogenesis in mice. Proc Natl Acad Sci U S A 2008;105(39):14976–80.

[73] Koubova J, Menke DB, Zhou Q, Capel B, Griswold MD, Page DC. Retinoic acid regulates sex-specific timing of meiotic initiation in mice. Proc Natl Acad Sci U S A 2006;103(8):2474–9.

[74] Handel MA, Eppig JJ, Schimenti JC. Applying "gold standards" to in-vitro-derived germ cells. Cell 2014; 157(6):1257–61.

[75] McLaren A, Lawson KA. How is the mouse germ-cell lineage established? Differentiation 2005;73(9–10):435–7.

[76] Sasaki H, Matsui Y. Epigenetic events in mammalian germ-cell development: reprogramming and beyond. Nat Rev Genet 2008;9(2):129–40.

[77] Yabuta Y, Kurimoto K, Ohinata Y, Seki Y, Saitou M. Gene expression dynamics during germline specification in mice identified by quantitative single-cell gene expression profiling. Biol Reprod 2006;75(5):705–16.

[78] Kurimoto K, Yamaji M, Seki Y, Saitou M. Specification of the germ cell lineage in mice: a process orchestrated by the PR-domain proteins, Blimp1 and Prdm14. Cell Cycle 2008;7(22):3514–8.

[79] Sato M, Kimura T, Kurokawa K, Fujita Y, Abe K, Masuhara M, et al. Identification of PGC7, a new gene expressed specifically in preimplantation embryos and germ cells. Mech Dev 2002;113(1):91–4.

[80] Tsuda M, Sasaoka Y, Kiso M, Abe K, Haraguchi S, Kobayashi S, et al. Conserved role of nanos proteins in germ cell development. Science 2003;301(5637):1239–41.

[81] Bourc'his D, Xu GL, Lin CS, Bollman B, Bestor TH. Dnmt3L and the establishment of maternal genomic imprints. Science 2001;294(5551):2536–9.

[82] Cooke HJ, Lee M, Kerr S, Ruggiu M. A murine homologue of the human DAZ gene is autosomal and expressed only in male and female gonads. Hum Mol Genet 1996;5(4):513–6.

[83] Fujiwara Y, Komiya T, Kawabata H, Sato M, Fujimoto H, Furusawa M, et al. Isolation of a DEAD-family protein gene that encodes a murine homolog of Drosophila vasa and its specific expression in germ cell lineage. Proc Natl Acad Sci U S A 1994;91(25):12258–62.

[84] Costoya JA, Hobbs RM, Barna M, Cattoretti G, Manova K, Sukhwani M, et al. Essential role of Plzf in maintenance of spermatogonial stem cells. Nat Genet 2004;36(6):653–9.

[85] Baltus AE, Menke DB, Hu YC, Goodheart ML, Carpenter AE, de Rooij DG, et al. In germ cells of mouse embryonic ovaries, the decision to enter meiosis precedes premeiotic DNA replication. Nat Genet 2006; 38(12):1430–4.

[86] Hu J, Chen YX, Wang D, Qi X, Li TG, Hao J, et al. Developmental expression and function of Bmp4 in spermatogenesis and in maintaining epididymal integrity. Dev Biol 2004;276(1):158–71.

[87] Puglisi R, Montanari M, Chiarella P, Stefanini M, Boitani C. Regulatory role of BMP2 and BMP7 in spermatogonia and Sertoli cell proliferation in the immature mouse. Eur J Endocrinol 2004;151(4):511–20.

[88] Mithraprabhu S, Mendis S, Meachem SJ, Tubino L, Matzuk MM, Brown CW, et al. Activin bioactivity affects germ cell differentiation in the postnatal mouse testis in vivo. Biol Reprod 2010;82(5):980–90.

[89] Clark AT, Bodnar MS, Fox M, Rodriquez RT, Abeyta MJ, Firpo MT, et al. Spontaneous differentiation of germ cells from human embryonic stem cells in vitro. Hum Mol Genet 2004;13(7):727–39.

[90] Thomson JA, Itskovitz-Eldor J, Shapiro SS, Waknitz MA, Swiergiel JJ, Marshall VS, et al. Embryonic stem cell lines derived from human blastocysts. Science 1998;282(5391):1145–7.

[91] Takahashi K, Yamanaka S. Induction of pluripotent stem cells from mouse embryonic and adult fibroblast cultures by defined factors. Cell 2006;126(4):663–76.

[92] Takahashi K, Tanabe K, Ohnuki M, Narita M, Ichisaka T, Tomoda K, et al. Induction of pluripotent stem cells from adult human fibroblasts by defined factors. Cell 2007;131(5):861–72.

[93] Hiller M, Liu C, Blumenthal PD, Gearhart JD, Kerr CL. Bone morphogenetic protein 4 mediates human embryonic germ cell derivation. Stem Cells Dev 2011;20(2):351–61.

[94] Kee K, Angeles VT, Flores M, Nguyen HN, Reijo Pera RA. Human DAZL, DAZ and BOULE genes modulate primordial germ-cell and haploid gamete formation. Nature 2009;462(7270):222–5.

[95] Eguizabal C, Montserrat N, Vassena R, Barragan M, Garreta E, Garcia-Quevedo L, et al. Complete meiosis from human induced pluripotent stem cells. Stem Cells 2011;29(8):1186–95.

[96] Kurimoto K, Yabuta Y, Hayashi K, Ohta H, Kiyonari H, Mitani T, et al. Quantitative dynamics of chromatin remodeling during germ cell specification from mouse embryonic stem cells. Cell Stem Cell 2015;16(5):517–32.

[97] Hua J, Pan S, Yang C, Dong W, Dou Z, Sidhu KS. Derivation of male germ cell-like lineage from human fetal bone marrow stem cells. Reprod BioMed Online 2009;19(1):99–105.

[98] Drusenheimer N, Wulf G, Nolte J, Lee JH, Dev A, Dressel R, et al. Putative human male germ cells from bone marrow stem cells. Soc Reprod Fertil Suppl 2007;63:69–76.

[99] Huang P, Lin LM, Wu XY, Tang QL, Feng XY, Lin GY, et al. Differentiation of human umbilical cord Wharton's jelly-derived mesenchymal stem cells into germ-like cells in vitro. J Cell Biochem 2010;109(4):747–54.

Developmental Genetics of the Female Reproductive Tract

Kavita Narang*,†, Zebulun S. Cope*,†, Jose M. Teixeira*

*Michigan State University, College of Human Medicine, Department of Obstetrics, Gynecology and Reproductive Biology, Grand Rapids, MI, United States †Spectrum Health, Butterworth Hospital, Women's Center, Department of Obstetrics and Gynecology, Grand Rapids, MI, United States

OUTLINE

INTRODUCTION

The development of the female reproductive tract during embryogenesis is a complex and dynamic process that involves a series of events; all of which are critical for proper tissue and

Human Reproductive and Prenatal Genetics
https://doi.org/10.1016/B978-0-12-813570-9.00006-1

organ differentiation, sex determination, and ultimately continuation of the species. Any deviation from this complex orchestration of signaling pathways invites risk for errors and abnormalities. In mammals, the bipotential urogenital ridges develop from swellings of the intermediate mesoderm during the embryonic to fetal transition by still unknown mechanisms. They also contain the primitive gonads and reproductive ducts, which at this stage are identical in the genotypic male and female. In the absence of a Y chromosome, as is the case in females, the Müllerian (paramesonephric) ducts differentiate to form the uterus, fallopian tubes, cervix, and interior portion of the vagina. In the presence of a Y chromosome, as in males, the sex-determining region of the Y chromosome (*SRY*) gene will direct the bipotential gonads to differentiate into testes, which produce testosterone and induce the Wolffian (mesonephric) ducts to differentiate into the rete testes, efferent ducts, epididymides, ductus deferens, and seminal vesicles. Anti-Mullerian Hormone (AMH) production during early testes differentiation will induce apoptosis in the Müllerian ducts, which would otherwise lead to males with Persistent Müllerian Duct Syndrome, a rare form of pseudohermaphroditism. Molecular genetics and experimental embryology in animal models have made it possible to understand some of the intricate pathways behind this fascinating stage of our development. This chapter will first elaborate some of the details of these pathways and then describe how errors in these pathways contribute to congenital anomalies of the female reproductive tract.

EMBRYOGENESIS OF THE FEMALE REPRODUCTIVE TRACT

In the embryo, the primordial urogenital ridges, which consist of coelomic epithelium and underlying mesenchymal cells that will differentiate into the urinary and genital systems, begin their development from the intermediate mesoderm along the posterior wall of the abdominal cavity. The urinary system includes the kidneys, ureters, urinary bladder, and urethra and will not be discussed further within this chapter. The genital system forms the gonads, the ovaries or testes, and the Müllerian duct or Wolffian duct. This chapter will focus on embryogenesis of the genital system, specifically the female reproductive tract.

The study of animal models has led to a greater understanding of the signaling pathways and molecular mechanisms involved in Müllerian duct and Wolffian duct formation and differentiation. Most studies use mouse models that are readily amenable to genetic analysis or chicken embryos that allow experimental manipulation of the developing ducts [1]. In humans, the initial stages of gonadal development begin during the fifth week of gestation. The gonads are indifferent or bipotential at this stage and are being populated by primordial germ cells migrating from the area where the allantois and the yolk sac meet. The Müllerian ducts and Wolffian ducts are lateral to the bipotential gonads and will differentiate into female and male reproductive tracts, respectively, depending on the chromosomal sex of the gonads. If the embryo is XX, the gonads will differentiate into ovaries and the Müllerian ducts will persist and differentiate into the uterus, fallopian tubes, cervix, and vagina. Alternatively, if the embryo is XY, the gonads differentiate into the testes, the Müllerian ducts will regress, and the Wolffian ducts will develop into the male reproductive tract (Fig. 6.1).

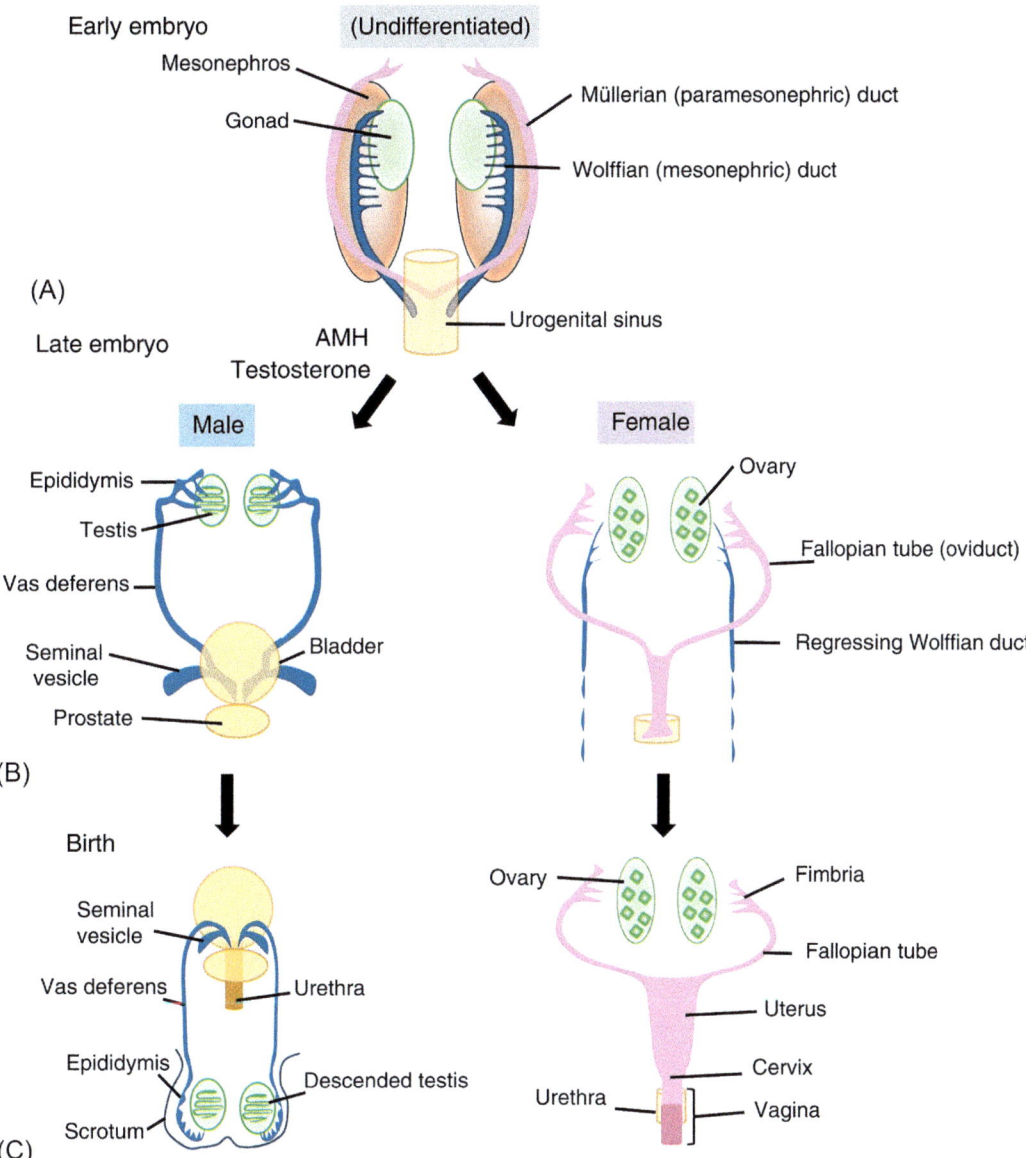

FIG. 6.1 Schematic illustration of mammalian reproductive tract and gonadal differentiation. (A) The indifferent gonads with embryonic Müllerian and Wolffian ducts at 5 weeks gestation in humans. (B) Gonadal differentiation begins and the production of AMH and testosterone in the testes direct Müllerian duct regression and Wolffian duct differentiation, respectively, which is completed around 10 weeks. (C) At birth, gonadal and ductal differentiation is complete [1].

Development of Ovaries

Gonads develop lateral to the mesonephros, which is the primordial structure for the urinary system. The absence of a Y chromosome is responsible for differentiation of gonads into ovaries, historically referred to as the "default gonadal differentiation pathway" but now well known to require activation of a variety of transcriptional programs [2–4,4a]. This occurs at around 5–7 weeks gestation in humans and requires expression of autosomal genes described in more detail below for normal ovarian differentiation (Fig. 6.2). The ovary contains an outer cortex and inner medulla. Cortical cords develop from the outer cortex and run medially into the ovary. They host several million proliferating primordial germ cells, which are also known as oogonia. At birth, two million remaining oogonia will have begun meiosis as primary oocytes and become arrested in meiotic prophase. Once ovarian organogenesis is complete, it separates from the regressing mesonephros and remains suspended by the mesovarium.

To facilitate this process, three important genes, *Wnt4*, *Fgf9*, and *Sox9*, are involved. *Wnt4* belongs to the Wingless-Type MMTV Integration Site Family of genes and has been extensively observed in the development of the female reproductive system. *Fgf9* is expressed in Sertoli cells, the male germ cells, and plays a role in mesonephric cell migration and testes development. *SOX9*, an autosomal member of the HMG-box protein superfamily, is the master regulator of Sertoli cell differentiation and ultimately male development [5]. *Wnt4* is expressed at similar levels in the bipotential gonad. As seen in Fig. 6.2, the absence of the *SRY* gene in XX gonads upregulates *Wnt4* expression. This subsequently stabilizes beta-

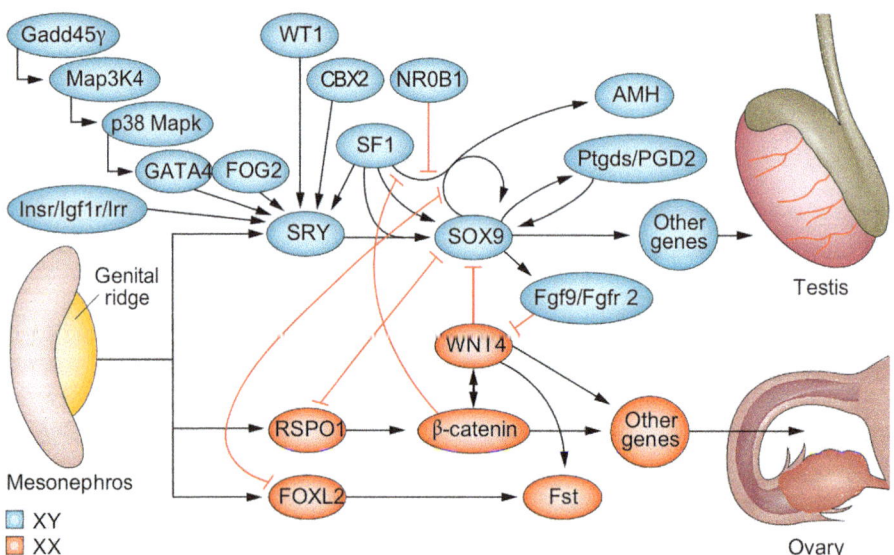

FIG. 6.2 Diagram depicting genes involved in the pathway for gonadal differentiation. In the presence of a Y chromosome, the SRY and SOX9 genes are expressed and trigger production of anti-Mullerian hormone and testosterone from Sertoli and Leydig cells, respectively. This results in the formation of the male genital tract. In the absence of the Y chromosome in females, the *WNT4* gene is expressed and causes oogenesis [4a].

catenin and silences *FGF9* and *SOX9*, resulting in the formation of the ovary [5]. Please refer to Chapter 2 for an in-depth description of gonadogenesis.

Development of Müllerian Ducts

The progression from primitive Müllerian ducts within urogenital ridges to mature and functional reproductive tract organs in females is a dynamic process that can be categorized into three main stages: Müllerian duct formation, Wolffian duct regression, and Müllerian duct differentiation [1].

Formation

Formation is the first stage of this developmental cascade and can be described in three important phases: initiation, invagination, and elongation [6] (Fig. 6.3).

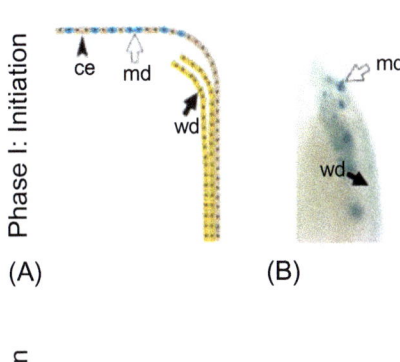

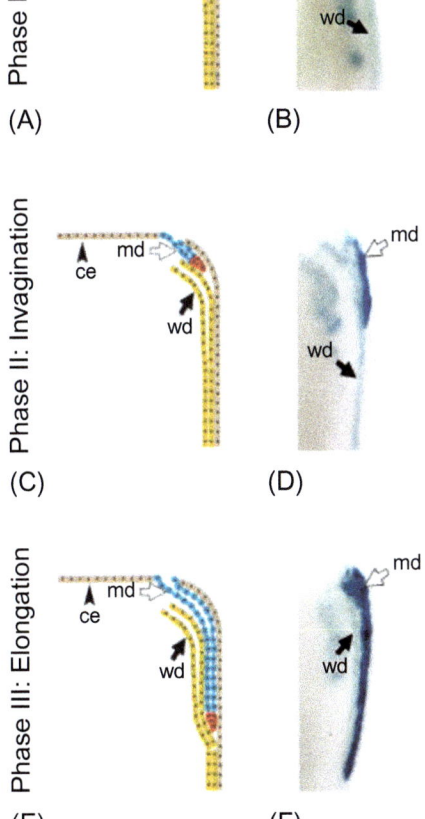

FIG. 6.3 A three-phase model for Müllerian duct development. In the first phase, cells of the coelomic epithelium are specified to become Müllerian duct cells (A, B). After specification, the second phase begins and these cells invaginate caudally toward the Wolffian duct (C, D). Once the Müllerian duct comes into contact with the Wolffian duct, the third phase begins (E, F) and the Müllerian duct elongates caudally, following the path of the Wolffian duct toward the urogenital sinus. *Blue cells*, mesoepithelial Müllerian duct cells; *red cells*, proliferating Müllerian duct precursor cells; *brown cells*, coelomic epithelial cells; *yellow cells*, Wolffian epithelial cells; *ce*, coelomic epithelium; *MD*, Müllerian duct; *WD*, Wolffian duct [6].

INITIATION

This phase can be described as the formation of primitive, stoma-like placode structures of thickened coelomic epithelial cells at the anterior end of the developing urogenital ridge near the Wolffian duct [7, 8]. These meso-epithelial Müllerian duct precursor cells express *Lhx1*, which is necessary for Müllerian duct invagination and is not dependent on whether the Wolffian duct is present [9,10]. *Pax2* appears to be expressed in the Müllerian duct precursor cells earlier than *Lhx1*, and reduction of *Pax2* expression also lowered *Lhx1* expression, which suggests that PAX2 regulates the expression of *Lhx1* [10]. The same group showed that expression of bone morphogenetic proteins 2, 4, and 7 (BMP2, 4, 7) was observed in the Müllerian duct precursor cells. Inhibition of their signaling potential by overexpression of noggin, an extracellular inhibitor of BMP receptor binding, inhibited placode thickening and *Pax2/Lhx1* expression, suggesting that BMP signaling is also important for this early stage of Müllerian duct formation [10]. They also showed that inhibition of fibroblast growth factor receptor (FGFR) signaling was critical for *Lhx1* expression and Müllerian duct formation [10]. These studies have demonstrated that sequential actions of BMP/PAX2 and FGF/LHX1 signaling are critical for the specification of these cells to become Müllerian duct precursors. However, earlier studies of the *Pax2* and *Wnt4* knockout mice suggested that the Müllerian duct precursor cells could still be detected and thus that the absence of either gene does not affect this early developmental stage of Müllerian duct formation [11, 12].

INVAGINATION

Once specified, the Müllerian duct precursor cells undergo significant changes in their morphology and begin to proliferate and invaginate caudally. The Müllerian duct-specified cells occupy the space between the Wolffian duct and the coelomic epithelium. Studies in chick embryos showed that this process is triggered by apical constriction of the duct, represented by elevated N-Cadherin (a cell-cell adhesion protein) prior to invagination [10]. This process is controlled by FGF, a fibroblast growth factor important for tissue repair, signaling through the Ras/ERK pathway [1, 8]. Inhibition of the expression of Rac1, a Rho family of GTPase important for many cytoskeletal functions, appears to be required for invagination as well [10] and is likely to be important for maintaining the meso-epithelial character of the Müllerian duct cells. In chick studies, inhibition of *Pax2* expression completely blocked invagination [10]. Conversely, in the mouse, Pax2-null embryos that survived long enough for reproductive tract analysis showed that the anterior portion of the Müllerian duct was still present [11]. Mice lacking *Wnt4* also form this rudimentary Müllerian duct structure [12]. Once in contact with or close proximity to the Wolffian duct cells, the Müllerian duct precursor cells begin the next phase of development, which is elongation.

ELONGATION

After invagination is complete, the nascent Müllerian duct epithelial cells proliferate in a cranio-caudal fashion to form a tube that runs alongside the Wolffian duct and ultimately fuses with the urogenital sinus, where proliferation probably ends [8]. In the absence of a Wolffian duct, as is the case for mice deleted for *Pax2* [11] or *Emx2* [13] and possibly *Lhx1* [9], elongation does not occur and the Müllerian duct does not form. This is in contrast to initiation and invagination that proceed independently of the Wolffian duct [9, 10]. Some of the *Wnt* genes are also crucial for both elongation and subsequent differentiation into the

Müllerian duct-derived tissues. Expression of *Wnt4* is essential to the elongation process [12] because it appears to direct cellular migration and proliferation of the Müllerian duct at this time [14]. During elongation, the highly proliferative cells located at the caudal region of the Müllerian duct start to extend pseudopodia and transition from epithelium to mesenchymal cells by a process controlled by DMTR1 [15]. It has also been postulated that the Wolffian duct secretes chemoattractants or morphogens to guide cell proliferation in a caudal direction. One of the proteins identified in this process is WNT9B, as demonstrated by the failure of duct elongation in *Wnt9b* null mouse embryos [16]. Epithelium to mesenchyme transition begins with basement membrane breakdown [17] followed by continuous proliferation to produce the Müllerian duct-derived mesenchyme that later differentiates into the myometrium and endometrial stroma by a process that appears to also be controlled by DMTR1 [18]. Mouse embryos homozygously deleted for the *Hoxa13* gene, which is described in more detail below, lack the caudal end of the Müllerian duct [19], suggesting that expression of this homeobox gene is needed for complete elongation. Once elongation is complete, the mesoepithelial cells will differentiate into the Müllerian duct mesenchyme and subsequently the different stroma cell types of the female reproductive tract.

Sex Determination and Regression of the Wolffian Duct

At about 8 weeks gestation, the process of Müllerian duct formation is complete, and the fetus will have bilateral Müllerian ducts, Wolffian ducts, and differentiating gonads. At this stage, the urogenital ridges of both sexes are no longer identical because the gonads in males will have begun to form visible testicular cords, spermatogenic veins, and hormone-secreting Sertoli and Leydig cells. The gonads in females form ovaries containing oocytes.

Between the eighth and ninth weeks of gestation, the sex-determining region on the Y chromosome, *SRY*, gene which is located on the short arm of the Y chromosome, triggers the synthesis and secretion of AMH (aka Müllerian inhibiting substance or MIS) from Sertoli cells [20, 21]. During the ninth week, the interstitial Leydig cells will begin secreting testosterone. AMH production by the Sertoli cell is activated by SOX9, a transcription factor from the *SRY*-related HMG box gene group, and inhibits the development of the Müllerian ducts into the fallopian tubes, uterus, cervix, and the interior part of the vagina [21–24]. Mouse studies have shown that increasing levels of AMH cause regression of the Müllerian duct by apoptosis [25], as seen by proportional increase in levels of Caspase-3 and BAX, an apoptosis promoter [26, 27]. Following regression of the Müllerian duct, testosterone secreted by Leydig cells induces stabilization and differentiation of Wolffian ducts into the male reproductive tract organs, the epididymides, vasa deferens, and seminal vesicles [28]. Another enzyme called 5 alpha reductase produced by the testes converts testosterone into dihydrotestosterone, which is responsible for the development of external male genitalia. In the absence of a Y chromosome, and consequently production of *SRY*, the bipotential gonads will differentiate into ovaries. This creates an environment devoid of AMH and testosterone, which allows ongoing development of the Müllerian ducts into the uterus, fallopian tubes, cervix, and vagina. This all transpires by 12 weeks of gestation [29]. Expression of MSH homeobox 2 *(Msx2)* will trigger caudal regression of the Wolffian duct by apoptosis [30], a process that requires the activity of the orphan nuclear receptor, COUP-TFII *(Nr2f2)* [31]. Remnants of the Wolffian duct can persist near the female vagina, cervix, or between the layers of the broad ligament of the uterus in up to one-third of women and are not usually associated with clinical symptoms [32] (Fig. 6.4).

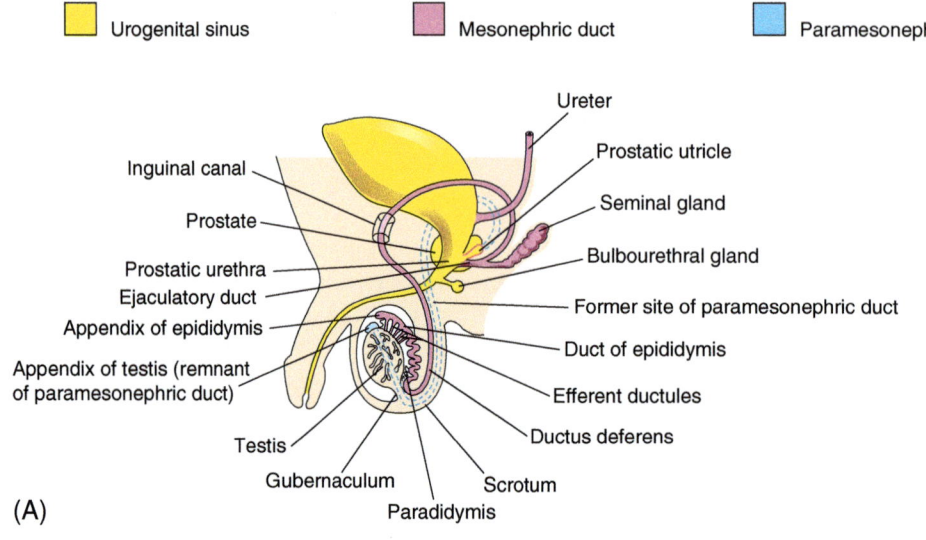

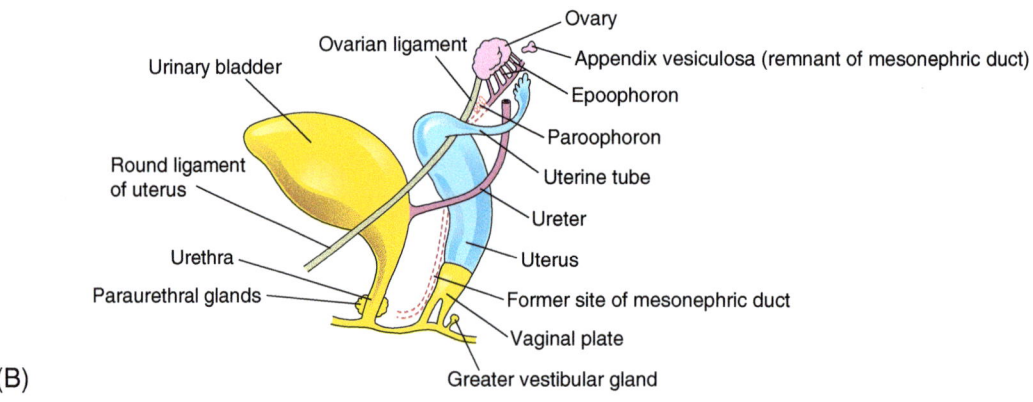

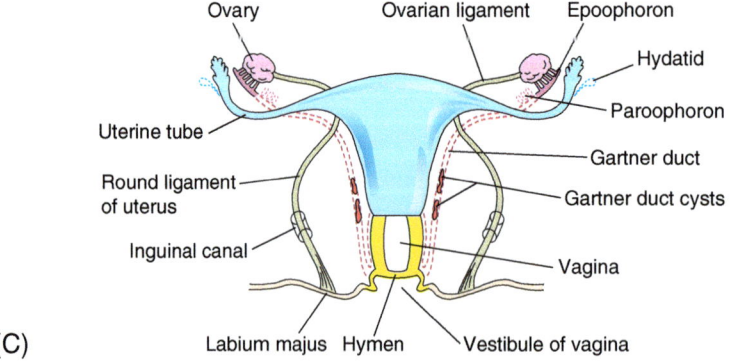

FIG. 6.4 (A) This schematic diagram summarizes the final structures formed from the Wolffian duct. (B) and (C) Lateral and frontal views of the final outcomes of the Müllerian duct forming the female reproductive tract, respectively [29].

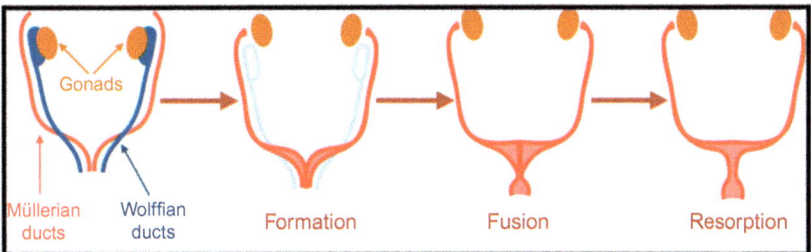

FIG. 6.5 This schematic diagram shows how the bilateral Müllerian ducts fuse in the midline, starting at the caudal region. Once fusion is complete, the septum in the midline is resorbed by the process of apoptosis, resulting in a single uterine cavity. *Used with permission from http://www.howardisms.com/obgyn/four-tips-for-correctly-diagnosing-uterine-anomalies/.*

Development of Uterus, Fallopian Tubes, and Upper Vagina From Müllerian Ducts

Once regression of the Wolffian duct is completed, the Müllerian ducts continue to develop through stages of fusion, canalization, and septal resorption to form the uterus, bilateral fallopian tubes, and interior portion of the vagina (Fig. 6.5) [15].

Fusion

This stage is characterized by the fusion of both Müllerian ducts. The elongated duct can be divided into cranial, horizontal, and caudal regions. The cranial region appears funneled in shape and remains open and separated; this area becomes the fimbriae or infundibula of the fallopian tubes. The horizontal and caudal regions begin to migrate medially and extend caudally. They fuse at the midline to form a tubular structure with the walls of the two tubes forming a medial septum. Patients with Hand-Foot-Genital Syndrome present with hypodactyly and incomplete fusion of the Müllerian duct, usually bicornuate or didelphic uteri. Mutations in the *HOXA13* gene have been linked to patients with this syndrome [33]. HOXA13 is one of a cluster of homeobox genes important for embryonic cell fate in many organs along the anterior-posterior axis, including the Müllerian duct. Disruption of one of these genes often leads to homeotic transformation, whereby some organs will have characteristics of more distal organs. In mice, analyses of Hox gene expression in the female reproductive tract show *Hoxa9* is expressed in the oviduct, *Hoxa10* in the uterus, *Hoxa11* in the uterus and cervix, and *Hoxa13* in the cervix and internal vagina [33a].

Canalization

Canalization of the Müllerian duct results in two channels with a midline dividing septum. This process occurs in synchrony with the next process of septal resorption.

Septal Resorption

Once fused and canalized, the septum separating the two ducts begins to resorb in a craniocaudal fashion, giving rise to a single Y-shaped tubular structure—the uterus and fallopian tubes. This occurs at approximately 20 weeks gestation [34].

FALLOPIAN TUBE ANOMALIES

The fallopian tubes are formed from the cranial nonfused part of the Müllerian ducts. They are usually 10–12 cm in length and can be divided into four parts from medial to lateral: intramural, isthmic, ampullary, and fimbrial. They function to transport ova released from the ovaries for fertilization and subsequently transport the embryo to the uterus. Most anomalies associated with fallopian tubes are seen in conjunction with the aforementioned Müllerian duct anomalies. However, another rare anomaly of the fallopian tube significant to gynecologists and infertility specialists is accessory fallopian tubes. They are nonpatent, cylindrical tubes attached to a functional fallopian tube. Although rare, they can contribute to infertility and ectopic pregnancy if they pick up a releasing ovum before the functional fallopian tube does [35].

UTERINE ANOMALIES

Patients with Müllerian duct anomalies represent a clinically fascinating and challenging subset of the population encountered by obstetricians and gynecologists. Overall, the ductal anomalies occur at an incidence of ~1.9%–4% in the general female population with a higher incidence, 6.3%–10%, being reported in the infertile female subpopulation [36, 37]. The clinical implications pertinent to women affected by these failures of normal development reveal themselves by the increased risk of infertility, pregnancy loss, preterm labor, fetal malpresentation, and retained placenta seen within this subset. The impact expands to include measures of quality of life as these patients have been found to be at increased risk of significant distress and depression secondary to concerns regarding their bodies as well as the ability to procreate and engage in "normal" sexual activities [36, 38, 39]. The exact clinical presentation of each of these abnormalities relates directly to their timing of disruption within the precise temporospatial development of the uterus, cervix, and upper vagina. Yet, despite some drastic phenotypic alterations, a large majority of these patients are not formally diagnosed until early adolescence as they present for evaluations of primary amenorrhea in the setting of normal secondary sexual characteristics. This is in part because the ovaries undergo a pathway largely independent of the development of the Müllerian duct and their hormonal influence into puberty is unaltered in the course of these anomalies. Although most occurrences appear sporadic in nature, familial inheritance has been observed, implicating a possible genetic etiology [40]. The desire to fully understand the mechanisms of disruption is rooted not solely in a desire for knowledge, but also in an understanding that may uncover the key to treatment.

Recently, efforts to evolve the terminology as it pertains to individuals affected by complications of reproductive organ development have examined nomenclature and definitions to better stratify anomaly processes and phenotypic presentations. Multiple pediatric endocrine societies throughout the world have determined that classifications such as "intersex," "hermaphrodite," and "pseudohermaphrodite" are perceived as not only pejorative by the patients to whom they are prescribed but also as confusing and inexact to practitioners [41]. A 2006 gathering of international endocrine experts produced consensus nomenclature and definitions. Chief among the outcomes of this meeting was the term "disorders of sex development" (DSD). DSD is defined as congenital conditions in which development of chromosomal, gonadal, or anatomic sex is atypical. Although the Müllerian anomalies that follow indeed fit that definition, they do not comprise the entirety of DSD conditions as defined [42].

Classification of Mullerian Duct Anomalies

The American Society of Reproductive Medicine (ASRM) outlines an important classification for Müllerian duct anomalies (Fig. 6.6). Most but not all anomalies will fit into one of these classifications [43–45]. It is important to note that although these Müllerian duct anomalies will be discussed here in sections as outlined by ASRM, currently no universally accepted classification system exists. This point may not be solely arbitrary as diagnosis of a ductal anomaly may result in treatment or management plans that increase patient distress and morbidity without any perceivable increase in desired treatment outcomes [45]. Notwithstanding these caveats, here are the anomalies presented by way of ASRM classifications:

Class I—Agenesis/Hypoplasia

This classification denotes anomalies that present as either complete agenesis or hypoplasia of the vagina, cervix, fallopian tubes, uterine fundus, or a combination thereof. The most notable form has three commonly used names: Mayer-Rokitansky-Kuster-Hauser (MRKH) Syndrome, Vaginal agenesis, and Müllerian agenesis. The incidence of this particular syndrome is ~1 in 5000 females [46]. Although central to the diagnosis is an anomaly of the

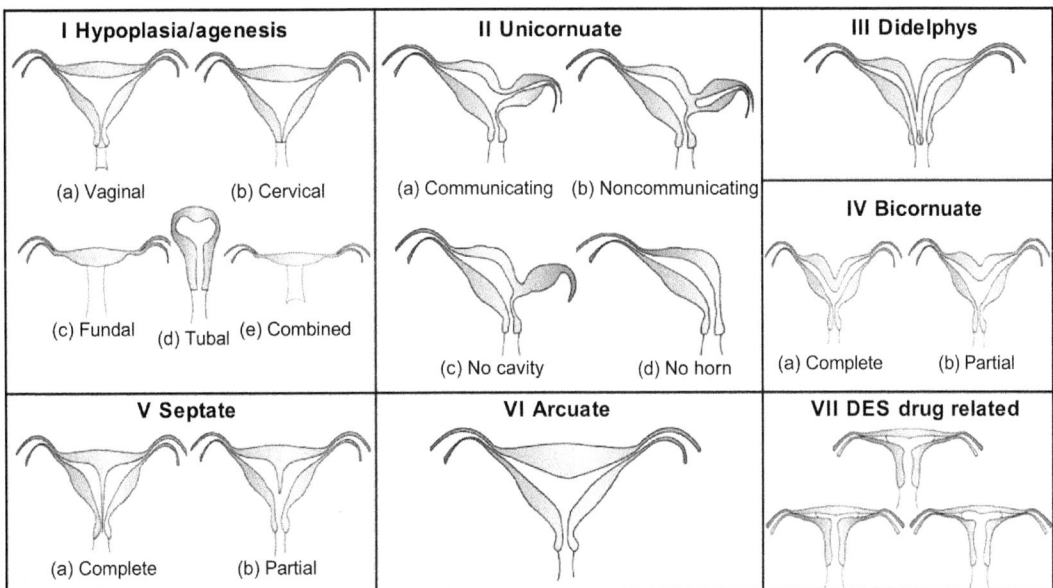

FIG. 6.6 Classification of Mullerian duct anomalies according to the American Society of Reproductive Medicine [43a].

vagina, uterus, fallopian tubes, and/or cervix, extraductal anomalies can be found in up to 53% of patients with this condition [47]. This point is further illustrated by a proposed, although not universally accepted, classification system first discussed by Schmid-Tannwald and Hauser in 1977, expanded upon by Duncan et al. in 1979, and reiterated by Oppelt et al. in 2006 [47]. In typical MRKH syndrome, anomalies are confined to the uterus, cervix, or vagina; however, with atypical MRKH syndrome, anomalies of the uterus, cervix, and/or vagina are associated with anomalies of either the ovary or renal system. Additionally, there is MURCS Syndrome (Müllerian agenesis, renal agenesis, and cervicothoracic somite abnormalities), which represents uterine, cervical, and/or vaginal malformation along with renal malformation, skeletal malformation, and/or cardiac malformation.

The clinical evaluation, of course, begins with a detailed history and physical. The physical exam will often narrow the differential diagnosis, especially if the patient is encountered well into adolescence. As a diagnosis of primary amenorrhea is often elicited from the patient's history, a physical exam will often reveal evidence of an obstructed vaginal opening without evidence of hematocolposis, which serves to differentiate between MRKH and a atransvaginal septum or an imperforate hymen. The presence of axillary and pubic hair will serve to differentiate between Androgen Insensitivity Syndrome and MRKH as patients with the former typically have no or scant hair in those regions. Afterward, transabdominal and translabial ultrasound (US) should occur to further define the degree of the anomaly. Given that up to 29% of affected patients will have a renal anomaly, the US evaluation of the kidneys should be routine. Additionally, a spine radiograph should also be routine as up to 32% of patients may have a skeletal abnormality, such as vertebral arch defects, sacralization of L5, or scoliosis, even if asymptomatic. Other associations to simply be aware of are a small increase in hearing impairment in MRKH patients and a rare association with VATER/VACTERL anomalies (vertebral, anorectal, cardiovascular, tracheoesophageal fistula, esophageal atresia, renal anomalies, and renal defects) [48, 49]. Because rudimentary Müllerian structures may be observed in up to 90% of patients, strong consideration of MRI should be considered for evaluation of endometrial activity (Fig. 6.7) [51]. If an MKRH patient reports cyclical or chronic abdominal pain, the MRI may help evaluate for active endometrial tissue. Beyond an MRI, a laparoscopy is not outside the warranted realm of evaluation, especially in patients who maintain persistent abdominal pain [36].

FIG. 6.7 MRI image showing absence of uterus between bladder and rectum. *b*, bladder; *r*, rectum. *Arrow* shows absence of uterus in its anatomic position [50].

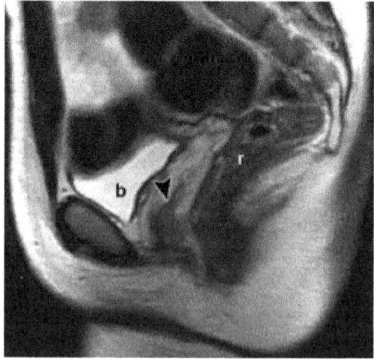

Following a full evaluation for diagnosis, the management strategy should include not only surgery but also psychosocial counseling. The anatomical address focuses on vaginal elongation via dilation as the first-line approach. Upwards of 90% of patients will be able to achieve functional success with this management plan alone [52, 53]. Even if the surgical undertaking of creating a neovagina is deemed appropriate, the maintenance of development of the neovagina would still include dilation therapy or vaginal intercourse to maintain length and diameter.

With advancements in radiologic medicine, the diagnosis has become more accurate; yet very little is known about the molecular etiology of MRKH. This has become a focus of recent research and several studies have been published on possible genetic contributions. Brucker (2017) identified mutations in *ESR1* and *OXTR* as potential gene candidates associated with the manifestation of this syndrome. Several other studies report the homebox genes (*HOXA10* and *HOXA13*) as significant contributors [54]. The most frequently discussed signaling pathway in regards to Müllerian agenesis, however, has been the WNT pathway, particularly in genetically modified mouse models with *WNT4* being discussed as a possible causative gene in humans. However, even this most notable gene has conflicting data in regards to causation [55]. The genes *HNF1B* and *LHX1* have also been implicated in association with MRKH. These genes reside on 17q12, which itself has been noted to display significant copy number variants (CNVs) in affected patients [46]. Despite numerous studies seeking to find causative genes or inheritance patterns, there is no definitive conclusive evidence of a specific gene or genes involved with MRKH. To date, the most likely mode of inheritance proposed, if there indeed is one, is a de-novo autosomal dominant or multifactorial/polygenic mode of inheritance [56]. As previously stated, conclusive evidence of associated genes regarding MRKH is lacking. For the Müllerian anomalies that follow, associations with the aforementioned genes have been postulated without definitive evidence of causation [57].

Class II—Unicornuate

This is defined as the complete or near complete absence of one of the Müllerian ducts, resulting in four possible outcomes: (i) absent rudimentary horn, (ii) nonfunctional rudimentary horn (noncavitary), (iii) cavity communicating rudimentary horn, and (iv) cavitary noncommunicating rudimentary horn [50]. This manifestation of the anomaly is attributed to what is called a "lateral fusion defect;" more specifically it is an asymmetric lateral fusion defect. This defect is overall the most common form of Müllerian anomaly and has been estimated to occur at an incidence of ~1 in 4020 women. The abnormality results from the failure of formation of one Müllerian duct, migration of a duct, abnormal fusion of the ducts, or absorption of the intervening septum. The anatomical manifestations range from a fully formed unicornuate uterus in isolation to various formations of the contralateral aspect of the uterus. The affected part of the uterus can represent all partially formed variations of the uterus with or without communication to the "normally formed" aspect of the uterus and/or cervix (See Fig. 6.8). Clinically, this may present as cyclical abdominal pain, especially around the time of adolescence once menarche begins if the functioning endometrial tissue does not have an outlet for proper evacuation of menses. A unicornuate uterus can also be completely asymptomatic.

The implications in regards to pregnancy do reveal a tendency to poorer outcomes, as evidenced by women with infertility having a higher likelihood of having this anomaly.

FIG. 6.8 (A) Schematic diagram showing a single uterine horn. (B) 3D ultrasound of a patient with a unicornuate uterus [50a].

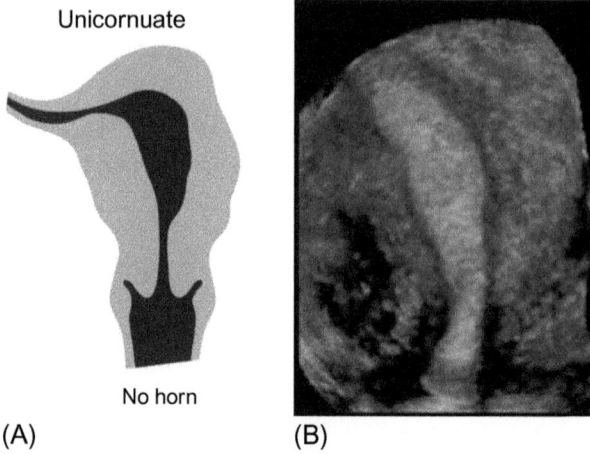

Unicornuate

No horn

(A)   (B)

A literature review found that affected woman had pregnancy outcomes of ectopic pregnancy at an incidence of 2.7% of the time, preterm delivery in 20.1% of cases, and intrauterine fetal demise (IUFD) in 3.8% of pregnancies [58]. This is compared to an overall pregnancy outcome in the general population of 0.6%–2% for ectopic [59, 60], 9.6% for preterm delivery, and 5.96 IUFDs per 1000 live births [61].

Class III—Didelphys

Uterine didelphys denotes in its truest sense a double uterus. The most frequently seen accompanying duplication is bicollis, which simply denotes two cervices. Additionally, accompanying anatomical duplication may be found in the genital tract by way of duplication of the vagina or vulva, of the urinary tract with duplication of the urethra and bladder, and even occasional duplication of the GI tract by way of duplication of the anus. There appears to be an association of this anomaly with the concomitant appearance of both hemivagina and ipsilateral renal agenesis [62]. In cases where concomitant hemivagina and renal agenesis are noted, there tends to be a prevalence of the anomaly to the right side [63].

The etiology stems from complete midline nonfusion of the Müllerian duct, resulting in two functional horns (hemiuteri) and cervices (Fig. 6.9); this is another example of a lateral fusion defect. Because fusion occurs in the caudal to cranial direction, the vagina is often fused and presents as a single vagina. In the case of poor fusion in the vagina, patients may present with a vaginal septum [50].

Class IV—Bicornuate

A bicornuate uterus is usually defined by the presence of a significant external fundal indention in an otherwise normal appearing uterus. That degree of indention is normally defined as 1 cm or more, according to American Society for Reproductive Medicine (ASRM).

A bicornuate uterus occurs when there is partial nonfusion of the Müllerian ducts at the level of the fundus. It is characterized by the presence of a cleft >1 cm in the external contour

Didelphus

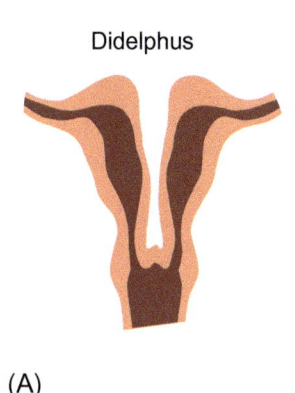

(A)

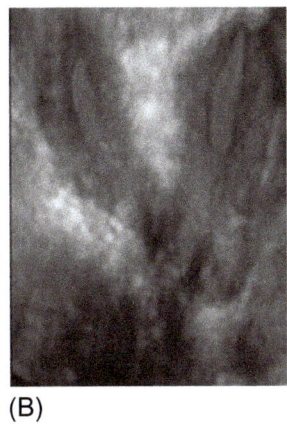

(B)

FIG. 6.9 (A) Schematic diagram showing uterine didelphys with two noncommunicating uterine horns with two cervices. (B) 3D ultrasound of patient showing the same [50a].

Bicornuate

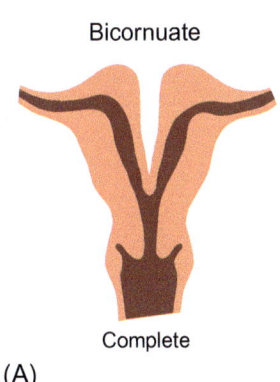

Complete

(A)

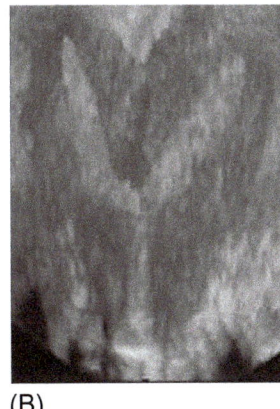

(B)

FIG. 6.10 (A) Schematic diagram showing bicornuate uterus with two uterine horns and a single cervix. (B) 3D ultrasound of patient showing the same [50a].

of the uterine fundus (Fig. 6.10). Subclassification into complete or partial categories depends on septum length. Complete uterine septa that extend either to the internal or external cervical os are known as bicornuate unicollis uterus and bicornuate bicollis uterus, respectively. When the septum is confined to the fundal region, it is considered a partial bicornuate uterus [50].

Class V—Septate

By ASRM standards, the definition of a septate uterus is the presence of fundal uterine tissue with a depth from the interstitial (endometrial fundal) line to the apex of the indentation >1.5 cm AND an angle of the indentation <90 degrees. A septate uterus occurs when there is failure of resorption of the septum after fusion of the Müllerian ducts (Fig. 6.11). It may be fibrous or muscular in nature and is important clinically due to its association with recurrent

FIG. 6.11 (A) Schematic diagram showing a septate uterus showing a fibrous septum from the uterine fundus into the uterine cavity. (B) Hysterosalpingography of patient showing the same [50a].

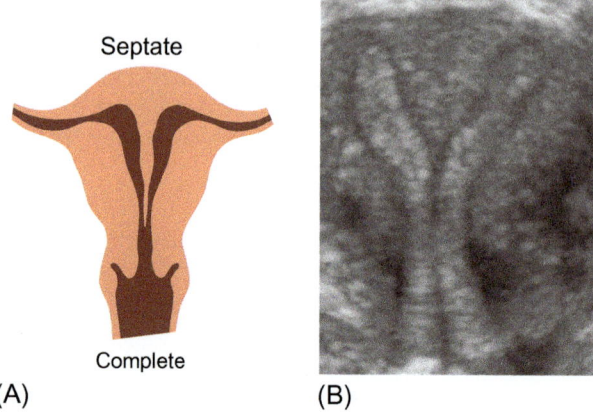

(A)　　　　(B)

pregnancy loss. Multiple studies have illustrated this association, including one in particular that noted a miscarriage rate of 41% in women with septums compared to a miscarriage rate of 11.9% in women who had surgical resection of their septums [64]. However, it has also been noted through multiple studies that women with and without septums have overall comparable fertility rates [65].

Class VI—Arcuate

By ASRM standards, the definition of an arcuate uterus is depth from the interstitial line to the apex of the indentation <1 cm AND angle of the indentation >90 degrees (Fig. 6.12). An arcuate uterus occurs when there is near complete resorption of the uterovaginal septum. Due to minimal clinical implications associated with this anomaly, it is often considered an anatomic variant.

FIG. 6.12 (A) Schematic diagram showing an arcuate uterus with an indentation of <1 cm from the uterine fundus into the cavity. (B) Hysterosalpingography of patient showing the same [50a].

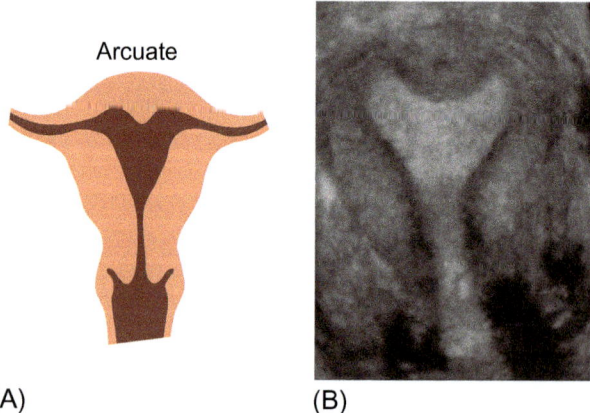

(A)　　　　(B)

DES

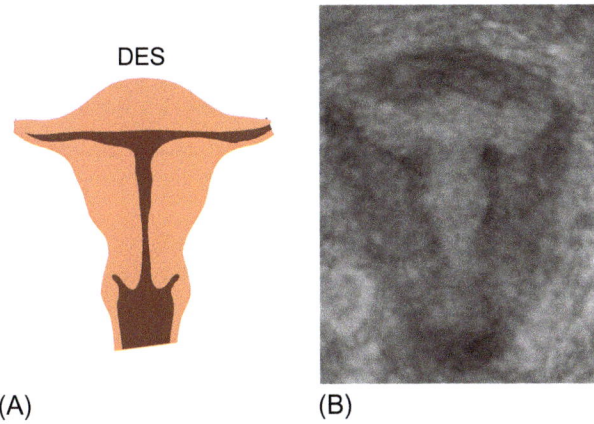

(A)   (B)

FIG. 6.13 (A) Schematic diagram showing a T-shaped uterus as expected in patients with DES exposure. (B) Hysterosalpingography of patient showing the same [50a].

Class VII—Diethylstilbestrol Related

Classically, prenatal exposure to diethylstilbestrol (DES), a potent estrogen, can result in a T-shaped uterus (Fig. 6.13)—although varying degrees of uterine hypoplasia have been described. DES is a now infamous medication prescribed to pregnant women from 1945 to 1971 to prevent miscarriage, preterm labor, and morning sickness. Clinical observation found that 15% of females born to women exposed to DES had uterine abnormalities and were at increased risk of clear cell vaginal cancer, prompting the United States Food and Drug Administration to urge physicians not to prescribe DES to pregnant women in 1971 and to ban it outright in 2000. The overall effects of DES proved to be quite devastating over time. Along with the anatomical anomaly that tended to be present in "DES daughters," the US National Cancer Institute revealed that affected individuals had increased risk of infertility, spontaneous abortion, ectopic pregnancy, preeclampsia, stillbirth, preterm delivery, early menopause, and breast cancer beyond the age of 40 by about double the rate in the unaffected population [66].

In mouse studies, DES was found to transiently disrupt developmental signals that permanently changed the expression of TRP63, whose normal function is to induce mesenchymal cells to either become columnar epithelium, which is normal for uterine tissue, or squamous epithelium, which is normal for cervicovaginal tissue. When TRP63 is disrupted, the cervicovaginal tissue differentiates into columnar cells [67]. This DES-induced cervicovaginal adenosis is thought to be the precursor to the clear-cell vaginal and/or cervical adenocarcinoma that is more prevalent in DES daughters [68]. In other mouse studies, the anatomic appearance of the female reproductive tract appeared similar to mice exposed to DES and those lacking WNT7a, thus insinuating that DES may affect that gene directly [69]. DES-exposed patients also are known to have reduced fertility. This phenomenon is not solely an effect of anatomical distortion, as illustrated in mouse studies examining HOXA10 and HOXA11. Both genes have been shown to be instrumental in regard to both embryo implantation and correct anatomical formation of the uterus. DES has been shown to repress their expression within the Müllerian duct, resulting in the aforementioned abnormalities [70, 71].

Other Mullerian Duct Anomalies

Although most Müllerian duct anomalies can be classified under ASRM's classification as discussed, other rare anomalies that have been studied include Hand-Foot-Genital Syndrome, Klippel-Fiel Syndrome, and Holt-Oram Syndrome.

Hand-Foot-Genital Syndrome

Hand-Foot-Genital Syndrome (HFGS) is a rarely encountered autosomal dominant inherited condition defined by a combination of urogenital defects and limb malformations. The clinical manifestations in female patients typically reveal degrees of incomplete Müllerian fusion as evidenced by vaginal septums, potential of bicollis, ectopic ureteral orifices, bilateral abnormally short thumbs and great toes, fifth finger clinodactaly, and hypoplastic thenar eminences. Pathogenic variants of the gene *HOXA13* have been known to cause HFGS; however, the genetic association with Müllerian anomalies has not been clearly illustrated [72, 73].

Klippel-Feil Syndrome

Klippel-Feil Syndrome is a heterogeneously inherited condition that occurs in about 1 in 40,000 newborns; it is characterized by congenital fusion of two or more cervical spine vertebrae. A "classic triad" is often associated with Klippel-Feil, consisting of cervical fusion, short neck, and a low hairline. Klippel-Feil has been shown to have an association with both vaginal agenesis and MRKH [73, 74]. Although mutations in MEOX1, GDF3, and GDF6 genes have been identified to cause some cases of Klippel-Fiel syndrome, the association between specific gene mutations and their effects on urogenital abnormalities as it pertains to these patients has not been illustrated [75].

Holt-Oram Syndrome

Holt-Oram Syndrome (HOS) is an autosomal dominant syndrome characterized by skeletal abnormalities of the upper extremities, with at least one abnormality affecting the bones of the wrist. Also, 75% of individuals have been shown to have concomitant cardiac conditions. Affected individuals may have abnormal variations of the thumbs ranging from complete absence to elongation, various presentations of the radius bone ranging from partial to complete absence, and abnormalities of the clavicle and scapula are also frequent. Associations between HOS and MRKH have been discussed in the literature. Although, HOS is known to be caused by mutations of the TBX5 gene with an AD inheritance pattern; sporadic mutations have also been illustrated in this syndrome that affects ~1 in 100,000 individuals. No clear genetic mutations have been elucidated in regard to HOS and its association with MRKH [76–78].

Evaluation of Müllerian Duct Anomalies

As with the diagnosis of most other female pelvic organ diseases and abnormalities, transvaginal ultrasound is the initial tool of choice. However, given that method's limitation for accurate diagnosis of some diseases, MRI and hysterosalpingogram can also be considered as tools of evaluation. Detailed below are how different imaging modalities can play an important role in accurately diagnosing Müllerian duct anomalies [79].

Transvaginal Ultrasonograpy and Saline Infusion Sonohysterography (TVUS With SIS)

Images are obtained by infusion of saline into the uterine cavity under direct transvaginal ultrasound guidance. This is an excellent tool for outlining the shape and contour of the intrauterine cavity. Modern three-dimensional transvaginal ultrasound can generate reconstructed images and increase the sensitivity and specificity of the diagnosis.

Hysterosalpingography (HSG)

This imaging modality is completed by slow injection of contrast medium via a catheter into the uterine cavity. A serial time lapse X-ray is performed as dye is being injected. This provides a clear image of the fill within the cavity and information in regard to the shape and patency of the fallopian tubes. This test can be performed in conjunction with saline infusion sonohysterography to improve diagnostic accuracy (Fig. 6.14).

Magnetic Resonance Imaging

Magnetic resonance imaging (MRI) is becoming widely used in the diagnosis of several diseases. It is the ideal imaging modality for the diagnosis of Müllerian duct anomalies as it can outline the external contour of the uterus and internal uterine lining.

Hysteroscopy

This is considered the most invasive procedure from among those previously listed. It requires utilization of a small camera passed through the cervical canal and into the uterine cavity under light or deep sedation. It allows a clear, real-time survey of the intrauterine cavity and provides the opportunity for concurrent treatment if the pathology is identified, for example, resection of a uterine septum in patients with a septate uterus.

Müllerian duct anomalies can be both complex and varied in their presentations. All these imaging modalities have their advantages and limitations and can be used to complement

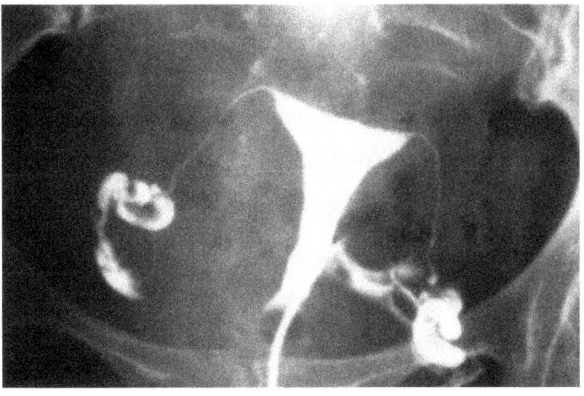

FIG. 6.14 Normal HSG outlining normal fundus and uterine contour. *From Baramki TA. Hysterosalpingography. Fertil Steril 2005;83(6):1595–606.*

each other in order to most accurately define the extent of the anomaly. The thorough and complete evaluation then allows for the most effective treatment for affected patients.

CERVICAL ANOMALIES

The cervix is formed from the caudal fusion of the paired Müllerian ducts, as discussed earlier. Hence, anomalies of the cervix usually coexist with Müllerian duct anomalies. The normal cervix is fusiform in shape and ~3 cm in length. It is longer than the uterine body in prepubertal girls, but this ratio reverses after puberty. In contrast to the muscular tissue of the uterus, the cervical stroma contains collagen and is elastic in nature. Congenital cervical anomalies can be classified into three categories: duplication, agenesis, or DES exposure [80]. Duplicated cervices occur as a result of nonfusion of Müllerian ducts. These cervices may sometime be obstructed and result in hematometra (blood-filled uterus) or pyometra (abscess-filled uterus). This requires surgical intervention as part of treatment. Cervical agenesis is a rare condition and can be seen as part of MRKH syndrome. Patients with complete cervical agenesis will also have poor development of the upper vagina, given their shared embryonic development. Incomplete cervical agenesis is also known as cervical hypoplasia and patients may have a normal upper vagina [81]. These patients are usually diagnosed at the time of menarche (start of menstruation), presenting with cyclic pelvic pain or primary amenorrhea. Congenital cervical anomalies can be seen in up to 20% of women with DES exposure. These women are at an increased risk of clear cell carcinoma of the vagina and should be followed closely. Another rare congenital cervical lesion is mesonephric cysts, formed from microscopic remnants of the Wolffian (mesonephric) ducts within the cervical stroma. Patients are usually asymptomatic, in which case treatment is not warranted [32].

DEVELOPMENT OF THE VAGINA

There has been a lot of controversy on the development of the vagina over the past century. In 1933, Koff studied the morphological organization of the vagina and explained that the upper two-thirds of the vagina is derived from the Müllerian duct and the lower one-third from the sinovaginal bulb (part of the urogenital system) [82]. However, this concept has been questioned due to the difference in the epithelium, disease progression, and functional properties of the uterus and vagina [83]. Several genetic and molecular studies have been conducted to better understand the embryonic origin of the vagina. In 2009, Cai Y reported that the entire length of the vagina originates from Müllerian duct-derived tissue that expresses the protein BMP4, which reshapes the intermediate mesoderm-derived Müllerian duct into the primordial vagina. This results in formation of the stratified squamous epithelium of the vagina [84].

In 2010, Kurita published another study to elucidate the origin of the vagina [83]. He used fluorescent labeled epithelial cells of the Müllerian duct, Wolffian duct, and urogenital sinus in mouse embryos by crossing tdTomato-EGFP dual-reporter transgenic mice with transgenic mouse lines that express Cre-recombinase in each type of epithelium. The vagina in newborn mice consisted of a fused Müllerian duct in the cranial region and sinovaginal tissue from the urogenital sinus in the caudal region. However, the Müllerian vagina continued to extend caudally and by postpartum day 7, it formed the entire length of the vagina. The sinovaginal tissue remained only at the junction between the vagina and vulva [83]. In 2012, Fritsch reported the genes involved in this process. His findings showed that three components—the urogenital sinus, the Müllerian duct, and the Wolffian duct-contributed. Cells containing p63 (a marker for squamous epithelium) were provided by the urogenital sinus and transported to the fused Müllerian duct via the Wolffian duct. As the Müllerian duct elongates caudally, it incorporates these cells that form the squamous epithelium of the vagina [85]. These studies demonstrate that the vagina is derived from the modified Müllerian duct and certain diseases of the Müllerian duct may also cause vaginal disease or anomalies.

Vaginal Anomalies

Given its dual origin, vaginal anomalies can be part of Müllerian duct anomalies or isolated developmental defects. MRKH, the most common vaginal anomaly, is associated with Müllerian duct anomalies and is classified as a Type I Müllerian anomaly under ASRM classification [44, 45]. This is discussed in more detail in "Uterine Anomalies" section. Isolated developmental defects of the vagina present as vaginal atresia and are seen in syndromes such as Bardet-Biedl Syndrome (BBS), McKusick-Kaufman Syndrome (MKKS), Fraser Syndrome, and Winters Syndrome. These are rare conditions with an incidence anywhere between 1:4000 and 1:10,000 [86]. The specific molecular mechanisms leading to vaginal atresia have not been clearly elucidated. However, proposed mechanisms include inappropriate production of AMH in the female embryonic gonads, regional absence or deficiency of estrogen receptors in the lower Müllerian duct structure, arrest of Müllerian duct development by a teratogenic agent, a mesenchymal inductive defect, or sporadic gene mutations [87]. Vaginal atresia is characterized by the absence of a hymen or by the absence of the vagina extending to the cervix. A physical exam reveals normal Müllerian structures, including the uterus and cervix, but the vagina is replaced with fibrous tissue.

BBS and MKKS are two autosomal recessive syndromes associated with vaginal agenesis and can be characterized by the presence of congenital heart malformations, postaxial polydactyl, and hydrometrocolpos, which is a fluid-filled dilated vagina and uterus caused by outflow tract obstruction secondary to the absence of the vagina [88, 89]. Patients with BBS also have visual impairment, developmental delays, and obesity. The MKKS gene involved in chaperone protein folding, processing, and assembly has been mapped to chromosome 20p12 and is responsible for the MKKS phenotype [88]. This gene, along with 10 other genes in the BBS gene family, are associated with the BBS phenotype [89].

CONCLUSION

The development of the female reproductive tract is a dynamic process. Recent advances have elucidated more of the individual genetics and the workings of their programmed proteins in regard to their effects on development. While our current understanding has been greatly enhanced over the past few years, the future holds not only further discovery but also the possibility to scientifically alter the presence and effects of the genes of female reproductive development. This could lead to earlier diagnoses and the potential application of groundbreaking gene therapies in the management of Müllerian duct anomalies. The moral and ethical implications of elucidating the genes involved with anatomical sex development cannot be overlooked as we advance to treatment. It must be considered that, as genetic research may make it possible for the steps of binary phenotypic sex development to become clearer, the debate regarding gender, sexual orientation, and sexual identity may, in effect, become exponentially more complex.

References

[1] Roly ZY, Backhouse B, Cutting A, Tan TY, Sinclair AH, Ayers KL, Major AT, Smith CA. The cell biology and molecular genetics of Mullerian duct development. Wiley Interdiscip Rev Dev Biol 2018;7(3):e310.

[2] Eicher EM, Washburn LL. Genetic control of primary sex determination in mice. Annu Rev Genet 1986;20:327–60.

[3] Nef S, Schaad O, Stallings NR, Cederroth CR, Pitetti JL, Schaer G, Malki S, Dubois-Dauphin M, Boizet-Bonhoure B, Descombes P, Parker KL, Vassalli JD. Gene expression during sex determination reveals a robust female genetic program at the onset of ovarian development. Dev Biol 2005;287:361–77.

[4] Yao HH. The pathway to femaleness: current knowledge on embryonic development of the ovary. Mol Cell Endocrinol 2005;230:87–93.

[4a] Ono M, Harley VR. Disorders of sex development: new genes, new concepts. Nat Rev Endocrinol 2012;9:71–91.

[5] Mullen RD, Behringer RR. Molecular genetics of Mullerian duct formation, regression and differentiation. Sex Dev 2014;8:281–96.

[6] Orvis GD, Behringer RR. Cellular mechanisms of Mullerian duct formation in the mouse. Dev Biol 2007;306:493–504.

[7] Jacob M, Konrad K, Jacob HJ. Early development of the mullerian duct in avian embryos with reference to the human. An ultrastructural and immunohistochemical study. Cells Tissues Organs 1999;164:63–81.

[8] Kobayashi A, Behringer RR. Developmental genetics of the female reproductive tract in mammals. Nat Rev Genet 2003;4:969–80.

[9] Kobayashi A, Shawlot W, Kania A, Behringer RR. Requirement of Lim1 for female reproductive tract development. Development 2004;131:539–49.

[10] Atsuta Y, Takahashi Y. Early formation of the Mullerian duct is regulated by sequential actions of BMP/Pax2 and FGF/Lim1 signaling. Development 2016;143:3549–59.

[11] Torres M, Gomez-Pardo E, Dressler GR, Gruss P. Pax-2 controls multiple steps of urogenital development. Development 1995;121:4057–65.

[12] Vainio S, Heikkila M, Kispert A, Chin N, McMahon AP. Female development in mammals is regulated by Wnt-4 signalling. Nature 1999;397:405–9.

[13] Miyamoto N, Yoshida M, Kuratani S, Matsuo I, Aizawa S. Defects of urogenital development in mice lacking Emx2. Development 1997;124:1653–64.

[14] Prunskaite-Hyyrylainen R, Skovorodkin I, Xu Q, Miihalainen I, Shan J, Vainio SJ. Wnt4 coordinates directional cell migration and extension of the Mullerian duct essential for ontogenesis of the female reproductive tract. Hum Mol Genet 2016;25:1059–73.

[15] Guioli S, Sekido R, Lovell-Badge R. The origin of the Mullerian duct in chick and mouse. Dev Biol 2007;302:389–98.

[16] Carroll TJ, Park JS, Hayashi S, Majumdar A, McMahon AP. Wnt9b plays a central role in the regulation of mesenchymal to epithelial transitions underlying organogenesis of the mammalian urogenital system. Dev Cell 2005;9:283–92.

[17] Nakaya Y, Sukowati EW, Wu Y, Sheng G. RhoA and microtubule dynamics control cell-basement membrane interaction in EMT during gastrulation. Nat Cell Biol 2008;10:765–75.

[18] Ayers KL, Cutting AD, Roeszler KN, Sinclair AH, Smith CA. DMRT1 is required for Mullerian duct formation in the chicken embryo. Dev Biol 2015;400:224–36.

[19] Warot X, Fromental-Ramain C, Fraulob V, Chambon P, Dolle P. Gene dosage-dependent effects of the Hoxa-13 and Hoxd-13 mutations on morphogenesis of the terminal parts of the digestive and urogenital tracts. Development 1997;124:4781–91.

[20] Warne GL, Kanumakala S. Molecular endocrinology of sex differentiation. Semin Reprod Med 2002;20:169–80.

[21] Teixeira J, Maheswaran S, Donahoe PK. Mullerian inhibiting substance: an instructive developmental hormone with diagnostic and possible therapeutic applications. Endocr Rev 2001;22:657–74.

[22] Josso N, Belville C, di Clemente N, Picard JY. AMH and AMH receptor defects in persistent Mullerian duct syndrome. Hum Reprod Update 2005;11:351–6.

[23] Behringer RR, Finegold MJ, Cate RL. Mullerian-inhibiting substance function during mammalian sexual development. Cell 1994;79:415–25.

[24] Mishina Y, Rey R, Finegold MJ, Matzuk MM, Josso N, Cate RL, Behringer RR. Genetic analysis of the Mullerian-inhibiting substance signal transduction pathway in mammalian sexual differentiation. Genes Dev 1996;10:2577–87.

[25] Catlin EA, Tonnu VC, Ebb RG, Pacheco BA, Manganaro TF, Ezzell RM, Donahoe PK, Teixeira J. Mullerian inhibiting substance inhibits branching morphogenesis and induces apoptosis in fetal rat lung. Endocrinology 1997;138:790–6.

[26] Rehman ZU, Worku T, Davis JS, Talpur HS, Bhattarai D, Kadariya I, Hua G, Cao J, Dad R, Farmanullah, Hussain T, Yang L. Role and mechanism of AMH in the regulation of Sertoli cells in mice. J Steroid Biochem Mol Biol 2017;174:133–40.

[27] Apte SS, Mattei MG, Olsen BR. Mapping of the human BAX gene to chromosome 19q13.3-q13.4 and isolation of a novel alternatively spliced transcript, BAX delta. Genomics 1995;26:592–4.

[28] Sinisi AA, Pasquali D, Notaro A, Bellastella A. Sexual differentiation. J Endocrinol Investig 2003;26:23–8.

[29] Moore KL, Persaud TVN, Torchia MG. The developing human: clinically oriented embryology. 10th ed. Philadelphia, PA: Elsevier; 2015.

[30] Yin Y, Lin C, Ma L. MSX2 promotes vaginal epithelial differentiation and wolffian duct regression and dampens the vaginal response to diethylstilbestrol. Mol Endocrinol 2006;20:1535–46.

[31] Zhao F, Franco HL, Rodriguez KF, Brown PR, Tsai MJ, Tsai SY, Yao HH. Elimination of the male reproductive tract in the female embryo is promoted by COUP-TFII in mice. Science 2017;357:717–20.

[32] Howitt BE, Nucci MR. Mesonephric proliferations of the female genital tract. Pathology 2018;50:141–50.

[33] Mortlock DP, Innis JW. Mutation of HOXA13 in hand-foot-genital syndrome. Nat Genet 1997;15:179–80.

[33a] Taylor HS, Vanden Heuvel GB, Igarashi P. A conserved Hox axis in the mouse and human female reproductive system: late establishment and persistent adult expression of the Hoxa cluster genes. Biol Reprod 1997;57:1338–45.

[34] Connell M, Owen C, Segars J. Genetic syndromes and genes involved in the development of the female reproductive tract: a possible role for gene therapy. J Genet Syndr Gene Ther 2013;4:127.

[35] Gandhi KR, Siddiqui AU, Wabale RN, Daimi SR. The accessory fallopian tube: A rare anomaly. J Hum Reprod Sci 2012;5:293–4.

[36] Committee on Adolescent Health Care. ACOG Committee Opinion No. 728: Mullerian agenesis: diagnosis, management, and treatment. Obstet Gynecol 2018;131:e35–42.

[37] Ashton D, Amin HK, Richart RM, Neuwirth RS. The incidence of asymptomatic uterine anomalies in women undergoing transcervical tubal sterilization. Obstet Gynecol 1988;72:28–30.

[38] Gueniche K, Yi MK, Nataf N. And god created woman? The link between female sexuality and the mother-daughter relationship in Mayer-Rokitansky-Kuster-Hauser syndrome in adolescents. Bull Menn Clin 2014;78:57–69.

[39] Labus LD, Djordjevic ML, Stanojevic DS, Bizic MR, Stojanovic BZ, Cavic TM. Rectosigmoid vaginoplasty in patients with vaginal agenesis: sexual and psychosocial outcomes. Sex Health 2011;8:427–30.

[40] Jacquinet A, Millar D, Lehman A. Etiologies of uterine malformations. Am J Med Genet A 2016;170:2141–72.

[41] Conn J, Gillam L, Conway GS. Revealing the diagnosis of androgen insensitivity syndrome in adulthood. BMJ 2005;331:628–30.

[42] Hughes IA, Houk C, Ahmed SF, Lee PA, Group LEC. Consensus statement on management of intersex disorders. Arch Dis Child 2006;91:554–63.

[43] van der Veen NM, Brouns JF, Doornbos JP, van Wijngaarden WJ. Misoprostol and termination of pregnancy: is there a need for ultrasound screening in a general population to assess the risk for adverse outcome in cases of uterine anomaly? Arch Gynecol Obstet 2011;283:1–5.

[43a] The American Fertility Society classifications of adnexal adhesions, distal tubal occlusion, tubal occlusion secondary to tubal ligation, tubal pregnancies, Mullerian anomalies and intrauterine adhesions. Fertil Steril 1988;49:944–55.

[44] Buttram Jr. VC, Gibbons WE. Mullerian anomalies: a proposed classification (an analysis of 144 cases). Fertil Steril 1979;32:40–6.

[45] Ludwin A, Ludwin I. Comparison of the ESHRE–ESGE and ASRM classifications of Müllerian duct anomalies in everyday practice. Hum Reprod 2015;30:569–80.

[46] Fontana L, Gentilin B, Fedele L, Gervasini C, Miozzo M. Genetics of Mayer-Rokitansky-Kuster-Hauser (MRKH) syndrome. Clin Genet 2017;91:233–46.

[47] Oppelt P, Renner SP, Kellermann A, Brucker S, Hauser GA, Ludwig KS, Strissel PL, Strick R, Wallwiener D, Beckmann MW. Clinical aspects of Mayer-Rokitansky-Kuester-Hauser syndrome: recommendations for clinical diagnosis and staging. Hum Reprod 2006;21:792–7.

[48] Kapczuk K, Iwaniec K, Friebe Z, Kedzia W. Congenital malformations and other comorbidities in 125 women with Mayer-Rokitansky-Kuster-Hauser syndrome. Eur J Obstet Gynecol Reprod Biol 2016;207:45–9.

[49] Breech L. Gynecologic concerns in patients with anorectal malformations. Semin Pediatr Surg 2010;19:139–45.

[50] Behr SC, Courtier JL, Qayyum A. Imaging of Müllerian duct anomalies. Radiographics 2012;32:E233–50.

[50a] Bermejo C, Martinez Ten P, Cantarero R, Diaz D, Perez Pedregosa J, Barron E, Labrador E, Ruiz Lopez L. Three-dimensional ultrasound in the diagnosis of Müllerian duct anomalies and concordance with magnetic resonance imaging. Ultrasound Obstet Gynecol 2010;35(5):593–601.

[51] Preibsch H, Rall K, Wietek BM, Brucker SY, Staebler A, Claussen CD, Siegmann-Luz KC. Clinical value of magnetic resonance imaging in patients with Mayer-Rokitansky-Kuster-Hauser (MRKH) syndrome: diagnosis of associated malformations, uterine rudiments and intrauterine endometrium. Eur Radiol 2014;24:1621–7.

[52] Roberts CP, Haber MJ, Rock JA. Vaginal creation for mullerian agenesis. Am J Obstet Gynecol 2001;185:1349–52 [discussion 1352–1353].

[53] Edmonds DK, Rose GL, Lipton MG, Quek J. Mayer-Rokitansky-Kuster-Hauser syndrome: a review of 245 consecutive cases managed by a multidisciplinary approach with vaginal dilators. Fertil Steril 2012;97:686–90.

[54] Zhu Y, Cheng Z, Wang J, Liu B, Cheng L, Chen B, Cao Y, Wang B. A novel mutation of HOXA11 in a patient with septate uterus. Orphanet J Rare Dis 2017;12:178.

[55] Biason-Lauber A, Konrad D, Navratil F, Schoenle EJ. A WNT4 mutation associated with Mullerian-duct regression and virilization in a 46, XX woman. N Engl J Med 2004;351:792–8.

[56] Herlin M, Højland AT, Petersen MB. Familial occurrence of Mayer–Rokitansky–Küster–Hauser syndrome: a case report and review of the literature. Am J Med Genet A 2014;164:2276–86.

[57] Liatsikos SA, Grimbizis GF, Georgiou I, Papadopoulos N, Lazaros L, Bontis JN, Tarlatzis BC. HOX A10 and HOX A11 mutation scan in congenital malformations of the female genital tract. Reprod BioMed Online 2010;21:126–32.

[58] Reichman D, Laufer MR, Robinson BK. Pregnancy outcomes in unicornuate uteri: a review. Fertil Steril 2009;91:1886–94.

[59] Van Den Eeden SK, Shan J, Bruce C, Glasser M. Ectopic pregnancy rate and treatment utilization in a large managed care organization. Obstet Gynecol 2005;105:1052–7.

[60] Hoover KW, Tao G, Kent CK. Trends in the diagnosis and treatment of ectopic pregnancy in the United States. Obstet Gynecol 2010;115:495–502.

[61] Gregory EC, MacDorman MF, Martin JA. Trends in fetal and perinatal mortality in the United States, 2006–2012. NCHS Data Brief 2014;1–8.

[62] Kim MS, Nam SY, Lee G. Complete septate uterus, obstructed hemivagina, and ipsilateral adnexal and renal agenesis in pregnancy. Obstet Gynecol Sci 2014;57:310–3.

[63] Attar R, Yildirim G, Inan Y, Kuzilkale O, Karateke A. Uterus didelphys with an obstructed unilateral vagina and ipsilateral renal agenesis: a rare cause of dysmenorrhoea. J Turk Ger Gynecol Assoc 2013;14:242–5.

[64] Kupesic S, Kurjak A, Skenderovic S, Bjelos D. Screening for uterine abnormalities by three-dimensional ultrasound improves perinatal outcome. J Perinat Med 2002;30:9–17.

[65] Chan YY, Jayaprakasan K, Zamora J, Thornton JG, Raine-Fenning N, Coomarasamy A. The prevalence of congenital uterine anomalies in unselected and high-risk populations: a systematic review. Hum Reprod Update 2011;17:761–71.

[66] Hoover RN, Hyer M, Pfeiffer RM, Adam E, Bond B, Cheville AL, Colton T, Hartge P, Hatch EE, Herbst AL, Karlan BY, Kaufman R, Noller KL, Palmer JR, Robboy SJ, Saal RC, Strohnitter W, Titus-Ernstoff L, Troisi R. Adverse health outcomes in women exposed in utero to diethylstilbestrol. N Engl J Med 2011;365:1304–14.

[67] Kurita T, Mills AA, Cunha GR. Roles of p63 in the diethylstilbestrol-induced cervicovaginal adenosis. Development 2004;131:1639–49.

[68] Robboy SJ, Szyfelbein WM, Goellner JR, Kaufman RH, Taft PD, Richard RM, Gaffey TA, Prat J, Virata R, Hatab PA, McGorray SP, Noller KL, Townsend D, Labarthe D, Barnes AB. Dysplasia and cytologic findings in 4,589 young women enrolled in diethylstilbestrol-adenosis (DESAD) project. Am J Obstet Gynecol 1981;140:579–86.

[69] Miller C, Degenhardt K, Sassoon DA. Fetal exposure to DES results in de-regulation of Wnt7a during uterine morphogenesis. Nat Genet 1998;20:228–30.

[70] Ma L, Benson GV, Lim H, Dey SK, Maas RL. Abdominal B (AbdB) Hoxa genes: regulation in adult uterus by estrogen and progesterone and repression in mullerian duct by the synthetic estrogen diethylstilbestrol (DES). Dev Biol 1998;197:141–54.

[71] Li S, Ma L, Chiang T, Burow M, Newbold RR, Negishi M, Barrett JC, McLachlan JA. Promoter CpG methylation of Hox-a10 and Hox-a11 in mouse uterus not altered upon neonatal diethylstilbestrol exposure. Mol Carcinog 2001;32:213–9.

[72] Innis JW. Hand-foot-genital syndrome. In: Adam MP, Ardinger HH, Pagon RA, Wallace SE, LJH B, Stephens K, Amemiya A, editors. GeneReviews. Seattle, WA: University of Washington, Seattle University of Washington; 1993.

[73] Kimberley N, Hutson JM, Southwell BR, Grover SR. Vaginal agenesis, the hymen, and associated anomalies. J Pediatr Adolesc Gynecol 2012;25:54–8.

[74] Willemsen WNP. Combination of the Mayer-Rokitansky-Küster and Klippel-Feil syndrome—a case report and literature review. Eur J Obstet Gynecol Reprod Biol 1982;13:229–35.

[75] Bayrakli F, Guclu B, Yakicier C, Balaban H, Kartal U, Erguner B, Sagiroglu MS, Yuksel S, Ozturk AR, Kazanci B, Ozum U, Kars HZ. Mutation in MEOX1 gene causes a recessive Klippel-Feil syndrome subtype. BMC Genet 2013;14:95.

[76] Basson CT, Cowley GS, Solomon SD, Weissman B, Poznanski AK, Traill TA, Seidman JG, Seidman CE. The clinical and genetic spectrum of the Holt-Oram syndrome (heart-hand syndrome). N Engl J Med 1994;330:885–91.

[77] Fakih MH, Williamson HO, Seymour EQ, Pai S. Concurrence of the Holt-Oram syndrome and the Rokitansky-Kuster-Hauser syndrome. A case report. J Reprod Med 1987;32:549–50.

[78] Ulrich U, Schrickel J, Dorn C, Richter O, Lewalter T, Luderitz B, Rhiem K. Mayer-von Rokitansky-Kuster-Hauser syndrome in association with a hitherto undescribed variant of the Holt-Oram syndrome with an aorto-pulmonary window. Hum Reprod 2004;19:1201–3.

[79] Fukunaga T, Fujii S, Inoue C, Mukuda N, Murakami A, Tanabe Y, Harada T, Ogawa T. The spectrum of imaging appearances of mullerian duct anomalies: focus on MR imaging. Jpn J Radiol 2017;35:697–706.

[80] Choussein S, Nasioudis D, Schizas D, Economopoulos KP. Mullerian dysgenesis: a critical review of the literature. Arch Gynecol Obstet 2017;295:1369–81.

[81] Niver DH, Barrette G, Jewelewicz R. Congenital atresia of the uterine cervix and vagina: three cases. Fertil Steril 1980;33:25–9.

[82] Koff AK. Development of the vagina in the human fetus. Contrib Embryol 1933;24:59–91.

[83] Kurita T. Developmental origin of vaginal epithelium. Differentiation 2010;80:99–105.

[84] Cai Y. Revisiting old vaginal topics: conversion of the Mullerian vagina and origin of the "sinus" vagina. Int J Dev Biol 2009;53:925–34.

[85] Fritsch H, Richter E, Adam N. Molecular characteristics and alterations during early development of the human vagina. J Anat 2012;220:363–71.

[86] Evans TN, Poland ML, Boving RL. Vaginal malformations. Am J Obstet Gynecol 1981;141:910–20.

[87] Rock JA. Te Linde's operative gynecology. In: Rock JA, Thomson J, editors. Surgery for anomalies of the Mullerian ducts. 8th ed. Philadelphia, PA: Lippincott-Raven; 1997.

[88] David A, Bitoun P, Lacombe D, Lambert JC, Nivelon A, Vigneron J, Verloes A. Hydrometrocolpos and polydactyly: A common neonatal presentation of Bardet-Biedl and McKusick-Kaufman syndromes. J Med Genet 1999;36:599–603.

[89] Forsythe E, Beales PL. Bardet-Biedl syndrome. In: Adam MP, Ardinger HH, Pagon RA, Wallace SE, LJH B, Stephens K, Amemiya A, editors. GeneReviews. Seattle, WA: University of Washington, Seattle University of Washington; 1993.

The Molecular Genetics of Oogenesis

Fan Zhai*, Xinyi Ma*, Liying Yan*, Jie Qiao*,†

*Center for Reproductive Medicine, Department of Obstetrics and Gynecology, Peking University
Third Hospital, Beijing, China †Fellow of the Chinese Academy of Engineering, Beijing, China

INTRODUCTION

The production of functional gametes has a central biological importance and is essential for the propagation of the sexually reproducing species. Oogenesis is the creation of a female gamete (also known as an ovum or oocyte) from a primordial germ cell (PGC). This is a prolonged and carefully regulated developmental process. It involves the differentiation from a diploid germ cell into a haploid oocyte, followed by maturation to a developmentally competent oocyte capable of fertilization. Unlike male gametes that supply nuclear genetic material to the newly formed zygote upon fertilization, an oocyte also contributes mitochondrial genome and cytoplasmic components as well as fundamental molecular cues that govern early embryonic development.

The journey toward successful oogenesis begins within the ovarian niche and relies on the paracrine and junctional interactions between an oocyte and the surrounding somatic cells destined to form an ovarian follicle. This developmental pathway involves the coordinated progression through mitotic divisions, followed by the periods of precisely timed active meiotic division and intermittent meiotic arrest. During oogenesis, periods of transcriptional activity coincide with active cell growth and alternate with the stages of oocyte quiescence where there is transcriptional silencing, cell growth arrest, and little metabolic activity. Multiple genetic and environmental factors can perturb any of the differentiation events occurring during oocyte development. The presence of the crucial checkpoints allows the flawless progression of the oocyte through the process of division, growth, and maturation. Both the nuclear and mitochondrial genome are implicated in the oocyte-specific unique life cycle and play an important role in oocyte maturation and embryo development.

Over the last decade, substantial progress in the basic aspects of oocyte research has been made in elucidating the factors that regulate oocyte and follicle development. The findings delivered from the animal models and in vitro approaches contributed to understanding the molecular basis of the events that govern the complex cell transition during oogenesis and the acquisition of oocyte developmental competence. In this chapter, we discuss the key aspects of human oogenesis with an emphasis on the underlying cellular and molecular mechanisms and the genetic determinants of oocyte maturation.

OOGENESIS: AN OVERVIEW

Oogenesis involves three key phases: proliferation, growth, and maturation, during which PGCs progress to primary oocytes, secondary oocytes, and then to mature ootids [1]. During the proliferation phase (oocytogenesis), the PGCs migrate to the gonadal ridges from the hindgut by crossing the dorsal mesentery. Migration and proliferation of PGCs is controlled by bone morphogenetic proteins (BMPs) and activin, the members of the transforming growth factor beta (TGFβ) family, and a number of germ cell-derived transcription factors [2, 3]. BMP2 and BMP4 increase and BMP7 decrease the number of PGCs in mice ovaries while activin increases the number of PGCs in humans. The factors that are vital for survival, migration, and proliferation of PGCs include SOX17 [4], BLIMP1, PRDM14, etc. In the female gonads, PGCs differentiate into oogonia and small diploid cells [5, 6]. Oogonia undergo massive proliferation by mitotic divisions from ~6 to 8 weeks gestation [7]. Owing to incomplete division of the cytoplasm (incomplete cytokinesis) during mitosis, dividing oogonia form germ cell cysts that will develop into the ovarian follicle [8]. The number of oogonia increases only during embryonic life and no germ cells forms after birth.

The growth phase of oogenesis starts when oogonia stop division, enlarge to form primary oocytes, and initiate meiosis. The first meiotic division in primary oocytes begins around ~11–20 weeks of embryonic life before birth (Fig. 7.1). RNA binding proteins, including DAZL and BOLL, are involved in different stages of human meiotic division [9].

Subsequently, oocyte is arrested in the diplotene stage during the prophase of the first meiosis and enters prolonged resting phase for decades, called dictyate (Fig. 7.1) [8]. Initiation of meiosis in the human and mouse ovary involves STRA8 (stimulated by retinoic acid gene 8) signaling, which dependents on retinoic acid [8]. Simultaneously, germ cell cysts break and

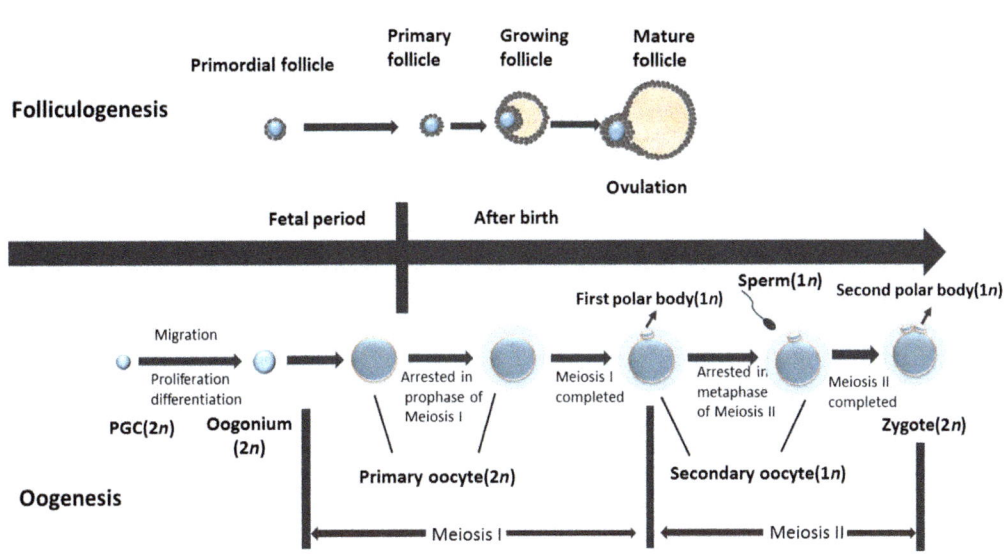

FIG. 7.1 A description of human oogenesis and corresponding folliculogenesis. PGC migrates to the gonadal ridge during proliferation phase and then differentiate into oogonium. After that, oogonium stops division and progresses into meiosis for the first time and arrests in prophase of Meiosis I until puberty, and forms the primary oocyte. Meiosis resumes during ovulation and M II begins immediately after M I is completed, and it halts at the metaphase stage of Meiosis II until fertilization, after which the second polar body is created. The letter "n" and "$2n$" represents the number of chromosome set in the cells of different stages. The process of folliculogenesis is shown corresponding to different stages of development oogenesis.

ovarian stromal (pregranulosa) cells surround the primary oocyte to form the single-layer primordial follicles. In humans, follicle formation takes place during the second trimester of fetal development and completes before birth [6]. Animal models revealed involvement of estrogen and its receptors as well as the TGFβ superfamily, particularly GDF9 and BMP15, in the process of follicle cyst breakdown and primordial follicle formation. However, the factors and signaling pathways in humans remain unclear. During formation of the primordial follicles, a vast number of oocytes undergo apoptosis. Different pro- and antiapoptotic proteins have been implicated in the germ cell fate during primordial follicle formation. This includes antiapoptotic proteins, such as the B cell lymphoma-leukemia (BCL) protein family members BCL2 and BCLX, and the proapoptotic protein BAX as well as caspase (casp)-regulated apoptotic pathways. It has been demonstrated that the absence of BCL2 or BCLX results in a reduced number of primordial follicles whereas the loss of BAX or casp2 activity leads to an increase in the primordial follicle pool [3].

Primordial follicles present at birth determine the reservoir of germ cells available during the female reproductive life. Thus, mutations in any of the genes involved in PGC migration and survival or regulation of follicle formation and apoptosis can lead to depletion of the germ cell pool and, in turn, affect a woman's reproductive span.

Primary oocytes within primordial follicles remain dormant until puberty and then undergo activation as the cohorts of continuously recruited follicles. Primordial follicles give rise to the primary follicles in which oocytes enlarge, and flattened granulosa cells develop into cuboidal follicular cells. Activation of the primordial follicles is a highly regulated dynamic process. PTEN/PI3K signaling has been identified as crucial for the transition from primordial to primary follicles. PTEN (phosphatase and tensin homolog deleted on chromosome 10) leads to the inactivation of PI3K (phosphatidylinositol 3 kinase), a serine-threonine protein kinase that stimulates cell proliferation and suppresses expression of the proapoptosis factor FOXO3 (forkhead box O3). Research on the mouse and human ovary demonstrated that manipulation of the PTEN/PI3K pathway by inhibition of PTEN and activation of PI3K induces primordial follicle activation [10]. Another signaling pathway mediated by the Tsc/mTORC1 (tuberous clerosus complex 1 and mammalian target of rapamycin complex 1) regulates activation of primordial follicles [11]. Other vital factors involved in the process include the KIT ligand and its receptor, neurotrophins, the transcriptional factors Nobox and Sohln1, and the members of the TGFβ family, the BMP4, BMP7, and the anti-Mullerian hormone (AMH); however, most data are derived from animal studies [3].

As follicles mature, primary oocytes form zona pellucida and cuboidal somatic cells gain additional layers, which designates progression to the secondary and antral stages. Only a small number of oocytes progress through maturation in every menstrual cycle. Several factors produced by oocytes and their surrounding somatic granulosa and theca cells are important for the progression from primary to secondary follicles and the corresponding stages of oogenesis. These include GDF9 and BMP15 from the TGFβ family, the TATA-binding protein 2 (TBP2), and the TATA box binding protein (TBP)-associated factor (TAF4B) [12]. At early stages, ovarian follicles develop in FSH-independent fashion under the control of locally secreted factors and become FSH-dependent when they reach the antral stage. Just before ovulation, the primary oocyte completes the first meiotic division, which results in the formation of the haploid secondary oocyte with extrusion of the first polar body, a small nonfunctional cell that subsequently degenerates. The second meiotic division begins immediately at ovulation and completes if fertilization occurs, leading to the creation of the haploid ootid with extrusion of the second polar body [13].

The maturation phase occurs through a gradual sequential process during which the oocyte develops into a mature ootid. The acquisition of the developmental competence by oocyte takes place concomitantly with folliculogenesis and relies on bidirectional communication with the neighboring somatic cells (Fig. 7.1). Among the factors that orchestrate oocyte maturation, activation of NOTCH signaling in gonadal somatic cells by ligands from adjacent germ cells (GCs) [14] highlights the role of NOTCH signaling in oocyte-controlled proliferation and differentiation of GCs. Moreover, members of the TGF-β superfamily, the oocyte-derived factors GDF9 and BMP15, have been shown to be involved in oocyte maturation, GC proliferation, and cholesterol biosynthesis [15]. In addition, GC-derived signaling as well as KITLG and its receptor KIT, previously implicated in paracrine signaling, were expressed in GCs and oocytes, respectively, and play crucial roles in primordial follicle activation [16].

NUCLEAR DNA (NDNA) INHERITANCE

The nuclear DNA (nDNA) is inherited from both parents (half of which is derived from an oocyte and the other half from a sperm). During the process of oogenesis, the ploidy status, which defines the number of sets of chromosomes, undergoes dynamic change. Germ cells undergo mitotic divisions, which maintain the diploid chromosomal set, followed by two rounds of meiosis to produce genetically distinct haploid gametes with one set of chromosomes [17]. After fertilization, two haploid gametes give rise to the diploid zygote (Fig. 7.1). Thus, a highly regulated mechanism of both types of cell division is critical for adequate gamete development and for producing a zygote with the correct number of chromosomes. Mitosis and meiosis are integral parts of the cellular life cycle. Both processes are essential for maintaining genomic stability, adequate gamete development, and a correct number of chromosomes in the zygote. Aberrations occurring in either mitosis or meiosis may result in failed oocyte maturation, chromosomal structural or numerical abnormalities, which can subsequently lead to adverse pregnancy outcomes and genetic diseases in offspring.

Mitotic Division

Mitotic division takes place when PGCs migrate to the female gonads and differentiate into oogonia [6]. Mitosis involves a nuclear division that produces two genetically identical daughter cells through the stages of prophase, prometaphase, metaphase, anaphase, and telophase. Mitosis is a reasonably short stage of the cell cycle, preceded by the long interphase. In preparation for the division, the cell synthesizes mRNA and proteins during gap 1 (G1-pase) of the interphase. Then the cell progresses to the synthesis (S-phase) of the interphase, during which the whole genome encoded within double DNA strands is replicated. Subsequently, rapid growth and protein synthesis occur during gap 2 (G2-phase) of the interphase, which ends with the onset of prophase, a first phase of mitosis (M-phase).

Prophase involves condensation and separation of the duplicated DNA into sister chromatids, each of which corresponds to one set of the duplicated genome. Nuclear chromatin becomes visible in the light microscope as chromosomes. During prometaphase of mitosis, the nuclear membrane dissolves and spindle fibers attach to the chromosomal centromeres. This is followed by alignment of sister chromatids along the middle of the cell in metaphase (metaphase plate). Chromosomes then segregate during anaphase to generate two genetically identical nuclei, receiving one copy of each chromosome. Afterward, chromosomes disperse in telophase and spindle fibers gradually disappear. The process is accompanied by cytokinesis, which results in the division of cytoplasm around each nucleus and the production of two identical daughter cells. All the cell cycle phases round sequentially under a strict surveillance of the checkpoint proteins to ensure an adequate cycle progression.

Mitosis in the vertebrate is activated by cyclin-dependent kinase Cdk1 (also known as Cdc2), which is dependent on cyclin B. Other proteins and protein kinases involved in the regulation of mitosis include Cdc25C, Wee1, and Myt1 [18]. An important mitotic checkpoint (also known as the spindle assembly checkpoint) involves a complex signaling cascade that ensures uninterrupted chromosome segregation, and arrests aberrant mitotic division via inhibition of the anaphase-promoting complex/cyclosome (APC/C), an E3 ubiquitin ligase. Defects in this signaling complex have been linked with tumor formation [19]. Additional, a less common mitotic derangement is tripolar mitosis, which results in the production of three daughter cells with uneven distribution of genetic material [20].

Meiosis Division

Meiosis is a specialized cell division during which the number of chromosomes is reduced by half, creating genetically distinct haploid cells. Analogous to mitosis, the event preceding mitotic division involves DNA replication during the S-phase of the interphase, which leads to the production of two identical sister chromatids attached at a centromere. The G2 phase is not present in meiosis. Instead, DNA replication is followed by two rounds of cell division, known as meiosis I and meiosis II.

Meiosis I and II entail four stages: prophase, metaphase, anaphase, and telophase. During the prophase of meiosis I, homologous chromosomes pair with each other and undergo genetic recombination (crossovers) when they exchange genetic information and produce unique genetic combinations. Meiotic prophase is subdivided into five stages based on the appearance of chromosomes: leptotene, zygotene, pachytene, diplotene, and diakinesis [3]. The oocyte maturation suspends at the diplotene stage of prophase I and enters a resting stage called dictyate or germinal vesicle (GV) arrest, until meiosis resumption is triggered by the luteinizing hormone (LH) at ovulation [21]. Resumed meiosis I progresses through the metaphase I when homologous chromosome pairs move together along the metaphase plate, followed by segregation of the homologous chromosomes in which a pair of sister chromatids remain together in anaphase I. It ends when the chromosomes arrive at the opposite poles in telophase I. Upon completion of meiosis I, the primary oocyte divides into a larger secondary oocyte and extrudes a smaller polar body to discard half the genetic material. Meiosis I results in two haploid cells, each with a single set of chromosomes (half the number of the original parent cell chromosomes), although each chromosome contains a pair of sister chromatids.

Meiosis II starts after meiosis I without DNA replication. The process is similar to mitosis and involves equational segregation of sister chromatids after degradation of cohesin, a protein complex holding sister chromatids at the centromere. As a result, the number of DNA copy halves while the number of chromosomes remains the same. Following meiosis II, the secondary oocyte divides into an oocyte and a second polar body. Notably, in female gametogenesis, only one cell develops into an oocyte and the other meiotic products are eliminated by the extrusion of polar bodies. However, in males four daughter cells produced by meiotic division form sperm.

Meiosis has important biological implications. Compared to male meiosis, female meiosis is more prone to errors, resulting in aneuploidy, which increases with a woman's age. For instance, in 35-year-old women, aneuploidy is observed in about 20% of oocytes, but reaches 60% around menopause [22]. Although aneuploid oocytes can still be fertilized, the embryo can hardly be viable. Aneuploidy is one of the leading causes of reproductive failure and the most frequent genetic cause of developmental disabilities [23, 24].

In contrast to mitotic division, where incorrect segregation affects only a fraction of the cells resulting in mosaicism, the chromosome missegregation in meiosis could affect all the cells. During meiotic division, cells should equally share the chromosome, but sometimes, the whole pair of chromosomes or bivalent end up in one cell while the other one gets nothing. Incorrect chromosome segregation results in aneuploidy, carrying an abnormal number of chromosomes [24]. The aneuploid oocyte can still function, fertilize, and generate an aneuploid embryo. The one copy of autosomes is lethal in humans, leading to spontaneous abortion. Missing a sex chromosome (monosomy) or the addition of a chromosome copy (trisomy) could be viable, but the offspring's health is compromised, including trisomy 21 (Down syndrome), trisomy 18 (Edwards syndrome), trisomy 13 (Patau syndrome), and monosomy X (Turner syndrome). Trisomy 21 (47, XX/XY, +21), one of the most common viable aneuploidies in humans, results from the incorrect segregation of chromosome 21 in meiosis

I (65%) or meiosis II (23%) [25]. Nondisjunction during meiosis is the most common cause of the aneuploidies, which can also occasionally result from a chromosomal rearrangement (e.g., Robertsonian translocation or a balanced sex chromosome translocation) [26].

Nuclear Mechanics During Gametogenesis: Chromosome Replication and Interactions

Chromosomal interactions during meiosis will produce combinations of traits that come from both parents, yet are different from either parent. Precise mechanisms of nuclear polarization, genetic recombination, chromosome alignment, chromosome segregation, and spindle assembly and checkpoint (SAC) guard oogenesis against the misdistribution of genetic material in oocytes.

Nuclear Polarization and the Chromosomal Bouquet

Homolog pairing complexes (synapsis) occurring in early prophase are essential for the process of DNA recombination. This involves nuclear polarization formation during the prophase preceding the next two instances of meiosis. It is formed through the unique intertelomeres connections, involving a SUN domain protein, a KASH domain protein, and microtubules, which locate in the inner nuclear envelope (NE), the outer NE, and cytoplasm, respectively. At the onset of the homolog pairing, chromosomes rearrange into continuous leptotene strands. In the zygotene of prophase I, the tails (telomeres) of chromosomes scatter around the nucleus, tightly congregating in a restricted area on the nuclear periphery through the Shelterin protein complex and then attaching to the NE via a neighboring SUN1 protein [27]. Chromosome movement during leptotene is facilitated by the SUN/KASH complex, which binds the telomere to the inner side of NE and connects with perinuclear microtubules by the dynein motor protein in the outer NE membrane [28].

This arrangement resembles the stems of flowers gathered in a bouquet and requires various motor proteins and microtubules [28]. The exact mechanism by which telomeres attach to SUN1 in the human oocyte remains unclear. This process is regulated by cyclin-dependent kinase Cdk2 through phosphorylation of the SUN1 N-terminal domain, which in turn is activated by RingoA, an atypical noncyclin Cdk activator protein. The experiments using in vivo mutant and protein domain deletion analyses have demonstrated that the telomere-SUN1 association involves the Majin, Terb1, Terb2, and Trf1 proteins that comprise the Shelterin complex [29–31]. Bouquet formation stabilizes the interaction between homologs and occurs independently of synapsis formation and homologous chromosome recombination [32]. Mutation in NDJ1, a protein involved in the formation of the bouquet, has been associated with defects in the meiosis-specific distribution of telomeres [33]. Small chromosomes do not form bouquets, probably because they have less chance to encounter their homolog chromosomes.

Genetic Recombination

Homologous chromosome recombination occurs in meiosis and plays an important role in genetic diversity, and is also a key determinant in the unique gene profile of an individual.

This naturally occurring process has important biological implications in oogenesis, which occurs through DNA repair. The process spans the pachytene, leptotene, and zygotene stages of prophase I and requires the homolog pairing of chromosomes through synapsis complexes.

DNA recombination begins when a topoisomerase SPO11, a type 2 topoisomerase, generates DNA double-strand breaks (DSBs) [34]. DSBs are confined to the specific chromosome regions (hotspots) that contain 13-base pair sequence motifs and are regulated by a number of proteins (e.g., methyl transferases) and histone modifications [e.g., histone 3 lysine 4 trimethylation] [35].

Then, MRN (Mre11-Rad50-Nbs1), a heterotrimeric protein complex, performs resection from $5'$ to $3'$ end of DNA strands at DSB and creates two overhanging $3'$ ends. This is followed by unwinding the double-stranded DNA by Sgs1 helicase and slicing the single strand by nucleases. The exchange between two homologous chromosomes involves the activity of the protein complex RPA-Rad51 and the DNA polymerase to create a cross structure called the Holliday junction. DNA is then repaired either through the DSBR (double-strand break repair) or SDSA (synthesis-dependent strand annealing) pathway that results in crossover and noncrossover products, respectively [36]. A recent single-cell whole-genome sequencing analysis of human polar bodies and oocyte pronuclei from the same individual revealed that the occurrence of crossover in a human oocyte occurs around 43 times ($\sim$2 per chromosome), which is 1.7 times more frequent than in sperm. In this research, a weak chromatid interference was found. This implies that one crossover between the sister chromatid only slightly reduces the probability for the additional nearby crossover, and indicates that genetic information in the oocyte is being exchanged in dynamic fashion [37]. The recombination probability varies depending on the proximity to the transcription start site (TSS) with underrepresentation of crossovers around TSS [37].

Chromosome Alignment and Segregation

Chromosome alignment and segregation occur after the spindle structure is built. As we know, in order to ensure accurate segregation, chromosomes must align in the correct patterns in both mitosis and meiosis.

In mitosis, kinetochores could capture and stabilize microtubules, which assemble the spindle, by the formation of kinetochore fibers (K-fiber). When K-fibers connect kinetochores and opposite spindle poles, the chromosome will transport to the equator of the spindle. During meiosis, however, the kinetochores are not responsible for holding the connections with microtubules [38]. Despite the absence of typical K-fibers, chromosomes could also align on the metaphase plate, owing to the pushing forces of a series of microtubule motors along the chromosome arms. It has been revealed that multiple microtubules exist between the two chromosome arms during prometaphase. Thus, chromosome alignment in oocytes involves interactions among chromokinesins, Aurora B/C Kinase proteins, and microtubules that "push" chromosomes toward the spindle equator [39]. Once chromosomes have reached the spindle equator, they oscillate along the spindle axis while attempting biorientation.

To achieve the formation of a haploid gamete, homologous chromosomes are segregated during the first meiotic division and sister chromatids are segregated during the second division. The ability of chromosomes to remove cohesion is essential for adequate chromosome segregation. The removal of cohesion takes place in a stepwise pattern after chromosomes move to the spindle equator, cohesion in chromosome arms during meiosis I, and cohesion in the centromere region during meiosis II. During meiosis I, sister chromatids of meiotic chromosomes are held together with the chromosome arms and centromeres by the meiosis-specific protein complexes and cohesins comprising Rec8. The kinase Plk1 marks Rec8 at chromatid arms by phosphorylation, which allows Rec8 cleavage by Separase and release of cohesion between sister chromatid arms. Cohesins in the centromere undergo dephosphorylation by Shugoshin (Sgo), PP2A, and Bub1. During meiosis II, in the absence of Emi2, a component of CSF (cytostatic factor), the Separase is released to cleave Rec8 at the centromeres of sister chromatids for chromosome separation [40]. Once biorientation and attachment to opposite spindle poles is achieved, the oocyte may get into anaphase II.

Spindle Assembly and Checkpoint (SAC)

A spindle assembly checkpoint (SAC) is involved in the regulation of chromosome segregation in meiosis. The main components of the SAC protein complex include Mad1, Mad2, BubR1, Bub1, and Bub3 [41]. The SAC ensures that all the microtubules from the spindle poles connect with the kinetochores and the chromosomes are aligned to the equatorial plate before the onset of anaphase. The checkpoint halts cell cycle progression in the cases of chromosome misalignment and anaphase will not commence until the defects are corrected, inducing a metaphase arrest [42]. SAC ensures progression to anaphase when all chromosomes are successfully attached to the bipolar spindle [43]. Unattached kinetochores can inhibit the activation of an E3 ubiquitin ligase, also known as an anaphase-promoting complex/cyclosome (APC/C) [44]. If the bipolar pattern of the kinetochores is flawless, the SAC will not be activated, and Securin and Cyclin B are degraded through the ubiquitin pathway by APC/C. This results in the activation of Separase and allows a metaphase-to-anaphase transition [43, 45]. In addition, the SAC can be activated by a distortion of the kinetochore due to lack of tension and destabilization (an intrakinetochore stretch) when the kinetochores are not attached to the neither poles. Wrong attachments of kinetochores escaping SAC control may lead to inappropriate chromosome alignment and segregation errors. Missegregations of chromosomes during meiosis cause aneuploidy [46]. Fertilization of aneuploid oocytes in human may lead to spontaneous abortion in early pregnancy, or may also lead to the occurrence of chromosomal disorders after the baby is born [47]. One of the most common viable aneuploidies is trisomy 21 because of the missegregation of chromosome 21 in female meiosis [48].

SAC is also involved in Rec8-related meiosis I regulation, ensuring the cohesion of sister chromatids during meiosis I and guarding against aberrant acentrosomal spindle formation in the oocyte [49, 50].

MITOCHONDRIAL INHERITANCE

Mitochondria are semiautonomous energy-supplying organelles that are essential for cellular survival and development. Mitochondria play a vital role in numerous cellular processes, including adenosine-5′-triphosphate (ATP) generation, cell signaling, apoptosis, and calcium (Ca^{2+}) homeostasis. Mitochondria are the major source of reactive oxygen species (ROS) and are involved in oxidative stress-induced aging of cells [51]. The number of mitochondria per cell varies with the cell type and fluctuates under different physiological conditions. Mitochondria are one of the most abundant cellular organelles in the mature oocyte [52]. During oogenesis, mitochondria exhibit unique features and undergo multiple alterations in structure, number, and activity, all of which influence energy provision throughout development of the oocyte and early embryo.

The Role of Mitochondria in Oogenesis

Cellular processes involved in oocyte development, maturation, and fertilization have high energy requirements. In a maturing oocyte, preimplantation embryo glycolysis is limited and mitochondria provide the primary source of ATP to meet cellular energy demands. A reduced ATP level may limit the energy supply for the oocytes or preimplantation embryos [53]. In addition, mitochondria are involved in intracellular calcium homeostasis and are an important source of calcium in the oocyte [54]. Ca^{2+} has a key function in cytoplasmic maturation; it is also involved in the energy production before fertilization and is important for embryonic cell-cycle transition and embryonic axis establishment [55]. Thus, the mitochondrial content and functional integrity within the oocyte are of critical importance for the establishment of developmental competence and sustaining the earliest stages of life in the preimplantation embryo [56]. Structural, spatial, and genetic abnormalities leading to mitochondrial dysfunction are strongly associated with infertility, poor reproductive performance, and reduced embryonic survival, owing to the mitochondrial influence on oocyte metabolism and development [57, 58]. In contrast to the somatic cells, mitochondria in growing and maturing oocytes are smaller, have a spherical or oval shape, contain less cristae, display a denser matrix, and presumably exhibit lower activity [59]. Limited mitochondrial activity in oocytes has been proposed as a defensive mechanism that protects mtDNA from harmful ROS production [52, 60]. Mitochondrial ultrastructure changes in the embryo and from the blastocyst stage take on an appearance of that observed in somatic cells, which coincides with increased metabolic activity [52].

Mitochondrial DNA (mtDNA)

The mammalian mitochondrial genome, around 16.5 kb, is packaged as double-stranded circular DNA molecules (mtDNA) that lack introns and compose 37 genes [61]. Mitochondrial gene products include 13 polypeptides that constitute the mitochondrial oxidative phosphorylation (OXPHOS) respiratory chain complex, 22 transfer RNAs (tRNAs), and two ribosomal RNAs (rRNAs), all of which are indispensable for mitochondrial function. Most of the mitochondrially localized proteins that comprise mitoproteome (around 1500 proteins) are encoded by nDNA and imported into mitochondria by various transport mechanisms [62]. Thus, the cooperation between the nuclear and mitochondrial genome is essential to ensure mitochondrial function [63, 64]. There are approximately 100,000–800,000 mtDNA copies in the human oocyte, which vary between individuals and between different physiological states [65–67]. Each mitochondrion in the oocyte contain only 1–2 copies of the mitochondrial genome [65]. Developmentally competent oocytes have higher mtDNA content compared to poor quality or aging oocytes [68], whereas a low mtDNA copy number has been linked with fertilization failure and decreased oocyte developmental competence [64, 66, 67]. However, a correlation between mtDNA quantity and reproductive outcomes has not been demonstrated by other investigators, hence the optimal amount of mtDNA required for normal oocyte development remains unclear [69].

Transmission of mtDNA in the Female Germ Line

In most mammalian species, mitochondria are inherited down the maternal line [70]. Rare cases of mixed biparental mitochondrial inheritance have been observed in several species and interspecies crosses, but this seems to be an exception from a general rule. Sperm mitochondria are eliminated within a few days after fertilization in two- to eight-cell stage embryos. This varies between the species but always occurs before activation of the embryonic genome [71]. It has been suggested that paternal mitochondria in the developing embryo are targeted by the proteolytic marker ubiquitin and are destroyed via a ubiquitin-dependent mechanism [72]. Although the exact reason for sperm mitochondria elimination remains unclear, it has been hypothesized that excessive production of reactive oxygen species (ROS) during spermatogenesis damages sperm mtDNA, which is subsequently sacrificed and removed from the embryo [72]. On contrast, oocyte mitochondria are bioenergetically less active, which presumably protects oocyte mtDNA and allows transmission to subsequent generations [60]. It has been also proposed that elimination of paternal mitochondrial material is an evolutionary mechanism to avoid lethal genome conflict between these essential organelles in the developing embryo [73].

MtDNA Replication and Transcription

During oogenesis, mitochondria undergo a dramatic increase in number, from several hundred in the PGC to approximately 10,000 in the primary oocyte and the hundreds of thousands in the mature oocyte [52]. Throughout the transmission of mtDNA molecules along the cell line, only a subset of them shift from PGC to the primary oocyte, followed by rapid replication of these templates. As a result, certain sequence variants populate the maturing oocyte with potential transmission to future progenies. Such a phenomenon of selected mtDNA segregation, called "the mitochondrial genetic bottleneck," is likely implicated in transgenerational transmission of mitochondrial mutations, although the underlying mechanism and a biological rationale for the process are not entirely understood.

MtDNA replication occurs by the formation of a displacement loop (D-loop) and starts at the origin of replication on heavy strand (H-strand). The leading strand is then synthetized by the DNA-dependent RNA synthesis. The synthesis of the other strand, the light strand (L-strand), starts when the replication of the H-strand reaches two-thirds of the molecule length and uncovers the L-strand origin of replication.

Mitochondrial replication and transcription rely on interplay between the nuclear and mitochondrial genomes. The mitochondrial RNA polymerase and transcription factor TFAM are responsible for maintenance of the mtDNA copy number in mammals, and are regulated by the pathways involving PGC-1α, the mitochondrial transcription specificity factors TFB1M and TFB2M, and the nuclear respiratory factors NRF-1 and NRF-2. Other factors and enzymes—mitochondrial DNA polymerase γ (POLG), mitochondrial single-stranded DNA-binding protein (mtSSBP), and TWINKLE, a mitochondrial-specific helicase—are also important players in the mitochondrial replication process [74]. MtDNA replication is temporary inactivated in the grown oocyte and resumes at postimplantation [75].

Thus, the mitochondrial content of the mature oocyte must be sufficient to meet the energy requirements of fertilization and early embryonic development until mitochondrial replication is reinitiated. The mitochondrial genome, however, becomes transcriptionally active in the two-cell preimplantation embryo, and the precise mechanisms responsible for activation and silencing of mitochondrial replication and transcription are not entirely understood. It is clear, however, that genomic silencing in mitochondria concomitantly with the increased energy requirements make the oocyte more susceptible to mitochondrial dysfunction, particularly if mitochondria are abnormally distributed during cell division.

Mitochondrial Distribution and Dynamics

Mitochondria show a dynamic morphology and form complex interconnected networks [76]. Frequent fission and fusion events remodel mitochondrial interconnections within the cell, which seems to be important for the maintenance of mitochondrial bioenergetics and mitochondrial distribution [77].

Homogeneousness and heterogeneousness describe two different patterns of mitochondrial distribution in the oocyte cytoplasm. Homogeneous distribution of mitochondria is commonly observed at the germinal vesicle stage [78]. Heterogeneous distribution occurs during meiosis in the metaphase I and II oocyte when mitochondria aggregate around the nucleus [79]. Homogenously distributed mitochondria are dominate in the later stages of oocyte maturation, which indicates poor oocyte and embryo quality. This suggests that the distribution pattern of mitochondria may be used to estimate cytoplasmic maturation [65, 68].

Only a limited number of mtDNA copies is transmitted from PGCs to primary oocytes. Asymmetric division of the oocyte during meiosis results in a dramatic reorganization of mitochondria during the first and second meiotic division. Upon initiation of meiosis I, mitochondria accumulate around the MI spindle and accompany the spindle during its migration with uneven segregation into the daughter cells. In mice, mitochondria undergo several cycles of dispersal and aggregation until extrusion of the first polar body [80]. Analogously, mitochondria accumulate around the MII spindle during the second meiotic division and undergo redistribution with extrusion of the second polar body [81]. Mitochondrial redistribution has a crucial impact on mitochondrial content of the oocyte [82]. Abnormal distribution of mitochondria during the first or second meiosis or at the pronucleus stage has been linked to a reduced capacity for embryo development, suggesting that mitochondrial redistributions are a feature of oocyte competency and successful early embryogenesis [83–85]. The precise mechanism by which mitochondrial redistributions contribute to developmental competence remains unclear. It has been proposed that the aggregation of mitochondria in specific intracellular locations concentrates ATP supply or Ca^{2+} buffering in areas of localized high demand [86]. This supposedly reduces mitochondrial activity, and in turn ROS production, while maintaining sufficiently high energy production. However, experimental data in oocytes to support this assumption are lacking.

Mitochondrial redistribution within the oocyte is mediated by the microtubule network of cytoskeleton and the motor proteins dynein and kinesin that influence mitochondrial movement to the inner cytoplasm [79, 86–88]. Inappropriate formation of the cytoplasmic microtubules can lead to the arrest of oocyte development or deranged embryogenesis due to the abnormal ATP distribution [83]. Aberrations in cytoplasmic microtubule organization and mitochondrial transport have been observed in in vitro matured oocytes, which possibly explains the lower developmental capacity of the in vitro matured oocytes compared to those matured in vivo [87].

Mitochondrial distribution is also regulated by the proteins involved in the process of fusion (Drp1 and Fis1) and fission (Mfn1, Mfn2, and Opa1) that ensure mitochondrial motility. Mouse knockout models have demonstrated that mutations in these genes result in embryonic arrest [89–91]. The mitochondrial adapter proteins Trak and Miro, a novel family of Rho GTPases, have been recently demonstrated to play an important role in establishing mitochondria redistribution in mice oocytes by interacting with microtubular components [92].

Mitochondrial Aberrations Leading to Disease

Mitochondrial DNA lacks protective histones and has a limited DNA repair capacity [93]. The proximity of mtDNA to the site of ROS production, and the relative inefficiency of the mtDNA repair mechanisms, make mtDNA more susceptible to DNA damage, resulting in mutations and genomic instability. Although the mutation rate is highly variable across the mitochondrial genome, it is at least two orders of magnitude higher than that for nDNA. Accumulation of mtDNA mutations may eventually lead to apoptosis, progression of several diseases, or in organismal aging [94]. The majority of the mitochondrial proteins are encoded by nuclear genes. Therefore, mitochondrial dysfunction leading to diseases can also be caused by mutations in nDNA [95].

The type and severity of mitochondrial diseases are related to mutation type and mutation load in the cell, which determine the ATP amount generated by mitochondria and the ability to meet the energy demand of a cell [96]. Mitochondrial diseases can present at childhood or adulthood and predominantly affect cells with high-energy demand such as skeletal muscles, kidneys, the heart, and the central nervous system [97]. Point mutations of mtDNA and deletions of large-scale mtDNA in any of the mitochondria-encoding genes can cause mtDNA disease, resulting in different phenotypes from nonsyndromic sensorineural deafness to serious systemic conditions such as opthalmoplegeia (CPEO), Kearns-Sayre Syndrome, and Pearson's marrow-pancreas syndrome [98]. Nuclear mitochondrial diseases are caused by nDNA mutations in the genes encoding for the components of OXPHOS energy-generating machinery in mitochondria. These mutations are commonly inherited in the autosomal pattern and often manifest early in life, for example cardiomyopathy and Leigh syndrome [99]. In addition, mitochondrial dysfunction can affect spindle formation and the segregation of chromosomes during maturation, leading to numerical chromosomal abnormalities from nondisjunction events [100]. This is also one of the proposed mechanisms for age-related aneuploidy or for oocyte damage caused by excessive oxidative stress [101].

SUMMARY

The oocyte is one of the long-living cells in the human body that develops through a remarkably coordinated process from differentiation in the fetal ovary to fertilization decades later. Oogenesis, folliculogenesis, and oocyte maturation involve complex interactions between oocytes and their surrounding somatic cells and are driven by interplay between cytoplasmic and nuclear compartments of the oocyte.

Over the last several decades, substantial progress has been made in elucidating the molecular mechanisms governing oocyte growth and maturation. The crucial roles of a large number of proteins expressed during oocyte development have been confirmed through the evaluation of genetic products of these proteins by using knockout approaches or targeted deletions. Nonetheless, despite fundamental discoveries made in understanding oocyte biology, many questions remain. Regulation of the checkpoints that ensure flawless progress of the oocyte through development and acquisition by oocyte developmental competence are just coming into focus. Owing to the scarcity of the human oocyte and preimplantation embryos for research, the evidence is mainly derived from animal studies, of which the translational value to human application is not entirely clear. Impressive advances in molecular biology techniques, including proteomic approaches, genome-wide profiling, and single-cell high resolution analyses, are expected to generate exciting discoveries. Achieving a more integrated understanding of the regulatory processes involved in follicle development and oocyte maturation will hopefully lead to refinement of the methodologies for in vitro fertilization techniques and for developing new treatment modalities for patients with reproductive disorders.

Acknowledgments

We thank Dr. Victoria Nisenblat for critical reading and editing of the manuscript. This work was supported by National Natural Science Funds of China (81521002, 31522034).

References

[1] Shibuya H, Watanabe Y. The meiosis-specific modification of mammalian telomeres. Cell Cycle 2014; 13(13):2024–8.

[2] de Souza GB, et al. Bovine ovarian stem cells differentiate into germ cells and oocyte-like structures after culture in vitro. Reprod Domest Anim 2017;52(2):243–50.

[3] Sánchez F, Smitz J. Molecular control of oogenesis. Biochim Biophys Acta (BBA) Mol Basis Dis 2012; 1822(12):1896–912.

[4] Irie N, et al. SOX17 is a critical specifier of human primordial germ cell fate. Cell 2015;160(1–2):253–68.

[5] Bowles J, Koopman P. Sex determination in mammalian germ cells: extrinsic versus intrinsic factors. Reproduction 2010;139(6):943–58.

[6] Pepling ME. From primordial germ cell to primordial follicle: mammalian female germ cell development. Genesis 2006;44(12):622–32.

[7] Elkouby YM. All in one-integrating cell polarity, meiosis, mitosis and mechanical forces in early oocyte differentiation in vertebrates. Int J Dev Biol 2017;61(3–4–5):179–93.

[8] Baltus AE, et al. In germ cells of mouse embryonic ovaries, the decision to enter meiosis precedes premeiotic DNA replication. Nat Genet 2006;38(12):1430.

[9] He J, et al. A developmental stage-specific switch from DAZL to BOLL occurs during fetal oogenesis in humans, but not mice. PLoS ONE 2013;8(9):e73996.

[10] Li J, et al. Activation of dormant ovarian follicles to generate mature eggs. Proc Natl Acad Sci 2010; 107(22):10280–4.

[11] Adhikari D, Liu K. mTOR signaling in the control of activation of primordial follicles. Cell Cycle 2010; 9(9):1673–4.

[12] Gazdag E, et al. TBP2 is essential for germ cell development by regulating transcription and chromatin condensation in the oocyte. Genes Dev 2009;23(18):2210–23.

[13] MacLennan M, et al. Oocyte development, meiosis and aneuploidy. Semin Cell Dev Biol 2015;45:68–76.

[14] Li L, et al. Single-cell RNA-Seq analysis maps development of human germline cells and gonadal niche interactions. Cell Stem Cell 2017;20(6):858–873.e4.

[15] Su Y, et al. Oocyte regulation of metabolic cooperativity between mouse cumulus cells and oocytes: BMP15 and GDF9 control cholesterol biosynthesis in cumulus cells. Development 2008;135(1):111–21.

[16] Thomas FH, Vanderhyden BC. Oocyte-granulosa cell interactions during mouse follicular development: regulation of kit ligand expression and its role in oocyte growth. Reprod Biol Endocrinol 2006;4(1):19.

[17] Petronczki M, Siomos MF, Nasmyth K. Un ménage à quatre: the molecular biology of chromosome segregation in meiosis. Cell 2003;112(4):423.

[18] Gleyzer N, Vercauteren K, Scarpulla RC. Control of mitochondrial transcription specificity factors (TFB1M and TFB2M) by nuclear respiratory factors (NRF-1 and NRF-2) and PGC-1 family coactivators. Mol Cell Biol 2005; 25(4):1354–66.

[19] Kops GJ, Weaver BA, Cleveland DW. On the road to cancer: aneuploidy and the mitotic checkpoint. Nat Rev Cancer 2004;5(10):773.

[20] Kalatova B, et al. Tripolar mitosis in human cells and embryos: occurrence, pathophysiology and medical implications. Acta Histochem 2015;117(1):111–25.

[21] Hunt PA, Hassold TJ. Human female meiosis: what makes a good egg go bad? Trends Genet 2008;24(2):86–93.

[22] Kuliev A, et al. Meiosis errors in over 20,000 oocytes studied in the practice of preimplantation aneuploidy testing. Reprod BioMed Online 2011;22(1):2–8.

[23] Eichenlaub-Ritter U, et al. Exposure of mouse oocytes to bisphenol A causes meiotic arrest but not aneuploidy. Mutat Res Genet Toxicol Environ Mutagen 2008;651(1):82–92.

[24] Hassold T, Hunt P. To err (meiotically) is human: the genesis of human aneuploidy. Nat Rev Genet 2001;2(4):280.

[25] Herbert M, et al. Meiosis and maternal aging: insights from aneuploid oocytes and trisomy births. Cold Spring Harb Perspect Biol 2015;7(4)a017970.

[26] De Souza E, Alberman E, Morris JK. Down syndrome and paternal age, a new analysis of case-control data collected in the 1960s. Am J Med Genet A 2009;149(6):1205–8.

[27] Hiraoka Y, Dernburg AF. The SUN rises on meiotic chromosome dynamics. Dev Cell 2009;17(5):598–605.

[28] Starr DA, Fridolfsson HN. Interactions between nuclei and the cytoskeleton are mediated by SUN-KASH nuclear-envelope bridges. Annu Rev Cell Dev Biol 2010;26:421–44.

[29] Mikolcevic P, et al. Essential role of the Cdk2 activator RingoA in meiotic telomere tethering to the nuclear envelope. Nat Commun 2016;7:11084.

[30] Viera A, et al. CDK2 regulates nuclear envelope protein dynamics and telomere attachment in mouse meiotic prophase. J Cell Sci 2015;128(1):88–99.

[31] Shibuya H, Ishiguro K, Watanabe Y. The TRF1-binding protein TERB1 promotes chromosome movement and telomere rigidity in meiosis. Nat Cell Biol 2014;16(2):145–56.

[32] Scherthan H. A bouquet makes ends meet. Nat Rev Mol Cell Biol 2001;2(8):621–7.

[33] Conrad MN, et al. MPS3 mediates meiotic bouquet formation in *Saccharomyces cerevisiae*. Proc Natl Acad Sci U S A 2007;104(21):8863–8.

[34] Keeney S, Giroux CN, Kleckner N. Meiosis-specific DNA double-strand breaks are catalyzed by Spo11, a member of a widely conserved protein family. Cell 1997;88(3):375–84.

[35] Munoz-Fuentes V, Di Rienzo A, Vila C. Prdm9, a major determinant of meiotic recombination hotspots, is not functional in dogs and their wild relatives, wolves and coyotes. PLoS ONE 2011;6(11):e25498.

[36] Shibuya H, et al. MAJIN links telomeric DNA to the nuclear membrane by exchanging telomere cap. Cell 2015;163(5):1252–66.

[37] Hou Y, et al. Genome analyses of single human oocytes. Cell 2013;155(7):1492–506.

[38] Kaltschmidt JA, et al. Rotation and asymmetry of the mitotic spindle direct asymmetric cell division in the developing central nervous system. Nat Cell Biol 2000;2(1):7–12.

[39] Antonio C, et al. Xkid, a chromokinesin required for chromosome alignment on the metaphase plate. Cell 2000;102(4):425–35.

[40] Vogt E, et al. Spindle formation, chromosome segregation and the spindle checkpoint in mammalian oocytes and susceptibility to meiotic error. Mutat Res 2008;651(1–2):14–29.

[41] Sun SC, Kim NH. Spindle assembly checkpoint and its regulators in meiosis. Hum Reprod Update 2012; 18(1):60–72.

[42] London N, Biggins S. Signalling dynamics in the spindle checkpoint response. Nat Rev Mol Cell Biol 2014; 15(11):736–47.

[43] Musacchio A, Salmon ED. The spindle-assembly checkpoint in space and time. Nat Rev Mol Cell Biol 2007; 8(5):379–93.

[44] Peters JM. The anaphase promoting complex/cyclosome: a machine designed to destroy. Nat Rev Mol Cell Biol 2006;7(9):644–56.

[45] Lara-Gonzalez P, Taylor SS. Cohesion fatigue explains why pharmacological inhibition of the APC/C induces a spindle checkpoint-dependent mitotic arrest. PLoS ONE 2012;7(11):e49041.

[46] Arguello-Miranda O, et al. Casein kinase 1 coordinates cohesin cleavage, gametogenesis, and exit from M phase in meiosis II. Dev Cell 2017;40(1):37–52.

[47] Menasha J, et al. Incidence and spectrum of chromosome abnormalities in spontaneous abortions: new insights from a 12-year study. Genet Med 2005;7(4):251–63.

[48] Nagaoka SI, Hassold TJ, Hunt PA. Human aneuploidy: mechanisms and new insights into an age-old problem. Nat Rev Genet 2012;13(7):493–504.

[49] Nagaoka SI, et al. Oocyte-specific differences in cell-cycle control create an innate susceptibility to meiotic errors. Curr Biol 2011;21(8):651–7.

[50] Hayashi A, et al. Reconstruction of the kinetochore during meiosis in fission yeast *Schizosaccharomyces pombe*. Mol Biol Cell 2006;17(12):5173–84.

[51] Ott M, et al. Mitochondria, oxidative stress and cell death. Apoptosis 2007;12(5):913–22.

[52] Sathananthan AH, Trounson AO. Mitochondrial morphology during preimplantational human embryogenesis. Hum Reprod 2000;15(Suppl. 2):148–59.

[53] Thouas GA, et al. Mitochondrial dysfunction in mouse oocytes results in preimplantation embryo arrest in vitro. Biol Reprod 2004;71(6):1936–42.

[54] Krisher RL. The effect of oocyte quality on development. J Anim Sci 2004;82(Suppl. 13):E14–23.

[55] Whitaker M. Calcium signalling in early embryos. Philos Trans R Soc Lond B Biol Sci 2008;363(1495):1401–18.

[56] Van Blerkom J. Mitochondria as regulatory forces in oocytes, preimplantation embryos and stem cells. Reprod BioMed Online 2008;16(4):553–69.

[57] Van Blerkom J. Mitochondria in human oogenesis and preimplantation embryogenesis: engines of metabolism, ionic regulation and developmental competence. Reproduction 2004;128(3):269–80.

[58] Tilly JL, Sinclair DA. Germline energetics, aging, and female infertility. Cell Metab 2013;17(6):838–50.

[59] Motta PM, et al. Mitochondrial morphology in human fetal and adult female germ cells. Hum Reprod 2000; 15(Suppl. 2):129–47.

[60] Leese HJ, et al. Metabolism of the viable mammalian embryo: quietness revisited. Mol Hum Reprod 2008; 14(12):667–72.

[61] Schatten H, Sun Q, Prather R. The impact of mitochondrial function/dysfunction on IVF and new treatment possibilities for infertility. Reprod Biol Endocrinol 2014;12(1):111.

[62] Meisinger C, Sickmann A, Pfanner N. The mitochondrial proteome: from inventory to function. Cell 2008;134(1):22–4.

[63] Ma H, et al. Incompatibility between nuclear and mitochondrial genomes contributes to an interspecies reproductive barrier. Cell Metab 2016;24(2):283–94.

[64] Zhang C, Montooth KL, Calvi BR. Incompatibility between mitochondrial and nuclear genomes during oogenesis results in ovarian failure and embryonic lethality. Development 2017;144(13):2490–503.

[65] Pikó L, Taylor KD. Amounts of mitochondrial DNA and abundance of some mitochondrial gene transcripts in early mouse embryos. Dev Biol 1987;123(2):364–74.

[66] Reynier P, et al. Mitochondrial DNA content affects the fertilizability of human oocytes. Mol Hum Reprod 2001;7(5):425–9.

[67] Santos TA, El Shourbagy S, John JCS. Mitochondrial content reflects oocyte variability and fertilization outcome. Fertil Steril 2006;85(3):584–91.

[68] López-Lluch G, et al. Mitochondrial biogenesis and healthy aging. Exp Gerontol 2008;43(9):813–9.

[69] Barritt JA, et al. Quantification of human ooplasmic mitochondria. Reprod BioMed Online 2002;4(3):243–7.

[70] Giles RE, et al. Maternal inheritance of human mitochondrial DNA. Proc Natl Acad Sci 1980;77(11):6715–9.

[71] St. John JC, et al. Mitochondrial DNA transmission, replication and inheritance: a journey from the gamete through the embryo and into offspring and embryonic stem cells. Hum Reprod Update 2010;16(5):488–509.

[72] Sutovsky P, et al. Degradation of paternal mitochondria after fertilization: implications for heteroplasmy, assisted reproductive technologies and mtDNA inheritance. Reprod BioMed Online 2004;8(1):24–33.

[73] Hurst LD. Intragenomic conflict as an evolutionary force. Proc Biol Sci 1992;248(1322):135–40.

[74] Montier LLC, Deng JJ, Bai Y. Number matters: control of mammalian mitochondrial DNA copy number. J Genet Genomics 2009;36(3):125–31.

[75] Shoubridge EA, Wai T. Mitochondrial DNA and the mammalian oocyte. Curr Top Dev Biol 2007;77:87–111.

[76] Bereiter-Hahn J, et al. Structural implications of mitochondrial dynamics. Biotechnol J 2008;3(6):765–80.

[77] Westermann B. Mitochondrial fusion and fission in cell life and death. Nat Rev Mol Cell Biol 2010;11(12):872–84.

[78] Nishi Y, et al. Change of the mitochondrial distribution in mouse ooplasm during in vitro maturation. J Nippon Med Sch 2003;70(5):408–15.

[79] Torner H, et al. Mitochondrial aggregation patterns and activity in porcine oocytes and apoptosis in surrounding cumulus cells depends on the stage of pre-ovulatory maturation. Theriogenology 2004; 61(9):1675–89.

[80] Yu Y, et al. Redistribution of mitochondria leads to bursts of ATP production during spontaneous mouse oocyte maturation. J Cell Physiol 2010;224(3):672–80.

[81] Calarco PG. Polarization of mitochondria in the unfertilized mouse oocyte. Dev Genet 1995;16(1):36–43.

[82] Wang LY, et al. Mitochondrial functions on oocytes and preimplantation embryos. J Zhejiang Univ Sci B 2009; 10(7):483–92.

[83] Nagai S, et al. Correlation of abnormal mitochondrial distribution in mouse oocytes with reduced developmental competence. Tohoku J Exp Med 2006;210(2):137–44.

[84] Van Blerkom J, Davis P, Alexander S. Differential mitochondrial distribution in human pronuclear embryos leads to disproportionate inheritance between blastomeres: relationship to microtubular organization, ATP content and competence. Hum Reprod 2000;15(12):2621–33.

[85] Wilding M, et al. Mitochondrial aggregation patterns and activity in human oocytes and preimplantation embryos. Hum Reprod 2001;16(5):909–17.

[86] Sun QY, et al. Translocation of active mitochondria during pig oocyte maturation, fertilization and early embryo development in vitro. Reproduction 2001;122(1):155–63.

[87] Brevini TA, et al. Role of adenosine triphosphate, active mitochondria, and microtubules in the acquisition of developmental competence of parthenogenetically activated pig oocytes. Biol Reprod 2005;72(5):1218–23.

[88] Liu S, et al. Dynamic modulation of cytoskeleton during in vitro maturation in human oocytes. Am J Obstet Gynecol 2010;203(2):151.e1–7.

[89] Chen H, et al. Mitofusins Mfn1 and Mfn2 coordinately regulate mitochondrial fusion and are essential for embryonic development. J Cell Biol 2003;160(2):189–200.

[90] Davies VJ, et al. Opa1 deficiency in a mouse model of autosomal dominant optic atrophy impairs mitochondrial morphology, optic nerve structure and visual function. Hum Mol Genet 2007;16(11):1307–18.

[91] Wakabayashi J, et al. The dynamin-related GTPase Drp1 is required for embryonic and brain development in mice. J Cell Biol 2009;186(6):805–16.

[92] Dalton C. Mitochondrial dynamics during mouse oocyte maturation. UCL (University College London); 2011.

[93] Kujoth GC, et al. Mitochondrial DNA mutations, oxidative stress, and apoptosis in mammalian aging. Science 2005;309(5733):481–4.

[94] Chan DC. Mitochondria: dynamic organelles in disease, aging, and development. Cell 2006;125(7):1241–52.

[95] Alston CL, et al. The genetics and pathology of mitochondrial disease. J Pathol 2017;241(2):236–50.

[96] Johns DR. Seminars in medicine of the Beth Israel Hospital, Boston. Mitochondrial DNA and disease. N Engl J Med 1995;333(10):638–44.

[97] McFarland R, Taylor RW, Turnbull DM. Mitochondrial disease–its impact, etiology, and pathology. Curr Top Dev Biol 2007;77:113–55.

[98] Haas RH, et al. Mitochondrial disease: a practical approach for primary care physicians. Pediatrics 2007; 120(6):1326–33.

[99] Holt IJ, Harding AE, Morgan-Hughes JA. Deletions of muscle mitochondrial DNA in patients with mitochondrial myopathies. Nature 1988;331(6158):717–9.

[100] Eichenlaub-Ritter U. Genetics of oocyte ageing. Maturitas 1998;30(2):143–69.

[101] Zhang X, et al. Deficit of mitochondria-derived ATP during oxidative stress impairs mouse MII oocyte spindles. Cell Res 2006;16(10):841–50.

Epigenetic Control of Oocyte Development

Maria M. Viveiros*,†, Rabindranath De La Fuente*,†

*Department of Physiology and Pharmacology, College of Veterinary Medicine, University of Georgia, Athens, GA, United States †Regenerative Biosciences Center (RBC), University of Georgia, Athens, GA, United States

INTRODUCTION

The establishment of epigenetic modifications in the oocyte genome occurs during a critical window of postnatal oocyte development. Oocyte growth and differentiation ensue from a limited pool of primordial follicles established shortly after birth that are maintained in a protracted meiotic arrest at the diplotene stage [1, 2]. The duration of meiotic arrest varies according to the species. For example, human primordial oocytes may require up to 4 months to differentiate into fully grown preovulatory oocytes. In contrast, mouse oocytes require only 3 weeks to complete this process [3]. During this period, mouse oocytes will undergo an approximate 300-fold increase in volume [1, 2]. Oocyte growth takes place within the ovarian follicle and is strictly dependent on the establishment of a complex bidirectional communication with cumulus granulosa cells. This communication is essential for the synthesis and storage of maternal mRNA, the accumulation of cell cycle regulatory proteins as well as cytoplasmic organelles, and formation of the zona pellucida [3, 4]. Granulosa cells are known to modulate the process of global transcription and chromatin remodeling in the oocyte genome [5, 6], play an active role in the transport of RNA molecules to the oocyte through the establishment of functional gap junctions [7], and to regulate the translation of maternal mRNA stores [8]. The signaling pathways by which granulosa cells modulate transcription and chromatin remodeling in the oocyte genome are not clear at present. However, functional differentiation of chromatin structure through the establishment of epigenetic modifications is critical for the control of transcription, the establishment of maternal-specific DNA methylation marks or genomic imprinting, and the maintenance of chromosome stability. Therefore, regulation of the epigenetic landscape in the oocyte genome is essential for acquisition of both meiotic and developmental competence [9, 10].

EPIGENETIC MODIFICATIONS IN THE OOCYTE GENOME

Epigenetic maturation involves a series of chromatin modifications that affects the patterns of gene expression without exerting any change at the level of the DNA sequence. These changes can occur on a genome-wide basis as well as at specific chromosomal domains such as centromeres and telomeres, or may affect chromatin composition and structure at the single gene level [10]. Epigenetic modifications have been recently defined as a structural adaptation of chromosomal regions to register, signal, or perpetuate altered activity states. This definition involves a number of transient modifications that are essential for DNA repair or that may be unique to a specific cell cycle stage in addition to stable changes inherited through mitotic or meiotic cell division [11]. Thus, epigenetic modifications may induce transient or heritable changes in gene expression through DNA methylation; histone posttranslational modifications such as histone methylation, acetylation, phosphorylation, and poly (ADP) ribosylation; or the incorporation of histone variants [12]. Notably, elegant genome-wide histone methylation studies and DNA methylome analysis have recently revealed the presence of a growing list of epigenetic modifications that exhibit a germ cell-specific function in order to remodel the genome generation after generation, and to fulfill the unique requirements of gamete formation and accurate chromosome segregation during meiosis [10, 13–16].

ESTABLISHMENT OF DNA METHYLATION DURING OOCYTE GROWTH

During fetal development, primordial germ cells undergo a process of genomic reprogramming during which most DNA methylation marks from the previous generation are erased [17]. In female germ cells, the establishment of maternal-specific DNA methylation

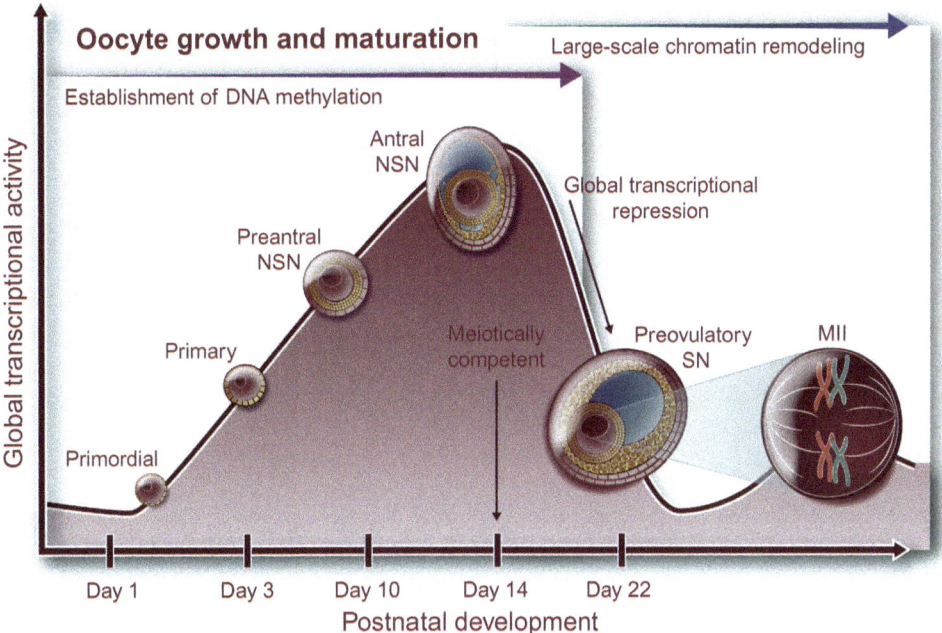

FIG. 8.1   Epigenetic maturation during oocyte growth. Global transcriptional activity increases during the early stages of oocyte growth, reaching a peak in oocytes from early antral follicles on days 14–16 of postnatal development. These oocytes exhibit a nonsurrounded nucleolus (NSN) configuration. Global transcriptional repression takes place on days 18–21 in preovulatory oocytes that exhibit the surrounded nucleolus (SN) configuration. Large-scale chromatin remodeling induces the SN configuration and is essential for meiotic centromere function as well as chromosome segregation. Establishment of DNA methylation takes place on a locus-by-locus basis beginning on day 10 of postnatal development.

marks or de novo methylation takes place during postnatal oocyte growth. This process begins when the oocyte reaches approximately 50 µM in diameter and is present within preantral follicles on day 10 of postnatal development (Fig. 8.1) [18–21]. De novo methylation takes place on a locus-by-locus basis and is completed in the fully grown preovulatory oocyte contained in the graafian follicle [19, 21]. This results in the establishment of a sex-specific mark or genomic imprint that is strictly required for normal preimplantation development [22–25].

Differences in the patterns of DNA methylation between the paternal and maternal genomes are critical to regulate allele-specific gene expression in the developing embryo and thus constitute the basis of genomic imprinting in mammals [26–28]. Imprinted genes are required for placental differentiation, regulation of fetal growth, postnatal development, and maternal behavior in mice [29–31]. Aberrant DNA methylation at specific imprinting control regions (ICRs) is associated with several human syndromes such as Beckwith-Wiedemann-Syndrome (BWS). BWS is characterized by increased birth weight and length, midline abdominal wall defects, and neonatal hypoglycemia; it arises as a consequence of demethylation of the *KvDMR1* region within the KCNQ1OT gene [32].

Notably, oocytes can mature and progress through meiosis as well as become fertilized in the absence of DNA methylation at maternally imprinted genes. However, the patterns of DNA methylation established during oogenesis play a critical role in the regulation of gene expression during subsequent embryo development. DNMT3A/DNMT3L knockout embryos die shortly after implantation due to the absence of maternal methylation imprints, abnormal allelic transcription, and impaired trophoblast development [33]. DNA methylation during oogenesis may also have a profound effect in regulating transgenerational epigenetic inheritance [33]. Also of clinical importance, assisted reproductive techniques such as embryo transfer and in vitro culture of oocytes and embryos may be associated with DNA methylation changes at several imprinted genes, such as H19, Snrpn, Igf2, Kcnq1ot1, Cdkn1c, Kcnq1, Mknr3, Ascl2, Zim1, and Peg3. This may pose an increased risk for imprinting disorders in the offspring [31, 34, 35].

The primary mechanisms regulating the establishment and maintenance of oocyte de novo methylation are not fully understood. However, functional interaction of the DNA methyltransferase Dnmt3a and the DNA methyltransferase-like (Dnmt3L) [26, 36] is essential for this process [18, 37–39]. Initiation of oocyte growth is sufficient to trigger transcriptional activation of the maintenance methylase Dnmt1 as well as Dnmt3L. Loading of Dnmt1 to the nucleus also seems to be an oocyte autonomous process mediated by stem cell factor-dependent stimulation of the phosphatidyl inositol 3 kinase (PI3K) pathway [40]. De novo methylation patterns at maternally imprinted loci are established by the functional interaction of Dnmt3a with Dnmt3L in the growing oocyte [41] where Dnmt3L is essential for directing the targeting of Dnmt3a2 to specific imprinted control regions (ICRs) [41, 42]. ICRs or differentially methylated regions (DMRs) are located within the 5′ flanking regions of imprinted genes. Both histone modifications as well as transcription at these regulatory sequences dictate the placing and timing of methylation marks [34, 43].

The mechanism for specific targeting of this protein complex to ICRs at specific loci is not clear. However, recent studies elegantly demonstrated that the patterns of histone methylation at ICRs might be critical for this process. Indeed, the first example of a functional relationship between histone methylation and DNA methylation during oogenesis was demonstrated after the analysis of methylation patterns in the oocytes of mice deficient for the histone demethylase (KDM1). Functional ablation of KDM1 in growing oocytes results in overexpression of histone H3 di-methylation at lysine 4 (H3K4me2) and disrupts the establishment of maternal imprints in four out of seven imprinted loci. This indicates that removal of the H3K4Me2 by KDM1 is required for the establishment of proper DNA methylation patterns of maternally imprinted genes [44]. The precise role of KDM1B in this process remains to be determined. However, it has been suggested that KDM1B may induce de-methylation of histone H3 at lysine 4 (H3K4) to generate an open chromatin configuration that facilitates the recruitment of the DNA methylation machinery to specific imprinted loci [44]. Additional factors, such as the KRAB zinc finger protein Zfp57, may exhibit a direct interaction with DNA to establish the maternal imprinting signal at the Snrpn imprinted region. This may occur through a potential interaction with the chromatin binding protein KAP1 (also known as TRIM28), which is known to facilitate an interaction with the histone methyltransferase SETDB1 and DNMT3A [45–47], although the mechanisms of the Zfp57 function are not fully understood at present [45].

Perhaps one of the most intriguing advancements toward unraveling the mechanisms responsible for the establishment of maternal imprinting is the potential role that transcription across imprinted control regions may play in the establishment and maintenance of de novo methylation during oogenesis [48]. In these studies, transcripts originating from promoters within a differentially methylated region (DMR) of the GNAS locus as well as histone modifications were required for the establishment of maternal imprints [48]. Whether this is a universal mechanism that is functional at all known ICRs or whether different mechanisms operate at the specific ICR of different maternally imprinted genes is not currently known. Genome-wide chromatin immunoprecipitation (ChIP-seq) provided additional evidence that H3K4me2 interferes with DNA methylation and must first be removed by KDM1B to establish DNA methylation at oocyte gDMRs, while permissive H3K36me3 marks accumulate in growing oocytes as de novo methylation is established [49]. This model implies that different histone modifications are likely required at different genomic locations [33, 48, 50]. The precise mechanisms by which transcription regulates chromatin remodeling remain to be determined. The same applies to the functional interactions of several chromatin-binding proteins that are critical for the maintenance of DNA methylation patterns in oocytes, such as the maintenance DNA methylase UHRF1 [51], Sall4 [52], a zinc finger protein, the transcription factor TRIM28 [46], and the histone H3.3 chaperone HIRA [53].

Powerful strategies such as genome-wide DNA methylation at single nucleotide resolution have begun to shed important mechanistic insight into this critical process [21]. Current studies indicate that both transcription and histone modifications across germ line differentially methylated regions (gDMRs) regulate the placing and time of establishment of de novo methylation marks at defined loci [33, 48, 50]. Importantly, the effect of transcription on the establishment of DNA methylation seems to be unique to the oocyte [33]. Genome-wide DNA methylome analysis also revealed striking new evidence for the presence of unique patterns of DNA methylation that extend beyond imprinted genes in the oocyte genome, consistent with the presence of germ-cell specific mechanisms for epigenetic regulation [43, 49, 53, 54]. For example, in contrast to somatic cells, almost two-thirds of all DNA methylation occurs in a non-CpG context. Instead, it is enriched at gene bodies, resulting in the formation of large-scale hypermethylated and hypomethylated domains in which transcription regulates the establishment of almost 90% of the oocyte methylome. Thus, DNA methylation occurs predominantly at expressed gene bodies in fully grown oocytes [50, 55, 56]. Importantly, whereas promoter DNA methylation has been negatively associated with gene expression in many cell types, gene body methylation is positively correlated with gene expression in the mammalian oocyte genome [50, 55, 56].

Another intriguing aspect of the oocyte genome is the recently described broad genomic distribution patterns for histone H3 tri-methylated at lysine 4 (H3K4me3). For example, in ES cells and somatic cells, H3K4me3 is preferentially enriched at transcription start sites of active genes, where it is detected as a narrow high peak of H3K4me3 following ChIP-seq analysis. By contrast, in fully grown oocytes as well as during the oocyte to embryo transition, the H3K4me3 mark is present in large genomic regions spanning >10,000 base pairs and is found distant from the transcription start sites. This unique pattern has been termed broad or noncanonical H3K4me3 to emphasize its difference with the clearly delineated peaks of H3K4me3 associated with transcription start sites in somatic and ES cells [57–59].

The mechanisms regulating broad H3K4me3 patterns and their effects on transcription or DNA methylation at specific compartments in the oocyte genome are not fully understood. However, oocyte-conditional deletion of the histone methyltransferase (MLL2), which is known to regulate the nuclear localization of H3K4me3 [60], revealed that MLL2 may regulate the broad distribution patterns of H3K4me3 over large genomic regions in fully grown oocytes [61]. Importantly, H3K4me3 in the nongrowing oocyte is restricted to transcribed gene promoters and is dependent on transcription. In the fully grown oocyte, the accumulation of H3K4me3 becomes independent of transcription, spreading throughout thousands of bivalent domains and broad distal domains. Only the noncanonical distribution of H3K4me3 is regulated by MLL2 function. This suggests the presence of two independent mechanisms of H3K4me3 deposition in the oocyte genome, with MLL2 specifically targeting unmethylated CpG rich regions in a transcription independent manner [61]. Additional factors may be involved in the regulation of broad H3K4me3 patterns as the zinc finger protein SALL4 has also been recently described to affect H3K4me3 patterns during oocyte growth by inducing and overexpressing the histone de-methylase KDM5B with a subsequent reduction of H3K4me3 levels [52].

Two intriguing observations provide potential clues regarding the biological significance of broad H3K4me3 genome deposition. First, overexpression of KDM5B reduced broad H3K4me3 distribution and reactivated global transcription in preovulatory oocytes [57], which are known to silence transcription in preparation for meiosis [10]. Loss of MLL2 also prevented transcriptional repression in the oocyte, suggesting that MLL2-dependent deposition of broad H3K4me3 domains may be required for global transcriptional silencing [60, 61]. Intriguingly, removal of broad H3K4me3 domains is also required for normal zygotic genome activation, suggesting that this may be an important epigenetic mechanism for regulation of global transcriptional activity during the oocyte to embryo transition [57–59].

HISTONE MODIFICATIONS DURING OOCYTE GROWTH

The mammalian oocyte nucleus is subject to dynamic chromatin modifications at several developmental transitions in preparation for meiosis. Histone deposition and its posttranslational modifications directly regulate chromatin structure and function [62]. Either by changing the primary organization of the nucleosome or by inducing changes in large-scale chromatin remodeling of entire chromosomal domains, histone posttranslational modifications have the potential to regulate chromatin condensation and establish either a transcriptionally permissive or repressive chromatin environment [63, 64]. The levels of both Histone H3 and H4 methylation as well as H3 and H4 acetylation increase during oocyte growth, concomitantly with the levels of expression of their regulatory histone methyl transferases and histone acetyl transferases [65]. Yet, we have only a limited mechanistic understanding concerning the functional interaction of most of these modifications in the oocyte genome. The mechanisms regulating this "meiotic histone code" are only beginning to be unraveled through the generation of transgenic mouse models with oocyte conditional deletions and the use of genome-wide transcriptome and chromatin immunoprecipitation (ChIP-seq) analyses.

These are powerful strategies that will accelerate our understanding of how the meiotic histone code or changes in large-scale chromatin structure regulate genome function during oogenesis and the maternal inheritance of epigenetic states to the early embryo.

Chromatin structure during oogenesis is highly dynamic and requires continuous deposition of the histone variant H3.3 by its chaperone molecule HIRA, which is essential for chromatin homeostasis, transcriptional regulation, de novo methylation in the oocyte, and female fertility [53]. Depletion of HIRA in primordial oocytes causes extensive oocyte death due to the lack of H3/H4 deposition and the accumulation of DNA damage. The resulting chromatin displayed altered structure and increased DNAseI sensitivity [53]. Notably, a systematic analysis of genome-wide histone modifications using ChIP-sequencing revealed that histone modifications are established before the onset of de novo methylation and thus have a direct impact on the subsequent acquisition of oocyte-specific DNA methylation patterns [49]. Histone modifications shape the oocyte methylome by directing the methylation machinery to specific genomic regions. For example, histone methylation at lysine 36 (H3K36me3) can recruit DNMT3A in vitro, suggesting that a similar mechanism may function in vivo [33]. Notably, a new type of noncanonical imprinting mechanism mediated by the inheritance of oocyte-specific H3K27me3 was recently described. This mechanism is critical for the regulation of a subset of genes whose imprinting is regulated through a DNA methylation-independent mechanism [66]. Other histone modifications are essential to regulate entire chromosomal domains. For example, histone tri-methylation at lysine 9 (H3K9me3) plays a critical role in regulating the function of large-scale chromosomal domains such as centromeres and telomeres. Together with H4K20me3, this is essential for the formation of constitutive heterochromatin, the regulation of large scale chromatin remodeling, and the maintenance of chromosome stability [10].

LARGE-SCALE CHROMATIN REMODELING IN THE OOCYTE GENOME

Large-scale chromatin remodeling is defined as a series of genome-wide changes in nuclear architecture that take place in response to environmental or differentiation stimuli and can be recognized at the level of specific chromosomes or chromosomal domains [67]. Mammalian meiosis involves two sequential cell divisions without an intervening S-phase in order to reduce the number of chromosomes and generate mature haploid gametes. During oocyte meiosis-I, one set of homologous chromosomes is extruded into the first polar body and the oocyte arrests at metaphase-II. Subsequently, and in response to fertilization, sister chromatids become separated during meiosis-II, resulting in the formation of a second polar body [68, 69]. Unique structural and functional properties are required to orchestrate the accurate segregation of homologous chromosomes during meiosis-I as well as of sister chromatids during meiosis-II. For example, in addition to complex interchromosomal interactions, such as centromeric cohesion and formation of chiasmata, meiotic division also requires specialized spindle microtubule attachments with paired centromeres at homologous chromosomes during metaphase-I or with individualized centromeric domains of sister chromatids at metaphase-II [70, 71]. Therefore, mammalian oocytes undergo striking chromatin remodeling events in preparation for meiotic division with far-reaching implications for the individualization of chromosome bivalents and the maintenance of chromosome stability. Importantly, these chromatin-remodeling events are essential for the acquisition of both meiotic and developmental competence [9, 10, 72, 73].

Growing oocytes exhibit decondensed chromatin in a nonsurrounded nucleolus (NSN) configuration [74–76] and show high levels of transcriptional activity [5]. Subsequent oocyte growth and differentiation leads to a striking developmentally programmed nuclear re-organization with progressive chromatin condensation, and the formation of a perinucleolar heterochromatin rim described as the surrounded nucleolus (SN) configuration in fully grown preovulatory oocytes [5, 75, 77]. This configuration is present in both human and mouse oocytes [74, 75, 78]. Concomitantly with these large-scale chromatin-remodeling events, oocytes also undergo a remarkable transition from high levels of transcriptional activity in NSN oocytes to global transcriptional repression at the SN stage (Fig. 8.1). Both processes, global chromatin remodeling and transcriptional repression, are of critical importance for the timely progression of meiotic maturation [10, 74, 79, 80] and the acquisition of developmental potential. Notably, despite being temporally linked, large-scale chromatin remodeling and global transcriptional silencing in the oocyte genome are autonomous developmental processes under the control of distinct cellular pathways [9]. However, the specialized nuclear architecture acquired during the transition into the SN configuration is of direct functional significance for meiotic progression, centromere function, and accurate chromosome segregation [9, 10].

HETEROCHROMATIN FORMATION AND CHROMOSOME STABILITY DURING OOCYTE MEIOSIS

Pericentric heterochromatin is required to support the highly specialized centromeric interactions regulating homologous chromosome segregation during metaphase-I as well as the prevention of premature sister chromatid separation before the end of metaphase-II [70]. Mammalian centromeres consist of repetitive major satellite DNA sequences that are necessary but not sufficient for centromere formation, suggesting that centromeres are under epigenetic regulation [81, 82]. Tri-methylation of H3K9 is a constitutive component of pericentric heterochromatin required for centromere structure and function and the maintenance of a transcriptionally repressive chromatin environment. Importantly, H3K9me3 and H4K20me3 act as docking sites for additional chromatin binding proteins such as heterochromatin protein 1 (HP1) and the chromatin remodeling protein ATRX [83–87]. Centromere structure and function also entails the incorporation of histone variants, such as CENP-A, to induce the formation of a higher-order chromatin structure through histone deacetylation and large-scale chromatin remodeling [81, 88–91].

ATRX belongs to a large protein family of ATP-dependent chromatin-remodeling proteins and binds pericentric heterochromatin domains in human and mouse somatic cells. ATRX is essential for the establishment of proper DNA methylation patterns at repetitive sequences within the human genome [92–95]. In mouse oocytes, ATRX binds to pericentric heterochromatin of metaphase-I and metaphase-II chromosomes and is involved in mediating chromosome alignment at the meiotic spindle [96]. Pericentric heterochromatin integrity is essential to coordinate sister centromere cohesion during mitosis [97]. For example, chromatin remodeling complexes such as SNF2h are essential to load the cohesin subunit Rad21 onto the centromeres in human mitotic cells [98]. Moreover, loss of HP1 from pericentric heterochromatin in mouse somatic cells deficient for the SUV39 histone methyltransferase severely affects cohesion between sister chromatids [99, 100]. Thus, pericentric heterochromatin plays a major role in centromere cohesion in mitotic cells.

Loss of maternal ATRX function in mouse oocytes contributes to chromosome misalignment at the metaphase-II spindle and pericentromeric DNA breaks. This leads to severe meiotic centromere instability, the formation of aneuploid embryos, and female subfertility [73, 96, 101]. Interestingly, ATRX deficiency at the pericentric heterochromatin domains results in abnormal axial chromatid condensation and is associated with reduced levels of histone H3 phosphorylation (H3S10ph), known to be an epigenetic mark associated with proper chromosome condensation. In addition, loss of ATRX function results in failure to recruit the transcriptional regulator DAXX to pericentric heterochromatin at the GV stage [73]. Taken together, these findings support a crucial role of ATRX in the formation and molecular composition of pericentric heterochromatin in the oocyte genome, a process that is essential to mediate the complex chromatin remodeling events during meiosis in mammalian oocytes. Intriguingly, ATRX is a novel component of the epigenetic landscape at the inactive X chromosome in somatic cells and during imprinted X chromosome inactivation in trophoblast stem cells [102]. These findings suggest that ATRX is an important component of both facultative and constitutive heterochromatin and underscore a critical role for this chromatin-remodeling factor in heterochromatin formation in mammalian germ cells. Notably, the chromosomal defects observed in ATRX-deficient oocytes provide direct evidence that epigenetic modifications in mammalian oocytes are essential for the maintenance of chromosome stability at metaphase-II and during the oocyte-to-embryo transition [101]. Elucidating the molecular mechanisms through which ATRX regulates chromosome condensation and phosphorylation of histone H3 (H3S10Ph) at pericentric heterochromatin will be an essential step toward understanding critical epigenetic factors contributing to the onset of aneuploidy in the female gamete.

CHROMOSOME-MICROTUBULE INTERACTIONS AT THE KINETOCHORE

The fidelity of chromosome segregation during mitotic and meiotic division relies on correctly oriented and stable chromosome-microtubule attachments. Improper interactions can result in chromosome segregation errors that lead to aneuploidy in gametes and developing embryos upon fertilization [103]. Spindle microtubules bind to the chromosomes specifically at the kinetochore, a specialized region of the centromere [104]. During mitosis, stable bioriented (amphitelic) attachments occur when the kinetochore from each sister chromatid is bound to microtubules from opposing spindle poles. Disruptions in this process can result in erroneous attachments such that two kinetochores may bind to the same spindle pole (syntelic attachment), or a single kinetochore binds to both spindle poles (merotelic attachment). Similar attachment errors occur during meiotic division in mouse and human oocytes [103, 105, 106]. Moreover, chromosome-microtubule interactions are more complex during meiosis-I as homologous chromosome separation necessitates coorientation—the attachment and cosegregation of sister chromatids of each homologue to the same spindle pole. Despite these unique facets, our understanding of the mechanisms that regulate kinetochore-microtubule interactions during meiosis is based largely on somatic cell studies.

The kinetochore is a large multiprotein complex characterized by a trilayer structure. The inner kinetochore plate directly contacts centromeric heterochromatin while the outer plate and corona layer contain proteins required for microtubule interaction. Specific aspects of the centromere mark it as the exclusive site for kinetochore formation, including (i) essential epigenetic marks such as nucleosomes containing the histone H3 variant Centromere Protein A (CENP-A) that specify the centromere and are necessary for kinetochore assembly, and (ii) histone modifications that affect chromatin structure [83, 107, 108]. In contrast, satellite DNA sequences associated with the centromere are not required for kinetochore assembly [109, 110]. CENP-A forms a centromere-specific nucleosome with histone H4 and H2A/B [111, 112]. In vertebrates, a group of 16 proteins comprises the Constitutive Centromere Associated Network (CCAN) that is targeted to CENP-A nucleosomes on centromeric chromatin [111–114]. Other epigenetic features, including histone H4 Lys20 monomethylation and histone H4 Lys5 and Lys12 acetylation in the CENP-A nucleosome, also contribute to CCAN assembly on the centromere and CENP-A deposition, respectively [115, 116]. Conversely, CCAN protein components such as CENP-C further stabilize CENP-A at the centromere [117]. The CCAN complex localizes to the centromere throughout the cell cycle. Importantly, CCAN association with CENP-A nucleosomes establishes and maintains centromeric chromatin, leading to kinetochore formation [111, 112, 118].

At the onset of mitosis, another essential protein network is recruited to the CCAN, which is essential for microtubule binding. The KMN network is composed of three complexes, KNL1, Mis12, and Ndc80 [111, 112, 119]. Studies demonstrate that the NDC80 complex directly binds to microtubules and is necessary for the formation of kinetochore-microtubule attachments. Depletion of the NDC80 complex proteins results in microtubule attachment defects during mitosis [104, 119]. Other recognized proteins with microtubule binding activity include the dimeric kinetochore kinesin CENP-E [120, 121] and the Ska1 complex [122]. Whether these mechanisms are fully conserved in gametes is not known. Mouse oocytes do express CENP-E at the kinetochores and the inhibition of CENP-E function leads to MI-arrest and unstable kinetochore-microtubule interactions [121, 123]. CENP-C is also expressed at kinetochores in both maternal and paternal chromosomes during the first cleavage division of the one-cell embryo [101]. Notably, the first two divisions in the early embryo kinetochore domains are significantly larger in the maternal chromosomes, demonstrating a gradual and unique conversion from meiotic to mitotic kinetochores during the oocyte-to-embryo transition [101].

Key proteins at the outer kinetochore function to ensure correct microtubule attachment to the chromosomes. More specifically, these proteins detect and correct inappropriate chromosome-microtubule interactions by promoting the stability of appropriate bioriented microtubule attachments and eliminating (destabilizing) incorrect attachments. Pivotal studies established that tension across the centromere is essential for stabilizing chromosome-microtubule interactions [124, 125]. In contrast, phosphorylation of kinetochore substrates by the conserved Ipl1/Aurora B kinase selectively eliminates incorrect attachments by promoting the turnover of microtubule attachments in the absence of tension. In mitotic cells, Aurora B regulates kinetochore microtubule attachments by modulating the microtubule binding activity of essential KNM-network proteins such as Hec1/Ndc8 [126–128]. In germ cells, Aurora C kinase is reportedly the primary isoform expressed and plays an important role in regulating chromosome-microtubule interactions during meiotic division [129]. Inhibition of Aurora C activity in mouse oocytes promotes abnormal kinetochore-microtubule attachments as well as premature chromosome segregation and cytokinesis failure during meiosis-I [130]. Moreover, analysis of Aurora C null female mice [131] reveals both meiotic defects in oocytes as well as cell division errors in early embryos [132]. Studies support a vital role for Aurora C as well as the Aurora B function. Notably, Aurora B transcripts are significantly upregulated in ATRX-deficient embryos characterized by centromeric instability in the early embryo [101].

Detection of chromosome-microtubule attachment errors normally activates the spindle assembly checkpoint (SAC), which inhibits anaphase promoting complex APC/C^{cdc20} activity to delay anaphase onset and allow for error correction [133, 134]. Essential SAC proteins, including Mad2, BubR1, and Bub3, function as an important surveillance mechanism and localize to the outer kinetochore until stable chromosome-microtubule interactions are established. Studies support the expression and function of these key SAC proteins in mammalian oocytes [135–138]. High-resolution live cell imaging analyses also reveal that homologous chromosome coorientation during meiosis-I is error prone in oocytes, as the majority of chromosomes undergo error correction of kinetochore-microtubule attachments prior to anaphase onset and completion of the first meiotic division [139]. However, the high rates of aneuploidy reported in mammalian gametes and embryos [103] indicate that attachment errors can potentially elude detection and/or correction by SAC. Analysis of oocytes from $Sycp3^{-/-}$ mice revealed bioriented univalents during meiosis-I and loss of Mad2 at the kinetochores, indicating that SAC is satisfied [140]. Moreover, prematurely separated single chromatids in $Sycp3^{-/-}$ MII oocytes establish merotelic (bidirectional) kinetochore-microtubule attachments [141], leading to equal tension that potentially escaping SAC-mediated arrest. These merotelic attachments can promote aneuploidy, and suggest that in mammalian oocytes, the kinetochore may be organized into multiple domains that enable bidirectional attachment by chromatids [141]. Yet, other studies with $Mlh1$ mutant mice demonstrate that metaphase alignment and bidirectional attachment of all chromosomes is not required for anaphase onset in oocytes [142]. It has been proposed that in contrast to mitosis, anaphase onset during meiotic division likely requires bipolar attachments of a "critical mass," but not all chromosomes [142]. This concept is also supported by studies in which meiotic spindle organization and stability are disrupted in mouse and human oocytes [105, 106, 143]. Targeted ablation or loss of essential microtubule organizing center (MTOC)-associated proteins such as NEDD1 and pericentrin (Pcnt) leads to significant defects in meiotic spindle organization with a high incidence of chromosome attachment errors at metaphase-I. Despite the presence of erroneous attachments, the majority of oocytes progressed to metaphase-II and showed high rates of aneuploidy [105, 106, 143]. These studies demonstrate that SAC was satisfied such that anaphase onset was initiated despite the presence of inappropriate kinetochore-microtubule attachments. Why SAC is seemingly less effective in the detection and/or correction of attachment errors in the mammalian oocyte is not known. Nevertheless, the limited effectiveness of this mechanism poses a major risk in contributing to aneuploidy in gametes and embryos.

ENVIRONMENTAL AND HORMONAL CAUSES OF OOCYTE ANEUPLOIDY

Numerical chromosome abnormalities in the human embryo commonly originate from errors in oocyte chromosome segregation during meiotic division. In spite of its clinical importance, the underlying mechanisms of oocyte aneuploidy are not fully understood. Advanced maternal age is widely recognized as the most critical factor for an increased risk of aneuploidy [69, 144, 145]. The incidence of chromosomal defects in clinically recognized pregnancies is about 2%–3% women less than 30 years old. However, there is a striking increase to >35% of aneuploid oocytes in women over 40 [69]. In addition to chromosomal abnormalities, spindle defects and mitochondrial dysfunction have all been implicated in the age-related declines in fertility [146]. Transcriptome analysis of oocytes from young and old mice revealed that aging is associated with a deregulation of mechanisms controlling the spindle assembly checkpoint, kinetochore function, spindle formation, and chromatin-remodeling processes [147]. Additional studies revealed that loss of centromere cohesion is a major factor in the etiology of aneuploidy during reproductive senescence [14, 148–151]. Importantly, oocyte aging also leads to a reduction in the amount of the alpha-kleisin protein Rec8 and an increase in the interkinetochore distance, suggesting a gradual loss of cohesion [149, 150]. Moreover, aged oocytes also exhibit a significant decrease in chromosomal Shugoshin 2 [150], a protein crucial to preventing cleavage of centromeric cohesins during metaphase-I and, hence, premature sister chromatid separation [152, 153].

Oocytes obtained from aged female mice show a significant reduction of *Atrx* mRNA and protein levels [147], suggesting that loss of ATRX function during reproductive senescence may contribute to the onset of aneuploidy in the female gamete. Notably, the most common chromosome segregation defects in ATRX-deficient oocytes, namely nondisjunction and premature centromere separation [73], are also the most common types of aneuploidy in oocytes from women of advanced reproductive age [145]. This underscores the importance of ATRX deficient ova as a model to determine the epigenetic mechanisms involved in the onset of aneuploidy during reproductive senescence.

Reproductive senescence may affect the epigenetic composition of important chromatin domains in the oocyte. For example, oocytes obtained from older mice exhibit high levels of acetylated histone H4 on lysine 8 and lysine 12 at the metaphase-II stage of meiosis [154, 155]. In addition, oocytes obtained from women of advanced reproductive age exhibit hyperacetylation of lysine 12 in histone H4 and abnormal chromosome alignment [156]. Therefore, disruption of global histone acetylation patterns impairs chromosome segregation, leading to aneuploidy and infertility. In addition to changes in global histone acetylation and methylation levels, reproductive aging has also been linked to a significantly altered expression of several DNA methyltransferases [157].

Establishment of epigenetic modifications during gametogenesis and meiotic transitions can be disrupted by both hormonal and environmental factors [158–160]. For example, high levels of follicle-stimulating hormone (FSH) are associated with increased nondisjunction in both human and mouse oocytes [161]. Importantly, hormonal changes during the decade preceding reproductive senescence in women are characterized by increasing FSH serum concentrations [161]. Ovarian hyperstimulation and suboptimal oocyte maturation conditions during assisted reproductive technologies may have additional effects on oocyte quality as the establishment of epigenetic modifications is extremely susceptible during oocyte growth and maturation, potentially contributing to oocyte aneuploidy. Moreover, loss of methylation on the maternal allele of the KvDMR1 locus was associated with biallelic expression of the Kcnq1ot1 gene, a common characteristic of Beckwith-Wiedemann Syndrome in human patients. Endometriosis is a clinical condition characterized by chromic inflammation, a hyperestrogenic hormonal environment, and progesterone resistance that affect 5%–10% of women of reproductive age in the United States. This condition causes infertility in 50% of patients [162]. Importantly, the use of an established experimental baboon model revealed striking changes in the epigenetic landscape of both the ovary and the oocyte genome following 15 months of disease progression [163]. Notably, dramatic changes in the expression levels of key chromatin-remodeling proteins were observed. Most prominently, the arginine methyl transferases CARM1, PRMT2, and PRMT8 and the downregulation of CARM1 protein expression was also observed in fully grown preovulatory oocytes of the affected females following 15 months of endometriosis [163]. Notably, CARM1 is essential for maintenance of pluripotency in the cleavage stage embryo [164], suggesting that the altered hormonal environment present during endometriosis has the potential to directly impact both the ovarian and oocyte epigenetic landscape as well as oocyte quality and developmental potential.

CONCLUSIONS

Epigenetic modifications in the oocyte genome are essential for transcriptional regulation, establishment of maternal-specific DNA methylation marks, and maintenance of chromosome stability. The epigenetic landscape of the oocyte genome is regulated by histone post-translational modifications, deposition of histone variants, and chromatin-remodeling proteins. Epigenetic modifications may be established on a locus-by-locus basis and modulate the local chromatin environment at regulatory sequences of maternally imprinted genes. On the other hand, these changes may occur over entire chromosomal domains in order to regulate large-scale chromatin structure. Oocytes can mature, progress through meiosis, and become fertilized in the absence of DNA methylation at maternally imprinted genes. However, the patterns of DNA methylation established during oogenesis play a critical role in the regulation of gene expression during subsequent embryo development. In contrast, even subtle defects in the regulation of a large-scale chromatin structure during oocyte growth have a direct impact on the epigenetic control of centromere function, chromosome segregation, and oocyte aneuploidy. Genome-wide methylone analysis and transcriptome studies have provided critical molecular insight into the ontogeny and primary mechanisms of DNA methylation. Germ-cell specific mechanisms regulate the unique nature of the oocyte methylome and the recently discovered broad H3K4me3 patterns in fully grown oocytes. Transgenic mouse models with oocyte-conditional deletions of chromatin-remodeling proteins will be critical to gain a mechanistic understanding of the epigenetic origins of chromosome instability, genomic reprogramming, and the control of gene expression during the mammalian oocyte-to-embryo transition. In turn, these studies will provide invaluable information to understand the effect of environmental or hormonal disruption of chromosome stability in the human oocyte and preimplantation embryo.

Acknowledgments

We thank Brad Gilleland (Educational Resources, College of Veterinary Medicine, University of Georgia) for the preparation of Fig. 8.1. Funding support provided by NSF through the Research Center for Cell Manufacturing CMaT, the Regenerative Engineering and Medicine Center, and a USDA-NIFA Animal Health Capacity Pilot Grant to Dr. De La Fuente. Funding support was also provided by the NIH (HD086528 and HD0713330) to Dr. Viveiros.

References

[1] Reddy P, Liu L, Adhikari D, Jagarlamudi K, Rajareddy S, Shen Y, et al. Oocyte-specific deletion of Pten causes premature activation of the primordial follicle pool. Science 2008;319(5863):611.

[2] Liu K, Rajareddy S, Liu L, Jagarlamudi K, Boman K, Selstam G, et al. Control of mammalian oocyte growth and early follicular development by the oocyte PI3 kinase pathway: new roles for an old timer. Dev Biol 2006;299(1):1–11.

[3] Clarke HJ. Regulation of germ cell development by intercellular signaling in the mammalian ovarian follicle. Wiley Interdiscip Rev Dev Biol 2018;7(1)e294.

[4] Matzuk MM, Burns KH, Viveiros MM, Eppig JJ. Intercellular communication in the mammalian ovary: oocytes carry the conversation. Science 2002;296(5576):2178–80.

[5] De La Fuente R, Eppig JJ. Transcriptional activity of the mouse oocyte genome: companion granulosa cells modulate transcription and chromatin remodeling. Dev Biol 2001;229(1):224–36.

[6] Liu H, Aoki F. Transcriptional activity associated with meiotic competence in fully grown mouse GV oocytes. Zygote 2002;10(4):327–32.

[7] Macaulay AD, Gilbert I, Caballero J, Barreto R, Fournier E, Tossou P, et al. The gametic synapse: RNA transfer to the bovine oocyte. Biol Reprod 2014;91(4):90. 1-12-90, 1-12.

[8] Chen J, Torcia S, Xie F, Lin C-J, Cakmak H, Franciosi F, et al. Somatic cells regulate maternal mRNA translation and developmental competence of mouse oocytes. Nat Cell Biol 2013;15:1415.

[9] De La Fuente R, Viveiros M, Burns K, Adashi E, Matzuk M, Eppig J. Major chromatin remodeling in the germinal vesicle (GV) of mammalian oocytes is dispensable for global transcriptional silencing but required for centromeric heterochromatin function. Dev Biol 2004;275(2):447–58.

[10] De La Fuente R. Chromatin modifications in the germinal vesicle (GV) of mammalian oocytes. Dev Biol 2006;292(1):1–12.

[11] Bird A. Perceptions of epigenetics. Nature 2007;447:396–8.

[12] Soshnev Alexey A, Josefowicz Steven Z, Allis CD. Greater than the sum of parts: complexity of the dynamic epigenome. Mol Cell 2016;62(5):681–94.

[13] Kimmins S, Sassone-Corsi P. Chromatin remodelling and epigenetic features of germ cells. Nature 2005;434(7033):583–9.

[14] Revenkova E, Eijpe M, Heyting C, Hodges C, Hunt P, Liebe B, et al. Cohesin SMC1 beta is required for meiotic chromosome dynamics, sister chromatid cohesion and DNA recombination. Nat Cell Biol 2004;6(6):555–62.

[15] Ivanovska I, Orr-Weaver TL. Histone modifications and the chromatin scaffold for meiotic chromosome architecture. Cell Cycle 2006;5(18):2064–71.

[16] De La Fuente R. Histone deacetylation: establishing a meiotic histone code. Cell Cycle 2014;13(6):879–80.

[17] Auclair G, Weber M. Mechanisms of DNA methylation and demethylation in mammals. Biochimie 2012;94(11):2202–11.

[18] Morgan HD, Santos F, Green K, Dean W, Reik W. Epigenetic reprogramming in mammals. Hum Mol Genet 2005;14(1):R47–58.

[19] Lucifero D, Mann MR, Bartolomei MS, Trasler JM. Gene-specific timing and epigenetic memory in oocyte imprinting. Hum Mol Genet 2004;13(8):839–49.

[20] Fedoriw AM, Stein P, Svoboda P, Schultz RM, Bartolomei MS. Transgenic RNAi reveals essential function for CTCF in H19 gene imprinting. Science 2004;303(5655):238–40.

[21] Smallwood SA, Tomizawa S, Krueger F, Ruf N, Carli N, Segonds-Pichon A, et al. Dynamic CpG island methylation landscape in oocytes and preimplantation embryos. Nat Genet 2011;43(8):811–4.

[22] Barton SC, Surani M, Norris ML. Role of paternal and maternal genomes in mouse development. Nature 1984;311(5984):374–6.

[23] Surani M, Barton SC, Norris ML. Development of reconstituted mouse eggs suggests imprinting of the genome during gametogenesis. Nature 1984;308(5959):548–50.

[24] McGrath J, Solter D. Completion of mouse embryogenesis requires both the maternal and paternal genomes. Cell 1984;37(1):179–83.

[25] Obata Y, Kaneko-Ishino T, Koide T, Takai Y, Ueda T, Domeki I, Shiroishi T, Ishino F, Kono T. Disruption of primary imprinting during oocyte growth leads to the modified expression of imprinted genes during embryogenesis. Development 1998;125(8):1553–60.

[26] Bourc'his D, Xu GL, Lin CS, Bollman B, Bestor TH. Dnmt3L and the establishment of maternal genomic imprints. Science 2001;294(5551):2536–9.

[27] Kaneda M, Okano M, Hata K, Sado T, Tsujimoto N, Li E, et al. Essential role for de novo DNA methyltransferase Dnmt3a in paternal and maternal imprinting. Nature 2004;429(6994):900–3.

[28] Bestor TH, Bourc'his D. Transposon silencing and imprint establishment in mammalian germ cells. Cold Spring Harb Symp Quant Biol 2004;69:381–7.

[29] Tilghman S. The sins of the fathers and mothers: genomic imprinting in mammalian development. Cell 1999;96(2):185–93.

[30] Moore T. Genetic conflict, genomic imprinting and establishment of the epigenotype in relation to growth. Reproduction 2001;122(2):185–93.

[31] Kelly TL, Trasler JM. Reproductive epigenetics. Clin Genet 2004;65(4):247–60.

[32] Higashimoto K, Soejima H, Saito T, Okumura K, Mukai T. Imprinting disruption of the CDKN1C/KCNQ1OT1 domain: the molecular mechanisms causing Beckwith-Wiedemann syndrome and cancer. Cytogenet Genome Res 2006;113(1–4):306–12.

[33] Stewart KR, Veselovska L, Kelsey G. Establishment and functions of DNA methylation in the germline. Epigenomics 2016;8(10):1399–413.

[34] Weaver JR, Bartolomei MS. Chromatin regulators of genomic imprinting. Biochim Biophys Acta (BBA) Gene Regul Mech 2014;1839(3):169–77.

[35] Trasler JM. Gamete imprinting: setting epigenetic patterns for the next generation. Reprod Fertil Dev 2006;18(1–2):63–9.

[36] Bourc'his D, Bestor TH. Origins of extreme sexual dimorphism in genomic imprinting. Cytogenet Genome Res 2006;113(1–4):36–40.

[37] Surani M. Reprogramming of genome function through epigenetic inheritance. Nature 2001;414(6859):122–8.

[38] Surani MA, Hayashi K, Hajkova P. Genetic and epigenetic regulators of pluripotency. Cell 2007;128(4):747–62.

[39] Ferguson-Smith AC, Surani MA. Imprinting and the epigenetic asymmetry between parental genomes. Science 2001;293(5532):1086–9.

[40] Auclair G, Guibert S, Bender A, Weber M. Ontogeny of CpG island methylation and specificity of DNMT3 methyltransferases during embryonic development in the mouse. Genome Biol 2014;15:545.

[41] Kaneda M, Hirasawa R, Chiba H, Okano M, Li E, Sasaki H. Genetic evidence for Dnmt3a-dependent imprinting during oocyte growth obtained by conditional knockout with Zp3-Cre and complete exclusion of Dnmt3b by chimera formation. Genes Cells 2009;15(3):169–79.

[42] Nimura K, Ishida C, Koriyama H, Hata K, Yamanaka S, Li E, et al. Dnmt3a2 targets endogenous Dnmt3L to ES cell chromatin and induces regional DNA methylation. Genes Cells 2006;11(10):1225–37.

[43] Kelsey G, Feil R. New insights into establishment and maintenance of DNA methylation imprints in mammals. Philos Trans R Soc Lond B Biol Sci 2012;368(1609).

[44] Ciccone DN, Su H, Hevi S, Gay F, Lei H, Bajko J. KDM1B is a histone H3K4 demethylase required to establish maternal genomic imprints. Nature 2009;461:415–8.

[45] Li X, Ito M, Zhou F, Youngson N, Zuo X, Leder P, et al. A maternal-zygotic effect gene, Zfp57, maintains both maternal and paternal imprints. Dev Cell 2008;15(4):547–57.

[46] Messerschmidt DM, de Vries W, Ito M, Solter D, Ferguson-Smith A, Knowles BB. Trim28 is required for epigenetic stability during mouse oocyte to embryo transition. Science 2012;335(6075):1499–502.

[47] Quenneville S, Verde G, Corsinotti A, Kapopoulou A, Jakobsson J, Offner S, et al. In embryonic stem cells, ZFP57/KAP1 recognize a methylated hexanucleotide to affect chromatin and DNA methylation of imprinting control regions. Mol Cell 2011;44(3):361–72.

[48] Chotalia M, Smallwood SA, Ruf N, Dawson C, Lucifero D, Frontera M, et al. Transcription is required for establishment of germline methylation marks at imprinted genes. Genes Dev 2009;23(1):105–17.

I. REPRODUCTIVE TRACT DEVELOPMENT AND GAMETOGENESIS

[49] Stewart KR, Veselovska L, Kim J, Huang J, Saadeh H, Tomizawa S-i, et al. Dynamic changes in histone modifications precede de novo DNA methylation in oocytes. Genes Dev 2015;29(23):2449–62.

[50] Veselovska L, Smallwood S, Saadeh H, Stewart K, Krueger F, Maupetit-Mehouas S, et al. Deep sequencing and de novo assembly of the mouse oocyte transcriptome define the contribution of transcription to the DNA methylation landscape. Genome Biol 2015;16(1):209.

[51] Maenohara S, Unoki M, Toh H, Ohishi H, Sharif J, Koseki H, et al. Role of UHRF1 in de novo DNA methylation in oocytes and maintenance methylation in preimplantation embryos. PLoS Genet 2017;13(10): e1007042.

[52] Xu K, Chen X, Yang H, Xu Y, He Y, Wang C, et al. Maternal Sall4 is indispensable for epigenetic maturation of mouse oocytes. J Biol Chem 2017;292(5):1798–807.

[53] Nashun B, Hill Peter WS, Smallwood Sebastien A, Dharmalingam G, Amouroux R, Clark Stephen J, et al. Continuous histone replacement by Hira is essential for normal transcriptional regulation and de novo DNA methylation during mouse oogenesis. Mol Cell 2015;.

[54] Smith ZD, Chan MM, Humm KC, Karnik R, Mekhoubad S, Regev A, et al. DNA methylation dynamics of the human preimplantation embryo. Nature 2014;511(7511):611–5.

[55] Kobayashi H, Sakurai T, Imai M, Takahashi N, Fukuda A, Yayoi O. Contribution of intragenic DNA methylation in mouse gametic DNA methylomes to establish oocyte-specific heritable marks. PLoS Genet 2012;8: e1002440.

[56] Shirane K, Toh H, Kobayashi H, Miura F, Chiba H, Ito T. Mouse oocyte methylomes at base resolution reveal genome-wide accumulation of non-CpG methylation and role of DNA methyltransferases. PLoS Genet 2013;9: e1003439.

[57] Zhang B, Zheng H, Huang B, Li W, Xiang Y, Peng X, et al. Allelic reprogramming of the histone modification H3K4me3 in early mammalian development. Nature 2016;537:553.

[58] Dahl JA, Jung I, Aanes H, Greggains GD, Manaf A, Lerdrup M, et al. Broad histone H3K4me3 domains in mouse oocytes modulate maternal-to-zygotic transition. Nature 2016;537:548.

[59] Liu X, Wang C, Liu W, Li J, Li C, Kou X, et al. Distinct features of H3K4me3 and H3K27me3 chromatin domains in pre-implantation embryos. Nature 2016;537:558.

[60] Andreu-Vieyra CV, Chen R, Agno JE, Glaser S, Anastassiadis K, Stewart AF, et al. MLL2 is required in oocytes for bulk histone 3 lysine 4 trimethylation and transcriptional silencing. PLoS Biol 2010;8(8): e1000453.

[61] Hanna CW, Taudt A, Huang J, Gahurova L, Kranz A, Andrews S, et al. MLL2 conveys transcription-independent H3K4 trimethylation in oocytes. Nat Struct Mol Biol 2018;25(1):73–82.

[62] Kouzarides T. Chromatin modifications and their function. Cell 2007;128:.

[63] Cheung P, Allis CD, Sassone-Corsi P. Signaling to chromatin through histone modifications. Cell 2000;103(2):263–71.

[64] Margueron R, Trojer P, Reinberg D. The key to development: interpreting the histone code? Curr Opin Genet Dev 2005;15(2):163–76.

[65] Kageyama S-i, Liu H, Kaneko N, Ooga M, Nagata M, Aoki F. Alterations in epigenetic modifications during oocyte growth in mice. Reproduction 2007;133(1):85–94.

[66] Inoue A, Jiang L, Lu F, Suzuki T, Zhang Y. Maternal H3K27me3 controls DNA methylation-independent imprinting. Nature 2017;547:419.

[67] Berger S, Felsenfeld G. Chromatin goes global. Mol Cell 2001;8(2):263–8.

[68] Eppig JJ, Viveiros MM, Marin Bivens C, De La Fuente R. Regulation of mammalian oocyte maturation. In: Leung PCK, Adashi EY, editors. The ovary. 2nd ed. Elsevier, Academic Press; 2004. p. 113–29.

[69] Hassold T, Hunt P. To err (meiotically) is human: the genesis of human aneuploidy. Nat Rev Genet 2001;2:280–91.

[70] Petronczki M, Siomos M, Nasmyth K. Un menage a quatre: the molecular biology of chromosome segregation in meiosis. Cell 2003;112(4):423–40.

[71] Page S, Hawley R. Chromosome choreography: the meiotic ballet. Science 2003;301(5634):785–9.

[72] Yang F, Baumann C, De La Fuente R. Persistence of histone H2AX phosphorylation after meiotic chromosome synapsis and abnormal centromere cohesion in poly (ADP-ribose) polymerase (Parp-1) null oocytes. Dev Biol 2009;331:326–38.

[73] Baumann C, Viveiros MM, De La Fuente R. Loss of maternal ATRX results in centromere instability and aneuploidy in the mammalian oocyte and pre-implantation embryo. PLoS Genet 2010;6(9): e1001137.

[74] Wickramasinghe D, Ebert KM, Albertini DF. Meiotic competence acquisition is associated with the appearance of M-phase characteristics in growing mouse oocytes. Dev Biol 1991;143:162–72.

[75] Miyara FMC, Dumont-Hassan M, Meur AL, Cohen-Bacrie P, Aubriot FX, Glissant A, Nathan C, Douard S, Stanovici A, Debey P. Chromatin configuration and transcriptional control in human and mouse oocytes. Mol Reprod Dev 2003;64(4):458–70.

[76] Pesty A, Miyara F, Debey P, Lefevre B, Poirot C. Multiparameter assessment of mouse oogenesis during follicular growth in vitro. Mol Hum Reprod 2007;13(1):3–9.

[77] Longo F, Garagna S, Merico V, Orlandini G, Gatti R, Scandroglio R, et al. Nuclear localization of NORs and centromeres in mouse oocytes during folliculogenesis. Mol Reprod Dev 2003;66(3):279–90.

[78] Zuccotti M, Piccinelli A, Rossi PG, Garagna S, Redi CA. Chromatin organization during mouse oocyte growth. Mol Reprod Dev 1995;(4):479–85.

[79] Debey P, Szöllösi M, Szöllösi D, Vautier D, Girousse A, Besombes D. Competent mouse oocytes isolated from antral follicles exhibit different chromatin organization and follow different maturation dynamics. Mol Reprod Dev 1993;36(1):59–74.

[80] Schramm RD, Tennier MT, Boatman DE, Bavister BD. Chromatin configurations and meiotic competence of oocytes are related to follicular diameter in nonstimulated rhesus monkeys. Biol Reprod 1993;48(2):349–56.

[81] Karpen GH, Allshire RC. The case for epigenetic effects on centromere identity and function. Trends Genet 1997;13(12):489–96.

[82] Dillon N, Festenstein R. Unravelling heterochromatin: competition between positive and negative factors regulates accessibility. Trends Genet 2002;18(5):252–8.

[83] Howman EV, Fowler KJ, Newson AJ, Redward S, MacDonald AC, Kalitsis P, et al. Early disruption of centromeric chromatin organization in centromere protein a (Cenpa) null mice. Proc Natl Acad Sci U S A 2000;97(3):1148–53.

[84] Lachner M, O'Carroll D, Rea S, Mechtler K, Jenuwein T. Methylation of histone H3 lysine 9 creates a binding site for HP1 proteins. Nature 2001;410(6824):116–20.

[85] Bannister AJ, Zegerman P, Partridge JF, Miska EA, Thomas JO, Allshire RC, et al. Selective recognition of methylated lysine 9 on histone H3 by the HP1 chromo domain. Nature 2001;410(6824):120–4.

[86] Kourmouli N, Sun YM, van der Sar S, Singh PB, Brown JP. Epigenetic regulation of mammalian pericentric heterochromatin in vivo by HP1. Biochem Biophys Res Commun 2005;337(3):901–7.

[87] Schotta G, Lachner M, Sarma K, Ebert A, Sengupta R, Reuter G, et al. A silencing pathway to induce H3-K9 and H4-K20 trimethylation at constitutive heterochromatin. Genes Dev 2004;18(11):1251–62.

[88] Pluta AF, Mackay AM, Ainsztein AM, Goldberg IG, Earnshaw WC. The centromere: hub of chromosomal activities. Science 1995;270(5242):1591–4.

[89] Murphy TD, Karpen GH. Centromeres take flight: alpha satellite and the quest for the human centromere. Cell 1998;93(3):317–20.

[90] Ahmad K, Henikoff S. Centromeres are specialized replication domains in heterochromatin. J Cell Biol 2001;153(1):101–10.

[91] Wiens GR, Sorger PK. Centromeric chromatin and epigenetic effects in kinetochore assembly. Cell 1998;93(3):313–6.

[92] McDowell TL, Gibbons RJ, Sutherland H, O'Rourke DM, Bickmore WA, Pombo A, et al. Localization of a putative transcriptional regulator (ATRX) at pericentromeric heterochromatin and the short arms of acrocentric chromosomes. Proc Natl Acad Sci U S A 1999;96(24):13983–8.

[93] Gibbons RJ, Higgs DR. Molecular-clinical spectrum of the ATR-X syndrome. Am J Med Genet 2000;97(3):204–12.

[94] Picketts DJ, Tastan AO, Higgs DR, Gibbons RJ. Comparison of the human and murine ATRX gene identifies highly conserved, functionally important domains. Mamm Genome 1998;9:400–3.

[95] Gibbons R, Bachoo S, Picketts D, Aftimos S, Asenbauer B, Bergoffen J, et al. Mutations in transcriptional regulator ATRX establish the functional significance of a PHD-like domain. Nat Genet 1997;17(2):146–8.

[96] De La Fuente R, Viveiros M, Wigglesworth K, Eppig J. ATRX, a member of the SNF2 family of helicase/ATPases, is required for chromosome alignment and meiotic spindle organization in metaphase II stage mouse oocytes. Dev Biol 2004;272:1–14.

[97] Guenatri M, Bailly D, Maison C, Almouzni G. Mouse centric and pericentric satellite repeats form distinct functional heterochromatin. J Cell Biol 2004;166(4):493–505.

[98] Hakimi M, Bochar D, Schmiesing J, Dong Y, Barak O, Speicher D, et al. A chromatin remodelling complex that loads cohesin onto human chromosomes. Nature 2002;418(6901):994–8.

[99] Peters AH, O'Carroll D, Scherthan H, Mechtler K, Sauer S, Schofer C, et al. Loss of the Suv39 h histone methyltransferases impairs mammalian heterochromatin and genome stability. Cell 2001;107:323–37.

[100] Maison C, Bailly D, Peters AH, Quivy JP, Roche D, Taddei A, et al. Higher-order structure in pericentric heterochromatin involves a distinct pattern of histone modification and an RNA component. Nat Genet 2002;30(3):329–34.

[101] De La Fuente R, Baumann C, Viveiros MM. ATRX contributes to epigenetic asymmetry and silencing of major satellite transcripts in the maternal genome of the mouse embryo. Development 2015;142(10):1806–17.

[102] Baumann C, De La Fuente R. ATRX marks the inactive X chromosome (Xi) in somatic cells and during imprinted X chromosome inactivation in trophoblast stem cells. Chromosoma 2009;118(2):209–22.

[103] Nagaoka SI, Hassold TJ, Hunt PA. Human aneuploidy: mechanisms and new insights into an age-old problem. Nat Rev Genet 2012;13(7):493–504.

[104] Cheeseman IM, Desai A. Molecular architecture of the kinetochore-microtubule interface. Nat Rev Mol Cell Biol 2008;9(1):33–46.

[105] Holubcová Z, Blayney M, Elder K, Schuh M. Error-prone chromosome-mediated spindle assembly favors chromosome segregation defects in human oocytes. Science 2015;348(6239):1143–7.

[106] Baumann C, Wang X, Yang L, Viveiros MM. Error-prone meiotic division and subfertility in mice with oocyte-conditional knockdown of pericentrin. J Cell Sci 2017;130(7):1251.

[107] Oegema K, Desai A, Rybina S, Kirkham M, Hyman AA. Functional analysis of kinetochore assembly in *Caenorhabditis elegans*. J Cell Biol 2001;153(6):1209–26.

[108] Black BE, Brock MA, Bedard S, Woods VL, Cleveland DW. An epigenetic mark generated by the incorporation of CENP-A into centromeric nucleosomes. Proc Natl Acad Sci U S A 2007;104(12):5008–13.

[109] Amor DJ, Choo KHA. Neocentromeres: role in human disease, evolution, and centromere study. Am J Hum Genet 2002;71(4):695–714.

[110] Bassett EA, Wood S, Salimian KJ, Ajith S, Foltz DR, Black BE. Epigenetic centromere specification directs aurora B accumulation but is insufficient to efficiently correct mitotic errors. J Cell Biol 2010;190(2):177–85.

[111] McKinley KL, Cheeseman IM. The molecular basis for centromere identity and function. Nat Rev Mol Cell Biol 2015;17:16.

[112] Fukagawa T, Earnshaw William C. The centromere: chromatin foundation for the kinetochore machinery. Dev Cell 2014;30(5):496–508.

[113] Foltz DR, Jansen LET, Black BE, Bailey AO, Yates JR, Cleveland DW. The human CENP-A centromeric nucleosome-associated complex. Nat Cell Biol 2006;8(5):458–69.

[114] Hori T, Amano M, Suzuki A, Backer CB, Welburn JP, Dong Y, et al. CCAN makes multiple contacts with centromeric DNA to provide distinct pathways to the outer kinetochore. Cell 2008;135(6):1039–52.

[115] Hori T, Shang W-H, Toyoda A, Misu S, Monma N, Ikeo K, et al. Histone H4 Lys 20 Monomethylation of the CENP-A nucleosome is essential for kinetochore assembly. Dev Cell 2014;29(6):740–9.

[116] Shang W-H, Hori T, Westhorpe FG, Godek KM, Toyoda A, Misu S, et al. Acetylation of histone H4 lysine 5 and 12 is required for CENP-A deposition into centromeres. Nat Commun 2016;7:13465.

[117] Falk SJ, Guo LY, Sekulic N, Smoak EM, Mani T, Logsdon GA, et al. CENP-C reshapes and stabilizes CENP-A nucleosomes at the centromere. Science 2015;348(6235):699–703.

[118] Pesenti ME, Weir JR, Musacchio A. Progress in the structural and functional characterization of kinetochores. Curr Opin Struct Biol 2016;37:152–63.

[119] Cheeseman IM, Chappie JS, Wilson-Kubalek EM, Desai A. The conserved KMN network constitutes the core microtubule-binding site of the kinetochore. Cell 2006;127(5):983–97.

[120] Wood KW, Sakowicz R, Goldstein LSB, Cleveland DW. CENP-E is a plus end-directed kinetochore motor required for metaphase chromosome alignment. Cell 1997;91(3):357–66.

[121] Duesbery NS, Choi T, Brown KD, Wood KW, Resau J, Fukasawa K, et al. CENP-E is an essential kinetochore motor in maturing oocytes and is masked during Mos-dependent, cell cycle arrest at metaphase II. Proc Nat Acad Sci U S A 1997;94(17):9165–70.

[122] Welburn JPI, Grishchuk EL, Backer CB, Wilson-Kubalek EM, Yates JR, Cheeseman IM. The human kinetochore Ska1 complex facilitates microtubule depolymerization-coupled motility. Dev Cell 2009;16(3):374–85.

[123] Gui L, Homer H. Spindle assembly checkpoint signalling is uncoupled from chromosomal position in mouse oocytes. Development 2012;139(11):1941–6.

[124] Nicklas RB, Ward SC. Elements of error correction in mitosis: microtubule capture, release, and tension. J Cell Biol 1994;126(5):1241–53.

[125] Nicklas RB, Waters JC, Salmon ED, Ward SC. Checkpoint signals in grasshopper meiosis are sensitive to microtubule attachment, but tension is still essential. J Cell Sci 2001;114(23):4173–83.

[126] Cheeseman IM, Anderson S, Jwa M, Green EM, J-S K, Yates III JR, et al. Phospho-regulation of kinetochore-microtubule attachments by the aurora kinase Ipl1p. Cell 2002;111(2):163–72.

[127] DeLuca JG, Gall WE, Ciferri C, Cimini D, Musacchio A, Salmon ED. Kinetochore microtubule dynamics and attachment stability are regulated by Hec1. Cell 2006;127(5):969–82.

[128] Welburn JPI, Vleugel M, Liu D, Yates JR, Lampson MA, Fukagawa T, et al. Aurora B phosphorylates spatially distinct targets to differentially regulate the kinetochore-microtubule interface. Mol Cell 2010;38(3):383–92.

[129] Yang K-T, Li S-K, Chang C-C, Tang C-JC, Lin Y-N, Lee S-C, et al. Aurora-C kinase deficiency causes cytokinesis failure in meiosis-I and production of large polyploid oocytes in mice. Mol Biol Cell 2010;21(14):2371–83.

[130] Chang Y-C, Chen Y-J, Wu C-H, Wu Y-C, Yen T-C, Ouyang P. Characterization of centrosomal proteins Cep55 and pericentrin in intercellular bridges of mouse testes. J Cell Biochem 2010;109(6):1274–85.

[131] Godmann M, Auger V, Ferraroni-Aguiar V, Sauro AD, Sette C, Behr R, et al. Dynamic regulation of histone H3 methylation at lysine 4 in mammalian spermatogenesis. Biol Reprod 2007;77(5):754–64.

[132] Schindler K, Davydenko O, Fram B, Lampson MA, Schultz RM. Maternally recruited aurora C kinase is more stable than aurora B to support mouse oocyte maturation and early development. Proc Natl Acad Sci U S A 2012;109(33):E2215–22.

[133] Musacchio A. The molecular biology of spindle assembly checkpoint signaling dynamics. Curr Biol 2017;25(20):R1002–18.

[134] Marston AL, Wassmann K. Multiple duties for spindle assembly checkpoint kinases in meiosis. Front Cell Dev Biol 2017;5:109.

[135] Homer HA, McDougall A, Levasseur M, Yallop K, Murdoch AP, Herbert M. Mad2 prevents aneuploidy and premature proteolysis of cyclin B and securin during meiosis I in mouse oocytes. Genes Dev 2005;19:202–7.

[136] Niault T, Hached K, Sotillo R, Sorger PK, Maro B, Benezra R, et al. Changing Mad2 levels affects chromosome segregation and spindle assembly checkpoint control in female mouse meiosis-I. PLoS ONE 2007;2(11):e1165.

[137] Leland S, Nagarajan P, Polyzos A, Thomas S, Samaan G, Donnell R, et al. Heterozygosity for a Bub1 mutation causes female-specific germ cell aneuploidy in mice. Proc Natl Acad Sci U S A 2009;106(31):12776–81.

[138] McGuinness BE, Anger M, Kouznetsova A, Gil-Bernabè AM, Helmhart W, Kudo NR, et al. Regulation of APC/C activity in oocytes by a Bub1-dependent spindle assembly checkpoint. Curr Biol 2009;19(5):369–80.

[139] Kitajima TS, Ohsugi M, Ellenberg J. Complete kinetochore tracking reveals error-prone homologous chromosome biorientation in mammalian oocytes. Cell 2011;146(4):568–81.

[140] Kouznetsova A, Lister L, Nordenskjold M, Herbert M, Hoog C. Bi-orientation of achiasmatic chromosomes in meiosis I oocytes contributes to aneuploidy in mice. Nat Genet 2007;39(8):966–8.

[141] Kouznetsova A, Hernández-Hernández A, Höög C. Merotelic attachments allow alignment and stabilization of chromatids in meiosis II oocytes. Nat Commun 2014;5.

[142] Nagaoka S, Hodges CA, Albertini DF, Hunt PA. Oocyte-specific differences in cell-cycle control create an innate susceptibility to meiotic errors. Curr Biol 2011;21(8):651–7.

[143] Ma W, Baumann C, Viveiros MM. NEDD1 is crucial for meiotic spindle stability and accurate chromosome segregation in mammalian oocytes. Dev Biol 2010;339:439–50.

[144] Hunt PA, Hassold TJ. Human female meiosis: what makes a good egg go bad? Trends Genet 2008;24(2):86–93.

[145] Vialard F, Petit C, Bergere M, Gomes DM, Martel-Petit V, Lombroso R, et al. Evidence of a high proportion of premature unbalanced separation of sister chromatids in the first polar bodies of women of advanced age. Hum Reprod 2006;21(5):1172–8.

[146] Pellestor F, Anahory T, Lefort G, Puechberty J, Liehr T, Hédon B, et al. Complex chromosomal rearrangements: origin and meiotic behavior. Hum Reprod Update 2011;17(4):476–94.

[147] Pan H, Ma P, Zhu W, Schultz RM. Age-associated increase in aneuploidy and changes in gene expression in mouse eggs. Dev Biol 2008;316(2):397–407.

[148] Hodges CA, Revenkova E, Jessberger R, Hassold TJ, Hunt PA. SMC1beta-deficient female mice provide evidence that cohesins are a missing link in age-related nondisjunction. Nat Genet 2005;37(12):1351–5.

[149] Chiang T, Duncan FE, Schindler K, Schultz RM, Lampson MA. Evidence that weakened centromere cohesion is a leading cause of age-related aneuploidy in oocytes. Curr Biol 2010;20(17):1522–8.

[150] Lister LM, Kouznetsova A, Hyslop LA, Kalleas D, Pace SL, Barel JC, et al. Age-related meiotic segregation errors in mammalian oocytes are preceded by depletion of Cohesin and Sgo2. Curr Biol 2010;20(17):1511–21.

[151] Revenkova E, Herrmann K, Adelfalk C, Jessberger R. Oocyte cohesin expression restricted to predictyate stages provides full fertility and prevents aneuploidy. Curr Biol 2010;20(17):1529–33.

[152] Llano E, Gomez R, Gutierrez-Caballero C, Herran Y, Sanchez-Martin M, Vazquez-Quinones L, et al. Shugoshin-2 is essential for the completion of meiosis but not for mitotic cell division in mice. Genes Dev 2008;22(17):2400–13.

[153] Lee J, Kitajima TS, Tanno Y, Yoshida K, Morita T, Miyano T, et al. Unified mode of centromeric protection by shugoshin in mammalian oocytes and somatic cells. Nat Cell Biol 2008;10(1):42–52.

[154] Akiyama T, Nagata M, Aoki F. Inadequate histone deacetylation during oocyte meiosis causes aneuploidy and embryo death in mice. Proc Natl Acad Sci U S A 2006;103(19):7339–44.

[155] Suo L, Meng Q-G, Pei Y, Yan C-L, Fu X-W, Bunch TD, et al. Changes in acetylation on lysine 12 of histone H4 (acH4K12) of murine oocytes during maternal aging may affect fertilization and subsequent embryo development. Fertil Steril 2010;93(3):945–51.

[156] van den Berg IM, Eleveld C, van der Hoeven M, Birnie E, Steegers EAP, Galjaard RJ, et al. Defective deacetylation of histone 4 K12 in human oocytes is associated with advanced maternal age and chromosome misalignment. Hum Reprod 2011;.

[157] Hamatani T, Falco G, Carter MG, Akutsu H, Stagg CA, Sharov AA, et al. Age-associated alteration of gene expression patterns in mouse oocytes. Hum Mol Genet 2004;13(19):2263–78.

[158] Susiarjo M, Hassold TJ, Freeman E, Hunt PA. Bisphenol A exposure in utero disrupts early oogenesis in the mouse. PLoS Genet 2007;3(1)e5.

[159] Dolinoy DC, Weidman JR, Waterland RA, Jirtle RL. Maternal genistein alters coat color and protects Avy mouse offspring from obesity by modifying the fetal epigenome. Environ Health Perspect 2006;114(4):567–72.

[160] Jirtle RL, Skinner MK. Environmental epigenomics and disease susceptibility. Nat Rev Genet 2007;8(4):253–62.

[161] Roberts R, Iatropoulou A, Ciantar D, Stark J, Becker DL, Franks S, et al. Follicle-stimulating hormone affects metaphase I chromosome alignment and increases aneuploidy in mouse oocytes matured in vitro. Biol Reprod 2005;72(1):107–18.

[162] Afshar Y, Hastings J, Roqueiro D, Jeong J-W, Giudice LC, Fazleabas AT. Changes in eutopic endometrial gene expression during the progression of experimental endometriosis in the baboon, *Papio Anubis*. Biol Reprod 2013;88(44):1–9.

[163] Baumann C, Olson M, Wang K, Fazleabas A, De La Fuente R. Arginine methyltransferases mediate an epigenetic ovarian response to endometriosis. Reproduction 2015;150(4):297–310.

[164] Torres-Padilla ME, Parfitt DE, Kouzarides T, Zernicka-Goetz M. Histone arginine methylation regulates pluripotency in the early mouse embryo. Nature 2007;445(7124):214–8.

RNA Binding Protein Networks and Translational Regulation in Oocytes

Xuan G. Luong, Marco Conti

Department of OBGYN-RS, University of California, San Francisco, San Francisco, CA,
United States

Abbreviations

GV   germinal vesicle
GVBD   germinal vesicle breakdown
KD   knockdown
KO   knockout

Human Reproductive and Prenatal Genetics
https://doi.org/10.1016/B978-0-12-813570-9.00009-7

MI   metaphase I
MII   metaphase II
RBP   RNA-binding protein

INTRODUCTION

Distinct developmental strategies underlie the unique function of the germline to transmit genetic information to the progeny. During embryo development, the germline is among the first to be specified in mammals, and in some species, the cytoplasm to be inherited by germ cells is defined as early as at the first embryonic cleavage. Indeed, the cytoplasm of gametes imparts unique properties as extensive epigenetic remodeling occurs in the germline during embryo development. An additional, remarkable strategy adopted by germ cells from the most diverse organisms is the reliance on posttranscriptional controls to regulate gene expression. This is best highlighted by the developmental transitions occurring in fully grown mammalian oocytes. In the oocytes of preovulatory follicles, the chromatin becomes highly condensed and transcription decreases to undetectable levels. As a consequence of this remodeling, all the events associated with the final stages of oocyte maturation and early embryogenesis are driven by posttranscriptional regulation of gene expression. These final stages of oogenesis are critical in determining whether the oocyte will acquire the ability to complete maturation and sustain embryo development, a property termed developmental competence. Although it has been difficult to define the key events required for acquisition of developmental competence, the generation of a healthy female gamete depends on at least two major events. The first event is the coordinated development of somatic and germ cells in the ovarian follicle, which depends on hormonal and molecular signals regulating the exchange of information between the two cell types. The second is proper maturation of the oocyte nuclear and cytoplasmic compartments, which occur during the final stages of gamete development and are major determinants of cell fitness. While some of the molecular constituents involved in these processes have been identified, a comprehensive view of the machinery and interactions required for acquisition of developmental competence is still lacking. The assembly of this machinery is thought to be highly dependent on a coordinated program of maternal mRNA translation, which is dominant during oocyte maturation, as transcription is greatly reduced.

This chapter will summarize the major cellular and molecular events throughout meiotic maturation and describe how these processes support the ability of the oocyte to mature and sustain embryo development. Emphasis will be placed on the network of RNA-binding proteins and their involvement in coordinating a program of maternal mRNA translation and eventual degradation. Because of the logistical and ethical issues associated with conducting experiments using human samples, much of what is known has been gathered using model organisms such as frogs and mice. As some of the major aspects of translation regulation are probably conserved among vertebrates, we believe this knowledge may be relevant to the human system. Elucidating the molecular details of oocyte translational control is crucial in addressing female infertility, defects in embryonic development, and other human pathologies linked to improper gene expression, such as cancer. Discovery of new biological features resulting from this knowledge will make it possible to develop biomarkers to better predict oocyte quality and to select the best oocytes for fertility treatment.

OOCYTE MEIOTIC MATURATION

Oocyte Growth, Meiotic Arrest, and Cell-Cycle Reentry

In mammals, the female germ cell enters meiosis during the fetal stage and remains arrested in the diplotene stage of prophase (GV) for months or years, depending on the species [1]. During this "dormant" postnatal period, oocytes are small and enclosed within primordial follicles, formed by a layer of pregranulosa cells. Upon activation, a subpopulation of oocytes undergoes significant growth in size, decondensation of the chromatin, and activation of transcription [2, 3]. This results in the accumulation of large pools of maternal mRNA [4]. Toward the end of this growth phase, the chromatin condenses and there is progressive silencing of transcription. The ability to resume meiosis is termed meiotic competence, and is achieved during the late stages of folliculogenesis [5, 6]. Additionally, microtubule organizing centers (MTOCs), important for spindle assembly, are formed during this time [7]. Although the functional machinery is present, meiotic-competent oocytes do not resume meiosis unless the surrounding follicular cells are stimulated by luteinizing hormone (LH) or if the oocytes are removed from the follicular environment (spontaneous maturation) [8, 9]. Signals from the surrounding follicle prevent the oocyte from entering meiosis prematurely, ensuring their differentiation is complete and developmental competence is acquired. The events during this last stage of differentiation and requirements for developmental competence are less well characterized; it has been proposed that reorganization of the nucleus and cytoplasm must occur in a coordinate fashion [10].

The signaling pathways and second messengers controlling meiotic arrest have been described using the mouse model. However, some of the major components and functions have also been detected in other species as well as in human oocytes [8], suggesting that similar regulatory circuits are present among species. In both mouse and human GV-arrested oocytes, the G protein-coupled receptor GPR3 is constitutively active, leading to stimulation of oocyte adenylyl cyclase [11, 12]. The activated adenylyl cyclase produces cAMP at concentrations in the oocyte sufficient to stimulate PKA activity. PKAs phosphorylate key cell cycle components that prevent activation of the metaphase-promoting factor (MPF), a cyclin-dependent kinase 1 (CDK1)-cyclin complex, resulting in meiotic arrest [13]. Furthermore, cAMP degradation via the phosphodiesterase PDE3A is blocked by exogenous cGMP that is synthesized by surrounding granulosa cells and transferred to the oocytes through gap junctions [14, 15]. This arrangement confirms the early proposal of the presence of an oocyte maturation inhibitor [16], and explains why, upon removal from the follicular compartment, both mouse and human oocytes will undergo spontaneous maturation.

LH-induced oocyte reentry into the meiotic cell cycle within an intact follicle is dependent on a decrease in cGMP levels in the mural granulosa cells. This major decrease of cGMP in the follicular compartment also causes export of cGMP from the oocyte, activation of PDE3A, and a rapid degradation of cAMP within the gamete [17]. As a result, PKA is inactivated, two key regulators of MPF—CDC25 and WEE2—become dephosphorylated, and MPF is activated, allowing the oocyte to reenter the cell cycle [13]. Other events that lead to the overall persistent decrease in cGMP at ovulation include dephosphorylation of NPR2 [18], release of epidermal-like growth factors (e.g., amphiregulin (AREG)) [15, 19], activation of PDE5 [20], inhibition of CNP production [21], and termination of the gap junction communication among granulosa cells [22].

Nuclear Maturation

Reentry into the meiotic cell cycle is followed by germinal vesicle breakdown (GVBD), and progression through the first meiotic division is marked by extrusion of the first polar body (PBI). The oocyte then enters meiosis II and remains arrested in metaphase II (MII) until fertilization. The time in which maturation occurs varies depending on species.

In mouse oocytes, GVBD is observed approximately 1–3 hours after hormone stimulation [23] while the process takes about 6–8 hours in bovine oocytes [24] and 18–24 hours in porcine oocytes [25]. In humans, it is difficult to ascertain the exact timing of GVBD in vivo, but in vitro time microscopy of human GV oocytes revealed that GVBD may occur in anywhere from 5 to 19.8 hours [26]. This prolonged period to reach GVBD in humans may be due to the fact that several cell cycle components (e.g., cyclins) must be synthesized de novo [27]. If this hypothesis is correct, the decreased cAMP concentration required for cell cycle reentry may somehow lead to the translation of a subset of maternal mRNAs in humans and other species with prolonged intervals between hormone stimulation and GVBD. In contrast, de novo protein synthesis is not required for GVBD in mouse oocytes [28, 29].

In human oocytes, chromatin condensation is observed within the first few hours after GVBD, and the chromosomes may serve as initiation sites for microtubule nucleation followed by progressive spindle assembly [30]. This is contrary to the widely accepted view that spindle formation is driven by the organization of MTOCs, as observed in mice [31]. The stage between GVBD and MI is known as prometaphase and is sustained for several hours to ensure proper spindle assembly [31]. After nucleation, the microtubules extend and attach to the kinetochores of homologous chromosomes, and the organization of the stable, bioriented meiotic spindle is concomitant with metaphase progression. During this time, the spindle assembly checkpoint (SAC) is active to ensure proper spindle assembly. Once the MI spindles are formed, the SAC is inactivated and the cell is allowed to progress to anaphase I.

At anaphase I, the bivalent homologous chromosomes separate and one homolog set is extruded with the first polar body at telophase I. Shugoshin is a protein that prevents dissociation of sister chromatids via recruitment of a phosphatase to block cohesion phosphorylation, stabilizing the centromeric interaction of sister kinetochores [32]. The oocyte then assembles the MII spindle with the chromosomes now aligned on the equatorial plate. The cell remains arrested in this state until fertilization, when the pairs of sister chromatids segregate and one set is extruded with the second polar body.

Remarkably, these cellular events take place during a period when gene transcription is virtually nonexistent, suggesting that coordinated translation of preexisting maternal mRNA directs these transitions. In order to generate a developmentally competent oocyte, proper chromatin condensation, spindle assembly, and chromosome translocation must occur. Errors at any number of these steps may result in impaired cell cycle progression or an oocyte that may be fertilized, but then is unable to sustain embryo development. For example, the increased incidence of aneuploidy in aged oocytes is thought to be due to compromised chromosome cohesion, resulting in improper segregation [33, 34]. Although the exact mechanisms leading to homolog instability are unclear, a hypothesis from our lab is that a faulty translation program in aged oocytes leads to incorrect levels of components important in spindle assembly and chromosome alignment.

Cytoplasmic Maturation

Concurrent with nuclear maturation is considerable reorganization of the ooplasm. This remodeling has been extensively reviewed and includes coordinated trafficking of vesicles, mitochondria, the Golgi apparatus, and the endoplasmic reticulum [35, 36]. Another phenomenon associated with redistribution of the oocyte organelles is cytoplasmic streaming [37]. This property is likely due to progressive cycles of cytoskeleton polymerization and depolymerization. Functionally, this streaming is thought to be responsible for movement of the spindle to the surface of the oocyte [38]. This view has been challenged by the presence of early asymmetry or polarization in the oocyte with an eccentric nucleus, which is then followed by assembly of the spindle chromosome complex [36]. Regardless, this substantial cytoplasmic remodeling has been associated with the acquisition of oocyte developmental competence. Balbiani bodies, which are composed of mitochondria clusters and nuage material, are dynamic structures formed in the oocytes early in development that are thought to be involved in mRNA storage, stabilization, and transport [39, 40]. Later during development, an extensive cytoplasmic lattice develops; this structure may be involved in ribosome storage and translation of maternal mRNAs [41]. Inactivation of genes coding for the building blocks of this lattice (e.g., *Padi6* and *Mater*) leads to impaired developmental competence and defective embryo development [42, 43]. Furthermore, mutation of PADI6 in humans is associated with female infertility due to early embryo arrest [44].

COMPONENTS OF THE TRANSLATIONAL PROGRAM IN OOCYTES

During the oocyte growth phase, transcription is highly active and a large number of maternal mRNAs accumulate in the oocyte. However, once the oocytes are fully grown, transcription ceases and all subsequent transitions up to the two- to eight-cell embryo are dependent on the unmasking, translation, and degradation of these maternal mRNAs. Thus, gene expression is regulated exclusively in the cytoplasm of the oocyte, also known as the ooplasm. This program of regulated translation is essential for generating the molecular machinery required for fertilization, reprogramming of the zygote to totipotency, and embryo development, further supporting the view that embryogenesis starts during oogenesis.

Translation of mRNA relies on the following consecutive steps: initiation, when eukaryotic initiation factors assemble the machinery required for translation, including the ribosome; elongation, when information encoded by the mRNA is decoded into a polypeptide through ribosome scanning of the mRNA strand; and termination, when the polypeptides and ribosomes are released from the mRNA. The mRNA itself contains *cis*-acting elements that function as signals to direct each of these transitions. In addition, translation or repression of a message is largely dependent on the function of a large class of *trans*-acting proteins interacting with mRNAs. The canonical RBPs usually recognize *cis*-acting elements, many of which are clustered in the 3′ UTR of the message, via binding domains. Some of the well-characterized domains include the RNA recognition motif (RRM) domain, the K homology (KH) domain, the zinc finger (Znf) domain, and the double-stranded RNA-binding domain (RBD) [45]. The affinity of RBPs for nucleic acid is usually low; the sequences they recognize are variable and are often degenerated. These properties generate a broad array of affinities, and usually, several RBPs interact with one another and are recruited to a message in a synchronous fashion to stabilize the protein-RNA complex [46].

With the development of novel genome-wide techniques, the list of proteins interacting with mRNA has expanded tremendously, and >1000 RBPs have been identified [47]. Of the >200 RBPs found in the oocyte, we will focus mostly on the RBPs whose functions are established (Table 9.1). We will also discuss their physiological function throughout meiosis and during the oocyte-to-zygote transition. We should also mention that mRNAs coding for >400 additional noncanonical mRNA-interacting proteins have been detected in the oocyte. The functions of these atypical RNA interactors are mostly unknown.

TABLE 9.1　Canonical RNA-Binding Proteins in the Oocyte

Gene Symbol	Full Gene Name	RNA-Binding Domain	Consensus Sequence	Phenotypes	Reference
CPEB1	Cytoplasmic polyadenylation element binding protein 1	RRMx2, Znf	UUUUA(A)U	KO: embryonic oocyte development suspended at the pachytene stage of prophase I; adult females are sterile	Tay and Richter [48]
Cpeb4	Cytoplasmic polyadenylation element binding protein 4	RRMx2, Znf	UUUUA(A)U	KD: oocytes show impaired MI to MII transition and no first polar body extrusion	Igea and Méndez [49]
DAZL	Deleted in azoospermia-like	RRMx1	UUUG/CUUU	KD: oocytes show decreased translation during late meiosis and improper spindle assembly; KO: females are sterile	Chen et al. [50] and McNeilly et al. [50a]
ELAVL2	ELAV like RNA binding protein 2	RRMx3	$(AUUUA)_n$	Overexpression: improper chromosome segregation in the oocytes; KD: increased translation and decreased developmental competence	Mai Nguyen et al. [51]
EPABP	Embryonic poly (A)-binding protein	RRMx4	mRNA poly(A) tail	KO: immature oocytes had impaired growth and no transcriptional silencing; fully grown oocytes failed to activate translational; female adults are infertile	Guzeloglu-Kayisli et al. [52] and Lowther and Mehlmann [53]
MSI2	Musashi homolog 2 (Drosophila)	RRMx2	$G/AU_{1-3}AGU$	KO: females showed impaired folliculogenesis and a decrease in the number of MII oocytes	Sutherland et al. [54]
MSY2	Y-box-binding protein 2	CSDx1	UCCAUCA	KO: females are infertile, show impaired folliculogenesis, and loss of oocytes	Yang et al. [55]
PUM1	Pumilio 1 (Drosophila)	PUMx1	$UGUAX_{2-4}UA$	KO: oocytes showed delayed meiosis; females had diminished ovarian reserves	Mak et al. [56]

TABLE 9.1 Canonical RNA-Binding Proteins in the Oocyte

Gene Symbol	Full Gene Name	RNA-Binding Domain	Consensus Sequence	Phenotypes	Reference
PUM2	Pumilio 2 (Drosophila)	PUMx1	$UGUAX_{2-4}UA$	Increases in translation during maturation	Chen et al. [50]
ZAR1	Zygote arrest 1	PHD	A/UUUA/GUCU	KO: fertilized oocytes do not progress past the one-cell stage and the few blastocysts show impaired ZGA	Wu et al. [57]
ZAR1L	Zar1-like	PHD	A/UUUA/GUCU	KD: two-cell stage embryonic arrest	Hu et al. [58]
ZFP36L2	mRNA decay activator protein ZFP36L2	Znf_CCCHx2	$(AUUUA)_n$	KO in C57BL/6NTac mouse strain: embryonice arrest at the two-cell stage; KO in F3 mouse strain: females anovulous and oocytes did not mature	Ramos et al. [59] and Ball et al. [60]

Poly(A) Tail Length Determines the Rate of Translation of Maternal mRNAs

One of the several key mechanisms involved in the activation of translation is the regulation of polyadenylation at the mRNA 3′ end. Once the maternal messages are transcribed in the nucleus, a cap is added to the 5′ UTR using the modified nucleotide, 7-methylguanosine. These nuclear messages have long poly(A) tails and the nuclear cap-binding complex (CBC) is initially bound to the 5′ cap, which marks the mRNA for translocation to the ooplasm. Upon export, the 3′ poly(A) tail is deadenylated to about 20–40 residues. The message is stabilized and stored, and translation is repressed via the binding of several RBPs (e.g., Cpeb1, Elavl2, and Msy2). With meiotic resumption, a series of events (detailed below) leads to additional polyadenylation of the 3′ end (150–200 adenosine monophosphates) that then allows various RBPs to bind and create a bridge between the 3′ UTR and the 5′ cap (now bound by eukaryotic initiation factor (eIF) 4F). The eIF4F, composed of three distinct protein subunits (eIF4A, eIF4E, and eIF4G), facilitates the formation of the ribosomal translation preinitiation complex. The activated message is then recruited for translation.

Studied extensively in frog oocytes, the poly(A) tail length of a subset of maternal mRNAs increases at the time of translation activation, which in this animal model, precedes GVBD [61]. This process is confirmed in other species, including the cow and mouse, although the timing of translation activation varies [62, 63]. Genome-wide analysis confirms that poly(A) tail length dictates mRNA stabilization and translation specifically in the female gamete, as this function is not apparent in somatic cells [64].

Additional Posttranscriptional mRNA Modifications

Posttranscriptional addition of nontemplated nucleotides to the 3′ ends of RNA molecules or methylation had been observed years ago, but only recently has the functional consequences of these modifications been revealed.

An increasing number of terminal nucleotidyl transferases (TUTases) that uridylate various RNA classes have been identified. ZCCHC11/TUT4 and ZCCHC6/TUT7 uridylate maternal mRNAs with short poly(A) tails (less than ∼25 nucleotides) [65]. The significance of this posttranscriptional modification has been explored by loss-of-function studies in the mouse. Double-knockout (KO) mice for *Tut4/7* are infertile, although ovulation is not affected. Oocytes from these mice are either not fertilized or the few that are fertilized do not progress beyond the two-cell stage. Thus, *Tut4/7*-null oocytes are incompetent to support embryo development. At the molecular level, GV oocytes depleted of these TUTases showed altered gene expression with a larger number of upregulated genes, consistent with the role of uridylation in mRNA stability. Of note, PABPC1 (see below) antagonizes uridylation of polyadenylated mRNAs, contributing to the specificity of TUTases for short (A)-tails [66].

Epitranscriptomics or mRNA epigenetics is a field that has recently gained considerable attention as an increasing number of reports describe alternative posttranscriptional modifications of mRNA [67–69], similar to mechanisms reported for DNA. Recent findings indicate that m6A-methylation of RNA is a major determinant of splicing, alternative polyadenylation (APA), and mRNA stability. RNA methylation involves a writer complex (METTL3 and METTL14) [70] associated with an adapter (WTAP) [71], erasers (FTO and ALKBH5) [72], and five readers (YTHDC1, YTHDC2, YTHDF1, YTHDF2, and YTHDF3). For example, m6A-modifed mRNA is recognized by YTHDF2, which binds to destabilize the message [73]. Mutagenesis of these genes in the oocyte results in improper mRNA degradation and frequent female infertility [74]. Although the field is in its infancy, it is emerging that the machinery required for imparting and detecting RNA methylation is indispensable for proper oocyte maturation and nuclear reprogramming. With the exception of one study, no data are available for humans; patients with the homozygotic loss-of-function mutation of *FTO*, which encodes a dioxygenase, showed retarded growth, microcephaly, psychomotor delay, and general brain deficits [75].

RNA-Binding Proteins in the Oocyte

Poly(A)-Binding Proteins (PABPs)

Poly(A) binding proteins (PABP 1–5) are a family of RRM-containing proteins involved in multiple levels of mRNA regulation, including translational initiation, termination, stability, and mRNA-specific degradation [76, 77]. PABPs are both nuclear and cytoplasmic and are major components of mRNA-protein complexes. Given their property of binding to the ubiquitous poly(A) tails, PABPs serve various "global" functions in posttranscriptional regulation. The best-characterized role is to enhance translational initiation through interaction with eIF4G of the translational initiation complex and increased recruitment of ribosomal subunits [78]. This dual-binding to the poly(A) tail and interaction with the 5′ cap stabilizes the closed-loop conformation of a message, enhancing translation. Regarding mRNA turnover, PABPs are thought to protect the poly(A) tail from deadenylation. They have been shown to inhibit nonsense-mediated mRNA decay, again supporting mRNA stability [79]. In specific cases, however, PABPC1 recruits a deadenylase to mRNA, thus initiating the destabilization process [80]. Although PABPs should function globally in translational regulation given the ubiquitous presence of mRNA poly(A) tails, these proteins interact with numerous other RBPs that recognize specific motifs (see below). These interactions convey specificity of PABPs to maternal mRNA regulation.

A specific form of PABP, termed embryonic PABP (EPABP, encoded by *Pabpc1l*), is expressed in the oocytes [52]. Mice deficient in EPABPs are infertile; oocytes are dysmorphic and do not complete maturation. This phenotype is thought to be due to failure to translate maternal mRNAs that are essential for oocyte development. Injection of *Pabpc1l* into *Pabpc1l*-null GV oocytes did not rescue maturation, confirming a role of this RBP earlier during oogenesis. A later study showed that KO-*Epabp* oocytes were unable to silence transcription during growth, with growth also being impaired [53]. Interestingly, microinjection of *Epab* mRNA into these preantral follicle-enclosed KO oocytes resulted in rescue of the phenotype.

In addition to interacting with the eIF4F initiation complex, PABPs also interact with PAIP proteins (PAIP1, PAIP2A, PAIP2B), which are expressed in the oocyte. PAIP1 stabilizes the interaction of PABP with the eIF4F complex whereas PAIP2A and PAIP2B function as binding competitors of PABP to the 5′ cap [81, 82]. *Paip2*-KO resulted in impaired translation during spermatogenesis and male infertility [83], but surprisingly, there were no apparent reproductive defects in females.

Dual Function of CPE-Binding Proteins (CPEBs) and Associated Complexes

Several *Cpeb* paralogs that recognize the cytoplasmic polyadenylation element (CPE), [UUUUA(A)U], have been identified in mammals, including CPEB 1–4 [84]. *Cpeb1*-KO male and female mice are sterile [48]. Specifically, CPEB1 binds to mRNAs, and, along with other proteins, assembles macromolecular complexes to either prevent or promote polyadenylation, and therefore, regulate translation. A prototypical complex, mainly identified in frog oocytes, includes a scaffold protein (Symplekin) [85]; the cleavage and polyadenylation factor (Cpsf) [86], which recognizes the polyadenylation signal (PAS); and a repressor of the cap complex (Maskin) [87]. In addition, a cytoplasmic poly(A) polymerase (Gld2) [88] and a deadenylase (Parn) form the complex. The competing activities of Gld2 and Parn determine the degree of polyadenylation at the 3′ end of the target mRNA. In the GV state, this complex binds to stabilize the message and repress translation. However, upon meiotic resumption, CPEB1 is phosphorylated (by Aurora A in frogs and IAK1/Eg2, ERK1/2, or CDK1 in mice) [86, 89–91], promoting rearrangement of the protein subunits. Also, Maskin is released to allow formation of the translation initiation complex, Parn is expelled, and Gld2 may then act to increase poly(A) tail length.

This dual role of Cpeb1 is thought to be dictated by a combinatorial code of CPEs distributed on the 3′ UTR of maternal mRNAs in frog oocytes [92]. CPE-mediated translational repression requires at least two CPEs, with 10–12 nucleotides between the elements proposed as the optimal distance for repression. CPE-mediated cytoplasmic polyadenylation and translational activation, on the other hand, require a single consensus CPE and the distance between the CPE and PAS modulates the extent of Cpeb1 function, with an optimal distance of 25 nucleotides. A clear relationship between CPE position and repression/activation could not be defined in mouse maternal mRNAs [50].

In frogs, the Cpeb1 protein is present in oocytes starting in GV, with a marked decrease after entry into MI. This is due to a second wave of Cpeb1 phosphorylation via polo-like kinase 1 (Plx1) and Cdk1/Cdc2 that leads to the degradation of Cpeb1 via the ubiquitin-proteasome pathway [93]. This partial degradation may be necessary for the polyadenylation of mRNAs that are initially repressed by Cpeb1. Similar findings have been reported in the mouse [91]. Interestingly, a very rapid shift in the mobility of CPEB1 in mouse oocytes within 1 hour of reentry into the cell cycle suggests an initial phosphorylation [91]. This is followed by a supershift in mobility between 2 and 4 hours after meiotic reentry. This latter shift is dependent on CDK1 and PLK1 activity, as it is completely abolished by inhibitors of these two enzymes. Furthermore, in frogs, Cpeb1 has been shown to upregulate the translation of Cpeb4 later in meiosis. Cpeb4 activity may be necessary to sustain translational activation once Cpeb1 levels have decreased by anaphase I [49].

Deleted in Azoospermia (DAZ) Proteins

DAZ, deleted-in azoospermia like (Dazl), and boule-like (Boll) are homologues that comprise a family of RBPs with essential roles in primordial germ cell (PGC) differentiation [94]. While Boll is the ancestral gene, *Dazl* is found in vertebrates and *DAZ* is limited to humans and old world monkeys [95]. Both male and female mice deficient in DAZL are sterile while *DAZL* polymorphisms in women are associated with infertility and premature ovarian failure [96]. In mice, DAZL expression is maintained throughout late meiosis until the zygote stage. During this period, studies utilizing knockdown (KD) strategies indicate that this RBP regulates the translation of messages important for spindle assembly, the MI-to-MII transition, and the degradation of maternal mRNA [50].

DAZL recognizes the sequence [UUUG/CUUU], and indeed, this motif is found in the 3′ UTRs of many oocyte-regulated genes. However, its binding is rather promiscuous because DAZL may also recognize a poly(U) sequence, with only a threefold decrease in affinity [97]. Thus, DAZL is expected to interact with a large number of RNAs in the oocyte. This possibility has been tested by RNA-IP in both mouse and human fetal PGCs [98, 99]. In mature mouse oocytes, DAZL RIP-Chip indicated specific binding to almost 1000 mRNAs (Conti, unpublished data). The presence of two or more DAZL-binding elements was associated with transcripts that increased in the polysome fraction during maturation to MII. Using a frog reconstitution system, Dazl was found to bind to the 3′ UTR of maternal transcripts to recruit Pabps, increasing end-to-end complex formation and, thus, leading to enhanced polysome recruitment [99]. This Dazl-mediated process is polyadenylation-independent and provides an alternative method of activating translation.

In females, known targets of DAZL include *Tex19.1*, cell cycle regulators (*Bub1b, Cdc20*, and *Tpx2*), and chromatin remodelers (*Arid1a* and *Smarca5*) [50, 99]. Other targets, observed predominantly in the male, include *Sycp3*, mouse vasa homolog (*Vasa/Mvh*), and *Tpx1* [98]. As previously mentioned, the KD of DAZL in fully grown mouse oocytes displayed reduced translation activation and improper spindle assembly while most did not progress to MII and were not fertilizable, providing additional evidence that translation is required for developmental competence [50]. Additionally, DAZL may control the synthesis of other proteins, such as BTG4, that function in the degradation of stored maternal mRNA later during maturation in preparation for fertilization and zygotic genome activation (ZGA) [90].

Zygote Arrest (ZAR) Proteins

ZAR1 is a lesser-studied protein initially identified as a maternal effect gene in the mouse; its expression during oocyte development is key for the initiation of embryogenesis [57]. Detected predominantly in the cytoplasm of fully grown GV oocytes, ZAR1 is present throughout maturation and is found at low levels until the one-cell zygote. Its expression was absent in two-cell stage embryos. The majority of zygotes derived from *Zar1* KO oocytes were arrested at the one-cell stage with the maternal and paternal genomes remaining separate entities. These cells were able to incorporate BrdU, suggesting entry into the S phase, and that arrest was specifically in the M phase. The few degenerate *Zar1*-KO embryos displayed decreased levels of transcription-requiring complex (TRC) proteins (15% that of controls), suggesting impaired ZGA. The study concluded that ZAR1 may be a regulator of transcription. Since then, new evidence has emerged supporting the idea that Zar1 and its paralog, Zar1l, are RBPs that control translation during frog and zebrafish gametogenesis [100, 101]. Both proteins bind to RNA via an atypical plant homeodomain (PHD) zinc finger domain found at the C-terminus of the protein. Expression of a dominant-negative mutant of ZAR1L that only contains the PHD domain caused two-cell stage embryonic arrest in the mouse [58]. In frog oocytes, both RBPs recognize and bind to the translational control sequence (TCS), [A/UUUA/GUCU], in the 3′ UTRs of important cell cycle regulators, *wee1* and *mos*, to suppress translation during oocyte growth and the initial phases of meiotic resumption [102]. Like Cpeb1, Zar1 may play dual roles in translation regulation. The TCS confers mRNA translational repression in immature oocytes and early pattern cytoplasmic polyadenylation and translational activation in progesterone-stimulated oocytes [103].

Pumilio (PUM) Proteins

Pumilios (PUM) are RBPs that are members of the PUF family of proteins and are considered repressors of translation, possibly via message destabilization. They have been shown to be critical for germ cell development in the worm [104]. Most recently, PUM1 has been implicated in oocyte developmental competence [56]. PUM proteins recognize the PUM-binding element (PBE), [UGUAX$_{2-4}$UA], and the Nanos response element (NRE), [UGUA]. While *Pum1* mRNA was barely detectable in mouse oocytes, *Pum2* mRNA translation increased during oocyte maturation [50]. In GV frog oocytes, Pum2 binds and represses the translation of *ringo*, an atypical Cdk1 activator, until progesterone stimulation, when Pum2 dissociated from the *ringo* mRNA, resulting in *ringo* translation [105]. Ringo may then activate the free catalytic Cdk subunits, triggering phosphorylation of several targets, including Musashi, which may go on to bind and increase the translation of several downstream targets (described below) [106].

Musashi (MSI) Proteins

Musashi (MSI) is an RBP believed to activate translation by recognizing its binding element (MBE), [G/AU$_{1-3}$AGU], and has been shown to be involved in the translation activation of *mos* and *ccnb5* prior to GVBD in frog oocytes [107, 108]. At present, little is known about its function in rodent and human oocytes. Confirming the role of MSI in translation activation, the MBE was found to be enriched in mouse maternal mRNAs whose translation is activated during oocyte meiosis [50]. Although both MSI1 and MSI2 are expressed in mouse GV oocytes, MSI2 expression persists through MII while MSI1 expression is decreased [54]. Furthermore, KO-*Msi2* female mice showed decreased ovary size, impaired folliculogenesis, and fewer MII oocytes.

DNA (cytosine-5)-methyltransferase is a protein responsible for methylation in the early embryo and is present in the maturing oocyte. The 3′ UTR of its gene, *Dnmt1*, contains an

MBE and CPE [109]. Its message is polyadenylated and translated early during maturation. Injection of the mouse *Dnmt1* 3′ UTR with a mutant MBE into frog oocytes showed no polyadenylation as compared to the mouse *Dnmt1* WT 3′ UTR, which was properly polyadenylated prior to GVBD. Mutation of the CPE, however, did not prevent polyadenylation, although there was a decreasing trend in *Dnmt1* mRNA poly(A) tail length. This suggested that cytoplasmic polyadenylation of *Dnmt1* depends on MSI function and an MBE in the 3′ UTR.

AU-Rich Element (ARE)-Binding Proteins

A group of RNA-binding proteins abundant in somatic cells is comprised of proteins that recognize the AU-rich sequence $[AUUUA]_n$. These proteins are diverse and include members of the ELAVL family [110] and zinc finger proteins. It has been estimated that these RBPs bind to 6%–7% of mRNAs and are involved in translation repression and mRNA destabilization. Numerous studies have defined their role in immune cells and transformed cells [111]. However, little is known about the function of these proteins in oocytes. Details about the few observations on their function during gamete development will be discussed in later sections of this chapter.

Other RNA-Binding Proteins

Over the years, numerous studies have described the expression of several additional RBPs in the oocytes, even though exploration of their function during maturation has been limited. The Staufen family has two members, Staufen1 (Stau1) and Staufen 2 (Stau2), both expressed in the oocyte [112]. KD of Stau2 in the oocyte leads to disruption of the spindle, a finding that is consistent with the known interaction of these RBPs with microtubules and their involvement in mRNA transport [113]. Oocyte mitochondrial abnormalities have been associated with altered expression of the fragile X mental retardation 1 (FMR1) protein [114]. Studies of the brain indicate that FMR1 is an RBP that may be involved in mRNA trafficking from the nucleus to the cytoplasm. A complex composed of plasminogen activator inhibitor 1 RNA-binding protein (SERBP1) and Spindlin-1 (SPIN1) has been reported in oocytes, and disruption of this complex prevents oocyte maturation [115]. Deleted in azoospermia (DAZ)-associated protein 1 (DAZAP1/Prrp) is a widely expressed RBP. In addition to pleiotropic effects, the KO of its gene mostly affects spermatogenesis in mice [116], but resulting female sterility has also been observed [117]. Finally, Filia is a component of a subcortical maternal complex (SCMC) in mouse oocytes. Analysis of the crystal structures of the N-terminal domain of Filia indicates that it contains an anomalous KH-domain capable of interacting with RNA [118]. Filia-null oocytes show defects in spindle assembly, chromosome instability, and euploidy in early cleavage, causing delayed embryo development [119].

BIOLOGICAL SIGNIFICANCE OF THE TRANSLATION PROGRAM IN OOCYTES

Genome-wide analysis of mRNA in maturing mouse oocytes indicate two major classes of transcripts whose translation shows significant opposing regulation (defined by significant fold-change in ribosome association from GV to MI or MII) and a third class (class I) that

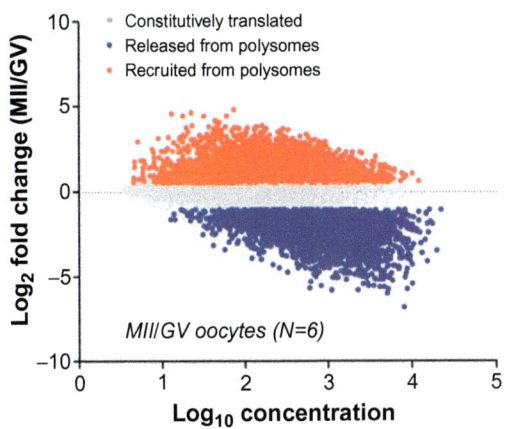

FIG. 9.1   Genome-wide analysis of the translation program during mouse oocyte maturation. Mouse oocytes expressing an HA tag on the large ribosomal subunit were used in this experiment. After a brief preincubation in the presence of the PDE inhibitor, cilostamide, oocytes were allowed to mature after transfer in fresh medium without cilostamide. Oocytes were collected at the time indicted in the abscissa. Oocytes lysates were immunoprecipitated with HA antibodies (RiboTag) and the IP pellet was used for RNA extraction and cDNA library preparation. RNA-Seq was performed and data analysis followed a published workflow. The data represent the mean +/− range of two biological replicates. The differential pattern of translation was confirmed by RiboTag-IP/qPCR.

P value/ fold change	Up (Class III)	Down (Class II)	Unchanged (Class I)
P = .05, 2x	1767	2554	11909
P = .01, 2x	1320	1816	13086
P = .001, 2x	619	716	14845

is constitutively translated (i.e., where ribosome loading remains unchanged throughout maturation) (Fig. 9.1) [50]. Class II transcripts are released from the polysomes, suggesting translation downregulation, while class III transcripts are recruited to the polysomes, suggesting increased translation.

An initial assessment of the mechanisms underlying these divergent behaviors of maternal mRNAs comes from the unbiased search for enriched motifs in the 3′ UTR of regulated transcripts. The most abundant motif enriched in the 3′ UTRs of class III transcripts matched closely to a consensus CPE element recognized by the CPEBs [50]. The presence of two or more CPEs was associated with transcripts that increased in the polysome fraction during maturation to MII. Additional consensus sequences enriched in transcripts undergoing similar regulation bind to DAZL, PUM, or MSI. It should be pointed out that, even within a single class of transcripts, temporal differences were observed in polysome recruitment, opening the possibility that divergent mechanisms of translation activation/repression are operating during meiosis. Although numerous studies have explored the transcriptome in human oocytes, message association with polysomes and translation have not been investigated.

RBPs Work Cooperatively to Regulate Translation

Further analysis of the mechanism of translational activation of candidate transcripts points to cooperative roles of RBPs. For instance, CPEB1 and DAZL work in concert to activate the translation of proteins important to maturation and late-stage maternal RNA degradation in the oocyte. CPEB1 can bind the 3′ UTR of *Dazl* to increase its translation, and DAZL may bind to its own 3′ UTR to self upregulate translation when DAZL levels are increased and

CPEB1 levels are decreased [120]. TEX19.1 is a protein expressed during spermatogenesis and oogenesis and thought to be involved in transposon repression [121]. We have reported that TEX19.1 protein levels increase almost 10-fold during oocyte maturation [120]. Detailed analysis of the translational regulation responsible for the accumulation of this protein showed that ribosome loading on the *Tex19.1* mRNA is dependent on both CPEB1 and DAZL. Mutagenesis of the consensus binding sites in the 3′ UTR of this mRNA documented that the two RBPs act synergistically to support the immense increase in translation. In addition to *Dazl* itself and *Tex19.1*, other genes whose translations are coregulated by both RBPs include *Btg4*, *Bub1b*, *Sycp3*, and *Tpx2* [90].

As previously mentioned, Cpeb1 regulates the translation of Cpeb4 in frog oocytes [49]. Cpeb1 binds to the 3′ UTR of *Cpeb4* and activates its translation. Cpeb4 in turn binds and activates the translation of mRNAs containing CPEs. This mechanism provides a positive feedback loop that temporally extends CPE-mediated translational activation. Because it has been shown that CPEB4 is activated by MAPK [122], this arrangement also allows for translational activation driven by different kinases functioning at different times during the meiotic cell cycle.

As described earlier, the TCS (proposed binding element for Zar1 and Zar1l) confers mRNA translational repression in immature frog oocytes and cytoplasmic polyadenylation and translational activation in progesterone-stimulated oocytes [103]. The presence of CPEs in frog WT *wee1* 3′ UTR suppresses TCS-directed early translational activation and defers its translation to a later time. The CPEs, however, did not interfere with TCS-directed repression in immature oocytes—another example of multiple RBPs functioning to regulate the timing and degree of translation of a message.

In frogs, the 3′ UTRs of several cyclin mRNAs were found to contain PBEs proximate to consensus and nonconsensus CPEs [92]. Interestingly, when the translation of various forms of the *ccnb1* 3′ UTRs was studied with a tagged luciferase reporter, the presence of PBE was shown to have a much stronger role in translation activation than in repression of the message. A PBE with three CPEs increased translation repression by twofold, but a PBE alone or associated with less than three CPE(s) had no effect on repression. A PBE in combination with a single consensus CPE increased translation. Moreover, requirement for the PBE for translational activation became more stringent when a nonconsensus CPE (CPE2) was present, and the impact of PBE on translation was negligible when present with two CPEs. The presence of PBE probably stabilizes Cpeb binding to the mRNA, with its effects highest with a weaker CPE and lowest in the presence of a possible Cpeb1 dimer binding to two proximal CPEs. Mutating the PBE did not affect polyadenylation, even though translational activation was compromised, indicating that Pum-dependent stabilization of Cpeb-binding could trigger a different mechanism of translation activation.

In frogs, Pum2 binds to the PBE found in the 3′ UTR of dormant *ringo* transcripts and forms a complex with Dazl and ePAB to suppress *ringo* translation [105]. Upon oocyte maturation, Pum2 is released from the transcript and dissociates from Dazl and ePAB, resulting in *ringo* translation. The expression of Ringo is upstream of Cpeb1 phosphorylation and was required for Cpeb-mediated cytoplasmic polyadenylation. Human PUM2 is expressed predominantly in embryonic stem cells and germ cells and has been shown to also colocalize with DAZ and DAZL in male germ cells [123, 124].

In *Xenopus* oocytes, Gld2, the previously described adenylase, interacts with Msi and contributes to cytoplasmic polyadenylation of target mRNAs prior to GVBD [125]. Furthermore, the enzyme was found to be associated with both Msi proteins in immature and progesterone-stimulated oocytes. Oocytes coinjected with Msi antisense oligonucleotides and a mutant Msi with a deleted Gld2-binding domain displayed impaired GVBD; KD of Gld2 synthesis resulted in blocked cell cycle progression and decreased Msi-dependent mRNA cytoplasmic polyadenylation. Conversely, overexpression of Gld2 and Msi led to a synergistic increase of *c-mos* polyadenylation and accelerated oocyte maturation. Furthermore, mouse MSI has been shown to interact with GLD2, suggesting that its function in translation activation via cytoplasmic polyadenylation may be conserved across vertebrate species. Both overexpression and KD of GLD2 in mouse oocytes resulted in delayed or blocked meiotic maturation [126].

A recent study has shown that Msi1 alone is not sufficient to induce cytoplasmic polyadenylation in *Xenopus* oocytes [127]. It is, in fact, conformational change of the mRNA induced upon Msi1 binding that results in exposure of proximal CPEs and allows Cpeb1 to bind, resulting in increased polyadenylation. Cpeb1 and Msi1 RNA-IP followed by microarray hybridization revealed 491 potential mutual targets of Cpeb1 and Msi1. *c-mos* was confirmed to be a known target of the two RBPs and a novel mutual target, *net1*, was identified.

These relationships provide evidence that multiple RBPs interact to orchestrate waves of translation activation throughout oocyte maturation. Current studies are focusing on maternal messages whose 3′ UTR contains binding sites for multiple RBPs in efforts to elucidate the exact mechanisms by which these proteins interact and alter translation.

Mechanisms Defining Distinct Temporal Patterns of mRNA Translation

An example of how distinct temporal patterns of translation in the mouse drive progression through the cell cycle is the divergent patterns of translation of Cyclin B proteins (Fig. 9.2). Mouse oocytes express three Cyclin Bs (CCNB1, B2, and B3). *Ccnb2* is translated at high levels in GV oocytes whereas *Ccnb1* translation is repressed, supporting the possibility that pre-MPF is mostly constituted of the CCNB2/CDK1 dimer. *Ccnb3* mRNA translation was constant or slightly increased during the first 4 hours of maturation, but then decreased and minimal translation was detected in MII oocytes.

While *Ccnb2* translation is apparently independent of CPEB1, *Ccnb1* repression and translational activation relied on CPEB1-mediated polyadenylation [91]. Another level of control required for these divergent time courses has been uncovered by investigating the ribosome loading onto the different *Ccnb* transcripts. *Ccnb1* mRNAs are heterogeneous at the 3′ UTR with the short, intermediate, and long 3′ UTRs defined by the presence of multiple PAS elements [128]. While the short form was constitutively translated, the intermediate and long forms were recruited to the polysome fraction after GVBD. This diversity in the 3′ UTR of *Ccnb1* has also been reported in bovine and porcine oocytes [129, 130]. Similarly, *Ccnb2* mRNAs exist with two distinct 3′ UTRs that engender distinct rates of translation (Daldello, personal communication). More importantly, widespread heterogeneity in the 3′ UTR of maternal mRNAs has been found using bioinformatics tools, reinforcing the idea that alternative polyadenylation usage is a critical component of the overall program of translation during meiotic maturation. Also, human

FIG. 9.2   **Pattern of ribosome loading on endogenous Cyclin B mRNAs during oocyte maturation.** Oocytes matured in vivo at the GV or MII stages were collected and oocyte extracts were fractionated on sucrose density gradients. The fraction corresponding to the sedimentation of polysomes was collected, pooled, and had RNA extracted. After linear amplification, cDNA was labeled and hybridized to Affymetrix chips. Intensities were recorded and normalized as described [50]. Two datasets were combined (GEO GSE46640 GSE35106) and analyzed for differences in signals between MII and GV oocytes. The cutoff was set to twofold differences and FDR < 0.05. In the table, significant different transcripts are reported using different FDR values. Class I are mRNAs not significantly different in polysomes from MII and GV oocytes; class II mRNAs significantly decreased on the polysomes between MII and GV; and class III are mRNAs significantly increased in the polysome fraction when comparing MII and GV oocytes. The data are the average of six biological replicates.

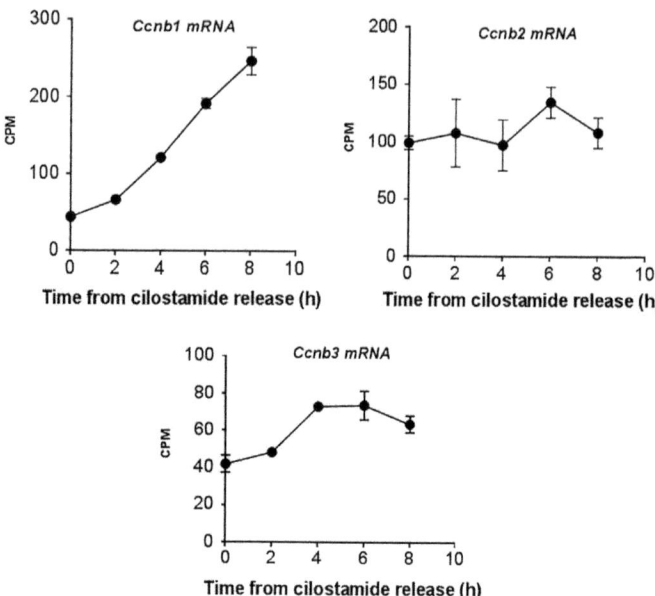

CCNB1 mRNA contains multiple PAS elements in the 3′ UTR, indicating that this mode of regulation is also relevant in the human gamete.

In frog oocytes, *ringo*, coding for an atypical Cdc2 activator, is translated early in the GV state and is followed by the translation of Cyclin B2 and B5 [105]. Later, after GVBD, other cyclins, such as B1 and B4, are translated when Cdc2 is already activated. A combinatorial code of CPE elements present in the 3′ UTR of these frog mRNAs is associated with their time-dependent translation [92]. Early or Cdc2-independent cytoplasmic polyadenylation is dictated by single or multiple CPEs distinct from the PAS. Late cytoplasmic polyadenylation, however, was associated with the presence of at least two CPEs, with one overlapping with the PAS. Furthermore, adding a PAS-overlapping CPE into the 3′ UTR of *ccnb5* shifted the translation pattern of this mRNA from early to late. As described previously, the shift from Cpeb1- to Cpeb4-driven polyadenylation after MI may serve as an additional mechanism for late translation regulation.

c-Mos, a Map kinase-kinase-kinase, is expressed exclusively in the oocyte and is required to activate MAP/ERK kinases during meiosis. *c-Mos*-KO oocytes show defects in MI spindle and leaky MII arrest [131]; *Mapk*-KO oocytes show similar phenotypes. Consistent with its divergent functions in amphibians and mammals, *c-Mos* mRNA translation follows distinct temporal patterns in the two models. In frog oocytes, protein translation is activated early in prophase whereas in the mouse, translation is activated later during prometaphase. In both species, a CPE element has been identified in the *c-Mos* 3′ UTR and functions with the PAS to promote polyadenylation and translation [132, 133]. Mutation of these binding elements inhibited translation. Furthermore, KD of the endogenous c-Mos via antisense oligonucleotides blocked oocyte maturation to MII. This phenotype was rescued by injection with WT *c-Mos* mRNA, but not with *c-Mos* with mutated CPEs. The human *MOS* 3′ UTR also contains a functional CPE, and endogenous *MOS* mRNA undergoes maturation-dependent cytoplasmic polyadenylation [134]. Interestingly, expression of a reporter tagged to the human *MOS* 3′ UTR injected into frog oocytes followed a delayed time course similar to that of *MOS* mRNA found in human oocytes. The authors explained this divergent temporal pattern of activation as the result of the different function of RBPs with Msi functioning in frogs and a late CPEB-mediated activation functioning in humans.

Preliminary RNA-Seq analysis of the translation program in mouse oocytes from 0 to 8 hours after cell cycle reentry confirms the previously reported three classes of differentially regulated maternal mRNA. Additionally, there may be two subpopulations of class III transcripts that show different timings of initial translation activation—an "early" group that undergoes initial activation at 2–4 hours and a "late" group whose activation is not observed until 6–8 hours after meiotic reentry (Luong, unpublished). Further exploration is needed to confirm the existence of these two distinct groups and to elucidate the mechanism behind their differential regulation.

These observations support the hypothesis that timing of translation is encoded in the 3′ UTR of the messages themselves. Future analysis of functional clustering among groups of transcripts showing differences in temporal regulation of translation would shed light on the physiological significance of this level of control.

Termination of Translation and Protein Degradation

We have described the importance of translation activation for oocyte maturation and acquisition of development competence, but the balance of protein synthesis and degradation is also critical to proper development. Although de novo protein synthesis is not required for GVBD in mouse oocytes, sufficient steady state protein levels are needed in the GV state to prepare the oocyte for subsequent cell cycle reentry. For instance, GV arrest may be partially due to the activity of the anaphase-promoting complex (APC) that is continuously degrading CCNB1 via ubiquitin conjugation in order to prevent accumulation of CDK1 activity [135]. The action of APC must be counterbalanced with the translation of *Ccnb1* in order to maintain requisite protein levels for the oocyte to resume meiosis. This equilibrium between Ccnb1 synthesis and degradation is also observed at MII arrest in frog oocytes [136].

In quiescent oocytes, ELAVL2, an oocyte-specific AU-rich element binding protein, and Y-box-binding protein 2 (MSY2), which can bind to both DNA and RNA, are proteins important for message stabilization and translation repression. ELAVL2 is thought to be a repressor of translation and is found at high levels in the growing oocyte with a nonsurrounded nucleolus, but the protein is dramatically decreased once the oocyte reaches the surrounded nucleolus configuration [51]. Overexpression of ELAVL2 resulted in improper chromosome segregation while premature KD of ELAVL2 was associated with increased translation and decreased development competence. Depletion of MSY2 in mice resulted in disruption of oocyte growth and maturation and fertility [55, 137]. It is proposed that MSY2 binds to protect maternal mRNAs from degradation, and its phosphorylation by CDK1 leads to loss of this protective function, resulting in message destabilization. Similar to ELAVL2, ZFP36L2, a zinc finger protein, binds to the ARE and decreases the stability of mRNAs. Female mice with deletions in this gene produced seemingly normal oocytes, but the resulting embryos were blocked at the two-cell stage [59]. However, in a later study, KO of *Zfp35l2* in a different mouse strain cause anovulation in females with oocytes that are unable to mature [60].

A critical feature of the oocyte-to-zygote transition is the progressive depletion of maternal mRNAs to allow the embryo to take over gene expression at ZGA. This implies that the messages switch from a stable state during oocyte growth to an unstable state due to activation of waves of mRNA degradation. Genome-wide analysis comparing GV and MII transcript levels confirms this significant decline, which is likely to be transcript-specific [50]. Several reports indicate that the messages coding for mitochondria and ribosome components were enriched among the mRNAs whose translation decreased during maturation and among transcripts that are destabilized [138]. Additional degradation events have been observed at the oocyte-to-embryo transition and up to the two-cell stage in mouse embryo development. At these times, exhausted maternal components are purged to accommodate expression of the zygotic genome [139, 140]. The program that determines the timing of translational termination and degradation is, with few exceptions, poorly defined. There is, however, evidence in the mouse that disruption of these waves of degradation obstructs correct oocyte and embryo development.

mRNA stabilization is dependent on modifications at the 5' and 3' ends of mRNAs. Decapping proteins, such as DCP1/2, dismantle the 5' cap and promote XRN-1-mediated exonucleases and exosome-complex-mediated hydrolysis [141]. *Dcp1/2* transcripts are recruited to the polysome and their proteins accumulate from GV to MII—a pattern associated with function in maternal mRNA destabilization during late oocyte maturation. Additionally, the two proteins undergo M-phase phosphorylation. Oocytes depleted of DCP1/2 do not undergo mRNA degradation and have altered ZGA.

PARN or CCR4-Not proteins are recruited to the 3' end and induce deadenylation of the poly(A) tail, causing mRNA destabilization. When translation of components of the CCR4-NOT complex, such as *Cnot7* and *Cnot6l*, was blocked, mRNA deadenylation was prevented [142]. B-cell translocation gene-4 (BTG-4) is a scaffold protein that recruits CNOT7 to the translation initiation factor (eIF4E) or polyadenylate-binding protein 2 (PABP2/PABPN1) [143, 144]. *Btg4*-depleted oocytes display mRNA stabilization and untimely polyadenylation. Additionally, these gametes spontaneously progress to anaphase II before fertilization. The failure to maintain MII arrest is probably due to their inability to accumulate the APC inhibitor, the endogenous meiotic inhibitor 2 (EMI2). Other studies showed that *Btg4*-KO female mice were infertile, only producing embryos that were blocked at the one- or two-cell stage [143]. As previously described, DAZL has been shown to bind and upregulate the translation of *Btg4*, resulting in degradation of stored maternal mRNA later during maturation in preparation for fertilization and ZGA.

MicroRNAs (miRNAs), which are known to inhibit translation and promote mRNA destabilization, are surprisingly inactive in mouse oocytes [145, 146]. Instead, mRNA degradation is due to small interfering RNA (siRNA), as supported by the *Dicer1*-KO phenotype [147].

All these findings are consistent in supporting the concept that the machinery required to destabilize maternal mRNAs is assembled during oocyte via concerted translation activation of a subset of mRNAs, and that this process is essential in conferring developmental competence. The only data available for humans are from transcriptomic analyses of oocytes with compromised developmental competence, and they indicate upregulation of a subset of maternal mRNAs. Given the infertility phenotypes associated with defective maternal mRNA destabilization, it is likely that a subgroup of infertile patients carries mutations in any of the genes described above.

Regulated Translation of Components Used During Preimplantation Development

Understanding translation is an approach to identifying cytoplasmic factors contributing to the acquisition of developmental competence during oocyte maturation. Indeed, gene ontology (GO) analysis of transcripts recruited to the polysome (class III) during mouse oocyte maturation showed enrichment in mRNAs coding for transcription machinery and proteins involved in chromatin remodeling [50]. This highlights the importance of the oocyte translation program in reprogramming the oocyte and sperm pronuclei to that of a totipotent cell and provides molecular correlates that confer the unique ability of the ooplasm to induce somatic nuclei reprogramming during somatic cell nuclear transfer.

A conspicuous number of mRNA coding for transcription factors is expressed in the oocytes and dynamically transferred to and from the actively translated pool during maturation [148]. Translation of polymerase II and III components, general transcription factors (GTF2A1, GTF2A2, GTF2B, GTF2H2, and GTF2H3), TATA box components (TFIIA1, TFIIH, TFIIH3, and components of the transcription factor TAFIID), are activated during oocyte maturation (class III). Additional coactivators and repressors of transcription in embryo development that are translationally regulated during oocyte maturation include members of the Krupple-like factor, GATA, Forkhead (FOX), PRMT7, and Glis families of proteins.

SIN3A is a proposed maternal factor involved in chromatin reconfiguration during embryo development. It serves as a scaffold in the histone deacetylase 1/2 (HDAC1/2) chromatin repressor complex, which keeps chromatin in a repressive conformation and is indispensable for the growth of pluripotent cells of the inner cell mass [149]. *Sin3a* is a class III mRNA and suppression of its translation during oocyte meiosis via siRNA resulted in suspended embryo development at the two- to four-cell transition. Notably, *Sin3a*-KO embryos are not viable past embryonic day 6.5 [150].

In a similar fashion, major changes in the translation of epigenetic factors and components of chromatin remodeling are observed during oocyte maturation. For instance, different histone-coding transcripts show different patterns of translation. Although several transcripts are constitutively associated or released from the polysomes (class I and II, respectively), the *H3.3* histone mRNA becomes progressively associated with ribosomes during maturation (class III), indicating a large increase in H3.3 protein synthesis. Specifically, H3.3 is important in chromatin decondensation and transcription in the zygote [151]. KD of this protein in oocytes with morpholino oligonucleotides resulted in no apparent phenotype during meiosis, but early embryo development was suspended at the two-cell stage [152]. Simultaneously, neither sperm nor oocyte chromatin decondensed and DNA replication was impaired. Hira is a chaperone required for H3.3 loading onto the chromatin and KO of this gene resulted in a phenocopy of the H3.3 KD [153]. A recent study has shown that methylation of H3.3 via the maternal protein METTL23 is important paternal genome reprogramming in the zygote [154].

Another class III transcript is *Tet3*, a member of the ten-eleven translocation (TET) family of enzymes. TET3 may induce epigenetic modifications of the paternal genome via generation of 5-hydroxymethylcytosine (5hmC) [155], which is a possible first step required for demethylation. Moreover, female mice with conditional KO of *Tet3* in the oocytes showed reduced fecundity and the heterozygous mutant embryos were developmentally abnormal. Conversely, DNA methyl transferases (*Dnmt1*, *Dnmt3a*, and *Dnmt3l*) show an inverse pattern of translation as compared to *Tet3*, with recruitment to the polysome decreasing from GV to MII (class II). The DNMT3A2-HDAC2 complex has been reported to be essential for genomic imprinting and stabilization in mouse oocytes [156]. Furthermore, aged female mice showed significantly decreased oocyte *Dnmt* expression, which may be a cause for aging-associated faulty embryo development [157].

The translation of several other epigenetic modulators has been shown to be regulated during oocyte meiosis. For instance, all the components of polycomb complex PRC2 (*Ezh1/2*, *Suz12*, and *Eed*) and some of the more heterogenous PRC1 become translationally activated during oocyte maturation (Luong, unpublished data). PRC2 is a repressive complex involved in the maintenance of the H3K27m3 methylation [158]. Numerous observations indicate that the maternally provided polycomb complex is required for gene repression in the embryo and its activation invariably results in failed embryogenesis [159]. Similarly, components of the SWItch/Sucrose Nonfermentable (SWI/SNF) chromatin-remodeling complex are present within the oocytes of various species [160, 161]. SMARCA4/BRG1 is a prototypical component of this complex and its depletion in mouse oocytes led to suspended embryo development and failed ZGA [162]. Different SWI/SNF components show divergent patterns of translation and the configuration of the complex itself is dynamic throughout oocyte maturation, but the physiological significance of these differences is still unclear [163].

These findings further support the notion that maternal mRNAs translated during oocyte maturation have a role later during embryo development.

Somatic Cells Promote Oocyte mRNA Translation

The importance of somatic cell-to-oocyte communication in meiotic arrest and resumption has been described above and is evident in denuded oocytes that have impaired fertilization and decreased embryo production [164]. Crosstalk between surrounding cumulus cells and the oocyte via gap junctions is required for acquiring developmental competence and may be acting through translation regulation. TPX2, a previously mentioned protein important in spindle assembly, is detected at the highest levels in oocytes matured in vivo while denuded oocytes matured in vitro have the lowest levels. Furthermore, studies have shown that stimulation by cumulus cells may promote accumulation of several proteins in the oocyte via activation of the PI3K-AKT-mTOR pathway, and reporter assays strongly suggest that increased translation is the reason for this accumulation [165]. mTOR is the kinase in two complexes, TORC 1 and TORC 2, that phosphorylates 4E-BP, a protein that inhibits the 5' cap complex. Upon phosphorylation, 4E-BP detaches from eIF4E, allowing formation of the 5' cap and binding of the small ribosomal subunit, and thereby increasing translation. This cumulus-driven translation is synergistic with cell-cycle dependent translation activation as reflected by increased endogenous protein levels and elevated luciferase activity mediated by the 3′ UTR of targets such as *Dazl*, *Il7*, and *Tpx2*. Additionally, *Pten*-KO oocytes, which result in constitutively active AKT, also produced elevated translation of similar targets. Remarkably, these oocytes seem to display greater fitness and developmental competence even in the absence of follicle-stimulating hormone or growth factors [166]. Some of these effects are global, but only a subset of maternal mRNAs are affected; specificity may be conferred by terminal oligopyrimidine (TOP) sequences present in the 5′ UTR of transcripts [167].

CONCLUSIONS AND PERSPECTIVES

Oogenesis engenders complex developmental transitions with the latest stages of oocyte maturation in the antral follicle being among the most elaborate. They prepare a germ cell to become a totipotent zygote that will support embryo development. Complex nuclear and cytoplasmic remodeling takes place during these final stages of oocyte development and is remarkably driven by gene expression regulated in the cytoplasm via translation of stored mRNAs. This reliance on posttranscriptional regulation is common in all species studied, from flies to frogs to humans. Many theories have been proposed to explain this unique property, including the requirement to establish an environment that supports nuclear reprogramming or is simply a consequence of the two rapid meiotic divisions. What is certain is that a full understanding of this program will provide an unprecedented view of the requirements for nuclear reprogramming and embryo development.

In this chapter, we characterized some major aspects of the translational regulations required for oocyte maturation. Novel concepts reviewed include the noncoding nucleotide modification of the mRNA (polyadenylation and uridylation) and the emerging field of RNA epigenomics. We have described the networks of RNA-binding proteins and macromolecular complexes that tailor translation through the different phases of meiosis and prepare the oocyte for fertilization and embryo development. Whenever possible, we have emphasized the differences present among species.

We have highlighted how little is known about the translation program in human oocytes. It will be important to overcome the logistic and ethical hurdles and apply the tools developed with model organisms to understand how translation is regulated in the oocytes of a woman. Although it will not be possible to obtain the same level of detail, we should be able to infer the processes taking place in humans from detailed information gathered from these model systems. In this regard, it will be important to identify the species with mechanisms most similar to that of humans.

In addition to addressing major biological questions, a thorough understanding of the oocyte translational program will be valuable for assisted reproductive technologies (ARTs). ARTs have made major progress in the past three decades and are becoming the treatment of choice for infertility. In Western countries, they also are having an impact on family planning. However, there is still room for improvement. A major field of study is the development of tools to distinguish a "good" egg, so that only the ones with the best developmental potential and the least genomic disruption are transferred. In this respect, a better understanding of the translation program in human oocytes may be beneficial on several levels. Proving that the translational program is correctly executed will assist in developing new paradigms to select a gamete to be transferred. It will also provide new strategies to manipulate the human gamete and to optimize culture conditions. As an example, a thorough understanding of the translation program underlying oocyte secretion during the periovulatory period may be used to identify biomarkers helpful in assessing the state of differentiation of an oocyte. While still in the initial stages, pharmacological therapies targeting RBPs are being developed for cancer treatment [168, 169]. Perhaps in the future, drugs will become available to correct aberrant translation in the oocyte and improve the fitness of the gamete to be fertilized and develop as an embryo.

References

[1] Von Stetina JR, Orr-Weaver TL. Developmental control of oocyte maturation and egg activation in metazoan models. Cold Spring Harb Perspect Biol 2011;3(10).

[2] De La Fuente R, Eppig JJ. Transcriptional activity of the mouse oocyte genome: companion granulosa cells modulate transcription and chromatin remodeling. Dev Biol 2001;229(1):224–36.

[3] Bouniol-Baly C, Hamraoui L, Guibert J, Beaujean N, Szollosi MS, Debey P. Differential transcriptional activity associated with chromatin configuration in fully grown mouse germinal vesicle oocytes. Biol Reprod 1999; 60(3):580–7.

[4] Clarke HJ. Posttranscriptional control of gene expression during mouse oogenesis. Results Probl Cell Differ 2012;55:1–21.

[5] Fair T. Follicular oocyte growth and acquisition of developmental competence. Anim Reprod Sci 2003; 78(3–4):203–16.

[6] Eppig JJ. Oocyte control of ovarian follicular development and function in mammals. Reproduction 2001; 122(6):829–38.

[7] Wickramasinghe D, Albertini DF. Centrosome phosphorylation and the developmental expression of meiotic competence in mouse oocytes. Dev Biol 1992;152(1):62–74.

[8] Conti M, Hsieh M, Zamah AM, Oh JS. Novel signaling mechanisms in the ovary during oocyte maturation and ovulation. Mol Cell Endocrinol 2012;356(1–2):65–73.

[9] Edwards RG. Maturation in vitro of mouse sheep cow pig rhesus monkey and human ovarian oocytes. Nature 1965;208(5008):349.

[10] Eppig JJ. Coordination of nuclear and cytoplasmic oocyte maturation in Eutherian mammals. Reprod Fertil Dev 1996;8(4):485–9.

[11] Hinckley M, Vaccari S, Horner K, Chen R, Conti M. The G-protein-coupled receptors Gpr3 and Gpr12 are involved in camp signaling and maintenance of meiotic arrest in rodent oocytes. Dev Biol 2005;287(2):249–61.

[12] Freudzon L, Norris RP, Hand AR, Tanaka S, Saeki Y, Jones TLZ, et al. Regulation of meiotic prophase arrest in mouse oocytes by Gpr3, a constitutive activator of the G(S) G protein. J Cell Biol 2005;171(2):255–65.

[13] Oh JS, Han SJ, Conti M. Wee1b, Myt1, and Cdc25 function in distinct compartments of the mouse oocyte to control meiotic resumption. J Cell Biol 2010;188(2):199–207.

[14] Norris RP, Ratzan WJ, Freudzon M, Mehlmann LM, Krall J, Movsesian MA, et al. Cyclic GMP from the somatic cells of the mouse ovarian follicle regulates cyclic amp and meiosis in the oocyte. Dev Biol 2009;331(2):419.

[15] Vaccari S, Weeks JL, Hsieh M, Menniti FS, Conti M. Cyclic GMP signaling is involved in the luteinizing hormone-dependent meiotic maturation of mouse oocytes. Biol Reprod 2009;81(3):595–604.

[16] Tsafriri A, Pomerantz SH. Oocyte maturation inhibitor. Clin Endocrinol Metab 1986;15(1):157–70.

[17] Shuhaibar LC, Egbert JR, Norris RP, Lampe PD, Nikolaev VO, Thunemann M, et al. Intercellular signaling via cyclic GMP diffusion through gap junctions restarts meiosis in mouse ovarian follicles. Proc Natl Acad Sci U S A 2015;112(17):5527–32.

[18] Shuhaibar LC, Egbert JR, Edmund AB, Uliasz TF, Dickey DM, Yee S-P, et al. Dephosphorylation of juxtamembrane serines and threonines of the Npr2 guanylyl cyclase is required for rapid resumption of oocyte meiosis in response to luteinizing hormone. Dev Biol 2016;409(1):194–201.

[19] Norris RP, Freudzon M, Nikolaev VO, Jaffe LA. Epidermal growth factor receptor kinase activity is required for gap junction closure and for part of the decrease in ovarian follicle cGMP in response to LH. Reproduction 2010;140(5):655–62.

[20] Egbert JR, Uliasz TF, Shuhaibar LC, Geerts A, Wunder F, Kleiman RJ, et al. Luteinizing hormone causes phosphorylation and activation of the cGMP phosphodiesterase PDE5 in rat ovarian follicles, contributing, together with PDE1 activity, to the resumption of meiosis. Biol Reprod 2016;94(5).

[21] Liu XQ, Xie F, Zamah AM, Cao BY, Conti M. Multiple pathways mediate luteinizing hormone regulation of cGMP signaling in the mouse ovarian follicle. Biol Reprod 2014;91(1).

[22] Gershon E, Plaks V, Dekel N. Gap junctions in the ovary: expression, localization and function. Mol Cell Endocrinol 2008;282(1–2):18–25.

[23] Hsieh M, Lee D, Panigone S, Homer K, Chen R, Theologis A, et al. Luteinizing hormone-dependent activation of the epidermal growth factor network is essential for ovulation. Mol Cell Biol 2007;27(5):1914–24.

[24] Sirard MA, Florman HM, Leibfriedrutledge ML, Barnes FL, Sims ML, First NL, Timing of nuclear progression and protein-synthesis necessary for meiotic maturation of bovine oocytes. Biol Reprod 1989;40(6):1257–63.

[25] Dieci C, Lodde V, Franciosi F, Lagutina I, Tessaro I, Modina SC, et al. The effect of cilostamide on gap junction communication dynamics, chromatin remodeling, and competence acquisition in pig oocytes following parthenogenetic activation and nuclear transfer. Biol Reprod 2013;89(3).

[26] Escrich L, Grau N, De Los Santos MJ, Romero JL, Pellicer A, Escriba MJ. The dynamics of in vitro maturation of germinal vesicle oocytes. Fertil Steril 2012;98(5):1147–51.

[27] Schultz GA, Gifford DJ, Mahadevan MM, Fleetham JA, Taylor PJ. Protein synthetic patterns in immature and mature human oocytes. Ann N Y Acad Sci 1988;541:237–47.

[28] Szollosi MS, Debey P, Szollosi D, Rime H, Vautier D. Chromatin behavior under influence of puromycin and 6-DMAP at different stages of mouse oocyte maturation. Chromosoma 1991;100(5):339–54.

[29] Ferrell JE. Building a cellular switch: more lessons from a good egg. Bioessays 1999;21(10):866–70.

[30] Holubcova Z, Blayney M, Elder K, Schuh M. Error-prone chromosome-mediated spindle assembly favors chromosome segregation defects in human oocytes. Obstet Gynecol Surv 2015;70(9):572–3.

[31] Schuh M, Ellenberg J. Self-organization of MTOCS replaces centrosome function during acentrosomal spindle assembly in live mouse oocytes. Cell 2007;130(3):484–98.

[32] Lee J, Kitajima TS, Tanno Y, Yoshida K, Morita T, Miyano T, et al. Unified mode of centromeric protection by Shugoshin in mammalian oocytes and somatic cells. Nat Cell Biol 2008;10(1). 42-U29.

[33] Webster A, Schuh M. Mechanisms of aneuploidy in human eggs. Trends Cell Biol 2017;27(1):55–68.

[34] Nagaoka SI, Hassold TJ, Hunt PA. Human aneuploidy: mechanisms and new insights into an age-old problem. Nat Rev Genet 2012;13(7):493–504.

[35] Li R, Albertini DF. The road to maturation: somatic cell interaction and self-organization of the mammalian oocyte. Nat Rev Mol Cell Biol 2013;14(3):141–52.

[36] Coticchio G, Dal Canto M, Renzini MM, Guglielmo MC, Brambillasca F, Turchi D, et al. Oocyte maturation: gamete-somatic cells interactions, meiotic resumption, cytoskeletal dynamics and cytoplasmic reorganization. Hum Reprod Update 2015;21(4):427–54.

[37] Niwayama R, Nagao H, Kitajima TS, Hufnagel L, Shinohara K, Higuchi T, et al. Bayesian inference of forces causing cytoplasmic streaming in *Caenorhabditis elegans* embryos and mouse oocytes. PLoS ONE 2016;11(7).

[38] Brunet S, Verlhac MH. Positioning to get out of meiosis: the asymmetry of division. Hum Reprod Update 2011;17(1):68–75.

[39] Bilinski SM, Kloc M, Tworzydlo W. Selection of mitochondria in female germline cells: is Balbiani body implicated in this process? J Assist Reprod Genet 2017;34(11):1405–12.

[40] Pepling ME, Wilhelm JE, O'hara AL, Gephardt GW, Spradling AC. Mouse oocytes within germ cell cysts and primordial follicles contain a Balbiani body. Proc Natl Acad Sci U S A 2007;104(1):187–92.

[41] Capco DG, Gallicano GI, Mcgaughey RW, Downing KH, Larabell CA. Cytoskeletal sheets of mammalian eggs and embryos - a lattice-like network of intermediate filaments. Cell Motil Cytoskeleton 1993;24(2):85–99.

[42] Esposito G, Vitale AM, Leijten FPJ, Strik AM, Koonen-Reemst A, Yurttas P, et al. Peptidylarginine deiminase (PAD) 6 is essential for oocyte cytoskeletal sheet formation and female fertility. Mol Cell Endocrinol 2007;273 (1–2):25–31.

[43] Kim B, Kan R, Anguish L, Nelson LM, Coonrod SA. Potential role for mater in cytoplasmic lattice formation in murine oocytes. PLoS ONE 2010;5(9).

[44] Xu Y, Shi YL, Fu J, Yu M, Feng RZ, Sang Q, et al. Mutations in PADI6 cause female infertility characterized by early embryonic arrest. Am J Hum Genet 2016;99(3):744–52.

[45] Ray D, Kazan H, Cook KB, Weirauch MT, Najafabadi HS, Li X, et al. A compendium of RNA-binding motifs for decoding gene regulation. Nature 2013;499(7457):172–7.

[46] Jankowsky E, Harris ME. Specificity and nonspecificity in RNA-protein interactions. Nat Rev Mol Cell Biol 2015;16(9):533–44.

[47] Castello A, Fischer B, Eichelbaum K, Horos R, Beckmann BM, Strein C, et al. Insights into RNA biology from an atlas of mammalian mRNA-binding proteins. Cell 2012;149(6):1393–406.

[48] Tay J, Richter JD. Germ cell differentiation and synaptonemal complex formation are disrupted in CPEB knock-out mice. Dev Cell 2001;1(2):201–13.

[49] Igea A, Mendez R. Meiosis requires a translational positive loop where CPEB1 ensues its replacement by CPEB4. EMBO J 2010;29(13):2182–93.

[50] Chen J, Melton C, Suh N, Oh JS, Horner K, Xie F, et al. Genome-wide analysis of translation reveals a critical role for deleted in azoospermia-like (Dazl) at the oocyte-to-zygote transition. Genes Dev 2011;25(7):755–66.

[50a] McNeilly JR, Saunders PTK, Taggart M, Cranfield M, Cooke HJ, McNeilley AS. Loss of oocytes in Dazl knock-out mice results in maintained ovarian steroidogenic function but altered gonadotropin secretion in adult animals. Endocrinology 2000;141(11):4284-94.

[51] Mai Nguyen C, Auriol J, Jegou B, Kontoyiannis DL, Turner JMA, De Rooij DG, et al. The RNA-binding protein ELAVL1/HuR is essential for mouse spermatogenesis, acting both at meiotic and postmeiotic stages. Mol Biol Cell 2011;22(16):2875–85.

[52] Guzeloglu-Kayisli O, Lalioti MD, Aydiner F, Sasson I, Ilbay O, Sakkas D, et al. Embryonic poly(A)-binding protein (EPAB) is required for oocyte maturation and female fertility in mice. Biochem J 2012;446:47–58.

[53] Lowther KM, Mehlmann LM. Embryonic poly(A)-binding protein is required during early stages of mouse oocyte development for chromatin organization, transcriptional silencing, and meiotic competence. Biol Reprod 2015;93(2).

[54] Sutherland JM, Sobinoff AP, Gunter KM, Fraser BA, Pye V, Bernstein IR, et al. Knockout of RNA binding protein MSI2 impairs follicle development in the mouse ovary: characterization of MSI1 and MSI2 during folliculogenesis. Biomolecules 2015;5(3):1228–44.

[55] Yang JX, Medvedev S, Yu JY, Tang LC, Agno JE, Matzuk MM, et al. Absence of the DNA-/RNA-binding protein MSY2 results in male and female infertility. Proc Natl Acad Sci U S A 2005;102(16):5755–60.

[56] Mak W, Fang CD, Holden T, Dratver MB, Lin HF. An important role of Pumilio 1 in regulating the development of the mammalian female germline. Biol Reprod 2016;94(6).

[57] Wu XM, Viveiros MM, Eppig JJ, Bai YC, Fitzpatrick SL, Matzuk MM. Zygote arrest 1 (Zar1) is a novel maternal-effect gene critical for the oocyte-to-embryo transition. Nat Genet 2003;33(2):187–91.

[58] Hu J, Wan F, Zhu X, Yuan Y, Ding M, Ga S. Mouse ZAR1-like (Xm_359149) colocalizes with mRNA processing components and its dominant-negative mutant caused two-cell-stage embryonic arrest. Dev Dyn 2010; 239(2):407–24.

[59] Ramos SBV, Stumpo DJ, Kennington EA, Phillips RS, Bock CB, Ribeiro-Neto F, et al. The CCCH tandem zinc-finger protein Zfp36l2 is crucial for female fertility and early embryonic development. Development 2004; 131(19):4883–93.

[60] Ball CB, Rodriguez KF, Stumpo DJ, Ribeiro-Neto F, Korach KS, Blackshear PJ, et al. The RNA-binding protein, ZFP36L2, influences ovulation and oocyte maturation. PLoS ONE 2014;9(5).

[61] Mendez R, Richter JD. Translational control by CPEB: a means to the end. Nat Rev Mol Cell Biol 2001;2(7):521–9.

[62] Tay J, Hodgman R, Richter JD. The control of cyclin B1 mRNA translation during mouse oocyte maturation. Dev Biol 2000;221(1):1–9.

[63] Reyes JM, Ross PJ. Cytoplasmic polyadenylation in mammalian oocyte maturation. Wiley Interdiscip Rev RNA 2016;7(1):71–89.

[64] Subtelny AO, Eichhorn SW, Chen GR, Sive H, Bartel DP. Poly(A)-tail profiling reveals an embryonic switch in translational control. Nature 2014;508(7494):66.

[65] Morgan M, Much C, Digiacomo M, Azzi C, Ivanova I, Vitsios DM, et al. mRNA 3′ uridylation and poly(A) tail length sculpt the mammalian maternal transcriptome. Nature 2017;548(7667):347.

[66] Lim J, Ha M, Chang H, Kwon SC, Simanshu DK, Patel DJ, et al. Uridylation by TUT4 and TUT7 marks mRNA for degradation. Cell 2014;159(6):1365–76.

[67] Meyer KD, Jaffrey SR. The dynamic epitranscriptome: N-6-methyladenosine and gene expression control. Nat Rev Mol Cell Biol 2014;15(5):313–26.

[68] Fu Y, Dominissini D, Rechavi G, He C. Gene expression regulation mediated through reversible m(6)A RNA methylation. Nat Rev Genet 2014;15(5):293–306.

[69] Saletore Y, Meyer K, Korlach J, Vilfan ID, Jaffrey S, Mason CE. The birth of the epitranscriptome: deciphering the function of RNA modifications. Genome Biol 2012;13(10).

[70] Liu J, Yue Y, Han D, Wang X, Fu Y, Zhang L, et al. A METTL3-METTL14 complex mediates mammalian nuclear RNA N-6-adenosine methylation. Nat Chem Biol 2014;10(2):93–5.

[71] Ping X-L, Sun B-F, Wang L, Xiao W, Yang X, Wang W-J, et al. Mammalian WTAP is a regulatory subunit of the RNA N6-methyladenosine methyltransferase. Cell Res 2014;24(2):177–89.

[72] Bartosovic M, Molares HC, Gregorova P, Hrossova D, Kudla G, Vanacova S. N6-methyladenosine demethylase FTO targets pre-mRNAs and regulates alternative splicing and 3′-end processing. Nucleic Acids Res 2017; 45(19):11356–70.

[73] Wang X, Lu ZK, Gomez A, Hon GC, Yue YN, Han DL, et al. N-6-methyladenosine-dependent regulation of messenger RNA stability. Nature 2014;505(7481):117.

[74] Ivanova I, Much C, Di Giacomo M, Azzi C, Morgan M, Moreira PN, et al. The RNA m(6)A reader YTHDF2 is essential for the post-transcriptional regulation of the maternal transcriptome and oocyte competence. Mol Cell 2017;67(6):1059.

[75] Boissel S, Reish O, Proulx K, Kawagoe-Takaki H, Sedgwick B, Yeo GSH, et al. Loss-of-function mutation in the dioxygenase-encoding FTO gene causes severe growth retardation and multiple malformations. Am J Hum Genet 2009;85(1):106–11.

[76] Kahvejian A, Svitkin YV, Sukarieh R, M'boutchou MN, Sonenberg N. Mammalian poly(A)-binding protein is a eukaryotic translation initiation factor, which acts via multiple mechanisms. Genes Dev 2005;19(1):104–13.

[77] RWP S, Gray NK. Poly(A)-binding protein (PABP): a common viral target. Biochem J 2010;426(1):11.

[78] Wakiyama M, Imataka H, Sonenberg N. Interaction of eLF4G with poly(A)-binding protein stimulates translation and is critical for Xenopus oocyte maturation. Curr Biol 2000;10(18):1147–50.

[79] Fatscher T, Boehm V, Weiche B, Gehring NH. The interaction of cytoplasmic poly(A)-binding protein with eukaryotic initiation factor 4G suppresses nonsense-mediated mRNA decay. RNA 2014;20(10):1579–92.

[80] Stupfler B, Birck C, Seraphin B, Mauxion F. BTG2 bridges PABPC1 RNA-binding domains and CAF1 deadenylase to control cell proliferation. Nat Commun 2016;7:.

[81] Martineau Y, Derry MC, Wang X, Yanagiya A, Berlanga JJ, Shyu A-B, et al. Poly(A)-binding protein-interacting protein 1 binds to eukaryotic translation initiation factor 3 to stimulate translation. Mol Cell Biol 2008; 28(21):6658–67.

[82] Khaleghpour K, Svitkin YV, Craig AW, Demaria CT, Deo RC, Burley SK, et al. Translational repression by a novel partner of human poly(A) binding protein, PAIP2. Mol Cell 2001;7(1):205–16.

[83] Yanagiya A, Delbes G, Svitkin YV, Robaire B, Sonenberg N. The poly(A)-binding protein partner Paip2a controls translation during late spermiogenesis in mice. J Clin Invest 2010;120(9):3389–400.

[84] Darnell JC, Richter JD. Cytoplasmic RNA-binding proteins and the control of complex brain function. Cold Spring Harb Perspect Biol 2012;4(8).

[85] Barnard DC, Ryan K, Manley JL, Richter JD. Symplekin and xGLD-2 are required for CPEB-mediated cytoplasmic polyadenylation. Cell 2004;119(5):641–51.

[86] Hodgman R, Tay J, Mendez R, Richter JD. CPEB phosphorylation and cytoplasmic polyadenylation are catalyzed by the kinase IAK1/Eg2 in maturing mouse oocytes. Development 2001;128(14):2815–22.

[87] Cao QP, Richter JD. Dissolution of the maskin-eIF4E complex by cytoplasmic polyadenylation and poly(A)-binding protein controls cyclin B1 mRNA translation and oocyte maturation. EMBO J 2002;21(14):3852–62.

[88] Kwak JE, Wang LT, Ballantyne S, Kimble J, Wickens M. Mammalian GLD-2 homologs are poly(A) polymerases. Proc Natl Acad Sci U S A 2004;101(13):4407–12.

[89] Sarkissian M, Mendez R, Richter JD. Progesterone and insulin stimulation of CPEB dependent polyadenylation is regulated by Aurora A and glycogen synthase kinase-3. Genes Dev 2004;18(1):48–61.

[90] Sha Q-Q, Dai X-X, Dang Y, Tang F, Liu J, Zhang Y-L, et al. A MAPK cascade couples maternal mRNA translation and degradation to meiotic cell cycle progression in mouse oocytes. Development 2017;144(3):452–63.

[91] Han SJ, Martins JPS, Yang Y, Kang MK, Daldello EM, Conti M. The translation of cyclin B1 and B2 is differentially regulated during mouse oocyte reentry into the meiotic cell cycle. Sci Rep 2017;7.

[92] Pique M, Lopez JM, Foissac S, Guigo R, Mendez R. A combinatorial code for CPE-mediated translational control. Cell 2008;132(3):434–48.

[93] Setoyama D, Yamashita M, Sagata N. Mechanism of degradation of CPEB during Xenopus oocyte maturation. Proc Natl Acad Sci U S A 2007;104(46):18001–6.

[94] Kee K, Angeles VT, Flores M, Nguyen HN, Pera RAR. Human DAZL, DAZ and BOULE genes modulate primordial germ-cell and haploid gamete formation. Nature 2009;462(7270). 222–U95.

[95] Tung JY, Luetjens CM, Wistuba J, Xu EY, Pera RAR, Gromoll J. Evolutionary comparison of the reproductive genes, DAZL and BOULE, in primates with and without DAZ. Dev Genes Evol 2006;216(3):158–68.

[96] Tung JY, Rosen MP, Nelson LM, Turek PJ, Witte JS, Cramer DW, et al. Novel missense mutations of the Deleted-in-AZoospermia-Like (DAZL) gene in infertile women and men. Reprod Biol Endocrinol 2006;4.

[97] Jenkins HT, Malkova B, Edwards TA. Kinked beta-strands mediate high-affinity recognition of mRNA targets by the germ-cell regulator DAZL. Proc Natl Acad Sci U S A 2011;108(45):18266–71.

[98] Rosario R, Smith RWP, Adams IR, Anderson RA. RNA immunoprecipitation identifies novel targets of DAZL in human foetal ovary. Mol Hum Reprod 2017;23(3):177–86.

[99] Collier B, Gorgoni B, Loveridge C, Cooke HJ, Gray NK. The DAZL family proteins are PABP-binding proteins that regulate translation in germ cells. EMBO J 2005;24(14):2656–66.

[100] Miao L, Yuan Y, Cheng F, Fang J, Zhou F, Ma W, et al. Translation repression by maternal RNA binding protein Zar1 is essential for early oogenesis in zebrafish. Development 2017;144(1):128–38.

[101] Charlesworth A, Yamamoto TM, Cook JM, Silva KD, Kotter CV, Carter GS, et al. *Xenopus laevis* zygote arrest 2 (zar2) encodes a zinc finger RNA-binding protein that binds to the translational control sequence in the maternal Wee1 mRNA and regulates translation. Dev Biol 2012;369(2):177–90.

[102] Yamamoto TM, Cook JM, Kotter CV, Khat T, Silva KD, Ferreyros M, et al. Zar1 represses translation in *Xenopus* oocytes and binds to the TCS in maternal mRNAs with different characteristics than Zar2. Biochim Biophys Acta Gene Regul Mech 2013;1829(10):1034–46.

[103] Wang YY, Charlesworth A, Byrd SM, Gregerson R, Macnicol MC, Macnicol AM. A novel mRNA 3′ untranslated region translational control sequence regulates *Xenopus* Wee1 mRNA translation. Dev Biol 2008;317(2):454–66.

[104] Kimble J, Crittenden SL. Controls of germline stem cells, entry into meiosis, and the sperm/oocyte decision in *Caenorhabditis elegans*. Annu Rev Cell Dev Biol 2007;23:405–33.

[105] Padmanabhan K, Richter JD. Regulated Pumilio-2 binding controls RINGO/Spy mRNA translation and CPEB activation. Genes Dev 2006;20(2):199–209.

[106] Arumugam K, Macnicol MC, Wang YY, Cragle CE, Tackett AJ, Hardy LL, et al. Ringo/cyclin-dependent kinase and mitogen-activated protein kinase signaling pathways regulate the activity of the cell fate determinant Musashi to promote cell cycle re-entry in *Xenopus* oocytes. J Biol Chem 2012;287(13):10639–49.

[107] Macnicol MC, Cragle CE, Macnicol AM. Context-dependent regulation of Musashi-mediated mRNA translation and cell cycle regulation. Cell Cycle 2011;10(1):39–44.

[108] Charlesworth A, Wilczynska A, Thampi P, Cox LL, Macnicol AM. Musashi regulates the temporal order of mRNA translation during *Xenopus* oocyte maturation. EMBO J 2006;25(12):2792–801.

[109] Rutledge CE, Lau HT, Mangan H, Hardy LL, Sunnotel O, Guo F, et al. Efficient translation of Dnmt1 requires cytoplasmic polyadenylation and Musashi binding elements. PLoS ONE 2014;9(2).

[110] Wiszniak SE, Dredge BK, Jensen KB. HuB (Elavl2) mRNA is restricted to the germ cells by post-transcriptional mechanisms including stabilisation of the message by DAZL. PLoS ONE 2011;6(6).

[111] Akamatsu W, Okano HJ, Osumi N, Inoue T, Nakamura S, Sakakibara SI, et al. Mammalian ELAV-like neuronal RNA-binding proteins HuB and HuC promote neuronal development in both the central and the peripheral nervous systems. Proc Natl Acad Sci U S A 1999;96(17):9885–90.

[112] Furic L, Maher-Laporte M, Desgroseillers LA. Genome-wide approach identifies distinct but overlapping subsets of cellular mRNAs associated with Staufen1- and Staufen2-containing ribonucleoprotein complexes. RNA 2008;14(2):324–35.

[113] Cao Y, Du J, Chen DD, Wang Q, Zhang NN, Liu XY, et al. RNA-binding protein Stau2 is important for spindle integrity and meiosis progression in mouse oocytes. Cell Cycle 2016;15(19):2608–18.

[114] Dioguardi CC, Uslu B, Haynes M, Kurus M, Gul M, Miao DQ, et al. Granulosa cell and oocyte mitochondrial abnormalities in a mouse model of fragile X primary ovarian insufficiency. Mol Hum Reprod 2016;22(6):384–96.

[115] Chew TG, Peaston A, Lim AK, Lorthongpanich C, Knowles BB, Solter D. A Tudor domain protein SPINDLIN1 interacts with the mRNA-binding protein SERBP1 and is involved in mouse oocyte meiotic resumption. PLoS ONE 2013;8(7).

[116] Smith RWP, Anderson RC, Smith JWS, Brook M, Richardson WA, Gray NK. DAZAP1, an RNA-binding protein required for development and spermatogenesis, can regulate mRNA translation. RNA 2011;17(7):1282–95.

[117] Hsu LCL, Chen HY, Lin YW, Chu WC, Lin MJ, Yan YT, et al. DAZAP1, an hnRNP protein, is required for normal growth and spermatogenesis in mice. RNA 2008;14(9):1814–22.

[118] Wang JK, Xu MY, Zhu K, Li L, Liu XQ. The N-terminus of FILIA forms an atypical KH domain with a unique extension involved in interaction with RNA. PLoS ONE 2012;7(1).

[119] Zheng P, Dean J. Role of Filia, a maternal effect gene, in maintaining euploidy during cleavage-stage mouse embryogenesis. Proc Natl Acad Sci U S A 2009;106(18):7473–8.

[120] Martins JPS, Liu X, Oke A, Arora R, Franciosi F, Viville S, et al. DAZL and CPEB1 regulate mRNA translation synergistically during oocyte maturation. J Cell Sci 2016;129(6):1271–82.

[121] Ollinger R, Childs AJ, Burgess HM, Speed RM, Lundegaard PR, Reynolds N, et al. Deletion of the pluripotency-associated Tex19.1 gene causes activation of endogenous retroviruses and defective spermatogenesis in mice. PLoS Genet 2008;4(9).

[122] Ortiz-Zapater E, Pineda D, Martinez-Bosch N, Fernandez-Miranda G, Iglesias M, Alameda F, et al. Key contribution of CPEB4-mediated translational control to cancer progression. Nat Med 2012;18(1):83–90.

[123] Moore FL, Jaruzelska J, Fox MS, Urano J, Firpo MT, Turek PJ, et al. Human Pumilio-2 is expressed in embryonic stem cells and germ cells and interacts with DAZ (Deleted in AZoospermia) and DAZ-like proteins. Proc Natl Acad Sci U S A 2003;100(2):538–43.

[124] Fox M, Urano J, Pera RAR. Identification and characterization of RNA sequences to which human PUMILIO-2 (PUM2) and deleted in Azoospermia-like (DAZL) bind. Genomics 2005;85(1):92–105.

[125] Cragle C, Macnicol AM. Musashi protein-directed translational activation of target mRNAs is mediated by the poly(A) polymerase, germ line development defective-2. J Biol Chem 2014;289(20):14239–51.

[126] Nakanishi T, Kubota H, Ishibashi N, Kumagai S, Watanabe H, Yamashita M, et al. Possible role of mouse poly(A) polymerase mGLD-2 during oocyte maturation. Dev Biol 2006;289(1):115–26.

[127] Weill L, Belloc E, Castellazzi CL, Mendez R. Musashi 1 regulates the timing and extent of meiotic mRNA translational activation by promoting the use of specific CPEs. Nat Struct Mol Biol 2017;24(8):672.

[128] Yang Y, Yang C-R, Han SJ, Daldello EM, Cho A, Martins JPS, et al. Maternal mRNAs with distinct 3′ UTRs define the temporal pattern of Ccnb1 synthesis during mouse oocyte meiotic maturation. Genes Dev 2017;31(13):1302–7.

[129] Tremblay K, Vigneault C, Mcgraw S, Sirard MA. Expression of cyclin B1 messenger RNA isoforms and initiation of cytoplasmic polyadenylation in the bovine oocyte. Biol Reprod 2005;72(4):1037–44.

[130] Zhang D-X, Cui X-S, Kim N-H. Molecular characterization and polyadenylation-regulated expression of cyclin B1 and Cdc2 in porcine oocytes and early parthenotes. Mol Reprod Dev 2010;77(1):38–50.

[131] Araki K, Naito K, Haraguchi S, Suzuki R, Yokoyama M, Inoue M, et al. Meiotic abnormalities of c-mos knockout mouse oocytes: activation after first meiosis or entrance into third meiotic metaphase. Biol Reprod 1996; 55(6):1315–24.

[132] Gebauer F, Xu WH, Cooper GM, Richter JD. Translational control by cytoplasmic polyadenylation of c-mos messenger-RNA is necessary for oocyte maturation in the mouse. EMBO J 1994;13(23):5712–20.

[133] Demoor CH, Richter JD. The Mos pathway regulates cytoplasmic polyadenylation in *Xenopus* oocytes. Mol Cell Biol 1997;17(11):6419–26.

[134] Prasad CK, Mahadevan M, Macnicol MC, Macnicol AM. Mos 3′ UTR regulatory differences underlie species-specific temporal patterns of Mos mRNA cytoplasmic polyadenylation and translational recruitment during oocyte maturation. Mol Reprod Dev 2008;75(8):1258–68.

[135] Reis A, Chang HY, Levasseur M, Jones KT. APC(Cdh1) activity in mouse oocytes prevents entry into the first meiotic division. Nat Cell Biol 2006;8(5):539–40.

[136] Yamamoto TM, Iwabuchi M, Ohsumi K, Kishimoto T. APC/C-Cdc20-mediated degradation of cyclin B participates in CSF arrest in unfertilized *Xenopus* eggs. Dev Biol 2005;279(2):345–55.

[137] Medvedev S, Pan H, Schultz RM. Absence of MSY2 in mouse oocytes perturbs oocyte growth and maturation, RNA stability, and the transcriptome. Biol Reprod 2011;85(3):575–83.

[138] Su YQ, Sugiura K, Woo Y, Wigglesworth K, Kamdar S, Affourtit J, et al. Selective degradation of transcripts during meiotic maturation of mouse oocytes. Dev Biol 2007;302(1):104–17.

[139] Svoboda P, Franke V, Schultz RM. Sculpting the transcriptome during the oocyte-to-embryo transition in mouse. Curr Top Dev Biol 2015;113:305–49.

[140] Yartseva V, Giraldez AJ. The maternal-to-zygotic transition during vertebrate development: a model for reprogramming. Curr Top Dev Biol 2015;113:191–232.

[141] Ma J, Flemr M, Strnad H, Svoboda P, Schultz RM. Maternally recruited DCP1A and DCP2 contribute to messenger RNA degradation during oocyte maturation and genome activation in mouse. Biol Reprod 2013;88(1).

[142] Ma J, Fukuda Y, Schultz RM. Mobilization of dormant Cnot7 mRNA promotes deadenylation of maternal transcripts during mouse oocyte maturation. Biol Reprod 2015;93(2):48.

[143] Yu C, Ji SY, Sha QQ, Dang YJ, Zhou JJ, Zhang YL, et al. BTG4 is a meiotic cell cycle-coupled maternal-zygotic transition licensing factor in oocytes. Nat Struct Mol Biol 2016;23(5):387–94.

[144] Pasternak M, Pfender S, Santhanam B, Schuh M. The BTG4 and CAF1 complex prevents the spontaneous activation of eggs by deadenylating maternal mRNAs. Open Biol 2016;6(9).

[145] Ma J, Flemr M, Stein P, Berninger P, Malik R, Zavolan M, et al. MicroRNA activity is suppressed in mouse oocytes. Curr Biol 2010;20(3):265–70.

[146] Suh N, Baehner L, Moltzahn F, Melton C, Shenoy A, Chen J, et al. MicroRNA function is globally suppressed in mouse oocytes and early embryos. Curr Biol 2010;20(3):271–7.

[147] Murchison EP, Stein P, Xuan Z, Pan H, Zhang MQ, Schultz RM, et al. Critical roles for dicer in the female germline. Genes Dev 2007;21(6):682–93.

[148] Kim K-H, Lee K-A. Maternal effect genes: findings and effects on mouse embryo development. Clin Exp Reprod Med 2014;41(2):47–61.

[149] Streubel G, Fitzpatrick DJ, Oliviero G, Scelfo A, Moran B, Das S, et al. Fam60a defines a variant Sin3a-Hdac complex in embryonic stem cells required for self-renewal. EMBO J 2017;36(15):2216–32.

[150] Dannenberg JH, David G, Zhong S, Van Der Torre J, Wong WH, Depinho RA. mSin3a corepressor regulates diverse transcriptional networks governing normal and neoplastic growth and survival. Genes Dev 2005; 19(13):1581–95.

[151] Wen DC, Banaszynski LA, Liu Y, Geng FQ, Noh KM, Xiang J, et al. Histone variant H3.3 is an essential maternal factor for oocyte reprogramming. Proc Natl Acad Sci U S A 2014;111(20):7325–30.

[152] Lin CJ, Conti M, Ramalho-Santos M. Histone variant H3.3 maintains a decondensed chromatin state essential for mouse preimplantation development. Development 2013;140(17):3624–34.

[153] Lin CJ, Koh FM, Wong P, Conti M, Ramalho-Santos M. Hira-mediated H3.3 incorporation is required for DNA replication and ribosomal RNA transcription in the mouse zygote. Dev Cell 2014;30(3):268–79.

[154] Hatanaka Y, Tsusaka T, Shimizu N, Morita K, Suzuki T, Machida S, et al. Histone H3 methylated at arginine 17 is essential for reprogramming the paternal genome in zygotes. Cell Rep 2017;20(12):2756–65.

[155] Gu TP, Guo F, Yang H, Wu HP, Xu GF, Liu W, et al. The role of Tet3 DNA dioxygenase in epigenetic reprogramming by oocytes. Nature 2011;477(7366). 606-U136.

[156] Ma PP, De Waal E, Weaver JR, Bartolomei MS, Schultz RM. A DNMT3A2-HDAC2 complex is essential for genomic imprinting and genome integrity in mouse oocytes. Cell Rep 2015;13(8):1552–60.

[157] Hamatani T, Falco G, Akutsu H, Stagg CA, Sharov AA, Dudekula DB, et al. Age-associated alteration of gene expression patterns in mouse oocytes. Hum Mol Genet 2004;13(19):2263–78.

[158] Margueron R, Reinberg D. The Polycomb complex PRC2 and its mark in life. Nature 2011;469(7330):343–9.

[159] Posfai E, Kunzmann R, Brochard V, Salvaing J, Cabuy E, Roloff TC, et al. Polycomb function during oogenesis is required for mouse embryonic development. Genes Dev 2012;26(9):920–32.

[160] Lisboa LA, Bordignon V, Seneda MM. Immunolocalization of BRG1-SWI/SNF protein during folliculogenesis in the porcine ovary. Zygote 2012;20(3):243–8.

[161] Zheng P, Patel B, Mcmenamin M, Paprocki AM, Schramm RD, Nagi NG, et al. Expression of genes encoding chromatin regulatory factors in developing rhesus monkey oocytes and preimplantation stage embryos: possible roles in genome activation. Biol Reprod 2004;70(5):1419–27.

[162] Bultman SJ, Gebuhr TC, Pan H, Svoboda P, Schultz RM, Magnuson T. Maternal BRG1 regulates zygotic genome activation in the mouse. Genes Dev 2006;20(13):1744–54.

[163] Ho L, Crabtree GR. Chromatin remodelling during development. Nature 2010;463(7280):474–84.

[164] Gilchrist RB, Ritter LJ, Armstrong DT. Oocyte-somatic cell interactions during follicle development in mammals. Anim Reprod Sci 2004;82–83:431–46.

[165] Chen J, Torcia S, Xie F, Lin CJ, Cakmak H, Franciosi F, et al. Somatic cells regulate maternal mRNA translation and developmental competence of mouse oocytes. Nat Cell Biol 2013;15(12):1415.

[166] Franciosi F, Manandhar S, Conti M. FSH regulates mRNA translation in mouse oocytes and promotes developmental competence. Endocrinology 2016;157(2):872–82.

[167] Susor A, Jansova D, Cerna R, Danylevska A, Anger M, Toralova T, et al. Temporal and spatial regulation of translation in the mammalian oocyte via the mTOR-eIF4F pathway. Nat Commun 2015;6.

[168] Lan L, Appelman C, Smith AR, Yu J, Larsen S, Marquez RT, et al. Natural product (−)-gossypol inhibits colon cancer cell growth by targeting RNA-binding protein Musashi-1. Mol Oncol 2015;9(7):1406–20.

[169] Van Kouwenhove M, Kedde M, Agami R. MicroRNA regulation by RNA-binding proteins and its implications for cancer. Nat Rev Cancer 2011;11(9):644–56.

Translational Regulation of Gene Expression During Oogenesis and Preimplantation Embryo Development

Ecem Esencan, Emre Seli

Department of Obstetrics, Gynecology, and Reproductive Sciences, Yale School of Medicine, New Haven, CT, United States

INTRODUCTION

Oogenesis portrays a series of developmental steps that enable the oocyte to mature and become competent for fertilization. During the embryogenesis of most metazoans, primordial germ cells (PGCs) migrate to gonads from their extragonadal origin and replicate by mitosis. In the female fetus, replicated PGCs (now termed oogonia) differentiate into oocytes and enter meiosis to become primary oocytes [1]. All primary oocytes become arrested at the diplotene stage of the first meiotic prophase, a state that lasts until the female reaches reproductive maturity [2, 3]. At this point, oocytes are surrounded by a single layer of squamous granulosa cells forming primordial follicles [4].

The duration of the first meiotic arrest, dictyate, is determined by the species and can range from a few weeks to a few decades [1]. In *Xenopus* ovaries, primary oocytes start to resume meiosis 4 weeks after the formation of the embryo. This happens in mice after they reach the age of 4–6 weeks whereas in humans, primary oocytes may remain arrested in prophase I for up to 50 years [5]. This arrest allows the oocyte to transcribe and store significant amounts of messenger RNA (mRNA), which will later be translated into proteins that mediate gene expression during meiotic resumption, oocyte maturation, fertilization, and early embryogenesis while the oocyte and the embryo remain transcriptionally silent [1].

Meiotic resumption and subsequent oocyte maturation are hormonally induced and begin to occur once the female reaches sexual maturation. To mimic this, in in vitro *Xenopus laevis* models, progesterone can be used to release the oocyte from the first meiotic arrest [6, 7]. Whereas in in vivo models of *Xenopus*, mice, and humans, pituitary gonadotropins are responsible for major cytoplasmic and nuclear changes in the oocyte, which culminate in oocyte maturation [8, 9]. At a molecular level, meiosis resumption requires the activation of the maturation-promotion factor (MPF), which is a heterodimer formed by cyclin-dependent kinase 1 (CDK1 or CDC2) and cyclin B1 [10–12]. During the first meiotic arrest, MPF is inactive because CDK1 is phosphorylated at the Thr^{14} and Thr^{15} sites [13–16]. CDK activation kinase (CAK) complex mediates phosphorylation of Thr^{16} and CDC25B dephosphorylates the Thr^{14} and Thr^{15} sites of CDK1, resulting in activation of MPF and resumption of first meiosis [17, 18].

In mice and humans, significant growth and differentiation of the follicle that encloses the oocyte precede meiotic reactivation and maturation. In every estrous/menstrual cycle, a number of primordial follicles become activated to form a primary follicle with only one layer of cuboidal granulosa cells and, later, a secondary follicle that has multiple layers of cuboidal granulosa cells surrounding the oocyte. Formation of a fluid-filled cavity (antrum) within the follicle results in an antral follicle, characterized by the differentiation of granulosa cells into cumulus cells that surround the oocyte and mural granulosa cells that line the antral cavity. Follicle growth beyond the antral stage requires the pituitary FSH [19]. Once the follicle reaches the late antral stage (Graafian follicle), the LH surge stimulates the growing oocyte to resume meiosis. With completion of the first meiotic division, the primary oocyte asymmetrically divides into two haploid daughter cells, one called the secondary oocyte and the other the first polar body. LH also promotes a series of changes in the somatic compartment that result in ovulation. Subsequently, both follicle and oocyte maturation are achieved as the secondary oocyte reaches and later arrests in the metaphase of the second meiotic division, waiting to be ovulated from the antral follicle and be fertilized.

The suppression of transcription, which starts upon meiotic reactivation, persists during oocyte maturation, fertilization, and early cleavage divisions of the embryo. Reactivation of embryonic genome transcription is called zygotic genome activation (ZGA) and its timing varies between species [1]. In the *Xenopus* embryo, ZGA takes place when the embryo reaches a state of approximately 4000 cells, which corresponds to 12 consecutive mitotic divisions [20,

21]. However, in mouse and human embryos, ZGA occurs earlier, at the two-cell and four- to eight-cell stages, respectively [22–24]. Until the ZGA, maternally derived transcripts accumulated in the oocyte are the main mediators of gene expression. This requires transcript synthesis and regulation to be precise and efficient.

FACTORS DETERMINING OOCYTE TRANSCRIPTOME

Transcription is the initial step of gene expression in a metazoan cell, where a short segment of DNA is used as a template to reproduce the information it carries in the form of mRNA. Transcription is divided into three phases: initiation, elongation, and termination. In metazoan cells, synthesis of mRNA starts with binding RNA polymerase II to the promoter region on the DNA, upstream of the specific gene to be transcribed. In the elongation stage, the coding region of the DNA becomes unwound and ribonucleotides covalently bind to each other to construct the initial form of mRNA, called the precursor mRNA (pre-mRNA) in the nucleus. The chain continues to elongate on its 3′ end until RNA polymerase II reaches the stop signal.

Once the pre-mRNA is transcribed, it needs to undergo further processing in the nucleus before reaching its mature form and translocating to the cytoplasm. The first step of maturation occurs during transcription when 7-methylguanosine, which is also called 5′ cap, binds to the 5′ end of the pre-mRNA. It guides the splicing of exons, assists cytoplasmic translocation of newly transcribed mRNA with the help of RNA transport proteins, protects the mRNA against enzymatic degradation, and also binds to the cap binding protein, which is the first step to initiate translation in the cytoplasm [25, 26]. The second step of pre-mRNA maturation is cleavage, which occurs at the cleavage site located in the 3′ untranslated region (3′ UTR) at the 3′ end of the mRNA. It is regulated by a multiprotein complex called the cleavage and polyadenylation specificity factor (CPSF) [27, 28]. CPSF specifically binds to a consensus sequence, an AAUUAAA hexamer on pre-mRNA [1]. Cleavage is followed by the addition of a poly(A) tail that consists of approximately 250 adenosine nucleotides at the 3′ end of pre-mRNA with the help of poly(A) polymerase (PAP). During its elongation, the tail is bound with a poly(A) binding protein called poly(A)-binding protein nuclear 1 (PABPN1); this enhances the poly(A) tail expansion and protects it from exonucleases. When mRNA translocates to the cytoplasm, PABPN1 gets replaced with its cytoplasmic counterpart, poly(A)-binding protein cytoplasmic 1 (PABPC1), which also functions to shield poly(A) tail against deadenylation [1]. In the last step of nuclear processing, pre-mRNA goes through splicing, during which noncoding intron regions are removed with spliceosomes and the remaining exons that code for the protein of interest are ligated back to each other. Finally, with the help of the 5′ cap and RNA transport proteins, mature mRNA is transported to the cytoplasm, where it would be used as a template to be translated into the required proteins [1]. In somatic cells, this process constantly takes place based on the necessities of the cells.

In comparison to somatic cells, transcription machinery shows some differences in the oocyte and the early embryo. During oogenesis, immature oocytes are transcriptionally active and produce large amounts of a variety of mRNAs. These mRNAs are stored in cytoplasm, either concentrated at a specific region or dispersed into the entire cytoplasm as part of ribonuclease protein (mRNP) complexes. Transcriptional activity ceases as the oocyte undergoes maturation and posttranscriptional processes mediate meiotic resumption, fertilization, and embryonic genome reprogramming. Short sequence motifs in the 3′ UTR regions of the oocyte and early embryo mRNAs are essential in this process [29].

During oocyte maturation, existing mRNAs bind to MSY2, an RNA-binding protein, and with the help of low levels of RNA degradation activity, remain stable. Right after fertilization, MSY2 becomes phosphorylated, leading to unstable maternal mRNAs susceptible to degradation [30]. At the time when maternal mRNA degradation starts, ZGA is initiated. ZGA takes place in two distinct steps, minor and major ZGA. In mice, minor ZGA is weak and inefficient with a minimal contribution to zygotic gene expression. It ensues right before embryo cleavage, specifically at the male pronucleus, and leads to the translation of small polypeptides, which are temporally active during the two-cell embryo stage. Major ZGA in which there is a striking gene expression profile reprogramming occurs in the two-cell embryo stage in mice and the four- and eight-cell embryo stage in humans. Consequently, as the embryo undergoes cleavage divisions, the control of cellular processes is transferred from the maternal mRNAs and proteins to the zygotic genome [29].

ESSENTIAL REGULATORY COMPONENTS OF TRANSLATION IN SOMATIC CELLS

Translation is the last step of gene expression and takes place in the cytoplasm. It enables the decoding of the nucleotide sequence in mRNAs into polypeptide chains with specific functions. mRNAs are read in ribosomes where every third nucleotide, termed a codon, on the transcript translates into a specific amino acid. Amino acids carried by tRNAs are added until the termination codon is reached. The translated polypeptide then undergoes posttranslational modification before it becomes a functional protein. Translation occurs continuously in an energy-efficient manner in a somatic cell and the growing oocyte, according to the needs of the cell [1].

The translation machinery is very complex and requires multiple essential components for both initiation and continuation. One of them is the poly(A) tail length, which determines whether an mRNA is going to be translated. Moreover, it is imperative that the mRNA has a circular shape that enables the 3′ end to connect with the bound proteins on the 5′ end. Circular mRNA also helps to recycle ribosomes once translation is completed. Translation requires quite sophisticated complexes such as a cap binding complex and a preinitiation complex. A cap binding complex is essential for translation initiation; it attaches to the 5′ cap of mRNA and incorporates a cap binding protein (eIF4E), an RNA helicase (eIF4A), and a scaffolding protein (eIF4G) [31]. eIF4E association with the cap is the rate-limiting step of the cap-dependent initiation of translation [32]. Activation of translation via phosphorylation of eIF4E can be modulated by growth factors [33–36]. Also, eIF4E availability can be regulated with other eIF4E-binding proteins (eIF4E-BPs) competing with eIF4G [35]. eIF4G is crucial as it binds to eIF4E and eIF4A, forms a bridge between the mRNA and ribosome, and also binds to PABPC1 to assist the translation of polyadenylated mRNA through circularization. eIF4G mutations lead to disarray of translation regulation and circular mRNA formation [37–40]. The translation preinitiation complex consists of eIF1A and two other complexes called the

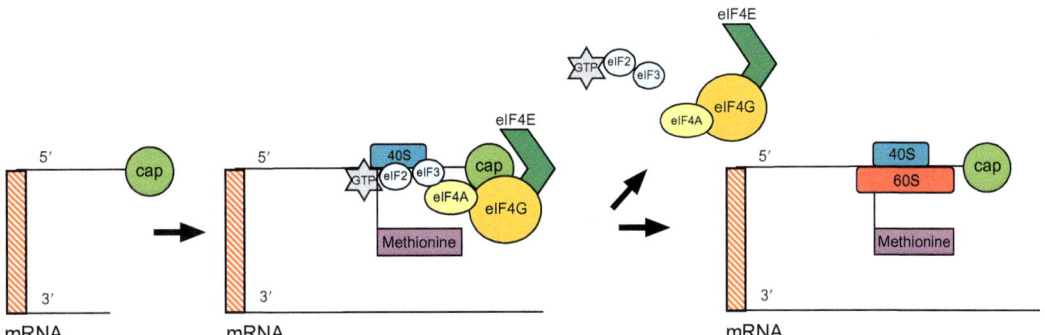

FIG. 10.1   Translation initiation. In order for translation to take place, mRNAs of interest should be bound with two different complexes. The first is the cap-binding complex consisting of cap-binding protein eIF4E, RNA helicase eIF4A, and scaffolding protein eIF4G. The second one is the preinitiation complex including the 40S subunit of ribosome, eIF3, and attached to it, a ternary complex comprising eIF2, GTP, and methionine tRNA. All together, the cap-binding complex, the preinitiation complex and mRNA form the 48S initiation complex. The 40S ribosomal subunit in the 48S initiation complex scans mRNA in the direction of the 5′ end to the 3′ end until it encounters the AUG sequence, which is also called the start codon. When 40S reaches the start codon, bound proteins are released and the 60S subunit of ribosome is recruited. Together with 40S, they scan and translate the mRNA into protein. *Modified from Friend K, Seli E. Molecular biology of the gamete. Stem Cells in Human Reproduction; 2009. p. 22 [chapter 3].*

43S complex and the ternary complex. The 43S complex is made up of 40S small ribosomal units and an eIF3 complex is attached to it. The ternary complex includes eIF2, GTP, and a tRNA carrying the initiator amino acid, methionine. eIF2 can be found in a phosphorylated form when the cell is trying to downregulate protein synthesis.

When the translation preinitiation complex binds to the cap binding complex and mRNA, the 48S initiation complex is formed; this is a rate-limiting step of translation initiation. Together with attached proteins, 40S ribosomal subunit, which is a component of 48S initiation complex, scans the mRNA of interest in the direction 5′- toward the 3′-end until it finds the start codon, AUG. After that, attached proteins are released and the 60S ribosomal subunit incorporates into 40S to form the 80S ribosomal unit, which is the functional form of ribosome (Fig. 10.1). Subsequently, translation is initiated and it continues to read the mRNA until the 80S unit comes across one of the three stop codons. Then, the amino acid chain is released for further modification.

TRANSLATION REGULATORY PATHWAYS IN OOCYTES AND EARLY PREIMPLANTATION EMBRYOS

Compared to somatic cells, the regulation of translation is more complicated and tightly controlled in the oocyte [41–45]. Because the mature oocyte and early embryo are transcriptionally silent, gene expression and production of vital functional proteins depend on maternally derived mRNAs synthesized during the first meiotic arrest. The activation and suppression of these mRNAs are regulated temporally and spatially by orchestrating cis-acting elements on the mRNAs and trans-acting factors bound to them, and may be cytoplasmic polyadenylation-dependent or -independent [46].

Polyadenylation-Independent Induction of Early Translation in the Oocyte

Rapid inducer of G2/M progression in oocytes/Speedy (RINGO/Spy) was first identified in *Xenopus* oocytes. It is one of the key proteins involved in translational activation at the time of oocyte maturation stimulation. The RINGO/Spy amount is very scant in immature oocytes, as its mRNA is kept in a dormant state without being translated [47]. Translational suppression of *RINGO/Spy* mRNA is mediated by Pumilio-2 (PUM2), which binds to the pumilio-binding element (PBE) on the 3′ UTR of *RINGO/Spy* mRNA, inhibiting its translation. PUM2 also binds to deleted in azospermia-like (DAZL) and embryonic poly(A) binding protein (EPAB). DAZL and EPAB act as co-repressors of translation when they are bound with PUM2. Upon progesterone-mediated induction of maturation in in vitro *Xenopus* models, PUM2 detaches from PBE, DAZL, and EPAB. This triggers the DAZL-EPAB complex to activate RINGO/Spy translation. The effect of the DAZL-EPAB complex seems to be vital in translation regulation as *Dazl* knockout mice models are demonstrated to be infertile in both genders [48].

RINGO/Spy recruits cyclin-dependent kinase (CDK), which in turn initiates the phosphorylation of several targets such as Aurora A/Eg2 and Musashi [49]. Eg2 is responsible for phosphorylation of CPEB in translation whereas phosphorylated Musashi induces the cytoplasmic polyadenylation and subsequent translation initiation of several early and essential transcripts, including *c-Mos* and *Cyclin B5* [50, 51]. c-MOS is responsible for the activation of the MAPK signaling pathway ensuing CPEB1 activation through phosphorylation at a different site than Eg2 [52], leading to derepression and translation of CPE-containing mRNAs, including *cyclin B1* and *Wee1* [53–55]. In the immature oocyte cytoplasm, *RINGO/Spy* mRNA is in a latent form, bound by an atypical poly(A) polymerase GLD2, CPEB1, and polyA specific ribonuclease (PARN). When the oocyte is stimulated by progesterone, CPEB1 gets phosphorylated and PARN is expelled, resulting in polyadenylation of CPE-containing transcripts by GLD2 [56, 57].

Translational Activation of Gene Expression in the Oocyte by Cytoplasmic Polyadenylation

mRNAs synthesized in the oocyte prior to the initiation of oocyte maturation undergo deadenylation in the cytoplasm and bind a group of proteins that protects them from being degraded and allows them to be stored in a silent state until their translational activation by cytoplasmic polyadenylation [58].

Cytoplasmic polyadenylation of previously deadenylated transcripts is the best-studied pathway of translational regulation. It was first identified in sea urchin eggs where, immediately after fertilization, there is a twofold increase in the number of polyadenylated mRNAs [59, 60]. Subsequently, this phenomenon was also observed in *Xenopus* and mice oocytes, and it was concluded that the poly(A) tail of certain mRNAs is elongated in the oocyte cytoplasm, resulting in their translation [61]. Usually, long poly(A) tails consist of 80–500 adenosine nucleotides whereas mRNAs with short tails have only 20–50 adenosine nucleotides; in *Xenopus* oocytes, mRNAs with long poly(A) tails (following cytoplasmic polyadenylation) include transcripts of cell cycle regulators, including *cyclin*, *Cdk2*, *Wee1*, and *Aurora A* [62].

Cytoplasmic polyadenylation is strictly controlled through several mechanisms that collectively regulate temporal expression of specific transcripts. Among those, the cytoplasmic polyadenylation element (CPE) is the best characterized translation regulatory sequence in the metazoan. CPE is located at the 3′ UTR, near the consensus cleavage signal and 20–30 nucleotides upstream of the polyadenylation signal, AAUAAA. As CPE gets closer to the polyadenylation signal on an mRNA, the efficiency of translational activation increases. CPE is explicit in polyadenylated mRNAs such as *cyclin* and other cell cycle regulators [63]; it is rich in uracil nucleotide with a consensus sequence $UUUUUA_{1-2}U$, and interacts with CPE-binding protein (CPEB). CPEB has a zinc finger and two RNA recognition motifs (RRMs) and plays a vital role in *Xenopus* and mammalian oogenesis [62]. Introduction of an antibody against CPEB or an excess CPE sequence into *Xenopus* oocytes leads to inhibition of polyadenylation and maturational arrest. CPEB global knockout mice are infertile in both sexes and display disruption of meiotic progression at the pachytene due to the inhibited translation of key regulatory mRNAs, including that of synaptonemal complex protein 3 (SCP3) [64]. Zp3-mediated conditional CPEB knockout in oocytes following the primary follicle stage also results in infertility, which this time is associated with oocyte detachment from the cumulus/granulosa cell layer, and spindle and nuclear anomalies.

CPE-CPEB interaction is the common target of various regulatory pathways. In immature *Xenopus* oocytes, inactive CPE-containing mRNAs are bound by Maskin, which is also attached to eIF4E (located at the Cap), impeding its interaction with eIF4G and inhibiting the eIF4G-mediated 43S translation initiation complex. When oogenesis is stimulated with progesterone, serine/threonine kinase Aurora A/Eg2, which is an Aurora family kinase, becomes activated as a result of phosphorylation. Eg2 transcript is also subjected to cytoplasmic polyadenylation, forming a positive feedback loop in translation activation. Activated Eg2 in turn leads to phosphorylation of Serine[174] on CPEB, which is attached to Maskin [65–67]. Phosphorylated CPEB stabilizes and activates CPSF located at the AAUAAA sequence at the 3′ end and it attaches to GLD2, a PAP. GLD2 recruits additional adenosine nucleotides and elongates the poly(A) tail by cytoplasmic polyadenylation and a poly(A)-binding protein (PABP) is recruited. It binds to eIF4G, helping it to dissociate from Maskin and its inhibitory effect is disrupted. Scaffolding factor eIF4G also binds to eIF4E to initiate translation [46] (Fig. 10.2).

The elimination of Maskin's inhibitory effect on translation upon CPEB activation also requires the presence of a PABP. However, the ubiquitous somatic cytoplasmic PABCP1 is absent from the oocyte. Instead, oocytes and pre-ZGA early embryos contain an embryonic poly(A)-binding protein (EPAB), which plays a key role in cytoplasmic polyadenylation and oocyte maturation.

EPAB's Role in Translational Regulation

EPAB is the predominant PABP in *Xenopus*, mouse, and human oocytes as well as early embryos [10, 68–71]. EPAB prevents deadenylation of mRNAs in *Xenopus* oocytes and is required for cytoplasmic polyadenylation and translational activation of key regulators of oocyte maturation [68, 70, 72, 73]. Phosphorylation of *Xenopus* EPAB at a four-residue cluster (Ser460, Ser461, Ser464, Thr465) is required for cytoplasmic polyadenylation and oocyte maturation but not EPAB's ability to stimulate translation. Importantly, in *Xenopus*, EPAB is part of both polyadenylation-dependent (CPEB1-SYMPK-CPSF) and independent (DAZL-PUM2) complexes that regulate translation [72, 74].

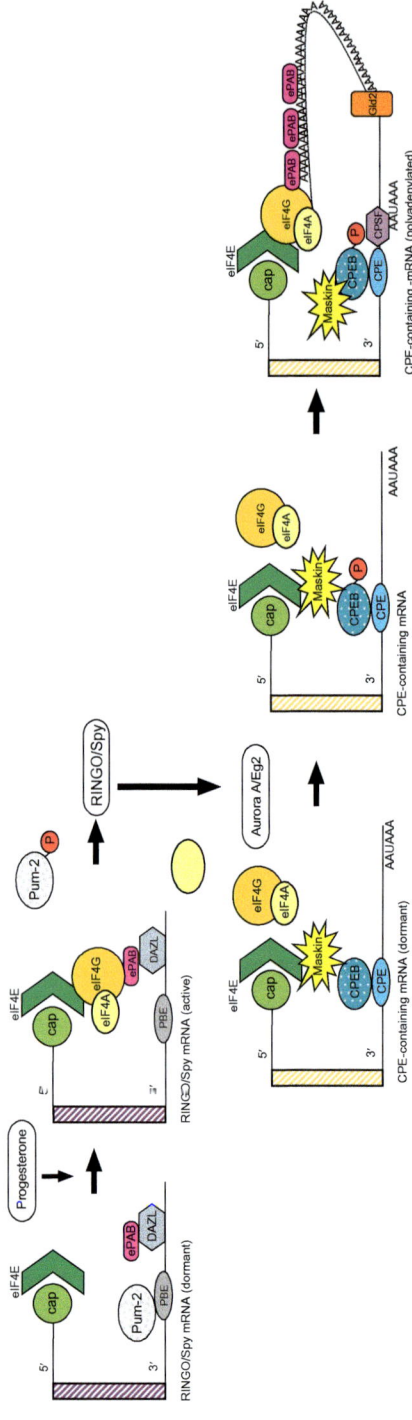

FIG. 10.2 Translation regulation. In immature oocytes, RINGO/Spy mRNA is found in a dormant form with bound Pumilio-2 at the PBE region and ePAB and DAZL at the 3′ UTR. With the progesterone stimulation, Pumilio-2 becomes phosphorylated and detaches from PBE, leading to preclusion of its suppressive effect on the mRNA. The cap-binding protein eIF4E recruits eIF4A-bound eIF4G and initiates translation of the RINGO/Spy protein. This protein, in turn, activates Aurora A/Eg2 to phosphorylate CPEB attached to CPE-containing mRNAs, which are also bound by another repressor, Maskin. Maskin inhibits association of eIF4G with eIF4E, impeding translation initiation. With the phosphorylation of CPEB, CPSF is recruited to bind the consensus sequence, AAUUAAA, and it augments binding of Gld2, which polyadenylates the 3′ tail. EPAB, a PABP, attaches to the formed poly(A) tail and eIF4G, enabling Maskin to dissociate from eIF4E. As a result, eIF4G associates with eIF4E and initiates translation of the CPE-containing mRNA. *Modified from Friend K, Seli E. Molecular biology of the gamete. Stem Cells in Human Reproduction; 2009. p. 22 [chapter 3].*

In mice, *Epab* is solely expressed in oocytes and one- and two-cell embryos until ZGA, when it becomes replaced by *Pabpc1* [69, 72]. *Epab* knockout female mice are infertile and cannot generate embryos or mature oocytes in vivo or in vitro [10]. In addition, late antral follicles in *Epab* knockout mice ovaries exhibit impaired cumulus expansion and significant decrease in ovulation [10], associated with impaired responsiveness of *Epab* knockout cumulus and granulosa cells to LH and EGF signaling [75]. Importantly, oocyte maturation can be restored by microinjection of *Epab* mRNA into *Epab* knockout follicle-enclosed oocytes (FEOs) at the preantral stage, but not into denuded, GV-stage oocytes [10, 76, 77]. Consistent with this finding, EPAB is required in oocyte-somatic cell communication in preantral stages of mouse folliculogenesis [77], suggesting that EPAB plays a role in oocyte development prior to suppression of transcription.

Cap-Independent Translation Initiation

Initiation of translation in a minority of maternal mRNAs can be accomplished without cap dependency. While these pathways are not well characterized, the best-identified modulator of cap-independent translation initiation is internal ribosome entry sites (IREs) [32].

Translational Quiescence by Deadenylation, Regulatory Proteins, and Others

Deadenylation

In somatic cells, deadenylated transcripts are subject to degradation by either decapping or exonucleases such as XPN [30, 78]. On the other hand, in oocytes and early embryos, deadenylated transcripts are translationally silenced and either they are stored in a dormant state or their degradation is initiated. Cytoplasmic deadenylation is as crucial as polyadenylation in regulating gene expression during early development, and it requires meticulous control.

The embryo deadenylation element (EDEN) is the best-characterized regulatory element for deadenylation in *Xenopus* embryos. A consensus sequence of EDEN is composed of uracil repeats with purine nucleotides in addition to adenosine-uracil sequences, and it was first identified in fertilized *Xenopus* oocyte *c-Mos* and *Aurora A* transcripts [79]. For EDEN to induce deadenylation, it should bind an EDEN-binding protein (EDEN-BP), which incorporates three RRMs [80, 81]. In in vitro human models, CUG-BP, which is closely related to EDEN-BP, can replace the EDEN-BP and induce deadenylation of transcripts [80, 82, 83].

Also, it is important to note that deadenylation doesn't always halt the translation of mRNAs completely. In *Xenopus* oocytes, *histone B4* mRNAs are commonly found with short poly(A) tails, even though the translated protein amounts are increased along oogenesis [84]. This brings attention to the existence of other regulatory pathways in translation repression.

Deadenylated transcripts might be further eliminated by decapping. During oocyte maturation, CPE-incorporated transcripts of cardinal RNA-decapping proteins such as DCP1 and DCP2 are also activated. The decapping process can be enhanced by EDC1 and EDC2 and is followed by 5' to 3' mRNA degradation, with the help of exonucleases such as XPN [85–88]. In addition to them, *PAN2* and *C-not* are also critical mRNAs in deadenylation machinery [30].

Regulatory Proteins

Maskin plays a significant role in the translational suppression of CPE-containing transcripts. As mentioned above, Maskin interacts simultaneously with both CPEB and eIF4E [89] and inhibits the assembly of the eIF4G-mediated 43S translation initiation complex on CPE-containing mRNAs [90]. Because the eIF4E amount is the rate-limiting step of translation initiation, its availability for eIF4G is crucial. Maskin is therefore an eIF4E-binding protein, which inhibits cap-dependent translation. *Cyclin B1* mRNA carries CPE and is prone to translation suppression by Maskin. When Maskin levels are decreased, the level of cyclin B1 in *Xenopus* oocytes is noted to increase [53, 91]. Maskin is not conserved in mammalians but CPE-related translation regulation is observed in mammalian oocytes, where the eIF4E-binding protein1 (eIF4E-BP1) is the most commonly encountered and studied eIF4E-BP [92]. Its phosphorylation is required for disassembly from eIF4E and is modulated by the mTOR pathway, specifically by mTORc1 [93]. Rapamycin, which is a known mTOR inhibitor, causes inhibition of eIF4E-BP1 phosphorylation in mouse oocytes, leading to asymmetric meiotic division due to decreased synthesis of proteins important in spindle formation and functioning in meiosis I and II [94, 95]. Abnormal spindle morphology results in misalignment of chromosomes, leading to aneuploidy [96].

There are additional pathways of translation suppression involving CPE. Two of those proteins that bind to CPEB, called Pumilio1/2 (PUM1/2) and RNA helicase Xp54, are involved in translational regulation in *Xenopus* oocytes [97, 98]. PUM2 is bound to PBE at the 3' UTR region of mRNAs and induces translation repression, preventing polyadenylation of the short tail. Increased PUM2 levels repress *cyclin B1* expression whereas an anti-Pumilio antibody injection increases cyclin B1 levels [99]. The inhibitory effect of PUM1/2 is considered to be a consequence of its interaction with CCR4-POP2-NOT deadenylase [100]. DAZL and EPAB are also considered to be corepressors in immature oocytes because they are interacting with PUM2 [101]. DAZL associates with uracil-rich parts of the 3' UTR in transcripts and its action is independent from polyadenylation [102]. PUM2 can detach from PBE only after progesterone induction; it enables DAZL and EPAB to activate translation initiation [103]. The suppressive effect of Xp54 was implied to be inhibited by poly(A) tail, leading to the activation of translation in *Xenopus* oocytes [104].

CPEB is also bound to PARN, which is a known deadenylase, ensuring poly(A) tail shortening of maternally derived mRNAs stored in the oocyte cytoplasm [72, 105–107]. It is expelled after CPEB phosphorylation, which simultaneously allows recruitment of GLD2 for polyadenylation [52]. Hyperphosphorylation of eIF4E-binding inhibitory proteins can impede their interaction with eIF4E and increase eIF4E availability for eIF4G [108, 109]. Although it is poorly understood, eIF4E activation has also been shown to be important in the mRNA export from the nucleus. HOXA9, which promotes eIF4E activity, has been shown to also stimulate the translocation of some transcripts [110].

Transcripts synthetized during the first meiotic arrest remain stable during oogenesis as a result of bound MSY2 and the low efficiency of RNA degradation machinery. The degradation of maternal transcripts starts with the completion of meiosis I and the entire maternally derived mRNA pool is degraded as the early embryo goes through major ZGA. As mentioned above, with maturation of oocytes, CDK1 induces phosphorylation of MSY2 and maternal mRNA stability shifts to instability, making them more vulnerable to degradation.

A decrease in maturation-related mRNA degradation was observed in oocytes when CDK1 was inhibited with roscovitine [30]. In an immature oocyte, MSY2 comprises approximately 2% of total protein content [111]. Due to instability in their posttranscriptional gene expression, MSY-knockout female mice are described as infertile and to have almost 25% fewer mRNAs compared to their wild-type counterparts [112, 113].

Small Noncoding RNAs

mRNA translational repression can also be mediated by small RNAs that are tissue-specific and 20–30 nucleotides long [30]. The three best-characterized small RNAs that are noted to increase in mammalian oocytes are microRNA (miRNA), short interfering RNA (siRNA), and PIWI-associated RNA (pi-RNA) [114].

miRNAs are 21–22 nucleotide long, evolutionarily conserved, noncoding mRNAs that repress expression of targeted genes through partial or full sequence complementarity, resulting in the degradation of target mRNAs and/or inhibition of their translation [115–117]. Precursor molecules, which correspond either to transcripts of independent miRNA genes or the introns of protein-coding mRNAs, are used to generate miRNAs; the biogenesis of miRNAs is a multistep mechanism. First, in the nucleus, primary miRNA transcripts (pri-miRNAs) are transcribed by RNA polymerase III. They are then processed by the Drosha/DGCR8 complex into precursor miRNAs (pre-miRNAs) which consist of 60–70 nucleotides. Pre-miRNAs are then transported into the cytoplasm by exportin 5, where they are cleaved by DICER to yield shorter, ~21-bp long double-stranded mature miRNAs [118]. These duplexes are then unwound upon incorporation with Argonaute (AGO) proteins 1–4 into RNA-induced silencing complex (RISC). Consequently, the single-stranded miRNA mediates the suppression or cleavage of target mRNAs through partial or full sequence complementation within the 3′ UTR of targets [119]. The recognition by miRNA is mediated mainly by complementary binding of the "seed" region, which is the first eight residues at the 5′ end of the miRNA to the 3′ UTR of target mRNA; six or seven Watson-Crick matches between the target 3′ UTR and the seed region result in successful silencing of mRNA expression [120]. The DICER enzyme is required for biogenesis of both miRNAs and siRNAs. It has been demonstrated to be essential for normal oocyte [121] and zygote [122] development, using constitutive and conditional Dicer-knockout mice, respectively. Compared to a wild-type, DICER-deficient mice have a similar number in generated and ovulated oocytes; however, its deletion leads to a catastrophe in meiosis I resulting from disorganized spindle formation and subsequent chromosomal anomalies [122, 123]. In addition, Giraldez et al. showed that in zebrafish, miR-430 is expressed at the onset of maternal-to-zygotic transition (MZT) and regulates morphogenesis during early embryo development. With the help of a microarray approach and in vivo validation, they demonstrated miR-430′s direct regulatory role on several hundred maternally expressed transcripts that accumulate in the absence of miR-430. miR-430 was also shown to accelerate deadenylation of maternal mRNAs, a possible mechanism enabling miRNA-mediated clearance of maternal mRNAs in the embryo [124]. Combined, these findings point to the conclusion that miRNAs play a key role in oocyte and preimplantation embryo development. However, in 2010, two independent studies [125, 126] presented data suggesting that miRNA activity is actually suppressed in fully grown mouse oocytes and early embryos. In addition, unlike the case for DICER-deficient mice oocytes that had significant changes in their mRNA content, oocytes of mice deficient of DGCR8, an RNA-binding protein specifically required for miRNA processing but not for siRNAs, have chromosomally normal oocyte and preimplantation embryo development and their mRNA profiles remain unaffected, even though they have no mature miRNAs [126]. These results concluded that siRNAs are crucial in mouse oocytes and the early embryo whereas miRNAs that are essential in translation regulation in somatic cells do not play a significant role in mouse oocytes. While miRNA-related translation repression was found not to be required in mouse oocytes, miRNAs are considered to be one of the primary regulators of transcript decapping in mammals [127]. It has also been demonstrated that miRNA expression changes with aging in human ovarian follicles, effecting the cumulus gene expression profile, which is essential for successful oocyte maturation [128].

siRNAs that are active only in mammalian oocytes bypass nuclear Drosha/DGCR8 processing and directly get cleaved by DICER in the cytoplasm to reach their mature form before associating with the endonucleolytic AGO2 protein to form RISC [129–131]. Similar to miRNAs, siRNAs also recognize their targets by Watson-Crick base pairing. However, they require full complementarity for target cleavage as opposed to miRNAs, which can repress transcripts even with partial complementarity. Among the AGO protein family, AGO2 is unique in having an endonucleolytic property and is the only one that incorporates siRNA to result in transcript cleavage [130]. Similar to *Dicer* knockout mice, female mice with catalytically inactive knock-in in *Ago2* demonstrated disrupted meiotic maturation with spindle formation defects leading to severe chromosomal segregation errors, even though their oogenesis and hormonal response were normal. This emphasizes the importance of siRNAs in translation regulation in mouse oocytes (especially when compared to evolutionarily lower organisms such as zebrafish or Drosophilia) [132]. Even though it doesn't affect the phenotype, it has been shown that miRNA levels diminished in *Ago2*-null oocytes, which can be explained by the decrease in miRNA stability and the turnover in the absence of AGO2 [130]. Baumann et al. also established that microinjection of siRNAs into mice oocytes can target specific mRNAs to diminish the amount of proteins of interest [133].

Similar to siRNAs, piRNAs also act through RNA interference. They are maternally inherited and are identified to be exclusive to germ cells. However, their function could have been only identified in spermatogenesis and male fertility [114].

mRNA Localization as a Determinant of Spatial and Temporal Translational Regulation

Spatial segregation of transcripts seems to play a fundamental role in regulation of translation and in distribution of genetic material [107]. Transcripts are dispersed in the cytoplasm in an irregular and asymmetric pattern, possibly making translation regulation more efficient by decreasing the energy needed to actively transport proteins. Locally synthetized proteins also have different structural and functional properties compared to the dynamically transported ones. They are generally involved in protein-protein interaction and are more exposed to posttranslational modification [107]. Besides, transcript localization at this step is crucial as they are going to be distributed into two different cells. During meiosis I, spindles travel from the central plate to the cortex for a successful asymmetrical division of cellular components and in order to form the first polar body [134].

Recent data suggest that some transcripts are localized to the nucleus. This could result either from retention in the nucleus or active transport from the cytoplasm. These transcripts are especially clustered very close to chromosomes and spindles, which signifies their importance in functions related to proper spindle formation [96, 107]. They are generally found in RNPs consisting of mRNA and RNA-binding proteins [107, 135]. Those include functional protein-encoding transcripts along with RNPs composed of long noncoding RNAs (lncRNA) such as *Xist1, Neat1*, and *Malat1* [136]. It is postulated that retention or import of these mRNAs allows temporal and spatial regulation of dynamic processes of meiosis, especially spindle assembly, as these transcripts become translated upon nuclear envelope breakdown [100].

TRANSLATIONAL REGULATION IN THE EARLY EMBRYO

Right after fertilization, an early embryo's gene expression is dependent on the maternally stored mRNAs transferred through the oocyte. Functional proteins translated from maternal transcripts enable completion of meiotic maturation, embryonic genome activation, and degradation of maternally derived transcripts.

Our knowledge about how translation is regulated in the early embryo (prior to ZGA) is not as comprehensive as the oocytes' due to the insufficient molecular studies. The small number of embryos that can be reproduced from every estrous cycle and the size of the embryos are limiting factors. Big data approaches using RNA sequencing and consensus sequence/binding protein identification can be used to overcome this problem and extend our knowledge on preimplantation embryos [137].

Similar to the translational regulation of gene expression observed in the oocyte, CPE-dependent cytoplasmic polyadenylation and other not well-identified translational pathways also seem to play an important role during early embryogenesis in *Xenopus* and mammals [138, 139]. As the mature mammalian oocyte is fertilized, calcium levels increase and activate calmodulin kinase II. This enzyme promotes the degradation of cyclin B1 and c-MOS through the ubiquitin proteasome pathway [29]. A change in the cytoplasmic environment also leads to induction of the reprogramming process [85].

Gene expression profiles change dramatically during transition from oocyte to embryo. Compared to mature human oocytes, 1-cell embryos have 79 upregulated and 70 downregulated genes expressed in their cytoplasm [137]. In a mature oocyte, cell communication and stimulus response genes are overexpressed. In comparison, these genes are mostly downregulated in the early embryo while development, metabolism, and cellular physiological process genes such as the ones regulating cell division and chromosome segregation are upregulated [140]. Maternal mRNAs encoding the continuously needed genes such as house-keeping genes or ones that are only needed in the oocyte as oocyte-specific transcription factors such as *Figla, Nobox, Lhx8, Sohlh1/2*, and *Foxo3a* or transcripts needed for embryonic gene activation are also degraded in a nonsequential way [141].

The aim of ZGA is to delete the identity of the oocyte from the embryo completely while building the zygotic totipotent identity. Maternal transcript deletion occurs in three recognizable waves [142]. After resumption of the first meiotic division and extrusion of the first polar body, the translation of dormant mRNAs of degradation machinery including *Dcp2* and *C-not* starts. During this first wave, mRNAs with short 3′ UTRs, including plenty of housekeeping genes and genes regulating basic metabolic functions, are targeted [142]. More stable transcripts rich in AU motifs and with longer poly(A) tails are not degraded at this phase [85]. The second wave occurs with fertilization and the last wave occurs along with cleavage and zygotic genome activation. The activating mechanism for these waves is unclear but considered to be initiated by change in the ratio between the cytoplasm and nucleus during rapid cleavage of the embryo [20]. Along with degradation of maternal identity, the embryo generates its own identity with minor and especially major ZGAs. Yan et al. applied next-generation sequencing in human embryos to identify the genomic profile and they described an increase in approximately 2500 genes during four- and eight-cell embryos, corresponding to major ZGA in humans [143].

CONCLUSION

Because the transcriptionally quiescent oocyte and early embryo can only modulate their gene expression through translational regulation, this process is essential to generating a viable embryo. The evolutionarily conserved cis-acting mRNA sites and trans-acting regulators in this process are carefully orchestrated to achieve timely repression and activation of specific genes.

Cytoplasmic polyadenylation and deadenylation are the most well-known end result of regulatory pathways of translation. In addition, there are other regulators such as RNA localization that control temporal and spatial translation of stored transcripts. There are still many unknown effectors of translational regulation waiting to be discovered.

References

[1] Friend K, Seli E. Molecular biology of the gamete. In: Simón C, Pellicer A, editors. Stem cells in human reproduction: basic science and therapeutic potential. 2nd ed. Boca Raton: CRC Press; 2009. p. 22–34.
[2] Page AW, Orr-Weaver TL. Stopping and starting the meiotic cell cycle. Curr Opin Genet Dev 1997;7(1):23–31.
[3] Sagata N. Meiotic metaphase arrest in animal oocytes: its mechanisms and biological significance. Trends Cell Biol 1996;6(1):22–8.
[4] Virant-Klun I, et al. Gene expression profiling of human oocytes developed and matured in vivo or in vitro. Biomed Res Int 2013;2013:879489.
[5] Rasar MA, Hammes SR. The physiology of the Xenopus laevis ovary. Methods Mol Biol 2006;322:17–30.
[6] Bayaa M, et al. The classical progesterone receptor mediates Xenopus oocyte maturation through a nongenomic mechanism. Proc Natl Acad Sci U S A 2000;97(23):12607–12.
[7] Tian J, et al. Identification of XPR-1, a progesterone receptor required for Xenopus oocyte activation. Proc Natl Acad Sci U S A 2000;97(26):14358–63.
[8] Faiman C, Ryan RJ. Serum follicle-stimulating hormone and luteinizing hormone concentrations during the menstrual cycle as determined by radioimmunoassays. J Clin Endocrinol Metab 1967;27(12):1711–6.
[9] Rao AJ, et al. The role of FSH and LH in the initiation of ovulation in rats and hamsters: a study using rabbit antisera to ovine FSH and LH. J Reprod Fertil 1974;37(2):323–30.
[10] Guzeloglu-Kayisli O, et al. Embryonic poly(A)-binding protein (EPAB) is required for oocyte maturation and female fertility in mice. Biochem J 2012;446(1):47–58.
[11] Masui Y, Markert CL. Cytoplasmic control of nuclear behavior during meiotic maturation of frog oocytes. J Exp Zool 1971;177(2):129–45.
[12] Gautier J, et al. Cyclin is a component of maturation-promoting factor from Xenopus. Cell 1990;60(3):487–94.

[13] Lundgren K, et al. mik1 and wee1 cooperate in the inhibitory tyrosine phosphorylation of cdc2. Cell 1991; 64(6):1111–22.

[14] Parker LL, Piwnica-Worms H. Inactivation of the p34cdc2-cyclin B complex by the human WEE1 tyrosine kinase. Science 1992;257(5078):1955–7.

[15] Mueller PR, et al. Myt1: a membrane-associated inhibitory kinase that phosphorylates Cdc2 on both threonine-14 and tyrosine-15. Science 1995;270(5233):86–90.

[16] Wells NJ, et al. The C-terminal domain of the Cdc2 inhibitory kinase Myt1 interacts with Cdc2 complexes and is required for inhibition of G(2)/M progression. J Cell Sci 1999;112(Pt 19):3361–71.

[17] Fesquet D, et al. The MO15 gene encodes the catalytic subunit of a protein kinase that activates cdc2 and other cyclin-dependent kinases (CDKs) through phosphorylation of Thr161 and its homologues. EMBO J 1993; 12(8):3111–21.

[18] Solomon MJ, Harper JW, Shuttleworth J. CAK, the p34cdc2 activating kinase, contains a protein identical or closely related to p40MO15. EMBO J 1993;12(8):3133–42.

[19] Oktem O, Urman B. Understanding follicle growth in vivo. Hum Reprod 2010;25(12):2944–54.

[20] Newport J, Kirschner M. A major developmental transition in early *Xenopus* embryos: I. Characterization and timing of cellular changes at the midblastula stage. Cell 1982;30(3):675–86.

[21] Newport J, Kirschner M. A major developmental transition in early *Xenopus* embryos: II. Control of the onset of transcription. Cell 1982;30(3):687–96.

[22] Clegg KB, Piko L. RNA synthesis and cytoplasmic polyadenylation in the one-cell mouse embryo. Nature 1982;295(5847):343–4.

[23] Flach G, et al. The transition from maternal to embryonic control in the 2-cell mouse embryo. EMBO J 1982; 1(6):681–6.

[24] Braude P, Bolton V, Moore S. Human gene expression first occurs between the four- and eight-cell stages of preimplantation development. Nature 1988;332(6163):459–61.

[25] Izaurralde E, et al. A nuclear cap binding protein complex involved in pre-mRNA splicing. Cell 1994; 78(4):657–68.

[26] Flaherty SM, et al. Participation of the nuclear cap binding complex in pre-mRNA 3′ processing. Proc Natl Acad Sci U S A 1997;94(22):11893–8.

[27] Zhao J, Hyman L, Moore C. Formation of mRNA 3′ ends in eukaryotes: mechanism, regulation, and interrelationships with other steps in mRNA synthesis. Microbiol Mol Biol Rev 1999;63(2):405–45.

[28] Proudfoot N. New perspectives on connecting messenger RNA 3′ end formation to transcription. Curr Opin Cell Biol 2004;16(3):272–8.

[29] Hamatani T, et al. Global gene expression profiling of preimplantation embryos. Hum Cell 2006;19(3):98–117.

[30] Svoboda P, Franke V, Schultz RM. Sculpting the transcriptome during the oocyte-to-embryo transition in mouse. Curr Top Dev Biol 2015;113:305–49.

[31] Dowling RJ, et al. mTORC1-mediated cell proliferation, but not cell growth, controlled by the 4E-BPs. Science 2010;328(5982):1172–6.

[32] Gebauer F, Hentze MW. Molecular mechanisms of translational control. Nat Rev Mol Cell Biol 2004; 5(10):827–35.

[33] Jones RM, et al. An essential E box in the promoter of the gene encoding the mRNA cap-binding protein (eukaryotic initiation factor 4E) is a target for activation by c-myc. Mol Cell Biol 1996;16(9):4754–64.

[34] Rosenwald IB, et al. Elevated levels of cyclin D1 protein in response to increased expression of eukaryotic initiation factor 4E. Mol Cell Biol 1993;13(12):7358–63.

[35] Gingras AC, Raught B, Sonenberg N. eIF4 initiation factors: effectors of mRNA recruitment to ribosomes and regulators of translation. Annu Rev Biochem 1999;68:913–63.

[36] Rhoads RE. Regulation of eukaryotic protein synthesis by initiation factors. J Biol Chem 1993;268(5):3017–20.

[37] Kessler SH, Sachs AB. RNA recognition motif 2 of yeast Pab1p is required for its functional interaction with eukaryotic translation initiation factor 4G. Mol Cell Biol 1998;18(1):51–7.

[38] Tarun Jr. SZ, et al. Translation initiation factor eIF4G mediates in vitro poly(A) tail-dependent translation. Proc Natl Acad Sci U S A 1997;94(17):9046–51.

[39] Tarun Jr. SZ, Sachs AB. A common function for mRNA 5′ and 3′ ends in translation initiation in yeast. Genes Dev 1995;9(23):2997–3007.

[40] Tarun Jr. SZ, Sachs AB. Binding of eukaryotic translation initiation factor 4E (eIF4E) to eIF4G represses translation of uncapped mRNA. Mol Cell Biol 1997;17(12):6876–86.

[41] Clemens MJ, Bommer UA. Translational control: the cancer connection. Int J Biochem Cell Biol 1999;31(1):1–23.

[42] Hake LE, Richter JD. Translational regulation of maternal mRNA. Biochim Biophys Acta 1997;1332(1):M31–8.

[43] Morris DR. Growth control of translation in mammalian cells. Prog Nucleic Acid Res Mol Biol 1995;51:339–63.

[44] Pain VM. Initiation of protein synthesis in eukaryotic cells. Eur J Biochem 1996;236(3):747–71.

[45] Saffman EE, Lasko P. Germline development in vertebrates and invertebrates. Cell Mol Life Sci 1999; 55(8–9):1141–63.

[46] Piccioni F, Zappavigna V, Verrotti AC. Translational regulation during oogenesis and early development: the cap-poly(A) tail relationship. C R Biol 2005;328(10 – 11):863–81.

[47] Ferby I, et al. A novel p34(cdc2)-binding and activating protein that is necessary and sufficient to trigger G (2)/M progression in *Xenopus* oocytes. Genes Dev 1999;13(16):2177–89.

[48] Rosario R, Adams IR, Anderson RA. Is there a role for DAZL in human female fertility? Mol Hum Reprod 2016;22(6):377–83.

[49] Arumugam K, et al. Ringo/cyclin-dependent kinase and mitogen-activated protein kinase signaling pathways regulate the activity of the cell fate determinant Musashi to promote cell cycle re-entry in *Xenopus* oocytes. J Biol Chem 2012;287(13):10639–49.

[50] Arumugam K, et al. Enforcing temporal control of maternal mRNA translation during oocyte cell-cycle progression. EMBO J 2010;29(2):387–97.

[51] Charlesworth A, et al. Musashi regulates the temporal order of mRNA translation during *Xenopus* oocyte maturation. EMBO J 2006;25(12):2792–801.

[52] Keady BT, et al. MAPK interacts with XGef and is required for CPEB activation during meiosis in *Xenopus* oocytes. J Cell Sci 2007;120(Pt 6):1093–103.

[53] Cao Q, Richter JD. Dissolution of the maskin-eIF4E complex by cytoplasmic polyadenylation and poly(A)-binding protein controls cyclin B1 mRNA translation and oocyte maturation. EMBO J 2002;21(14):3852–62.

[54] Charlesworth A, Welk J, MacNicol AM. The temporal control of Wee1 mRNA translation during *Xenopus* oocyte maturation is regulated by cytoplasmic polyadenylation elements within the 3′-untranslated region. Dev Biol 2000;227(2):706–19.

[55] de Moor CH, Richter JD. Cytoplasmic polyadenylation elements mediate masking and unmasking of cyclin B1 mRNA. EMBO J 1999;18(8):2294–303.

[56] Barnard DC, et al. Symplekin and xGLD-2 are required for CPEB-mediated cytoplasmic polyadenylation. Cell 2004;119(5):641–51.

[57] Kim JH, Richter JD. Opposing polymerase-deadenylase activities regulate cytoplasmic polyadenylation. Mol Cell 2006;24(2):173–83.

[58] Kim JH, Richter JD. Measuring CPEB-mediated cytoplasmic polyadenylation-deadenylation in *Xenopus laevis* oocytes and egg extracts. Methods Enzymol 2008;448:119–38.

[59] Slater DW, Slater I, Gillespie D. Post-fertilization synthesis of polyadenylic acid in sea urchin embryos. Nature 1972;240(5380):333–7.

[60] Wilt FH. Polyadenylation of maternal RNA of sea urchin eggs after fertilization. Proc Natl Acad Sci U S A 1973;70(8):2345–9.

[61] Lim J, et al. mTAIL-seq reveals dynamic poly(A) tail regulation in oocyte-to-embryo development. Genes Dev 2016;30(14):1671–82.

[62] de Moor CH, Meijer H, Lissenden S. Mechanisms of translational control by the 3′ UTR in development and differentiation. Semin Cell Dev Biol 2005;16(1):49–58.

[63] Mendez R, Richter JD. Translational control by CPEB: a means to the end. Nat Rev Mol Cell Biol 2001;2(7):521–9.

[64] Tay J, Richter JD. Germ cell differentiation and synaptonemal complex formation are disrupted in CPEB knock-out mice. Dev Cell 2001;1(2):201–13.

[65] Sarkissian M, Mendez R, Richter JD. Progesterone and insulin stimulation of CPEB-dependent polyadenylation is regulated by Aurora A and glycogen synthase kinase-3. Genes Dev 2004;18(1):48–61.

[66] Mendez R, et al. Phosphorylation of CPE binding factor by Eg2 regulates translation of c-mos mRNA. Nature 2000;404(6775):302–7.

[67] Mendez R, et al. Phosphorylation of CPEB by Eg2 mediates the recruitment of CPSF into an active cytoplasmic polyadenylation complex. Mol Cell 2000;6(5):1253–9.

[68] Voeltz GK, et al. A novel embryonic poly(A) binding protein, ePAB, regulates mRNA deadenylation in Xenopus egg extracts. Genes Dev 2001;15(6):774–88.

[69] Seli E, et al. An embryonic poly(A)-binding protein (ePAB) is expressed in mouse oocytes and early preimplantation embryos. Proc Natl Acad Sci U S A 2005;102(2):367–72.

[70] Wilkie GS, et al. Embryonic poly(A)-binding protein stimulates translation in germ cells. Mol Cell Biol 2005; 25(5):2060–71.

[71] Guzeloglu-Kayisli O, et al. Identification and characterization of human embryonic poly(A) binding protein (EPAB). Mol Hum Reprod 2008;14(10):581–8.

[72] Kim JH, Richter JD. RINGO/cdk1 and CPEB mediate poly(A) tail stabilization and translational regulation by ePAB. Genes Dev 2007;21(20):2571–9.

[73] Friend K, et al. Embryonic poly(A)-binding protein (ePAB) phosphorylation is required for Xenopus oocyte maturation. Biochem J 2012;445(1):93–100.

[74] Pushpa K, Kumar GA, Subramaniam K. Translational control of germ cell decisions. Results Probl Cell Differ 2017;59:175–200.

[75] Yang CR, et al. Embryonic poly(A)-binding protein (EPAB) is required for granulosa cell EGF signaling and cumulus expansion in female mice. Endocrinology 2016;157(1):405–16.

[76] Lowther KM, Mehlmann LM. Embryonic poly(A)-binding protein is required during early stages of mouse oocyte development for chromatin organization, transcriptional silencing, and meiotic competence. Biol Reprod 2015;93(2):43.

[77] Lowther KM, et al. Embryonic poly(A)-binding protein is required at the preantral stage of mouse folliculogenesis for oocyte-somatic communication. Biol Reprod 2017;96(2):341–51.

[78] Wilusz CJ, Wormington M, Peltz SW. The cap-to-tail guide to mRNA turnover. Nat Rev Mol Cell Biol 2001; 2(4):237–46.

[79] Sagata N, et al. The c-mos proto-oncogene product is a cytostatic factor responsible for meiotic arrest in vertebrate eggs. Nature 1989;342(6249):512–8.

[80] Paillard L, et al. c-Jun ARE targets mRNA deadenylation by an EDEN-BP (embryo deadenylation element-binding protein)-dependent pathway. J Biol Chem 2002;277(5):3232–5.

[81] Paillard L, et al. EDEN and EDEN-BP, a cis element and an associated factor that mediate sequence-specific mRNA deadenylation in Xenopus embryos. EMBO J 1998;17(1):278–87.

[82] Delaunay J, et al. The Drosophila Bruno paralogue Bru-3 specifically binds the EDEN translational repression element. Nucleic Acids Res 2004;32(10):3070–82.

[83] Paillard L, Legagneux V, Beverley Osborne H. A functional deadenylation assay identifies human CUG-BP as a deadenylation factor. Biol Cell 2003;95(2):107–13.

[84] de Moor CH, Richter JD. The Mos pathway regulates cytoplasmic polyadenylation in Xenopus oocytes. Mol Cell Biol 1997;17(11):6419–26.

[85] Ma J, et al. Maternally recruited DCP1A and DCP2 contribute to messenger RNA degradation during oocyte maturation and genome activation in mouse. Biol Reprod 2013;88(1):11.

[86] Coller JM, et al. The DEAD box helicase, Dhh1p, functions in mRNA decapping and interacts with both the decapping and deadenylase complexes. RNA 2001;7(12):1717–27.

[87] Schwartz D, Decker CJ, Parker R. The enhancer of decapping proteins, Edc1p and Edc2p, bind RNA and stimulate the activity of the decapping enzyme. RNA 2003;9(2):239–51.

[88] Tharun S, Parker R. Targeting an mRNA for decapping: displacement of translation factors and association of the Lsm1p-7p complex on deadenylated yeast mRNAs. Mol Cell 2001;8(5):1075–83.

[89] Stebbins-Boaz B, et al. Maskin is a CPEB-associated factor that transiently interacts with elF-4E. Mol Cell 1999; 4(6):1017–27.

[90] Richter JD, Theurkauf WE. Development. The message is in the translation. Science 2001;293(5527):60–2.

[91] Groisman I, et al. Translational control of the embryonic cell cycle. Cell 2002;109(4):473–83.

[92] Jansova D, et al. Regulation of 4E-BP1 activity in the mammalian oocyte. Cell Cycle 2017;16(10):927–39.

[93] Scheper GC, Proud CG. Does phosphorylation of the cap-binding protein eIF4E play a role in translation initiation? Eur J Biochem 2002;269(22):5350–9.

[94] Mayer S, Wrenzycki C, Tomek W. Inactivation of mTor arrests bovine oocytes in the metaphase-I stage, despite reversible inhibition of 4E-BP1 phosphorylation. Mol Reprod Dev 2014;81(4):363–75.

[95] Lee SE, et al. mTOR is required for asymmetric division through small GTPases in mouse oocytes. Mol Reprod Dev 2012;79(5):356–66.

[96] Susor A, et al. Temporal and spatial regulation of translation in the mammalian oocyte via the mTOR-eIF4F pathway. Nat Commun 2015;6:6078.

[97] Minshall N, Thom G, Standart N. A conserved role of a DEAD box helicase in mRNA masking. RNA 2001; 7(12):1728–42.

[98] Nakahata S, et al. Biochemical identification of *Xenopus* Pumilio as a sequence-specific cyclin B1 mRNA-binding protein that physically interacts with a Nanos homolog, Xcat-2, and a cytoplasmic polyadenylation element-binding protein. J Biol Chem 2001;276(24):20945–53.

[99] Nakahata S, et al. Involvement of *Xenopus* Pumilio in the translational regulation that is specific to cyclin B1 mRNA during oocyte maturation. Mech Dev 2003;120(8):865–80.

[100] Susor A, Kubelka M. Translational regulation in the mammalian oocyte. Results Probl Cell Differ 2017;63:257–95.

[101] Chen HH, et al. DAZL limits pluripotency, differentiation, and apoptosis in developing primordial germ cells. Stem Cell Reports 2014;3(5):892–904.

[102] Collier B, et al. The DAZL family proteins are PABP-binding proteins that regulate translation in germ cells. EMBO J 2005;24(14):2656–66.

[103] Vasudevan S, Seli E, Steitz JA. Metazoan oocyte and early embryo development program: a progression through translation regulatory cascades. Genes Dev 2006;20(2):138–46.

[104] Minshall N, Standart N. The active form of Xp54 RNA helicase in translational repression is an RNA-mediated oligomer. Nucleic Acids Res 2004;32(4):1325–34.

[105] Kwak JE, et al. Mammalian GLD-2 homologs are poly(A) polymerases. Proc Natl Acad Sci U S A 2004; 101(13):4407–12.

[106] Korner CG, et al. The deadenylating nuclease (DAN) is involved in poly(A) tail removal during the meiotic maturation of *Xenopus* oocytes. EMBO J 1998;17(18):5427–37.

[107] Susor A, et al. Translation in the mammalian oocyte in space and time. Cell Tissue Res 2016;363(1):69–84.

[108] Khaleghpour K, et al. Translational homeostasis: eukaryotic translation initiation factor 4E control of 4E-binding protein 1 and p70 S6 kinase activities. Mol Cell Biol 1999;19(6):4302–10.

[109] Pause A, et al. Insulin-dependent stimulation of protein synthesis by phosphorylation of a regulator of 5′-cap function. Nature 1994;371(6500):762–7.

[110] Topisirovic I, et al. Eukaryotic translation initiation factor 4E activity is modulated by HOXA9 at multiple levels. Mol Cell Biol 2005;25(3):1100–12.

[111] Roovers EF, et al. Piwi proteins and piRNAs in mammalian oocytes and early embryos. Cell Rep 2015; 10(12):2069–82.

[112] Yang J, et al. Absence of the DNA-/RNA-binding protein MSY2 results in male and female infertility. Proc Natl Acad Sci U S A 2005;102(16):5755–60.

[113] Medvedev S, Pan H, Schultz RM. Absence of MSY2 in mouse oocytes perturbs oocyte growth and maturation, RNA stability, and the transcriptome. Biol Reprod 2011;85(3):575–83.

[114] Garcia-Lopez J, et al. Global characterization and target identification of piRNAs and endo-siRNAs in mouse gametes and zygotes. Biochim Biophys Acta 2014;1839(6):463–75.

[115] Ambros V. The functions of animal microRNAs. Nature 2004;431:350–5.

[116] Bartel DP. MicroRNAs: genomics, biogenesis, mechanism, and function. Cell 2004;116:281–97.

[117] Pillai RS, Bhattacharyya SN, Filipowicz W. Repression of protein synthesis by miRNAs: how many mechanisms? Trends Cell Biol 2007;17:118–26.

[118] Adur MK, Hale BJ, Ross JW. Detection of miRNA in mammalian oocytes and embryos. Methods Mol Biol 2017;1605:63–81.

[119] Chen PY, Meister G. microRNA-guided posttranscriptional gene regulation. Biol Chem 2005;386:1205–18.

[120] Lewis BP, Burge CB, Bartel DP. Conserved seed pairing, often flanked by adenosines, indicate that thousands of human genes are microRNA targets. Cell 2005;120:15–20.

[121] Bernstein E, et al. Dicer is essential for mouse development. Nat Genet 2003;35:215–7.

[122] Tang F, et al. Maternal microRNAs are essential for mouse zygotic development. Genes Dev 2007;21(6):644–8.

[123] Murchison EP, et al. Critical roles for Dicer in the female germline. Genes Dev 2007;21(6):682–93.

[124] Giraldez AJ, et al. Zebrafish MiR-430 promotes deadenylation and clearance of maternal mRNAs. Science 2006;312:75–9.

[125] Ma J, et al. MicroRNA activity is suppressed in mouse oocytes. Curr Biol 2010;20(3):265–70.

[126] Suh N, et al. MicroRNA function is globally suppressed in mouse oocytes and early embryos. Curr Biol 2010; 20(3):271–7.

[127] Fabian MR, Sonenberg N, Filipowicz W. Regulation of mRNA translation and stability by microRNAs. Annu Rev Biochem 2010;79:351–79.

[128] Russell DL, et al. Bidirectional communication between cumulus cells and the oocyte: old hands and new players? Theriogenology 2016;86(1):62–8.

[129] Svoboda P. Long and small noncoding RNAs during oocyte-to-embryo transition in mammals. Biochem Soc Trans 2017;45(5):1117–24.

[130] Stein P, et al. Essential role for endogenous siRNAs during meiosis in mouse oocytes. PLoS Genet 2015;11(2) e1005013.

[131] Bernstein E, et al. Role for a bidentate ribonuclease in the initiation step of RNA interference. Nature 2001; 409(6818):363–6.

[132] Dallaire A, Simard MJ. The implication of microRNAs and endo-siRNAs in animal germline and early development. Dev Biol 2016;416(1):18–25.

[133] Baumann C, Viveiros MM. Meiotic spindle assessment in mouse oocytes by siRNA-mediated silencing. J Vis Exp 2015;(104).

[134] Kusch J, Liakopoulos D, Barral Y. Spindle asymmetry: a compass for the cell. Trends Cell Biol 2003;13(11):562–9.

[135] VerMilyea MD, et al. Transcriptome asymmetry within mouse zygotes but not between early embryonic sister blastomeres. EMBO J 2011;30(9):1841–51.

[136] Chen LL, Carmichael GG. Altered nuclear retention of mRNAs containing inverted repeats in human embryonic stem cells: functional role of a nuclear noncoding RNA. Mol Cell 2009;35(4):467–78.

[137] Xue Z, et al. Genetic programs in human and mouse early embryos revealed by single-cell RNA sequencing. Nature 2013;500(7464):593–7.

[138] Groisman I, et al. CPEB, maskin, and cyclin B1 mRNA at the mitotic apparatus: implications for local translational control of cell division. Cell 2000;103(3):435–47.

[139] Sousa Martins JP, et al. DAZL and CPEB1 regulate mRNA translation synergistically during oocyte maturation. J Cell Sci 2016;129(6):1271–82.

[140] Evsikov AV, et al. Cracking the egg: molecular dynamics and evolutionary aspects of the transition from the fully grown oocyte to embryo. Genes Dev 2006;20(19):2713–27.

[141] Abe K, et al. The first murine zygotic transcription is promiscuous and uncoupled from splicing and 3′ processing. EMBO J 2015;34(11):1523–37.

[142] Hamatani T, et al. Dynamics of global gene expression changes during mouse preimplantation development. Dev Cell 2004;6(1):117–31.

[143] Yan L, et al. Single-cell RNA-Seq profiling of human preimplantation embryos and embryonic stem cells. Nat Struct Mol Biol 2013;20(9):1131–9.

11

MicroRNAs in Gametes and Preimplantation Embryos: Clinical Implications

Allison Tscherner, Leanne Stalker, Jonathan LaMarre

Department of Biomedical Sciences, University of Guelph, Guelph, ON, Canada

Abbreviations

aCGH	array comparative genomic hybridization
AGO	argonaute
ART	assisted reproductive technologies
ATM	ataxia telangiectasia mutated
BSA	bovine serum albumin
cAMP	cyclic adenosine monophosphate
CL	corpus luteum
COC	cumulus-oocyte complex
DGCR8	DiGeorge syndrome critical region 8
eCG	equine chorionic gonadotropin
endo-siRNA	endogenous short interfering ribonucleic acid
FIGLA	factor in the germline alpha
FSH	follicle stimulating hormone
GDP	guanosine diphosphate
GTP	guanosine triphosphate
GV	germinal vesicle
HAS2	hyaluronan synthase 2
hCG	human chorionic gonadotropin
IVF	in vitro fertilization
IVM	in vitro maturation
LH	luteinizing hormone
MAPK	mitogen-activated protein kinase
MII	metaphase II
miRISC	microRNA-induced silencing complex
miRNA	micro ribonucleic acid
mRNA	messenger ribonucleic acid
MSK1	mitogen- and stress-activated protein kinase
MZT	maternal-zygotic transition
NGS	next generation sequencing
NOBOX	newborn ovary homeobox protein
Pcna	proliferating cell nuclear antigen
PCOS	polycystic ovarian syndrome
PGC	primordial germ cells
PGS	preimplantation genetic screening
piRNA	Piwi-interacting ribonucleic acid
pre-miRNA	precursor microRNA
pri-miRNA	primary microRNA
TGF-β	transforming growth factor beta
TRBP	TAR RNA-binding protein
TZP	transzonal projections
UTR	untranslated region
ZP	zona pellucida

INTRODUCTION

The advent of assisted reproductive technologies (ART) has profoundly influenced clinical approaches to infertility, with approximately 1.5 million human embryo transfers and 500,000 births annually employing these approaches [1]. In vitro fertilization (IVF), the central technique in modern ART, has become a broadly employed clinical practice that has enjoyed widespread success, although a number of important questions and issues remain [1]. This is particularly true with respect to the accurate and timely identification of gametes and embryos that have optimal potential to experience successful development, pregnancy, birth, and postnatal health (reviewed in [2, 3]).

Improvements in the success of ART outcomes in specific patient groups have resulted from innovative approaches to embryo selection that aim to eliminate embryos with lower developmental potential due to aneuploidy, a common cause of ART failure [4, 5]. Very recently, preimplantation genetic screening (PGS) of DNA from trophectoderm biopsies using next-generation sequencing (NGS) [6] has proven to be a powerful and effective tool in this regard. The basic principle underlying these approaches is that changes in DNA sequence or abundance resulting from chromosomal abnormalities severely limit developmental potential and negatively impact fertility. Because technological improvements have allowed the safe, rapid, and sensitive detection of such changes, this approach can be employed to avoid transferring embryos that are unlikely to develop successfully. However, there remains considerable disagreement over when and how these techniques should be employed [7], and further refinement will likely be necessary before widespread implementation can occur.

In addition to enabling a detailed and sensitive examination of genetic (i.e., DNA-based) changes in embryo genomes in both experimental and diagnostic contexts, NGS and other molecular techniques have facilitated comprehensive studies of the mammalian embryo transcriptome—the specific RNAs that are transcribed from DNA at different developmental stages and under different conditions in vitro and in vivo [8–13]. Studies in the bovine model in particular provide compelling evidence that the oocyte and embryonic transcriptome reflect not only the underlying characteristics of the DNA from which they are derived, but also the influence of different environmental factors on the expression of the appropriate genetic sequences. In addition to the extensive complement of mRNAs identified in oocyte and embryonic transcriptomes, a number of recent studies and reviews have highlighted the presence and potential roles of small RNAs (micro (mi)RNAs, Piwi-interacting (pi)RNAs) during preimplantation development [14–20].

While the embryonic transcriptome is profoundly important for normal development, one of the most unique properties of embryos is that they are usually considered to be "transcriptionally silent" during the earliest developmental phases [21, 22]. In this period, the embryo relies instead on transcripts that are expressed and stored during oogenesis and folliculogenesis. This transcriptional legacy dictates the cohort of RNAs present during the period before the zygotic genome becomes actively transcribed [23–26]. The active destruction of these transcripts during the critical maternal-to-zygotic transition actually relies in many species on small RNA pathways that utilize individual miRNAs to target the transcripts for degradation [14, 25, 26]. Because the entire oocyte transcriptome, and much of that in the early embryo after fertilization, is generated during the formation and maturation of female gametes, our discussion of some clinically relevant small RNA pathways in the embryo will begin with a review of folliculogenesis, highlighting key processes in which small RNAs have been implicated. This will be followed by an introduction to the biology of miRNAs and their roles as regulators of cellular function in the context of gametes and embryos. The chapter will conclude with a discussion of the clinical potential of miRNAs as biomarkers of gamete quality and embryo health.

OVERVIEW OF FOLLICULOGENESIS AND OOCYTE MATURATION IN MAMMALS

Oogenesis and Folliculogenesis

Oogenesis and folliculogenesis are intimately related processes from their inception. Under the influence of many factors, oogenesis begins with the migration of primordial germ cells (PGCs) and colonization of the developing fetal gonad where these cells proliferate mitotically, giving rise to oogonia (reviewed in [27]). Between weeks 10–12 of fetal development in humans, oogonial proliferation with incomplete cytokinesis results in the formation of germ cell nests, surrounded by elongated cells characteristic of granulosa cells [28]. After developmental week 12, few dividing cells remain; most germ cells have entered meiosis, which identifies them as oocytes, while pregranulosa cells have elongated considerably and extend around and between individual oocytes to form the earliest primordial follicles [28].

Once meiosis is initiated, oocytes enter a lengthy dormant stage, arrested at the diplotene stage of meiosis I (dictyotene stage), until final maturation and ovulation, or until the follicle becomes atretic [29]. During this period, which can exceed 50 years in humans, the large population of nongrowing primordial follicles represents the source of developing follicles throughout the individual's reproductive life (reviewed in [30]). Cohorts of primordial follicles are continuously recruited from this pool as the oocyte begins its growth phase and the single layer of squamous primitive granulosa cells becomes cuboidal and proliferative, forming a primary follicle [31]. Some members of this cohort undergo development to the multilayered secondary follicle stage, which can occur in the absence of gonadotropins, until a cavity forms and begins to fill with follicular fluid, creating an early antral follicle [32]. Development and differentiation past this stage are dependent primarily on the gonadotropins follicle stimulating hormone (FSH) and luteinizing hormone (LH). A rise in FSH creates the basis for follicle selection during which one follicle becomes a dominant preovulatory follicle, inhibiting the growth and development of others from the cohort (the roles of gonadotropins in this process are reviewed in [33]). Over the course of folliculogenesis, a follicle that becomes dominant will grow from a size of approximately 40 μm to greater than 20 mm [34], and the oocyte itself will grow from approximately 35 to 120 μm in diameter, a size that is reached by few other mammalian cells [35].

Gonadotropin-Dependent Follicle Growth and Concomitant Maturation of the Cumulus-Oocyte Complex

Antrum formation marks a critical transition for both the follicle and the oocyte itself. Oocytes residing within preantral follicles are unable to progress from their arrest at the dictyotene stage of meiosis, which is attributed to an insufficiency in critical factors necessary to drive meiotic progression (reviewed in [32]). Oocytes in antral follicles become meiotically competent and will resume meiosis spontaneously if they are removed from the follicle [36, 37]. Importantly, however, achievement of meiosis resumption alone is insufficient to support fertilization and embryo development. Further oocyte maturation that comprises a series of nuclear and cytoplasmic events including transcriptional changes, imprinting, cortical granule release, histone exchange, and changes in microtubule organization (many events summarized in [38, 39]) is additionally required. Most of these required events require somatic cell support, particularly from cumulus cells: an essential group of specialized granulosa cells that differentiate at the antral stage of folliculogenesis [32]. The removal of oocytes from their surrounding cumulus cells before maturation significantly affects nuclear and cytoplasmic maturation as well as the rate of development to the blastocyst stage if subsequent fertilization occurs [40–44].

The relationship between the oocyte and its cumulus oophorus is complex. During the final stages of maturation, cumulus cells propagate external endocrine and biochemical signals to promote meiotic progression and the achievement of competency in the oocyte while responding to oocyte cues that regulate a number of cumulus cell functions [45–48]. Dramatic functional changes in cumulus cells are induced by the preovulatory surge of gonadotropins in vivo and directly by FSH in vitro [49, 50], preparing the follicle for ovulation. A remarkable conformational change occurs as cumulus cells secrete and become embedded in a matrix of hyaluronic acid to cause cumulus expansion and separation from the follicle wall [51]. Meanwhile, biochemical changes initiate a drop in oocyte cyclic adenosine monophosphate (cAMP), leading to the resumption of meiosis [52]. These essential cumulus-oocyte communications occur through several mechanisms, including the transfer of small molecules via gap junctions [53] and by transzonal projections (TZPs), granulosa cell extensions that traverse the glycoprotein matrix of the zona pellucida (ZP) to make direct contact with the oocyte [54]. Paracrine signaling by oocyte and cumulus cell secreted factors also play a substantial role in the concomitant development of the cumulus cells and the oocyte (cumulus-oocyte complex (COC)) [47, 55–57].

Many factors present in this microenvironment influence cumulus cell gene expression, which in turn results in the functional changes necessary for extracellular matrix synthesis, cumulus expansion, and the readiness of the COC for fertilization [58–60]. A number of cumulus cell ribonucleic acid transcripts are induced and rapidly turned over during the final maturation period, and suboptimal levels of these transcripts may ultimately reflect a reduced state of oocyte competency [61]. In vitro maturation (IVM) of COCs extracted from small follicles is a technique that has advantages for certain populations of patients, but can result in an intrinsic reduction of oocyte developmental potential (reviewed by [62]). It has been observed that IVM results in a divergent population of cumulus cell transcripts when compared to COCs matured in vivo, and these gene expression differences may contribute to the adverse effects of IVM. The transcript profile of cumulus cells derived from human COCs matured in vitro shows reduced expression of cumulus expansion factors such as *amphiregulin, prostaglandin-endoperoxide synthase 2*, and *hyaluronan synthase 2* compared to COCs matured in vivo [63]. In the bovine model, the cumulus cell transcriptome from COCs matured in vitro is enriched for genes involved with stress responses while the abundance of key genes required for cumulus expansion and oocyte maturation (such as *TNF alpha induced protein 6, Inhibin beta A, Follistatin*) is lower than in vivo matured cumulus cells [64]. Together, these data suggest that cumulus cell gene expression changes are strongly linked with developmental potential. Transcriptome analysis has shown a relationship between the expression of specific cumulus cell genes and oocyte competence, as indicated by nuclear maturation, cleavage and blastocyst rates, and oocyte apoptosis [65–68]. Cumulus cell gene expression is therefore a rich potential resource for the identification of specific factors that are highly correlated with developmental potential. A panel of eight differentially expressed cumulus cell genes has shown prognostic value as an indicator of early cleavage in human zygotes, which is a useful clinical parameter of embryo viability [69].

Prior to the antral stage of folliculogenesis, the growing oocyte is highly transcriptionally active [70]. Once the antral cavity forms, the oocyte has reached full size, and chromatin remodeling in preparation for the resumption of meiosis is accompanied by a progressive shutdown of transcriptional machinery whereupon the oocyte is considered to be essentially transcriptionally silent during the germinal vesicle (GV) to metaphase II (MII) transition [21, 70]. If the oocyte becomes fertilized, the early embryo will remain transcriptionally silent until the maternal-to-zygotic transition and activation of the embryonic genome (discussed in further detail below), and the early embryo will rely on proteins and RNA transcripts stored in the oocyte to perform essential processes. TZPs appear to play important roles in delivering these molecules to the oocyte [71, 72]. Our understanding of the roles and importance of TZPs has recently been improved substantially by Macaulay et al. [73], who demonstrated that, in addition to small molecules, TZPs traffic small and large (>200 nucleotide) RNA transcripts from cumulus cells to the oocyte in preparation for maturation. This supports a model where the transcriptionally silent oocyte relies to some extent on the transcriptional activity of the somatic cells with which it maintains direct contact. RNA sequencing analysis of bovine TZPs show the presence of many mRNA sequences as well as small and regulatory RNAs [72] that appear to provide essential support to the oocyte as it prepares for fertilization.

The Maternal to Zygotic Transition: Destruction of Maternal Transcripts and Activation of the Embryonic Genome

Immediately after fertilization, the embryonic genome remains transcriptionally quiescent, and a relatively rapid series of cell cycles dominates the first hours of embryo development. Transcription resumes around the 4- to 8-cell stage in humans [22] and the 8- to 16-cell stage in bovine embryos [74], at which time the embryonic genome is considered to be activated. This critical process is apparent through changes in the pattern of expressed polypeptides [22], visualization of nucleolus organizer regions [74], sensitivity of transcripts [23] and peptides [22] to α-amanitin treatment, RNA sequencing, and RNA sequencing of incompletely spliced nascent transcripts [24].

From the onset of transcriptional quiescence to the resumption of de novo transcription in the cleavage-stage embryo, the maturing oocyte and subsequent zygote rely on RNA transcripts of maternal origin. This pool of maternal RNA influences the two distinctly different processes of oocyte maturation and early embryo development and, as the requirements of the embryo change, a general degradation of maternal transcripts occurs (extensively studied in many models and summarized in [14, 25, 26, 75]). Microarray analysis has shown that a substantial fraction of the maternal transcriptome from the oocyte is degraded at the maternal-zygotic transition (MZT) in mammalian embryos [76], with a decline in the expression of many genes necessary for oogenesis, oocyte maturation, and fertilization [77–79]. Importantly, the persistence of maternal effect genes can induce developmental arrest [80], and the timely degradation of mRNA is a highly conserved, critical checkpoint in embryo development. Numerous molecular pathways are involved in the clearance of maternal mRNA at the MZT. Several RNA binding proteins, including Argonaute, Smaug, and Pumilio, recruit the enzymes and factors that are necessary to direct deadenylation and degradation of mRNA (extensively reviewed in [81, 82]). Considerable evidence supports a model whereby small noncoding RNAs, particularly miRNAs, that are present in the zygote are important participants in the decay of maternal mRNAs across multiple species [15, 16, 78, 83].

MicroRNA BIOSYNTHESIS, MECHANISM OF ACTION, AND REGULATION

MicroRNAs: Ubiquitous Regulators of Cellular Gene Expression

MicroRNAs are an important class of small noncoding RNAs that act as effectors of posttranscriptional gene silencing and have emerged as potent regulators of gene expression in a vast array of biological systems and cellular contexts. Genes encoding miRNAs are highly conserved among eukaryotes and participate broadly in development, physiology, and pathology by influencing specific biological processes such as proliferation, differentiation, signal transduction, and apoptosis (extensively reviewed in [84] and with detailed examples of adult-specific physiology [85]). These small RNAs typically act by repressing the activity of "target" genes through antisense base-pairing interactions with the 3' untranslated region (UTR) of the target mRNA. Although active miRNA molecules are typically ~22 nucleotides long, they bind targets via Watson-Crick base pairing involving primarily an essential 7 nucleotide seed sequence [86]. Considering this short pairing requirement, one miRNA has the potential to bind hundreds of target 3'UTRs, and any 3'UTR can theoretically be targeted by hundreds of different miRNAs. Based on in silico analysis, miRNAs were initially implicated as regulators of ~30% of expressed mammalian genes [86], a figure that has since been extended to more than 60% [87]. Given the potential for these molecules to profoundly alter cellular gene expression and function, the mechanisms by which miRNAs are themselves regulated during the complex sequential biosynthesis process (see Fig. 11.1, [88]) remain very active and important areas of research.

Transcriptional Regulation of microRNAs

Genes encoding miRNAs are transcribed as long primary (pri-miRNA) transcripts that form local hairpin structures containing mature miRNA sequences [89]. This is most frequently mediated by RNA polymerase II, and the resulting transcripts contain 5'cap structures and poly(A) tails, structurally resembling mRNA transcripts and demonstrating polymerase II "gene-like" expression control [90]. MicroRNA-encoding genes are found throughout the genome, with 40%–70% of vertebrate miRNA genes mapping to introns of protein-coding and noncoding units, and the remainder arising from intergenic regions [91]. A common, though not universal, feature of miRNA genes is their organization into clusters of several miRNAs [92], which are transcribed as single polycistronic transcriptional units [89] containing multiple hairpin structures and controlled by common regulatory sequences. While polycistronic miRNAs from the same parent transcript are often concurrently expressed at similar levels, several miRNA clusters are transcribed as a single pri-miRNA but

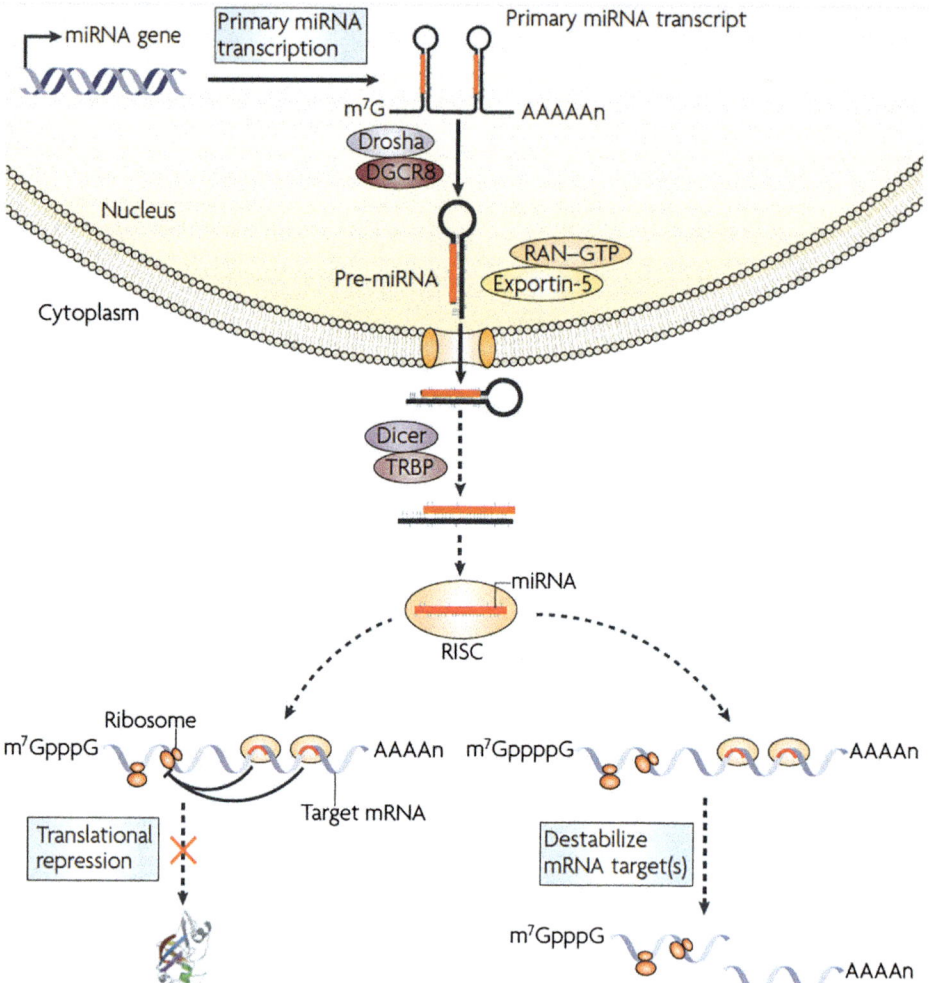

FIG. 11.1   MiRNA biosynthesis and mechanism of action. MiRNAs are initially transcribed as primary-miRNA transcripts that undergo sequential processing steps until they are loaded in their mature form into an RNA-induced silencing complex (RISC). In participation with RISC, miRNAs pair with mRNA targets and initiate translational repression or target mRNA destabilization. See references in text. *Reprinted from Lodish HF, Zhou B, Liu G, Chen C-Z. Micromanagement of the immune system by microRNAs. Nat Rev Immunol 2008, with permission from RightsLink Permissions Springer Customer Service Centre GmbH: Springer,* ©*2008.*

expressed at different levels as mature miRNA molecules [93], suggesting processing differences downstream of transcription.

From a regulatory standpoint, the genomic location of any specific miRNA gene is one key factor determining its expression profile. Mono- and polycistronic pri-miRNAs that are present in genomic regions distinct from other known transcripts are controlled by miRNA-specific promoters. Individual or clustered pri-miRNAs that occur within introns of genes can be produced as a by-product of transcription of the host gene or be independently controlled by their own miRNA-specific promoters that lie within the host gene (reviewed in [94]). In these instances, miRNA expression can occur independently of host gene expression. Regardless of their location, the transcription factors and chromatin features that regulate pri-miRNA transcription appear essentially the same as those that regulate coding transcript expression.

MicroRNA Processing and Export

Once transcribed, nascent pri-miRNAs undergo multiple processing events before reaching a mature, active state. The initiation of miRNA processing is executed in the nucleus by the class II RNase III enzyme Drosha [95], in conjunction with the RNA binding protein DiGeorge Syndrome Critical Region 8 (DGCR8) [96, 97] that together form the minimal constituents of the "Microprocessor Complex" [97, 98]. DGCR8 recognizes pri-miRNA substrates [99] while Drosha creates a double strand break to release a ~70 nucleotide hairpin precursor (pre-) miRNA.

Purified Drosha and DGCR8 are capable of performing pri-miRNA processing in vitro. However, Drosha-containing complexes isolated from human cells are much larger than the combination of Drosha plus DGCR8. The larger complex contains a number of additional RNA-associated proteins such as DEAD-box RNA helicases, which appear to influence pri-miRNA processing in vivo [98]. Importantly, these auxiliary proteins associated with the Microprocessor are a major source of regulatory control that may act to facilitate or suppress Microprocessor activity and influence miRNA abundance in specific cellular contexts (reviewed in [100]). In addition to the influence of auxiliary proteins, a posttranscriptional cross-regulatory circuit between Drosha and DGCR8 constrains their own availability, creating an additional level of pri-miRNA processing control [101].

After initial processing by the Microprocessor, the resulting pre-miRNAs are bound by the nucleo-cytoplasmic transport factor Exportin 5, which binds nuclear pre-miRNA cargo in the presence of Ran-guanosine triphosphate (GTP) [102], and releases this cargo upon hydrolysis of Ran-GTP to Ran-guanosine diphosphate (GDP) in the cytoplasm [103]. Once in the cytoplasm, pre-miRNAs are cleaved by the RNase III enzyme Dicer [104, 105] into a ~22 nucleotide double-stranded RNA duplex that, in association with the double-stranded RNA binding protein (TAR RNA-binding protein: TRBP), is recruited to an Argonaute protein [106], where one strand becomes incorporated into an miRNA-induced silencing complex (miRISC).

MicroRNA Target Binding and Destabilization

MiRISCs are multisubunit ribonucleoprotein complexes that mediate miRNA-driven gene silencing. The core of the miRISC consists of the complementary miRNA guide strand that provides sequence specificity to the silencing complex, and an Argonaute from the AGO clade that presents the miRNA seed sequence for target scanning and coordinates the gene silencing events that follow [107]. AGO-directed silencing of mRNA targets can occur by translational repression or deadenylation and decay [108, 109], reviewed by [110]. In some circumstances, target destruction occurs directly by mRNA cleavage when a catalytically active AGO is present [111]. The most common mechanisms of gene silencing are translational repression and deadenylation followed by degradation (reviewed in [112]).

MicroRNAs IN REPRODUCTION

MicroRNAs in the Ovary

The ovary is a highly dynamic organ that undergoes constant remodeling, wound healing, and regular responses to endocrine signals (reviewed in [113, 114]). MiRNAs are abundant in the ovary, are expressed in unique patterns specific to the different cell types, and their expression fluctuates actively throughout the estrus cycle [115] and reproductive aging [116]. Because of these unique profiles, it is widely postulated that miRNAs play important roles in ovarian biology. Profiling of miRNAs expressed in the ovaries of model species (mouse, cow, sheep, and pig) suggests that miRNAs from the let-7 family, miR-21, miR-99a, miR-125b, miR-126, miR-143, miR-145, and miR-199b are generally the most abundantly expressed [117]. Analysis of the fetal bovine ovary, where miRNAs represented the significant majority of all small RNAs detected, reveals that miR-99a and miR-125b are the most abundant miRNAs, and that a group including miR-99a, miR-10b, miR-199a, miR-424, miR-100, miR-455, and miR-214 showed 10-fold higher relative abundance when compared to other bovine tissues [118], suggesting a potential relationship with ovarian development. In the adult bovine ovary, the let-7 family, miR-21, miR-23b, miR-24, miR-27a, miR-126, and miR-143, were detected frequently [119]. This study also demonstrated some important differences between the expression patterns of the fetal and adult ovary, particularly the expression of miR-29a that was detected in follicular cells but not in the fetal ovary and that fluctuated with the ovarian cycle. Dynamic changes in miRNA expression throughout the follicular and luteal phases of the estrus cycle have since been confirmed, including the induction of miR-29a in the early luteal phase of the cycle [120], supporting a role for this miRNA in follicular recruitment, dominance, and the suppression of subordinate follicles. While it is difficult to make direct comparisons between miRNA profiles generated in different studies and with different sequencing technologies, the common features between different datasets and model organisms support conservation of function in the ovary.

Functional categorization of the miRNAs most abundant in the adult ovary indicate that the pathways enriched for targeting by these miRNAs include those known to be important for ovarian function [119]. In many instances, molecular targets that are known to cause functional changes during specific phases of the estrous cycle correlate inversely with differentially expressed miRNAs, suggesting that cyclic miRNAs actively participate in cyclic changes in ovarian function [120]. These studies suggest that the gonadotropins FSH and LH are regulators of miRNA and that miRNA-mediated changes in gene expression participate in their effects. In support of this, FSH supplementation of cultured granulosa cells induces rapid changes in miRNA expression; 31 miRNAs including miR-23b, miR-29a, and miR-30d were significantly altered within 12 hours of FSH treatment [121]. In vitro, mir-21 is strongly induced within hours of equine chorionic gonadotropin (eCG) or human chorionic gonadotropin (hCG) [122] treatment. In vivo, ovulation-inducing doses of hCG elicit changes in 13 known miRNAs within 4 hours of treatment [123], suggesting that circulating hormones strongly influence miRNA induction in the preovulatory and periovulatory periods.

Changes in miRNA expression during ovarian development, puberty, throughout the estrous cycle, and in response to gonadotropins in vitro and in vivo clearly implicate miRNAs as active regulators of ovarian function. While many studies have predicted targets of miRNA action in the ovary based on gene ontology and pathway analysis, validation of these targets and definitive evidence of specific miRNA-mediated gene silencing in the ovary is lacking. Unfortunately, most of the studies demonstrating the functional role of miRNAs in ovarian function have utilized gene knockouts of Dicer in the mouse and demonstrate that oocyte-specific Dicer knockout results in both depletion of almost all miRNAs and developmental failure at the one- to two-cell stage [17], and that more generalized knockout of Dicer impairs ovarian function [124]. While these results originally suggested essential functions for miRNA in the ovary, evidence has since been compiled that Dicer-dependent endogenous short interfering RNAs (endo-siRNAs) appear to play a much larger role in the mouse ovary than initially suspected [125].

Insights into the roles of miRNAs in ovarian biology have often been obtained through detailed examination of their mRNA targets and their downstream functions. A direct functional role for miRNAs is revealed by studies on the regulation of proliferating cell nuclear antigen (Pcna), which influences primordial follicle assembly by promoting oocyte loss in the fetal and neonatal ovary [126]. Mir-376a directly binds the *Pcna* 3'UTR and promotes oocyte survival by suppressing *Pcna*. As a result, miR-376a overexpression increases the number of primordial follicles and decreases oocyte apoptosis in the fetal mouse ovary [127]. Other studies have shown that MiR-145 is expressed in neonatal mouse ovaries, and the administration of miR-145 antagonists significantly depletes primordial follicle numbers while increasing the number of growing follicles [128].

MicroRNAs also influence apoptosis in the adult ovary. MiR-26b expression positively correlates with progression of follicular atresia and acts as a direct regulator of granulosa cell apoptosis in vitro by targeting ataxia telangiectasia mutated (ATM) protein, increasing the number of DNA strand breaks and apoptosis [129]. Subsequent studies have demonstrated that hyaluronan synthase 2 (HAS2) is also a direct target of miR-26b, through which it alters the HAS2-HA-CD44-Caspase-3 apoptosis pathway in granulosa cells [130]. Inhibition of miR-21 in the mouse ovary has also been linked to initiation of apoptosis in the follicle, although the direct molecular targets are presently unknown [122].

Steroidogenesis in the ovary is subject to miRNA-mediated control, as demonstrated by the miR-378-aromatase-estradiol axis. The aromatase (*Cyp19a1*) 3'UTR is a direct target of mir-378, and overexpression of this miRNA in granulosa cells decreases aromatase protein and estradiol production [131]. MiR-224 is expressed in follicles at various stages of development, and is involved in TGF-β mediated granulosa cell proliferation and follicle maturation [132]. Treatment of mural granulosa cells with TGF-β or miR-224 mimics increases proliferation and significantly increases both *Cyp19a1* and estrogen release [132]. Together, these findings demonstrate how a single miRNA may act on several targets to regulate multiple aspects of follicular maturation, including steroidogenesis, granulosa cell proliferation, and inhibition of atresia.

After an ovulatory event, theca and granulosa cells from the remains of the preovulatory follicle undergo a series of morphological and biochemical changes to form the corpus luteum

(CL) [133]. There is limited but growing knowledge concerning the population of miRNAs present in the CL. Stage-specific miRNA profiling studies demonstrate that miRNAs, including let-7c, miR-21, and miR-126, undergo dynamic changes from luteinization to late-cycle CL and CL regression (reviewed by [134]). Importantly, these profiling studies have revealed that a small subset of miRNAs diverge between regressing CLs of nonpregnancy and the CL of pregnancy, suggesting they may be specifically involved in the rescue of the CL from luteolysis that is essential for the maintenance of pregnancy [135].

MicroRNAs in the COC: Implications for Oocyte Maturation

The cumulus-oocyte complex is the product of ovulation and, as discussed previously, one crucial component of oocyte maturation is the bidirectional communication and transit of small molecules between the gamete and its somatic support cells in order to prepare the oocyte for a fertilization event, providing the embryo with the materials needed to undergo the first cell division stages [71, 73, 136]. Considering the combined importance of gene expression changes in the cumulus cells and the accumulation of developmentally essential RNAs in the oocyte, the assortment of miRNAs present in the COC during final maturation has come under intense scrutiny. MicroRNAs show differential enrichment between the cumulus and oocyte compartments of the bovine COC, and this enrichment changes over the course of in vitro maturation. Thirty-nine miRNAs, including miR-183, miR-18b, miR-96, miR-150, and miR-205, are enriched in immature oocytes compared to their surrounding cumulus cells while 45 miRNAs, including miR-317, miR-146a, miR-18b, and miR-222, are more abundant in mature oocytes relative to cumulus cells [137].

The different miRNA expression profiles observed between the germ cell and somatic compartments of the COC suggest that they may be actively regulated in each compartment, implying specific functional importance for oocyte maturation and cumulus expansion. Despite many dramatic changes in mRNA and miRNA levels, the number of experimentally validated miRNA-target relationships in the oocyte remains limited. Storage of oocyte RNA often involves localization to specific bodies or cytoplasmic granules in the form of ribonucleoproteins in order to repress translation and protect these transcripts from decay until the time of translational activation [138]. In these compartments, dormant RNAs are bound to a complex of 3'UTR binding proteins that mask miRNA binding sites and prevent miRNA-mediated gene silencing (reviewed by [139]). Our incomplete understanding of miRNA-mediated gene regulation, combined with the technical challenges of experimentally manipulating miRNA levels in immature oocytes in order to validate their targets, limits our knowledge concerning the overall importance of miRNAs in oocyte maturation. This issue has been further complicated by the realization that the mouse has significant limitations as a translational model for miRNA dynamics in the oocyte of many other species (including humans), as miRNA expression is uniquely suppressed in mouse oocytes and embryos [140]. However, some examples of miRNA-mediated gene regulation in the COC have been characterized in other species. Mir-130b has been detected in immature and mature bovine oocytes [141] and cumulus cells during in vitro maturation, where it has been shown to promote cumulus cell viability and proliferation while acting as a positive regulator of glucose metabolism [142]. In the oocyte, miR-130b inhibition reduced polar body extrusion rates and rates of development to the morula and blastocyst stages when fertilized and cultured in vitro. The effects of miR-130b reported in this study appear to result from subtle changes in the miR-130b targets SMAD5 and mitogen- and stress-activated protein kinase [MSK1], which are key mediators of the TGF-β and mitogen-activated protein kinase (MAPK) signaling pathways, respectively [142]. In the porcine model of in vitro maturation, miR-21 increases significantly in the COC during maturation, and the addition of miR-21 inhibitors to IVM culture media reduces the number of oocytes that reach metaphase II [143]. While it contributes to steroidogenesis in granulosa cells, miR-378 expression decreases in cumulus cells over the course of maturation, and forced overexpression of miR-378 in the porcine COC impairs cumulus expansion and is detrimental to oocyte nuclear maturation and development to the blastocyst stage [144].

MicroRNAs in the Male Gamete: Mature Sperm Contain microRNAs That May Influence Embryo Development

This review is primarily focused on miRNAs in the oocyte and preimplantation embryo. However, the male gamete transcriptome, particularly the small RNAs within it, may provide significant contributions to development and is briefly discussed here in that context. The formation and differentiation of the male gamete through the process of spermatogenesis rivals oogenesis in the female in its complexity, and is beyond the scope of this review. Importantly, the current view of sperm contributions to the zygote has expanded beyond the exclusive delivery of the paternal genome, and sperm are now thought to provide organelles (reviewed by [145]) and RNAs [146, 147] that are critical to subsequent embryo development. Some RNAs present in sperm and early embryos are absent from nonfertilized oocytes, suggesting a unique role for these RNAs in postfertilization events [148].

The transcriptome in spermatozoa includes both long (transfer RNA, ribosomal RNA, and mRNA) and small RNAs including miRNAs, endogenous short interfering RNA (endo-siRNAs), and piRNAs, where miRNAs make up ~7% of the small RNA transcriptome and piRNAs make up ~17% [146, 149]. In addition to their potential regulatory roles in early embryonic development, miRNAs are present in the testis and profiling has revealed populations of miRNAs that are either preferentially or exclusively expressed in this tissue [150]. These miRNAs may be required for the completion of spermatogenesis, which is supported by the observation that loss of Drosha in testes results in severe disruptions to this process [151]. A survey of human sperm has uncovered a unique set of primary miRNA transcripts, including *pri-miR-181c*, that are present in sperm cells yet essentially absent from the testis, and are not observable in their short, mature form in the sperm [152]. It has been suggested that these miRNA precursors may be delivered to the oocyte upon fertilization, and are cleaved into their mature form only in the embryo, where they begin targeting mRNAs [152]. Evidence supporting a functional role for sperm miRNAs has been obtained in the mouse, where miR-34b and -34c are present in mouse zygotes but not oocytes or parthenotes, and appear to be required for the first zygotic cleavage event [147]. In addition to the functional roles of sperm-borne miRNAs in the embryo, miRNA profiles of sperm may be a useful parameter that is correlated with male fertility [153]. The miRNA profile in seminal plasma is useful in diagnosing infertility in human patients [153], and high versus low fertility status bulls have been shown to exhibit different miRNA signatures [154, 155]. While the quantitative contribution of small RNAs in sperm to the entire population present in the zygote and beyond is small and somewhat speculative, it remains possible that small paternal RNAs in sperm have the potential to influence development and reflect developmental potential.

MicroRNAs in the Preimplantation Embryo

The diverse and potent regulatory functions of miRNAs for many different genes strongly suggest that miRNA-mediated control of gene expression has strong potential to influence early developmental outcomes. This importance is particularly evident in model species such as zebrafish, where miR-430 alone has been shown to control the decay of hundreds of maternal transcripts during the maternal-zygotic transition [15]. Almost all these targets are of maternal origin, and it was demonstrated that maternal but not zygotic transcripts are enriched for miR-430 binding sites. In the absence of miR-430, these transcripts fail to degrade [15]. MiR-427, the miR-430 ortholog in *Xenopus*, similarly increases at the MZT to $\sim 10^9$ copies per embryo, and is responsible for deadenylation and clearance of maternal mRNA [83]. To date it has been difficult to determine whether the action of miR-430 and its orthologs are conserved processes in preimplantation development in other species; evidence suggesting that comparable single miRNA mediated transcript clearance occurs in mammals has not been reported. Notably, however, murine miR-290 contains the same seed sequence as miR-427 and miR-430, and it is specifically expressed in the preimplantation embryo, where it is required for embryo survival [156].

While no comparable broad maternal transcript targeting by any specific miRNAs has been demonstrated in mammals examined to date, proof for the principle that miRNA-mediated gene silencing may play more specific roles in the bovine embryo has been demonstrated by repression of the maternal effect transcription factor newborn ovary homeobox protein (NOBOX) by miR-196a at the 8- to 16-cell stage [157]. Another oocyte-specific transcription factor known as factor in the germline alpha (FIGLA) is subject to degradation by miR-212 during the MZT in bovine embryos [78], further confirming that miRNAs are involved in mRNA degradation at the MZT. In addition to these two examples, miR-21 and miR-130a are also strongly induced at the eight-cell stage in bovine embryos [16]; however their specific targets remain unknown and their relevance to human embryos has not been established.

Some miRNAs peak during the MZT and are thought to play important roles in degradation of maternal mRNA while other miRNA species are abundant in blastocysts, where their principal roles may be related to embryo implantation and subsequent placental development. The expression of miRNAs in the endometrium and the functions of miRNAs in embryo-endometrium communication, receptivity, implantation, decidualization, and early gestation are beyond the scope of this review and have been well described elsewhere [158, 159]. It is, however, interesting to note that miRNAs expressed in the blastocyst are known to act as signals that can affect implantation and the establishment of pregnancy. MiR-661 is secreted by human blastocysts that failed to implant after culture in vitro, and this secreted miRNA is taken up by cultured human endometrial epithelial cells where it reduces adhesion by targeting the poliovirus receptor-related 1 (*PVRL1*) gene product [160]. Conversely, miR-30d is specifically expressed in the endometrium at the window of implantation, where it is secreted into endometrial fluid and can be taken up by murine blastocysts, facilitating the process of embryo adhesion [161]. While limited in number, these few studies collectively suggest that miRNAs participate in embryo-endometrium cross talk and can act as positive or negative regulators of embryo adhesion. The possibility that miRNAs expressed in the blastocyst can regulate gene expression in adjacent tissues to improve or impede embryo attachment highlights the importance of factors that control miRNA expression in the embryo, and any potential relationships between blastocyst miRNA expression and pregnancy success/failure.

Potential Utility of MicroRNAs as Noninvasive Markers of Oocyte and Embryo Developmental Potential

In vitro embryo production strategies for both clinical (human) and agricultural (domestic animal) sectors continue to grapple with a specific fundamental challenge: how to select, from a cohort of embryos, the single embryo that is most likely to result in a successful pregnancy yielding healthy offspring. In clinical settings particularly, this challenge is further complicated by the suboptimal embryo quality that often accompanies the specific patient population. The most common strategy for selection relies heavily on the morphological assessment of embryos or oocytes, an effective technique that can nevertheless be undermined by the subjectivity and experience of the assessor or by the imperfect relationship between morphologic criteria and underlying embryo quality. Clearly, poor morphology is highly correlated with reduced developmental competence and chromosomal abnormalities; however, chromosomal abnormalities and other problems do occur in embryos with good morphological criteria [162]. Overall, morphological assessments are still somewhat ineffective in identifying all the characteristics most strongly associated with either implantation or developmental potential.

One ancillary approach to improve embryo assessment that has recently emerged is the examination of spent embryo culture medium for the presence of proteins, low molecular weight molecules, metabolites, and secreted nucleic acids as indirect indicators of embryo health (reviewed by [2, 163]). The protein secretome and the metabolome were the first to be investigated as potential biomarkers, but factors such as cost and a requirement for highly specialized equipment to perform the relevant assays has hampered their routine use in clinical settings (reviewed by [164, 165]). The quantification and profiling of secreted nucleic acids, including genomic and mitochondrial DNA [166, 167] as well as miRNAs by quantitative PCR based-approaches, are more sensitive, more widely available, and less costly than the quantitative detection of metabolites. Due to these advantages and the very high potential for the nucleic acid profile to reflect many different aspects of embryo health, this approach, particularly the detection of miRNAs, has recently emerged as a promising direction in embryo assessment.

Several inherent properties of miRNAs support their use as biomarkers in media and biological fluids. Levels of specific miRNAs fluctuate in a pattern reflecting the state of the cell or tissue, regulating processes such as cell growth and apoptosis, as described above, or responding to conditions such as acute or chronic inflammation [168] or oxidative stress [169]. While some miRNAs are ubiquitously expressed [170, 171], numerous others exhibit expression patterns that are restricted to specific cell-types or developmental stages [172]. Furthermore, they have been widely implicated in cell-fate decisions [173] and organogenesis [174]. Based on these characteristics, the profiles and relative levels of expression of different small RNAs can, at least theoretically, be integrated to form a "signature" or "fingerprint" that may reflect the underlying physiological state and developmental potential of the embryo. In other health contexts, miRNA signatures can be used to distinguish between healthy and disease states. This signature is often detectable in biological fluids [175], which makes them useful as components of many different diagnostic strategies.

In biological fluids, miRNAs are frequently present in extracellular vesicles known as exosomes [176], but can also be found outside these structures when stabilized by binding to carrier proteins such as lipoproteins [177] or AGO proteins [178, 179]. These physical associations, combined with the inherent stability of short-length RNAs, protects miRNAs from many endogenous and exogenous nucleases and confer some resistance to changes in pH and temperature changes (including freezing) [175, 180], enhancing their specific utility as markers. A spectrum of miRNAs has been confirmed in saliva, semen, tears, urine, and amniotic fluid, where a range of ~200–500 different miRNAs is concurrently expressed, each at different levels [171]. An miRNA signature has also been established for follicular fluid, and there exists a strong correlation between two specific miRNAs and a clinical diagnosis of polycystic ovarian syndrome (PCOS), where fluid from the follicles of PCOS patients contains significantly fewer copies of miR-132 and miR-320 [181]. In a follow-up study conducted by the same group, follicular fluid miR-320 levels positively correlated with Day 3 embryo cell number and quality [182]. An additional independent study correlated follicular overexpression of miR-9, miR-18b, miR-32, miR-34c, and miR-135a with a diagnosis of PCOS [183].

As discussed in previous sections, miRNAs are present in cleavage-stage embryos and blastocysts. Technical advances in detection methods have made it possible to quantify miRNA expression in individual embryos, and the abundance of specific miRNA genes can reflect chromosomal content. In one study of human embryos, the abundance of miR-518d-5p was found to be significantly higher in male embryos, and a group of miRNAs—including miR-141 and miR-27b—was significantly more abundant in euploid embryos [184]. Other studies have shown that a low abundance of blastocyst let-7a and miR-24 correlate with a diagnosis of male factor infertility in the father, and a panel of altered miRNAs including a low abundance of miR-93 correlates with a diagnosis of PCOS in the mother [185]. Interestingly, a subsequent study found that elevated levels of miR-93 correlated with advanced age of maternal patients [186]. Collectively, these observations suggest that specific miRNA abundance may provide valuable information relating to embryo sex, ploidy, and parental characteristics, with the obvious limitation that the embryo needs to be destroyed to perform the analysis. These preliminary studies do, however, help establish the molecular basis for the concept that, like other cell types, embryos may secrete miRNAs into their culture media, and that this miRNA profile could serve as an indirect indicator of ploidy, embryo health, and developmental potential (Fig. 11.2). One such example is miR-191, which has been detected at elevated levels in the spent media of blastocysts that were morphologically of good quality but confirmed to be aneuploid by trophectoderm biopsy and array comparative

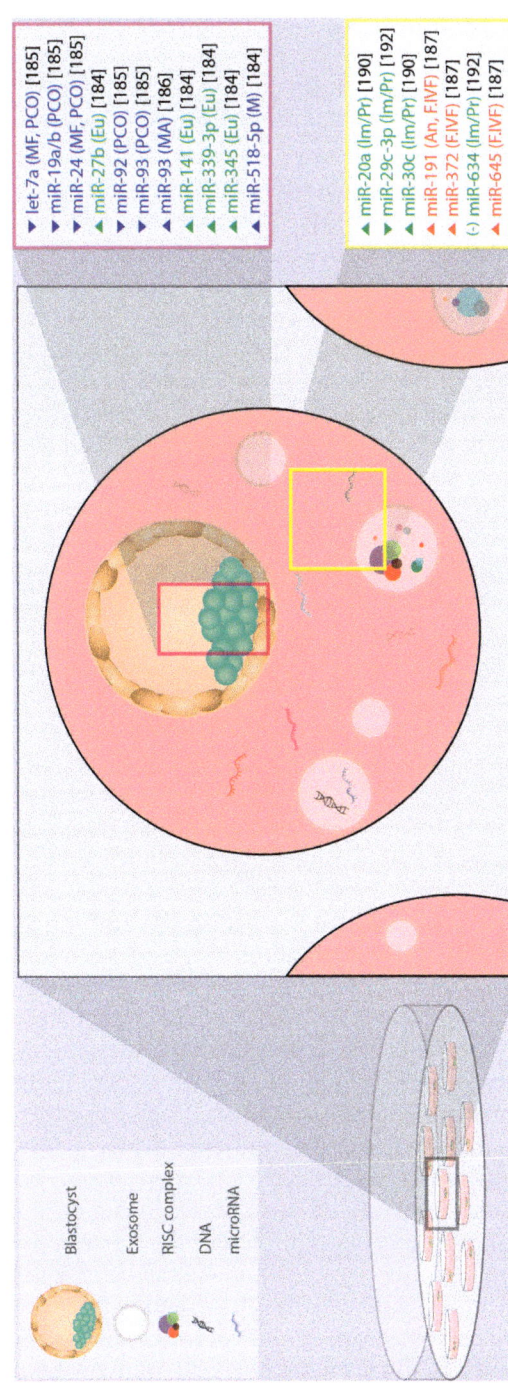

FIG. 11.2 MiRNA expression in blastocysts and spent culture media as potential indicators of embryo quality and IVF outcome. Changes in miRNA abundance that have been correlated significantly with chromosomal makeup, parental characteristics, and implantation potential of human blastocysts cultured in vitro: combined evidence from multiple studies. *Red text:* miRNA correlates with poor embryo quality or negative IVF outcome, *green text:* high embryo quality or positive IVF outcome, *blue text:* no relationship to IVF outcome. ▲, elevated miRNA abundance; ▼, reduced miRNA abundance; (−), miRNA not detected; *An,* aneuploid; *Eu,* euploid; *F.IVF,* failed IVF cycle; *Im/Pr,* implantation/embryo transfer resulting in positive pregnancy; *M,* male embryo; *MA,* maternal age; *MF,* male factor infertility diagnosis of parent; *PCO,* polycystic ovaries diagnosis of parent; [] indicates corresponding reference in text.

genomic hybridization (aCGH). The same study also found that elevated miR-191 correlated with a failed pregnancy outcome (biochemical pregnancy, spontaneous abortion, or implantation failure) [187].

One of the most highly expressed miRNAs in human blastocysts is miR-372 [184]. The miR-290-295 cluster is the murine ortholog of human miR-372, which is specifically expressed in early embryos and adult germ cells. Deletion of this cluster results in decreased embryo viability, germ cell defects, and impaired female fertility [156], suggesting that miR-372 plays functional roles in early embryo development. miR-372 is more abundant in the spent media from failed-IVF cycle blastocysts compared to those that resulted in live births [187]. Together, the finding that miR-372 is abundant in high quality blastocysts while its presence in spent culture media correlates with IVF failure suggests a mechanism where the release of developmentally important miRNAs into culture medium may be an early indicator of developmental arrest. Additional studies will be required to confirm this finding and the underlying concept. The principle that miRNAs are functional biomarkers in this regard has been validated to some extent in other species. Elevated miR-191 has been detected in culture media conditioned by degenerate bovine embryos (failed to develop from the morula to blastocyst stage), compared to media conditioned by blastocysts [188]. *Bos taurus* and *Homo sapiens* miR-191 have identical sequences [91] and in human IVF, elevated miR-191 in culture media is correlated with aneuploidy and failed pregnancy, suggesting that this miRNA may act as one specific indicator of poor developmental potential across species. miR-25 is also detectable in degenerate bovine embryos and is found in culture media derived from bovine blastocysts and degenerates, as well as culture media derived from human blastocysts [189]. Whether a connection exists between miR-25 abundance and developmental potential remains to be determined. One study has evaluated differences in the spent media miRNA profile from a cohort of euploid blastocysts only, and determined that an miRNA signature with elevated miR-20a and miR-30c correlates with implantation success when compared to euploid morphologically good embryos that failed to implant [190]. These results are particularly encouraging because they suggest that a secreted miRNA profile may provide unique prognostic data that could be a useful addition to the information provided by PGS, the most widely employed tool to characterize chromosomal status in the embryo [191].

While these early studies performed at single academic IVF centers have shown promising results, it should be emphasized that multicenter cross-validation of results between studies has not yet been reported. Major challenges with the use of spent media miRNA as an indicator of embryo quality include the minute amounts of starting material and differences in IVF protocols and reagents available internationally, which have the potential to undermine the accuracy and reproducibility of any given "signature" at different institutions. One recent study has attempted to address some of these issues by reproducing the same spent media miRNA profile using completely different RNA extraction, preparation, and quantification methods; a subset of miRNAs was successfully cross-validated using independent methods [190]. The standardization of detection protocols should improve the reproducibility of spent media studies conducted at different clinics. However, concerns have arisen from the findings that common media supplements contain miRNAs that readily pass through 0.22 μm filters [178] used in the sterilization of protein supplements for cell culture. For example, several miRNAs have been detected in embryo culture media *before* exposure to embryos, introduced by the addition of human serum albumin [187, 190] or bovine serum albumin (BSA) [189]. Rosenbluth et al. determined that as many as 9 out of 11 miRNAs detected in spent media were also present in media-alone controls, and Capalbo et al. [190] consistently amplified a group of 14 miRNAs in control media, deemed to be false-positives. The issue becomes most apparent when these independent studies are reviewed together; a major finding from Capalbo et al. [190] was the correlation between high spent-media miR-30c and blastocyst implantation potential, which was found to be solely present in embryo conditioned media and absent in controls, but which was identified as a false-positive by Rosenbluth et al. [187]. Because media supplements appear to be a major source of false-positive miRNA signals, standardization of reagents would be essential for these types of analyses to become clinically applicable. Interestingly, different results for miR-30c have been reported when different detection methods are employed; Abu-Halima et al. found that elevated miR-30c in spent blastocyst media correlated significantly with a positive pregnancy outcome using miRNA microarray, but a decrease in miR-30c was found in samples associated with pregnancy by qRT-PCR [192]. Together, these combined data suggest that miR-30c may not be an ideal candidate miRNA indicator of embryo quality or pregnancy. However, further studies will be required to determine whether the quantification of this gene has clinical utility.

In addition to the presence of miRNAs in the media prior to embryo culture, it is important to consider a number of other technical factors that have the potential to reduce their diagnostic utility. When considering the number of miRNA molecules that can potentially be secreted from a single embryo into a 20 µL drop of media over a 4- to 5-day culture period, and how some fraction of the RNA is lost during sample preparation, it is perhaps not surprising that procedural differences combined with the use of different media supplements has resulted in differing miRNA signatures in spent media between independent studies.

One interesting technique that has recently emerged and which may circumvent this is the use of blastocoel fluid as a rich source of nucleic acids from the embryo. Initial studies of blastocoel fluid focused on DNA markers and the potential for embryo ploidy assessment. Tobler et al. obtained molecular karyotypes from genomic DNA isolated from blastocoel fluid that was aspirated from blastocysts by micropipette [193]. While this study was designed using research embryos unsuitable for transfer, blastocoel fluid is expelled from blastocysts undergoing routine laser-induced collapse before vitrification [194]. Theoretically, the aspiration or expulsion of blastocoel fluid into media or physiologic solutions is likely to enrich the sample with RNA and facilitate higher reproducibility of subsequent miRNA profiling. However, it remains unknown at present whether miRNAs can be effectively and reproducibly collected in laser collapsed blastocyst spent culture media. The population of miRNAs detected in spent blastocyst culture medium by Capalbo et al. [190] suggests that 97% originates from the trophectoderm, and therefore the profile of miRNA in the medium is predominantly a signature of trophectoderm biology. In contrast, because the fluid within the blastocoel contacts both the inner cell mass and the trophectoderm, this fluid, or media enriched with it, may provide a signature that more accurately reflects the status of cells in both trophectoderm and the inner cell mass. Clearly, such a signature would be likely to have greater prognostic value for pregnancy outcome. Regardless of the techniques employed, or the samples at which they are directed, it is important to emphasize that an miRNA signature alone is unlikely to represent the single definitive assessment method for embryo health under current technical and conceptual constraints. It is more probable that such a signature could be used in conjunction with a number of other assessment strategies to provide a more comprehensive and accurate picture of the health and developmental potential of in vitro embryo development.

CONCLUSIONS

MicroRNAs play well-established roles in the development of the ovary and testes and also participate in the interactions between the embryo and the endometrium that ultimately determine whether successful pregnancy will be established. Somatic cells in the gonads constantly respond to environmental signals and changing cellular dynamics, highly active processes that require continuous turnover of RNA and protein that are in turn highly subject to posttranscriptional regulation by miRNAs. In contrast, the functional roles of miRNAs in germ cells and early embryos have been more difficult to elucidate, for reasons involving protected storage of dormant mRNAs and the influence of other pathways that may direct maternal mRNA degradation in the vertebrate embryo. While a small number of individual studies have established specific roles for miRNAs in the mammalian oocyte and embryo, their ultimate roles may be directed at fine-tuning gene expression rather than acting as major regulatory switches. MiRNA activity often results in subtle or mild changes in the expression of genes that they target, making an absolute functional importance more difficult to definitively establish. However, the patterns of expression and secretion of miRNAs appear to vary rapidly and consistently in response to changes in cellular function and to extracellular factors that elicit such changes. When the expression patterns of multiple independent miRNAs are examined simultaneously using sequencing and bioinformatics strategies, clear indications of the overall health status of the embryo and its developmental competence are likely to emerge. By way of analogy, much like metadata derived from electronic communication often reflects ongoing behavior, the "epigenomic metadata" derived from miRNAs that developing embryos generate for intra- and extracellular communication during development is likely to provide a useful yet incomplete picture of their health status and developmental potential. Because it will likely be easier to quantitate miRNA expression and identify patterns associated with success than to fully characterize their function, their use as biomarkers may well precede a comprehensive understanding of miRNA function in this context. However, ongoing exploration of the regulation and consequences of miRNA expression in oocytes, follicular cells, and embryos should ultimately enable us to develop a more comprehensive picture of the factors that contribute to developmental success, starting from the earliest stages of follicular and oocyte growth.

Acknowledgments

The authors wish to thank Graham Gilchrist, Stewart Russell, Moysés Miranda, Pavneesh Madan, Allan King, Monica Antenos, Don Reiger, Monica Mezezi, Mike Cecchi, and Michael Neal for helpful discussions. Jennifer Kao's efforts in preparing the illustrations are highly appreciated. J. LaMarre has acted as a consultant for LifeGlobal LLC in areas related to the work described. The Natural Sciences and Engineering Research Council (NSERC: Canada), Ontario Ministry of Agriculture, Food and Rural Affairs (OMAFRA) and the Ontario Centres of Excellence (OCE) have provided funding for studies in the LaMarre laboratory in the area of miRNAs and reproduction.

References

[1] Dyer S, Chambers GM, de Mouzon J, Nygren KG, Zegers-Hochschild F, Mansour R, et al. International Committee for Monitoring Assisted Reproductive Technologies world report: assisted reproductive technology 2008, 2009 and 2010. Hum Reprod 2016;31(7):1588–609.

[2] Perkel KJ, Tscherner A, Merrill C, Lamarre J, Madan P. The ART of selecting the best embryo: a review of early embryonic mortality and bovine embryo viability assessment methods. Mol Reprod Dev 2015;82:822–38.

[3] Daughtry BL, Chavez SL. Chromosomal instability in mammalian pre-implantation embryos: potential causes, detection methods, and clinical consequences. Cell Tissue Res 2016;363(1):201–25.

[4] Smith SE, Toledo AA, Massey JB, Kort HI. Simultaneous detection of chromosomes X, Y, 13, 18, and 21 by fluorescence in situ hybridization in blastomeres obtained from preimplantation embryos. J Assist Reprod Genet 1998;15(5):314–9.

[5] Vanneste E, Voet T, Le Caignec C, Ampe M, Konings P, Melotte C, et al. Chromosome instability is common in human cleavage-stage embryos. Nat Med 2009;15(5):577–83.

[6] Fiorentino F, Bono S, Biricik A, Nuccitelli A, Cotroneo E, Cottone G, et al. Application of next-generation sequencing technology for comprehensive aneuploidy screening of blastocysts in clinical preimplantation genetic screening cycles. Hum Reprod 2014;29(12):2802–13.

[7] Sermon K, Capalbo A, Cohen J, Coonen E, De Rycke M, De Vos A, et al. The why, the how and the when of PGS 2.0: current practices and expert opinions of fertility specialists, molecular biologists, and embryologists. Mol Hum Reprod 2016;22(8):845–57.

[8] Vassena R, Boué S, González-Roca E, Aran B, Auer H, Veiga A, et al. Waves of early transcriptional activation and pluripotency program initiation during human preimplantation development. Development 2011;138(17):3699–709.

[9] Giritharan G, Li MW, Di Sebastiano F, Esteban FJ, Horcajadas JA, Lloyd KC, et al. Effect of ICSI on gene expression and development of mouse preimplantation embryos. Hum Reprod 2010;25(12):3012–24.

[10] Milazzotto MP, Goissis MD, Chitwood JL, Annes K, Soares CA, Ispada J, et al. Early cleavages influence the molecular and the metabolic pattern of individually cultured bovine blastocysts. Mol Reprod Dev 2016;83(4):324–36.

[11] Wang N, Li CY, Zhu HB, Hao HS, Wang HY, Yan CL, et al. Effect of vitrification on the mRNA transcriptome of bovine oocytes. Reprod Domest Anim 2017;52(4):531–41.

[12] Heras S, De Coninck DI, Van Poucke M, Goossens K, Bogado Pascottini O, Van Nieuwerburgh F, et al. Suboptimal culture conditions induce more deviations in gene expression in male than female bovine blastocysts. BMC Genomics 2016;1772.

[13] Sirard MA. Factors affecting oocyte and embryo transcriptomes. Reprod Domest Anim 2012;47(Suppl. 4):148–55.

[14] Giraldez AJ. microRNAs, the cell's Nepenthe: clearing the past during the maternal-to-zygotic transition and cellular reprogramming. Curr Opin Genet Dev 2010;20(4):369–75.

[15] Giraldez AJ, Mishima Y, Rihel J, Grocock RJ, Van Dongen S, Inoue K, et al. Zebrafish MiR-430 promotes deadenylation and clearance of maternal mRNAs. Science 2006;312(5770):75–9.

[16] Mondou E, Dufort I, Gohin M, Fournier E, Sirard MA. Analysis of microRNAs and their precursors in bovine early embryonic development. Mol Hum Reprod 2012;18(9):425–34.

[17] Tang F, Kaneda M, O'Carroll D, Hajkova P, Barton SC, Sun YA, et al. Maternal microRNAs are essential for mouse zygotic development. Genes Dev 2007;21(6):644–8.

[18] Uhde K, van Tol H, Stout T, Roelen B. MicroRNA expression in bovine cumulus cells in relation to oocyte quality. ncRNA 2017;3(1):12.

[19] Russell S, Patel M, Gilchrist G, Stalker L, Gillis D, Rosenkranz D, et al. Bovine piRNA-like RNAs are associated with both transposable elements and mRNAs. Reproduction 2017;153(3):305–18.

[20] Roovers EF, Rosenkranz D, Mahdipour M, Han CT, He N, Chuva de Sousa Lopes SM, et al. Piwi proteins and piRNAs in mammalian oocytes and early embryos. Cell Rep 2015;10(12):2069–82.

[21] Crozet N, Kanka J, Motlik J, Fulka J. Nucleolar fine structure and RNA synthesis in bovine oocytes from antral follicles. Gamete Res 1986;14:65–73.

[22] Braude P, Bolton V, Moore S. Human gene expression first occurs between the four- and eight-cell stages of preimplantation development. Nature 1988;332(6163):459–61.

[23] Kanka J, Kepkova K, Nemcova L. Gene expression during minor genome activation in preimplantation bovine development. Theriogenology 2009;72(4):572–83.

[24] Graf A, Krebs S, Zakhartchenko V, Schwalb B, Blum H, Wolf E. Fine mapping of genome activation in bovine embryos by RNA sequencing. Proc Natl Acad Sci U S A 2014;111:4139–44.

[25] Schier AF. The maternal-zygotic transition: death and birth of RNAs. Science 2007;316(5823):406–7.

[26] Stitzel ML, Seydoux G. Regulation of the oocyte-to-zygote transition. Science 2007;316(5823):407–8.

[27] Sánchez F, Smitz J. Molecular control of oogenesis. Biochim Biophys Acta 2012;1822(12):1896–912.

I. REPRODUCTIVE TRACT DEVELOPMENT AND GAMETOGENESIS

[28] Gondos B, Bhiraleus P, Hobel CJ. Ultrastructural observations on germ cells in human fetal ovaries. Am J Obstet Gynecol 1971;110(5):644–52.

[29] Borum K. Oogenesis in the mouse. A study of the meiotic prophase. Exp Cell Res 1961;24:495–507.

[30] Hsueh AJ, Billig H, Tsafriri A. Ovarian follicle atresia: a hormonally controlled apoptotic process. Endocr Rev 1994;15(6):707–24.

[31] Eppig JJ. Intercommunication between mammalian oocytes and companion somatic cells. Bioessays 1991; 13(11):569–74.

[32] Eppig JJ. Oocyte control of ovarian follicular development and function in mammals. Reproduction 2001; 122(6):829–38.

[33] Palermo R. Differential actions of FSH and LH during folliculogenesis. Reprod BioMed Online 2007;15(3):326–37.

[34] Kerin JF, Edmonds DK, Warnes GM, Cox LW, Seamark RF, Matthews CD, et al. Morphological and functional relations of Graafian follicle growth to ovulation in women using ultrasonic, laparoscopic and biochemical measurements. Br J Obstet Gynaecol 1981;88(2):81–90.

[35] Picton H, Briggs D, Gosden R. The molecular basis of oocyte growth and development. Mol Cell Endocrinol 1998;145(1–2):27–37.

[36] Pincus G, Enzmann EV. The comparative behavior of mammalian eggs in vivo and in vitro: I. The activation of ovarian eggs. J Exp Med 1935;62(5):665–75.

[37] Dekel N, Beers WH. Rat oocyte maturation in vitro: relief of cyclic AMP inhibition by gonadotropins. Proc Natl Acad Sci U S A 1978;75(9):4369–73.

[38] Albertini DF, Sanfins A, Combelles CM. Origins and manifestations of oocyte maturation competencies. Reprod BioMed Online 2003;6(4):410–5.

[39] Li R, Albertini DF. The road to maturation: somatic cell interaction and self-organization of the mammalian oocyte. Nat Rev Mol Cell Biol 2013;14(3):141–52.

[40] Fukui Y, Sakuma Y. Maturation of bovine oocytes cultured in vitro: relation to ovarian activity, follicular size and the presence or absence of cumulus cells. Biol Reprod 1980;22(3):669–73.

[41] Vanderhyden BC, Armstrong DT. Role of cumulus cells and serum on the in vitro maturation, fertilization, and subsequent development of rat oocytes. Biol Reprod 1989;40(4):720–8.

[42] Dey SR, Deb GK, Ha AN, Lee JI, Bang JI, Lee KL, et al. Coculturing denuded oocytes during the in vitro maturation of bovine cumulus oocyte complexes exerts a synergistic effect on embryo development. Theriogenology 2012;77(6):1064–77.

[43] Tatemoto H, Horiuchi T, Terada T. Effects of cycloheximide on chromatin condensations and germinal vesicle breakdown (GVBD) of cumulus-enclosed and denuded oocytes in cattle. Theriogenology 1994;42(7):1141–8.

[44] Zhang L, Jiang S, Wozniak PJ, Yang X, Godke RA. Cumulus cell function during bovine oocyte maturation, fertilization, and embryo development in vitro. Mol Reprod Dev 1995;40(3):338–44.

[45] Buccione R, Vanderhyden BC, Caron PJ, Eppig JJ. FSH-induced expansion of the mouse cumulus oophorus in vitro is dependent upon a specific factor(s) secreted by the oocyte. Dev Biol 1990;138(1):16–25.

[46] Hussein TS, Froiland DA, Amato F, Thompson JG, Gilchrist RB. Oocytes prevent cumulus cell apoptosis by maintaining a morphogenic paracrine gradient of bone morphogenetic proteins. J Cell Sci 2005;118 (Pt 22):5257–68.

[47] Gilchrist RB, Ritter LJ, Myllymaa S, Kaivo-Oja N, Dragovic RA, Hickey TE, et al. Molecular basis of oocyte-paracrine signalling that promotes granulosa cell proliferation. J Cell Sci 2006;119(Pt 18):3811–21.

[48] Matzuk MM, Burns KH, Viveiros MM, Eppig JJ. Intercellular communication in the mammalian ovary: oocytes carry the conversation. Science 2002;296(5576):2178–80.

[49] Eppig JJ, Ward-Bailey PF. Sulfated glycosaminoglycans inhibit hyaluronic acid synthesizing activity in mouse cumuli oophori. Exp Cell Res 1984;150(2):459–65.

[50] Chen L, Russell PT, Larsen WJ. Sequential effects of follicle-stimulating hormone and luteinizing hormone on mouse cumulus expansion in vitro. Biol Reprod 1994;51(2):290–5.

[51] Eppig JJ. Regulation of cumulus oophorus expansion by gonadotropins in vivo and in vitro. Biol Reprod 1980;23(3):545–52.

[52] Norris RP, Ratzan WJ, Freudzon M, Mehlmann LM, Krall J, Movsesian MA, et al. Cyclic GMP from the surrounding somatic cells regulates cyclic AMP and meiosis in the mouse oocyte. Development 2009; 136(11):1869–78.

[53] Simon AM, Goodenough DA, Li E, Paul DL. Female infertility in mice lacking connexin 37. Nature 1997;385 (6616):525–9.

[54] Suzuki H, Jeong BS, Yang X. Dynamic changes of cumulus-oocyte cell communication during in vitro maturation of porcine oocytes. Biol Reprod 2000;63(3):723–9.

[55] Park J-Y, Su Y-Q, Ariga M, Law E, Jin SLC, Conti M. EGF-like growth factors as mediators of LH action in the ovulatory follicle. Science 2004;303(5658):682–4.

[56] Gilchrist RB, Ritter LJ. Differences in the participation of TGFB superfamily signalling pathways mediating porcine and murine cumulus cell expansion. Reproduction 2011;142(5):647–57.

[57] Kidder GM, Vanderhyden BC. Bidirectional communication between oocytes and follicle cells: ensuring oocyte developmental competence. Can J Physiol Pharmacol 2010;88(4):399–413.

[58] Fülöp C, Szanto S, Mukhopadhyay D, Bardos T, Kamath R, Rugg MS, et al. Impaired cumulus mucification and female sterility in tumor necrosis factor-induced protein-6 deficient mice. Development 2003;130(10): 2253–61.

[59] Salustri A, Garlanda C, Hirsch E, De Acetis M, Maccagno A, Bottazzi B, et al. PTX3 plays a key role in the organization of the cumulus oophorus extracellular matrix and in in vivo fertilization. Development 2004; 131(7):1577–86.

[60] Dragovic RA, Ritter LJ, Schulz SJ, Amato F, Thompson JG, Armstrong DT, et al. Oocyte-secreted factor activation of SMAD 2/3 signaling enables initiation of mouse cumulus cell expansion. Biol Reprod 2007;76 (5):848–57.

[61] McKenzie LJ, Pangas SA, Carson SA, Kovanci E, Cisneros P, Buster JE, et al. Human cumulus granulosa cell gene expression: a predictor of fertilization and embryo selection in women undergoing IVF. Hum Reprod 2004;19(12):2869–74.

[62] Smitz JE, Thompson JG, Gilchrist RB. The promise of in vitro maturation in assisted reproduction and fertility preservation. Semin Reprod Med 2011;29(1):24–37.

[63] Guzman L, Adriaenssens T, Ortega-Hrepich C, Albuz FK, Mateizel I, Devroey P, et al. Human antral follicles <6 mm: a comparison between in vivo maturation and in vitro maturation in non-hCG primed cycles using cumulus cell gene expression. Mol Hum Reprod 2013;19(1):7–16.

[64] Tesfaye D, Ghanem N, Carter F, Fair T, Sirard M-A, Hoelker M, et al. Gene expression profile of cumulus cells derived from cumulus–oocyte complexes matured either or. Reprod Fertil Dev 2009;21(3):451.

[65] Assidi M, Dufort I, Ali A, Hamel M, Algriany O, Dielemann S, et al. Identification of potential markers of oocyte competence expressed in bovine cumulus cells matured with follicle-stimulating hormone and/or phorbol myristate acetate in vitro. Biol Reprod 2008;79(2):209–22.

[66] Kussano NR, Leme LO, Guimarães AL, Franco MM, Dode MA. Molecular markers for oocyte competence in bovine cumulus cells. Theriogenology 2016;85(6):1167–76.

[67] Feuerstein P, Cadoret V, Dalbies-Tran R, Guerif F, Bidault R, Royere D. Gene expression in human cumulus cells: one approach to oocyte competence. Hum Reprod 2007;22(12):3069–77.

[68] Dieci C, Lodde V, Labreque R, Dufort I, Tessaro I, Sirard MA, et al. Differences in cumulus cell gene expression indicate the benefit of a pre-maturation step to improvein-vitro bovine embryo production. Mol Hum Reprod 2016;22:882–97.

[69] van Montfoort AP, Geraedts JP, Dumoulin JC, Stassen AP, Evers JL, Ayoubi TA. Differential gene expression in cumulus cells as a prognostic indicator of embryo viability: a microarray analysis. Mol Hum Reprod 2008; 14(3):157–68.

[70] Fair T, Hyttel P, Greve T. Bovine oocyte diameter in relation to maturational competence and transcriptional activity. Mol Reprod Dev 1995;42(4):437–42.

[71] Albertini DF, Combelles CM, Benecchi E, Carabatsos MJ. Cellular basis for paracrine regulation of ovarian follicle development. Reproduction 2001;121(5):647–53.

[72] Macaulay AD, Gilbert I, Caballero J, Barreto R, Fournier E, Tossou P, et al. The gametic synapse: RNA transfer to the bovine oocyte. Biol Reprod 2014;91:1–12.

[73] Macaulay AD, Gilbert I, Scantland S, Fournier E, Ashkar F, Bastien A, et al. Cumulus cell transcripts transit to the bovine oocyte in preparation for maturation. Biol Reprod 2016;94:1–11.

[74] King WA, Niar A, Chartrain I, Betteridge KJ, Guay P. Nucleolus organizer regions and nucleoli in preattachment bovine embryos. J Reprod Fertil 1988;82(1):87–95.

[75] Barckmann B, Simonelig M. Control of maternal mRNA stability in germ cells and early embryos. Biochim Biophys Acta 2013;1829(6–7):714–24.

[76] Hamatani T, Carter MG, Sharov AA, Ko MS. Dynamics of global gene expression changes during mouse pre-implantation development. Dev Cell 2004;6(1):117–31.

[77] Pennetier S, Uzbekova S, Perreau C, Papillier P, Mermillod P, Dalbiès-Tran R. Spatio-temporal expression of the germ cell marker genes MATER, ZAR1, GDF9, BMP15, and VASA in adult bovine tissues, oocytes, and preimplantation embryos. Biol Reprod 2004;71(4):1359–66.

[78] Tripurani SK, Wee G, Lee KB, Smith GW, Wang L, Jianboyao M. MicroRNA-212 post-transcriptionally regulates oocyte-specific basic-helix-loop-helix transcription factor, factor in the germline alpha (FIGLA), during bovine early embryogenesis. PLoS ONE 2013;8(9)e76114.

[79] Wang QT, Piotrowska K, Ciemerych MA, Milenkovic L, Scott MP, Davis RW, et al. A genome-wide study of gene activity reveals developmental signaling pathways in the preimplantation mouse embryo. Dev Cell 2004; 6(1):133–44.

[80] Sagata N, Watanabe N, Vande Woude GF, Ikawa Y. The c-mos proto-oncogene product is a cytostatic factor responsible for meiotic arrest in vertebrate eggs. Nature 1989;342(6249):512–8.

[81] Yartseva V, Giraldez AJ. The maternal-to-zygotic transition during vertebrate development: a model for reprogramming. Curr Top Dev Biol 2015;113:191–232.

[82] Izaurralde E. Elucidating the temporal order of silencing. EMBO Rep 2012;13(8):662–3.

[83] Lund E, Liu M, Hartley RS, Sheets MD, Dahlberg JE. Deadenylation of maternal mRNAs mediated by miR-427 in *Xenopus laevis* embryos. RNA 2009;15(12):2351–63.

[84] Flynt AS, Lai EC. Biological principles of microRNA-mediated regulation: shared themes amid diversity. Nat Rev Genet 2008;9(11):831–42.

[85] Sun K, Lai EC. Adult-specific functions of animal microRNAs. Nat Rev Genet 2013;14(8):535–48.

[86] Lewis BP, Burge CB, Bartel DP. Conserved seed pairing, often flanked by adenosines, indicates that thousands of human genes are microRNA targets. Cell 2005;120(1):15–20.

[87] Friedman RC, Farh KK, Burge CB, Bartel DP. Most mammalian mRNAs are conserved targets of microRNAs. Genome Res 2009;19(1):92–105.

[88] Lodish HF, Zhou B, Liu G, Chen C-Z. Micromanagement of the immune system by microRNAs. Nat Rev Immunol 2008;8(3):120–30.

[89] Lee Y, Jeon K, Lee JT, Kim S, Kim VN. MicroRNA maturation: stepwise processing and subcellular localization. EMBO J 2002;21(17):4663–70.

[90] Lee Y, Kim M, Han J, Yeom KH, Lee S, Baek SH, et al. MicroRNA genes are transcribed by RNA polymerase II. EMBO J 2004;23(20):4051–60.

[91] Griffiths-Jones S, Saini HK, van Dongen S, Enright AJ. miRBase: tools for microRNA genomics. Nucleic Acids Res 2008;36(Database issue):D154–8.

[92] Lagos-Quintana M, Rauhut R, Lendeckel W, Tuschl T. Identification of novel genes coding for small expressed RNAs. Science 2001;294(5543):853–8.

[93] Yu J, Wang F, Yang GH, Wang FL, Ma YN, Du ZW, et al. Human microRNA clusters: genomic organization and expression profile in leukemia cell lines. Biochem Biophys Res Commun 2006;349(1):59–68.

[94] Olena AF, Patton JG. Genomic organization of microRNAs. J Cell Physiol 2010;222(3):540–5.

[95] Lee Y, Ahn C, Han J, Choi H, Kim J, Yim J, et al. The nuclear RNase III Drosha initiates microRNA processing. Nature 2003;425(6956):415–9.

[96] Landthaler M, Yalcin A, Tuschl T. The human DiGeorge syndrome critical region gene 8 and its *D. melanogaster* homolog are required for miRNA biogenesis. Curr Biol 2004;14(23):2162–7.

[97] Denli AM, Tops BB, Plasterk RH, Ketting RF, Hannon GJ. Processing of primary microRNAs by the microprocessor complex. Nature 2004;432(7014):231–5.

[98] Gregory RI, Yan KP, Amuthan G, Chendrimada T, Doratotaj B, Cooch N, et al. The microprocessor complex mediates the genesis of microRNAs. Nature 2004;432(7014):235–40.

[99] Han J, Lee Y, Yeom KH, Nam JW, Heo I, Rhee JK, et al. Molecular basis for the recognition of primary microRNAs by the Drosha-DGCR8 complex. Cell 2006;125(5):887–901.

[100] Ha M, Kim VN. Regulation of microRNA biogenesis. Nat Rev Mol Cell Biol 2014;15:509–24.

[101] Han J, Pedersen JS, Kwon SC, Belair CD, Kim YK, Yeom KH, et al. Posttranscriptional crossregulation between Drosha and DGCR8. Cell 2009;136(1):75–84.

[102] Yi R, Qin Y, Macara IG, Cullen BR. Exportin-5 mediates the nuclear export of pre-microRNAs and short hairpin RNAs. Genes Dev 2003;17(24):3011–6.

[103] Brownawell AM, Macara IG. Exportin-5, a novel karyopherin, mediates nuclear export of double-stranded RNA binding proteins. J Cell Biol 2002;156(1):53–64.

[104] Bernstein E, Caudy AA, Hammond SM, Hannon GJ. Role for a bidentate ribonuclease in the initiation step of RNA interference. Nature 2001;409(6818):363–6.

[105] Hutvágner G, McLachlan J, Pasquinelli AE, Bálint E, Tuschl T, Zamore PD. A cellular function for the RNA-interference enzyme Dicer in the maturation of the let-7 small temporal RNA. Science 2001;293(5531): 834–8.

[106] Chendrimada TP, Gregory RI, Kumaraswamy E, Norman J, Cooch N, Nishikura K, et al. TRBP recruits the Dicer complex to Ago2 for microRNA processing and gene silencing. Nature 2005;436(7051):740–4.

[107] Hammond SM, Boettcher S, Caudy AA, Kobayashi R, Hannon GJ. Argonaute2, a link between genetic and biochemical analyses of RNAi. Science 2001;293(5532):1146–50.

[108] Wu L, Fan J, Belasco JG. MicroRNAs direct rapid deadenylation of mRNA. Proc Natl Acad Sci U S A 2006; 103(11):4034–9.

[109] Djuranovic S, Nahvi A, Green R. miRNA-mediated gene silencing by translational repression followed by mRNA deadenylation and decay. Science 2012;336(6078):237–40.

[110] Meister G. Argonaute proteins: functional insights and emerging roles. Nat Rev Genet 2013;14(7):447–59.

[111] Azuma-Mukai A, Oguri H, Mituyama T, Qian ZR, Asai K, Siomi H, et al. Characterization of endogenous human Argonautes and their miRNA partners in RNA silencing. Proc Natl Acad Sci U S A 2008;105(23):7964–9.

[112] Jonas S, Izaurralde E. Towards a molecular understanding of microRNA-mediated gene silencing. Nat Rev Genet 2015;16(7):421–33.

[113] Richards JS, Pangas SA. The ovary: basic biology and clinical implications. J Clin Invest 2010;120(4):963–72.

[114] Fraser HM. Regulation of the ovarian follicular vasculature. Reprod Biol Endocrinol 2006;4:18.

[115] McBride D, Carré W, Sontakke SD, Hogg CO, Law A, Donadeu FX, et al. Identification of miRNAs associated with the follicular-luteal transition in the ruminant ovary. Reproduction 2012;144(2):221–33.

[116] Diez-Fraile A, Lammens T, Tilleman K, Witkowski W, Verhasselt B, De Sutter P, et al. Age-associated differential microRNA levels in human follicular fluid reveal pathways potentially determining fertility and success of in vitro fertilization. Hum Fertil (Camb) 2014;17(2):90–8.

[117] Hossain MM, Sohel MM, Schellander K, Tesfaye D. Characterization and importance of microRNAs in mammalian gonadal functions. Cell Tissue Res 2012;349(3):679–90.

[118] Tripurani SK, Xiao C, Salem M, Yao J. Cloning and analysis of fetal ovary microRNAs in cattle. Anim Reprod Sci 2010;120(1–4):16–22.

[119] Hossain MM, Ghanem N, Hoelker M, Rings F, Phatsara C, Tholen E, et al. Identification and characterization of miRNAs expressed in the bovine ovary. BMC Genomics 2009;10:443.

[120] Salilew-Wondim D, Ahmad I, Gebremedhn S, Sahadevan S, Hossain MD, Rings F, et al. The expression pattern of microRNAs in granulosa cells of subordinate and dominant follicles during the early luteal phase of the bovine estrous cycle. PLoS ONE 2014;9:e106795.

[121] Yao N, Yang BQ, Liu Y, Tan XY, Lu CL, Yuan XH, et al. Follicle-stimulating hormone regulation of microRNA expression on progesterone production in cultured rat granulosa cells. Endocrine 2010;38(2):158–66.

[122] Carletti MZ, Fiedler SD, Christenson LK. MicroRNA 21 blocks apoptosis in mouse periovulatory granulosa cells. Biol Reprod 2010;83(2):286–95.

[123] Fiedler SD, Carletti MZ, Hong X, Christenson LK. Hormonal regulation of MicroRNA expression in periovulatory mouse mural granulosa cells. Biol Reprod 2008;79(6):1030–7.

[124] Otsuka M, Zheng M, Hayashi M, Lee JD, Yoshino O, Lin S, et al. Impaired microRNA processing causes corpus luteum insufficiency and infertility in mice. J Clin Invest 2008;118(5):1944–54.

[125] Watanabe T, Totoki Y, Toyoda A, Kaneda M, Kuramochi-Miyagawa S, Obata Y, et al. Endogenous siRNAs from naturally formed dsRNAs regulate transcripts in mouse oocytes. Nature 2008;453(7194):539–43.

[126] Xu B, Hua J, Zhang Y, Jiang X, Zhang H, Ma T, et al. Proliferating cell nuclear antigen (PCNA) regulates primordial follicle assembly by promoting apoptosis of oocytes in fetal and neonatal mouse ovaries. PLoS ONE 2011;6(1):e16046.

[127] Zhang H, Jiang X, Zhang Y, Xu B, Hua J, Ma T, et al. microRNA 376a regulates follicle assembly by targeting Pcna in fetal and neonatal mouse ovaries. Reproduction 2014;148(1):43–54.

[128] Yang S, Wang S, Luo A, Ding T, Lai Z, Shen W, et al. Expression patterns and regulatory functions of microRNAs during the initiation of primordial follicle development in the neonatal mouse ovary. Biol Reprod 2013;89(5):1–11.

[129] Lin F, Li R, Pan ZX, Zhou B, Yu DB, Wang XG, et al. miR-26b promotes granulosa cell apoptosis by targeting ATM during follicular atresia in porcine ovary. PLoS ONE 2012;7(6)e38640.

[130] Liu J, Tu F, Yao W, Li X, Xie Z, Liu H, et al. Conserved miR-26b enhances ovarian granulosa cell apoptosis through HAS2-HA-CD44-Caspase-3 pathway by targeting HAS2. Sci Rep 2016;18:21197.

[131] Xu S, Linher-Melville K, Yang BB, Wu D, Li J. Micro-RNA378 (miR-378) regulates ovarian estradiol production by targeting aromatase. Endocrinology 2011;152(10):3941–51.

[132] Yao G, Yin M, Lian J, Tian H, Liu L, Li X, et al. MicroRNA-224 is involved in transforming growth factor-beta-mediated mouse granulosa cell proliferation and granulosa cell function by targeting Smad4. Mol Endocrinol 2010;24(3):540–51.

[133] Schams D, Berisha B. Regulation of corpus luteum function in cattle–an overview. Reprod Domest Anim 2004;39(4):241–51.

[134] Maalouf SW, Liu WS, Pate JL. MicroRNA in ovarian function. Cell Tissue Res 2016;363(1):7–18.

[135] Maalouf SW, Liu WS, Albert I, Pate JL. Regulating life or death: potential role of microRNA in rescue of the corpus luteum. Mol Cell Endocrinol 2014;398(1–2):78–88.

[136] Albertini DF, Barrett SL. Oocyte-somatic cell communication. Reprod Suppl 2003;61:49–54.

[137] Abd El Naby WS, Hagos TH, Hossain MM, Salilew-Wondim D, Gad AY, Rings F, et al. Expression analysis of regulatory microRNAs in bovine cumulus oocyte complex and preimplantation embryos. Zygote 2013;21(1):31–51.

[138] Flemr M, Ma J, Schultz RM, Svoboda P. P-body loss is concomitant with formation of a messenger RNA storage domain in mouse oocytes. Biol Reprod 2010;82(5):1008–17.

[139] Mendez R, Richter JD. Translational control by CPEB: a means to the end. Nat Rev Mol Cell Biol 2001;2(7):521–9.

[140] Suh N, Baehner L, Moltzahn F, Melton C, Shenoy A, Chen J, et al. MicroRNA function is globally suppressed in mouse oocytes and early embryos. Curr Biol 2010;20(3):271–7.

[141] Tesfaye D, Worku D, Rings F, Phatsara C, Tholen E, Schellander K, et al. Identification and expression profiling of microRNAs during bovine oocyte maturation using heterologous approach. Mol Reprod Dev 2009; 76(7):665–77.

[142] Sinha PB, Tesfaye D, Rings F, Hossien M, Hoelker M, Held E, et al. MicroRNA-130b is involved in bovine granulosa and cumulus cells function, oocyte maturation and blastocyst formation. J Ovarian Res 2017;10(1):37.

[143] Wright EC, Hale BJ, Yang CX, Njoka JG, Ross JW. MicroRNA-21 and PDCD4 expression during in vitro oocyte maturation in pigs. Reprod Biol Endocrinol 2016;14:21.

[144] Pan B, Toms D, Shen W, Li J. MicroRNA-378 regulates oocyte maturation via the suppression of aromatase in porcine cumulus cells. Am J Physiol Endocrinol Metab 2015;308(6):E525–34.

[145] Sun QY, Schatten H. Centrosome inheritance after fertilization and nuclear transfer in mammals. Adv Exp Med Biol 2007;591:58–71.

[146] Krawetz SA, Kruger A, Lalancette C, Tagett R, Anton E, Draghici S, et al. A survey of small RNAs in human sperm. Hum Reprod 2011;26(12):3401–12.

[147] Liu WM, Pang RT, Chiu PC, Wong BP, Lao K, Lee KF, et al. Sperm-borne microRNA-34c is required for the first cleavage division in mouse. Proc Natl Acad Sci U S A 2012;109(2):490–4.

[148] Ostermeier GC, Miller D, Huntriss JD, Diamond MP, Krawetz SA. Reproductive biology: delivering spermatozoan RNA to the oocyte. Nature 2004;429(6988):154.

[149] Grivna ST, Beyret E, Wang Z, Lin H. A novel class of small RNAs in mouse spermatogenic cells. Genes Dev 2006;20(13):1709–14.

[150] Ro S, Park C, Sanders KM, McCarrey JR, Yan W. Cloning and expression profiling of testis-expressed microRNAs. Dev Biol 2007;311(2):592–602.

[151] Wu Q, Song R, Ortogero N, Zheng H, Evanoff R, Small CL, et al. The RNase III enzyme DROSHA is essential for microRNA production and spermatogenesis. J Biol Chem 2012;287(30):25173–90.

[152] Sendler E, Johnson GD, Mao S, Goodrich RJ, Diamond MP, Hauser R, et al. Stability, delivery and functions of human sperm RNAs at fertilization. Nucleic Acids Res 2013;41(7):4104–17.

[153] Wang C, Yang C, Chen X, Yao B, Yang C, Zhu C, et al. Altered profile of seminal plasma microRNAs in the molecular diagnosis of male infertility. Clin Chem 2011;57(12):1722–31.

[154] Govindaraju A, Uzun A, Robertson L, Atli MO, Kaya A, Topper E, et al. Dynamics of microRNAs in bull spermatozoa. Reprod Biol Endocrinol 2012;10:82.

[155] Fagerlind M, Stålhammar H, Olsson B, Klinga-Levan K. Expression of miRNAs in bull spermatozoa correlates with fertility rates. Reprod Domest Anim 2015;50(4):587–94.

[156] Medeiros LA, Dennis LM, Gill ME, Houbaviy H, Markoulaki S, Fu D, et al. Mir-290-295 deficiency in mice results in partially penetrant embryonic lethality and germ cell defects. Proc Natl Acad Sci U S A 2011; 108(34):14163–8.

[157] Tripurani SK, Lee KB, Wee G, Smith GW, Yao J. MicroRNA-196a regulates bovine newborn ovary homeobox gene (NOBOX) expression during early embryogenesis. BMC Dev Biol 2011;11:25.

[158] Dior UP, Kogan L, Chill HH, Eizenberg N, Simon A, Revel A. Emerging roles of microRNA in the embryo-endometrium cross talk. Semin Reprod Med 2014;32(5):402–9.

[159] Galliano D, Pellicer A. MicroRNA and implantation. Fertil Steril 2014;101(6):1531–44.

[160] Cuman C, Van Sinderen M, Gantier MP, Rainczuk K, Sorby K, Rombauts L, et al. Human blastocyst secreted microRNA regulate endometrial epithelial cell adhesion. EBioMedicine 2015;2(10):1528–35.

[161] Vilella F, Moreno-Moya JM, Balaguer N, Grasso A, Herrero M, Martínez S, et al. Hsa-miR-30d, secreted by the human endometrium, is taken up by the pre-implantation embryo and might modify its transcriptome. Development 2015;142(18):3210–21.

[162] Munné S, Alikani M, Tomkin G, Grifo J, Cohen J. Embryo morphology, developmental rates, and maternal age are correlated with chromosome abnormalities. Fertil Steril 1995;64(2):382–91.

[163] Rødgaard T, Heegaard PM, Callesen H. Non-invasive assessment of in-vitro embryo quality to improve transfer success. Reprod BioMed Online 2015;31(5):585–92.

[164] Botros L, Sakkas D, Seli E. Metabolomics and its application for non-invasive embryo assessment in IVF. Mol Hum Reprod 2008;14(12):679–90.

[165] Seli E, Robert C, Sirard MA. OMICS in assisted reproduction: possibilities and pitfalls. Mol Hum Reprod 2010;16(8):513–30.

[166] Stigliani S, Anserini P, Venturini PL, Scaruffi P. Mitochondrial DNA content in embryo culture medium is significantly associated with human embryo fragmentation. Hum Reprod 2013;28(10):2652–60.

[167] Stigliani S, Persico L, Lagazio C, Anserini P, Venturini PL, Scaruffi P. Mitochondrial DNA in Day 3 embryo culture medium is a novel, non-invasive biomarker of blastocyst potential and implantation outcome. Mol Hum Reprod 2014;20(12):1238–46.

[168] Viennois E, Zhao Y, Han MK, Xiao B, Zhang M, Prasad M, et al. Serum miRNA signature diagnoses and discriminates murine colitis subtypes and predicts ulcerative colitis in humans. Sci Rep 2017;7:2520.

[169] Mateescu B, Batista L, Cardon M, Gruosso T, de Feraudy Y, Mariani O, et al. miR-141 and miR-200a act on ovarian tumorigenesis by controlling oxidative stress response. Nat Med 2011;17(12):1627–35.

[170] Liang Y, Ridzon D, Wong L, Chen C. Characterization of microRNA expression profiles in normal human tissues. BMC Genomics 2007;8:166.

[171] Weber JA, Baxter DH, Zhang S, Huang DY, Huang KH, Lee MJ, et al. The microRNA spectrum in 12 body fluids. Clin Chem 2010;56(11):1733–41.

[172] Subramanyam D, Lamouille S, Judson RL, Liu JY, Bucay N, Derynck R, et al. Multiple targets of miR-302 and miR-372 promote reprogramming of human fibroblasts to induced pluripotent stem cells. Nat Biotechnol 2011;29(5):443–8.

[173] Barroso-delJesus A, Lucena-Aguilar G, Sanchez L, Ligero G, Gutierrez-Aranda I, Menendez P. The nodal inhibitor lefty is negatively modulated by the microRNA miR-302 in human embryonic stem cells. FASEB J 2011;25(5):1497–508.

[174] Ribeiro AO, Schoof CR, Izzotti A, Pereira LV, Vasques LR. MicroRNAs: modulators of cell identity, and their applications in tissue engineering. Microrna 2014;3(1):45–53.

[175] Chen X, Ba Y, Ma L, Cai X, Yin Y, Wang K, et al. Characterization of microRNAs in serum: a novel class of biomarkers for diagnosis of cancer and other diseases. Cell Res 2008;18(10):997–1006.

[176] Valadi H, Ekstrom K, Bossios A, Sjostrand M, Lee JJ, Lotvall JO. Exosome-mediated transfer of mRNAs and microRNAs is a novel mechanism of genetic exchange between cells. Nat Cell Biol 2007;9(6):654–9.

[177] Vickers KC, Palmisano BT, Shoucri BM, Shamburek RD, Remaley AT. MicroRNAs are transported in plasma and delivered to recipient cells by high-density lipoproteins. Nat Cell Biol 2011;13(4):423–33.

[178] Turchinovich A, Weiz L, Langheinz A, Burwinkel B. Characterization of extracellular circulating microRNA. Nucleic Acids Res 2011;39(16):7223–33.

[179] Creemers EE, Tijsen AJ, Pinto YM. Circulating microRNAs: novel biomarkers and extracellular communicators in cardiovascular disease? Circ Res 2012;110(3):483–95.

[180] Mitchell PS, Parkin RK, Kroh EM, Fritz BR, Wyman SK, Pogosova-Agadjanyan EL, et al. Circulating microRNAs as stable blood-based markers for cancer detection. Proc Natl Acad Sci U S A 2008;105(30):10513–8.

[181] Sang Q, Yao Z, Wang H, Feng R, Wang H, Zhao X, et al. Identification of microRNAs in human follicular fluid: characterization of microRNAs that govern steroidogenesis in vitro and are associated with polycystic ovary syndrome in vivo. J Clin Endocrinol Metab 2013;98(7):3068–79.

[182] Feng R, Sang Q, Zhu Y, Fu W, Liu M, Xu Y, et al. MiRNA-320 in the human follicular fluid is associated with embryo quality in vivo and affects mouse embryonic development in vitro. Sci Rep 2015;5:8689.

[183] Roth LW, McCallie B, Alvero R, Schoolcraft WB, Minjarez D, Katz-Jaffe MG. Altered microRNA and gene expression in the follicular fluid of women with polycystic ovary syndrome. J Assist Reprod Genet 2014; 31(3):355–62.

[184] Rosenbluth EM, Shelton DN, Sparks AE, Devor E, Christenson L, Van Voorhis BJ. MicroRNA expression in the human blastocyst. Fertil Steril 2013;99(3):855–861.e3.

[185] McCallie B, Schoolcraft WB, Katz-Jaffe MG. Aberration of blastocyst microRNA expression is associated with human infertility. Fertil Steril 2010;93(7):2374–82.

[186] McCallie BR, Parks JC, Strieby AL, Schoolcraft WB, Katz-Jaffe MG. Human blastocysts exhibit unique microrna profiles in relation to maternal age and chromosome constitution. J Assist Reprod Genet 2014;31(7):913–9.

[187] Rosenbluth EM, Shelton DN, Wells LM, Sparks AE, Van Voorhis BJ. Human embryos secrete microRNAs into culture media–a potential biomarker for implantation. Fertil Steril 2014;101(5):1493–500.

[188] Kropp J, Khatib H. Characterization of microRNA in bovine in vitro culture media associated with embryo quality and development. J Dairy Sci 2015;98(9):6552–63.

[189] Kropp J, Salih SM, Khatib H. Expression of microRNAs in bovine and human pre-implantation embryo culture media. Front Genet 2014;5:91.

[190] Capalbo A, Ubaldi FM, Cimadomo D, Noli L, Khalaf Y, Farcomeni A, et al. MicroRNAs in spent blastocyst culture medium are derived from trophectoderm cells and can be explored for human embryo reproductive competence assessment. Fertil Steril 2016;105(1):225–235.e1.

[191] Palini S, De Stefani S, Primiterra M, Galluzzi L. Pre-implantation genetic diagnosis and screening: now and the future. Gynecol Endocrinol 2015;31(10):755–9.

[192] Abu-Halima M, Häusler S, Backes C, Fehlmann T, Staib C, Nestel S, et al. Micro-ribonucleic acids and extracellular vesicles repertoire in the spent culture media is altered in women undergoing in vitro fertilization. Sci Rep 2017;7(1):13525.

[193] Tobler KJ, Zhao Y, Ross R, Benner AT, Xu X, Du L, et al. Blastocoel fluid from differentiated blastocysts harbors embryonic genomic material capable of a whole-genome deoxyribonucleic acid amplification and comprehensive chromosome microarray analysis. Fertil Steril 2015;104(2):418–25.

[194] Van Landuyt L, Polyzos NP, De Munck N, Blockeel C, Van de Velde H, Verheyen G. A prospective randomized controlled trial investigating the effect of artificial shrinkage (collapse) on the implantation potential of vitrified blastocysts. Hum Reprod 2015;30(11):2509–18.

EMBRYO IMPLANTATION, PLACENTA DEVELOPMENT AND PREGNANCY

Transcriptomics of the Human Endometrium and Embryo Implantation

Jose Miravet-Valenciano, María Ruiz-Alonso*,
Carlos Simón[†,‡,§,¶]*

*IGENOMIX, Valencia, Spain [†]University of Valencia/INCLIVA, Valencia, Spain [‡]Igenomix S.L,
Valencia, Spain [§]Department of Ob/Gyn, Stanford University, Stanford, CA, United States
[¶]Department of Ob/Gyn, Baylor College of Medicine, Houston, TX, United States

OUTLINE

Human Reproductive and Prenatal Genetics
https://doi.org/10.1016/B978-0-12-813570-9.00012-7

Acronyms and Abbreviations

ART	assisted reproductive technology
BMI	body mass index
DET	deferred embryo transfer
DNA	deoxyribonucleic acid
EF	endometrial fluid
ERA	endometrial receptivity analysis
Eth	endometrial thickness
FET	fresh embryo transfer
HRT	hormonal replacement therapy
IR	implantation rates
IVF	in vitro fertilisation
LH	luteinizing hormone
LIF	leukemia inhibitory factor
miRNA	micro-RNA
MMPs	matrix metalloproteinases
mRNA	messenger RNA
NGS	next-generation sequencing
NPV	negative predictive value
NR	nonreceptive
OPR	ongoing pregnancy rate
P	progesterone
PCA	principal component analysis
pET	personalized embryo transfer
PPV	positive predictive value
PR	pregnancy rate
R	receptive
RIF	recurrent implantation failure
RNA	Ribonucleic acid
NK	natural killer cells
WOI	window of implantation

INTRODUCTION

Despite careful embryo selection, two of every three in vitro fertilization (IVF) cycles fail to result in pregnancy, and more than 8 of every 10 transferred embryos fail to implant [1], making reproduction in humans an inefficient process. The key to successful implantation and subsequent invasion and decidualization is synchronization. The embryo must not only evolve to the blastocyst stage, but the endometrium must also achieve a specific receptive status and cross-talk between the embryo and endometrium must occur during a specific period known as the window of implantation (WOI). The embryo itself is thought to be responsible for two-thirds of implantation failures while inadequate uterine receptivity has been estimated to contribute to one-third [2, 3]. Leading up to the WOI, the endometrium must undergo a highly regulated and coordinated gene expression regulation during the whole endometrial cycle.

During the menstrual cycle, the endometrium undergoes physiological and morphological changes specific to each phase. These changes are initiated by responses of the different endometrial cell types to ovarian steroid hormones and paracrine-secreted molecules from neighboring cells. The menstrual cycle can be divided into the proliferative, the secretory, and the menstrual phases. During the proliferative phase, estrogen levels begin to rise, leading to a proliferation of stromal cells and glands and to the elongation of spiral arteries. This phase corresponds to the follicular phase in the ovary and ends with ovulation. The secretory phase is the time between ovulation and menstruation, and in its beginning, a rise in progesterone levels occurs. Progesterone receptors (PR-A and PR-B) and estrogen receptors (ERα and ERβ) are differentially expressed in different cell types in the endometrium. The secretory phase corresponds to the luteal phase in the ovary. In this phase, the endometrium acquires the receptive phenotype for a short, self-limited period, allowing for the implantation of the blastocyst. This state of receptivity, the WOI, occurs in the midsecretory phase. A successful interaction between a healthy embryo and a receptive endometrium will lead the embryo to appose, attach, penetrate, and invade the immune barrier-free endometrium and gain access to essential nutrients required for proper growth and development. In the absence of embryo implantation, the progesterone and estrogen levels decrease, the spiral arteries suffer vasoconstriction, and the endometrium involutions lead to menstrual bleeding and the starting of a new cycle.

Morphologically, the endometrium is divided into a functional and a basal layer. The functional layer occupies two thirds of the endometrial thickness and presents four cellular compartments: the luminal epithelium, the glandular epithelium, the stroma, and the vascular compartment. It is responsible for proliferation, secretion, and tissue degeneration. The functional layer develops from the basal layer and is responsible for tissue regeneration. The basal layer is adjacent to the myometrium, and the tissue below the functional layer is not shed at any time during the menstrual cycle.

Desynchronization between the embryo and the endometrium has been described in one out of four patients with recurrent implantation failure (RIF) [4]. RIF is generally defined as failure to achieve a clinical pregnancy after transferring at least four good-quality embryos in at least three fresh or frozen cycles [5]. Indeed, implantation failure is a major challenge for assisted reproductive technology (ART). However, it is possible to perform a personalized embryo transfer (pET) if the embryo is transferred according to the personal WOI of every patient. After correcting the window of implantation, pET normalizes clinical results. Patients suffering from RIF of endometrial origin have found a chance to achieve pregnancy thanks to pET because previous transfers have always been performed on the same day, regardless of their endometrial receptivity status. The success of this technique highlights the importance to fertility of the endometrial component as well as the complex molecular language exchanged between the embryo and the endometrium.

Early histological dating methods [6, 7] such as tissue histology have been reassessed following the publication of randomized studies challenging their accuracy, reproducibility, and functional relevance [8, 9]. This has encouraged further investigation and application of new technologies used to diagnose endometrial receptivity. To this end, the application of "-omics" high-throughput analyses, including genomics, epigenomics, proteomics, secretomics, metabolomics, and especially transcriptomics, has revolutionized our knowledge of the human endometrium and its involvement in fertility following the completion of the Human Genome Project [10].

The aim of this chapter is to review recent publications describing the transcriptomics of the human endometrium and its use as an objective tool in diagnosing endometrial receptivity and personalizing embryo transfer according to endometrial status.

TRANSCRIPTOMICS

Transcriptomics allow for gene expression characterization at the messenger RNA (mRNA) level of a population, leading to a sample-specific molecular profile correlating with the underlying biology. The main objective of transcriptomic studies is to identify the genes differentially expressed among predefined classes of samples to develop a mathematical function based on the expression profiles that can accurately predict the biologic group, diagnostic category, or prognostic stage [11]. Gene expression profiling has been used for tumor classification and to assess functions that might be reversed, such as the progression of cancer, heart disease, neuropsychiatric disorders, or factors associated with infertility [11a]. Machine-learning algorithms have been implemented as predictive models for the microarray signatures [11b] used in diagnostic and prognostic prediction [11c].

In the last decade, a fair number of publications describing the transcriptomes of the human endometrium, based on microarray studies, have validated the suitability of this technology in determining personalized endometrial factors in reproductive medicine. Additionally, several consortia have produced guidelines on the quality of microarray analyses and developed accurate and reproducible multivariate expression-based gene prediction models, known as "predictors" or "classifiers" [11d, 55, 56]. The predictor strategy is as follows: if the differences between the classes are the result of measurable differences in gene expression levels, these differences can be identified and used to assign the classes for a new profiling set. These tools help guarantee the efficacy of the clinical applicability of transcriptomics to phenotype diseases and thereafter their application to personalized medicine by stratifying patients and subphenotype diseases, improving diagnoses, and allowing for the personalization of treatments. The most extensive application of transcriptomic predictors in reproduction has been the assessment of the endometrial factor, a complex and a multifactorial contributor to infertility and, until recently, the black box of reproductive medicine.

TRANSCRIPTOMICS OF THE HUMAN ENDOMETRIUM

Analysis of endometrial transcriptome patterns in physiological and pathophysiological conditions has been to date the most commonly applied "-omics" technique utilized in the human endometrium. Other techniques such as genomics, proteomics, lipidomics, metabolomics, and epigenomics have failed to produce a tool that can be brought into clinical practice with sufficient reliability.

For more than 65 years, histologic evaluation has been considered the standard for clinical diagnosis based on morphological observations. The limitations of this method (including subjectivity, the need to perform the dating on a functional and healthy endometrium, and the incapacity to discriminate between fertile and infertile women [8]) underscore a need to understand the genetic mechanisms underlying the observed histological changes. The possibility of classifying the endometrium using transcriptomic profiles offers a powerful tool in clinical applications and is independent of the specific functional meaning of the transcriptomic signature [12].

Endometrial transcriptomics examines mRNA expression throughout the menstrual cycle using high-throughput techniques and massive data analysis. In the last decade, the preferred technique has been microarray-based gene expression. Microarray techniques have revealed the existence of specific expression profiles for each phase of the menstrual cycle [11,13,14,14a,14b,15,15a] as well as different profiles for the different types of cycles studied (reviewed in Martinez-Conejero et al. [15b]; Ruiz-Alonso et al. [11]). Specific expression profiles are also observed in the case of endometriosis [15c,16] and endometrial cancer [17]. This allows for the classification of the endometrium according to its transcriptomic profile [18] to overcome subjectivity in the morphological dating of the endometrium. Nonetheless, microarray strategies have some limitations that lie in the differences in experimental design, in the type of array used, the timing, the conditions of endometrial sampling, sample selection criteria, annotation versions used, the pipelines used for data processing, and the absence of consistent standards for data presentation [11,14b,18a] (Fig. 12.1).

Notably, most research in human endometrial transcriptomics is focused on finding a receptive transcriptomic signature and on comparing the expression profile during the WOI to other phases to develop a tool that assesses endometrial receptivity or identifies abnormalities in the endometrium.

From Histologic Evaluation to Molecular Medicine

Even though microarray technology had been successfully used to search for a molecular signature of receptivity [19–22], Ponnampalam et al. were the first to propose the transcriptomic characterization of the human endometrium throughout the menstrual cycle and use it as an alternative to histological dating [13]. In this study, samples were grouped in accordance with the endometrial cycle via hierarchical clustering. In the first arm were the proliferative, early secretory, and menstrual phases while the second arm contained the samples in the midsecretory and the late secretory phases. This group also found differences in the clustering based on the expression profiles of some samples and how they were classified by the histological criteria. Interestingly, two years later, Talbi et al. released a similar study [23] in which they reported a principal component analysis (PCA) detecting four clusters of samples corresponding to the predominant proliferative, early secretory, midsecretory, and late secretory phases. Despite differences in the sets of genes they used in this study for PCA and hierarchical clustering, the same clusters were found by both methods. Furthermore, six samples histologically dated as "ambiguous" were categorized in the same phase in both PCA and hieratical clustering, suggesting the existence of well-defined transcriptomic profiles for each menstrual phase that can be used to accurately

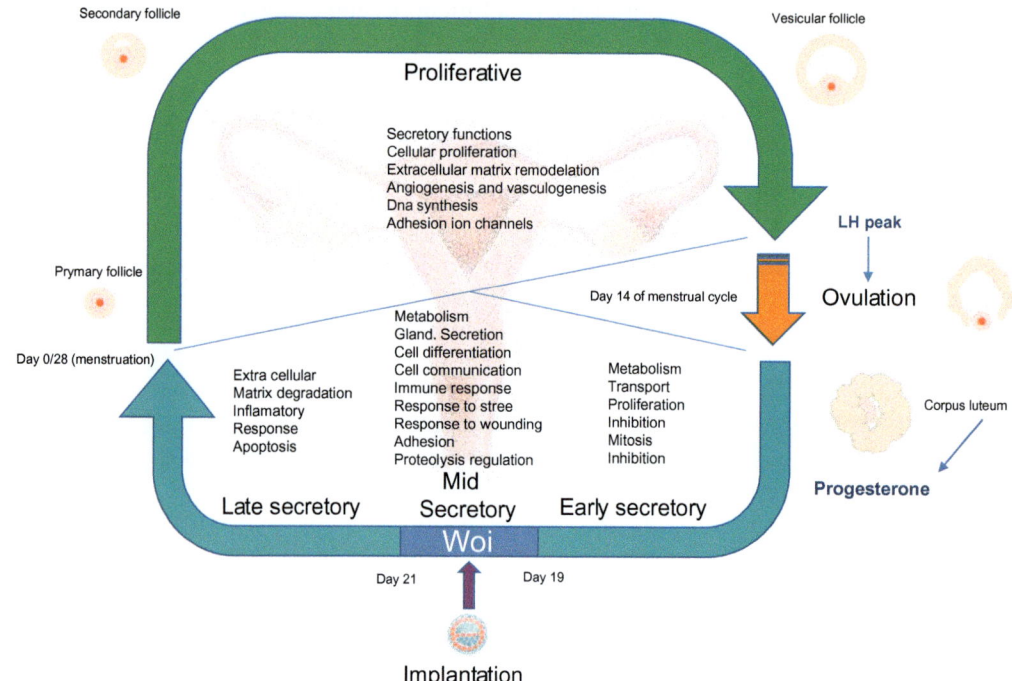

FIG. 12.1 Maturation of the endometrium throughout the menstrual cycle and its relationship to the biological processes implied in each endometrial stage.

catalogue endometria at different stages, further encouraging the transition from anatomical classification of the human endometrium to molecular methods.

Transcriptomics in the Menstrual Cycle

Changes in gene expression, as evidenced by transcriptomic studies, are useful to reveal the most crucial functions of each of the phases that make up an endometrial cycle. Although the most studied is the secretory phase because it is at this stage when the endometrium reaches its state of receptivity, every phase has its peculiarities.

Transcriptomics in Menstruation

Menstruation occurs on day 1 of the cycle, with an approximate duration of 3–5 days. This phase includes loss of the functional layer of the endometrium due to a drop in the estrogen levels [24], and it is one of the stages where major changes occur at the transcriptomic level. In contrast, it is the least studied from the gene expression perspective.

The main processes involved in menstruation are apoptosis and tissue breakdown. Thus, it is not surprising that the analysis of temporal gene expression patterns reveals a peak in the expression of genes related to apoptosis, inflammation, signal transduction, transcription, and DNA repair, including:

- Natural cytotoxicity-triggering receptor 3 (NCR3), which is involved in the inflammatory response in recognition of no-HLA ligands by natural killer cells [25]. This was reported for the first time by Ponnampalam et al. Its expression is up-regulated toward the late secretory phase, sustained in the menstrual phase, and down-regulated in the proliferative phase [26].
- Wnt family members: WNT5A and WNT7A may be implicated in endometrial gland development, as shown in mice [27]. Their high expression level during the menstrual phase lowers in the proliferative phase.
- Matrix metalloproteinase (MMP) family members MMP1, MMP3, and MMP10 are involved in tissue desquamation and show an increase in expression toward the late secretory phase before dropping back to the late proliferative phase.

Transcriptomics in the Proliferative Phase

The proliferative phase starts with the end of menstruation and lasts until ovulation (days 12–14 of the cycle). During this phase, estrogen levels increase, leading to stromal and glandular proliferation and to the elongation of spiral arteries to create a new functional layer [24]. The processes that occur during the proliferative phase typically involve cellular proliferation and differentiation, extracellular matrix remodeling, angiogenesis, and vasculogenesis [23, 28–30].

Interestingly, 64% of the genes that are differentially expressed between the menstrual and proliferative phases are down-regulated in the proliferative phase, showing a global decrease in gene expression in the transition from both phases. Among the remarkable genes, the following should be considered:

- MMPs are repressed in this phase, except MMP26, whose expression is activated and maintained until the WOI, suggesting that could be implicated in both tissue remodeling and blastocyst invasion [26].
- The trefoil family of peptides (TFF), whose up-regulation in the proliferative versus secretory phases makes them candidates as participants in endometrium regeneration after the menstrual phase [19].
- The HOX family genes, whose overexpression may be due to increased estrogen levels that inhibit the WNT5A and WNT7A genes to allow the expression of HOXA10 and HOXA11. The HOX gene expression completes glandular development and differentiation initiated by WNT5A and WNT7A [26].
- DNA replication licensing members (MCM2, MCM3, MCM4, MCM5, CM6, MCM7, and origin recognition complex proteins) suggest that, in addition to cell cycle regulation, DNA replication licensing is likely to be another key regulatory point in the hormonal regulation of epithelial cell proliferation in the human endometrium [28].
- Genes related to mitotic activity and tissue remodeling occurring in this phase due to the dramatic changes suffered by the endometrium in the predominant processes during the transition from the proliferative phase to the early secretory phase. These changes include cell motility, cell communication, cell adhesioncellular signaling, cellular cycle regulation, cellular division, cellular proliferation, collagen metabolism, extracellular matrix

regulation, and signal transduction and regulation of enzymes and ion channels. The described genes include Olfactomedin 1, *SOX4*, *MMP11*, *tPA (PLAT)*, *ADAM12*, retinoic acid, members of the IGF and TGF family, *FGF1*, *HGF*, *FGFR3* and *1.2*, and specific activator protein-kinase activities [23].

Transcriptomics in the Secretory Phase

The secretory phase takes place between ovulation and the onset of menstruation in a new cycle (days 28–30) and comprises the early, mid, and late secretory phases. This phase is the most studied because the WOI takes place during the midsecretory phase. The WOI opens on either day 7 after the surge of luteinizing hormone that indicates impending ovulation, or on day 5 after progesterone (P) production or administration. However, as will be described later, the WOI can be displaced and open two days before or after this day.

The main objective of the transcriptomic profile in this phase has been to find a molecular signature that is characteristic of a receptive endometrium to gain insight into the implantation phenomenon. Most studies have focused on comparing samples from this phase to the rest and applying variations in the many other parameters used, such as number of patients, taking samples from the same patient or not, using pools of samples, the days or types of cycles when samples are taken, or types of data analysis [14, 19–22, 31–35]. Despite a lack of consensus on the genes making up this transcriptomic signature, these studies are complementary to each other and shed light on the complexity of endometrial receptivity and the molecules involved in the implantation process. They all agree on the existence of a transcriptomic profile that is typical of the WOI and suggest that transcriptional activation must occur to achieve a state of receptivity. This is because most genes are up-regulated compared to their expression in the prereceptive phase. This process is known as "the transcriptional awakening process" [14].

The secretory phase is subdivided into three phases:

1. Early secretory phase. The predominant processes in this phase are related to cell metabolism, transport, germ cell migration, and negative cell proliferation regulation. This phase is highly biosynthetically active because the endometrium is preparing for embryo implantation. Among the up-regulated genes, the following should be highlighted:
   - *MSX1* (HOX family genes) correlates temporally and spatially with critical morphogenesis events [34].
   - *17βHSD* regulates oestradiol availability in the endometrium [23].
   - *PIP5K1B* regulates a wide range of cellular processes [34].
   - *MUC1* maintains cell surface hydration by lubricating and protecting cells against microorganisms and enzymes [36]
2. Midsecretory phase. The main processes carried out during this phase are related to a high active metabolism, secretion, and an increased innate immune response, stress response, and response to wounding. The progesterone peak that occurs in this phase provokes a number of gene expression changes that coincide with the time that the endometrium becomes receptive.

There are some genes whose expression is repressed during the transition from the early to the midsecretory phases. They are mainly involved in cell cycle regulation, ion binding, transport of signaling proteins, and members of the family of immunomodulators. There is an overexpression of genes related to implantation preparation, such as those involved in processes related to cell adhesion, metabolism, response to external stimuli, signaling, immune response (including stress response, defense response, humoral immune response, innate immune response, and response to wounding), cell communication, and negative regulation of proliferation and development [23, 32]. Among the most interesting genes that are up-regulated in this phase are:

- *LIF*: a cytokine that induces proliferation and has been shown to play a key role in the adhesion and invasion stages during implantation due to an anchor effect in the trophoblast [37].
- Osteopontin (*SPP1*): a structural glycoprotein of the extracellular matrix that mediates cellular adhesion and migration during implantation, is regulated by progesterone, and shows its peak of expression in endometrial epithelial cells [38, 39].
- *CXCL14*: a gene involved in the activation of the innate immune response that stimulates the chemotaxis of natural killer cells for clustering around epithelial glands. Due to its strong overexpression in this stage, it is thought to be the major recruiter of monocytes during the WOI [40]
- IL15 is also related to the immune system, plays important roles in uNK cell proliferation and differentiation, and is involved in the recruitment of peripheral blood cells CD16-NK [41, 42].
- Glycodelin also participates in immune response regulation, promoting the opening of the WOI by lowering maternal immunological responses to the implanting embryo [15].
- Other overexpressed genes related to the protection of the endometrium or the embryo, such as metallothioneins and GPXs (involved in protecting the cell from free radicals and heavy metals); decay accelerating factor, *DAF*, (protects the embryo from the maternal complement-mediated attack); and *GADD45* (preventing anticipated implantation and blastocyst invasion) [15, 43, 44].
- An overexpression of genes related to the humoral and cellular immune response has also been reported in the transition between the midsecretory and the late secretory phases [23].

3. Late secretory phase. The main processes that take place in this phase include extracellular matrix degradation, inflammatory response, and apoptosis. None of them favor implantation. This implies a closure of the WOI and a return to the nonreceptive endometrial phenotype.

The following gene regulation must be highlighted:
- The genes that are repressed are mostly involved in autophagy (*GABARAPL1*, *GABARAPL3*, and *DKK1*). This may be because autophagic degradation is involved in the mechanism responsible for the formation of new blood vessels [35]
- Up-regulated genes in this phase include *SOX4*, *ADAMTS5*, *GNG4*, Integrin α2, prolactin receptor, *MMPs*, *ADAMS*, serine proteases (*PLAU* and *PLAT*), *EBAF*, endothelin receptor B (*EDNRB*), and *MMP10*, which are involved in desquamation, tissue breakdown, and inflammatory mediation [23, 31]. The drop in P levels that occurs in this phase mediates all these processes as well as the overexpression of inflammatory mediators [45].

- Regarding immune activation, there is an overexpression of Fc receptors, MHC molecules, NK molecules, and T-cell-secreted molecules that corresponds to an innate and adaptive immunity response [23].

According to the gene regulation observed, the predominant activities that occur in the endometrium in this phase correspond to preparation for scaling in the next menstrual phase, which is when the process starts again.

miRNA Expression in the Endometrium

MicroRNAs (miRNAs) are small noncoding RNA molecules of 21–23 nucleotides that regulate the expression of other genes through the complementary binding sites in their mRNA targets. MicroRNAs can lead to the degradation or repression of their targets, regulating genes linked to processes such as development, cell differentiation, apoptosis, transduction signals, organogenesis, and cell proliferation, among others [46]. While most miRNAs are located within the cell, some miRNAs, commonly known as circulating miRNAs, have also been found in the extracellular environment in vivo and in vitro [47].

Studies of the expression of endometrial miRNAs have revealed the existence of cycle phase-specific miRNA expression profiles [48]. These studies have focused either on the expression during the WOI [28] or comparing the expression in natural cycles and ovarian stimulation cycles [49]. Studies have also been done to profile miRNA expression in women with implantation failure and compare that to women with proven fertility [50].

Kuokkannen et al. found that the miRNAs up-regulated in the midsecretory phase have target genes involved in DNA replication licensing and cell cycle regulators, which is consistent with the suppression of cell proliferation during the secretory phase. Among these up-regulated miRNA are *MIR31, MIR29B, MIR29C, MIR30B, MIR30D*, and *MIR210*, whose targets are key regulators of the cell cycle: cyclins and their cyclin-dependent kinase (CDK) partners. The overexpression of these miRNAs in the midsecretory phase is consistent with the repression of some of their target genes, but not all. This study suggests that these targets could be regulated at the translation level without any effect on gene expression.

Shah et al. [49] found different miRNA expression profiles between the early secretory and the midsecretory phases in natural cycles. Bioinformatics analysis revealed that these miRNAs regulate many genes, some of which are known to be differentially expressed during the WOI. They conclude that the differential expression of miRNAs may play a significant role in establishing and maintaining endometrial receptivity.

The molecular composition of the endometrial fluid (EF) was recently analyzed for potential markers of endometrial receptivity. The EF is a complex biological fluid secreted by the endometrial glands. It provides nutrients for embryonic survival during the peri-implantation period and constitutes the microenvironment in which the endometrial-blastocyst dialogue occurs before implantation. A study analyzing the miRNA content of the EF throughout the menstrual cycle identified a set of 27 miRNAs differentially expressed during the WOI, with miR-30d being the most up-regulated. This miRNA was shown to be internalized by trophectodermal cells of the blastocyst, modulating blastocyst adhesion [50a]. These results confirm the role of miRNAs in endometrial receptivity and their potential as biomarkers.

DIAGNOSIS OF ENDOMETRIAL RECEPTIVITY BASED ON TRANSCRIPTOMICS

In 2011, the endometrial receptivity analysis (ERA) test was launched, taking advantage of evidence accumulated in the last 15 years regarding transcriptomics and the genes involved in receptivity. The ERA is a microarray-based machine-learning predictive model used to diagnose human endometrial receptivity status. It does so by comparing the transcriptomic profile of a test sample with those of control samples taken seven days after the luteinising hormone peak (LH+7) in a natural cycle or five days after P administration (P+5) in a hormone replacement therapy (HRT) cycle. It is composed of a set of select genes that defines the transcriptomic signature of receptivity of the human endometrium during the implantation window [32] and describes the personalized WOI of a patient regardless of endometrial histology [4]. ERA is currently performed using NGS technology.

To design this diagnostic tool, 238 genes with differential expression were selected (rate of change >3) from the analysis of previously published data. This group of genes was coupled to a computational predictor and an algorithm that can diagnose the personalized endometrial WOI of a given patient regardless of endometrial histology. The predictor was trained with gene expression profiles obtained from samples at different stages of the menstrual cycle (proliferative, prereceptive, receptive, and postreceptive) to classify a test sample according to the gene expression values obtained with the array. The specificity and sensitivity of this array is 0.88 and 0.99, respectively. In addition, its efficacy and consistency have been demonstrated [32]. In this work, the Kappa concordance index shows a much better efficiency of the molecular diagnosis compared to the histological data. Further, a consistency analysis, for which endometrial biopsies were taken from the same woman two to three years apart in the same type and day of the cycle, showed the same molecular diagnosis given by the computational predictor of the ERA as well as high similarity in the analysis of paired samples by PCA and hierarchical grouping. This study reaffirms the existence of specific and consistent expression patterns in the secretory phase of the menstrual cycle.

The ERA test is the only known molecular tool whose clinical applicability has been proven for its use in reproductive medicine. The value of this test lies in its ability to diagnose the individual status of the window of implantation in a patient and in the possibility of a pET [4]. Many of the failures of implantation of endometrial origin are not due to an irreversible alteration of the endometrium, but to a displacement of the implantation window along with the absence of an adequate diagnostic system.

The analysis of the ERA consists of the identification of possible displacements of the window of implantation to find the moment of specific receptivity for each patient, which supposes a totally innovative concept in the assessment of endometrial receptivity. ERA has been validated in a prospective multicenter clinical study in which implantation and pregnancy rates were analyzed for patients with RIF who have carried out a personalized embryo transfer [4]. This study focused on a group of patients less than 40 years old, with RIF and at least three previous failed cycles, and a control group consisting of patients with one or no prior IVF treatment. RIF and control patients underwent ERA-based endometrial receptivity diagnosis using an endometrial biopsy obtained either on day LH+7 in a natural cycle or on day P+5 in an HRT cycle [4]. The diagnostic ratios of receptive/nonresponsive were 74/26 for the

RIF group and 88/12 for the control group, demonstrating the existence of an endometrial alteration in terms of displacement of the window of implantation in patients with RIF. Their inability to implant could be attributed to an endometrial factor. In the case of patients with nonreceptive diagnosis, the expression profile of each patient was analyzed and 84% were found to have a prereceptive profile, implying a delay in the implantation window. This displacement was validated with a second ERA test performed one or two days after the first test (according to each specific case). In cases in which an advancement of the implantation window was observed, it was validated by advancing the biopsy sample by one or two days. In 15 of the 18 cases of displacement analyzed, a receptivity profile was obtained and, following the indications of the test, the personalized embryo transfer was applied, taking place in the same type and day of the cycle in which that diagnosis was obtained. Using this strategy, a 50.0% pregnancy rate and a 38.5% implantation rate were achieved. These rates are similar to those of patients who had a receptive result at their first biopsy, at 51.7% and 33.9%, respectively [4].

Until now, the primary factor guiding the timing of embryo transfer in ART has been the stage of embryo development because it was generally accepted that the timing and duration of the WOI were constant in all women. However, this study has shown that it is possible to identify the status of the endometrium using the transcriptomic profile of a select group of genes to identify a delayed or advanced WOI. Thus, we now have the ability to diagnose displacement of the WOI and to personalize embryo transfer in each patient if necessary [4]. This highlights the need to synchronize embryonic and endometrial development, personalizing the timing of embryo transfer.

A prospective clinical trial using ERA at the expected moment of receptivity found that 26% of patients with RIF but only 12% of control patients (with one or no previous failed cycles) presented a displaced WOI. Interestingly, when pET, guided by the ERA prediction of WOI, was performed on RIF patients, the implantation and pregnancy rates increased to the levels observed in the control group. This implies that in those patients the endometrial factor was responsible for the failed cycles. Thus, the diagnostic value of ERA was proven in clinical practice to improve pregnancy outcomes in women with RIF [4]. This study demonstrates that RIF is not an endometrial dysfunction, but rather a desynchronization between embryo and endometrium. This initial study has been further validated by the report of a clinical case of successful pET after seven previous failed IVF attempts. This case report was complemented by a pilot study of 17 patients undergoing oocyte donation who experienced multiple failed implantations with routine embryo transfer, but were subsequently treated with pET after determining their pWOI resulting in normalization of their reproductive outcome [51].

The clinical efficiency of pET has also been assessed according to its specificity and sensitivity. Following a similar protocol to the pilot study, the clinical outcome of pET was analyzed in a group of 205 receptive patients and compared to frozen embryo transfer on a day after the determination of nonreceptive status in 52 patients. Differences in implantation rate (IR) and pregnancy rate (PR) between the groups were similar to those found in previous research, with a 23% PR and 13% IR in FET versus 60% and 45% when pET was performed. To calculate specificity and sensitivity, the positive condition was considered to be nonreceptive and the negative considered to be receptive. After analysis, a specificity of 0.91 and a sensitivity of 0.33 was obtained due to the multifactorial character of the implantation process.

TABLE 12.1 Clinical Outcome and Efficiency of Embryo Transfer According to ERA Status

Clinical Outcome	NR (52)	R (205)
IR First attempt	13% (12/90)	45% (161/355)
IR Total attempts	10% (17/174)	41% (182/441)
PR First attempt	23% (12/52)	60% (123/205)
PR Total attempts	17% (17/100)	55% (140/253)
OPR First attempt	0% (0/12)	74 % (91/123)
OPR Total attempts	0% (0/100)	74% (103/140)
Clinical efficiency	Positive (52)	Negative (205)
True	40	123
False	12	82
Sensitivity (TP/TP+FN)	0.33	
Specificity (TN/TN+FP)	0.91	
PPV (TP/TP+FN)	0.77	
NPV (TN/TN+FN)	0.60	

IR, implantation rate; PR, pregnancy rate; OPR, ongoing pregnancy rate; PPV, positive predictive value; NPV, negative predictive value; NR, nonreceptive; R, receptive.

The positive predictive value obtained was 0.77 while the negative predictive value was 0.60 (16). Clinical outcomes were more similar after increasing the number of patients to 400 receptive and 100 nonreceptive cases: 20% PR and 12% IR was observed in FET versus 58% and 45% when pET was performed. Data obtained from this study are shown in Table 12.1.

An international randomized controlled study is underway to perform endometrial assessment during fertility screening at the beginning of reproductive care (The ERA as a diagnostic guide for personalized embryo transfer. ClinicalTrials.gov Identifier: NCT01954758). An ERA RCT consortium was created to include 28 clinics worldwide. This randomized study included patients undergoing transfer at the blastocyst stage (day 5 or day 6) in their first IVF/ICSI cycle with a body mass index (BMI) between 18.5 and 30 who were younger than 37 and had a normal ovarian reserve. Patients with any pathology affecting the endometrial cavity were operated previously to participate in the study. Patients with recurrent pregnancy loss and/or severe male factor were excluded.

The study consists of three arms comparing fresh embryo transfer under stimulation protocol, frozen embryo transfer at P+5 in HRT cycles, and pET guided by ERA with frozen embryos in HRT cycles. At the midpoint of recruitment, results show significant differences between PR for pET arm (85.7%) versus fresh embryo transfer (FET) (61.7%) and deferred embryo transfer (DET) (60.8%). Although not yet significant, there are also differences in IR (47.8% for pET, 35.3% for FET, and 41.4% for DET) and in ongoing pregnancy rate (OPR) per embryo transfer (55.1% for pET, 43.3% for FET, and 44.6% for DET). These interim results, shown in Table 12.2, were published in the American Society of Reproductive Medicine

TABLE 12.2 Interim Clinical Outcome of a Randomized Controlled Study Comparing Fresh Embryo Transfer, Deferred Embryo Transfer, and Personalized Embryo Transfer Guided by ERA

Variable	FET	DET	pET
Recruited (n)	117	122	117
Embryo transfer (n)	60	74	49
Implantation rate (%)	36/102 (35.3)	53/128 (41.4)	43/90 (47.8)
Pregnancy rate/ET (%)	37/60 (61.7)[a]	45/74 (60.8)[a]	42/49 (85.7)[a]
Pregnancy loss (%)	11/37 (29.7)	12/45 (26.7)	15/42 (35.7)
Biochemical pregnancy (%)	8/37 (21.6)	3/45 (6.7)	5/42 (11.9)
Ectopic pregnancies (%)	1/37 (2.7)	0/45 (0.0)	1/42 (2.4)
Clinical miscarriages (%)	2/37 (5.4)	9/45 (20.0)	9/42 (21.4)
Ongoing pregnancy/et (%)	26/60 (43.3)	33/74 (44.6)	27/49 (55.1)
Twins (%)	8/28 (28.6)	11/42 (26.2)	7/36 (19.4)
Singleton (%)	20/28 (71.4)	31/42 (73.8)	29/36 (80.6)

[a] Statistical comparisons between groups were performed using Chi-square test (*P< 0.05).

(ASRM) 2016 scientific congress [52] and show that 14% of patients have a displaced WOI whose correction would likely result in an effective cost-benefit strategy at the first clinical appointment.

Since 2017, ERA has been performed by NGS, and the transcriptomic subclassification refinement released by Diaz-Gimeno et al. [52a] has been applied for this version of the test. This study provides new insight into transcriptomic stratification and improves the predictive accuracy of the gene signature during the WOI, making it possible to detect the best transcriptome associated with successful pregnancy. Furthermore, they have discovered a signature that may help prevent biochemical pregnancies of endometrial origin and their findings accurately define and provide clinical meaning for the personalized variability among menstrual cycle WOIs.

Other studies have attempted to describe the transcriptomic profile of endometrial receptivity (review by Gómez et al. [52b]). A recent metaanalysis found that 57 genes, including genes present in the ERA (i.e., SPP1, ANXA4, CLDN4, DPP4, GPX3, MAOA, and PAEP), were identified as potential receptivity biomarkers in multiple studies and are the most representative panel for predicting the WOI [52c]. However, these findings have not been translated to the clinic.

FACTORS INFLUENCING ENDOMETRIAL TRANSCRIPTOMICS DURING RECEPTIVITY

Endometriosis

Endometriosis is an estrogen-dependent disorder that typically affects women of reproductive age, impacting their physical, mental, and social well-being. Endometriosis, a benign condition historically related to infertility, is characterized by the presence of endometrial tissue outside the uterine cavity suffered by an estimated 10% of women [53]. Symptoms range from practically nonexistent to severe chronic pelvic pain, dysmenorrhea, and cyclic urinary or bowel complaints.

Endometriosis does not affect embryo implantation in ovum recipients, meaning that the endometrial factor is not responsible for infertility in these patients. However, disruption of endometrial genes may be responsible for endometriosis. To uncover the role of endometrial receptivity in endometriosis-associated infertility, ERA results arising from endometrial samples from 17 infertile patients with different stages of endometriosis (stages I–IV) based on the revised staging system of the American Society for Reproductive Medicine [54] were compared to the results of samples from five healthy patients [16]. Endometrial biopsies were taken from each woman at day 18–20 of a natural cycle, according to Noyes criteria [6]. Interestingly, the results showed clustering of samples that was not due to the endometriosis stage, but rather to the day of the cycle on which the samples were taken. Depending on the gene expression profiling, the ERA showed no over- or under expression in the 238 genes of the customized array when comparing the four endometriosis stages and controls. Alternatively, comparisons between samples taken on different days of the menstrual cycle revealed that only 13 genes were differentially regulated: *ARG2*, *CLDN4*, *HRASLS3*, *MAOA*, *EFNA1*, *RPRM*, *DEFB1*, *S100P*, *KRT7*, *BCL6*, *RARRES3*, *GDF15*, and *GABARAPL1*.

This pilot study concluded that the transcriptomic profile of endometrial receptivity was similar in patients with or without endometriosis, regardless of the stage of the disease, indicating that endometriosis does not affect endometrial receptivity.

Obesity

According to the WHO, obesity is defined as abnormal or excessive fat accumulation that may impair health as a result of a variety of associated pathologies. It is an increasingly prevalent health burden worldwide that has nearly tripled since 1975. In the field of reproduction, maternal obesity is linked to infertility. Obesity is associated with menstrual disorders, hirsutism, obstetric complications, anovulatory problems, a lower implantation potential, increased miscarriage rates, and longer time topregnancy in both spontaneous pregnancies and ART cycles. The poorer reproductive outcomes in obese patients may involve the egg, the embryo, the endometrium, or a combination.

To clarify the role of endometrial receptivity in obesity-associated infertility, Comstock et al. analyzed the endometrial transcriptomic profile during the window of implantation in infertile obese women [54a]. To this aim, the ERA test was performed in infertile patients stratified by body mass index (BMI) as normal-weight (BMI 18.5-24.9 kg/m^2), overweight (BMI 25.0-29.9 kg/m^2), or obesity class I (BMI 30.0-34.9 kg/m^2), II (BMI 35.0-39.9 kg/m^2), or III (BMI over 40.0 kg/m^2). This study found that the *XCL1*, *XCL2*, *HMHA1*, *S100A1*, *KLRC1*, *COTL1*, *COL16A1*, *KRT7*, and *MFAP5* genes are significantly dysregulated during the window of implantation in the receptive endometrium of obese patients. On the other hand, *COL16A1*, *COTL1*, *HMHA1*, *KRCL1*, *XCL1*, and *XCL2* genes were down-regulated and *KRT7*, *MFAP5*, and *S100A1* genes were up-regulated in the endometrium of obese patients. Thus, this study demonstrated that the transcriptomic profile of the endometrium during the WOI is altered in obese patients. The displacement of endometrial receptivity was more pronounced as BMI increased according to the proportion of nonreceptive patients and the levels of specific dysregulated genes (either up- or down-regulated), providing evidence that an altered endometrial gene expression may contribute to infertility in obese patients.

Ethnicity

Race or ethnicity may also influence endometrial receptivity. The ERA test has been applied to patients of different ethnicities to elucidate whether differential genomic variants affecting ethnicity could influence the WOI. To this end, the transcriptomic profiles of endometrial biopsies of infertile women from the following ethnicities were compared: African ($n=19$), Arabian ($n=20$), Caucasian ($n=1003$), East Asian ($n=209$), Hispanic ($n=45$), and South Asian ($n=1,244$). Most race/ethnic groups exhibited similar proportions of receptive and nonreceptive women. However, African women were distinct. In this group, there was a significant decrease in receptive patients with an apparent shift to a prereceptive phenotype. A potential reason for this is that the African women involved in this study had significantly higher BMIs than the women from other groups, which could bias the results obtained from the ERA test. Further studies are required to clarify whether race or ethnicity influences the acquisition and/or prediction of endometrial receptivity (Valbuena and Simon, unpublished data).

Endometrial Thickness

Many studies have investigated the association between endometrial thickness (Eth) and receptivity, concluding that a thickness of 6 mm before progesterone administration in HRT cycles indicates a receptive endometrium. To corroborate these results using a reliable molecular method, ERA analysis was performed in retrospective endometrial samples classified by endometrial thickness at ultrasound into three groups: atrophic endometrium (Eth < 6 mm), normal endometrium (Eth 6–12 mm), and hypertrophic endometrium (Eth > 12 mm).

Endometrial thickness was measured the day before progesterone supplementation in HRT cycles prior to ovum donation or frozen embryo transfer and ERA was performed during the WOI after five days of progesterone administration. A normal ratio of receptive and nonreceptive profiles was observed in samples from endometria with either normal or increased thicknesses. However, in samples from atrophic endometria, the percentage of nonreceptive profiles was significantly increased and accounted for more than 50% of the analyzed samples. Specifically, patients with atrophic endometrium presented a transcriptomic profile consistent with the prereceptive status, suggesting that the WOI in those patients is delayed due to insufficient growth of the endometrium (Valbuena and Simon, unpublished data).

Limitations of the ERA Test

Despite the refinement of the ERA as a predictor of endometrial receptivity based on transcriptomic profile during the WOI, certain limitations must be considered:

1. Transcriptomic analysis uses mRNA, a highly sensitive genetic material that can be degraded if strict considerations such as ranges of temperature for shipment and storage, use of RNAse inhibitors, and sterile conditions are not applied.
2. Endometrial biopsies could present difficulties during collection, resulting in a few samples that are not suitable to process because it is not possible to obtain enough RNA, or it is degraded. Furthermore, taking a biopsy necessitates causing pain to the patients.
3. Because ERA has only been tested during HRT and natural cycles, it is not possible to extrapolate to controlled ovarian stimulation cycles.
4. The duration of the current protocol limits embryo transfer to the same cycle in which the biopsy is taken and it is only possible to perform pET with frozen embryos from the same patient or fresh embryos in ovum donation cycles.
5. Finally, ERA diagnoses the receptivity status of the endometrium at the transcriptional level. Nevertheless, other factors must be taking into account regarding implantation success. For example, an altered uterine microbiome may impair the clinical results of an otherwise receptive endometrium. In addition, a chromosomally normal embryo must be transferred to improve the chance of implantation.

CONCLUSIONS

Despite careful embryo selection, pregnancy and birth rates resulting from assisted reproductive technologies remain lower than optimal. Among the multiple factors implied in effective IVF treatment, the primary limiting factor is successful embryo implantation. Implantation failures are caused primarily by poor endometrial receptivity, defects in the embryo, and diseases or disorders in the endometrium. It is currently accepted that two-thirds of these implantation failures have their origin in low endometrial receptivity or in a defective endometrium-embryo dialogue. At this point, it is not surprising that the key to human reproduction requires the perfect synchrony between a healthy embryo and the endometrium. Otherwise, implantation will never occur.

The human endometrium undergoes cyclic physiological and morphological changes in response to several factors, including estrogen and progesterone, that induce the biological processes according to endometrial gene expression. It only becomes receptive to blastocyst implantation during the WOI, which can be displaced forward or backward from the standard day or can be extremely narrow. Unless the blastocyst or the day 3 embryo is in the perfect state of development and is transferred during the WOI, it will be wasted, so timing is absolutely critical for achieving pregnancy.

This chapter describes the variation in gene expression profiles among different phases of the cycle along with diverse conditions of the endometrium. The functional genomics of endometrial receptivity have been extensively investigated to find transcriptomic markers of endometrial receptivity during the implantation window, with the vision of using this information in diagnosing endometrial receptivity. This advance implies the substitution of other classic biochemical and morphological markers whose effectiveness has been frequently questioned. The ERA has become the gold standard for the diagnosis of WOI displacement in patients with RIF based on the transcriptomic profile of endometrial samples and has been used for clinical and academic research in endometrial receptivity. Currently, our group is validating a noninvasive test to provide consistent results and make it easier for clinicians to obtain samples and avoid unnecessary pain and discomfort to the patients.

References

[1] Kovalevsky G, Patrizio P. High rates of embryo wastage with use of assisted reproductive technology: a look at the trends between 1995 and 2001 in the United States. Fertil Steril 2005;84(2):325–30.

[2] Cha J, Vilella F, Dey SK, Simón C. Molecular interplay in successful implantation. In: Ten critical topics in reproductive medicine. 2013. p. 44–8.

[3] Macklon NS, Stouffer RL, Giudice LC, Fauser BC. The science behind 25 years of ovarian stimulation for in vitro fertilization. Endocr Rev 2006;27(2):170–207.

[4] Ruiz-Alonso M, Blesa D, Díaz-Gimeno P, Gómez E, Fernández-Sánchez M, Carranza F, et al. The endometrial receptivity array for diagnosis and personalized embryo transfer as a treatment for patients with repeated implantation failure. Fertil Steril 2013;100(3):818–24.

[5] Coughlan C, Ledger W, Wang Q, Liu F, Demirol A, Gurgan T, et al. Recurrent implantation failure: definition and management. Reprod BioMed Online 2014;28:14–38. https://dx.doi.org/10.1016/j.rbmo.2013.08.011.

[6] Noyes RW, Hertig AT, Rock J. Dating the endometrial biopsy. Obstet Gynecol Surv 1950;5(4):561–4.

[7] Noyes RW, Hertig AT, Rock J. Dating the endometrial biopsy. Am J Obstet Gynecol 1975;122(2):262.

[8] Coutifaris C, Myers ER, Guzick DS, Diamond MP, Carson SA, Legro RS, et al. Histological dating of timed endometrial biopsy tissue is not related to fertility status. Fertil Steril 2004;82(5):1264–72.

[9] Murray MJ, Meyer WR, Zaino RJ, Lessey BA, Novotny DB, Ireland K, et al. A critical analysis of the accuracy, reproducibility, and clinical utility of histologic endometrial dating in fertile women. Fertil Steril 2004;81(5):1333–43.

[10] Venter JC, Adams MD, Myers EW, Li PW, Mural RJ, Sutton GG, et al. The sequence of the human genome. Science 2001;291(5507):1304–51.

[11] Ruiz-Alonso M, Blesa D, Simón C. The genomics of the human endometrium. Biochim Biophys Acta 2012;1822(12):1931–42.

[11a] Quackenbush J. Microarray analysis and tumor classification. N Engl J Med 2006;354(23):2463–72.

[11b] Medina I, Montaner D, Tárraga J, Dopazo J. Prophet, a web-based tool for class prediction using microarray data. Bioinformatics 2006;23(3):390–1.

[11c] Simon R. Using DNA microarrays for diagnostic and prognostic prediction. Expert Rev Mol Diagn 2003;3(5):587–95.

[11d] Brazma A, Hingamp P, Quackenbush J, Sherlock G, Spellman P, Stoeckert C, Gaasterland T. Minimum information about a microarray experiment (MIAME)—toward standards for microarray data. Nat Genet 2001;29(4):365.

[12] Shi W, Bessarabova M, Dosymbekov D, Dezso Z, Nikolskaya T, Dudoladova M, et al. Functional analysis of multiple genomic signatures demonstrates that classification algorithms choose phenotype-related genes. Pharmacogenomics J 2010;10(4):310–23.

[13] Ponnampalam AP, Weston GC, Trajstman AC, Susil B, Rogers PA. Molecular classification of human endometrial cycle stages by transcriptional profiling. Mol Hum Reprod 2004;10(12):879–93.

[14] Horcajadas JA, Riesewijk A, Domínguez F, Cervero ANA, Pellicer A, Simón C. Determinants of endometrial receptivity. Ann N Y Acad Sci 2004;1034(1):166–75.

[14a] Giudice LC. Application of functional genomics to primate endometrium: insights into biological processes. Reprod Biol Endocrinol 2006;4(1):S4.

[14b] Horcajadas JA, Pellicer A, Simon C. Wide genomic analysis of human endometrial receptivity: new times, new opportunities. Hum Reprod Update 2007;13(1):77–86.

[15] Aghajanova L, Simón C, Horcajadas JA. Are favorite molecules of endometrial receptivity still in favor? Expert Rev Obstet Gynecol 2008;3(4):487–501.

[15a] Garrido-Gómez T, Ruiz-Alonso M, Blesa D, Diaz-Gimeno P, Vilella F, Simón C. Profiling the gene signature of endometrial receptivity: clinical results. Fertil Steril 2013;99(4):1078–85.

[15b] Martínez-Conejero JA, Simón C, Pellicer A, Horcajadas JA. Is ovarian stimulation detrimental to the endometrium? Reprod Biomed Online 2007;15(1):45–50.

[15c] Sherwin JRA, Sharkey AM, Mihalyi A, Simsa P, Catalano RD, D'hooghe T. Global gene analysis of late secretory phase, eutopic endometrium does not provide the basis for a minimally invasive test of endometriosis. Human Reprod 2008;23(5):1063–8.

[16] Garcia-Velasco JA, Fassbender A, Ruiz-Alonso M, Blesa D, Thomas DH, Simon C. Is endometrial receptivity transcriptomics affected in women with endometriosis? A pilot study. Reprod BioMed Online 2015;31(5):647–54.

[17] Habermann JK, Bündgen NK, Gemoll T, Hautaniemi S, Lundgren C, Wangsa D, … Jörnvall H. Genomic instability influences the transcriptome and proteome in endometrial cancer subtypes. Mol Cancer 2011;10(1):132.

[18] Wilcox AJ, Baird DD, Weinberg CR. Time of implantation of the conceptus and loss of pregnancy. N Engl J Med 1999;340:1796–9. https://dx.doi.org/10.1056/NEJM199906103402304.

[18a] Ulbrich SE, Groebner AE, Bauersachs S. Transcriptional profiling to address molecular determinants of endometrial receptivity—lessons from studies in livestock species. Methods 2013;59(1):108–15.

[19] Borthwick JM, Charnock-Jones DS, Tom BD, Hull ML, Teirney R, Phillips SC, Smith SK. Determination of the transcript profile of human endometrium. Mol Hum Reprod 2003;9(1):19–33.

[20] Carson DD, Lagow E, Thathiah A, Al-Shami R, Farach-Carson MC, Vernon M, et al. Changes in gene expression during the early to mid-luteal (receptive phase) transition in human endometrium detected by high-density microarray screening. Mol Hum Reprod 2002;8(9):871–9.

[21] Kao LC, Tulac S, Lobo SA, Imani B, Yang JP, Germeyer A, et al. Global gene profiling in human endometrium during the window of implantation. Endocrinology 2002;143(6):2119–38.

[22] Riesewijk A, Martín J, van Os R, Horcajadas JA, Polman J, Pellicer A, et al. Gene expression profiling of human endometrial receptivity on days LH+ 2 versus LH+ 7 by microarray technology. Mol Hum Reprod 2003;9(5):253–64.

[23] Talbi S, Hamilton AE, Vo KC, Tulac S, Overgaard MT, Dosiou C, et al. Molecular phenotyping of human endometrium distinguishes menstrual cycle phases and underlying biological processes in normo-ovulatory women. Endocrinology 2006;147(3):1097–121.

[24] Hawkins SM, Matzuk MM. The menstrual cycle. Ann N Y Acad Sci 2008;1135(1):10–8.

[25] Pende D, Parolini S, Pessino A, Sivori S, Augugliaro R, Morelli L, et al. Identification and molecular characterization of NKp30, a novel triggering receptor involved in natural cytotoxicity mediated by human natural killer cells. J Exp Med 1999;190(10):1505–16.

[26] Punyadeera C, Dassen H, Klomp J, Dunselman G, Kamps R, Dijcks F, et al. Estrogen-modulated gene expression in the human endometrium. Cell Mol Life Sci 2005;62(2):239–50.

[27] Miller C, Sassoon DA. Wnt-7a maintains appropriate uterine patterning during the development of the mouse female reproductive tract. Development 1998;125(16):3201–11.

[28] Kuokkanen S, Chen B, Ojalvo L, Benard L, Santoro N, Pollard JW. Genomic profiling of microRNAs and messenger RNAs reveals hormonal regulation in microRNA expression in human endometrium. Biol Reprod 2010;82(4):791–801.

[29] Maruyama T, Yoshimura Y. Molecular and cellular mechanisms for differentiation and regeneration of the uterine endometrium. Endocr J 2008;55(5):795–810.

[30] Simmen FA, Simmen RC. Orchestrating the menstrual cycle: discerning the music from the noise. Endocrinology 2006;147(3):1094–6.

[31] Critchley HO, Robertson KA, Forster T, Henderson TA, Williams AR, Ghazal P. Gene expression profiling of mid to late secretory phase endometrial biopsies from women with menstrual complaint. Am J Obstet Gynecol 2006;195(2):406–14.

[32] Díaz-Gimeno P, Horcajadas JA, Martínez-Conejero JA, Esteban FJ, Alamá P, Pellicer A, Simón C. A genomic diagnostic tool for human endometrial receptivity based on the transcriptomic signature. Fertil Steril 2011;95(1):50–60.

[33] Haouzi D, Assou S, Mahmoud K, Tondeur S, Rème T, Hedon B, et al. Gene expression profile of human endometrial receptivity: comparison between natural and stimulated cycles for the same patients. Hum Reprod 2009;24(6):1436–45.

[34] Mirkin S, Arslan M, Churikov D, Corica A, Diaz JI, Williams S, et al. In search of candidate genes critically expressed in the human endometrium during the window of implantation. Hum Reprod 2005;20(8):2104–17.

[35] Tseng LH, Chen I, Chen MY, Yan H, Wang CN, Lee CL. Genome-based expression profiling as a single standardized microarray platform for the diagnosis of endometrial disorder: an array of 126-gene model. Fertil Steril 2010;94(1):114–9.

[36] Brayman M, Thathiah A, Carson DD. MUC1: a multifunctional cell surface component of reproductive tissue epithelia. Reprod Biol Endocrinol 2004;2(1):4.

[37] Dimitriadis E, Nie G, Hannan NJ, Paiva P, Salamonsen LA. Local regulation of implantation at the human fetal-maternal interface. Int J Dev Biol 2009;54(2-3):313–22.

[38] Apparao KBC, Murray MJ, Fritz MA, Meyer WR, Chambers AF, Truong PR, Lessey BA. Osteopontin and its receptor αvβ3 integrin are coexpressed in the human endometrium during the menstrual cycle but regulated differentially. J Clin Endocrinol Metab 2001;86(10):4991–5000.

[39] Lessey BA. Two pathways of progesterone action in the human endometrium: implications for implantation and contraception. Steroids 2003;68(10):809–15.

[40] Mokhtar NM, Cheng CW, Cook E, Bielby H, Smith SK, Charnock-Jones DS. Progestin regulates chemokine (CXC motif) ligand 14 transcript level in human endometrium. Mol Hum Reprod 2009;16(3):170–7.

[41] Kitaya K, Yamaguchi T, Honjo H. Central role of interleukin-15 in postovulatory recruitment of peripheral blood CD16 (−) natural killer cells into human endometrium. J Clin Endocrinol Metab 2005;90(5):2932–40.

[42] Okada H, Nakajima T, Sanezumi M, Ikuta A, Yasuda K, Kanzaki H. Progesterone enhances interleukin-15 production in human endometrial stromal cells in vitro. J Clin Endocrinol Metab 2000;85(12):4765–70.

[43] Franchi A, Zaret J, Zhang X, Bocca S, Oehninger S. Expression of immunomodulatory genes, their protein products and specific ligands/receptors during the window of implantation in the human endometrium. Mol Hum Reprod 2008;14(7):413–21.

[44] Francis J, Rai R, Sebire NJ, El-Gaddal S, Fernandes MS, Jindal P, et al. Impaired expression of endometrial differentiation markers and complement regulatory proteins in patients with recurrent pregnancy loss associated with antiphospholipid syndrome. Mol Hum Reprod 2006;12(7):435–42.

[45] Critchley HO, Kelly RW, Brenner RM, Baird DT. The endocrinology of menstruation—a role for the immune system. Clin Endocrinol 2001;55(6):701–10.

[46] Bartel DP. MicroRNAs: genomics, biogenesis, mechanism, and function. Cell 2004;116(2):281–97.

[47] Sohel MH. Extracellular/circulating microRNAs: release mechanisms, functions and challenges. Achiev Life Sci 2016;10(2):175–86.

[48] Lessey BA. Fine tuning of endometrial function by estrogen and progesterone through microRNAs. Biol Reprod 2010;82(4):653–5.

[49] Sha AG, Liu JL, Jiang XM, Ren JZ, Ma CH, Lei W, et al. Genome-wide identification of micro-ribonucleic acids associated with human endometrial receptivity in natural and stimulated cycles by deep sequencing. Fertil Steril 2011;96(1):150–5.

[50] Revel A, Achache H, Stevens J, Smith Y, Reich R. MicroRNAs are associated with human embryo implantation defects. Hum Reprod 2011;26(10):2830–40.

[50a] Vilella F, Moreno-Moya JM, Balaguer N, Grasso A, Herrero M, Martínez S, Simón C. Hsa-miR-30d, secreted by the human endometrium, is taken up by the pre-implantation embryo and might modify its transcriptome. Development 2015;142(18):3210–21.

[51] Ruiz-Alonso M, Galindo N, Pellicer A, Simón C. What a difference two days make;"personalized" embryo transfer (pET) paradigm: a case report and pilot study. Hum Reprod 2014;29(6):1244–7.

[52] Simon C, Vladimirov IK, Cortes GC, Ortega I, Cabanillas S, Vidal C, et al. Prospective, randomized study of the endometrial receptivity analysis (ERA) test in the infertility work-up to guide personalized embryo transfer versus fresh transfer or deferred embryo transfer. Fertil Steril 2016;106(3):e46–7.

[52a] Díaz-Gimeno P, Ruiz-Alonso M, Sebastian-Leon P, Pellicer A, Valbuena D, Simón C. Window of implantation transcriptomic stratification reveals different endometrial subsignatures associated with live birth and biochemical pregnancy. Fertil Steril 2017;108(4):703–10.

[52b] Gómez E, Ruíz-Alonso M, Miravet J, Simón C. Human endometrial transcriptomics: implications for embryonic implantation. Cold Spring Harb Perspect Med 2015;a022996.

[52c] Altmäe S, Koel M, Võsa U, Adler P, Suhorutšenko M, Laisk-Podar T, Aghajanova L. Meta-signature of human endometrial receptivity: a meta-analysis and validation study of transcriptomic biomarkers. Sci Rep 2017;7(1):10077.

[53] Meuleman C, Vandenabeele B, Fieuws S, Spiessens C, Timmerman D, D'Hooghe T. High prevalence of endometriosis in infertile women with normal ovulation and normospermic partners. Fertil Steril 2009;92(1):68–74.

[54] Canis M, Donnez JG, Guzick DS, Halme JK, Rock JA, Schenken RS, Vernon MW. Revised american society for reproductive medicine classification of endometriosis: 1996. Fertil Steril 1997;67(5):817–21.

[54a] Comstock IA, Diaz-Gimeno P, Cabanillas S, Bellver J, Sebastian-Leon P, Shah M, Simon C. Does an increased body mass index affect endometrial gene expression patterns in infertile patients? A functional genomics analysis. Fertil Steril 2017;107(3):740–8.

[55] MAQC, 2006: Shi L, Reid LH, Jones WD, Shippy R, Warrington JA, Baker SC, Luo Y. The MicroArray Quality Control (MAQC) project shows inter and intraplatform reproducibility of gene expression measurements. Nature biotechnology 2006;24(9):1151.

[56] MAQC, 2010: Shi L, Campbell G, Jones WD, Campagne F, Wen Z, Walker SJ, Shaughnessy Jr JD. The MicroArray Quality Control (MAQC)-II study of common practices for the development and validation of microarray-based predictive models. Nature biotechnology 2010;28(8):827.

Further Reading

Achache H, Revel A. Endometrial receptivity markers, the journey to successful embryo implantation. Hum Reprod Update 2006;12:731–46. https://dx.doi.org/10.1093/humupd/dml004.

Glossary

Biomarker	A characteristic that is objectively measured and evaluated as an indicator of normal biological processes, pathogenic processes, or pharmacologic responses to a therapeutic intervention.
Transcriptome	The messenger RNA (mRNA) molecules expressed by an organism.
Endometrial fluid	Body fluid inside the uterine cavity that contains the molecules secreted from the endometrial glands into the uterine lumen. This fluid can be obtained by aspiration of the cavity by transcervical introduction of a thin and flexible catheter.
Endometrial receptivity	A period within the menstrual cycle in which the endometrium acquires a functional status able to promote and support blastocyst implantation.
Endometrial receptivity analysis	A personalized genetic test, developed and patented in 2009 by IGENOMIX, to diagnose the endometrial receptivity state within the window of implantation.
Genome-wide association studies	Studies to identify DNA markers of variants across the genome related to a trait or disease. This type of study has proven useful for the prevention, diagnosis, and treatment of numerous diseases.
Next-generation sequencing	Modern technology that allows the sequencing of DNA and RNA much more quickly and cheaply than the previously used Sanger sequencing method, revolutionizing the study of genomics and molecular biology.
"-Omics" sciences	General term for a broad discipline of science and engineering that analyzes different molecules or processes at a large scale as well as the interactions of biological information objects in various "-omes," for example, genome, transcriptome, proteome, lipidome, secretome, etc.
Personalized embryo transfer	Novel therapeutic strategy used in patients with a displaced window of implantation in which the embryo is transferred, guided by the endometrial timing.
Recurrent implantation failure	Clinical condition that prevents the final success of assisted reproduction. It is defined as three or more IVF cycles in which one or two morphologically high-grade embryos were transferred.
Window of implantation	Period during the menstrual cycle in which the endometrium is ready to receive the embryo and start the pregnancy. This occurs around days 19–21 in each menstrual cycle of a fertile woman.

Epigenetic Modifications in the Human Placenta

Wendy P. Robinson*,†, Maria S. Peñaherrera*,†, Chaini Konwar*,†, Victor Yuan*,†, Samantha L. Wilson*,†

*Department of Medical Genetics, University of British Columbia, Vancouver, BC, Canada †BC Children's Hospital Research, Vancouver, BC, Canada

EPIGENETIC FEATURES OF THE PLACENTA

Overview of Epigenetics

Epigenetic modifications include DNA methylation (DNAm), histone methylation and acetylation, and subsets of noncoding RNAs that interact with DNA to regulate gene transcription. These marks work together to affect chromatin packaging and accessibility of the DNA to the binding of proteins, including transcription factors, DNA polymerase,

Human Reproductive and Prenatal Genetics
https://doi.org/10.1016/B978-0-12-813570-9.00013-9

insulator proteins, and activator proteins. In general, suppressive marks, that is, those that prevent protein binding, include DNAm and histone H3K9 or H3K27 di- and tri-methylation while examples of activation marks are H3K4 methylation, or H3K9 or H3K27 acetylation [1]. However, it is important to note that active gene transcription can itself alter chromatin conformation and affect these marks. Furthermore, as epigenetic marks can be transmitted through cell division, they can in some cases represent a cellular memory of past processes rather than current gene expression. In other words, while epigenetic information is related to gene expression, there is not a one-to-one correspondence and the information contained in these epigenetic marks is complex.

The study of epigenetic modifications in the placenta is still in its infancy. DNAm has been the most widely investigated epigenetic mark in the placenta due to the ease of assaying it and the availability of technologies such as DNAm microarrays or bisulfite sequencing that allow for a survey of DNAm across the genome. DNAm involves the addition of a methyl group onto a DNA base, most commonly cytosine residues in the context of a CpG dinucleotide. While DNAm can occur on other nucleotide bases, this is rare outside of embryonic cells. There are ~28 million CpG sites in the genome, most of which are methylated. However, some CpG dense regions called CpG islands are typically associated with gene promoters or other protein-binding regions and tend to be unmethylated [2]. DNAm at island-associated gene promoters is typically associated with gene repression. However, absence of DNAm at promoters cannot alone lead to gene transcription, as specific transcription factors, enhancer proteins, and various cofactors are also needed and can modulate expression levels. Gene body regions tend to exhibit increased methylation when a gene is active, although many sites of DNAm within a gene show little clear functional relationship to expression or to each other. 5-Hydroxymethyl cytosine is much less common than methyl-cytosine and represents a transition step in the process of demethylation as well as possibly playing a role in gene regulation. It is found more commonly in germ cells undergoing reprogramming and embryonic cells. In the placenta, 5hMC represents only 1%–3% of all cytosine methylation, but is found at functionally relevant regions and may play a role in regulation of a subset of genes [3].

Unique Features of the Placental Epigenome

The placenta shows many unique epigenetic features as compared to somatic tissues (Table 13.1). On average there is a lower level of DNAm, but this is not randomly distributed. In fact, the majority of the placental genome shows high levels of DNAm similar to that observed in somatic cells. However, it also includes large blocks (>100 kb in length) of lower methylation, referred to as partially methylated domains (PMDs), that uniquely cover 37% of the placental genome. The PMDs are relatively gene-poor and those genes contained within them tend to be tissue specific and highly methylated despite being found in a region of relatively low methylation [4, 5]. The PMDs are also associated with a relative increase in H3K9me3 and H3K27me3 marks. The function of PMDs is unknown, but their presence suggests a high-level organization of the placental genome that is fundamentally different from that observed in other cell or tissue types and does not reflect simply a passive loss of methylation due to failure to maintain it [6].

TABLE 13.1   Epigenetic Features of the Placenta

DNA methylation	• Global hypomethylation • Presence of partially methylated domains (PMDs) • Reduced methylation at some retrotransposable elements • Hypomethylation of X-chromosome gene promoters in females • High number of maternally imprinted loci
Pseudomalignancy	• Increased DNAm at tumor suppressor genes • Loss of DNAm at genes promoting invasion
Highly variable	• Intraplacental (site-to-site) variability • Interplacental variability • Methylation allelic polymorphism • Polymorphic imprinting
Developmental changes	• Cell type specific epigenetic differences • Alterations associated with hypoxia exposure and syncytialization • Increased average DNAm with gestational age
Histone features	• High H3K27 methylation and low histone H2A and/or H4 phosphorylation in trophectoderm; higher H3K9me3 and H3K27me3 in PMDs
ncRNA	• Many placental specific miRNAs, for example, the imprinted C19MC cluster

There is also on average less methylation at several families of retrotransposable elements, including L1s and HERVs. This has been linked to the greater use of retroviral promoters in placental gene expression [7]. However, within these element families, methylation levels can be widely variable [7, 8]. The hypomethylation of retrotransposable elements is independent of location relative to PMDs. While the presence of PMDs and retrotransposable element hypomethylation are relatively stable with gestational age, there is an increase in genome-wide DNAm [9] and a reduction in the extent of PMDs with increasing gestational age [10]. These changes may be associated with oxygenation linked to maternal blood perfusion in the late first trimester as well as changes to placental function throughout development [10]. Some of the most differentially methylated genes across gestation are involved in immune regulation [9]. which may be due to a combination of functional changes in immune modulation combined with changes in cell composition.

While the placental epigenome differs strikingly from somatic tissues, many of these unique features (hypomethylation, presence of PMDs, etc.) are shared with cancer cells [11]. Similar to cancer, the placenta is highly invasive, proliferates under hypoxic conditions, evades host immune defenses, and establishes its own blood supply. In parallel, there is a tendency toward methylation of tumor suppressor genes and activation of oncogenes in the human placenta. For example, there is placental specific hypermethylation of *RASSF1* as well as methylation of *APC*, *SFRP2*, *WIF1*, and *EN1*, which are negative regulators of Wnt/β-catenin signaling, a key pathway contributing to human tumor progression [10–12]. Understanding what limits trophoblast invasion during pregnancy may provide key insights into the differences that are unique to malignant cells.

A High Degree of Intra- and Interplacental Epigenetic Variability

The placenta is also unusual in the polymorphic and variable nature of epigenetic marks [13]. This may allow the placenta to be more responsive to signals from mother and fetus, to adapt as needs change, and to buffer the environment to which the fetus is exposed. This variability is important to consider for: (1) adequately capturing a sample that is representative of the whole placenta when obtaining a biopsy for clinical or research purposes, and (2) interpreting the biological meaning of changes associated with pathology, particularly if these are small relative to the range of variation within healthy control placentas.

Intraplacental variation reflects that the 50–60 villous trees that comprise the bulk of the placenta each develop from only small number of precursors cells. As a consequence, epigenetic changes arising early and present in only a subset of precursor cells may be propagated throughout single "trees," but do not spread to other villus trees [13, 14]. One also expects a greater degree of heterogeneity at CpG sites that are methylated in a cell-specific manner due to local variation in cell composition. By investigating the patterns of DNAm by site and depth within a placenta, one can gain some insights into when such variations arise [15]. For example, the *APC* gene promoter can show bimodal patterns of DNAm within a single placenta due to variable monoallelic/biallelic DNAm [14], likely reflecting that these marks are set early in development. In contrast, DNAm variation for the *H19/IGF2* imprinting control region is more consistent with an origin after formation of primary villi, possibly in response to local variation in nutrient supply [15]. While this intraplacental epigenetic heterogeneity could theoretically result in regions of the placenta that are more/less healthy, the placenta can generally provide adequate support for the growing fetus, even in the presence of localized regions of pathology [13].

Interindividual variation in epigenetic marks can also arise due to a combination of genetic and environmental influences as well as stochastic effects. There is a greater variability in DNAm levels within PMDs than in the remaining highly methylated domains [16]. In addition to *APC*, mentioned above, many other functionally important genes in the placenta have been reported to be variably monoallelically methylated and expressed, including *WNT2*, *TUSC3*, *EPHB4*, and *LEP* [11, 17–19]. This can contribute to within and between placenta variation. Widespread polymorphic placental specific imprinting has also been reported by several groups [20–22] (discussed further below). While some of this variation may be nonfunctional, as only a subset of epigenetic changes affects gene expression, the implications of this variability for placental health will be important to clarify.

The Epigenetic Profiles of Placental Cell Types

In the early blastocyst, the majority of gamete-derived epigenetic marks have been erased [23] (except for a small subset of largely oocyte-derived marks discussed relative to imprinting below) (Fig. 13.1). The trophectoderm will develop into one of the layers of the chorion as well as the villous trophoblast, which retains this low methylated state. Hypomethylation of chorionic villi as a whole reflects that trophoblast is the predominant cell type within this tissue [24]. In the mouse, trophoblast hypomethylation is associated with down-regulation of the DNA methyltransferases, Dnmt3a, Dnmt3b, and Dnmt1a [25]. However, the role of DNMTs in human trophoblast is less clear [26]. In contrast, the inner cell mass undergoes

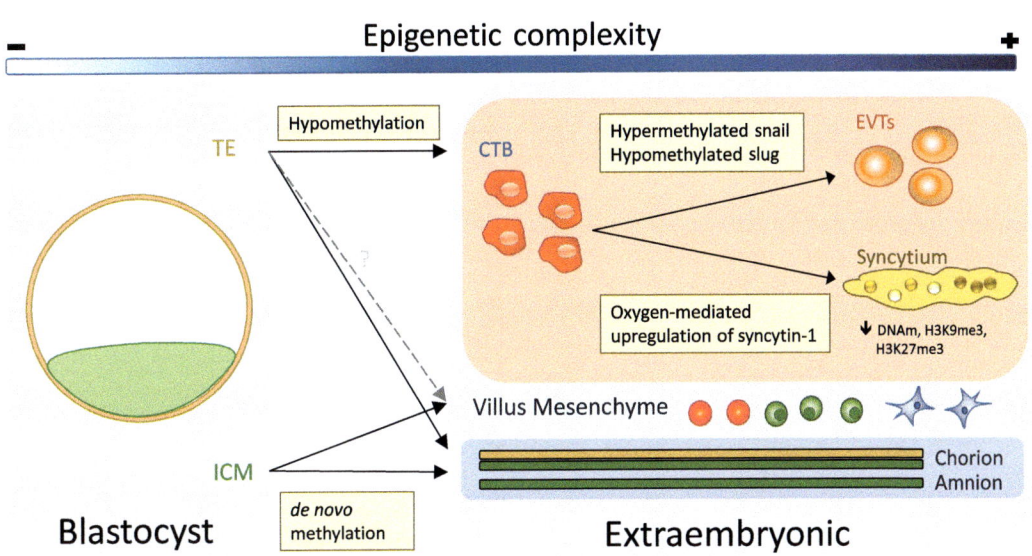

FIG. 13.1 The relationship of epigenetic changes to placental cell development.

de novo global DNAm. These cells are the origin of the fetal tissues, but also contribute to primitive endoderm, from which placental endothelial cells are derived, and to the extraembryonic mesoderm, from which villus stroma is derived. In addition to differences in DNAm, the early trophectoderm shows higher H3K27 methylation as well as lower levels of histone H2A and/or H4 phosphorylation [27]. Further differentiation processes in both trophoblast and other lineages are associated with additional DNAm changes and histone modifications at specific regulatory loci [28].

The tree-like structures of chorionic villi that compose the placenta consist of an outer layer of stem-like cytotrophoblasts (CTBs), which fuse together to form syncytial trophoblast (STB). This syncytium facilitates maternal-fetal exchange and produce hormones to support pregnancy, such as progesterone, leptin, human chorionic gonadotropin, and human placental lactogen. Transcriptional and epigenetic events control cytotrophoblast differentiation and are triggered by pregnancy-associated levels in oxygen content [29, 30]. Under low oxygen conditions, CTB fusion is impaired, and certain STB-specific CpG sites become hypermethylated [31]. For example, syncytin-1, which is critical for STB formation, is downregulated in response to hypoxia through DNMT3A-dependent hypermethylation [32]. Histone deacetylase complexes (HDAC) and histone acetyltransferases (HAT) are also involved through their effects on the transcription factor GCMa, which regulates syncytin-1 expression [33].

The terminally differentiated syncytium contains nuclei derived from the fusion of CTBs at various stages in gestation. As a result, nuclei in the syncytium are heterogeneous in their age and structure. Nuclei from recently fused CTBs tend to contain a transcriptionally active euchromatic structure whereas older heterochromatic nuclei form condensed structures called syncytial knots [34]. The majority of STB nuclei are transcriptionally active with the majority of the nuclei lacking repressive epigenetic marks, such as DNAm, H3K9me3, and

H3K27me3 [35]. However, there is also an enrichment of H4K20me3, a repressive chromatin mark, that is suggested to potentially originate from the older nuclei.

CTBs can also differentiate into extravillous trophoblasts (EVTs), which exhibit strikingly different transcriptional, epigenetic, and morphological characteristics compared to STBs. During this process, expression of cell adhesion molecules and polarity are lost and expression of mesenchymal markers, migratory, and invasive properties are gained [36]. Similar to cancer cells, EVTs down-regulate E-cadherin by DNA methylation and up-regulate many proteases (e.g., *MMP9*), phosphatidylinositol 3′-kinase (*PI3K*), and several autocrine or paracrine loops (e.g., involving *EGF* and *VEGF*) [37] as they transition into a highly invasive phenotype. Disruption of DNAm by *DNMT* inhibitor 5-aza-dC results in loss of invasive and migratory properties in EVT cells; up-regulation of epithelial markers, such as E-Cadherin, beta-Catenin, and Cadherin18; and down-regulation of mesenchymal genes, such as *VIM* and *CDH2* [38]. Transcription factors essential to the initiation of this transition process, *SNAIL* and *SLUG*, are also differentially methylated between CTBs and EVTs [39].

Histone acetylation is also essential for trophoblast differentiation, and is thought to be controlled by placental-utero oxygen content through the activity of hypoxia-inducible factor 1 (HIF-1) and HDACs [30]. The HDAC inhibitor, trichostatin-A, results in increased histone acetylation and overexpression of tumor suppressor gene *SERPINB5*, leading to reduced EVT motility and invasion [40]. In contrast, acetylation of histones H2A and H2B by the CREB-binding protein (CBP) acetyltransferases inhibits the epithelial-mesenchyme transition and promotes epithelial characteristics in mouse studies [41]. A higher-order chromatin structure mediated by histone modifications in combination with a locus repeat structure was demonstrated to play a role in STB-specific expression within the human growth hormone gene cluster [42], illustrating that the DNA sequence itself can affect chromatin structure.

Our knowledge of cell-specific epigenetics is confined primarily to trophoblastic populations. However, additional placental cell populations are functionally important and may also be implicated in disease. Chorionic villus cultures yield the predominant growth of fibroblast cells, which have a DNAm profile that is much more similar to the fetal membranes than to whole placental villi [43]. Placental-specific macrophages, Hofbauer cells (HBs), lie in the mesenchymal core of the chorionic villi and in comparison to maternal and fetal macrophages, show hypermethylation of many immune response-related and classical macrophage-activation associated genes, possibly as a result of high expression of DNMT1, DNMT3A, and DNMT3B [44].

Our current understanding of pathological diseases that affect the placenta is contributed predominantly from bulk tissue profiling studies, which suffer from the limitation that cellular heterogeneity can result in the dilution of cell-specific signals and the creation of spurious ones. More targeted studies investigating cell-specific signatures will contribute to our understanding of changes associated with normal placental functioning as well as of abnormal ones.

MONOALLELIC GENE INACTIVATION

X-Chromosome Inactivation in the Placenta

Epigenetic inactivation of one of the two copies of a gene can result in monoallelic gene expression and is involved in X-chromosome inactivation (XCI), genomic imprinting, and nonimprinted allelic inactivation. XCI is a unique regulatory mechanism that results in dosage compensation of X-linked genes between males and female mammals via the epigenetic inactivation of one of the X chromosomes in females. While preferential inactivation of the paternally derived X chromosome is observed in the extraembryonic tissues of rodents, XCI is random in the human placenta and in both mouse and human somatic tissues [15, 45]. As XCI is initiated in the blastocyst and the inactive X is stably maintained through any subsequent somatic cell divisions, females are generally mosaics for X chromosome gene expression. In the placenta, this results in a "patchiness" in terms of which X is inactivated at differing sampling locations, as cells within a villus tree are clonally descended from only one or a few precursor cells [15].

Genes that are subject to XCI exhibit methylation of the CpG island promoter regions on the inactive X chromosome in somatic tissues, leading to partial methylation in females at these sites that are typically low methylated in males [46, 47]. Outside the promoter regions, the active X chromosome generally has higher methylation than the inactive X. In the human placenta, however, the differences between the single X in males and the Xs in females are less pronounced. In particular, X-linked CpG Island promoters in females show ~27% lower methylation in the placenta as compared to the blood [46]. Additionally, the placenta, relative to blood, shows decreased methylation in highly methylated X-linked intragenic and intergenic regions, consistent with the overall hypomethylation of the placental genome [8, 46].

The X chromosome bears a number of genes important for trophoblast development. In mice, the loss of XCI has been implicated in placental malformations, growth retardation, failed development, and lethality [48, 49]. It is not yet clear whether altered XCI leads to placental disease in humans. A case of an X/autosome translocation in a female showed the expected nonrandom XCI in the somatic tissues and amnion but mostly random XCI in the placenta, suggesting that the selective forces acting in somatic tissues in such cases might be either weaker or absent in the human placenta [50]. The numerous X-linked genes involved in placentogenesis, the early unequal gene expression by the sex chromosomes in males and females, and the unique characteristics of placenta-specific epigenetic processes are among the factors that may play a role in the placental contribution to the origins of sexual dimorphism in health and disease [51]. While sex-specific DNAm at the autosomal loci has been well demonstrated in the placentas in animal models, this is a vastly unexplored and understudied research area in humans.

Genomic Imprinting

Normal placental growth, development, physiology, and morphology are dependent on the correct establishment and maintenance of genomic imprinting [52]. Imprinted genes are those that are monoallelically expressed based on their parent of origin. These patterns of expression tend to be species- and tissue-specific and in some cases also developmentally regulated [53]. The selective inactivation of one of the parental alleles can be achieved by differential DNAm, allele specific histone modifications, antisense ncRNA mediated silencing, and long-range chromatin interactions [52]. An evolutionary link between imprinting and placentation has been suggested with imprinted genes in the placenta playing key roles in trophoblast differentiation and placental development [52]. Many of the classic imprinted loci, that is, conserved between mouse and human, have been associated with disorders such as Beckwith-Wiedemann syndrome and Silver Russell syndrome that are linked to altered growth of the placenta and fetus [54]. These two syndromes are associated with opposite alterations at the imprinting control regions for *H19/IGF2* and/or *KCNQ1OT1* at 11p15, with placentomegaly and fetal overgrowth associated with Beckwith-Wiedemann syndrome and small placenta with hypoplastic villi plus fetal growth restriction for Silver-Russell syndrome. Alterations of this same region have also been seen in placentas with features of placental mesenchymal dysplasia [55].

Imprinted genes are typically associated with differentially methylated regions (DMRs) that are established in either the germline and maintained after fertilization or during embryonic development as secondary or somatic DMRs. While sperm-derived methylation marks are largely lost by the blastocyst stage, there is incomplete erasure in the maternal genome such that hundreds or even thousands of imprinted DMRs are retained in the placenta [20–22]. The majority of these are placenta-specific and not conserved in the mouse or other mammals with the exception of primates [20–22, 56]. More than 30 of these placental-specific DMRs have confirmed paternal expression in the placenta [21]. While the function of many of the associated genes is unclear, many of these genes play vital roles in placental development. For example, imprinted monoallelic (paternal) expression confined to the placenta is observed for *DNMT1* and *AIM1* [20, 57]. There are also two large microRNA clusters that show imprinted expression that is exclusively or predominantly found in placenta: a large paternally expressed microRNA cluster on chromosome 19q13.41 (C19MC) that has a role in protection against viral infections, and a maternally expressed cluster on chromosome 14 (C14MC) [58]. The expression of these miRNAs changes throughout gestation has been suggested to be altered in preterm birth and is readily detected in maternal serum [59]. Most of these placental-specific DMRs show intermediate levels of methylation (transient imprinting) in chorion but loss of methylation in amnion, likely reflecting the partial trophoblast contribution to the chorion and later origin of the amniotic epithelial layer [20]. They also show enrichment for zinc finger protein 57 (ZFP57) binding sites and for H3K9me3. In mice, di/trimethylation of histone 3 lysine 9 (H3K9me2/3) and ZFP57 protects imprinted DMRs from demethylation during embryonic reprogramming [60].

Surprisingly, about half of placental-specific imprinted DMRs exhibit polymorphic imprinted methylation (also referred to as methylation allelic polymorphism) [19]. That is, some placental samples will exhibit monoallelic methylation while the others show loss of imprinting at the given DMR [20–22]. This has been demonstrated to lead to polymorphic monoallelic/biallelic expression for a number of genes, including *LIN28B*, *NTM*, and *MAGI2* [21, 52, 61]. Interestingly, some genes, for example, *IGF2R* and the closely linked *SLC22A2*, show polymorphic imprinted expression that appears to be independent of DNAm [52]. This pervasive polymorphic imprinting in the human placenta is largely independent of gestational age and cell composition [20].

Evaluation of DNAm at imprinted genes can be used in the diagnosis of genomic imbalances of the parental contributions, as occurs in (i) complete hydatidiform moles (CHM) that may be androgenetic or carry mutations in *NLPR7*, which affects maintenance of maternal imprints; (ii) triploidy, which may be digynic or diandric; and (iii) placental mesenchymal dysplasia, which is typically due to androgenetic/biparental cell chimerism, and presents with a distinct placental pathology characterized by placentomegaly and grape-like vesicles that resemble a molar pregnancy but may have a normal fetus [43, 62]. Nonetheless, it is important to note that the methylation level at some imprinted DMRs changes with gestational age and can vary by cell/tissue type [53]. These relationships are complex, though. For example, at two imprinted genes important for placental function, *MEST* and *CDKN1C*, some differentially methylated CpGs change the DNAm level with gestational age while others associated with the same genes are stable.

Methylation Allelic Polymorphism

Methylation allelic polymorphism (MAP), or an "on-off" type of DNA methylation (0% or 50% corresponding to biallelic/monoallelic expression patterns) that cannot be explained by a nearby genetic polymorphism, has been reported in the placenta for multiple genes, including *WNT2*, *TUSC3*, *EPHB4*, *LEP*, and *CGB5* [17, 19, 63]. Variable monoallelic DNA methylation may be a way to fine-tune gene expression and increase placental adaptability. While loss of monoallelic methylation may be linked to abnormal placental function, its role in healthy and/or diseased placentas has not yet been well defined [43].

ALTERED DNAM AND PLACENTAL PATHOLOGY (TABLE 13.2)

Miscarriage

Approximately 20% of clinically recognized pregnancies end in miscarriage, the majority due to large-scale chromosomal abnormalities. Only a few studies have investigated whether epigenetic differences are found in the placentas from such spontaneously aborted specimens. Several types of chromosomal trisomies have been associated with DNAm changes, but these may largely be a consequence of widespread disruption in gene expression [64, 65]. Triploidy is also associated with many epigenetic changes, which is in part accounted for by an imbalance of imprinted regions [20]. Among chromosomally normal miscarriages, data is sparse. Nonetheless, several studies have reported more outliers in DNAm at imprinted regions in spontaneous miscarriage, suggesting that epimutations might contribute [66, 67]. Overall, lower expression of *DNMT1* and lower global DNAm was reported in placental villi from early miscarriages as compared to elective terminations [68]. However, it is extremely challenging to study epigenetics in the context of miscarriage due to the lack of well-matched controls, heterogeneity in etiology, low quality of tissue obtained, and the difficulty in establishing an accurate developmental age of the specimen.

Spontaneous Preterm Birth

Spontaneous preterm birth (sPTB) in the present context is defined as any noniatrogenic cause of delivery at <37 weeks gestation, including both preterm rupture of membranes and spontaneous preterm labor. It is heterogeneous in etiology but is frequently associated with signs of chronic and/or acute inflammation in the decidua, placenta, and/or the fetal membranes [69]. Maternal chronic disease (e.g. obesity, diabetes), genetic variants in the

TABLE 13.2 Examples of Altered Placental DNAm and Pathology/Exposures

Pathology	
Miscarriage	• Widespread alterations with triploidy and trisomy • Increased outlier methylation at imprinted genes in chromosomally normal miscarriage
Preterm birth	• Inconsistent findings, heterogeneous in etiology
Preeclampsia	• Widespread changes in early but not late onset PE; hypomethylation at gene enhancer regions, particularly those associated with angiogenesis. May partly reflect syncytialtrophoblast proliferation
Intrauterine growth restriction	• No consistent changes globally once cases associated with preeclampsia are excluded. Some reports of altered DNA methylation or expression at some imprinted genes
Exposure	
Maternal Smoking	• DNAm changes at *CYP1A1, HTR2A* and *AHRR*
Maternal Diet	• Global and site-specific changes to DNA and histone methylation associated with choline intake
Maternal Health	• Global placental hypermethylation associated with gestational diabetes
Assisted Reproduction	• Evidence for an effect in humans is weak and studies inconsistent

placenta and/or the mother, and/or lifestyle factors (e.g., smoking, poor nutrition) are all risk factors, and may together alter the balance between the pro- and antiinflammatory signaling pathways in the placenta, making it more susceptible to inflammation associated with sPTB [70]. As many of the maternal risk factors for sPTB have been linked to epigenetic alterations in both mother and baby, it is possible to identify associated epigenetic inflammatory signatures in the mother that are predictive of sPTB [70]. Such inflammation-associated signatures might be best detected in maternal DNA, but placental-derived products circulating in the maternal blood could also reflect changes due to inflammation and infection.

Identifying changes to the placenta specifically associated with PTB is difficult as the epigenetic profile of the placenta changes with gestational age, and hence term controls are not a proper match for comparison. Furthermore, given that the etiology and presentation of sPTB is heterogeneous, the associated features have not typically been defined in the epigenetic studies performed as yet. Additionally, iatrogenic cases (delivered because of preeclampsia or other indication) are sometimes included. A systematic review of placental epigenetic modifications associated with sPTB also highlighted the lack of standardized procedures for tissue collection, case-control ascertainment, clinical data collection, quality control procedures, and analytical approaches among previous publications [71]. Additional high-quality studies are needed before the extent of epigenetic changes associated with sPTB can be fully assessed.

Preeclampsia

Preeclampsia (PE) is associated with a number of pathological changes in the placenta, including intervillous fibrin deposition, infarcts, increased syncytial knots, and other signs of accelerated villus maturation. These changes are more common in PE cases with early onset (diagnosis made <34 weeks). There have now been multiple studies reporting widespread hypomethylation in placentas associated with PE [72–77], particularly early onset cases. DNAm profiling can be used to refine the clinical diagnoses, with late onset PE cases diagnosed between 34 and 36 weeks with cooccurring IUGR showing strong similarities to early onset cases while the remaining late onset cases cluster with controls [77]. Although the exact repertoire of significant hits associated with early onset PE varies between studies, the altered sites tend to be located in enhancer rather than promoter regions [73]. The linked genes are reported to be involved in cell proliferation, apoptosis, cell adhesion, cell structure, transport, and immune response. Specific genes associated with these DNAm changes include those shown to have altered gene expression in PE, such as FLT1, INHBA, and WNT2. Interestingly, these results appear to be placenta-specific with hypermethylation reported in the fetal cord blood, along with differing gene involvement from the same pregnancies that exhibit hypomethylation in the placenta [72]. Using a targeted approach, increased DNAm and H3K9me3 was observed for VEGF and JUN promoters in placentas associated with PE [78]. These genes correspondingly show decreased expression in PE, which likely plays an important role in the decreased vascular remodeling. Altered DNAm at genes involved in cortisol signaling and steroid regulation has also been observed [79]. In addition, LEP (encoding leptin), which frequently shows nonimprinted monoallelic expression in the placenta, showed evidence of higher biallelic expression in PE, indicating that loss of the monoallelic expression may be detrimental [17].

Based on a selection of DNAm sites that change with gestational age, it was also shown that placentas associated with PE show DNAm profiles consistent with accelerated aging [80]. This may reflect the increased syncytial knots and other pathological features associated with advanced villus maturation. However, telomere length, another aging marker, does not appear to be significantly altered [81]. The indication for accelerated aging in PE from DNAm but not telomere data may be explained by previous observations that, while these two marks both correlate with chronological age, they are representing different aging processes. For example, as nuclei in syncytial knots are more heterochromatic than those in a healthy syncytialtrophoblast, one would expect altered epigenetic profiles to be associated with increased knots, but this would likely be independent of telomere length as these cells are not dividing.

As DNAm varies widely between cell or tissue types, the observed changes may reflect altered cell composition in PE as well as changes to gene methylation/expression within a cell type, possibly as a result of hypoxia or reoxygenation stress. In fact, many sites hypomethylated in PE overlap those that become hypomethylated as cytotrophoblast differentiates to syncytialtrophoblast [73], suggesting that a subset of changes could be attributable to syncytialtrophoblast proliferation. In contrast, hypoxia tends to result in higher, rather than lower, methylation at many of these same sites [31]. As changes in cell type composition may be either causal or consequential to pathology, the nature of the epigenetic changes in whole tissue are important to explore [82]. Larger sample sizes, consistent diagnostic criteria, and use of validation cohorts should help narrow down the key epigenetic changes most characteristic of PE. It is also still unclear what the earliest epigenetic changes associated with PE are, but all could be considered for the identification of predictive biomarkers.

Fetal Growth Restriction

Poor fetal growth can be due to a wide variety of causes including, but not limited to, maternal stress exposure, poor nutrition, smoking, chromosome abnormalities confined to the placenta, placental dysfunction associated with PE, and/or placental inflammation/infection associated with sPTB. Unexplained poor fetal growth outside these other conditions can be due to inherited genetic variation, epigenetic variation, or undefined events related to early implantation and placentation that are critical to maintain nutrient supply to the fetus. In fact, genetic variation is a major contributor to birth weight, with maternal factors, possibly due to mitochondrial inheritance and/or imprinting, playing a stronger role [83].

Searches for epigenetic causes of poor fetal growth have largely focused on the role of imprinted genes because many appear to affect placental and fetal growth patterns [27]. However, outside of rare changes associated with known imprinting disorders, the evidence for imprinting defects having a significant contribution to growth restriction is unclear. Interestingly, a number of studies have reported altered expression of imprinted genes [52], but this might not be mediated by epigenetic changes specifically at the imprinting control regions commonly investigated. Altered placental DNAm at the imprinting control region for *IGF2/H19* has been reported, but others have not found this association. In one study, altered expression of the imprinted genes *PLAGL1* and *HYMAI* was observed in IUGR-associated placentas compared to normal placentas, but DNAm at the associated promoters was not altered [84]. Altered DNAm or differential expression has also been reported for *PHLDA2, NNAT, PEG10, CDKNC, MEST, MEG, GNAS, PLAGL1, GRB10, OSBPL5,* and *ZNF331* [43, 54]. Gestational age changes have not always been considered in these studies. While changes to imprinted gene expression in the context of fetal growth could be a cause of reduced growth, it may also be a consequence of reduced fetal demand, that is, if the fetus is restricted in growth for independent reasons, it may signal to the placenta to reduce nutrient transport.

Using a genome-wide approach and limiting cases to a gestational age at birth >36 weeks, 22 methylation changes in the placenta were linked to small for gestational age and/or IUGR [85]. A positive correlation between LINE-1 and AluYb8 methylation level and birthweight percentile has also been reported [86]. Another approach to identify altered placental epigenetics related to fetal growth is to compare twin pregnancies that are discordant for fetal growth. Significant DNAm changes at eight genes were reported in one study, with further analysis of the data demonstrating significant enrichment of changes linked to gene pathways associated with lipid metabolism and cadherin and Wnt signaling [87]. While other smaller studies have also reported altered DNA methylation associated with fetal growth, there is not yet a consensus on the top altered sites. This is likely due to a combination of general issues surrounding the reproducibility of large-scale data analysis, population differences, and differences in study design, including patient ascertainment.

EPIGENETIC ASSOCIATIONS WITH MATERNAL EXPOSURES

Maternal diet and health indices (diabetes, obesity, smoking) are well established to affect fetal growth and development. Specific examples in mouse models have clearly demonstrated that gene expression changes can result from exposure to such factors as dietary folate or genistein [88, 89]. There has therefore been widespread interest in identifying epigenetic changes in the human placenta that could be indicators of prenatal exposures [27].

Extreme nutrient deprivation conditions such as the Dutch Famine (1944–45) provide strong human evidence to illustrate the effects of in utero malnutrition on adult outcomes. While altered DNAm has been reported at select genes after such periconceptional famine exposure [90], data on placentas is unavailable. Maternal choline intake during the third trimester of pregnancy was associated with global and site-specific DNAm and histone methylation [91]. Of interest, increased promoter methylation of the placental corticotropin-releasing hormone gene (CRH) and reduced placental CRH expression was noted. This CRH variation could affect fetal HPA axis reactivity.

The most robust evidence for altered placental DNA methylation is associated with maternal smoking. The most reproduced finding is altered DNAm at the aryl-hydrocarbon receptor repressor (AHRR) gene, which has also been shown to persist in blood postnatally [92]. Altered methylation and expression have also been reported for CYP1A1 [93] and HTR2A [90]. A methylation change at a CpG site near LEKR1 was linked to self-reported and cotinine-based measures of smoke exposure as well as potentially having a causal link to low birthweight [94]. These studies collectively suggest that methylation changes in specific genes could have functional consequences.

Maternal health conditions such as diabetes and obesity are associated with increased risk of adverse pregnancy outcomes and a few studies have suggested altered DNAm in the associated placentas [95, 96]. The altered sites associated with maternal diabetes tended to be hypermethylated and showed enrichment for a number of pathways, including endocytosis, MAPK signaling, and cell adhesion, among others. A global increase in DNAm in placentas from diabetic mothers was also reported based on a liquid chromatography–mass spectrometry approach [97]. As with all cases of maternal disease, changes present in these placentas could be due either to the presence of the disease or the treatments for the disease.

The effect of assisted reproduction on placental and offspring epigenetic programming has been an area of much investigation. There is substantial research in mouse models that shows that loss of methylation at imprinted loci is associated with a variety of embryo culture and transfer techniques [27]. The placenta appears to be more affected than embryonic cells and this effect has been linked to suboptimal oxygen levels during in vitro culture [98]. As trophectoderm cells are directly exposed to culture conditions, it is possible that the effects are largely limited to these cells [52]. However, the evidence for abnormal imprinting after assisted reproduction in humans is not strong [52]. Furthermore, although there are reports of subtle alterations in placental DNAm after ART, based on genome-wide studies, the results often contradict and are still inconclusive at this point [99].

Although the study of human placental epigenetic patterns associated with environmental exposures is progressing at a rapid rate, mechanisms of fetal programming still remain elusive. Despite compelling reports from mice models, there are several challenges in translating findings from experimental animal models to the human population. Furthermore, few epigenetic changes have been well validated in human placental studies. This is a growing field that will surely yield many interesting findings in the future.

CLINICAL APPLICATIONS

Our understanding of epigenetics in placental development is still in its infancy, but shows much promise for impacting the clinical assessment of fetal health (Fig. 13.2). One application is to improve our definition of placental pathology and distinguish normal variation from abnormal findings. An illustration of this is a study of DNAm profiling of preeclampsia and IUGR placentas whereby DNAm at a limited number of CpG sites could cluster placentas into at least three groups: severe preeclampsia (diagnosed <34 weeks or <36 weeks with IUGR), control, and a mixed group suggestive of preterm birth associated inflammation [77]. This is similar to findings using gene expression profiling and a larger sample size, whereby five distinct placental phenotypic clusters could be identified [100]. Cell-specific epigenetic profiles can potentially be used to quantify changes in cell composition such as those associated with inflammation. This is extremely important because conditions such as PE, IUGR, and sPTB are heterogeneous in nature. Thus, predicting these prior to the onset of symptoms is likely to be dependent on each distinct underlying cause.

Defining epigenetic profiles linked to specific pathologies can then be used to develop novel predictive tests to predict poor outcomes. In theory, it might be possible to detect epigenetic abnormalities such as imprinting errors in preimplantation embryos, although the application of this is likely limited. Evaluation of imprinting errors in chorionic villus is also feasible, but could be error-prone if there is a discrepancy between placental and fetal findings. Evaluating placental health via maternal blood during pregnancy is a viable application. Placental DNAm at delivery has been linked to altered protein levels in maternal blood earlier in pregnancy. For example, reduced promoter DNAm at *INHBA* at delivery was inversely associated with INHA measurements in maternal serum in the second trimester as was also the case for placental *FN1* DNAm relative to third trimester maternal serum FN1 [101].

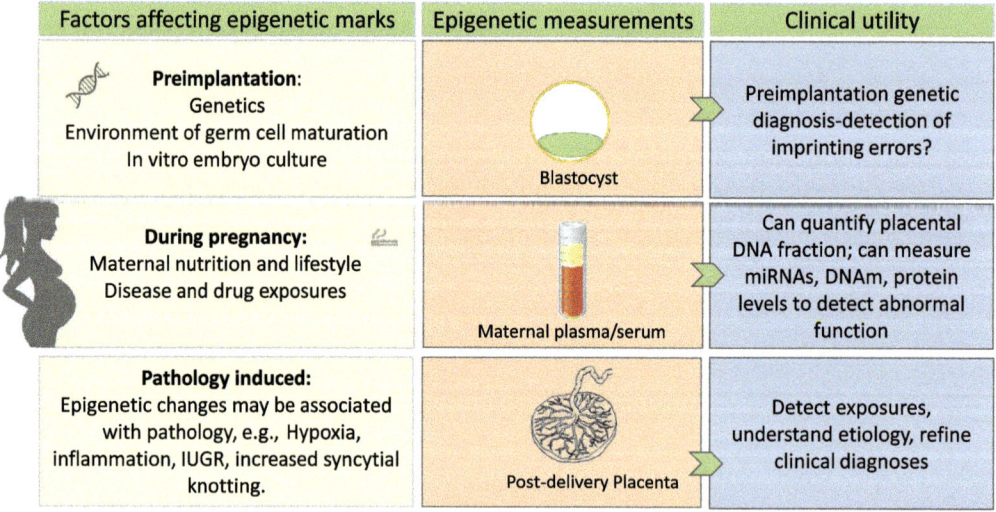

FIG. 13.2   Factors that influence placental epigenetics and their clinical utility.

INHBA and *FN1* DNAm was associated with gestational-age corrected birth weight in the same study. Similarly, low DNAm of *LEP* in the placenta is associated with increased maternal serum leptin levels and linked to early onset PE [17]. RNA derived from the placenta can be detected in maternal blood and has also been evaluated for prediction of pregnancy complications [102, 103]. These results suggest that other DNAm changes associated with adverse outcomes might identify potential RNA and protein serum markers, although many factors, including native maternal expression levels, can impact these measurements [101].

Direct measurement of methylated placental DNA in maternal plasma is possible when methylation is limited to the placenta and not found in maternal blood. For example the tumor suppressor gene *RASSF1*, which is specifically methylated in placenta, has been used to quantify placental DNA in maternal plasma and is suggested to be useful to predict preeclampsia. While significant, the sensitivity of this approach is low [103]. More promise appears with the detection of trophoblast specific microRNAs such as those from the imprinted C19MC, a number of which have shown altered levels in the blood of women who ultimately developed preeclampsia [103].

There are many avenues to explore to further our knowledge of the epigenetics of the human placenta. Human placentas are genetically diverse and also exposed to a wide range of maternal environments, both of which interact to influence epigenetic marks such that the role of the environment depends on genetic makeup. Placental/fetal sex is also an underexplored variable that clearly impacts placental response to environmental exposures. Developing a fundamental understanding of these processes will be important to help us evaluate how factors such as embryo manipulation in assisted reproduction, maternal drug exposures, and maternal diseases impact the developing fetus, and then in turn develop approaches to improve pregnancy outcomes and newborn health.

References

[1] Bannister AJ, Kouzarides T. Regulation of chromatin by histone modifications. Cell Res 2011;21(3):381–95.

[2] Deaton AM, Bird A. CpG islands and the regulation of transcription. Genes Dev 2011;25(10):1010–22.

[3] Green BB, Houseman EA, Johnson KC, et al. Hydroxymethylation is uniquely distributed within term placenta, and is associated with gene expression. FASEB J 2016;30(8):2874–84.

[4] Schroeder DI, Blair JD, Lott P, et al. The human placenta methylome. Proc Natl Acad Sci U S A 2013; 110(15):6037–42.

[5] Schroeder DI, LaSalle JM. How has the study of the human placenta aided our understanding of partially methylated genes? Epigenomics 2013;5:645–54.

[6] de Goede OM, Lavoie PM, Robinson WP. Characterizing the hypomethylated DNA methylation profile of nucleated red blood cells from cord blood. Epigenomics 2016;8:1481–94.

[7] Reiss D, Zhang Y, Mager DL. Widely variable endogenous retroviral methylation levels in human placenta. Nucleic Acids Res 2007;35(14):4743–54.

[8] Price EM, Cotton AM, Peñaherrera MS, McFadden DE, Kobor MS, Robinson W. Different measures of "genome-wide" DNA methylation exhibit unique properties in placental and somatic tissues. Epigenetics 2012;7(6):652–63.

[9] Novakovic B, Yuen RK, Gordon L, et al. Evidence for widespread changes in promoter methylation profile in human placenta in response to increasing gestational age and environmental/stochastic factors. BMC Genomics 2011;12:529.

[10] Nordor AV, Nehar-Belaid D, Richon S, et al. The early pregnancy placenta foreshadows DNA methylation alterations of solid tumors. Epigenetics 2017;12:793–803.

[11] Novakovic B, Saffery R. Placental pseudo-malignancy from a DNA methylation perspective: unanswered questions and future directions. Front Genet 2013;4:285.

[12] Chiu RW, Chim SS, Wong IH, et al. Hypermethylation of RASSF1A in human and rhesus placentas. Am J Pathol 2007;170(3):941–50.

[13] Yuen R, Robinson W. Review: a high capacity of the human placenta for genetic and epigenetic variation: implications for assessing pregnancy outcome. Placenta 2011;32:S136–41.

[14] Avila L, Yuen R, Diego-Alvarez D, Peñaherrera M, Jiang R, Robinson W. Evaluating DNA methylation and gene expression variability in the human term placenta. Placenta 2010;31(12):1070–7.

[15] Penaherrera MS, Jiang R, Avila L, Yuen RK, Brown CJ, Robinson WP. Patterns of placental development evaluated by X chromosome inactivation profiling provide a basis to evaluate the origin of epigenetic variation. Hum Reprod 2012;27(6):1745–53.

[16] Schroeder DI, Schmidt RJ, Crary-Dooley FK, et al. Placental methylome analysis from a prospective autism study. Mol Autism 2016;7(1):51.

[17] Hogg K, Blair JD, von Dadelszen P, Robinson WP. Hypomethylation of the LEP gene in placenta and elevated maternal leptin concentration in early onset preeclampsia. Mol Cell Endocrinol 2013;367(1):64–73.

[18] Wong N, Novakovic B, Weinrich B, et al. Methylation of the adenomatous polyposis coli (APC) gene in human placenta and hypermethylation in choriocarcinoma cells. Cancer Lett 2008;268(1):56–62.

[19] Yuen RK, Avila L, Peñaherrera MS, et al. Human placental-specific epipolymorphism and its association with adverse pregnancy outcomes. PLoS One 2009;4(10):e7389.

[20] Hanna CW, Peñaherrera MS, Saadeh H, et al. Pervasive polymorphic imprinted methylation in the human placenta. Genome Res 2016;26:756–67.

[21] Sanchez-Delgado M, Vidal E, Medrano J, et al. Human oocyte-derived methylation differences persist in the placenta revealing widespread transient imprinting. PLoS Genet 2016;12(11):e1006427.

[22] Hamada H, Okae H, Toh H, et al. Allele-specific methylome and transcriptome analysis reveals widespread imprinting in the human placenta. Am J Hum Genet 2016;99(5):1045–58.

[23] Cantone I, Fisher AG. Epigenetic programming and reprogramming during development. Nat Struct Mol Biol 2013;20(3):282–9.

[24] Grigoriu A, Ferreira JC, Choufani S, Baczyk D, Kingdom J, Weksberg R. Cell specific patterns of methylation in the human placenta. Epigenetics 2011;6(3):368–79.

[25] Oda M, Oxley D, Dean W, Reik W. Regulation of lineage specific DNA hypomethylation in mouse trophectoderm. PLoS One 2013;8(6)e68846.

[26] Novakovic B, Wong NC, Sibson M, et al. DNA methylation-mediated down-regulation of DNA methyltransferase-1 (DNMT1) is coincident with, but not essential for, global hypomethylation in human placenta. J Biol Chem 2010;285(13):9583–93.

[27] Nelissen EC, van Montfoort AP, Dumoulin JC, Evers JL. Epigenetics and the placenta. Hum Reprod Update 2010;17(3):397–417.

[28] Hemberger M. Epigenetic landscape required for placental development. Cell Mol Life Sci 2007;64(18):2422–36.

[29] Soncin F, Natale D, Parast MM. Signaling pathways in mouse and human trophoblast differentiation: a comparative review. Cell Mol Life Sci 2015;72(7):1291–302.

[30] Maltepe E, Krampitz GW, Okazaki KM, et al. Hypoxia-inducible factor-dependent histone deacetylase activity determines stem cell fate in the placenta. Development 2005;132(15):3393–403.

[31] Yuen RK, Chen B, Blair JD, Robinson WP, Nelson DM. Hypoxia alters the epigenetic profile in cultured human placental trophoblasts. Epigenetics 2013;8(2):192–202.

[32] Ruebner M, Strissel PL, Ekici AB, et al. Reduced syncytin-1 expression levels in placental syndromes correlates with epigenetic hypermethylation of the ERVW-1 promoter region. PLoS One 2013;8(2):e56145.

[33] Chuang H, Chang C, Chang G, Yao T, Chen H. Histone deacetylase 3 binds to and regulates the GCMa transcription factor. Nucleic Acids Res 2006;34(5):1459–69.

[34] Burton GJ, Jones CJ. Syncytial knots, sprouts, apoptosis, and trophoblast deportation from the human placenta. Taiwan J Obstet Gynecol 2009;48(1):28–37.

[35] Fogarty NM, Burton G, Ferguson-Smith A. Different epigenetic states define syncytiotrophoblast and cytotrophoblast nuclei in the trophoblast of the human placenta. Placenta 2015;36(8):796–802.

[36] E Davies J, Pollheimer J, Yong HE, et al. Epithelial-mesenchymal transition during extravillous trophoblast differentiation. Cell Adhes Migr 2016;10(3):310–21.

[37] Ferretti C, Bruni L, Dangles-Marie V, Pecking A, Bellet D. Molecular circuits shared by placental and cancer cells, and their implications in the proliferative, invasive and migratory capacities of trophoblasts. Hum Reprod Update 2007;13(2):121–41.

[38] Chen Y, Wang K, Leach R. 5-aza-dC treatment induces mesenchymal-to-epithelial transition in 1st trimester trophoblast cell line HTR8/SVneo. Biochem Biophys Res Commun 2013;432(1):116–22.

[39] Chen Y, Wang K, Qian C, Leach R. DNA methylation is associated with transcription of snail and slug genes. Biochem Biophys Res Commun 2013;430(3):1083–90.

[40] Dokras A, Gardner L, Kirschmann D, Seftor E, Hendrix M. The tumour suppressor gene maspin is differentially regulated in cytotrophoblasts during human placental development. Placenta 2002;23(4):274–80.

[41] Abell AN, Jordan NV, Huang W, et al. MAP3K4/CBP-regulated H2B acetylation controls epithelial-mesenchymal transition in trophoblast stem cells. Cell Stem Cell 2011;8(5):525–37.

[42] Tsai Y, Cooke NE, Liebhaber SA. Tissue specific CTCF occupancy and boundary function at the human growth hormone locus. Nucleic Acids Res 2014;42(8):4906–21.

[43] Robinson WP, Price EM. The human placental methylome. Cold Spring Harb Perspect Med 2015;5(5):a023044.

[44] Kim SY, Romero R, Tarca AL, et al. Methylome of fetal and maternal monocytes and macrophages at the Feto-Maternal interface. Am J Reprod Immunol 2012;68(1):8–27.

[45] Mello d, Moreira JC, Araujo d, Souza ES, Stabellini R, et al. Random X inactivation and extensive mosaicism in human placenta revealed by analysis of allele-specific gene expression along the X chromosome. PLoS One 2010;5(6):e10947.

[46] Cotton AM, Avila L, Penaherrera MS, Affleck JG, Robinson WP, Brown CJ. Inactive X chromosome-specific reduction in placental DNA methylation. Hum Mol Genet 2009;18(19):3544–52.

[47] Cotton AM, Price EM, Jones MJ, Balaton BP, Kobor MS, Brown CJ. Landscape of DNA methylation on the X chromosome reflects CpG density, functional chromatin state and X-chromosome inactivation. Hum Mol Genet 2015;24(6):1528–39.

[48] McGraw S, Oakes CC, Martel J, et al. Loss of DNMT1o disrupts imprinted X chromosome inactivation and accentuates placental defects in females. PLoS Genet 2013;9(11)e1003873.

[49] Hemberger M. The role of the X chromosome in mammalian extra embryonic development. Cytogenet Genome Res 2002;99(1-4):210–7.

[50] Penaherrera M, Ma S, Ho Yuen B, Brown C, Robinson W. X-chromosome inactivation (XCI) patterns in placental tissues of a paternally derived bal t (X; 20) case. Am J Med Genet A 2003;118(1):29–34.

[51] Gabory A, Roseboom TJ, Moore T, Moore LG, Junien C. Placental contribution to the origins of sexual dimorphism in health and diseases: sex chromosomes and epigenetics. Biol Sex Differ 2013;4(1):5.

[52] Monk D. Genomic imprinting in the human placenta. Obstet Gynecol 2015;213(4):S152–62.

[53] Yuen RK, Jiang R, Peñaherrera MS, McFadden DE, Robinson WP. Genome-wide mapping of imprinted differentially methylated regions by DNA methylation profiling of human placentas from triploidies. Epigenetics Chromatin 2011;4(1):10.

[54] Hanna C, Robinson W. Placentation and genomic imprinting. In: Neumova AK, Taketo T, editors. Epigenetics in Human Reproduction and Development. World Scientific Publishing Co.; 2016. p. 159–84

[55] Robinson W, Slee J, Smith N, et al. Placental mesenchymal dysplasia associated with fetal overgrowth and mosaic deletion of the maternal copy of 11p15. 5. Am J Med Genet A 2007;143(15):1752–9.

[56] Court F, Tayama C, Romanelli V, et al. Genome-wide parent-of-origin DNA methylation analysis reveals the intricacies of human imprinting and suggests a germline methylation-independent mechanism of establishment. Genome Res 2014;24(4):554–69.

[57] Das R, Lee YK, Strogantsev R, et al. DNMT1 and AIM1 imprinting in human placenta revealed through a genome-wide screen for allele-specific DNA methylation. BMC Genomics 2013;14(1):685.

[58] Ouyang Y, Mouillet J, Coyne C, Sadovsky Y. Review: placenta-specific microRNAs in exosomes–good things come in nano-packages. Placenta 2014;35:S69–73.

[59] Morales-Prieto D, Chaiwangyen W, Ospina-Prieto S, et al. MicroRNA expression profiles of trophoblastic cells. Placenta 2012;33(9):725–34.

[60] Nakamura T, Liu Y, Nakashima H, et al. PGC7 binds histone H3K9me2 to protect against conversion of 5mC to 5hmC in early embryos. Nature 2012;486(7403):415–9.

[61] Barbaux S, Gascoin-Lachambre G, Buffat C, et al. A genome-wide approach reveals novel imprinted genes expressed in the human placenta. Epigenetics 2012;7(9):1079–90.

[62] Bourque D, Penaherrera M, Yuen R, Van Allen M, McFadden D, Robinson W. The utility of quantitative methylation assays at imprinted genes for the diagnosis of fetal and placental disorders. Clin Genet 2011; 79(2):169–75.

[63] Uusküla L, Rull K, Nagirnaja L, Laan M. Methylation allelic polymorphism (MAP) in chorionic gonadotropin β5 (CGB5) and its association with pregnancy success. J Clin Endocrinol Metab 2011;96(1):E199–207.

[64] Blair J, Langlois S, McFadden D, Robinson W. Overlapping DNA methylation profile between placentas with trisomy 16 and early-onset preeclampsia. Placenta 2014;35(3):216–22.

[65] Hatt L, Aagaard MM, Bach C, et al. Microarray-based analysis of methylation of 1st trimester trisomic placentas from down syndrome, edwards syndrome and patau syndrome. PLoS One 2016;11(8):e0160319.

[66] Hanna CW, McFadden DE, Robinson WP. DNA methylation profiling of placental villi from karyotypically normal miscarriage and recurrent miscarriage. Am J Pathol 2013;182(6):2276–84.

[67] Pliushch G, Schneider E, Weise D, et al. Extreme methylation values of imprinted genes in human abortions and stillbirths. Am J Pathol 2010;176(3):1084–90.

[68] Yin L, Zhang Y, Lv P, et al. Insufficient maintenance DNA methylation is associated with abnormal embryonic development. BMC Med 2012;10(1):26.

[69] Bastek JA, Gómez LM, Elovitz MA. The role of inflammation and infection in preterm birth. Clin Perinatol 2011;38(3):385–406.

[70] Olson DM, Severson EM, Verstraeten BS, Ng JW, McCreary JK, Metz GA. Allostatic load and preterm birth. Int J Mol Sci 2015;16(12):29856–74.

[71] Toure DM, ElRayes W, Barnes-Josiah D, Hartman T, Klinkebiel D, Baccaglini L. Epigenetic modifications of human placenta associated with preterm birth: a systematic review. J Matern Fetal Neonatal Med 2017;1–12.

[72] Yeung KR, Chiu CL, Pidsley R, Makris A, Hennessy A, Lind JM. DNA methylation profiles in preeclampsia and healthy control placentas. Am J Physiol Heart Circ Physiol 2016;310(10):H1295–303.

[73] Blair JD, Yuen RK, Lim BK, McFadden DE, von Dadelszen P, Robinson WP. Widespread DNA hypomethylation at gene enhancer regions in placentas associated with early-onset preeclampsia. Mol Hum Reprod 2013;19(10):697–708.

[74] Anton L, Brown AG, Bartolomei MS, Elovitz MA. Differential methylation of genes associated with cell adhesion in preeclamptic placentas. PLoS One 2014;9(6):e100148.

[75] Herzog EM, Eggink AJ, Willemsen SP, et al. Early-and late-onset preeclampsia and the tissue-specific epigenome of the placenta and newborn. Placenta 2017;58:122–32.

[76] Chu T, Bunce K, Shaw P, et al. Comprehensive analysis of preeclampsia-associated DNA methylation in the placenta. PLoS One 2014;9(9):e107318.

[77] Wilson SL, Leavey K, Cox B, Robinson WP. Mining DNA methylation alterations towards a classification of placental pathologies. Hum Mol Genet 2017;.

[78] Rahat B, Najar RA, Hamid A, Bagga R, Kaur J. The role of aberrant methylation of trophoblastic stem cell origin in the pathogenesis and diagnosis of placental disorders. Prenat Diagn 2017;37(2):133–43.

[79] Hogg K, Blair JD, McFadden DE, von Dadelszen P, Robinson WP. Early onset preeclampsia is associated with altered DNA methylation of cortisol-signalling and steroidogenic genes in the placenta. PLoS One 2013; 8:e62969.

[80] Mayne BT, Leemaqz SY, Smith AK, Breen J, Roberts CT, Bianco-Miotto T. Accelerated placental aging in early onset preeclampsia pregnancies identified by DNA methylation. Epigenomics 2017;9(3):279–89.

[81] Wilson SL, Liu Y, Robinson WP. Placental telomere length decline with gestational age differs by sex and TERT, DNMT1, and DNMT3A DNA methylation. Placenta 2016;48:26–33.

[82] Lappalainen T, Greally JM. Associating cellular epigenetic models with human phenotypes. Nat Rev Genet 2017;18(7):441–51.

[83] Saenger P, Czernichow P, Hughes I, Reiter EO. Small for gestational age: short stature and beyond. Endocr Rev 2007;28(2):219–51.

[84] Iglesias-Platas I, Martin-Trujillo A, Petazzi P, Guillaumet-Adkins A, Esteller M, Monk D. Altered expression of the imprinted transcription factor PLAGL1 deregulates a network of genes in the human IUGR placenta. Hum Mol Genet 2014;23(23):6275–85.

[85] Banister CE, Koestler DC, Maccani MA, Padbury JF, Houseman EA, Marsit CJ. Infant growth restriction is associated with distinct patterns of DNA methylation in human placentas. Epigenetics 2011;6(7):920–7.

[86] Wilhelm-Benartzi CS, Houseman EA, Maccani MA, et al. In utero exposures, infant growth, and DNA methylation of repetitive elements and developmentally related genes in human placenta. Environ Health Perspect 2012;120(2):296–302.

[87] Roifman M, Choufani S, Turinsky AL, et al. Genome-wide placental DNA methylation analysis of severely growth-discordant monochorionic twins reveals novel epigenetic targets for intrauterine growth restriction. Clin Epigenetics 2016;8(1):70.

[88] Dolinoy DC, Weidman JR, Waterland RA, Jirtle RL. Maternal genistein alters coat color and protects avy mouse offspring from obesity by modifying the fetal epigenome. Environ Health Perspect 2006;114(4):567–72.

[89] Waterland RA, Jirtle RL. Transposable elements: targets for early nutritional effects on epigenetic gene regulation. Mol Cell Biol 2003;23(15):5293–300.

[90] Tobi EW, Goeman JJ, Monajemi R, et al. DNA methylation signatures link prenatal famine exposure to growth and metabolism. Nat Commun 2014;5:5592.

[91] Jiang X, Yan J, West AA. Maternal choline intake alters the epigenetic state of fetal cortisol-regulating genes in humans. FASEB J 2012;26(8):3563–74.

[92] Januar V, Desoye G, Novakovic B, Cvitic S, Saffery R. Epigenetic regulation of human placental function and pregnancy outcome: considerations for causal inference. Obstet Gynecol 2015;213(4):S182–96.

[93] Suter M, Abramovici A, Showalter L, et al. In utero tobacco exposure epigenetically modifies placental CYP1A1 expression. Metab Clin Exp 2010;59(10):1481–90.

[94] Morales E, Vilahur N, Salas LA, et al. Genome-wide DNA methylation study in human placenta identifies novel loci associated with maternal smoking during pregnancy. Int J Epidemiol 2016;45(5):1644–55.

[95] Finer S, Mathews C, Lowe R, et al. Maternal gestational diabetes is associated with genome-wide DNA methylation variation in placenta and cord blood of exposed offspring. Hum Mol Genet 2015;24(11):3021–9.

[96] Binder AM, LaRocca J, Lesseur C, Marsit CJ, Michels KB. Epigenome-wide and transcriptome-wide analyses reveal gestational diabetes is associated with alterations in the human leukocyte antigen complex. Clin Epigenetics 2015;7(1):79.

[97] Reichetzeder C, Putra SD, Pfab T, et al. Increased global placental DNA methylation levels are associated with gestational diabetes. Clin Epigenetics 2016;8(1):82.

[98] de Waal E, Mak W, Calhoun S, et al. In vitro culture increases the frequency of stochastic epigenetic errors at imprinted genes in placental tissues from mouse concepti produced through assisted reproductive technologies. Biol Reprod 2014;90(2):22.

[99] Choux C, Carmignac V, Bruno C, Sagot P, Vaiman D, Fauque P. The placenta: phenotypic and epigenetic modifications induced by assisted reproductive technologies throughout pregnancy. Clin Epigenetics 2015;7(1):87.

[100] Leavey K, Bainbridge SA, Cox BJ. Large scale aggregate microarray analysis reveals three distinct molecular subclasses of human preeclampsia. PLoS One 2015;10(2):e0116508.

[101] Wilson SL, Blair JD, Hogg K, Langlois S, von Dadelszen P, Robinson WP. Placental DNA methylation at term reflects maternal serum levels of INHA and FN1, but not PAPPA, early in pregnancy. BMC Med Genet 2015;16(1):1.

[102] Manokhina I, Wilson SL, Robinson WP. Noninvasive nucleic acid–based approaches to monitor placental health and predict pregnancy-related complications. Obstet Gynecol 2015;213(4):S197–206.

[103] Manokhina I, del Gobbo GF, Konwar C, Wilson SL, Robinson WP. Placental biomarkers for assessing fetal health. Hum Mol Genet 2017;26:R237–45.

microRNAs in Pregnancy: Implications for Basic Research and Clinical Management

Ming Liu*, Xiaotao Bian*,‡, Hao Wang†, Yan-Ling Wang*,‡

*State Key Laboratory of Stem Cell and Reproductive Biology, Institute of Zoology, Chinese Academy of Sciences, Beijing, China †Medical Research Center, Peking University Third Hospital, Beijing, China ‡University of Chinese Academy of Sciences, Beijing, China

OUTLINE

Human Reproductive and Prenatal Genetics
https://doi.org/10.1016/B978-0-12-813570-9.00014-0

Abbreviations

miRNA	microRNA
PCR	polymerase chain reaction
3′-UTR	3′-untranslated region
RISC	RNA-induced silencing complex
C19MC	chromosome 19 microRNA cluster
C14MC	chromosome 14 microRNA cluster
PLAP	placental alkaline phosphatase
PRKG1	protein kinase, cGMP-dependent, type I
IGF1/2	insulin-like growth factor 1/2
PHLDA2	pleckstrin homology like domain family a member 2
PLAGL1	PLAG1 like zinc finger 1
HIF-1	hypoxia-induced-factor 1
HRE	hypoxia response element
TLR	toll-like receptor
AGO2	argonaute 2
aPL	antiphospholipid antibody
EMT	epithelial to mesenchymal transition
MET	mesenchymal to epithelial transition
IVF	in vitro fertilization
MMP	matrix metalloproteinase
dMSC	decidua-derived mesenchymal stem cell
VEGF	vascular endothelial growth factor
PGF2α	prostaglandin F2α
COX-2	cyclooxygenase-2
NF-κB	nuclear factor-κB
CX43	connexin-43
OXTR	oxytocin receptor
TNF-α	tumor necrosis factor-α
CXCL12	C-X-C motif chemokine ligand 12
T2D	type 2 diabetes
LPD	low-protein diet
HFD	high-fat diet
MAPK1	mitogen-activated protein kinase 1
FGR	fetal growth restriction
GDM	gestational diabetes mellitus
RSA	recurrent spontaneous abortion
SNP	single nucleotide polymorphisms

INTRODUCTION

MicroRNAs (miRNAs) are single-stranded noncoding regulatory RNA molecules in length with 19–25 nucleotide (nt). The first miRNA, lin-4, was found to participate in the embryonic development of *C. elegans* in 1993. Since then, accumulating numbers of miRNAs have been identified in plants, animals, protists, and viruses, with the exception of bacteria. miRNAs are highly evolutionarily conserved and generally regulate gene expression in the posttranscription level. In humans, they can interact with more than 30% of genes in a wide variety of cell types [1]. miRNAs play a role in almost all cellular events, including cell proliferation, differentiation, migration, apoptosis, metabolism, stress response, etc.

Distinct types of cells may express common miRNAs or display their unique miRNA profiles. The expression patterns of miRNAs in certain cell types or tissues usually change under different physiological or pathological conditions. Therefore, it has been suggested that miRNAs not only take part in the regulation of the physiological process, but are also involved in the etiology of various diseases, which may have a wide range of clinical applications.

During pregnancy, the expression pattern of miRNAs changes significantly compared with the nonpregnant stage, indicating that miRNAs are probably essential parts for pregnancy regulation. Abundant miRNAs are expressed in the uterus and placenta, participating in the regulation of embryonic implantation, placentation, fetal development, and maternal adaptation. Dysregulation of miRNAs has been demonstrated in association with various pregnancy complications.

BASIC KNOWLEDGE OF miRNAs

The Biosynthesis of miRNAs

The biosynthesis of miRNAs is a complex multistep process. Two members of the RNase III family, Drosha and Dicer, participate in the process in the nucleus and cytoplasm, respectively. In mammals, miRNAs are usually synthesized through a canonical approach, which begins with the RNA polymerase II-dependent transcription of the primary-miRNA (pri-miRNA) from the miRNA-encoding gene. These long-chain transcripts fold into a stem-loop structure containing the sequences of mature miRNA. The next step is processed by a microprocessor complex, a protein complex containing Drosha and double-stranded RNA binding protein DGCR8. The pri-miRNA is cleaved by the microprocessor complex into the precursor of miRNA (pre-miRNA), which is 65–80 nt in length with an irregular hairpin structure and a 30 nt sequence out of the hairpin structure. Pre-miRNA is subjected to active transportation to cytoplasm by RanGTPase-dependent Exportin 5 (Exp5), and is further processed by Dicer with the assistance of the RNA-binding protein TPBP. Pre-miRNA is cleaved at one of the two arms in the hairpin structure to release double-stranded miRNA. The small molecule is loaded to the RNA-induced silence complex (RISC) containing the Argonaute (Ago) protein core, and is eventually unwound into single strands. One of them with a less stable pair of 2–4 bases in the 5′ end, called the guide strand, binds to RISC and remains as mature miRNA. The remainder, called the passenger strand, will be degraded through unknown mechanisms.

More than 80% of miRNA-encoding genes are located in the introns of longer primary transcripts that are either protein-encoding mRNAs or mRNA-like transcripts. These pri-miRNAs are usually transcribed in company with their host genes. On the other hand, those intergenic miRNA-encoding genes are transcribed independently by their own promoter. In addition, there are alternative pathways of miRNA biosynthesis, and these nonclassical mammalian miRNAs usually exist in low abundance. Some rare miRNAs, named mirtron, originate from the short hairpin structure of mRNA introns. They bypass the Drosha-mediated cleavage and are released by splicesomes, linearized by lariat debranching, and fold into pre-miRNA-like hairpin structures. Other kinds of small RNAs, such as small nucleolar RNA (snoRNA) and short hairpin RNA (shRNA), can also bypass the microprocessing complex to produce miRNA. Some miRNAs, such as miR-451, are synthesized by relying on the shear activity of Ago2 instead of Dicer [2].

Posttranscriptional Function of miRNAs

It is generally accepted that the major working mechanism of miRNA is posttranscriptional suppression on the target genes. An miRNA-RISC complex with the combination of miRNA and Argonaute protein cores, DP103, MOV10, and others, can be complementarily base paired with the binding site of the target mRNA in its 3′-untranslated region (3′-UTR), leading to instability of the mRNA or blockage of its translation by deadenylation. Additionally, it is reported that some miRNAs can also directly bind to the coding sequence or 5′-UTR of the target mRNAs to inhibit or activate the expression of the targets.

In animals, the pairing between most of the miRNAs and their target genes is often incomplete, with only 2–7 bases in the 5′ end of miRNAs as seed regions for the complementary pair. However, miRNAs in plants need to completely pair with the binding site in the target genes to promote the degradation of mRNA. miRNAs construct a complex network with their target genes. One simple miRNA has the ability to modulate the expression of multiple target genes while one gene can be regulated by many miRNAs. Notably, the suppression on the target genes by miRNAs is generally mild, being no more than half of the target protein level [1].

THE SOURCES AND CHARACTERS OF miRNAs DURING PREGNANCY

Placenta-Derived miRNAs and Placenta-Specific miRNAs

The placenta is an abundant source of miRNAs, with more than 600 miRNAs being found in this organ. The majority of placenta-derived miRNAs belong to three placental-specific miRNA clusters: C19MC, C14MC, and the miR-371-3 gene cluster [3]. C19MC, the chromosome 19 microRNA cluster, is the largest human miRNA gene cluster with a length of approximately 100 kb. It is a primate-specific miRNA cluster, located in chromosome 19q13.41 of human genome and consisting of 46 genes that encode 59 mature miRNAs. C19MC is imprinted with only the paternally inherited allele expressed, and the members of this cluster are principally expressed in the placenta as well as in the testis, embryonic stem cells, and some tumors [4]. It is demonstrated that the abnormal expression of C19MC miRNAs is associated with adverse pregnant outcomes and antiviral defenses during maternal-fetal infections. C14MC is a maternally expressed imprinted gene family that is located in chromosome 14q32, spanning 40 kb and encoding 52 miRNAs. Some members of this cluster are specifically expressed in the placenta. Cluster miR-371-3 locates in chromosome 19 of human genome, being 1050 bp upstream of C19MC, and encodes miR-371-3p, miR-372, miR-373-3p, miR-371-5p, miR-373-5p, and miR-371b-3p. All three clusters exhibit a gestation-dependent expression pattern during pregnancy. The expressions of C19MC and miR-371-3 members increase while the levels of C14MC members decrease from the first to the third trimester of pregnancy [3].

The Sfmbt2 miRNA clusters are located in the Sfmbt2 gene. It is highly expressed in the placenta of Old World rodents, including mice and rats, and shows imprinted expression. The Sfmbt2 miRNA cluster is one of the largest miRNA clusters in mice. The miRBase database records that it is located in intron 10 of the Sfmbt2 gene, containing 72 miRNA precursor sequences. The Sfmbt2 gene is paternally expressed in the mouse placenta and is essential for trophoblast maintenance, as confirmed by knockout mice with paternal deletion of the gene. Inoue et al. generated the 53 kb knockout mice with a large-scale deletion of the entire Sfmbt2 miRNA region to uncover the functions of these small RNAs [5]. These miRNAs are paternally expressed in the placenta and promote spongiotrophoblast cell proliferation by suppressing target genes with cell proliferation regulators or tumor-repressors, in cooperation with the host Sfmbt2 gene under the same imprinted placental expression control.

In addition to these placental-specific miRNAs, the placenta also produces other miRNAs with relatively low levels. Many of them present time-specific patterns at different stages of placentation. For example, the members of the miR-17-92 gene cluster are significantly highly expressed in the first trimester while the levels of the let-7 family, the miR-34 family, the miR-29 gene cluster, miR-195, and miR-181c increase significantly in the third trimester [6].

Extracellular Vesicles as Cargo for miRNAs

During pregnancy, a large number of placenta-produced miRNAs can be released to the extracellular environment and transferred to maternal circulation. These circulating miRNAs exist in at least two forms: vesicle and nonvesicle. They are resistant to the endogenous ribonuclease in circulation. The miRNAs in nonvesicle form can bind to specific lipoproteins, including Ago2, high-density lipoprotein (HDL), and nucleophosmin, to form a complex. Those in vesicle form are packaged into the phospholipid bilayer of extracellular vesicles (EVs). The EVs are classified into three types based on their sizes and synthesis approaches: exosomes (40–100 nm), microvesicles (100–1000 nm), and apoptotic blebs (1000–5000 nm), respectively. Among them, the exosomes are proved to mediate local and long-distance extracellular communication, and they may be potentially less-invasive biomarkers for abnormal placentation. Most of the placenta-derived EVs originate from syncytiotrophoblasts, the terminal differentiated multinucleated trophoblast layer covered in the outermost of the floating villi. The placenta-derived exosomes can be released to the maternal circulation as early as the sixth week of gestation. They bring placental alkaline phosphatase (PLAP) and can be specifically collected for subsequent analysis. The levels of bioactive exosomes in the maternal plasma increase by 50-fold during pregnancy, and rapidly decrease to the basal level after delivery [7, 8].

It has been proved that the placenta-derived exosomes contain miRNAs. Luo et al. found that exosomes from villous trophoblast containing trophoblast-specific miRNAs such as miR-517a were released to maternal circulation [9]. Donker et al. demonstrated that C19MC members were the most abundant miRNAs in the exosomes released by the placenta [10]. The miRNAs packaged in the trophoblast exosomes participate in the maternal-fetal crosstalk. In normal pregnancy, the number of placenta-derived exosomes elevated significantly along with the gestational age [11]. The miR-517a that was packaged in the trophoblast exosomes was inversely correlated with the expression of PRKG1 in NK cells [12]. The data seems to indicate that trophoblast exosomes may work somehow like hormones to mediate the maternal-fetal communication. It has been proved that the C19MC members in exosomes induced autophagy in nonplacental recipient cells, enhancing antiviral defenses [13]. Separating the placenta-derived or maternal exosomes and measuring the packaged miRNAs has been suggested as a new way to monitor the pregnancy process and diagnose pregnancy complications.

REGULATION OF miRNA EXPRESSION IN THE PLACENTA

The expression of miRNAs in the placenta can be regulated by multiple factors, including hormones, oxygen tension, inflammatory factors, etc. Large amounts of evidence has been derived from the analysis of clinical samples and in vitro cell models. The in-depth molecular mechanisms remain less investigated.

Oxygen Tension

Oxygen tension is an important factor to regulate trophoblast cell fate. At the early stage of gestation, especially before the uterine spiral arteries are well remodeled, the embryo/placenta is in a physiological low oxygen condition. Such a hypoxic environment was demonstrated to actively promote the rapid proliferation in trophoblast cells. The increase in oxygen tension at the fetomaternal interface could induce the differentiation of trophoblast cells toward various pathways. In the case of preeclampsia, the aberrantly hypoxic stress due to the insufficient remodeling of spiral arteries was well recognized as the critical pathological cause leading to excessive growth and insufficient invasion of the trophoblast cells. Although several proteins involved in miRNA biosynthesis, such as Drosha, Exp5, Ago2, and DP103, are not affected by hypoxia, Dicer has been shown to be down-regulated under hypoxia. Accumulated data revealed a number of miRNAs being differentially expressed upon hypoxic stimulation in placental trophoblasts. Among them, miR-210 is the well-studied hypoxia-responsive gene. It is transcriptionally activated by hypoxia-induced-factor 1 (HIF-1) under a hypoxia condition via the hypoxia response element (HRE) on the upstream sequence of its coding gene [14]. Other miRNAs that can be up-regulated by hypoxia in trophoblast cells include miR-30b-5p, miR-30d-5p, miR-34a, miR-93, miR-135b, miR-149, miR-203, miR-299-5p, miR-300, miR-365, miR-424, miR-503, miR-517a/b, miR-517c, miR-518c, miR-620, and miR-3074-3p. Hypoxia can also suppress the expression of several miRNAs in trophoblast cells, such as miR-7-5p, miR-33b-3p, miR-663a, miR-720, miR-1260b, miR-1280, miR-3656, and miR-4417 [15, 16]. The regulatory mechanism of oxygen tension on miRNA expression needs to be further investigated.

Inflammatory Factors

It is well known that the fetomaternal interface during normal pregnancy is under a mildly inflammatory status. The inflammatory signaling can stimulate or repress certain miRNA expressions, which further participate in the innate or adaptive immune response. Evidence revealed that Toll-like receptors (TLRs) or IFN-α treatment could significantly reduce the protein level of Dicer while IFN-γ induced Dicer expression in trophoblast cells, indicating the potential of these immune factors to regulate miRNA synthesis in the placenta.

Studies on the inflammation-associated pregnancy complications may give some hints. For instance, Ackerman and collegues identified seven differentially expressed miRNAs—miR-887-3p, miR-154-5p, miR-376b-5p, miR-376c-3p, miR-500a-5p, miR-133a-3p, and miR-223-3p—in villous trophoblasts derived from intraamniotic infection-induced preterm birth patients when compared with the noninfection spontaneous preterm birth individuals [17]. In the preeclamptic placenta, the miR-15b expression was increased. In human trophoblast cell models, Yang et al. proved that the inflammatory stimuli LPS could increase miR-15b expression and subsequently inhibit AGO2 expression. AGO2 acted not only as a major effector protein in the core of RISC, but also as a transcriptional regulator of miR-15b by occupying the promoter, indicating the roles of the miR-15b-AGO2 feedback loop in trophoblast function [18]. Our previous study revealed that miR-210 expression was evidently enhanced by hypoxia and the inflammatory factor TNFα, indicating the complex causes of the abnormal overexpression of miR-210 in the preeclamptic placenta [19, 20].

TLR-mediated pathways may participate in inflammatory regulation on the placental miRNAs. TLR3 activation could mimic the preeclampsia-like symptoms in mice, including pregnancy-specific hypertension, endothelial dysfunction, and proteinuria. In the placenta of this mouse model, the activation of TLR3 significantly elevated miR-210 expression [21]. In human trophoblast cells, TLRs could mediate the up-regulation of miR-23a, miR-329, and let-7c in response to LPS treatment [22]. TLR4 was found to be involved in the effect of the antiphospholipid antibody (aPL) to elevate cellular and exosomal miR-146-5p, miR-146a-3p, and miR-210 expression [23].

A common downstream signal molecule of the inflammatory factors is NF-κB. It could mediate the response of multiple miRNAs, including let-7, miR-9, miR-146a, and miR-224 to various proinflammatory factors in various cell types [24–27], but it remains unclear whether the same regulation occurs in the placenta. There is evidence showing that TNF-α induced miR-155 in HUVECs via the transcriptional regulation of NF-κB [28], and TLR3 enhanced miR-210 via HIF-1α and NF-κBp50 [21].

Besides, the inflammation-elicited miRNAs regulation was also found in the decidual tissue. For instance, IL-1b treatment in human decidual cell cultures caused differential expression of six miRNAs, including miR-146a, miR-155, etc. [29].

THE FUNCTION OF miRNAs DURING PREGNANCY

There has been much evidence indicating the participation of miRNAs in the regulation of various processes of pregnancy, including embryonic implantation, placental development, pregnant immunetolerance, placental endocrine, and parturition. Many studies reported the differentially expressed miRNAs in pregnancy-associated diseases, and the functional investigation of these miRNAs using in vitro cell models or in vivo animal models, which revealed the possible working mechanisms of the deregulated miRNAs in the occurrence or the development of the pregnant disorders. However, one puzzle is that gene deletion of certain miRNAs in mice did not cause obvious pregnancy defects, although the knockout of Dicer resulted in fetal lethality. It is most likely that miRNAs may commit fine-tune regulation of the target genes. In addition, the multiplex network among miRNAs and their targets may be a redundant way to ensure the miRNA-regulated cell events. Therefore, the dysregulation of miRNAs may account for the pathological changes observed in the pregnant complications, but the knockout of a certain miRNA gene in mice cannot reflect severe developmental problems.

Role of miRNAs in Uterine Endometrium and Embryonic Implantation

Embryo implantation is a critical beginning step of pregnancy that involves blastocyst apposition, attachment to the uterine epithelium, and decidualization of the uterine stroma. Decidualization is regulated by multiple factors, and miRNAs have been implicated as important players. Drosha expression in a mouse uterus presents spatiotemporal features during early pregnancy. A high level of Drosha was observed in decidual stromal cells at the implantation window and the artificially induced decidua. With the stromal cell culture model, it was confirmed that the Drosha expression gradually increased along the progression of decidualization.

Many miRNAs exhibit spatial and/or temporal change patterns in uterine endometrium, and may regulate various cell events in the uterus. The expression of Let-7a in mice endometrial epithelial cells increased from E0.5 to E3.5, and it could influence embryo implantation by targeting Mucin 1 (Muc1) [30]. miR-21 expression could be regulated by the implantation of blastocyst, being intensive in the stromal cells of the implantation site at E4.5. Two miRNAs with spatiotemporal expression in the uterus, miR-101a and miR-199a, were involved in embryonic implantation by targeting cyclooxygenase-2 [31]. Microarray analysis of miRNA profiles at the implantation sites and the interimplantation sites revealed 30 up-regulated and 42 down-regulated miRNAs in the endometrium of the implantation sites at E4.5 [32]. An RNA sequencing study also found more than 60 differential miRNAs in a mouse uterus under delayed and activated implantation [33].

In the human endometrium, several miRNAs varied along with the menstrual cycle. Altmäe et al. found the up-regulation of miR-30b and miR-30d and the down-regulation of miR-494 and miR-923 in a receptive endometrium [34]. Another 12 miRNAs were found increasing in the endometrium at the midsecretory phase, and they were involved in the regulation of the cell cycle [35]. The miR-200 family and miR-205 were found to promote the mesenchymal-to-epithelial transition (MET) of endometrial cells by targeting transcriptional repressors ZEB1 and ZEB2, and therefore upregulate the expression of E-cadherin [36].

The dysregulation of miRNAs in the endometrium or embryo indicated a failure in embryonic implantation or early pregnancy. A comparative study between the stimulated cycles and the natural cycles revealed 22 differential miRNAs in the human endometrium. Half these miRNAs exhibited a progesterone response element in their promotor regions, indicating the altered endometrial receptivity during the stimulated cycles. A microRNA chip study demonstrated the association of miR-451 with poor pregnancy rates of IVF [37]. In the IVF patients with repeated implantation failure, 13 significantly altered miRNAs were identified when compared with fertile women [38]. These miRNAs were involved in the regulation of cell adhesion, the cell cycle, and the Wnt signaling pathway.

Role of miRNAs in Placental Trophoblast Cell Differentiation

In humans, the cytotrophoblast (CTB) progenitor cells derived from the trophectoderm of the blastocyst give rise to the placental trophoblast subtypes through two general pathways: villous trophoblasts and invasive extravillous trophoblasts. In the villous pathway, mononucleated CTB cells fuse into multinucleated STBs, forming the syncytial layer that covers the placental villous tree. These cells are principally involved in the exchange of gases, nutrients, and waste across the maternal-fetal interface [39]. The syncytial layer is the major place to produce pregnancy-related hormones, such as human chorionic gonadotropin (hCG) and human placental lactogen (hPL) [40]. Additionally, STBs are in direct contact with the maternal blood and are therefore required to exhibit a degree of immune tolerance. This tolerance is achieved in large part through their lack of expression of the classical class I human leucocyte antigens (HLA) [41]. In the extravillous pathway, CTB cells proliferate to form trophoblast cell columns of the anchoring villi [42, 43]. Some trophoblast cells, called extravillous trophoblasts (EVTs), can detach from the columns and migrate into the decidua. A subset of these, interstitial trophoblasts (iEVTs), migrate into the deep layer of the decidua and even into the inner third of the myometrium, thereby anchoring the fetus to the mother. Another subset of EVTs, endovascular trophoblasts (enEVT), acquires endothelial-like characteristics and penetrates the uterine spiral arteries to replace maternal endothelial cells. In this way, the uterine spiral arteries are remodeled into low-resistance, high-capacity uteroplacental arteries that provide increased blood flow toward the placenta to meet the requirements of the growing fetus [44].

Trophoblast differentiation during placental development is precisely regulated by environmental factors, such as oxygen tension within the maternal-fetal interface, and by various hormones and growth factors. miRNAs have been demonstrated to participate in the modulation of various trophoblast cell events. The dysregulation of trophoblast activities is in tight association with the development of preeclampsia, a pregnancy-associated disorder characterized by hypertension and proteinurea [43, 45, 46].

miRNAs Regulate Syncytiolization and Hormone Production in STBs

It is generally accepted that the formation of STBs is a fusogenic event. Glial cells missing-1 (GCM-1) has been recognized as a principal transcription factor to induce the expression of fusion peptide synctytin-1, which triggers the fusion process of CTB cells. A positive feedback loop composed of Gcm1 and Fzd5 was critical for normal initiation of branching in the chorionic plate [47]. The STBs are responsible for the production of hormones to maintain pregnancy, such as human chorionic gonadotropin (hCG), progesterone, estradiol, testosterone, etc. [48].

A member of C19MC, miR-515-5p, being transcriptionally regulated by c-MYC, exhibited decreased expression along the trophoblast cell fusion process. This small RNA was found negatively regulating GCM1 and FZD5, which are critical for trophoblast cell fusion, and suppressing CYP19A1 (encoding aromatase) expression to hamper estrogen production [49]. A similar working mechanism was found in the miR-17-92 cluster and the miR-106a-363 cluster. Two members of these clusters, miR-106a and miR-19b, which were also transcriptionally regulated by c-MYC, could target and decrease the expression of CYP19A1 and GCM1. Our recent study demonstrated that miR-22, a testosterone-induced miRNA, targeted ERα and subsequently inhibited aromatase expression and estradiol production.

In preeclamptic placenta, a significant up-regulation of miR-22 and increased testosterone and reduced estradiol production were observed when compared with controls. The study therefore reveals a mechanism underlying the balanced production of androgen and estrogen modulated by miR-22 in human placenta [50]. Some other studies found the progesterone production could be enhanced by miR-96 [51], and miR-130b-3p inhibited PGC-1α to affect mitochondrial biogenesis signaling in syncytiatrophoblast cells [52].

miRNAs Regulate Trophoblast Cell Invasion, Proliferation, Apoptosis

A number of differentially expression miRNAs have been identified in preeclamptic placenta, and many of them were demonstrated to be involved in regulating trophoblast cell invasion. For instance, the C19MC cluster was unregulated in preeclamptic placenta, and two members of this cluster, miR-517a/b and miR-517c, impeded trophoblast cell invasion and promoted the expression of antiangiogenic sFLT1 [53]. However, miR-518d could strengthen the trophoblast cell abilities of migration and invasion [54]. miR-210 was the most up-regulated miRNA in preeclamptic placenta [55], and it inhibited trophoblast cell migration and invasion by targeting Ephrin-A3 and Homeobox-A9 [56]. Other miRNAs that were reported to suppress trophoblast cell invasion included miR-137, which targeted ERRα [57]; miR-193b-3p, which blocked TGF-β2 [58]; miR-346 and miR-582-3p, which inhibited EG-VEGF; MiR-29b and miR-519d-3p, which targeted MMP-2; and miR-204, which inhibited MMP-9 [59–61]. LPS stimulation-induced miR-15b reduced trophoblast cell invasion and endothelial tube formation via targeting AGO2 [18].

Interestingly, our previous studies revealed that several downregulated miRNAs in preeclamptic placenta could target Activin/Nodal signaling in trophoblast cells, resulting in an excessive activation of this signaling pathway and aberrant trophoblast cell functions [62]. Targeting of Nodal by miR-378-5p, ActRIIA and ActRIIB by miR-195, ALK5 and ALK7 by miR-376c, and Smad2 by miR-18a resulted in enhanced trophoblast cell invasion or reduced cell apoptosis. The lowered expression of these miRNAs may therefore contribute to the shallow trophoblast cell invasion and excessive trophoblast cell apoptosis in preeclampsia. Another down-regulated miRNA in preeclamptic placentas, miR-675, could target Nodal modulator 1 (NOMO1) and regulate trophoblast cell proliferation [63].

miR-23a and miR-128a, which were up-regulated in preeclamptic placenta, could induce trophoblast cell apoptosis via targeting XIAP and Bax. The down-regulated miR-101 in preeclamptic placenta may contribute to ER stress-induced trophoblast cell apoptosis by targeting ERp44 [64]. The other differential miRNAs in preeclamptic placentas that were associated with cell apoptosis included miR-155, miR-21, and miR-122 [65].

There are studies indicating that miRNAs derived from other cells at the fetomaternal interface can regulate trophoblast cell function. An example is miR-494, which was up-regulated in preeclamptic decidua-derived mesenchymal stem cells (dMSCs). It could target CDK6 and CCND1 to arrest G1/S transition, and suppress the migration of trophoblast cells and the tube formation of endothelial cells [66]. miR-495 was significantly increased in umbilical cord tissue and mesenchymal stem cells in severe preeclamptic patients. This small RNA arrested the cell cycle in S phase, promoted apoptosis and inhibited migration by targeting Bmi-1, and inhibited trophoblast cell invasion and tube formation of HUVEC in a paracrine way [67].

Role of miRNAs in Parturition

Steroid hormones progesterone (P4) and 17β-estradiol (E2) play critical roles in regulating uterine excitability (quiescence or contractility) throughout gestation, and they usually work in opposite ways. Uterine quiescence is maintained by increased levels of P4, which act via progesterone receptor (PR) and partially interact with inflammatory transcription factor, nuclear factor-κB (NF-κB), to suppress the expression of contraction-related genes such as cyclooxygenase-2 (COX-2) and the prostaglandin F2α (PGF2α) receptor. Uterine contraction is initiated with increased levels of E2, enhanced estrogen receptor (ERα) activity, and withdrawal of PR activity during late gestation. Activation of ERα promotes parturition by enhancing the expression of genes encoding contraction-related proteins, primarily including COX-2, connexin-43 (CX43), and oxytocin receptor (OXTR).

Multiple miRNAs have been demonstrated to be functional in the regulation of contractile and proinflammatory genes in the myometrium of humans and mice, such as the miR-200 family [68, 69], the miR-199a/214 cluster [70], and the miR-181a family [71]. During late gestation and labor, members of the miR-200 family were up-regulated, leading to the down-regulation of their direct targets, zinc finger E-box binding homeobox proteins ZEB1 and ZEB2, and subsequently the suppression of P4 signaling and the expression of OXTR and CX43 [68]. One member, miR-200a, could also target STAT5b, which is a transcriptional repressor of the progesterone-metabolizing enzyme 20α-hydroxysteroid dehydrogenase (20α-HSD) [69]. On the contrary, the miR-199a/214 cluster and the miR-181a family were significantly decreased in the myometrium of humans and mice during late gestation and labor. The former caused a significant increase in the target gene COX-2, an important enzyme in the synthesis of proinflammatory prostaglandins [70], and the latter led to the enhanced expression of several proinflammatory factors, including TNF-α, ERα, and c-Fos [71]. These miRNAs and their targets were suggested to play reciprocal roles in regulating uterine quiescence/contractility throughout the entire pregnancy.

Role of miRNAs in Immune Tolerance During Pregnancy

An immune-tolerant microenvironment at the fetomaternal interface was essential for the appropriate maternal adaptation to the semiallograft fetus. A complex crosstalk occurs among multiple types of immune cells and trophoblast cells at the fetomaternal interface to alter their characteristics and functions toward an immune-tolerant state. Accumulating evidence reveals the roles of miRNAs to regulate the release of inflammatory factors from trophoblast cells. A differentially expressed miRNA in early onset severe preeclamptic placenta, miR-125, enhanced IL-8 production by targeting SGPL1 in trophoblast cells [72]. miR-30a could inhibit TAB3 and therefore suppress IL-1β-induced activation of NF-κB and JNK signaling. Its up-regulation in preeclamptic placentas decreased the production of IL-6, IL-8, and COX2. miR-329 targeted NF-κB and interfered with IL-6 expression in trophoblasts [22]. These miRNAs may participate in the immune dysregulation in the microenvironment of preeclampsia [73].

Hypoxia has been well accepted as an essential factor to initiate the pathological change in the preeclamptic placenta. Several miRNAs are significantly induced by hypoxia in trophoblast cells. miR-210 is a well-proved hypoxia-induced miRNA, and it is evidently up-regulated in preeclampsia placentas. A study from Kopriva et al. revealed miR-210 could target STAT6 to reduce IL-4 expression in the mouse placenta and in human primary cytotrophoblast cells [21]. Interestingly, miR-210 could also be induced by TNFα, TLR3, and NF-κB [19, 21], suggesting a complex network of immunoregulation and hypoxia that may be mediated by miR-210. Other examples of hypoxia-induced miRNA in trophoblast cells include miR-135b and miR-365, which could influence the expression of CXCL12 and HLA-G, respectively [15, 16].

The functions of immune cells in the decidua can be affected by miRNAs. The production of IL-17 in CD4$^+$ T cells was enhanced by miR-155, whose level was higher in preeclamptic plasma [74]. Some miRNAs, such as miR-34a-3p/5p, miR-141-3p/5p, and miR-24, regulated the function of decidual NK (dNK) cells, and their aberrant expressions were in association with recurrent spontaneous abortion (RSA) [75]. In addition, miRNAs from the placenta may be transported to the maternal peripheral blood NK (pNK) cells and commit functions. For instance, a recent study revealed that the level of some members in C19MC were found in pNK cells, and their levels were significantly higher at the third trimester than the first trimester and rapidly decreased after delivery [76]. It is most likely that the placental miRNAs, especially those being packaged in exosomes, may work like hormones to modulate the actions of multiple target cells, either within the fetomaternal interface or in the maternal organs.

Maternal Diet-Induced miRNAs and Metabolic Health in the Offspring

More and more epidemiological evidence emphasizes that unbalanced maternal nutrition status, particularly in the cases of type 2 diabetes (T2D) and obesity, is in tight association with the risk of metabolic diseases in their offspring in adulthood. Studies in mouse models have demonstrated the involvement of miRNAs in these health problems. A study from Zheng et al. using a mouse model of maternal low-protein diet (LPD) during pregnancy identified a group of differentially expressed miRNAs in the offspring from the LPD mother [77]. The putative targets of these miRNAs were mapped into seven inflammation-related pathways. It was suggested that these differential miRNAs may lead to impaired insulin secretion and glucose intolerance via regulating chronic low-grade inflammation. Alejandro et al. revealed that a maternal low-protein diet could influence insulin secretion and glucose homeostasis by altering miRNA and mTOR expression in the offspring [78]. Insulin levels, β cell fraction, and mTOR signaling were all decreased in newborns of the mothers that had been fed with LPD throughout pregnancy. In islets of these offspring, reduced mTOR and increased expression of a subset of miRNAs were observed, and a blockade of miR-199a-3p and miR-342 could restore mTOR and insulin secretion to normal levels [78]. In addition, in rodents at pregnancy or with obesity, compensatory β cell mass expansion was found associated with changes in the expression of several islet miRNAs, particularly miR-338-3p [79].

A maternal high-fat diet (HFD) during pregnancy or lactation has been found to be associated with insulin resistance, hepatic lipid accumulation, and increased serum cytokine levels in the offspring until adulthood. In a mouse model of a maternal high-fat diet (HFD), decreased miR-122 and increased miR-370 expressions were found in recently weaned offspring of the dams fed with HFD [80]. miR-122 is a liver-specific miRNA that has been known to participate in multiple metabolic processes, including cholesterol synthesis, fatty acid synthesis, and oxidation. Its repression could lead to hepatic insulin resistance [81].

miR-370 was shown to directly down-regulate carnitine palmitoyltransferase 1α (CPT 1α) and modulate fatty acid oxidation [82]. The alteration of these miRNAs in the offspring from the HFD mother significantly increased their risk of ectopic lipid accumulation and obesity and may result in more serious metabolic disorders in the long term [80]. In the livers of offspring from high-calorie diet (HCD) dams, down-regulation of miR-615-5p, miR-3079-5p, miR-124*, and miR-101b* and upregulation of miR-143* were discovered. The target genes of these miRNAs, such as TNF-α and mitogen-activated protein kinase 1 (MAPK1), are inflammatory process-associated [83]. It was suggested that the dysregulation of miRNAs was a potential cause for the high risk of obesity and diabetes in HCD dams.

CIRCULATING miRNA AS DIAGNOSTIC OR PREDICTIVE BIOMARKERS FOR PREGNANCY-ASSOCIATED DISORDERS

It has been demonstrated that the miRNA levels in plasma from pregnant women are much higher than those from nonpregnant women and their levels correspond to the stage of pregnancy [7, 84–86]. Plasma miRNA levels do not appear to be affected by incubation temperature or pH or even by RNase A treatment [87, 88]. In addition, many of the cell-free miRNAs are found enclosed in extracellular vesicles like exosomes or bound by miRNA-binding proteins such as Ago2 and high-density lipoprotein, making them relative stable in maternal circulation. The circulating miRNAs are therefore very promising to serve as unique markers for monitoring pregnancy outcomes. There have been increasing data describing the changes of miRNAs in the maternal circulation of various pregnant disorders.

Preeclampsia

Preeclampsia is the leading cause of maternal morbidity, mortality, and premature delivery, affecting approximately 2%–7% of pregnancies [89]. Much effort has been put into the clinical and basic research on this complication. However, the only treatment available for this disease is still premature delivery/termination of pregnancy. The major challenge is to effectively predict the condition early before the onset of the clinical signs, and develop preventive and therapeutic strategies that will minimize the burden of the disease.

Reports from various groups demonstrated the alteration of several circulating miRNAs in preeclamptic patients at early to mid-gestation. Compared with the normal pregnant controls, preeclamptic patients exhibited increased circulating levels of miR-210, miR-155, miR-942, miR-885-5p, miR-206, miR-21, miR-215, miR-650, miR-29a, miR-133b, miR-148a-3p, and members of C19MC (including miR-518b, miR-1323, miR-516b, miR-516a-5p, miR-517-5p, miR-517-3p, miR-525-5p, miR-515-5p, miR-520h, miR-520a-5p, miR-520a-3p, miR-526a and miR-526b), and declined levels of miR-146a-5p, miR-199a-5p, miR-221-3p, miR-18a and miR-19b1, miR-195, miR-376c, miR-223, miR-1229, and miR-1267 [56, 90–95]. It is suggested that the combination of a different set of miRNAs may improve the specificity and sensitivity of the predictive value. For example, a combination of seven differential miRNAs, the up-regulated miR-21, miR-155, miR-210, miR-215, miR-650, and the down-regulated miR-18a and miR-19b1, were suggested to distinguish preeclamptic patients from healthy pregnant women [92]. An addition of four miRNAs, higher expressed miR-518b and miR-29a and lower expressed miR-144 and miR-15b, would improve the specificity to distinguish severe pre-eclampsia from mild preeclampsia. An miRNA panel comprising miR-1, miR-133b, miR-148a-3p, miR-199a, miR-223, miR-1229, and miR-1267 was also evaluated for the prediction of adverse pregnancy outcomes as early as the first trimester [95].

Fetal Growth Restriction

Fetal growth restriction (FGR), defined as fetal weight below the 10th percentile for gestational age, is often complicated by other pregnant symptoms, such as preeclampsia. Some of the differential miRNAs are shared by FGR and preeclampsia pregnancy, such as miR-210, miR-21, miR-199a-5p, miR-195, miR-199, miR-516a-5p, miR-517-5p, miR-517-3p, miR-518b, miR-520a-3p, miR-520h, miR-525, and miR-526a. The expression of placental specific C19MC miRNAs in maternal circulating was associated with the risk of FGR. Members of C19MC miRNAs in first-trimester plasma were observed to be up-regulated in pregnancy, later developing pre-eclampsia and FGR, including miR-516a-5p, miR-517-5p, miR-517-3p, miR-518b, miR-520a-3p, miR-520h, miR-525, and miR-526a [96, 97]. In addition, FGR pregnancies were associated with the reduced level of miR-17-5p, miR-146a-5p, miR-221-3p, miR-574-3p, miR-16, miR-100, miR-122, miR-125b, miR-126-3p, miR-143-3p, miR-195, miR-199a, miR-221-3p, miR-342-3p, and miR-574-3p in maternal circulation [98]. Lower levels of miR-20b-5p, miR-942-5p, miR-324-3p, miR-223-5p, and miR-127-3p were reported in association with higher odds for a small-for-gestational-age (SGA) infant. Some hypoxia-regulated miRNAs such as miR-210, miR-21, miR-424, miR-199a, miR-20b and miR-373 were significantly increased in severe preterm FGR compared with gestation-matched controls [99]. Interestingly, some reports have demonstrated that the overall level of hypoxia-regulated miRNAs was elevated in the plasma of FGR patients, suggesting the essential contribution of hypoxia to the occurrence of FGR.

Gestational Diabetes Mellitus and Macrosomia Babies

Higher seral levels of miR-183-5p, miR-200b-3p, miR-125-5p, and miR-1290 were observed in gestational diabetes mellitus (GDM) cases than in normal pregnancies at the first trimester [100]. At the 16th–19th gestational week, serum miR-29a, miR-132, and miR-222 were down-regulated while miR-16-5p, miR-17-5p, miR-19a-3p, miR-19b-3p, and miR-20a-5p were up-regulated in GDM women compared with healthy pregnancies [101, 102]. At early to mid-gestation, the associations of miR-155-5p, miR-21-3p, miR-146b-5p, miR-223-3p, miR-517-5p, and miR-29a-3p with GDM were observed only in women bearing male fetuses, and miR-21-3p and miR-210-3p was associated with overweight but not lean GDM women [103].

For the GDM women bearing macrosomia babies, the circulating level of miR-143 declined while miR-21 increased [104]. miR-194 and miR-376a decreased in the early second-trimester serum of nondiabetic macrosomia [105].

Preterm Birth

Currently, limited diagnostic or predictive miRNA candidates have been reported for preterm birth. Gray et al. reported that the levels of miR-302b, miR-1253, and a clustering of miR-548 miRNA were down-regulated while that of miR-223 was increased in the plasma of women who later experienced a spontaneous preterm birth [106]. Winger et al. measured the expression of eight miRNAs (miR-148a, miR-301a, miR-671, miR-181a, miR-210, miR-1267, miR-223, and miR-340) in peripheral blood mononuclear cells at the first trimester, and discussed the potential clinical utility of these miRNAs as predictive markers for preterm birth [107]. However, many other studies failed to identify differential circulating miRNAs in preterm birth, and it has been suggested that miRNAs in maternal circulation may be unlikely as useful predictive biomarkers for preterm birth.

Recurrent Spontaneous Abortion

Evidence for circulating miRNAs as diagnostic or predictive biomarkers for RSA is limited. Elevated levels of miR-320b, miR-146b-5p, miR-221-3p, and miR-559 and declined miR-101-3p were found in RSA patients, and these miRNAs targeted multiple functional genes involved in immune, apoptosis, and angiogenic functions. More evidence was obtained regarding the genetic polymorphism of miRNA-encoding genes in RSA patients. A mutation site (+29A>G) on the encoding gene of miR-125a increased the risk for RSA. This mutation could disturb the pattern of the miR-125 targetome, which would regulate embryonic development, cell proliferation, migration, and invasion [108]. SNPs in pri-miR-10a (rs3809783) and pri-miR-125a (rs41275794 and rs12976445) in the mother and the mutations miR-146a C>G, miR-149 T>C, miR-196a2 T>C, and miR-499 A>G in the fetus have been reported as possible risk factors for RSA [59, 109].

Some Concerns in Measuring Circulating miRNAs

An increasing number of studies demonstrate the potential of circulating miRNAs to predict pregnant syndromes. However, clinical translation is a long way away. Some concerns must be taken into consideration. First, the abundance of many miRNAs in the plasma or serum is extremely low. A highly sensitive and accurate measuring method needs to be developed. Improvement of techniques is an urgent task to be undertaken for enhancing the identification of plasma miRNA biomarkers. The second concern is how to control the quality of circulating miRNAs. It has been reported that miRNAs isolated from plasma or serum often give different results. Kim et al. demonstrated that some highly expressed miRNAs in plasma decreased severely in serum [110], and the type of anticoagulant also affected the results of plasma miRNA isolation. It was also shown that the time interval between blood collection and centrifugation was significantly associated with the results of circulating miRNA measurement, indicating that the hemolysis may affect the quality of circulating miRNAs [111]. In addition, because some miRNAs are enriched in microvesicles or platelets, the speed of centrifugation will interfere with the measuring results [112]. The placenta-derived vesicles and exosomes specifically express alkaline phosphatase, making it possible to specifically separate these components for further measurement. The third concern is the high-throughput platform for circulating miRNA measurement, which should be less time-consuming, more easily handled, and less costly to meet clinical requirements.

SUMMARY AND PERSPECTIVES

In recent years, emerging evidence points to the important roles of miRNAs in pregnancy establishment and maintenance. However, the working mechanisms of these small RNAs in various cell behaviors at the fetomaternal interface remain poorly understood. Gene deficiency in a certain miRNA-encoding gene in the mouse has led to a restricted phenotype, making it even more difficult to elucidate the physiological effect of miRNAs in vivo. A possible explanation might be the redundant network among miRNAs and their target genes. The miRNAs in one cluster or family usually work synergistically or antagonistically, and gene knockout of a cluster or a family may help in understanding their roles in an animal model. In addition, aberrant expression of miRNAs in the decidua or the placenta derived from the pregnant complications suggests a link between the deregulation of miRNAs and the diseases, but the precise mechanisms through which these miRNAs contribute to the pathogenesis have not yet been elucidated. Clarifying how the expression of the essential or predominant miRNAs are regulated by various factors and signaling molecules, and identifying the gene networks regulated by those miRNAs, will be a novel step in understanding the regulatory mechanisms in normal and compromised pregnancies. Furthermore, the detection of miRNAs in maternal plasma raises the exciting possibility of using them as novel biomarkers to monitor the pregnancy outcomes. The improvement of technique to specifically and sensitively detect circulating miRNAs is an urgent task for enhancing the translation of bench research into effective clinical application.

References

[1] Baek D, Villen J, Shin C, Camargo FD, Gygi SP, Bartel DP. The impact of microRNAs on protein output. Nature 2008;455(7209):64–71.
[2] Cifuentes D, Xue H, Taylor DW, Patnode H, Mishima Y, Cheloufi S, et al. A novel miRNA processing pathway independent of Dicer requires Argonaute2 catalytic activity. Science 2010;328(5986):1694–8.
[3] Morales-Prieto DM, Chaiwangyen W, Ospina-Prieto S, Schneider U, Herrmann J, Gruhn B, et al. MicroRNA expression profiles of trophoblastic cells. Placenta 2012;33(9):725–34.
[4] Noguer-Dance M, Abu-Amero S, Al-Khtib M, Lefevre A, Coullin P, Moore GE, et al. The primate-specific microRNA gene cluster (C19MC) is imprinted in the placenta. Hum Mol Genet 2010;19(18):3566–82.
[5] Inoue K, Hirose M, Inoue H, Hatanaka Y, Honda A, Hasegawa A, et al. The rodent-specific microRNA cluster within the Sfmbt2 gene is imprinted and essential for placental development. Cell Rep 2017;19(5):949–56.
[6] Gu Y, Sun J, Groome LJ, Wang Y. Differential miRNA expression profiles between the first and third trimester human placentas. Am J Physiol Endocrinol Metab 2013;304(8):E836–43.
[7] Chim SS, Shing TK, Hung EC, Leung TY, Lau TK, Chiu RW, et al. Detection and characterization of placental microRNAs in maternal plasma. Clin Chem 2008;54(3):482–90.
[8] Salomon C, Torres MJ, Kobayashi M, Scholz-Romero K, Sobrevia L, Dobierzewska A, et al. A gestational profile of placental exosomes in maternal plasma and their effects on endothelial cell migration. PLoS One 2014;9(6): e98667.
[9] Luo SS, Ishibashi O, Ishikawa G, Ishikawa T, Katayama A, Mishima T, et al. Human villous trophoblasts express and secrete placenta-specific microRNAs into maternal circulation via exosomes. Biol Reprod 2009;81(4):717–29.
[10] Donker RB, Mouillet JF, Chu T, Hubel CA, Stolz DB, Morelli AE, et al. The expression profile of C19MC microRNAs in primary human trophoblast cells and exosomes. Mol Hum Reprod 2012;18(8):417–24.

[11] Sarker S, Scholz-Romero K, Perez A, Illanes SE, Mitchell MD, Rice GE, et al. Placenta-derived exosomes continuously increase in maternal circulation over the first trimester of pregnancy. J Transl Med 2014;12:204.

[12] Kambe S, Yoshitake H, Yuge K, Ishida Y, Ali MM, Takizawa T, et al. Human exosomal placenta-associated miR-517a-3p modulates the expression of PRKG1 mRNA in Jurkat cells. Biol Reprod 2014;91(5):129.

[13] Delorme-Axford E, Donker RB, Mouillet JF, Chu T, Bayer A, Ouyang Y, et al. Human placental trophoblasts confer viral resistance to recipient cells. Proc Natl Acad Sci U S A 2013;110(29):12048–53.

[14] Chan YC, Banerjee J, Choi SY, Sen CK. miR-210: the master hypoxamir. Microcirculation 2012;19(3):215–23.

[15] Tamaru S, Mizuno Y, Tochigi H, Kajihara T, Okazaki Y, Okagaki R, et al. MicroRNA-135b suppresses extravillous trophoblast-derived HTR-8/SVneo cell invasion by directly down regulating CXCL12 under low oxygen conditions. Biochem Biophys Res Commun 2015;461(2):421–6.

[16] Mori A, Nishi H, Sasaki T, Nagamitsu Y, Kawaguchi R, Okamoto A, et al. HLA-G expression is regulated by miR-365 in trophoblasts under hypoxic conditions. Placenta 2016;45:37–41.

[17] WEt A, Buhimschi IA, Eidem HR, Rinker DC, Rokas A, Rood K, et al. Comprehensive RNA profiling of villous trophoblast and decidua basalis in pregnancies complicated by preterm birth following intra-amniotic infection. Placenta 2016;44:23–33.

[18] Yang M, Chen Y, Chen L, Wang K, Pan T, Liu X, et al. miR-15b-AGO2 play a critical role in HTR8/SVneo invasion and in a model of angiogenesis defects related to inflammation. Placenta 2016;41:62–73.

[19] Luo R, Shao X, Xu P, Liu Y, Wang Y, Zhao Y, et al. MicroRNA-210 contributes to preeclampsia by downregulating potassium channel modulatory factor 1. Hypertension 2014;64(4):839–45.

[20] Xu P, Zhao Y, Liu M, Wang Y, Wang H, Li YX, et al. Variations of microRNAs in human placentas and plasma from preeclamptic pregnancy. Hypertension 2014;63(6):1276–84.

[21] Kopriva SE, Chiasson VL, Mitchell BM, Chatterjee P. TLR3-induced placental miR-210 down-regulates the STAT6/interleukin-4 pathway. PLoS One 2013;8(7):e67760.

[22] Garg M, Potter JA, Abrahams VM. Identification of microRNAs that regulate TLR2-mediated trophoblast apoptosis and inhibition of IL-6 mRNA. PLoS One 2013;8(10):e77249.

[23] Gysler SM, Mulla MJ, Guerra M, Brosens JJ, Salmon JE, Chamley LW, et al. Antiphospholipid antibody-induced miR-146a-3p drives trophoblast interleukin-8 secretion through activation of Toll-like receptor 8. Mol Hum Reprod 2016;22(7):465–74.

[24] Bazzoni F, Rossato M, Fabbri M, Gaudiosi D, Mirolo M, Mori L, et al. Induction and regulatory function of miR-9 in human monocytes and neutrophils exposed to proinflammatory signals. Proc Natl Acad Sci U S A 2009;106(13):5282–7.

[25] Iliopoulos D, Hirsch HA, Struhl K. An epigenetic switch involving NF-kappaB, Lin28, Let-7 MicroRNA, and IL6 links inflammation to cell transformation. Cell 2009;139(4):693–706.

[26] Scisciani C, Vossio S, Guerrieri F, Schinzari V, De Iaco R. D'Onorio de Meo P, et al., Transcriptional regulation of miR-224 upregulated in human HCCs by NFkappaB inflammatory pathways. J Hepatol 2012;56(4):855–61.

[27] Taganov KD, Boldin MP, Chang KJ, Baltimore D. NF-kappaB-dependent induction of microRNA miR-146, an inhibitor targeted to signaling proteins of innate immune responses. Proc Natl Acad Sci U S A 2006;103(33):12481–6.

[28] Kim J, Lee KS, Kim JH, Lee DK, Park M, Choi S, et al. Aspirin prevents TNF-alpha-induced endothelial cell dysfunction by regulating the NF-kappaB-dependent miR-155/eNOS pathway: role of a miR-155/eNOS axis in preeclampsia. Free Radic Biol Med 2017;104:185–98.

[29] Ibrahim SA, WEt A, Summerfield TL, Lockwood CJ, Schatz F, Kniss DA. Inflammatory gene networks in term human decidual cells define a potential signature for cytokine-mediated parturition. Am J Obstet Gynecol 2016;214(2):284.e1–284.e47.

[30] Inyawilert W, Fu TY, Lin CT, Tang PC. Let-7-mediated suppression of mucin 1 expression in the mouse uterus during embryo implantation. J Reprod Dev 2015;61(2):138–44.

[31] Chakrabarty A, Tranguch S, Daikoku T, Jensen K, Furneaux H, Dey SK. MicroRNA regulation of cyclooxygenase-2 during embryo implantation. Proc Natl Acad Sci U S A 2007;104(38):15144–9.

[32] Geng Y, He J, Ding Y, Chen X, Zhou Y, Liu S, et al. The differential expression of microRNAs between implantation sites and interimplantation sites in early pregnancy in mice and their potential functions. Reprod Sci 2014;21(10):1296–306.

[33] Su RW, Lei W, Liu JL, Zhang ZR, Jia B, Feng XH, et al. The integrative analysis of microRNA and mRNA expression in mouse uterus under delayed implantation and activation. PLoS One 2010;5(11):e15513.

[34] Altmae S, Martinez-Conejero JA, Esteban FJ, Ruiz-Alonso M, Stavreus-Evers A, Horcajadas JA, et al. MicroRNAs miR-30b, miR-30d, and miR-494 regulate human endometrial receptivity. Reprod Sci 2013;20(3):308–17.

[35] Kuokkanen S, Chen B, Ojalvo L, Benard L, Santoro N, Pollard JW. Genomic profiling of microRNAs and messenger RNAs reveals hormonal regulation in microRNA expression in human endometrium. Biol Reprod 2010;82(4):791–801.

[36] Gregory PA, Bert AG, Paterson EL, Barry SC, Tsykin A, Farshid G, et al. The miR-200 family and miR-205 regulate epithelial to mesenchymal transition by targeting ZEB1 and SIP1. Nat Cell Biol 2008;10(5):593–601.

[37] Li R, Qiao J, Wang L, Li L, Zhen X, Liu P, et al. MicroRNA array and microarray evaluation of endometrial receptivity in patients with high serum progesterone levels on the day of hCG administration. Reprod Biol Endocrinol 2011;9:29.

[38] Revel A, Achache H, Stevens J, Smith Y, Reich R. MicroRNAs are associated with human embryo implantation defects. Hum Reprod 2011;26(10):2830–40.

[39] Kaufmann P, Black S, Huppertz B. Endovascular trophoblast invasion: implications for the pathogenesis of intrauterine growth retardation and preeclampsia. Biol Reprod 2003;69(1):1–7.

[40] Graham CH, Lala PK. Mechanisms of placental invasion of the uterus and their control. Biochem Cell Biol 1992;70(10–11):867–74.

[41] Nakamura O. Children's immunology, what can we learn from animal studies (1): Decidual cells induce specific immune system of feto-maternal interface. J Toxicol Sci 2009;34(Suppl. 2):SP331–9.

[42] Knofler M. Critical growth factors and signalling pathways controlling human trophoblast invasion. Int J Dev Biol 2010;54(2–3):269–80.

[43] Red-Horse K, Zhou Y, Genbacev O, Prakobphol A, Foulk R, McMaster M, et al. Trophoblast differentiation during embryo implantation and formation of the maternal-fetal interface. J Clin Invest 2004;114(6):744–54.

[44] Lyall F. Mechanisms regulating cytotrophoblast invasion in normal pregnancy and pre-eclampsia. Aust N Z J Obstet Gynaecol 2006;46(4):266–73.

[45] LP R, Redmer DA. Angiogenesis in the placenta. Biol Reprod 2001;64(4):1033–40.

[46] Rossant J, Cross JC. Placental development: lessons from mouse mutants. Nat Rev Genet 2001;2(7):538–48.

[47] Lu J, Zhang S, Nakano H, Simmons DG, Wang S, Kong S, et al. A positive feedback loop involving Gcm1 and Fzd5 directs chorionic branching morphogenesis in the placenta. PLoS Biol 2013;11(4)e1001536.

[48] Zbella EA, Ilekis J, Scommegna A, Benveniste R. Competitive studies with dehydroepiandrosterone sulfate and 16 alpha-hydroxydehydroepiandrosterone sulfate in cultured human choriocarcinoma JEG-3 cells: effect on estrone, 17 beta-estradiol, and estriol secretion. J Clin Endocrinol Metab 1986;63(3):751–7.

[49] Zhang M, Muralimanoharan S, Wortman AC, Mendelson CR. Primate-specific miR-515 family members inhibit key genes in human trophoblast differentiation and are upregulated in preeclampsia. Proc Natl Acad Sci U S A 2016;113(45):E7069–76.

[50] Shao X, Liu Y, Liu M, Wang Y, Yan L, Wang H, et al. Testosterone represses estrogen signaling by upregulating miR-22: a mechanism for imbalanced steroid hormone production in preeclampsia. Hypertension 2017;69(4):721–30.

[51] Mohammed BT, Sontakke SD, Ioannidis J, Duncan WC, Donadeu FX. The adequate corpus luteum: miR-96 promotes luteal cell survival and progesterone production. J Clin Endocrinol Metab 2017;102:2188–98.

[52] Jiang L, Long A, Tan L, Hong M, Wu J, Cai L, et al. Elevated microRNA-520g in pre-eclampsia inhibits migration and invasion of trophoblasts. Placenta 2017;51:70–5.

[53] Anton L, Olarerin-George AO, Hogenesch JB, Elovitz MA. Placental expression of miR-517a/b and miR-517c contributes to trophoblast dysfunction and preeclampsia. PLoS One 2015;10(3):e0122707.

[54] Xie L, Sadovsky Y. The function of miR-519d in cell migration, invasion, and proliferation suggests a role in early placentation. Placenta 2016;48:34–7.

[55] Pineles BL, Romero R, Montenegro D, Tarca AL, Han YM, Kim YM, et al. Distinct subsets of microRNAs are expressed differentially in the human placentas of patients with preeclampsia. Am J Obstet Gynecol 2007;196(3):261.e1–6.

[56] Zhang Y, Fei M, Xue G, Zhou Q, Jia Y, Li L, et al. Elevated levels of hypoxia-inducible microRNA-210 in pre-eclampsia: new insights into molecular mechanisms for the disease. J Cell Mol Med 2012;16(2):249–59.

[57] Lu TM, Lu W, Zhao LJ. MicroRNA-137 affects proliferation and migration of placenta trophoblast cells in preeclampsia by targeting ERRalpha. Reprod Sci 2016;24(1):85–96.

[58] Zhou X, Li Q, Xu J, Zhang X, Zhang H, Xiang Y, et al. The aberrantly expressed miR-193b-3p contributes to preeclampsia through regulating transforming growth factor-beta signaling. Sci Rep 2016;6:19910.

[59] Hu Y, Liu CM, Qi L, He TZ, Shi-Guo L, Hao CJ, et al. Two common SNPs in pri-miR-125a alter the mature miRNA expression and associate with recurrent pregnancy loss in a Han-Chinese population. RNA Biol 2011;8(5):861–72.

[60] Ma L, Zhang XQ, Zhou DX, Cui Y, Deng LL, Yang T, et al. Feasibility of urinary microRNA profiling detection in intrahepatic cholestasis of pregnancy and its potential as a noninvasive biomarker. Sci Rep 2016;6:31535.

[61] Rao ZZ, Zhang XW, Ding YL, Yang MY. miR-148a-mediated estrogen-induced cholestasis in intrahepatic cholestasis of pregnancy: role of PXR/MRP3. PLoS One 2017;12(6):e0178702.

[62] Ji L, Brkic J, Liu M, Fu G, Peng C, Wang YL. Placental trophoblast cell differentiation: physiological regulation and pathological relevance to preeclampsia. Mol Asp Med 2013;34(5):981–1023.

[63] Gao WL, Liu M, Yang Y, Yang H, Liao Q, Bai Y, et al. The imprinted H19 gene regulates human placental trophoblast cell proliferation via encoding miR-675 that targets Nodal Modulator 1 (NOMO1). RNA Biol 2012;9(7):1002–10.

[64] Zou Y, Jiang Z, Yu X, Zhang Y, Sun M, Wang W, et al. MiR-101 regulates apoptosis of trophoblast HTR-8/SVneo cells by targeting endoplasmic reticulum (ER) protein 44 during preeclampsia. J Hum Hypertens 2014;28 (10):610–6.

[65] Lasabova Z, Vazan M, Zibolenova J, Svecova I. Overexpression of miR-21 and miR-122 in preeclamptic placentas. Neuro Endocrinol Lett 2015;36(7):695–9.

[66] Chen S, Zhao G, Miao H, Tang R, Song Y, Hu Y, et al. MicroRNA-494 inhibits the growth and angiogenesis-regulating potential of mesenchymal stem cells. FEBS Lett 2015;589(6):710–7.

[67] Li X, Song Y, Liu D, Zhao J, Xu J, Ren J, et al. MiR-495 promotes senescence of mesenchymal stem cells by targeting Bmi-1. Cell Physiol Biochem 2017;42(2):780–96.

[68] Renthal NE, Chen CC, Williams KC, Gerard RD, Prange-Kiel J, Mendelson CR. miR-200 family and targets, ZEB1 and ZEB2, modulate uterine quiescence and contractility during pregnancy and labor. Proc Natl Acad Sci U S A 2010;107(48):20828–33.

[69] Williams KC, Renthal NE, Condon JC, Gerard RD, Mendelson CR. MicroRNA-200a serves a key role in the decline of progesterone receptor function leading to term and preterm labor. Proc Natl Acad Sci U S A 2012;109(19):7529–34.

[70] Williams KC, Renthal NE, Gerard RD, Mendelson CR. The microRNA (miR)-199a/214 cluster mediates opposing effects of progesterone and estrogen on uterine contractility during pregnancy and labor. Mol Endocrinol 2012;26(11):1857–67.

[71] Gao L, Wang G, Liu WN, Kinser H, Franco HL, Mendelson CR. Reciprocal feedback between miR-181a and E2/ ERalpha in myometrium enhances inflammation leading to labor. J Clin Endocrinol Metab 2016;101 (10):3646–56.

[72] Yang W, Wang A, Zhao C, Li Q, Pan Z, Han X, et al. miR-125b enhances IL-8 production in early-onset severe preeclampsia by targeting sphingosine-1-phosphate lyase 1. PLoS One 2016;11(12)e0166940.

[73] Hu E, Ding L, Miao H, Liu F, Liu D, Dou H, et al. MiR-30a attenuates immunosuppressive functions of IL-1beta-elicited mesenchymal stem cells via targeting TAB3. FEBS Lett 2015;589(24 Pt B):3899–907.

[74] Yang X, Zhang J, Ding Y. Association of microRNA-155, interleukin 17A, and proteinuria in preeclampsia. Medicine (Baltimore) 2017;96(18):e6509.

[75] Li D, Li J. Association of miR-34a-3p/5p, miR-141-3p/5p, and miR-24 in decidual natural killer cells with unexplained recurrent spontaneous abortion. Med Sci Monit 2016;22:922–9.

[76] Ishida Y, Zhao D, Ohkuchi A, Kuwata T, Yoshitake H, Yuge K, et al. Maternal peripheral blood natural killer cells incorporate placenta-associated microRNAs during pregnancy. Int J Mol Med 2015;35(6): 1511–24.

[77] Zheng J, Xiao X, Zhang Q, Wang T, Yu M, Xu J. Maternal low-protein diet modulates glucose metabolism and hepatic microRNAs expression in the early life of offspring dagger. Nutrients 2017;9(3).

[78] Alejandro EU, Gregg B, Wallen T, Kumusoglu D, Meister D, Chen A, et al. Maternal diet-induced microRNAs and mTOR underlie beta cell dysfunction in offspring. J Clin Invest 2014;124(10):4395–410.

[79] Jacovetti C, Abderrahmani A, Parnaud G, Jonas JC, Peyot ML, Cornu M, et al. MicroRNAs contribute to compensatory beta cell expansion during pregnancy and obesity. J Clin Invest 2012;122(10):3541–51.

[80] Benatti RO, Melo AM, Borges FO, Ignacio-Souza LM, Simino LA, Milanski M, et al. Maternal high-fat diet consumption modulates hepatic lipid metabolism and microRNA-122 (miR-122) and microRNA-370 (miR-370) expression in offspring. Br J Nutr 2014;111(12):2112–22.

[81] Yang YM, Seo SY, Kim TH, Kim SG. Decrease of microRNA-122 causes hepatic insulin resistance by inducing protein tyrosine phosphatase 1B, which is reversed by licorice flavonoid. Hepatology 2012;56 (6):2209–20.

[82] Iliopoulos D, Drosatos K, Hiyama Y, Goldberg IJ, Zannis VI. MicroRNA-370 controls the expression of microRNA-122 and Cpt1alpha and affects lipid metabolism. J Lipid Res 2010;51(6):1513–23.

[83] Zheng J, Zhang Q, Mul JD, Yu M, Xu J, Qi C, et al. Maternal high-calorie diet is associated with altered hepatic microRNA expression and impaired metabolic health in offspring at weaning age. Endocrine 2016;54(1):70–80.

[84] Gilad S, Meiri E, Yogev Y, Benjamin S, Lebanony D, Yerushalmi N, et al. Serum microRNAs are promising novel biomarkers. PLoS One 2008;3(9)e3148.

[85] Miura K, Miura S, Yamasaki K, Higashijima A, Kinoshita A, Yoshiura K, et al. Identification of pregnancy-associated microRNAs in maternal plasma. Clin Chem 2010;56(11):1767–71.

[86] Morales-Prieto DM, Schleussner E, UR M. Reduction in miR-141 is induced by leukemia inhibitory factor and inhibits proliferation in choriocarcinoma cell line JEG-3. Am J Reprod Immunol 2011;66(Suppl. 1):57–62.

[87] Chen X, Ba Y, Ma L, Cai X, Yin Y, Wang K, et al. Characterization of microRNAs in serum: a novel class of biomarkers for diagnosis of cancer and other diseases. Cell Res 2008;18(10):997–1006.

[88] Mitchell PS, Parkin RK, Kroh EM, Fritz BR, Wyman SK, Pogosova-Agadjanyan EL, et al. Circulating microRNAs as stable blood-based markers for cancer detection. Proc Natl Acad Sci U S A 2008;105(30):10513–8.

[89] Cartwright JE, Fraser R, Leslie K, Wallace AE, James JL. Remodelling at the maternal-fetal interface: relevance to human pregnancy disorders. Reproduction 2010;140(6):803–13.

[90] Akehurst C, Small HY, Sharafetdinova L, Forrest R, Beattie W, Brown CE, et al. Differential expression of microRNA-206 and its target genes in preeclampsia. J Hypertens 2015;33(10):2068–74.

[91] Hromadnikova I, Kotlabova K, Hympanova L, Krofta L. Gestational hypertension, preeclampsia and intrauterine growth restriction induce dysregulation of cardiovascular and cerebrovascular disease associated microRNAs in maternal whole peripheral blood. Thromb Res 2016;137:126–40.

[92] Jairajpuri DS, Malalla ZH, Mahmood N, Almawi WY. Circulating microRNA expression as predictor of preeclampsia and its severity. Gene 2017;627:543–8.

[93] Miura K, Higashijima A, Murakami Y, Tsukamoto O, Hasegawa Y, Abe S, et al. Circulating chromosome 19 miRNA cluster microRNAs in pregnant women with severe pre-eclampsia. J Obstet Gynaecol Res 2015;41(10):1526–32.

[94] Sandrim VC, Luizon MR, Palei AC, Tanus-Santos JE, Cavalli RC. Circulating microRNA expression profiles in pre-eclampsia: evidence of increased miR-885-5p levels. BJOG 2016;123(13):2120–8.

[95] Winger EE, Reed JL, Ji X. First-trimester maternal cell microRNA is a superior pregnancy marker to immunological testing for predicting adverse pregnancy outcome. J Reprod Immunol 2015;110:22–35.

[96] Hromadnikova I, Kotlabova K, Ivankova K, Krofta L. First trimester screening of circulating C19MC microRNAs and the evaluation of their potential to predict the onset of preeclampsia and IUGR. PLoS One 2017;12(2)e0171756.

[97] Hromadnikova I, Kotlabova K, Ondrackova M, Kestlerova A, Novotna V, Hympanova L, et al. Circulating C19MC microRNAs in preeclampsia, gestational hypertension, and fetal growth restriction. Mediat Inflamm 2013;2013:186041.

[98] Hromadnikova I, Kotlabova K, Hympanova L, Krofta L. Cardiovascular and cerebrovascular disease associated microRNAs are dysregulated in placental tissues affected with gestational hypertension, preeclampsia and in trauterine growth restriction. PLoS One 2015;10(9)e0138383.

[99] Whitehead CL, Teh WT, Walker SP, Leung C, Larmour L, Tong S. Circulating MicroRNAs in maternal blood as potential biomarkers for fetal hypoxia in-utero. PLoS One 2013;8(11)e78487.

[100] Lamadrid-Romero M, Solis KH, Cruz-Resendiz MS, Perez JE, Diaz NF, Flores-Herrera H, et al. Central nervous system development-related microRNAs levels increase in the serum of gestational diabetic women during the first trimester of pregnancy. Neurosci Res 2018;130:8–22.

[101] Cao YL, Jia YJ, Xing BH, Shi DD, Dong XJ. Plasma microRNA-16-5p, -17-5p and -20a-5p: novel diagnostic biomarkers for gestational diabetes mellitus. J Obstet Gynaecol Res 2017;43(6):974–81.

[102] Zhu Y, Tian F, Li H, Zhou Y, Lu J, Ge Q. Profiling maternal plasma microRNA expression in early pregnancy to predict gestational diabetes mellitus. Int J Gynaecol Obstet 2015;130(1):49–53.

[103] Wander PL, Boyko EJ, Hevner K, Parikh VJ, Tadesse MG, Sorensen TK, et al. Circulating early- and mid-pregnancy microRNAs and risk of gestational diabetes. Diabetes Res Clin Pract 2017;132:1–9.

[104] Zhang JT, Cai QY, Ji SS, Zhang HX, Wang YH, Yan HT, et al. Decreased miR-143 and increased miR-21 placental expression levels are associated with macrosomia. Mol Med Rep 2016;13(4):3273–80.

[105] Hu L, Han J, Zheng F, Ma H, Chen J, Jiang Y, et al. Early second-trimester serum microRNAs as potential biomarker for nondiabetic macrosomia. Biomed Res Int 2014;2014:394125.

[106] Gray C, McCowan LM, Patel R, Taylor RS, Vickers MH. Maternal plasma miRNAs as biomarkers during mid-pregnancy to predict later spontaneous preterm birth: a pilot study. Sci Rep 2017;7(1):815.

[107] Winger EE, Reed JL, Ji X. Early first trimester peripheral blood cell microRNA predicts risk of preterm delivery in pregnant women: proof of concept. PLoS One 2017;12(7):e0180124.

[108] Hu Y, Huo ZH, Liu CM, Liu SG, Zhang N, Yin KL, et al. Functional study of one nucleotide mutation in pri-miR-125a coding region which related to recurrent pregnancy loss. PLoS One 2014;9(12):e114781.

[109] Li Y, Wang XQ, Zhang L, Lv XD, Su X, Tian S, et al. A SNP in pri-miR-10a is associated with recurrent spontaneous abortion in a Han-Chinese population. Oncotarget 2016;7(7):8208–22.

[110] Kim DJ, Linnstaedt S, Palma J, Park JC, Ntrivalas E, Kwak-Kim JY, et al. Plasma components affect accuracy of circulating cancer-related microRNA quantitation. J Mol Diagn 2012;14(1):71–80.

[111] Page K, Guttery DS, Zahra N, Primrose L, Elshaw SR, Pringle JH, et al. Influence of plasma processing on recovery and analysis of circulating nucleic acids. PLoS One 2013;8(10):e77963.

[112] Cheng HH, Yi HS, Kim Y, Kroh EM, Chien JW, Eaton KD, et al. Plasma processing conditions substantially influence circulating microRNA biomarker levels. PLoS One 2013;8(6):e64795.

Glossary

Trophoblast cells   Cells from the outer layer of a blastocyst (trophectoderm), which give rise to the major component of the placenta.

Cytotrophoblast cells   Mononucleated trophoblast cells located at the inner layer of placental villi.

Syncytiotrophoblasts   Multinucleated trophoblast cells located at the outer layer of placental villi.

Genetics and Genomics of Preterm Birth

Lubna Nadeem*, Stephen J. Lye*,†,‡, Oksana Shynlova*,‡

*Lunenfeld-Tanenbaum Research Institute at Sinai Health System, Toronto, ON, Canada
†Department of Physiology, University of Toronto, Toronto, ON, Canada ‡Department of
Obstetrics and Gynecology, University of Toronto, Toronto, ON, Canada

OUTLINE

Abbreviations

20α HSD	20 alpha hydroxysteroid dehydrogenase
20α-OHP	20α-dihydroprogesterone
ADRB2	β2-adrenergic receptor
AP-1	activator protein-1
AR	androgen receptor

Human Reproductive and Prenatal Genetics
https://doi.org/10.1016/B978-0-12-813570-9.00015-2

BMI	body mass index
BV	bacterial vaginosis
C19MC	chromosome 19 microRNA cluster
CAPs	contraction-associated proteins
CCL2	C-C motif chemokine ligand 2
CDX2	caudal type homeobox 2
COX2	cyclooxygenase 2
CpG	cytosine-phosphate-Guanine
CRH	corticotropin releasing hormone
CRHR1	corticotropin-releasing-hormone receptor 1
Cx43	connexin 43
CXCL	C-X-C motif chemokine ligand
CYP2E1	cytochrome P450, family 2, subfamily E, poly-peptide 1
dbGaP	database of genotypes and phenotypes
DHCR7	7-dehyrocholesterol reductase
DHFR	dihydrofolate reductase
DNBC	Danish National Birth Cohort
DRD2	dopamine receptor D2
ECM	extracellular matrix
ENPP1	ectonucleotide pyrophosphatase
EP1	prostaglandin E_2 receptor 1
EVs	extracellular vesicles
FokI	flavobacterium okeanokoites
FP	prostaglandin F receptor
GENEVA	Gene Environment Association Studies Initiative
GPN	genomic and proteomic network
GWAS	genome wide association
HDL	high-density lipoproteins
HPC	hydroxyprogesterone caproate
IGF1R	insulin-like growth factor 1 receptor
IGFBP3	insulin-like growth factor binding protein 3
IL	interleukins
IL2RG	IL-2 receptor γ subunit
IVF	in vitro fertilization
LDL	low-density lipoproteins
MELAS	mitochondrial encephalomyopathy, lactic acidosis, and stroke-like episodes
miRNA	micro RNA
MMPs	matrix metalloproteinases
MnSOD	manganese superoxide dismutase
MoBa	The Norwegian Mother and Child Cohort Study
mtDNA	mitochondrial genome
MT-TL1	mitochondrially encoded tRNA leucine 1
NCHS	National Center for Health Statistics
NF-κB	nuclear factor kappa-light-chain-enhancer of activated B cells
OHD	hydroxyvitamin D
OS	oxidative stress
OTR	oxytocin receptor
P4	progesterone
PGE2	prostaglandin E2
PGF2α	prostaglandin F2α
PGI2	prostacyclin
PGs	prostaglandins
pPROM	preterm premature rupture of membranes
PREs	progesterone response elements
PRKCA	protein kinase C-α

PROC	protein C
PROCR	protein C receptor
PRs	progesterone receptors
PTB	preterm birth
PTGER	prostaglandin E receptor
PTGES	prostaglandin E synthase
PTGFR	prostaglandin F receptor
PTH-rP	parathyroid hormone-related peptide
PTL	preterm labor
PUFA	polyunsaturated fatty acids
RNASET2	ribonuclease T2
ROS	reactive oxygen species
RVA	rare variant analysis
SNP	single nucleotide polymorphism
SPs	surfactant proteins
STAT5b	signal transducer and activator of transcription 5B
TFPI	tissue factor pathway inhibitor
TFs	transcription factors
THBD	thrombomodulin
TL	term labor
TNFα	tumor necrosis factor
TRAF2	TNF receptor-associated factor 2
VDR	vitamin D receptor
VEGF	vascular endothelial growth factor
WHO	World Health Organization

THE PROBLEM OF PRETERM BIRTH

Preterm birth (PTB, birth at <37 completed weeks of gestation) is a significant worldwide problem of clinical obstetrics. It is the single largest cause of perinatal mortality and morbidity in industrialized countries. Statistics suggest that 60%–80% of infant deaths without congenital anomalies are related to PTB, a total of 1 million infant deaths annually [1]. Perhaps even more disturbing is the observation that 47% of surviving very preterm babies (weight at birth 500–1000 g) have significant health impairments. They are at increased risk of neurosensory and respiratory illnesses that demand neonatal intensive care [2]. PTB is responsible for the majority of newborn morbidity (intraventricular hemorrhage, necrotizing enterocolitis, respiratory distress syndrome, sepsis, bronchopulmonary dysplasia, retinopathy of prematurity, periventricular leukomalacia, and patent ductus arteriosus), and is associated with cerebral palsy and other long-term health sequelae, including cognitive impairment, blindness, deafness, and respiratory illness [3]. In particular, Lipper et al. found that 72% of children who weighed <800 g at birth and reached the age of 7 years required rehabilitation or early intervention services due to neurological or developmental abnormalities [4]. While most studies have focused on very preterm infants born <32 weeks gestation, infants born between 32 and 36 weeks gestation are also at higher relative risk for death during infancy than babies born at term gestations [5]. Infants born with such complications are further challenged with the educational disadvantages and adverse health outcomes during their school age and adulthood [6, 7]. In addition to the infants, mothers of the preterm infants are also at a greater risk for mental illness, cardiovascular disease, type II diabetes, and breast cancer later in their lives.

Besides the medical and emotional burden, PTB poses enormous economic liability upon the families, communities, and the health care system [8]. The average hospital charges for a premature infant in a Canadian neonatal intensive care unit is estimated at $1500 per day while total neonatal care for preterm infants approaches $4 billion per year in the United States. These figures do not consider the long-term costs in caring for individuals born with PTB-associated disability, nor do they consider the long-term impact of PTB on health in later life.

Epidemiology of PTB

Globally, PTB presents one of the greatest challenges in perinatal care. Over the past two decades, the rate of PTB has been escalating steadily and alarmingly. The World Health Organization recently estimated that 14.9 million babies worldwide were born preterm in 2010, with a PTB rate ranging between 5% and 18%, depending upon the country (incidence rate of 9.3%.) Approximately 85% of this burden is confined to Africa and South Asia where >10 million babies are born preterm, and where infant and child mortality is highest. About 0.5 million PTBs occur in Europe and Northern America, and 0.8 million babies are born preterm in Latin America and the Caribbean. In the United States, the PTB rate is on a rise, which results in about 0.4 million babies being born prematurely every year [9]. Similar increases in the incidence of PTB have been reported in Canada and Australia [2]. With the exception of France and Finland, no developed country has reported a decrease in the rate of PTB. In the neonatal period, preterm infants are 40 times more likely to die than term infants. Almost all (98%) newborn deaths occur in the poorest countries of the world. In developed countries, the neonatal mortality rate is 5 per 1000 live births or lower, compared to rates hovering around 42 per 1000 live births in the least developed nations [10].

Clinical Presentation of PTB

PTB can be classified mainly into three groups: (1) medically indicated (iatrogenic) PTB, (2) spontaneous (idiopathic) PTB with intact membranes, and (3) spontaneous PTB with preterm premature rupture of the membranes (pPROM). In particular, about 30%–35% of the PTBs are iatrogenic, 40%–45% are idiopathic without pPROM, and 25%–30% are idiopathic with pPROM [11]. Universally, spontaneous PTB is further sub-classified into three groups based on the gestational age at delivery: late preterm (32–36 weeks), moderate preterm (28–31 weeks), and very preterm (<28 weeks). More than one risk factor is generally associated with all three groups of PTB, suggesting etiologic heterogeneity of this syndrome.

Risk Factors and Etiology for PTB

Many association studies have been conducted to identify the possible risk factors for PTB (both exogenous and endogenous), which include maternal, paternal, fetal, environmental, social, genetic, and hormonal factors (Fig. 15.1). Identification of risk factors aids in categorizing women into a priority group of "threatened preterm labor (PTL)" in order to provide them with risk-specific treatment in a timely manner. In addition, risk factor identification

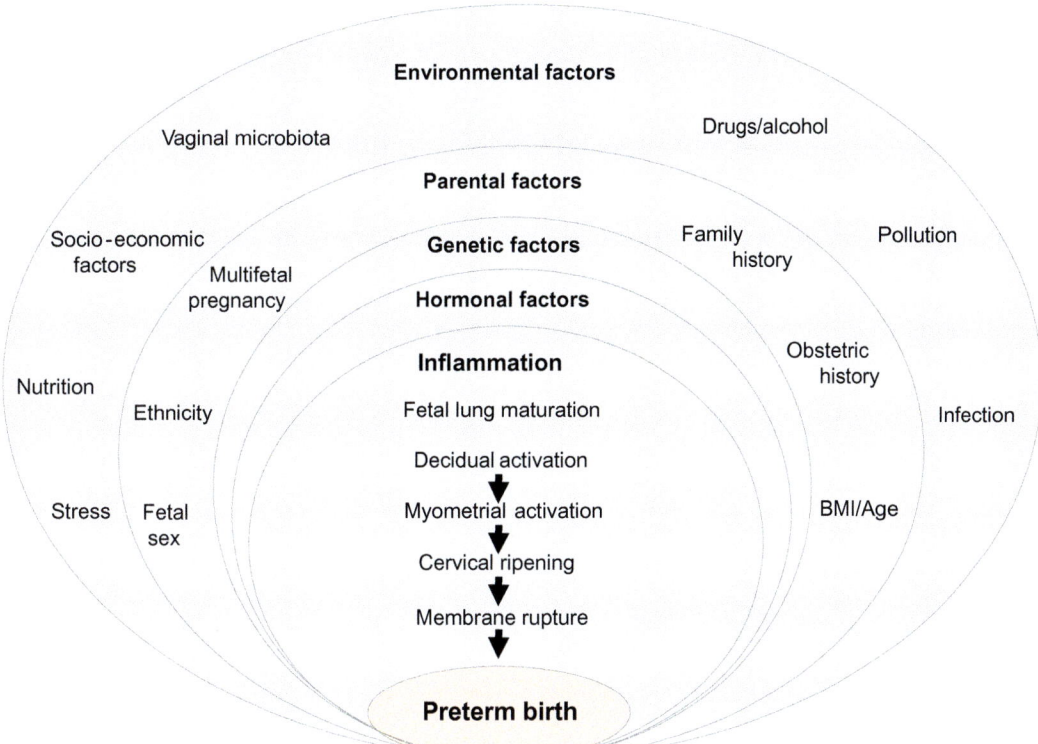

FIG. 15.1 Currently known risk factors for preterm birth (both exogenous and endogenous).

provides insight about the possible molecular mechanisms that may be involved in PTB induction.

Maternal Factors

a) Family History

Family studies suggest that genetic factors may contribute to about 40% of PTB. Predisposition to PTB occurs in a matrilineal fashion across generations. Porter et al. reported that the risk of PTB was significantly higher in mothers born preterm than in those who were not [12]. Also, the risk of PTB is higher if the subject's mother, full sister, or maternal half-sister has had a PTB. However, no risk is imposed if family from the paternal side or the paternal half-sister has had cases of PTB suggestive of epigenetic imprinting.

b) Previous Obstetric History/Medical Procedures

Previous spontaneous and missed miscarriages have been associated with pPROM and PTB [13]. Interpregnancy intervals are also considered important with respect to the PTB rate: an interval of less than 12 months is considered to be a risk factor for the subsequent pregnancy [14]. An amniocentesis performed in the second trimester of pregnancy, induced abortion, or termination before 20 weeks is associated with pPROM and PTB (14–24 weeks) in the subsequent pregnancy [15].

c) Ethnicity/Racial Predisposition

In 2016, the National Center for Health Statistics (NCHS) estimated that African-American women have a two- to sixfold higher PTB risk compared to Caucasians [16]. Moreover, in a group of women with prior spontaneous PTB, the incidence of recurrent PTB is higher in black women compared to white women (8.2% vs. 13.4% for very preterm and moderate preterm delivery) [17]. In addition, a US study of computerized stress testing showed a twofold increase in PTB risk for black military women as compared to white women due to their higher cardiac reactivity [18]. The prevalence of bacterial vaginosis (BV, a PTB-associated factor), is also higher in black women compared to white [19]. Moreover, African-American women are reported to have a diverse vaginal microbiome that lacks the dominance of the Lactobacillus species. Therfore, they are more prone to intrauterine infections during pregnancy compared to the white population [20]. In Canada, aboriginal women have higher rates of PTB as compared to nonaboriginal women due to factors such as low weight gain during pregnancy, high stress levels, and inadequate prenatal care [21].

d) BMI and Maternal Age

Maternal obesity and malnutrition are both associated with adverse pregnancy outcomes. However, the predictive value of BMI, maternal weight gain during pregnancy, and maternal height are considered to be poor predictors of PTB due to the heterogeneity of results among various studies. Prepregnancy BMI and weight gain during pregnancy evaluated for the late PTB (32–36 weeks) and very preterm/moderate PTB (20–31 weeks) showed a strong association between very low weight gain and very preterm PTB [22]. Physical activity is another important factor that has been evaluated with respect to the occurrence of PTB. In particular, a study concluded that leisure time exercise in the first and second trimester of gestation proves beneficial and presents a protective effect against PTB while intense physical load (climbing stairs more than 10 times a day) increased the risk of PTB [23]. Association between PTB and young maternal age (>18 years and between 18 and 19 years) was assessed with the optimal reference age (between 25 and 29 years) in Brazil where the prevalence of teenage pregnancy is high (29%). The risk of PTB was found to be higher in women younger than 18 but not for women between 18 and 19 [24] or 20 and 49 [17].

Paternal Factors

Risk of PTB increases when the paternal age is greater than 45 years. A comparison between women aged 20–29 years with older partners versus younger ones showed a strong association of early PTB with advanced age of the father [25]. Similarly, increased association was identified between high paternal/maternal age difference and PTB when the comparison was made between older women with young partners versus average age/young couples. However, this association was found only in white women [26]. Importantly, regardless of maternal race, paternal black race was also associated with increased odds of PTB [27]. These studies suggest a paternal contribution to fetal/placental genotype that ultimately influences the risk of a preterm delivery.

Fetal Factors/Multifetal Pregnancies

Fetal sex is an important factor that has been associated with the incidence of PTB. Most studies that include fetal sex in the analysis (IVF studies) showed a strong association of male fetuses with PTB [28]. Multifetal pregnancy alone accounts for 10% of all PTB. Within this group, twins resulting from assisted reproductive technologies have shown to be at a higher risk of PTB compared to naturally conceived twins [29].

Environmental Factors

a) Infections and Vaginal Microbiome

Intrauterine infection during pregnancy most commonly occurs by bacterial ascension through the vagina, which is host to a vast microbiome. Vaginal Lactobacilli populations contribute to gynecological health by modulating the immune response to decrease pathogenic adherence, producing bacteriocins that lower pathogen growth and secreting acidic compounds such as lactic acid and H_2O_2 to lower the pH of the vaginal fluids, which is intolerant to pathogens. With pregnancy, the vaginal flora becomes much more stable with a Lactobacilli dominance, which is shown to have a protective role against amniotic infections [30]. The absence of Lactobacillus species in the vaginal microbiota during pregnancy has been linked to intrauterine infection (chorioamnionitis) and PTB [20]. Importantly, studies have demonstrated that the third trimester vaginal microbiome in women who delivered preterm (idiopathic or infectious PTB) was different from women who delivered at full term [31]. Detection of bacterial vaginosis (particularly early in gestation) and changes in vaginal flora after antibiotic treatments of Escherichia coli or *Klebsiella pneumoniae* were associated with PTB. Clinical trials of probiotic Lactobacilli treatment, however, have shown a small lowering of the sPTB rate with a weak dose-dependent effect [32]. Importantly, oral pathogens such as a chronic periodontal infection can also lead to PTB [33]. Periodontitis may act as a distant reservoir of microbes and inflammatory mediators and contribute to the induction of PTB. However, several intervention studies on the incidence of PTB reported contradictory results with no clear beneficial effect of periodontal treatment [34].

b) Socioeconomic Factors/Stress

Importantly, increased psychological stress is associated with higher prevalence of BV and increased incidence of PTB [35]. Physiological changes induced by stress include the production of glucocorticoids and uterotonic prostaglandins (PGs), which may trigger the initiation of preterm labor (PTL) [36]. In particular, it was reported that a hypothalamic neuropeptide corticotropin releasing hormone (CRH) is a major mediator of stress-induced PTB [37]. CRH can induce PTL by stimulating the synthesis and production of uterotonic PGF2α and PGE2, which leads to the activation of the uterine muscle (myometrium) and the initiation of labor contractions. Importantly, maternal plasma CRH levels rise exponentially with advancing gestation and reach peak during term labor (TL) [38]. Therefore, stress causes a premature increase in the CRH plasma level comparable to that during TL, which can stimulate PTL [39].

c) Drugs/Substance Usage

The association of heavy maternal alcohol consumption during pregnancy and various adverse birth outcomes has been well established [40]. A recent systematic review and meta-analysis indicates a nonlinear association between maternal alcohol consumption and the risks of low birth weight, PTB, and babies that are small for gestational age. It was reported that for mothers who consumed more than three alcoholic drinks per day, the risk of having a preterm born child increased by 23% [41]. Marijuana is the illicit drug most commonly used during pregnancy. The self-reported prevalence of marijuana use during pregnancy ranges from 2% to 5%. A retrospective cohort study and a recent meta-analysis both found that the risk of PTB among only marijuana users was not associated with an increased risk of PTB, but concomitant marijuana and tobacco use doubled the likelihood of PTB in pregnant users as compared to pregnant women not using either substance [42]. Thus, concurrent tobacco use may be an important mediator for some adverse pregnancy outcomes among marijuana users. Research indicates that 11% of women in the United States smoke during pregnancy, which is thought to be a cause of 5% of all PTBs. In addition, new studies show that antidepressants, once thought to be safe to take during pregnancy, may cause PTB and the risk increases when antidepressants are used during the third trimester. For instance, sumatripan (commonly used in Denmark for the treatment of headaches and migraines) was found associated with an increased risk of PTB [43].

d) Nutrition

Consumption or deficiency of specific nutrients during pregnancy may influence gestation length and the risk of PTB. For instance, it was reported that the preconception intake of multivitamins reduces the risk for both early and late PTB [44] while suboptimal levels of vitamins B12 and B6 were associated with PTB [45]. Moreover, in the women who delivered preterm, decreased serum selenium concentrations were reported during early gestation compared to controls [46]. Some food items have been reported to be associated with a reduced risk of PTB, specifically milk and yoghurt products enriched with probiotic *Lactobacillus* [47]. Also, two large cohort studies found that Mediterranean-type dietary patterns, characterized by a high intake of fruit, vegetables, olive oil, and fish, were associated with a reduced risk of preterm delivery [48]. While the role of polyunsaturated fatty acids (PUFA, precursors of Omega-3 and Omega-6) in modulating the level of PGs has been described in the literature, there are contradicting results on this matter. One report proposed the use of Omega-3 as a secondary prophylaxis for PTB due to its pregnancy-delaying effect (4–7 days) [49] while another report found no association between Omega-3-rich seafood intake with the length of gestation or risk of PTB [50].

A low maternal level of Vitamin D (25 hydroxyvitamin D/25-OHD) is considered a risk factor for PTB [51] as 25-OHD can reduce oxidative stress in the placenta and has a protective role against inflammation (reviewed in Ref. [52]). In pregnant women, low blood [25-OHD] <50 nmol/L is considered a "deficiency" while <75 nmol/L is an "insufficiency" [52]. Two meta-analyses of observational studies and randomized controlled trials concluded that maternal blood 25-OHD deficiency (<50 vs. >50 nmol/L) is associated with an increased risk of PTB while insufficiency is not a threat [52]. A recent clinical trial of vitamin D supplementation showed a significant reduction in the PTB risk when [25-OHD] >100 nmol/L in maternal blood; however, similar studies failed to reproduce these results.

e) Pollution

The chemical composition of air pollution and tobacco smoke are quite similar and while tobacco use has established associations with PTB (as explained elsewhere in this chapter), air pollution becomes a likely risk factor for PTB. Air pollution has a role in modulating the epigenetic processes such as DNA methylation and histone modifications and is associated with inflammation. DNA methylation of CpG sites has been explored in various tissues from women who experience PTB (compared to the term controls), including the placenta, myometrium, and cervix; however, a direct link between DNA methylation and PTB has not been established. Importantly, a significant association between increased DNA methylation and infection has been unraveled, suggesting an indirect association of air pollution with PTB [53].

While there are no peer-reviewed reports of an association between domestic water fluoridation and an increased risk of PTB, some reports indicate that elevated fluoride levels in pregnant women may be linked to iron-deficiency anemia, which potentially might cause preterm delivery and low-birth weight infants, disruption of gut flora, thyroid disorders, preeclampsia, and placental and vascular calcification [54].

Hormonal Factors

Endocrine disorders such as diabetes and hypothyroidism have been associated with pPROM and PTB [55, 56]. The role of the sex steroid hormone progesterone is discussed in great detail in Section "Regulation of Parturition" of this chapter.

Genetic Factors

An important role for genetic factors in the etiology of PTB has been documented for a long time. This is described in depth later in this chapter (see Section "Genetics of PTB").

REGULATION OF PARTURITION

Although the initial triggers may differ, the physiology of human term and preterm parturition share the common pathways of cervical ripening, myometrial/decidual activation, and fetal membrane rupture [57]. The timing of labor onset is determined by the hormonal cross-talk between maternal and fetal systems. Animal studies suggest that the fetal genome ultimately regulates the timing of parturition. Two distinct yet integrated pathways–an endocrine cascade comprising the fetal hypothalamic-pituitary-adrenal-placental axis and a mechanical pathway in which fetal growth imposes tension on the uterine wall–both play a key role in labor initiation by inducing molecular changes within the uterine muscle (myometrium). These changes include the expression of a cassette of contraction-associated proteins (CAPs), that is, the gap junction protein connexin43 (Cx43/*GJA1*); receptors for oxytocin and prostaglandins (*OTR*, *EP1-4* and *FP*); and Na$^+$ and Ca^{2+} ion channels that control the excitability of the myometrium [58]; as well as extracellular matrix (ECM) proteins [59], cell-matrix adhesion complexes [60], *COX2*, and proinflammatory genes (*NF-κB2, CCL2, IL-1, CXCL8*, etc.).

Myometrial contractions mark the beginning of parturition whether it is term or preterm. During the course of gestation, the myometrium undergoes a remarkable transformation with labor representing the terminal step in this process of smooth muscle differentiation. In the first trimester of pregnancy, myometrial cells go through an initial phase of proliferation.

During the second/third trimester, a phase of growth occurs involving cellular hypertrophy and interstitial matrix elaboration. Near term is a contractile phase when the elevated expression of CAPs increases the contractile potential while a higher expression of cell-adhesion complexes induces adherence to uterine ECM. The final phase is achieved during labor, when highly contractile smooth muscle cells are connected and synchronized with each other to form a myometrial syncytium. Thus, activated myometrium is transformed into a strong contractile muscle capable of responding optimally to the increased production of uterine agonists (oxytocin and stimulatory PGs), and generating rhythmic forceful labor contractions to expel the fetus from the uterus and deliver the placenta [61]. The final contractile/labor phases of myometrial transformation are associated with the up-regulation of the expression of proinflammatory cytokines and cellular adhesion molecules and an infiltration of leukocytes into the uterine muscle, inducing a localized myometrial inflammation. This physiological inflammation is a critical element in the initiation of labor.

Increasing evidence suggests that the triggers of inflammation and the signals for the termination of gestation can come from the fetus [62] as it reaches maturity. Numerous research results support this notion: (1) placenta produces CRH, and its secretion increases exponentially with advancing gestation, reaching its peak concentration before term parturition [41]. Placental CRH from maternal plasma stimulates the fetal pituitary gland to increase the production of corticotropin, which induces the synthesis and release of cortisol from the fetal adrenal gland [63]. (2) Fetal cortisol augments the maturation of fetal lungs, which secrete surfactant proteins (SPs: SP-A, SP-D) [64] into the amniotic fluid via fetal breathing movements. The synthesis of SP-A by the fetal lung is initiated during late gestation and reaches its peak near parturition [65]. SP-A stimulates the synthesis of PGE in the human amnion [66] and inflammation that leads to the onset of labor. (3) Fetal cortisol also induces the expression of COX2 in the placenta, resulting in the increased production of PGs [67], which boosts myometrial contractility. In addition, it was suggested that CRH can decrease progesterone synthesis, therefore inducing functional progesterone withdrawal and labor onset.

Role of Progesterone in Pregnancy and Labor

During pregnancy, factors such as ovarian hormone progesterone (P4), relaxin, prostacyclin (PGI2), parathyroid hormone-related peptide (PTH-rP), nitric oxide, and CRH [68] keep the myometrium in a state of quiescence by suppressing the proinflammatory and procontractile genes. Among these factors, P4 alone is proven strong enough to suppress all labor-associated changes because the treatment of pregnant rodents with P4 at term delays myometrial contractions and parturition [69]. In all viviparous species examined to date, P4 is essential for the establishment and maintenance of pregnancy. Studies suggested that disruption of P4 signaling by the specific antagonist RU486 at any stage of pregnancy can induce myometrial contractions and PTL in mice [70], rats [59], and women [71]. In most mammals, there is a systemic P4 withdrawal at term resulting in myometrial activation, which triggers labor onset. Paradoxically, in humans, circulating P4 levels remain elevated at term and yet labor occurs normally, which points toward a mechanism of local P4 withdrawal. We have recently discovered that in term human myometrium, the expression and activity of a P4-metabolizing enzyme 20alpha-hydroxysteroid dehydrogenase (20α-HSD) is increased, thereby converting P4 into an inactive metabolite [44]. Despite high circulating hormone levels, the local decrease in P4 concentration in the human myometrium results in "functional P4 withdrawal," up-regulation of CAP genes, and the onset of labor [72]. P4 is also known to be a regulator of the maternal immune system during pregnancy. In a mouse model of infection-induced PTL, pretreatment with medroxyprogesterone acetate down-regulated the expression of Cx43, Ptgs2, Tnfα, and Il1b (induced by infection) and delayed preterm delivery [73]. In humans, the local P4 application of 17alpha hydroxyprogesterone caproate (HPC) was proposed as a treatment for women at risk for PTB. In 2003 in a large double-blind trial, HPC emerged as a promising treatment that significantly reduced the risk of PTB (reviewed in Ref. [74]), but in 2017, a subsequent prospective analysis revealed no benefit of this drug for the prevention of recurrent PTB [74]. In 2016, a double-blind, randomized, placebo-controlled trial of vaginal P4 prophylaxis in high-risk women (because of previous spontaneous PTB) was not associated with a reduced risk of PTB or composite neonatal adverse outcomes [75]. However, a 2017 meta-analysis showed that the administration of vaginal P4 to asymptomatic women with a twin gestation and a sonographic short cervix in midtrimester reduces the risk of PTB as well as neonatal mortality and morbidity [76].

Progesterone acts through its receptors (PRs). There are two major isoforms of PRs: PRA and PRB, which are identical proteins except for a short deletion in PRA. PRB is the full-length P4 receptor consisting of three transcriptional activation domains and is responsible for the majority of P4-related functions. Both PRs have equal binding affinity for P4 and, upon ligand binding, become transcriptionally active, that is, they dimerize, interact with other transcriptional coactivators, and bind to the specific DNA elements in the gene promoters, called progesterone response elements (PREs). In the uterus, both PR isoforms are expressed and the ratio of PRA to PRB significantly increases in the murine [77] and human myometrium during labor [78]. Throughout gestation, P4-PR interaction regulates myometrial contractility by suppressing the transcription of labor genes, including those directly regulating contractility (CAPs) and proinflammatory genes. The molecular mechanism of this suppression involves direct interaction of PRs with the AP-1 family of transcription factors, comprising the Jun and Fos proteins. We have found that the expression and localization of the Jun and Fos genes and proteins in the myometrium is regulated across gestation [79].

Physiologic Inflammation and Labor Onset

A successful pregnancy (implantation, early placentation, uterovascular remodeling to delivery) requires an intricate interaction with the maternal immune system. Pregnancy is associated with marked changes in the maternal immune system. Disruption of immune function during pregnancy has been proposed to underlie some pregnancy complications, including PTB and preeclampsia [80]. Thus, term and preterm labor are associated with induction of an inflammatory cascade within uterine tissues, leading to the production of proinflammatory cytokines, PGs, and CAPs that contribute to labor and delivery. Numerous studies indicate that physiological uterine inflammation is induced at term by mechanical and endocrine factors, causing the expression and release of proinflammatory cytokines and chemokines (IL-1β, IL-6, CCL2, and CXCL8 and its receptors, etc.) by different uterine compartments (i.e. myometrium, decidua, cervix, fetal membranes), followed by massive infiltration of inflammatory cells, predominantly neutrophils and macrophages [81]. These immune cells are a rich source of proinflammatory cytokines (IL1β, IL6, and TNFα) and PGs, contributing to their tissue-specific expression. Cytokines mediated the activation of the proinflammatory transcription factors such as NF-κB [82], which subsequently induced biochemical and molecular changes within the uterus [83]. Numerous studies have documented that each uterine compartment synthesizes a specific repertoire of proinflammatory cytokines and chemokines, contributing to the inflammatory cascade. Recent gene microarray analyses have revealed the induction of multiple genes and core inflammatory signaling pathways in both the myometrium and cervix with the onset of term labor, with *CXCL8* (the major chemoattractant for neutrophils) being the most highly expressed gene.

Preterm Labor Initiation

Depending upon the population, about half of spontaneous PTB is attributed to intrauterine or systemic infection and half to idiopathic causes for which, despite significant research effort, the underlying mechanisms remain largely elusive. One factor that appears to be a common element in both infectious and idiopathic PTL is the presence of an inflammatory state. Numerous studies support a role for inflammation in term labor and it is suggested that premature activation of inflammatory responses likely underlies the initiation of PTL.

Each PTB likely has a unique biological signature due to a specific combination of individual risk factors. Systemic maternal infection (i.e., pneumonia, pyelonephritis, malaria, typhoid fever, etc.) has been associated with PTL and PTB; however, the frequency of these conditions in developed countries is relatively low. In contrast, intrauterine infection/inflammation is often associated with PTL and has been traditionally considered an acute complication of pregnancy. Most studies conducted to determine the presence of infection in patients with PTL have focused on microbial invasion of the amniotic cavity (defined as a positive amniotic fluid culture for microorganisms when the fluid is retrieved by transabdominal amniocentesis). It has been estimated that intrauterine infection is responsible for 25%–40% of spontaneous PTL [11]. Moreover, the incidence of histological chorioamnionitis is inversely related to gestational age at birth [84]. Infection is the only pathologic process for which a firm causal link with PTL has been established and for which a defined molecular pathophysiology is known [85]. The evidence in support of this link includes: (1) subclinical intrauterine infection (absence of clinical signs and symptoms) is associated with PTL and delivery [86]; (2) intrauterine infection or systemic administration of microbial products to pregnant animals can result in PTL and delivery [73] while antibiotic treatment of ascending intrauterine infections in experimental models of chorioamnionitis can prevent PTL [87]; and (3) patients with elevated concentrations of proinflammatory cytokines (IL-6 [88], TNFα [89]), and matrix metalloproteases (MMPs [90]) in amniotic fluid at the time of midtrimester amniocentesis (clinical inflammation) are at risk for subsequent PTL and delivery [85]. Unfortunately, prevention and treatment of infection-associated PTB remains an unsolved problem. Antimicrobial therapy has been associated with a reduction in the PTL rate in women with asymptomatic bacteriuria [91]. However, randomized clinical trials of antibiotic treatment of patients with BV have contradictory results. A beneficial effect was demonstrated only in patients with a history of PTL [92, 93] while other studies in patients without previous PTL yielded negative results. Similarly, antibiotic therapy of patients with asymptomatic *Trichomonas vaginalis* [94], *Streptococcus agalactia* [95], and *Ureaplasma urealyticum* [96] did not prevent PTL. Importantly, accumulating evidence suggests that intraamniotic infection may be a chronic condition. In addition, an allergic reaction can also trigger preterm delivery [97].

Genomics of PTB

Systematic review and meta-analysis by Eidem and colleagues characterized transcriptomic changes related to spontaneous PTB within gestational tissues including the placenta, decidua, myometrium, maternal blood, cervix, fetal membranes (chorion and amnion), umbilical cord, fetal blood, and basal plate [98]. It was reported that only 18% of the preterm studies focused on spontaneous onset of labor, which is responsible for 45% of all PTB cases. There are limited microarray-based analyses of maternal peripheral leukocyte in relation to spontaneous PTL [25]. Recently, our group reported gene expression signatures obtained from total maternal blood leukocytes of women in threatened PTL that were predictive of delivery within 48 h [24]. We also used next-generation sequencing to characterize the transcriptome in whole blood leukocytes and peripheral monocytes of women undergoing spontaneous PTL compared to healthy pregnant women that subsequently delivered at full term [99]. RNA sequencing is a highly sensitive method that enables the discovery of subtle changes in gene expression directly related to active preterm parturition. In this analysis, we identified significant differences in the expression of 262 genes in the peripheral monocytes and 184 genes in the whole blood of women who were in spontaneous PTL compared to pregnant women of the same gestational age not undergoing labor, with 43 of these genes differentially expressed in both white blood cells and peripheral monocytes [99].

microRNA and Regulation of Labor Onset

The rapid transition of the myometrium at term from a quiescent to a highly contractile state suggests that the labor genes may also be regulated at the posttranscriptional level, possibly through the actions of microRNAs (miRNAs). MiRNAs, often called "micromanagers of gene expression," are highly conserved small noncoding RNAs comprising of 18–25 nucleotides. They recognize a target mRNA through a complementary sequence within the 3'-untranslated region [100]. MiRNAs mediate their action by (1) degradation of their target mRNAs, (2) repression, or (3) stimulation of target gene translation [101]. There are more than 1000 miRNAs identified so far that regulate 50% of all human genes [102]. They are implicated in numerous physiologic and pathologic cellular processes, including human reproduction, and therefore have been extensively studied for their role in PTBs.

Several studies examined the expression profile of miRNAs in serum and uterine tissues from women undergoing term or preterm labor (i.e., in the placenta, chorioamniotic membranes, myometrium, and cervix). In one study, miRNAs (that modulate inflammatory and tissue remodeling genes) from cervical tissue were implicated in the etiology of PTB through disruption of the epithelial barrier that might allow the inflammatory factors to enter the cervix and cause premature ripening [53]. The miRNA cluster C19MC on chromosome 19 is the largest human miRNA gene cluster. In the placenta, it is regulated by genomic imprinting and is paternally inherited. Montenegro and colleagues analyzed 455 miRNAs from the chorioamniotic membranes and reported differential expression of 39 miRNAs (mostly down-regulated) in preterm compared to term laboring women while no difference was detected at membranes from term laboring versus nonlaboring women [103]. Another study reported differences in the expression of 32 miRNAs in amnion from laboring and nonlaboring women; 31 of those were also down-regulated [104]. In particular, they demonstrated that miR-143 targets prostaglandin-endoperoxidase synthase 2 (PTGS2) in amnion mesenchymal cells and hypothesized that the decrease in this miRNA may be responsible for the induction of PTGS2 preceding parturition. A recent study demonstrated a significant increase in the expression of miR-200a in the myometrium of women in labor as compared to not in labor. miR-200a targets STAT5b, a transcriptional repressor of the P4-metabolizing enzyme 20a-HSD. This results in enzyme up-regulation and subsequent decline in P4, potentially contributing to a "functional" P4 withdrawal [105].

Besides their regulatory roles in different gestational tissues and cells, miRNAs can also be detected in various body fluids. These cell-free circulating miRNAs are either bound to RNA-binding proteins or lipoproteins such as HDL or LDL or they are encapsulated into extracellular vesicles (EVs) to escape degradation by RNAses. Circulating miRNAs have already been used to identify markers for disease diagnosis or prognosis. Several studies report changes of specific miRNA concentrations in the serum or plasma of patients in PTL, but the results are inconsistent or even contradictory and are influenced by the sample used (serum vs. plasma), the low RNA concentration in samples, and the lack of robust measurement technology. We used next-generation sequencing as a platform for miRNA analysis in whole plasma, EV, and EV-depleted plasma of women in spontaneous PTL and found a number of miRNAs in EVs that may be potential biomarkers for PTL [106]. We proposed that changes in the levels of these miRNAs may reflect pathological changes of the placenta associated with PTL.

GENETICS OF PTB

Several observations strongly support a role for genetic variation in the etiology of PTB. First, evidence from Swedish and Australian twin studies suggests a genetic predisposition of PTB with heritability estimates ranging from 20% to 40% [107]. Second, data collected in the United Kingdom suggest that there is an association between PTB and ethnicity/race as suggested by the higher rates (odds ratio ~1.5) of PTB in Asians versus whites [108]. African-Americans also have a higher PTB rate, even when socioeconomic factors (income and access to healthcare) are carefully controlled [109]. This suggests a genetic origin for susceptibility, but the source remains obscure. Interestingly, recent African migrants have a lower risk of PTB as compared to African-Americans and Canadians of African descent [110, 111]. Third, the leading risk factor for PTB is a previous PTB, with the greatest impact occurring when the gestational age of the prior PTB was less than 27 weeks. The risk of PTB also increases with the number of prior PTBs, with the most recent PTB being the most predictive [112]. Finally, as mentioned previously, there is a matrilineality of risk wherein maternal and matrilineally related histories of PTB are predictive (pregnant women who were born preterm and/or have sisters born preterm have an increased risk of PTB), but paternal and patrilineally related histories are not [113].

Candidate Genes Analysis

Investigators have searched for associations between single nucleotide polymorphisms (SNPs) and PTB. An SNP is a site in the DNA in which a single base-pair or nucleotide varies from person to person. While the majority of the SNPs are of no biological consequence, a fraction of SNPs have functional significance and are the basis for the physical diversity found among humans. SNPs can influence the promoter activity of DNA and pre-mRNA conformation, which may result in altered amino acid sequences and protein functions. Thus, SNPs can play a direct or indirect role in phenotypic expression.

Several groups have explored different candidate genes, looking for SNPs in groups of patients with PTB. Gene polymorphisms have been identified in association with the spontaneous pPROM and PTB [114]. The numbers of SNPs investigated in relation to PTB are still relatively low and they tend to be variants associated with diseases that could potentially be related to PTB. For example, SNPs have been identified in the coding regions of genes related to inflammation (i.e., cytokines and cytokine receptors). SNPs within the proinflammatory cytokine TNFa [115] and protein kinase C-α (PRKCA) [116] have been identified in black women in association with PTB. Interestingly, this study reported that the risk of PTB for mothers carrying the TNF polymorphism is amplified in the presence of BV [115]. Moreover, TNF haplotypes were associated with spontaneous PTB in both African-American and white women [117]. Several of these studies have been able to identify genetic associations with PTB. However, the number of patients enrolled in these studies is relatively small; therefore, the findings are not well replicated, and the results do not always translate across ethnic populations.

Despite the clear role for prostaglandins in the regulation of myometrial contractility, no polymorphisms in PG receptors (PTGER and PTGFR) or prostaglandin E synthase (PTGES) have been associated with PTB. Similarly, no associations with PTB were found for SNPs within genes encoding the β2-adrenergic receptor (ADRB2) and the dopamine receptor D2 (DRD2), which promote smooth muscle relaxation. Importantly, two SNPs—Flavobacterium okeanokoites (FokI, rs2228570 A>G) and Caudal Type Homeobox 2 (Cdx-2, rs11568820 T>C)—in the Vitamin D receptor (VDR) were associated with PTB in the Brazilian population. Both SNPs caused significant changes in *VDR* mRNA levels and in protein structure [118]. The importance of the polymorphisms is strengthened by the important role of the Vitamin D pathway in immune response modulation and the proven association of its deficiency with PTB [52].

Placentation is a critical process for a successful pregnancy and defects in this process can lead to major pregnancy complications, including susceptibility to PTB. Studies have therefore searched for genetic associations with placental dysfunction. One small cohort study reported a polymorphism in the vascular endothelial growth factor (VEGF); however, this was not replicated in larger cohorts. Polymorphism in genes related to thrombosis, including thrombomodulin (THBD), factor II, VII, XIII, F13A1, F2, F7, protein C (PROC), protein

C receptor (PROCR), and tissue factor pathway inhibitor (TFPI), have also been studied. Polymorphisms were identified in Factor VII in mothers, while polymorphisms were identified in Factor XIII in infants; both were independently associated with PTB [119].

Thus, despite considerable effort, a candidate gene approach to identifying a genetic basis for PTB has shown limited success. This is likely a challenging task due to multiple physiological and environmental factors acting as contributors (i.e., stress, infections, smoking etc.) in addition to the fact that pregnancy involves two genomes (maternal and fetal) [120].

Genome-Wide Association

Genome-wide association (GWAS) studies have been used to identify biomarkers for disease susceptibility, potential targets for therapeutics, or to provide information about the molecular mechanisms/pathways involved in disease pathology [121]. GWAS are based on a case-control design in which the whole genome is analyzed with hundreds of thousands of DNA markers in large nonrelated population cohorts presenting with a common phenotype. GWAS is considered to be a robust, unbiased, and hypothesis-free methodology; however, it is not problem free. For example, confounding can arise from population stratification when genetically heterogeneous populations are analyzed together without further adjustment, the need for large sample sizes to detect minor effect alleles, and inflated false-positive association rates arising from a thousand tests that are an integral part of any such study.

Several datasets are publicly available through the Database of Genotypes and Phenotypes (dbGaP). These include the Danish National Birth Cohort (DNBC, a large dataset including 1000 PTB mother-child pairs along with 1000 control pairs) [122], two datasets from the Gene Environment Association Studies initiative (GENEVA, including DNBC and an African-American cohort), a Genomic and Proteomic Network (GPN) cohort [123], and another African-American cohort collected by Boston Medical Center [124]. Using these cohorts, a comparison of mother-infant genomes from early PTB versus normal TL control pairs did not detect PTB associations with SNP from the maternal genome while the SNPs in ribonuclease T2 (*RNASET2*) and in the extended major histocompatibility complex (immune response-related genes) in the fetal genome were associated with PTB [125]. Unfortunately, subsequent studies involving independent fetal cohorts did not confirm this association. Another study using the DNBC cohort did not find any significant fetal autosomal genomic associations with PTB and suggested that the family-based linkage studies may prove more useful for the PTB risk assessment [126]. A nested study that used the data from the GENEVA cohort reported no association between PTB and candidate genes that was previously reported in the literature. However, using gene set enrichment analyses, they identified 30 pathways related to inflammation and metabolic disorders that were associated with PTB [127].

It is suggested that the impact of genomic analysis of variants can be improved with the use of "admixture mapping" that analyzes the ancestry of interbred populations. Admixture mapping is performed to correlate the ancestry at genetic loci with a phenotype and/or disease [128]. Admixture mapping, using subjects from African descent, resulted in the identification (in African-American women) of a region on chromosome 7 (7q21–7q22) that was associated with PTB. Importantly, this region carries genes related to metabolism, inflammation, collagen, and calcium regulation [129].

Rare Variant Analysis (RVA)

Although GWAS datasets present a powerful tool for the analysis of genetic variants, some limitations such as variations in sample collection, analysis methods, inadequate statistical power due to sample size, population-based heterogeneity in the frequencies of gene polymorphism, and the complex etiology of PTB make it a less-effective approach for the analysis of genomic associations. Attention has been diverted toward identifying rare gene variants associated with PTB using high-throughput sequencing of whole genomes and exomes rather than GWAS, which focuses on common variants.

One such study evaluated 33 genes among 257 families with a history of PTB, performed parametric and nonparametric analyses on 99 SNPs, and identified several rare variants within (1) the maternal genome, such as ectonucleotide pyrophosphatase/phosphodiesterase 1 (ENPP1), insulin-like growth factor binding protein 3 (IGFBP3), 7-dehyrocholesterol reductase (DHCR7), and TNF receptor-associated factor 2 (TRAF2); and (2) the fetal genome, such as corticotropin-releasing-hormone receptor 1 (CRHR1) and cytochrome P450, family 2, subfamily E, poly-peptide 1 (CYP2E1), that were moderately associated with PTB. However, further DNA sequence analysis for CRHR1 and TRAF2 did not reveal any potential causative mutations or variants [130].

Family-Based Designs/Genome-Wide Linkage Studies

Family-based designs are more reliable in determining the genetic association of diseases because they increase the control over population-based heterogeneity, the power to detect associations, and to provide information on the effects of allele origin and transmission of the disease phenotype [131]. Linkage analysis is a technique to identify genetic effects and relies on the fact that, if family members affected by the disease share a specific area of a chromosome not shared by unaffected members, then the gene or genes predisposing to the disease is likely to be on or near that area.

In family-based linkage studies, parametric linkage analysis is performed and genetic factors from a mother who gave birth to a preterm baby or a fetus that was born preterm are evaluated. A study in Finland that used multiplex families with later extension to nuclear families, followed by case-control study and haplotype segregation analysis, found the insulin-like growth factor 1 receptor (IGF1R) to be a potential susceptibility gene in the fetal genome that can result in PTB [132]. In another study using linkage analysis and the same set of Finnish families, two X-chromosomal linked genes were associated with PTB: (1) the androgen receptor (AR) located at Xq12, and (2) the IL-2 Receptor γ subunit (IL2RG) located at Xq13 [133]. A study within a Mexican-American population used linkage analysis on 1439 subjects to identify another PTB susceptibility linkage region on 18q21.33-q23 [134].

It is likely that the PTB syndrome, as with other multifactorial complex diseases, is caused by the combined effects of multiple genes and nongenetic environmental factors. For instance, an increased risk of PTB was found in the presence of low dietary folate and a deletion of the gene encoding dihydrofolate reductase (the enzyme that converts the folic acid to the reduced forms used for cell division) [135].

However, in addition to the usual issues associated with diseases of multifactorial etiology, PTB presents a unique problem, namely interactions between the maternal and fetal genomes.

For example, a recent study of glutathione S-transferase T1 showed that null genotypes in both mothers and fetuses increased the risk of PTB whether or not the mothers smoked during the third trimester [136]. However, if women smoked, only one individual needed to have the null genotype. These data show that the genotypes of both mother and fetus are important and that both can interact with the environment to affect risk. Therefore, it is likely that sequence variation alone is not sufficient to predict the risk of disease susceptibility, particularly in homeostatic organisms such as humans.

Mitochondrial Genetics

The mitochondrial genome (mtDNA) is extremely small compared to the nuclear genome. However, mtDNA mutations make a significant contribution to inherited diseases. Given that African-American women have an excess of PTB and a well-known matrilineality of PTB risk, a mitochondrially inherited factor is reasonably implied because mitochondrial inheritance is matrilineal and most African-Americans have a similar mtDNA sequence.

Mitochondria are a major source of intracellular reactive oxygen species (ROS). Excessive ROS production can cause oxidative stress (OS) leading to oxidation of macromolecules, cellular aging, and death. A healthy body produces antioxidants to remove the excessive ROS content by converting it to H_2O and thus maintaining intracellular homeostasis. During pregnancy, OS is high due to the increased mitochondrial content contributed by the placenta. Infection is also a source of increased OS, which induces the activity of matrix metalloproteinases (MMPs). In the case of amniotic infection, high levels of OS can cause damage to the amnion via degradation of collagens by MMPs [137] and decreased antioxidant capacity [138], subsequently resulting in PPROM and PTB.

OS can also affect the length of telomeres (a nucleoprotein structure located at the ends of chromosomes) that are subject to shortening during each cycle of cell division. Telomere length has been reported to be shorter in the fetuses from pregnancy complicated by PPROM versus PTB, implicating a role for OS [139]. A mitochondrial protein manganese superoxide dismutase (MnSOD) scavenges the intracellular ROS, thereby offering protection against OS-induced damages. Term labor is associated with a 2.6-fold increase in the expression levels of the MnSOD gene while its mutation is implicated in the PTB [140].

Women carrying pathogenic mtDNA mutations (m.3243A>G mutation in the *MT-TL1* gene causing MELAS syndrome) who have only mild mitochondrial disease manifestation themselves had a 25.3% incidence of PTB [141]. mtDNA mutations (A4917G, G10398A, and T4216C) in this population were analyzed for association between smoking (as a source of OS) and PTB; however, only marginal association of A4917G and T416C was found [142]. Lastly, a well-performed case-control association study of mtDNA variants was conducted using the DNBC cohort and a Finish-Norwegian cohort (MoBa) where mitochondrial SNPs were analyzed for their association with PTB. However, this study did not uncover any clear linkage between mitochondrial polymorphisms and PTB [143].

It was recently proposed that the impact of mitochondrial polymorphisms is dependent upon the interaction with an individual's nuclear inheritance. The misalignment between the mtDNA and nuclear DNA occurs as mitochondrial protein-coding polymorphisms exist in the context of nuclear polymorphisms to which they are poorly aligned [144]. This misalignment is more common in African-Americans due to a history of admixture and would be inherited in a matrilineal fashion. It remains to be determined whether women who have higher degrees of misalignment between their nuclear and mitochondrial genome are more susceptible to PTB.

CONCLUSION REMARKS

Despite considerable research, no therapies have been shown to be effective across populations in preventing PTB and improving infant outcomes; clearly, new approaches are required [145]. While the analysis of genetic polymorphisms linked to disease is a powerful tool, it has so far provided limited actionable information with respect to the etiology of PTB. Identification of women at risk for delivering preterm based on their genetic predisposition will potentially assist the development and implementation of preventive measures to improve the clinical management of PTB. It will not, however, fully resolve this critical public health problem until we develop a better understanding of how genotype is influenced by environmental factors. PTB is a multifactorial complex syndrome caused by multiple gene interactions and nongenetic environmental factors. It is clear that the genotypes of both mother and fetus are equally important and that both can interact with the environment to affect risk. The search for a genetic contribution to PTB has also been complicated by methodologic challenges such as variations in inclusion/exclusion criteria, population size, and racial differences, which can significantly influence the interpretation of the results.

Progress in PTB prevention will ultimately require a greater understanding of the molecular mechanisms underlying an individual woman's risk of PTB so that treatments can be individualized. For example, which patients should be candidates for progesterone prophylaxis, for immunomodulators, for antibiotics, or for cervical cerclage? The remarkable advances in the speed and cost effectiveness of large-scale, high-throughput genetic screening will make individual genotyping feasible in the near future. We now need to develop tools to accurately and deeply phenotype the multiple diseases that comprise PTB in order to harness this technology to individualize care options that can prevent PTB and improve outcomes for mother and baby.

References

[1] Goldenberg RL. The management of preterm labor. Obstet Gynecol 2002;100(5 Pt 1):1020–37.
[2] Brostrom EB, Akre O, Katz-Salamon M, Jaraj D, Kaijser M. Obstructive pulmonary disease in old age among individuals born preterm. Eur J Epidemiol 2013;28(1):79–85.
[3] Ward RM, Beachy JC. Neonatal complications following preterm birth. BJOG 2003;110(Suppl. 20):8–16.
[4] Lipper EG, Ross GS, Auld PA, Glassman MB. Survival and outcome of infants weighing less than 800 grams at birth. Am J Obstet Gynecol 1990;163(1 Pt 1):146–50.
[5] Kramer MS, Demissie K, Yang H, Platt RW, Sauve R, Liston R. The contribution of mild and moderate preterm birth to infant mortality. Fetal and Infant Health Study Group of the Canadian Perinatal Surveillance System. JAMA 2000;284(7):843–9.
[6] Hack M, Flannery DJ, Schluchter M, Cartar L, Borawski E, Klein N. Outcomes in young adulthood for very-low-birth-weight infants. N Engl J Med 2002;346(3):149–57.

[7] Hack M, Taylor HG, Drotar D, Schluchter M, Cartar L, Andreias L, et al. Chronic conditions, functional limitations, and special health care needs of school-aged children born with extremely low-birth-weight in the 1990s. JAMA 2005;294(3):318–25.

[8] Petrou S. Economic consequences of preterm birth and low birthweight. BJOG 2003;110(Suppl. 20):17–23.

[9] Martin JA, Park MM, Sutton PD. Births: preliminary data for 2001. Natl Vital Stat Rep 2002;50(10):1–20.

[10] MacDorman MF, Matthews TJ, Mohangoo AD, Zeitlin J. International comparisons of infant mortality and related factors: United States and Europe, 2010. Natl Vital Stat Rep 2014;63(5):1–6.

[11] Goldenberg RL, Culhane JF, Iams JD, Romero R. Epidemiology and causes of preterm birth. Lancet 2008; 371(9606):75–84.

[12] Porter TF, Fraser AM, Hunter CY, Ward RH, Varner MW. The risk of preterm birth across generations. Obstet Gynecol 1997;90(1):63–7.

[13] Buchmayer SM, Sparen P, Cnattingius S. Previous pregnancy loss: risks related to severity of preterm delivery. Am J Obstet Gynecol 2004;191(4):1225–31.

[14] Hsieh TT, Chen SF, Shau WY, Hsieh CC, Hsu JJ, Hung TH. The impact of interpregnancy interval and previous preterm birth on the subsequent risk of preterm birth. J Soc Gynecol Investig 2005;12(3):202–7.

[15] Kilpatrick SJ, Patil R, Connell J, Nichols J, Studee L. Risk factors for previable premature rupture of membranes or advanced cervical dilation: a case control study. Am J Obstet Gynecol 2006;194(4):1168–74. discussion 74-5.

[16] DeFranco EA, Hall ES, Muglia LJ. Racial disparity in previable birth. Am J Obstet Gynecol 2016;214(3):394.e1–7.

[17] Adams MM, Elam-Evans LD, Wilson HG, Gilbertz DA. Rates of and factors associated with recurrence of preterm delivery. JAMA 2000;283(12):1591–6.

[18] Hatch M, Berkowitz G, Janevic T, Sloan R, Lapinski R, James T, et al. Race, cardiovascular reactivity, and preterm delivery among active-duty military women. Epidemiology 2006;17(2):178–82.

[19] Royce RA, Jackson TP, Thorp Jr. JM, Hillier SL, Rabe LK, Pastore LM, et al. Race/ethnicity, vaginal flora patterns, and pH during pregnancy. Sex Transm Dis 1999;26(2):96–102.

[20] Fettweis JM, Brooks JP, Serrano MG, Sheth NU, Girerd PH, Edwards DJ, et al. Differences in vaginal microbiome in African American women versus women of European ancestry. Microbiology 2014;160:2272–82. Pt 10.

[21] Heaman MI, Blanchard JF, Gupton AL, Moffatt ME, Currie RF. Risk factors for spontaneous preterm birth among Aboriginal and non-Aboriginal women in Manitoba. Paediatr Perinat Epidemiol 2005;19(3):181–93.

[22] Dietz PM, Callaghan WM, Cogswell ME, Morrow B, Ferre C, Schieve LA. Combined effects of prepregnancy body mass index and weight gain during pregnancy on the risk of preterm delivery. Epidemiology 2006; 17(2):170–7.

[23] Misra DP, Strobino DM, Stashinko EE, Nagey DA, Nanda J. Effects of physical activity on preterm birth. Am J Epidemiol 1998;147(7):628–35.

[24] da Silva AA, Simoes VM, Barbieri MA, Bettiol H, Lamy-Filho F, Coimbra LC, et al. Young maternal age and preterm birth. Paediatr Perinat Epidemiol 2003;17(4):332–9.

[25] Astolfi P, De Pasquale A, Zonta LA. Paternal age and preterm birth in Italy, 1990 to 1998. Epidemiology 2006; 17(2):218–21.

[26] Kinzler WL, Ananth CV, Smulian JC, Vintzileos AM. Parental age difference and adverse perinatal outcomes in the United States. Paediatr Perinat Epidemiol 2002;16(4):320–7.

[27] Simhan HN, Krohn MA. Paternal race and preterm birth. Am J Obstet Gynecol 2008;198(6):644.c1 6.

[28] Zeitlin J, Saurel-Cubizolles MJ, De Mouzon J, Rivera L, Ancel PY, Blondel B, et al. Fetal sex and preterm birth: are males at greater risk? Hum Reprod 2002;17(10):2762–8.

[29] Verstraelen H, Goetgeluk S, Derom C, Vansteelandt S, Derom R, Goetghebeur E, et al. Preterm birth in twins after subfertility treatment: population based cohort study. BMJ 2005;331(7526):1173.

[30] Witkin SS. The vaginal microbiome, vaginal anti-microbial defence mechanisms and the clinical challenge of reducing infection-related preterm birth. BJOG 2015;122(2):213–8.

[31] Kindinger LM, Bennett PR, Lee YS, Marchesi JR, Smith A, Cacciatore S, et al. The interaction between vaginal microbiota, cervical length, and vaginal progesterone treatment for preterm birth risk. Microbiome 2017;5(1):6.

[32] Myhre R, Brantsaeter AL, Myking S, Gjessing HK, Sengpiel V, Meltzer HM, et al. Intake of probiotic food and risk of spontaneous preterm delivery. Am J Clin Nutr 2011;93(1):151–7.

[33] Pozo E, Mesa F, Ikram MH, Puertas A, Torrecillas-Martinez L, Ortega-Oller I, et al. Preterm birth and/or low birth weight are associated with periodontal disease and the increased placental immunohistochemical expression of inflammatory markers. Histol Histopathol 2016;31(2):231–7.

[34] Teshome A, Yitayeh A. Relationship between periodontal disease and preterm low birth weight: systematic review. Pan Afr Med J 2016;24:215.

[35] Klebanoff MA, Hillier SL, Nugent RP, MacPherson CA, Hauth JC, Carey JC, et al. Is bacterial vaginosis a stronger risk factor for preterm birth when it is diagnosed earlier in gestation? Am J Obstet Gynecol 2005; 192(2):470–7.

[36] Kalantaridou SN, Zoumakis E, Makrigiannakis A, Lavasidis LG, Vrekoussis T, Chrousos GP. Corticotropin-releasing hormone, stress and human reproduction: an update. J Reprod Immunol 2010;85(1):33–9.

[37] Brooks AN, Challis JR. Regulation of the hypothalamic-pituitary-adrenal axis in birth. Can J Physiol Pharmacol 1988;66(8):1106–12.

[38] Chen Y, Holzman C, Chung H, Senagore P, Talge NM, Siler-Khodr T. Levels of maternal serum corticotropin-releasing hormone (CRH) at midpregnancy in relation to maternal characteristics. Psychoneuroendocrinology 2010;35(6):820–32.

[39] You X, Liu J, Xu C, Liu W, Zhu X, Li Y, et al. Corticotropin-releasing hormone (CRH) promotes inflammation in human pregnant myometrium: the evidence of CRH initiating parturition? J Clin Endocrinol Metab 2014;99(2): E199–208.

[40] Henderson J, Kesmodel U, Gray R. Systematic review of the fetal effects of prenatal binge-drinking. J Epidemiol Community Health 2007;61(12):1069–73.

[41] Patra J, Bakker R, Irving H, Jaddoe VW, Malini S, Rehm J. Dose-response relationship between alcohol consumption before and during pregnancy and the risks of low birthweight, preterm birth and small for gestational age (SGA)-a systematic review and meta-analyses. BJOG 2011;118(12):1411–21.

[42] American College of Obstetricians and Gynecologists Committee on Obstetric Practice. Committee Opinion No. 637: Marijuana use during pregnancy and lactation. Obstet Gynecol 2015;126(1):234–8.

[43] Olesen C, Steffensen FH, Sorensen HT, Nielsen GL, Olsen J. Pregnancy outcome following prescription for sumatriptan. Headache 2000;40(1):20–4.

[44] Vahratian A, Siega-Riz AM, Savitz DA, Thorp Jr. JM. Multivitamin use and the risk of preterm birth. Am J Epidemiol 2004;160(9):886–92.

[45] Ronnenberg AG, Goldman MB, Chen D, Aitken IW, Willett WC, Selhub J, et al. Preconception homocysteine and B vitamin status and birth outcomes in Chinese women. Am J Clin Nutr 2002;76(6):1385–91.

[46] Rayman MP, Wijnen H, Vader H, Kooistra L, Pop V. Maternal selenium status during early gestation and risk for preterm birth. CMAJ 2011;183(5):549–55.

[47] Stojanovic N, Plecas D, Plesinac S. Normal vaginal flora, disorders and application of probiotics in pregnancy. Arch Gynecol Obstet 2012;286(2):325–32.

[48] Mikkelsen TB, Osterdal ML, Knudsen VK, Haugen M, Meltzer HM, Bakketeig L, et al. Association between a Mediterranean-type diet and risk of preterm birth among Danish women: a prospective cohort study. Acta Obstet Gynecol Scand 2008;87(3):325–30.

[49] Facchinetti F, Fazzio M, Venturini P. Polyunsaturated fatty acids and risk of preterm delivery. Eur Rev Med Pharmacol Sci 2005;9(1):41–8.

[50] Oken E, Kleinman KP, Olsen SF, Rich-Edwards JW, Gillman MW. Associations of seafood and elongated n-3 fatty acid intake with fetal growth and length of gestation: results from a US pregnancy cohort. Am J Epidemiol 2004;160(8):774–83.

[51] Qin LL, Lu FG, Yang SH, Xu HL, Luo BA. Does maternal vitamin D deficiency increase the risk of preterm birth: a meta-analysis of observational studies. Nutrients 2016;8(5).

[52] Zhou SS, Tao YH, Huang K, Zhu BB, Tao FB. Vitamin D and risk of preterm birth: up-to-date meta-analysis of randomized controlled trials and observational studies. J Obstet Gynaecol Res 2017;43(2):247–56.

[53] Lin VW, Baccarelli AA, Burris HH. Epigenetics-a potential mediator between air pollution and preterm birth. Environ Epigenet 2016;2(1).

[54] MacArthur J. Pregnancy and Fluoride Do Not Mix: Prenatal Fluoride and Premature Birth. Preeclampsia, Autism: CreateSpace Independent Publishing Platform; 2016.

[55] Rosenberg TJ, Garbers S, Lipkind H, Chiasson MA. Maternal obesity and diabetes as risk factors for adverse pregnancy outcomes: differences among 4 racial/ethnic groups. Am J Public Health 2005;95(9):1545–51.

[56] Cleary-Goldman J, Malone FD, Lambert-Messerlian G, Sullivan L, Canick J, Porter TF, et al. Maternal thyroid hypofunction and pregnancy outcome. Obstet Gynecol 2008;112(1):85–92.

[57] Heng YJ, Liong S, Permezel M, Rice GE, Di Quinzio MK, Georgiou HM. Human cervicovaginal fluid biomarkers to predict term and preterm labor. Front Physiol 2015;6:151.

[58] Shynlova O, Tsui P, Dorogin A, Chow M, Lye SJ. Expression and localization of alpha-smooth muscle and gamma-actins in the pregnant rat myometrium. Biol Reprod 2005;73(4):773–80.

[59] Shynlova O, Mitchell JA, Tsampalieros A, Langille BL, Lye SJ. Progesterone and gravidity differentially regulate expression of extracellular matrix components in the pregnant rat myometrium. Biol Reprod 2004; 70(4):986–92.

[60] MacPhee DJ, Mostachfi H, Han R, Lye SJ, Post M, Caniggia I. Focal adhesion kinase is a key mediator of human trophoblast development. Lab Invest 2001;81(11):1469–83.

[61] Shynlova O, Lee YH, Srikhajon K, Lye SJ. Physiologic uterine inflammation and labor onset: integration of endocrine and mechanical signals. Reprod Sci 2013;20(2):154–67.

[62] Torricelli M, Giovannelli A, Leucci E, De Falco G, Reis FM, Imperatore A, et al. Labor (term and preterm) is associated with changes in the placental mRNA expression of corticotrophin-releasing factor. Reprod Sci 2007;14(3):241–5.

[63] Goland RS, Jozak S, Warren WB, Conwell IM, Stark RI, Tropper PJ. Elevated levels of umbilical cord plasma corticotropin-releasing hormone in growth-retarded fetuses. J Clin Endocrinol Metab 1993; 77(5):1174–9.

[64] Kuroki Y, Takahashi M, Nishitani C. Pulmonary collectins in innate immunity of the lung. Cell Microbiol 2007; 9(8):1871–9.

[65] Mendelson CR, Boggaram V. Hormonal and developmental regulation of pulmonary surfactant synthesis in fetal lung. Baillieres Clin Endocrinol Metab 1990;4(2):351–78.

[66] Lopez Bernal A, Newman GE, Phizackerley PJ, Turnbull AC. Surfactant stimulates prostaglandin E production in human amnion. Br J Obstet Gynaecol 1988;95(10):1013–7.

[67] Liggins GC, Fairclough RJ, Grieves SA, Kendall JZ, Knox BS. The mechanism of initiation of parturition in the ewe. Recent Prog Horm Res 1973;29:111–59.

[68] Challis JRG, Lye SJ, Gibb W, Whittle W, Patel F, Alfaidy N. Understanding preterm labor. Ann N Y Acad Sci 2001;943(1):225–34.

[69] Kokubu K, Hondo E, Sakaguchi N, Sagara E, Kiso Y. Differentiation and elimination of uterine natural killer cells in delayed implantation and parturition mice. J Reprod Dev 2005;51(6):773–6.

[70] Shynlova O, Nedd-Roderique T, Li Y, Dorogin A, Lye SJ. Myometrial immune cells contribute to term parturition, preterm labour and post-partum involution in mice. J Cell Mol Med 2013;17(1):90–102.

[71] Maria B, Stampf F, Goepp A, Ulmann A. Termination of early pregnancy by a single dose of mifepristone (RU 486), a progesterone antagonist. Eur J Obstet Gynecol Reprod Biol 1988;28(3):249–55.

[72] Nadeem L, Shynlova O, Matysiak-Zablocki E, Mesiano S, Dong X, Lye S. Molecular evidence of functional progesterone withdrawal in human myometrium. Nat Commun 2016;7:11565.

[73] Elovitz MA, Mrinalini C. Animal models of preterm birth. Trends Endocrinol Metab 2004;15(10):479–87.

[74] Nelson DB, McIntire DD, McDonald J, Gard J, Turrichi P, Leveno KJ. 17-alpha Hydroxyprogesterone caproate did not reduce the rate of recurrent preterm birth in a prospective cohort study. Am J Obstet Gynecol 2017; 216(6):600.e1–9.

[75] Norman JE, Marlow N, Messow CM, Shennan A, Bennett PR, Thornton S, et al. Vaginal progesterone prophylaxis for preterm birth (the OPPTIMUM study): a multicentre, randomised, double-blind trial. Lancet 2016; 387(10033):2106–16.

[76] Romero R, Conde-Agudelo A, El-Refaie W, Rode L, Brizot ML, Cetingoz E, et al. Vaginal progesterone decreases preterm birth and neonatal morbidity and mortality in women with a twin gestation and a short cervix: an updated meta-analysis of individual patient data. Ultrasound Obstet Gynecol 2017; 49(3):303–14.

[77] Zeng Z, Velarde MC, Simmen FA, Simmen RC. Delayed parturition and altered myometrial progesterone receptor isoform A expression in mice null for Kruppel-like factor 9. Biol Reprod 2008;78(6):1029–37.

[78] Mesiano S, Chan EC, Fitter JT, Kwek K, Yeo G, Smith R. Progesterone withdrawal and estrogen activation in human parturition are coordinated by progesterone receptor A expression in the myometrium. J Clin Endocrinol Metab 2002;87(6):2924–30.

[79] Nadeem L, Farine T, Dorogin A, Matysiak-Zablocki E, Shynlova O, Lye S. Differential expression of myometrial AP-1 proteins during gestation and labour. J Cell Mol Med 2018;22:452–71.

[80] Griffin C. Probiotics in obstetrics and gynaecology. Aust N Z J Obstet Gynaecol 2015;55(3):201–9.

[81] Osman I, Young A, Ledingham MA, Thomson AJ, Jordan F, Greer IA, et al. Leukocyte density and proinflammatory cytokine expression in human fetal membranes, decidua, cervix and myometrium before and during labour at term. Mol Hum Reprod 2003;9(1):41–5.

[82] Condon JC, Hardy DB, Kovaric K, Mendelson CR. Up-regulation of the progesterone receptor (PR)-C isoform in laboring myometrium by activation of nuclear factor-kappaB may contribute to the onset of labor through inhibition of PR function. Mol Endocrinol 2006;20(4):764–75.

[83] Shynlova O, Lee Y-H, Srikhajon K, Lye SJ. Physiologic uterine inflammation and labor onset. Reprod Sci 2013; 20(2):154–67.

[84] Kim CJ, Romero R, Kusanovic JP, Yoo W, Dong Z, Topping V, et al. The frequency, clinical significance, and pathological features of chronic chorioamnionitis: a lesion associated with spontaneous preterm birth. Mod Pathol 2010;23(7):1000–11.

[85] Romero R, Espinoza J, Goncalves LF, Kusanovic JP, Friel LA, Nien JK. Inflammation in preterm and term labour and delivery. Semin Fetal Neonatal Med 2006;11(5):317–26.

[86] Gomez R, Ghezzi F, Romero R, Munoz H, Tolosa JE, Rojas I. Premature labor and intra-amniotic infection. Clinical aspects and role of the cytokines in diagnosis and pathophysiology. Clin Perinatol 1995;22(2):281–342.

[87] Fidel P, Ghezzi F, Romero R, Chaiworapongsa T, Espinoza J, Cutright J, et al. The effect of antibiotic therapy on intrauterine infection-induced preterm parturition in rabbits. J Matern Fetal Neonatal Med 2003;14(1):57–64.

[88] Ghidini A, Jenkins CB, Spong CY, Pezzullo JC, Salafia CM, Eglinton GS. Elevated amniotic fluid interleukin-6 levels during the early second trimester are associated with greater risk of subsequent preterm delivery. Am J Reprod Immunol 1997;37(3):227–31.

[89] Spong CY, Scherer DM, Ghidini A, Pezzullo JC, Salafia CM, Eglinton GS. Midtrimester amniotic fluid tumor necrosis factor-alpha does not predict small-for-gestational-age infants. Am J Reprod Immunol 1997;37(3):236–9.

[90] Park KH, Chaiworapongsa T, Kim YM, Espinoza J, Yoshimatsu J, Edwin S, et al. Matrix metalloproteinase 3 in parturition, premature rupture of the membranes, and microbial invasion of the amniotic cavity. J Perinat Med 2003;31(1):12–22.

[91] Goncalves LF, Chaiworapongsa T, Romero R. Intrauterine infection and prematurity. Ment Retard Dev Disabil Res Rev 2002;8(1):3–13.

[92] Hauth JC, Goldenberg RL, Andrews WW, DuBard MB, Copper RL. Reduced incidence of preterm delivery with metronidazole and erythromycin in women with bacterial vaginosis. N Engl J Med 1995;333(26):1732–6.

[93] Mercer BM, Arheart KL. Antimicrobial therapy in expectant management of preterm premature rupture of the membranes. Lancet 1995;346(8985):1271–9.

[94] Klebanoff MA, Carey JC, Hauth JC, Hillier SL, Nugent RP, Thom EA, et al. Failure of metronidazole to prevent preterm delivery among pregnant women with asymptomatic Trichomonas vaginalis infection. N Engl J Med 2001;345(7):487–93.

[95] Klebanoff MA, Regan JA, Rao AV, Nugent RP, Blackwelder WC, Eschenbach DA, et al. Outcome of the Vaginal Infections and Prematurity Study: results of a clinical trial of erythromycin among pregnant women colonized with group B streptococci. Am J Obstet Gynecol 1995;172(5):1540–5.

[96] Eschenbach DA, Nugent RP, Rao AV, Cotch MF, Gibbs RS, Lipscomb KA, et al. A randomized placebo-controlled trial of erythromycin for the treatment of Ureaplasma urealyticum to prevent premature delivery. The Vaginal Infections and Prematurity Study Group. Am J Obstet Gynecol 1991;164(3):734–42.

[97] Witkin SS. Allergy and preterm birth. Am J Obstet Gynecol 2007;196(1):e27. Author reply e-8.

[98] Eidem HR, WEt A, McGary KL, Abbot P, Rokas A. Gestational tissue transcriptomics in term and preterm human pregnancies: a systematic review and meta-analysis. BMC Med Genet 2015;8:27.

[99] Paquette ASO, Kibschull M, Price ND, Lye SJ. Comparative analysis of gene expression in maternal peripheral blood and monocytes during spontaneous preterm labor. Am J Obstet Gynecol 2018;218:345.e1–345.e30.

[100] Tay Y, Kats L, Salmena L, Weiss D, Tan SM, Ala U, et al. Coding-independent regulation of the tumor suppressor PTEN by competing endogenous mRNAs. Cell 2011;147(2):344–57.

[101] Vasudevan S. Posttranscriptional upregulation by microRNAs. Wiley Interdiscip Rev RNA 2012;3(3):311–30.

[102] Rodriguez A, Griffiths-Jones S, Ashurst JL, Bradley A. Identification of mammalian microRNA host genes and transcription units. Genome Res 2004;14(10A):1902–10.

[103] Montenegro D, Romero R, Kim SS, Tarca AL, Draghici S, Kusanovic JP, et al. Expression patterns of microRNAs in the chorioamniotic membranes: a role for microRNAs in human pregnancy and parturition. J Pathol 2009; 217(1):113–21.

[104] Kim SY, Romero R, Tarca AL, Bhatti G, Lee J, Chaiworapongsa T, et al. miR-143 regulation of prostaglandin-endoperoxidase synthase 2 in the amnion: implications for human parturition at term. PLoS One 2011;6:e24131.

[105] Williams KC, Renthal NE, Condon JC, Gerard RD, Mendelson CR. MicroRNA-200a serves a key role in the decline of progesterone receptor function leading to term and preterm labor. Proc Natl Acad Sci U S A 2012;109(19):7529–34.

[106] Fallen S, Baxter D, Wu X, KimT-K SO, Lee M, Scherler K, Lye S, HoodL WK. Extracellular vesicle RNAs reflect placenta dysfunction and are a biomarker source for preterm labor. J Cell Mol Med 2018;22:2760–73.

[107] Varner MW, Esplin MS. Current understanding of genetic factors in preterm birth. BJOG 2005;112(Suppl. 1):28–31.

[108] Patel RR, Steer P, Doyle P, Little MP, Elliott P. Does gestation vary by ethnic group? A London-based study of over 122,000 pregnancies with spontaneous onset of labour. Int J Epidemiol 2004;33(1):107–13.

[109] Goldenberg RL, Cliver SP, Mulvihill FX, Hickey CA, Hoffman HJ, Klerman LV, et al. Medical, psychosocial, and behavioral risk factors do not explain the increased risk for low birth weight among black women. Am J Obstet Gynecol 1996;175(5):1317–24.

[110] Elo IT, Vang Z, Culhane JF. Variation in birth outcomes by mother's country of birth among non-Hispanic black women in the United States. Matern Child Health J 2014;18(10):2371–81.

[111] Urquia ML, Frank JW, Moineddin R, Glazier RH. Immigrants' duration of residence and adverse birth outcomes: a population-based study. BJOG 2010;117(5):591–601.

[112] Esplin MS, O'Brien E, Fraser A, Kerber RA, Clark E, Simonsen SE, et al. Estimating recurrence of spontaneous preterm delivery. Obstet Gynecol 2008;112(3):516–23.

[113] Svensson AC, Sandin S, Cnattingius S, Reilly M, Pawitan Y, Hultman CM, et al. Maternal effects for preterm birth: a genetic epidemiologic study of 630,000 families. Am J Epidemiol 2009;170(11):1365–72.

[114] Crider KS, Whitehead N, Buus RM. Genetic variation associated with preterm birth: a HuGE review. Genet Med 2005;7(9):593–604.

[115] Macones GA, Parry S, Elkousy M, Clothier B, Ural SH, Strauss 3rd JF. A polymorphism in the promoter region of TNF and bacterial vaginosis: preliminary evidence of gene-environment interaction in the etiology of spontaneous preterm birth. Am J Obstet Gynecol 2004;190(6):1504–8. Discussion 3A.

[116] Frey HA, Stout MJ, Pearson LN, Tuuli MG, Cahill AG, Strauss 3rd JF, et al. Genetic variation associated with preterm birth in African-American women. Am J Obstet Gynecol 2016;215(2):235.e1–8.

[117] Engel SA, Erichsen HC, Savitz DA, Thorp J, Chanock SJ, Olshan AF. Risk of spontaneous preterm birth is associated with common proinflammatory cytokine polymorphisms. Epidemiology 2005;16(4):469–77.

[118] Javorski NR, Lima CAD, Silva LVC, Crovella S, de Azevedo Silva J. Vitamin D receptor (VDR) polymorphisms are associated to spontaneous preterm birth and maternal aspects. Gene 2018;642:58–63.

[119] Hartel C, von Otte S, Koch J, Ahrens P, Kattner E, Segerer H, et al. Polymorphisms of haemostasis genes as risk factors for preterm delivery. Thromb Haemost 2005;94(1):88–92.

[120] Dolan SM, Hollegaard MV, Merialdi M, Betran AP, Allen T, Abelow C, et al. Synopsis of preterm birth genetic association studies: the preterm birth genetics knowledge base (PTBGene). Public Health Genomics 2010;13 (7–8):514–23.

[121] Manolio TA. Genomewide association studies and assessment of the risk of disease. N Engl J Med 2010; 363(2):166–76.

[122] Falah N, McElroy J, Snegovskikh V, Lockwood CJ, Norwitz E, Murray JC, et al. Investigation of genetic risk factors for chronic adult diseases for association with preterm birth. Hum Genet 2013;132(1):57–67.

[123] Zhang J, Zhou J, Xu B, Chen C, Shi W. Different expressions of TLRs and related factors in peripheral blood of preterm infants. Int J Clin Exp Med 2015;8(3):4108–14.

[124] Nachman RM, Mao G, Zhang X, Hong X, Chen Z, Soria CS, et al. Intrauterine inflammation and maternal exposure to ambient PM2.5 during preconception and specific periods of pregnancy. the Boston Birth Cohort. Environ Health Perspect 2016;124(10):1608–15.

[125] Zhang H, Baldwin DA, Bukowski RK, Parry S, Xu Y, Song C, et al. A genome-wide association study of early spontaneous preterm delivery. Genet Epidemiol 2015;39(3):217–26.

[126] Wu W, Clark E, Manuck T, Esplin M, Varner M, Jorde L. A Genome-Wide Association Study of spontaneous preterm birth in a European population. F1000Research 2013;2:255 [version 1; referees: 2 approved with reservations].

[127] Uzun A, Dewan AT, Istrail S, Padbury JF. Pathway-based genetic analysis of preterm birth. Genomics 2013;101 (3):163–70.

[128] Shriner D. Overview of admixture mapping. Curr Protoc Hum Genet 2013; Editorial board, Haines JL [et al]. Chapter 1:Unit 1.23.

[129] Manuck TA, Lai Y, Meis PJ, Sibai B, Spong CY, Rouse DJ, et al. Admixture mapping to identify spontaneous preterm birth susceptibility loci in African Americans. Obstet Gynecol 2011;117(5):1078–84.

[130] Bream EN, Leppellere CR, Cooper ME, Dagle JM, Merrill DC, Christensen K, et al. Candidate gene linkage approach to identify DNA variants that predispose to preterm birth. Pediatr Res 2013;73(2):135–41.

[131] Ott J, Kamatani Y, Lathrop M. Family-based designs for genome-wide association studies. Nat Rev Genet 2011;12(7):465–74.

[132] Haataja R, Karjalainen MK, Luukkonen A, Teramo K, Puttonen H, Ojaniemi M, et al. Mapping a new spontaneous preterm birth susceptibility gene, IGF1R, using linkage, haplotype sharing, and association analysis. PLoS Genet 2011;7(2)e1001293.

[133] Karjalainen MK, Huusko JM, Ulvila J, Sotkasiira J, Luukkonen A, Teramo K, et al. A potential novel spontaneous preterm birth gene, AR, identified by linkage and association analysis of X chromosomal markers. PLoS One 2012;7(12)e51378.

[134] Chittoor G, Farook VS, Puppala S, Fowler SP, Schneider J, Dyer TD, et al. Localization of a major susceptibility locus influencing preterm birth. Mol Hum Reprod 2013;19(10):687–96.

[135] Johnson WG, Scholl TO, Spychala JR, Buyske S, Stenroos ES, Chen X. Common dihydrofolate reductase 19-base pair deletion allele: a novel risk factor for preterm delivery. Am J Clin Nutr 2005;81(3):664–8.

[136] Nukui T, Day RD, Sims CS, Ness RB, Romkes M. Maternal/newborn GSTT1 null genotype contributes to risk of preterm, low birthweight infants. Pharmacogenetics 2004;14(9):569–76.

[137] Woods Jr. JR. Reactive oxygen species and preterm premature rupture of membranes—a review. Placenta 2001;22(Suppl. A):S38–44.

[138] Kacerovsky M, Tothova L, Menon R, Vlkova B, Musilova I, Hornychova H, et al. Amniotic fluid markers of oxidative stress in pregnancies complicated by preterm prelabor rupture of membranes. J Matern Fetal Neonatal Med 2014;27:1–10.

[139] Menon R, Yu J, Basanta-Henry P, Brou L, Berga SL, Fortunato SJ, et al. Short fetal leukocyte telomere length and preterm prelabor rupture of the membranes. PLoS One 2012;7(2)e31136.

[140] Than NG, Romero R, Tarca AL, Draghici S, Erez O, Chaiworapongsa T, et al. Mitochondrial manganese superoxide dismutase mRNA expression in human chorioamniotic membranes and its association with labor, inflammation, and infection. J Matern Fetal Neona 2009;22(11):1000–13.

[141] de Laat P, Fleuren LH, Bekker MN, Smeitink JA, Janssen MC. Obstetric complications in carriers of the m.3243A>G mutation, a retrospective cohort study on maternal and fetal outcome. Mitochondrion 2015;25:98–103.

[142] Velez DR, Menon R, Simhan H, Fortunato S, Canter JA, Williams SM. Mitochondrial DNA variant A4917G, smoking and spontaneous preterm birth. Mitochondrion 2008;8(2):130–5.

[143] Alleman BW, Myking S, Ryckman KK, Myhre R, Feingold E, Feenstra B, et al. No observed association for mitochondrial SNPs with preterm delivery and related outcomes. Pediatr Res 2012;72(5):539–44.

[144] Falk MJ, Sondheimer N. Mitochondrial genetic diseases. Curr Opin Pediatr 2010;22(6):711–6.

[145] Jain J, Gyamfi-Bannerman C. Future directions in preterm birth research. Semin Fetal Neonatal Med 2016; 21(2):129–32.

Glossary

Agonist	A chemical, often an analogue of a hormone that binds to its receptor and activates the receptor to elicit a biological response.
Amniocentesis	A procedure that involves drawing a sample of amniotic fluid from the amniotic cavity to perform genetic testing on fetal DNA or to detect an intrauterine infection.
Antagonist	A chemical, often an analogue of a hormone that binds to its receptor but elicits no biological response and can block hormonal response.
Bacteriuria	The presence of bacteria in urine.
Caucasian	The term Caucasian is used to refer to white people whose ancestry can be traced to Europe.
Cellular hypertrophy	An increase in the size of cells.
Cerebral Palsy	A neurological disorder that affects the part of the brain that controls muscle movements.

Chorioamnionitis	Chorioamnionitis or intraamniotic infection (IAI) is an inflammation of the fetal membranes (amnion and chorion) due to a bacterial infection.
Haplotypes	A group of alleles of different or closely linked genes on a single chromosome that is inherited together from a single parent.
Iatrogenic PTB	Preterm birth as a result of a medical procedure or treatment.
Idiopathic	Spontaneous disease for which the causes are unknown.
Matrilineal transmission	The transmission of disease from mother to daughter.
Myocytes	Muscle cells that form muscle tissue. The specialized forms are cardiac, skeletal, and smooth muscle myocytes.
Myometrium	Uterine muscle, the middle layer of the uterine wall, consisting mainly of uterine smooth muscle cells
Neonate	A newborn child.
Polymorphism	The occurrence of two or more forms/phenotypes.
Preterm birth	Parturition before the completion of 37 weeks of pregnancy.
Randomized control trial	It is a type of study that aims to reduce bias. In such studies, the participants are assigned by chance to separate groups; neither the researchers nor the participants can choose to allocate people in groups.

The Intergenerational Effects on Fetal Programming

He-Feng Huang[*,†], *Guo-Lian Ding*[*,†], *Xin-Mei Liu*[*,†],
Zi-Ru Jiang[*,†]

[*]Institute of Embryo-Fetal Original Adult Disease and Shanghai Key Laboratory for Reproductive Medicine, School of Medicine, Shanghai Jiao Tong University, Shanghai, People's Republic of China [†]The International Peace Maternity and Child Health Hospital, School of Medicine, Shanghai Jiao Tong University, Shanghai, People's Republic of China

Abbreviations

5caC	5-carboxylcytosine
5fC	5-formylcytosine
5hmC	5-hydroxymethylcytosine
5mC	5 carbon of cytosine residues
ART	assisted reproductive technology
BNST	bed nucleus of stria terminalis
CpGs	cytosine-phosphate-guanine dinucleotides
DMR	differentially methylated region

Human Reproductive and Prenatal Genetics
https://doi.org/10.1016/B978-0-12-813570-9.00016-4

Dnmts	DNA methyltransferases
endo-**siRNAs**	endogenous small interfering
HFD	high-fat diet
HPA	hypothalamic-pituitary-adrenal
IGF2	insulin-like growth factor 2
JHDM2A	JmjC-domain-containing-histone demethylase 2A
miRNAs	microRNAs
PGCs	primordial germ cells
piRNAs	Piwi-interacting RNAs
PVN	paraventricular nucleus
RISC	RNA-induced silencing complexes
SN	surrounded nucleolus
sncRNAs	small noncoding RNAs
sRNA	small RNA
Tet	ten-eleven translocation
tRNAs	transfer RNAs
tsRNAs	transfer RNA-derived small RNAs

GAMETE AND EMBRYO-FETAL ORIGIN OF DISEASES

At the beginning of this chapter, we have to mention the famous Barker hypothesis, also known as the thrifty phenotype hypothesis. This is a hypothesis proposed in 1990 by the British epidemiologist David Barker that intrauterine growth retardation, low birth weight, and premature birth have a causal relationship to the origins of hypertension, coronary heart disease, and noninsulin-dependent diabetes in middle age. Barker's hypothesis is derived from a historical cohort study that revealed a significant association between the occurrence of hypertension and coronary heart disease in middle age and premature birth or low birth weight. The proposition that a baby's nourishment in utero and during infancy determines subsequent development is a major determinant of health and disease later in life.

Actually, earlier evidence indicated that early postnatal events increase the risk of some chronic diseases in later life. In 1934, a landmark paper published in the *Lancet* found that death rates from all causes in the United Kingdom and Sweden decreased with each successive year-of-birth cohort between 1751 and 1930, indicating that the health of the child is determined by the environmental conditions existing during early life (0–15 years), and that the health of the adult is largely determined by the physical constitution of the child [1]. In 1977, Forsdahl et al. [2, 3] discovered poverty in childhood followed by prosperity in later life is a risk factor for arteriosclerotic heart disease.

Studies in the UK a decade later shifted the focus back to prenatal rather than postnatal events. In 1989, Barker et al. [4] examined relationships between postneonatal mortality for the period 1911–25 and later adult mortality in 1968–78. They found that regional differences in stroke and coronary heart disease mortality were predicted by birth weight. Barker subsequently showed that lower birth weights, and weight at one year, were associated with an increased risk of death from stroke and coronary heart disease in adults. This proposed that the roots of cardiovascular disease lay in the effects of poverty on the mother and undernutrition in fetal life and early infancy [5]. Subsequent studies in the UK, Europe, the United States, and China have confirmed these findings and showed that it is restricted fetal growth rather than preterm delivery that carries the risk of later adult diseases [6]. These observations have been collectively termed the "Barker hypothesis."

The embryonic and fetal periods are clearly vulnerable to environmental factors, and acquired changes can persist transgenerationally despite the lack of continued exposure. A possible explanation of such an outcome is epigenetic regulation of the human genome. In 2010, Motrenko [7] claimed the embryo-fetal origin of diseases, proposing that the abnormal development of the early embryo could induce poor health status after birth. Thus, it has been hypothesized that the adaptive responses of the gamete/embryo reacting with adverse factors (culture systems and manipulations in assisted reproductive technology (ART), endocrine disrupting chemicals, toxins, etc.) make it extremely susceptible for permanent damage on organ function and structure. This can then induce chronic adult diseases (such as diabetes, impaired glucose intolerance, and insulin resistance) related to the epigenetic reprogramming and development of individuals. What could be a reason for worry is the fact that such changes in the offspring can be passed on to subsequent generations, producing transgenerational altered epigenetic reprogramming.

Most human physiological systems and organs begin to develop early in gestation but become fully mature only after birth. A relatively long gestation and period of postnatal maturation allow for prolonged pre- and postnatal interactions with the environment. The primary determinants of fetal growth are genes, the integrity of the fetoplacental unit, and the appropriate endocrine environment that is largely represented by insulin action and the insulin-like growth factor system [8, 9]. Normal fetal growth and development take place in two phases, the embryonic and fetal phases. The embryonic phase consists of the proliferation, organization, and differentiation of the embryo whereas the fetal phase describes the continued growth and functional maturation of different tissues and organs [8, 9]. Embryonic and fetal periods are clearly vulnerable to environmental factors. The acquired changes can exist persistently, even transgenerationally, despite the lack of continued adverse exposure. One possible explanation is the epigenetic regulation of the human genome where changes in gene expression or cellular phenotype result from mechanisms other than changes in the underlying DNA [10]. Passing such changes to offspring may result in transgenerational epigenetic reprogramming with transmission of adverse traits and characteristics to offspring. Besides embryonic and fetal periods, the period of gametogenesis is an important and more vulnerable developmental stage with the programming and reprogramming process.

This chapter introduces epigenetic modification in germ cells and the growing body of evidence from epidemiological observations and clinical and experimental animal studies that supports the intergenerational effects on fetal programming associated with the gamete and embryo-fetal origins of the diseases.

EPIGENETIC PATTERNS IN MALE GERM CELLS

Epigenetics is the study of stable heritable changes in gene functions that do not involve DNA sequence changes, referring to nonsequence-based mechanisms that control gene expression. The paternal epigenome plays an important role in the developing embryo, which is not limited to nucleosome retention data. To date, three main mechanisms—DNA methylation, histone modifications, and RNA-associated silencing—have been associated with epigenetic silencing of gene expression [11]. The sperm epigenetic program is unique and tailored to meet the needs of this highly specialized cell. Chromatin changes in sperm contribute to virtually every function that the male gamete must perform throughout spermatogenesis and in the mature cell [12]. But the requisite replacement of canonical histones with sperm-specific protamine proteins has called into question the utility of the paternal epigenome in embryonic development [13]. The protamination of sperm chromatin provides the compaction necessary for safe delivery to the oocyte, but removes histones that are capable of eliciting gene activation or silencing via tail modifications [14]. In effect, protamination removes a potentially informative epigenetic layer from the paternal chromatin, leading to the previously held belief that sperm are incompetent to drive epigenetic changes in the embryo and suggesting that their utility is found only in the delivery of an undamaged DNA blueprint [15].

DNA methylation, one of the best-characterized DNA modifications associated with the modulation of gene activity, is a common regulatory mark found on the 5 carbon of cytosine residues (5mC) at cytosine-phosphate-guanine dinucleotides (CpGs), which exert strong epigenetic regulation in many cell types [16]. In mammals, maternal and paternal alleles of most genes are expressed at similar levels, but some genes behave differently, depending on their parent of origin. These genes are called imprinted genes and are regulated by DNA methylation in the differentially methylated region (DMR) during gametogenesis. Nonimprinted genes acquire their methylation similarly to imprinted genes; however, after fertilization, both the maternal and paternal genomes become demethylated while imprinted genes retain their methylation status [17]. Some repeat sequences appear to escape demethylation completely during gametogenesis, and retain a high proportion of their initial methylation marking during preimplantation development [18]. The founding cells of the germline, the primordial germ cells (PGCs), are thought to carry full complements of parental methylation profiles when they begin migrating toward the genital ridge [19]. Upon entry into the genital ridge, they undergo extensive genome-wide demethylation [20]. Early studies in mice employing methylation-sensitive restriction enzymes as well as Southern blot and PCR approaches indicated that PGCs have completely demethylated genomes by 13.5 days of gestation [21–23]. A number of imprinted genes, including Peg3, Kcnq1ot1, Snrpn, H19, Rasgrf1, and Gtl2 as well as nonimprinted genes such as α-actin become demethylated between 10.5 and 13.5 days of gestation [24]. However, certain sequences (at least some repetitive elements) appear to be treated differently: IAP, LINE-1, and minor satellite sequences are only subject to partial demethylation whereas most imprinted and single-copy genes become demethylated [25, 26]. Rapid, and possibly active, genome-wide erasure of methylation patterns takes place between 10.5 and 12.5 days of gestation, leaving PGCs of both sexes in an equivalent epigenetic state by embryonic day 13.5 [27, 28]. Following demethylation in PGCs, male and female gametes acquire sex- and sequence-specific genomic methylation patterns. For nonimprinted genes and repeat sequences, DNA methylation can be assessed directly. For imprinted genes, determination of DNA methylation status and assessment of mono- or biallelic expression of the genes of interest in the resulting embryos are necessary. The second genome-wide demethylation occurs in the early embryo. Marks established on imprinted genes and some repeat sequences must be faithfully maintained during preimplantation development at a time when the methylation of nonimprinted sequences is lost.

Protamination creates a highly condensed nuclear structure that helps to enable proper motility and protects DNA from damage. Although incorporation of these unique sperm-specific proteins results in a quiescent chromatin structure, some regions retain histones and their associated modifications. Recent studies have found this nucleosome retention is programmatic, and not due to a result of random distribution [15]. In theory, this selective retention in sperm may allow for targeted gene activation or silencing in the embryo. Multiple histone variants found in sperm are essential during spermatogenesis as well as in the mature spermatozoa. Among these, the important nuclear proteins are histone 2A and B (H2A and H2B), histone 3 (H3), histone 4 (H4), and the testes variant (tH2B) [29]. Recent studies implicate aberrant histone methylation and/or acetylation in the mature sperm in various forms of infertility. The loss-of-function mutation of JmjC-domain-containing-histone demethylase 2A (JHDM2A), an enzyme with known H3K9 demethylase activity, reveals decreased transcription of transition protein 1 and P1 during spermatogenesis [30]. Additional studies demonstrated that varying degrees of infertility, including sterility, are correlated with perturbations in histone methylation.

As a terminally differentiated cell, the ejaculated spermatozoon is exquisitely specialized for delivering the paternal genome to the egg. The presence of RNA in the sperm nucleus is paradoxical if one assumes that it serves no function [31]. The selective retention of mRNAs and siRNAs when most cytoplasmic RNA is lost to the residual body (normally destroyed during sperm preparation) during remodeling argues against passive trapping, as does the evidence that sperm RNA can support protein synthesis de novo during capacitation [31]. Based on a heterologous model system, the spermatozoon delivers its RNA cargo to the oocyte [32]. One study showed that a c-Kit-derived heritable effect on hair color in the mouse was strongly influenced by the presence of aberrant levels of "scrambled" noncoding c-Kit RNA transferred by the spermatozoa of the affected individual [33]. The concept of active RNA-dependent translation in capacitating sperm is supported by reports showing that specific sperm RNAs are "consumed" during manual processing that supports capacitation of normal viable sperm in vitro [34]. An effect of sperm RNA on non-Mendelian inheritance of coat color in mice has also been reported [33]. The mechanism appeared to be through an RNA-mediated epigenetic and heritable paramutation effect. RNA transcripts colocalize with nucleosome-bound chromatin near the nuclear envelope in the mature sperm, as is the case with the insulin-like growth factor 2 (IGF2) locus [35]. Spermatozoal RNA transcripts are capable of inhibiting the protamination process and maintaining a histone-bound chromatin structure.

EPIGENETIC REGULATION DURING OOGENESIS AND OOCYTE MATURATION

The molecular changes that occur during oogenesis are important for oocytes to acquire developmental competence. In addition to transcription factors binding to promoters, regulation of transcription may be achieved through epigenetic mechanisms. Epigenetic mechanisms in oocytes and early embryos include chromatin remodeling, DNA methylation, histone modification, and noncoding RNAs [36].

During mammalian oocyte growth, the nucleus of the oocyte arrests at the diplotene stage, termed the germinal vesicle (GV), and undergoes chromatin remodeling for control of gene expression. GV oocytes from murine antral follicles divide into two groups, according to the chromatin distribution in the nucleus: (I) surrounded nucleolus (SN) oocytes, with rather condensed chromatin surrounding the nucleolus; and (II) nonsurrounded nucleolus (NSN) oocytes, with more dispersed chromatin not surrounding the nucleolus [37]. During oocyte growth and maturation, the GV chromatin configuration varies in different species. In mice, oocytes in preantral follicles with a diameter between ~10–40 μm have NSN configuration. SN oocytes can only be found in antral follicles and the proportion of SN configuration increases as the size of the oocytes becomes larger. SN oocytes are silent during transcription while NSN oocytes are actively transcribing. Transcriptional repression is associated with meiotic competence in fully grown GV oocytes [38]. The percentage of oocytes that resume meiosis is higher in SN oocytes than in NSN oocytes. After fertilization, NSN oocytes cannot develop beyond the two-cell stage while a proportion of SN oocytes can develop to blastocyst [39]. Thus, the chromatin configuration is highly related to oocyte developmental competence.

Up to now, >100 imprinted genes have been identified in mammals and most of them are maternally imprinted. In female germlines, DNA methylation is erased during the differentiation of primordial germ cells, and de novo DNA methylation initiates asynchronously during the growth phase of the diplotene-arrested oocyte [40]. DNA methylation is established in an oocyte in a size-dependent manner, and the maternal methylation imprints become fully established by the fully grown oocyte stages [41]. DNA methyltransferases (Dnmts) are a family of enzymes that catalyzes the transfer of a methyl group to DNA and every member fulfills different functions respectively during DNA methylation. Dnmt3a and Dnmt3b are responsible for establishing de novo CpG methylation while Dnmt1 maintains the methylation pattern during chromosome replication. The activity of Dnmt3a and Dnmt3b is catalyzed by a related protein, Dnmt3L. Dnmt3L is highly expressed in germ cells and forms a complex with Dnmt3a and Dnmt3b [40]. Although Dnmt3b is dispensable for the establishment of maternal imprints, Dnmt3a and Dnmt3L are both necessary to establish maternal imprints in growing oocytes. Dnmt1o, the oocyte-specific isoform of Dnmt1s methyltransferase, is produced in oocytes and maintains the CpG methylation in oocytes and preimplantation embryos [42].

In addition to DNA methylation, histone modification plays an important role in controlling gene expression in gametes and early embryos. The nucleosome is the fundamental building component of chromatin. It is composed of 147 base pairs of genomic DNA and an octamer of two subunits of each of the core histones H2A, H2B, H3, and H4. The amino-terminal portion of the histone protein contains a flexible and highly basic tail region that is subject to various posttranslational modifications, including acetylation, phosphorylation, methylation, ADP-ribosylation, ubiquitination, sumoylation, biotinylation, and proline isomerization [43]. Histone acetylation is associated with enhanced transcriptional activity whereas histone deacetylation is correlated with repression of gene expression. The acetylation of H3 and H4 is more extensively studied than H2A and H2B. The level of acetylation on histone H3 and H4 increases during oocyte growth and generally all the lysine residues are acetylated in fully grown GV oocytes. However, with the resumption of meiosis, deacetylation will take place in several lysine residues and reach its peak in MII oocytes. Histone (de)acetylation is related to chromatin remodeling during oocyte growth and is necessary for the binding of a chromatin remodeling protein to the centromeric heterochromatin, an essential step for the correct alignment of the chromosomes [42]. Although all core histones contain phosphor-acceptor sites, the phosphorylation of serine 10 and 28 residue on histone H3 (H3/Ser10ph and H3/Ser28ph) is the most extensively characterized. However, studies on the distribution and expression of H3/Ser10ph and H3/Ser28ph during oocyte maturation are discordant. The phosphorylation level of both H3/Ser10ph and H3/Ser28ph increases as oocytes proceed to the MI stage, but the distribution patterns are different between them. Although phosphorylation of H3/Ser10ph correlates with chromosome condensation in mitotic cells, recent studies demonstrated that there are no relationships between H3/Ser10ph phosphorylation and chromosome condensation. However, H3/Ser28ph may be associated with chromosome condensation in oocytes [43].

In contrast to acetylation and phosphorylation, histone methylation is relatively stable during oocyte maturation. The main methylation sites are the basic amino acid side chains of lysine (K) and arginine (R) residues. Histone methylation may contribute to the establishment and maintenance of an imprinted pattern of gene expression together with DNA methylation [42,43].

From the fully grown oocyte stage until zygotic genome activation (ZGA), the genome is transcriptionally silent. During this period, all mRNA regulation must occur posttranscriptionally. Transcripts expressed by oocytes will support the maturation, fertilization, and early stages of embryonic development. By the MII stage, more than half the mRNA stored in oocytes will be degraded. Small noncoding RNAs are implicated in the elimination of maternal mRNAs [44]. Small noncoding RNAs range in size from 18 to 32 nucleotides (nt) in length and play a critical role in posttranscriptional regulation. Three major classes of small, noncoding RNAs have been identified in mammals: microRNAs (miRNAs), endogenous small interfering (endo-siRNAs), and Piwi-interacting RNAs (piRNAs). There are two subclasses of miRNAs, canonical and noncanonical miRNAs. Canonical miRNAs will be processed by the Drosha-Dgcr8 complex to form premiRNAs transported from the nucleus into the cytoplasm while noncanonical miRNAs can bypass the Drosha-Dgcr8 step [45]. Both miRNAs and endo-siRNAs involve RNase III enzyme Dicer processing with Dicer products being assembled into ribonucleoprotein complexes called RNA-induced silencing complexes (RISC). RISC binds the target RNA and silences gene expression by cleaving the target RNA [46]. The key components of RISC are proteins of the Argonaute (Ago) family. In mammals, four Ago proteins function in miRNA repression but only Ago2 functions in siRNA repression. Ago2 is maternally expressed and plays an essential role in the degradation of maternal mRNAs. In contrast, piRNAs do not require Dicer processing and are expressed predominantly in the germlines of mammals. They are able to interact with the piwi proteins, a distinct family of the Argonaute family [45]. Previous studies demonstrated that miRNAs, such as Let-7, Mir22, Mir16-1 and Mir29, are present in oocytes. Furthermore, mRNA profiling and bioinformatic analyses show that many targets for the expressed miRNAs are also present in oocytes. However, there is little miRNA function in mature oocytes [47] and the mRNA levels of oocytes from Dgcr8 mutant mice are unchanged [48]. These results suggest that miRNA function is down-regulated during oocyte development. In contrast, endo-siRNAs may play an essential role in oocyte meiosis. Dicer loss in oocytes shows hundreds of misregulated transcripts and results in meiosis arrest with abnormal spindles and severe chromosome congression defects [49]. Ago2-deficient oocytes have similar phenotypes [50]. These results suggest a role for the endo-siRNA in regulating oocyte meiosis. piRNAs have been identified in growing oocytes, but deletion of Piwi proteins has no oocyte phenotype [45]. The role of piRNAs in oogenesis remains to be defined.

DYNAMIC EPIGENETIC MODIFICATION

Although a fertilized egg inherits a haploid set of chromosomes from both the egg and the sperm, in mammals these maternal and paternal gametes do not contribute equal genetic functions to the developing diploid embryo. Functional differences exist between the two sets of parental chromosomes due to "genomic imprinting." Imprinting is a particularly important genetic mechanism in mammals, and may influence the transfer of nutrients to the fetus and the newborn from the mother. That differentially marks the maternally and paternally inherited chromosome homologues and results in particular genes being expressed or repressed in response to this parent-specific modification. The effect of the imprint on gene activity is to allow the expression of some imprinted genes from either the maternally inherited chromosome or others from the paternally inherited chromosome. Aberrant imprinting can have profound effects on mammalian embryonic development and cause human disease.

Parental imprinting may be a functional reflection of different patterns of DNA methylation during gametogenesis in males and females. The methylation pattern that a zygote inherits is transmitted to all daughter cells after mitosis, even in postnatal life. When gametes enter meiosis, the original imprints on the chromosomes are erased and new imprints are made. Sperm chromosomes receive a paternal imprint and egg chromosomes receive a maternal imprint. As mentioned previously, imprints are "established" during the development of germ cells into sperm or eggs, and are necessary for normal fetal development. After fertilization, they are "maintained" as chromosomes duplicate and segregate in the developing organism. In the germ cells of the new organism, imprints are "erased" at an early stage. This is followed by establishment again at a later stage of germ-cell development, thus completing the imprinting cycle. In somatic cells, imprints are maintained and modified during development.

The resetting of the imprint is a critical portion of epigenetic reprogramming in germ cells. For most imprints, current evidence indicates that there might be two stages for this resetting process, the step of "erasure" and the later step of "establishment." Several studies have now demonstrated that the erasure of at least methylation imprints occurs in the germline and is completed by embryonic day 12–13 (E12–13) in both sexes after the PGCs enter the gonadal ridge [51]. Indeed, in female embryo germ cells-somatic hybrid cells there were striking changes in the methylation of the somatic nucleus, resulting in the demethylation of several imprinted and nonimprinted genes; whether this demethylation is active or passive is not known [52]. The evidence so far indicates that all methylation imprints are probably erased at this stage. This is important because it implies that imprints inherited from a parent with the same sex as the developing embryo are erased and are unlikely to persist unchanged. Methylation imprints are still present and may be functionally intact before the erasure stage. In addition to methylation imprints, differential replication of DNA is also apparently erased in both germlines. In the female germline this coincides with demethylation, but in the male germline it occurs substantially later, after birth [53]. After erasure of germline methylation imprints, differentiating germ cell genomes must become maternalized or paternalized depending on germ cell sex, and this must occur before the onset of meiosis. In the female germline, imprints are reestablished in growing oocytes. Various imprinted genes receive an "imprinting mark" asynchronously at particular stages during oocyte meiotic prophase I, during the transition from primordial to antral follicles [25, 54]. In the process of imprinting establishment, de novo methylation begins in both germlines at late fetal stages and continues after birth. It is not fully understood which enzymes are responsible for de novo methylation in germ cells. Dnmt1 and its germ cell specific isoforms, Dnmt3a or Dnmt3b, are candidates that are required for de novo methylation in postimplantation embryos [55]. DNA methylation in mammals occurs in the dinucleotide CpG. Allele-specific methylation patterns are maintained due to the presence of repetitive sequence regions near DMRs. Methyl groups can be introduced into unmethylated DNA by the de novo methylation enzymes Dnmt3a and Dnmt3b (and perhaps others). Once the imprints are introduced into the parental germlines, maintained in the early embryo, and fully matured during differentiation, they need to be "read." Reading means the conversion of methylation or chromatin imprints into differential gene expression. Differential gene expression occurs largely at the level of transcription, although posttranscriptional mechanisms may also exist.

As mentioned previously, during germ cell and early embryonic development, the most sensitive and vulnerable period of epigenetic reprogramming, exposure to an adverse environment leads to abnormal methylation and, possibly, long-term health problems. DNA methylation, a major epigenetic mechanism for gene silencing, is recognized to be responsible for the stability of gene expression status. The majority of CpGs in mammalian genomes are methylated. Dnmt 3A and 3B are essential for de novo methylation, and Dnmt1 maintains methylation patterns during cell division [56]. Establishment and maintenance of cell type-specific DNA methylation patterns are dependent on both methylation and demethylation. DNA demethylation is the process of the removal of a methyl group from nucleotides in DNA, which can be passive or active. It has been generally understood that passive DNA demethylation occurs by a reduction in activity or absence of Dnmts whereas the mechanism of active DNA demethylation has been controversial in recent decades.

Recently, three enzyme families have been implicated in active DNA demethylation via DNA repair. The first is the ten-eleven translocation (Tet) family of enzymes. 5mC can be hydroxylated by Tet to form 5-hydroxymethylcytosine (5hmC), which can be further oxidized to 5-formylcytosine (5fC) and 5-carboxylcytosine (5caC). The second family is the AID/APOBEC family. 5mC (or 5hmC) can be deaminated by AID/APOBEC family members to form 5-methyluracil (5mU) or 5-hydroxymethyluracil (5hmU). The third is the UDG family of base excision repair glycosylases. TDG and SMUG1 replace these intermediates (i.e., 5mU, 5hmU, or 5caC), culminating in cytosine replacement and DNA demethylation [57–59]. Environment and nutrition have been confirmed to affect epigenetic modification. Although it has been documented that hyperglycemia induces demethylation of specific cytosines throughout the genome [60, 61], whether the demethylation could be persistent and the mechanism involved need to be further investigated.

STUDIES ON INTERGENERATIONAL TRANSMISSION

Epigenetic modifications such as DNA methylation and histone modification play critical roles during embryogenesis. However, the knowledge of how much epigenetic information in gametes can be transferred to the next generation is still unclear.

Previous studies involving animal models revealed that poor nutrition during pregnancy can significantly affect the health of F1 offspring and even F2 offspring. Female guinea pigs that were undernourished during pregnancy produced F1 male offspring with cardiovascular defects. Although F1 females were not tested for these effects, they were mated without further dietary manipulation to produce the F2 offspring that had similar cardiovascular anomalies as the F1 males [62]. In rodent models, if F0 females were fed low-protein diets during pregnancy, they would give rise to F1 offspring with reduced birth weight, increased insulin sensitivity, high cholesterol, high body fat content, hypertension, and altered kidney morphology [63–65]. These abnormal phenotypes were transmitted to the F2 generation despite a normal diet being given to both the F1 and F2 offspring. Some studies have even shown that all the F1, F2, and F3 generations derived from F0 female mice that were fed a protein-restricted diet maintained low serum insulin levels and had altered pancreatic islet morphology [66]. Because the F3 offspring were not directly exposed to maternal protein restriction, it is possible that this intergenerational transmission may be epigenetically inherited through the germline.

In addition to maternal exposure, the offspring of prenatally undernourished fathers, but not mothers, were heavier and more obese than the offspring of parents with a normal diet [67, 68]. Mice fed a low-protein diet passed on a high-cholesterol phenotype with gene expression differences and modest DNA methylation differences to their paternal offspring [69]. Recent studies in rodent models have focused on nutritional effects transmitted via the paternal lineage. Such paternal-lineage risk is likely to be conferred via sperm, although whether this is via alterations in chromatin, small RNAs, or other agents is currently unclear.

For epigenetic epidemiology, the findings linking over- or undernutrition during pregnancy to epigenetic changes in offspring naturally raise the opportunity for epigenomic surveys in humans to potentially identify targets for therapeutic intervention. Indeed, there are numerous studies in humans examining the relationship between fetal nutrient availability and epigenetic modifications in the offspring. At present, many of these are confounded by small sample size, cellular heterogeneity of the tissues examined, and lack of validation. For example, most DNA methylation assays are performed in total peripheral blood monocytes where the unique methylation profiles of the various cellular lineages complicate interpretation of the data. Despite these issues, multiple studies in diverse populations report changes in DNA methylation associated with low birth weight and/or altered nutrient availability. Thus, it is likely that an adverse in utero milieu does indeed induce epigenetic modifications in the offspring, but whether these modifications have biological relevance remains to be determined. The field of "epigenetic epidemiology" remains an active and growing field of investigation, and researchers anticipate exciting advances in this area in the coming years [70].

Adverse prenatal environments can promote metabolic disease in offspring and subsequent generations. Animal models and epidemiological data implicate epigenetic inheritance, but the mechanisms remain unknown. In an intergenerational developmental programming model affecting the F2 mouse metabolism, Radford et al. [71] demonstrated that the in utero nutritional environment of F1 embryos alters the germline DNA methylome of F1 adult males in a locus-specific manner. Differentially methylated regions are hypomethylated and enriched in nucleosome-retaining regions. A substantial fraction is resistant to early embryo methylation reprogramming, potentially impacting F2 development.

Differential methylation is not maintained in F2 tissues, yet locus-specific expression is perturbed. Thus, in utero nutritional exposure during critical windows of germ cell development can impact the male germline methylome, associated with metabolic disease in offspring.

Evidence from animal studies and human famines suggests that starvation may affect the health of the progeny of famished individuals. However, it is not clear whether starvation affects only immediate offspring or has lasting effects; it is also unclear how such epigenetic information is inherited. Small RNA-induced gene silencing can persist over several generations via transgenerationally inherited small RNA molecules in *Caenorhabditis elegans*, but all known transgenerational silencing responses are directed against foreign DNA introduced into the organism. Rechavi et al. [72] found that starvation-induced developmental arrest, a natural and drastic environmental change, leads to the generation of small RNAs that are inherited through at least three consecutive generations. These small endogenous transgenerationally transmitted RNAs target genes with roles in nutrition. They defined genes that are essential for this multigenerational effect. Moreover, we show that the F3 offspring of starved animals show an increased lifespan, corroborating the notion of a transgenerational memory of past conditions. This is the first study that demonstrates that an endogenous response could trigger small RNA inheritance. The targets of these inherited small RNAs include sets of genes involved in nutrition. This epigenetic response is compromised in rde-4 mutants, and transgenerational transmission of the starvation-induced small RNAs depends on the germline-expressed nuclear argonaute HRDE-1.

To investigate the potential mechanisms by which paternal stress may contribute to offspring hypothalamic-pituitary-adrenal (HPA) axis dysregulation, Rodgers et al. [73] exposed mice to 6 weeks of chronic stress prior to breeding. As epidemiological studies support variation in paternal germ cell susceptibility to reprogramming across the lifespan, male stress exposure occurred either throughout puberty or in adulthood. Remarkably, offspring of sires from both paternal stress groups displayed significantly reduced HPA axis stress responsivity. Gene set enrichment analyses in offspring stress-regulating brain regions, the paraventricular nucleus (PVN) and the bed nucleus of stria terminalis (BNST), revealed global pattern changes in transcription suggestive of epigenetic reprogramming and consistent with altered offspring stress responsivity, including increased expression of glucocorticoid-responsive genes in the PVN. In examining potential epigenetic mechanisms of germ cell transmission, they found robust changes in sperm miRNA (miR) content where nine specific miRs were significantly increased in both paternal stress groups. These results demonstrate that paternal experience across the lifespan can induce germ cell epigenetic reprogramming and impact offspring HPA stress axis regulation, and may therefore offer novel insight into factors influencing neuropsychiatric disease risk.

In a recent study, through zygote microinjection of nine specific sperm miRs previously identified in our paternal stress mouse model, Rodgers et al. [74] further demonstrated that sperm miRs function to reduce maternal mRNA stores in early zygotes, ultimately reprogramming gene expression in the offspring hypothalamus and recapitulating the offspring stress dysregulation phenotype. Small noncoding RNAs (sncRNAs) are potential mediators of gene-environment interactions that can relay signals from the environment to the genome and exert regulatory functions on gene activity. Gapp et al. [75] reported that environmental conditions involving traumatic stress in early life in mice altered miRNAs expression as well as behavioral and metabolic responses in the progeny. Several miRNAs were affected in the serum and brain of both, the traumatized animals and their progeny when adults, but also in the sperm of traumatized males. Injection of sperm RNAs from these males into fertilized wild-type oocytes reproduced the behavioral and metabolic alterations in the resulting offspring. These results strongly suggest that sncRNAs are sensitive to environmental factors in early life and contribute to the inheritance of trauma-induced phenotypes across generations.

RESEARCH PROGRESS IN FETAL PROGRAMMING

Our knowledge of how offspring inherit the DNA methylome from parents is limited. Jiang et al. [76] generated nine single-base resolution DNA methylomes, including zebrafish gametes and early embryos. The oocyte methylome is significantly hypomethylated compared to sperm. Strikingly, the paternal DNA methylation pattern is maintained throughout early embryogenesis. The maternal DNA methylation pattern is maintained until the 16-cell stage. Then, the oocyte methylome is gradually discarded through cell division and is progressively reprogrammed to a pattern similar to that of the sperm methylome. The passive demethylation rate and the de novo methylation rate are similar in the maternal DNA. By the midblastula stage, the embryo's methylome is virtually identical to the sperm methylome. Moreover, inheritance of the sperm methylome facilitates the epigenetic regulation of embryogenesis. Therefore, besides DNA sequences, sperm DNA methylome is also inherited in zebrafish early embryos.

The reprogramming of parental methylomes is essential for embryonic development. In mammals, paternal 5mCs have been proposed to be actively converted to oxidized bases. These paternal oxidized bases and maternal 5mCs are believed to be passively diluted by cell divisions. By generating single-base resolution, allele-specific DNA methylomes from mouse gametes, early embryos, and PGCs as well as single base-resolution maps of oxidized cytosine bases for early embryos, Wang et al. [77] reported the existence of 5hmC and 5fC in both maternal and paternal genomes and found that 5mC or its oxidized derivatives, at the majority of demethylated CpGs, are converted to unmodified cytosines independent of passive dilution from gametes to four-cell embryos. Therefore, paternal methylome and at least a significant proportion of maternal methylome go through active demethylation during embryonic development. Additionally, all the known imprinting control regions (ICRs) were classified into germline or somatic ICRs. This study refines the knowledge of the inheritance and reprogramming of parental methylomes in mammals and also provides a powerful resource for future developmental studies.

Resetting the epigenome in human primordial germ cells (hPGCs) is critical for development. Tang et al. [78] showed that the transcriptional program of hPGCs is distinct from that in mice, with coexpression of somatic specifiers and naive pluripotency genes TFCP2L1 and KLF4. This unique gene regulatory network, established by SOX17 and BLIMP1, drives comprehensive germline DNA demethylation by repressing DNA methylation pathways and activating TET-mediated hydroxymethylation. Base-resolution methylome analysis reveals progressive DNA demethylation to basal levels in week 5–7 in vivo hPGCs. Concurrently, hPGCs undergo chromatin reorganization, X reactivation, and imprint erasure. Despite global hypomethylation, evolutionarily young and potentially hazardous retroelements such as SVA remain methylated. Remarkably, some loci associated with metabolic and neurological disorders are also resistant to DNA demethylation, revealing the potential for transgenerational epigenetic inheritance that may have phenotypic consequences. This study provided comprehensive insight on the early human germline transcriptional network and epigenetic reprogramming that subsequently impacts human development and disease.

Global DNA demethylation in humans is a fundamental process that occurs in preimplantation embryos and reversion to naive ground state pluripotent stem cells (PSCs). However, the extent of DNA methylation reprogramming in human germline cells is unknown. Gkountela et al. [79] performed whole-genome bisulfite sequencing (WGBS) and RNA-sequencing (RNA-seq) of human prenatal germline cells from 53 to 137 days of development. They discovered that the transcriptome and methylome of the human germline is distinct from both human PSCs and the inner cell mass (ICM) of human blastocysts. Using this resource to monitor the outcome of global DNA demethylation with reversion of primed PSCs to the naive ground state, we uncovered hotspots of ultralow methylation at transposons that are protected from demethylation in the germline and ICM. Taken together, the human germline serves as a valuable in vivo tool for monitoring the epigenome of cells that have emerged from a global DNA demethylation event.

Germ cells are vital for transmitting genetic information from one generation to the next and for maintaining the continuation of species. Guo et al. [80] analyzed the transcriptomes of 233 individual male and female human PGCs from 15 embryos at between 4 and 19 weeks of gestation as well as 86 neighboring somatic cells from 13 of these embryos using a single-cell RNA sequencing (RNA-seq) method. Furthermore, they analyzed the DNA methylomes of both male and female human gonadal PGCs as well as the neighboring somatic cells of 11 of these embryos at between 7 and 19 weeks of gestation using whole-genome bisulfite sequencing (WGBS). For the first time, the transcriptome of human PGCs of migrating and gonadal stages was analyzed at single-cell and single-base resolutions. Human PGCs show unique transcription patterns involving the simultaneous expression of both pluripotency genes and germline-specific genes, with a subset of them displaying developmental stage-specific features. They found that, during the 7 weeks of development in the 4–11 week embryos, the transcriptomes of the PGCs were stable in general with only several hundreds of genes changing their expression significantly. By contrast, the global DNA methylation was drastically decreased to 7.8% in the 11-week male PGCs and 6.0% in the 10-week female PGCs during this developmental period. This is very similar to the patterns in mice. How the PGCs maintain a relatively stable RNA expression pattern when the DNA methylation is globally removed warrants further analysis. The work paves the way for deciphering the complex epigenetic reprogramming of the germline with the aim of restoring totipotency in fertilized oocytes.

There may be a combination of molecular mechanisms underlying paternal transgenerational epigenetic inheritance involving changes in histone states and/or RNA in sperm. The function of sperm histones and their modifications in embryonic development, offspring health, and epigenetic inheritance is unknown. By overexpressing the human KDM1A histone lysine 4 demethylase during mouse spermatogenesis, Siklenka et al. [81] generated a mouse model producing spermatozoa with reduced H3K4me2 within the CpG islands of genes implicated in development, and studied the development and fitness of the offspring. Male transgenic offspring were bred with C57BL/6 females, generating the experimental heterozygous transgenic (TG) and nontransgenic (nonTG) brothers. Each generation from TG and non-TG animals (F1 to F3 in the transgenerational studies) was bred with C57BL/6 females, and the offspring (pups from generations F1 to F4) were analyzed for intergenerational and transgenerational effects. They found that KDM1A overexpression in one generation severely impaired the development and survivability of offspring. These defects lasted for two subsequent generations in the absence of KDM1A germline expression. They characterized histone and DNA methylation states in the sperm of TG and non-TG sires. Overexpression of KDM1A was associated with a specific loss of H3K4me2 at >2300 genes, including many developmental regulatory genes. Unlike in other examples of paternal transgenerational inheritance, we observed no changes in sperm DNA methylation associated with primarily CpG-enriched regions. Instead, they measured robust and analogous changes in sperm RNA content of TG and non-TG descendants as well as in their offspring at the two-cell stage. These changes in expression and the phenotypic abnormalities observed in offspring correlated with altered histone methylation levels at genes in sperm. This study demonstrated that KDM1A activity during sperm development has major developmental consequences for offspring and implicates histone methylation and sperm RNA as potential mediators of transgenerational inheritance. The data emphasized the complexity of transgenerational epigenetic inheritance likely involving multiple molecular factors, including the establishment of chromatin states in spermatogenesis and sperm-borne RNA. These findings demonstrated the potential of histone methylation as a molecular mechanism underlying paternal epigenetic inheritance.

Increasing evidence indicates that metabolic disorders in offspring can result from the father's diet, but the mechanism remains unclear. In a paternal mouse model given a high-fat diet (HFD), Chen et al. [82] showed that a subset of sperm transfer RNA-derived small RNAs (tsRNAs), mainly from 5′ transfer RNA halves and ranging in size from 30 to 34 nucleotides, exhibited changes in expression profiles and RNA modifications. Injection of sperm tsRNA fractions from HFD males into normal zygotes generated metabolic disorders in the F1 offspring and altered gene expression of metabolic pathways in early embryos and islets of F1 offspring, which was unrelated to DNA methylation at CpG-enriched regions. Hence, sperm tsRNAs represent a paternal epigenetic factor that may mediate the intergenerational inheritance of diet-induced metabolic disorders. Sharma et al. [83] also found that protein restriction in mice affects small RNA (sRNA) levels in mature sperm, with decreased let-7 levels and increased amounts of 5′ fragments of glycine transfer RNAs (tRNAs). In testicular sperm, tRNA fragments are scarce but increase in abundance as sperm mature in the epididymis. Epididymosomes (vesicles that fuse with sperm during epididymal transit) carry RNA payloads matching those of mature sperm and can deliver RNAs to immature sperm in vitro. Functionally, tRNA-glycine-GCC fragments repress genes associated with the endogenous retroelement MERVL in both embryonic stem cells and embryos. The results shed light on sRNA biogenesis and its dietary regulation during posttesticular sperm maturation, and they also link tRNA fragments to regulation of endogenous retro elements active in the preimplantation embryo.

Gametes carry parental genetic material to the next generation. Stress-induced epigenetic changes in the germline can be inherited and can have a profound impact on offspring development. However, the molecular mechanisms and consequences of transgenerational epigenetic inheritance are poorly understood. Zenk et al. [84] found that Drosophila oocytes transmit the repressive histone mark H3K27me3 to their offspring. Maternal contribution of the histone methyltransferase Enhancer of zeste, the enzymatic component of Polycomb repressive complex 2, is required for active propagation of H3K27me3 during early embryogenesis. H3K27me3 in the early embryo prevents aberrant accumulation of the active histone mark H3K27ac at regulatory regions and precocious activation of lineage-specific genes at zygotic genome activation. Disruption of the germline-inherited Polycomb epigenetic memory causes embryonic lethality that cannot be rescued by late zygotic reestablishment of H3K27me3. Thus, maternally inherited H3K27me3, propagated in the early embryo, regulates the activation of enhancers and lineage-specific genes during development.

Hyperandrogenism is a common endocrine disorder that happens in 5%–10% of reproductive-aged women. Previous animal studies suggest that intrauterine exposure to elevated androgen levels could impair glucose tolerance in offspring. Whether this is true in humans has not been determined. Tian et al. [85] measured basal blood androgen levels in pregestational women and then performed glucose tolerance tests in their children at the age of 5 years. A total of 156 children from 147 women with hyperandrogenism (testosterone >2.4 nmol/L or dehydroepiandrosterone >8.8 μmol/L) were tested while 1060 children from 969 women served as control. They found that maternal hyperandrogenism significantly increased the chance, by nearly fourfold, of developing prediabetes in offspring children (fasting glucose 5.6–6.9 mmol/L). In addition, children born to the mothers with pregestational hyperandrogenism have significantly higher glucose and insulin levels during oral glucose tolerance tests, suggesting the development of insulin resistance. To explore the underlying molecular mechanism, the authors checked DNA methylation at imprinted genes previously implicated in the pathogenesis of diabetes [86]. They found that children born to mothers with pregestational hyperandrogenism have lower DNA methylation levels and higher gene expression levels in two imprinted genes, Igf2 and Grb10, in blood lymphocytes. The authors went on to show that the lower DNA methylation levels and higher gene expression levels of Igf2 occurred in both F0 oocytes and F1 pancreatic islets in a rat model of pregestational hyperandrogenism, which further supports a gamete-mediated mechanism in this maternal effect. These findings also suggest that pancreatic islets are a major target tissue contributing to the glucose intolerance phenotype in F1 offspring.

CONCLUSIONS

The combined epidemiological, clinical, and animal studies clearly demonstrate that the intrauterine environment influences both the growth and development of the fetus and the subsequent development of adult diseases. During gametogenesis and embryonic development, there are some specific critical windows. The stimulus or insult in these critical periods may have long-lasting consequences on tissue or organ function postnatally. Fetal programming provides a strong candidate mechanism in the pathogenesis of the fetal origins of adult disease. The association between low birth weight and subsequent disease has been well documented; however, it is also clear that fetal programming effects may not be limited to the first generation. There is some evidence from human studies that nongenetic effects may be associated with the intergenerational transmission of disease risk. However, any impact of such intergenerational effects will result from the interaction of genes and the pre- and postnatal environment, and will also occur against a heterogeneous background of genetic susceptibility. The increased epigenetic epidemiology about the surveys of epigenetic marks in children who experience adverse intrauterine environments will enlarge our knowledge regarding the mechanisms responsible for long-term metabolic reprogramming. Animal models of intergenerational transmission of disease will be invaluable for studying the environmental effects over multiple generations and will eventually help to shed some light on the mechanisms involved. Prevention of metabolic abnormalities will, of course, be one of the key goals for future efforts, as epigenetic marks provide promising candidates for therapeutic intervention and research efforts should be focused in this area. Policies aimed at improving the health of one generation, in particular those directed at improving maternal, fetal, and infant health, may have important benefits for a number of succeeding generations.

References

[1] Kermack WO, McKendrick AG, McKinlay PL. Death-rates in Great Britain and Sweden. Some general regularities and their significance. Int J Epidemiol 2001;30(4):678–83.

[2] Forsdahl A. Are poor living conditions in childhood and adolescence an important risk factor for arteriosclerotic heart disease? Br J Prev Soc Med 1977;31(2):91–5.

[3] Forsdahl A. Living conditions in childhood and subsequent development of risk factors for arteriosclerotic heart disease. The cardiovascular survey in Finnmark 1974–75. J Epidemiol Community Health 1978;32 (1):34–7.

[4] Barker DJ, Osmond C, Law CM. The intrauterine and early postnatal origins of cardiovascular disease and chronic bronchitis. J Epidemiol Community Health 1989;43(3):237–40.

[5] Barker DJ, Winter PD, Osmond C, Margetts B, Simmonds SJ. Weight in infancy and death from ischaemic heart disease. Lancet 1989;2(8663):577–80.

[6] Paneth N, Susser M. Early origin of coronary heart disease (the "Barker hypothesis"). BMJ 1995;310(6977):411–2.

[7] Motrenko T. Embryo-fetal origin of diseases—new approach on epigenetic reprogramming. Arch Perinat Med 2010;6:.

[8] Gluckman PD, Harding JE. Nutritional and hormonal regulation of fetal growth—evolving concepts. Acta Paediatr Suppl 1994;399:60–3.

[9] Kanaka-Gantenbein C, Mastorakos G, Chrousos GP. Endocrine-related causes and consequences of intrauterine growth retardation. Ann N Y Acad Sci 2003;997:150–7.

[10] Canani RB, Costanzo MD, Leone L, Bedogni G, Brambilla P, Cianfarani S, et al. Epigenetic mechanisms elicited by nutrition in early life. Nutr Res Rev 2011;24(2):198–205.

[11] Egger G, Liang G, Aparicio A, Jones PA. Epigenetics in human disease and prospects for epigenetic therapy Nature 2004;429(6990):457–63.

[12] Jenkins TG, Carrell DT. The sperm epigenome and potential implications for the developing embryo. Reproduction 2012;143(6):727–34.

[13] Orsi GA, Couble P, Loppin B. Epigenetic and replacement roles of histone variant H3.3 in reproduction and development. Int J Dev Biol 2009;53(2–3):231–43.

[14] Strahl BD, Allis CD. The language of covalent histone modifications. Nature 2000;403(6765):41–5.

[15] Hammoud SS, Nix DA, Zhang H, Purwar J, Carrell DT, Cairns BR. Distinctive chromatin in human sperm packages genes for embryo development. Nature 2009;460(7254):473–8.

[16] Portela A, Esteller M. Epigenetic modifications and human disease. Nat Biotechnol 2010;28(10):1057–68.

[17] Reik W. Stability and flexibility of epigenetic gene regulation in mammalian development. Nature 2007;447 (7143):425–32.

[18] Lucifero D, Chaillet JR, Trasler JM. Potential significance of genomic imprinting defects for reproduction and assisted reproductive technology. Hum Reprod Update 2004;10(1):3–18.

[19] Shiota K. DNA methylation profiles of CpG islands for cellular differentiation and development in mammals. Cytogenet Genome Res 2004;105(2–4):325–34.

[20] Kafri T, Gao X, Razin A. Mechanistic aspects of genome-wide demethylation in the preimplantation mouse embryo. Proc Natl Acad Sci U S A 1993;90(22):10558–62.

[21] Brandeis M, Ariel M, Cedar H. Dynamics of DNA methylation during development. Bioessays 1993;15(11):709–13.

[22] Chaillet JR, Vogt TF, Beier DR, Leder P. Parental-specific methylation of an imprinted transgene is established during gametogenesis and progressively changes during embryogenesis. Cell 1991;66(1):77–83.

[23] Kafri T, Ariel M, Brandeis M, Shemer R, Urven L, McCarrey J, et al. Developmental pattern of gene-specific DNA methylation in the mouse embryo and germ line. Genes Dev 1992;6(5):705–14.

[24] Li E. Chromatin modification and epigenetic reprogramming in mammalian development. Nat Rev Genet 2002;3(9):662–73.

[25] Hajkova P, Erhardt S, Lane N, Haaf T, El-Maarri O, Reik W, et al. Epigenetic reprogramming in mouse primordial germ cells. Mech Dev 2002;117(1–2):15–23.

[26] Lane N, Dean W, Erhardt S, Hajkova P, Surani A, Walter J, et al. Resistance of IAPs to methylation reprogramming may provide a mechanism for epigenetic inheritance in the mouse. Genesis 2003;35(2):88–93.

[27] Szabo PE, Mann JR. Biallelic expression of imprinted genes in the mouse germ line: implications for erasure, establishment, and mechanisms of genomic imprinting. Genes Dev 1995;9(15):1857–68.

[28] Szabo PE, Hubner K, Scholer H, Mann JR. Allele-specific expression of imprinted genes in mouse migratory primordial germ cells. Mech Dev 2002;115(1–2):157–60.

[29] Gatewood JM, Cook GR, Balhorn R, Schmid CW, Bradbury EM. Isolation of four core histones from human sperm chromatin representing a minor subset of somatic histones. J Biol Chem 1990;265(33):20662–6.

[30] Okada Y, Scott G, Ray MK, Mishina Y, Zhang Y. Histone demethylase JHDM2A is critical for Tnp1 and Prm1 transcription and spermatogenesis. Nature 2007;450(7166):119–23.

[31] Millar MR, Sharpe RM, Maguire SM, Gaughan J, West AP, Saunders PT. Localization of mRNAs by in-situ hybridization to the residual body at stages IX-X of the cycle of the rat seminiferous epithelium: fact or artefact? Int J Androl 1994;17(3):149–60.

[32] Ostermeier GC, Dix DJ, Miller D, Khatri P, Krawetz SA. Spermatozoal RNA profiles of normal fertile men. Lancet 2002;360(9335):772–7.

[33] Rassoulzadegan M, Grandjean V, Gounon P, Vincent S, Gillot I, Cuzin F. RNA-mediated non-Mendelian inheritance of an epigenetic change in the mouse. Nature 2006;441(7092):469–74.

[34] Lambard S, Galeraud-Denis I, Martin G, Levy R, Chocat A, Carreau S. Analysis and significance of mRNA in human ejaculated sperm from normozoospermic donors: relationship to sperm motility and capacitation. Mol Hum Reprod 2004;10(7):535–41.

[35] Boissonnas CC, Abdalaoui HE, Haelewyn V, Fauque P, Dupont JM, Gut I, et al. Specific epigenetic alterations of IGF2-H19 locus in spermatozoa from infertile men. Eur J Hum Genet 2010;18(1):73–80.

[36] Prather RS, Ross JW, Isom SC, Green JA. Transcriptional, post-transcriptional and epigenetic control of porcine oocyte maturation and embryogenesis. Soc Reprod Fertil Suppl 2009;66:165–76.

[37] Tan JH, Wang HL, Sun XS, Liu Y, Sui HS, Zhang J. Chromatin configurations in the germinal vesicle of mammalian oocytes. Mol Hum Reprod 2009;15(1):1–9.

[38] Liu H, Aoki F. Transcriptional activity associated with meiotic competence in fully grown mouse GV oocytes. Zygote 2002;10(4):327–32.

[39] Zuccotti M, Ponce RH, Boiani M, Guizzardi S, Govoni P, Scandroglio R, et al. The analysis of chromatin organisation allows selection of mouse antral oocytes competent for development to blastocyst. Zygote 2002;10(1):73–8.

[40] Li Y, Sasaki H. Genomic imprinting in mammals: its life cycle, molecular mechanisms and reprogramming. Cell Res 2011;21(3):466–73.

[41] Hiura H, Obata Y, Komiyama J, Shirai M, Kono T. Oocyte growth-dependent progression of maternal imprinting in mice. Genes Cells 2006;11(4):353–61.

[42] Zuccotti M, Merico V, Cecconi S, Redi CA, Garagna S. What does it take to make a developmentally competent mammalian egg? Hum Reprod Update 2011;17(4):525–40.

[43] Gu L, Wang Q, Sun QY. Histone modifications during mammalian oocyte maturation: dynamics, regulation and functions. Cell Cycle 2010;9(10):1942–50.

[44] Rana TM. Illuminating the silence: understanding the structure and function of small RNAs. Nat Rev Mol Cell Biol 2007;8(1):23–36.

[45] Suh N, Blelloch R. Small RNAs in early mammalian development: from gametes to gastrulation. Development 2011;138(9):1653–61.

[46] Evans WS, Griffin ML, Yankov VI. The pituitary gonadotroph: dynamics of gonadotropin release. In: Adashi EY, Rock JA, Rosenwaks Z, editors. Reproductive endocrinology, surgery, and technology. Philadelphia: Lippincott-Raven; 1996. p. 181–210.

[47] Ma J, Flemr M, Stein P, Berninger P, Malik R, Zavolan M, et al. MicroRNA activity is suppressed in mouse oocytes. Curr Biol 2010;20(3):265–70.

[48] Suh N, Baehner L, Moltzahn F, Melton C, Shenoy A, Chen J, et al. MicroRNA function is globally suppressed in mouse oocytes and early embryos. Curr Biol 2010;20(3):271–7.

[49] Murchison EP, Stein P, Xuan Z, Pan H, Zhang MQ, Schultz RM, et al. Critical roles for Dicer in the female germline. Genes Dev 2007;21(6):682–93.

[50] Kaneda M, Tang F, O'Carroll D, Lao K, Surani MA. Essential role for Argonaute2 protein in mouse oogenesis. Epigenetics Chromatin 2009;2(1):9.

[51] Allegrucci C, Thurston A, Lucas E, Young L. Epigenetics and the germline. Reproduction 2005;129(2):137–49.

[52] Tada M, Tada T, Lefebvre L, Barton SC, Surani MA. Embryonic germ cells induce epigenetic reprogramming of somatic nucleus in hybrid cells. EMBO J 1997;16(21):6510–20.

[53] Simon I, Tenzen T, Reubinoff BE, Hillman D, McCarrey JR, Cedar H. Asynchronous replication of imprinted genes is established in the gametes and maintained during development. Nature 1999;401(6756):929–32.

[54] Obata Y, Kono T. Maternal primary imprinting is established at a specific time for each gene throughout oocyte growth. J Biol Chem 2002;277(7):5285–9.

[55] Okano M, Bell DW, Haber DA, Li E. DNA methyltransferases Dnmt3a and Dnmt3b are essential for de novo methylation and mammalian development. Cell 1999;99(3):247–57.

[56] Jaenisch R, Bird A. Epigenetic regulation of gene expression: how the genome integrates intrinsic and environmental signals. Nat Genet 2003;33(Suppl):245–54.

[57] Bhutani N, Burns DM, Blau HM. DNA demethylation dynamics. Cell 2011;146(6):866–72.

[58] Williams K, Christensen J, Pedersen MT, Johansen JV, Cloos PA, Rappsilber J, et al. TET1 and hydroxymethyl-cytosine in transcription and DNA methylation fidelity. Nature 2011;473(7347):343–8.

[59] He YF, Li BZ, Li Z, Liu P, Wang Y, Tang Q, et al. Tet-mediated formation of 5-carboxylcytosine and its excision by TDG in mammalian DNA. Science 2011;333(6047):1303–7.

[60] Volkmar M, Dedeurwaerder S, Cunha DA, Ndlovu MN, Defrance M, Deplus R, et al. DNA methylation profiling identifies epigenetic dysregulation in pancreatic islets from type 2 diabetic patients. EMBO J 2012;31(6):1405–26.

[61] Pirola L, Balcerczyk A, Tothill RW, Haviv I, Kaspi A, Lunke S, et al. Genome-wide analysis distinguishes hyperglycemia regulated epigenetic signatures of primary vascular cells. Genome Res 2011;21(10):1601–15.

[62] Bertram C, Khan O, Ohri S, Phillips DI, Matthews SG, Hanson MA. Transgenerational effects of prenatal nutrient restriction on cardiovascular and hypothalamic-pituitary-adrenal function. J Physiol 2008;586(8):2217–29.

[63] Peixoto-Silva N, Frantz ED, Mandarim-de-Lacerda CA, Pinheiro-Mulder A. Maternal protein restriction in mice causes adverse metabolic and hypothalamic effects in the F1 and F2 generations. Br J Nutr 2011;106(9):1364–73.

[64] Zambrano E, Martinez-Samayoa PM, Bautista CJ, Deas M, Guillen L, Rodriguez-Gonzalez GL, et al. Sex differences in transgenerational alterations of growth and metabolism in progeny (F2) of female offspring (F1) of rats fed a low protein diet during pregnancy and lactation. J Physiol 2005;566(Pt 1):225–36.

[65] Harrison M, Langley-Evans SC. Intergenerational programming of impaired nephrogenesis and hypertension in rats following maternal protein restriction during pregnancy. Br J Nutr 2009;101(7):1020–30.

[66] Frantz ED, Aguila MB, Pinheiro-Mulder Ada R, Mandarim-de-Lacerda CA. Transgenerational endocrine pancreatic adaptation in mice from maternal protein restriction in utero. Mech Ageing Dev 2011;132(3):110–6.

[67] Painter RC, Osmond C, Gluckman P, Hanson M, Phillips DI, Roseboom TJ. Transgenerational effects of prenatal exposure to the Dutch famine on neonatal adiposity and health in later life. BJOG 2008;115(10):1243–9.

[68] Veenendaal MV, Painter RC, de Rooij SR, Bossuyt PM, van der Post JA, Gluckman PD, et al. Transgenerational effects of prenatal exposure to the 1944–45 Dutch famine. BJOG 2013;120(5):548–53.

[69] Carone BR, Fauquier L, Habib N, Shea JM, Hart CE, Li R, et al. Paternally induced transgenerational environmental reprogramming of metabolic gene expression in mammals. Cell 2010;143(7):1084–96.

[70] Rando OJ, Simmons RA. I'm eating for two: parental dietary effects on offspring metabolism. Cell 2015;161(1):93–105.

[71] Radford EJ, Ito M, Shi H, Corish JA, Yamazawa K, Isganaitis E, et al. In utero effects. In utero undernourishment perturbs the adult sperm methylome and intergenerational metabolism. Science 2014;345(6198):1255903.

[72] Rechavi O, Houri-Ze'evi L, Anava S, Goh WSS, Kerk SY, Hannon GJ, et al. Starvation-induced transgenerational inheritance of small RNAs in *C. elegans*. Cell 2014;158(2):277–87.

[73] Rodgers AB, Morgan CP, Bronson SL, Revello S, Bale TL. Paternal stress exposure alters sperm microRNA content and reprograms offspring HPA stress axis regulation. J Neurosci 2013;33(21):9003–12.

[74] Rodgers AB, Morgan CP, Leu NA, Bale TL. Transgenerational epigenetic programming via sperm microRNA recapitulates effects of paternal stress. Proc Natl Acad Sci U S A 2015;112(44):13699–704.

[75] Gapp K, Jawaid A, Sarkies P, Bohacek J, Pelczar P, Prados J, et al. Implication of sperm RNAs in transgenerational inheritance of the effects of early trauma in mice. Nat Neurosci 2014;17(5):667–9.

[76] Jiang L, Zhang J, Wang JJ, Wang L, Zhang L, Li G, et al. Sperm, but not oocyte, DNA methylome is inherited by zebrafish early embryos. Cell 2013;153(4):773–84.

[77] Wang L, Zhang J, Duan J, Gao X, Zhu W, Lu X, et al. Programming and inheritance of parental DNA methylomes in mammals. Cell 2014;157(4):979–91.

[78] Tang WW, Dietmann S, Irie N, Leitch HG, Floros VI, Bradshaw CR, et al. A unique gene regulatory network resets the human germline epigenome for development. Cell 2015;161(6):1453–67.

[79] Gkountela S, Zhang KX, Shafiq TA, Liao WW, Hargan-Calvopina J, Chen PY, et al. DNA demethylation dynamics in the human prenatal germline. Cell 2015;161(6):1425–36.

[80] Guo F, Yan L, Guo H, Li L, Hu B, Zhao Y, et al. The transcriptome and DNA methylome landscapes of human primordial germ cells. Cell 2015;161(6):1437–52.

[81] Siklenka K, Erkek S, Godmann M, Lambrot R, McGraw S, Lafleur C, et al. Disruption of histone methylation in developing sperm impairs offspring health transgenerationally. Science 2015;350(6261):aab2006.

[82] Chen Q, Yan M, Cao Z, Li X, Zhang Y, Shi J, et al. Sperm tsRNAs contribute to intergenerational inheritance of an acquired metabolic disorder. Science 2016;351(6271):397–400.

[83] Sharma U, Conine CC, Shea JM, Boskovic A, Derr AG, Bing XY, et al. Biogenesis and function of tRNA fragments during sperm maturation and fertilization in mammals. Science 2016;351(6271):391–6.

[84] Zenk F, Loeser E, Schiavo R, Kilpert F, Bogdanovic O, Iovino N. Germ line-inherited H3K27me3 restricts enhancer function during maternal-to-zygotic transition. Science 2017;357(6347):212–6.

[85] Tian S, Lin XH, Xiong YM, Liu ME, Yu TT, Lv M, et al. Prevalence of prediabetes risk in offspring born to mothers with hyperandrogenism. EBioMedicine 2017;16:275–83.

[86] Ding GL, Wang FF, Shu J, Tian S, Jiang Y, Zhang D, et al. Transgenerational glucose intolerance with Igf2/H19 epigenetic alterations in mouse islet induced by intrauterine hyperglycemia. Diabetes 2012;61(5):1133–42.

INFERTILITY AND ASSISTED REPRODUCTIVE TECHNOLOGY

Genetic Testing in Male Infertility

Alberto Ferlin, Savina Dipresa, Carlo Foresta

Department of Medicine, Unit of Andrology and Reproductive Medicine, University of Padova, Padova, Italy

INTRODUCTION

In Western countries, approximately 15% of couples of reproductive age are infertile as defined by the inability to conceive after one year of unprotected intercourse. A male factor is responsible, alone or in combination with female factors, in about half the cases.

Male infertility represents one of the clearest examples of a complex phenotype with a substantial genetic basis. Several risk factors and causes might affect male fertility, including lifestyle, diabetes, obesity, hormonal diseases, testicular trauma, cryptorchidism, varicocele, genitourinary infections, ejaculatory disorders, and chemo/radio or surgical therapies.

Identifiable genetic abnormalities contribute to 15%–20% of the most severe forms of male infertility (azoospermia) while the majority (30%–60%) of infertile males do not receive a clear diagnosis; therefore, they are reported as idiopathic with a strong suspicion of genetic underpinnings. This is particularly evident in cases of infertility or repeated assisted reproduction failure with normal semen parameters.

Among known genetic causes of male infertility, there are chromosomal abnormalities (identifiable by karyotype analysis), Y-microdeletions (identifiable by molecular techniques, especially PCR), and X-linked and autosomal gene mutations (identifiable by molecular diagnostic techniques) that influence at different levels many physiological processes involved in male reproduction, such as hormonal homeostasis, spermatogenesis, and sperm quality. Genetic association studies, gene mutation screening, animal models, and basic research in the last few years have clearly demonstrated the high prevalence of genetic causes of spermatogenic impairment and male infertility. They have also provided interesting results on new possible genetic factors involved in male fertility and as prognostic markers of the fertilizing potential of human spermatozoa. Rearrangements, namely copy number variations (CNVs) in the sexual chromosomes (identifiable by specific array CGH or array SNP), have been associated with an increased risk for spermatogenic impairment. Recently, epigenetic mechanisms were also taken into account.

Studies on the genetics of male infertility are intrinsically difficult and complicated because this condition is heterogeneous, it has multiple causes, it can be classified in different ways (clinical, semen analysis, fatherhood), and it results from intricate gene-environment interactions that are difficult to determine [1].

Although a relatively small number of genetic tests are available in clinical practice and are currently recommended in the diagnostic workup of the infertile male [2–4], new developments in this field will probably suggest in the near future the possible translation of new genetic markers into clinical practice. Therefore, clinicians should be aware of their significance and presumed role in achieving a more precise diagnosis of distinct spermatogenic alterations, sperm defects, and prognostic information. Genetic tests should routinely be included in the diagnostic workup of infertile men. Different guidelines have been proposed to correctly use these tests before entering an assisted reproduction program, including selecting the tests on the basis of the initial clinical evaluation, personal and familial history, physical examination, semen and hormonal data, and ultrasound analysis of the male reproductive tract [2, 3, 5].

These tests allow the appropriate assistance for infertile couples, as they explore the cause of the infertility and assess the risk of the couple to transmit their genetic characteristics [6]. In fact, if natural selection prevents the transmission of mutations causing infertility, it is also true that this protective mechanism might be overcome by assisted reproduction techniques. Consequently, the identification of genetic factors of male infertility is important for the appropriate assistance of infertile couples.

Cytogenetic testing (karyotype) and Y chromosome microdeletion analysis are recommended during the diagnostic workup of patients with nonobstructive azoospermia and severe oligozoospermia due to testicular failure, whereas analysis of the cystic fibrosis gene CFTR is recommended in cases of obstructive azoospermia [2, 3, 5]. Selected cases would merit analysis of other genes, such as the androgen receptor (AR), genes responsible for hypogonadotropic hypogonadism (HH), genes involved in testicular descent and cryptorchidism (INSL3, RXFP2), or genes responsible for complete asthenozoospermia or globozoospermia.

The following is a more extensive illustration of the genetic causes of male infertility for a better understanding of the proposed genetic testing.

CHROMOSOMAL ABNORMALITIES

Chromosomal abnormalities are present in about 5% of men with severe oligospermia defined as a sperm concentration <5 million sperm per milliliter, with an increase to about 15% in azoospermic males. Therefore, chromosomal analysis is not routinely performed, except for men with severe oligospermia or azoospermia, situations of recurrent pregnancy loss, or unexplained repeat failures of assisted reproductive techniques [7]. In such cases, a chromosomal analysis, that is, a karyotype, is obtained to evaluate men with insufficient spermatogenesis. The prevalence of karyotype anomalies is directly related to the severity of spermatogenic impairment, but also normozoospermic men might be carriers of chromosomal disorders [8, 9]. Chromosomal abnormalities include both numerical and structural aberrations of chromosomes that might involve the sex chromosomes and the autosomes and that can be homogeneous or in mosaicisms. Aneuploidies are defined by an abnormal number of chromosomes either more or less than the euploid state (46,XY or 46,XX). Most common aneuploidies involve the gonosomes (X and Y chromosomes); examples include Klinefelter Syndrome (KS) (47, XXY) and mixed gonadal dysgenesis (MGD) (45,X/46,XY) with the latter being a mosaic of two different chromosomal numbers. *Klinefelter* syndrome represents the most frequent karyotype abnormality detected in infertile subjects, with a prevalence of 1%–2%. The prevalence of KS in azoospermic males is 10%–12%, and in severely oligozoospermic men, it is 0.6% [10, 11]. The genetic characteristic is the extra X chromosome: approximately 80%–90% of KS men have a 47,XXY karyotype. Karyotypes among the remaining 10%–20% include mosaicisms of two different genetic lines such as 47,XXY/46,XY, or 47,iXq,Y in which iXq refers to an isochromosome where the X chromosome is structurally abnormal with the chromosomal arms being mirrors of the "q" long arm of the X chromosome rather than a normal long (q) and short (p) arm; or 48,XXYY, etc. [12]. The mechanism for the additional X chromosome is because of nondisjunction where the sex chromosomes fail to separate. This event may occur during oogenesis in meiosis I (50%) or meiosis II (10%) or during meiosis I (40%) in spermatogenesis. Infrequently (3%), nondysjunction occurs during early embryogenesis in the fertilized egg [12].

Variations in X inactivation and KS results in reduced androgen production, increased estrogen-to-testosterone ratios, and variable sensitivity of the androgen receptor (AR) that may be associated with different CAG repeats. Phenotypically, individuals with KS demonstrate substantial variability in the severity of classic features and worsen as they age. Varying degrees of genes escaping from these extra X chromosome(s) may influence the heterogeneity of the phenotype. Infertility is the prevailing symptom of most KS patients, althout in azoospermic KS, a patient's residual spermatogenesis can be found both in classic forms and in mosaic (46,XY/47,XXY) forms. Classically, KS men are of tall stature, have gynecomastia, gynoid hips, propensity toward obesity, sparse body hair, are hypotrophic, and have firm testicles (<4 ccs) [12].

Less commonly, KS patients may have undervirilized features such as micropenis, undescended testicles, bifid scrotum, and hypospadias [13]. They are typically identified to have hypergonadotropic hypogonadism (primary hypogonadism), with 65%–85% of KS men with total testosterone levels <12 nmol/L [13]. Thus, they express signs and symptoms of hypogonadism such as osteoporosis, reduced libido, erectile dysfunction, and infertility.

Testicles of men with KS typically demonstrate progressive hyalinization, fibrosis, and degeneration of germ cells and sertoli cells, most often resulting in sertoli cell only (SCO) syndrome. This is thought to worsen immediately after puberty [14]. The degeneration of spermatogenesis appears to occur immediately after puberty and seems to remain stable thereafter in life. Psychosocial, language, and speech abilities may also be affected, and deficits have been shown to correlate to increasing supernumary of the X chromosome. Fertility options are generally dependent upon in vitro fertilization intracytoplasmic sperm injection (IVF-ICSI); most commonly, men with KS are azoospermic and require testicular sperm extraction (TESE). Although strong evidence exists that 47,XXY can produce sperm, it is also possible in rare patients that spermatogenesis may be due to focal mosaicism of the normal karyotype, with germ cells that are able to progress through meiosis, mitosis, and spermiation [15, 16]. The sperm retrieved are considered safe for IVF [17], but do carry a higher diploid sperm incidence of 6.3% [7].

The karyotype *47,XYY* is the second most frequent aneuploidy of the sex chromosomes following KS, with an incidence of 0.1%. The mechanism of aneuploidy resulting in 47, XYY is because of nondisjunction at meiosis II during spermatogenesis, resulting in the additional Y chromosome in 84% of instances [18] or postzygotic mitotic error in 16% of instances [19]. Most patients are not diagnosed until later in life due to a typically normal phenotype. Fertility is variable with semen analyses ranging from azoospermia to normal sperm counts [20, 21]. Testicular biopsy findings among azoospermic men most commonly demonstrate SCO or early or late maturation arrest. FSH levels are typically elevated in response to impaired spermatogenesis, and testosterone levels are generally normal or may be elevated [22]. Although some studies by fluorescent in situ hybridization (FISH) showed an increase in aneuploidy in sperm samples, an increased risk of aneuploidy in the offspring has not been documented.

A *46,XX* karyotype might be rarely found in infertile patients with male phenotype, both with and without detection of the SRY gene (in these cases testes formation is guaranteed by another gene, autosomal or X-linked, involved in sex determination). 46,XX male subjects are invariably azoospermic with testicular atrophy [23].

Men with *XY,X mixed gonadal dysgenesis* (MGD) may present as children with ambiguous or abnormal genitalia or as adults with infertility, gonadal failure, or short stature [24]. The phenotype is variable and has been shown to generally follow gonadal development and location. Gonads may develop into testicles, or undifferentiated streaks, and may be located in the respective scrotum, intraabdominally, or along the path of descent in the inguinal canals. Individuals with bilateral scrotal testicles typically present as a male with short stature and gonadal failure. Those with one scrotal testicle and an intraabdominal streak tend to present with sexual ambiguity as an undervirilized infant. Individuals who present with a streak and an intraabdominal testicle tend to present as in infant with sexual ambiguity resembling female clitoromegaly, and those with bilateral intraabdominal streaks tend to present as a phenotypic female infant. Most have azoospermia, with one report of successful mTESE sperm extraction [25]. There are limited reports of oligozoospermia among individuals with MGD [26]. Testicular biopsies describe disorganized cytoarchitecture, hyalinization, and atrophy of the seminiferous tubules [27]. Most have testicular failure with approximately 45% requiring testosterone replacement when they are adults [28]. MGD is also associated with cardiac and renal malformations, gonadal blastomas, and germ cell tumors. Genetically, those with MGD typically have a karyotype consisting of a mosaicism of 45,X/46,XY; however, variants and ring mosaicisms have also been described such as 45,X/46,X(r)Y [26]. These abnormalities are because of chromosomal misaggregation secondary to anaphase lag or chromosomal rearrangement during early embryonic mitosis.

Hundreds of genes distributed on the Y chromosome, the X chromosome, and the autosomes are necessary for normal sexual development, for the formation and the descent of the testes, and for spermatogenesis and fertility [29]. Autosome aberrations detected by karyotype anaylsis include translocations, inversions, rings, isochromosomes, and other structural abnormalities such as an extra satellite marker chromosome and clinical syndromes including partial deletions or duplications (if larger than 10 Mb).

Further alterations of the sex chromosomes include *structural aberrations of the Y* chromosome, such as deletions, rings, isochromosomes, inversions, and translocations. Among these, the most frequent alterations are translocations between the Y chromosome and autosomal chromosomes. The frequency in the general population is 0.02%, but the frequency increases to 0.2% in oligozoospermic patients and 0.9% in subjects undergoing intracytoplasmic sperm injection (ICSI) [30]. Deletions of genetic material from the Y chromosome may result in a microdeletion within the Y chromosome that cannot be detected on karyotype (Y microdeletion); this is the cause of up to 10% of men with nonobstructive azoospermia. These are further discussed in the sections below.

Chromosomal *translocations* can affect male fertility and pregnancy outcome and, in particular, males with reciprocal translocations have a high rate of unbalanced spermatozoa due to meiotic segregation errors [31]. Therefore, genetic counseling and analysis of chromosomal constitution in the sperm of translocation carriers is mandatory in order to assess the risk of transmission of unbalanced forms. From a quantitative point of view, translocations may be balanced or unbalanced. Reciprocal translocations involve an exchange of genetic material between two or more chromosomes. They are the most common chromosomal structural anomalies in humans and are 10 times more common among infertile men [32]. The most frequent translocations in infertile patients are robertsonian translocations or centric fusions [33]. Robertsonian translocations (RTs) occur with an incidence of 0.9% of men with severe male factor infertility [34]. These occur among acrocentric chromosomes such as chromosomes 13, 14, 15, 21, and 22; the long arms fuse, resulting in the loss of genetic material among the chromosomal short arms (final complement of 45 chromosomes). The most common RTs include t(13q;14q) and t(14q;21q). These men tend to be phenotypically normal; however, they may demonstrate impaired spermatogenesis with increased rates of sperm aneuploidy among those sperm produced [35].

Autosomal *inversions* can be considered intrachromosomal reciprocal translocations without loss of genetic material. Inversions relevant to male infertility include that of chromosome 9. This inversion accounts for up to 3%–5% of male infertility and results in a variable phenotype including normospermia, oligospermia, azoospermia, and asthenospermia [36, 37].

In conclusion, karyotype analysis should be performed in azoospermic and severely oligozoospermic men with primary testiculopathy, and in all cases (including normozoospermia) before assisted reproduction techniques (ART) and/or after repeated ART failure [2].

Y-CHROMOSOME MICRODELETIONS

In recent years, the high frequency of genetic alterations of the Y chromosome have been found in male infertile patients. Y chromosome microdeletions have been extensively studied because of the recognition that Yq has factors important for spermatogenesis: the azoospermia factor (AZF) region of the euchromatin long arm of the Y chromosome. Y chromosome microdeletions are clinically important because they are associated with severe male infertility, and the likelihood of treatment success can be determined by the location of the deletion. Microdeletions in the Y chromosome long arm (Yq) represent the most frequent molecular genetic cause of severe infertility, being detected in 10%–15% of nonobstructive azoospermic and in 5%–10% of severely oligozoospermic patients [38]. Although the genetic pathways and mechanisms of spermatogenic impairment are still largely unknown, three regions, referred to as "azoospermic factors" (AZFa, b and c from proximal to distal), have been identified on Yq [39], but different classifications have also been proposed [40]. The AZF loci harbor 14 protein-coding genes critical for spermatogenesis [41]. Each of these regions may be deleted independently or in combination. The six classic forms of AZF deletions and their corresponding phenotype, in order of decreasing severity, include: AZFabc (SCO), AZFa (SCO), AZFbc (SCO/maturation arrest), AZFb (maturation arrest), AZFc (severe oligospermia to azoospermia), and partial AZFc (normal spermatogenesis to azoospermia) [42]. The most frequent microdeletion involves the AZFc region (about 60%–70% of the deletions) and produces the loss of several genes. Different degrees of spermatogenetic alterations might be found but in general, most patients with this alteration have sperm in the ejaculate or in the testes [43]. The AZFa region spans 1100 kb, and harbors two protein-coding genes: USP9Y and DBY [42]. The DBY gene encodes RNA helicase [44] and has been demonstrated to play a major role in spermatogenesis [45]. These deletions are because of intrachromosomal recombination between flanking repeating genetic sequences or palindromes [46]. Complete deletions of AZFa are rare, accounting for only 3% of Y microdeletions. They carry the poorest prognosis with azoospermia in all men. Histology typically demonstrates SCO, with no previous reports identifying sperm with mTESE. However, partial AZFa deletions have been reported; among these cases, USP9Y is deleted in isolation. The clinical phenotypes were azoospermia and severe oligospermia, with histology demonstrating hypospermatogenesis [47, 48]. The AZFb region contains seven protein-encoding genes experimentally shown to be implicated in spermatogenesis. These include EIF1AY, RPS4Y2, and SMCY located in the X-degenerate euchromatin, and HSFY, XKRY, PRY, and RBMY located in the ampliconic regions. Deletions of the AZFb region are large (4.96e6.92 Mbs) [43], and thought to be due to homologous recombination (HR) and nonhomologous recombination (NHR) among other mechanisms yet to be described [43]. AZFb deletions account for 15% of Y microdeletions. Invariably, complete deletions result in azoospermia and SCO or early maturation arrest histology. The AZFc region spans 4.5 Mb of euchromatin and contains five protein-encoding genes shown to be implicated in spermatogenesis: BPY2, CDY, DAZ, CSPG4LY, and GOLGAZLY [41]. The DAZ family of four genes has been most prolifically studied. This family encodes RNA-binding proteins expressed exclusively in germ cells [44] and exists in palindromic sequences DAZ1/2 and DAZ3/4. Deletions among this region are thought to occur through HR or NHR. Most frequently, deletions occur through HR and involve ampliconic regions b2/b4, b1/b3, b2/b3, and gr/g, resulting in the deletion of several genes, including DAZ genes. NHR accounts for deletions of ampliconic regions P3a, P3b, and P3c. AZFc deletions are the most common Y microdeletions because this is composed of amplicons, which are particularly susceptible to deletions by the above methods [49]. AZFc deletions account for 60% of all clinically relevant Y microdeletions [50]. Up to 70% of men with AZFc deletions have sperm in the ejaculate, typically <1 million sperm per milliliter for these cryptospermic men with AZFc deletions. In men with azoospermia and AZFc deletions, mTESE can be used to harvest sperm from the testicle in 50%–60% of cases. Thus, fertility potential is present in some patients with AZFc deletions; however, Y microdeletions are passed to male offspring. Reports indicate that offspring with Y microdeletions typically have impaired spermatogenesis because their only source of Y chromosome-associated genes is from the partially deleted Y chromosome. Furthermore, combinations of deletions in various AZF regions may occur. Combined AZFb e AZFc deletions are the most common because the two regions overlap and share 1.5 Mb [51]. These deletions do not extend to the same proximal extent as isolated AZFb deletions. Specifically, CDY1 gene is composed of two copies; one copy exists in the AZFc region while the other is in the area of AZFb overlap [43]. This combination of deletions accounts for approximately 13% of Y microdeletions. Patient phenotypes demonstrate SCO or maturation arrest; sperm retrieval attempts are uniformly unsuccessful. In summary, Y microdeletions occur frequently and are indicated among patients with azoospermia and severe oligospermia. Men with complete AZFa and b deletions do not produce sperm in our experience, and successful surgical sperm retrieval has not been described. Furthermore, AZFc deletions are associated with severe oligospermia or azoospermia at presentation. These couples will likely require IVF-ICSI. Deletions of these regions are identified by molecular techniques, especially PCR, following specific guidelines [52]. Y chromosome microdeletion analysis is not indicated when sperm concentration is above 5 million/mL [4, 53]. ART techniques allow the transmission of Yq microdeletions, and male offspring of men with this genetic alteration will therefore also carry the deletion and will possess an impairment of spermatogenesis [54]. Moreover, Yq microdeletions could produce a higher percentage of sperm with sex chromosome aneuploidies [55]. Genetic counseling should be provided before assisted reproduction as the Y microdeletion will be transmitted to the male offspring, with variable but adverse effects on spermatogenesis.

CNVs (COPY NUMBER VARIATIONS)

By applying a genome-wide array for single-nucleotide polymorphism (SNP) and CNV, some authors identified a number of rare autosomal deletions, X-linked CNVs, and Y-linked duplications that increased the risk of male infertility.

Among them, a deletion in the *DMRT1* gene was particularly attractive as a genetic cause of male infertility, even if with a very low frequency. DMRT1 is a testis-specific transcriptional regulator required for testicular differentiation maping on chromosome 9p24.3, a region involved in XY gonadal dysgenesis.

Other studies provide strong evidence for the involvement of X-chromosome CNVs in spermatogenic impairment. The Krausz group reported studies performed with X chromosome specific array CGH and recurrent deletion CNVs on Xq. The authors found that two CNVs (CNV64 and CNV69) that represent risk factors for spermatogenic impairment are present in patients and in controls, whereas another CNV (CNV67) containing the gene MAGEA9 is present exclusively in patients [56].

Rearrangements other than Y chromosome microdeletions in the long arm of this chromosome, namely copy number variations (CNVs) in the AZFc region (mainly the *gr/gr deletion*), have been associated with an increased risk for spermatogenic impairment [57], suggesting that gr/gr deletions could be included among the genetic tests in the workup of infertile men. The analysis of gr/gr deletions is made with a quite simple PCR technique [4].

X-LINKED GENE MUTATIONS

More than 3000 differentially expressed genes have been associated with defective spermatogenesis; however, fewer than 0.01% of these genes have been further explored [58]. The main conditions associated with the X chromosome are represented by mutations in the KAL1 gene causing the Kallmann syndrome and mutations in the androgen receptor (AR) gene causing varying degrees of androgen insensitivity.

Kallmann syndrome has a frequency of 1:10.000 and is characterized clinically by isolated hypogonadotropic hypogonadism (HH) (low serum level of testosterone, luteinizing hormone, and follicle-stimulating hormone) and anosmia [59]. Beyond HH and midline defects such as cleft palate, anosmia, or hyposmia, other clinical features associated with KLS including infertility, tall stature, cryptorchidism, unilateral renal agenesis, and neurogenic deafness [60]. The most frequent genetic alteration is a mutation in the X-linked gene KAL1 (14% of familial cases and 11% of sporadic cases) encoding the protein anosmin-1, which is a cell adhesion protein of the extracellular matrix. During embryogenesis, anosmin-1 is required for the organized migration of both olfactory axons and GnRH neurons from the olfactory placode through the cribriform plate and into the preoptic area of the hypothalamus. As such, defects result in anosmia and GnRH deficiency. GnRH deficiency results in a subsequent lack of the pulsatile luteinizing hormone (LH) that is required for normal gonadal function [61]. There are further genes involved in hypogonadotropic hypogonadism mapping on the chromosome other than the X chromosome (see below).

Mutations in the *AR* gene have been described in 2%–3% of azoospermic and severely oligozoospermic men and might be suggested by high testosterone and luteinizing hormone levels [62–64]. Mutations in the AR gene cause a variety of defects known as androgen insensitivity syndrome, with phenotypes ranging from a female appearance of external genital in the complete forms to various kinds of genital abnormalities in the partial forms and to isolated male infertility in milder forms [62]. Another interesting gene is *TEX11* (Testis-expressed 11), which is located on chromosome Xq13.2. It encodes a protein that contributes to repairing double-strand DNA breaks and regulates recombination and homologous chromosome synapses [65]. It is exclusively expressed among spermatocytes and spermatids in men with normal fertility potential. However, among men with TEX11 mutations, they present with azoospermia and testicular biopsy demonstrates maturation arrest.

AUTOSOMAL GENE MUTATIONS

Another genetic analysis that could be considered in the infertile male, depending on clinical suspicion, is the mutation analysis of the CFTR gene. Cystic fibrosis (CF) is an autosomal recessive disease due to the presence of severe mutations in both copies of the CFTR gene encoding a salt homeostasis anion channel in epithelial cells. Mutations in this gene are very frequent in the general population (1/25 subjects is a carrier in western countries) [66] and patients with CFTR mutations are frequently candidates for in vitro fertilization. Among men who carry a mild form of the CFTR gene mutation, they may be phenotypically normal or present with abnormal Wolfian duct formation (the caput of the epididymis with failure of development of the remaining epididymis and vas deferens), congenital unilateral or bilateral absence of the vas deferens (CBAVD) presenting as obstructive azoospermia, mild oligozoospermia or even normozoospermia nonpatent vas deferens or blind ending vas deferens, hypoplasia or absence of the seminal vesicles, epididymal obstruction or ejaculatory duct obstruction. CBAVD carries a prevalence of 1% among infertile patients and 25% of men with primary obstructive azoospermia [67]. Among men presenting with CBAVD, 78% carry at least one identifiable CFTR mutation, and 46% carry two mutations (compound heterozygotes) [67]. Screening for cystic fibrosis transmembrane conductance regulator (CFTR) gene mutations (including the 5T allele) is strongly recommended in infertile individuals with bilateral or unilateral congenital absence of vas deferens [2, 3, 5]. Because of the high prevalence of CFTR mutations in the general population, screening for CFTR gene mutations is frequently recommended by international guidelines on both partners before assisted reproduction techniques are employed to have an informed discussion regarding risk of CF to the offspring. Men with vasal agenesis should have an abdominal ultrasound to evaluate for renal agenesis. These men are amenable to surgical sperm retrieval from the epididymal remnant through percutaneous epididymal sperm aspiration (PESA) or the approach that we prefer to obtain optimal sperm quality microsurgical epididymal sperm aspiration (MESA). The sperm is amenable for use in IVF-ICSI, with appropriate genetic counseling and consideration of preimplantation genetic testing.

Other genetic abnormalities of this group are rarer and their role in the pathogenesis of infertility is not yet fully understood. Numerous autosomal genes have been suggested to play a role in male reproductive function, but mutations in only a few of them have routine clinical importance, namely the *INSL3* (insulin-like factor 3) gene and the *RXFP2* gene (relaxin family peptide receptor 2, formerly known as LGR8, leucine-rich repeat-containing G-protein coupled receptor 8) in patients with a history of cryptorchidism [68]. INSL3 is a member of the relaxin-like hormone family produced by the Leydig cells and is a major determinant for the transabdominal phase of testicular descent [69, 70]. Besides the role in testicular descent, INSL3 seems to cause endocrine and paracrine actions in adults, and deficiency of this hormone might represent a sign of functional hypogonadism [71]. Mutations in INSL3 and its receptor RXFP2 represent a cause or risk factor for cryptorchidism [72], although their role in infertile men without a history of cryptorchidism is still unknown.

Some authors have reported a possible association between mutations in *dynein genes* and isolated asthenozoospermia with consequent implications for male infertility. Mutations in dynein genes cause primary ciliary dyskinesia (PCD). Three of the several dynein arm subunits have been found to be mutated in individual patients and/or in several families affected by PCD. These genes encode for three proteins: DNAI1 (axonemal dynein, intermediate chain 1) [73], DNAH5 (axonemal dynein, heavy chain 5) [74], and DNAH11 (axonemal dynein, heavy chain 11) [75].

Globozoospermia is a rare autosomal recessive condition characterized by sperm cells with a round head and no acrosome. Approximately 70% of the cases are caused by mutations in the DPY19L2 gene involved in the development of the acrosome and elongation of the sperm head. Without an acrosome, the abnormal sperm are unable to get through the outer membrane of an egg cell. Genetic analysis of infertile globozoospermic patients identified causative mutations also in PICK1 (protein interacting with CK1) and in SPATA16 (spermatogenesis associated 16) [76].

Genes other than the X-linked gene KAL1 are known to be involved in *hypogonadotropic hypogonadism*. The FGFR1 gene (8p12) encodes the FGFR1 receptor implicated in olfactory and GnRH neuron development [77]. Aberrant FGFR1 functioning has been associated with midline defects, cleft palate, dental agenesis, and malformations of extremities observed in some cases of KLS [78, 79]. Secondary testicular failure due to HH occurs because of a lack of stimulation of the testicle to induce testosterone production and subsequent spermatogenesis. Other genes involved are GnRH (8p21–8p11.2), GnRHR (4q21.2), FGF8 (10q24), HS6ST1 (2q21), PROKR2 (20p13) and PROK2 (3p21), KISS1R (19p13.3), KISS1 (1q32), TACR3 (4q25), TAC3 (12q13–q21), and CHD7 (8q12.1) [80].

NEW GENETIC MARKERS IN AUTOSOMES

Glutathione S-transferases (*GSTs*) represent an important superfamily of metabolic enzymes that are important for protecting cells against oxidative stress. Recent data suggested that the GSTM1 null phenotype is slightly associated with male infertility in Asians and whites, although the GSTT1 null phenotype is slightly associated only in Asians.

Methylenetetrahydrofolate reductase (*MTHFR*) is an important enzyme of the folate pathway that maintains the methyl pool required for regulatory functions and conversion of homocysteine to methionine. A number of case-control association studies have been performed in recent years with contrasting results, with some evidence suggesting c.1793G > A polymorphism as being protective against infertility.

Also, interesting results have been found on the association between polymorphisms in telomerase reverse transcriptase (TERT) and telomerase-associated protein 1 (TEP1) and male infertility [81].

Studies in recent years have led to the evaluation of new potential loci for male fertility, including the following *candidates genes*: USP8 (ubiquitin specific peptidase 8, on 15q21.2), a gene encoding an essential deubiquitinating enzyme that has a role in acrosome assembly; UBD (ubiquitin B, on 6p22.1) and EPSTI1 (epithelial stromal interaction 1, on 13q14.11), which have a role in innate immunity; and LRRC32 (leucine rich repeat containing 32, on 11q13.5), which encodes a latent transforming growth factor-b receptor of regulatory T cells [82].

SPERM GENETICS

Sperm DNA analysis can help in defining some forms of idiopathic infertility. A number of methods have been proposed to assess sperm quality and function as well as DNA integrity, such as protamination and DNA packaging, DNA fragmentation, chromosome aneuploidy, mitochondrial function, apoptosis, and telomere length.

In human spermatozoa, the protamination process (the substitution of the nuclear protein histones with protamines during the transition from spermatids to mature sperm) is not complete and a fraction of DNA (10%–15%) remains bound to histones. There are different assays (aniline blue, toluidine blue, chromomycin A3) that could indicate protamine deficiency or aberrant chromatin packing [83–85].

The integrity of sperm DNA is of vital importance for normal sperm function and embryo development [86, 87]. The integrity of sperm DNA can be evaluated with different techniques: the acridine orange staining assay, the sperm chromatin structure assay (SCSA), the COMET assay, the Spem Chromatin Dispersion (HALO) test, the Terminal deoxynucleotidyl transferase-mediated dUTP Nick End Labelling assay (TUNEL) test, γH2AX evaluation, and the in situ nick translation test. The term "DNA fragmentation" is often used inappropriately and the different tests proposed to evaluate it actually assess different DNA damage. Generally, the estimate of DNA fragmentation is expressed as the percentage of sperm with DNA damage with respect to the total number of sperm. Most of them are not quantitative, that is, they do not measure how much of the sperm DNA in each cell is damaged (the ideal test should obviously be a method directly assessing the amount of DNA damaged in a cell rather than one giving only the percentage of sperm with some fragmentation, irrespective of the amount of DNA damaged). Some of them also do not distinguish DNA fragmentation occurring in a single DNA helix from that occurring in both helixes. DNA damage might directly affect sperm function, especially if it occurs in the coding part of DNA. However, it might be relatively harmless if there is little damage along the DNA and located in introns or regions of DNA not important for DNA functions (none of the tests are able to distinguish these different conditions).

The technology of FISH (Fluorescence in situ Hybridization) analysis using chromosome-specific DNA probes is able to detect numerical chromosomal abnormalities in decondensed sperm. It offers the possibility to obtain data on aneuploidy frequencies in large populations of sperm. Usually, five-color FISH is performed to detect aneuploidies for the sex chromosomes and chromosomes 13, 18, and 21, which represent the most common aneuploidies detected at birth in humans [88]. Normal values (percentage of sperm with specific aneuploidy) are still debatable. Furthermore, chromosomal aneuploidies in general are responsible for a great deal of pregnancy loss and assisted reproduction failure [89].

In addition, the molecular karyotype on single human sperm by array-CGH showed that 8% of sperm from nor-mozoospermic men is unbalanced [89].

The usefulness of these methods in the evaluation of male infertility, as routine investigation and as prognostic markers for fertility, is still debatable due to many factors (different methods have been proposed, standardization has not been reached for some of these methods, normal values are yet under investigation, prospective studies have not been performed). Nevertheless, they have the potential to provide information in selected forms of male infertility and they can be used as additional, second-level sperm tests after standard semen analysis in idiopathic and nonidiopathic cases of oligozoospermia. Furthermore, they can give additional information on sperm function and fertilizing ability in infertile males with normal standard semen parameters and in couples experiencing repeated abortion, implantation failure, and recurrent early pregnancy loss during assisted reproduction techniques.

Nowadays, one of the major areas of interest in male reproduction is the role of sperm **telomeres** in spermatogenesis. Telomeres confer stability on the chromosome and preserve genomic stability, and they undergo progressive shortening with each cell division. Telomere length is maintained by telomerase, which is highly expressed in germ cells, and this is probably the reason why telomere length in sperm increases with age in contrast with somatic cells [90–92]. A study on sperm and leukocyte telomere length in a group of young men (18–19 year old students) with different semen parameters found that sperm telomere length is related to sperm count. Also, it is lower in oligozoospermic than in normozoospermic men and it is related to parents' age at conception [93].

Recent studies analyzing the *expression of some genes in sperm* have shown the possibility of identifying genetic markers of the fertilizing potential of human spermatozoa in cases in which standard semen analysis does not detect any abnormality [94].

SPERM EPIGENETICS

Epigenetic biomarkers are extremely interesting and potentially important in elucidating the cause of idiopathic male infertility. Evidence for a link between sperm epigenetic modifications and male infertility in humans is a recent demonstration. Epigenetic events are crucial for proper sperm function. They are mainly characterized by reorganization and condensation of the sperm genome during postmeiotic maturation, DNA methylation loss or gain on the global level and on imprinted genes, histone acetylation, and histone-to-protamine transition [95]. Furthermore, hypermethylation of the cAMP response element modulator (CREM) and the mesoderm-specific transcript (MEST) have been shown to have a negative impact on semen parameters and fertility [96, 97]. Noncoding and microRNAs have also been an area of focus in identifying regulators of spermatogenesis and biomarkers for fertile and infertile men. Deeper investigations are needed to identify the clinical utility of these markers.

References

[1] Ferlin A. New genetic markers for male fertility. Asian J Androl 2012;14:807–8.
[2] Foresta C, Ferlin A, Gianaroli L, Dallapiccola B. Guidelines for the appropriate use of genetic tests in infertile couples. Eur J Hum Genet 2002;10:303–12.
[3] Jungwirth A, Giwercman A, Tournaye H, Diemer T, Kopa Z, Dohle G, et al. European Association of Urology guidelines on male infertility: the 2012 update. Eur Urol 2012;62:324–32.
[4] Krausz C, Hoefsloot L, Simoni M, Tuttelmann F. EAA/EMQN best practice guidelines for molecular diagnosis of Y-chromosomal microdeletions: state-of-the-art 2013. Andrology 2014;2:5–19.
[5] Stahl PJ, Schlegel PN. Genetic evaluation of the azoospermic or severely oligozoospermic male. Curr Opin Obstet Gynecol 2012;24:221–8.

[6] Ferlin A, Raicu F, Gatta V, Zuccarello D, Palka G, Foresta C. Male infertility: role of genetic background. Reprod Biomed Online 2007;14:734–45.

[7] Piomboni P, Stendardi A, Gambera L. Chromosomal aberrations and aneuploidies of spermatozoa. Adv Exp Med Biol 2014;791:27–52.

[8] Elghezal H, Hidar S, Braham R, Denguezli W, Ajina M, Saâd A. Chromosome abnormalities in one thousand infertile males with nonobstructive sperm disorders. Fertil Steril 2006;86:1792–5.

[9] Dul EC, Van Ravenswaaij-Arts CM, Groen H, Van Echten-Arends J, Land JA. Who should be screened for chromosomal abnormalities before ICSI treatment? Hum Reprod 2010;25:2673–7.

[10] Groth KA, Skakkebæk A, Høst C, Gravholt CH, Bojesen A. Clinical review: Klinefelter syndrome—a clinical update. J Clin Endocrinol Metab 2013;98:20–30.

[11] Paduch DA, Bolyakov A, Cohen P, Travis A. Reproduction in men with Klinefelter syndrome: the past, the present, and the future. Semin Reprod Med 2009;27:137–48.

[12] Bonomi M, Rochira V, Pasquali D, Balercia G, Jannini EA, Ferlin A. Klinefelter syndrome (KS): genetics, clinical phenotype and hypogonadism. J Endocrinol Invest 2017;40(2):123–34.

[13] Lanfranco F, Kamischke A, Zitzmann M, Nieschlag E. Klinefelter's syndrome. Lancet 2004;364(9430):273–83.

[14] Wosnitzer MS, Paduch DA. Endocrinological issues and hormonal manipulation in children and men with Klinefelter syndrome. Am J Med Genet C Semin Med Genet 2013;163C(1):16–26.

[15] Lenz P, Luetjens CM, Kamischke A, Kuhnert B, Kennerknecht I, Nieschlag E. Mosaic status in lymphocytes of infertile men with or without Klinefelter syndrome. Hum Reprod 2005;20(5):1248–55.

[16] Foresta C, Galeazzi C, Bettella A, Marin P, Rossato M, Garolla A, et al. Analysis of meiosis in intratesticular germ cells from subjects affected by classic Klinefelter's syndrome. J Clin Endocrinol Metab 1999;84(10):3807–10.

[17] Madureira C, Cunha M, Sousa M, Neto AP, Pinho MJ, Viana P, et al. Treatment by testicular sperm extraction and intracytoplasmic sperm injection of 65 azoospermic patients with non-mosaic Klinefelter syndrome with birth of 17 healthy children. Andrology 2014;2(4):623–31.

[18] Kim IW, Khadilkar AC, Ko EY, Sabanegh ES. 47,XYY syndrome and male infertility. Rev Urol 2013;15(4):188–96.

[19] Rives N, Simeon N, Milazzo JP, Barthélémy C, Macé B. Meiotic segregation of sex chromosomes in mosaic and non-mosaic XYY males: case reports and review of the literature. Int J Androl 2003;26(4):242–9.

[20] Rives N, Milazzo JP, Miraux L, North MO, Sibert L, Macé B. From spermatocytes to spermatozoa in an infertile XYY male. Int J Androl 2005;28(5):304–10.

[21] Abdel-Razic MM, Abdel-Hamid IA, El Sobky ES. Non mosaic 47,XYY syndrome presenting with male infertility: case series. Andrologia 2012;44(3):200–4.

[22] Bardsley MZ, Kowal K, Levy C, Gosek A, Ayari N, Tartaglia N, et al. 47,XYY syndrome: clinical phenotype and timing of ascertainment. J Pediatr 2013;163(4):1085–94.

[23] Abusheikha N, Lass A, Brinsden P. XX males without SRY gene and with infertility. Hum Reprod 2001;16:717–8.

[24] Layman LC, Tho SP, Clark AD, Kulharya A, McDonough PG. Phenotypic spectrum of 45,X/46,XY males with a ring Y chromosome and bilaterally descended testes. Fertil Steril 2009;91(3):791–7.

[25] Flannigan RK, Chow V, Ma S, Yuzpe A. 45,X/46,XY mixed gonadal dysgenesis: a case of successful sperm extraction. Can Urol Assoc J 2014;8(1e2):E108–10.

[26] Arnedo N, Nogues C, Bosch M, Templado C. Mitotic and meiotic behaviour of a naturally transmitted ring Y chromosome: reproductive risk evaluation. Hum Reprod 2005;20(2):462–8.

[27] Hsieh MH, Hollander A, Lamb DJ, Turek PJ. The genetic and phenotypic basis of infertility in men with pediatric urologic disorders. Urology 2010;76(1):25–31.

[28] Martinerie L, Morel Y, Gay CL, Pienkowski C, de Kerdanet M, Cabrol S, et al. Impaired puberty, fertility, and final stature in 45,X/46,XY mixed gonadal dysgenetic patients raised as boys. Eur J Endocrinol 2012;166(4):687–94.

[29] Aston KI, Conrad DF. A review of genome-wide approaches to study the genetic basis for spermatogenic defects. Methods Mol Biol 2013;927:397–410.

[30] Ferlin A, Tessari A, Ganz F, Marchina E, Barlati S, Garolla A, et al. Association of partial AZFc region deletions with spermatogenic impairment and male infertility. J Med Genet 2005;42:209–13.

[31] Nishikawa N, Sato T, Suzumori N, Sonta S, Suzumori K. Meiotic segregation analysis in male translocation carriers by using fluorescent in situ hybridization. Int J Androl 2008;31:60–6.

[32] Estop AM, Cieply KM, Aston CE. The meiotic segregation pattern of a reciprocal translocation t(10;12)(q26.1; p13.3) by fluorescence in situ hybridization sperm analysis. Eur J Hum Genet 1997;5(2):78–82.

[33] Ogur G, Van Assche E, Vegetti W, Verheyen G, Tournaye H, Bonduelle M, et al. Chromosomal segregation in spermatozoa of 14 Robertsonian translocation carriers. Mol Hum Reprod 2006;12:209–15.

[34] Mau-Holzmann UA. Somatic chromosomal abnormalities in infertile men and women. Cytogenet Genome Res 2005;111(3e4):317–36.

[35] Keymolen K, Van Berkel K, Vorsselmans A, Staessen C, Liebaers I. Pregnancy outcome in carriers of Robertsonian translocations. Am J Med Genet A 2011;155A(10):2381–5.

[36] Dana M, Stoian V. Association of pericentric inversion of chromosome 9 and infertility in romanian population. Maedica (Buchar) 2012;7(1):25–9.

[37] Mozdarani H, Meybodi AM, Karimi H. Impact of pericentric inversion of Chromosome 9 [inv (9) (p11q12)] on infertility. Indian J Hum Genet 2007;13(1):26–9.

[38] Ferlin A, Arredi B, Speltra E, Cazzadore C, Selice R, Garolla A, et al. Molecular and clinical characterization of Y chromosome microdeletions in infertile men: a 10-year experience in Italy. J Clin Endocrinol Metab 2007;92:762–70.

[39] Premi S, Srivastava J, Epplen JT, Ali S. AZFc region of the Y chromosome shows singular structural organization. Chromosome Res 2010;18:419–30.

[40] Sun K, Chen XF, Zhu XB, Hu HL, Zhang W, Shao FM, et al. A new molecular diagnostic approach to assess Y chromosome microdeletions in infertile men. J Int Med Res 2012;40:237–48.

[41] Vogt PH. AZF deletions and Y chromosomal haplogroups: history and update based on sequence. Hum Reprod Update 2005;11(4):319–36.

[42] Sadeghi-Nejad H, Farrokhi F. Genetics of azoospermia: current knowledge, clinical implications, and future directions. Part II: Y chromosome microdeletions. Urol J 2007;4(4):192–206.

[43] Repping S, Skaletsky H, Lange J, Silber S, Van Der Veen F, Oates RD, et al. Recombination between palindromes P5 and P1 on the human Y chromosome causes massive deletions and spermatogenic failure. Am J Hum Genet 2002;71:906–22.

[44] Kleiman SE, Yogev L, Hauser R, Botchan A, Maymon BB-S, Paz G, et al. Expression profile of AZF genes in testicular biopsies of azoospermic men. Hum Reprod 2007;22(1):151–8.

[45] Foresta C, Ferlin A, Moro E. Deletion and expression analysis of AZFa genes on the human Y chromosome revealed a major role for DBY in male infertility. Hum Mol Genet 2000;9(8):1161–9.

[46] Blanco P, Shlumukova M, Sargent CA, Jobling MA, Affara M, Hurles ME. Divergent outcomes of intrachromosomal recombination on the human Y chromosome: male infertility and recurrent polymorphism. J Med Genet 2000;37(10):752–8.

[47] Brown GM, Furlong RA, Sargent CA, Erickson RP, Longepie g MM, et al. Characterisation of the coding sequence and fine mapping of the human DFFRY gene and comparative expression analysis and mapping to the Sxrb interval of the mouse Y chromosome of the Dffry gene. Hum Mol Genet 1998;7(1):97–107.

[48] Sun C, Skaletsky H, Birren B, Devon K, Tang Z, Silber S, et al. An azoospermic man with a de novo point mutation in the Y-chromosomal gene USP9Y. Nat Genet 1999;23(4):429–32.

[49] Wu B, Lu NX, Xia YK, Gu AH, Lu CC, Wang W, et al. A frequent Y chromosome b2/b3 subdeletion shows strong association with male infertility in Han-Chinese population. Hum Reprod 2007;22(4):1107–13.

[50] Asero P, Calogero AE, Condorelli RA, Mongioi l VE, Lanzafame F, et al. Relevance of genetic investigation in male infertility. J Endocrinol Invest 2014;37(5):415–27.

[51] Stahl PJ, Masson P, Mielnik A, Marean MB, Schlegel PN, Paduch DA. A decade of experience emphasizes that testing for Y microdeletions is essential in American men with azoospermia and severe oligozoospermia. Fertil Steril 2010;94(5):1753e6.

[52] Vogt PH, Bender U. Human Y chromosome microdeletion analysis by PCR multiplex protocols identifying only clinically relevant AZF microdeletions. Methods Mol Biol 2013;927:187–204.

[53] Foresta C, Moro E, Ferlin A. Y chromosome microdeletions and alterations of spermatogenesis. Endocr Rev 2001;22:226–39.

[54] Vogt PH. Genomic heterogeneity and instability of the AZF locus on the human Y chromosome. Mol Cell Endocrinol 2004;224:1–9.

[55] Patsalis PC, Skordis N, Sismani C, Kousoulidou L, Koumbaris G, Eftychi C, et al. Identification of high frequency of Y chromosome deletions in patients with sex chromosome mosaicism and correlation with the clinical phenotype and Y-chromosome instability. Am J Med Genet A 2005;135:145–9.

[56] Krausz C, Giachini C, Lo Giacco D, Daguin F, Chianese C, Ars E, et al. High resolution X chromosome-specific array-CGH detects new CNVs in infertile males. PLoS ONE 2012;7:e44887.

[57] Stouffs K, Lissens W, Tournaye H, Haentjens P. What about gr/gr deletions and male infertility? Systematic review and meta-analysis. Hum Reprod Update 2011;17:197–209.

[58] Schultz N, Hamra FK, Garbers DL. A multitude of genes expressed solely in meiotic or postmeiotic spermatogenic cells offers a myriad of contraceptive targets. Proc Natl Acad Sci U S A 2003;100(21):12201–6.

[59] Zhang S, H X, Wang T, Liu G, Liu J. The KAL1 pVal610Ile mutation is a recessive mutation causing Kallmann syndrome. Fertil Steril 2013;99:1720–3.

[60] Silveira LF, MacColl GS, Bouloux PM. Hypogonadotropic hypogonadism. Semin Reprod Med 2002;20(4):327–38.

[61] Oliveira LM, Seminara SB, Beranova M, Hayes FJ, Valkenburgh SB, Schipani E, et al. The importance of autosomal genes in Kallmann syndrome: genotype-phenotype correlations and neuroendocrine characteristics. J Clin Endocrinol Metab 2001;86(4):1532–8.

[62] Thomas Jr. PS, Fraley GS, Damian V, Woodke LB, Zapata F, Sopher BL, et al. Loss of endogenous androgen receptor protein accelerates motor neuron degeneration and accentuates androgen insensitivity in a mouse model of X-linked spinal and bulbar muscular atrophy. Hum Mol Genet 2006;15:2225–38.

[63] Katagiri Y, Neri QV, Takeuchi T, Moy F, Sills ES, Palermo GD. Androgen receptor CAG polymorphism (Xq11-12) status and human spermatogenesis: a prospective analysis of infertile males and their offspring conceived by intracytoplasmic sperm injection. Int J Mol Med 2006;18:405–13.

[64] Gottlieb B, Lombroso R, Beitel LK, Trifiro TA. Molecular pathology of the androgen receptor in male (in)fertility. Reprod Biomed Online 2005;10:42–8.

[65] Adelman CA, Petrini JH. ZIP4H (TEX11) deficiency in the mouse impairs meiotic double strand break repair and the regulation of crossing over. PLoS Genet 2008;4(3)e1000042.

[66] Strausbaugh SD, Davis PB. Cystic fibrosis: a review of epidemiology and pathobiology. Clin Chest Med 2007;28:279–88.

[67] Yu J, Chen Z, Ni Y, Li Z. CFTR mutations in men with congenital bilateral absence of the vas deferens (CBAVD): asystemic review and meta-analysis. Hum Reprod 2012;27(1):25–35.

[68] Ferlin A, Zuccarello D, Zuccarello B, Chirico MR, Zanon GF, Foresta C. Genetic alterations associated with cryptorchidism. JAMA 2008;300:2271–6.

[69] Huang Z, Rivas B, Agoulnik AI. Insulin-like 3 signaling is important for testicular descent but dispensable for spermatogenesis and germ cell survival in adult mice. Biol Reprod 2012;87:143.

[70] Kumagai J, Hsu SY, Matsumi H, Roh JS, Fu P, Wade JD, et al. INSL3/Leydig insulin-like peptide activates the LGR8 receptor important in testis descent. J Biol Chem 2002;277:31283–6.

[71] Ivell R, Kotula-Balak M, Glynn D, Heng K, Anand-Ivell R. Relaxin family peptides in the male reproductive system–a critical appraisal. Mol Hum Reprod 2011;17:71–84.

[72] Ferlin A, Zuccarello D, Garolla A, Selice R, Vinanzi C, Ganz F, et al. Mutations in INSL3 and RXFP2 genes in cryptorchid boys. Ann N Y Acad Sci 2009;1160:213–4.

[73] Zariwala MA, Leigh MW, Ceppa F, Kennedy MP, Noone PG, Carson JL, et al. Mutations of DNAI1 in primary ciliary dyskinesia: evidence of founder effect in a common mutation. Am J Respir Crit Care Med 2006;174:858–66.

[74] Hornef N, Olbrich H, Horvath J, Zariwala MA, Fliegauf M, Loges NT, et al. DNAH5 mutations are a common cause of primary ciliary dyskinesia with outer dynein arm defects. Am J Respir Crit Care Med 2006;174:120–6.

[75] Schwabe GC, Hoffmann K, Loges NT, Birker D, Rossier C, de Santi MM, et al. Primary ciliary dyskinesia associated with normal axoneme ultrastructure is caused by DNAH11 mutations. Hum Mutat 2008;29:289–98.

[76] Fujihara Y, Oji A, Larasati T, Kojima-Kita K, Ikawa M. Human globozoospermia-related gene spata16 is required for sperm formation revealed by CRISPR/Cas9-mediated mouse models. Int J Mol Sci 2017;18(10). https://dx.doi.org/10.3390/ijms18102208.

[77] Neto FT, Bach PV, Najari BB, Li PS, Goldstein M. Genetics of male infertility. Curr Urol Rep 2016;17(10):70.

[78] Sarfati J, Bouvattier C, Bry-Gauillard H, Cartes A, Bouligand J, Young J. Kallmann syndrome with FGFR1 and KAL1 mutations detected during fetal life. Orphanet J Rare Dis 2015;10:71.

[79] Villanueva C, Jacobson-Dickman E, Xu C, Manouvreir S, Dwyer AA, Sykiotis GP. Congenital hypogonadotropic hypogonadism with split hand/foot mal-formation: a clinical entity with a high frequency of FGFR1 mutations. Genet Med 2015;17(8):651–9.

[80] Bonomi M, Libri DV, Guizzardi F, Guarducci E, Maiolo E, Pignatti E, et al. Idiopathic Central Hypogonadism Study group of the Italian Societies of Endocrinology and Pediatric Endocrinology and Diabetes. New understandings of the genetic basis of isolated idiopathic central hypogonadism. Asian J Androl 2012;14:49–56.

[81] Yan L, Wu S, Zhang S, Ji G, Gu A. Genetic variants in telomerase reverse transcriptase (TERT) and telomerase-associated protein 1 (TEP1) and the risk of male infertility. Gene 2014;534:139–43.

[82] Kosova G, Scott NM, Niederberger C, Prins GS, Ober C. Genome-wide association study identifies candidate genes for male fertility traits in humans. Am J Hum Genet 2012;90:950–61.

[83] Dadoune JP, Mayaux MJ, Guihard-Moscato ML. Correlation between defects in chromatin condensation of human spermatozoa stained by aniline blue and semen characteristics. Andrologia 1988;20:211–7.

[84] Erenpreisa J, Freivalds T, Slaidina M, Krampe R, Butikova J, Ivanov A, et al. Toluidine blue test for sperm DNA integrity and elaboration of image cytometry algorithm. Cytometry 2003;52:19–27.

[85] Manicardi GC, Bianchi PG, Pantano S, Azzoni P, Bizzaro D, Bianchi U, et al. Presence of endogenous nicks in DNA of ejaculated human spermatozoa and it relationship to chromomycin A3 accessibility. Biol Reprod 1995;52:864–7.

[86] Ferlin A, Foresta C. New genetic markers for male infertility. Curr Opin Obstet Gynecol 2014;26:193–8.

[87] Bisht S, Faiq M, Tolahunase M. Oxidative stress and male infertility. Nat Rev Urol 2017;14:470–85.

[88] Hassold T, Hunt P. To err (meiotically) is human: the genesis of human aneuploidy. Nat Rev Genet 2001;2:280–91.

[89] Templado C, Vidal F, Estop A. Aneuploidy in human spermatozoa. Cytogenet Genome Res 2011;133:91–9.

[90] Baird DM, Britt-Compton B, Rowson J, Amso NN, Gregory L, Kipling D. Telomere instability in the male germline. Hum Mol Genet 2006;15:45–51.

[91] Kimura M, Cherkas LF, Kato BS, Demissie S, Hjelmborg JB, Brimacombe M, et al. Offspring's leukocyte telomere length, paternal age, and telomere elongation in sperm. PLoS Genet 2008;4–37.

[92] Aston KI, Hunt SC, Susser E, Kimura M, Factor-Litvak P, Carrell D, et al. Divergence of sperm and leukocyte age-dependent telomere dynamics: implications for male driven evolution of telomere length in humans. Mol Hum Reprod 2012;18:517–22.

[93] Ferlin A, Rampazzo E, Rocca MS, Keppel S, Frigo AC, De Rossi A, et al. In young men sperm telomere length is related to sperm number and parental age. Hum Reprod 2013;28:3370–6.

[94] Bonache S, Mata A, Ramos MD, Bassas L, Larriba S. Sperm gene expression profile is related to pregnancy rate after insemination and is predictive of low fecundity in normozoospermic men. Hum Reprod 2012;27:1556–67.

[95] Boissonnas CC, Jouannet P, Jammes H. Epigenetic disorders and male subfertility. Fertil Steril 2013;99:624–31.

[96] Rajender S, Avery K, Agarwal A. Epigenetics, spermatogenesis and male infertility. Mutat Res 2011;727(3):62–71.

[97] Kobayashi H, Sato A, Otsu E, Hiura H, Tomatsu C, Utsunomiya T, et al. Aberrant DNA methylation of imprinted loci in sperm from oligospermic patients. Hum Mol Genet 2007;16(21):2542–51.

Further Reading

Patassini C, Garolla A, Bottacin A, Menegazzo M, Speltra E, Foresta C, et al. Molecular karyotyping of human & single sperm by array-comparative genomic hybridization. PLoS ONE 2013;8:e60922.

Genetics and Genomics of Endometriosis [☆]

Linda C. Giudice[*], Richard O. Burney[†], Christian Becker[‡], Stacey Missmer[§,¶,‖], Grant Montgomery[#], Nilufer Rahmioglu[**], Peter A.W. Rogers[††], Krina Zondervan[**,‡‡]

[*]Center for Reproductive Sciences, Center for Reproductive Health, University of California, San Francisco, San Francisco, CA, United States [†]Department of Clinical Investigation, Madigan Army Medical Center, Tacoma, WA, United States [‡]Nuffield Department of Obstetrics and Gynaecology, University of Oxford, Women's Centre, John Radcliffe Hospital, Oxford, United Kingdom [§]Department of Epidemiology, Harvard T.H. Chan School of Public Health, Boston, MA, United States [¶]Boston Center for Endometriosis, Boston Children's and Brigham and Women's Hospitals, Boston, MA, United States [‖]Department of Obstetrics, Gynecology and Reproductive Biology, Michigan State University, East Lansing, MI, United States [#]Institute for Molecular Bioscience, The University of Queensland, Brisbane, QLD, Australia [**]Wellcome Trust Center for Human Genetics, University of Oxford, Oxford, United Kingdom [††]Department of Obstetrics and Gynaecology, University of Melbourne, Royal Women's Hospital, Parkville, VIC, Australia [‡‡]Nuffield Department of Obstetrics and Gynaecology, John Radcliffe Hospital, University of Oxford, Oxford, United Kingdom

[☆] Dedication: The authors would like to dedicate this chapter to Ronald Blatt, MD, PhD, a pioneer in endometriosis research who passed away in 2017. His curiosity and investigations into endometriosis and adenomyosis have inspired generations of students and researchers in the field, and he will be greatly missed. His legacy, however, will continue through the efforts of generations of researchers studying endometriosis and mechanisms, genetics, and genomics underlying this and other disorders that intersect with this complex disease.

Abbreviations

ADP	adenosine diphosphate ribose
ASRM	American Society for Reproductive Medicine
BMI	body mass index
CGH	comparative genomic hybridization
ChIP	chromatin immunoprecipitation
COX2	cyclooxygenase 2
CpG	cytosine-phosphate-guanine sequence
Cyp19A1	aromatase
DNA	deoxyribonucleic acid
E_1	estrone
E_2	estradiol
EAOC	endometriosis-associated ovarian cancer
EBAF	endometrial bleeding factor
ENCODE	encyclopedia of DNA elements
eQTLs	expression quantitative trait loci
ER	estrogen receptor
ERFFI1	ErbB receptor feedback inhibitor 1
ERα	estrogen receptor alpha
ERβ	estrogen receptor beta
ESR1	ERα gene
ESR2	ERβ gene
EZH2	enhancer of zeste homologue
GnRH	gonadotropin-releasing hormone
GTEx	genotype-tissue expression
GWAS	genome-wide association studies
GWAS	genome-wide association studies
H	histone
H3	acetylated histone 3
H3K27	H3 lysine 27

H3K4	H3 lysine 4
H3K9	H3 lysine 9
H4	acetylated histone 4
HapMap	haplotype map
HAT	histone acetylase
HDAC	histone deacetylases
HDMT	histone demethyltransferases
HSD17B2	17β-hydroxysteroid dehydrogenase 2
IL	interleukin
K	lysine
K-ras	Kirsten rat sarcoma
LOH	loss of heterozygosity
LSD1	lysine-specific demthylase 1
MCP-1	monocyte chemotactic protein 1
MeCP2	methyl-CpG-binding domain protein
MeDIP	methylated DNA immunoprecipitation
miR	microRNA
MMPs	matrix metalloproteinases
MRI	magnetic resonance imaging
NGF	nerve growth factor
NIH	National Institutes of Health
P$_4$	progesterone
PCR	polymerase chain reaction
PGE	prostaglandin E
PGE2	prostaglandin E2
PGR	PR gene
PR	progesterone receptor
PRA	progesterone receptor A
PRB	progesterone receptor B
Pten	Phosphate and tensin homologue
RA	retinoic acid
rASRM	revised American Society for Reproductive Medicine
RNA	ribonucleic acid
SF1	steroidogenic factor 1
SIRT1	sirtuin 1
SNPs	single nucleotide polymorphisms
SOPs	standard operating procedures
TIMPs	tissue inhibitors of MMPs
TNF-α	tumor necrosis factor α
TSS	transcription start sites
UTR	untranslated region
VEGF	vascular endothelial growth factor
VPA	valproic acid
WERF	World Endometriosis Foundation
WERF-EPHect	WERF Biobanking and Phenome Harmonization Project

INTRODUCTION

Endometriosis Epidemiology, Pathogenesis, Pathophysiology

Endometriosis is a common, estrogen-dependent, inflammatory disease that is associated with a high prevalence of pelvic pain and reduced fertility in women and whose pathogenesis and pathophysiology remain challenges for clinicians, researchers, and those affected [1, 2]. It is characterized by endometrial-like tissue outside the uterus, primarily on the pelvic peritoneum, ovaries, and rectovaginal septum, and rarely on the lung, pericardium, and brain, that responds to cycling ovarian-derived steroid hormones, including hormone withdrawal at the time of menses [1]. Endometriosis affects 6%–10% of women of reproductive age (~176 million women and teen girls globally), 35%–50% of women and teens with dysmenorrhea and chronic pelvic pain, and up to 50% of women with infertility [3–5]. Higher risk is associated with early menarche, short menstrual cycle length, increased menstrual flow, decreased parity, lean body mass, positive family history, and in utero exposure to estrogenic substances [1, 5]. Women with endometriosis have higher risks of other chronic medical conditions, including ovarian and breast cancer, cutaneous melanoma, asthma, some autoimmune and atopic disorders, and cardiovascular disease [6]. Overall, this complex disease has a major impact on the quality of life of those affected and their families [7] and on the health economy, totaling $69 billion in the United States in 2009 for disease-associated diagnostic evaluations and treatments for pain and infertility [8].

Pelvic endometriosis, dependent on estradiol (E_2) for growth, largely derives (Fig. 18.1) from retrograde menstruation through the fallopian tubes of shed steroid hormone-sensitive eutopic (within the uterus) endometrial cells and tissues that attach to the peritoneum, establish a blood supply, and elicit an inflammatory response [1]. This is accompanied by neuroangiogenesis, fibrosis, adhesion formation, and anatomic distortion and results in pain and infertility. In women with this disease, the eutopic endometrium exhibits multiple abnormalities that likely contribute to disease establishment and maintenance outside the uterus [2]. Normally, the endometrium responds throughout the cycle to the proliferative effects of E_2, and, after ovulation, the antiproliferative differentiating effects of progesterone (P_4) to prepare for pregnancy. These steroid hormones act via their cognate nuclear receptors, ERα, ERβ, PRA, and PRB, and coregulators [9]. However, in women with endometriosis, the eutopic endometrium is abnormal and has a proinflammatory proliferative phenotype and displays resistance to P_4 action [10–13]. While retrograde menses explains the physical displacement of endometrial fragments into the peritoneal cavity, additional steps are needed for disease pathogenesis and pathophysiology. The 90% prevalence of retrograde menstruation in women and nearly 10% prevalence of the disease suggest hereditary or acquired properties of the eutopic endometrium, hereditary or acquired abnormalities of the peritoneum, and/or defective immune clearance of sloughed endometrium predisposing to disease pathogenesis [2]. Distal disease is believed to derive from Mullerian transformation or lymphatic dissemination [2].

The genetic and genomic predispositions for developing endometriosis in the pelvis or at other sites or of developing associated medical conditions are not well understood. However, increasing data reveal important roles for heritable (genetic) and environmental/lifestyle (epigenetic and exposome) risk factors, offering opportunities for developing diagnostics, risk stratification, and potential risk modification of environment and lifestyle contributors. The genetic and epigenetic basis of endometriosis is considered herein.

Endometriosis Diagnosis, Staging, Treatment

Definitive diagnosis of endometriosis is by surgical (laparoscopic) visualization of the disease and imaging (ultrasound or MRI) of ovarian endometriomas and deep-infiltrating

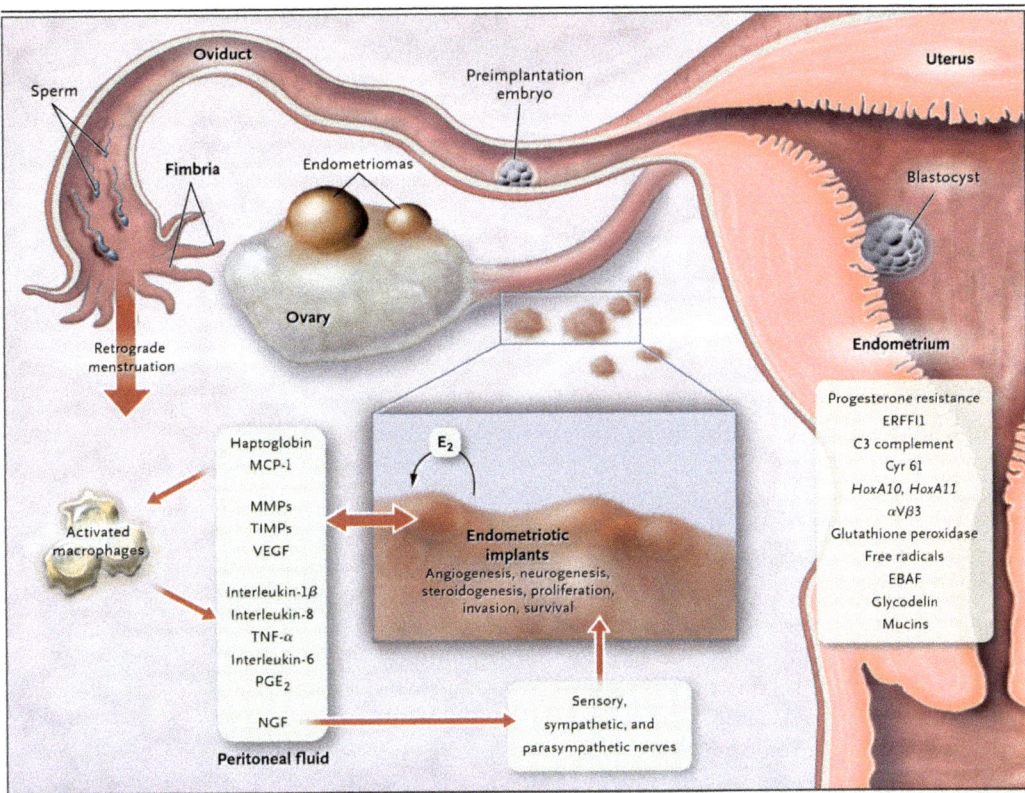

FIG. 18.1 Pathophysiology of pain and infertility associated with endometriosis. Retrograde transplanted endometrial tissue and cells attach to peritoneal surfaces, establish a blood supply, and invade nearby structures. They are infiltrated by sensory, sympathetic, and parasympathetic nerves and elicit an inflammatory response. Endometriotic implants secrete estradiol (E2) as well as prostaglandin E2 (PGE2), monocyte chemotactic protein 1 [MCP-1]) that attracts macrophages, neurotrophic peptides (nerve growth factor [NGF]), enzymes for tissue remodeling (matrix metalloproteinases [MMPs]), tissue inhibitors of MMPs (TIMPs), and proangiogenic substances (e.g., vascular endothelial growth factor (VEGF) and interleukin (IL-8) Lesions secrete haptoglobin, which decreases macrophage adhesion and phagocytic function. Lesions and activated macrophages, which are abundant in the peritoneal fluid in women with endometriosis, also secrete proinflammatory cytokines (IL-1β, IL-8, IL-6, and tumor necrosis factor α [TNF-α]). Local (and systemic) estradiol can stimulate lesion production of PGE2, which can activate pain fibers, enhance neuronal invasion of lesions by stimulating production of NGF and other neurotrophins, and promote sprouting of nociceptors that contribute to persistent inflammatory pain and inhibit neuronal apoptosis. Endometrial bleeding factor (EBAF) is misexpressed and may contribute to uterine bleeding. Infertility results from toxic effects of the inflammatory process on gametes and embryos, compromised fimbrial function, and the eutopic endometrium that is resistant to the action of progesterone and is inhospitable to embryonic implantation. ERFFI1 (ErbB receptor feedback inhibitor 1) is constitutively expressed and there is excess mitogenic signaling. *From Giudice LC. Clinical practice. Endometriosis. N Engl J Med 2010;362(25):2389–98, with permission.*

disease [14]. The World Endometriosis Research Foundation (WERF) Global Study on Women's Health revealed a delay of diagnosis of 7 years after the onset of symptoms, with an average of 1 year from symptom onset to the first medical consultation and 6 years between the first consultation and ultimate diagnosis by a gynecologist [7]. This long delay is due largely to the nonspecific character of many of the associated symptoms, the normalization of pain symptoms, the lack of familiarity among some health care providers about the disorder, and the need for surgery to make the diagnosis [7]. A minor contributor is uncertainty about the symptoms and reluctance on the part of patients to see physicians for fear of a catastrophic diagnosis [15]. Thus, endometriosis represents an area of major unmet personal, clinical, and societal need for which novel methods of noninvasive diagnosis (biomarkers) urgently need to be identified. Genetic and genomic approaches afford an opportunity to move the field forward in this regard.

Endometriosis is commonly staged using the revised American Society for Reproductive Medicine (ASRM) staging system [16]. It is based on the extent (number and sizes) of lesions, the presence or absence of adhesions, and ovarian and/or peritoneal involvement [16]. Higher stages occur with ovarian cysts and extensive adhesions, but do not correlate well with symptoms or predicting treatment response [16]. In contrast, deeply infiltrating disease, which does not contribute to ASRM stage scoring, correlates with pain severity [17]. Current medical therapies for endometriosis-related pain focus on hormonal treatments to minimize or eliminate menstruation, inhibit ovulation, oppose the action of estradiol, inhibit the synthesis of estradiol, and minimize disease burden [1, 17]. These therapies include contraceptive steroids (combined oral contraceptive pills, vaginal ring, patch, progestin injections, implants, and intrauterine devices), danazol (an immunosuppressive androgen), gonadotropin-releasing hormone (GnRH) analogues, and aromatase inhibitors [1, 17]. Metaanalyses have demonstrated essentially equivalent efficacy of these treatments for pain management, although they are abandoned in about 80% of women within 1–2 years of initiation due to waning efficacy and/or intolerable side effects [1, 17]. Surgical approaches reduce disease burden and offer the opportunity to ablate adhesive disease with improved pain relief, although about 50% of women have a recurrence of symptoms severe enough to warrant repeat surgery within 5 years of the index procedure. Clinically, medical therapies can be administered prior to or adjunctively postoperatively for pain management and disease suppression [1, 17].

Infertility associated with endometriosis derives from poor oocyte quality and an adverse, proinflammatory endometrial environment [1, 18–20], with lower implantation rates in advanced versus early stage disease [21]. Infertile women with endometriosis, who attain a singleton pregnancy either spontaneously or through fertility therapies, are at higher risk of miscarriage and adverse pregnancy outcomes [22]. Most of the latter have placenta and implantation-based etiologies attributable to abnormal endometrium prepregnancy and likely in early gestation In women with disease [23]. Medical therapies for pain are counterproductive for fertility as they inhibit conception potential, and surgery is sometimes needed to restore anatomic normalcy in the pelvises of women with disease and infertility [1]. Assisted reproductive technologies result in reasonable pregnancy rates, particularly for women with stage I–II disease [16]. Even with spontaneous conceptions, these patients are at higher risk of pregnancy complications [22, 24, 25].

Endometriosis Genomics, Clinical Phenotyping, and Biospecimen Standardization

Which patients will respond to specific medical and surgical therapies for pain relief or successful pregnancy and who is at risk for developing scarring, fibrosis, disease recurrence, symptom recurrence, poor pregnancy outcomes, and associated chronic medical conditions are currently unpredictable. Integrated analysis of different types of molecular data provide an unprecedented opportunity to improve human health broadly by refining current knowledge about various disease diagnostics and therapeutics. Since the millennium, the Human Genome Project, the International SNP Consortium, the HapMap and 1000 Genomes Projects, the ENCODE Project, the NIH Roadmap Epigenomics Program, the Genotype-Tissue Expression (GTEx) Project, and other major efforts have resulted in the development and application of multiple technologies that allow high-throughput analyses of genomic, epigenomic, and transcriptomic data. Coupled with powerful bioinformatics tools, unbiased genome-wide approaches are providing opportunities to generate novel insights into the pathophysiology of specific disease states and ultimately to develop new, patient-specific treatments. Added to these are innovations in gene editing and other approaches to modify cellular behavior.

The field is poised for pursuing advanced genetic and genomic approaches to develop new targeted, personalized therapies for pain relief and fertility in women with endometriosis and innovative diagnostic and prognostic tools. Critical to these "big data" approaches are deep clinical phenotyping and standardization of specimen collection, processing, and storage across all groups studying this disorder. The World Endometriosis Foundation (WERF) Biobanking and Phenome Harmonization Project (WERF-EPHect: https://endometriosisfoundation.org/ephect/), a collaboration of 34 academic centers and three industry partners, has developed freely available data collection tools and established global standard operating procedures (SOPs) for tissue and fluid collection and storage and standardized phenotypic and surgical data collection instruments that are essential for meaningful study design and accurate data interpretation in the field of endometriosis and endometrium-related disorders [26–29].

ENDOMETRIOSIS AS A GENETIC DISEASE

Heritability of Endometriosis

A considerable heritable component to endometriosis risk has been established through both family- and population-based studies [[30, 31], reviews]. The risk for first-degree relatives of women with severe endometriosis ranges between 2- and 15-fold higher than for relatives of unaffected women [[31], review]. Studies of monozygotic twins have demonstrated high concordance for histologically confirmed endometriosis [32], and large twin studies have estimated the heritable component to be about 50% (0.51 [33] and 0.47 [34]). Genetic linkage (powerful in their design for finding rare variants in "major genes" responsible for the majority of familial disease) and candidate gene approaches have been disappointing, but not unexpected, given that endometriosis is a multifactorial, complex trait ([30, 31], reviews).

Genome-Wide Association Studies

More recently, genome-wide association studies (GWAS), using high-throughput genotyping technologies and extensive bioinformatics analyses, have transformed the understanding of genetic contributions to complex diseases. GWAS has been extraordinarily successful in identifying DNA sequence variants associated with human diseases and phenotypes, with thousands of risk-loci identified across hundreds of traits (http://www.ebi.ac.uk/gwas; https://www.genome.gov/GWAStudies). To date, a total of seven GWA endometriosis studies have been conducted in populations of Japanese and European ancestry with robust analyses and replication data across populations and different ethnic groups [35–41]. Until 2016, 12 independent single nucleotide polymorphisms (SNPs) for endometriosis had been identified: rs10965235 in *CDKN2BAS* on 9p21.3 [36]; rs1519761 on 2q23.3 [39]; rs7521902 near *WNT4* on 1p36.12; r13391619 in *GREB1* on 2p25.1; rs4141819 on 2p14; rs7739264 near *ID4* on 6p22.3; rs12700667 on 7p15.2 in an intergenic region and 99 kb upstream of microRNA miRNA-148a and 290 kb upstream of *NFEL2L3* on chromosome 7; rs1537377 near *CDKN2B-AS-1* on *9p21.3; rs10859871* near *VEZT* on 12q22 [37, 38]; rs17773813 near *KDR* on 4q12; rs519664 in t/tC39B on 9p22 [40]; and rs6542095 near *IL1A* [41]. The most recent metaanalysis of 11 GWA case-control data sets (17,045 endometriosis cases and 191,596 controls) replicated nine of the previously reported loci and identified five novel loci significantly associated with endometriosis risk and involving sex steroid hormone pathways and metabolism (*FN1, CCDC170, ESR1, SYNE1, FSHB*) [42]. A Manhattan plot for GWAS for endometriosis based on all 17,045 endometriosis cases is shown in Fig. 18.2, and Table 18.1 shows the summary of the GWA metaanalysis results for 14 genome-wide significant loci.

Although some of the identified variants in candidate loci are intergenic, seven genes (*WNT4, GREB1, FN1, CCDC170, ESR1, SYNE1, FSHB*) have been implicated with biological

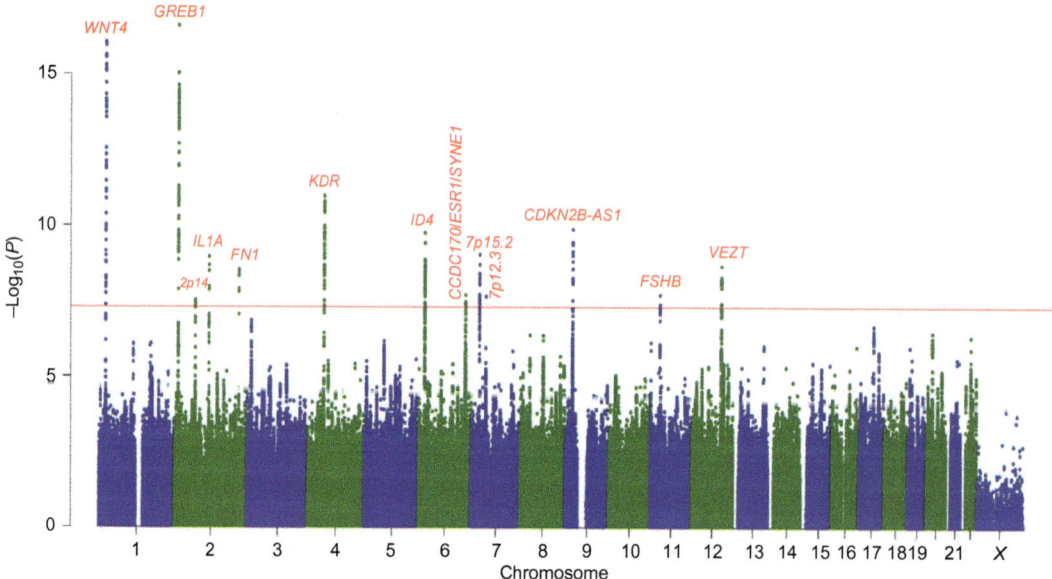

FIG. 18.2   Manhattan plot for GWAS with endometriosis. Data are based on GWA metaanalysis of all endometriosis cases. The horizontal axis shows the chromosome position and the vertical axis shows the significance of tested markers combined in a fixed-effects metaanalysis. Markers that reached genome-wide significance ($P < 5 \times 10^{-8}$) are highlighted. *From Sapkota Y, Steinthorsdottir V, Morris AP, Fassbender A, Rahmioglu N, De Vivo I, et al. Meta-analysis identifies five novel loci associated with endometriosis highlighting key genes involved in hormone metabolism. Nat Commun 2017;8:15539, with permission.*

TABLE 18.1 Summary of the GWA Metaanalysis for 14 Genome-Wide Significant Loci

Chr	SNP	Position (bp)	RA	OA	RAF_{EUR}	RAF_{JPT}	Metaanalysis (All)		Metaanalysis (Grade B)		Associated Gene/ Cytoband
							OR (95% CI)	P Value	OR (95% CI)	P Value	
Previously reported loci											
1	rs12037376	22,462,111	A	G	0.17	0.58	1.16 (1.12–1.19)	8.87×10^{-17}	1.28 (1.18–1.36)	2.69×10^{-9}	WNT4/1p36.l2
2	rs11674184	11,721,535	T	G	0.61	0.54	1.13 (1.10–1.15)	2.67×10^{-17}	1.18 (1.10–1.24)	1.94×10^{-6}	GREB1/2p25.1
2	rs6546324	67,856,490	A	C	0.31	0.21	1.08 (1.05–1.11)	3.01×10^{-8}	1.19 (1.11–1.26)	3.71×10^{-7}	ETAA1/2p14
2	rs10167914	113,563,361	G	A	0.30	0.75	1.12 (1.08–1.15)	1.10×10^{-9}	1.15 (1.07–1.21)	7.59×10^{-5}	IL1A/2q13
4	rs1903068	56,008,477	A	G	0.68	0.88	1.11 (1.07–1.13)	1.04×10^{-11}	1.33 (1.24–1.40)	2.58×10^{-15}	KDR/4q12
6	rs760794	19,790,560	T	C	0.43	0.71	1.09 (1.06–1.12)	1.79×10^{-10}	1.17 (1.10–1.24)	8.74×10^{-7}	ID4/6p22.3
7	rs12700667	25,901,639	A	G	0.74	0.20	1.10 (1.07–1.13)	9.08×10^{-10}	1.28 (1.19–1.36)	6.69×10^{-11}	7p15.2
9	rs1537377	22,169,700	C	T	0.40	0.39	1.09 (1.06–1.12)	1.33×10^{-10}	1.21 (1.13–1.27)	6.31×10^{-9}	CDKN2B-AS1/9p21.3
12	rs4762326	95,668,951	T	C	0.47	0.48	1.08 (1.05–1.11)	2.20×10^{-9}	1.15 (1.08–1.21)	1.08×10^{-5}	VEZT/12q22
Novel loci											
2	rs1250241	216,295,312	T	A	0.29	0.06	1.06 (1.03–1.09)	6.20×10^{-5}	1.23 (1.15–1.30)	2.99×10^{-9}	FN1/2q35
6	rs1971256	151,816,011	C	T	0.20	0.35	1.09 (1.06–1.13)	3.74×10^{-8}	1.28 (1.19–1.36)	1.50×10^{-10}	CCDC170/6q25.1
6	rs71575922	152,554,014	G	C	0.16	–	1.11 (1.07–1.15)	2.02×10^{-8}	1.35 (1.24–1.43)	2.87×10^{-12}	SYNE1/6q25.1
7	rs74491657	46,947,633	G	A	0.91	0.78	1.08 (1.03–1.13)	1.23×10^{-3}	1.46 (1.28–1.59)	2.24×10^{-8}	7p12.3
11	rs74485684	30,242,287	T	C	0.84	0.98	1.11 (1.07–1.15)	2.00×10^{-8}	1.26 (1.15–1.35)	7.77×10^{-7}	FSHB/11p14.1

Chr, chromosome; *SNP*, single-nucleotide polymorphism; genomic position is shown relative to GRCh37 (hg19); *GWA*, genome-wide association; *RA*, risk allele; *OA*, other allele; *OR*, odds ratio with respect to RA; *CI*, confidence interval; RAF_{EUR}, average risk allele frequency in European samples; RAF_{JPT}, average risk allele frequency in Japanese samples.

From Sapkota Y, Steinthorsdottir V, Morris AP, Fassbender A, Rahmioglu N, De Vivo I, et al. Meta-analysis identifies five novel loci associated with endometriosis highlighting key genes involved in hormone metabolism. Nat Commun 2017;8:15539 with permission.

plausibility [30, 35–39, 42, 43]. A particular pathway implicated in the GWAS involves WNT/β-catenin signaling, which is essential to the development of the female reproductive tract and a key player in the monthly regeneration of the endometrium. Some SNPs are associated with other traits and diseases in other GWA studies. For example, bone mineral density is associated in or near *WNT4*; diseases including glaucoma, glioma, intracranial aneurysm, and coronary artery disease are associated with variants in the region of *CDKN2B-AS1*; and *VEZT* is associated with adverse response to chemotherapy [44]. Follow-up of an intergenic endometriosis locus on chromosome 7p15.2 that was independently associated with fat distribution (waist-to-hip ratio, adjusted for BMI) identified WNT/β-catenin signaling as the shared pathway between the traits [45]. Moreover, a recent pathway analysis of GWAS data uncovered a novel variant in the mitogen-activated protein kinase *MAP3K4* gene and revealed multiple pathways statistically enriched for genetic association signals. This included MAPK-related pathways in stage I/II disease, extracellular matrix/fibrosis in stage III/IV disease, and interleukin-signaling, apoptosis-signaling, and GnRH-signaling pathways overall [46].

Consistently, GWA studies have demonstrated a differential genetic burden associated with varying degrees of endometriosis disease stage [16], and the strongest effect sizes of association are observed with more advanced stage disease [37, 47]. It is unclear what underlying distinct molecular phenotypes are captured by these broad, surgically defined subtypes, although statistical enrichment and pathways analyses [46, 48] are beginning to address these phenotypes. Moreover, further refinement of the data is anticipated with increased utilization of the WERF-EPHect tools by investigators across the globe. Currently identified genome-wide significantly associated variants explain 5% of heritability [42, 48], and combined analyses of all common germline variants account for about 26% of the risk for endometriosis [37, 49]—underscoring the need for larger GWAS metaanalyses to mine the remaining 21% heritability of this disease due to common variation.

Genetic Control of Gene Expression

The question arises whether an individual's genotype can influence gene expression and thus function within a given tissue. Indeed, numerous studies have demonstrated that the expression levels of many genes are under genetic control (GTEx Consortium 2017 [50]), and these genetic effects (expression quantitative trait loci (eQTLs)) can affect variation in gene expression in various tissues among individuals. A recent study, using genome-wide genotyping and endometrial gene expression data across the menstrual cycle, identified eQTLs for 198 unique genes. After separating the analysis into ubiquitously or inconsistently expressed genes across the cycle, 1851 genes were differentially expressed and 1037 genes were turned on or off at different cycle stages [51]. Thus, genotype affects gene expression in eutopic the, and significant differences in global endometrial gene expression are observed in women with versus without endometriosis [11, 13, 18]. Whether genetic variants in endometriosis women play a role in the different expression of specific genes in the eutopic endometrium and whether these predispose to establishment of endometriotic lesions remain to be determined. It should be noted that genome-wide analyses of gene expression in the eutopic endometrium in women with versus without endometriosis differ among studies [11, 13, 18, 52], possibly due to disease stages, heterogeneity of disease, different platforms for analysis, different data analytical approaches, associated pain, infertility or both, different definitions of the control population, individual heterogeneity (genetic, epigenetic), failure to control accurately for cycle stage, relatively small sample sizes, medications, stress, and other factors or exposures. Thus, patient phenotyping and data analytics with large sample sizes are key for understanding the differences observed in the literature and when considering the interplay of genetics and the regulation of gene expression in endometriosis lesions and the eutopic endometrium of women with endometriosis.

GWAS variants associated with endometriosis are most commonly located within introns or in intergenic regions, implying they affect endometriosis risk through genetic *regulation* of gene expression rather than through direct protein coding effects. Analysis of results for genes with significant eQTLs in the endometrium identified one genome-wide significant *cis*-eQTL in the endometrium for the long noncoding RNA *LINC00339/HSPC157* at chromosome 1p36.12 that overlapped genomic regions associated with endometriosis [51]. The small sample size of eQTL studies in the endometrium (whole tissue analysis) is a major limitation [53]. Results from the GTEx project show >70% of eQTLs are shared among multiple tissues (GTEx Consortium 2017 [50]). Signals for three eQTLs observed in blood samples overlap two genomic regions associated with endometriosis at chromosome 1p36 [54] and chromosome 12 [55]. Genetic effects on *VEZT* in the endometrium show the same magnitude and direction of effect as the results for blood samples, but the results were not genome-wide significant [55]. The top SNP at chromosome 1p36 significantly affects expression of *LINC00339* in both the blood and endometrium [51, 54] and the expression of cell division control protein 42 (*CDC42*) in the blood. Genetic effects on the expression of *CDC42* in endometrial tissue showed effects in the same direction and of similar magnitude, but the effects were not significant [54]. However, analysis of chromatin interactions on chromosome 1p36 provides evidence that both *LINC00339* and *CDC42*, but not *WNT4*, are target genes for the genetic effects on endometriosis risk in this region [54]. These findings underpin the necessity to conduct functional studies integrating genetic data with gene expression and epigenetic profiling to identify the genes affected by GWAS variants rather than conclusions based on the proximity of genes to the signal.

Structural Genomic Alterations in Endometriosis

Endometriosis lesions (ovarian endometrioma, deep infiltrating disease, superficial peritoneal lesions) are comprised of endometrial epithelium, stromal fibroblasts, vascular components, and immune cells; they are heterogeneous in their location and gross appearance [2]. While histologic analyses demonstrate that these lesions are benign, they display some characteristics of malignancy, including cellular invasiveness, neoangiogenesis, enhanced proliferation and cell migration, resistance to apoptosis, distant metastases, and recurrence [56, 57]. Cytogenetic studies of endometriosis lesions have demonstrated the loss of heterozygosity (LOH) [58, 59] and chromosomal aneuploidy and copy number abnormalities [60–63], similar to malignancies. Comparative genomic hybridization (CGH) microarray analysis of laser-dissected glandular epithelial cells in lesions demonstrated that lesions cluster by anatomic location (peritoneal or ovarian) [63], underscoring heterogeneity across lesion types and commonalities of abnormalities within specific lesion types. Genomic alterations

(LOH, chromosome loss/gain, translocations, and copy numbers) have been demonstrated in epithelial cells of the *eutopic* endometrium of women with endometriosis, several of which are common to aberrations in lesions [62, 64]. These data suggest overall that preexisting genomic alterations in this tissue within the uterus may confer a survival advantage to subsequently sloughed endometrial cells in establishing endometriotic implants, compared with no disease in the majority of women who have retrograde menstruation and who have a normal eutopic endometrium [64]. Alternatively, the environment within the pelvis, including inflammation, heme- and free iron-induced oxidative stress, and steroid hormones may promote genomic abnormalities in sloughed tissue and the establishment of disease and potentially the oncogenic potential of this tissue [56].

Endometriosis and Ovarian Cancer

Epidemiologic and Genetic Associations

Among the 21 epidemiological studies that have investigated endometriosis in relation to ovarian cancer risk, 20 reported a positive association (including 16 reporting statistically significant findings) [6]. The pooled analysis Ovarian Cancer Association Consortium (OCAC) quantified 50% greater odds among women with endometriosis (RR: 1.46, 95% CI: 1.31–1.63 [6]. A recent metaanalysis that focused on endometriosis-associated ovarian cancer (EAOC) reported a 30%–80% greater risk that varied with study design [65]. EAOC occurs in ovarian endometriomas but not in deep infiltrating disease or superficial peritoneal lesions [56]. Epidemiologic and histologic data demonstrate that the majority of EAOCs are endometrioid and clear cell carcinomas and, to a lesser extent, low grade serous ovarian cancer [6, 56, 57]. A recent study investigated whether the epidemiologic association between endometriosis and ovarian cancer subtypes could be attributable to shared genetic susceptibility loci [66]. Indeed, the strongest genetic correlation with endometriosis was with ovarian clear cell carcinoma (0.51, 95% CI: 0.18–0.84), and similar correlations with endometriosis were found with endometrioid (0.48, 95% CI: 0.07–0.89) and low-grade serous (0.40, 95% CI: 0.05–0.75) carcinomas [66]. High-grade serous ovarian carcinoma (largely derived from fallopian tube epithelium), despite no epidemiologic association with endometriosis, showed a genetic correlation, albeit weaker than other subtypes (0.25, 95% CI: 0.11–0.39) [66]. Of critical importance are definitive epidemiologic studies quantifying the risk among women with endometriomas relative to other endometriosis macrophenotypes as well as the lesion microphenome to solidify the role of malignant transformation [6].

Somatic Mutations in Endometriosis and Ovarian Cancer

A mouse model of endometriosis has provided insight into the pathogenesis of peritoneal endometriosis as well as EAOC [67] (Fig. 18.3) [68]. The conditional activation of the K-ras oncogene in endometrial cells deposited into the peritoneum resulted in histologically confirmed peritoneal endometriotic implants in about 50% of mice within 8 months [67]. On the other hand, similar activation of the K-ras oncogene in peritoneal cells showed no progression to endometriosis, suggesting an endometrial origin of peritoneal disease. Furthermore, Cre-activation of *Kras* in the mouse ovarian surface epithelium or inactivation of *Pten* was sufficient for the formation of ovarian endometriotic lesions, but a second "hit" was required

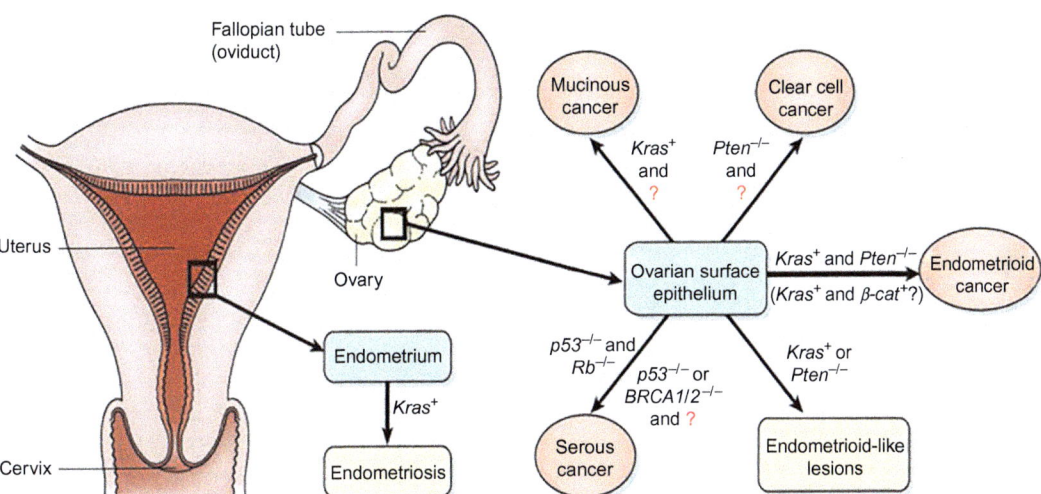

FIG. 18.3   Proposed genetic alterations in the development of epithelial ovarian cancers and endometriosis. Normal tissue is in a green rectangle, a benign pathologic disease is placed in a yellow rectangle, and cancerous lesions are enclosed by a red oval. The "+" symbol denotes activating mutations in oncogenes while "−/−" denotes inactivating mutations in tumor suppressor genes. A question mark indicates that the second genetic hit is unknown. *From Matzuk MM. Gynecologic diseases get their genes. Nat Med 2005;11(1):24–6, with permission.*

(i.e., both *Kras* activation and *Pten* inactivation) for development of invasive and metastatic ovarian endometrioid adenocarcinoma [67, 68] (Fig. 18.3).

In humans, several cancer-associated ("driver") somatic mutations have been identified in endometriomas and deep infiltrating endometriosis, although malignant transformation, to date, has only been reported in the former [56]. Histologic studies of endometriomas support a continuum from benign → atypia (cytologic atypia and hyperplasia) → cancer [56], and several studies have identified that endometriomas and multifocal lesions are monoclonal [56, 69]. Driver mutations alone or in combination include activation of the oncogene *KRAS* [70], silencing of the tumor suppressor genes *TP53* [71] and *PTEN* [72], suppression of *ARIDIA 1* (the AT-rich interactive domain 1A [SWI-like] gene) coding for BAF250a [73], and activation of *PIK3C*, *PPP2RIA* (An74), and *CTTNB1* [74]. A recent study using exome sequencing of deep infiltrating endometriosis revealed that the number of somatic mutations in each lesion was variable and lesions were found in 19 of 24 patients (79%), although only 26% carried known somatic cancer driver mutations in *ARID1A*, *PIK3CA*, *KRAS*, or *PPP2R1A*; the somatic mutations were in the epithelial compartment of the lesions [75]. Overall, these data raise several important questions, including whether these and other specific mutations are shared among other types of endometriosis lesions and cell types, if they contribute to disease severity and progression, if they are also observed in the eutopic endometrium of women with disease, if they contribute to variable responses observed in medical therapies, and what the "second hit" is to progress to malignancy or not. Indeed, the cellular and molecular changes preceding deletions and mutations in the above genes resulting in developing EAOC are not well understood and comprise an area of active investigation for prognosis and potentially targeted therapeutics.

ENDOMETRIOSIS AS AN EPIGENETIC DISEASE

Epigenetics has been proposed as a unifying principle in the pathogenesis of complex traits and diseases [76] and includes epigenetic modifications of DNA and histones and microRN As (Fig. 18.4) participate in regulating steroid hormone-responses of various tissues and in inflammatory processes [77]. As endometriosis is an estrogen-dependent, progesterone-resistant, inflammatory disorder, it is not surprising that epigenetic modifications of DNA and histones and expression of various microRNAs have been studied in the eutopic endometrium of women with endometriosis (and without disease) and in endometriotic lesions and their cellular components in vitro. An early study revealed nuclear receptor and coreceptor signaling and chromatin remodeling pathways in the human endometrium that implicated the epigenome in the steroid hormone response of this tissue and with abnormalities in the endometrium of women with endometriosis [78]. Indeed, accumulating evidence over the past decade suggests that various epigenetic aberrations in lesions and the eutopic endometrium play key roles in the pathogenesis and pathophysiology of endometriosis. They also have implications for possible preventive strategies through lifestyle/exposome modifications and also innovative therapies to address pain and infertility associated with the disease [18, 79, 80]. These are summarized below.

DNA-Methylation

Targeted Gene Approaches

Targeted approaches have provided strong evidence for hypomethylation and hypermethylation of select genes involved in steroid hormone action, inflammation, and cell adhesion in endometriotic lesions and in the eutopic endometrium of women with disease [18, 80]. The promoters of *ESR2*, steroidogenic factor 1 (*SF-1*), aromatase (*Cyp19A1*), cyclooxygenase 2 (*COX2*), and *syncytin-1* genes are hypomethylated in endometriotic lesions and promoters of *PGR*, *HOXA10*, and *E*-Cadherin genes are hypermethylated with concomitant up- and down-regulation, respectively, of their corresponding gene products [18, 80–84]. Fig. 18.5 shows a schema of the activities and consequences of some of these epigenetic

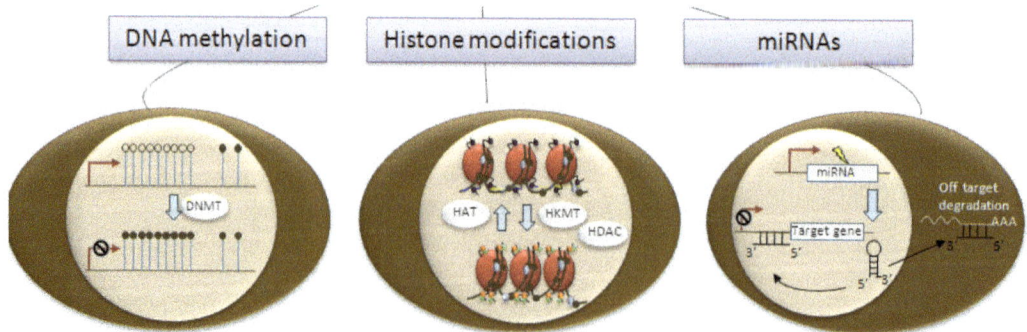

FIG. 18.4   Epigenetic mechanisms (see text). Abbreviations: *DNMT*, DNA methyltransferase; *HAT*, histone acetyl transferase; *HDAC*, histone deacetylase; *HKMT*, histone lysine methyl transferase; *mi RNA*, microRNA.

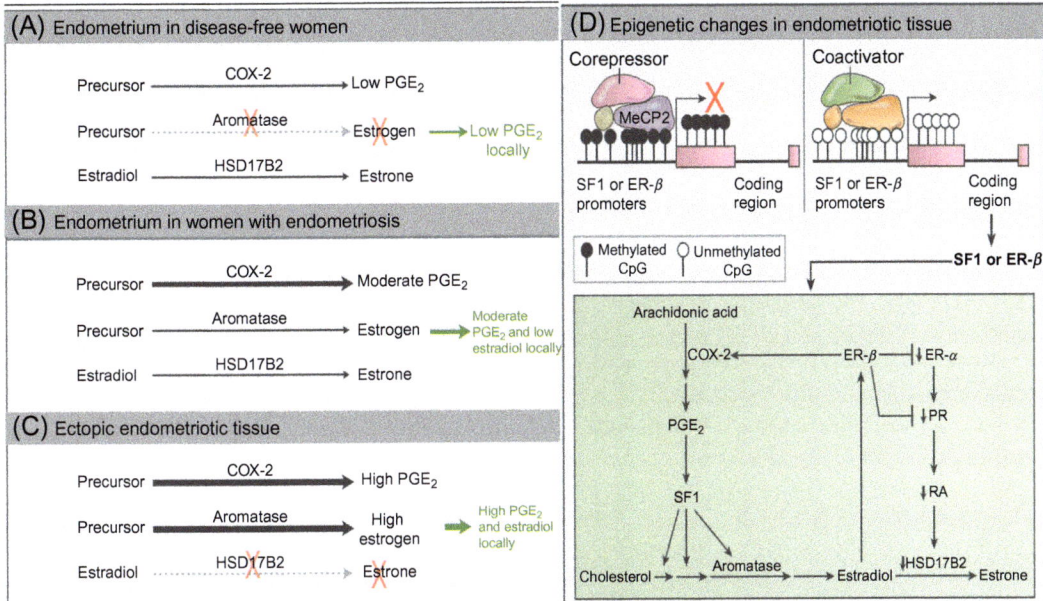

FIG. 18.5 Epigenetic changes in endometriotic tissues resulting in altered steroidogenesis and enhanced prostaglandin production and comparison of prostaglandins and estrogens in the eutopic endometrium of women without and with endometriosis and in endometriotic tissue (see text). Increasing enzyme activity is denoted by increasing thickness of arrows. *COX-2*, cyclooxygenase-2; *CpG*, a cytosine-phosphate-guanine sequence; *HSD17B2*, 17β-hydroxysteroid dehydrogenase 2; *MeCP2*, methyl-CpG-binding domain protein 2; *RA*, retinoic acid. *From Bulun SE. Endometriosis. N Engl J Med 2009;360(3):268–79, with permission.*

changes and their relevance to the pathophysiology of endometriosis. In a normal eutopic endometrium (Fig. 18.5, panel A), the activity of COX-2, and thus prostaglandin (PGE) 2 production, is low, and E_2 is not produced locally due to the absence of aromatase. During the luteal phase, the P_4-dependent 17β-hydroxysteroid dehydrogenase 2 (HSD17B2) enzyme catalyzes conversion of E_2 to estrone (E_1), a less potent estrogen. However, in the endometrium of women with endometriosis (Fig. 18.5, panel B), there is a subtle increase in COX-2 and detectable aromatase activity [18]. In ectopic endometriotic tissue (Fig. 18.5, panel C), molecular abnormalities include high COX-2 and aromatase levels, the latter resulting in markedly increased lesion E_2 levels. Moreover, E_2 is not metabolized due to deficient HSD17B2 activity [18]. Normally, the promoters of the two nuclear receptors, SF1 and ESR2 (ER-β), are heavily methylated and thus silenced in eutopic endometrial stromal cells (Fig. 18.5, panel D). However, in *endometriotic* stromal cells, a lack of promoter methylation results in promoter activation and pathologic overexpression of SF1and ESR2 (ER β) (Fig. 18.5, panel D). SF1 mediates PGE2-dependent induction of E_2, and ER-β suppresses ER-α and PRs, leading to P_4-resistance and deficient inactivation of E_2. Thus, in endometriotic tissue, E_2 and PGE2 are produced in large amounts and they enhance cell survival and inflammation in endometriotic lesions and in the pelvic environment.

Clinically, increased PGE2 in eutopic endometrium and endometriotic tissues causes severe dysmenorrhea and pain. Within the past decade, aromatase inhibitors have been introduced in the clinical management of endometriosis-associated pelvic pain, as they inhibit lesion (and ovarian) E_2 synthesis and have been shown to markedly decrease lesion size and pelvic pain in women refractory to other medical therapies [85]. P_4-resistance of lesions, attributable to lower ERα levels and hypermethylation of PGR [18], results in variable responses to progestins for pain relief [17]. Selective progesterone-receptor modulators with mixed agonist and antagonist properties have shown promise to reduce endometriosis-related pelvic pain compared to conventional progestins [86], and their clinical application is an active area of therapeutic development.

Genome-Wide Analyses

Genome-wide DNA-methylation analyses of endometriotic lesions have given additional insight into the pathophysiology of endometriosis. For example, Borghese et al. [87], using methylated DNA immunoprecipitation (MeDIP) arrays to profile specific (~25,000) promoters, reported similar methylation between endometriosis lesions and the eutopic endometrium, with lesion methylation variation occurring at discrete loci across the genome, supporting eutopic tissue as the origin of ectopic lesions and potentially local modifiers of lesions in situ. Bulun and colleagues [88] reported genome-wide DNA methylation analysis of eutopic endometrial stromal cells from women without disease and ectopic stromal cells from endometriomas, using the Illumina 450 K bead array and gene expression using large-scale arrays. They found similar DNA methylation in eutopic and ectopic stromal cells, although with extensive differences intragenically and at sites distal to classic CpG islands where methylation was typically negatively correlated with gene expression. They further found an epigenetic switch for GATA factor expression in endometriosis. Recently, Yotova and colleagues [89], studying endometrioma and eutopic endometrial stromal fibroblasts and using whole genome methylation bead arrays, described differential methylation of genes encoding transporters (SLC22A23), signaling components (BDNF, DAPK1, ROR1, and WNT5A), and transcription factors known to be dysregulated in endometriotic lesions or the eutopic endometrium of women with disease and involved with the P_4-resistant phenotype (GATA family members, HAND2, HOXA cluster, NR5A1, OSR2, and TBX3). RNA-Seq revealed a subset of the differentially methylated genes that was differentially expressed, with the likelihood of differential expression increasing with the extent of methylation and location in enhancer elements. Treatment of the stromal fibroblasts with 5aza-dC (a hypomethylating drug) led to activation of DAPK1 and SLC22A23 and repression of HAND2, JAZF1, OSR2, and ROR1 mRNA expression [89]. Importantly, these data, along with those described above, underscore a consistent epigenetic signature in endometriosis stromal cells with specific transcriptional and signaling pathways that could be pursued as therapeutic targets.

Genome-wide DNA methylation analysis of whole human endometrial tissue has revealed that the DNA methylome is distinct in different cycle phases, with the greatest changes in the secretory (P_4-dominant) phase in women without endometriosis [53, 90, 91]. Profiles differ in the endometrium of women with endometriosis [91–93], particularly in the secretory phase [93]. While differentially methylated loci across the cycle in controls are located in both CpG and nonCpG islands, in the eutopic endometrium of women with severe endometriosis, the majority of methylated loci, either in phase comparisons or across the cycle in disease, are located nearly exclusively in CpG islands [90, 93]. The biological significance of this remains to be determined. For many genes, changes in DNA methylation in the eutopic endometrium of women without and with endometriosis are associated with specific gene expression changes, and in disease, expression differences are relevant to endometrial function and dysfunction [53, 90, 93]. For example, these involve genes in cell cycle regulation (e.g., *CDKN2B*), inflammation and immune response (e.g., *PLEK, CCL3, BST2*), and with potential roles in disease pathophysiology affecting diverse processes such as cell adhesion, cell migration, oxidative stress protection, and genes involved in transcription function and regulation [93].

Of note, age is significantly associated with DNA methylation patterns in the eutopic and ectopic endometrium [53]. Rahmioglu and colleagues [53] partitioned variation in DNA methylation and RNA expression profiles in the endometrium and endometriotic disease tissue into between-individual, within-tissue (cellular heterogeneity), and technical variation categories. They found that interindividual variation is overwhelmingly in the top 10%–50% variable DNA methylation/RNA expression sites in relation to within-tissue and technical variation. Moreover, differential methylation analysis comparing eutopic and ectopic endometrium from the same women revealed many significant differences that are enriched for WNT signaling, angiogenesis, cadherin signaling, and gonadotropin-releasing-hormone-receptor pathways [53]. These data overall underscore the importance of deep phenotyping of patient characteristics. As the endometrium is a heterogeneous tissue of multiple cell types, accounting for cell type heterogeneity in samples is increasingly challenging. Single-cell approaches are anticipated to mitigate some of these challenges and, as additional data are derived from GWA studies and DNA methylation analyses, the question arises whether genome-wide genetic loci are associated with DNA methylation profiles in the eutopic endometrium and gene expression. This is an active area of investigation.

Histone Modifications

There is increasing evidence that histone modifications are involved in endometriosis pathophysiology [79, 84, 94] and that disease phenotypes may be modified by targeting specific histone marks in a variety of conditions [95], specifically in endometriosis [96–99]. Covalent histone modifications include methylation, phosphorylation, acetylation, sumoylation, ubiquitination, citrullination, and ADP-ribosylation. They play a major role in chromatin dynamics and activating or silencing transcriptional activity [100]. Histone modifications are derived from the action of specific enzymes, including histone deacetylases (HDACs), histone acetylases (HATs), and histone demethyltransferases (HDMTs); histone acetylation is generally associated with gene activation and histone methylation with gene silencing [101].

Histone Modifications in Endometriosis Lesions

Global and lysine (K)-specific acetylation and methylation studies have revealed that endometriotic lesions are globally hypoacetylated at H3 (but not H4), have lower levels of H3K9ac and H4K16ac, and are hypermethylated at H3K4, H3K9, and H3K27, compared to the control eutopic endometrium [102]. Moreover, chromatin immunoprecipitation (ChIP)-polymerase chain reaction (PCR) studies of histone acetylation of promoter regions of candidate genes known to be down- and up-regulated in endometriosis lesions versus the normal eutopic endometrium, respectively, revealed hypoacetylation of H3H4 within promoters of *HOXA19, ESR1, CDH1,* and *P21,* and enrichment of H3 and H4 acetylation in the *SF1* promoter versus the normal eutopic endometrium [102]. These studies underscore a complex interplay of histone modifications and DNA methylation (see above) in the regulation of gene expression in endometriosis pathophysiology and, while promising, also pose challenges for targeted therapeutic development.

Immunoreactive lysine-specific demthylase 1 (LSD1, which demthylates mono- and di-methylated lysines in H3 lysines 4 and 9 (H3K4 H3K9)), sirtuin 1 (SIRT1, a HDAC), and the enhancer of zeste homologue (EZH2, a histone methyltransferase that methylates H3K27) have been reported to be elevated in endometriomas compared to the normal eutopic endometrium [99]. However, a recent study found no difference in trimethylation of H3 in endometriosis lesions (mixture of endometriomas, peritoneal disease, and deep infiltrating lesions), although EZH2 expressed in endometriotic epithelial cells was higher as assessed by immunohistochemistry and was up-regulated by progesterone [103]. HDAC I, but not other class I HDACs (HDACII, HDACIII), is significantly increased in endometriotic lesions compared to the normal eutopic endometrium in the proliferative phase, but without differences among lesion types, and significantly correlate with ERα, ERβ, and PR levels [104]. These results are consistent with reduced acetylation levels in histone H3 and H4 [97] and elevated HDAC I in another study that included lesions in different phases of the cycle [105]. Cycle phase is an important driver, as HDACs are hormone-dependent in the normal eutopic endometrium [78], and thus studies on their regulation or dysregulation in ectopic disease warrant focus on this potential confounder.

Therapeutic Potential

With regard to potential therapeutic applications, attention has focused on HDAC inhibitors, which share antiproliferative, antiangiogenic, antiinflammatory, and antinociceptive properties; suppress cell invasion and uterine contractility; and induce cellular apoptosis ([99], review). Both in vitro and in vivo studies have been conducted with select HDAC inhibitors and show promising results for therapeutic applications in women with endometriosis-associated pain although off-target effects may limit clinical application. For example, Trichostatin A, a class I HDAC inhibitor, attenuates invasiveness and activates E-cadherin expression of an immortalized endometriotic cell line in vitro [106]. Additionally, Trichostatin A and valproic acid (VPA, another HDAC inhibitor) induce cell cycle arrest and p21 expression in this endometriotic cell line [79]. Acetylated histones 3 and 4 (H3, H4, respectively) are decreased in cultured endometrioma stromal fibroblasts (vs. normal endometrium-derived fibroblasts), and treatment of these cells with the HDAC inhibitors VPA, apicidin, and suberoyl anilide bishydroxamine results in accumulation of acetylated histones in the chromatin and specifically in the promoters of p16, p21, p27, and cycle check-point kinase 2 genes, resulting in suppression of cell proliferation, induction of cell cycle arrest, and cellular apoptosis [97]. In a mouse model of endometriosis, Trichostatin A reduces lesion growth and hyperalgesia [107] and in a rat model of endometriosis, VPA and progestins inhibit lesion growth and also reduce hyperalgesia [98]. With regard to women, a pilot study of VPA administered for three months to three patients with endometriosis, adenomyosis, and severe dysmenorrhea revealed complete pain relief, a 33% mean reduction in uterine size, and disappearance or reduction of tender palpable cul-de-sac endometriotic nodules [98]. While these preclinical findings are exciting and the clinical pilot study is promising, prospective randomized controlled trials are necessary to assess safety and efficacy and validate these observations more broadly. Furthermore, specificity of response is important, as is consideration of side effects and major and minor teratogenic effects of VPA, the latter of which are especially concerning, particularly in women with symptomatic endometriosis who are in their reproductive years.

microRNAs

Regulation of gene transcription has increasingly become more complex as more and more molecular and genomic players are identified along with continuing progress on elucidation of their mechanisms of action. microRNAs (miRNAs) are relative newcomers, having been discovered only about 15 years ago [108]. Since that time, they have enjoyed widespread study in multiple organisms and diseases, with their origins, functions, and mechanisms of action enabled tremendously by deep sequencing technologies. miRNAs are short noncoding RNAs (about 22 nucleotides in length) that regulate gene expression posttranscriptionally by targeting mRNAs for cleavage or translational repression. They generally bind via their "seed sequences" to the 3'-UTR (untranslated region) of their target mRNAs and repress protein production by destabilizing the mRNA and translational silencing [109].

microRNAs in Endometriosis Lesions and the Eutopic Endometrium

Over the past 8 years, miRNAs have been widely studied in endometriosis lesions, the peripheral blood of affected women and controls in search of biomarkers of disease, and in the eutopic endometrium of women with and without endometriosis. Excellent recent reviews are provided by Mari-Alexandre [110] and Nothwick [111]. More than 100 miRNAs have been found to be over- or underexpressed in endometriotic lesions compared to the eutopic endometrium (Table 18.2) [111–116], although replication of results between studies is a major issue. Of note, a few studies have also found marked differences in miRNA expression between the eutopic endometrium of women with and without disease, including miR-34c, miR-9, miR-34b, miR-488-5p, and miR629-3p, involved in inflammation [110, 117, 118]. Given the multiplicity of targets in lesions and the eutopic endometrium, trying to understand the significance of these data has been challenging, and thus, focus has turned to candidate miRNAs with biological plausibility in terms of disease pathophysiology. These include analyses of specific miRNAs governing mRNAs involved in cell proliferation, angiogenesis, cell mobility, and cell invasion, and complementary functional studies of knockdown or overexpression to assess cell behavior and as proof of concept [111]. Relevant to angiogenesis are miR-15a-5p, miR-15b, miR16, miR-17-5p, miR-20a, miR-21, miR125a, miR126, miR-221, miR-222, and miR199a. Relevant to cell adhesion, migration, proliferation, and survival are miR-15a-5p miR-142-3p, miR145, miR183, miR-199a, miR451, and 483-5p, and relevant to the inflammatory phenotype are miR451 and miR629-3p [110, 111]. There is considerable lack of agreement among studies, likely due to numerous factors including quality of samples, deep clinical phenotyping of subjects, cycle phase differences, and tissue heterogeneity (numbers and distribution of cell types in lesions and the eutopic endometrium) [119]. Saare and colleagues recently summarized the shortcomings of miRNA studies in endometriosis research and called for pure cell isolates for analysis to minimize heterogeneity [119]. They further identified three major

TABLE 18.2 Over- and Underexpressed Endometriotic Lesion miRNA Identified by Array or Deep Sequencing Analysis

miRNA	Level of Expression	Reference
miR-145, miR-143, miR-99a, miR-99b, **miR126, miR100,** miR-125b, **miR-150**, miR-125a, miR-223, miR-194, miR-365, **miR29c, miR-1**	Increased (lesion/eutopic)	[112]
miR-200a, miR-141, miR-200b, miR-142-3p, miR-424, miR-34c, **miR-20a, miR-196b**	Decreased (lesion/eutopic)	[112]
miR-1, miR-100, miR-101, **miR-126**, miR-130a, **miR-143, R-145**, miR-148a, **miR-150**, miR-186, miR-196b, miR-199a, miR-202, miR-221, miR-28, miR-299-5p, miR-29b, **miR-29c**, miR-30e-3p, miR-30e-5p, miR-34, miR-365, miR-368, miR-376, miR-379, miR-411, miR-493-5p, miR-99a	Increased (lesion/eutopic)	[113]
miR-106a, miR-106b, miR-130b, miR-132, miR-17-5p, miR-182, **miR-183, miR-196b, miR-200a, miR-200b, miR-200c, miR-20a, miR-25**, miR-425-5p, miR-486, miR-503, miR-638, miR-663, miR-671, miR-768-3p, miR-768-5p, miR-93	Decreased (lesion/eutopic)	[113]
let7d, let7i, miR-202, 193a-3p, **miR-29c**, miR-708, miR-509-3-5p, miR-574-3-p, miR-193a-5p, miR-451, miR-485-3p, **miR-100**, miR-720	Increased (lesion/eutopic)	[114]
let7a, b, c, e, f, g, h, miR-504, **miR-141**, miR-429, miR-203, miR-10a, **miR-200a**, miR-873, **miR-200b, miR-200c**, miR-449b, miR-375, miR-34c-5p	Decreased (lesion/eutopic)	[114]
miRPlus-F1038, miR-1915, miR-637, miR-518e, miR-519a, miR-519b-5p, miR-519c-5p, miR-522, miR-523, miR-574-5p, miR-615-3p, miR-1909, miR-224, miR-133b, miR-622, miR-628-3p, ebv-miR-BHRF1-2, miRPlus-F1215, miRPlus-F1221, miR-1470, miR-1469, miR-520d-5p, miR-551b, miR-361-3p, miR-941, miRPlus-F1223, miR-202, miR663b, miRPlus-F1042, miR-381, miR-412	Increased (lesion/eutopic)	[115]
miR-203, miR-425, **miR-183**, miR-92a, **miR-196b**, miR-363, let-7i, miRPlus-E1031, **miR-200b**, miRPlus-F1231, miR-215, miR-362-3p, miR-342-3p, **miR-200c**, miR-93, miR-24-1, **miR-25**, miR-106b	Decreased (lesion/eutopic)	[115]
miR-106-5p, miR-142, miR-145-5p, miR-16-5p, miR-181a, miR-205, miR-663	Increased (lesion/eutopic)	[116]
miR-126, miR-146a-5p, miR-148a, miR-150, miR-15a-5p, miR-19b-1-5p, **miR-200b, miR200c**, miR-423, miR-675	Decreased (lesion/eutopic)	[116]

Adapted from Nothnick WB. MicroRNAs and endometriosis: distinguishing drivers from passengers in disease pathogenesis. Semin Reprod Med 2017;35(2);173–80, with permission.

miRNA families whose select members are most frequently differentially expressed in endometriotic lesions versus the eutopic endometrium [119]. These include miR200b, a member of the miR200 family involved in cell migration and epithelial-to-mesenchymal transition; miR145 (145-3p and miR-145-5p), involved in apoptosis of cancer cells [111];

and miR196b, located near the *Hox* gene cluster. The latter has also been shown to regulate numerous members of the Hox gene family, including those relevant to female reproductive tract patterning and whose abnormal regulation (e.g., *HoxA10*) has been demonstrated in response to progesterone in endometriosis lesions and the eutopic endometrium of women with versus without disease [120].

Circulating microRNAs

A few studies have evaluated circulating miRNAs as potential noninvasive biomarkers of endometriosis in an effort to bypass the current gold standard of surgical diagnosis of the disease [110]. In 2013, a small study of 10 women with endometriosis and 10 controls without disease identified specific miRNAs in serum that were subsequently validated by q-RT-PCR in a separate, larger cohort of 60 women with endometriosis and 25 controls [121]. They found that miR-199a and miR-122 levels were significantly higher and miR-145, miR-141, miR-542-3p, and miR9 were significantly lower in women with endometriosis. Subsequently, another group studying 23 women with disease and 23 controls found plasma miR-17-5p, miR-20a, and miR-22 levels down-regulated in women with disease and that these "biomarkers" could discriminate between disease and no disease [122]. Subsequently, Taylor and colleagues, quantifying areas under the curve, found that combined levels of let-7b, let-7d, and let-7f in the proliferative phase gave the best result to distinguish women with and without endometriosis [123]. Another group found that down-regulation of serum miR-200a-3p, miR-200b-3p, and mir141-3p demonstrated the greatest accuracy to discriminate between those with and without disease [124]. The most recent contribution to this rapidly evolving field is the finding of elevated serum miR-451a in women and baboons with endometriosis [125]. Interestingly, these studies overall describe unique and nonoverlapping biomarkers for endometriosis, underscoring the importance of considering endometriosis case phenotyping, control characteristics and selection, modifiers such as menstrual cycle dependence or independence, differences among technology platforms, analytical tools and protocols employed, and whether there are informative stratifying subgroups defined by disease burden, stage, pain, infertility, women's characteristics such as age, body size, parity, and lifestyle factors. Large, prospective, multicenter replication studies with diverse participant populations are important in moving this field forward, and with robust consideration of these confounders, mediators, and modifiers so far, circulating miRNAs offer the most promising noninvasive diagnostic for endometriosis. This could expand to the predicting stage of disease, response to therapies, risk of disease recurrence and malignant potential, and prognostic indicators for response to treatments.

Epigenetic and Exposome Contributions

Associations between genetic loci and exposome factors (often referred to as gene x environment interactions) have been documented in many complex diseases. The exposome encompasses modifiable and innate exposures, including anthropometrics, diet, environmental chemicals, parity, resilience, and menstrual cycle characteristics, among others. While it is well confirmed that epigenetics vary within the same woman and same biological sample type, as described above, the rate of epigenetic changes, their causes, and the magnitude of change attributable to each cause are not well defined. Confirmatory data will influence the validity of study designs with respect to sample timing, phenomic data collection, inference from single versus multiple samples from participants, and disease-specific hypotheses with respect to identification of modifiable risk factors. Recent studies have shown that interindividual differences in characteristics, lifestyle factors, or environmental exposures are associated with DNA methylation and that DNA methylation profiles can predict trait liability independent of genetic profiles [126]. For example, body mass index (BMI) is moderately heritable and methylation profiles predict BMI independent of genetic profiles. In contrast, height has high heritability, and methylation profiles contribute little variation to this trait, consistent with the large genetic contribution to height [126]. Thus, methylation profiles reflect a portion of the pathophysiologic processes by which the exposome can affect complex diseases and have the potential to significantly improve complex-trait prediction over and above that of genetic predictors [126]. While most of the data suggesting environmental influences on epigenetic marks have focused on DNA methylation loci, it is likely that in the future, as more data accumulate from emerging high-throughput technologies, other associations between the exposome and epigenetic profiles will similarly reveal environmental influences on their phenotypes.

Making valid, replicable inferences regarding genetic and epigenetic regulation of subtype-specific endometriosis pathogenesis requires access to large numbers of deeply phenotyped patients and study participants; temporally defined personal characteristics, behavior, and lifestyle (e.g., exogenous hormone use, cigarette smoking status, environmental toxin exposures, anthropometry, parity); clinical symptoms and course (e.g., symptom profile, comorbidities, treatment and response history); and serial biologic fluids and tissue (ectopic and eutopic endometrial) samples from women with and without the disease. Furthermore, with environment/lifestyle influences, there is potential to initiate modifiable approaches and potentially decrease manifestations of pain and infertility of those affected with disease and decrease risk during development, adolescence, and through adulthood.

CONCLUSIONS

Overall, studies on the genomics, genetics, and epigenetics of endometriosis have provided tremendous insights into the pathogenesis and pathophysiology of endometriosis. As technologies develop and with increasingly detailed clinical phenotyping and biospecimen standardization, the promise to elucidate this complex trait, its distinct phenotypes and comorbidities, noninvasive diagnostics, and targeted therapeutics is increasingly within our grasp. Elucidating mechanisms underlying the diverse manifestations of endometriosis is an important goal for the future with the hope of improving the quality of life of women and adolescent girls affected by this challenging and debilitating disorder.

Acknowledgment

The authors acknowledge the support of NIH NICHD R01 for coauthors LCG, CB, SM, GM, NR, PAWR, and KZ and Allison Power for her expert administrative support for this chapter.

References

[1] Giudice LC. Clinical practice: endometriosis. N Engl J Med 2010;362(25):2389–98.

[2] Burney RO, Giudice LC. Pathogenesis and pathophysiology of endometriosis. Fertil Steril 2012;98(3):511–9.

[3] Goldstein DP, deCholnoky C, Emans SJ, Leventhal JM. Laparoscopy in the diagnosis and management of pelvic pain in adolescents. J Reprod Med 1980;24(6):251–6.

[4] Eskenazi B, Warner ML. Epidemiology of endometriosis. Obstet Gynecol Clin North Am 1997;24(2):235–58.

[5] Missmer SA. Epidemiological and clinical risk factors for endometriosis. In: Farland LV, Shah DK, Kvaskoff M, Zondervan K, D'Hooghe T, editors. Biomarkers for Endometriosis. New York: Springer Science; 2017.

[6] Kvaskoff M, Mu F, Terry KL, Harris HR, Poole EM, Farland L, et al. Endometriosis: a high-risk population for major chronic diseases? Hum Reprod Update 2015;21(4):500–16.

[7] Nnoaham KE, Hummelshoj L, Webster P, d'Hooghe T, de Cicco Nardone F, de Cicco Nardone C, et al. Impact of endometriosis on quality of life and work productivity: a mutlicenter study across ten countries. Fertil Steril 2001;96(2):366–73.

[8] Simoens S, Dunselman G, Dirksen C, Hummelshoj L, Bokor A, Brandes I, et al. The burden of endometriosis: costs and quality of life of women with endometriosis and treated in referral centres. Hum Reprod 2012; 27(5):1292–9.

[9] Zheng Y, Murphy L. Regulation of steroid hormone receptors and coregulators during the cell cycle highlights potential novel function in addition to roles as transcription factors. Nucl Recept Signal 2016;14:e001.

[10] Klemmt PA, Carver JG, Kennedy SH, Koninckx PR, Mardon HJ. Stromal cells from endometriotic lesions and endometrium from women with endometriosis have reduced decidualization capacity. Fertil Steril 2006;85 (3):564–72.

[11] Burney RO, Talbi S, Hamilton AE, Vo KC, Nyegaard M, Giudice LC, et al. Gene expression analysis of endometrium reveals progesterone resistance and candidate susceptibility genes in women with endometriosis. Endocrinology 2007;148(8):3814–26.

[12] Aghajanova L, Tatsumi K, Horcajadas JA, Zamah AM, Esteban FJ, Herndon CN, Conti M, Giudice LC. Unique transcriptome, pathways, and networks in the human endometrial fibroblast response to progesterone in endometriosis. Biol Reprod 2011;84(4):801–15.

[13] Tamaresis JS, Irwin JC, Goldfien GA, Rabban JT, Burney RO, Giudice LC, et al. Molecular classification of endometriosis and disease stage using high-dimensional genomic data. Endocrinology 2014;155(12):4986–99.

[14] Brosens I, Puttemans P, Campo R, Gordts S, Kinkel K. Diagnosis of endometriosis: pelvic endoscopy and imaging techniques. Best Pract Res Clin Obstet Gynaecol 2004;18(2):285–303.

[15] Ballard K, Lowton K, Wright J. What's the delay? A qualitative study of women's experiences of reaching a diagnosis of endometriosis. Fertil Steril 2006;86(5):1296–301.

[16] American Society for Reproductive Medicine. Revised American Society for Reproductive Medicine classification of endometriosis: 1996. Fertil Steril 1997;67(5):817–21.

[17] Vercellini P, Vigano P, Somigliana E, Fedele L. Endometriosis; pathogenesis and treatment. Nat Rev Endocrinol 2014;10(5):261–75.

[18] Bulun SE. Endometriosis. N Engl J Med 2009;360(3):268–79.

[19] Lessey BA, Lebovic DI, Taylor RN. Eutopic endometrium in women with endometriosis: ground zero for the study of implantation defects. Semin Reprod Med 2013;31(2):109–24.

[20] Lessey BA, Kim JJ. Endometrial receptivity in the eutopic endometrium of women with endometriosis: it is affected, and let me show you why. Fertil Steril 2017;108(1):19–27.

[21] Kuivasaari P, Hippelainen M, Anttila M, Heinonen S. Effect of endometriosis on IVF/ICSI outcome: stage III/IV endometriosis worsens cumulative pregnancy and live-born rates. Hum Reprod 2005;20(11):3130–5.

[22] Luke B, Gopal D, Cabral H, Stern JE, Diop H. Pregnancy, birth, and infant outcomes by maternal fertility status: the Massachusetts outcomes study of assisted reproductive technology. Am J Obstet Gynecol 2017;217(3) 327: e1-327:e14.

[23] Vannuccini S, Clifton VL, Fraser IS, Taylor HS, Critchley H, Giudice LC, et al. Infertility and reproductive disorders: impact of hormonal and inflammatory mechanisms on pregnancy outcome. Hum Reprod Update 2016;22(1):104–15.

[24] Saraswat L, Ayansina DT, Cooper KG, Bhattacharya S, Miligkos D, Horne AW, et al. Pregnancy outcomes in women with endometriosis: a national record linkage study. Br J Obstet Gynaecol 2017;124(3):444–52.

[25] Glavind MT, Forman A, Arendt LH, Nielsen K, Henriksen TB. Endometriosis and pregnancy complications: a Danish cohort study. Fertil Steril 2017;107(1):160–6.

[26] Becker CM, Laufer MR, Stratton P, Hummelshoj L, Missmer SA, Zondervan KT, et al. World endometriosis research foundation endometriosis phenome and biobanking harmonization project: I. Surgical phenotype data collection in endometriosis research. Fertil Steril 2014;102(5):1213–22.

[27] Vitonis AF, Vincent K, Rahmioglu N, Fassbender A, Buck Louis GM, Giudice LC, et al. World endometriosis research foundation endometriosis phenome and biobanking harmonization project: II. Clinical and covariate phenotype data collection in endometriosis research. Fertil Steril 2014;102(5):1223–32.

[28] Rahmioglu N, Fassbender A, Vitonis AF, Tworoger SS, Hummelshoj L, D'Hooghe TM, et al. World endometriosis research foundation endometriosis phenome and biobanking harmonization project: III. Fluid biospecimen collection, processing, and storage in endometriosis research. Fertil Steril 2014;102(5):1233–43.

[29] Fassbender A, Rahmioglu N, Vitonis AF, Viganò P, Giudice LC, D'Hooghe TM, et al. World endometriosis research foundation endometriosis phenome and biobanking harmonisation project: IV. Tissue collection, processing, and storage in endometriosis research. Fertil Steril 2014;102(5):1244–53.

[30] Fung JN, Rogers PA, Montgomery GW. Identifying the biological basis of GWAS hits for endometriosis. Biol Reprod 2015;92(4):1–12.

[31] Rahmioglu N, Montgomery GW, Zondervan KT. Genetics of endometriosis. Womens Health 2016;11(5):577–86.

[32] Hadfield RM, Mardon HJ, Barlow DH, Kennedy SH. Endometriosis in monozygotic twins. Fertil Steril 1997;68(5):941–2.

[33] Treloar SA, O'Connor DT, O'Connor VM, Martin NG. Genetic influences on endometriosis in an Australian twin sample. Fertil Steril 1999;71(4):701–10.

[34] Saha R, Pettersson HJ, Svedberg P, Olovsson M, Bergqvist A, Marions L, et al. Heritability of endometriosis. Fertil Steril 2015;104(4):947–52.

[35] Adachi S, Tajima A, Quan J, Haino K, Yoshihara K, Masuzaki H, et al. Metaanalysis of genome-wide association scans for genetic susceptibility to endometriosis in Japanese population. J Hum Genet 2010;55(12):816–22.

[36] Uno S, Zembutsu H, Hirasawa A, Takahashi A, Kubo M, Akahane T, et al. A genome-wide association study identified genetic variants in the CDKN2BAS locus associated with endometriosis in Japanese. Nat Genet 2010;42(8):707–10.

[37] Painter JN, Anderson CA, Nyholt DR, Macgregor S, Lin J, Lee SH, et al. Genome-wide association study identifies a locus at 7p15.2 associated with endometriosis. Nat Genet 2011;43(1):51–4.

[38] Nyholt DR, Low SK, Anderson CA, Painter JN, Uno S, Morris AP, et al. Genome-wide association meta-analysis identifies new endometriosis risk loci. Nat Genet 2012;44(12):1355–9.

[39] Albertsen HM, Chettier R, Farrington P, Ward K. Genome-wide association study link novel loci to endometriosis. PLoS One 2013;8(3)e58257.

[40] Steinthorsdottir V, Thorleifsson G, Aradottir K, Feenstra B, Sigurdsson A, Stefansdottir L, et al. Common variants upstream of KDR encoding VEGFR2 and in TTC39B associate with endometriosis. Nat Commun 2016;7:12350.

[41] Sapkota Y, Low SK, Attia J, Gordon SD, Henders AK, Holliday EG, et al. Association between endometriosis and the interleukin 1A (IL1A) locus. Hum Reprod 2015;30(1):239–48.

[42] Sapkota Y, Steinthorsdottir V, Morris AP, Fassbender A, Rahmioglu N, De Vivo I, et al. Meta-analysis identifies five novel loci associated with endometriosis highlighting key genes involved in hormone metabolism. Nat Commun 2017;8:15539.

[43] Treloar SA, Wicks J, Nyholt DR, Montgomery GW, Bahlo M, Smith V, et al. Genomewide linkage study in 1,176 affected sister pair families identifies a significant susceptibility locus for endometriosis on chromosome 10q26. Am J Hum Genet 2005;77(3):365–76.

[44] Rahmioglu N, Nyholt DR, Morris AP, Missmer SA, Montgomery GW, Zondervan KT. Genetic variants underlying risk of endometriosis: insights from meta-analysis of eight genome-wide association and replication datasets. Hum Reprod Update 2014;20(5):702–16.

[45] Rahmioglu N, MacGregor S, Drong AW, Hedman AK, Harris HR, Randall JC, et al. Genome-wide enrichment analysis between endometriosis and obesity related traits reveals novel susceptibility loci. Hum Mol Genet 2015;24(4):1185–99.

[46] Uimari O, Rahmioglu N, Nyholt DR, Vincent K, Missmer SA, Becker C, et al. Genome-wide genetic analyses highlight mitogen-activated protein kinase (MAPK) signaling in the pathogenesis of endometriosis. Hum Reprod 2017;32(4):780–93.

[47] Sapkota Y, Attia J, Gordon SD, Henders AK, Holliday EG, Rahmioglu N, et al. Genetic burden associated with varying degrees of disease severity in endometriosis. Mol Hum Reprod 2015;21(7):594–602.

[48] Zondervan KT, Rahmioglu N, Morris AP, Nyholt DR, Montgomery GW, Becker CM, et al. Beyond endometriosis genome-wide association study: from genomics to phenomics to the patient. Semin Reprod Med 2016;34(4):242–54.

[49] Lee SH, Harold D, Nyholt DR, ANZGene Consortium, International Endogene Consortium, et al. Estimation and partitioning of polygenic variation captured by common SNPs for Alzheimer's disease, multiple sclerosis and endometriosis. Hum Mol Genet 2013;22(4):832–41.

[50] GTEx Consortium, Laboratory, Data Analysis and Coordinating Center (LDACC)—Analysis Working Group, Statistical Methods group—Analysis Working Group, Enhancing GTEx (eGTEx) groups; NIH Common Fund, NIH/NCI, et al. Genetic effects on gene expression across human tissues. Nature 2017;550 (7675):204–13.

[51] Fung JN, Girling JE, Lukowski SW, Sapkota Y, Wallace L, Joldsworth-Carson SJ, et al. The genetic regulation of transcription in human endometrial tissue. Hum Reprod 2017;32(4):893–904.

[52] Fassbender A, Verbeeck N, Bornigen D, Kyama CM, Bokor A, et al. Combined mRNA microarray and proteomic analysis of eutopic endometrium of women with and without endometriosis. Hum Reprod 2012;27 (7):2020–9.

[53] Rahmioglu N, Drong AW, Lockstone H, Tapmeier T, Hellner K, Saare M, et al. Variability of genome-wide DNA methylation and mRNA expression profiles in reproductive and endocrine disease related tissues. Epigenetics 2017;3:1–12.

[54] Powell JE, Fung JN, Shakhbazov K, Sapkota Y, Cloonan N, Hemani G, et al. Endometriosis risk alleles at 1p36.12 act through inverse regulation of CDC42 and LINC00339. Hum Mol Genet 2016;25(22):5046–58.

[55] Holdsworth-Carson SJ, Fung JN, Luong HT, Sapkota Y, Bowdler LM, Wallace L, et al. Endometrial vezatin and its association with endometriosis risk. Hum Reprod 2016;31(5):999–1013.

[56] Munksgaard PS, Blaakaer J. The association between endometriosis and ovarian cancer: a review of histological, genetic and molecular alterations. Gynecol Oncol 2012;124(1):164–9.

[57] Worley MJ, Welch WR, Berkowitz RS, Ng SW. Endometriosis-associated ovarian cancer: a review of pathogenesis. Int J Mol Sci 2013;14(3):5367–79.

[58] Jiang X, Hitchcock A, Bryan EJ, Watson RH, Englefield P, Thomas EJ, Campbell IG. Microsatellite analysis of endometriosis reveals loss of heterozygosity at candidate ovarian tumor suppressor gene loci. Cancer Res 1996;56(15):3534–9.

[59] Sato N, Tsunoda H, Nishida M, Morishita Y, Takimoto Y, Kubo T, et al. Loss of heterozygosity on 10q23.3 and mutation of the tumor suppressor gene PTEN in benign endometrial cyst of the ovary: possible sequence progression from benign endometrial cyst to endometrioid carcinoma and clear cell carcinoma of the ovary. Cancer Res 2000;60(24):7052–6.

[60] Ballouk F, Ross JS, Wolf BC. Ovarian endometriotic cysts. An analysis of cytologic atypia and DNA ploidy patterns. Am J Clin Pathol 1994;102(4):415–9.

[61] Shin JC, Ross HL, Elias S, Nguyen DD, Mitchell-Leef D, Simpson JL, et al. Detection of chromosomal aneuploidy in endometriosis in endometriosis by multi-color fluorescence in situ hybridization (FISH). Hum Genet 1997;100(3–4):401–6.

[62] Gogusev J, Bouquet de Joliniere J, Telvis L, Doussau M, du Manoir S, Stojkoski A, et al. Detection of DNA copy number changes in human endometriosis by comparative genomic hybridization. Hum Genet 1999;105 (5):444–51.

[63] Wu Y, Strawn E, Basir Z, Wang Y, Halverson G, Jailwala P, et al. Genomic alterations in ectopic and eutopic endometria of women with endometriosis. Gynecol Obstet Invest 2006;62(3):148–59.

[64] Guo SW, Wu Y, Strawn E, Basir Z, Wang Y, Halverson G, et al. Genomic alterations in the endometrium may be a proximate cause for endometriosis. Eur J Obstet Gynecol Reprod Biol 2004;116(1):89–99.

[65] Kim HS, Kim TH, Chung HH, Song YS. Risk and prognosis of ovarian cancer in women with endometriosis: a meta-analysis. Br J Cancer 2014;110(7):1878–90.

[66] Lu Y, Cuellar-Partida G, Painter JN, Nyholt DR, Australian Ovarian Cancer Study, International Endogene Consortium (IEC), et al. Shared genetics underlying epidemiological association between endometriosis and ovarian cancer. Hum Mol Genet 2015;24(20):5955–64.

[67] Dinulescu DM, Ince TA, Quade BJ, Shafer SA, Crowley D, Jacks T. Role of K-ras and Pten in the development of mouse models of endometriosis and endometrioid ovarian cancer. Nat Med 2005;11(1):63–70.

[68] Matzuk MM. Gynecologic diseases get their genes. Nat Med 2005;11(1):24–6.

[69] Anglesio MS, Bashashati A, Wang YK, Senz J, Ha G, Yang W, et al. Multifocal endometriotic lesions associated with cancer are clonal and carry a high mutation burden. J Pathol 2015;236(2):201–9.

[70] Amemiya S, Sekizawa A, Otsuka J, Tachikawa T, Saito H, Okai T. Malignant transformation of endometriosis and genetic alterations of K-ras and microsatellite instability. Int J Gynecol Obstet 2004;86(3):371–6.

[71] Sainz de la Cuesta R, Izquierdo M, Canamero M, Granizo JJ, Manzarbeitia F. Increased prevalence of p53 overexpression from typical endometriosis to atypical endometriosis and ovarian cancer associated with endometriosis. Eur J Obstet Gynecol Reprod Biol 2004;113(1):87–93.

[72] Govatati S, Kodati VL, Deenadayal M, Chakravarty B, Shivaji S, Bhanoori M. Mutations in the PTEN tumor gene and risk of endometriosis: a case-control study. Hum Reprod 2014;29(2):324–36.

[73] Wiegand KC, Shah SP, Al-Agha OM, Zhao Y, Tse K, Zeng T, et al. *ARID1A* mutations in endometriosis-associated ovarian carcinomas. N Engl J Med 2010;363(16):1532–43.

[74] McConechy MK, Ding J, Senz J, Yang W, Melnyk N, Tone AA, et al. Ovarian and endometrial endometrioid carcinomas have distinct CTNNB1 and PTEN mutation profiles. Mod Pathol 2014;27(1):128–34.

[75] Anglesio MS, Papadopoulos N, Ayhan A, Nazeran TM, Noe M, Horlings HM, et al. Cancer-associated mutations in endometriosis without cancer. N Engl J Med 2017;376(19):1835–48.

[76] Petronis A. Epigenetics as a unifying principle in the aetiology of complex traits and diseases. Nature 2010;465 (7299):721–7.

[77] Martinez-Arguelles DB, Papadopoulos V. Epigenetic regulation of the expression of genes involved in steroid hormone biosynthesis and action. Steroids 2010;75(7):467–76.

[78] Zelenko Z, Aghajanova L, Irwin JC, Giudice LC. Nuclear receptor, coregulatory signaling, and chromatin remodeling pathways suggest involvement of the epigenome in the steroid hormone response of endometrium and abnormalities in endometriosis. Reprod Sci 2011;19(2):152–62.

[79] Guo SW. Epigenetics of endometriosis. Mol Hum Reprod 2009;15(10):587–607.

[80] Koukoura O, Sifakis S, Spandidos D. DNA methylation in endometriosis (review). Mol Med Rep 2016;13 (4):2939–48.

[81] Wu Y, Halverson G, Basir Z, Strawn E, Yan P, Guo SW. Aberrant methylation at HOXA10 may be responsible for its aberrant expression in the endometrium of patients with endometriosis. Am J Obstet Gynecol 2005;193 (2):371–80.

[82] Xue Q, Lin Z, Yin P, Milad MP, Cheng YH, Confino E, et al. Transcriptional activation of steroidogenic factor-1 by hypomethylation of the 5′ CpG island in endometriosis. J Clin Endocrinol Metabol 2007;92(8):3261–7.

[83] Izawa M, Taniguchi F, Harada T. Epigenetics in endometriosis. In: Harada T, editor. Chapter 8. Endometriosis: pathogenesis and treatment. Japan: Springer; 2014.

[84] Forte A, Cipollaro M, Galderisi U. Genetic, epigenetic and stem cell alterations in endometriosis: new insights and potential therapeutic perspectives. Clin Sci 2014;126(2):123–38.

[85] Attar E, Bulun SE. Aromatase inhibitors: the next generation of therapeutics for endometriosis? Fertil Steril 2006;85(5):1307–18.

[86] Chwalisz K, Perez MC, Demanno D, Winkel C, Schubert G, Elger W. Selective progesterone receptor modulator development and use in the treatment of leiomyomata and endometriosis. Endocr Rev 2005;26(3):423–38.

[87] Borghese B, Barbaux S, Mondon F, Santulli P, Pierre G, Vinci G, et al. Research resource: genome-wide profiling of methylated promoters in endometriosis reveals a subtelomeric location of hypermethylation. Mol Endocrinol 2010;24(9):1872–85.

[88] Dyson MT, Roqueiro D, Monsivais D, Ercan CM, Pavone ME, Brooks DC, et al. Genome-wide DNA methylation analysis predicts an epigenetic switch for GATA factor expression in endometriosis. PLoS Genet 2014;10(3): e1004158.

[89] Yotova I, Hsu E, Do C, Gaba A, Sczabolcs M, Dekan S, et al. Epigenetic alterations affecting transcription factors and signaling pathways in stromal cells of endometriosis. PLoS ONE 2017;12:e0170859.

[90] Houshdaran S, Zelenko Z, Irwin JC, Giudice LC. Human endometrial DNA methylome is cycle-dependent and is associated with gene expression regulation. Mol Endocrinol 2014;28(7):1118–35.

[91] Saare M, Modhukur V, Suhorutschenko M, Rajashekar B, Rekker K, Soritsa D, et al. The influence of menstrual cycle and endometriosis on endometrial methylome. Clin Epigenetics 2016;8:2–10.

[92] Naqvi H, Ilagan Y, Krikun G, Taylor HS. Altered genome-wide methylation in endometriosis. Reprod Sci 2014;21(10):1237–43.

[93] Houshdaran S, Nezhat CR, Vo KC, Zelenko Z, Irwin JC, Giudice LC. Aberrant endometrial DNA methylome and associated gene expression in endometriosis. Biol Reprod 2016;95(5):93.

[94] Bannister AJ, Kouzarides T. Regulation of chromatin by histone modifications. Cell Res 2011;21(3):381–95.

[95] Ferguson LR, Tatham AL, Lin Z, Denny WA. Epigenetic regulation of gene expression as an anticancer drug. Curr Cancer Drug Targets 2011;11(2):199–212.

[96] Wu Y, Guo SW. Histone deacetylase inhibitors trichostatin A and valproic acid induce cell cycle arrest and p21 expression in immortalized human endometrial stromal cells. Eur J Obstet Gynecol Reprod Biol 2008;137 (2):198–203.

[97] Kawano Y, Nasu K, Li H, Tsuno A, Abe W, Takai N, et al. Application of the histone deacetylase inhibitors for the treatment of endometriosis: histone modifications as pathogenesis and novel therapeutic target. Hum Reprod 2011;26(9):2486–98.

[98] Liu M, Liu X, Zhang Y, Guo SW. Valproic acid and progestin inhibit lesion growth and reduce hyeralgesia in experimentally induced endometriosis in rats. Reprod Sci 2012;19(4):360–73.

[99] Li X, Liu X, Guo SW. Histone deacetylase inhibitors as therapeutics for endometriosis. Expert Rev Obstet Gynecol 2012;7(5):451–66.

[100] Jenuwein T, Allis CD. Translating the histone code. Science 2001;293(10):1074–80.

[101] Fischer JJ, Toedling J, Krueger T, Schueler M, Huber W, Sperling S. Combinatorial effects of four histone modifications in transcription and differentiation. Genomics 2008;91(1):41–51.

[102] Monteiro JB, Colon-Diaz M, Garcia M, Gutierrez S, Colon M, Seto E, et al. Endometriosis is characterized by a distinct pattern of histone 3 and histone 4 lysine modifications. Reprod Sci 2014;21(3):305–18.

[103] Colon-Caraballo M, Monteiro JB, Flores I. H3K27me3 is an epigenetic mark of relevance in endometriosis. Reprod Sci 2015;22(9):1134–42.

[104] Samartzis EP, Noske A, Samartzis N, Fink D, Imesch P. The expression of histone deacetylase 1, but not other class I histone deacetylases, is significantly increased in endometriosis. Reprod Sci 2013;20(12):1416–22.

[105] Colon-Diaz M, Baez-Vega P, Garcia M, Ruiz A, Monteiro JB, Fourquet J, Bayona M, Alvarez-Garriga C, Achille A, Seto E, Flores I. HDAC I and HDAC II are differentially expressed in endometriosis. Reprod Sci 2012;19(5):483–92.

[106] Wu Y, Starzinski-Powitz A, Guo SW. Trichostatin A, a histone deacetylase inhibitor, attenuates invasiveness and reactivates E-cadherin expression in immortalized endometriotic cells. Reprod Sci 2007;14(4):374–82.

[107] Lu Y, Nie J, Liu X, Zheng Y, Guo SW. Trichostatin A, a histonedeacetylase inhibitor, reduces lesion growth and hyperagesia in experimentally-induced endometriosis in mice. Hum Reprod 2010;25(4):1014–25.

[108] Lee RC, Feinbaum RL, Ambros V. The *C. elegans* heterochronic gene lin-4 encodes small RNAs with antisense complementarity to lin-14. Cell 1993;75(5):843–54.

[109] Bartel DP. microRNAs: genomics, biogenesis, mechanism, and functions. Cell 2004;116(10):281–97.

[110] Mari-Alexandre J, Sanchez-Izquierdo D, Gilabert-Estelles J, Barcelo-Molina M, Braza-Boils A, Sandoval J. miRNAs regulation and its role as biomarkers in endometriosis. Int J Mol Sci 2016;17(1):93–110.

[111] Nothnick WB. MicroRNAs and endometriosis: distinguishing drivers from passengers in disease pathogenesis. Semin Reprod Med 2017;35(2):173–80.

[112] Ohlsson Teague EM, Van der Hoek KH, Van der Hoek MB, et al. MicroRNA-regulated pathways associated with endometriosis. Mol Endocrinol 2009;23(2):265–75.

[113] Filigheddu N, Gregnanin I, Porporato PE, et al. Differential expression of microRNAs between eutopic and ectopic endometrium in ovarian endometriosis. J Biomed Biotechnol 2010;2010:369549.

[114] Hawkins SM, Creighton CJ, Han DY, et al. Functional microRNA involved in endometriosis. Mol Endocrinol 2011;25(5):821–32.

[115] Shi XY, Gu L, Chen J, Guo XR, Shi YL. Downregulation of miR-183 inhibits apoptosis and enhances the invasive potential of endometrial stromal cells in endometriosis. Int J Mol Med 2014;33(1):59–67.

[116] Yang RQ, Teng H, Xu XH, et al. Microarray analysis of microRNA deregulation and angiogenesis-related proteins in endometriosis. Genet Mol Res 2016;15(2).

[117] Burney RO, Hamilton AE, Aghajanova L, Vo KC, Nezhat CN, Lessey BA, Giudice LC. MicroRNA expression profiling of eutopic secretory endometrium in women with versus without endometriosis. Mol Hum Reprod 2009;15(10):625–31.

[118] Laundanski P, Charkiewicz R, Kuzmicki M, Szamatowicz J, Charkiewicz A, Niklinski J. MicroRNAs expression profiling of eutopic proliferative endometrium in women with ovarian endometriosis. Reprod Biol Endocrinol 2013;11:78–84.

[119] Saare M, Rekker KJ, Laisk-Podar T, Rahmioglu N, Zondervan K, Salumets A, et al. Challenges in endometriosis miRNA studies—from tissue heterogeneity to disease-specific miRNAs. Biochim Biophys Acta 2017;1863 (9):2282–92.

[120] Taylor HS, Bagot C, Kardana A, Olive D, Arici A. HOX gene expression is altered in the endometrium of women with endometriosis. Hum Reprod 1999;14(5):1328–31.

[121] Jia SZ, Yang Y, Lang J, Sun P, Leng J. Plasma miR-17-5p, miR-20a and miR-22 are down-regulated in women with endometriosis. Hum Reprod 2013;28:322–30.

[122] Wang WT, Zhao YN, Han BW, Hong SJ, Chen YQ. Circulating microRNAs identified in a genome-wide serum microRNA expression analysis as non-invasive biomarkers for endometriosis. J Clin Endocrinol Metabol 2013;98(1):281–9.

[123] Cho S, Mutlu L, Grechukhina O, Taylor HS. Circulating microRNs as potential biomarkers for endometriosis. Fertil Steril 2015;103(5):1252–60.

[124] Rekker K, Saare M, Roost AM, Kaart T, Soritsa D, Karro H, et al. Circulating miR-200 family micro-RNAs have altered plasma levels in patients with endometriosis and vary with blood collection time. Fertil Steril 2015;104 (4):938–46.

[125] Nothnick WB, Falcone T, Joshi N, Fazleabas AT, Graham A. Serum miR-451a levels are significantly elevated in women with endometriosis and recapitulated in baboons (*Papio anubis*) with experimentally-induced disease. Reprod Sci 2016;24(8):1195–202.

[126] Shah S, Bonder MJ, Marioni RE, Zhu Z, McRae AF, Zhernakova A, et al. Improving phenotypic prediction by combining genetic and epigenetic associations. Am J Hum Genet 2015;97(1):75–85.

Further Reading

Liu X, Guo SW. A pilot study on the off-label use of valproic acid to treat adenomyosis. Fertil Steril 2008;89(1):246–50.

Glossary

Endometrium	tissue lining the uterus
Ectopic endometrium	endometrial-like tissue outside the uterus
Eutopic endometrium	endometrial tissue within the uterus
Exposome	measure of all the exposures of an individual during development and throughout the life course and how those exposures relate to health (https://www.cdc.gov/niosh/topics/exposome/default.html)
Phenome	totality of all phenotypic traits

Genetics and Genomics of Primary Ovarian Insufficiency

Elena J. Tucker[*,†]*, Sylvie Jaillard*[*,‡,§]*, Andrew H. Sinclair*[*,†]

[*]Murdoch Children's Research Institute, Royal Children's Hospital, Melbourne, VIC, Australia
[†]Department of Paediatrics, University of Melbourne, Melbourne, VIC, Australia [‡]Rennes University Hospital, Cytogenetics and Cell Biology Department, Rennes, France [§]INSERM U1085-IRSET, Rennes 1 University, Rennes, France

Abbreviations

BPES blepharophimosis/ptosis/epicanthus inversus syndrome
FMR1 fragile X mental retardation 1
FSH follicle stimulating hormone

LH	leutinizing hormone
NGS	next-generation sequencing
PEO	progressive external opthalmoplegia
PGC	primordial germ cell
POI	primary ovarian insufficiency
SMC	structural maintenance of chromosomes

INTRODUCTION

Primary ovarian insufficiency (POI) is a form of infertility affecting 1 in 10,000 women by age 20, 1 in 1000 women by age 30, and 1 in 100 women by age 40 [1]. POI is diagnosed in women up to age 40 who have irregular or absent menstruation as well as elevated follicle-stimulating hormone (FSH), measured by repeated blood tests spaced greater than one month apart (https://www.eshre.eu/Guidelines-and-Legal/Guidelines/Management-of-premature-ovarian-insufficiency.aspx) [1, 2]. POI is sometimes referred to as premature ovarian failure or premature menopause. These terms are misleading because many women diagnosed with POI have some evidence of residual ovarian function and spontaneous pregnancy is possible, although rare [3–6]. ESHRE consensus recommends the term "insufficiency" instead of "failure" [7]. An American consensus meeting decided on "primary ovarian insufficiency" as the preferred terminology [8, 9]. Although infertility is a pertinent feature of the condition, POI is in fact a chronic illness with numerous emotional and medical sequalae. A diagnosis of POI means a woman must face the loss of reproductive potential, years of hormone replacement therapy, and an increased risk of osteoporosis, cardiovascular disease, and earlier mortality in general [10, 11]. POI can be secondary to medical interventions such as ovarian surgery, cytotoxic cancer therapy, infections, and autoimmune diseases, although it is often idiopathic. It can be an isolated condition or part of a known syndrome. Both isolated and syndromic POI often have a genetic basis, with more than 50 genes in which variants can be causative and many other genes that are implicated [12]. These genes can affect various processes such as gonadal development, meiosis, DNA repair, follicle recruitment and progression, steroidogenesis, apoptosis, immune regulation, mitochondrial function, and metabolism.

THE GENETIC BASIS OF POI

Many studies have demonstrated the heritability of POI, supportive of its genetic basis. For example, the largest described POI cohort with 675 patients revealed that 18% had at least one affected relative [12]. Similarly, a study of 71 women with idiopathic POI showed 31% had at least one affected family member [13]. Many studies have shown that menopausal age, like POI, has a significant genetic basis. For example, a strong relationship between mother and daughter menopausal age was revealed by a survey of 551 women aged 45–54 (recruited at an osteoporosis screening program) [14]. In this study, 75% of the patients with POI also reported early maternal menopause [14]. Also supporting the genetic basis of POI is the fact that there are many genes that cause POI in a monogenic manner (Table 19.1, Fig. 19.1). Many of these cause POI as part of a syndrome such as Perrault syndrome, blepharophimosis/ptosis/epicanthus inversus syndrome (BPES), ataxia telegiectasia, or Turner's syndrome, but some genes can also be responsible for isolated POI. Genes considered clearly causative are those

TABLE 19.1 Human Genes in Which Variants Are Known to Cause Premature Ovarian Insufficiency

Gene	Role of the Gene	Inheritance	Phenotype
AARS2	Charges mitochondrial tRNA-ala with alanine during mitochondrial translation	AR	Leukodystrophy + POI
AIRE	Putative transcription factor required to regulate immunity	AR	Autoimmune polyglandular syndrome, type 1 + POI
ANTXR1	Extracellular matrix regulation	AR	GAPO syndrome + POI
ATM	Role in cellular responses to genomic damage	AR	Ataxia telengiecstasia + POI
BLM	DNA helicase, regulates homologous recombination	AR	Bloom syndrome + POI
BMP15	Member of transforming growth factor-beta superfamily, regulates folliculogenesis	XLD	POI
BMPR1B	Transmembrane serine/threonine kinases involved in endochondral bone formation and embryogenesis	AR	Acromesomelic chondrodysplasia + POI
C10ORF2	mtDNA helicase, required for mtDNA replication and maintenance	AR	Perrault syndrome + POI
PMM2	Enzyme necessary for the synthesis of GDP-mannose	AR	Congenital disorder of glycosylation + POI
CLPP	Component of a mitochondrial ATP-dependent proteolytic complex, required for unfolded protein response	AR	Perrault syndrome + POI
CYP11B1	Steroid 11-beta-hydroxylase required for conversion of 11-deoxycortisol to cortisol and 11-deoxycorticosterone to corticosterone	AR	Nonclassical congenital adrenal hyperplasia + POI
CYP17A1	Monooxygenase, key enzyme in the steroidogenic pathway that produces progestins, mineralocorticoids, glucocorticoids, androgens, and estrogens	AR	POI
CYP19A1	Monooxygenase, key enzyme responsible for the biosynthesis of estrogens	AR	POI, fetal masculinization
DMC1	Required for meiotic homologous recombination	AR	POI (azoospermia in males)
EIF2B2	GTP exchange factor that is essential for protein synthesis	AR	Ovarioleukodystrophy + POI
EIF2B4	GTP exchange factor that is essential for protein synthesis	AR	Ovarioleukodystrophy + POI
EIF2B5	GTP exchange factor that is essential for protein synthesis	AR	Ovarioleukodystrophy + POI
EIF4ENIF1	Nucleocytoplasmic shuttling protein responsible for the nuclear import of EIF4E	AD	POI
ERCC6	Putative role in cellular responses to genomic damage	AD	POI

Continued

TABLE 19.1   Human Genes in Which Variants Are Known to Cause Premature Ovarian Insufficiency—cont'd

Gene	Role of the Gene	Inheritance	Phenotype
FANCA	Required for S phase of the growth cycle and after exposure to DNA crosslinking agents	AR	Fanconi anemia + POI
FANCC	Required for S phase of the growth cycle and after exposure to DNA crosslinking agents	AR	Fanconi anemia + POI
FANCG	Required for S phase of the growth cycle and after exposure to DNA crosslinking agents	AR	Fanconi anemia + POI
FANCM	Interacts with the FA core complex in DNA damage repair	AR	POI
FIGLA	Transcription factor required for folliculogenesis	AR	POI
FOXL2	Crucial role in ovarian development and maintenance and female fertility	AD	BPES type 1 + POI
FMR1	RNA-binding protein that associates with polyribosomes, putative role in translation	XLD (premutation)	POI
FSHR	Receptor for FSH, required for folliculogenesis	AR	POI
GALT	Responsible for the conversion of galactose-1-phosphate (gal-1P) and UDP-glucose into glucose-1-phosphate and UDP-galactose	AR	Galactosemia + POI
GDF9	Growth differentiation factor required for folliculogenesis	AR, AD	POI
GNAS	Signalling molecule that couples FSHR and LHCGR to their downstream effector	AD	Pseudo-hypoparathyroidism + POI
HARS2	Charges mitochondrial tRNA-his with histidine during mitochondrial translation	AR	Perrault syndrome + POI
HAX1	Required for maintaining the inner mitochondrial membrane potential	AR	Severe congenital neutropenia + POI
HFM1	Required for normal progression of homologous recombination and proper synapsis between homologous chromosomes	AR	POI
HSD17B4	Involved in fatty acid β-oxidation and steroid metabolism	AR	Perrault syndrome + POI
LARS2	Charges mitochondrial tRNA-leu with leucine during mitochondrial translation	AR	Perrault syndrome + POI
LMNA	Structural protein component of the nuclear lamina, a protein network underlying the inner nuclear membrane	AD	Cardiomyopathy + POI
LRPPRC	Required for mitochondrial mRNA stability and transport	AR	Leigh syndrome + POI
MCM8	Required for homologous recombination-mediated repair of double-strand breaks	AR	POI

TABLE 19.1 Human Genes in Which Variants Are Known to Cause Premature Ovarian Insufficiency—cont'd

Gene	Role of the Gene	Inheritance	Phenotype
MCM9	Required for homologous recombination-mediated repair of double-strand breaks	AR	POI
MRE11A	Required for double strand break repair	AR	Ataxia-telangiectasia-like disorder + POI
MSH4	Meiotic (post replicative) DNA mismatch repair	AR	POI
MSH5	Meiotic (post replicative) DNA mismatch repair	AR	POI
NANOS3	Suppression of apoptosis	AR	POI
NBN	Required for the repair of double strand breaks	AR	Nijmegen breakage syndrome + POI, infertility
NOBOX	Homeobox gene involved in regulation of oocyte-specific genes	AD, AR	POI
NOG	Signaling molecule involved in promoting somite patterning during embryogenesis	AD	Proximal symphalangism + POI
NR5A1	Transcription factor required for sex development, steroidogenesis and fertility	AD	POI (DSD in males)
NSMCE2	E3-SUMO ligase, required for tolerance of replicative stress	AR	Primordial dwarfism + POI
NUP107	Nucleoporin protein involved in transport between cytoplasm and nucleus, putative role in meiosis/mitosis progression and/or DNA damage repair	AR	POI (XX gonadal dysgenesis)
PGRMC1	Binds and activates P450 proteins for steroidogenesis, responsible for antiapoptotic action of progesterone	AD	POI
POF1B	Binds nonmuscle actin filaments, possible role in germ cell division	XLR	POI
POLG	Enzyme that synthesizes new mtDNA and corrects mtDNA errors	AR, AD	Progressive external opthalmoplegia and parkinonism + POI
POLR2C	Subunit of RNA polymerase II	AD	POI (+autoimmunity?)
PSMC3IP	Role in meiotic recombination, coactivator of nuclear hormone receptor-dependent transcription	AR	POI, XX ovarian dysgenesis
RCBTB1	Ubiquitination, likely regulates oxidative stress	AR	POI, retinal dystrophy, intellectual disability
RECQL4	DNA helicase that unwinds double-stranded DNA, involved in DNA repair	AR	Rothmund-Thomson syndrome + POI
SGO2	Maintains the integrity of the cohesion complex, required for meiosis	AR	POI

Continued

TABLE 19.1　Human Genes in Which Variants Are Known to Cause Premature Ovarian Insufficiency—cont'd

Gene	Role of the Gene	Inheritance	Phenotype
SOHLH1	Gonadal transcription factor required for folliculogenesis	AR	POI
SPIDR	Homologous recombination repair gene	AR	POI
STAG3	Subunit of cohesin, required for proper pairing and segregation of chromosomes during meiosis	AR	POI
STAR	Required for conversion of cholesterol to pregnenolone	AR	Congenital lipoid adrenal hyperplasia + POI
SYCE1	Part of the synaptonemal complex that physically links homologous chromosomes	AR	POI
TRIM37	Peroxisomal membrane protein with ubiquitin E3 ligase activity	AR	Mulibrey nanism + POI
WRN	DNA helicase that unwinds double-stranded DNA, involved in DNA repair	AR	Werner syndrome + POI
XRCC4	Nonhomologous end-joining DNA repair	AR	Dwarfism, metabolic syndrome + POI

Gene symbol, role, inheritance pattern and patient phenotype are indicated. *AD*, autosomal dominant; *AR*, autosomal recessive; *DSD*, disorder of sex development; *POI*, premature ovarian insufficiency; *XLD*, X-linked dominant; *XLR*, X-linked recessive.
Adapted from Tucker EJ, Grover SR, Bachelot A, Touraine P, Sinclair AH. Premature ovarian insufficiency: new perspectives on genetic cause and phenotypic spectrum. Endocr Rev 2016;37(6):609–35.

for which pathogenic variants are found in multiple unrelated individuals, those that segregate with POI in large families, and/or those that have functional validation in cell lines or animal models. In addition to these genes known to cause POI in a monogenic manner, there are other candidates revealed by studies in animal models or genome-wide association studies (GWAS). The genes responsible for POI have roles that include development, hormone signaling, meiosis, DNA damage repair, immunity, and metabolism. Despite the clear evidence for the genetic basis of POI and the many known genes involved in its etiology, POI is a heterogeneous condition and the causative gene is only identified in a minority of cases. The majority of cases that remain unexplained indicate that our knowledge of the genetic basis of the condition is incomplete.

DEVELOPMENTAL GENES

The ovary and its constituent oocytes develop during embryogenesis. The absence of a Y chromosome in females allows the bipotential gonad to develop into an ovary with granulosa cells surrounding the germ cells that have migrated from the hindgut. Some genes known to underlie POI are involved in this gonadogenesis. *NR5A1*, for example, encodes the steroidogenic factor-1, which is a transcription factor that cooperates with male- and female-specific factors to direct the development of the adrenal and reproductive system [15].

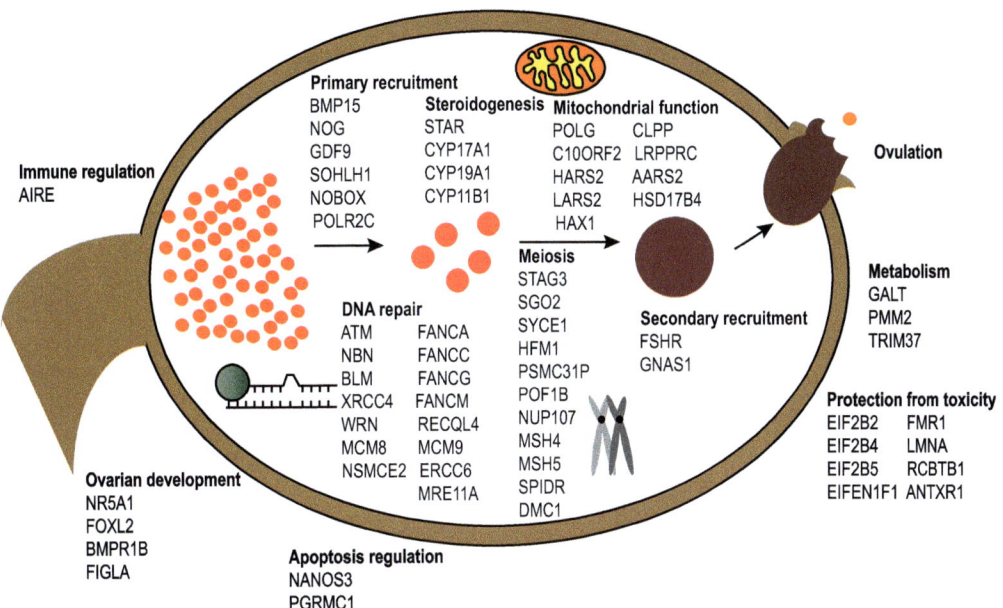

FIG. 19.1 An ovary with vital processes for its functioning and associated POI genes, indicated. Developmental genes ensure the proper structure of the ovary and its oocyte pool is established. Genes involved in apoptosis and genes that prevent build-up of toxic factors aid oocyte survival. Efficient metabolism and mitochondrial function are required for the amount of energy needed for ovarian function. Signaling molecules and hormones direct primary and secondary recruitment of follicles. Steroidogenesis is responsible for hormone production. Many genes are required for proper meiotic cell division and for efficient DNA repair to ensure the integrity and survival of oocytes.

Mutations in *NR5A1* can cause a range of ovarian anomalies in females, including 46, XX gonadal dysgenesis, and POI [16].

FOXL2 is another gene required for gonadogenesis, but it is also required for the postnatal maintenance of ovaries [17]. It is expressed in developing eyelids as well as ovarian follicles. Human mutations are associated with BPES, a dominant syndrome characterized by eyelid abnormalities with POI (Type 1) or without POI (Type 2) [18].

BMPR1B is a receptor for growth differentiation factors such as GDF5; it is important for gonadal and skeletal development. Mutations in *BMPR1B* have been found to cause acromesomelic chondrodysplasia with absent or underdeveloped ovaries and POI [19].

MEIOTIC GENES

Proper development of the ovaries is necessary for the protection, support, and development of the oocytes they contain. Oocytes are generated via gametogenesis. After migrating to the ovary from the hindgut, primordial germ cells (PGCs) undergo rapid cell division to generate the primordial pool. The maximum number of germ cells is achieved at ∼20 weeks gestation, after which oocytes begin depleting (at peak the pool is ∼7 million but whittles down to 1–2 million at birth). Meiosis commences prenatally but arrests at prophase I until menarche. Each menstrual cycle promotes the resumption of meiosis and the maturation of one (or less frequently, two) oocytes each month. Given the important role of meiosis in the development of oocytes, it is not surprising that meiotic genes can harbor variants causing POI.

During meiosis, sister chromatids are tethered by the cohesion complex and homologous recombination occurs between paired chromosomes via the synaptonemal complex. *STAG3* encodes a cohesin protein, and variants within this gene can cause POI [20]. *SGO2* encodes a protein required to maintain the integrity of the cohesion complex, and variants within this gene have similarly been shown responsible for POI [21]. Variants in *SYCE1*, which encodes a component of the synaptonemal complex, have also been found responsible for POI [22, 23]. Other genes required for homologous recombination such as *HFM1* and *PSMC3IP* have recently been found to harbor mutations causing POI in humans [24, 25]. Other genes believed important for meiosis in which variants can cause POI include *POF1B* and *NUP107*. *POF1B* encodes a myosin-like protein that can bind nonmuscle actin filaments and is therefore believed to be involved in germ cell division [26]. *NUP107* encodes a nuclear pore complex protein and the related genes in *Drosophila melanogaster* are required for mitotic and meiotic division [27]. *MSH4* and *MSH5* are DNA mismatch repair genes required for meiotic recombination via postreplicative DNA repair, and causative variants have recently been found in these genes in POI patients [28, 29]. Similarly, the homologous recombination repair gene, *SPIDR*, has been found to harbor variants causing POI [30].

DNA REPAIR GENES

Because oocytes have such a dependence on accurate cell division, they must also have efficient DNA damage repair mechanisms. As indicated above, DNA repair genes are also critical for the repair of double strand breaks as part of meiotic homologous recombination. Many genes involved in the DNA damage repair pathways have been found to have causative POI-related variants. Because DNA damage repair is a ubiquitous process, many of the variants in DNA damage repair genes cause a syndromic phenotype with POI as only a minor feature. ATM is a cell-cycle checkpoint kinase needed for the cellular response to DNA damage. Pathogenic variants in *ATM* cause ataxia telangiectasia, a chromosomal instability disorder that manifests with cerebellar degeneration, immunodeficiency, and cancer susceptibility as well as gonadal abnormalities such as POI [31]. Other chromosomal instability disorders associated with POI and increased cancer risk, immunodeficiency, and/or premature aging include Nijmegen Breakage syndrome caused by pathogenic variants in *NBN*; Bloom syndrome caused by pathogenic variants in *BLM*; Werner syndrome caused by pathogenic variants in *WRN*; Fanconi anemia caused by pathogenic variants in genes such as *FANCA*, *FANCC*, and *FANCG*; and Rothmund-Thomson syndrome caused by pathogenic variants in *RECQL4* [32–36]. E3-SUMO ligase, encoded for by the *NSMCE2* gene, is part of a structural maintenance of chromosomes (SMC) complex believed to be required for tolerance of replicative stress. Pathogenic variants in *NSMCE2* cause a syndromic presentation of primordial dwarfism, insulin resistance, and POI [37]. Despite DNA repair gene variants often manifesting in syndromic phenotypes, some have been described in milder phenotypes or isolated POI. *MCM8* and *MCM9* encode minichromosome maintenance proteins that are required for homologous recombination and the repair of double-stranded DNA breaks.

Pathogenic variants in *MCM8* have been found to cause POI with hypothyroidism [38] whereas mutations in *MCM9* have been found to cause POI associated with short stature [39], isolated POI [40], or POI associated with colorectal cancer [41]. Pathogenic variants in the *CSB-PGBD3* fusion gene (*ERCC6*), which encodes a protein involved in transcription-coupled DNA repair, also cause POI [42].

APOPTOTIC GENES

Germ cell apoptosis is an important part of ovarian functioning, needed to eliminate cells that have failed to replicate properly [43]. Variants in genes involved in apoptosis may therefore sometimes be responsible for POI. Pathogenic variants in *NANOS3*, which encodes a protein that represses apoptosis of PGCs, have been associated with POI [44]. Progesterone also acts to repress apoptosis in ovarian cells and pathogenic variants of the gene-encoding progesterone receptor membrane component 1, *PGRMC1*, have been found in POI patients [45].

GENES PROTECTING FROM TOXICITY

Because oocyte integrity is so important, they are sensitive to toxicity. A number of genes, when disrupted, appear to cause abnormal mRNA or protein levels that are toxic to oocytes. This includes pathogenic variants in genes encoding components of the eukaryotic translation initiation factor 2B, *eIF2B2*, *eIF2B4*, and *eIF2B5* [46]. Patients with pathogenic variants in these genes usually present with ovarioleukodystrophy, characterized by POI and progressive leukodystrophy. This eukaryotic translation initiation factor modulates protein translation under cellular stress, thereby preventing the accumulation of denatured proteins that can cause follicular atresia [46]. Similarly, pathogenic variants in *RCBTB1*, another gene believed to be involved in the stress response, are associated with POI along with intellectual disability, goiter, and retinal dystrophy [47]. Pathogenic variants in *eIF4EN1F1* can also cause POI. This gene is involved in translation repression. It is proposed that follicular atresia results from the toxicity of increased mRNA levels [48]; however *eIF4EN1F1* is also proposed to have a role in meiotic progression so it may be considered a meiotic gene [49]. The *ANTXR1* gene product regulates the extracellular matrix. Pathogenic variants in this gene cause GAPO syndrome with growth retardation, alopecia, pseudoanodontia, and progressive optic atrophy as characteristic features. Female individuals also experience POI due to ovarian damage that results from the toxic accumulation of extracellular connective tissue matrix [50].

One of the most common genetic contributors to POI is premutation of the *Fragile X Mental Retardation 1* (*FMR1*) gene. This gene is believed to explain up to 13% of familial cases and 3% of sporadic cases [51], and is the only gene currently recommended for testing in the diagnostic setting [7]. The normal allele contains 5–44 CGG repeats in the 5′ untranslated region. Expansion of this region to 200 or more repeats leads to transcriptional silencing and Fragile X mental retardation whereas the "premutation" expansion to 55–199 repeats leads to elevated *FMR1* mRNA that may be responsible for ovarian damage via mRNA-mediated toxicity [52].

Within the ovary, the follicle basal lamina is a specialized barrier of extracellular matrix that protects follicles through all stages of their maturation. *LMNA* encodes a component of the follicle basal membrane, Laminin A. Mutations in *LMNA* disrupt this protective layer around follicles, making them more susceptible to damage, and can cause POI with associated cardiomyopathy and dysmorphic features [53].

GENES REQUIRED FOR FOLLICLE DEVELOPMENT

Every month, a subset of the follicular pool is recruited for progression (primary recruitment). One follicle then completes further maturation and ovulation (secondary recruitment). Many factors are involved in recruitment, in particular growth factors and neuropeptides for primary recruitment and the hypothalamus-pituitary-ovarian axis for secondary recruitment. An example is BMP15, which promotes follicular growth and maturation. Studies in sheep support a role for this gene in folliculogenesis, with variants causing increased fertility when heterozygous and sterility when homozygous [54]. In humans, pathogenic variants in *BMP15* have been described in X-linked dominant POI [55].

NOG encodes a protein that interacts with BMP4 and BMP7, which are factors known to regulate folliculogenesis. Mutations in *NOG* are associated with proximal symphalangism with POI [56]. Mouse studies indicate that *Gdf9* is an important gene for folliculogenesis beyond the primary stage, and some variants have been identified in the *GDF9* gene in humans that may contribute to the development of POI [57–59].

NOBOX is an oocyte-specific homeobox gene that regulates the transcription of *GDF9* and *BMP15*, among other factors. Pathogenic variants in *NOBOX* have been found to cause POI [60]. Most *NOBOX* variants described have been heterozygous in affected patients, and pathogenicity has been supported by in vitro functional studies [61]. More recently, however, homozygous *NOBOX* variants have been described that cause POI with autosomal recessive inheritance [62].

Variants causing POI have also been identified in the gene encoding transcription factor, *SOHLH1* [63]. SOHLH1 is required for early folliculogenesis and possibly ovarian development also [63].

GENES REQUIRED FOR HORMONAL SIGNALING

After a pool of follicles undergoes primary recruitment, secondary recruitment commences. This stage of follicular growth relies heavily on hormones such as FSH, which promotes the development of antral follicles, and the leutinizing hormone (LH), which stimulates the oocyte in the dominant follicle to complete meiosis and ovulation. Variants in the genes encoding subunits of these hormones or their receptors can underlie cases of POI. Variants in *FSHR*, which encodes the FSH receptor, can cause POI of variable severity from hypoplastic ovaries with no ovarian function to partial ovarian impairment [64–66]. Furthermore, variants in *GNAS*, which encodes the intracellular factor that couples FSHR and LHCGR to their enzymatic effector, cause POI as part of pseudohypoparathyroidism [67].

STEROIDOGENIC GENES

Steroidogenesis is another key hormonal process that needs to be intact for proper ovarian function. Estrogen is an important hormone required for ovarian function, and is produced via steroidogenesis. Steroidogenesis begins in the theca cells where cholesterol is converted to androgens and is completed by granulosa cells that convert androgens to estrogens. Factors involved in steroidogenesis can cause POI. For example, POI-causing pathogenic variants can occur in *STAR*. This gene encodes the STeroidogenic Acute Regulatory protein, which is responsible for the transport of cholesterol into mitochondria, the site of steroidogenesis. Pathogenic variants in this gene usually cause lipoid congenital adrenal hyperplasia (CAH) with adrenal failure and potential neonatal death. However, women with mild nonclassical CAH may survive into adulthood but develop POI due to the toxicity of lipoid accumulation [68]. Enzymes directly involved in the biochemical pathway that converts cholesterol to estrogen can harbor POI-causing variants. Variants in the *CYP17A1* and *CYP19A1* genes, for example, encoding enzymes for pregnenolone/progesterone hydroxylation and estrogen aromatization, respectively, can cause POI [69–72].

MITOCHONDRIAL GENES

Mitochondria are not only important for ovarian functioning because of their role in steroidogenesis, but also because they are responsible for generating the majority of cellular energy (ovaries have a particularly high energy demand) and for minimizing damage via oxidative stress, to which oocytes are particularly susceptible. A number of studies have demonstrated that women with POI do indeed have elevated markers of oxidative stress [73, 74]. Because energy is a ubiquitous requirement of human cells, however, patients with pathogenic variants in mitochondrial genes usually have multisystem involvement and POI is a symptom of a more severe syndrome. POI has been described as a feature of mitochondrial diseases caused by variants in many different nuclear genes encoding mitochondrial proteins. For example, pathogenic variants in *POLG*, which encodes a factor required for the replication of mitochondrial DNA, can cause POI as part of progressive external ophthalmoplegia (PEO), characterized by blindness and myopathy [75, 76]. Similarly, patients with pathogenic variants in another gene required for mitochondrial DNA replication, *C10orf2*, present with syndromic POI. In this case, patients develop hearing loss as part of Perrault syndrome [77]. Mitochondrial tRNA synthetases are required for mitochondrial protein translation, and pathogenic variants in a number of these have been described in patients with syndromic POI: Perrault syndrome for patients with *LARS2* [78] or *HARS2* [79] variants, and adult-onset encephalopathy for patients with *AARS2* pathogenic variants [80]. Pathogenic variants in *CLPP*, which encodes a mitochondrial protease required for the unfolded-protein response, also cause Perrault syndrome, including ovarian dysfunction [81]. Another mitochondrial disease, Leigh syndrome, can be caused by mutations in the *LRPPRC* gene that modulates mitochondrial gene expression. Patients with mutations in this gene have a severe and progressive neurological condition that often results in death. However, surviving women exhibit POI with primary amenorrhea and arrested puberty [82].

METABOLIC GENES

Proper metabolism is required for ovarian function, for several reasons. The metabolism of substrates generates energy for the labor-intensive function of the ovaries while also breaking down potentially toxic factors and synthesizing necessary factors for ovarian function. Some of the genes known to underlie cases of POI are involved in the metabolism. The *GALT* gene, for example, encodes an enzyme required for galactose metabolism. The majority of women with pathogenic variants in this gene exhibit POI because galactose accumulates to toxic levels, causing follicular atresia [83]. Another metabolic gene in which variants can cause POI is *PMM2* (*CDG1*), which encodes phosphomannomutase, an enzyme required for the synthesis of glycoprotein. Patients present with neurological symptoms of varying severity and POI [84].

GENES AFFECTING IMMUNE REGULATION

POI can be a symptom of a number of autoimmune diseases such as systemic lupus erythematosus, Hashimoto's thyroiditis, or Addison's disease [85, 86]. Individuals with autoimmune oophoritis often have autoantibodies directed at ovarian, steroidogenic, and/or adrenocortical tissue, which leads to ovarian inflammation [85, 86]. Autoimmune diseases tend to run in families and often have a genetic basis. One such autoimmune disease that is associated with POI is autoimmune polyendocrine syndrome type 1 (APS-1), caused by pathogenic variants in the gene encoding the autoimmune regulator, AIRE. Affected individuals have adrenal, thyroid, and gonadal dysfunction as well as candidiasis. Although other genes associated with autoimmune POI in a monogenic fashion have yet to be discovered, susceptibility loci have been identified [87] and a number of genes involved in immune regulation have shown significant association with menopausal age.

CHROMOSOMAL CAUSES

Chromosomal aberrations are another genetic cause of POI. Turner syndrome, which results from loss of an X chromosome (45,X) complete or partial, causes short stature and rapid apoptosis of fetal oocytes as well as cardiovascular, lymphatic and renal system anomalies [88]. This oocyte depletion could result from the reduced dosage of a critical gene [88] or from impaired meiosis due to the lack of a homologous X chromosome pair [89]. In contrast, an *extra* X chromosome, as in trisomy X (47, XXX), usually does not impair fertility, but has been reported in occasional patients with POI [90, 91].

POI can also be associated with other chromosomal abnormalities such as terminal deletions or translocations of the X chromosome [92]. Women with balanced X-autosome translocations are a rare and clinically heterogeneous group of patients whose most common phenotype is POI, with consequences depending on the location of the breakpoints and on the X-chromosome inactivation pattern [93]. The common occurrence of chromosomal aberrations affecting Xq13-28 (with POF1 in Xq26-Xq28 and POF2 in Xq13-Xq21) and Xp13-11 suggests that these may be critical regions for normal ovarian development [94]. Consistent with these regions being critical for ovarian function, they contain at least two known POI-related genes, *POF1B* and *FMR1*. In chromosomal rearrangements, the genes within DNA breakpoints are candidates that may be responsible for POI. Mapping of chromosome X breakpoints is a way to identify POI candidate genes but many of the genes disrupted have been discarded as candidate genes, suggesting a long-range position effect altering the expression of genes flanking the breakpoints on the X chromosome, or on the autosomes involved in the rearrangement or a meiotic arrest. In many cases with chromosomal aberrations, POI indeed likely results from impaired meiosis due to unpaired chromosomes.

THE MISSING HERITABILITY OF POI

There are clearly a myriad of genes required for ovarian function and many genes that may be disrupted, resulting in POI. These known genes, however, collectively explain only a small proportion of patients. The majority of POI patients never receive a genetic diagnosis. Genetic investigations are usually limited to chromosome and *FMR1* analysis. So what causes POI in the remaining cases, particularly those that clearly have a familial nature?

The diversity of processes required for ovarian function and the number of factors required means that many more genes may cause POI in a monogenic manner. Studies in mice and *in vitro* have demonstrated hundreds of additional candidate POI genes that are yet to be confirmed as disease-causing in humans [12]. In the past, candidate genes had to be investigated sequentially via Sanger sequencing, limiting the rate at which discoveries could be made. Now that next-generation sequencing (NGS) is widespread, reducing time and economic constraints, it is likely that the rate of monogenic POI variant discovery will accelerate.

Many genes have been reported in association with POI, but lack evidence of causality based on current guidelines [95]. Variants in these genes have merely been identified via Sanger sequencing of candidate genes in POI patients, or have been found in association with POI in underpowered GWAS studies. The functional impact of variants is not established nor have the variants in these genes been found to segregate with POI in multiple families. Again, the widespread use of NGS may confirm some of these genes as truly causative, if multiple patients with related variants and associated phenotypes are discovered.

In addition to monogenic POI, however, it is likely that POI, in some cases, results from multifactorial or oligogenic inheritance, with more than one gene variant acting in concert to impair ovarian function. Several studies have indicated an interaction between multiple genes and POI susceptibility [96, 97], and studies of double and triple knockout mice have demonstrated multigenic inheritance of POI [98, 99]. Although yet to be established, NGS of large numbers of candidate genes in large cohorts of POI patients will likely provide insight into this type of inheritance in human POI.

CHANGING DIAGNOSTIC PRACTICE

Although current guidelines recommend only karyotype analysis and investigation of *FMR1* triplet repeat, the ease and cost at which NGS can now be performed mean it is likely to enter the clinical setting for the management of POI patients. NGS will establish more genetic diagnoses that can enable personalized treatments and may give insight into likely disease progression or associated symptoms that may develop. For example, in rare cases POI may develop prior to inevitable neurodegeneration or may precede cancer in patients with chromosomal instability disorders [12, 100]. Such insights would change patient management and could improve outcomes for patients, such as earlier detection of cancer. The sensitivity of diagnoses such as these means that genetic analysis should only be performed with careful consultation and with the involvement of genetic counselors. Identifying a genetic cause of POI in an individual also enables screening of family members who may be affected. The identification of presymptomatic patients may allow strategies to mitigate the onset of infertility, such as cryopreservation of oocytes before their depletion.

CONCLUSION/SUMMARY

POI is a condition with significant emotional, medical, and economic burdens. For the 1 in 100 women affected, the loss of reproductive potential, the associated comorbidities, and the need for ongoing hormone therapy can be devastating. There is a clear genetic basis to POI, however, it is a heterogeneous condition with more than 50 causative variants and many other candidate genes involved. The known causative genes are involved in varying processes required for ovarian function, including ovarian development, meiosis, DNA repair, protection from toxicity, steroidogenesis, hormone signaling, immune function, mitochondrial function, and metabolism. Our understanding of POI genetics is far from complete with the cause for the majority of cases completely unknown. The missing heritability could be explained by monogenic causes that are yet to be discovered, complex inheritance that can be difficult to prove, or interaction of genes with the environment. As genetic investigations of POI patients become more intensive and more common, we are likely to learn a lot more about the genetics of this complex condition. More clear genotype:phenotype relationships may emerge and prognostic insight may be gained. Women may seek genetic diagnoses because this can enable personalized treatment and counseling, better understanding of potential comorbidities and reproductive potential, and the screening of family members who may benefit from preemptive therapy. Genetic studies for patients with POI should be done with the involvement of genetic counselors because some diagnoses may come with unexpected implications. The better understanding of POI genetics that is likely to occur in the near future ought to provide new insights into the processes required for ovarian function and hopefully lead to better therapies and treatments for affected women.

Funding

This work was supported by a Peter Doherty Early Career Fellowship (1054432, EJT), an NHMRC program grant (1074258, AS), and a fellowship (1062854, AS) from the Australian National Health and Medical Research Council, and the Victorian government's Operational Infrastructure Support Program.

References

[1] Coulam CB, Adamson SC, Annegers JF. Incidence of premature ovarian failure. Obstet Gynecol 1986;67(4): 604–6.

[2] Fabre S, Pierre A, Pisselet C, Mulsant P, Lecerf F, Pohl J, et al. The Booroola mutation in sheep is associated with an alteration of the bone morphogenetic protein receptor-IB functionality. J Endocrinol 2003;177(3):435–44.

[3] Bidet M, Bachelot A, Bissauge E, Golmard JL, Gricourt S, Dulon J, et al. Resumption of ovarian function and pregnancies in 358 patients with premature ovarian failure. J Clin Endocrinol Metab 2011;96(12):3864–72.

[4] Bidet M, Bachelot A, Touraine P. Premature ovarian failure: predictability of intermittent ovarian function and response to ovulation induction agents. Curr Opin Obstet Gynecol 2008;20(4):416–20.

[5] Bachelot A, Nicolas C, Bidet M, Dulon J, Leban M, Golmard JL, et al. Long-term outcome of ovarian function in women with intermittent premature ovarian insufficiency. Clin Endocrinol (Oxf) 2016;.

[6] Nelson LM, Covington SN, Rebar RW. An update: spontaneous premature ovarian failure is not an early menopause. Fertil Steril 2005;83(5):1327–32.

[7] ESHRE. Management of women with premature ovarian insufficiency. In: Guideline of the European society of human reproduction and embryology; 2015.

[8] Cooper AR, Baker VL, Sterling EW, Ryan ME, Woodruff TK, Nelson LM. The time is now for a new approach to primary ovarian insufficiency. Fertil Steril 2011;95(6):1890–7.

[9] Nelson LM. Clinical practice. Primary ovarian insufficiency. N Engl J Med 2009;360(6):606–14.

[10] Tao XY, Zuo AZ, Wang JQ, Tao FB. Effect of primary ovarian insufficiency and early natural menopause on mortality: a meta-analysis. Climacteric: J Int Menopause Soc 2016;19(1):27–36.

[11] Eastell R. Management of osteoporosis due to ovarian failure. Med Pediatr Oncol 2003;41(3):222–7.

[12] Tucker EJ, Grover SR, Bachelot A, Touraine P, Sinclair AH. Premature ovarian insufficiency: new perspectives on genetic cause and phenotypic spectrum. Endocr Rev 2016;37(6):609–35.

[13] Vegetti W, Grazia Tibiletti M, Testa G, de Lauretis Y, Alagna F, Castoldi E, et al. Inheritance in idiopathic premature ovarian failure: analysis of 71 cases. Hum Reprod 1998;13(7):1796–800.

[14] Torgerson DJ, Thomas RE, Reid DM. Mothers and daughters menopausal ages: is there a link? Eur J Obstet Gynecol Reprod Biol 1997;74(1):63–6.

[15] Lin L, Achermann JC. Steroidogenic factor-1 (SF-1, Ad4BP, NR5A1) and disorders of testis development. Sex Dev: Genet Mol Biol Evol Endocrinol Embryol Pathol Sex Determ Differ 2008;2(4–5):200–9.

[16] Lourenco D, Brauner R, Lin L, De Perdigo A, Weryha G, Muresan M, et al. Mutations in NR5A1 associated with ovarian insufficiency. N Engl J Med 2009;360(12):1200–10.

[17] Uhlenhaut NH, Jakob S, Anlag K, Eisenberger T, Sekido R, Kress J, et al. Somatic sex reprogramming of adult ovaries to testes by FOXL2 ablation. Cell 2009;139(6):1130–42.

[18] Crisponi L, Deiana M, Loi A, Chiappe F, Uda M, Amati P, et al. The putative forkhead transcription factor FOXL2 is mutated in blepharophimosis/ptosis/epicanthus inversus syndrome. Nat Genet 2001;27(2):159–66.

[19] Demirhan O, Turkmen S, Schwabe GC, Soyupak S, Akgul E, Tastemir D, et al. A homozygous BMPR1B mutation causes a new subtype of acromesomelic chondrodysplasia with genital anomalies. J Med Genet 2005;42(4):314–7.

[20] Caburet S, Arboleda VA, Llano E, Overbeek PA, Barbero JL, Oka K, et al. Mutant cohesin in premature ovarian failure. N Engl J Med 2014;370(10):943–9.

[21] Llano E, Gomez R, Gutierrez-Caballero C, Herran Y, Sanchez-Martin M, Vazquez-Quinones L, et al. Shugoshin-2 is essential for the completion of meiosis but not for mitotic cell division in mice. Genes Dev 2008;22(17):2400–13.

[22] Bolcun-Filas E, Hall E, Speed R, Taggart M, Grey C, de Massy B, et al. Mutation of the mouse Syce1 gene disrupts synapsis and suggests a link between synaptonemal complex structural components and DNA repair. PLoS Genet 2009;5(2):e1000393.

[23] de Vries L, Behar DM, Smirin-Yosef P, Lagovsky I, Tzur S, Basel-Vanagaite L. Exome sequencing reveals SYCE1 mutation associated with autosomal recessive primary ovarian insufficiency. J Clin Endocrinol Metab 2014;99(10):E2129–32.

[24] Wang J, Zhang W, Jiang H, Wu BL. Primary ovarian insufficiency C. Mutations in HFM1 in recessive primary ovarian insufficiency. N Engl J Med 2014;370(10):972–4.

[25] Zangen D, Kaufman Y, Zeligson S, Perlberg S, Fridman H, Kanaan M, et al. XX ovarian dysgenesis is caused by a PSMC3IP/HOP2 mutation that abolishes coactivation of estrogen-driven transcription. Am J Hum Genet 2011;89(4):572–9.

[26] Lacombe A, Lee H, Zahed L, Choucair M, Muller JM, Nelson SF, et al. Disruption of POF1B binding to nonmuscle actin filaments is associated with premature ovarian failure. Am J Hum Genet 2006;79(1):113–9.

[27] Weinberg-Shukron A, Renbaum P, Kalifa R, Zeligson S, Ben-Neriah Z, Dreifuss A, et al. A mutation in the nucleoporin-107 gene causes XX gonadal dysgenesis. J Clin Invest 2015;.

[28] Carlosama C, Elzaiat M, Patino LC, Mateus HE, Veitia RA, Laissue P. A homozygous donor splice-site mutation in the meiotic gene MSH4 causes primary ovarian insufficiency. Hum Mol Genet 2017;26(16):3161–6.

[29] Guo T, Zhao S, Zhao S, Chen M, Li G, Jiao X, et al. Mutations in MSH5 in primary ovarian insufficiency. Hum Mol Genet 2017;26(8):1452–7.

[30] Smirin-Yosef P, Zuckerman-Levin N, Tzur S, Granot Y, Cohen L, Sachsenweger J, et al. A Biallelic mutation in the homologous recombination repair gene SPIDR is associated with human gonadal dysgenesis. J Clin Endocrinol Metab 2017;102(2):681–8.

[31] Miller ME, Chatten J. Ovarian changes in ataxia telangiectasia. Acta Paediatr Scand 1967;56(5):559–61.

[32] Arora H, Chacon AH, Choudhary S, McLeod MP, Meshkov L, Nouri K, et al. Bloom syndrome. Int J Dermatol 2014;53(7):798–802.

[33] Rossi ML, Ghosh AK, Bohr VA. Roles of Werner syndrome protein in protection of genome integrity. DNA Repair 2010;9(3):331–44.

[34] Siitonen HA, Sotkasiira J, Biervliet M, Benmansour A, Capri Y, Cormier-Daire V, et al. The mutation spectrum in RECQL4 diseases. Eur J Human Genet 2009;17(2):151–8.

[35] Chrzanowska KH, Gregorek H, Dembowska-Baginska B, Kalina MA, Digweed M. Nijmegen breakage syndrome (NBS). Orphanet J Rare Dis 2012;7:13.

[36] Giri N, Batista DL, Alter BP, Stratakis CA. Endocrine abnormalities in patients with Fanconi anemia. J Clin Endocrinol Metab 2007;92(7):2624–31.

[37] Payne F, Colnaghi R, Rocha N, Seth A, Harris J, Carpenter G, et al. Hypomorphism in human NSMCE2 linked to primordial dwarfism and insulin resistance. J Clin Invest 2014;124(9):4028–38.

[38] AlAsiri S, Basit S, Wood-Trageser MA, Yatsenko SA, Jeffries EP, Surti U, et al. Exome sequencing reveals MCM8 mutation underlies ovarian failure and chromosomal instability. J Clin Invest 2015;125(1):258–62.

[39] Wood-Trageser MA, Gurbuz F, Yatsenko SA, Jeffries EP, Kotan LD, Surti U, et al. MCM9 mutations are associated with ovarian failure, short stature, and chromosomal instability. Am J Hum Genet 2014;95(6):754–62.

[40] Fauchereau F, Shalev S, Chervinsky E, Fruchter RB, Legois B, Fellous M, et al. A non-sense MCM9 mutation in a familial case of primary ovarian insufficiency. Clin Genet 2016;.

[41] Goldberg Y, Halpern N, Hubert A, Adler SN, Cohen S, Plesser-Duvdevani M, et al. Mutated MCM9 is associated with predisposition to hereditary mixed polyposis and colorectal cancer in addition to primary ovarian failure. Cancer Genet 2015;208(12):621–4.

[42] Qin Y, Guo T, Li G, Tang TS, Zhao S, Jiao X, et al. CSB-PGBD3 mutations cause premature ovarian failure. PLoS Genet 2015;11(7):e1005419.

[43] Morita Y, Tilly JL. Oocyte apoptosis: like sand through an hourglass. Dev Biol 1999;213(1):1–17.

[44] Santos MG, Machado AZ, Martins CN, Domenice S, Costa EM, Nishi MY, et al. Homozygous inactivating mutation in NANOS3 in two sisters with primary ovarian insufficiency. Biomed Res Int 2014;2014:787465.

[45] Mansouri MR, Schuster J, Badhai J, Stattin EL, Losel R, Wehling M, et al. Alterations in the expression, structure and function of progesterone receptor membrane component-1 (PGRMC1) in premature ovarian failure. Hum Mol Genet 2008;17(23):3776–83.

[46] Fogli A, Rodriguez D, Eymard-Pierre E, Bouhour F, Labauge P, Meaney BF, et al. Ovarian failure related to eukaryotic initiation factor 2B mutations. Am J Hum Genet 2003;72(6):1544–50.

[47] Coppieters F, Ascari G, Dannhausen K, Nikopoulos K, Peelman F, Karlstetter M, et al. Isolated and syndromic retinal dystrophy caused by Biallelic mutations in RCBTB1, a gene implicated in ubiquitination. Am J Hum Genet 2016;99(2):470–80.

[48] Kasippillai T, MacArthur DG, Kirby A, Thomas B, Lambalk CB, Daly MJ, et al. Mutations in eIF4ENIF1 are associated with primary ovarian insufficiency. J Clin Endocrinol Metab 2013;98(9):E1534–9.

[49] Pfender S, Kuznetsov V, Pasternak M, Tischer T, Santhanam B, Schuh M. Live imaging RNAi screen reveals genes essential for meiosis in mammalian oocytes. Nature 2015;524(7564):239–42.

[50] Benetti-Pinto CL, Ferreira V, Andrade L, Yela DA, De Mello MP. GAPO syndrome: a new syndromic cause of premature ovarian insufficiency. Climacteric: J Int Menopause Soc 2016;19(6):594–8.

[51] Conway GS, Payne NN, Webb J, Murray A, Jacobs PA. Fragile X premutation screening in women with premature ovarian failure. Hum Reprod 1998;13(5):1184–7.

[52] Elizur SE, Lebovitz O, Derech-Haim S, Dratviman-Storobinsky O, Feldman B, Dor J, et al. Elevated levels of FMR1 mRNA in granulosa cells are associated with low ovarian reserve in FMR1 premutation carriers. PLoS ONE 2014;9(8)e105121.

[53] McPherson E, Turner L, Zador I, Reynolds K, Macgregor D, Giampietro PF. Ovarian failure and dilated cardiomyopathy due to a novel lamin mutation. Am J Med Genet A 2009;149A(4):567–72.

[54] Hanrahan JP, Gregan SM, Mulsant P, Mullen M, Davis GH, Powell R, et al. Mutations in the genes for oocyte-derived growth factors GDF9 and BMP15 are associated with both increased ovulation rate and sterility in Cambridge and Belclare sheep (Ovis aries). Biol Reprod 2004;70(4):900–9.

[55] Di Pasquale E, Beck-Peccoz P, Persani L. Hypergonadotropic ovarian failure associated with an inherited mutation of human bone morphogenetic protein-15 (BMP15) gene. Am J Hum Genet 2004;75(1):106–11.

[56] Kosaki K, Sato S, Hasegawa T, Matsuo N, Suzuki T, Ogata T. Premature ovarian failure in a female with proximal symphalangism and Noggin mutation. Fertil Steril 2004;81(4):1137–9.

[57] Dixit H, Rao LK, Padmalatha V, Kanakavalli M, Deenadayal M, Gupta N, et al. Mutational screening of the coding region of growth differentiation factor 9 gene in Indian women with ovarian failure. Menopause 2005; 12(6):749–54.

[58] Kovanci E, Rohozinski J, Simpson JL, Heard MJ, Bishop CE, Carson SA. Growth differentiating factor-9 mutations may be associated with premature ovarian failure. Fertil Steril 2007;87(1):143–6.

[59] Zhao H, Qin Y, Kovanci E, Simpson JL, Chen ZJ, Rajkovic A. Analyses of GDF9 mutation in 100 Chinese women with premature ovarian failure. Fertil Steril 2007;88(5):1474–6.

[60] Qin Y, Choi Y, Zhao H, Simpson JL, Chen ZJ, Rajkovic A. NOBOX homeobox mutation causes premature ovarian failure. Am J Hum Genet 2007;81(3):576–81.

[61] Bouilly J, Bachelot A, Broutin I, Touraine P, Binart N. Novel NOBOX loss-of-function mutations account for 6.2% of cases in a large primary ovarian insufficiency cohort. Hum Mutat 2011;32(10):1108–13.

[62] Li L, Wang B, Zhang W, Chen B, Luo M, Wang J, et al. A homozygous NOBOX truncating variant causes defective transcriptional activation and leads to primary ovarian insufficiency. Hum Reprod 2017;32(1):248–55.

[63] Bayram Y, Gulsuner S, Guran T, Abaci A, Yesil G, Gulsuner HU, et al. Homozygous loss-of-function mutations in SOHLH1 in patients with non-syndromic hypergonadotropic hypogonadism. J Clin Endocrinol Metab 2015. jc20151150.

[64] Aittomaki K, Lucena JL, Pakarinen P, Sistonen P, Tapanainen J, Gromoll J, et al. Mutation in the follicle-stimulating hormone receptor gene causes hereditary hypergonadotropic ovarian failure. Cell 1995;82 (6):959–68.

[65] Touraine P, Beau I, Gougeon A, Meduri G, Desroches A, Pichard C, et al. New natural inactivating mutations of the follicle-stimulating hormone receptor: correlations between receptor function and phenotype. Mol Endocrinol 1999;13(11):1844–54.

[66] Beau I, Touraine P, Meduri G, Gougeon A, Desroches A, Matuchansky C, et al. A novel phenotype related to partial loss of function mutations of the follicle stimulating hormone receptor. J Clin Invest 1998;102(7):1352–9.

[67] Mantovani G, Spada A. Mutations in the Gs alpha gene causing hormone resistance. Best Pract Res Clin Endocrinol Metab 2006;20(4):501–13.

[68] Bose HS, Pescovitz OH, Miller WL. Spontaneous feminization in a 46,XX female patient with congenital lipoid adrenal hyperplasia due to a homozygous frameshift mutation in the steroidogenic acute regulatory protein. J Clin Endocrinol Metab 1997;82(5):1511–5.

[69] Miura K, Yasuda K, Yanase T, Yamakita N, Sasano H, Nawata H, et al. Mutation of cytochrome P-45017 alpha gene (CYP17) in a Japanese patient previously reported as having glucocorticoid-responsive

hyperaldosteronism: with a review of Japanese patients with mutations of CYP17. J Clin Endocrinol Metab 1996;81(10):3797–801.

[70] Simsek E, Ozdemir I, Lin L, Achermann JC. Isolated 17,20-lyase (desmolase) deficiency in a 46,XX female presenting with delayed puberty. Fertil Steril 2005;83(5):1548–51.

[71] Conte FA, Grumbach MM, Ito Y, Fisher CR, Simpson ER. A syndrome of female pseudohermaphrodism, hypergonadotropic hypogonadism, and multicystic ovaries associated with missense mutations in the gene encoding aromatase (P450arom). J Clin Endocrinol Metab 1994;78(6):1287–92.

[72] Ito Y, Fisher CR, Conte FA, Grumbach MM, Simpson ER. Molecular basis of aromatase deficiency in an adult female with sexual infantilism and polycystic ovaries. Proc Natl Acad Sci U S A 1993;90(24):11673–7.

[73] Tokmak A, Yildirim G, Sarikaya E, Cinar M, Bogdaycioglu N, Yilmaz FM, et al. Increased oxidative stress markers may be a promising indicator of risk for primary ovarian insufficiency: a cross-sectional case control study. Rev Bras Ginecol Obstet: Rev Feder Bras Soc Ginecol Obstetr 2015;37(9):411–6.

[74] Venkatesh S, Kumar M, Sharma A, Kriplani A, Ammini AC, Talwar P, et al. Oxidative stress and ATPase6 mutation is associated with primary ovarian insufficiency. Arch Gynecol Obstet 2010;282(3):313–8.

[75] Pagnamenta AT, Taanman JW, Wilson CJ, Anderson NE, Marotta R, Duncan AJ, et al. Dominant inheritance of premature ovarian failure associated with mutant mitochondrial DNA polymerase gamma. Hum Reprod 2006;21(10):2467–73.

[76] Luoma P, Melberg A, Rinne JO, Kaukonen JA, Nupponen NN, Chalmers RM, et al. Parkinsonism, premature menopause, and mitochondrial DNA polymerase gamma mutations: clinical and molecular genetic study. Lancet 2004;364(9437):875–82.

[77] Morino H, Pierce SB, Matsuda Y, Walsh T, Ohsawa R, Newby M, et al. Mutations in Twinkle primase-helicase cause Perrault syndrome with neurologic features. Neurology 2014;83(22):2054–61.

[78] Pierce SB, Gersak K, Michaelson-Cohen R, Walsh T, Lee MK, Malach D, et al. Mutations in LARS2, encoding mitochondrial leucyl-tRNA synthetase, lead to premature ovarian failure and hearing loss in Perrault syndrome. Am J Hum Genet 2013;92(4):614–20.

[79] Pierce SB, Chisholm KM, Lynch ED, Lee MK, Walsh T, Opitz JM, et al. Mutations in mitochondrial histidyl tRNA synthetase HARS2 cause ovarian dysgenesis and sensorineural hearing loss of Perrault syndrome. Proc Natl Acad Sci U S A 2011;108(16):6543–8.

[80] Dallabona C, Diodato D, Kevelam SH, Haack TB, Wong LJ, Salomons GS, et al. Novel (ovario) leukodystrophy related to AARS2 mutations. Neurology 2014;82(23):2063–71.

[81] Jenkinson EM, Rehman AU, Walsh T, Clayton-Smith J, Lee K, Morell RJ, et al. Perrault syndrome is caused by recessive mutations in CLPP, encoding a mitochondrial ATP-dependent chambered protease. Am J Hum Genet 2013;92(4):605–13.

[82] Ghaddhab C, Morin C, Brunel-Guitton C, Mitchell GA, Van Vliet G, Huot C. Premature ovarian failure in French Canadian Leigh syndrome. J Pediatr 2017;184:227–9: [e1].

[83] Banerjee S, Chakraborty P, Saha P, Bandyopadhyay SA, Banerjee S, Kabir SN. Ovotoxic effects of galactose involve attenuation of follicle-stimulating hormone bioactivity and up-regulation of granulosa cell p53 expression. PLoS ONE 2012;7(2):e30709.

[84] Matthijs G, Schollen E, Pardon E, Veiga-Da-Cunha M, Jaeken J, Cassiman JJ, et al. Mutations in PMM2, a phosphomannomutase gene on chromosome 16p13, in carbohydrate-deficient glycoprotein type I syndrome (Jaeken syndrome). Nat Genet 1997;16(1):88–92.

[85] Silva CA, Yamakami LY, Aikawa NE, Araujo DB, Carvalho JF, Bonfa E. Autoimmune primary ovarian insufficiency. Autoimmun Rev 2014;13(4–5):427–30.

[86] La Marca A, Brozzetti A, Sighinolfi G, Marzotti S, Volpe A, Falorni A. Primary ovarian insufficiency: autoimmune causes. Curr Opin Obstet Gynecol 2010;22(4):277–82.

[87] Kahaly GJ. Polyglandular autoimmune syndromes. Eur J Endocrinol/Eur Feder Endocr Soc 2009;161(1):11–20.

[88] Zinn AR, Page DC, Fisher EM. Turner syndrome: the case of the missing sex chromosome. Trends Genet 1993; 9(3):90–3.

[89] Burgoyne PS, Baker TG. Perinatal oocyte loss in XO mice and its implications for the aetiology of gonadal dysgenesis in XO women. J Reprod Fertil 1985;75(2):633–45.

[90] Villanueva AL, Rebar RW. Triple-X syndrome and premature ovarian failure. Obstet Gynecol 1983; 62(3 Suppl):70s–73s.

[91] Tungphaisal S, Jinorose U. True 47,XXX in a patient with premature ovarian failure: the first reported case in Thailand. J Med Assoc Thailand = Chotmaihet Thangphaet 1992;75(11):661–5.

[92] Simpson JL, Rajkovic A. Ovarian differentiation and gonadal failure. Am J Med Genet 1999;89(4):186–200.

[93] Moyses-Oliveira M, Guilherme Rdos S, Dantas AG, Ueta R, Perez AB, Haidar M, et al. Genetic mechanisms leading to primary amenorrhea in balanced X-autosome translocations. Fertil Steril 2015;103(5):1289–96: e2.

[94] Therman E, Laxova R, Susman B. The critical region on the human Xq. Hum Genet 1990;85(5):455–61.

[95] MacArthur DG, Manolio TA, Dimmock DP, Rehm HL, Shendure J, Abecasis GR, et al. Guidelines for investigating causality of sequence variants in human disease. Nature 2014;508(7497):469–76.

[96] Pyun JA, Kim S, Cha DH, Kwack K. Epistasis between polymorphisms in TSHB and ADAMTS16 is associated with premature ovarian failure. Menopause 2014;21(8):890–5.

[97] Pyun JA, Kim S, Cha DH, Kwack K. Epistasis between IGF2R and ADAMTS19 polymorphisms associates with premature ovarian failure. Hum Reprod 2013;28(11):3146–54.

[98] Pangas SA, Li X, Umans L, Zwijsen A, Huylebroeck D, Gutierrez C, et al. Conditional deletion of Smad1 and Smad5 in somatic cells of male and female gonads leads to metastatic tumor development in mice. Mol Cell Biol 2008;28(1):248–57.

[99] Li Q, Pangas SA, Jorgez CJ, Graff JM, Weinstein M, Matzuk MM. Redundant roles of SMAD2 and SMAD3 in ovarian granulosa cells in vivo. Mol Cell Biol 2008;28(23):7001–11.

[100] Tucker EJ, Grover SR, Robevska G, van den Bergen J, Hanna C, Sinclair AH. Identification of variants in pleiotropic genes causing "isolated" premature ovarian insufficiency: implications for medical practice. Eur J Hum Genet 2018; https://dx.doi.org/10.1038/s41431-018-0140-4.

Genetics of Polycystic Ovary Syndrome

Tristan Hardy[*,†], *Robert J. Norman*[‡,§,¶,‖]

[*]Genetics and Molecular Pathology, SA Pathology, Adelaide, SA, Australia [†]Repromed, Dulwich, SA, Australia [‡]University of Adelaide, Adelaide, SA, Australia [§]Fertility SA, Adelaide, SA, Australia [¶]Royal Adelaide Hospital, Adelaide, SA, Australia [‖]NHMRC Centre for Research Excellence in Polycystic Ovary Syndrome, Adelaide, SA, Australia

Abbreviations

AMH	antiMullerian hormone
BMI	body mass index
FSH	follicle stimulating hormone
GWAS	genome-wide association studies
hCG	human chorionic gonadotropin
LH	luteinizing hormone
NIH	National Institutes of Health
PCOS	polycystic ovary syndrome

Human Reproductive and Prenatal Genetics
https://doi.org/10.1016/B978-0-12-813570-9.00020-6

SNP   single nucleotide polymorphism
T1DM   type 1 diabetes mellitus
T2DM   type 2 diabetes mellitus
WES   whole exome sequencing
WGS   whole genome sequencing

INTRODUCTION

Polycystic ovary syndrome (PCOS) is the most common endocrine disorder affecting women in their reproductive phase of life. The phenotype of PCOS is heterogeneous, manifesting in its more pronounced form primarily as chronic anovulation and/or hyperandrogenism. As a result of the clinical heterogeneity of the disorder, a variety of diagnostic criteria have been proposed that attempt to capture the salient elements of the phenotype.

It has been evident for some time that PCOS is a highly heritable condition, and numerous studies have been performed attempting to dissect the underlying genetic architecture of the condition. As a polygenic disorder involving a range of endocrine and metabolic pathways, it has been difficult to isolate individual genetic variants with evidence of high impact using the candidate gene approach. Genome-wide analytical techniques such as genome-wide association studies (GWAS) have identified numerous loci of interest, many of which have been confirmed across different populations, suggesting the impact of nearby genes on the underlying pathogenesis of the condition. However, these loci explain only a fraction of the estimated heritability of the condition, and it is likely that further studies using whole exome or whole genome sequencing (WES/WGS) will be required to identify additional genetic variants of high impact.

This chapter focuses on genomic approaches to the identification of important loci in PCOS pathogenesis, reviewing in detail the current GWAS studies and the emerging studies using WES/WGS.

DEFINING PCOS

Several well-known criteria for the diagnosis of PCOS have been enumerated [1], the most commonly used being the Rotterdam criteria that require two out of three of oligo/anovulation, clinical or biochemical hyperandrogenism, and polycystic ovarian morphology [2]. The variety of diagnostic criteria and the clinical heterogeneity of the condition has made it difficult to study the underlying genetic architecture of PCOS, as accurate phenotyping is a key component of any genetic study. Some studies have used polycystic ovaries alone to define the condition whereas others have demanded hyperandrogenism and menstrual disturbances to be included in any genetic examination. Regardless of the clinical criteria used, it has been evident to researchers for some time that the features of PCOS appear to have a familial relationship, suggesting a genetic component to the condition [3].

HERITABILITY OF PCOS

Initial studies of the heritability of PCOS concluded that the disorder was most likely a dominantly inherited trait with low penetrance and expressivity [4], although more recently these features have been thought to be consistent with a polygenic model of inheritance, along with the majority of complex phenotypic traits [5]. Cooper et al. [3] were the first to systematically analyze the potential genetic contributions to PCOS, ruling out a chromosomal origin of the disorder and suggesting that the pattern of symptoms (predominantly menstrual irregularities and hirsutism) in mothers and sisters of affected probands was most likely consistent with an autosomal dominant mode of inheritance. Expansion of inclusion criteria to patients with polycystic ovarian morphology on ultrasound similarly concluded that the pattern of features in family members was consistent with an autosomal dominant mode of transmission [6]. Subsequent studies confirmed that metabolic features of the disorder were also prevalent in family members, including male relatives. Hyperinsulinemia and hypertriglyceridemia affected 69% and 56% of family members overall, and polycystic ovaries were present in 74% of female relatives while male pattern baldness was found in 88% of male relatives [7]. Legro et al. [8] similarly demonstrated that sisters of probands with PCOS had an increased risk of both PCOS (22%) and hyperandrogenemia with regular menstrual cycles (24%).

Twin studies have been extremely effective in establishing the heritability of multiple conditions, and initial studies with small numbers concluded that PCOS was not likely to result from variants within a single gene but was either polygenic or multifactorial [9]. The most definitive twin family study analyzed 3205 females (comprising 1332 monozygotic twins, 680 dizygotic twins, 474 individuals from dizygotic opposite sex pairs, and 719 nontwin sisters), according to the Rotterdam criteria [10]. Monozygotic twin pairs had significantly higher correlation of the features of PCOS than dizygotic or nontwin sisters; the relative risk of oligomenorrhea was 0.67 versus 0.07, acne 0.78 versus 0.44, and hirsutism 0.86 versus 0.28 [10]. Overall, the risk of PCOS in monozygotic twin sisters was 0.71 (95% CI 0.43 to 0.88) versus 0.38 (95% CI 0.00 to 0.66) in dizygotic twin or nontwin sisters. After modeling the three variables of oligomenorrhea, acne, and hirsutism in an independent pathway model, the authors concluded that PCOS was affected predominantly by genetic variance (79%) and unique environmental influence (21%), with no role for shared environmental influence [10]. Although this and many other studies pointed to the strongly heritable nature of PCOS, the establishment of heritability gave no clues as to the putative genes involved.

CANDIDATE GENE STUDIES

The majority of genetic studies in PCOS have used the candidate gene approach [11], partly due to investigator-driven hypotheses regarding the pathogenesis of PCOS, but also as a reflection of the technology available for genetic analysis at the time of the studies. Overall, a recent database included information on 241 genes and 114 SNPs that have been associated with PCOS [12], demonstrating the variety of genetic variants that have been identified in the literature. Within these, a few promising candidate genes have been identified and replicated in a number of studies, although their overall contribution to the pathogenesis of PCOS remains unclear. Generally there is a history of identifying gene relationships that do not stand the test of time.

The only candidate gene that has been subsequently validated by GWAS studies is the insulin receptor (INSR) [13]. Tucci et al. [14] identified the polymorphic variant (D19S884) upstream of INSR within the fibrillin 3 gene (FBN3) as a potential risk factor for PCOS, but could not validate whether the susceptibility locus was in the insulin gene receptor itself or within the gene region. The D19S884 marker was subsequently analyzed in a larger group of 367 families [15] and 453 families [16], with both authors concluding that there was a significant association with the transmission of PCOS. These findings were replicated in a study of Han Chinese patients with PCOS [17], although a smaller Spanish study concluded that there was no relationship between the D19S884 marker and PCOS [18].

Given the importance of insulin resistance to the overall phenotype of PCOS, there has been ongoing investigation into the role of polymorphisms in the insulin receptor gene itself with seven studies finding SNPs within INSR contributing to PCOS risk, one of which was eventually confirmed in a large study (rs2252673) [13]. Researchers have also postulated that the strong association between PCOS and obesity may imply a relationship between obesity-susceptibility variants and PCOS. Of these, the most studied is that of the fat mass and obesity associated (FTO) gene, in particular, the rs9939609 polymorphism [19]. Meta-analysis of five studies covering 5010 PCOS patients and 5300 controls demonstrated that rs9939609 was significantly different between groups, suggesting that the A allele was a risk factor for PCOS susceptibility in both Asian and Caucasian subgroups (OR 1.43 and OR 1.33, respectively) [19].

The genetic basis of hyperandrogenism in PCOS has led to a number of studies exploring variants in the enzymes involved in the steroidogenic pathway [6] and sensitivity to androgen signaling [20]. Early studies of the polymorphic trinucleotide repeat (CAG) in the androgen receptor gene suggested that infertile women with PCOS had a greater number of repeats compared with fertile control patients [21]. However, later meta-analysis of the role of CAG repeats within the androgen receptor gene concluded in favor of an association of the short repeat group with hyperandrogenism (56.25% vs. 29.14%, $P < 0.001$) [22]. Furthermore, in a discovery cohort of 354 PCOS and 161 control patients, SNPs within genes in the androgen receptor signaling pathway were analyzed (HSPA1A, HSPA8, ST13, STIP1, PTGES3, FKBP4, BAG1, and STUB1) and two SNPs in FKBP4 were associated with a reduced odds ratio of PCOS. However, only one of these (rs4409904) was confirmed in a subsequent replication cohort (397 cases and 306 controls) [20]. A recent study also suggested that alternative splice variants of the androgen receptor may be associated with abnormal folliculogenesis and hyperandrogenemia in patients with PCOS [23]. Studies concurrently examining multiple genes in the steroidogenic pathway along with those involved with gonadotropin action, obesity, energy regulation, and insulin action have failed to demonstrate a compelling relationship between steroidogenic enzyme pathway variants and PCOS, but have suggested that variants in gonadotropin genes (Follistatin) may be associated with PCOS [24].

An inflammatory phenotype associated with PCOS and metabolic alterations has also been proposed, the genetic basis of which may be linked to polymorphisms in inflammatory cytokine genes such as TNF-alpha, IL-6, and IL-1 beta [25]. All reports examining polymorphisms in these genes were subjected to meta-analysis, with 14 studies eligible for inclusion. For *TNF-alpha*, there were 802 cases and 802 controls in total, with no evidence of significance of the -308 G/A polymorphism with regard to PCOS or obesity. For *IL-6*, 351 cases and 464 controls were identified, examining the -174 *G/C* polymorphism and finding no significant association overall; however, using an allelic model led to some evidence of significance. Similarly, no obvious association between *IL-1 beta* polymorphism (-511C/T) was found, suggesting overall that previous studies using small numbers of cases and controls found associations that disappeared on wider meta-analysis [25].

Overall, despite many promising leads, the candidate gene approach is unsuited to the analysis of a polygenic condition such as PCOS, due to the extremely low likelihood of finding a high-impact variant in a single gene when comparing only small numbers of individuals with and without the disorder. In addition, previous studies have tested only for variants that are known in the population, and are therefore unlikely to discover new variants with high genetic impact. The transition to GWAS studies has proven useful for defining risk loci in PCOS due to the emergence of new genetic technologies.

GENOME-WIDE ASSOCIATION STUDIES

Improvements in the throughput of genetic analysis technologies have allowed the application of genome-wide techniques to analyze hundreds of thousands to millions of variants per individual, in significantly larger groups of cases and controls. There have been five major GWAS studies focusing on PCOS that have identified a number of promising genetic loci, many of which would not have been predicted by a hypothesis-driven candidate gene approach. The findings of these studies are worth reviewing in detail as they present the most up-to-date analysis of our current understanding of the heritability of PCOS.

Chen et al. [26] published the first GWAS study on PCOS in the Han Chinese population, analyzing 744 PCOS cases and 895 controls in a discovery set and further validating the susceptibility loci in 2840 PCOS cases and 5012 controls. The case definition in this study followed the 2003 Rotterdam Criteria requiring two of three criteria to be present: oligo- and/or anovulation, evidence of clinical and/or biochemical hyperandrogenism, and polycystic ovarian morphology, with the exclusion of other causes of oligomenorrhea or hyperandrogenism. Single nucleotide polymorphisms (SNPs) were analyzed using Affymetrix SNP 6.0 chips, with a total of 611,633 SNPs subjected to analysis over the population. In the discovery set, four distinct regions with 29 SNPs showed strong evidence of association: 2p16.3, 2p21, 5q14.3, and 9q33.3. Two independent replication sets from slightly different populations (northern Han Chinese, and southern and central Han Chinese) were used to validate the regions identified in the discovery set, with 28 of the 29 SNPs showing genome-wide significance, and 1 SNP representing the 5q14.3 region being removed from analysis. The three leading SNPs identified in the study were representative of each of the loci: rs13405728 at 2p16.3, rs13429458 at 2p21, and rs2479106 at 9q33.3. These regions remained significant when logistic regression analysis was performed to remove the effects of age and body mass index (BMI). Also, the regions did not overlap with those identified on previous GWAS studies analyzing BMI, suggesting an independent association with PCOS rather than metabolic dysfunction.

Analyzing nearby genes (± 500 kb) and linkage disequilibrium (LD) blocks involving the most significant SNPs, Chen et al. [26] identified a number of candidate genes of interest in the pathogenesis of PCOS. At 2p16.3, the most significant nearby genes were the TFIIA-alpha and beta-like factor (*GTF2A1L*) as well as the luteinizing hormone and human chorionic gonadotropin gene receptor (*LHCGR*), both of which have a documented role in human reproduction. *GTF2A1L* had previously been reported to be a potential cause of human infertility, being expressed during germ cell development and playing a role in the testis [26]. The *LHCGR* gene has a central role in determining the sensitivity of granulosa cells to the luteinizing hormone, with previous studies suggesting inactivating mutations may result in LH resistance (manifesting as increased LH levels, enlarged ovaries, and oligomenorrhea) and activating mutations resulting in hyperandrogenism without infertility [26]. The association of SNP rs13405728 in this region did not significantly differ according to the BMI of the cases and controls, suggesting that the SNP independently predicted PCOS as defined by the Rotterdam criteria through a mechanism separate to the metabolic associations of PCOS. The authors also noted that the *FSHR* gene encoding the FSH receptor was nearby (211 kb downstream of rs13405728), although this location was beyond a strong recombination hotspot. Targeted analysis of 65 intragenic SNPs within *FSHR* demonstrated 13 with a PCA-adjusted *P*-value between 2×10^{-3} and 4×10^{-4}, lower than that required for genome-wide significance but potentially significant in combination, thus not ruling out a role for variants in the FSH receptor in susceptibility to PCOS.

The 2p21 region comprised the majority of significant SNPs (21 out of 28), spanning 304 kb and located within two different LD blocks [26]. The most significant independent associations in this region involved the *THADA* gene, which was originally identified in thyroid adenomas but may have a role in susceptibility to insulin resistance/type II diabetes mellitus (T2DM). Subjects with PCOS carrying these SNPs did not, however, have an increased risk of insulin resistance in this study population. The 9q33.3 region contained six significant SNPs spanning 42.3 kb within *DENND1A* (differentially expressed in normal and neoplastic cells domain containing 1A) [26]. This protein is a negative regulator of endoplasmic reticulum aminopeptidase 1 (ERAP1), which has previously been associated with PCOS and obesity [27].

A follow-up study from the same group [28] studied a larger cohort of Han Chinese (8226 cases and 7578 controls) in order to confirm the three previously reported loci, and independently identified eight new PCOS association signals. In addition, the authors identified an independent signal at the previously reported 2p16.3 region, located within *FSHR*. This study used the Affymetrix Axiom array and combined the data from genotyped and imputed SNPs from the Chen et al. [26] study to maximize statistical power. The study analyzed 2254 PCOS cases defined by the Rotterdam criteria and 3001 controls in the discovery set meta-analysis and replicated the findings in two independent sets of 1908/6318 cases and 1913/5665 controls. The previously identified regions 2p16.3, 2p21, and 9q33.3 again reached genome-wide significance and a further 19 new regions were identified in the discovery meta-analysis stage. The 19 new regions and variants in *FSHR* were analyzed in the replication set, of which eight remained significant (2p.16.3/*FSHR*, 9q22.32, 11q22.1, 12q13.2, 12q14.3, 16q12.1, 19p13.3, and 20q13.2), the majority containing candidate genes related to hormones, insulin resistance, and organ growth.

The most significant SNP at 9q22.32 was located in an intron of *C9orf3* that encodes for a protein within the zinc aminopeptidase family that had previously been associated with erectile dysfunction following radiotherapy for prostate cancer, alongside *FSHR* variants that have also been strongly associated with erectile dysfunction. A SNP within *YAP1* at 11q22.1 (rs1894116) was posited to have a role in altering the expression of genes associated with cell proliferation and organ size control, although the function within the ovary is not defined. The most significant SNP at 12q13.2 was located between *RAB5B* and *SUOX*, which had previously been associated with type 1 diabetes mellitus (T1DM). Another SNP in this region (rs2292239) located within the *ERBB3* gene showed evidence of association with PCOS, and identified a family of genes that have been identified in subsequent studies. The nearby region of 12q14.3 had an independent signal from a SNP within *HMGA2*, which encodes a protein involved in DNA transcription regulation and had previously been associated with T2DM.

The 19p13.3 region identified as associated with PCOS was significant for containing the insulin receptor gene (*INSR*), with a signal from an intronic SNP showing genome-wide significance (rs2059807). This ties in strongly with the common understanding of one of the primary pathogenetic mechanisms of PCOS being insulin resistance, as mutations in the *INSR* gene have previously been shown to be associated with severe hyperinsulinemia and insulin resistance. In addition, variants within the *FSHR* gene at 2p16.3 that had previously failed to meet genome-wide significance in the Chen et al. [26] study were identified as significant variants, independent from the previously identified signals at *LHCGR*. The addition of these regions to the GWAS loci known to be associated with PCOS added further weight to the most common explanations regarding the pathogenesis of PCOS. However, the strength of GWAS studies is in identifying regions that would not have been identified based on a hypothesis-driven approach. Shi et al. [28] added other candidate genes at 16q12.1 (*TOX3*) and 20q13.2 (located between *SUMO1P1* and *ZNF217*) that do not have roles in growth and metabolism at the endocrine level, but are related to cell growth and DNA modification.

A GWAS involving Korean women with PCOS by Rotterdam or NIH criteria subsequently identified a novel gene associated with PCOS in combination with obesity [29]. In this study, the PCOS cohort consisted of 774 patients with 967 controls and genotyping was performed using the Illumina HumanOmni1 Quad v1. In total, 619,339 SNPs were analyzed for association with PCOS. Three PCOS-associated SNPs were identified at 12p12.2 (rs10841843, rs6487237, rs7485509) that were related to *GYS2*, a glycogen synthase gene with a role in glycogen storage disease 0 (OMIM 138571). The variants in this gene were subjected to further analysis in relation to the BMI of the subjects, identifying an additional 14 significant variants. However, the initial three SNPs remained significant after adjustment for BMI, suggesting an independent risk for PCOS rather than being driven by an association with BMI. The authors then proceeded to analyze a cohort of childhood obesity (482 children) and gestational diabetes mellitus (1710 women), suggesting that these variants had a pleiotropic effect on obesity-related conditions across the lifespan. The authors analyzed seven of the SNPs reported by Chen et al. [26] at 2p16.3, 2p21, and 9q33.3 and concluded that six (with the exception of rs2479106 in *DENND1A*) had a *P*-value between 2×10^{-2} and 8×10^{-4} that failed to reach genome-wide significance. The authors also analyzed variants in *FSHR*, concluding that there was an association in this cohort of Korean women (*P*-value between 2.2×10^{-3} and 5.9×10^{-4}).

Lee et al. [30] published a further GWAS study in a population of Korean women with PCOS, according to the Rotterdam criteria. The discovery cohort consisted of 1000 PCOS cases and 1000 controls, with a replication study of 249 cases and 778 controls. Genotyping was performed using the HumanOmni1-Quad v1 array with 636,870 SNPs analyzed for association with PCOS. The discovery stage revealed 56 SNPs over 24 regions, of which 21 were analyzed in the replication stage using TaqMan technology. Twelve SNPs demonstrated the same direction of effect in the discovery and replication phases, of which one remained significant in decreasing the risk of PCOS for those carrying the variant rs10505648 at 8q24.2. This signal was located 487 kb upstream of *KHDRBS3,* which may be related telomerase activity, a pathogenetic mechanism explored in PCOS by Li et al. [31]. Several moderate associations with PCOS were shown in other loci. However, the small size of the replication cohort may have limited the ability of the study to detect an association. The authors were only able to analyze 10 of the 11 PCOS loci identified in the Chen et al. [26] and Shi et al. [28] studies due to a different genotyping method, and found that seven were associated with a consistent direction of effect at statistical significance. A follow-up pathway analysis utilizing the same dataset also identified variants in oocyte meiosis as the top-ranking pathway associated with PCOS [32].

Day et al. [33] have reported the largest GWAS study to date in PCOS with analysis of 5184 self-reported cases of PCOS with European ancestry and 82,759 controls and a replication set of clinically validated cases (1875 according to Rotterdam criteria and 861 according to NIH criteria) and 181,645 controls. Six independent signals were identified at genome-wide significance, of which four were novel and two were located within previously reported genes (*YAP1* and *THADA*). All six were independently associated with PCOS risk and were not associated with BMI, suggesting an impact on PCOS-specific pathways. The strongest novel signal was an intronic variant in *ERBB4/HER4,* with signals in *ERBB3/HER3* and *ERRB2/HER2* also nearing genome-wide significance. All those are members of the epidermal growth factor receptor family and may have a role in mediating LH-induced steroidogenesis in the ovary. Another variant rs11031006 near *FSHB* (encoding the FSH-specific beta subunit) reached genome-wide significance, adding further evidence to the importance of the FSH hormone or receptor gene variants in modulating the process of follicular maturation. The *FSHR* variant reported in the Shi et al. [28] study was detected but only weakly associated with PCOS in this population of Caucasian European women.

Other novel signals were near *RAD50* (a dsDNA break repair gene) and *KRR1* (a ribosome assembly factor) [33]. Similar to the previous Korean GWAS studies [29, 30], this European study found associations between 10 of the 11 reported variants in the Han Chinese population [28], of which six were significant but not at the genome-wide level. Two genes previously reported (*YAP1* and *THADA*) in the Han Chinese cohorts were identified as significant loci but with novel variants, and a further variant in *DENND1A* was not confirmed in the replication phase of the study. The authors also report an association between PCOS risk and delayed menopause in this cohort, suggesting a possible evolutionary mechanism for the persistence of PCOS risk alleles in the population.

Hayes et al. [34] undertook a discovery GWAS in 984 PCOS cases defined by NIH criteria and 2694 controls, followed by two replication studies in 1799/217 cases and 1231/1335 controls. Three loci associated with PCOS at genome-wide significance were identified: 8p32.1,

11p14.1, and 9q22.32. Of these, 8p32.1 was entirely novel and was near the *GATA4/NEIL2* genes, which have a role in steroidogenesis and the repair of DNA damage, respectively, as well as *FDFT1*, which is involved in the cholesterol-biosynthesis pathway. The 9q22.32 locus included the previously reported C9orf3 gene, with the most strongly associated SNP in this study (rs10993397) being independent of the association with rs3802457 reported in Shi et al. [28]. The 11p14.1 locus identified the rs11031006 variant in the *FSHB* gene reported in Day et al. [33] and in a quantitative trait analysis, demonstrated that this variant was strongly associated with LH levels. Again, seven of the 11 loci identified in the Shi et al. [28] study were confirmed, although the 8q24.2 locus identified by Lee et al. [30] was not found to be significant.

Overall, the contribution of the GWAS methodology to the understanding of PCOS has been moderate (Table 20.1). The recurrence of variants in the regions of gonadotropin or gonadotropin receptor genes across multiple studies confirms the importance of the hypothalamic-pituitary-ovarian axis in the development of PCOS. In addition, multiple regions with genes related to metabolism and cellular proliferation have been identified, in keeping with the hypotheses regarding insulin resistance, obesity, and the development of PCOS. Novel variants in genes such as *EGFR, DENND1A, THADA,* and C9orf3 have also appeared in diverse populations, suggesting a potential role for hitherto unexplored pathways and the possibility of new pharmacological interventions for PCOS. However, it must be remembered that the overall contribution of each of these variants and their effect on PCOS risk is very small, accounting for <10% of the overall ~80% heritability of the condition [35]. In addition, the variants identified are only effective in highlighting genomic regions of interest, and do not in themselves represent pathological variants. However, the confirmation of multiple loci across studies in different populations and using different diagnostic criteria has been a positive impact of the GWAS study era [13].

TABLE 20.1 Pathways and Genes Identified in PCOS GWAS Studies

Gene Family	Genes	Evidence From GWAS Studies
Gonadotropin function	*FSHR, FSHB, LHCGR*	+++
Insulin function, obesity	*INSR, THADA, DENND1A, RAB5B/SUOX*	+++
Steroidogenesis	*GATA4, FDFT1*	++
Epidermal growth factor receptors	*ERRB3/HER3, ERRB4/HER4, ERRB2/HER2*	++
Cell growth, DNA repair, telomerase activity	*C9orf3, YAP1, NEIL2, RAD50, KRR1, KHDRBS3, TOX3, SUMO1P1, ZNF217, HMGA2*	++
Reproductive tract development	*GTF2A1L*	+
Glycogen storage	*GYS2*	+

POST-GWAS STUDIES

The GWAS studies detailed above have provided a significant resource for researchers attempting to understand the pathogenesis of PCOS as well as those attempting to find or replicate population-specific variants. Numerous studies using the candidate gene approach have attempted to confirm the findings of the GWAS studies above in different populations. Goodarzi et al. [36] investigated the three loci from the original Chen et al. [26] GWAS in a population of European-derived PCOS patients defined by NIH criteria (939/535 cases and 957/845 controls across two cohorts). Variants in *DENND1A* and *THADA* were associated with PCOS; however, there was no evidence for association of the *LHCGR* variation. An Icelandic study replicated the 9q33.3 signals found in the Chen et al. [26] study, and was associated with the risk of hyperandrogenism in women without PCOS [37].

Studies involving the gonadotropin polypeptides and receptors have similarly confirmed involvement across multiple populations. Capalbo et al. [38] studied a Sardinian population with 198 PCOS patients and 187 controls, demonstrating a 2.0/2.7-fold risk with the heterozygous/homozygous S312N variant at the *LHCGR* locus. Almawi et al. [39] genotyped *FSHR* and *LHCGR* in 203 women with PCOS and 211 controls in a Bahraini population, demonstrating novel SNPs in both genes (rs7371084, rs4953616, and rs11692782) associated with PCOS risk. Ha et al. [40] demonstrated that variants in the *LHCGR* gene were associated with PCOS in Hui Chinese women, a separate ethnic group to the original Han Chinese study. However, numbers were small (151 cases and 99 controls) and the variants in *DENND1A* and *THADA* that had been identified in Chen et al. [26] were not found to be associated with PCOS in this population. Another study examined the novel loci identified in the European populations [33, 34] in a case-control cohort of Han Chinese, with 1500 PCOS cases and 1220 controls [41]. Marker SNPs in *ERBB4* again reached significance in the Chinese population, including after adjustment for BMI.

In addition, functional studies have added to our understanding of these risk loci and their differential expression in patients with and without PCOS. Jones et al. [42] studied methylation and mRNA expression in the regions surrounding the 11 risk loci identified in the Shi et al. [28] study, generating functional maps of adipose tissue gene expression in subjects with and without PCOS. Their finding of *LHCGR* overexpression in the adipose tissue of subjects with PCOS correlated with previous studies of granulosa and theca cell expression [43] and again suggests that variants in *LHCGR* identified in GWAS studies may have a key role in the pathogenesis of PCOS. Subjects with PCOS had lower *INSR* expression in adipose tissue, although cumulus cells of obese PCOS subjects had overexpressed *INSR*, suggesting that the sensitivity of ovarian tissue to insulin may differ from that in the periphery, allowing continuation of ovarian steroidogenesis.

Another study examining gene expression and DNA methylation in adipose tissue demonstrated significant expression differences in multiple genes, including those previously linked to PCOS (*RAB5B*) and those associated with insulin resistance, adipocyte size, and hyperandrogenism. However, this study did not find differences in *LHCGR* or *INSR* expression [44]. A study using differential in-gel electrophoresis analysis and mass spectroscopy of ovarian tissue in women with PCOS showed 18 differentially expressed proteins, including progesterone receptor component 1 and retinol-binding protein 1 (PGRMC1 and RBP1), which may serve as potential biomarkers to aid in identification of cases [45].

A functional study involving RNA sequencing of adipose tissue in subjects with PCOS before and after treatment with metformin demonstrated that one of the variants identified in the Han Chinese population (*THADA*, rs12478601) was related to a greater response to metformin [46]. The study also confirmed the relationship between variants near the *FSHB* gene and LH levels and the LH:FSH ratio. The findings suggest that susceptibility loci may not only have the ability to predict the likelihood of developing PCOS, but may also guide treatment based on pharmacogenomics.

Epigenetic studies have also attempted to explain features of PCOS and the association with metabolic syndrome. Zhao et al. [47] studied *PPARGC1A* promoter methylation and mitochondrial DNA (mtDNA) content in peripheral blood leukocytes of women with PCOS and concluded that there was a significant association with increased *PPARGC1A* promoter methylation and decreased mtDNA content with increasing metabolic risk. The contribution of dynamic changes in genomic expression to the phenotype of PCOS requires significant further exploration.

GENOMIC APPROACHES

The emergence of genomic technologies capable of whole exome and whole genome sequencing greatly increases the potential for the identification of pathogenic variants associated with PCOS. This allows the analysis of rare sequence variants rather than identification of associated variants (which by inclusion on a SNP array are by definition common population variants). A recent study by Gorsic et al. [48] used whole genome sequencing in 80 patients with PCOS according to NIH criteria and compared to 24 controls without PCOS. Three rare, putative functional coding variants were identified in the *AMH* gene in five of the women with PCOS, and were not identified in controls. Targeted resequencing of a larger PCOS cohort (643 cases and 153 controls) identified 21 additional rare coding variants in *AMH*, 18 of which were found only in patients with PCOS. Of these, 17 of the variants were shown to decrease AMH activity in a functional assay. The variants that were found in the control group did not have a functional impact. The findings also contrasted with previous metaanalyses of common variants in the AMH gene [49], suggesting that rare but strongly deleterious mutations may have a role in the pathogenesis of PCOS, demonstrating the potential for sequencing technologies in future gene discovery.

Another aspect of genome-wide analysis in future clinical practice will be the prospective identification of individuals at risk of PCOS based on known risk loci. A study by Lee et al. [50] analyzed a cohort of 862 PCOS and 860 control patients and assigned a genetic risk score based on the 11 susceptibility loci identified by Shi et al. [28]. The risk score was calculated by the addition of the number of susceptibility loci and was significantly higher on average in women with PCOS rather than controls. The odds ratio of having PCOS in the highest quartile of risk scores (>12) was 6.28 ($P < 0.001$), suggesting that the current loci had substantial predictive power. The genetic risk score for menopause may also be a surrogate marker for PCOS, with variants predicting late menopause also having an association with PCOS [51].

Cui et al. [51] analyzed the relative contribution of each of the SNPs identified in the Han Chinese population to the individual clinical features of PCOS, rather than PCOS as a syndrome. Individuals with oligo-anovulation only (746), hyperandrogenism only (278), and polycystic ovarian morphology only (536) were compared to 1790 healthy controls. Individual SNPs were identified to have a unique relationship with each phenotype, with the exception of rs4385527 in *C9orf3*, which had a relationship with all three phenotypes. Variants in *LHCGR* and *INSR* were associated with oligo-anovulation and variants in *THADA and DENND1A* with polycystic ovarian morphology. The identification of individuals at risk of PCOS using genotypic data may allow early lifestyle intervention and is a key goal of the genomic era, alongside understanding the underlying pathogenesis of the condition and the development of new therapeutic agents.

CONCLUSION

In many ways, PCOS represents an evolutionary puzzle: why is it such a common disorder, if the ultimate effect is a reduction in reproductive fitness? Many potential reasons for the prevalence of PCOS have been identified, including the beneficial impact of hyperandrogenemia in improving muscle strength and fitness in women, the potential for insulin resistance in previous fasting periods, the lengthened period of fertility associated with delayed menopause, and the reduced exposure to pregnancy events that likely had a high mortality in the evolutionary past [52]. It seems likely that the variants affecting our susceptibility to PCOS arose in prehistoric times, prior to the migration out of Africa [53].

Ultimately, it is the complexity of mammalian female reproductive endocrinology that predisposes one to a risk of suboptimal functioning across multiple pathways, and it is likely that individual susceptibility to PCOS will have contributions from multiple genes [54]. Although the role of key pathways such as insulin resistance and gonadotropin secretion and receptor function have been confirmed by genetic studies, there are a number of pathways with clear signals but that need further functional studies to understand their role in pathogenesis. In addition, the overall heritability of PCOS remains unexplained by current studies, with analysis of the 11 loci predicted in the Han Chinese population estimated to explain 2.4% of the variance in risk of PCOS [55]. The application of whole exome/genome sequencing technologies has the potential to identify informative variants at the level of both the individual and the population, promising significant advances in our understanding of PCOS.

References

[1] Azziz R, Carmina E, Chen Z, Dunaif A, Laven JSE, Legro RS, et al. Polycystic ovary syndrome. Nat Rev: Disease Primers 2016;2:16057.
[2] Rotterdam ESHRE/ASRM-Sponsored PCOS Consensus Workshop Group. Revised 2003 consensus on diagnostic criteria and long-term health risks related to polycystic ovary syndrome. Fertil Steril 2004;81(1):19–25.
[3] Cooper HE, Spellacy WN, Prem KA, Cohen WD. Hereditary factors in the Stein-Leventhal syndrome. Am J Obstet Gynaecol 1968;100(3):371–87.
[4] Amato P, Simpson JL. The genetics of polycystic ovary syndrome. Best Pract Res Clin Obstet Gynaecol 2004;18 (5):707–18.
[5] De Leo V, Musacchio MC, Cappelli V, Massaro MG, Morgante G, Petraglia F. Genetic, hormonal and metabolic aspects of PCOS: an update. Reprod Biol Endocrinol 2016;14:38.

[6] Franks S, Gharani N, Waterworth D, Batty S, White D, Williamson R, et al. The genetic basis of polycystic ovary syndrome. Hum Reprod 1997;12(12):2641–8.

[7] Norman RJ, Masters S, Hague W. Hyperinsulinemia is common in family members of women with polycystic ovary syndrome. Fertil Steril 1996;66(6):942–7.

[8] Legro RS, Driscoll D, Strauss JF, Fox J, Dunaif A. Evidence for a genetic basis for hyperandrogenemia in polycystic ovary syndrome. Proc Natl Acad Sci U S A 1998;95:14956–60.

[9] Jahanfar S, Seppala M, Eden JA, Nguyen TV, Warren P. A twin study of polycystic ovary syndrome. Fertil Steril 1995;63(3):478–86.

[10] Vink JM, Sadrzadeh S, Lambalk CB, Boomsma DI. Heritability of polycystic ovary syndrome in a Dutch twin-family study. J Clin Endocrinol Metab 2006;91(6):2100–4.

[11] Dumesic DA, Oberfield SE, Stener-Victorin E, Marshall JC, Laven JS, Legro RS. Scientific statement on the diagnostic criteria, epidemiology, pathophysiology, and molecular genetics of polycystic ovary syndrome. Endocr Rev 2015;36(5):487–525.

[12] Joseph S, Barai RS, Bhujbalrao R, Idicula-Thomas S. PCOSKB: a KnowledgeBase on genes, diseases, ontology terms and biochemical pathways associated with polycystic ovary syndrome. Nucleic Acids Res 2016;44: D1032–5.

[13] Jones MR, Goodarzi MO. Genetic determinants of polycystic ovary syndrome: progress and future directions. Fertil Steril 2016;106(1):25–32.

[14] Tucci S, Futterweit W, Conception ES, Greenberg DA, Villanueva R, Davies TF, et al. Evidence for association of polycystic ovary syndrome in caucasian women with a marker at the insulin receptor gene locus. J Clin Endocrinol Metab 2001;86(1):446–9.

[15] Urbanek M, Woodroff A, Ewens KG, Diamanti-Kandarakis E, Legro RS, Strauss JF, et al. Candidate gene region for polycystic ovary syndrome on chromosome 19p13.2. J Clin Endocrinol Metab 2005;90(12):6623–9.

[16] Ewens KG, Stewart DR, Ankener W, Urbanek M, McAllister JM, Chen C, et al. Family-based analysis of candidate genes for polycystic ovary syndrome. J Clin Endocrinol Metab 2010;95:2306–15.

[17] Xie G, Xu P, Che Y, Xia Y, Cao Y, Wang W, et al. Microsatellite polymorphism in the fibrillin 3 gene and susceptibility to PCOS: a case-control study and meta-analysis. Reprod Biomed Online 2013;26:168–74.

[18] Villuendas G, Escobar-Morreale HF, Tosi F, Sancho J, Moghetti P, San Millan JL. Association between the D19S884 marker at the insulin receptor gene locus and polycystic ovary syndrome. Fertil Steril 2003;79(1):219–20.

[19] Liu AL, Xie HJ, Xie HY, Liu J, Yin J, Hu JS, et al. Association between fat mass and obesity associated (FTO) gene rs9939609 A/T polymorphism and polycystic ovary syndrome: a systematic review and meta-analysis. BMC Med Genet 2017;18:89.

[20] Ketefian A, Jones MR, Krauss RM, Chen YI, Legro RS, Azziz R, et al. Association study of androgen signaling pathway genes in polycystic ovary syndrome. Fertil Steril 2016;105(2):467–73.

[21] Hickey T, Chandy A, Norman RJ. The androgen receptor CAG repeat polymorphism and X-chromosome inactivation in Australian Caucasian women with infertility related to polycystic ovary syndrome. J Clin Endocrinol Metab 2002;87(1):161–5.

[22] Peng CY, Xie HJ, Guo ZF, Nie YL, Chen J, Zhou JM, et al. The association between androgen receptor gene CAG polymorphism and polycystic ovary syndrome: a case-control study and meta-analysis. J Assist Reprod Genet 2014;31:1211–9.

[23] Wang F, Pan J, Liu Y, Meng Q, Lv P, Qu F, et al. Alternative splicing of the androgen receptor in polycystic ovary syndrome. Proc Natl Acad Sci U S A 2015;112(15):4743–8.

[24] Urbanek M, Legro RS, Driscoll D, Azziz R, Ehrmann DA, Norman RJ, et al. Thirty-seven candidate genes for polycystic ovary syndrome: strongest evidence for linkage is with follistatin. Proc Natl Acad Sci U S A 1999;96:8573–8.

[25] Guo R, Zheng Y, Yang J, Zheng N. Association of TNF-alpha, IL-6 and Il-1beta gene polymorphisms with polycystic ovary syndrome: a meta-analysis. BMC Genet 2015;16:5.

[26] Chen ZJ, Zhao H, He L, Shi Y, Qin Y, Shi Y, et al. Genome-wide association study identifies susceptibility loci for polycystic ovary syndrome on chromosome 2p16.3, 2p21 and 9q33.3. Nat Genet 2011;43(1):55–9.

[27] Olszanecka-Glinianowicz M, Banas M, Zahorska-Markiewicz B, Janowska J, Kocelak P, Madej P, et al. Is the polycystic ovary syndrome associated with chronic inflammation per se? Eur J Obstet Gynecol Reprod Biol 2007;133:197–202.

[28] Shi Y, Zhao H, Cao Y, Yang D, Li Z, Zhang B, et al. Genome-wide association study identifies eight new risk loci for polycystic ovary syndrome. Nat Genet 2012;44(9):1020–5.

[29] Hwang JY, Lee EJ, Go MJ, Sung YA, Lee HJ, Kwak SH, et al. Genome-wide association study identifies *GYS2* as a novel genetic factor for polycystic ovary syndrome through obesity-related condition. J Hum Genet 2012;57:660–4.

[30] Lee H, Oh JY, Sung YA, Chung H, Kim HL, Kim GS, et al. Genome-wide association study identified new susceptibility loci for polycystic ovary syndrome. Hum Reprod 2015;30(3):723–31.

[31] Li Q, Du J, Feng R, Xu Y, Wang H, Sang Q, et al. A possible new mechanism in the pathophysiology of polycystic ovary syndrome (PCOS): the discovery that leukocyte telomere length is strongly associated with PCOS. J Clin Endocrinol Metab 2014;99(2):E234–40.

[32] Shim U, Kim HN, Lee H, Oh JY, Sung YA, Kim HL. Pathway analysis based on a genome-wide association study of polycystic ovary syndrome. PLoS ONE 2015;10(8):e0136609.

[33] Day FR, Hinds DA, Tung JY, Stolk L, Styrkarsdottir U, Saxena R, et al. Causal mechanisms and balancing selection inferred from genetic association with polycystic ovary syndrome. Nat Commun 2015;6:8464.

[34] Hayes MG, Urbanek M, Ehrmann DA, Armstrong LL, Lee JY, Sisk R, et al. Genome-wide association of polycystic ovarian syndrome implicates alterations in gonadotropin secretion in European ancestry populations. Nat Commun 2015;6:7502.

[35] Dunaif A. Perspectives in polycystic ovary syndrome: From hair to eternity. J Clin Endocrinol Metab 2016;101 (3):759–68.

[36] Goodarzi MO, Jones MR, Li X, Chua AK, Garcia O, Chen YI, et al. Replication of association of *DENND1A* and *THADA* variants with polycystic ovary syndrome in European cohorts. J Med Genet 2012;49(2):90–5.

[37] Welt CK, Styrkarsdottir U, Ehrmann DA, Thorleifsson G, Arason G, Gudmundsson JA, et al. Variants in *DENND1A* are associated with polycystic ovary syndrome in women of European ancestry. J Clin Endocrinol Metab 2012;97(7):E1342–7.

[38] Capalbo A, Sagnella F, Apa R, Fulghesu AM, Lanzones A, Morciano A, et al. The 312N variant of the luteinizing hormone/choriogonadotropin receptor gene (*LHCGR*) confers up to 2.7-fold increased risk of polycystic ovary syndrome in a Sardinian population. Clin Endocrinol (Oxf) 2012;77:113–9.

[39] Almawi WY, Hubail B, Arekat DZ, Al-Farsi SM, Al-Kindi SK, Arekat MR, et al. Luteinizing hormone/choriogonadotropin receptor and follicle stimulating hormone receptor gene variants in polycystic ovary syndrome. J Assist Reprod Genet 2015;32:607–14.

[40] Ha L, Shi Y, Zhao J, Li T, Chen ZJ. Association study between polycystic ovarian syndrome and the susceptibility genes polymorphisms in hui Chinese women. PLoS ONE 2015;10(5)e0126505.

[41] Peng Y, Zhang W, Yang P, Tian Y, Su S, Zhang C, et al. *ERBB4* confers risk for polycystic ovary syndrome in Han Chinese. Sci Rep 2017;7:42000.

[42] Jones MR, Brower MA, Xu N, Cui J, Mengesha E, Chen YI, et al. Systems genetics reveals the functional context of PCOS loci and identifies genetic and molecular mechanisms of disease heterogeneity. PLoS Genet 2015;11(8) e1005455.

[43] Wang P, Zhao H, Li T, Zhang W, Wu K, Li M, et al. Hypomethylation of the LH/choriogonadotropin receptor promoter region is a potential mechanism underlying susceptibility to polycystic ovary syndrome. Endocrinology 2014;155:1445–52.

[44] Kokosar M, Benrick A, Perfilyev A, Fornes R, Nilsson E, Maliqueo M, et al. Epigenetic and transcriptional alterations in human adipose tissue of polycystic ovary syndrome. Sci Rep 2016;6:22883.

[45] Li L, Zhang J, Deng Q, Li J, Li Z, Xiao Y, et al. Proteomic profiling for identification of novel biomarkers differentially expressed in human ovaries from polycystic ovary syndrome patients. PLoS ONE 2016;11(11):e0164538.

[46] Pau CT, Mosbruger T, Saxena R, Welt CK. Phenotype and tissue expression as a function of genetic risk in polycystic ovary syndrome. PLoS ONE 2017;12(1):e0168870.

[47] Zhao H, Zhao Y, Ren Y, Li M, Li T, Li R, et al. Epigenetic regulation of an adverse metabolic phenotype in polycystic ovary syndrome: the impact of the leukocyte methylation of *PPARGC1A* promoter. Fertil Steril 2017;107:467–74.

[48] Gorsic LK, Kosova G, Werstein B, Sisk R, Legro RS, Hayes MG, et al. Pathogenic anti-Mullerian hormone variants in polycystic ovary syndrome. J Clin Endocrinol Metab 2017;102(8):2862–72.

[49] Pabalan N, Montagna E, Singian E, Tabangay L, Jarjanazi H, Barbosa CP, et al. Associations of polymorphisms in anti-mullerian hormone (*AMH* Ile49Ser) and its type II receptor *AMHRII* -482 A > G on reproductive outcomes and polycystic ovary syndrome: a systematic review and meta-analysis. Cell Physiol Biochem 2016;39:2249–61.

[50] Lee H, Oh JY, Sung YA, Chung HW. A genetic risk score is associated with polycystic ovary syndrome-related traits. Hum Reprod 2016;31(1):209–15.

[51] Cui L, Li G, Zhong W, Bian Y, Su S, Sheng Y, et al. Polycystic ovary syndrome susceptibility single nucleotide polymorphisms in women with a single PCOS clinical feature. Hum Reprod 2015;30(3):732–6.
[52] Fessler DM, Natterson-Horowitz B, Azziz R. Evolutionary determinants of polycystic ovary syndrome: part 2. Fertil Steril 2016;106:42–7.
[53] Casarini L, Brigante G. The polycystic ovary syndrome evolutionary paradox: a genome-wide association studies-based, in silico evolutionary explanation. J Clin Endocrinol Metab 2014;99(11):E2412–20.
[54] Joshi N, Chan JL. Female genomics: infertility and overall health. Semin Reprod Med 2017;35:217–24.
[55] Xu Y, Li Z, Ai F, Chen J, Xing Q, Zhou P, et al. Systematic evaluation of genetic variants for polycystic ovary syndrome in a Chinese population. PLoS ONE 2015;10(10)e0140695.

Further Reading

Saxena R, Bjonnes AC, Georgopoulos NA, Koika V, Panidis D, Welt CK. Gene variants associated with age at menopause are also associated with polycystic ovary syndrome, gonadotrophins and ovarian volume. Hum Reprod 2015;30(7):1697–703.

Glossary

Oligo/anovulation   Failure to achieve ovulation on a regular basis/failure to achieve ovulation; manifests as menstrual disturbance and subfertility.

Hyperandrogenism   Elevated levels of androgens measured either by clinical means (hirsutism, acne) or biochemical tests (increased testosterone, elevated free androgen index).

Polycystic ovarian morphology   Appearance of polycystic ovaries on transvaginal ultrasound, various definitions including increased number of peripheral small follicles and increased ovarian volume overall.

21

Genetics and Genomics of Recurrent Pregnancy Loss

Laura Kasak*, Kristiina Rull*,†, Maris Laan*

*Institute of Biomedicine and Translational Medicine, University of Tartu, Tartu, Estonia
†Department of Obstetrics and Gynecology, Women's Clinic of Tartu University Hospital, Tartu, Estonia

OUTLINE

Abbreviations

3′UTR	3′ untranslated region
aCGH	array comparative genomic hybridization
ALOX15	arachidonate 15-lipoxygenase
AP2	activating protein 2
ART	assisted reproductive technology
ASRM	American Society for Reproductive Medicine
ATF4	activating transcription factor 4
AURKB	aurora kinase B
bp	base pair
C3	complement component 3
CD74	CD74 molecule, major histocompatibility complex, class II invariant chain
CGB	chorionic gonadotropin beta
CNV	copy number variant
C-section	caesarean section
CTNNA3	catenin alpha 3
CYP	cytochrome P450
DNMT1	DNA methyltransferase 1
DYNC2H1	dynein cytoplasmic 2 heavy chain 1
E2F	transcription factor E2F
EGR1	early growth response 1
F2	coagulation factor II/prothrombin
F5	coagulation factor V/factor V Leiden
FISH	fluorescent in situ hybridization
G9aMT	methyl transferase G9a
GOLPH3	golgi phosphoprotein 3
GPX4	glutathione peroxidase 4
GSTM1	glutathione S-transferase mu 1
GSTT1	glutathione S-transferase theta 1
H3-K9	methylated lysine (K) residue at position 9 of histone H3
HCG	human chorionic gonadotropin
HIST1H1B	histone cluster 1 H1 family member B
HIST1H4A	histone cluster 1 H4 family member A
HLA	human leucocyte antigen
ICAM1	intercellular adhesion molecule 1
IFNγ	interferon gamma
IL	interleukin
IVF	in vitro fertilization
JAK2	Janus kinase 2

kb	kilo base pairs $=1000$ base pairs
KIR	killer-cell immunoglobulin-like receptors
KLF4	Kruppel like factor 4, transcription factor
LAPTM5	lysosomal protein transmembrane 5
MAP K3	mitogen-activated protein kinase 3
Mb	mega base pairs $=1,000,000$ base pairs
miRNA, miR	microRNA
MLPA	multiplex ligation-dependent probe amplification
mRNA	messenger RNA
MSR1	macrophage scavenger receptor 1
MTHFR	methylenetetrahydrofolate reductase
MTRR	5-methyltetrahydrofolate-homocysteine methyltransferase reductase
ncRNA	noncoding RNA
NGS	next-generation sequencing
NIPT	noninvasive prenatal testing
NOS3	nitric oxide synthase 3
NUP98	nucleoporin 98
PAPPA	pregnancy-associated plasma protein A
PCR	polymerase chain reaction
PDLIM1	PDZ and LIM domain protein 1
PDZD2	PDZ domain containing 2 gene
PGD	preimplantation genetic diagnosis
PGS	preimplantation genetic screening
PHLDA2	pleckstrin homology like domain family A member 2
PLK2	polo-like kinase 2
POC	products of conception
QF-PCR	quantitative fluorescent polymerase chain reaction
RNA-Seq	RNA sequencing
RPL	recurrent pregnancy loss
RT-qPCR	reverse transcription quantitative PCR
SDF	sperm DNA fragmentation
SLC16A2	solute carrier family 16 member 2
snoRNA	small nucleolar RNA
SNP	single nucleotide polymorphism
snRNA	small nuclear RNA
SNV	single nucleotide variant
SREBP	sterol regulatory element binding protein
SYCP3	synaptonemal complex protein 3
TGFB1	transforming growth factor, beta 1
TIMP2	tissue metallopeptidase inhibitor 2
TNF	tumor necrosis factor
TP53	tumor protein 53
TRAIL (*alias* TNFSF10)	TNF-related apoptosis-inducing ligand (TNF superfamily member 10)
VEGFA	vascular endothelial growth factor A
WES	whole exome sequencing
WGS	whole genome sequencing
XCI	skewed X-chromosome inactivation
ZF5	transcription factor ZF5
ZFP36	ring finger protein ZFP36

INTRODUCTION TO RECURRENT PREGNANCY LOSS

Recurrent Pregnancy Loss—Multiplicity of Definitions

Recurrent pregnancy loss (RPL) is a stressful condition that affects 1%–5% of couples aiming at childbirth [1]. Such patients should be considered as a high-risk obstetric population not only in the context of RPL management, but because of the special clinical attention also required during subsequent pregnancies. RPL mothers show a tendency toward preterm birth, a higher incidence of C-section, and postpartum hemorrhage. Indicative of suboptimal placental function, there is increased risk to develop second-trimester abortions, late-onset preeclampsia, placenta accreta/increta/percreta, fetal oligohydramnios, and early onset growth restriction [2].

RPL is a unique diagnostic entity as only recurrent independent events form the basis for clinical diagnosis. In Europe, RPL diagnosis is traditionally assigned to the couple or female with at least three pregnancy losses before the 20th gestational week [3], whereas American guidelines also include cases with two miscarriages [4]. The prevalence of risk factors associated with pregnancy losses as well as the prognosis for subsequent pregnancy are similar among couples with two and three previous losses [1]. Thus, equal clinical management should be offered for these couples regardless of the number and pattern (consecutive or nonconsecutive) of previous losses.

Research addressing risk factors and pathogenic mechanisms behind recurrent pregnancy loss classifies RPL cases into several clinical subgroups (Table 21.1). For example, discrimination of couples with primary (no previous live birth), secondary (losses after birth(s)), or even tertiary (losses between live births) RPL is used. Some studies include patients with self-reported losses and biochemical pregnancies (pregnancies documented only by a positive urine or serum hCG test) while other investigations have included only cases confirmed

TABLE 21.1 Subclassification Options for Diagnosis of Recurrent Pregnancy Loss

Parameter for Recurrent Pregnancy Loss Characterization	Subclassifications
Number of previous losses	Two or three
Gestational age at losses	Early = embryonic loss (<10 gestational weeks) Late = fetal loss (>10 gestational weeks)
Visualization or confirmation of losses	Nonvisualized = biochemical pregnancy Visualized/histologically confirmed = miscarriage
Series	Consecutive Nonconsecutive = interspersed with live births
Type	Primary = no live births Secondary = after live birth(s) Tertiary = between live birth(s)
Localization of pregnancies	Intrauterine Extrauterine, including unknown location
Partnership during losses	The same or a different partner

by histological or sonographic examination. There is no clear consensus on whether only women with pregnancy losses are patients or whether a couple can be considered as an RPL case, especially as the genetic causes are among the most prevalent ones (see below). As pregnancy loss is a psychologically devastating condition, the change of cohabiting partners is quite frequent among these patients, which further complicates the diagnostic workup and discovery of causes behind the pregnancy losses.

Although international consensus is still to be reached in the clinical practice and guidelines in defining RPL, there is an increasing support to initiate a specific clinical workup already after the second pregnancy loss of a couple.

Known Causes Behind Recurrent Pregnancy Loss

Accepting the 15% pregnancy loss rate, it can be calculated that the incidence of RPL (≥ 3 pregnancy losses) by chance would be about 0.35% [5]. Yet, the observed incidence of RPL in several populations is three times higher, indicative of predisposing factors [5, 6]. It is largely recognized that chromosomal abnormalities in the products of conception (POC) are the most common cause of all pregnancy losses, accounting for 70% of sporadic losses and approximately 40% of recurrent losses [7, 8] (see below). However, among RPL cases, the fraction of aborted fetuses with a normal karyotype increases almost linearly with the accumulating total number of miscarriages, pointing to other major causes that may lead to repeated loss of a chromosomally normal fetus (Fig. 21.1). There are several well-acknowledged risk factors and contributing clinical conditions that cause or predispose to RPL [3, 4, 8]. Inability to develop immunotolerance, unfavorable milieu to ensure suitable environment for implantation, early embryonic development due to anatomic and endocrine problems, attack by various known embryotoxic agents (e.g., infections), etc., may impair implantation and placental development while triggering biological mechanisms leading to expulsion of the fetus from the maternal organism (Fig. 21.2). Consequently, the current clinical workup offered to couples suffering from RPL includes testing for the presence of antiphospholipid antibodies, endocrine and uterine anatomical abnormalities, and parental karyotype aberrations. In some cases, testing of thrombophilia markers (e.g., Leiden V mutation) and additional immunological factors (e.g., thyroid peroxidase antibodies, antinuclear antibodies, and tissue transglutaminase antibodies) may be included in the clinical protocol. Today's consensus opinion is that gross chromosomal abnormalities and submicroscopic structural variants in either of the partners are responsible for ~10% of RPL cases [10, 11, 24] (Fig. 21.2). The main female-related factors contributing to RPL are anatomic abnormalities, including uterine malformations, fibroids, and endometrial polyps (10%–20%) and endocrine disorders such as subclinical hypo- or hyperthyreosis, hyperprolactinemia, and menstrual disorders (10%–20%). In 10%–20% of cases, various parental immunological factors such as antiphospholipid syndrome of the female partner are considered as critical contributors to RPL [9]. Infections in the genital tract in either of the partners are responsible for 0.5%–5% of RPL cases. However, due to variable diagnostic criteria applied in different centers to define the RPL couples, the reported prevalence of acknowledged risk factors for RPL varies to a great extent and represents only an approximate estimate. In clinical practice, a typical RPL couple is diagnosed simultaneously with several potential parental contributing factors, and the sole causal factor

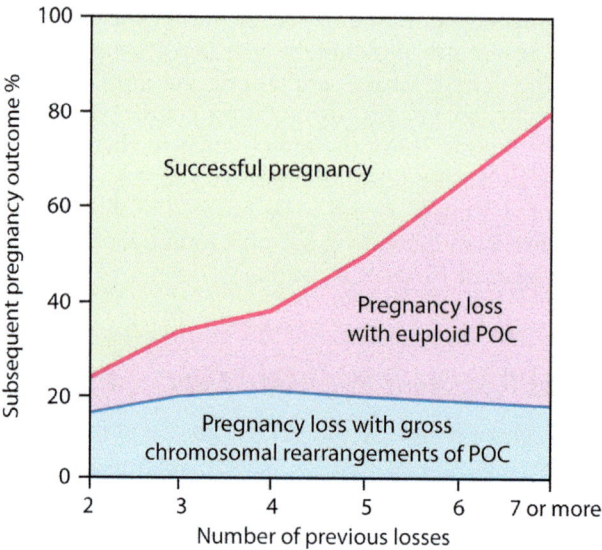

FIG. 21.1 Pregnancy outcome in RPL cases is dependent on the cumulative number of previous pregnancy losses and karyotypes of the products of conception (POC). A higher number of previous miscarriages is a negative predictor to reach a successful outcome in subsequent pregnancies. For RPL couples with seven and more pregnancy losses, the success rate in delivering a healthy newborn is only about 20%. As a main cause for pregnancy loss, an abnormal karyotype of the POC is a positive predictor for success in the next gestation as it has no direct effect on the viability of subsequent conceptuses. Notably, the overall proportion of POCs with aberrant karyotypes is not dependent on the number of previous pregnancy losses of the couple. In contrast, the proportion of pregnancy losses with normal karyotype of the POC is significantly increased with the number of previous miscarriages of the RPL couple. This points to some systematic (genetic or nongenetic) error during early development of the conceptus and/or an unfavorable environment at the maternal-fetal interface preventing the establishment of a viable pregnancy. *Modified from reference Ogasawara M, Aoki K, Okada S, Suzumori K. Embryonic karyotype of abortuses in relation to the number of previous miscarriages. Fertil Steril 2000;73(2):300–4.*

cannot be assigned. Even if the partner(s) carry gross chromosomal abnormalities, the main cause leading to the particular pregnancy loss may be different. After testing an RPL couple for all currently known parental risk factors, the proportion of idiopathic RPL represents 50%–70% of the cases [4]. However, when the clinical workup incorporates (molecular) karyotyping of the POC and information on the fetal chromosomal rearrangements as a causative factor for a miscarriage, approximately 25% of the RPL cases remain truly unexplained [11] (Fig. 21.2). There is a high probability that these families carry genetic variants in either or both of the partners predisposing to abnormal early embryonic development, impaired pregnancy establishment, and their series of pregnancy losses. Lifestyle (such as smoking and alcohol) and environmental factors may play an additional undefined role in the pathogenesis of pregnancy loss, but it remains challenging to estimate their effect.

Regarding the predisposing factors, the primary and secondary RPL cases may be considered as two different clinical entities. Immunological factors may play a greater role in women with losses after a live birth than in women who have never had a successful pregnancy [12, 13]. Maternal endocrine disorders or anatomic causes are more frequent among patients with primary RPL [14, 15]. Another study showed that couples with primary compared to

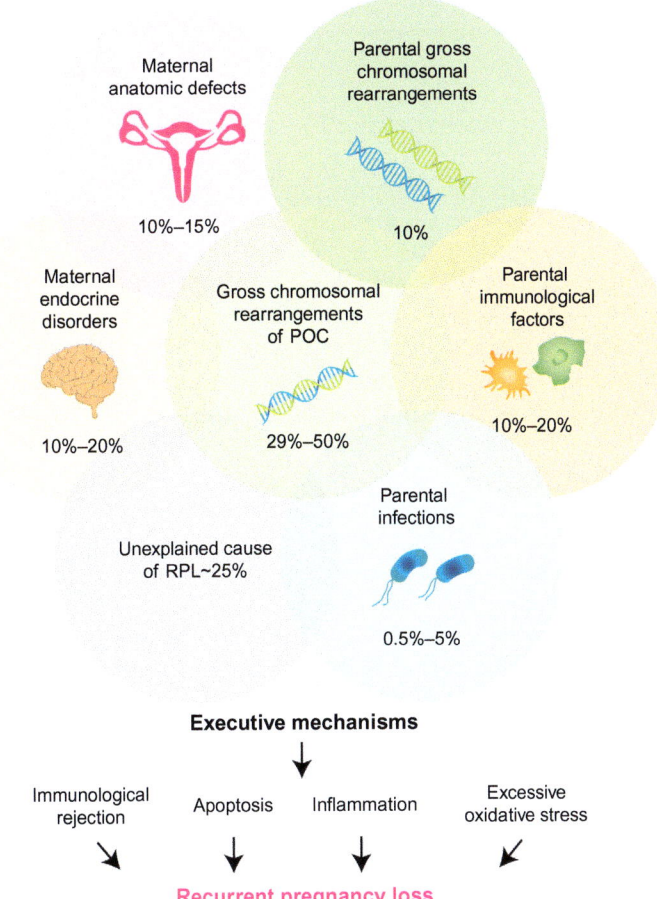

FIG. 21.2 Known causes for recurrent pregnancy loss (RPL). Known parental causative factors for RPL explain 30%–50% of the clinical cases. These include gross chromosomal rearrangements (aneuploidy, microdeletions/duplications, translocations, extensive CNV load, etc.), infections and immunological factors in either of the partners, maternal endocrine disturbances (hypo- or hyperthyreosis, hyperprolactinemia, and menstrual disorders), and anatomical defects (uterine malformations, fibroids, and endometrial polyps). Frequently, an RPL couple is diagnosed simultaneously with several potential parental contributing factors and the sole causal factor cannot be explicitly assigned. Whereas parental causative factors remain more or less constant across different gestations, each conceptus is different with its unique genome. In total, 29%–50% of the conceptuses of RPL couples have chromosomal abnormalities, indicating major disturbances in the core processes of genome replication, repair, and mitosis as well as cellular proliferation and differentiation during early development. Several parental risk factors may predispose to the chromosomal disturbances of the POC. Irrespective of the cause, the executive mechanisms initially defining the fate of the conceptus, initiating and/or accompanying already triggered pregnancy loss, are thought to be universal. Of all RPL cases, approximately 25% remain truly unexplained.

secondary RPL have lower aneuploidy rates and a higher proportion of morphologically normal embryos, indicative of other causes than defective meiosis [16]. Irrespective of the cause for the clinical condition, the prognosis of live birth rates in subsequent pregnancies appears to be similar for the primary and secondary RPL patients [15, 17].

ROLE OF CHROMOSOMAL REARRANGEMENTS IN RPL

Analysis of Parental Genomes

Today, a standard clinical evaluation of RPL cases includes parental karyotyping. The couples suffering from RPL carry major chromosomal abnormalities (mostly structural abnormalities) five to six times more frequently than a random subset of the general adult population [18]. Overall carrier frequency of chromosomal aberrations does not differ between the couples with consecutive miscarriages and RPL cases with nonconsecutive pregnancy losses interspersed with successful gestations [19]. Among all RPL couples, the proportion of cases with gross chromosomal abnormalities in either or both of the parents is 2%–4%, including asymptomatic carriers with balanced translocations or inversions [20, 21]. Although >70% of aborted conceptuses of these couples are with an abnormal karyotype, an appropriate clinical management results in the cumulative live birth rate of up to 60% across these RPL families [11, 21]. Due to the development of prenatal diagnostic methods, such as amniocentesis and chorionic villus sampling, karyotyping of the fetus has been offered for RPL couples with structural genomic abnormalities since the late 1980s [22].

Until recently, there has been insufficient attention to submicroscopic chromosomal variations that are often also referred to as DNA copy number variants (CNVs), in susceptibility to RPL. CNVs represent deletions, duplications, and complex rearrangements of chromosomal regions, and their size ranges from a few hundred bp to several Mb (Fig. 21.3A). The effect of these structural genomic variants on pregnancy failure depends on their size, type, and chromosomal localization as well as the number of involved genes critical to early pregnancy. Pericentromeric abnormalities were associated with RPL >10 years ago [23, 24]. Two recent independent studies have shown an increased prevalence of large pericentromeric and subtelomeric CNVs in the genomes of female and male partners of RPL couples [10, 25]. In addition, a positive correlation between an overall burden of CNVs and miscarriage risk has been reported [25, 26].

The critical role of CNV burden in the parental genomes of RPL couples underlines the importance of including microarray-based analysis of submicroscopic structural abnormalities in clinical management. Identification of all chromosomal disturbances in one or both of the parents possibly predisposing to RPL would enhance genetic counseling about the chances of transmission of the identified abnormality and the couples' risks for additional pregnancy losses as well as future reproductive choices.

Analysis of Products of Conception

Testing the products of conception (POC) is important to assist in defining the cause behind the pregnancy losses and improve counseling for couples about prognosis in a following gestation. Immediately after the chromosomal banding protocols were established in the early 1970s, cytogenetic methods were applied to study the karyotype of the aborted fetuses. A comprehensive set involving data from 1498 karyotyped spontaneous human abortions (< 12 gestational weeks) revealed that 61% had chromosomal abnormalities [7]. Since then, numerous studies focusing on aborted fetuses have demonstrated convincingly that gross

Structural variants are deletions, duplications or complex genomic rearrangements

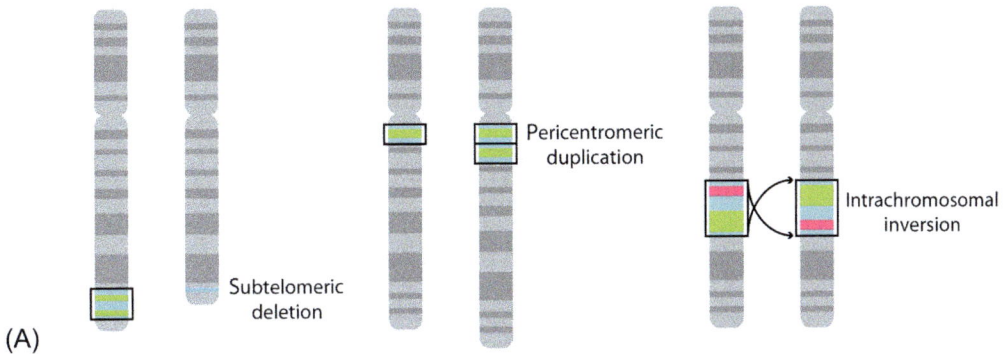

(A)

62 kb duplication of the genomic region involving the *PDZD2* and *GOLPH3* genes is enriched among RPL cases

RPL cases have substantially less placental CNVs compared to normal pregnancy

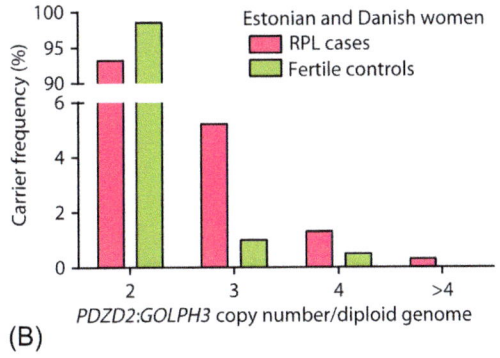

(B)

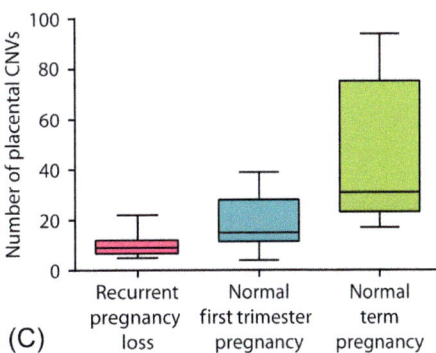

(C)

FIG. 21.3   Parental and placental copy number variants (CNVs) in predisposition to recurrent pregnancy loss (RPL). (A) Schematic illustration of some typical submicroscopic chromosomal rearrangements in RPL couples, potentially interfering with normal chromosome segregation, disrupting functional genomic loci, or leading to pathogenic gene copy number dosage, deletion, or duplication. (B) Copy number distribution and carrier frequency of the *PDZD2:GOLPH3* duplication (5p13.3; 61.6 kb) in Estonian and Danish female RPL patients ($n = 309$) compared to fertile women ($n = 205$). Normally, the genome harbors one copy of this segment on each chromosome and in total, there are two copies per diploid genome. Carriers of *PDZD2:GOLPH3* duplication on one or both chromosomes have three or more copies of this genomic segment per genome. The *PDZD2:GOLPH3* duplication carrier frequency was significantly enriched among RPL female patients. (C) Placental genomes of normal pregnancies (first trimester, $n = 9$; term, $n = 8$) harbor a substantial load of somatic CNVs that emerges during early gestation and accumulates over gestation. In contrast, a limited number of somatic rearrangements were detected in the placental samples of RPL cases ($n = 10$). *Part (B) was assembled using published data from reference Nagirnaja L, Palta P, Kasak L, Rull K, Christiansen OB, Nielsen HS, et al. Structural genomic variation as risk factor for idiopathic recurrent miscarriage. Human Mutat 2014;35 (8):972–82 and part (C) based on data reported in reference Kasak L, Rull K, Sõber S, Laan M. Copy number variation profile in the placental and parental genomes of recurrent pregnancy loss families. Sci Rep 2017;7:45327.*

chromosomal aberrations are the most common cause for both sporadic miscarriages (pooled prevalence from 13 studies, 45%, 95% CI: 38%–52%), and RPL (six studies 39%, 29%–50%) [27, 28]. It is well established that errors in mitosis and meiosis are very common among all women. It has been estimated that ~50% of oocytes are aneuploid [29] and ~30% of conceptuses are lost already before missed menstruation [30]. However, in a subset of RPL cases, an extremely increased rate of impaired chromosome segregation may point to systematic errors in the meiotic process.

Twenty years ago, the Euro-Team Early Pregnancy protocol for RPL (1995) recommended that if both partners of a couple experiencing two spontaneous miscarriages have a normal karyotype, cytogenetic investigation of the POC material should be performed in order to provide prognostic information and to assess the efficacy of treatment [6,31]. However, routine genetic testing of the POC has not been fully introduced into clinical practice until now due to challenges in obtaining the biological material suitable for culturing in laboratory conditions that is a prerequisite for karyotyping. Additional complexities in analyzing the genome of the POC are possible maternal tissue contamination, placental mosaicism, and polyploidy. Conventional karyotyping, the gold standard to detect chromosome abnormalities in pregnancy loss samples, determines the number, form, and size of chromosomes throughout the whole genome. Unfortunately, karyotyping of the POC fails for at least 20% of samples due to culture failure, maternal contamination, or poor quality of chromosomal preparations [32]. In addition, it is a laborious and time-consuming procedure.

The first methodological innovation was the development of targeted molecular assays to detect aneuploidy from uncultured material, such as fluorescent in situ hybridization (FISH), multiplex ligation-dependent probe amplification (MLPA), and quantitative fluorescent PCR (QF-PCR) (Table 21.2) [27]. Current advanced technologies, array comparative genomic hybridization (aCGH), and SNP microarrays enable genome-wide high-resolution profiling of microdeletions/duplications rapidly and cost effectively with a high diagnostic yield and low failure rate (Table 21.2) [27, 28, 33–36]. Compared to conventional karyotyping, microarray-based analysis reveals up to 10% additional pathogenic chromosomal rearrangements [27, 28, 35, 37, 38]. Similarly to RPL parental genomes, also the genomes of miscarried POCs are shown to be enriched for large pericentromeric and subtelomeric CNVs [27, 39]. However, the limitation of microarray-based analysis is the failure to detect balanced rearrangements and very low-level mosaicism (<10%–15%) [40]. An efficient laboratory protocol combining QF-PCR and aCGH has already replaced standard cytogenetic analysis in testing of miscarriage tissues in several clinical centers [36, 41]. The most advanced available technology, next-generation sequencing (NGS), exhibits not only single base pair resolution but also the highest sensitivity to detect mosaicism. A study using the NGS platform on embryos previously deemed euploid demonstrated that 32.6% and 5.2% of all analyzed cases were, in fact, mosaic and polyploid, respectively [42].

In case of a euploid POC, a complete clinical workup has to be assigned for the RPL couple [34–36]. Detection of a genomic structural abnormality in the POC as a causal factor for miscarriage excludes the need for additional diagnostic tests, relieving stress while saving money and time. An example is a recent report of a euploid male (XY) POC with maternally inherited 3.5 Mb X-chromosomal microdeletion encompassing 13 developmentally important genes [35]. The couple was counseled to consider preimplantation genetic diagnosis in order to avoid

TABLE 21.2 Overview of Methods Used for Genetic Testing

Method [Resolution]	Advantages	Disadvantages
Karyotyping Microscope [3–5 Mb]	Gold standard High specificity	High cost Time-consuming Microdeletion/duplication syndromes not detectable Chance of culture failure or maternal contamination
aCGH and SNP-microarrays [500 bp]	Fresh or preserved material High resolution Detection of microdeletion/ duplication syndromes Detects maternal contamination Quick results	Balanced rearrangements not detectable
WES, WGS [1 bp]	Highest resolution Detects all kinds of abnormalities Detects mosaicism Very low false positive rate	High cost
FISH Microscope [10 kb-1 Mb]	Performed on interphase cells, eliminates the need for culture Detects polyploidy and aneuploidy Quick results	Limited resolution Detects abnormalities only where probes are designed Moderate false positive rate
QF-PCR [200 bp]	Detection of maternal contamination Locus-specific method, high sensitivity to target Quick results	Detects abnormalities only where the test is designed (usually chromosome 13, 18, and 21)
MLPA [50 bp]	Targets a large number of small DNA sequences at once Low cost Quick results	Mosaicism, polyploidy and balanced rearrangements not detectable Maternal contamination not detectable

aCGH, array comparative genomic hybridization; *FISH*, fluoresence in situ hybridization; *MLPA*, multiplex ligation-dependent probe amplification; *QF-PCR*, quantitative fluorescence PCR; *WES*, whole exome sequencing; *WGS*, whole genome sequencing.
Modified from van den Berg MM, van Maarle MC, van Wely M, Goddijn M. Genetics of early miscarriage. Biochim Biophys Acta 2012;1822 (12):1951–9.

transmission of the deletion-carrying chromosome. In several cases in clinical decision-making, it would be informative to consider chromosomal profiling of the POC not only from index gestation, but also from the previous miscarriages of the couple, if available.

NOVEL MOLECULAR AND CLINICAL INSIGHTS FROM MICROARRAY-BASED STUDIES

Novel Molecular Insights From the Analysis of CNVs in RPL Families

Availability of innovative high-resolution methodological approaches and attention to a more detailed clinical history of RPL families have brought several novel molecular insights regarding the role of submicroscopic genomic rearrangements in early pregnancy success.

Analysis of the genes underlying the CNVs detected in parental genomes and POC samples has shown an enrichment of immune system-related pathways, for example, allograft rejection, complement cascade, antigen binding, and presentation [25, 26, 43]. In addition, CNVs detected in RPL tissue material compared to normal pregnancies were shown to lack enrichment in genes regulated by transcription factors critical in early embryonic development and regulation of basic cellular processes, such as *E2F*, *ZF5*, *KLF4*, *AP2*, and *SREBP* [44]. If disrupted by a CNV, the resulting haploinsufficiency of genes regulated by these essential factors could lead to developmental defects. In perspective, these pathways represent a great source for new biomarkers for the evaluation of fetal health and the early detection of pregnancy complications.

A few recurrent CNVs have been identified as associated with RPL risk, highlighting novel genes and mechanisms possibly implicated in RPL development. A 60 kb duplication encompassing the *GOLPH3* and *PDZD2* genes was detected with fivefold higher prevalence among RPL females compared to fertile women (Fig. 21.3B) [26]. GOLPH3 has a regulatory role in Golgi trafficking and has been associated with tumor progression. The analysis of CNV profiles of miscarriage material has also identified loci disrupted in several POC samples. For example, RPL chorionic villi harbor deletions within the *CTNNA3* gene [10, 45] encoding a cell-cell adhesion molecule, shown to restrain the invasive properties of trophoblastic cells and possibly linked to preeclampsia [46]. Deletions in RPL samples were also shown to encompass the *NUP98* gene with a distinct role in human embryonic stem cell differentiation [47]. Two cases have been reported with duplications interrupting the *MTRR* gene, encoding an enzyme involved in folate metabolism, and associated with preeclampsia, spontaneous abortion, and idiopathic RPL in several independent studies [10]. A duplication disrupting the *TIMP2* gene was shown to directly reduce its placental expression levels [48]. Notably, *TIMP2* is mainly expressed in the placenta and functions in extracellular matrix homeostasis.

Compared to other somatic tissues, a specific feature of the placental genome is an excessive load of somatic genomic rearrangements [49, 50]. The phenomenon appears to be a common hallmark of human and murine placental genomes in normal gestations. The accumulated evidence has led to the hypothesis that similar to cancer genomes, these placental de novo chromosomal rearrangements encompassing genes involved in cellular proliferation and trophoblast invasion may facilitate successful implantation [49]. Consistent with this hypothesis, the analyzed placental genomes of RPL cases exhibited twofold fewer CNVs compared to uncomplicated first-trimester pregnancies with similar gestational age (first-trimester elective terminations) and this difference mainly arose from the lower number of de novo duplications (Fig. 21.3C) [10]. It supports the idea of impaired promotion of somatic structural variation in placental genomes as an additional potential contributor to pregnancy loss. Importantly, a reduced number of somatic CNVs was also reported for term placentas of several late pregnancy complications, such as preeclampsia, gestational diabetes, fetal growth restriction, and macrosomia [49].

Hypothesis of "Unfavorable Genomes": Link Between Miscarriage Risk and Developmental Disorders in Live Born Offspring of the RPL Families

In 1993, Khoury and Erickson showed that stillbirths and spontaneous abortions have higher rates of birth defects than live born infants [51]. It has been well established already that a large fraction of early aborted embryos from RPL cases have developmental defects, irrespective of the presence or absence of karyotype abnormalities [16, 52]. These include severe disturbances of growth and morphogenesis (e.g., embryonic growth disorganization) that have a low likelihood of survival after the first half of the pregnancy, but also defects that are observed among live born infants, for example, craniofacial, limb, neural tube, or heart defects. As the RPL families also show a higher prevalence of developmental defects in their live born children, it is indicative of some systematic errors in the early development of their conceptuses. The risk of serious birth defects was estimated to be 4.2% for infants of women with three or more pregnancy losses compared to 2.5% for the cases with no prior miscarriage [51]. It has also been reported that among the mothers with a *spina bifida* offspring, 24.4% reported a history of pregnancy loss, suggesting certain systematic impairment in embryogenesis [53]. A recent large-scale study of 242,187 newborns also demonstrated association between maternal history of RPL (5% of the study group) and long-term neurological morbidity, developmental, and movement disorders of the offspring [54].

It was recently suggested that a fraction of RPL cases may carry "unfavorable" genomes predisposed to severe developmental disturbances in their conceptuses [10]. Detailed analysis of clinical histories of families with parental chromosomal microdeletions or microduplications has revealed multiple fetal malformations in several pregnancy losses as well as in their live born children. In these families, an aberrant CNV profile in one or both of the partners apparently represents a risk factor for both pregnancy loss and fetal developmental disturbances in gestations reaching delivery, most probably executed via major disturbances in chromosome segregation in early development. A large and/or extensive number of CNVs, especially in pericentromeric and subtelomeric regions [10, 23, 39], may affect correct chromosome pairing in meiosis and mitosis, thereby causing further genomic instability and reduced potential to deliver a viable and healthy offspring [55]. Notably, the effect of genomic structural variants in maternal and paternal genomes is similar in regard to the couple's risk to RPL [56].

As an example of an unfavorable genome variant, 15q11.1-q11.2 microduplication was detected in 30% of the POC samples from euploid spontaneous abortions, compared to 2.4% in the general population [39]. An independent study reported an RPL family with paternal 500-kb microduplication at 15q11.2, between recurrent breakpoints for chromosomal rearrangements, termed as BP1-BP2 [10]. The reproductive history of this family showed variable pregnancy outcomes, including a high number of miscarriages, one healthy newborn, and one offspring with severe developmental delay and intellectual disability. Although 40% of the 15q11.2 BP1-BP2 microduplication carriers suffer from delayed development and speech, autism, and other neurobehavioral problems, phenotypically normal carriers have been identified in several instances, like the male partner in this clinical case. However, the carrier status of this large CNV appears to disturb the establishment of a healthy conceptus in the next generation.

As an interesting observation from a recent study [10], two couples diagnosed with primary RPL eventually achieved successful pregnancies, but with a delivery of two macrosomic newborns in both cases (from 4.5 to >5 kg). This suggests that in some conceptuses, compensatory mechanisms (e.g., through epigenetic programming) may be activated that enable to overcome deficiencies in early fetal and placental development, and to maintain the pregnancy in RPL cases with "unfavorable genomes". However, this may lead to placental malfunction and, in turn, to fetal overgrowth.

Further retrospective and prospective studies are needed to provide knowledge on how the combined effect of the genomic composition of RPL couples and their POCs may modulate the recurrence risk of subsequent miscarriages and predisposition to severe developmental disorders in live born offspring. In addition to large structural variants, pathogenic germline single nucleotide variants (SNVs) in critical developmental genes may also predispose to recurrent cases of aberrant embryonic development [57]. The accumulating knowledge on parental "unfavorable genomes" is expected to have clinical implications for pregnancy counseling in individual RPL cases and, in general, pathogenic implications related to birth defect etiology.

ROLE OF SINGLE NUCLEOTIDE VARIANTS

Design of Genetic Association Studies of RPL

Apart from karyotype analysis with the main objective being the detection of large-scale sporadic genomic rearrangements predisposing to miscarriage, studies on parentally inherited genetic risk factors for RPL have been challenging to design, perform, and interpret. What is the reported evidence for familial predisposition to RPL? The first evidence for familial aggregation of RPL was reported in 1988, claiming that 16% of women with primary RPL had a family history of RPL and a smaller number of siblings compared to controls with no pregnancy loss [58]. A few years later, an independent study showed a two- to sevenfold increased prevalence of pregnancy losses among first-degree blood relatives of women suffering from RPL [59]. Population-based register studies have shown that the overall frequency of miscarriage among the siblings of idiopathic RPL patients is approximately doubled compared to the general population [60].

The focus of genetic research in RPL has been on the discovery of DNA variants, which may directly or indirectly predispose to or cause miscarriage and have potential implications in diagnostics and clinical management. There are two broad options to investigate the inheritable component of human disease—family-based linkage studies and genetic association studies comparing unrelated cases and controls. A genome-wide linkage scan using sibling pairs with unexplained RPL suggested four genomic regions in chromosomes 3, 6, 9, and 11 [60]. However, no risk genes or mutations have been fine-mapped for any of these loci, apparently due to the heterogeneity of the genetic component. RPL may have a monogenic cause in individual families, but the susceptibility genes and variants are probably different from family to family, causing genetic heterogeneity and disturbing classical linkage analysis in pedigrees [60]. As the existence of a universal high-penetrance genetic mutation causing RPL is not probable, alternative approaches for the genetic analysis of RPL families, such as whole exome sequencing (WES) with high information content, are expected to lead to a higher success rate compared to linkage studies.

For multifactorial disorders with adult onset such as RPL, case-control genetic association studies represent an alternative option. So far, most of the case-control studies addressing the genetic predisposition to RPL have involved affected females and, less frequently, both partners of RPL couples. Women with one or more live births have been typically included as a control group. However, delaying childbirth until older reproductive age, the use of contraceptive methods after birth of one or two planned children, and a change of cohabitant partners are frequent limitations for recruitment of an ideal control group. Ideally, the controls should be exposed to a similar number of pregnancies at the same age as the affected RPL women. Thus, it is challenging to design and conduct a case control study with unbiased methodology. Even if it can be done, the study would still be underpowered due to a limited number of patients available for recruitment.

In addition, each pregnancy and each conceptus has a unique genome with its distinctive epigenetic settings, modulated by the efficiency of established maternal-fetal interface, maternal nutrition and health, and other nongenetic confounders. To study the genetic component of RPL, the most optimal approach would be the genetic analysis of the entire RPL family, incorporating both partners of a clinically well-described RPL couple as well as their aborted POCs and live born children. This would allow directly addressing the combined effect of paternally and maternally inherited genetic factors on both successful and failed fetal development and placental function.

Association Studies of RPL Point to a Moderate Role of Common SNVs

Genetic association studies of RPL have been mostly designed as hypothesis-based candidate gene analyses using either Sanger sequencing or genotyping of targeted single nucleotide variants (SNVs). A typical genetic association study of RPL has investigated female patients. However, the definition of RPL and the genetic models used in association testing have been extremely variable across studies [61–63]. Over 100 genes and approximately 500 SNVs have been examined in the research literature. The majority of the case-control studies have targeted genes involved in maternal physiological response to the pregnancy or previously linked to the executive mechanisms leading to abortion. The placental proteins encoded by the fetal genome have a direct influence on pregnancy success and have also been investigated as suitable candidates for genetic studies of RPL. Thus, the most frequently analyzed genes in RPL case-control association studies conducted to this date are involved in immune tolerance (e.g., *HLA*- and *KIR*-genes) and inflammation (e.g., *IFNγ*, *TNF*, *TGFB1*, *IL*-genes), blood coagulation (e.g., *MTHFR*, *F2*, *F5*, *VEGFA*), and placental development and function (e.g., *PAPPA*, *TP53*, *NOS3*, *hCGbeta* locus). SNVs in genes related to hormonal and detoxification systems have seldom been analyzed (e.g., *GSTT1*, *GSTM1*, *CYP*-genes).

Three recently published large systematic reviews came to a consensus conclusion that the majority of nearly 500 common SNVs analyzed have no or only modestly increased risk to develop RPL [61–63]. Almost two-thirds of these variants have so far been reported by only one study and therefore most likely represent sporadic associations arising by chance. Two metaanalyses published in 2017, targeting 124/36 SNVs in 73/16 genes, reached nominal significant P-values (<0.05) for 43%–58% of SNVs in 37/13 genes, respectively [62, 63]. However, the majority of the estimated odds ratios ranged from 0.5 to 2.0 and most of the reported associations did not remain statistically significant after more stringent inclusion criteria of patients and controls. Only five common variants in four genes (*FII*, *FV*, *MTHFR*, *NOS3*), which had been reported in at least five independent studies, were significantly associated with RPL in both metaanalyses. These include three of the most popular clinically tested variants in RPL patients: thrombophilia-associated factor V (Leiden factor) mutation (rs6025), factor II (prothrombin) G20210A mutation (rs1799963), and *MTHFR* C667T variant (rs1801133). As up to 30% of women with uncomplicated obstetric histories also carry one of these variants, their prognostic value in RPL cases is still questionable and routine testing of these SNVs in clinical practice is no longer recommended [34, 64]. The positive associations detected in metaanalyses may indicate publication bias as studies with positive findings are more likely to be made publicly available and a large amount of research with nonsignificant associations is left unpublished. Additionally, there is significant variation between studies in the definition of RPL as well as in selection criteria for patient and control groups, all of which may lead to controversial results. Also, research is conducted in diverse populations, making it difficult to compare the outcomes.

MicroRNAs, key posttranscriptional regulators of gene expression, have become a newly discovered focus area in the search for risk factors for RPL. For example, the expression of *HLA-G* implicated in RPL was shown to be modulated by an SNV localized within the 3′ UTR region (*HLA-G*+3142C>G). This variant demonstrated allele-specific affinity to microRNAs miR-148a, miR-148b, and miR-152, and consequently a differential rate of mRNA degradation and translation suppression processes [65]. However, although parental SNVs in genes encoding miRNAs have been linked with predisposition to RPL, the estimated effect sizes are modest and there is lack of replication of claimed associations in independent clinical sample sets (e.g., [66–68]). Furthermore, applying NGS-based microRNA sequencing to profile the whole miRNome in chorionic villi from RPL cases did not detect any large disturbances in miRNA expression compared to elective first-trimester terminations [44].

Currently, none of the studied SNVs have been proven to display sufficient diagnostic and prognostic value with additional benefit in the clinical management of RPL couples. One scenario is that suboptimal functioning of a single gene can be overcome by compensatory mechanisms. Thus, testing the carrier status of SNVs with a moderate risk effect provides only limited clinical information. An alternative option is that the genetic variant is rare and the conducted clinical follow-up studies have not had adequate statistical power. An example is a missense variant V617F in the *JAK2* gene that has an allele frequency of 0.2% among normal primigravidas compared to 1.06% in RPL cases (odds ratio: 5.33) [69]. The third likely scenario is that SNVs with a substantial effect on RPL predisposition either on the maternal/paternal or fetal side have not yet been discovered. Chromosomal abnormalities are the most common cause of pregnancy loss. However, variants within genes involved in chromosome segregation, recombination, and cell division have been rarely studied in association with RPL (e.g., *AURKB*, *PLK4*, *SYCP3* genes [57, 70, 71]). Functional variants in genes responsible for embryonic and placental development, gene transcription, and regulation of biological processes represent an additional attractive source for future genetic association studies of RPL [10, 44].

Performing a genome-wide association study for RPL would be an ideal approach, but its design is challenging due to the reasons highlighted above. The development of guidelines for genetic association studies in RPL is desperately needed to make comparisons between research findings straightforward.

Exome Sequencing—A Potential Tool to Uncover Mutations With Strong Effect

Whole-exome sequencing (WES) is a state-of-the-art method that captures all genetic variants within the protein-coding region of the genome; this also includes exon-intron boundaries potentially harboring splice site mutations. Mutations with pathogenic effect on the encoded protein represent both missense and loss-of-function variants. To date, only two WES studies have been performed in relation to RPL. Qiao et al. targeted seven miscarriage samples derived from four idiopathic RPL families [72]. In two families, compound heterozygous missense mutations were discovered in both miscarriages affecting *ALOX15* and *DYNC2H1* genes, respectively. Both genes are important in early fetal development. Quintero-Ronderos et al. applied WES on 49 unrelated women affected by RPL and identified 27 potentially deleterious coding variants in 22 genes [73]. Similar to the pathways detected in the CNV analysis of parental genomes and POC (see above), these genes are also involved in biological processes related to immunological function modulation, angiogenesis, coagulation, cell adhesion, and proliferation.

Further studies still have to uncover whether idiopathic RPL families carry individual mutations with strong effect or a pool of deleterious mutations jointly contributing to RPL. Without such data, it is impossible to estimate the potential role of WES in molecular diagnostics and clinical decision-making for RPL families.

GENETIC TESTING IN RPL MANAGEMENT: CURRENT GUIDELINES AND DISCUSSION POINTS

According to current recommendations, clinical management of RPL couples includes parental karyotyping as the first priority genetic test [3, 4]. As findings of large chromosomal abnormalities are rare (2%–4% of couples [20, 21]) and the cumulative live birth rate among carriers is high without any interventions, the added value of this conventional approach is fairly limited. Therefore, a broader concept combined with a more personalized selection of genetic tests should be applied according to the couples' needs (Table 21.3). The emerging critical burden of CNVs in the RPL parental and POC genomes indicates a necessity to include high-resolution sensitive microarray-based analysis of submicroscopic rearrangements into routine clinical management to improve counseling about the risks for additional pregnancy losses and future reproductive choices (see details above). To decrease the proportion of unexplained RPL cases, genetic analysis of the POC should be performed already after a second pregnancy loss. Improved clarity in the diagnosis will enable the obstetrician to offer better targeted psychological support to the affected couple, if needed, and plan the next steps in clinical management faster. For example, detection of CNVs or unbalanced translocations with uncertain clinical significance warrants an extended genetic analysis of both parents and also any live born children in order to improve the accuracy of prognostic estimates for future gestations and to possibly opt for additional preconception and prenatal molecular diagnostics. The use of preimplantation genetic diagnosis (PGD) may be offered in such cases. However, for RPL couples with a euploid miscarried POC, broader clinical analyses (e.g., hormonal, immunological, anatomical, and andrological assessment as well as lifestyle factors) are relevant.

IVF-preimplantation genetic screening (PGS), which allows selection of embryos for transfer based on genetic criteria, is increasingly used for unexplained RPL patients due to the prevalence of fetal aneuploidies in early miscarriages. The primary goal of IVF-PGS is to

TABLE 21.3   Genetic Analysis of Recurrent Pregnancy Loss Cases

Characteristics, Clinical Value, and Options of Genetic Testing	Analyzed Genomic Material	
	Maternal and Paternal Blood	Products of Conception
Time of testing	Any time	Immediately at pregnancy loss
Sampling procedure	Minimally invasive	Invasive surgical procedure
Contamination	None	Often by maternal cells
Role of other factors[a]	Strong	Limited
Effect of genetic factor	Identical for all conceptuses	Unique for every case
Clinical information	Recurrence risk assessment	Detects affected mechanisms
Relevance to index loss	Moderate to strong	May be causal
Prognostic value for next pregnancies	Moderate to strong	Low
Current universally applicable genetic tests for RPL cases	Conventional karyotyping using cultured cells Aneuploidy tests for single chromosomes using genomic DNA (MLPA, QF-PCR) or uncultured cells (FISH)[b] Profiling of genome-wide chromosomal aberrations using genomic DNA analysis on microarrays (SNP or aCGH)	
Genetic tests in families[c]	Exome sequencing[d]	
Targeted epigenetic tests	XCI in maternal tissues[e] SDF index[e]	Epigenetic profiling of targeted loci, for example, imprinted genes[d]
Perspective genetic tests in clinical practice	*All RPL cases:* Targeted gene panels for sequencing or SNP genotyping of known RPL loci; provides also CNV information[f] *Single families:* WGS providing both high-resolution information on chromosomal rearrangements and single gene mutations	

[a] *Role of modifying factors for the penetrance of genetic variants, such as parental age, life style, general health history and index pregnancy course.*
[b] *QF-PCR is preferred in clinical practice due to easy test, high and reliable success rate, low cost.*
[c] *Definition includes parents, miscarriage conceptuses and live born children; relevant to the families with the history of adverse pregnancy outcomes.*
[d] *Currently only basic research.*
[e] *Diagnostic/prognostic value is still under debate as the research data is controversial.*
[f] *Currently not yet established due to limited data, but with high future potential due to straightforward standardization, cost-effectiveness and clear interpretation.*
aCGH, *array comparative genomic hybridization;* CNV, *copy number variant;* FISH, *fluorescent in situ hybridization;* MLPA, *multiplex ligation-dependent probe amplification;* QF-PCR, *quantitative fluorescent PCR;* SDF, *sperm DNA fragmentation;* SNP, *single nucleotide polymorphism;* WGS, *whole-genome sequencing;* XCI, *skewed X-chromosome inactivation.*

decrease the number of pregnancy losses and increase the live birth rate. However, the benefit of this procedure in idiopathic RPL couples is unconvincing in terms of the clinical outcome and cost effectiveness. Two recent studies have shown that the IVF-PGS strategy has a 3.5 month longer time to pregnancy, a similar or even lower live birth rate, and a cost that is 100-fold more expensive compared to expectant management [74, 75]. IVF-PGS has a lower clinical miscarriage rate, which could decrease the emotional burden associated with miscarriages. However, there is limited data on the emotional impact of a failed IVF cycle. In addition, there is not enough information yet if time to live birth (the ultimate goal) could be decreased compared to expectant management. IVF-PGS applying aCGH resulted in an ongoing pregnancy (≥20 gestational weeks) rate of 70% in the most optimistic studies [76]. Without this invasive management, the probability of a successful spontaneous pregnancy ending with a live birth of a healthy newborn within 5 years after the first consultation is also near 70% among RPL couples [77]. Therefore, due to lack of evidence, the current guideline of the ASRM still recommends expectant management with supportive care for patients with idiopathic RPL [2]. However, PGD may be indicated in a small percentage of RPL couples with known single-gene disorders or translocations.

Due to inconsistent outcomes and low clinical value, current guidelines have dropped [4] or are dropping [3] the recommendation to test common SNVs associated in published research with the risk to RPL. However, in the era of WES and WGS, more data on familial rare mutations are expected to emerge during the next 5–10 years. Identifying the core genes enriched in mutations among RPL patients would allow assembling a relevant clinical gene panel for routine application in the management of RPL patients. Similar actions are already ongoing in other areas of reproductive medicine, such as the development of a male infertility gene panel [78]. The development of a diagnostic/prognostic gene panel for cost-effective screening of SNVs predisposing to pregnancy losses should be an ultimate long-term aim to improve the quality of personalized counseling of RPL couples.

APPLICATION OF "OMICS" TOOLS FOR RPL RESEARCH

With rapidly evolving technological advances in various fields of "omics," there are several appealing perspectives in the biomolecular research of RPL with two major objectives. First, understanding genome and cellular level alterations behind failed implantation, fetal lethality, and miscarriage will provide improved knowledge about the pathogenesis as well as molecular mechanisms and biological pathways triggering or being involved in a failed pregnancy. This may assist in identifying potential therapeutic targets to prevent pregnancy loss in clinical cases with a euploid POC and a possible maternal causative factor for RPL. Second, pinpointing candidate genes with aberrant function in RPL represents a valuable source for generating mouse models to investigate in vivo treatment options for RPL, for screening genetic variants predisposing to pregnancy failure, and for developing biomarkers to monitor early pregnancies in high-risk cases (e.g., pregnancies achieved through ART approaches).

As the genomics of RPL were already addressed in detail above, in this chapter the term "omics" refers mainly to studies targeting the epigenome, transcriptome, and proteome of RPL cases. Other advanced "omics" technologies have not been systematically applied to investigate relevant tissues in RPL patients. Further complexity arises from alternative study targets for the "omics"-level analysis in RPL cases. For example, it is relevant to perform gene, protein or microRNA expression analysis of either placental chorionic villi, maternal endometrial, or blood samples. Similarly, epigenetic profiles of parental, fetal, and placental genomes can be equally informative in the context of repeated pregnancy loss. Future perspectives in investigating the molecular pathogenesis behind RPL rely on integrating alternative genetic and genomic approaches (DNA variation, gene expression, epigenetic regulation) in studies of individual genes as well as whole-genome analysis (Fig. 21.4A).

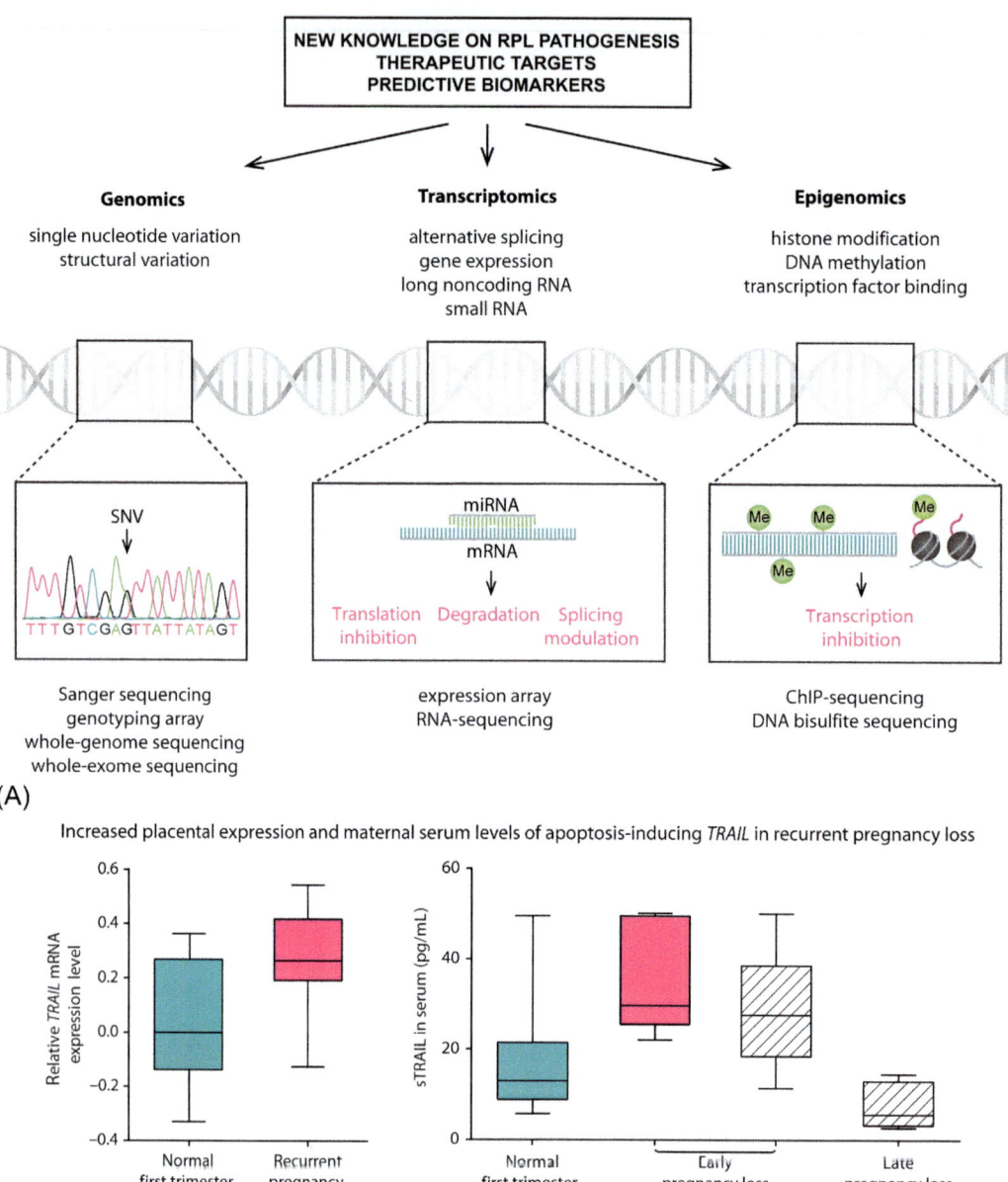

FIG. 21.4 Utilization of "omics" in recurrent pregnancy loss (RPL) research. (A) Application of genomics, transcriptomics, and epigenomics in RPL research is expected to provide improved knowledge about the pathogenesis, molecular mechanisms, and biological pathways behind pregnancy loss. Pinpointing candidate genes with aberrant function in RPL represents a valuable source to identify therapeutic targets, to generate mouse models for in vivo studies of RPL treatment, to screen genetic risk variants for RPL, and to develop biomarkers for monitoring early pregnancies in high risk cases. (B) TNF-related apoptosis-inducing ligand TRAIL was identified as a biomarker for RPL using microarray analysis of placental gene expression [105]. RT-qPCR analysis has confirmed significantly increased placental expression of *TRAIL* in RPL ($n=13$) compared to normal first-trimester pregnancy ($n=23$, left graph). The encoded soluble protein sTRAIL has revealed its potential as a predictive maternal serum marker for early pregnancy loss during the first trimester of gestation. Compared to women with uncomplicated pregnancies ($n=35$), circulating sTRAIL was significantly elevated in pregnancy loss cases at the miscarriage event ($n=13$) and also prospectively 2–50 days before clinical symptoms ($n=5$; right graph). sTRAIL serum levels were not indicative of late miscarriage ($n=4$). *Part B was assembled using published data from reference Rull K, Tomberg K, Kõks S, Männik J, Mõls M, Sirotkina M, et al. Increased placental expression and maternal serum levels of apoptosis-inducing TRAIL in recurrent miscarriage. Placenta 2013;34(2):141–8.*

ABERRANT EPIGENOME AND RISK TO RPL

Gross Epigenetic Disturbances in Parental Genomes

The diagnosis of RPL has been linked to systematic modifications in parental epigenomes, such as skewed X-chromosome inactivation (XCI) in the maternal genome and impaired sperm chromatin integrity on the paternal side. So far, the issue of skewed XCI and RPL has remained controversial and has not been included in the clinical genetic work-up of female RPL patients. Although seminal studies suggested an increased prevalence of skewed XCI among women with idiopathic RPL compared with controls [79–81], independent follow-up studies showed that skewed XCI is not primarily associated with RPL [82, 83], but rather with higher maternal age [83–85]. A metaanalysis across 12 studies reached the conclusion that only extremely skewed XCI ($\geq$90%) represents a risk factor for idiopathic RPL among women with three or more pregnancy losses [86].

Another currently debated issue is the contribution of sperm DNA fragmentation (SDF) to the risk of experiencing RPL. A high SDF index represents an indirect marker of an aberrantly packed sperm epigenome. There is solid scientific evidence that increased SDF is associated with impaired reproductive capacity and it has also been detected with increased prevalence among male partners of RPL couples compared to controls [87, 88]. However, the clinical potential of the SDF index as a predictive factor for the risk to develop RPL is doubtful [89, 90]. Furthermore, SDF was shown to depend on abstinence time and several conducted studies have not taken this into account when claiming the association with RPL [91].

Aberrant Epigenome of the POC

Epigenetic (re)programming is a tightly regulated process that is critical in early embryonic and trophoblast development and in preparing the endometrium for pregnancy [92–94]. Disturbed placental epigenetic regulation may cause abnormal trophoblastic invasion, which may contribute to the risk of pregnancy loss, fetal growth restriction, and preeclampsia [95]. Of note, the placental epigenome has distinct features such as general genome-wide hypomethylation compared to fetal tissues, parent-of-origin determined gene expression of several critical developmental genes, polymorphic imprinting of selected loci, and critical dynamic changes and fine tuning of its epigenetic profile over the gestation [96]. In addition, a recent study on mice suggested that the oocyte-derived DNA methylation may actually control multiple differentiation-related and physiological processes in the trophoblast via both imprinting-dependent and imprinting-independent mechanisms [94]. Studies aiming to uncover the role of aberrant epigenome programming in RPL cases are highly needed. However, there are clinical and technical challenges in collecting biomaterials suitable for epigenetic profiling of miscarried early human pregnancies. Thus, the published data on the epigenomes of human POC or placental samples of pregnancy losses are limited. Due to straightforward analysis and interpretation of data, the most frequently analyzed epigenetic mark is the methylation status of CpG dinucleotides, either in targeted gene promoters or using genome-wide profiling. The few conducted studies on RPL tissues have shown that aberrant DNA methylation at specific loci as well as global epigenetic dysregulation may contribute to or accompany the pregnancies destined for miscarriage [97–101]. For example, some RPL placentas have revealed polymorphic imprinting and monoallelic expression of the maternal allele of the HCG-beta encoding *CGB5* gene [98]. HCG is essential in human implantation and its biallelic expression is required for a viable pregnancy. Chorionic villi of RPL cases also exhibit an enrichment of outliers for DNA hemimethylation status of imprinted loci, normally epigenetically silenced in a parent-of-origin specific manner [97, 100]. On a genome-wide scale, early fetal losses have been associated with global DNA hypomethylation in chorionic villi and reduced overall expression of DNMT1, an enzyme for DNA methylation maintenance [99]. Interestingly, a recent study on blastocysts with a monosomy karyotype reported hypomethylation across the extant chromosome in a hemizygous status, accompanied with impaired cellular expression of DNMTs and developmental genes [101]. It can be speculated that aberrant epigenomic programming of the POC may itself represent a red flag contributing to the overall compromised implantation potential of a nonviable conceptus.

Consistent with the data on impaired epigenetic programming of the POC, decidual/endometrial tissue of unexplained RPL cases also shows a reduced expression level of critical epigenome "writer" enzymes, methyl transferase (G9aMT), and methylated histone (H3-K9) [102]. A recent study demonstrated that RPL is strongly associated with a specific lack of DNA methylation at distinct genomic motifs within CA-nucleotide rich regions in human endometrial stromal cells [103]. As DNA methylation in non-CpG sites is a hallmark of cellular multipotency, hypomethylation of CA-rich regions is predicted to lead to uterine stem cell deficiency and enhanced cellular senescence.

Although basic research on the RPL epigenomes has brought several novel insights, the niche of epigenetic testing of the POC in the clinical management and counseling of RPL couples is still to be established.

IMPAIRED GENE EXPRESSION IN RPL PLACENTAS

Profile of Placental Transcriptome in RPL

Disturbances in gene expression and its regulation at the maternal-fetal interface may contribute to abnormal implantation and placental development and, consequently, fetal viability. Thus, the analysis of systematic aberrations in RPL transcriptomes and proteomes may detect biological mechanisms involved in the initiation or promotion of pregnancy loss. Due to challenges in obtaining high-quality clinical samples, there are limited studies investigating transcriptome-wide gene expression signatures in biomaterials collected from RPL cases.

Only two studies have been published that utilized gene expression microarrays to analyze placental samples from RPL compared to controls. Neither study detected differentially expressed genes with highly significant statistical support. Analysis of decidual tissue identified 155 genes showing at least two-fold increased or decreased expression in RPL patients compared to uncomplicated pregnancies [104]. A parallel study identified only 27 genes with altered expression in the RPL placental tissue [105]. Both seminal reports highlighted an enrichment of dysregulated genes involved in cell communication and signaling, inflammatory, and immune response.

Despite the modest statistical support due to the small number of heterogeneous clinical samples, capturing gene expression profiles and biological pathways involved in failed pregnancy may assist in pinpointing novel biomarkers that are potentially applicable in clinical conditions for the benefit of RPL patients. Apoptosis-inducing ligand *TRAIL* has been identified as one of the most upregulated genes in RPL placentas [105]. Measurement of the level of encoded soluble protein sTRAIL in the maternal serum has revealed its potential as a predictive biomarker for early pregnancy outcome in two independent studies [105, 106] (Fig. 21.4B).

RNA sequencing (RNA-Seq), a state-of-the-art methodology, provides substantially higher resolution and statistical power to capture the entire transcriptome compared to microarray-based analysis. RNA-Seq is capable of precise measurement of the level of all protein-coding and noncoding (ncRNA) transcripts. So far, only one study has been published using RNA-Seq to analyze the transcriptome of chorionic villi from RPL cases in comparison with uncomplicated but electively terminated pregnancies [44]. The study revealed several novel insights into the pathogenesis behind RPL. The majority of differentially expressed genes in RPL chorionic villi were shown to be down-regulated compared to a normal viable pregnancy. RPL samples exhibited substantially decreased transcript levels of histones, regulatory ncRNAs, and genes involved in telomere, spliceosome, ribosomal, and mitochondrial and intracellular signaling functions (Fig. 21.5A,B). These genes are explicitly involved in the basic machinery required for replication and chromatin integrity, transcription, RNA processing, the maintenance of mitochondria, and other essential genome functions required in the process of rapid cellular proliferation and differentiation, critical in early placental development. In contrast, genes with increased transcript levels in RPL chorionic villi represent mostly protein encoding loci, which are involved in the protein processing/transport, fetal development, cell adhesion, immune response, and transcriptional regulation (Fig. 21.5A). Several of these genes overlapped with loci linked to pregnancy complications and/or placental malfunction in other studies (Fig. 21.5C). Among them, the genes *ATF4*, *C3*, *PHLDA2*, *GPX4*, and *ICAM1* have been previously suggested to be implicated in the pathogenesis of pregnancy loss: *GPX4* in preeclampsia, *PHLDA2* and *SLC16A2* in fetal growth restriction, and *CD74* and *LAPTM5* in preterm premature rupture of membranes (details and references in [44]). Some of the short-listed up-regulated genes (*EGR1*, *PDLIM1*, *MAPK3*) are known to function in trophoblast invasion, proliferation, and syncytialization processes. The overall emerging picture of the transcriptome profile in the RPL placentas suggests the malfunction of basic nuclear processes and the potential organismal feedback mechanisms to rescue the fetus.

As an unexpected outcome of this comprehensive transcriptome profiling, the gene ontology analysis revealed that a large fraction (~2/3) of differentially expressed genes in RPL placentas possess binding sites for E2F transcription factors. E2F transcription factors have a key role in regulating the placental transcriptome to guarantee proper cellular proliferation, endoreplication in trophoblastic cells, placental development, and fetal viability [107]. For a conceptus destined to miscarriage, the E2F TF-family represents a potential key coordinator in reprogramming the placental genome toward gradually stopping the maintenance of basic nuclear and cellular functions [44].

Functional categories of genes with elevated (left) and reduced (right) expression in RPL

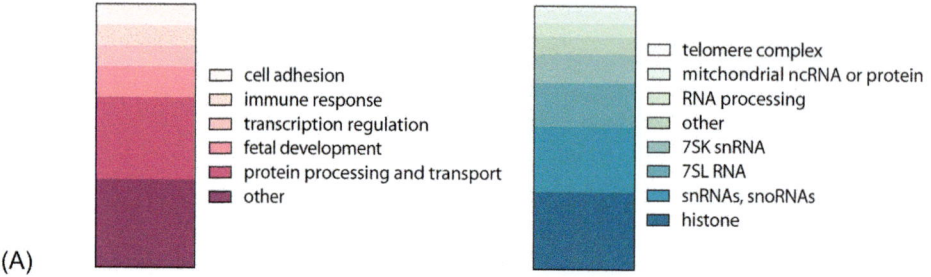

☐ cell adhesion
☐ immune response
☐ transcription regulation
☐ fetal development
☐ protein processing and transport
☐ other

☐ telomere complex
☐ mitchondrial ncRNA or protein
☐ RNA processing
☐ other
☐ 7SK snRNA
☐ 7SL RNA
☐ snRNAs, snoRNAs
☐ histone

(A)

Significantly decreased expression of *HIST1H4A* and *HIST1H1B* in recurrent pregnancy loss placental tissue

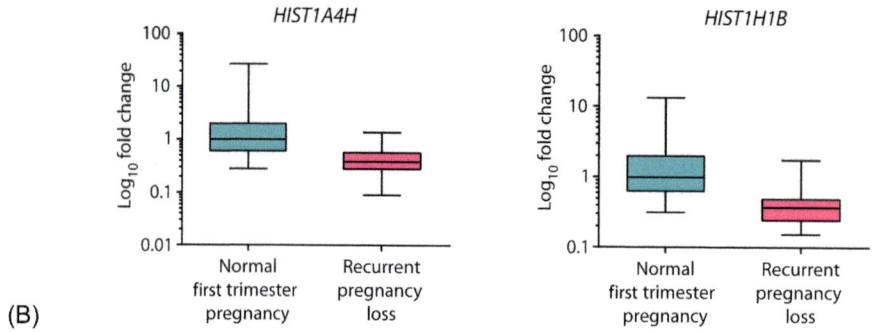

(B)

Genes with prior evidence for the involvement in placental function and pathology

Gene	Link to placental development/RPL
ICAM1	involved in inflammatory processes, higher placental expression in RPL
PHLDA2	regulates extraembryonic energy stores, associated with pregnancy loss and FGR
CD74	involved in inflammatory processes, downregulation in placental macrophages in preeclampsia
C3	complement activation, associated with RPL
GPX4	antioxidant, downregulated in preeclampsia
LAPTM5	preterm premature rupture of membranes
PLTP	maternal-fetal cholesterol transfer, up-regulated in GDM
SCL16A2	thyroid hormone transporter, modulates trophoblast cell invasion and viability, associated with FGR
MAPK3	trophoblast growth and invasion
EGR1	involved in embryo implantation, decidualization, and placental angiogenesis
PDLIM1	involved in differentiation of trophoblast cells
ZFP36	global post-transcriptional regulator of inflammation, associated with RPL
ATF4	decreased expression in spontaneous abortion

(C)

FIG. 21.5 Genes with significantly altered expression levels in chorionic villus samples from miscarried gestations of RPL couples compared to normal first-trimester pregnancies. Chorionic villus samples from miscarried gestations of RPL couples and uncomplicated but electively terminated first-trimester pregnancies have been analyzed for their transcriptome profile using RNA-Seq [44]. (A) Differential expression profiling detected 51 (27%) transcripts with increased and 138 (73%) with decreased expression in RPL compared to controls. Up-regulated loci in RPL samples are involved in the protein processing/transport, fetal development, cell adhesion, immune response, and transcriptional regulation (pink column, left) whereas decreased transcript levels were reported for histones, regulatory RNAs, and genes involved in telomere, spliceosome, ribosomal, mitochondrial, and intracellular signaling functions (blue column, right). (B) Examples of loci with significantly decreased expression in RPL chorionic villi, histone-encoding *HIST1H4A* and *HIST1H1B* genes. RT-qPCR analysis compared RPL placental tissue ($n=14$) with normal first-trimester pregnancy ($n=24$). (C) Genes with significantly increased transcript levels in RPL chorionic villi have been implicated in placental function and various adverse pregnancy outcomes, pointing to potentially shared molecular mechanisms. *Figure was assembled using published data from reference Sõber S, Rull K, Reiman M, Ilisson P, Mattila P, Laan M. RNA sequencing of chorionic villi from recurrent pregnancy loss patients reveals impaired function of basic nuclear and cellular machinery. Sci Rep 2016;6:38439.*

Proteomic Analysis of RPL Patient Biomaterials

Proteomics represent an "omics" field analyzing the profile of proteins at a targeted organism, organ, tissue, or cell. Proteomic analysis in tissues responding to an ongoing miscarriage event represents an alternative approach to identifying differentially expressed biomarkers [108]. Consistent with the transcriptome profiling studies, the proteomic analysis of chorionic villi from RPL cases revealed increased protein levels of apoptosis-related loci and decreased expression of angiogenesis proteins. Analysis of the follicular fluids and serum samples of RPL patients showed abnormal levels of immune response, inflammation, and coagulation regulating proteins, such as complement component C3c chain E, fibrinogen, antithrombin, and angiotensinogen [109, 110].

CONCLUSION

During the past 10 years, the availability of advanced genomic analysis tools has led to major novel insights in the causes of RPL. Today we can conclude that in a large proportion of cases, RPL is to be considered a disorder that is caused by intrinsic factors determined by the genome of the RPL partners and/or their products of conception (POC). These factors jointly determine the "unfavorable genomes" of these RPL cases, unable to result in a viable newborn in the majority of their conceptuses. The accumulated evidence shows that about 30%–50% of the POC of RPL couples have either gross chromosomal abnormalities or other critical structural genomic rearrangements. For one of 10 RPL couples, these genomic abnormalities are directly inherited from the parents. In the rest of the cases, genomic rearrangements of the POC have been initiated de novo either in oogenesis or spermatogenesis, or alternatively during early embryonic development. Abnormal chromosome segregation, increased probability of somatic mutations and structural variants, impaired DNA replication, and repair of the occurred errors refer to overall genomic instability in a major fraction of RPL conceptuses. This concept is supported by recent transcriptome analysis of RPL chorionic villi, clearly pointing to the deficit of components responsible for the maintenance of basic nuclear and cellular machinery during early rapid cellular proliferation and differentiation [44].

Although there is solid data about the high frequency of aberrant genomes among the conceptuses of RPL, there is little knowledge on possible genetic and nongenetic factors driving these events, apart from the aberrant parental karyotype. Genes involved in early human development and their mutations, either inherited from the parents or generated in the germline, could explain the emerging link between RPL and human developmental disorders. More pedigree-based studies utilizing novel NGS-based genome analysis are needed to uncover these mostly rare and private mutations in RPL families. This will be achieved only through dense collaboration between obstetricians managing the couple's recurrent pregnancy losses, pediatricians monitoring the offspring with developmental defects, and geneticists focusing on uncovering the pathogenic variant(s) behind this complex clinical phenotype profile in the family. This knowledge is expected to improve patient counseling and management not only to reach a successful pregnancy, but also to inform about the risks to deliver a child with some health complications. A further line of synergies to bring advanced understanding of genetic causes behind recurrently failed pregnancies may arise from joint actions with reproductive medicine experts involved in male infertility management. There is growing evidence on genes implicated in both impaired spermatogenesis and risk to RPL. One example is the *SYCP3* gene that is involved in the recombination and segregation of meiotic chromosomes, and harbors mutations causing azoospermia [111] or predisposing to RPL [57]. Another example of a locus with reported mutations leading to either male infertility or RPL is *AURKC*, contributing to proper cytokinesis and chromosome segregation [70, 112]. Ideally, families with mutations in reproduction-related genes with such pleiotropic effects would be managed jointly by both obstetricians and andrologists. An alternative line of genetic research and clinical management in RPL cases could be targeted to loci implicated in broad physiological functions, which can indirectly cause RPL if disturbed. For example, it is well established that genetic factors also play an important role in endocrine or immune-system related disorders, which are acknowledged risk factors for RPL.

Taken together, on the clinical side more cooperation is needed between complementary clinical experts to manage the families with RPL. As genetic causes of the majority of RPL families are expected to be private or rare in the general population, one should revisit the strategies of genetic analysis to maximize the number of solved cases. Currently, WES (and possibly in coming years also WGS) is underutilized, but is encouraged to be incorporated into the molecular diagnostic workup of relevant RPL cases. Example clinical cases warranted to WES/WGS based diagnostics are idiopathic RPL families with developmentally disabled children or pedigrees exhibiting anamnesis of both, recurrent pregnancy failure and infertility. Utilization of state-of-the-art "omics" tools to analyze biomaterial from RPL cases is currently only in its initiation phase, but it exhibits the potential to uncover novel molecular mechanisms and key biological triggers behind RPL. In addition, studies are needed to fill the knowledge gaps in the epigenetic programming in pregnancy success and failure. In perspective, this may assist in developing novel treatment options targeting critical biological markers and switches in early pregnancy establishment.

References

[1] Branch DW, Gibson M, Silver RM. Clinical practice. Recurrent miscarriage. N Engl J Med 2010;363(18):1740–7.
[2] Yang Y, Luo Y, Yuan J, Tang Y, Xiong L, Xu M, et al. Association between maternal, fetal and paternal MTHFR gene C677T and A1298C polymorphisms and risk of recurrent pregnancy loss: a comprehensive evaluation. Arch Gynecol Obstet 2016;293(6):1197–211.
[3] Royal College of obstetrics and gynecology green-top guideline no 17. The investigation and treatment of couples with recurrent first-trimester and second-trimester miscarriage; 2011.
[4] Practice Committee of the American Society for Reproductive M. Evaluation and treatment of recurrent pregnancy loss: a committee opinion. Fertil Steril 2012;98(5):1103–11.
[5] Bricker L, Farquharson RG. Types of pregnancy loss in recurrent miscarriage: implications for research and clinical practice. Hum Reprod 2002;17(5):1345–50.
[6] Berry CW, Brambati B, Eskes TK, Exalto N, Fox H, Geraedts JP, et al. The euro-team early pregnancy (ETEP) protocol for recurrent miscarriage. Hum Reprod 1995;10(6):1516–20.
[7] Boue J, Bou A, Lazar P. Retrospective and prospective epidemiological studies of 1500 karyotyped spontaneous human abortions. Teratology 1975;12(1):11–26.
[8] Ogasawara M, Aoki K, Okada S, Suzumori K. Embryonic karyotype of abortuses in relation to the number of previous miscarriages. Fertil Steril 2000;73(2):300–4.

[9] El Hachem H, Crepaux V, May-Panloup P, Descamps P, Legendre G, Bouet PE. Recurrent pregnancy loss: current perspectives. Int J Women's Health 2017;9:331–45.

[10] Kasak L, Rull K, Sõber S, Laan M. Copy number variation profile in the placental and parental genomes of recurrent pregnancy loss families. Sci Rep 2017;7:45327.

[11] Sugiura-Ogasawara M, Ozaki Y, Katano K, Suzumori N, Kitaori T, Mizutani E. Abnormal embryonic karyotype is the most frequent cause of recurrent miscarriage. Hum Reprod 2012;27(8):2297–303.

[12] Christiansen OB, Pedersen B, Nielsen HS, Nybo Andersen AM. Impact of the sex of first child on the prognosis in secondary recurrent miscarriage. Hum Reprod 2004;19(12):2946–51.

[13] Piosik ZM, Goegebeur Y, Klitkou L, Steffensen R, Christiansen OB. Plasma TNF-alpha levels are higher in early pregnancy in patients with secondary compared with primary recurrent miscarriage. Am J Reprod Immunol 2013;70(5):347–58.

[14] Jaslow CR, Kutteh WH. Effect of prior birth and miscarriage frequency on the prevalence of acquired and congenital uterine anomalies in women with recurrent miscarriage: a cross-sectional study. Fertil Steril 2013;99(7):1916–22:e1.

[15] Shapira E, Ratzon R, Shoham-Vardi I, Serjienko R, Mazor M, Bashiri A. Primary vs. secondary recurrent pregnancy loss—epidemiological characteristics, etiology, and next pregnancy outcome. J Perinat Med 2012;40(4):389–96.

[16] Feichtinger M, Wallner E, Hartmann B, Reiner A, Philipp T. Transcervical embryoscopic and cytogenetic findings reveal distinctive differences in primary and secondary recurrent pregnancy loss. Fertil Steril 2017;107(1):144–9.

[17] Brigham SA, Conlon C, Farquharson RG. A longitudinal study of pregnancy outcome following idiopathic recurrent miscarriage. Hum Reprod 1999;14(11):2868–71.

[18] Tharapel AT, Tharapel SA, Bannerman RM. Recurrent pregnancy losses and parental chromosome abnormalities: a review. Br J Obstet Gynaecol 1985;92(9):899–914.

[19] van den Boogaard E, Kaandorp SP, Franssen MT, Mol BW, Leschot NJ, Wouters CH, et al. Consecutive or non-consecutive recurrent miscarriage: is there any difference in carrier status? Hum Reprod 2010;25(6):1411–4.

[20] Franssen MT, Korevaar JC, van der Veen F, Leschot NJ, Bossuyt PM, Goddijn M. Reproductive outcome after chromosome analysis in couples with two or more miscarriages: index [corrected]-control study. Br Med J (Clin Res Ed) 2006;332(7544):759–63.

[21] Flynn H, Yan J, Saravelos SH, Li TC. Comparison of reproductive outcome, including the pattern of loss, between couples with chromosomal abnormalities and those with unexplained repeated miscarriages. J Obstet Gynaecol Res 2014;40(1):109–16.

[22] Drugan A, Koppitch 3rd FC, Williams 3rd JC, Johnson MP, Moghissi KS, Evans MI. Prenatal genetic diagnosis following recurrent early pregnancy loss. Obstet Gynecol 1990;75(3):381–4. Pt 1.

[23] Cockwell AE, Jacobs PA, Beal SJ, Crolla JA. A study of cryptic terminal chromosome rearrangements in recurrent miscarriage couples detects unsuspected acrocentric pericentromeric abnormalities. Hum Genet 2003;112(3):298–302.

[24] De la Fuente-Cortes BE, Cerda-Flores RM, Davila-Rodriguez MI, Garcia-Vielma C, De la Rosa Alvarado RM, Cortes-Gutierrez EI. Chromosomal abnormalities and polymorphic variants in couples with repeated miscarriage in Mexico. Reprod Biomed Online 2009;18(4):543–8.

[25] Karim S, Jamal HS, Rouzi A, Ardawi MSM, Schulten HJ, Mirza Z, et al. Genomic answers for recurrent spontaneous abortion in Saudi Arabia: An array comparative genomic hybridization approach. Reprod Biol 2017;17(2):133–43.

[26] Nagirnaja L, Palta P, Kasak L, Rull K, Christiansen OB, Nielsen HS, et al. Structural genomic variation as risk factor for idiopathic recurrent miscarriage. Hum Mutat 2014;35(8):972–82.

[27] van den Berg MM, van Maarle MC, van Wely M, Goddijn M. Genetics of early miscarriage. Biochim Biophys Acta 2012;1822(12):1951–9.

[28] Rosenfeld JA, Tucker ME, Escobar LF, Neill NJ, Torchia BS, McDaniel LD, et al. Diagnostic utility of microarray testing in pregnancy loss. Ultrasound Obstet Gynecol 2015;46(4):478–86.

[29] Ottolini CS, Newnham LJ, Capalbo A, Natesan SA, Joshi HA, Cimadomo D, et al. Genome-wide maps of recombination and chromosome segregation in human oocytes and embryos show selection for maternal recombination rates. Nat Genet 2015;47(7):727–35.

[30] Hertig AT, Rock J, Adams EC, Menkin MC. Thirty-four fertilized human ova, good, bad and indifferent, recovered from 210 women of known fertility; a study of biologic wastage in early human pregnancy. Pediatrics 1959;23(1 Part 2):202–11.

III. INFERTILITY AND ASSISTED REPRODUCTIVE TECHNOLOGY

[31] Cowchock FS, Gibas Z, Jackson LG. Chromosome errors as a cause of spontaneous abortion: the relative importance of maternal age and obstetric history. Fertil Steril 1993;59(5):1011–4.

[32] Shearer BM, Thorland EC, Carlson AW, Jalal SM, Ketterling RP. Reflex fluorescent in situ hybridization testing for unsuccessful product of conception cultures: a retrospective analysis of 5555 samples attempted by conventional cytogenetics and fluorescent in situ hybridization. Genet Med 2011;13(6):545–52.

[33] Robberecht C, Schuddinck V, Fryns JP, Vermeesch JR. Diagnosis of miscarriages by molecular karyotyping: benefits and pitfalls. Genet Med 2009;11(9):646–54.

[34] Kutteh WH. Novel strategies for the management of recurrent pregnancy loss. Semin Reprod Med 2015; 33(3):161–8.

[35] Thorne J, Craffey A, Nulsen 3rd JC. Detection of an inherited deletion in products of conception in a patient with recurrent losses and normal karyotype. Obstet Gynecol 2017;130(1):126–9.

[36] Donaghue C, Davies N, Ahn JW, Thomas H, Ogilvie CM, Mann K. Efficient and cost-effective genetic analysis of products of conception and fetal tissues using a QF-PCR/array CGH strategy; five years of data. Mol Cytogen 2017;10:12.

[37] Shah MS, Cinnioglu C, Maisenbacher M, Comstock I, Kort J, Lathi RB. Comparison of cytogenetics and molecular karyotyping for chromosome testing of miscarriage specimens. Fertil Steril 2017;107(4):1028–33.

[38] Levy B, Sigurjonsson S, Pettersen B, Maisenbacher MK, Hall MP, Demko Z, et al. Genomic imbalance in products of conception: single-nucleotide polymorphism chromosomal microarray analysis. Obstet Gynecol 2014;124(2 Pt 1):202–9.

[39] Viaggi CD, Cavani S, Malacarne M, Floriddia F, Zerega G, Baldo C, et al. First-trimester euploid miscarriages analysed by array-CGH. J Appl Genet 2013;54(3):353–9.

[40] Sahoo T, Dzidic N, Strecker MN, Commander S, Travis MK, Doherty C, et al. Comprehensive genetic analysis of pregnancy loss by chromosomal microarrays: outcomes, benefits, and challenges. Genet Med 2017;19(1):83–9.

[41] Wou K, Hyun Y, Chitayat D, Vlasschaert M, Chong K, Wasim S, et al. Analysis of tissue from products of conception and perinatal losses using QF-PCR and microarray: a three-year retrospective study resulting in an efficient protocol. Eur J Med Genet 2016;59(8):417–24.

[42] Maxwell SM, Colls P, Hodes-Wertz B, McCulloh DH, McCaffrey C, Wells D, et al. Why do euploid embryos miscarry? A case-control study comparing the rate of aneuploidy within presumed euploid embryos that resulted in miscarriage or live birth using next-generation sequencing. Fertil Steril 2016;106(6):1414–9.e5.

[43] Bagheri H, Mercier E, Qiao Y, Stephenson MD, Rajcan-Separovic E. Genomic characteristics of miscarriage copy number variants. Mol Hum Reprod 2015;21(8):655–61.

[44] Sõber S, Rull K, Reiman M, Ilisson P, Mattila P, Laan M. RNA sequencing of chorionic villi from recurrent pregnancy loss patients reveals impaired function of basic nuclear and cellular machinery. Sci Rep 2016;6:38439.

[45] Rajcan-Separovic E, Diego-Alvarez D, Robinson WP, Tyson C, Qiao Y, Harvard C, et al. Identification of copy number variants in miscarriages from couples with idiopathic recurrent pregnancy loss. Hum Reprod 2010; 25(11):2913–22.

[46] van Dijk M, van Bezu J, van Abel D, Dunk C, Blankenstein MA, Oudejans CB, et al. The STOX1 genotype associated with pre-eclampsia leads to a reduction of trophoblast invasion by alpha-T-catenin upregulation. Hum Mol Genet 2010;19(13):2658–67.

[47] Liang Y, Franks TM, Marchetto MC, Gage FH, Hetzer MW. Dynamic association of NUP98 with the human genome. PLoS Genet 2013;9(2)e1003308.

[48] Wen J, Hanna CW, Martell S, Leung PC, Lewis SM, Robinson WP, et al. Functional consequences of copy number variants in miscarriage. Mol Cytogen 2015;8:6.

[49] Kasak L, Rull K, Vaas P, Teesalu P, Laan M. Extensive load of somatic CNVs in the human placenta. Sci Rep 2015;5:8342.

[50] Hannibal RL, Chuong EB, Rivera-Mulia JC, Gilbert DM, Valouev A, Baker JC. Copy number variation is a fundamental aspect of the placental genome. PLoS Genet 2014;10(5):e1004290.

[51] Khoury MJ, Erickson JD. Recurrent pregnancy loss as an indicator for increased risk of birth defects: A population-based case-control study. Paediatr Perinat Epidemiol 1993;7(4):404–16.

[52] Philipp T, Philipp K, Reiner A, Beer F, Kalousek DK. Embryoscopic and cytogenetic analysis of 233 missed abortions: factors involved in the pathogenesis of developmental defects of early failed pregnancies. Hum Reprod 2003;18(8):1724–32.

[53] Farley TL. A reproductive history of mothers with spina bifida offspring—a new look at old issues. Cerebrospinal Fluid Res 2006;3:10.

[54] Walfisch A, Wainstock T, Segal I, Landau D, Sheiner E. Maternal history of recurrent pregnancy loss increases the risk for long term paediatric neurological morbidity of the offspring. Am J Obstet Gynecol Suppl 2017;2017: S474.

[55] Liu P, Yuan B, Carvalho CM, Wuster A, Walter K, Zhang L, et al. An organismal CNV mutator phenotype restricted to early human development. Cell 2017;168(5):830–42:e7.

[56] Neusser M, Rogenhofer N, Durl S, Ochsenkuhn R, Trottmann M, Jurinovic V, et al. Increased chromosome 16 disomy rates in human spermatozoa and recurrent spontaneous abortions. Fertil Steril 2015;104(5): 1130–7:e1–e10.

[57] Bolor H, Mori T, Nishiyama S, Ito Y, Hosoba E, Inagaki H, et al. Mutations of the SYCP3 gene in women with recurrent pregnancy loss. Am J Hum Genet 2009;84(1):14–20.

[58] Johnson PM, Chia KV, Risk JM, Barnes RM, Woodrow JC. Immunological and immunogenetic investigation of recurrent spontaneous abortion. Dis Markers 1988;6(3):163–71.

[59] Christiansen OB. A fresh look at the causes and treatments of recurrent miscarriage, especially its immunological aspects. Hum Reprod Update 1996;2(4):271–93.

[60] Kolte AM, Nielsen HS, Moltke I, Degn B, Pedersen B, Sunde L, et al. A genome-wide scan in affected sibling pairs with idiopathic recurrent miscarriage suggests genetic linkage. Mol Hum Reprod 2011;17(6):379–85.

[61] Rull K, Nagirnaja L, Laan M. Genetics of recurrent miscarriage: challenges, current knowledge, future directions. Front Genet 2012;3:34.

[62] Shi X, Xie X, Jia Y, Li S. Maternal genetic polymorphisms and unexplained recurrent miscarriage: a systematic review and meta-analysis. Clin Genet 2017;91(2):265–84.

[63] Pereza N, Ostojic S, Kapovic M, Peterlin B. Systematic review and meta-analysis of genetic association studies in idiopathic recurrent spontaneous abortion. Fertil Steril 2017;107(1):150–9.

[64] Bradley LA, Palomaki GE, Bienstock J, Varga E, Scott JA. Can factor V Leiden and prothrombin G20210A testing in women with recurrent pregnancy loss result in improved pregnancy outcomes? Results from a targeted evidence-based review. Genet Med 2012;14(1):39–50.

[65] Tan Z, Randall G, Fan J, Camoretti-Mercado B, Brockman-Schneider R, Pan L, et al. Allele-specific targeting of microRNAs to HLA-G and risk of asthma. Am J Hum Genet 2007;81(4):829–34.

[66] Su X, Hu Y, Li Y, Cao JL, Wang XQ, Ma X, et al. The polymorphism of rs6505162 in the MIR423 coding region and recurrent pregnancy loss. Reproduction 2015;150(1):65–76.

[67] Rah H, Chung KW, Ko KH, Kim ES, Kim JO, Sakong JH, et al. miR-27a and miR-449b polymorphisms associated with a risk of idiopathic recurrent pregnancy loss. PLoS ONE 2017;12(5):e0177160.

[68] Amin-Beidokhti M, Mirfakhraie R, Zare-Karizi S, Karamoddin F. The role of parental microRNA alleles in recurrent pregnancy loss: an association study. Reprod Biomed Online 2017;34(3):325–30.

[69] Mercier E, Lissalde-Lavigne G, Gris JC. JAK2 V617F mutation in unexplained loss of first pregnancy. N Engl J Med 2007;357(19):1984–5.

[70] Lopez-Carrasco A, Oltra S, Monfort S, Mayo S, Rosello M, Martinez F, et al. Mutation screening of AURKB and SYCP3 in patients with reproductive problems. Mol Hum Reprod 2013;19(2):102–8.

[71] McCoy RC, Demko Z, Ryan A, Banjevic M, Hill M, Sigurjonsson S, et al. Common variants spanning PLK4 are associated with mitotic-origin aneuploidy in human embryos. Science 2015;348(6231):235–8.

[72] Qiao Y, Wen J, Tang F, Martell S, Shomer N, Leung PC, et al. Whole exome sequencing in recurrent early pregnancy loss. Mol Hum Reprod 2016;22(5):364–72.

[73] Quintero-Ronderos P, Mercier E, Fukuda M, Gonzalez R, Suarez CF, Patarroyo MA, et al. Novel genes and mutations in patients affected by recurrent pregnancy loss. PLoS ONE 2017;12(10):e0186149.

[74] Murugappan G, Shahine LK, Perfetto CO, Hickok LR, Lathi RB. Intent to treat analysis of in vitro fertilization and preimplantation genetic screening versus expectant management in patients with recurrent pregnancy loss. Hum Reprod 2016;31(8):1668–74.

[75] Murugappan G, Ohno MS, Lathi RB. Cost-effectiveness analysis of preimplantation genetic screening and in vitro fertilization versus expectant management in patients with unexplained recurrent pregnancy loss. Fertil Steril 2015;103(5):1215–20.

[76] Yang Z, Liu J, Collins GS, Salem SA, Liu X, Lyle SS, et al. Selection of single blastocysts for fresh transfer via standard morphology assessment alone and with array CGH for good prognosis IVF patients: results from a randomized pilot study. Mol Cytogen 2012;5(1):24.

[77] Lund M, Kamper-Jorgensen M, Nielsen HS, Lidegaard O, Andersen AM, Christiansen OB. Prognosis for live birth in women with recurrent miscarriage: what is the best measure of success? Obstet Gynecol 2012; 119(1):37–43.

[78] Oud MS, Ramos L, O'Bryan MK, McLachlan RI, Okutman O, Viville S, et al. Validation and application of a novel integrated genetic screening method to a cohort of 1,112 men with idiopathic azoospermia or severe oligozoospermia. Hum Mutat 2017;38(11):1592–605.

[79] Lanasa MC, Hogge WA, Kubik C, Blancato J, Hoffman EP. Highly skewed X-chromosome inactivation is associated with idiopathic recurrent spontaneous abortion. Am J Hum Genet 1999;65(1):252–4.

[80] Sangha KK, Stephenson MD, Brown CJ, Robinson WP. Extremely skewed X-chromosome inactivation is increased in women with recurrent spontaneous abortion. Am J Hum Genet 1999;65(3):913–7.

[81] Uehara S, Hashiyada M, Sato K, Sato Y, Fujimori K, Okamura K. Preferential X-chromosome inactivation in women with idiopathic recurrent pregnancy loss. Fertil Steril 2001;76(5):908–14.

[82] Pasquier E, Bohec C, De Saint Martin L, Le Marechal C, Le Martelot MT, Roche S, et al. Strong evidence that skewed X-chromosome inactivation is not associated with recurrent pregnancy loss: an incident paired case control study. Hum Reprod 2007;22(11):2829–33.

[83] Hogge WA, Prosen TL, Lanasa MC, Huber HA, Reeves MF. Recurrent spontaneous abortion and skewed X-inactivation: is there an association? Am J Obstet Gynecol 2007;196(4):384.

[84] Hatakeyama C, Anderson CL, Beever CL, Penaherrera MS, Brown CJ, Robinson WP. The dynamics of X-inactivation skewing as women age. Clin Genet 2004;66(4):327–32.

[85] Knudsen GP, Pedersen J, Klingenberg O, Lygren I, Orstavik KH. Increased skewing of X chromosome inactivation with age in both blood and buccal cells. Cytogenet Genome Res 2007;116(1–2):24–8.

[86] Sui Y, Chen Q, Sun X. Association of skewed X chromosome inactivation and idiopathic recurrent spontaneous abortion: a systematic review and meta-analysis. Reprod Biomed Online 2015;31(2):140–8.

[87] Bareh GM, Jacoby E, Binkley P, Chang TC, Schenken RS, Robinson RD. Sperm deoxyribonucleic acid fragmentation assessment in normozoospermic male partners of couples with unexplained recurrent pregnancy loss: a prospective study. Fertil Steril 2016;105(2):329–36.

[88] Robinson L, Gallos ID, Conner SJ, Rajkhowa M, Miller D, Lewis S, et al. The effect of sperm DNA fragmentation on miscarriage rates: a systematic review and meta-analysis. Hum Reprod 2012;27(10):2908–17.

[89] Carlini T, Paoli D, Pelloni M, Faja F, Dal Lago A, Lombardo F, et al. Sperm DNA fragmentation in Italian couples with recurrent pregnancy loss. Reprod Biomed Online 2017;34(1):58–65.

[90] Bronet F, Martinez E, Gaytan M, Linan A, Cernuda D, Ariza M, et al. Sperm DNA fragmentation index does not correlate with the sperm or embryo aneuploidy rate in recurrent miscarriage or implantation failure patients. Hum Reprod 2012;27(7):1922–9.

[91] Pons I, Cercas R, Villas C, Brana C, Fernandez-Shaw S. One abstinence day decreases sperm DNA fragmentation in 90% of selected patients. J Assist Reprod Genet 2013;30(9):1211–8.

[92] Tang WW, Kobayashi T, Irie N, Dietmann S, Surani MA. Specification and epigenetic programming of the human germ line. Nat Rev Genet 2016;17(10):585–600.

[93] Messerschmidt DM, Knowles BB, Solter D. DNA methylation dynamics during epigenetic reprogramming in the germline and preimplantation embryos. Genes Dev 2014;28(8):812–28.

[94] Branco MR, King M, Perez-Garcia V, Bogutz AB, Caley M, Fineberg E, et al. Maternal DNA methylation regulates early trophoblast development. Dev Cell 2016;36(2):152–63.

[95] Choux C, Carmignac V, Bruno C, Sagot P, Vaiman D, Fauque P. The placenta: phenotypic and epigenetic modifications induced by assisted reproductive technologies throughout pregnancy. Clin Epigenetics 2015;7:87.

[96] Novakovic B, Saffery R. DNA methylation profiling highlights the unique nature of the human placental epigenome. Epigenomics 2010;2(5):627–38.

[97] Zechner U, Pliushch G, Schneider E, El Hajj N, Tresch A, Shufaro Y, et al. Quantitative methylation analysis of developmentally important genes in human pregnancy losses after ART and spontaneous conception. Mol Hum Reprod 2009;16(9):704–13.

[98] Uuskula L, Rull K, Nagirnaja L, Laan M. Methylation allelic polymorphism (MAP) in chorionic gonadotropin beta5 (CGB5) and its association with pregnancy success. J Clin Endocrinol Metabol 2011;96(1):E199–207.

[99] Yin LJ, Zhang Y, Lv PP, He WH, Wu YT, Liu AX, et al. Insufficient maintenance DNA methylation is associated with abnormal embryonic development. BMC Med 2012;10:26.

[100] Hanna CW, McFadden DE, Robinson WP. DNA methylation profiling of placental villi from karyotypically normal miscarriage and recurrent miscarriage. Am J Pathol 2013;182(6):2276–84.

[101] McCallie BR, Parks JC, Patton AL, Griffin DK, Schoolcraft WB, Katz-Jaffe MG. Hypomethylation and genetic instability in monosomy blastocysts may contribute to decreased implantation potential. PLoS ONE 2016;11(7) e0159507.

[102] Fatima N, Ahmed SH, Salhan S, Rehman SM, Kaur J, Owais M, et al. Study of methyl transferase (G9aMT) and methylated histone (H3-K9) expressions in unexplained recurrent spontaneous abortion (URSA) and normal early pregnancy. Mol Hum Reprod 2011;17(11):693–701.

[103] Lucas ES, Dyer NP, Murakami K, Lee YH, Chan YW, Grimaldi G, et al. Loss of endometrial plasticity in recurrent pregnancy loss. Stem Cells 2016;34(2):346–56.

[104] Krieg SA, Fan X, Hong Y, Sang QX, Giaccia A, Westphal LM, et al. Global alteration in gene expression profiles of deciduas from women with idiopathic recurrent pregnancy loss. Mol Hum Reprod 2012;18(9):442–50.

[105] Rull K, Tomberg K, Kõks S, Männik J, Möls M, Sirotkina M, et al. Increased placental expression and maternal serum levels of apoptosis-inducing TRAIL in recurrent miscarriage. Placenta 2013;34(2):141–8.

[106] Agostinis C, Bulla R, Tisato V, De Seta F, Alberico S, Secchiero P, et al. Soluble TRAIL is elevated in recurrent miscarriage and inhibits the in vitro adhesion and migration of HTR8 trophoblastic cells. Hum Reprod 2012; 27(10):2941–7.

[107] Ouseph MM, Li J, Chen HZ, Pecot T, Wenzel P, Thompson JC, et al. Atypical E2F repressors and activators coordinate placental development. Dev Cell 2012;22(4):849–62.

[108] Baek KH, Lee EJ, Kim YS. Recurrent pregnancy loss: the key potential mechanisms. Trends Mol Med 2007; 13(7):310–7.

[109] Baek KH. Aberrant gene expression associated with recurrent pregnancy loss. Mol Hum Reprod 2004; 10(5):291–7.

[110] Kim MS, Gu BH, Song S, Choi BC, Cha DH, Baek KH. ITI-H4, as a biomarker in the serum of recurrent pregnancy loss (RPL) patients. Mol Biosyst 2011;7(5):1430–40.

[111] Miyamoto T, Hasuike S, Yogev L, Maduro MR, Ishikawa M, Westphal H, et al. Azoospermia in patients heterozygous for a mutation in SYCP3. Lancet 2003;362(9397):1714–9.

[112] Dieterich K, Soto Rifo R, Faure AK, Hennebicq S, Ben Amar B, Zahi M, et al. Homozygous mutation of AURKC yields large-headed polyploid spermatozoa and causes male infertility. Nat Genet 2007;39(5):661–5.

Glossary

Antiphopholipid syndrome (APS)	APA is an autoimmune hypercoagulable state caused by antiphospholipid antibodies. The disorder manifests clinically as recurrent venous or arterial thrombosis and/or fetal loss.
Array comparative genomic hybridization (array CGH)	Array CGH is a molecular cytogenetic technique for the detection of submicroscopic chromosomal rearrangements on a genome-wide scale using oligonucleotide microarrays. The analysis compares the measured ploidy of the DNA between the test sample and a reference sample.
Copy number variants (CNV)	CNV represent genome structural variations that lead to the changed dosage of rearranged genomic segment. Microdeletions and microduplications may lead to an altered number of gene copies, affecting their function.
DNA methylation	It is a process by which DNA molecules are modified by adding methyl groups to nucleotides. DNA methylation is mostly concentrated in CpG sites, where the cytosine is followed by a guanine nucleotide. In the human somatic cells, 70%–80% of CpGs are methylated. Non-CpG methylation refers to the condition when methyl groups are added to the cytosine, followed by adenine, thymine, or another cytosine (CpA, CpT, CpC). Non-CpG methylation is rare in humans and has been reported mostly in the brain and embryonic stem cells.
Epigenomics	It is an "omics" field aiming to profile the complete set of epigenetic modifications (DNA methylation, histone modification, chromatin remodeling, etc.) in the context of analyzed organism, organ, tissue, disease, etc. Epigenetic modifications alter gene expression but do not change the underlying DNA sequence. Epigenome defines the functional properties of the cell.

Genome structural variation	It refers to rearrangements in a particular genomic region compared to the reference genome. These chromosomal aberrations include deletions, duplications, inversions, insertions, translocations, and complex genomic rearrangements.
Imprinting	It is an epigenetic silencing of selected genes in the maternal/paternal germline, leading to the parent-of-origin-specific monoallelic expression of these loci in their conceptus.
microRNAs (miRNA)	miRNA are short 20–25 bp noncoding RNAs that regulate posttranscriptional gene expression levels through binding to transcripts, suppressing protein synthesis, and/or initiating mRNA degradation. MiRNomics refers to the "omics" field aiming to profile the entire set of microRNAs in a cell or tissue at a certain time and physiological condition.
"Omics"	It refers to research in molecular and genome biology that aims to profile and quantitate large pools of biological molecules, and link the collected data to the function and dynamics of a genome, cell, tissue, organ, or organism. The subcategories of omics is defined based on the study objects, such as genomics, transcriptomics, miRNomics, proteomics, metabolomics, etc.
Preimplantation genetic diagnosis (PGD)	PGD is a procedure used prior to implantation that tests for specific genetic abnormalities known to be heritable and present in one or both of the parents. This prevents certain genetic disorders from being passed on to the next generation.
Preimplantation genetic screening (PGS)	PGS uses a global genetic assessment of the embryo to detect a wide variety of genetic abnormalities over the genome. Only chromosomally normal embryos will be transferred in the attempt to achieve a successful pregnancy and a healthy child.
SNP genotyping array	It is a type of DNA microarray that is used to simultaneously determine genotypes for a large set of single nucleotide polymorphisms (SNPs) in the subject's genome.
Transcriptomics	It is an omics field aiming to profile the complete set of RNA transcripts (transcriptome) that are produced by the genome at a given time by an organ, tissue, or cell.

Human Genetics and Assisted Reproduction in Endometriosis

Chien-Wen Chen*, Endah Rahmawati[†], Guan-Lin Lai[‡],
Ya-Ching Chou[§,¶], Yun-Yi Ma[§,¶], Yi-Xuan Lee[¶],
Chii-Ruey Tzeng[§,¶]

[*]Department of Obstetrics and Gynecology, Shuang Ho Hospital, Taipei Medical University, New Taipei, Taiwan [†]Graduate Institute of Clinical Medicine, College of Medicine, Taipei Medical University, Taipei, Taiwan [‡]School of Public Health, College of Public health and Nutrition, Taipei Medical University, Taipei, Taiwan [§]Department of Obstetrics and Gynecology, College of Medicine, Taipei Medical University, Taipei, Taiwan [¶]Center for Reproductive Medicine, Taipei Medical University Hospital, Taipei, Taiwan

OUTLINE

Abbreviations

2-DIGE	two-dimensional gel electrophoresis
AFS	the American Fertility Society
ART	assisted reproductive technology
ASRM	the American Society for Reproductive Medicine
AUC	area under the curve
BBzP	butyl-benzyl phthalate
BzBP	benzyl butyl phthalate
CA125	cancer antigen 125
COH	controlled ovarian hyperstimulation
COX-2	cyclooxygenase-2
DBP	dibutyl phthalate
DEHP	di(2-ethylhexyl) phthalate
DIE	deep infiltrating endometriosis
DNA	deoxyribonucleic acid
DnBP	di-n-butyl phthalate
DnOP	di-n-octyl phthalate
EDCs	endocrine disrupting chemicals
EFI	endometriosis fertility index
ERs	estrogen receptors
EST	ethanol sclerotherapy
FSH	follicle-stimulating hormone
GnRH	gonadotropin-releasing hormone
GSTM1	glutathione S-transferase M 1
HPLC	high-performance liquid chromatography
IFγ	interferon gamma
IL	interleukin
IUI	intrauterine insemination
IVF	in vitro fertilization
LPO	lipid peroxide
MBzP	monobenzyl phthalate
MEHHP	mono(2-ethyl-5 hydroxyhexyl) phthalate
MEHP	mono(2-ethylhexyl) phthalate
MEOHP	mono(2-ethyl-5-oxhexyl) phthalate
MEP	monoethyl phthalate
miRNA	micro ribonucleic acid
MMP	matrix metalloproteinase
MnBP	mono-n-buty phthalate
MRI	magnetic resonance imaging
NF-κB	the nuclear factor-kappa B
NOD/SCID	nonobese diabetic/severe combined immunodeficiency
NSAIDs	nonsteroidal antiinflammatory drugs
OCPs	oral contraceptives
PPAR-γ	peroxisome proliferator-activated receptor-γ
ROS	reactive oxygen species
SELDI-TOF-MS	surface enhanced laser desorption/ionization time-of-flight mass spectrometry
SOD	superoxide dismutase
TNFα	tumor necrosis factor alpha
TVU	transvaginal ultrasound
VEGF	vascular endothelial growth factor

INTRODUCTION

This chapter will start with the introduction of the etiology and diagnosis of endometriosis, followed by a review of the association between endometriosis and environmental factors such as blood trace metals and phthalates. In view of the detrimental impacts of endometriosis on human reproduction, the efficacy of the various treatments, especially the assisted reproductive technologies, will be discussed as well. At the end of the chapter, another disease with an ambiguous relationship with endometriosis, namely adenomyosis, will be featured in a short section including treatments for infertility.

BIOMARKERS IN ENDOMETRIOSIS

Endometriosis is characterized by the implantation of endometrial tissue outside the uterine cavity. The pelvic peritoneum is the most frequently affected location, followed by the ovaries and the rectovaginal septum. In rare circumstances, the endometriotic foci are distant from the pelvis, for example the thorax [1]. The prevalence of endometriosis varies from 1% in the general population [2] to 21.4% in women suffering chronic pelvic pain [3]. The etiology and pathogenesis of endometriosis are still unclear. Formulated by John Sampson in 1927, retrograde menstruation is the theory that has the greatest consensus. However, retrograde menstruation can be observed in 90% of women, which is inconsistent with the incidence of endometriosis. In addition, retrograde menstruation fails to explain the endometriotic foci distant from the pelvis as well as endometriosis in male patients [4]. Other hypotheses of endometriosis include celomic metaplasia, aberrant placement of stem cells, the immunoescaping of endometriotic cells, and other genetic or environmental factors. Women with endometriosis may suffer from dysmenorrhea, chronic pelvic pain, and dyspareunia. An elevated risk of developing depression and anxiety disorders is reported to be associated with endometriosis [5]. Women with endometriosis-associated pelvic pain have a poorer quality of life and mental health as compared to those with asymptomatic endometriosis as well as the healthy controls [6].

Diagnosing endometriosis has always been challenging. In a survey conducted on 7025 women with endometriosis, 65% of the women were misdiagnosed and 46% had to see five doctors or more before they were correctly diagnosed [7]. At present, the gold standard to diagnose endometriosis is through laparoscopic inspection, following with histological confirmation. The diagnostic delay of endometriosis between the onset of symptoms and a diagnosis is about 6–11 years [8, 9]. The delay in diagnosis may be caused by several factors such as the use of oral contraceptives (OCPs) and nonsteroidal antiinflammatory drugs (NSAIDs), the reluctance of physicians to refer patients to gynecologists for a definitive diagnosis, patients being afraid to check with physicians for fear of a cancer diagnosis, and patients regarding the pain as only a natural occurrence experienced by every woman.

The diagnosis of endometriosis can be established according to the following findings.

(a) *Symptoms:* The association between endometriosis and symptoms such as dysmenorrhea, nonmenstrual pelvic pain, dyspareunia, and infertility is widely accepted but limited for diagnosis. A previous study reported that there was no correlation between the severity of endometriosis and the type or severity of pain symptoms [10]. Another symptom-based model study found menstrual dyschezia and a history of benign ovarian cysts predictive of both any and stage III/IV endometriosis, according to the American Fertility Society/the American Society for Reproductive Medicine (AFS/ASRM) classification [11]. While AFS/ASRM stage III/IV disease was predicted with good accuracy (area under the curve [AUC]=84.9, sensitivity 82.3% and specificity 75.8%), any-stage endometriosis was predicted relatively poorly (AUC=68.3, sensitivity 84.8% and specificity 43.5%) [11]. The latest review of patient-completed or symptom-based screening tools for endometriosis summarized that the existing tools had limitations in clinical application and were inadequate as diagnosing tools for endometriosis [12].

(b) *Physical examinations:* Routine vaginal examination alone may be insufficient to detect endometriosis prior to laparoscopy. This is reinforced by the fact that in many women with endometriosis, no abnormality is detected by clinical examinations.

(c) *Imaging studies:* Transvaginal ultrasound (TVU) is adequate to detect ovarian endometriotic cysts, but fails to identify peritoneal endometriosis, endometriosis-associated adhesions, and some locations of deep infiltrating endometriosis (DIE). While endometrioma and DIE could be detected by TVU or transrectal ultrasound, TVU is superior in many cases [13]. Endometriosis located in ureters, bladder, the rectosigmoid colon, and other pelvic masses are easily detected by magnetic resonance imaging (MRI) [13]. However, the overall diagnosing accuracy of imaging studies is limited because of low resolution for pelvic adhesions and superficial peritoneal implants [13].

(d) *Blood tests:* Numerous markers have been proven to be expressed in the peripheral blood of women with endometriosis when compared with controls, such as cancer antigen 125 (CA125), cytokines, angiogenic and growth factors, and migration-related factors. Numerous serum biomarkers have been investigated in the last decade but none of them have been applied clinically because of weak validation.

(e) *Laparoscopic inspection with histologic confirmation:* Laparoscopy is a surgical procedure with rare but significant potential risks for the patients. Although this method up to now is the gold standard for diagnosing endometriosis, this procedure is invasive and expensive.

The above-mentioned methods adopted in current clinical practice highlight the dilemma of diagnosing endometriosis between invasiveness and accuracy. There is an urgent demand for an accurate test to diagnose endometriosis noninvasively.

It has been reported that there are many biochemical factors expressed differently between endometrial tissue and endometriosis tissue. A panel of biomarkers was proven to diagnose endometriosis that was undetectable by ultrasound with a high sensitivity and an acceptable specificity [14]. Biomarkers for endometriosis were categorized into three groups: (1) blood biomarkers, including glycoprotein, immunological markers and inflammatory cytokines, oxidative stress, cell adhesion and invasion, angiogenesis, hormones, autoantibodies, microribonucleic acid (miRNA), proteomics, metabolomics, circulating cell-free deoxyribonucleic acid (DNA), and cell populations; (2) urine biomarkers, such as cytokeratin-19; and (3) endometrial biomarkers, including endometrial transcriptome, miRNA, endometrial proteome, and neuronal markers [15].

The proteomics are approached by using the two-dimensional gel electrophoresis (2-DIGE) or the latest surface-enhanced laser desorption/ionization time-of-flight mass spectrometry (SELDI-TOF-MS) in combination with bioinformatics analysis. In 2014, Signorile et al. proposed the analysis of the serum Zn-alpha2-glycoprotein level as a new noninvasive diagnostic test for endometriosis with a sensitivity of 69.4% and a specificity of 100% [16]. However, the application in a clinical setting is still hard because of poor mass accuracy and reproducibility.

The involvement of neuroangiogenesis in the pathogenesis of endometriosis led to the investigation of nerve fibers as biomarkers. Ovarian endometriosis was found to be innervated by sensory and sympathetic nerves with a particularly high density of the sympathetic nerves [17]. In contrast, the pan-neuronal protein marker, PGP 9.5, was found in only 31.1% of ovarian endometriosis cases and 19% in patients with no pain symptoms [18]. Possible explanations for these contrasting results could be technical variations in the performance of immunohistochemical analysis or possible misinterpretation of the spatial relationship between the nerve fibers and the lesions. Routine endometrial nerve fiber assessment was proved neither sensitive nor specific for the diagnosis of endometriosis [19].

Aside from early detection, biomarkers to identify the recurrence of endometriosis were also important. There is an overexpression of cyclooxygenase-2 (COX-2) in ovarian endometrioma, which may serve as a biomarker for recurrence [20]. COX-2 expression was mostly observed in glandular epithelial cells of ovarian endometriotic cysts whereas vascular endothelial growth factor (VEGF) expression was observed mainly in stromal cells. COX-2 and VEGF were closely correlated with each other, and both contributed to the angiogenesis of ovarian endometriosis [21].

THE ASSOCIATIONS OF BLOOD TRACE METALS AND ENDOMETRIOSIS

Excessive accumulation, deficiency, or an imbalance of trace metals may disturb cell functions, resulting in abnormal cell proliferation and malignant transformation or cell degeneration and apoptosis. Being essential in the catalytic process of antioxidant enzymes, the trace metals in human blood might be involved in both the immune reactions and the protective factors against oxidative damage. Reports of the relationship between blood trace metals and endometriosis are inconsistent and limited. Among infertile women, the blood levels of zinc, lead, and cadmium were found to be significantly different between endometriosis and nonendometriosis groups (Table 22.1). The associations between blood trace metal levels and endometriosis are shown in Table 22.2 [22].

There is a negative association between the blood zinc level and the presence of endometriosis [22]. A deficiency of zinc has been associated with increases in inflammatory cytokines and other markers of inflammation in vitro and in vivo. Golovine et al. suggested that the depletion of intracellular zinc increases the expression of the proinflammatory cytokines interleukin (IL)-6 and IL-8 via the nuclear factor-kappa B (NF-κB)-dependent pathway in cancer cells [23]. Featured with inflammatory responses and adhesions, endometriotic lesions secrete several proinflammatory molecules, such as IL-1 and IL-8, and tumor necrosis factor alpha (TNFα) or interferon gamma (IFγ), which in turn recruit macrophages and

TABLE 22.1　Comparisons of Blood Trace Metal Levels Between Infertile Women With and Without Endometriosis

Variables	Women With Endometriosis ($n = 68$)				Women Without Endometriosis ($n = 122$)				P^a
	Median	IQR	Mean[b]	95% CI	Median	IQR	Mean[b]	95% CI	
Zn (mg/L)	4.47	2.59–22.52	6.72	5.01–9.02	11.62	4.08–38.53	11.86	9.38–14.99	.003
Cu (mg/L)	0.43	0.21–0.67	0.39	0.32–0.48	0.51	0.28–0.90	0.48	0.41–0.56	.076
Mn (µg/L)	0.53	0.29–2.09	0.72	0.53–0.98	0.54	0.29–1.99	0.65	0.53–0.79	.610
Fe (mg/L)	1179.50	655.65–2084.00	1185.86	998.53–1408.32	1336.00	713.30–1881.00	1103.34	923.53–1318.14	.978
Hg (µg/L)	0.25	0.04–1.43	0.15	0.08–0.27	0.07	0.03–0.82	0.10	0.07–0.15	.200
Pb (µg/L)[c]	21.30	4.12–46.59	13.37	9.27–19.28	4.64	1.75–32.94	8.53	6.55–11.10	.029
Cd (µg/L)[d]	0.45	0.05–4.88	0.42	0.25–0.72	0.21	0.02–1.41	0.21	0.14–0.31	.036
Cr (µg/L)	0.45	0.19–1.52	0.51	0.36–0.74	0.55	0.25–1.29	0.56	0.45–0.69	.351

[a] *Mann-Whitney-Wilcoxon nonparametric test.*
[b] *Data are geometric means.*
[c] *The proportion under limit of detection is 25.8%; LOD = 2 µg/L.*
[d] *The proportion under limit of detection is 31.6%; LOD = 0.02 µg/L.*

Zn, *zinc;* Cu, *copper;* Mn, *manganese;* Fe, *iron;* Hg, *mercury;* Pb, *lead;* Cd, *cadmium;* Cr, *chromium;* IQR, *interquartile range;* 95% CI, *95% confidence interval.*

T lymphocytes to the peritoneum [24]. In an animal model, Gonzalez-Ramos et al. reported that NF-κB was activated constitutively in peritoneal endometriotic lesions and that inhibition of NF-κB reduced the expression of intercellular adhesion molecule-1 and cell proliferation, but increased the rate of apoptosis [25, 26]. Bao et al. also found that zinc deficiency caused excessive activation of the NF-κB pathway, leading to augmentation of the innate immune and acute-phase responses systemically [27].

Zinc is also required for the activity of some antioxidant enzymes and plays an important role in modulating redox homeostasis. For example, the cytoplasmic type of cellular superoxide dismutase (SOD) has been identified as a copper- and zinc-containing enzyme that acts as a scavenger of toxic superoxide radicals. There is increasing evidence that oxidative stress is an important pathogenic factor for endometriosis. Many reports have suggested that the peritoneal fluid, blood, or tissues of patients with endometriosis contain high concentrations of oxidative stress markers such as malondialdehyde, 8-hydroxy-2-deoxyguanosine, lipid peroxides (LPOs), and nitric oxide [28]. Furthermore, endogenous reactive oxygen species (ROS) have been suggested to act as the second messengers in controlling cellular proliferation. Endometriotic cells display a high level of endogenous oxidative stress that activates the mitogen-activated protein kinase signaling pathway that regulates cell proliferation [29].

On the other hand, blood lead may be associated with the presence of endometriosis [22]. Lead is recognized as a toxin that changes the erythrocyte membrane composition, reduces life expectancy, and inhibits hemoglobin synthesis. Introduced mainly from environmental

TABLE 22.2 Odds Ratios and 95% Confidence Intervals for the Associations Between Blood Trace Metal Levels and Endometriosis

Variables	Tertiles[a]	Women With Endometriosis		Women Without Endometriosis		OR (95% CI)	aOR[b] (95% CI)
		n	%	n	%		
Metal							
Zn (mg/L)	1st (<5.5)	37	54.4	41	33.6	1.00	1.00
	2nd (5.5–33.2)	17	25.0	40	32.8	0.47 (0.23–0.97)*	0.42 (0.20–0.92)*
	3rd (>33.2)	14	19.0	41	33.6	0.38 (0.18–0.80)*	0.39 (0.18–0.88)*
Cu (mg/L)	1st (<0.4)	27	39.7	41	33.6	1.00	1.00
	2nd (0.4–0.7)	27	39.7	40	32.8	1.03 (0.51–2.04)	0.99 (0.47–2.07)
	3rd (>0.7)	14	20.6	41	33.6	0.52 (0.24–1.13)	0.45 (0.20–1.02)
Mn (μg/L)	1st (<0.3)	27	39.7	48	39.3	1.00	1.00
	2nd (0.3–1.1)	13	19.1	34	27.9	0.68 (0.31–1.50)	0.68 (0.30–1.56)
	3rd (>1.1)	28	41.2	40	32.8	1.24 (0.63–2.44)	1.25 (0.61–2.57)
Fe (mg/L)	1st (<825.5)	25	36.5	41	33.6	1.00	1.00
	2nd (825.5–1506.0)	18	27.0	40	32.8	0.74 (0.35–1.56)	0.70 (0.32–1.54)
	3rd (>1506.0)	25	36.5	41	33.6	1.00 (0.49–2.02)	0.83 (0.39–1.76)
Metal (toxins)							
Hg (μg/L)	1st (<0.04)	23	33.8	42	34.4	1.00	1.00
	2nd (0.04–0.37)	14	20.6	40	32.8	0.64 (0.29–1.41)	0.66 (0.29–1.54)
	3rd (>0.37)	31	45.6	40	32.8	1.42 (0.71–2.83)	1.54 (0.72–3.29)
Pb (μg/L)	1st (<3.8)	14	20.6	41	33.6	1.00	1.00
	2nd (3.8–30.5)	28	41.2	49	40.2	1.67 (0.78–3.59)	1.73 (0.77–3.88)
	3rd (>30.5)	26	38.2	32	26.2	2.38 (1.07–5.28)*	2.59 (1.11–6.06)*
Cd (μg/L)	1st (<0.02)	17	25.0	43	35.3	1.00	1.00
	2nd (0.02–0.54)	21	31.9	39	32.0	1.36 (0.63–2.95)	1.03 (0.45–2.37)
	3rd (>0.54)	30	44.1	40	32.7	1.90 (0.91–3.95)	1.73 (0.78–3.92)
Cr (μg/L)	1st (<0.3)	27	39.7	41	33.6	1.00	1.00
	2nd (0.3–0.8)	20	29.4	41	33.6	0.74 (0.36–1.53)	0.78 (0.36–1.68)
	3rd (>0.8)	21	30.9	40	32.8	0.80 (0.39–1.63)	0.81 (0.38–1.73)

[a] Based on their distribution in women without endometriosis group.
[b] aOR values were adjusted for the effects of age, body fat percentage, educational level, age at menarche, and regularity of menstrual cycles.
* Statistically significant at $P < .05$.
OR, odds ratio; CI, confidence interval; Zn, zinc; Cu, copper; Mn, manganese; Fe, iron; Hg, mercury, Pb, lead; Cd, cadmium; Cr, chromium.

exposure, blood lead is involved in the formation of ROS and in the depletion of intrinsic antioxidant defense mechanisms in vivo and in vitro, which might be linked with the progression of endometriosis. Many reports have suggested that lead exposure results in an imbalance of redox homeostasis based on evidence from occupationally exposed subjects showing higher levels of superoxidative indicators and lower activities of antioxidative markers than controls [30], which leads to lower SOD activity and is associated with endometriosis. A study on rhesus monkeys reported a consistent association [31]. In a study recruiting from the Taiwanese population with a generally low blood lead level, women in a high-level group had a three-fold increased risk of infertility compared with a low-level group [32]. On the contrary, lead was proposed as an antiestrogenic factor that helped to protect against endometriosis based on the finding that women suffering from endometriotic diseases showed lower levels of blood lead than controls [33]. The exact relationship between lead levels and endometriosis still requires further investigation.

As a metalloestrogen, cadmium might induce oxidative stress by the overproduction of superoxide anions and by modifying zinc and copper homeostasis. However, the association between blood cadmium levels and the risk of endometriosis in population-based human studies remains uncertain. Iron levels in the peritoneal fluids and macrophages of women with endometriosis were significantly higher than those in women without endometriosis [34]. Iron accumulation in the peritoneal cavity would affect the iron turnover in patients with endometriosis in addition to increasing oxidative stress. Copper, manganese, mercury, and chromium are also important metals for living organisms. They might catalyze redox homeostasis by cosynthesis with proteins or in the production of cellular toxins.

THE ASSOCIATIONS OF PHTHALATES AND ENDOMETRIOSIS

Phthalates are a class of man-made chemicals commonly used in the manufacturing of a variety of consumer products, including construction materials, food packaging, and children's toys. Phthalates are also used as solvents in cosmetics, the coating of medications, and dietary supplements. Human exposure to phthalates is mainly through the ingestion of contaminated food as well as packaging materials and cosmetic products. Phthalates are complex chemicals that are rapidly metabolized and excreted in urine [35]. Phthalate metabolites are not accumulated within the body and are detected in more than 78% of the US population [36]. Increasing evidence shows that phthalates affect human fertility and reproductive health, mainly by interfering with or disrupting the endocrine activity and system. Accordingly, they were termed endocrine disruptors or endocrine disrupting chemicals (EDCs). In humans, phthalates could compromise testicular function and spermatogenesis [37]. In animal studies, exposure to di(2-ethylhexyl) phthalate (DEHP) and its metabolite, mono(2-ethylhexyl) phthalate (MEHP), resulted in anovulation, smaller preovulatory follicles or delayed ovulation, decreased serum progesterone levels, and increased serum follicle-stimulating hormone (FSH) levels in sexually mature rats [38, 39]. In pregnant rats and pseudopregnant rats, exposure to dibutyl phthalate (DBP) and benzyl butyl phthalate (BzBP) impairs implantation [40, 41]. Furthermore, di-*n*-butyl phthalate (DnBP) and butylbenzyl phthalate (BBzP) had estrogenic activity [42]. MEHP decreased aromatase transcripts and protein levels in a dose-dependent manner in granulosa cells [43]. In adult cycling rats, DEHP decreased serum estradiol concentrations while prolonging the estrous cycle and anovulation [43].

In 2003, Cobellis et al. used high-performance liquid chromatography (HPLC) to measure DEHP and MEPH and found that the plasma DEHP concentrations of endometriosis patients are significantly higher than healthy controls [44]. Endometriosis patients showed significantly higher levels of blood DnBP, BBzP, di-*n*-octyl phthalate (DnOP) and DEHP compared with women free from endometriosis [45]. Phthalates are rapidly metabolized and excreted in urine. Urinary phthalate metabolites are stable and used as a biomarker of phthalate exposure in the low exposure general population [37]. Using high-performance liquid chromatography tandem mass spectrometry to measure monoethyl phthalate (MEP), mono-*n*-buty phthalate (MnBP), monobenzyl phthalate (MBzP), MEHP, mono(2-ethyl-5 hydroxyhexyl) phthalate (MEHHP), and mono(2-ethyl-5-oxhexyl) phthalate (MEOHP) found no significant association of endometriosis and these phthalate metabolites in Japanese women [46]. Huang et al. uncovered that glutathione S-transferase M 1 (GSTM1, a major detoxification enzyme) null genotype and phthalate exposure are associated with adenomyosis and leiomyoma [47]. In the United States, the association of MBP and endometriosis was observed in self-reported histories of endometriosis patients [48]. Moreover, Kim et al. used liquid chromatography tandem mass spectrometry and found that MEHP and DEHP concentrations of plasma in women with endometriosis are significantly higher than control groups [49]. Buck Louis et al. demonstrated that MEHP was associated with endometriosis when the disease was restricted to comparison women with a normal pelvis after surgery [50]. In 2013, Upson et al. found that MBzP and MEP may increase endometriosis risk [51]. Kim et al. demonstrated that DEHP causes increases of matrix metalloproteinase (MMP)-2 and MMP-9 activity; it also induces cell invasion and increases the endometrial implant in nonobese diabetic/severe combined immunodeficiency (NOD/SCID) mice [52]. Urinary concentration MEHHP and MEOHP are associated with endometriosis in Korean women [52]. Therefore, these studies suggest that phthalates may play an important role in the pathogenesis of endometriosis.

THE EFFECTS OF PHTHALATES ON OOCYTE AND EMBRYO DEVELOPMENT

MEHP is the major and primary metabolite of DEHP. Increasing evidence has shown that the cytotoxicity of MEHP was 10- to 20-fold in comparison to its parent compound because of the bioactivity [53, 54]. DEHP, with the lipophilic feature caused by two 2-ethylhexanol branched chains, activates intracellular signal cascades while crossing the lipid membrane. In contrast, MEHP has only one 2-ethylhexanol branched chain and is consequently less lipophilic than DEHP. The structural differences between DEHP and MEHP lead to their distinct effects. It has been demonstrated that MEHP affects the signaling molecules located on the membrane rather than activating the intracellular molecules [55].

Adverse pregnancy outcomes were reported in pregnant women having high concentrations of MEHP in the urine or serum, including low birth weight and small head circumference of the newborn as well as shortened pregnancy duration and increased pregnancy loss

[56]. in vitro studies revealed that MEHP inhibited meiotic maturation of the oocyte and negatively modulated embryo development in the bovine [57] and mouse [58] with a dose-dependent effect. Both DEHP and MEHP inhibited follicle growth in mice by reducing estradiol production in antral follicles [59]. MEHP might affect the oocyte by interfering with estradiol activity via binding to estrogen receptors (ERs). Mu et al. showed that the injection of DEHP into neonatal mice resulted in reduced expression of ERa and ERb in the ovary [60].

MEHP tends to accumulate in adipose tissue, an important regulator of energy balance and glucose homeostasis. Results from both murine and human fat cell models indicated that MEHP promoted adipogenesis via activation of the peroxisome proliferator-activated receptor-γ (PPAR-γ) [61]. Moreover, MEHP also spatially altered the metabolic genes of early organogenesis-stage mouse conceptus cultured in vitro [62]. These results were in agreement with the induced obesity observed in mice with perinatal exposure to DEHP [63].

Phthalates are ubiquitous environmental contaminants that in general display low toxicity. Although high dose exposure of these compounds results in gonadal injury, the dosages used in studies are too high to reflect exposure levels in the real world. Besides, the toxicological consequences of phthalates could be complex considering the structural differences, distinct effects, and interactions between all the metabolites. Recent results of transgenerational studies suggest that phthalates induce epigenetic changes in the offspring and they may become more sensitive to subsequent exposure of phthalates than the parents. These issues are important in the designation of further studies.

ASSISTED REPRODUCTIVE TECHNOLOGY (ART) IN PRIMARY AND RECURRENT ENDOMETRIOSIS

The impact of endometriosis on natural conception is clearly demonstrated by the Endometriosis Fertility Index (EFI) [64]. With external validation, EFI is recommended by the World Endometriosis Society as one of the classification systems for women with endometriosis if future fertility is a concern. Regarding spontaneous pregnancy rates, surgical management is more effective than hormonal therapies, especially for women with AFS/ASRM stage I/II endometriosis [65]. In a randomized control trial recruiting women with AFS/ASRM stage I/II endometriosis, the live birth rate was 5.6 times higher (95% confidence interval (CI): 1.18–17.4) among those undergoing intrauterine insemination (IUI) with controlled ovarian hyperstimulation (COH) compared with those managed expectantly [66]. in vitro fertilization (IVF) is recommended for women with endometriosis accompanied by tubal dysfunction or male factor infertility as well as those who failed to conceive by other treatments [65]. Surgical removal of peritoneal endometriosis prior to IVF in women with AFS/ASRM stage I/II endometriosis should not be performed routinely because the benefit is not well established [67]. In addition, a cystectomy for women with endometrioma before IVF should be considered only when the accessibility of follicles is compromised [65].

As for the COH protocols, the use of prolonged down-regulation with gonadotropin-releasing hormone (GnRH) agonists for women with endometriosis prior to IVF was regarded as superior to other protocols for achieving pregnancy [68]. In prospective studies, women with severe endometriosis undergoing a prolonged GnRH agonist to achieve pituitary down-regulation before IVF had significantly higher pregnancy rates compared to those receiving only gonadotropins (25% vs. 3.9%) [69] or treated with a short protocol (34.7% vs. 10.7%) [70]. In one randomized study enrolling women with various stages of endometriosis, pretreatment with a GnRH agonist for 3 months before initiation of COH achieved a significantly higher ongoing pregnancy rate compared with those using a long protocol or a microdose flare regimen (80% vs. 53.85%) [71]. Despite the limitations of small sample sizes and high heterogeneity, most studies and a Cochrane review concluded that a prolonged down-regulation with a GnRH agonist before the start of COH was beneficial to the IVF outcomes for women with endometriosis [68–72].

The later-developed GnRH antagonist offers another choice for the COH protocols. One randomized prospective study evaluating women with AFS/ASRM stage I/II endometriosis undergoing IVF/ICSI found that the pregnancy outcomes were similar in those using a GnRH agonist and a GnRH antagonist protocol [73]. Regarding women receiving cystectomy for endometrioma, a retrospective study reported that the number of follicles on versus human chorionic gonadotropin injection day, versus duration of hyperstimulation, versus number of retrieved metaphase II oocytes, and versus total number of grade 1 embryos were significantly higher in the long GnRH agonist group than in the GnRH antagonist group, but there were no differences in pregnancy rates [74]. Another retrospective study also demonstrated comparable pregnancy rates after COH with either GnRH agonists or GnRH antagonists among patients being matched according to the propensity score [75]. The above studies indicate that GnRH antagonists might be as equally effective as GnRH agonists in determining the reproductive outcomes of IVF for women with endometriosis.

The high recurrence rate of endometriosis after surgery is frustrating. In general, 40%–45% of patients with endometriosis have a relapse of the disease within 5 years after the primary surgery [76] and the average two-year recurrence rate was calculated to be 19.1% [77]. Multiple factors influence the ovarian endometrioma's recurrence rate. Numerous epidemiological studies have identified the risk factors for recurrence, often given conflicting results because of different surgical procedures, various end-point measurements (recurrence of symptoms or cyst, improvement of fertility), different staging, and biases in the selection of patients for a study. When only the better powered and executed studies were considered, the total revised AFS score but not the revised AFS stage seems to be a risk factor for recurrence [78]. Previous medical treatment of endometriosis and larger cysts appeared to be associated with higher recurrence; in contrast, postoperative pregnancy was associated with a lower recurrence rate. Persistent progesterone exposure, the absence of menstruation, and the formation of corpus lutea can explain the protective effect of pregnancy for the recurrence of endometriomas.

Medication with ovulation-suppression agents has been demonstrated to have benefits in alleviating symptoms associated with endometriosis and extending the interval between surgery and disease recurrence. However, no evidence was found that such ovarian suppression agents as danazol, GnRH agonist, and progestins improve fertility in the treatment of endometriosis [65]. In the case of recurrence, reoperation is often considered the best treatment option. However, the consequences of endometriosis relapse on fecundability may be particularly detrimental, owing to ovarian damage caused by both recurrent disease and repeated surgical trauma. The role of preservation of ovarian function has been highlighted in infertility treatment. According to the guidelines of ASRM and the European Society of Human

Reproduction and Embryology (ESHRE) [65], repeated operations are not recommended because further operations had detrimental effects on the ovarian reserve.

As previously discussed, the ovarian suppression agents have not been shown to enhance pregnancy rates associated with natural or stimulated cycles in infertile women with endometriosis who are not undergoing ART. Sallam et al. studied the role of medical treatment before IVF and they reported that pretreatment with GnRH agonists at least 3 months prior to IVF increased the pregnancy rate [72]. Similar outcomes had been reported by others but the study design and inclusion criteria vary. In a prospective randomized trial, Rickes et al. assessed the role of prolonged GnRH agonist therapy for 6 months prior to either IVF or IUI after surgical treatment of endometriosis [79]. A statistically significant benefit was noted only among patients with more severe disease (AFS/ASRM stages III/IV) who subsequently underwent IVF [79].

Dicker et al. proposed that the use of transvaginal ultrasound-guided aspiration of ovarian endometriomas significantly improved the embryo-to-oocyte ratio in women with endometriosis undergoing an IVF treatment [80]. A more recent study compared three different treatments of endometriomas: ultrasound-guided aspiration, laparotomy/laparoscopy, or with no treatment at all. Subsequently, IVF was performed in all patients, and the observed fertilization rates were 67%, 57%, and 56%, respectively [81]. Ultrasound-guided transvaginal ethanol sclerotherapy (EST) has been widely practiced in Japan for ovarian endometriotic cysts. A prospective study [82] was reported to use EST for the treatment of recurrent endometriotic cysts before an IVF attempt. Ovarian cysts recurred in only 12.9% of cases, at a mean time of 10 months after EST, without harmful effects on ovarian reserve and ovarian response to stimulation. Consequently, clinical and cumulative pregnancy rates of the EST group were higher than those of women who underwent laparoscopic cystectomy. According to the authors, EST may be a good alternative to surgical management of recurrent endometriotic cysts before ART [82].

There is evidence that IUI following COH seems to be a better than expected management technique in infertile women with endometriosis. Notably, those studies were assessed in patients with minimal to mild endometriosis, and there is insufficient evidence to support COH/IUI in patients with severe endometriosis. According to the ESHRE recommended guidelines [65], IUI is only recommended in subfertile women with minimal to mild endometriosis, and IUI with COH should be considered within 6 months following surgery in the treatment of infertile women with AFS/ASRM stage I/II endometriosis.

IVF is currently the most efficient and successful treatment of endometriosis associated infertility. Patients with endometriosis can be referred for early infertility treatment, including IVF, to increase the chances of conception. Direct IVF should be considered if the women's age is more than 35 years and the duration of infertility is long [83].

Endometriosis is an estrogen-dependent disease because it is rarely observed before menarche and usually disappears after menopause. There might be concern that temporary and repeated exposure to high estradiol concentration during COH for IVF contributes to the recurrence of endometriosis. D'Hooghe et al. conducted a retrospective cohort study and reported that the cumulative recurrence rates were not increased after ovarian hyperstimulation for IVF [84]. Benaglia et al. studied 48 women with ovarian endometriomas undergoing IVF, comparing cyst diameter before and 3–6 months after IVF. The median size of endometrioma pre- and post-IVF was 20 mm and 21 mm, respectively. It was concluded that ovarian stimulation is not associated with increased endometriosis recurrence [85]. In conclusion, the temporary exposure to high estradiol levels in women during COH for IVF is not a major risk factor for recurrence of endometriosis.

Endometriosis may impact negatively on fertility and IVF outcomes through various mechanisms. However, IVF remains the most successful, but not the only, approach to overcome endometriosis-related infertility. The dilemma regarding the best approach to manage endometriosis-associated infertility remains unresolved. For the recurrent disease, the decision about the most appropriate treatment is even more complicated and remains controversial. Before choosing the most appropriate treatment, it is critical for clinicians to perform a thorough assessment of ovarian reserve, the severity of clinical symptoms, and all other factors that may influence clinical outcomes before initiating therapy. The purpose of reproductive surgery is to restore anatomy and to improve fertility, not to eradicate the disease while sacrificing the ovarian reserve. Surgery, especially for an ovary that was previously operated on, is not recommend by ASRM and ESHRE guidelines unless a suspicious malignancy is noted. Cyst aspiration may be considered if the mass hinders oocyte growth and collection.

ASSISTED REPRODUCTIVE TECHNOLOGY (ART) IN ADENOMYOSIS

Coined in 1925 by Frankl, the current definition of adenomyosis was provided in 1972 by Bird, who stated: "Adenomyosis may be defined as the benign invasion of endometrium into the myometrium, producing a diffusely enlarged uterus which microscopically exhibits ectopic non-neoplastic, endometrial glands and stroma surrounded by the hypertrophic and hyperplastic myometrium" [86]. This invading process of the endometrium is hypothesized to interfere with fertility because of the disrupted myometrial architecture and the resulting local inflammation. One study enrolling recipients of oocyte donation reported similar implantation rates but higher miscarriage rates in recipients with adenomyosis compared with matched controls [87]. The etiology and pathogenesis of adenomyosis are still unclear. A strong association between adenomyosis and endometriosis leads to adenomyosis-associated infertility being explained in similar physiopathological terms to those of endometriosis. However, adenomyosis and endometriosis exhibit different clinical profiles [88]. Whether adenomyosis and endometriosis share a common pathogenesis remains an ongoing debate. In the past, the diagnosis of adenomyosis was usually based on pathological examinations for women undergoing hysterectomy over the age of 40. Nowadays, the late first pregnancy of women as well as the enhanced accuracy of imaging technologies to diagnose adenomyosis preoperatively lead to the increased prevalence of adenomyosis in infertility clinics. Although the association between adenomyosis and infertility has not been well established, pooled data from IVF/ICSI displayed a negative impact of adenomyosis on pregnancy outcomes, owing to the reduced likelihood of clinical pregnancy (RR of 0.72, 95% CI: 0.55–0.95) and increased risk of early pregnancy loss (RR of 2.12, 95% CI: 1.20–3.75) [89].

Successful pregnancy following GnRH agonist treatment for women with adenomyosis had been reported in scattered cases since 1993 [90]. Pooled data from two retrospective studies adopting a GnRH long protocol to induce ovulation for IVF/ICSI did not show lower pregnancy rates in women with adenomyosis (RR of 1.05, 95% CI: 0.75–1.48) [89]. As for women with adenomyosis receiving frozen embryo transfer, higher pregnancy rates were reported in those pretreated with the GnRH agonist [91, 92]. The GnRH agonist treatment not only decreases the size of adenomyotic lesions [93] but also ameliorates endometrial implantation markers [94]. Experiences from the IVF clinic of Taipei Medical University Hospital showed the uterine volume of women with adenomyosis reduced almost by half ($42.27 \pm 17.31\%$) after GnRH agonist treatment. Among women with adenomyosis aged 38.61 ± 3.57 years, IVF with GnRH agonist pretreatment achieved a clinical pregnancy rate of 36.1% and an ongoing pregnancy rate of 27.8% (Table 22.3). Age was found to be an important factor determining the IVF outcomes based on the higher clinical pregnancy rates (56.3% vs. 20.0%, $P < .05$) and ongoing pregnancy rates (50.0% vs. 10.0% $P < .05$) among women aged 38 or less compared to those older than 38, which was mainly attributed to more high-quality embryos yielded by younger women (Table 22.4).

Cytoreductive surgery that completely removed adenomyotic lesions was also reported to be efficient, as the delivery rate achieved 50% in a review comprising younger women with

TABLE 22.3   IVF Outcomes for Women With Adenomyosis Pretreated With GnRH Agonist in Taipei Medical University Hospital (Unpublished Data)

Cycle number, n	36
Age, years	38.61 ± 3.57
AMH, ng/mL	2.63 ± 2.22
CA-125 (before GnRH agonist treatment), U/mL	153.14 ± 124.64
CA-125 (after GnRH agonist treatment), U/mL	29.27 ± 9.95
Uterine volume reduction[a], %	42.27 ± 17.31
Estrogen on the trigger day, pg/mL	2171 ± 1606
Progesterone on the trigger day, ng/mL	0.73 ± 0.37
Endometrial thickness on the trigger day, mm	10.64 ± 2.03
Number of oocyte retrieved, n	7.92 ± 5.28
Number of metaphase II oocyte, n	5.14 ± 4.33
Number of fertilized embryos, n	5.17 ± 3.34
Number of transferred embryos, n	2.56 ± 1.05
Culture day of transferred embryo, days	3.17 ± 0.97
Supernumerary embryos for vitrification, n	1.61 ± 2.77
Clinical pregnancy rate, n/total (%)	13/36 (36.1)
Implantation rate, n/total (%)	17/108 (15.7)
Abortion rate, n/total (%)	3/13 (23.1)
Ongoing pregnancy rate (>12 weeks), n/total (%)	10/36 (27.8)

[a] The percentage of uterine volume reduction was calculated as [volume(before pretreatment) − volume(after pretreatment)]/volume (before pretreatment).
All the values were presented as mean ± standard deviation or percentage.

TABLE 22.4 The Impact of Age on the IVF Outcomes for Women With Adenomyosis Pretreated With GnRH Agonist in Taipei Medical University Hospital (Unpublished Data)

	Age ≤ 38 (n = 16)	Age > 38 (n = 20)	P value
Age, years	35.69 ± 2.80	40.95 ± 2.09	<.001
AMH, ng/mL	3.30 ± 2.34	2.09 ± 2.02	NS
CA-125 (before GnRH agonist treatment), U/mL	123.62 ± 102.64	176.75 ± 137.77	NS
CA-125 (after GnRH agonist treatment), U/mL	25.25 ± 8.95	32.48 ± 9.73	.028
Uterine volume reduction[a], %	44.75 ± 14.49	40.46 ± 19.44	NS
Estrogen on the trigger day, pg/mL	2967 ± 1676	1535 ± 1254	.006
Progesterone on the trigger day, ng/mL	0.85 ± 0.38	0.63 ± 0.34	NS
Endometrial thickness on the trigger day, mm	11.06 ± 1.95	10.30 ± 2.08	NS
Number of oocyte retrieved, n	11.06 ± 5.30	5.40 ± 3.78	<.001
Number of metaphase II oocyte, n	7.25 ± 5.05	3.45 ± 2.76	.013
Number of fertilized embryos, n	6.69 ± 3.75	3.95 ± 2.44	.012
Number of transferred embryos, n	2.19 ± 0.65	2.85 ± 1.23	.047
Culture day of transferred embryo, days	3.59 ± 0.87	2.81 ± 0.98	.011
Supernumerary embryos for vitrification, n	2.82 ± 3.39	0.42 ± 1.23	.012
Clinical pregnancy rate, n/total (%)	9/16 (56.3)	4/20 (20.0)	.027
Implantation rate, n/total (%)	9/35 (25.7)	6/57 (10.5)	NS
Abortion rate, n/total (%)	1/9 (11.1)	2/6 (33.3)	NS
Ongoing pregnancy rate (>12 weeks), n/total (%)	8/16 (50.0)	2/20 (10.0)	.009

[a] *The percentage of uterine volume reduction was calculated as [volume(before pretreatment) − volume(after pretreatment)]/volume(before pretreatment).*

All the values were presented as mean ± standard deviation or percentage.

localized adenomyosis [95]. Another review including women with extensive adenomyosis at an older age demonstrated a delivery rate of 37% [96]. Based on a report that cytoreductive surgery failed to improve pregnancy rates in women with adenomyosis who were older than 40 [97], age should always be a concern before surgery. Extensive surgery could be associated with uterine rupture. The optimal wall thickness in balance with conception and prevention of rupture after cytoreductive surgery may range from 9 to 15 mm [98].

It is notable that women with poor ovarian reserve who are unable to afford a time-consuming pretreatment such as GnRH agonist or cytoreductive surgery may benefit from segmented IVF. For these women, oocytes could be collected using any of the protocols and the resulting embryos could be preserved by vitrification. The collection of oocytes can be repeated until a sufficient number of frozen embryos are obtained. The embryos can then be transferred following a long-term pituitary down-regulation by a GnRH agonist or/and cytoreductive surgery to improve the implantation environment.

CONCLUSION

Although a tool that diagnoses endometriosis both noninvasively and accurately is currently unavailable, a panel of biomarkers with high sensitivity and acceptable specificity is promising. The noninvasive early diagnosis of endometriosis may soon be achieved through a combination of the various diagnosing modalities. Environmental factors also appear to interact with endometriosis. While blood zinc displays a protective role against endometriosis, the relationship between blood lead and endometriosis remains unclear. Phthalates, influencing the human homeostasis by a group of metabolites, seem to be associated with endometriosis and impose detrimental effects on the development of oocytes and embryos. However, results from the past studies are not only inconclusive but also inapplicable in the real world. An aggregate effect of metabolites on endometriosis and human reproduction should be focused in the future. While managing endometriosis, preserving ovarian function should always be kept in mind if future fertility is a concern. Although endometriosis is highly associated with infertility, women with endometriosis are still able to conceive naturally. For women with endometriosis suffering infertility, IVF is currently the most efficient and successful treatment. The temporary exposure to high estradiol levels during COH for IVF is not a major risk factor for recurrence of endometriosis. In recurrent endometriomas, EST serves as a good alternative to the surgical management of recurrent endometriomas before ART. Although the association between adenomyosis and infertility has not been well established, the invading process of endometrium in adenomyosis is hypothesized to interfere with fertility because of the disrupted myometrial architecture and the resulting local inflammation. IVF following pretreatment with GnRH agonists or surgical removal of the endometriotic lesions are both efficient management techniques for infertile women with adenomyosis.

References

[1] Machairiotis N, Stylianaki A, Dryllis G, Zarogoulidis P, Kouroutou P, Tsiamis N, et al. Extrapelvic endometriosis: a rare entity or an under diagnosed condition? Diagn Pathol 2013;8:194.

[2] Eisenberg VH, Weil C, Chodick G, Shalev V. Epidemiology of endometriosis: a large population-based database study from a healthcare provider with 2 million members. BJOG 2018;125(1):55–62.

[3] Mowers EL, Lim CS, Skinner B, Mahnert N, Kamdar N, Morgan DM, et al. Prevalence of endometriosis during abdominal or laparoscopic hysterectomy for chronic pelvic pain. Obstet Gynecol 2016;127(6):1045–53.

[4] Lagana AS, Vitale SG, Salmeri FM, Triolo O, Ban Frangez H, Vrtacnik-Bokal E, et al. Unus pro omnibus, omnes pro uno: a novel, evidence-based, unifying theory for the pathogenesis of endometriosis. Med Hypotheses 2017;103:10–20.

[5] Chen LC, Hsu JW, Huang KL, Bai YM, Su TP, Li CT, et al. Risk of developing major depression and anxiety disorders among women with endometriosis: A longitudinal follow-up study. J Affect Disord 2016;190:282–5.

[6] Facchin F, Barbara G, Saita E, Mosconi P, Roberto A, Fedele L, et al. Impact of endometriosis on quality of life and mental health: pelvic pain makes the difference. J Psychosom Obstet Gynaecol 2015;36(4):135–41.

[7] European Endometriosis Alliance. Announcement of results from pan-European pain and quality of life study. Available from: http://endometriosis.org/news/support-awareness/european-endometriosis-alliance/; 2005 [cited 21.03.18].

[8] Husby GK, Haugen RS, Moen MH. Diagnostic delay in women with pain and endometriosis. Acta Obstet Gynecol Scand 2003;82(7):649–53.

[9] Nnoaham KE, Hummelshoj L, Webster P, d'Hooghe T, de Cicco Nardone F, de Cicco Nardone C, et al. Impact of endometriosis on quality of life and work productivity: a multicenter study across ten countries. Fertil Steril 2011;96(2):366–73: e8.

[10] Kennedy S, Bergqvist A, Chapron C, D'Hooghe T, Dunselman G, Greb R, et al. ESHRE guideline for the diagnosis and treatment of endometriosis. Hum Reprod 2005;20(10):2698–704.

[11] Nnoaham KE, Hummelshoj L, Kennedy SH, Jenkinson C, Zondervan KT, World Endometriosis Research Foundation Women's Health Symptom Survey Consortium. Developing symptom-based predictive models of endometriosis as a clinical screening tool: results from a multicenter study. Fertil Steril 2012;98(3):692–701: e5.

[12] Surrey E, Carter CM, Soliman AM, Khan S, DiBenedetti DB, Snabes MC. Patient-completed or symptom-based screening tools for endometriosis: a scoping review. Arch Gynecol Obstet 2017;296(2):153–65.

[13] Hsu AL, Khachikyan I, Stratton P. Invasive and noninvasive methods for the diagnosis of endometriosis. Clin Obstet Gynecol 2010;53(2):413–9.

[14] Vodolazkaia A, El-Aalamat Y, Popovic D, Mihalyi A, Bossuyt X, Kyama CM, et al. Evaluation of a panel of 28 biomarkers for the non-invasive diagnosis of endometriosis. Hum Reprod 2012;27(9):2698–711.

[15] Fassbender A, Burney RO, O DF, D'Hooghe T, Giudice L. Update on biomarkers for the detection of endometriosis. Biomed Res Int 2015;2015:130854.

[16] Kuessel L, Jaeger-Lansky A, Pateisky P, Rossberg N, Schulz A, Schmitz AA, et al. Cytokeratin-19 as a biomarker in urine and in serum for the diagnosis of endometriosis—a prospective study. Gynecol Endocrinol 2014;30 (1):38–41.

[17] Tokushige N, Russell P, Black K, Barrera H, Dubinovsky S, Markham R, et al. Nerve fibers in ovarian endometriomas. Fertil Steril 2010;94(5):1944–7.

[18] Zhang X, Yao H, Huang X, Lu B, Xu H, Zhou C. Nerve fibres in ovarian endometriotic lesions in women with ovarian endometriosis. Hum Reprod 2010;25(2):392–7.

[19] Leslie C, Ma T, McElhinney B, Leake R, Stewart CJ. Is the detection of endometrial nerve fibers useful in the diagnosis of endometriosis? Int J Gynecol Pathol 2013;32(2):149–55.

[20] Yuan L, Shen F, Lu Y, Liu X, Guo SW. Cyclooxygenase-2 overexpression in ovarian endometriomas is associated with higher risk of recurrence. Fertil Steril 2009;91(4 Suppl):1303–6.

[21] Ceyhan ST, Onguru O, Baser I, Gunhan O. Expression of cyclooxygenase-2 and vascular endothelial growth factor in ovarian endometriotic cysts and their relationship with angiogenesis. Fertil Steril 2008;90(4):988–93.

[22] Lai GL, Yeh CC, Yeh CY, Chen RY, Fu CL, Chen CH, et al. Decreased zinc and increased lead blood levels are associated with endometriosis in Asian Women. Reprod Toxicol 2017;74:77–84.

[23] Golovine K, Uzzo RG, Makhov P, Crispen PL, Kunkle D, Kolenko VM. Depletion of intracellular zinc increases expression of tumorigenic cytokines VEGF, IL-6 and IL-8 in prostate cancer cells via NF-κB-dependent pathway. Prostate 2008;68(13):1443–9.

[24] Giudice LC, Kao LC. Endometriosis. Lancet 2004;364(9447):1789–99.

[25] Gonzalez-Ramos R, Donnez J, Defrere S, Leclercq I, Squifflet J, Lousse JC, et al. Nuclear factor-kappa B is constitutively activated in peritoneal endometriosis. Mol Hum Reprod 2007;13(7):503–9.

[26] Gonzalez-Ramos R, Van Langendonckt A, Defrere S, Lousse JC, Mettlen M, Guillet A, et al. Agents blocking the nuclear factor-κB pathway are effective inhibitors of endometriosis in an in vivo experimental model. Gynecol Obstet Invest 2008;65(3):174–86.

[27] Bao S, Liu MJ, Lee B, Besecker B, Lai JP, Guttridge DC, et al. Zinc modulates the innate immune response in vivo to polymicrobial sepsis through regulation of NF-κB. Am J Physiol Lung Cell Mol Physiol 2010;298(6):L744–54.

[28] Carvalho LF, Samadder AN, Agarwal A, Fernandes LF, Abrao MS. Oxidative stress biomarkers in patients with endometriosis: systematic review. Arch Gynecol Obstet 2012;286(4):1033–40.

[29] Ngo C, Chereau C, Nicco C, Weill B, Chapron C, Batteux F. Reactive oxygen species controls endometriosis progression. Am J Pathol 2009;175(1):225–34.

[30] Kasperczyk S, Slowinska-Lozynska L, Kasperczyk A, Wielkoszynski T, Birkner E. The effect of occupational lead exposure on lipid peroxidation, protein carbonylation, and plasma viscosity. Toxicol Ind Health 2015; 31(12):1165–71.

[31] Krugner-Higby L, Rosenstein A, Handschke L, Luck M, Laughlin NK, Mahvi D, et al. Inguinal hernias, endometriosis, and other adverse outcomes in rhesus monkeys following lead exposure. Neurotoxicol Teratol 2003; 25(5):561–70.

[32] Chang SH, Cheng BH, Lee SL, Chuang HY, Yang CY, Sung FC, et al. Low blood lead concentration in association with infertility in women. Environ Res 2006;101(3):380–6.

[33] Heilier JF, Donnez J, Verougstraete V, Donnez O, Grandjean F, Haufroid V, et al. Cadmium, lead and endometriosis. Int Arch Occup Environ Health 2006;80(2):149–53.

[34] Lousse JC, Defrere S, Van Langendonckt A, Gras J, Gonzalez-Ramos R, Colette S, et al. Iron storage is significantly increased in peritoneal macrophages of endometriosis patients and correlates with iron overload in peritoneal fluid. Fertil Steril 2009;91(5):1668–75.

[35] Kay VR, Chambers C, Foster WG. Reproductive and developmental effects of phthalate diesters in females. Crit Rev Toxicol 2013;43(3):200–19.

[36] Silva MJ, Barr DB, Reidy JA, Malek NA, Hodge CC, Caudill SP, et al. Urinary levels of seven phthalate metabolites in the U.S. population from the National Health and Nutrition Examination Survey (NHANES) 1999–2000. Environ Health Perspect 2004;112(3):331–8.

[37] Hauser R, Calafat AM. Phthalates and human health. Occup Environ Med 2005;62(11):806–18.

[38] Lovekamp TN, Davis BJ. Mono-(2-ethylhexyl) phthalate suppresses aromatase transcript levels and estradiol production in cultured rat granulosa cells. Toxicol Appl Pharmacol 2001;172(3):217–24.

[39] Davis BJ, Maronpot RR, Heindel JJ. Di-(2-ethylhexyl) phthalate suppresses estradiol and ovulation in cycling rats. Toxicol Appl Pharmacol 1994;128(2):216–23.

[40] Ema M, Miyawaki E, Kawashima K. Effects of dibutyl phthalate on reproductive function in pregnant and pseudopregnant rats. Reprod Toxicol 2000;14(1):13–9.

[41] Ema M, Miyawaki E, Kawashima K. Reproductive effects of butyl benzyl phthalate in pregnant and pseudopregnant rats. Reprod Toxicol 1998;12(2):127–32.

[42] Soto AM, Sonnenschein C, Chung KL, Fernandez MF, Olea N, Serrano FO. The E-SCREEN assay as a tool to identify estrogens: an update on estrogenic environmental pollutants. Environ Health Perspect 1995;103(Suppl. 7):113–22.

[43] Lovekamp-Swan T, Davis BJ. Mechanisms of phthalate ester toxicity in the female reproductive system. Environ Health Perspect 2003;111(2):139–45.

[44] Cobellis L, Latini G, De Felice C, Razzi S, Paris I, Ruggieri F, et al. High plasma concentrations of di-(2-ethylhexyl)-phthalate in women with endometriosis. Hum Reprod 2003;18(7):1512–5.

[45] Reddy BS, Rozati R, Reddy BV, Raman NV. Association of phthalate esters with endometriosis in Indian women. BJOG 2006;113(5):515–20.

[46] Itoh H, Iwasaki M, Hanaoka T, Sasaki H, Tanaka T, Tsugane S. Urinary phthalate monoesters and endometriosis in infertile Japanese women. Sci Total Environ 2009;408(1):37–42.

[47] Huang PC, Tsai EM, Li WF, Liao PC, Chung MC, Wang YH, et al. Association between phthalate exposure and glutathione S-transferase M1 polymorphism in adenomyosis, leiomyoma and endometriosis. Hum Reprod 2010;25(4):986–94.

[48] Weuve J, Hauser R, Calafat AM, Missmer SA, Wise LA. Association of exposure to phthalates with endometriosis and uterine leiomyomata: findings from NHANES, 1999-2004. Environ Health Perspect 2010;118(6):825–32.

[49] Kim SH, Chun S, Jang JY, Chae HD, Kim CH, Kang BM. Increased plasma levels of phthalate esters in women with advanced-stage endometriosis: a prospective case-control study. Fertil Steril 2011;95(1):357–9.

[50] Buck Louis GM, Peterson CM, Chen Z, Croughan M, Sundaram R, Stanford J, et al. Bisphenol A and phthalates and endometriosis: the endometriosis: natural history, diagnosis and outcomes study. Fertil Steril 2013;100 (1):162–9: e1–e2.

[51] Upson K, Sathyanarayana S, De Roos AJ, Thompson ML, Scholes D, Dills R, et al. Phthalates and risk of endometriosis. Environ Res 2013;126:91–7.

[52] Kim SH, Cho S, Ihm HJ, Oh YS, Heo SH, Chun S, et al. Possible role of phthalate in the pathogenesis of endometriosis: in vitro, animal, and human data. J Clin Endocrinol Metab 2015;100(12):E1502–11.

[53] Yang G, Zhang W, Qin Q, Wang J, Zheng H, Xiong W, et al. Mono(2-ethylhexyl) phthalate induces apoptosis in p53-silenced L02 cells via activation of both mitochondrial and death receptor pathways. Environ Toxicol 2015;30(10):1178–91.

[54] Dees JH, Gazouli M, Papadopoulos V. Effect of mono-ethylhexyl phthalate on MA-10 Leydig tumor cells. Reprod Toxicol 2001;15(2):171–87.

[55] Gazouli M, Yao ZX, Boujrad N, Corton JC, Culty M, Papadopoulos V. Effect of peroxisome proliferators on Leydig cell peripheral-type benzodiazepine receptor gene expression, hormone-stimulated cholesterol transport, and steroidogenesis: role of the peroxisome proliferator-activator receptor alpha. Endocrinology 2002;143(7):2571–83.

[56] Adibi JJ, Hauser R, Williams PL, Whyatt RM, Calafat AM, Nelson H, et al. Maternal urinary metabolites of di-(2-Ethylhexyl) phthalate in relation to the timing of labor in a US multicenter pregnancy cohort study. Am J Epidemiol 2009;169(8):1015–24.

[57] Kalo D, Roth Z. Effects of mono(2-ethylhexyl)phthalate on cytoplasmic maturation of oocytes—the bovine model. Reprod Toxicol 2015;53:141–51.

[58] Absalan F, Saremy S, Mansori E, Taheri Moghadam M, Eftekhari Moghadam AR, Ghanavati R. Effects of mono-(2-ethylhexyl) phthalate and di-(2-ethylhexyl) phthalate administrations on oocyte meiotic maturation, apoptosis and gene quantification in mouse model. Cell J 2017;18(4):503–13.

[59] Dalman A, Eimani H, Sepehri H, Ashtiani SK, Valojerdi MR, Eftekhari-Yazdi P, et al. Effect of mono-(2-ethylhexyl) phthalate (MEHP) on resumption of meiosis, in vitro maturation and embryo development of immature mouse oocytes. Biofactors 2008;33(2):149–55.

[60] Mu X, Liao X, Chen X, Li Y, Wang M, Shen C, et al. DEHP exposure impairs mouse oocyte cyst breakdown and primordial follicle assembly through estrogen receptor-dependent and independent mechanisms. J Hazard Mater 2015;298:232–40.

[61] Campioli E, Batarseh A, Li J, Papadopoulos V. The endocrine disruptor mono-(2-ethylhexyl) phthalate affects the differentiation of human liposarcoma cells (SW 872). PLoS ONE 2011;6(12):e28750.

[62] Sant KE, Dolinoy DC, Jilek JL, Sartor MA, Harris C. Mono-2-ethylhexyl phthalate disrupts neurulation and modifies the embryonic redox environment and gene expression. Reprod Toxicol 2016;63:32–48.

[63] Hao C, Cheng X, Guo J, Xia H, Ma X. Perinatal exposure to diethyl-hexyl-phthalate induces obesity in mice. Front Biosci (Elite Ed) 2013;5:725–33.

[64] Adamson GD, Pasta DJ. Endometriosis fertility index: the new, validated endometriosis staging system. Fertil Steril 2010;94(5):1609–15.

[65] Dunselman GA, Vermeulen N, Becker C, Calhaz-Jorge C, D'Hooghe T, De Bie B, et al. ESHRE guideline: management of women with endometriosis. Hum Reprod 2014;29(3):400–12.

[66] Tummon IS, Asher LJ, Martin JS, Tulandi T. Randomized controlled trial of superovulation and insemination for infertility associated with minimal or mild endometriosis. Fertil Steril 1997;68(1):8–12.

[67] Opoien HK, Fedorcsak P, Byholm T, Tanbo T. Complete surgical removal of minimal and mild endometriosis improves outcome of subsequent IVF/ICSI treatment. Reprod Biomed Online 2011;23(3):389–95.

[68] Zikopoulos K, Kolibianakis EM, Devroey P. Ovarian stimulation for in vitro fertilization in patients with endometriosis. Acta Obstet Gynecol Scand 2004;83(7):651–5.

[69] Dicker D, Goldman JA, Levy T, Feldberg D, Ashkenazi J. The impact of long-term gonadotropin-releasing hormone analogue treatment on preclinical abortions in patients with severe endometriosis undergoing in vitro fertilization-embryo transfer. Fertil Steril 1992;57(3):597–600.

[70] Marcus SF, Edwards RG. High rates of pregnancy after long-term down-regulation of women with severe endometriosis. Am J Obstet Gynecol 1994;171(3):812–7.

[71] Surrey ES, Silverberg KM, Surrey MW, Schoolcraft WB. Effect of prolonged gonadotropin-releasing hormone agonist therapy on the outcome of in vitro fertilization-embryo transfer in patients with endometriosis. Fertil Steril 2002;78(4):699–704.

[72] Sallam HN, Garcia-Velasco JA, Dias S, Arici A. Long-term pituitary down-regulation before in vitro fertilization (IVF) for women with endometriosis. Cochrane Database Syst Rev 2006;1:CD004635.

[73] Pabuccu R, Onalan G, Kaya C. GnRH agonist and antagonist protocols for stage I-II endometriosis and endometrioma in in vitro fertilization/intracytoplasmic sperm injection cycles. Fertil Steril 2007;88(4):832–9.

[74] Bastu E, Yasa C, Dural O, Mutlu MF, Celik C, Ugurlucan FG, et al. Comparison of ovulation induction protocols after endometrioma resection. JSLS 2014;18(3).

[75] Rodriguez-Purata J, Coroleu B, Tur R, Carrasco B, Rodriguez I, Barri PN. Endometriosis and IVF: are agonists really better? Analysis of 1180 cycles with the propensity score matching. Gynecol Endocrinol 2013;29(9):859–62.

[76] Garry R. The effectiveness of laparoscopic excision of endometriosis. Curr Opin Obstet Gynecol 2004;16(4):299–303.

[77] Guo SW. Recurrence of endometriosis and its control. Hum Reprod Update 2009;15(4):441–61.

[78] Vercellini P, Fedele L, Aimi G, De Giorgi O, Consonni D, Crosignani PG. Reproductive performance, pain recurrence and disease relapse after conservative surgical treatment for endometriosis: the predictive value of the current classification system. Hum Reprod 2006;21(10):2679–85.

[79] Rickes D, Nickel I, Kropf S, Kleinstein J. Increased pregnancy rates after ultralong postoperative therapy with gonadotropin-releasing hormone analogs in patients with endometriosis. Fertil Steril 2002;78(4):757–62.

[80] Dicker D, Goldman JA, Feldberg D, Ashkenazi J, Levy T. Transvaginal ultrasonic needle-guided aspiration of endometriotic cysts before ovulation induction for in vitro fertilization. J in vitro Fert Embryo Transf 1991; 8(5):286–9.

[81] Suganuma N, Wakahara Y, Ishida D, Asano M, Kitagawa T, Katsumata Y, et al. Pretreatment for ovarian endometrial cyst before in vitro fertilization. Gynecol Obstet Invest 2002;54(Suppl. 1):36–40. discussion 1–2.

[82] Yazbeck C, Madelenat P, Ayel JP, Jacquesson L, Bontoux LM, Solal P, et al. Ethanol sclerotherapy: a treatment option for ovarian endometriomas before ovarian stimulation. Reprod Biomed Online 2009;19(1):121–5.

[83] Practice Committee of the American Society for Reproductive Medicine. Endometriosis and infertility: a committee opinion. Fertil Steril 2012;98(3):591–8.

[84] D'Hooghe TM, Denys B, Spiessens C, Meuleman C, Debrock S. Is the endometriosis recurrence rate increased after ovarian hyperstimulation? Fertil Steril 2006;86(2):283–90.

[85] Benaglia L, Somigliana E, Vighi V, Nicolosi AE, Iemmello R, Ragni G. Is the dimension of ovarian endometriomas significantly modified by IVF-ICSI cycles? Reprod Biomed Online 2009;18(3):401–6.

[86] Benagiano G, Brosens I. History of adenomyosis. Best Pract Res Clin Obstet Gynaecol 2006;20(4):449–63.

[87] Martinez-Conejero JA, Morgan M, Montesinos M, Fortuno S, Meseguer M, Simon C, et al. Adenomyosis does not affect implantation, but is associated with miscarriage in patients undergoing oocyte donation. Fertil Steril 2011;96(4):943–50.

[88] Templeman C, Marshall SF, Ursin G, Horn-Ross PL, Clarke CA, Allen M, et al. Adenomyosis and endometriosis in the California Teachers Study. Fertil Steril 2008;90(2):415–24.

[89] Vercellini P, Consonni D, Dridi D, Bracco B, Frattaruolo MP, Somigliana E. Uterine adenomyosis and in vitro fertilization outcome: a systematic review and meta-analysis. Hum Reprod 2014;29(5):964–77.

[90] Harada T, Khine YM, Kaponis A, Nikellis T, Decavalas G, Taniguchi F. The impact of adenomyosis on women's fertility. Obstet Gynecol Surv 2016;71(9):557–68.

[91] Niu Z, Chen Q, Sun Y, Feng Y. Long-term pituitary downregulation before frozen embryo transfer could improve pregnancy outcomes in women with adenomyosis. Gynecol Endocrinol 2013;29(12):1026–30.

[92] Park CW, Choi MH, Yang KM, Song IO. Pregnancy rate in women with adenomyosis undergoing fresh or frozen embryo transfer cycles following gonadotropin-releasing hormone agonist treatment. Clin Exp Reprod Med 2016;43(3):169–73.

[93] Imaoka I, Ascher SM, Sugimura K, Takahashi K, Li H, Cuomo F, et al. MR imaging of diffuse adenomyosis changes after GnRH analog therapy. J Magn Reson Imaging 2002;15(3):285–90.

[94] Khan KN, Kitajima M, Hiraki K, Fujishita A, Nakashima M, Masuzaki H. Decreased expression of human heat shock protein 70 in the endometria and pathological lesions of women with adenomyosis and uterine myoma after GnRH agonist therapy. Eur J Obstet Gynecol Reprod Biol 2015;187:6–13.

[95] Grimbizis GF, Mikos T, Tarlatzis B. Uterus-sparing operative treatment for adenomyosis. Fertil Steril 2014;101 (2):472–87.

[96] Dueholm M. Uterine adenomyosis and infertility, review of reproductive outcome after in vitro fertilization and surgery. Acta Obstet Gynecol Scand 2017;96(6):715–26.

[97] Kishi Y, Yabuta M, Taniguchi F. Who will benefit from uterus-sparing surgery in adenomyosis-associated subfertility? Fertil Steril 2014;102(3):802–7. e1.

[98] Otsubo Y, Nishida M, Arai Y, Ichikawa R, Taneichi A, Sakanaka M. Association of uterine wall thickness with pregnancy outcome following uterine sparing surgery for diffuse uterine adenomyosis. Aust N Z J Obstet Gynaecol 2016;56(1):88–91.

Uterine Transplantation

Mats Brännström

Department of Obstetrics and Gynecology, Sahlgrenska Academy, University of Gothenburg, Gothenburg, Sweden Stockholm IVF, Stockholm, Sweden

INTRODUCTION

Absolute uterine factor infertility (AUFI) has by definition been untreatable until recently, when uterus transplantation (UTx) proved to be an effective treatment based on the reports of births occurring in 2014, after transplantation of a uterus from an altruistic living donor [1] and from a mother [2].

Women with AUFI have either uterine absence (congenital/surgical) or abnormalities (anatomic/functional) that preclude implantation of an embryo or completion of a pregnancy.

The uterus can be absent from birth as part of the Mayer-Rokitansky-Kuster-Hauser (MRKH) syndrome, which affects around 1:4000 girls [3]. This represents only around 3% of women with AUFI but it has relevance to this topic because most of the UTx attempts performed so far worldwide have involved recipients with MRKH. MRKH syndrome is usually diagnosed during adolescence, when the female is investigated because of primary amenorrhea. A diagnostic workup will find absence of the vagina above the hymen and nonappearance of a proper uterus. The uterus is replaced by rudimentary tissue above the vaginal dimple and usually two smaller collections of myometrial tissue on the pelvic sidewall. A large proportion of women with MRKH have additional malformations in the urinary/renal system, with unilateral renal agenesis being the most prevalent comalformation. This may also have relevance in the UTx setting because a single kidney may have been a major underlying cause of preeclampsia [4] in the pregnancy of the first live birth after UTx [1].

The most prevalent cause of AUFI is uterine absence because of a hysterectomy, which is the most frequent major gynecological surgery procedure that women undergo. A hysterectomy during fertile age could be secondary to symptomatic/large leiomyoma, cervical cancer, endometrial cancer, or massive obstetric bleeding because of uterine atony, uterine rupture, or invasive placentation.

Anatomical uterine abnormalities that will prevent pregnancy are all cases of hypoplastic uterus and a minor portion of women with the unification defects of unicornuate and bicornuate uterus. While there is no difference in conception rate in bicornuate/unicornuate uteri, as compared to normal uteri, the rate of first-trimester miscarriage is considerably increased [5]. It should be pointed out that the risks of preterm labor and fetal malpresentation in women with any of these unification malformations are also greatly enhanced [5].

The presence of adenomyosis or radiation injury of the uterus, with secondary repeated miscarriage/implantation failure, are other causes of uterine factor infertility. Uterus transplantation may provide a treatment for these women as well as of women with no signs of uterine disease on radiology imaging but with repeated miscarriage/implantation failure in spite of the high quality of oocytes/embryos.

Intrauterine adhesions, most common after curettage or endometritis, are usually treatable by hysteroscopic resection. However, almost 70% of those with severe intrauterine adhesions of stage 3 and 4 remain infertile despite repeated hysteroscopy [6] and UTx will then be the only infertility treatment for these women.

The total prevalence of AUFI is estimated to be around 20,000 women of fertile age in a population of 100 million [7].

Previously, adoption and use of a gestational carrier were the only options to acquire motherhood for patients with AUFI. Adoption is not acceptable in all societies and if accepted, often excludes single mothers or lesbian couples. Gestational surrogacy may be practiced either as an altruistic or commercial arrangement. The procedure is not allowed in most countries worldwide due to legal, religious, and/or ethical reasons. Uterus transplantation would, in contrast to adoption and gestational surrogacy, provide full motherhood in terms of genetic, gestational, and legal aspects. Moreover, the typical risks of pregnancy (thromboembolism, hypertension, preeclampsia, diabetes, etc.) and those associated with delivery, such as pelvic floor dysfunction, are taken by the mother and not by a third person, as in gestational surrogacy. However, it should be acknowledged that in live donor UTx, there is also the risk for a third person donating her womb to the woman with AUFI.

Thirty-six UTx attempts have been performed from 2013 through Nov. 2017, with around half of them published as scientific reports (Table 23.1). Prior to that, two failed attempts were performed in settings with no research or surgical preparations for the procedure. The first

TABLE 23.1 Uterus Transplant Experience (as of Sep. 2017) Divided Into Published Cases (Publ.) and Those Communicated by Personal Communication (Pers.)

City, Nation	Publ./Pers.	Year	Donor Type	Surgery Donor	Birth/Preg
Jeddah, Saudi Arabia	Publ. [8]	2000	LD (×1)	Laparotomy	
Antalya, Turkey	Publ. [9, 10]	2011	DD (×1)	Laparotomy	
Gothenburg, Sweden	Publ. [1, 2, 11]	2013	LD (×9)	Laparotomy	8 births
	Pers.	2017	LD (×2)	RAL	
Xián, China	Publ. [12]				
Prague, Czech Republic	Pers.	2016	DD (×4), LD (×4)	Laparotomy	
Cleveland, United States	Publ. [13]	2016	DD (×1)	Laparotomy	
Sao Paulo, Brazil	Publ. [14]	2016	DD (×2)	Laparotomy	1 pregnant
Dallas, United States	Publ. [15]	2016–17	LD (×6), DD (×1)	Laparotomy	
Tubingen, Germany	Pers.	2016–17	LD (×2)	Laparotomy	
Belgrade, Serbia	Pers.	2017 (monozygotic twins)	LD (×1)	Laparotomy	1 pregnant
Guangzhou, China	Pers.	2017	LD (×1)	Laparoscopy	
Pune, India	Pers.	2017	LD (×2)	Laparoscopy	

human UTx case in the world was a live donor UTx performed in Saudi Arabia in 2000 [8]. A necrotic uterus was removed shortly after transplantation. The second UTx attempt in the world involved a deceased donor UTx procedure in Turkey in 2011 [9]. Two early miscarriages occurred around two years after UTx [10], in the latter case. However, successful UTx with delivery of a healthy child has not yet been reported from this case although 6 years have passed since UTx.

Modern animal-based research on UTx was initiated around the millennium shift. Several animal models have been used to optimize and safeguard the UTx procedure, prior to clinical introduction. Initial research included rodents, which was followed initially by domestic species, such as the sheep and pig, and subsequently by nonhuman primate models, such as the baboon and macaque. The models have been used to investigate aspects such as surgery, tolerability to ischemia, detection of rejection, immunosuppression, and fertility [16, 17]. This research-based approach follows the established Moore criteria [18] and IDEAL recommendations [19] for introduction of surgical innovations.

In this review article, the UTx animal research with fertility outcome as an endpoint as well as all published human cases from 2013 and later will be covered in detail.

ANIMAL UTx RESEARCH ON FERTILITY

Fertility aspects of UTx have been studied in several animal species and the experiments have included models with autologous, syngeneic, and allogeneic transplantations.

The syngeneic and autologous UTx models only test the results of UTx surgery with a uterus with an altered supply and outflow of blood and with changed ligamentous fixations or position of the uterus. Additional effects of immunosuppression and possible rejection episodes before and during pregnancy are tested in an allogeneic UTx model.

Fertility in Mouse UTx

The first demonstration of implantation in a true UTx setting was in the mouse, with the uterus transplanted into a heterotopic position and with the cervix of the uterine graft positioned intraabdominally [20]. The transplantation model was with vascular anastomoses of the caval vein and the aorta of the graft coupled end to side to the midabdominal parts of the aorta and vena cava of the recipient, using microsurgical skills to anastomose the vessels with 11-0 sutures. This was a syngeneic transplantation between inbred females of C57BL76xCBA/ca F-1 hybrids with no immunological rejection and consequently no need for immunosuppression. In the initial report of pregnancies after embryo transfer (ET), accomplished by transmyometrial approach through a small midline abdominal incision, only occasional early pregnancies were seen [20]. This low implantation rate was most likely due to that the uteri accumulated fluid within the cavity, secondary to clogging of mucous inside the canal of the intraabdominally positioned uterine cervix.

The mouse UTx model was later modified to avoid intrauterine fluid accumulation. This was accomplished by exteriorizing the uterine cervix as a cervical cutaneous stroma to allow for drainage of uterine/cervical mucous [21]. The native uterus was kept in the transplanted animals as internal control, concerning implantation rate and pregnancy rate. Three to six blastocysts were transferred (transmyometrially through a minimidline incision) into each of the grafted and native uteri. Pregnancy rate per uterus was similar in the transplanted uteri as compared to the native control uteri within the same animal and also in comparison to the uteri of nontransplanted animals with sham operations. The offspring were of normal birth weight and the growth trajectory up to adulthood followed the normal curves [21]. Both male and female offspring from transplanted uteri proved to have normal fertility.

In an additional study, the added effect of ischemic preservation between organ harvesting and transplantation was tested in the syngeneic mouse UTx model. Live offspring was demonstrated after cold ischemic conditions for 24 h but not for 48 h [22], indicating that the uterus is an organ that is greatly tolerable to ischemic conditions.

There exist no studies on fertility after allogeneic UTx in the mouse.

Fertility in Rat UTx

In the rat, with a considerably larger body size than the mouse, fertility was tested both after syngeneic and allogeneic UTx. In syngeneic UTx, inbred Lewis rats were used as both donors and recipients [23]. The model was with orthotopic UTx, after hemihysterectomy of the left uterine horn, and with anastomoses end to side between the right common iliacs of the graft and the recipient. The upper part of the right uterine was anastomosed to the tip of the uterine graft and a vaginal-vaginal end-to-end anastomosis was accomplished to allow for normal fertilization by spontaneous mating. Controls were with left-sided hemihysterectomy only. The pregnancy rate was similar in UTx animals as in controls, and there was no difference in pups per pregnancy. Growth trajectory, up to 60 days, was similar in offspring from animals of the UTx group and the sham-operated control group [23].

The first ever report of fertility after allogeneic UTx was a study exploring this in the rat model [24]. The uterus donors were of the Dark Agouti strain and the recipients were Lewis rats, with discordance of two major histocompatibility sites (RT1, RT2). In order to avoid rejection, tacrolimus immunosuppression was given via miniosmotic pumps The experiments were terminated by cesarean section at 2/3 through pregnancy, as predetermined in the ethics approval. The pregnancy rates (number of pregnant females/total number of females within group) were similar and around 60% in the UTx group and the control group that had been sham-operated and also received tacrolimus. Moreover, the median ranges of fetus per animal were similar in these two groups but lower than in the sham-operated nontacrolimus control group. This demonstration of pregnancy after allogeneic UTx was a central proof of concept of UTx as a possible future treatment of AUFI in humans.

In a follow-up study, the allogeneic combination of Lewis rats as uterus donors and Piebald-Virol-Glaxo rats as recipients was used with tacrolimus as maintenance immunosuppression [25]. The pregnancy rate was somewhat lower in the UTx group as compared to the two sham-operated control groups, with one of them also receiving tacrolimus. Birth weights of UTx offspring were not different from controls and the growth trajectory of the pups until postnatal week 16 was also unaltered in comparison to controls. This data indicated for the first time that allogeneic UTx may be regarded as safe in terms of perinatal outcome, at least in a rodent.

Fertility in Rabbit UTx

Only one study has examined fertility in the rabbit UTx model. Nine allogeneic UTx procedures were done in New Zealand White rabbits with proven fertility of both donors and recipients [26]. The uterus with the entire vascular tree, including uterine vessels and internal iliacs as well as the lower abdominal parts of the caval vein and aorta, was surgically isolated. The two anastomoses were aorto-aortic and cavo-caval end to side. Immunosuppression was with tacrolimus. Vitrified donor morulae-stage embryos were used and ET was done after ovulation-induction with hCG in one rabbit after a posttransplantation recovery period of 2 months. A total of 17 embryos were thawed and placed inside the two uterine horns during a laparotomy procedure. Ultrasound detected a fetal sac with a pregnancy with heartbeat 9 days after ET and this continued to grow for seven more days [26]. However, spontaneous abortion with fetal resorption then occurred. The cause of the pregnancy arrest could not clearly be identified, but did not seem to be related to diminished blood flow because blood flow in the internal iliac artery of the graft was normal at autopsy.

Fertility in Sheep UTx

Fertility in the sheep has been tested in a nonrejection setting, with the autologous UTx-model, and in an allogeneic UTx model. There are variations in the surgery of the models and with the autologous UTx model permitting natural conception.

The autologous UTx model was with uterine-tubal-ovarian transplantation and end-to-side vascular anastomoses of the uterine artery, utero-ovarian vein, and the ovarian artery, including an aortic patch, to the external iliacs [27]. Around three months after auto-UTx, five ewes were placed with rams for mating. This occurred in four of the five ewes and three of these conceived. Offspring of normal sizes were delivered by cesarean section, around two weeks before term (145 days). The offspring were not followed after birth.

The allogeneic sheep UTx model involved surgery with hysterectomy with short vascular pedicles of the uterine artery and vein, divided above the ureteric level [28]. The same surgery was done in parallel in the recipient, and the uteri could be shifted between the outbred sheep. Transplantation was by bilateral end-to-end anastomosis of the uterine arteries and veins as well as attachment of the vaginal rim of the graft to the vaginal vault of the recipient. Twelve transplanted ewes received maintenance immunosuppression by cyclosporine with the addition of prednisone during the first post-UTx week. Around 3 months after UTx, ET was performed in five ewes that received donor, single fresh cleavage-stage ET in three cases and frozen blastocyst ET in two cases. One of the two ewes that received the thawed blastocyst became pregnant but this proved to be an ectopic pregnancy. In the group that received fresh cleavage-state embryos, two out of three ewes became pregnant. One pregnancy ended in late miscarriage (day 105). The other resulted in live birth by scheduled cesarean section at day 138 of pregnancy [28]. The lamb weighed 3.5 kg. It showed signs of fetal respiratory distress and was initially ventilated with oxygen support. Importantly, this was the first and so far only live birth from a large animal undergoing allogeneic UTx.

Fertility in Macaque UTx Models

The first, and so far only, offspring reported in a nonhuman primate species was after autologous UTx in the macaque [29]. In that report, two cynomolgus macaques underwent autotransplantation with unilateral preservation of the oviduct and the ovary in both cases. Only one of these animals became pregnant and in that case anastomoses were between the uterine arteries and the external iliac arteries bilaterally and with venous outflow only on the side of the preserved ovary-oviductal compartment by a deep uterine vein and the ovarian vein. Anastomoses were end to side of the small vessels, with a diameter of 1–2.5 mm, to the external iliac vessels with use of 12-0 sutures. The complicated surgery is illustrated by the fact that the total surgical time of the animal with pregnancy was 13.5 h and with almost 5 h of warm ischemia to accomplish the four vascular connections. Menstruation resumed spontaneously and after three menstruations, natural mating occurred in this animal with an intact ovary-oviductal compartment with only a unilateral connection to the uterus. A viable intrauterine pregnancy was then confirmed at around gestational week five. The pregnancy was uneventful until day 143, when genital bleeding occurred [29]. Due to signs of partial placental abruption, a cesarean section was performed with delivery of a live offspring but with fetal respiratory distress. No attempts were done to secure further survival of the fetus. The demonstration of live offspring in this primate UTx situation was instrumental in the process of achieving ethics approval for human UTx studies.

There exist no studies on fertility after allogeneic UTx in a nonhuman primate species.

CLINICAL TRIALS

The human UTx attempts that have been performed within proper clinical trials and with published data are reported below. That will not include the first two cases, from 2000 [8] and 2011 [9, 10] that did not result in any live birth and that were not registered as clinical trials.

The Swedish Trial and Results

Nine LD UTx procedures were performed in Sweden in 2013 within an observational clinical trial [30]. Eight recipients had MRKH and one had lost her uterus through a hysterectomy because of cervical cancer. In-depth medical and psychological investigations were done on donors, recipients, and the partners of recipients [31]. All donors had been through normal pregnancies and none had a history of pre- or postterm delivery or repeated miscarriages. Uterus recovery, with dissection of the uterus with bilaterally vascular pedicles including segments of the internal iliacs, had durations of 10.5–13h [30]. The peroperative outcomes of the donors were favorable and no patient needed a blood transfusion. One donor acquired a ureteric-vaginal fistula, possibly due to thermic injury from diathermy, that presented 2 weeks after hysterectomy. The fistula was repaired 3 months post uterus donation with reimplantation of the ureter. That patient and all other donors were in good psychological and medical health at follow-up 1 year after surgery [32].

The recipient surgery was initiated prior to final graft retrieval and back-table preparation. The initial surgical preparations of the recipient before the uterus was finally removed from the recipient included dissection of the external iliacs and vaginal vault, with separation from the bladder and rectum. This latter surgical step was somewhat more cumbersome in patients with MRKH as opposed to the patient who had undergone hysterectomy, possibly because of a shorter vagina of the MRKH patients and variations in the anatomy of the uterine rudiment above the vaginal vault. After graft procurement in the donor and back-table preparation, the chilled and flushed uterus was positioned inside the pelvis. Anastomosis was by bilateral end-to-side anastomoses to the external iliacs of the uterine pedicles that included uterine vessels and the anterior iliac arteries as well as patches/segments of the internal iliac veins. Surgical duration was 4–5h and the patients stayed in the hospital for 4–9days. The immunosuppression regimen was induction with two perioperative doses of thymoglobulin plus methylprednisolone. From the day of surgery, tacrolimus and mycophenolate mofetil (MMF) were also given daily and oral glucocorticoids were given for 4 days [30]. After 8 months, MMF was discontinued if no or only one rejection episode had occurred during this period, but replaced with azathioprine in patients with several rejection episodes. The six-month outcome [30] was that seven out of nine uteri were still in place. Two uterine grafts were removed within the first 4 months. The causes were bilateral thrombotic occlusion of the uterine vessels in one case and persistent intrauterine infection, developing into an intrauterine abscess, in the other case [30].

During the first post-UTx year, uterine artery blood flow was within normal ranges in all seven patients [33]. Notably, protocol cervical biopsies revealed that five out of seven women had subclinical, mild rejection episodes during the first year but treatment with brief courses of corticosteroids or increments of tacrolimus reversed all episodes of rejection [11, 33].

The psychological outcome of recipients and partners was an overall optimism with only minor anxiety concerning graft survival during the first 3 months post-UTx [34]. According to the study protocol, single ETs were performed from around 12 months after UTx. The fifth woman to undergo UTx in the Swedish trial [30] became pregnant at her first ET with a cleavage state embryo and subsequently the first live birth after UTx took place in Sweden on Sept. 4, 2014 [1]. This was the first successful UTx procedure by definition because the ultimate goal of UTx is a healthy baby.

In the first successful UTx case [1], a rejection episode at gestational week 18 was diagnosed, which was effectively reversed by an intermittent increase in corticosteroids. From then on, the pregnancy was uneventful and she worked full time up until 31 full weeks and 5 days, when she acquired a headache and was admitted to the hospital due to preeclampsia. The following morning, a cesarean section was performed and a healthy boy with normal weight for gestational age (1775 g; −11%) was delivered.

The second UTx baby [2] was delivered in Nov. 2014 by an elective cesarean section that was planned for 35+0 but was brought forward 3 days due to cholestasis. Also, this pregnancy was accomplished at the initial ET, which in this case was of a blastocyst. The baby was of normal (+4%) weight. The uniqueness of this case is that the donor was the mother of the woman that delivered the child. Hence, the mother herself had been born from this uterus that now delivered her own child.

This first [1] and second [2] UTx child, as well as the six children [11] that were delivered in 2014–17, are healthy. Six out of seven women who have undergone ET attempts in this Swedish trial have now taken home their babies, and the clinical pregnancy rate is 7/7 with one recipient having had a miscarriage as late as gestational week 15. Therefore, this substantial efficiency of UTx at this initial experimental stage clearly indicates that UTx will have a future clinical role as an established treatment for AUFI.

In Sweden, a second UTx trial with the live donor concept has been initiated. Robotic-assisted laparoscopy is used for donor surgery and so far two cases have been completed, with a surgically favorable outcome in the donor and graft survival for several months in the recipient. The plan is to complete an additional six to eight cases in 2018 and to start ET attempts in the first two cases, also in 2018.

The Chinese Trial and Results

Uterus transplantation attempt number 12 in the word occurred in late 2015 in China [13]. The case utilized robotic-assisted laparoscopy for uterus retrieval in the 42-year-old premenopausal mother that donated the uterus to her 22-year-old daughter with MRKH. The surgery followed the general principles of that used in the Swedish trial [30], but with one major difference. The secured uterine outflow was not through the uterine veins but rather through the utero-ovarian veins. The reason for this is unclear, but it is stated in the paper [12] that the uterine veins were difficult to identify. The use of the utero-ovarian veins necessitated an oophorectomy in the donor, who may have been around a decade away from menopause. Naturally, this raises concerns of long-term medical consequences of the donor in relation to osteoporosis and cardiovascular disease. However, the surgical duration in the donor was substantially reduced due to avoidance of the complicated dissection of the uterine veins. Those are firmly attached to the ureters and with several communicating branches between the deep and superficial uterine veins [30]. The retrieval of the graft was through the vagina. The donor surgery was by laparotomy, with bilateral end-to-side anastomosis to the external iliac vessels [12]. It is unclear whether the length of the utero-ovarian veins, now with outflow through the larger vessels of the pelvis instead of the natural inflows of the caval vein and the left renal vein, will affect a future pregnancy when the uterine-corneal positions of the utero-ovarian veins move toward the upper abdomen simultaneously with the uterus increase in size during pregnancy. This may lead to stretching of these veins and compromised blood flow. The duration of the recipient surgery was twice [12] of that of the Swedish trial, indicating that the anastomosis of the utero-ovarian veins, with minimal thickness of the vascular walls, is far more difficult than when using a patch/segment of the internal iliac vein [30]. This prolonged warm ischemic time did not seem to influence the function of the uterus because the patient had spontaneous and regular menstruations from around 1.5 months after UTx. In the report [12] that presents the results up to 12 months after UTx, it is stated that 10 spontaneous menses had occurred and that ET attempts would start during the second posttransplantation year. The patient only experienced one rejection episode (after 2.5 weeks), which was diagnosed by clinical symptoms (low back ache, fatigue, and fever) and confirmed by increased CD4/CD8 ratio. The rejection was resolved by intravenous corticosteroid treatment for 3 days. We will await reports on the results of ET attempts.

The US Trials and Results

Two UTx trials have been initiated in the United States. The first involved a study on a deceased donor UTx and is presently underway at Cleveland Clinic. The first case, which unfortunately ended in graft removal after around 2 weeks, has been presented both in the media and in a scientific report [13]. The UTx procedure took place in Feb. 2016. There was a long cold ischemic time of more than 8 h because the graft had to be transported interstate. A fungal vaginal infection was present in the donor and this was not diagnosed at retrieval. This candida contamination later affected the vascular tree of the graft, and the graft was removed due to an infectious aneurysm of the internal iliac artery of the graft.

The second UTx trial in the United States was a live donor trial that was initiated in Dallas in Sep. 2016 [15]. The initial results of the first five attempts in Dallas have been reported. A similar laparotomy technique as in the Swedish trial [30] was used and with surgical durations of around 8 h for retrieval. The first three cases failed, due to vascular complications that were related to both inflow and outflow problems [15]. The grafts were removed during the initial 2 weeks in these three cases. In the other two cases, graft survival for 3 and 6 months has been reported [15].

The Brazilian Trial and Results

The third deceased UTx case in the world was done in Sao Paolo, Brazil, in Sep. 2016 [14]. It involved one UTx procedure from a deceased donor to a young woman with MRKH. The procurement process was deliberately prolonged to avoid vascular leakage on the back table and after vascular anastomosis in the recipient. The postoperative recovery has been fine with regular menstruations. As of Sep. 2017, a 20-week long pregnancy was reported. This is a promising result and may be the first live birth after deceased donor UTx.

CONCLUSION

Uterus transplantation is the first available treatment for AUFI. After meticulous research and team preparations before the first clinical UTx trial was launched in 2013, excellent pregnancy and take-home-baby rates have been reported. This is promising concerning the future for human UTx. Several new trials are under way on all continents. Uterus transplantation should stay at this experimental stage for several years. This will allow time to optimize the procedure further and to ensure that the procedure is safe concerning long-term medical and psychological effects, which, in the setting of a live donor UTx, includes donor, recipient, partner of recipient, and future children.

References

[1] Brännström M, Johannesson L, Bokström H, Kvarnström N, Mölne J, Dahm-Kähler P, et al. Live birth after uterus transplantation. Lancet 2015;385:607–16.

[2] Brännström M, Bokström M, Dahm-Kähler P, Diaz-Garcia C, Ekberg J, Enskog A, et al. One uterus bridging three generations; first live birth after mother-to-daughter uterus transplantation. Fertil Steril 2016;107:261–6.

[3] Oppelt P, Renner SP, Kellermann A, Brucker S, Hauser GA, Ludwig KS, et al. Clinical aspects of Mayer-Rokitansky-Kuester-Hauser syndrome: recommendations for clinical diagnosis and staging. Hum Reprod 2006;21:792–7.

[4] Brännström M, Diaz-Garcia C, Johannesson L, Dahm-Kähler P, Bokström H. Livebirth after uterus transplantation—authors' reply. Lancet 2015;385:2352–3.

[5] Chan YY, Jayaprakasan K, Tan A, Thornton JG, Coomarasamy A, Raine-Fenning NJ. Reproductive outcomes in women with congenital uterine anomalies: a systematic review. Ultrasound Obstet Gynecol 2011;38:371–81.

[6] Fernandez H, Al-Najjar F, Chauveaud-Lambling A, Frydman R, Gervaise A. Fertility after treatment of Asherman's syndrome stage 3 and 4. J Minim Invas Gynecol 2006;13:398–402.

[7] Sieunarine K, Zakaria FB, Boyle DC, Corless DJ, Noakes DE, Lindsay I, et al. Possibilities for fertility restoration: a new surgical technique. Int Surg 2005;90:249–56.

[8] Fageeh W, Raffa H, Jabbah H, Marzouki A. Transplantation of the human uterus. Int J Gynaecol Obstet 2002;76:245–51.

[9] Özkan Ö, Akar ME, Özkan Ö, Erdogan O, Hadimioglu N, Yilmaz M, et al. Preliminary results of the first human uterus transplantation from a multiorgan donor. Fertil Steril 2013;99:470–6.

[10] Erman Akar M, Özkan Ö, Aydinuraz B, Dirican K, Cincik M, Mendilcioglu I, et al. Clinical pregnancy after uterus transplantation. Fertil Steril 2013;100:1358–63.

[11] Mölne J, Broecker V, Ekberg J, Nilsson O, Dahm-Kähler P, Brännström M. Monitoring of human uterus transplantation with cervical biopsies: a provisional scoring system for rejection. Am J Transplant 2017;17:1628–36.

[12] Wei L, Xue T, Tao KS, Zhang G, Zhao GY, Yu SQ, et al. Modified human uterus transplantation using ovarian veins for venous drainage: the first report of surgically successful robotic-assisted uterus procurement and follow-up for 12 months. Fertil Steril 2017;108:346–56.

[13] Flyckt RL, Farrell RM, Perni UC, Tzakis AG, Falcone T. Deceased donor uterine transplantation; innovation and adaption. Obstet Gynecol 2016;128:837–42.

[14] Soares Jr. JM, Ejzenberg D, Andraus W, LA D'A, Baracat EC. First latin uterine transplantation: we can do it! Clinics (Sao Paulo) 2016;71:627–8.

[15] Testa G, Koon EC, Johannesson L, McKenna GJ, Anthony T, Klintmalm GB, et al. Living donor uterus transplantation: a single center's observations and lessons learned from early setbacks to technical success. Am J Transplant 2017;17:2901–10.

[16] Brännström M, Diaz-Garcia C, Hanafy A, Olausson M, Tzakis A. Uterus transplantation: animal research and human possibilities. Fertil Steril 2012;97:1269–76.

[17] Brännström M. The Swedish uterus transplantation project; the story behind the project. Acta Obstet Gynecol Scand 2015;94:675–9.

[18] Moore FD. Ethical problems special to surgery. Arch Surg 2000;135:14–6.

[19] McCulloch P, Altman DG, Campbell WB, Flum DR, Glasziou P, Marshall JC, et al. No surgical innovation without evaluation: the IDEAL recommendations. Lancet 2009;374:1105–12.

[20] Racho El-Akouri R, Kurlberg G, Dindelegan G, Mölne J, Wallin A, Brännström. Heterotopic uterine transplantation by vascular anastomosis in the mouse. J Endocrinol 2002;174:157–66.

[21] Racho El-Akouri R, Kurlberg G, Brännström M. Successful uterine transplantation in the mouse: pregnancy and postnatal development of offspring. Hum Reprod 2003;18:2018–23.

[22] Racho El-Akouri R, Wranning CA, Mölne J, Kurlberg G, Brännström M. Pregnancy in transplanted mouse uterus after long-term cold ischaemic preservation. Hum Reprod 2003;18:2024–30.

[23] Wranning CA, Akhi SN, Diaz-Garcia C, Brännström M. Pregnancy after syngeneic uterus transplantation and spontaneous mating in the rat. Hum Reprod 2011;26:553–8.

[24] Diaz-Garcia C, Akhi SN, Wallin A, Pellicer A, Brännström M. First report on fertility after allogeneic uterus transplantation. Acta Obstet Gynecol Scand 2010;89:1491–4.

[25] Diaz-Garcia C, Johannesson L, Shao R, Bilig H, Brännström M. Pregnancy after allogeneic uterus transplantation in the rat: perinatal outcome and growth trajectory. Fertil Steril 2014;102:1545–52.

[26] Saso S, Petts G, David AL, Thum MY, Chatterjee J, Vicente JS, et al. Achieving an early pregnancy following allogeneic uterine transplantation in a rabbit model. Eur J Obstet Gynecol Reprod Biol 2015;85:164–9.

[27] Wranning CA, Marcickiewicz J, Enskog A, Dahm-Kähler P, Hanafy A, Brännström M. Fertility after autologous ovine uterine-tubal-ovarian transplantation by vascular anastomosis to the external iliac vessels. Hum Reprod 2010;25:1973–9.

[28] Ramirez ER, Ramirez Nessetti DK, Nessetti MB, Khatamee M, Wolfson MR, Shaffer TH, et al. Pregnancy and outcome of uterine allotransplantation and assisted reproduction in sheep. J Minim Invasive Gynecol 2011;18:238–45.

[29] Mihara M, Kisu I, Hara H, Iida T, Araki J, Shim T, et al. Uterine autotransplantation in cynomolgus macaques: the first case of pregnancy and delivery. Hum Reprod 2012;27:2332–40.

[30] Brännström M, Johannesson L, Dahm-Kähler P, Enskog A, Mölne J, Kvarnström N, et al. The first clinical uterus transplantation trial: a six months report. Fertil Steril 2014;101:1228–36.

[31] Järvholm S, Johannesson L, Brännström M. Psychological aspects in pre-transplantation assessments of patients prior to entering the first uterus transplantation trial. Acta Obstet Gynecol Scand 2015;94:1035–8.

[32] Kvarnström N, Järvholm S, Johannesson L, Dahm-Kähler P, Olausson M, Brännström M. Live donors of the initial observational study of uterus transplantation-psychological and medical follow-up until 1 year after surgery in the 9 cases. Transplantation 2017;101:664–70.

[33] Johannesson L, Kvarnström N, Mölne J, Dahm-Kähler P, Enskog A, Diaz-Garcia C, et al. Uterus transplantation trial: 1-year outcome. Fertil Steril 2015;103:199–204.

[34] Järvholm S, Johannesson L, Clarke A, Brännström M. Uterus transplantation trial: psychological evaluation of recipients and partners during the post-transplantation year. Fertil Steril 2015;104:1010–5.

PART IV

PREIMPLANTATION/ PRENATAL GENETIC DIAGNOSIS AND SCREENING

24

Epidemiology and Genetics of Human Aneuploidy

Howard Cuckle*, Svetlana Arbuzova†

*Faculty of Medicine, Tel Aviv University, Israel †Center of Medical Genetics and Prenatal Diagnosis, Mariupol, Ukraine

O U T L I N E

INTRODUCTION

Aneuploidy is a common event in pregnancy with a wide spectrum of medical consequences from the lethal to the completely benign. Most affected zygotes abort spontaneously early in the first trimester; many of them do so even before there are clinical signs of pregnancy. Those that survive into the second trimester also experience high late intrauterine mortality and increased risk of infant death. Viability and clinical outcome vary according to the genotype and this chapter will concentrate on the more common forms of aneuploidy that are sufficiently viable to survive to term in relatively large numbers.

By far the most frequent of these is Down syndrome, with birth prevalence in the absence of prenatal diagnosis and therapeutic abortion of 1–2 per 1000 in developed countries. Consequently, it is considered first and more extensively than Edwards and Patau syndromes, which are an order of magnitude less frequent at birth and nonviable, and the sex chromosome aneuploidies that are common but relatively benign.

In recent decades, the advent of widespread prenatal screening and selective termination of affected pregnancies has obscured much of the epidemiological features observed at birth. Therefore, in this chapter considerable reliance is placed on facts established in a previous era. At the same time, prenatal screening has itself revealed new information on the natural history of these conditions. At the same time, advances in molecular biology and the assessment of early embryos for in vitro fertilization (IVF) allow the earliest stages of aneuploidy to be examined.

STATISTICAL CONSIDERATIONS

When critically evaluating published studies on the epidemiology of aneuploidy, care is needed to ensure that the results are not subject to confounding or bias.

An example of confounding arises from the strong association between maternal age and Down syndrome prevalence. As a consequence of this association, factors that are correlated with maternal age such as paternal age, smoking status, and parity need to be assessed after maternal age adjustment. When this is done crudely by dividing age into say two groups (e.g., <35 and 35 or more) and computing the factor within group confounding will be reduced but not completely eliminated. Ideally single year of age adjustment or, in case-control studies, age matching is required.

An example of potential bias is the advent of prenatal screening itself, which can lead to two possible effects. First, birth prevalence is now much lower than in the past and some studies combine results from prenatally diagnosed and Down syndrome cases reaching term. This requires adjustment for the interuterine fatality rate in prenatal cases when a termination of pregnancy has been carried out. This can be inaccurate unless the adjustment factors are based on follow-up of prenatal cases where termination was refused. Second, Down syndrome detected through screening is not representative of all screened cases because maternal age is more advanced on average and there may be an association between screening marker levels and morbidity.

NATURAL HISTORY OF DOWN SYNDROME

Prior to the era of widespread prenatal diagnosis and selective termination of pregnancy, Down syndrome was the most common known cause of severe mental disability. In addition, affected individuals are likely to have a number of medical conditions affecting the cardiac, digestive, and respiratory systems. From an epidemiological perspective, two other disorders are important because an excess has been reported in both the affected individuals and their mothers: Alzheimer's disease and thyroid disease.

Adults with Down syndrome may experience cognitive deficits due to pathological changes in the brain normally associated with Alzheimer's disease. However, although Alzheimer-like changes are common in the brains of young people with Down syndrome, it is not inevitable that they will develop the clinical disease, and when dementia does occur it is not until middle age. Dementia is more difficult to diagnose in individuals with mental disability but a study from New York state used operational definitions of dementia in terms of declining adaptive behavior [1]. Statewide information systems were used to investigate 2534 individuals with Down syndrome and more than 16,000 controls with other forms of mental disability. No excess in dementia was seen until age 50. Thereafter, depending on the criteria used, the relative risk compared with controls was 1.7–3.2 at age 51–60 and 2.7–8.3 at 61–70. In the oldest group, the prevalence of dementia was 50% using the most lenient criteria and 15% for the most severe.

Thyroid disease is common in Down syndrome individuals. In a recent study of 663 patients enrolled at multiple institutions across the United States, 197 were diagnosed with thyroid disease (30%) [2]. About two-thirds had hypothyroidism and one-quarter had subclinical hypothyroidism/hyper thyrotropinemia.

The life expectancy of those born with Down syndrome today is considerably greater than in the past due to more effective treatment of cardiac and other conditions. For example, a nationwide Danish study used actuarial analysis on 3530 Down syndrome individuals diagnosed since 1961 [3]. Those with complete trisomy 21 had an estimated survival rate to age 50 of 89%. Nevertheless, such very high rates are not found everywhere and survival is generally dependent on access to good health care, particularly to treat the associated conditions.

SUBTYPES OF DOWN SYNDROME

In most cases of Down syndrome, there is complete ("standard") trisomy 21. Among more than 5000 cases in the National Down Syndrome Cytogenetic Register (NDSCR) for England and Wales, the proportion was 95% while 4% had a Robertsonian translocation, mostly t(14;21) or t(21;21), and 1% were mosaic [4].

Standard trisomy 21 can be divided according to the origin of the additional chromosome 21. In the largest series of 724 affected individuals and their parents, 89% were maternal, 9% paternal, and 2% had postzygotic mitotic changes [5]. A further classification can be made according to the stage of meiosis when the error occurred (see "Genetics" below). Based on chromosome heterozygocity, in about three-quarters of the maternal and half of the paternal cases it was the first stage (MI) [6]. Finally, standard trisomy 21 can be divided according to recombination pattern: absence of chiasmata, distal recombination found in MI, and proximal recombination in the second stage of meiosis (MII) [7].

MATERNAL AGE

By far the most important risk factor for Down syndrome is maternal age; birth prevalence increases rapidly with age, particularly after age 30. Consequently, the mean maternal age in Down syndrome births is about 5 years greater than unaffected births.

Age-Specific Birth Prevalence

The best available estimate is obtained from combining data from published series of birth prevalence for individual years of age that were carried out before prenatal diagnosis became common. A metaanalysis included a total of 4000–5000 Down syndrome births and more than five million unaffected births, and a regression curve fitted the data well: prevalence $= 0.000627 + \exp(-16.2395 + 0.0286 \times \text{age})$ [8]. Subsequent metaanalyses differed in the number of series included, the method of pooling series, the type of regression equation, and the extent to which the maternal age range was restricted. None of the curves derived by the various metaanalyses differed substantially over the 15–45 year age range.

Older Women

Another curve has been published based on a series of 11,000 cases from NDSCR [9]. It differs significantly from the above metaanalyses for older women: higher at age 36–41 and considerably lower after 45. However, the results are subject to potentially strong bias. In the previous series, cases were collected before antenatal screening and prenatal diagnosis became widespread whereas 45% of Register cases were diagnosed prenatally and 82% of these ended in termination of pregnancy. Birth prevalence was estimated by assuming an intrauterine survival rate following prenatal diagnosis derived from studies of older women (see below). This rate may not be applicable to women having prenatal diagnosis because of biochemical and ultrasound screening. Not only are they younger, but extreme levels of all the screening markers are associated with nonviability and the average marker levels of screen detected cases vary with age.

Younger Women

The metaanalyses all applied regression curves that increase monotonically with maternal age. This would not be valid if, as has been claimed, the prevalence of Down syndrome is relatively high at extremely young ages. Examination of the observed single-year prevalence in the combined metaanalysis series does not support this claim. The prevalence was 0.00%, 0.06%, 0.07%, 0.06%, and 0.06%, respectively, at ages 15–19 compared with 0.06%, 0.07%, 0.06%, 0.07%, and 0.08% at 20–24 [8]. Error in recording maternal age is the most likely explanation for the apparent increased prevalence among very young ages in some studies. Down syndrome cases for which maternal age has been underrecorded will trend to make the curve J-shaped.

Contribution of Maternal Age

One simple epidemiological question is what proportion of Down syndrome births is due to maternal age? A simple way of estimating that is from the age-specific birth prevalence curve. For example, applying the curve to the distribution of maternities in England and Wales in 2012 yields an expected Down syndrome prevalence of 1.89 per 1000. The first term of the additive-exponential regression equation described above [8] is 0.000627 or 0.63 per 1000, which is the background prevalence regardless of age. Hence, in 2002 the proportion attributable to age is (1–0.63/1.89), or 67%. In an earlier decade when the average maternal age was lower, the proportion would be less but today with a rapidly increasing number of older mothers, the proportion is rising.

Maternal Age and Subtype

The mean maternal age is greater in maternally derived cases of Down syndrome than those in which there is a paternal error: for example, 31.5 and 28.2 years in one study [5]. But the mean maternal age does not differ according to the apparent meiotic stage of the maternal error (MI 31.3 and MII 32.1 years, in the same study) or the paternal error (MI 27.4 and MII 27.5 years). However, the proportion with pericentromeric and telomeric recombinations is less frequent with advancing maternal age. In a study of 400 Down syndrome cases, the rate of such recombinations was 78% among those aged <29 years, 34% for 29–34 years, and 19% for older women [10].

FETAL LOSS

Although Down syndrome is not associated with extremely high intrauterine lethality, a large proportion of recognized pregnancies with the disorder are not viable.

Studies of prenatal diagnosis are used to estimate fetal loss rates, either by comparing the observed number of cases with that expected from birth prevalence, given the maternal age distribution, or by follow-up of individuals declining termination of pregnancy, using direct or actuarial survival analysis. Published prevalence studies include a total of 341 Down syndrome cases diagnosed at chorionic villus sampling (CVS) and 1159 at amniocentesis [11]. There are three published follow-up series, including 110 cases diagnosed at amniocentesis [12] and a series of 126 cases from the NDSCR that has been analyzed according to the gestational age at prenatal diagnosis [13]. However, the register study is biased as some miscarriages may have occurred in women who did intend to have a termination, thus inflating the rates. An actuarial survival analysis of the register data has now been carried out [14] that overcomes the bias and is more data-efficient because all cases contribute to the estimate, not just those where termination was refused.

Actual and potential heterogeneity between the various studies precludes a grand metaanalysis to estimate the fetal loss rates. But an informal synthesis supports the conclusion that about one-half of Down syndrome pregnancies are lost after first trimester CVS and one-quarter after midtrimester amniocentesis.

Maternal-Age Specific Rates

The incidence of Down syndrome among pregnancies ending in miscarriage also increases with maternal age. The combined results of two large studies, in New York and Hawaii, include 3395 karyotyped miscarriages [15]. The mean age in 92 with trisomy 21 was 30.7 years compared with 27.0 in chromosomally normal miscarriages.

For the purposes of genetic counseling, there is a need to calculate the maternal age-specific risk of Down syndrome at the time of prenatal diagnosis by applying the overall fetal loss rate to the term risk. Formulae have been published for this purpose that permit the calculation to be done for individual weeks of gestation. However, the calculations assume that the fetal loss rates do not vary with maternal age. Because the studies used to calculate the overall rates are largely based on women over 35, this assumption can only readily be examined in older women. In the combined results of the prevalence studies, the estimated loss rate after CVS, using the birth prevalence curve in [8], was 45% for women aged 35–39 and 47% in those aged 40 or older. For amniocentesis, the rates were 28% and 21%, respectively. In the combined results of three follow up studies of women refusing termination following Down syndrome diagnosis, the mean maternal age for 29 pregnancies ending in fetal loss was 38.7 and for 70 live births it was 39.0; in 11 cases age was not available [12].

Thus the available data in older women do not contradict the assumed lack of correlation between age and Down syndrome viability. Nevertheless, in view of the strong correlation between age and miscarriage in the population generally [16], a wider range of maternal ages needs to be considered for a correlation to be dismissed. However, there is a problem with studies that include younger women who are having invasive prenatal diagnosis. Such women are selected for diagnosis because of abnormal biochemical and ultrasound marker levels that are associated with miscarriage.

Data from NDSCR have been used to examine this question. In one publication, among pregnancies not terminated following Down syndrome diagnosis, the gestation standardized mean age for cases diagnosed at 14–21 weeks was 37.5 for the 44 fetal deaths and 36.3 for 54 live births (from Table 1A in [13]). Subsequently, a very large series of 5177 prenatally diagnosed cases was included in an actuarial survival analysis [17]. This revealed a strong correlation with maternal age. The estimated fetal loss rate at age 25 was 23% from the time of CVS and 19% from amniocentesis. By comparison, at age 45 the rates are 44% and 33%, respectively.

In a study of 2014 IVF pregnancies, the miscarriage rate following ultrasound confirmation of fetal heart activity was 12% and was correlated with maternal age [18]. Of the 233 losses, cytogenetic analysis was available for 74 and this showed a higher frequency of abnormal karyotypes in women over 40 compared to younger women (82% vs. 65%).

CLEAVAGE-STAGE EMBRYOS AND GAMETES

Molecular studies have now been performed on large numbers of oocytes and sperm as part of assisted reproduction. Aneuploidy occurs more often in oocytes than sperm, although there are methodological problems in the interpretation of such data [19]. Some of the results on oocytes are described above in relation to free chromatids.

The largest study of oocytes that failed to fertilize after IVF included more than 1000 karyotypes and the aneuploidy rate was strongly related to maternal age: 5.6%, 7.2%, 7.4%, 15%, 34%, and 65% in those aged under 27, 27–30, 31–34, 35–38, 39–42, and 43–46 [20]. A similar effect was seen in both single chromatid and whole chromosome nondisjunction. The frequency of diploid sperm also increases with age but this is not due to disomy 21 [21].

In similar studies of preimplantation embryos, there is a strong relationship between chromosome abnormalities in early embryos and maternal age. In one study of 1255 embryos, the aneuploidy rate increased from 3.1% in women aged 20–34 years to 17% in those 40 or older [22]. In a similar study of 2058 embryos, many karyotyped using PGD, an increase with age was specifically seen for trisomy 21 [23].

PATERNAL AGE

Maternal and paternal ages are highly correlated with relatively little variability in the age difference between the two parents. Therefore, an extremely large number of affected couples would have to be investigated in order to discern any independent paternal age effect. Consequently, while some moderately sized studies of couples have reported evidence for an effect in births [24] and miscarriages [25], many others found no association with age. If a paternal effect does exist, it is more likely to be present in paternally derived cases. Paternal age has been examined in a series of 67 such cases [26]. The mean age was 29.5 years in 57 meiotic cases and 31.8 in 10 mitotics compared with 30.3 in controls.

Another possibility apart from sample size that might account for some of the conflicting results between studies is a synergistic effect whereby paternal age is not important unless there is advanced maternal age. This was suggested by statistical modeling applied to a large cohort of 3419 Down syndrome births in New York state [27].

Two studies with sufficient size to establish an effect even restricted to older couples have been published. However, they yielded opposite effects. The first was a study of French donor insemination centers, where there is a large paternal age difference between donors and recipients [28]. A statistically significant positive effect of donor age was reported. The second was a study of 757 Down syndrome pregnancies and almost two million maternities in Switzerland [29]. There was a highly significant two-fold increase in Down syndrome risk for couples where the father was the same age or younger than the mother. A negative paternal age effect at birth could be explained by an increase in the chance of miscarriage with advancing paternal age. There is evidence to support this. In the Jerusalem Perinatal Study of 92,408 births, women with partners aged 35–39 years or 40 years or older had a nearly three-fold increase in spontaneous abortions as compared with women conceiving with men aged younger than 25 years [30].

GENETIC FACTORS

Previous Down Syndrome

In a small proportion of couples, the index case will be shown to have arisen from a parental structural chromosome rearrangement. The recurrence risk in these couples can be quite high, depending on the specific parental genotype. The most frequent is a heterozygous Robertsonian balanced translocation and for female carriers the risk is great enough to dwarf the age-specific risk at most ages. For example, among 185 amniocenteses in such women, 15% of fetuses had a translocation [31]. In contrast, male carriers of a balanced translocation do not appear to have a high risk; all 70 amniotic fluid samples in the same study had a normal karyotype.

If a woman has had a previous pregnancy with Down syndrome and the additional chromosome 21 was noninherited, there is still an increased risk of recurrence. The increase has been estimated at three points in pregnancy. In an unpublished study of >2500 women who had first trimester invasive prenatal diagnosis because of a previous affected pregnancy, the Down syndrome incidence was 0.75% higher than that expected from the maternal-age distribution (Kypros Nicolaides, personal communication). Similarly, a metaanalysis of four second and trimester amniocentesis series totalling 4953 pregnancies found an excess of 0.54% [32]. A metaanalysis of 433 livebirths had five recurrences, an excess risk of 0.52% [33]. The weighted average of these rates, allowing for fetal losses, is 0.77% in the first trimester, 0.54% in the second, and 0.42% at term. Examination of the data suggests that the excess is similar at different ages so the excess can be added to the age-specific risk expressed as a probability. The recurrence risk is relatively large for young women but by the age of about 40 it is not materially different from the risk in women without a family history.

Data from NDSCR suggest that although there is a fixed excess risk, this is dependent on the maternal age for the previous pregnancy [34]. A total of 11,281 Down syndrome pregnancies were registered in women who had at least one previous pregnancy, including 95 with a previous Down syndrome. They estimated that the excess risk at midtrimester decreased from 0.62% when the previous occurrence was at age 20 to as low as 0.04% at age 40. Another study reported a similar trend among 1160 women with a previous Down syndrome pregnancy [35]. The results were broken down according to whether the index case occurred before or after maternal age 30. The estimated increase–multiplicative rather than excess— was 4.7-fold for those with a prior case at a young age and 1.6-fold if later.

Previous Miscarriage

It is not general practice to karyotype products of conception obtained following spontaneous abortion. Given the high fetal loss rate in aneuploid pregnancies, it can be expected that one or more previous miscarriages will increase the subsequent risk of Down syndrome and other aneuploidies, as seen for previous affected births. A study of 46,939 women having invasive prenatal diagnosis over a 20-year period has confirmed this [36]. There were 699 diagnoses of aneuploidy, 81% of which were trisomies 21, 18, and 13. After allowing for maternal age, the odds of aneuploidy for those with one, two, or three or more previous miscarriages were 1.21, 1.26, and 1.51, respectively; for the common trisomies, it was 1.24, 1.32, and 1.47.

Recurrent pregnancy loss has been investigated further in relation to sperm aneuploidy. A series of 24 partners of women who had unexplained recurrent losses were studied; they had at most one previous live birth [37]. More than 5000 sperm were analyzed for each partner and 2.8% had aneuploidy, compared with 1.5% and 1.2% in the general population and fertile controls, respectively. The rates of chromosome 21 aneuploidy were 0.47%, 0.28%, and 0.24% respectively.

Other Abnormalities

Those with a previous Down syndrome pregnancy also have an increased risk of other types of aneuploidy, with an estimated excess risk of 0.25% based on metaanalysis [32] and in one study, a 2.3-fold risk of trisomies 18, 13, or a sex-chromosome abnormality [35]. It may be relevant that in NDSCR, among Down syndrome births, the chance of a double aneuploidy including standard Down syndrome was of a similar magnitude with an excess risk of 0.38% (21/5447) [4].

One study found that women with a previous Down syndrome pregnancy are at increased risk of neural tube defects (NTDs) and *vice versa* [38]. Two series were included: 493 families from Israel who were at high NTD risk and 516 families from Ukraine at high risk of Down syndrome. In the previous NTD series, there was a 5.9-fold increased incidence of Down syndrome compared with that expected according to maternal age. In the previous Down syndrome series, there was a 5.1-fold increase in NTDs. However, the Latin American collaborative study of congenital malformation failed to confirm these results [39].

Multiple Recurrence

The recurrence of Down syndrome in older women may be due to chance alone, but in young women it is more likely to have a genetic cause. Apart from a parental structural chromosome rearrangement, mosaicism may be involved. For example, in one study of 13 families with recurrent free trisomy 21, five were shown to involve parental mosaicism [40].

Even when there is no obvious mosaicism, the possibility remains of low-level mosaicism, perhaps confined to the gonads, which may be revealed by the use of molecular techniques.

There are 14 case reports in the literature of families with either two Down syndrome cases or one Down and another aneuploidy in which there were different reproductive partners in the parental or grandparental generation [32]. In every case except one, recurrence was on the maternal side and the discrepant case was from a highly inbred population.

REPRODUCTIVE FACTORS

Parity

A number of studies have reported that women of higher parity are at increased Down syndrome risk while others have failed to confirm such an effect. Most either took no account of maternal age or allowed for this covariable by stratifying the data into broad age groups. However, given the exponential increase in risk after age 30, if stratification is too broad, residual confounding will remain. Only three studies controlled for single year of maternal age: two reported a significant association between Down syndrome risk and parity [41, 42] while the third did not [43].

An additional problem of interpretation in this area is that the acceptability of prenatal diagnosis and termination of pregnancies affected by Down syndrome declines with parity. Therefore, analyses that are restricted to births, excluding terminations, are biased toward a positive effect. It is noteworthy that of the three that were fully age-controlled, only the negative study included terminations [43]. In an attempt to overcome this bias, one of the positive studies performed a secondary analysis after excluding pregnancies where the birth certificate reported that amniocentesis had been carried out [42]. This resulted in such a reduction of the original effect that it was no longer statistically significant. Moreover, it is known that amniocentesis is only mentioned on a US birth certificate in about half the pregnancies where the procedure is performed. Had complete information been available, it is likely that the effect would have been reduced further. Thus there is no unbiased information confirming an association with parity.

Early Menopause

Three studies have reported reduced menopausal age in association with trisomy. In one study, among women aged 35 or more at the time of a Down syndrome diagnosis, menopause occurred on average 10.2 years after a Down syndrome birth compared with 12.8 years for age-matched controls [44]. Among women aged under 35 the mean age of menopause was 0.7 years earlier than the general average, based on the National Health Survey. In a similar study, the mean age of menopause based on actuarial analysis was 1.7 years earlier than controls in the older age group and 0.7 years in younger women [45]. The third study was carried out among women with trisomic miscarriages [46]. Menopausal age was 1.0 years earlier than for age-selected women with chromosomally normal losses or births. None of the differences was statistically significant apart from the difference in means for older women in one study [45]. Nevertheless, the consistency between studies suggests that the effect is real.

One group experiencing an extremely premature menopause is the subset of women with Turner syndrome who do ovulate. Apart from pregnancies achieved by assisted reproduction technologies, this can occur in cases of structural defects that spare the critical region on Xq13-16 or in mosaic Turner syndrome.

Some 232 spontaneous nondonor pregnancies in such women have been reported in the literature, from a systematic review [47] together with more recent series [48–51]. There were 22 cases of Turner syndrome (9.5%) and five with Down syndrome, a rate of 2.2% overall and 2.4% excluding Turner syndrome, which represents a very large excess.

Reduced Ovarian Mass

A case-control study was carried out among 189 women who had meiotic live-born Down syndrome cases and 329 controls [52]. There were six cases with reduced ovarian mass, a rate of 3.7%. In one case, there was congenital absence of an ovary, three had surgical removal (two for a cyst and one a teratoma), and two had surgical reduction of both ovaries (one endometriosis and one obstruction). One control had surgical removal of an ovary after a tubal pregnancy, a rate of 0.3%, so reduced ovarian mass was associated with a ninefold increase in Down syndrome risk.

Support for this comes from experimental work with inbred CBA mice that have a small number of oocytes that are completely depleted by the time ovulation ceases. A large series of mice was given a unilateral oophorectomy that caused increased ovulation in the contralateral ovary and an early menopause [53]. The oophorectomized mice had an increased rate of aneuploid embryos at all ages compared with untreated control animals. The treated mice also had earlier onset of irregular cycles, indicating premature reproductive aging.

Biochemical Markers of Ovarian Function

Four markers have been investigated in relation to aneuploidy: follicle stimulating hormone (FSH), estriol (E_2), inhibin B, and anti-Müllerian hormone (AMH). Two were carried out in women with aneuploid miscarriages and two specifically in Down syndrome pregnancies.

The first study was carried out in 59 women who had had ART and the pregnancy ended in a fetal loss; a karyotype on products of conception revealed that 26 had aneuploidy [54]. In pre-ART cycles, maternal FSH and E_2 levels were compared between the groups and in those with affected pregnancies, 58% had elevated levels of one or more markers compared with 27% in unaffected pregnancies. A similar study recruited 54 women following a trisomy miscarriage and 21 maternal age matched controls with euploid miscarriages [55]. In subsequent cycles, measurements were made of FSH, E_2, and inhibin B and the mean level was increased in cases except for E_2, which was reduced. None of the differences in these two studies was statistically significant, perhaps due to a small sample size.

A study of Down syndrome pregnancies in 118 women was carried out and in 102 controls [56]. Initially, over three cycles, FSH, E_2, and inhibin B were measured and subsequently

AMH [57]. There was a statistically significant increase in FSH among cases with 14% having elevated levels in at least one cycle compared with 5% in controls and a significant reduction in AMH—12% classified as low compared with 5%. There were small but nonsignificant elevations in E_2 and inhibin B. A study of maternal serum AMH was also carried out in stored samples originally taken for prenatal screening in 25 Down syndrome pregnancies and 125 age-matched controls. The median level of AMH in cases, adjusted for gestational age, was reduced by 17%, although this did not reach statistical significance [58].

Taking all four studies together, the results provide support, albeit weak, for there being impaired ovarian function in affected pregnancies. This is also consistent with the results concerning early menopause.

Sexual Practises

Several studies have claimed an association between reduced frequency of coitus and Down syndrome risk; for a review of this subject see [59]. Some of the evidence is direct, based on interviews of parents, but much of it is simply by inference. Thus infrequent coitus has been suggested to underlie occasional reports of increased Down syndrome frequency in illegitimate births, long marriage, long interval between births, and among Catholics, who are assumed to be using the ovulatory method of contraception.

A study in Jerusalem found a higher Down syndrome prevalence in orthodox Jewish couples compared with the nonreligious population [60]. Again, infrequent coitus was evoked to explain this result because orthodox Jews delay coitus until 7 days after the end of the menses, at which time there is a religious obligation to resume sexual activity.

GENERAL RISK FACTORS

Twins

On theoretical grounds, the overall birth prevalence of Down syndrome per twin *fetus* should be greater than the birth prevalence in singletons. First, in dizygous twins, concordance for Down syndrome might be expected to occur at the same rate as recurrence among siblings. Second, twinning increases with maternal age. The relative increase in prevalence can be estimated for a given population from the maternal age distribution, race, and the monozygous twining rate, which can be assumed to be independent of age.

There are five large studies in the literature that compare observed Down syndrome prevalence in twins and singletons [61–65]; however, they are subject to considerable bias. All but one [63] reported a large reduction in prevalence for twin fetuses. Two of the studies are based on national registration records: at birth and restricted to live births [61], or within 7 days of birth and a voluntary scheme with low compliance [62]. Affected twins are more likely to be stillborn and, in that period, karyotype confirmation of suspected Down syndrome live birth could take more than a week. The later studies were carried out in the era of prenatal screening and include prenatally diagnosed cases. Down syndrome screening detection rates are much lower in twins than singletons because the abnormal biochemical marker concentration in maternal serum produced by an affected twin will be masked by the normal concentration from an unaffected cotwin [66]. Consequently, because a large proportion of cases undetected by screening miscarry, the observed Down syndrome prevalence will be reduced compared to singletons.

Ethnic Origin

All the studies with single year of age prevalence rates used to estimate maternal age-specific risk of Down syndrome are almost entirely based on women of European origin. There are many individual reports of relatively high or low birth prevalence in different ethnic groups. Some are from countries without reliable systems for collecting information on the maternal date of birth but 36 studies covering 49 populations included sufficiently detailed and reliable age information to be entered into a metaanalysis [67]. An age-standardized index was computed, dividing the observed number of Down syndrome cases by the expected number obtained by applying the age-specific risk curve to the distribution of maternities. There were two groups with some evidence for rates greater than Europeans. These are those of Mexican and Central American origin in California (standardized indices 1.19 and 1.30 in two studies) and Jews of Asian or African origin in Israel (1.27). The standardized indices were markedly reduced in some populations, including three studies in Africans, but the authors conclude that this is likely to be due to incomplete ascertainment.

Smoking

Several early studies reported that smoking was less common in the mothers of infants with Down syndrome, but the latest metaanalysis of 17 published studies failed to find a significant association [68]. Smoking habits are subject to strong birth cohort effects so it is important to take full account of maternal age. Some of the early studies either did not take account of age or stratified the data using broad age bands, which may not be adequate. This was demonstrated in one study that found a relative risk of 0.87 with broad age grouping, 0.89 adjusting for additional variables, and 1.00 when age adjustment, together with the additional variables, was in single years [69].

One of the studies also categorized subjects according to the parental origin of the additional chromosome 21 and the timing of the error [70]. The overall estimate of the relative risk among all 285 Down syndrome pregnancies combined was 0.96. But when only maternally derived cases were considered, the reduction in risk was greater (0.84) and in those cases where the error occurred at MI, it was reduced further (0.72).

Maternal Thyroid Disease

Nine studies have shown that several years after the affected birth, mothers of children with Down syndrome have a high risk of either frank thyroid disease or elevated thyroid antibody titres. In a metaanalysis of these results, the risk of Down syndrome was increased

2.2-fold [71]. Because both thyroid disease and Down syndrome incidence increase with age, this could be an artefact but in those studies that adjusted the results for age, the effect persisted.

However, three studies have examined the association during the affected pregnancy itself and have failed to find a strong association. The first used stored maternal serum samples collected during second trimester prenatal screening and tested 77 from Down syndrome and 385 from age-matched controls [71]. Thyroid antibody levels were higher in cases but the difference did not reach statistical significance. Two studies examined first trimester maternal serum thyroid stimulating hormone (TSH) levels. In a study of samples taken prior to invasive prenatal diagnosis from 27 Down syndrome pregnancies and 115 controls, the median level was 16% lower in cases, allowing for gestation [72]. In the second study, TSH was measured in samples taken at the time of prenatal screening; 25 from Down syndrome pregnancies were tested retrospectively and 3592 prospectively from unaffected pregnancies [73]. Median TSH was reduced by 24% after allowing for gestation, which was statistically significant. Free thyroxine and free triiodothyronine were also measured and levels did not differ between cases and controls.

One possible explanation for the discrepancy between the results some years after the affected birth and during the affected pregnancy itself is the thyrotropic affect of human chorionic gonadotrophin (hCG). During the affected pregnancy, hCG levels are, on average, approximately double those in unaffected pregnancies. Thus in some cases the thyroid might be damaged by the affected pregnancy, presenting years later as thyroid disease.

OTHER COMMON ANEUPLOIDIES

Edwards and Patau Syndromes

Edwards syndrome is a lethal condition with about one-third dying in the neonatal period, one-half by two months, and only a few percent surviving the first year as severely mentally retarded individuals. Intrauterine fatality from the midtrimester to term is about two-thirds.

The maternal age-specific birth prevalence can be taken to be a fixed fraction of the corresponding prevalence for Down syndrome. When six series of routinely karyotyped neonates were combined [74], there were a total of seven cases of Edwards syndrome and 71 of Down syndrome, a relative frequency of 1/10. Combining data from nine regional congenital malformation registers, a total of 2254 Edwards syndrome cases were identified [17]. Only about one-fifth were live births, but applying interuterine survival curves to cases ending in termination of pregnancy and fetal losses, the estimated relative frequency was 1/8.

Patau syndrome is generally lethal but about 10% will survive for more than a year, albeit with profound developmental delay. It is associated with preeclampsia and in one study, the condition was present in six out of 25 cases with a further four having other forms of pregnancy-induced hypertension [75]. About half miscarry between midtrimester and term [33]. The best estimate of the birth prevalence relative to Down syndrome is 1/13 [17, 74].

Sex Chromosome Aneuploidy

This is more common than autosomal aneuploidy, both at birth and in fetal losses, and is more frequently of paternal origin [76]. Birth prevalence of the 45,X genotype, complete or mosaic, is 1 per 2500 females and the prevalence *decreases* with maternal age. The fetal loss rate is very high with only one-third surviving from midtrimester to birth [33], but the corresponding Turner syndrome phenotype among survivors is relatively benign. A recent analysis supports the conclusion that all 45,X fetuses that survive are cryptic mosaic [77]. Birth prevalence of 47,XXY is about 1 per 1000 males and prevalence increases with age, albeit less steeply than for autosomal aneuploidies. The Klinefelter syndrome phenotype that presents with hypogonadism and gynaecomastia is associated with a small reduction in intellectual capacity, although this might be biased by underascertainment. Males with 47,XYY and females with 47,XXX, neither of which have marked clinical signs but are associated with a moderate intellectual impairment, the former with a 1 per 1000 birth prevalence unrelated to age and the latter with a similar age-specific prevalence to 47,XXY.

GENETICS

When Does Aneuploidy Happen?

Apart from structural chromosome rearrangements, aneuploidy is generally a consequence of mal-segregation of chromosomes during oogenesis and spermatogenesis. This process includes three phases: multiplication, growth, and maturation. During the multiplication phase, there are several mitotic divisions of the germ cells that are followed by growth and finally two meioses. The primary diploid germ cell undergoes reductional division (MI) and then equational division (MII). Mal-segregation may occur during the mitotic and both meiotic divisions; if it is at mitosis, more than one mature gamete may have a chromosomal mutation.

Mitotic Errors and Germline Mosaicism

The most likely explanation of recurring nonstructural aneuploidy is the presence in a parent of cell lines arising from gametes of different genotypes ("germline mosaicism"). This occurs more frequently in women than men. A mutant pool of spermatozoa, if fertilization has not occurred, disappears from the reproductive cycle in a period of three months whereas mutant precursors of oocytes remain in the ovaries for life.

Ethical considerations have limited the determination of mitotic mosaicism in germinal cells. However, ovarian and testicular biopsies have been carried out in fetuses and very high percentages have been found in females while the frequency was low in males [78]. Based on these results, it has been suggested that most normal female fetuses are ovarian trisomy 21 mosaics and the increased prevalence of Down syndrome with maternal age is caused by differential selection of these cells during fetal and postnatal germ cell development.

Meiotic Errors

Mal-segregation of chromosomes or chromatids occurs during meiosis. Studies in aneuploid oocytes have shown that extra whole chromosomes are rarely found in comparison with additional free chromatids [79]. Most errors are believed to occur at MI, which arrests at the diplotene stage and resumes many years later at ovulation. However, this view has recently been challenged for trisomy 21, proposing that maternal germinal mosaicism is also involved and can even explain the maternal age effect [80] because normal oogonia enter meiosis earlier than those with mutations [81].

Uniparental Disomy (UPD)

The presence of two or more different chromosome complements in an individual developed from a single zygote is the most common type of the human mosaicism [6]. It has been reported for many types of chromosome abnormalities, including trisomy, monosomy, triploidy, deletions, duplications, rings, and other types of structural rearrangements. There is lower morbidity and mortality compared with nonmosaic aneuploidies of the same type. In UPD there are two copies of a whole or part of a chromosome from one parent with the corresponding parental contribution missing. Either nonidentical chromosomes are inherited at MI or a single chromosome is inherited and later duplicated at MII.

Sperm Aneuploidy

A review of published aneuploidy studies using florescent in situ hybridization (FISH) in spermatozoa from healthy donors estimated that the rate is about 0.1% per autosome, ranging from 0.03 per chromosome 8 to 0.5 per chromosome 22, and the overall disomy rate is 2.3% [82]. There was no age effect for the majority of chromosomes. An almost threefold increase of sex chromosome and chromosome 21 disomies has also been reported in infertile men compared with controls [83]. These findings suggest that the difference between maternal and paternal contributions to aneuploidy are not due to segregation errors but rather to more effective checkpoint mechanisms in spermatogenesis than in oogenesis. It was believed that germ cell aneuploidies caused by mal-segregation and meiotic checkpoint errors were more common in female meiosis. However, a recent FISH study on sperm from three donors found MII errors in 14% and MI errors in 2% [84]. It has been reported that, for mice spermatocytes, there are fewer recombinations in juveniles compared to adults [85]. This is consistent with the recently reported increased aneuploidy prevalence in young fathers, allowing for maternal age [29].

Why Does Aneuploidy Happen?

Traditional karyotyping has a resolution of about 3 Mb but the advent of modern technologies such as array-based molecular genetic analysis with comparative genomic hybridization and next-generation sequencing makes it possible to shed light on the molecular basis of aneuploidy. This is now being studied using whole genomics, transcriptomics, and proteomics studies, including single cell analysis, for example, blastomeres.

Cohesin, Shugoshin, and Spindle Assembly Checkpoint (SAC)

In order to fully understand aneuploidy, it is necessary to study the mechanisms ensuring the maintenance and segregation of chromosomes during germ cell division and maturation. Currently, a central focus of this research is the relative transcript levels of cohesin, shugoshin, and SAC genes. Correct chromosome segregation is maintained by the coordinated interaction of a variety of proteins that supports chromosome binding and reliable assembly of the spindle checkpoint. Cohesin multiprotein complexes consist of four subunits: two structural maintenance of chromosomes (SMC) proteins, an α-kleisin, and the substrate of the stromal antigen. The specific proteins maintain cohesion between replicated chromosomes until one of the subunits is targeted for elimination. It has been proposed that the maternal age effect is due to altered expression of SAC genes in oocytes from older women [86]. Differences in *SGOL1* and *BUB1* gene expression have been found between mature and immature oocytes. These findings suggest that higher *SGOL1* expression levels in mature oocytes could better protect against a premature separation of sister chromatids. The absence of *BUB1* transcripts in mature oocytes is frequently associated with reduced expression of either mitotic cohesins or shugoshins. These finding have led to speculation that reduced expression of *BUB1* could induce mitotic arrest in oocytes that fail to express a complete complement of cohesins and shugoshins, thereby reducing the number of developing aneuploid preimplantation embryos.

Mosaicism

Modern molecular techniques have revealed that cryptic chromosome 21 mosaicism is much more common than previously recognized. It has been suggested that a more stringent control of mitotic segregation during early germ cell embryonic development may lead to different degrees of mosaicism in ovaries compared with testicular samples [87].

Mosaic embryos result from postzygotic mitotic errors that are more frequent than germinal meiotic errors [88]. In studies of embryos undergoing preimplantation genetic screening (PGS), about 80%–100% are found to be mosaic. The origin of these errors is predominantly anaphase lag or chromosome mal-segregation [89]. The clinical significance of mosaicism detected by PGS is not well characterized because a normal fetus may result from self-correction or migration of mutant cells during embryo development to the placenta and decidua.

In the absence of somatic recombination, cases of chromosome 21 mosaicism originating from postzygotic mal-segregation in an original normal disomy 21 zygote would be expected to be isodisomic for two of the three chromosomes 21, making up this somatic acquired aneuploidy.

Translocations and Inversions

Individuals with balanced reciprocal translocation are typically healthy but are at increased risk of miscarriage, infertility, and birth with multiple congenital anomalies. These events are associated with partial aneuploidy of the embryo. An unbalanced chromosome complement with partial monosomy of one chromosome and partial trisomy of another results in highly variable phenotypes. In balanced translocations, the incidence of a partial aneuploidy in the embryo does not depend on the age or gender of the parent; it is associated with recurrent pregnancy loss [90].

Polymorphism

Polymorphic variants on chromosomes are considered normal because they involve heterochromatic regions and occur without apparent clinical significance. But the incidence is increased in infertile patients, couples with recurrent miscarriages, and in men with poor sperm quality. Recently, the frequency of polymorphic variants has been investigated in spermatozoa and specific polymorphisms were associated with aneuploidy [91]. The study also found that female polymorphism carriers had a higher risk of an aneuploid embryo. Parental single-nucleotide polymorphisms are also more common in aneuploidy. Mutant alleles are associated with impaired hormonal immunological systems or metabolism.

The ε4 allele of the apolipoprotein (apo) E gene is associated with Alzheimer's disease, both sporadic and familial. The allele frequency in parents of Down syndrome children has been investigated because of the excess risk of this disease in women with a previously affected pregnancy [92]. In a series of 188 Danish cases, there was no overall difference in the allele distribution compared to a control population [93]. However, a significantly increased frequency of the ε4 apoE allele was found in young mothers with apparent MII errors.

Abnormal folate and methyl metabolism can lead to DNA hypomethylation and abnormal segregation, which has prompted the investigation of maternal polymorphisms in genes involved in folate metabolism. The common $677C \rightarrow T$ polymorphism in the 5,10-methylene-tetrahydrofolate reductase (MTHFR) gene occurs more frequently than usual in the mothers of children with Down syndrome. A metaanalysis of 37 case-control studies provides the best estimate of risk [94]. Both homozygous (TT) and heterozygous (CT) configurations were more common in cases than controls. The results were consistent with a dominant model (TT or CT versus CC) estimated to increase Down syndrome risk by 23% and recessive (TT vs. TC or CC) with 21% increase.

An increased frequency of the $66A \rightarrow G$ polymorphism in methionine synthase reductase (MTRR) was found in two studies [95, 96]. Both studies also tested MTHFR alleles in the same women and found that a combination of the two mutations conferred a higher risk than either alone.

Other Molecular Findings

Regions of genome imprinting on uniparental pairs of some chromosomes can have phenotypic consequences. For example, UPD in chromosome 15 leads to Prader-Willi and Angelman syndromes. Chromosomal rearrangements resulting from mitotic or meiotic mechanisms differ both in their effect on embryonic development and in the risk of their recurrence in subsequent pregnancies [32]. UPD can arise from various mechanisms: trisomy rescue, with and without concomitant trisomy; monosomy rescue; and mitotic formation of a mosaic segmental UPD. Meiotic UPDs are different both in molecular mechanism and phenotype. In genetic counseling, it is important to know which mechanism was involved in order to predict recurrence risk and phenotype.

General Etiological Hypotheses

Among the various hypotheses proposed before the mid-2000s, maternal mitochondrial (mt)DNA damage [32] is of particular interest because it has received some recent confirmation using modern molecular genetic methods. During the different phases of oogenesis and embryogenesis, the energy required to support oocyte maturation, fertilization, and embryo development is provided by the mitochondria. Defects in mtDNA have been found in relation to abnormal oocyte maturation, fertilization, and implantation. Recently, it was shown that specific maternal mtDNA base changes are associated with a higher susceptibility to aneuploidy [97].

Recent publications have also lent support to other hypotheses suggested in the past related to ovarian aging, dysfunction, and reduced reserve [81]. It has been shown that the proportion of disomy 21 oocytes is greater in the aging ovary; one suggestion is that such oocytes lag behind during development and accumulate in the total oocyte pool over time. In a study of women undergoing IVF because of recurrent pregnancy loss and diminished ovarian reserve, 57% of blastocysts had aneuploidy and in 25% of cycles, all blastocysts were affected [98]. The aneuploidy rate increased with age but compared with controls, the excess was greatest in women under 38.

A completely new hypothesis might be formulated whereby the microbiome plays a critical role in chromosomal alterations during human gametogenesis. There has been recent interest in the role of gut microbial compositions in the formation, regulation, and maintenance of mental and physical health. In particular, it is known that the human microbiome strongly influences reproduction health and gametogenesis. The metabolites of bacterial proteolysis may act at amino acid catabolism, serving as signals to modulate the function of the female reproductive tract such as hormone secretion, ovulation, and endometrial receptivity.

CONCLUSIONS

The positive association between the common aneuploidies seen at birth and maternal age is well established and similar associations have been documented in early pregnancy and in the embryo. There are many hypotheses that might account for this strong effect but none of them is proven. In general, aneuploidy is due to mal-segregation of chromosomes during oogenesis and spermatogenesis, but the reason for this has not been established. Correct segregation occurs through the coordinated interaction of a variety of proteins that supports chromosome binding and reliable assembly of the spindle checkpoint. While it is likely that the breakdown over time of one of these proteins, cohesin, is involved, much of the process is unknown. Epidemiology provides clues to effects other than maternal age that might account for the larger proportion of aneuploidy cases not due to maternal age. Paternal age is difficult to study because parental ages are highly correlated but require more research. Reproductive factors such as reduced ovarian function, which is manifest, for example, in a relatively early menopause, are probably involved. The birth prevalence of Down syndrome, the most common serious aneuploidy, is declining because of more efficient prenatal diagnosis and termination of pregnancy. This should stimulate more etiological studies with the hope, currently distant, of a *primary* prevention strategy.

References

[1] Zigman WB, Schuf N, Sersen E, Silverman W. Prevalence of dementia in adults with and without Down syndrome. Am J Mental Retard 1995;100:403–12.

[2] Lavigne J, Sharr C, Elsharkawi I, Ozonoff A, Baumer N, Brasington C, et al. Thyroid dysfunction in patients with Down syndrome: results from a multi-institutional registry study. Am J Med Genet A 2017;173(6):1539–45.

[3] Zhu JL, Hasle H, Correa A, Schendel D, Friedman JM, Olsen J, et al. Survival among people with Down syndrome: a nationwide population-based study in Denmark. Genet Med 2013;15(1):64–9.

[4] Mutton D, Alberman E, Hook EB. Cytogenetic and epidemiological findings in Down syndrome, England and Wales 1989 to 1993. J Med Genet 1996;33:387–94.

[5] Hassold T, Sherman S. Down syndrome: genetic recombination and the origin of the extra chromosome 21. Clin Genet 2000;57:95–100.

[6] Hassold T, Hall H, Hunt P. The origin of human aneuploidy: where we have been, where we are going. Hum Mol Genet 2007;16:R203–8. Spec No. 2. [Review].

[7] Lamb NE, Feingold E, Savage A, Avramopoulos D, Freeman SB, Gu Y, et al. Characterization of susceptible chiasma configurations that increase the risk for maternal nondisjunction of chromosome 21. Hum Mol Genet 1997;6(9):1391–9.

[8] Cuckle HS, Wald NJ, Thompson SC. Estimating a women's risk of having a pregnancy associated with Down's syndrome using her age and serum alpha-fetoprotein level. Br J Obstet Gynaecol 1987;94:387–402.

[9] Morris JK, Mutton D, Alberman E. Revised estimates of the maternal age specific live birth prevalence of Down's syndrome. J Med Screen 2002;9:2–6.

[10] Lamb NE, Yu K, Shaffer J, Feingold E, Sherman SL. Association between maternal age and meiotic recombination for trisomy 21. Am J Hum Genet 2005;76:91–9.

[11] Bray IC, Wright DE. Estimating the spontaneous loss of Down syndrome fetuses between the time of chorionic villus sampling and livebirth. Prenat Diag 1998;18:1045–54.

[12] Hook EB, Topol BB, Cross PK. The natural history of cytogenetically abnormal fetuses detected at midtrimester amniocentesis which are not terminated electively: new data and estimates of the excess and relative risk of late fetal death associated with 47,+21 and some other abnormal karyotypes. Am J Hum Genet 1989;45:855–61.

[13] Hook EB, Mutton DE, Ide R, Alberman E, Bobrow M. The Natural History of Down Syndrome conceptuses diagnosed prenatally that are not electively terminated. Am J Hum Genet 1995;57:875–81.

[14] Morris JK, Wald NJ, Watt HC. Fetal loss in Down syndrome pregnancies. Prenat Diag 1999;19:142–5.

[15] Hassold T, Warburton D, Kline J, Stein Z. The relationship of maternal age and trisomy among trisomic spontaneous abortions. Am J Hum Genet 1984;36:1349–56.

[16] Nybo Anderson M-A, Wohlfahrt J, Christens P, Olsen J, Melbye M. Maternal age and fetal loss: population based register linkage study. Br Med J 2000;320:1708–12.

[17] Savva GM, Walker K, Morris JK. The maternal age-specific live birth prevalence of trisomies 13 and 18 compared to trisomy 21 (Down syndrome). Prenat Diagn 2010;30(1):57–64.

[18] Spandorfer SD, Davis OK, Barmat LI, Chung PH, Rosenwaks Z. Relationship between maternal age and aneuploidy in in vitro fertilization pregnancy loss. Fertil Steril 2004;81(4):1265–9.

[19] Eichenlaub-Ritter U. Genetics of oocyte ageing. Maturitas 1998;30:143–69.

[20] Pellestor F, Andreo B, Arnal F, Humeau C, Demaille J. Maternal aging and chromosomal abnormalities: new data drawn from in vitro unfertilized human oocytes. Hum Genet 2003;112:195–203.

[21] Bosch M, Rajmil O, Martinez-Pasarell O, Egozcue J, Templado C. Linear increase of diploidy in human sperm with age: a four-colour FISH study. Eur J Hum Genet 2001;9:533–8.

[22] Marquez C, Sandalinas M, Bahce M, Alikani M, Munne S. Chromosome abnormalities in 1255 cleavage-stage embryos. Reprod Biomed Online 2000;1:17–26.

[23] Munne S, Bahce M, Sandalinas M, Escudero T, Marquez C, Velilla E, et al. Differences in chromosome susceptibility to aneuploidy and survival to first trimester. Reprod Biomed Online 2004;8:81–90.

[24] Stene E, Stene J, Stengel-Rutkowski S. A reanalysis of the New York State prenatal diagnosis data on Down's syndrome and paternal age effects. Hum Genet 1987;77(4):299–302.

[25] Hatch M, Kline J, Levin B, Hutzler M, Warburton D. Paternal age and trisomy among spontaneous abortions. Hum Genet 1990;85(3):355–61.

[26] Savage AR, Petersen MB, Pettay D, Taft L, Allran K, Freeman SB, et al. Elucidating the mechanisms of paternal non-disjunction of chromosome 21 in humans. Hum Mol Genet 1998;7(8):1221–7.

[27] Fisch H, Hyun G, Golden R, Hensle TW, Olsson CA, Liberson GL. The influence of paternal age on Down syndrome. J Urol 2003;169:2275–8.

[28] Lansac J, Thepot F, Mayaux MJ, Czyglick F, Wack T, Selva J, et al. Pregnancy outcome after artificial insemination or IVF with frozen semen donor: a collaborative study of the French CECOS Federation on 21,597 pregnancies. Eur J Obstet Gynecol Reprod Biol 1997;74(2):223–8.

[29] Steiner B, Masood R, Rufibach K, Niedrist D, Kundert O, Riegel M, et al. An unexpected finding: younger fathers have a higher risk for offspring with chromosomal aneuploidies. Eur J Hum Genet 2015;23(4):466–72.

[30] Kleinhaus K, Perrin M, Friedlander Y, Paltiel O, Malaspina D, Harlap S. Paternal age and spontaneous abortion. Obstet Gynecol 2006;108(2):369–77.

[31] Boué A, Gallano P. A collaborative study of the segregation of inherited chromosome arrangements in 1356 prenatal diagnoses. Prenat Diagn 1984;4:45–67.

[32] Arbuzova S, Cuckle H, Mueller R, Sehmi I. Familial Down syndrome: evidence supporting cytoplasmic inheritance. Clin Genet 2001;60:456–62.

[33] Hook EBH. In: Brock DJH, Rodeck CH, Ferguson-Smith MA, editors. Prenatal diagnosis and screening. Edinburgh: Churchill Livingstone; 1992. p. 351–92.

[34] Morris JK, Mutton DE, Alberman E. Recurrences of free trisomy 21: analysis of data from the National Down Syndrome Cytogenetic Register. Prenat Diagn 2005;25(12):1120–8.

[35] Warburton D, Dallaire L, Thangavelu M, Ross L, Levin B, Kline J. Trisomy recurrence: a reconsideration based on North American data. Am J Hum Genet 2004;75:376–85.

[36] Bianco K, Caughey AB, Shaffer BL, Davis R, Norton ME. History of miscarriage and increased incidence of fetal aneuploidy in subsequent pregnancy. Obstet Gynecol 2006;107(5):1098–102.

[37] Carrell DT, Wilcox AL, Lowy L, Peterson CM, Jones KP, Erickson L, et al. Elevated sperm chromosome aneuploidy and apoptosis in patients with unexplained recurrent pregnancy loss. Obstet Gynecol 2003;101(6):1229–35.

[38] Barkai G, Arbuzova S, Berkenstadt M, Heifetz S, Cuckle H. Frequency of Down's syndrome and neural-tube defects in the same family. Lancet 2003;361(9366):1331–5.

[39] Amorim MR, Castilla EE, Orioli IM. Is there a familial link between Down's syndrome and neural tube defects? Population and familial survey. BMJ 2004;328(7431):84.

[40] Pangalos CG, Talbot Jr. CC, Lewis JG, Adelsberger PA, Petersen MB, Serre JL, et al. DNA polymorphism analysis in families with recurrence of free trisomy 21. Am J Hum Genet 1992;51(5):1015–27.

[41] Kallen K. Parity and Down syndrome. Am J Med Genet 1997;70(2):196–201.

[42] Doria-Rose VP, Kim HS, Augustine ET, Edwards KL. Parity and the risk of Down's syndrome. Am J Epidemiol 2003;158(6):503–8.

[43] Chan A, McCaul KA, Keane RJ, Haan EA. Effect of parity, gravidity, previous miscarriage, and age on risk of Down's syndrome: population based study. BMJ 1998;317(7163):923–4.

[44] Phillips OP, Cromwell S, Rivas M, Simpson JL. Elias S. Trisomy 21 and maternal age of menopause: does reproductive age rather than chronological age influence risk of nondisjunction? Hum Genet 1995;95:117–8.

[45] Bartmann AK, Araujo FM, Iannetta O, Paneto JC, Martelli L, Ramos ES. Down syndrome and precocious menopause. J Assist Reprod Genet 2005;22:129–31.

[46] Kline J, Kinney A, Levin B, Warburton D. Trisomic pregnancy and earlier age at menopause. Am J Hum Genet 2000;67(2):395–404.

[47] Tarani L, Lampariello S, Raguso G, Colloridi F, Pucarelli I, Pasquino AM, et al. Pregnancy in patients with Turner's syndrome: six new cases and review of literature. Gynecol Endocrinol 1998;12(2):83–7.

[48] Birkebaek NH, Cruger D, Hansen J, Nielsen J, Bruun-Petersen G. Fertility and pregnancy outcome in Danish women with Turner syndrome. Clin Genet 2002;61:35–9.

[49] Bryman I, Sylvén L, Berntorp K, Innala E, Bergström I, Hanson C, et al. Pregnancy rate and outcome in Swedish women with Turner syndrome. Fertil Steril 2011;95(8):2507–10.

[50] Hadnott TN, Gould HN, Gharib AM, Bondy CA. Outcomes of spontaneous and assisted pregnancies in Turner syndrome: the U.S. National Institutes of Health experience. Fertil Steril 2011;95(7):2251–6.

[51] Bernard V, Donadille B, Zenaty D, Courtillot C, Salenave S, Brac de la Perrière A, et al. CMERC Center for Rare Disease. Spontaneous fertility and pregnancy outcomes amongst 480 women with Turner syndrome. Hum Reprod 2016;31(4):782–8.

[52] Freeman SB, Yang Q, Allran K, Taft LF, Sherman SL. Women with a reduced ovarian complement may have an increased risk for a child with Down syndrome. Am J Hum Genet 2000;66(5):1680–3.

[53] Brook JD, Gosden RG, Chandley AC. Maternal ageing and aneuploid embryos—evidence from the mouse that biological and not chronological age is the important influence. Hum Genet 1984;66:41–5.

[54] Nasseri A, Mukherjee T, Grifo JA, Noyes N, Krey L, Copperman AB. Elevated day 3 serum follicle stimulating hormone and/or estradiol may predict fetal aneuploidy. Fertil Steril 1999;71(4):715–8.

[55] Kline J, Kinney A, Reuss ML, Kelly A, Levin B, Ferin M, et al. Trisomic pregnancy and the oocyte pool. Hum Reprod 2004;19:1633–43.

[56] van Montfrans JM, Dorland M, Oosterhuis GJ, van Vugt JM, Rekers-Mombarg LT, Lambalk CB. Increased concentrations of follicle-stimulating hormone in mothers of children with Down's syndrome. Lancet 1999;353:1853–4.

[57] van der Stroom EM, König TE, van Dulmen-den Broeder E, Elzinga WS, van Montfrans JM, Haadsma ML, et al. Early menopause in mothers of children with Down syndrome? Fertil Steril 2011;96(4):985–90.

[58] Seifer DB, MacLaughlin DT, Cuckle HS. Serum müllerian-inhibiting substance in Down syndrome pregnancies. Hum Reprod 2007;22(4):1017–10209.

[59] Martin-DeLeon PA, Williams MB. Sexual behaviour and Down syndrome: the biological mechanism. Am J Med Genet 1987;27:693–700.

[60] Sharav T. Aging gametes in relation to incidence, gender and twinning in Down syndrome. Am J Med Genet 1991;39(1):116–8.

[61] Windham GC, Bjerkedal T. Malformations in twins and their siblings, Norway, 1967-79. Acta Genet Med Gemellol (Roma) 1984;33(1):87–95.

[62] Doyle PE, Beral V, Botting B, Wale CJ. Congenital malformations in twins in England and Wales. J Epid Comm Health 1990;45:43–8.

[63] Garchet-Beaudron A, Dreux S, Leporrier N, Oury JF, Muller F. ABA Study Group; Clinical Study Group. Second-trimester Down syndrome maternal serum marker screening: a prospective study of 11 040 twin pregnancies. Prenat Diagn 2008;28(12):1105–9.

[64] Boyle B, Morris JK, McConkey R, Garne E, Loane M, Addor MC, et al. Prevalence and risk of Down syndrome in monozygotic and dizygotic multiple pregnancies in Europe: implications for prenatal screening. BJOG 2014;121(7):809–19. discussion 820.

[65] Sparks TN, Norton ME, Flessel M, Goldman S, Currier RJ. Observed rate of Down syndrome in twin pregnancies. Obstet Gynecol 2016;128(5):1127–33.

[66] Cuckle HS, Pergament E, Benn P. Multianalyte maternal serum screening for chromosomal abnormalities and neural tube defects. In: Milunsky A, Milunsky JM, editors. Genetic disorders and the fetus: diagnosis, prevention and treatment. 7th ed. Hoboken: Wiley-Blackwell; 2016. p. 483–540.

[67] Carothers AD, Hecht CA, Hook EB. International variation in reported livebirth prevalence rates of Down syndrome, adjusted for maternal age. J Med Genet 1999;36:386–93.

[68] Rudnicka AR, Wald NJ, Huttly W, Hackshaw AK. Influence of maternal smoking on the birth prevalence of Down syndrome and on second trimester screening performance. Prenat Diagn 2002;22(10):893–7.

[69] Chen C-L, Gilbert TJ, Daling JR. Maternal smoking and Down syndrome: the confounding effect of maternal age. Am J Epidemiol 1999;149(5):442–6.

[70] Yang Q, Sherman SL, Hassold TJ, Allran K, Taft L, Pettay D, et al. Risk factors for trisomy 21: maternal cigarette smoking and oral contraceptive use in a population-based case-control study. Genet Med 1999;1:80–8.

[71] Cuckle HS, Wald N, Stone R, Densem J, Haddow J, Knight G. Maternal serum thyroid antibodies in early pregnancy and fetal Down's syndrome. Prenat Diagn 1988;8(6):439–45.

[72] Weinans MJ, Pratt JJ, de Wolf BT, Mantingh A. First-trimester maternal serum human thyroid-stimulating hormone in chromosomally normal and Down syndrome pregnancies. Prenat Diagn 2001;21(9):723–5.

[73] Ashoor G, Maiz N, Cuckle H, Jawdat F, Nicolaides KH. Maternal thyroid function at 11–13 weeks of gestation in fetal trisomies 21 and 18. Prenat Diagn 2011;31(1):33–7.

[74] Hook EB, Hammerton JL. The frequency of chromosome abnormalities detected in consecutive newborn studies; differences between studies; results by sex and severity of phenotypic involvement. In: Hook EB, Porter IH, editors. Population cytogenetics: studies in humans. New York: Academic Press; 1977. p. 63–79.

[75] Toughy JF, James DK. Pre-eclampsia and trisomy 21. Brit J Obstet Gynaecol 1992;99:891–4.

[76] Templado C, Uroz L, Estop A. New insights on the origin and relevance of aneuploidy in human spermatozoa. Mol Hum Reprod 2013;19(10):634–43.

[77] Hook EB, Warburton D. Turner syndrome revisited: review of new data supports the hypothesis that all viable 45,X cases are cryptic mosaics with a rescue cell line, implying an origin by mitotic loss. Hum Genet 2014;133(4):417–24.

[78] Hultén MA, Jonasson J, Nordgren A, Iwarsson E. Germinal and somatic trisomy 21 mosaicism: how common is it, what are the implications for individual carriers and how does it come about? Curr Genomics 2010; 11(6):409–19.

[79] Pellestor F, Andréo B, Anahory T, Hamamah S. The occurrence of aneuploidy in human: lessons from the cytogenetic studies of human oocytes. Eur J Med Genet 2006;49(2):103–16.

[80] Hultén MA, Patel SD, Westgren M, Papadogiannakis N, Jonsson AM, Jonasson J, et al. On the paternal origin of trisomy 21 Down syndrome. Mol Cytogenet 2010;3:4.

[81] Katz-Jaffe MG1, Surrey ES, Minjarez DA, Gustofson RL, Stevens JM, Schoolcraft WB. Association of abnormal ovarian reserve parameters with a higher incidence of aneuploid blastocysts. Obstet Gynecol 2013;121(1):71–7.

[82] Templado C, Vidal F, Estop A. Aneuploidy in human spermatozoa. Cytogenet Genome Res 2011;133(2-4):91–9.

[83] Sarrate Z, Vidal F, Blanco J. Role of sperm fluorescent in situ hybridization studies in infertile patients: indications, study approach, and clinical relevance. Fertil Steril 2010;93:1892–902.

[84] Uroz L, Templado C. Meiotic non-disjunction mechanisms in human fertile males. Hum Reprod 2012;27:1518–24.

[85] Zelazowski MJ, Sandoval M, Paniker L, Hamilton HM, Han J, Gribbell MA, Kang R, Cole F. Age-dependent alterations in meiotic recombination cause chromosome segregation errors in spermatocytes. Cell 2017. https://dx.doi.org/10.1016/j.cell.2017.08.042. [Epub ahead of print].

[86] C1 D, Harvey AJ, Armant DR, Zelinski MB, Brenner CA. Expression profiles of cohesins, shugoshins and spindle assembly checkpoint genes in rhesus macaque oocytes predict their susceptibility for aneuploidy during embryonic development. Cell Cycle 2012;11(4):740–8.

[87] Iourov IY, Vorsanova SG, Yurov YB. Chromosomal mosaicism goes global. Mol Cytogenet 2008;1:26.

[88] Delhanty JD. The origins of genetic variation between individual human oocytes and embryos: implications for infertility. Hum Fertil 2013;16(4):241–5.

[89] Marin D, Scott Jr. RT, Treff NR. Preimplantation embryonic mosaicism: origin, consequences and the reliability of comprehensive chromosome screening. Curr Opin Obstet Gynecol 2017;29(3):168–74.

[90] Hryshchenko NV, Bychkova GM, Tavokina LV, Brovko AO, Graziano C, Soloviov OO, et al. Unbalanced translocations involving chromosome region 10q25.3q26.3 in patients with intellectual disability and complex phenotypes. Cytogenet Genome Res 2014;144(3):169–77.

[91] Morales R, Lledó B, Ortiz JA, Ten J, Llácer J, Bernabeu R. Chromosomal polymorphic variants increase aneuploidies in male gametes and embryos. Syst Biol Reprod Med 2016. https://dx.doi.org/10.1080/19396368.2016.1212949.

[92] Schupf N, Kapell D, Nightingale B, Lee JH, Mohlenhoff J, Bewley S, et al. Specificity of the fivefold increase in AD in mothers of adults with Down syndrome. Neurology 2001;57:979–84.

[93] Avramopoulos D, Mikkelsen M, Vassilopoulos D, Grigoriadou M, Petersen MB. Apolipoprotein E allele distribution in parents of Down's syndrome children. Lancet 1996;347:862–5.

[94] Kaur A, Kaur A. Maternal MTHFR polymorphism (677 C-T) and risk of Down's syndrome child: meta-analysis. J Genet 2016;95(3):505–13.

[95] Hobbs CA, Sherman SL, Ping Y, Hopkins SE, Torfs CP, Hine RJ, et al. Polymorphisms in genes involved in folate metabolism as maternal risk factors for Down syndrome. Am J Hum Genet 2000;67:623–30.

[96] O'Leary VB, Parle-McDermott A, Molloy AM, Kirke PN, Johnson Z, Conley M, et al. MTRR and MTHFR polymorphism: link to Down syndrome? Am J Med Genet 2002;107(2):151–5.

[97] Gianaroli L, Luiselli D, Crivello AM, Lang M, Ferraretti AP, De Fanti S, Magli MC, Romeo G. Mitochondrial DNA analysis and numerical chromosome condition in human oocytes and polar bodies. Mol Hum Reprod 2015;21(1):46–57.

[98] Shahine LK, Marshall L, Lamb JD, Hickok LR. Higher rates of aneuploidy in blastocysts and higher risk of no embryo transfer in recurrent pregnancy loss patients with diminished ovarian reserve undergoing in vitro fertilization. Fertil Steril 2016;106(5):1124–8.

Elizabeth A. Normand*, Ignatia B. Van den Veyver*,†

*Department of Molecular and Human Genetics, Baylor College of Medicine, Houston, TX,
United States †Department of Obstetrics and Gynecology, Baylor College of Medicine,
Houston, TX, United States

OUTLINE

Abbreviations

A	adenine
ACMG	American College of Medical Genetics and Genomics
ACOG	American College of Obstetricians and Gynecologists
C	cytosine
CAKUT	congenital anomalies of the kidneys and urinary tract
CCMG	Canadian College of Medical Geneticists
cff-DNA	cell-free fetal DNA
ClinVar	clinical variation database
CMA	chromosomal microarray analysis
CNV	copy number variant
CVS	chorionic villus sampling
DECIPHER	Database of Chromosomal Imbalance and Phenotype in Humans Using Ensembl Resources
DNA	deoxyribonucleic acid
ECS	expanded carrier screening
ES	exome sequencing
ExAc	exome aggregation consortium
FADS	fetal akinesia deformation sequence
G	guanine
GS	genome sequencing
HGMD	human gene mutation database
HPO	human phenotype ontology
IF	incidental finding
ISPD	International Society for Prenatal Diagnosis
IVF	in vitro fertilization
MPS	massively parallel sequencing
MRI	magnetic resonance imaging
NGS	next generation sequencing
PGD	preimplantation genetic diagnosis
PGS	preimplantation genetic screening
PQF	Perinatal Quality Foundation
SMFM	Society for Maternal-Fetal Medicine
T	thymine
VUS or VOUS	variant of uncertain significance

INTRODUCTION

Karyotype analysis and chromosomal microarray analysis (CMA) are currently the standard genetic tests when fetal structural anomalies are detected by prenatal ultrasound [1–3], which affects 3%–5% of pregnancies, or when there is another risk factor such as maternal age. G-banded karyotype analysis reveals a diagnosis in about 30% of these cases and CMA provides an additive diagnostic yield of 6%–7% to >10% when there are multiple anomalies [3, 4]. Because most chromosomal abnormalities detectable by karyotyping are also found by CMA [4], professional guidelines now recommend CMA as the first-line genetic test when an amniocentesis or chorionic villus sampling (CVS) is performed for these indications [5, 6]. While the introduction of CMA into prenatal genetic diagnosis is an important advance, it still leaves at least 60% of cases undiagnosed [7, 8]. CMA has also become a first-line diagnostic test in the pediatric setting with an incremental diagnostic yield over standard karyotyping of 15%–20% for children with birth defects, suspected genetic syndromes, or neurodevelopmental and cognitive disabilities [9]. Until recently, when karyotype and CMA did not reveal a diagnosis, patients were often subjected to multiple sequential gene-by-gene tests or small multigene panels for clinically suspected single-gene disorders based on the phenotype [10]. This "diagnostic odyssey" can be very costly and continue for years, and in many cases, a genetic etiology is never found [11, 12]. For prenatal genetic diagnosis where phenotypes are often not well defined and time is limited, such gene-by-gene testing is ineffective.

Thanks to recent advances in rapid high-throughput DNA sequencing or next-generation sequencing (NGS), knowledge about the genetic mutations that cause single-gene disorders has increased exponentially and continues to grow. NGS has also stimulated the development of new clinical tests to simultaneously sequence multiple genes, grouped in disease-specific gene panels, or as clinical diagnostic whole exome sequencing (ES) tests [13]. Diagnostic whole genome sequencing (GS) is also being explored [14]. While ES and GS are already widely used for disease gene discovery and diagnostic testing in the pediatric and adult population, where they have a diagnostic rate of ~25%–40% [15–18], their introduction into prenatal diagnosis and reproductive medicine is more recent [7, 8, 19–22]. While early data with fetal ES are very promising, its diagnostic yield, benefit, challenges, and clinical utility are not yet fully established [7, 8, 22].

NEXT-GENERATION SEQUENCING METHODS

Principles of Next-Generation Sequencing

In next-generation sequencing, a patient's genomic DNA is first sheared into small (50–400 nucleotides long) fragments and adapters are then linked to one or both (in paired-end sequencing) ends of these fragments (Fig. 29.1A). For gene panels or exome sequencing, where only a fraction of the entire genome is targeted, the fragments are denatured and hybridized to a set of DNA baits (Fig. 29.1B, C), but for genome sequencing, this "library" of fragments is then subjected to sequencing (Fig. 29.1D). The baits are short strands of DNA with a known sequence attached to magnetic beads, allowing them to be purified to enrich the library for the desired target regions. The final (enriched or unenriched) sequencing library is then immobilized on a solid platform, clonally amplified, denatured, and sequenced by synthesis of a new complementary strand. In this process, each nucleotide A, C, G, and T that is added to a clonally amplified fragment has a different fluorescent tag so that its insertion in the sequenced fragment can be detected (Fig. 29.1E) [10, 17, 23]. The sequencer can simultaneously (in massively parallel sequencing; MPS) identify the incorporated nucleotide for each of the clusters of clonally amplified fragments, which significantly speeds up the process and reduces cost. The identified bases of the sequenced overlapping fragments are then aligned using bioinformatics tools to generate a consensus sequence that is compared to the human reference sequence [17, 23] to determine if there are variants in the aligned sequence (Fig. 29.1E). The number of overlapped reads for each nucleotide position is called the read depth or sequencing depth and the accuracy of the aligned consensus sequence correlates with the number of reads. The fraction of all the sequenced bases for a particular assay that is adequately covered at sufficient depth is called coverage breadth. Because NGS is more error-prone than traditional Sanger sequencing, each nucleotide is ideally covered tens of times; low read depth can result in false negative and false positive results. The American College of Medical Genetics and Genomics (ACMG)

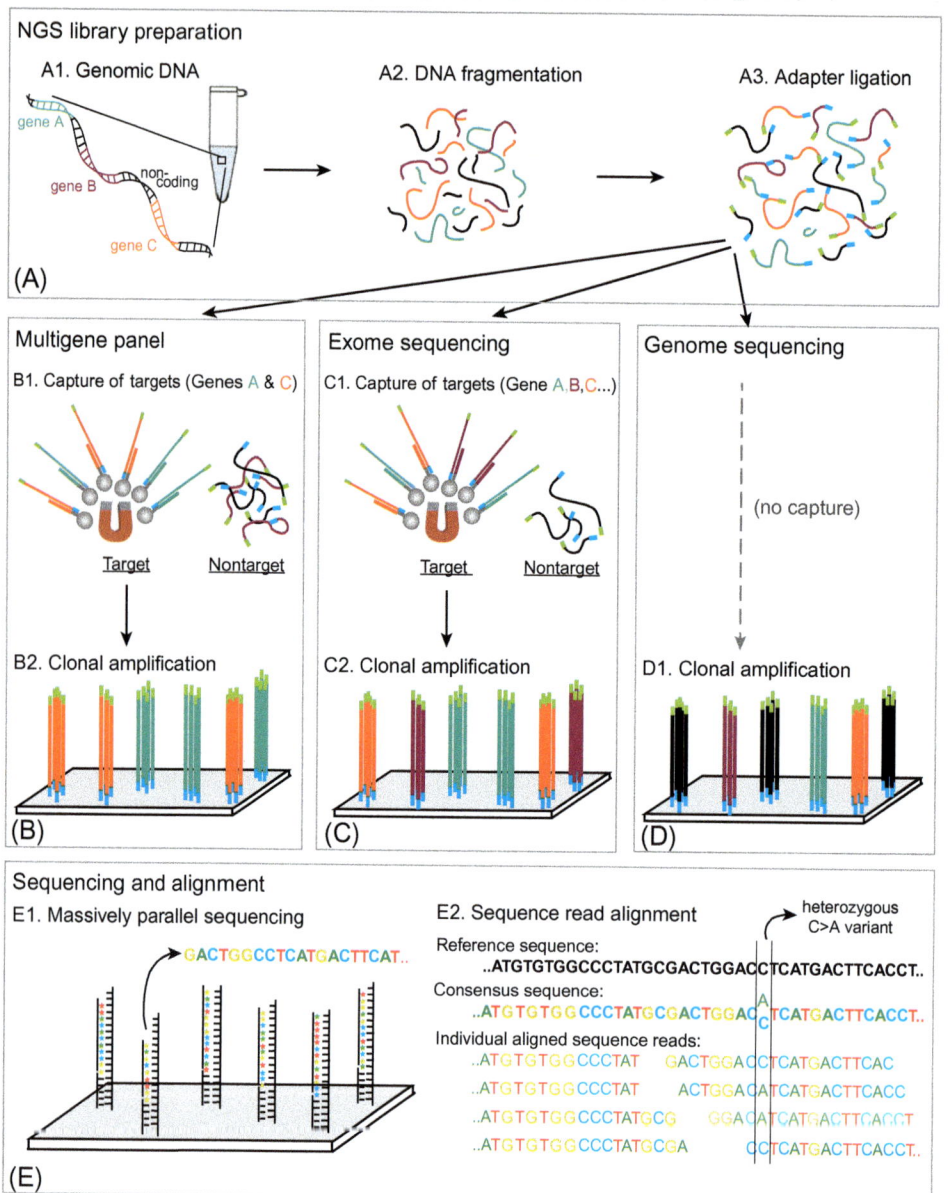

FIG. 29.1 Next-generation sequencing method. (A) A sequencing library is prepared by fragmenting genomic DNA and adding adapters (cyan and green). (B) For multigene panels, the coding regions of selected genes (e.g., teal, orange) are targeted by hybridizing to DNA baits attached to magnetic beads (silver). Nontarget regions (magenta, black) do not hybridize and are washed away. Individual DNA molecules are spatially separated, immobilized, and then clonally amplified. (C) For exome sequencing, the coding regions of all genes (e.g., teal, magenta, orange) are targeted by hybridizing to DNA baits attached to magnetic beads (silver), then clonally amplified. (D) For genome sequencing, the sequencing library does not undergo a capture step, but proceeds directly to the immobilization and clonal amplification step. (E) The immobilized DNA molecules are sequenced by synthesis using fluorescently tagged nucleotides. As each nucleotide is added during strand synthesis, its fluorescence is detected in real time by the sequencing machine and converted into a string of DNA bases, called a sequence read. The sequence reads from every DNA fragment are bioinformatically aligned to a reference genomic sequence to produce a consensus sequence. Any difference between the consensus sequence and the reference sequence represents putative genomic variants (arrow, heterozygous C > A variant). The number of overlapping sequence reads that align to any given genomic position is referred to as the coverage depth.

recommends that for diagnostic purposes, ES is done at an average depth of $\geq$100-fold and that $\geq$90%–95% of the sequenced regions are covered at least 10-fold [24].

Sequence Data Analysis and Interpretation

Each individual has millions of variants in their genome [25], and 20,000–50,000 in a sequenced exome [26], but only a minority of these cause disease. Organizations such as ClinGen and the ACMG have developed guidelines for laboratories for variant interpretation and reporting to facilitate more standardized and accurate interpretations [27, 28]. Based on these guidelines, a bioinformatics-supported interpretation process that considers a combination of parameters is applied, and these combined factors are scored by level of evidence to classify variants as pathogenic, likely pathogenic, likely benign, or benign. If there is not sufficient evidence to classify a variant into any of these categories, then it remains by default a variant of uncertain significance (VUS or VOUS) (Table 29.1) [13, 17, 27]. The frequency of the variant in the population is usually the first parameter that is considered. Variants that are common in healthy individuals (polymorphisms) are more likely to be benign whereas unique variants that have not been seen before are more likely disease-causing. The existence of other patients with the disease or an overlapping phenotype that has the same variant(s) strongly supports pathogenicity, although exceptions are known. Variants have been reclassified from pathogenic to benign and vice versa when new information becomes available. Another considered factor is the inheritance pattern in families. For inherited autosomal dominant conditions, a pathogenic or likely pathogenic variant should segregate with the dominantly inherited disease in the pedigree, whereas a "de novo" change is only found in the affected individual. For autosomal recessive disorders, the affected individual has two pathogenic or likely pathogenic variants in *trans*, one inherited from each "carrier" parent. Variants on the X chromosome can explain X-linked conditions. Another important parameter is the predicted molecular consequence of the variant. In most cases, variants that are predicted to lead to a complete absence of the gene product are very likely to be pathogenic (e.g., frameshift, nonsense, and canonical splice site variants as well as exonic deletions). On the other hand, missense variants, which result in the substitution of one amino acid for

TABLE 29.1 Classification of Sequence Variants

Type of mutation
Pathogenic
Likely pathogenic
Variant of uncertain significance
Likely benign
Benign
Location of mutation
In gene relevant to the phenotype
In gene causing a different disease

another, require additional lines of evidence to interpret. For these types of variants, computational algorithms can be used to predict whether the rare missense variant might disrupt the structure or function of the encoded protein [13, 17]. Data from animal models or functional assays can also be considered to refine the interpretation.

Professional societies such as the ACMG also encourage that research and diagnostic ES and GS data are shared with the scientific community in deidentified and protected databases [29]. This can significantly increase the evidence supporting variant classification, and consequently disease-gene associations and molecular diagnoses. Because scientific knowledge about disease-causing genetic variants is rapidly expanding, reinterpretation of exome sequencing results to include new information that could alter a prior result report is recommended. The frequency by which this should be done is still debated and must balance cost and practicality against the risk of missing a diagnosis or not revising an incorrect diagnosis. It is prudent to always consider reinterpreting a prior ES or GS result before it is used to inform reproductive testing and decision-making for a new pregnancy.

Challenges Unique to Sequence Variant Interpretation With Prenatal Testing

One of the key criteria that supports the final interpretation of variant pathogenicity is the correlation with a known phenotype. This is particularly challenging prenatally because prenatal phenotypic descriptions are limited by the resolution and accuracy of prenatal ultrasound and, more recently, fetal magnetic resonance imaging [8]. Other limiting factors include the timing and variability by which phenotypes become evident, the subtle nature of dysmorphology features beyond the resolution of prenatal imaging, and the inability to prenatally uncover certain components of a phenotype, such as metabolic abnormalities or developmental and behavioral disability. Furthermore, the scientific literature and commonly used databases and resources of benign and disease-causing variants, such as the exome aggregation consortium (ExAc) database, the genome aggregation database (gnomAD) [30], the clinical variation database (ClinVar) [31], the Human Gene Mutation Database (HGMD) [32], and the Database of Chromosomal Imbalance and Phenotype in Humans Using Ensembl Resources (DECIPHER) [33] are all biased toward postnatal, pediatric, and adult presentations of genetic diseases. These resources and the standard phenotyping nomenclature system, the Human Phenotype Ontology (HPO) [34], are also relatively depleted of information on prenatally lethal phenotypes because they are not ascertainable in the living population. These challenges highlight the need for dedicated efforts to curate fetal phenotypes and associated variants found in prenatal genome-wide sequencing [8].

Types of Mutations Detectable by Next-Generation Sequencing

Not all sequences are equally well covered with NGS (Table 29.2) because the technology has limitations that make it difficult for some mutations to be detected [13]. It performs well for point mutations, such as missense, nonsense, and frameshift variants, as well as for small insertion-deletions (indel) of a few nucleotides, but detecting larger structural chromosomal rearrangements or aneuploidy is more difficult, in particular from captured sequencing libraries used for exome and gene-panel sequencing. Newer approaches including paired-end sequencing and refined algorithms for copy number analysis from sequence data are

TABLE 29.2 Types of Mutations Detected by NGS

Well-Detected Variants	Poorly or Not Detected Variants
In unique genes and exons	In pseudogenes and repeated exons In highly homologous genes
Point mutations (missense, stop-gain, frameshift)	GC-rich sequences
Small insertions-deletions (indels)	Large rearrangements and aneuploidy[a]
Only in regions that can be sequenced with NGS	Repeat amplifications
Germline mutations	Low-level mosaic variants
Higher-level mosaic variants ($\geq$20%)	
Only in covered regions (exomes and panels)	In uncovered regions (exomes and panels)
In regions with sufficient depth	In regions with low coverage and depth
	Epigenetic modifications

[a] *Newer analysis algorithms and large insert sequencing are improving the detection of large copy number changes. Copy number changes are better detected by genome sequencing than by panel or exome sequencing.*

emerging [35] that may improve this, but until then, CMA and karyotype remain the methods of choice to detect larger chromosomal abnormalities [1, 2, 8, 10]. Because the consensus sequence is built from the alignment of smaller fragments, regions with high homology are challenging. These include repetitive sequences, short repeat amplifications (such as the CGG repeat expansion causing Fragile X syndrome), genes belonging to highly homologous gene families, and genes with homology to pseudoexons and pseudogenes. Because of the strong G—C chemical bond, regions with high CG content are difficult to capture and therefore often have low coverage in exome sequencing and gene panels. Mosaic mutations, present in only a subset of cells from which the sequenced DNA is prepared, are also challenging but can be detected providing parameters for variant calling are adapted and sequencing depth is adequate. Finally, uniparental disomy, that is, the inheritance of both pairs of entire chromosomes or large regions of DNA from the same parent, can be detected by trio exome and genome sequencing if haplotyping of identified polymorphic variants that integrates data from parental samples is performed. Some diagnostic laboratories complement a diagnostic exome sequencing test with a chromosomal microarray to detect deletions and regions of absence of homozygosity as well as complementary sequencing methods for clinically relevant regions that are difficult to sequence by NGS, but this is challenging within the time constraints for prenatal diagnosis.

DIAGNOSTIC NGS-BASED TESTS

Multigene Panels

Some phenotypes or structural abnormalities can be grouped in specific broader categories, associated with a few to up to hundreds of genes [10]. Examples include microcephaly or neuronal migration abnormalities, skeletal dysplasias, connective tissue disorders,

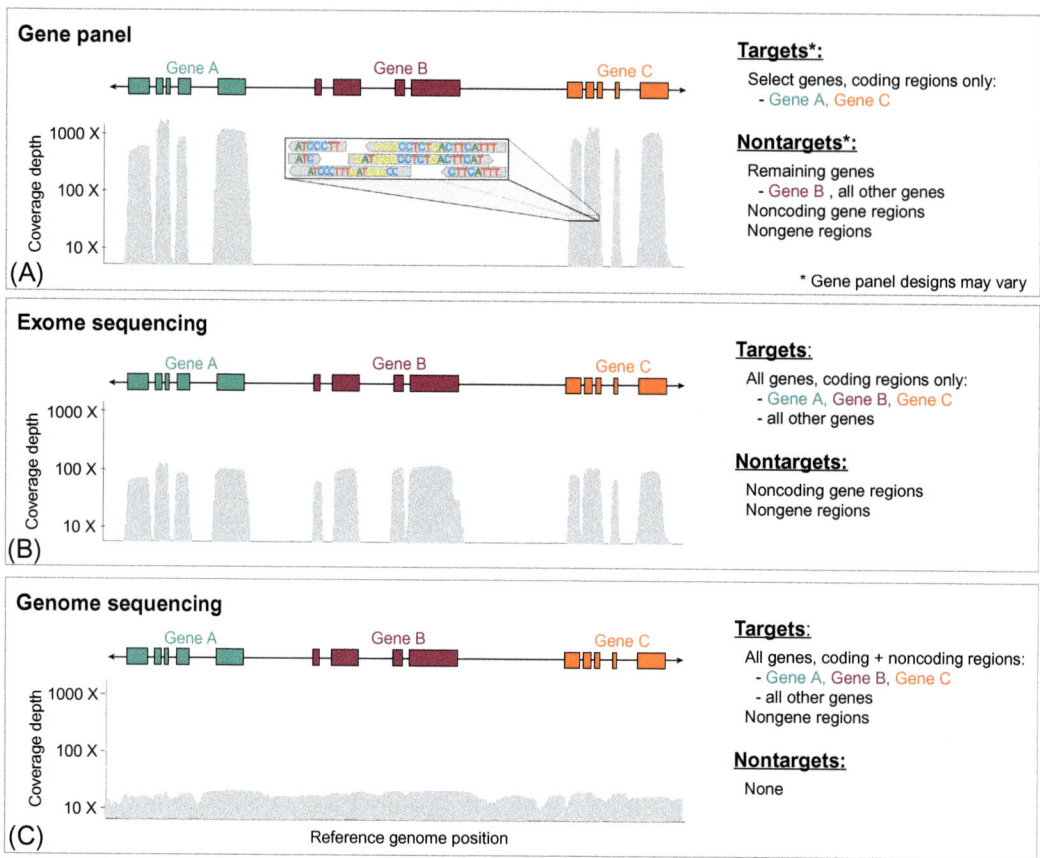

FIG. 29.2   Next-generation-based diagnostic tests. (A) Multigene panels target the coding regions of selected genes only (Gene A, C). Coverage depth, or the number of overlapping sequence reads at any particular genomic position, is graphed. For gene panels, most of the targeted regions have very high coverage depth ($>100\times$). Genes not included in the panel have no coverage (Gene B). (B) Exome sequencing targets the coding regions of all genes (Gene A, B, C, etc.). Coverage depth varies but most regions are covered $>10\times$ and many are covered around $100\times$. (C) Genome sequencing includes both coding and noncoding regions of the entire genome. Coverage depth is relatively uniform but is lower on average ($<100\times$) than gene panels or exome sequencing.

congenital heart defects, and intellectual disabilities [13, 36–38]. Testing with gene panels can be effective for disorders with genetic heterogeneity, whereby mutations in different genes cause the same phenotype, such as for Noonan syndrome [39], which prenatally can present with increased nuchal translucency, cystic hygroma, or certain congenital heart defects [40, 41]. Disease-specific panels (Fig. 29.2A) can be relatively fast and cost-effective when there is a high likelihood that a pathogenic variant in one of the genes on the chosen panel causes the phenotype. The sequencing depth of gene panels is typically higher than for exome or genome-wide sequencing. Interpretation is easier than for exomes because they contain fewer genes, which usually have been well characterized, resulting in a lower risk for detecting VUS and practically no risk for uncovering incidental or secondary findings (i.e., pathogenic variants unrelated to the indication for sequencing). Like any NGS-based assay, not all mutations can be detected due to technical limitations (see above). Thus, the diagnostic utility of such tests is affected by the more common mutation type responsible for a particular syndrome. Therefore, gene panels are often complemented by other assays, such as copy-number analysis for conditions in which a subset of affected individuals have copy number loss or gain of one or more exons or larger regions that include the gene, assays for triplet repeat amplifications (e.g., to detect the CGG-repeat amplification causing Fragile X syndrome), or Sanger sequencing for genes that are difficult to sequence by NGS. A major disadvantage of gene panels is that their gene content reflects current knowledge and the criteria for inclusion of genes in panels can vary, resulting in large differences between panels. With the current rapid pace of disease gene discoveries, panels can quickly become outdated, and each update can be time-consuming and must be revalidated. Finally, gene panels do not work well for patients with a nonspecific or atypical phenotype, which is particularly pertinent for prenatal diagnosis because gene panel designs are primarily informed by phenotypes observed after birth.

Exome Sequencing

For exome sequencing, the DNA baits are designed to capture all the coding exons and exon-intron boundaries of the approximately 20,000 known nuclear-encoded human genes, which make up the 1%–2% of the human genome that is estimated to contain at least 85% of all pathogenic sequence variants (Fig. 29.2B). It is often referred to as whole exome sequencing, but because of technical issues a typical exome sequencing library captures and provides interpretable results for about 95% of all exons, with typically an average sequencing depth that is lower than for gene panels, although this difference is narrowing with improving technology.

The major advantage of exome sequencing is that it is more unbiased and does not require prior detailed knowledge of the phenotype and possible associated genes. This is particularly helpful when there is high genetic heterogeneity; when phenotypes are very broad, nonspecific (e.g., intellectual disability or prenatal growth restriction), or do not fit a known genetic syndrome; or when limited phenotypic information is available. Fetal phenotypes of known conditions can be different than what is typically observed postnatally, suggesting that both research and diagnostic exome sequencing are particularly useful for structural fetal abnormalities, which is confirmed by early experiences [8]. At the same time, this is a disadvantage because interpretation of the clinical significance of a rare sequence variant requires correlation with prior knowledge about the phenotype. Thus, providers offering fetal exome sequencing should communicate all available phenotypic information with the sequencing laboratory for optimal result interpretation.

Because exome sequencing surveys nearly all coding genes, pathogenic variants in genes that cause a different disease, unrelated to the indication for testing, can be incidentally found. The ACMG has recommended that the diagnostic exome sequencing test include evaluation and reporting of "secondary findings" in 59 "actionable" genes for which preventive or therapeutic measures with health benefits can be offered to individuals with these secondary findings. Individuals undergoing sequencing can opt out of this analysis, and the guidelines do not include fetal samples [42–44].

Genome Sequencing

Genome sequencing is the most unbiased method to sequence the genome as it does not include a capture of specific targeted regions to prepare the library for sequencing (Fig. 29.2C). The resulting sequence data will include coding and noncoding regions, such as introns, promoters, and regulatory sequences. Thus, the amount of information that can be obtained is vastly larger than from exome sequencing, but challenges of classifying and interpreting variants are also amplified. It can often be difficult to prove by bioinformatics interpretation and complementary functional assays whether detected variants in the non-coding fraction of the genome are pathogenic. The sequencing depth in genome sequencing is lower but more uniform, which facilitates the detection of CNVs and has been shown to improve the detection of coding variants by 3% over exome sequencing. The cost of genome sequencing is currently still significantly higher than for exome sequencing, and it is not yet routinely included in clinical NGS-based diagnostic testing. However, as cost continues to fall, technology continues to advance, and our understanding of the noncoding genome improves, this is beginning to change.

Other Prenatal and Reproductive Applications That Use NGS Technology

Modern diagnostic assays currently applied for preconception, preimplantation, and pre-natal diagnosis and screening rely increasingly on NGS technologies. In a commonly used approach for noninvasive screening for fetal aneuploidy on cell-free fetal DNA, low-pass genome-wide sequencing is used and the number of fragments that are derived from each chromosome are counted to determine under- or overrepresentation of a particular chromosome [1]. In new noninvasive tests for screening and testing for single-gene disorders, panels of specific genes are sequenced from cell-free DNA and the presence of de novo or unique paternally inherited mutations as well as the dosage of maternally inherited mutations is determined (Zhang et al. in submission) [45]. Gene panel sequencing by NGS has also facilitated the expansion from targeted ethnicity-based reproductive carrier screening for a few selected conditions to development of pan-ethnic expanded carrier screening (ECS) as an acceptable alternate approach [46–48]. Finally, NGS is also becoming the cornerstone of the genetic assays used for preimplantation genetic testing of embryos for aneuploidy during *in vitro* fertilization (IVF) [49] as well as for the detection of specific point mutations with high specificity.

USE OF NEXT-GENERATION SEQUENCING IN REPRODUCTIVE GENETIC RESEARCH

Genome-wide sequencing is becoming the primary method of gene discovery for rare Mendelian disorders. It is also increasingly used for genome-wide association studies aiming to link specific variants to risks for polygenic and multifactorial disorders with gene-environment interactions. In the field of reproductive medicine, genome and exome sequencing are beginning to help elucidate causes and prognostic factors for gynecological cancers.

They are also adopted to search for the causes of fertility disorders and early pregnancy loss, and to identify maternal, paternal, and fetal genetic factors that contribute to preterm birth, preeclampsia, and fetal growth abnormalities.

Genetic Causes of Reproductive Disorders

An emerging area for genome-wide sequencing in reproductive medicine is to find the causes of unexplained recurrent pregnancy loss or unexplained infertility. Case reports or small series that used exome or genome sequencing to investigate recurrent pregnancy loss and recurrent chromosomal abnormalities [50–52] highlight this potential benefit. Genome-wide sequencing has also been used to find causes of infertility [53–55], primary gonadal failure [56], and primary ovarian insufficiency [57]. However, the clinical utility of introducing such tests into fertility care is not yet proven [58].

Causes of Birth Defects and Childhood Disorders

Research and clinical diagnostic uses of exome sequencing on fetuses and neonates with prenatally detected structural abnormalities has already resulted in new discoveries [20]. In an early example, Filges et al. discovered an autosomal recessive truncating mutation in *KIF14* in a family with recurrent intrauterine growth restriction and multiple abnormalities. The pathogenicity of this was supported by a mouse model, highlighting the importance of considering model organism data in interpretation of sequence variants. Other discoveries involve genetic defects that cause prenatal lethal [8, 59, 60] and severe in utero phenotypes. These include, as summarized in Westerfield et al. [20], fetal akinesia deformation sequence (FADS), arthrogryposis multiplex, and severe skeletal dysplasias. It has also been unexpectedly useful for renal and urogenital abnormalities with and without other abnormalities, for which prenatal CMA is often unrevealing. Alongside improved detail of fetal neuroimaging through the introduction of fetal magnetic resonance imaging (MRI), exome sequencing may yield new genes or new variants in known genes as a cause of fetal brain abnormalities that were previously not seen prenatally. Exome sequencing has also already resulted in expansion of new phenotypes for known disease genes to include previously unascertained different prenatal presentations [8, 20].

CLINICAL USE OF FETAL GENOME-WIDE SEQUENCING

Current Clinical Experience

In the pediatric population, diagnostic ES has a well-established diagnostic rate of at least 25%–30%, depending on the indication and targeted population [12, 15, 16, 61, 62], with higher detection rates in a neonatal population [18]. Initial reports on isolated cases or small series embedded in larger diagnostic exome experience reports highlighted the potential benefit of diagnostic exome sequencing for prenatal diagnosis in highly selected cases [15, 16, 50, 63–65]. A systematic investigation of the benefit of diagnostic ES for fetal abnormalities, by Carrs et al. reported a diagnostic rate of 10% on 30 cases with prenatally detected fetal anomalies and identified sequence variants of potential significance in another 17%, but none were from ongoing pregnancies. Alamillo et al. found mutations in four of seven prenatal cases [66], and Drury et al. described a 25% detection rate in a study of 24 fetuses with fetal abnormalities detected by ultrasound imaging [67]. In most cases in these series, the pregnancies had already ended when the exome sequencing was done on available samples. In a compilation of data from these and more recent studies, including unpublished work presented at national and international meetings, the diagnostic rate with exome sequencing for fetal anomalies had a wide range, from 6.2% to more than 80% [8]. This is likely related to the size of the cohort and the variability in selection of the tested population. In a recent study from our center (Normand et al. in submission), the overall diagnostic rate of exome sequencing for fetal anomalies was 32% and 35% in the subcohort of ongoing pregnancies. There was no significant difference in the detection rate between trio and proband-only sequencing, but the trio approach enhanced the speed and efficiency of result interpretation. In a more recent small series of 15 cases with fetal anomalies, a diagnosis or possible diagnosis was obtained for seven cases (47%) [68]. Lei et al. found pathogenic variants in four out of 30 (14%) fetuses with congenital anomalies of the kidneys and urinary tract (CAKUT), with or without other abnormalities that were sequenced after other tests were negative [69]. In the largest published series to date of sequencing done on a research basis for 196 fetuses with normal CMA and karyotype, that were selected from a group of >3900 with structural anomalies, the diagnostic rate of variants related to the presenting clinical phenotype was 24%, but the authors also reported a relatively high number of incidental findings (6.1%) and VUS (12.8%) [70].

Recent studies have also shown the effectiveness of exome sequencing for discovering new causes and obtaining a clinical diagnosis for prenatally lethal syndromes [50, 59, 71]. When there is high suspicion for autosomal recessive single-gene disorders, for example with parental consanguinity, exome sequencing with a focused analysis within regions of homozygosity, localized with concurrent or preceding SNP arrays, can be very effective [71]. Furthermore, one study found that a strategy of sequencing parental samples when fetal samples from prior affected pregnancies are not available and there is a high suspicion for recessive inheritance can have a diagnostic yield of 40% [72].

Taken together, these early studies all indicate that fetal genome-wide sequencing is a very promising strategy for the prenatal diagnosis of fetuses with structural abnormalities when standard testing by karyotype, CMA, and, in some cases, gene-specific analysis is uninformative. Research efforts with larger series to establish the clinical utility of fetal genome-wide sequencing for prenatal diagnosis are currently ongoing and planned. Such studies are needed to determine how best to implement it and explore in more detail the pitfalls and counseling difficulties that must be considered.

Does Fetal Genome-Wide Sequencing Affect Medical Management?

To demonstrate the clinical utility of a new test, it must be evaluated as to how its introduction into care impacts clinical management and patient outcomes. Experience in children and smaller prenatal series highlight cases where an exome result can alter management of a pregnancy or newborn [8, 12, 18, 73–75]. Prenatal diagnosis can result in early treatment for metabolic disorders, for example in urea cycle disorders where the initial hyperammonemia can be anticipated and treated early. It could be used to inform candidate selection for fetal therapy, for example in utero treatment with endotracheal occlusion for congenital diaphragmatic hernia, prenatal cardiac intervention for hypoplastic left heart syndrome, or for enrollment in trials of in utero stem cell transplantation [8].

The diagnostic information by itself can be valuable to parents faced with decisions after a prenatal diagnosis of fetal structural anomalies and can provide information on recurrence risk. It can also inform the management of future pregnancies with options of preimplantation or prenatal diagnosis, or the use of donor gametes to prevent affected pregnancies [20, 73].

GENETIC COUNSELING FOR GENOME-WIDE SEQUENCING

Genetic counseling for genome-wide sequencing is complex, lengthy, and especially challenging in the preconception and prenatal period, when a result can influence decisions regarding pregnancy management and continuation (Table 29.3). The information patients need to make informed decisions about testing is difficult to communicate in accessible terms under the stressful circumstances that follow the detection of congenital anomalies in the fetus. The types of findings, their significance, prognostic implications, and all management options should be conveyed accurately using clear language. This is often not accomplished in a single setting but is a process that begins before the test is performed and continues often well after the results are communicated. It is best done by professionals trained in genetics and with specific experience in prenatal and reproductive applications of genetic testing.

Pretest Education and Informed Consent

Exome and genome sequencing may reveal significant genetic information not only about the fetus, but also about other related family members. Therefore, it is important that both parents, where possible, receive the pretest education and participate in the informed consent process. The pregnant woman undergoing the diagnostic procedure, amniocentesis or CVS should be informed about their associated risks and give informed consent for the procedure itself. It is good practice that both parents provide consent for the actual sequencing of the fetal DNA as well as in trio sequencing approaches, each separately, for the sequencing of their own DNA. A recent position statement [22] summarized the topics to be covered in pretest education and counseling. These include generalized information about the nature of the test, the still fairly limited clinical experience with prenatal use of genome-wide sequencing, that testing is optional, the types of conditions it can detect, and the implications of uncovered genetic information for the family, including the potential to uncover close parentage (unknown consanguinity) and nonpaternity. It should also cover information on possible outcomes of the test, including best estimates of turn-around time to results (which with a rapid trio approach can be less than 3 weeks). The molecular diagnostic rate should be presented in the perspective of the clinical phenotype and be qualified to note that, until

TABLE 29.3   Genetic Counseling Components for Fetal Genome-Wide Sequencing

Pretest counseling

1. The types of results to be returned

2. Realistic expectations about diagnostic success

3. Possibility for variants of uncertain significance

4. Time frame to results

5. Possibility of no result or late result (including after birth)

6. Policy and choices for reporting incidental and secondary findings

7. Policy and choices for reporting adult-onset disorders

8. Policy and choices for reporting carrier status for recessive and X-linked disorders

9. Possible implications of results for parents and other family members

10. Possibility of discovering nonpaternity or close parentage

11. Information that reported results are based on current knowledge and strategies for data storage and reanalysis

12. Importance of data sharing

13. Expected cost

Posttest counseling

1. Confirm choices and policies discussed in pretest counseling

2. Disclose results and implications for prognosis for the fetus after birth

3. Discuss and make decisions about confirmatory testing and follow-up testing

4. Discuss and make decisions about pregnancy, delivery, and neonatal management

5. Discuss implications of results for parents and other family members

6. Provide emotional and social support

7. Facilitate multidisciplinary referral and care

now, data on the diagnostic rate of exome sequencing are still vague for prenatal diagnosis, ranging from 6% to >80%, depending on the indication and case selection [8]. The possibility of incidental and secondary findings and variants of uncertain significance should be communicated along with decisions about which class of variants will be reported and what the parental choices are about receiving such information [76, 77]. It is also important to convey in pretest counseling that reported results will be based on the knowledge current at the time of analysis as well as what the policies and options are for future reinterpretation of the results [22]. Cost of testing, insurance coverage, possible effects on future insurability, and the availability of posttest genetic counseling for result disclosure should also be addressed [78]. Sharing of variant information in public databases or research exchange services such as GeneMatcher can benefit the discovery of new disease genes and patients should be informed about that as well as provide consent for data sharing [79].

The categories of variants that are commonly reported include pathogenic and likely pathogenic variants in genes that are relevant to the fetal phenotype. In some situations, laboratories may report variants of uncertain significance, for example in autosomal recessive genes where there is one pathogenic variant and a second variant that is in trans (inherited from the other parent) and suspected to be pathogenic, but without confirmatory information. Other variants that may be reported are pathogenic or likely pathogenic variants in "medically actionable" genes for conditions that are incidentally uncovered, and for which the presence of the variant may result in a change in medical care that can improve the health of the sequenced fetus or one of the parents. Additionally, laboratories must develop standards for reporting categories of other medically important variants, such as carrier status for recessive or X-linked disorders, variants that cause adult-onset disorders, variants in cancer predisposition genes, or pharmacogenetic variants. How these are handled will be different in prenatal samples than in postnatal and adult samples, and approaches are not currently uniform. It is therefore important that the plan and decisions for these are discussed and documented during pretest counseling.

Posttest Counseling

Because of the complexity of the returned results from prenatal exome sequencing, all posttest counseling is ideally performed in person in a dedicated counseling session where all available information is conveyed (including negative results). This gives parents the opportunity to revisit their choices on the return of uncertain results and secondary findings as well as an opportunity to ask questions. Reproductive decisions based on the results and any changes in management of the pregnancy, perinatal care, and postnatal care must be discussed in posttest counseling, with the understanding that fully addressing these may require more than one encounter with a genetic professional and best involves a multidisciplinary team approach. In trio sequencing, results for the fetus as well as for the biological parents provide valuable information for future reproductive decisions and can facilitate the preparation for preimplantation testing or prenatal testing in subsequent pregnancies. The importance of any results for other family members and strategies to inform those who may benefit from knowing the results should also be discussed.

Attitudes and Perceptions of Women and Providers

A number of studies have investigated the understanding and perception of patients and providers of the benefits and potential pitfalls of genome-wide testing for prenatal diagnosis. In one survey, 83% of 186 parents with uncomplicated pregnancies felt that prenatal exome sequencing should be offered, but only 52% of them would accept amniocentesis to have the test [80]. A study on 15 women who had exome sequencing found that pretest genomic knowledge varied with socioeconomic status and educational background [68]. The perception about the chance for diagnostic information was higher in this study than what was provided in the pretest counseling (5.2/10 vs. 30%). A few studies and commentaries have explored guidance and the beliefs of parents and providers about the types of results that should be reported [19, 80–82]. One survey [82] found that providers were more conservative about result disclosure of uncertain and incidental findings, but patient representatives preferred to be informed about all possible results. A large survey of obstetricians in the United States highlighted the concerns by providers about lack of education along with the potential for overtreatment, high cost, increased parental anxiety, and disclosure of unexpected and undesirable results [83].

Ethical and Societal Aspects

When a new complex genetic test is introduced, such as NGS-based tests, the impact on society and ethical issues should be considered. Horn and Parker identified five areas they consider of particular concern: achieving valid consent; the management and feedback of information; the responsibilities of health professionals, priority setting, and resources; and the duties of providers and parents toward the future child [21]. The ethical implementation of new medical technologies must balance the guiding principles of autonomy, beneficence, nonmaleficence, and justice. Respect for autonomy validates the parents' role as the decision-makers for agreeing to undergo fetal testing. For parents to exert their autonomy, they have to understand the test and its implications, which imposes a duty for pretest counseling by knowledgeable providers that is adapted to their level of understanding [8, 21]. It must also cover all aspects, including disclosure of close parentage, nonpaternity, consanguinity, and the potential for actionable variants in parents when trio sequencing is done. This autonomy has to be balanced against the potential to deny the child autonomy about whether they wish to know genetic information [84]. This is particularly relevant for adult-onset disorders for which societies have recommended against presymptomatic testing in children. However, it is recognized that incidental detection of such findings can be complex [85–87].

Autonomy must also be balanced against the parental and provider responsibility of beneficence and nonmaleficence, the primary considerations that guide reporting of secondary (incidental) findings and VUS. The ACMG recommendation that 59 actionable genes should be analyzed in exome data to search for pathogenic variants (with an opt-out option) [43, 44] excludes fetal samples. What "actionable" means differs between prenatal and postnatal diagnosis. For example, an incidental discovery of a *BRCA1* pathogenic mutation in a fetus could result in a pregnancy termination decision. However, if found in children or adults, it will lead to a plan for regular screening and, if cancer is detected, surgical intervention or early treatment can be performed. Not all parents desire this information and the identification of such secondary findings also has ramifications for family members who were not part of the decision-making process regarding undergoing testing [7].

Whether and which VUS should be returned is also an ongoing debate that is not unique to prenatal genome-wide sequencing, but is amplified in this setting. In studies that investigate this for prenatal CMA, parents have indicated that, in some cases, the VUS information can be perceived as emotionally too difficult [88]. The interpretation of a variant as a VUS is changeable over time, which can be an argument in favor of disclosure (with future reanalysis), but this must be balanced with the emotional burden on families and with financial implications for the healthcare system [19, 73, 82]. A shared decision-making process between patients and providers has been recommended [20].

The cost effectiveness of fetal genome and exome sequencing has not yet been comprehensively evaluated [20], but must include the cost of counseling, follow-up testing, data storage, and reinterpretation, which together must be weighed against the cost to the healthcare system of an unknown diagnosis. The ethical principle of justice considers the equitable access to healthcare and medical resources. Fetal genome and exome sequencing is currently expensive and not reimbursed by all private health insurers or national health systems, who have to consider the equitable distribution of already limited resources [19, 73, 82]. Furthermore, in societies where termination of pregnancy is restricted, there is unequal access to all management options after a result is received.

Professional Guidelines

Both the ACMG and ACOG comment that fetal genome-wide sequencing should not be used routinely in ongoing pregnancies [89, 90] because its clinical utility has not yet been demonstrated. The more recent joint position statement of the ISPD, SMFM, and PQF [22], which is the first dedicated professional statement on fetal genome-wide sequencing, agrees with this recommendation. These organizations agree that while more research on benefits and pitfalls must be done, clinically offering prenatal genome-wide sequencing may be acceptable under specific circumstances if done by experienced teams. This could apply, for example, when standard testing fails to yield a diagnosis for a fetus with structural anomalies and the suspicion for a single-gene disorder is high, such as when there is a recurrent phenotype in subsequent pregnancies [22, 90].

CONCLUSION

Genome-wide sequencing has already become well integrated into pediatric genetic workups for children with birth defects or developmental and intellectual disabilities, where karyotype, CMA, and other tests, such as targeted gene sequencing, are uninformative. We do not yet fully know what the molecular diagnostic rate for exome sequencing and future genome sequencing will be for prenatal diagnosis, but early numbers support that it will be relatively similar to the 25%–40% seen in the pediatric population. This suggests that it has the potential of doubling the number of cases where a molecular diagnosis can be provided. It may also benefit the discovery of new genetic causes for prenatal phenotypes as well as for finding the cause of pregnancy failure and unexplained infertility. Studies on large cohorts for all these indications will be needed (some of which are ongoing) before the clinical utility of exome or genome sequencing can be accurately determined and comprehensive guidelines on its use developed. Until then, it should only be performed by experienced multidisciplinary teams. Currently, obtaining fetal genome-wide sequence data still requires performing a diagnostic invasive procedure to obtain fetal DNA, most commonly through amniocentesis or CVS. A number of reports have demonstrated the technical feasibility of noninvasive sequencing of the fetal genome from cell-free DNA circulating in the maternal plasma [91–93]. This is not currently ready for clinical application because it is both time and cost prohibitive, resulting from the inherent challenges associated with mixed fetal and maternal input DNA. In parallel, ongoing work on the isolation of intact fetal cells, mostly of trophoblastic origin, from maternal blood is holding great promise as a noninvasive source of fetal DNA that can be amplified and subjected to NGS to achieve diagnosis of single-gene disorders along with chromosomal copy number changes and aneuploidy [94–99].

References

[1] Van den Veyver IB. Recent advances in prenatal genetic screening and testing. F1000Res 2016;5:2591.

[2] Van den Veyver IB, Eng CM. Genome-wide sequencing for prenatal detection of fetal single-gene disorders. Cold Spring Harb Perspect Med 2015;5(10).

[3] Hillman SC, McMullan DJ, Hall G, Togneri FS, James N, Maher EJ, et al. Use of prenatal chromosomal micro-array: prospective cohort study and systematic review and meta-analysis. Ultrasound Obstet Gynecol 2013; 41(6):610–20.

[4] Wapner RJ, Martin CL, Levy B, Ballif BC, Eng CM, Zachary JM, et al. Chromosomal microarray versus karyotyping for prenatal diagnosis. N Engl J Med 2012;367(23):2175–84.

[5] ACOG Practice Bulletin No. 162. Summary: prenatal diagnostic testing for genetic disorders. Obstet Gynecol 2016;127(5):976–8.

[6] Society for Maternal-Fetal Medicine, Dugoff L, Norton ME, Kuller JA. The use of chromosomal microarray for prenatal diagnosis. Am J Obstet Gynecol 2016;215(4):B2–9.

[7] Hillman SC, Willams D, Carss KJ, McMullan DJ, Hurles ME, Kilby MD. Prenatal exome sequencing for fetuses with structural abnormalities: the next step. Ultrasound Obstet Gynecol 2015;45(1):4–9.

[8] Best S, Wou K, Vora N, Van den Veyver IB, Wapner R, Chitty LS. Promises, pitfalls and practicalities of prenatal whole exome sequencing. Prenat Diagn 2018;38(1):10–9.

[9] Miller DT, Adam MP, Aradhya S, Biesecker LG, Brothman AR, Carter NP, et al. Consensus statement: chromo-somal microarray is a first-tier clinical diagnostic test for individuals with developmental disabilities or congen-ital anomalies. Am J Hum Genet 2010;86(5):749–64.

[10] Normand EA, Alaimo JT, Van den Veyver IB. Exome and genome sequencing in reproductive medicine. Fertil Steril 2018;109(2):213–20.

[11] Bainbridge MN, Wiszniewski W, Murdock DR, Friedman J, Gonzaga-Jauregui C, Newsham I, et al. Whole-genome sequencing for optimized patient management. Sci Transl Med 2011;3(87):87re3.

[12] Sawyer SL, Hartley T, Dyment DA, Beaulieu CL, Schwartzentruber J, Smith A, et al. Utility of whole-exome sequencing for those near the end of the diagnostic odyssey: time to address gaps in care. Clin Genet 2016; 89(3):275–84.

[13] Xue Y, Ankala A, Wilcox WR, Hegde MR. Solving the molecular diagnostic testing conundrum for Mendelian disorders in the era of next-generation sequencing: single-gene, gene panel, or exome/genome sequencing. Genet Med 2015;17(6):444–51.

[14] Sun Y, Ruivenkamp CA, Hoffer MJ, Vrijenhoek T, Kriek M, van Asperen CJ, et al. Next-generation diagnostics: gene panel, exome, or whole genome? Hum Mutat 2015;36(6):648–55.

[15] Yang Y, Muzny DM, Reid JG, Bainbridge MN, Willis A, Ward PA, et al. Clinical whole-exome sequencing for the diagnosis of mendelian disorders. N Engl J Med 2013;369(16):1502–11.

[16] Yang Y, Muzny DM, Xia F, Niu Z, Person R, Ding Y, et al. Molecular findings among patients referred for clinical whole-exome sequencing. JAMA 2014;312(18):1870–9.

[17] Biesecker LG, Green RC. Diagnostic clinical genome and exome sequencing. N Engl J Med 2014;371(12):1170.

[18] Meng L, Pammi M, Saronwala A, Magoulas P, Ghazi AR, Vetrini F, et al. Use of exome sequencing for infants in intensive care units: ascertainment of severe single-gene disorders and effect on medical management. JAMA Pediatr 2017;e173438.

[19] Chitty LS, Friedman JM, Langlois S. Current controversies in prenatal diagnosis 2: should a fetal exome be used in the assessment of a dysmorphic or malformed fetus? Prenat Diagn 2016;36(1):15–9.

[20] Westerfield LE, Braxton AA, Walkiewicz M. Prenatal diagnostic exome sequencing: a review. Curr Genet Med Rep 2017; [Epub 2017/05/02], https://doi.org/10.1007/s40142-017-0120-y.

[21] Horn R, Parker M. Opening Pandora's box? Ethical issues in prenatal whole genome and exome sequencing. Prenat Diagn 2017;.

[22] International Society for Prenatal Diagnosis. Society for Maternal-Fetal Medicine, Perinatal Quality Foundation. Joint Position Statement from the International Society for Prenatal Diagnosis (ISPD), the Society for Maternal Fetal Medicine (SMFM), and the Perinatal Quality Foundation (PQF) on the use of genome-wide sequencing for fetal diagnosis. Prenat Diagn 2018;38(1):6–9.

[23] Bamshad MJ, Ng SB, Bigham AW, Tabor HK, Emond MJ, Nickerson DA, et al. Exome sequencing as a tool for Mendelian disease gene discovery. Nat Rev Genet 2011;12(11):745–55.

[24] Rehm HL, Bale SJ, Bayrak-Toydemir P, Berg JS, Brown KK, Deignan JL, et al. ACMG clinical laboratory standards for next-generation sequencing. Genet Med 2013;15(9):733–47.

[25] Genomes Project C, Auton A, Brooks LD, Durbin RM, Garrison EP, Kang HM, et al. A global reference for human genetic variation. Nature 2015;526(7571):68–74.

[26] Gonzaga-Jauregui C, Lupski JR, Gibbs RA. Human genome sequencing in health and disease. Annu Rev Med 2012;63:35–61.

[27] Richards S, Aziz N, Bale S, Bick D, Das S, Gastier-Foster J, et al. Standards and guidelines for the interpretation of sequence variants: a joint consensus recommendation of the American College of Medical Genetics and Genomics and the Association for Molecular Pathology. Genet Med 2015;17(5):405–24.

[28] Rehm HL, Berg JS, Brooks LD, Bustamante CD, Evans JP, Landrum MJ, et al. ClinGen—the clinical genome resource. N Engl J Med 2015;372(23):2235–42.

[29] ACMG Board of Directors. Laboratory and clinical genomic data sharing is crucial to improving genetic health care: a position statement of the American College of Medical Genetics and Genomics. Genet Med 2017; 19(7):721–2.

[30] Lek M, Karczewski KJ, Minikel EV, Samocha KE, Banks E, Fennell T, et al. Analysis of protein-coding genetic variation in 60,706 humans. Nature 2016;536(7616):285–91.

[31] Landrum MJ, Lee JM, Benson M, Brown G, Chao C, Chitipiralla S, et al. ClinVar: public archive of interpretations of clinically relevant variants. Nucleic Acids Res 2016;44(D1):D862–8.

[32] Stenson PD, Mort M, Ball EV, Evans K, Hayden M, Heywood S, et al. The Human Gene Mutation Database: towards a comprehensive repository of inherited mutation data for medical research, genetic diagnosis and next-generation sequencing studies. Hum Genet 2017;136(6):665–77.

[33] Firth HV, Richards SM, Bevan AP, Clayton S, Corpas M, Rajan D, et al. DECIPHER: database of chromosomal imbalance and phenotype in humans using Ensembl resources. Am J Hum Genet 2009;84(4):524–33.

[34] Robinson PN, Mundlos S. The human phenotype ontology. Clin Genet 2010;77(6):525–34.

[35] Gao J, Wan C, Zhang H, Li A, Zang Q, Ban R, et al. Anaconda: AN automated pipeline for somatic COpy number variation detection and annotation from tumor exome sequencing data. BMC Bioinform 2017;18(1):436.

[36] Paff T, Kooi IE, Moutaouakil Y, Riesebos E, Sistermans EA, JMA D, et al. Diagnostic yield of a targeted gene panel in primary ciliary dyskinesia patients, Hum Mutat 2018; [Epub 2018/01/25], https://doi.org/10.1002/humu.23403.

[37] Brett M, McPherson J, Zang ZJ, Lai A, Tan ES, Ng I, et al. Massively parallel sequencing of patients with intellectual disability, congenital anomalies and/or autism spectrum disorders with a targeted gene panel. PLoS ONE 2014;9(4):e93409.

[38] Sule G, Campeau PM, Zhang VW, Nagamani SC, Dawson BC, Grover M, et al. Next-generation sequencing for disorders of low and high bone mineral density. Osteoporos Int 2013;24(8):2253–9.

[39] Bhoj EJ, Yu Z, Guan Q, Ahrens-Nicklas R, Cao K, Luo M, et al. Phenotypic predictors and final diagnoses in patients referred for RASopathy testing by targeted next-generation sequencing. Genet Med 2017; 19(6):715–8.

[40] Ali MM, Chasen ST, Norton ME. Testing for Noonan syndrome after increased nuchal translucency. Prenat Diagn 2017;37(8):750–3.

[41] Alamillo CM, Fiddler M, Pergament E. Increased nuchal translucency in the presence of normal chromosomes: what's next? Curr Opin Obstet Gynecol 2012;24(2):102–8.

[42] Green RC, Berg JS, Grody WW, Kalia SS, Korf BR, Martin CL, et al. ACMG recommendations for reporting of incidental findings in clinical exome and genome sequencing. Genet Med 2013;15(7):565–74.

[43] ACMG Board of Directors. ACMG policy statement: updated recommendations regarding analysis and reporting of secondary findings in clinical genome-scale sequencing. Genet Med 2015;17(1):68–9.

[44] Kalia SS, Adelman K, Bale SJ, Chung WK, Eng C, Evans JP, et al. Recommendations for reporting of secondary findings in clinical exome and genome sequencing, 2016 update (ACMG SF v2.0): a policy statement of the American College of Medical Genetics and Genomics. Genet Med 2017;19(2):249–55.

[45] Hayward J, Chitty LS. Beyond screening for chromosomal abnormalities: advances in non-invasive diagnosis of single gene disorders and fetal exome sequencing, Semin Fetal Neonatal Med 2018; [Epub 2018/01/07], https://doi.org/10.1016/j.siny.2017.12.002.

[46] Haque IS, Lazarin GA, Kang HP, Evans EA, Goldberg JD, Wapner RJ. Modeled fetal risk of genetic diseases identified by expanded carrier screening. JAMA 2016;316(7):734–42.

[47] Edwards JG, Feldman G, Goldberg J, Gregg AR, Norton ME, Rose NC, et al. Expanded carrier screening in reproductive medicine-points to consider: a joint statement of the American College of Medical Genetics and Genomics, American College of Obstetricians and Gynecologists, National Society of Genetic Counselors, Perinatal Quality Foundation, and Society for Maternal-Fetal Medicine. Obstet Gynecol 2015;125(3):653–62.

[48] Lazarin GA, Haque IS, Nazareth S, Iori K, Patterson AS, Jacobson JL, et al. An empirical estimate of carrier frequencies for 400+ causal Mendelian variants: results from an ethnically diverse clinical sample of 23,453 individuals. Genet Med 2013;15(3):178–86.

[49] Brezina PR, Anchan R, Kearns WG. Preimplantation genetic testing for aneuploidy: what technology should you use and what are the differences? J Assist Reprod Genet 2016;33(7):823–32.

[50] Shamseldin HE, Swaid A, Alkuraya FS. Lifting the lid on unborn lethal Mendelian phenotypes through exome sequencing. Genet Med 2013;15(4):307–9.

[51] Filges I, Manokhina I, Penaherrera MS, McFadden DE, Louie K, Nosova E, et al. Recurrent triploidy due to a failure to complete maternal meiosis II: whole-exome sequencing reveals candidate variants. Mol Hum Reprod 2015;21(4):339–46.

[52] Qiao Y, Wen J, Tang F, Martell S, Shomer N, Leung PC, et al. Whole exome sequencing in recurrent early pregnancy loss. Mol Hum Reprod 2016;22(5):364–72.

[53] Chen T, Bian Y, Liu X, Zhao S, Wu K, Yan L, et al. A recurrent missense mutation in ZP3 causes empty follicle syndrome and female infertility. Am J Hum Genet 2017;101(3):459–65.

[54] Alazami AM, Awad SM, Coskun S, Al-Hassan S, Hijazi H, Abdulwahab FM, et al. TLE6 mutation causes the earliest known human embryonic lethality. Genome Biol 2015;16:240.

[55] Yariz KO, Walsh T, Uzak A, Spiliopoulos M, Duman D, Onalan G, et al. Inherited mutation of the luteinizing hormone/choriogonadotropin receptor (LHCGR) in empty follicle syndrome. Fertil Steril 2011;96(2):e125–30.

[56] Tenenbaum-Rakover Y, Weinberg-Shukron A, Renbaum P, Lobel O, Eideh H, Gulsuner S, et al. Minichromosome maintenance complex component 8 (MCM8) gene mutations result in primary gonadal failure. J Med Genet 2015;52(6):391–9.

[57] Gordon CM, Kanaoka T, Nelson LM. Update on primary ovarian insufficiency in adolescents. Curr Opin Pediatr 2015;27(4):511–9.

[58] Collins SC. Precision reproductive medicine: multigene panel testing for infertility risk assessment. J Assist Reprod Genet 2017;34(8):967–73.

[59] Filges I, Friedman JM. Exome sequencing for gene discovery in lethal fetal disorders—harnessing the value of extreme phenotypes. Prenat Diagn 2015;35(10):1005–9.

[60] Watson CM, Crinnion LA, Murphy H, Newbould M, Harrison SM, Lascelles C, et al. Deficiency of the myogenic factor MyoD causes a perinatally lethal fetal akinesia. J Med Genet 2016;53(4):264–9.

[61] Retterer K, Juusola J, Cho MT, Vitazka P, Millan F, Gibellini F, et al. Clinical application of whole-exome sequencing across clinical indications. Genet Med 2016;18(7):696–704.

[62] Trujillano D, Bertoli-Avella AM, Kumar Kandaswamy K, Weiss ME, Koster J, Marais A, et al. Clinical exome sequencing: results from 2819 samples reflecting 1000 families. Eur J Hum Genet 2017;25(2):176–82.

[63] Wang H, Sun Y, Wu W, Wei X, Lan Z, Xie J. A novel missense mutation of FGFR3 in a Chinese female and her fetus with Hypochondroplasia by next-generation sequencing. Clin Chim Acta 2013;423:62–5.

[64] Talkowski ME, Ordulu Z, Pillalamarri V, Benson CB, Blumenthal I, Connolly S, et al. Clinical diagnosis by whole-genome sequencing of a prenatal sample. N Engl J Med 2012;367(23):2226–32.

[65] Filges I, Nosova E, Bruder E, Tercanli S, Townsend K, Gibson WT, et al. Exome sequencing identifies mutations in KIF14 as a novel cause of an autosomal recessive lethal fetal ciliopathy phenotype. Clin Genet 2014;86(3):220–8.

[66] Alamillo CL, Powis Z, Farwell K, Shahmirzadi L, Weltmer EC, Turocy J, et al. Exome sequencing positively identified relevant alterations in more than half of cases with an indication of prenatal ultrasound anomalies. Prenat Diagn 2015;35(11):1073–8.

[67] Drury S, Williams H, Trump N, Boustred C, Gosgene LN, et al. Exome sequencing for prenatal diagnosis of fetuses with sonographic abnormalities. Prenat Diagn 2015;35(10):1010–7.

[68] Vora NL, Powell B, Brandt A, Strande N, Hardisty E, Gilmore K, et al. Prenatal exome sequencing in anomalous fetuses: new opportunities and challenges. Genet Med 2017;19(11):1207–16.

[69] Lei TY, Fu F, Li R, Wang D, Wang RY, Jing XY, et al. Whole-exome sequencing for prenatal diagnosis of fetuses with congenital anomalies of the kidney and urinary tract. Nephrol Dial Transplant 2017;32(10):1665–75.

[70] Fu F, Li R, Li Y, Nie ZQ, Lei TY, Wang D, et al. Whole exome sequencing as a diagnostic adjunct to clinical testing in a tertiary referral cohort of 3988 fetuses with structural abnormalities. Ultrasound Obstet Gynecol 2017; https://dx.doi.org/10.1002/uog.18915, [Epub ahead of print].

[71] Ellard S, Kivuva E, Turnpenny P, Stals K, Johnson M, Xie W, et al. An exome sequencing strategy to diagnose lethal autosomal recessive disorders. Eur J Hum Genet 2015;23(3):401–4.

[72] Stals KL, Wakeling M, Baptista J, Caswell R, Parrish A, Rankin J, et al. Diagnosis of lethal or prenatal-onset autosomal recessive disorders by parental exome sequencing. Prenat Diagn 2018;38(1):33–43.

[73] Westerfield LE, Stover SR, Mathur VS, Nassef SA, Carter TG, Yang Y, et al. Reproductive genetic counseling challenges associated with diagnostic exome sequencing in a large academic private reproductive genetic counseling practice. Prenat Diagn 2015;35(10):1022–9.

[74] Stark Z, Tan TY, Chong B, Brett GR, Yap P, Walsh M, et al. A prospective evaluation of whole-exome sequencing as a first-tier molecular test in infants with suspected monogenic disorders. Genet Med 2016;18(11):1090–6.

[75] Tan TY, Dillon OJ, Stark Z, Schofield D, Alam K, Shrestha R, et al. Diagnostic impact and cost-effectiveness of whole-exome sequencing for ambulant children with suspected monogenic conditions. JAMA Pediatr 2017; 171(9):855–62.

[76] Westerfield L, Darilek S, van den Veyver IB. Counseling challenges with variants of uncertain significance and incidental findings in prenatal genetic screening and diagnosis. J Clin Med 2014;3(3):1018–32.

[77] Abou Tayoun AN, Spinner NB, Rehm HL, Green RC, Bianchi DW. Prenatal DNA sequencing: clinical, counseling, and diagnostic laboratory considerations. Prenat Diagn 2018;38(1):26–32.

[78] Fonda Allen J, Stoll K, Bernhardt BA. Pre- and post-test genetic counseling for chromosomal and Mendelian disorders. Semin Perinatol 2016;40(1):44–55.

[79] Sobreira N, Schiettecatte F, Valle D, Hamosh A. GeneMatcher: a matching tool for connecting investigators with an interest in the same gene. Hum Mutat 2015;36(10):928–30.

[80] Kalynchuk EJ, Althouse A, Parker LS, Saller Jr. DN, Rajkovic A. Prenatal whole-exome sequencing: parental attitudes. Prenat Diagn 2015;35(10):1030–6.

[81] Pangalos C, Hagnefelt B, Lilakos K, Konialis C. First applications of a targeted exome sequencing approach in fetuses with ultrasound abnormalities reveals an important fraction of cases with associated gene defects. PeerJ 2016;4:e1955.

[82] Quinlan-Jones E, Kilby MD, Greenfield S, Parker M, McMullan D, Hurles ME, et al. Prenatal whole exome sequencing: the views of clinicians, scientists, genetic counsellors and patient representatives. Prenat Diagn 2016;36(10):935–41.

[83] Bayefsky MJ, White A, Wakim P, Hull SC, Wasserman D, Chen S, et al. Views of American OB/GYNs on the ethics of prenatal whole-genome sequencing. Prenat Diagn 2016;36(13):1250–6.

[84] Yurkiewicz IR, Korf BR, Lehmann LS. Prenatal whole-genome sequencing—is the quest to know a fetus's future ethical? N Engl J Med 2014;370(3):195–7.

[85] Burke W, Matheny Antommaria AH, Bennett R, Botkin J, Wright Clayton E, Henderson GE, et al. Recommendations for returning genomic incidental findings? We need to talk! Genet Med 2013;15(11):854–9.

[86] Mand C, Gillam L, Duncan RE, Delatycki MB. "It was the missing piece": adolescent experiences of predictive genetic testing for adult-onset conditions. Genet Med 2013;15(8):643–9.

[87] Anderson JA, Hayeems RZ, Shuman C, Szego MJ, Monfared N, Bowdin S, et al. Predictive genetic testing for adult-onset disorders in minors: a critical analysis of the arguments for and against the 2013 ACMG guidelines. Clin Genet 2015;87(4):301–10.

[88] Bernhardt BA, Roche MI, Perry DL, Scollon SR, Tomlinson AN, Skinner D. Experiences with obtaining informed consent for genomic sequencing. Am J Med Genet A 2015;167A(11):2635–46.

[89] ACOG Committee on Genetics and the Society for Maternal-Fetal Medicine. Committee opinion no. 682: microarrays and next-generation sequencing technology: the use of advanced genetic diagnostic tools in obstetrics and gynecology. Obstet Gynecol 2016;128(6):e262–8.

[90] ACMG Board of Directors. Points to consider in the clinical application of genomic sequencing. Genet Med 2012;14(8):759–61.

[91] Lo YM, Chan KC, Sun H, Chen EZ, Jiang P, Lun FM, et al. Maternal plasma DNA sequencing reveals the genome-wide genetic and mutational profile of the fetus. Sci Transl Med 2010;2(61):61ra91.

[92] Fan HC, Gu W, Wang J, Blumenfeld YJ, El-Sayed YY, Quake SR. Non-invasive prenatal measurement of the fetal genome. Nature 2012;487(7407):320–4.

[93] Kitzman JO, Snyder MW, Ventura M, Lewis AP, Qiu R, Simmons LE, et al. Noninvasive whole-genome sequencing of a human fetus. Sci Transl Med 2012;4(137):137ra76.

[94] Kolvraa S, Singh R, Normand EA, Qdaisat S, van den Veyver IB, Jackson L, et al. Genome-wide copy number analysis on DNA from fetal cells isolated from the blood of pregnant women. Prenat Diagn 2016;36(12):1127–34.

[95] Breman AM, Chow JC, U'Ren L, Normand EA, Qdaisat S, Zhao L, et al. Evidence for feasibility of fetal trophoblastic cell-based noninvasive prenatal testing. Prenat Diagn 2016;36(11):1009–19.

[96] Beaudet AL. Using fetal cells for prenatal diagnosis: history and recent progress. Am J Med Genet C Semin Med Genet 2016;172(2):123–7.

[97] Mouawia H, Saker A, Jais JP, Benachi A, Bussieres L, Lacour B, et al. Circulating trophoblastic cells provide genetic diagnosis in 63 fetuses at risk for cystic fibrosis or spinal muscular atrophy. Reprod Biomed Online 2012;25(5):508–20.

[98] Hatt L, Brinch M, Singh R, Moller K, Lauridsen RH, Uldbjerg N, et al. Characterization of fetal cells from the maternal circulation by microarray gene expression analysis—could the extravillous trophoblasts be a target for future cell-based non-invasive prenatal diagnosis? Fetal Diagn Ther 2014;35(3):218–27.

[99] Chen F, Liu P, Gu Y, Zhu Z, Nanisetti A, Lan Z, et al. Isolation and whole genome sequencing of fetal cells from maternal blood towards the ultimate non-invasive prenatal testing. Prenat Diagn 2017;37(13):1311–21.

Glossary

Amniocentesis	A diagnostic procedure typically performed after 15 weeks of pregnancy to obtain amniotic fluid and prepare cells or DNA for laboratory analysis.
Bait	The prepared DNA fragments of known sequence that are used to capture the complementary sequence from the collection of all fragments in a sequencing library in a capture assay (see also "capture").
Capture	The process by which a specific set of DNA fragments is enriched and purified from the total pool of DNA fragments (for multi gene-panel or exome sequencing) by hybridization (binding) to complementary baits (see also "bait").
Chorionic villus sampling	A diagnostic procedure performed between 10 and 13 weeks of pregnancy to obtain a small amount of placental villi and prepare cells or DNA for laboratory analysis.
Chromosomal microarray analysis	A method whereby DNA from a patient is hybridized to an array of small fragments of DNA (oligonucleotides), which allows the detection of extra or missing copies of DNA (deletions, duplications, aneuploidy) and in some platforms for the absence of heterozygosity (see below).
Clinical utility	The ability of a screening or diagnostic test to prevent or ameliorate adverse health outcomes such as mortality, morbidity, or disability through the adoption of efficacious treatments conditioned on test results [Grosse and Khoury; Genet Med 2006;8(7):448–50].
De novo sequence variant	A sequence variant in an individual's DNA that is not present in the DNA of either biological parent of that individual.
Exome sequencing	Next-generation sequencing of the 1%–2% of the genome that contains the coding exons of genes.
Genome sequencing	Next-generation sequencing of the genome.
Karyotype	An individual's collection of chromosomes, usually visually represented after laboratory methods are applied to prepare and stain the chromosomes.
Multi-gene panel test	A diagnostic or research test that uses a sequencing method (usually next-generation sequencing) to simultaneously sequence a defined number of genes known to cause similar diseases, syndromes, or phenotypes.
Next-generation sequencing	A method to efficiently sequence and annotate large amounts of DNA sequence in a massively parallel fashion (massively parallel sequencing).
Phenotype	The combination of observable characteristics of an individual.
Sequencing library	The complement of DNA fragments that has been prepared for sequencing.

Sequence variant A nucleotide (or a few nucleotides) in the sequenced DNA that is different from the expected nucleotide(s) at that position based on the known reference sequence.

Trio sequencing A method of sequencing whereby the DNA of an individual (or fetus) and that of their biological parents are sequenced and the sequence data from all three is integrated to facilitate the analysis and interpretation of the individual's sequencing results.

Variant of uncertain significance A sequence variant for which it is not known whether it is benign or can cause disease or an abnormal phenotype.

Chromosomal Microarrays and Exome Sequencing for Diagnosis of Fetal Abnormalities

Brynn Levy*, *Melissa Stosic*[†], *Jessica Giordano*[†], *Ronald Wapner*[†]

[*]Department of Pathology and Cell Biology, Columbia University Medical Center, New York, NY, United States [†]Department of Obstetrics and Gynecology, Columbia University Medical Center, New York, NY, United States

OUTLINE

INTRODUCTION

The longtime standard in prenatal cytogenetics has been karyotype analysis, which has been used to identify deletions and duplications 7–10 Mb or greater in size, most commonly aneuploidies such as Trisomies 21, 18, and 13 and sex chromosome aneuploidies. Historically, invasive prenatal diagnosis has been offered to women over the age of 35 as that is the age at which the risk of age-related aneuploidy equals the risk of miscarriage from invasive testing at 1/200 [1]. However, over time the risk of loss secondary to amniocentesis or chorionic villus sampling has been shown to be lower than previously described [2]. Simultaneously, the notion that only women 35 years of age and older are at risk for significant abnormalities has been disproven [3].

Major advances in genomic testing have been made over the past decade and these have now been applied to prenatal diagnosis. Chromosomal microarray analysis (CMA) became the first-tier diagnostic test in children with intellectual disabilities and/or structural abnormalities in 2010 [4]. By 2012, CMA was proven to have a 6% incremental yield over karyotype in anomalous fetuses and 1.7% in pregnancies without anomalies [3]. Today, CMA is often utilized as a primary test and is recommended in pregnancies with structural abnormalities [5]. Around the time CMA became the standard of care for fetuses with structural anomalies, the use of whole exome sequencing (WES) as a diagnostic tool expanded in pediatrics, allowing an even greater resolution of the genome. Because WES identifies a single gene disorder in 25%–30% of children undiagnosed by karyotype/CMA who are suspected to have a syndrome, clinical research in the utility of WES in the prenatal setting has emerged with a variety of unique challenges [6, 7]. While the incremental diagnostic rate of WES in fetuses with anomalies varies between reported series, approximately 14% of unselected structurally abnormal pregnancies with a normal karyotype and array will have a causative mutation identified by WES (Fig. 26.1, internal data).

For the past 50 years of prenatal diagnosis, we have been seeking to resolve the genomic etiology of fetal anomalies. With the introduction of each new genomic technique has come the ability to further identify a proportion of the causes contributing to fetal anomalies diagnosed in the second trimester (Fig. 26.1).

CHROMOSOMAL MICROARRAY ANALYSIS

Traditional G-banded karyotype has a resolution of approximately 7–10 million base pairs (MBs) when evaluating prenatal specimens such as amniotic fluid and chorionic villi. While karyotyping allows for the detection of all aneuploidies as well as large deletions and duplications, most chromosomal imbalances smaller than 7 MB are considered beyond the resolution of routine cytogenetic analysis. These submicroscopic imbalances are often referred to as copy number variants (CNVs), many of which are clinically insignificant and are found in apparently normal individuals [8–12]. However, certain CNVs are associated with genetic conditions that cause intellectual disability and/or birth defects [13]. Classic examples of these CNVs (aka microdeletions and microduplications) are the 22q11.2 deletion syndrome

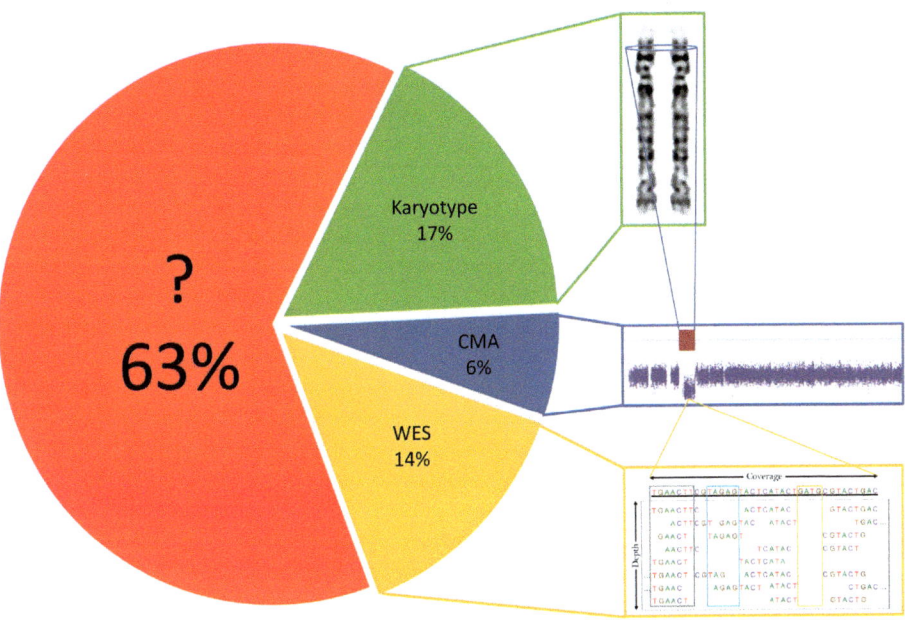

FIG. 26.1 Pie chart representing the contribution of various technologies in elucidating the etiology of fetal anomalies detected in the second trimester of pregnancy. The figure also illustrates the different resolution of each technology with karyotyping being able to discern imbalances only as large as individual bands (7–10 Mb), CMA increasing the diagnostic capabilities to the kilobase range, and WES taking it down to the individual base pair. The newer techniques generally add incremental diagnostic abilities as they are capable, to a large extent, of also discerning large copy number imbalances such as aneuploidy. The etiology of most fetal anomalies remains unresolved.

(DiGeorge syndrome) and the Prader-Willi/Angelman syndromes, which are typically caused by ~3 Mb and 4–6 Mb deletions, respectively. As such, these submicroscopic CNVs are also known as microdeletion/microduplication syndromes, particularly when they are associated with specific clinical syndromes [14, 15]. Chromosomal microarray analysis (CMA) has the ability to identify gains and losses in the kilobase range (roughly 100× the diagnostic resolution of standard karyotyping), and is therefore the preferred technique for identifying these microdeletion and microduplication syndromes. Table 26.1 lists many of the common microdeletion and microduplication syndromes, along with their cytogenetic location.

TYPES OF MICROARRAYS

Microarray technology can be performed utilizing either comparative genomic hybridization arrays (aCGH) or single nucleotide polymorphism (SNP) arrays; often, a combination of the two is used. In both methods, DNA probes from representative sequences across the human genome are bound to a slide or chip. Patient DNA is then fluorescently labeled and

TABLE 26.1 Common Microdeletion/Duplication Syndromes by Frequency

Condition; Cytogenetic Location	Incidence
16p11.2 duplication	1/1900
16p11.2 deletion	1/2300
16p13.11 deletion	1/2300
1q21.1 duplication	1/3300
22q11.2 deletion syndrome (DiGeorge, VCFS)	1/4000
22q11.2 duplication	1/4000
1p36 deletion syndrome	1/5000
Charcot-Marie-Tooth Type 1A; 17p12 duplication	1/5000–1/10,000
X-linked ichthyosis; Xp22.31 deletion	1/6000
Williams syndrome; 7q11.23 deletion	1/7500
7q11.23 duplication	1/7500
Prader-Willi syndrome; 15q11.2 paternal deletion	1/10,000
Angelman syndrome; 15q11.2 maternal deletion	1/12,000
17q12 deletion	1/14,500
Sotos syndrome; 5q35 deletion	1/15,000
Smith Magenis syndrome; 17p11.2 deletion	1/15,000–1/25,000
Cri-du-chat; 5p15 deletion	1/15,000–1/50,000
Koolen de Vries; 17q21 deletion	1/16,000
Potocki-Lupski syndrome; 17p11.2 duplication	1/20,000

hybridized to the DNA sequences present on the chip. Because the patient DNA is fluorescently tagged, the signal intensity can be measured to highlight the gain or loss of genetic material. The difference between aCGH and a SNP array is that aCGH necessitates the addition of control DNA alongside the patient DNA, whereas SNP arrays do not. In aCGH, the patient and control DNAs are labeled with different fluorescent colors, typically red and green, and the relative intensities are compared after a normalization process. The resulting data is then typically plotted as a Log2 ratio where a log2 of zero represents a normal DNA copy number of 2 (Fig. 26.2). An increased Log2 ratio indicates a DNA copy number gain and a decreased Log2 ratio indicates a DNA copy number loss (Fig. 26.2). SNPs are, by definition, variations between individuals at a single base pair site. SNP arrays function by adding fluorescently labeled patient DNA to a slide prepared with short oligonucleotide sequences, representing SNPs, across the genome. Unlike aCGH, SNP arrays work by measuring the absolute fluorescence probe intensities of the patient sample compared with the intensities of multiple normal controls that were independently hybridized (in silico comparison). The intensity of the signal

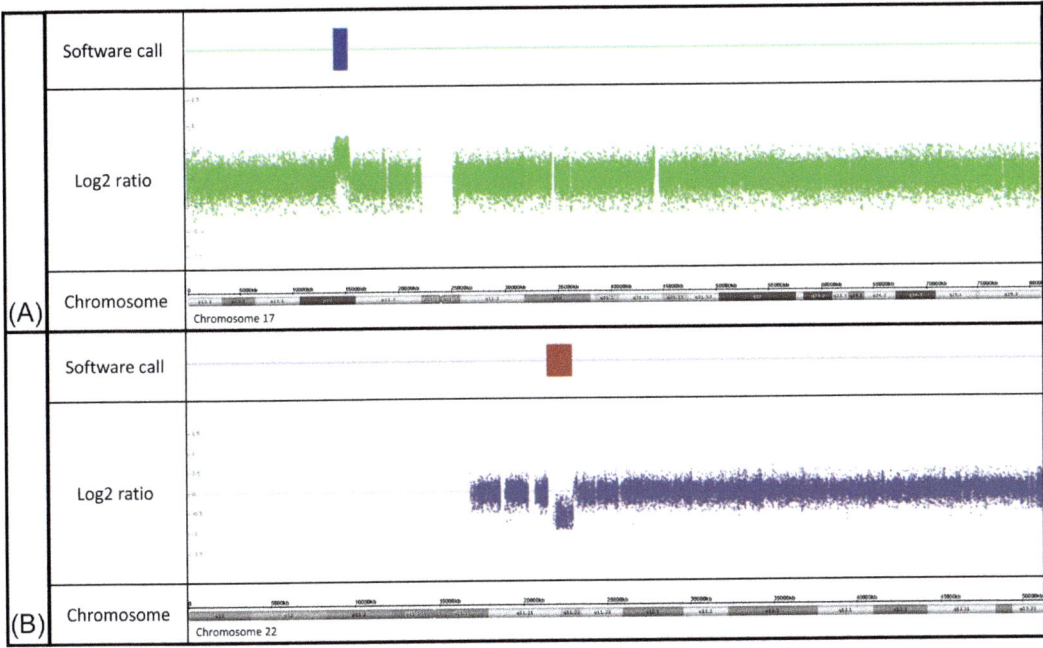

FIG. 26.2 Screenshot from the Affymetrix Chromosome Analysis Suite Software (Version 3.1) showing the Log2 ratios for a Microduplication and for a Microdeletion. (A) Fetus with a 1.34 Mb interstitial duplication in the short arm region (17p12) of chromosome 17, which is associated with a clinical diagnosis of Charcot-Marie-Tooth disease, type 1A demyelinating neuropathy (OMIM#118220) and is caused by a gain of the *PMP22* gene. The precise coordinates of the duplication correspond to chr17: 14,087,933-15,427,478 using Human Genome Build Hg19. The gene content within the duplicated region can be ascertained using the genome coordinates. A duplication is indicated in the software call panel by the presence of a blue bar. The duplication is identified by an increase in the Log2 ratio from zero as seen in the Log2 ratio panel. (B) Fetus with an interstitial deletion of the proximal long arm region (22q11.2) of chromosome 22, which is associated with a clinical diagnosis of DiGeorge syndrome (OMIM# 188400). The precise coordinates of the deletion correspond to chr22: 18,916,842-21,465,659 using Human Genome Build Hg19. The gene content within the deleted region can be ascertained using the genome coordinates. A deletion is indicated in the software call panel by the presence of a red bar. The deletion is identified by a decrease in the Log2 ratio from zero as seen in the Log2 ratio panel. The chromosome ideogram in the chromosome panels highlights the position (breakpoints) on the chromosome where copy number imbalances are present.

is measured at each SNP locus, providing both copy number and genotype information. Copy number data is also outputted as Log2 ratio plots, just like aCGH (Fig. 26.2). However, unlike aCGH, SNP microarrays can also yield additional clinically useful information that is extracted from the genotype plots generated from the SNPs. This includes uniparental disomy (UPD), triploidy, mosaicism, maternal cell contamination, parent of origin, consanguinity, and zygosity [16]. Triploidy cannot be determined from the Log2 ratio plots generated in either aCGH or SNP arrays. However, by assessing the SNP allele patterns (genotypes) on a SNP array, triploidy can easily be discerned (Fig. 26.3). In addition, long contiguous stretches of homozygosity (LCSH) observed on an SNP array often prove useful for further investigations, such as targeted genotyping or sequencing, as the LCSH region may contain a candidate recessive genetic condition that fits with the clinical phenotype [17].

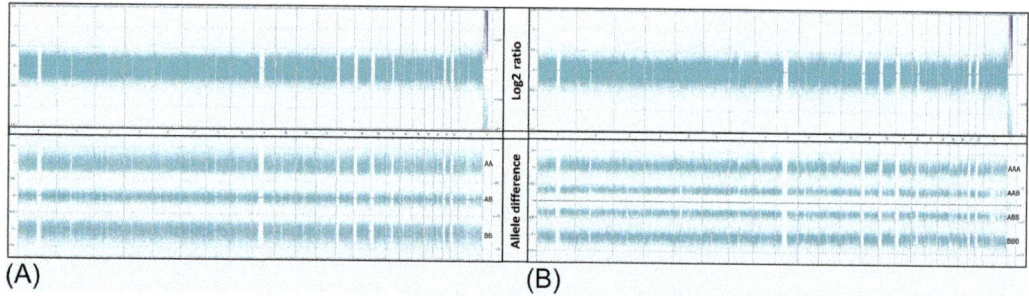

(A)   (B)

FIG. 26.3 Whole genome view showing the Log2 ratio and allele difference for every chromosome. (A) Normal Female Feus (46,XX). (B) Female Fetus with triploidy (69,XXX). Because the intensities of the probes are normalized, the Log2 ratios for a normal diploid sample and a triploid sample are indistinguishable (top panels), with both indicating a DNA copy number of 2. Genotypes extracted from the SNP data must be utilized to identify triploidy. The allele difference plots (bottom panels) show the various SNP genotypes for each SNP locus. In the presence of two chromosomes (normal diploid state), there are only three possible SNP combinations: AA, AB, and BB. A normal diploid female would show the characteristic three tracks for all the chromosomes. In the presence of three chromosomes (triploid state), there are four possible SNP combinations: AAA, AAB, ABB, and BBB, which results in four distinct tracks on the allele difference graph. The triploid 69,XXX fetus in this example displays the four characteristic tracks for every autosome as well as the X chromosome. Triploid fetuses with a 69,XXY constitution would show the four characteristic tracks for every autosome, three tracks representing the two X chromosomes, and two tracks representing the single Y chromosome.

MICROARRAY COVERAGE

CMA coverage can be whole genome or targeted, meaning that the array can be created to cover the entire genome at specific intervals or only specific regions associated with disease. The majority of microarrays utilized in the prenatal setting are designed to identify copy number gains or losses of any size in areas associated with known genetic conditions. They often also include broader coverage across the rest of the genome, called the backbone, at equally spaced intervals, which allows the detection of CNVs above a predetermined size, generally 1 Mb in prenatal studies [16].

REPORTING AND INTERPRETATION OF CMA RESULTS

The American College of Medical Genetics and Genomics (ACMGG), together with the Association for Molecular Pathology (AMP), published guidelines for the interpretation of variants and advocated a five-tier reporting system [18]: *pathogenic, likely pathogenic, uncertain significance, likely benign,* or *benign* [18]. Findings that are known to be associated with a specific condition are reported as pathogenic and those that are likely to cause abnormalities are reported as just that, likely pathogenic. Similarly, findings that are not disease causing are labeled as benign and those that are unlikely to cause disease are labeled as likely benign [18]. Generally, benign and likely benign findings are not reported. The remainder of the results is considered of uncertain significance and may be complex for the laboratory, clinician, and patient to interpret.

Clinical cytogeneticists use various tools, strategies, and guidelines to aid in the interpretation of CMA results. Public and in-house databases of normal individuals help to identify benign, common CNVs in the population, and online databases of clinically significant CNVs such as DECIPHER (https://decipher.sanger.ac.uk/), ISCA (http://dbsearch.clinicalgenome.org/search/), and Clinvar (https://www.ncbi.nlm.nih.gov/clinvar/) are tremendous resources to help characterize the clinical nature of pathogenic CNVs as well as assess evidence of pathogenicity. Evidence-based reports that show enrichment of candidate CNVs in large patient populations with like phenotypes compared to normal controls are highly desirable [13], and a single case report is not considered acceptable evidence for reporting an unknown CNV as pathogenic [18]. Variants of unknown significance (VOUS) present a major challenge as they may represent benign familial variants that produce no medical sequelae, or may be rare pathogenic changes that result in a clinical phenotype. The gene content, CNV size, and inheritance pattern can help to discern whether VOUS are more likely to be pathogenic or benign. Generally, a CNV that is inherited from a clinically normal parent is less likely to be detrimental. However, there are exceptions and these usually fall into the category of CNVs with incomplete penetrance and/or variable expressivity. In such cases, there is a spectrum of clinical outcomes (phenotypic heterogeneity) where parents could be normal or even mildly affected (i.e., at one end of the clinical spectrum) while their offspring present at the severe end of the clinical spectrum and manifest with the full-blown genetic syndrome [19–21]. Unfortunately, prospective and accurate predictions as to where on the spectrum a fetus will fall are not possible unless some aspects of the phenotype (such as congenital anomalies) are visible on an ultrasound scan.

PROFESSIONAL SOCIETY CMA GUIDELINES

Since 2007, recommendations have been in place that all women should be offered invasive prenatal diagnosis, regardless of maternal age [22, 23]. In the past, women who opted for testing would have a karyotype performed on the prenatal specimen. As of 2013, the American Congress of Obstetricians and Gynecologists (ACOG) and the Society for Maternal Fetal Medicine (SMFM) suggest that CMA be performed alongside or instead of karyotype in pregnancies where structural abnormalities have been identified, and that either CMA or karyotype is reasonable in pregnancies without ultrasound anomalies [5]. They also highlight that because the risk for CNVs is not age-dependent, CMA should not be limited to women 35 and older [5]. When prenatal CMA is ordered, both pre- and posttest counseling by a healthcare practitioner that has CMA experience such as a geneticist or genetic counselor are strongly suggested [24].

CLINICAL UTILITY OF CMA

In children with autism, developmental disability, and/or multiple congenital anomalies, CMA has replaced karyotype as the standard of care for diagnostic testing due to a 15% incremental yield in diagnosis [4, 25]. With a landmark study published in 2012, the benefit of

CMA over karyotype in the prenatal population became apparent [3]. To date, many studies have shown that CMA significantly increases the diagnostic yield in prenatal testing for both routine pregnancies and those with structural anomalies.

CMA IN ROUTINE PREGNANCIES

Maternal serum screening, which has been utilized for the past 30 years, has focused on detecting pregnancies affected with Down syndrome, which occurs in 1/700 live births. For pregnancies referred for invasive testing due to advanced maternal age, abnormal serum screening, maternal anxiety, and other routine indications, CMA has been show to diagnose a microdeletion or microduplication not seen by karyotype in 1%–2% of cases [3, 26–29]. Therefore, for the majority of routine pregnancies, the risk for a microdeletion or microduplication is higher than the risk for Down syndrome.

The variation in incremental yield over karyotype is partially a consequence of the change in categorization of results over time. For this reason, the risk of receiving results that are of uncertain significance has also changed, as shown in a recent reanalysis of a 2012 study where 2.5% of results were originally uncertain but when rereviewed, the rate was lowered to 0.9% [30]. A large metaanalysis of CMA for all indications reported that 1.4% of results were of uncertain clinical significance [31]. The risk of receiving uncertain results is one of the main limitations of microarray and is one of the reasons some women choose not to have testing, so the decrease in likelihood of this over time has made a tremendous impact. When results of uncertain significance are reported to patients, it is suggested that they meet with an expert to review the available information [24].

CMA IN PREGNANCIES WITH ULTRASOUND ANOMALIES

While many fetuses with a microdeletion or microduplication will not have any structural abnormalities, as shown by the risk in a routine pregnancy, there are some conditions that are known to be associated with certain anomalies, such as heart defects in the DiGeorge (22q11.2 deletion) syndrome [32]. Many times there is a large differential diagnosis for the findings seen on ultrasound, especially because the phenotype is often limited and important diagnostic clues such as intellectual disability or seizures are not yet visible. Therefore, because microarray covers a broader range of potential conditions, it is more appropriate than karyotype in fetuses with one or more structural anomalies [5, 33].

The ability of CMA to diagnose the underlying condition in pregnancies with findings on ultrasound has been well studied. The likelihood of finding a copy number variant in an anomalous fetus depends on the number of anomalies and the organ system(s) involved, but on average, 6%–7% will have a causative array finding identified [3, 26, 28, 31, 34]. The systems most likely to result in an abnormal CMA are the central nervous system as well as the urogenital, cardiac, renal, and skeletal systems [35–38]. Isolated cardiac and renal anomalies show a 10.6% and 15.0% incremental yield of CMA over karyotyping, respectively, and represent the highest diagnostic yield in fetuses with single organ system anomalies [39].

In the past, FISH studies have commonly been ordered to rule out DiGeorge (22q11.2 deletion) syndrome when fetal cardiac anomalies are discovered on ultrasound. Because 66.7% of patients with cardiac defects have CNVs other than the 22q11.2 deletion, CMA has proven to be more appropriate than targeted FISH when such an instance is encountered [39]. When multiple organ systems are affected, there is a 13.6% incidence of positive CMA results versus 5.1% when only a single organ system is affected [39].

BENEFITS AND LIMITATIONS OF CMA

The obvious benefit of CMA over karyotype is improved coverage of the genome. This comes into play not only in the diagnosis of microdeletion and microduplications syndromes, but also in identifying the chromosomal origin in cryptic karyotype results and in defining the breakpoints of karyotypic abnormalities [40]. For example, CMA may be used to search for small cytogenetically "invisible" submicroscopic deletions/duplications when an apparently balanced translocation is identified, or to delineate the borders of a visible deletion or duplication. CMA can also be designed to target specific regions or conditions and can be created to include very dense coverage over those areas. CMA is amenable to a wider range of sample types, including uncultured chorionic villi and uncultured amniocytes. As tissue culture failure is relatively common when studying stillbirth samples using routine cytogenetic techniques, CMA is the preferred method of analysis for these cases [41]. As discussed previously, SNP CMA can identify LCSH and therefore additional mechanisms for disease such as UPD. Lastly, CMA technology is objective rather than subjective [40].

One of the main limitations of CMA is that it only identifies abnormalities caused by gains or losses (imbalances) of DNA. Therefore, CMA cannot diagnose balanced rearrangements or provide the underlying mechanism of an imbalance. For example, the etiology of a gain observed by CMA could be a marker chromosome, an intrachromosomal direct duplication, or even an insertion of an additional chromosomal fragment into another chromosome [40]. Tetraploidy (and in the case of aCGH, triploidy) cannot be seen via CMA and low-level mosaicism may also be missed. As is also the case with other technologies, CMA is only as good as its coverage and cannot identify copy number variation outside the covered regions or below the resolution of the array used. In addition, routine clinical CMA will not detect genetic disease caused by sequence changes (mutation) and/or aberrant methylation.

PRENATAL WHOLE EXOME SEQUENCING (WES)

With ultrasound imaging becoming a routine part of prenatal care, an increasing number of fetal structural anomalies have been identified. Until recently, genetic evaluation of these cases was performed by karyotype with a diagnosis found in approximately 25%–35% of cases. Incorporation of CMA into standard practice increased the diagnostic yield by approximately 6%, still leaving the majority of families without a specific diagnosis [3]. This uncertainty as to the underlying etiology of the anomaly makes it difficult for families and providers to make medical management decisions because ultrasound anomalies can be associated with a lengthy differential diagnosis and uncertain prognosis. Although targeted gene panels are available to families when specific sonographic findings are involved (e.g., cystic hygroma and Noonan syndrome panels), for most cases they are of limited value because of the limited genes selected, the accuracy of the ultrasound phenotype, and a lack of knowledge regarding the prenatal phenotypes for many syndromes.

Next-generation sequencing such as whole exome sequencing (WES) and whole genome sequencing (WGS) allows for a more complete and higher resolution examination of the fetal DNA than karyotype or CMA and should provide more specific diagnoses. Postnatal studies to date suggest that WES can identify a single gene disorder in 25%–30% of structurally abnormal newborns unresolved by karyotype and CMA, suggesting that it is possibly valuable for prenatal evaluation [6, 7].

A recent systematic review [6] found 31 published case series and abstracts of prenatal WES and gave a diagnostic rate ranging from 6.2% to 80% (Table 26.2) [42–51]. The majority of these studies were relatively small case series with varying inclusion criteria and many were limited to cases with a high a priori suspicion for a Mendelian disease, which likely accounts for the wide diagnostic rates [42–51].

SEQUENCING TECHNOLOGY AND THE EXOME/GENOME

Although the human genome is composed of roughly three billion basepairs, only about 30 million are in coding regions and translated into functional proteins (~1%–1.5%). These essential coding basepairs are within exons that comprise an estimated 20,000 genes and the entire set of exons is called the exome. Because ~85% of Mendelian disease is caused by variants found in the exons, their analysis is most critical for clinical purposes [52]. Clinical interpretation of intronic changes is still limited, and because WES is less costly than WGS, which includes the introns, WES has been more readily used in the clinic and is the focus of this chapter.

A full description of WES techniques is beyond the scope of this chapter, but in general involves a series of steps in which DNA is prepared, exons are captured for analysis, sequencing of the exons is performed, and the results are analyzed by bioinformatic pipelines. To prepare the sample, DNA is extracted from the sample and fragmented. Various techniques are now available to select the DNA fragments of interest, which for WES are the exons. This collection of fragments referred to as the whole exome sequencing library is then sequenced. There are numerous commercially available sequencing platforms that have the ability to sequence millions of DNA fragments in parallel. Each platform has differences in depth of sequencing, sequencing run time, and reliability. The raw sequence data is then processed using base calling, read alignment (Fig. 26.4), variant calling, and variant annotation [53]. Notably, a greater depth of sequencing coverage allows for a more accurate variant call (Fig. 26.4).

WES creates an abundance of data, requiring careful curation using bioinformatics and specialized knowledge of clinical genetics. The technology identifies tens of thousands of variants, which are filtered using computational pipelines designed to identify those meeting certain preselected criteria. Some filtering factors may include the presence of the variant on a disease candidate list, the fit of the variant/s in the known inheritance model (de novo,

TABLE 26.2 Overview of Prenatal Whole Exome Case Series

First Author, Year	No. of Cases	Cohort Criteria	Method	Pathogenic Variants[a]
Aarabi, 2018	20	One or more major structural congenital anomaly detected by ultrasound	Trio	4/20 (20%)
Fu, 2017	196	Fetuses with structural abnormalities	147 Proband-only	34/147 (23.1%)
			49 Trio	13/49 (26.5%)
Lei, 2017	30	Fetuses with congenital anomalies of the kidney and urinary tract (CAKUT)[b]	23 Proband-only	3/23 (13%)
			7 Trio	1/7 (14%)
			Total	4/30 (13.3%)
Vora, 2017	15	Fetuses with multiple congenital anomalies highly suggestive of an underlying genetic disorder	Trio	7/15 (47%)
Yates, 2017	84	Fetuses with ultrasound abnormalities that resulted in fetal demise or pregnancy termination	29 Proband-only	4/29 (14%)
			45 Trios	11/45 (24%)
			6 Quads/4 Maternal Duos	2/10 (20%)
			Total	17/84 (20%)
Pangalos, 2016	14	Prenatal ultrasound abnormalities or malformations	Proband-only	6/14 (43%)
Alamillio, 2015	7	Multiple congenital anomalies on prenatal ultrasound	Trio	4/7 (57%)
Drury, 2015	24	Fetuses with an increased NT ($\geq$3.5 mm) or other ultrasound abnormality	14 Proband-only 10 Trio	5/24 (21%)
Carss, 2014	30	Structural abnormalities identified on prenatal ultrasound	Trio	3/30 (10%)
Yang, 2014	11	Terminated fetuses with anomalies	Trio	6/11 (54%)

[a] Not all series used ACMG variant interpretation guidelines; therefore pathogenicity is defined in some case series by the individual authors.
[b] Detection rate for isolated CAKUT: 2/22 (9.1%) and detection rate for CAKUT with other abnormalities: 2/8 (25%).

recessive x-linked), the mutation type (e.g., loss of function), the rarity of the variant, the presence or absence in control populations, in silico predictive scores of pathogenicity, and the assessment of a biological pathway. The use of a trio analysis in which both parents and the fetus are each sequenced and analyzed together greatly reduces the number of candidate variants because the inheritance of the variant is known. For example, if a variant is found in a proband and the disease mechanism is known as dominant, the index of suspicion is much higher if the variant is de novo than if the variant is inherited from a healthy parent. In the prenatal setting, WES trios are routinely used.

Laboratories are tasked with providing clinical geneticists and genetic counselors a list of candidate variants that is both manageable (not underfiltered) and does not exclude relevant

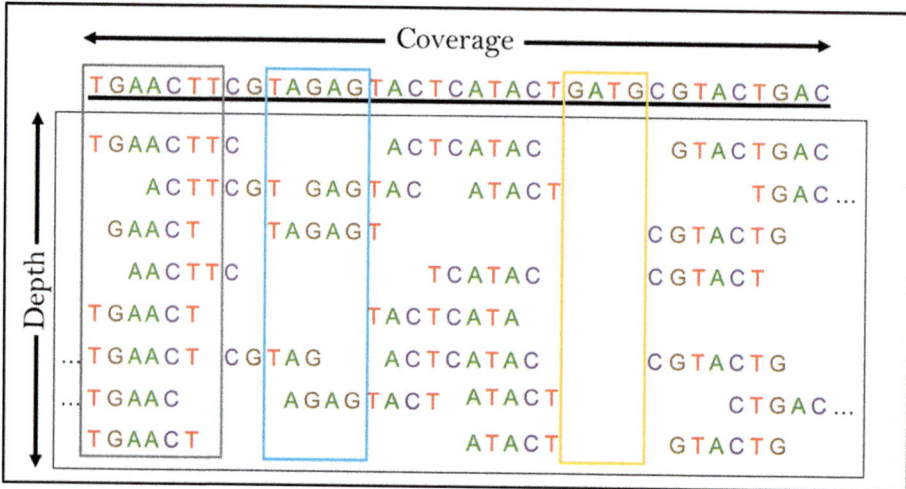

FIG. 26.4 Coverage and depth of sequencing. Individual reads are overlapped using bioinformatics tools to reconstruct the patient's genome using the reference genome as a guide. Sequence coverage refers to the extent (breadth) that sequencing reads are obtained from all representative regions of the reference genome. The depth of sequencing is also referred to as the redundancy with which reads are obtained from any given region. The greater the number of times a sequence of bases is read, the higher the depth of sequencing. Deep sequencing refers to the general concept of aiming for a high number of replicate reads of each region of a particular sequence. The reference sequence **T G A A C T T** represented in the gray box has more sequence reads (higher depth) than the reference sequence **T A G A G** represented in the blue box (lower depth). Because there are no sequence reads for the reference sequence **G A T G** represented in the yellow box, this region is said to have no coverage. The average coverage for a whole genome can be calculated from the length of the original genome, the number of reads, and the average read length (L). For example, a hypothetical genome with 2000 base pairs reconstructed from eight reads with an average length of 500 nucleotides will have 2× redundancy. If the eight reads only cover 1960 of the 2000 base pairs, then the breadth of coverage would be 98%.

causative variants (not overfiltered). Thus, the interpretation of WES requires a strong partnership between the molecular geneticists inside the laboratory and the clinical geneticist and genetic counselors. The final candidate variant list, usually composed of a limited number of variants, is usually sent to the clinical team, which determines the genotype-phenotype fit.

It is important to remember that WES has limitations because some genetic changes are not detectable by this technology. If there is a disruption in probe binding during exon enrichment, an exon may not be sequenced. WES also cannot detect epigenetic changes, nucleotide repeat expansions/contractions (e.g., Fragile X), pseudogene changes, deep intronic noncoding changes, variants in promoters, changes in nonnuclear DNA (i.e., mitochondrial), somatic mosaicisms (some detected), large CNVs found on CMA (also detectable using WGS), and balanced chromosomal changes (visible on karyotype).

CLINICAL INDICATIONS FOR WES

Pediatric guidelines recommend WES when an individual has multiple congenital birth defects or neurodevelopmental delay and no test has yielded a diagnosis. Data in the last 5 years has consistently shown about a 20%–30% diagnostic yield using these guidelines in the pediatric patient population [54, 55].

A joint statement from the International Society of Prenatal Diagnosis (ISPD), the Society of Maternal Fetal Medicine (SMFM), and the Perinatal Quality Foundation (PQF) recommended that fetal sequencing (WES/WGS) should be considered only in specific circumstances and preferably as part of a research protocol [56]. They recommended that the primary indication for fetal WES was the presence of fetal anomalies suggestive of a specific syndromic pattern or single-gene disorder [56]. They also indicated that WES/WGS could be considered for parents who have a history of an undiagnosed fetus for which no sample is available from the affected proband [56]. Although CMA is now deemed appropriate for any women choosing invasive prenatal diagnosis, the joint statement indicates that there is no evidence to support offering WES/WGS as part of routine diagnostic testing in the absence of anomalies [56].

The diagnostic yield using WES in the prenatal setting is significantly higher when multiple anomalies are present. Data is presently too limited to make clinical recommendations, but early studies suggest that involvement of specific organ systems may have a higher rate of abnormal results, with the CNS, renal, and skeletal systems having the highest yields. Whether this is related to a truly higher risk of pathogenic variants associated with these systems or simply is secondary to our present limited knowledge of fetal phenotypes involving other organs is uncertain. Additionally, the yield varies between the use of proband only or trio analysis, with a pathogenic variant more likely to be reported in trio analysis.

INTERPRETATION OF WES RESULTS

The American College of Medical Genetics and Genomics (ACMGG) and the Association for Molecular Pathology (AMP) have published a joint statement establishing standardization in the interpretation and classification of genetic variants [18]. Moreover, this recommendation describes a process for classifying variants into five categories based on criteria using typical types of variant evidence (e.g., population data, computational data, functional data, segregation data) [18]. Variants are categorized as *pathogenic, likely pathogenic, uncertain significance, likely benign,* or *benign* [18]. Most WES pipelines will filter out likely benign and benign variants, which are common in the population.

Interpretation of WES results generally requires a multidisciplinary team of clinical geneticists, genetic counselors, and molecular geneticists. For interpretation of prenatal cases, the inclusion of specialists with knowledge of prenatal imaging and dysmorphology is paramount. The accuracy of the ultrasound diagnosis and an understanding of its limitations are essential to accurate phenotype genotype correlation and variant interpretation.

One unique problem in the interpretation of prenatal WES results is the limited amount of information available to clinicians compared to that available in the postnatal period [42]. Prenatal imaging is excellent in identifying major structural anomalies, but may not reveal subtle dysmorphic features. Additionally, no information regarding neurocognitive development or epilepsy is available, which limits the clinician's ability to confidently interpret variants as having a clear genotype-phenotype match. Adding to this difficulty is that the prenatal phenotypes for many Mendelian disorders previously only described in their postnatal state are not well described. This limited data on potential prenatal phenotype expansion makes the interpretation of pathogenicity challenging and time consuming. For example, variants of

SCN2A had until recently been described as an epilepsy gene, until a few case reports described in utero CNS structural anomalies [57, 58]. At the time of this writing, this expanded phenotype remains unlisted in the Online Mendelian Inheritance in Man (OMIM).

REPORTING OF WES RESULTS

Presently, most centers take a conservative approach in reporting a finding as pathogenic. Given the limited prenatal genotype/phenotype data available, many variants found in prenatal WES do not meet the ACMGG/AMP guidelines for pathogenicity and are either not reported to patients or are reported as a variant of uncertain significance, creating an unclear path for clinical management. To augment the reporting process, clinicians monitor new publications and meeting abstracts, contact other researchers/clinicians known to be performing prenatal WES, and/or use web platforms such as Genematcher in hopes of identifying similar cases, allowing the reclassification of suspicious variants for patients [59].

Secondary and incidental findings in either the fetus or parents that are unrelated to the fetal diagnosis may also be identified through WES. Secondary findings appropriate for reporting are outlined in ACMGG guidelines and include known and likely pathogenic variants in 59 medically actionable genes [60]. These should only be reported if the patient is precounseled and opts for this information.

Predictive testing of a child is not typically recommended for adult-onset conditions, yet no specific guidance has been published regarding how to handle this in the fetal setting. This is usually reported at the discretion of the laboratory and clinical team. In addition, trio-based prenatal WES can uncover nonpaternity or parental consanguinity, which may also be unexpected to at least one of the parents.

CLINICAL IMPLICATIONS OF PRENATAL WES RESULTS

There is concern by some that the additional genetic information available by performing WES may lead to an increase in pregnancy terminations. In our experience, this has not been the case. In many cases, the certainty of a genetic diagnosis allows specific prognostic counseling as opposed to the uncertainty present when an anomaly with multiple potential outcomes is found. Once aware of the specific short- and long-term health issues facing the child, parents are better able to make an informed decision. Without this information, many parents elect pregnancy termination, fearing the worst outcomes.

Results from prenatal WES may allow for improved perinatal and neonatal management. Having a specific genetic diagnosis in a fetus with a structural anomaly can avoid the "diagnostic odyssey" that can occur when no diagnosis is known. This may save healthcare costs, reduce parental stress, and allow timely treatment when appropriate. Recent studies of acute neonatal sequencing in which results are available within 72h or less have demonstrated that in almost half the cases, this information has led to changes in management. These changes have included the initiation of new subspecialist care; additional or fewer diagnostic studies; changes in medication, diet, or major procedures; and the timely introduction of compassionate care [61, 62]. Similarly, it's anticipated that prenatal diagnosis will offer these same advantages and in addition can lead to altered fetal treatment.

Results from prenatal WES can also provide accurate recurrence risks, allowing couples to prepare for future pregnancies. If a potentially recurrent molecular diagnosis is made, the option for in vitro fertilization (IVF) with preimplantation genetic diagnosis (PGD), prenatal diagnosis via chorionic villus sampling (CVS), or amniocentesis is possible. If a de novo variant is found, couples can feel more reassured with conceiving a future unaffected pregnancy given the low recurrence risk (usually ~1% due to germline mosaicism).

PRE- AND POSTTEST COUNSELING FOR WES

The availability of pre- and posttest counseling, ideally by a genetic specialist, is integral to offering WES. The pretest counseling should include the purpose of the test (specifically, to identify a cause for the known phenotype), the limitations of negative results, and the possibility for a variant of uncertain significance. Additionally, preferences for reporting secondary and incidental findings should be discussed and documented. As with CMA, mentioning the potential for the test to reveal nonpaternity or relatedness is also recommended [33].

Given the rapidly expanding knowledge of variants and diseases, it is also essential to emphasize that the interpretation of the results may evolve with time and that negative and uncertain results will be reviewed on a regular basis with new findings reported to the patient. A long-term partnership between the patient and the clinician is crucial. The clinician should explain how a patient may be recontacted with new information and how a patient may recontact the provider if new information about the family history becomes available.

Posttest counseling depends on the specific results and context of the indication for testing. If a pathogenic variant related to the phenotype or a secondary/incidental finding is found, the risks to future pregnancies and family members must be described and appropriate referrals made. Prognosis and risk factors of the pathogenic variant should be described. If a variant of uncertain significance is found, the provider should explain the reasons for the uncertainty in the context of the indication, describe the planned methods to further elucidate pathogenicity, and ensure that there is a reliable way to recontact the family should interpretation change with time. If no reportable variant is found, the clinician should discuss the limitations of the technology, consider whethere there are alternative molecular methods that may yield a diagnosis given the specific indication, and review the possibility for reanalysis or new technologies in the future.

FUTURE DIRECTIONS

The use of whole exome sequencing for prenatal diagnosis is rapidly evolving and expanding. The experience to date has been limited to structural anomalies and more research is needed to standardize the pre- and posttest counseling, turnaround times, and variant interpretation. Although the cost for sequencing is decreasing, it remains substantial not only because of the technical complexity but also due to the time-consuming research done by a clinician.

A centralized database is needed to track and share the expanding fetal phenotypes and genotype-phenotype correlations, which may also improve turnaround times for variant interpretation. Further studies are also needed to determine the added benefits of an early molecular diagnosis in patient management and to quantify the patient experience. Insurance coverage for WES remains limited; however, it may expand if the benefits of prenatal WES become clear through research.

Given decreasing sequencing costs, testing will likely expand to whole genome sequencing in the near future, which will generate more data. This will initially make interpretation of results more complex, but is likely to ultimately identify previously unsuspected causes of birth defects.

CONCLUSION

Compared to the information offered by a karyotype, newer molecular techniques such as CMA and WES provide significantly greater analysis of the fetal genetic status. This additional information has led to an improved understanding of the etiology of many fetal structural anomalies and has expanded our understanding of the cause of many childhood disorders, including some cases of autism, learning disabilities, and dysmorphic phenotypes. At present, WES generates a significant amount of data that lacks sufficient evidence-based studies to facilitate appropriate diagnostic interpretation. As such, it is imperative that we use the experience gained from CMA to guide the implementation of WES, especially as it becomes more commonplace in the prenatal arena. At the onset of prenatal CMA, the nuances of a novel prenatal test quickly become apparent with the onslaught of variants of unknown clinical significance, ambiguous results, incidental findings, identification of nonpaternity, consanguinity and adult-onset conditions. These issues have now been discussed for more than 10 years and the lessons learned will certainly benefit the implementation of prenatal WES as similar hurdles are encountered.

As analysis of cell-free fetal DNA or the capture of fetal cells from the maternal circulation continues to improve, identification of fetal copy number variants or single base pair changes from a maternal blood sample will soon become feasible [63]. Additionally, there may come a time when WES/WGS is considered appropriate for the evaluation of nonanomalous pregnancies, allowing the genotype to be fully evaluated prior to the expression of the phenotype. Although this is already technically possible, intense discussion and investigation of its impact and ethics are required before it is routinely used [64].

References

[1] NICHD. Midtrimester amniocentesis for prenatal diagnosis. Safety and accuracy. JAMA 1976;236(13):1471–6.
[2] Akolekar R, Beta J, Picciarelli G, Ogilvie C, D'Antonio F. Procedure-related risk of miscarriage following amniocentesis and chorionic villus sampling: a systematic review and meta-analysis. Ultrasound Obstet Gynecol 2015;45(1):16–26.

[3] Wapner RJ, Martin CL, Levy B, Ballif BC, Eng CM, Zachary JM, et al. Chromosomal microarray versus karyotyping for prenatal diagnosis. N Engl J Med 2012;367(23):2175–84.

[4] Miller DT, Adam MP, Aradhya S, Biesecker LG, Brothman AR, Carter NP, et al. Consensus statement: chromosomal microarray is a first-tier clinical diagnostic test for individuals with developmental disabilities or congenital anomalies. Am J Hum Genet 2010;86(5):749–64.

[5] American College of O, Gynecologists Committee on G. Committee opinion No. 581: the use of chromosomal microarray analysis in prenatal diagnosis. Obstet Gynecol 2013;122(6):1374–7.

[6] Best S, Wou K, Vora N, Van der Veyver IB, Wapner R, Chitty LS. Promises, pitfalls and practicalities of prenatal whole exome sequencing. Prenat Diagn 2017.

[7] Yang Y, Muzny DM, Reid JG, Bainbridge MN, Willis A, Ward PA, et al. Clinical whole-exome sequencing for the diagnosis of mendelian disorders. N Engl J Med 2013;369(16):1502–11.

[8] Iafrate AJ, Feuk L, Rivera MN, Listewnik ML, Donahoe PK, Qi Y, et al. Detection of large-scale variation in the human genome. Nat Genet 2004;36(9):949–51.

[9] Redon R, Ishikawa S, Fitch KR, Feuk L, Perry GH, Andrews TD, et al. Global variation in copy number in the human genome. Nature 2006;444(7118):444–54.

[10] Sebat J, Lakshmi B, Troge J, Alexander J, Young J, Lundin P, et al. Large-scale copy number polymorphism in the human genome. Science 2004;305(5683):525–8.

[11] Sharp AJ, Locke DP, McGrath SD, Cheng Z, Bailey JA, Vallente RU, et al. Segmental duplications and copy-number variation in the human genome. Am J Hum Genet 2005;77(1):78–88.

[12] Tuzun E, Sharp AJ, Bailey JA, Kaul R, Morrison VA, Pertz LM, et al. Fine-scale structural variation of the human genome. Nat Genet 2005;37(7):727–32.

[13] Cooper GM, Coe BP, Girirajan S, Rosenfeld JA, Vu TH, Baker C, et al. A copy number variation morbidity map of developmental delay. Nat Genet 2011;43(9):838–46.

[14] Malcolm S. Microdeletion and microduplication syndromes. Prenat Diagn 1996;16(13):1213–9.

[15] Nevado J, Mergener R, Palomares-Bralo M, Souza KR, Vallespin E, Mena R, et al. New microdeletion and microduplication syndromes: a comprehensive review. Genet Mol Biol 2014;37(1 Suppl):210–9.

[16] Levy B, Wapner R. Prenatal diagnosis by chromosomal microarray analysis. Fertil Steril 2018;109(2):201–12.

[17] Alkuraya FS. Discovery of rare homozygous mutations from studies of consanguineous pedigrees. Curr Protoc Hum Genet 2012. Chapter 6:Unit 6 12.

[18] Richards S, Aziz N, Bale S, Bick D, Das S, Gastier-Foster J, et al. Standards and guidelines for the interpretation of sequence variants: a joint consensus recommendation of the American College of Medical Genetics and Genomics and the Association for Molecular Pathology. Genet Med 2015;17(5):405–24.

[19] Deak KL, Horn SR, Rehder CW. The evolving picture of microdeletion/microduplication syndromes in the age of microarray analysis: variable expressivity and genomic complexity. Clin Lab Med 2011;31(4):543–64 [viii].

[20] Girirajan S, Rosenfeld JA, Coe BP, Parikh S, Friedman N, Goldstein A, et al. Phenotypic heterogeneity of genomic disorders and rare copy-number variants. N Engl J Med 2012;367(14):1321–31.

[21] Veltman JA, Brunner HG. Understanding variable expressivity in microdeletion syndromes. Nat Genet 2010;42(3):192–3.

[22] ACOG Practice Bulletin No. 77. Screening for fetal chromosomal abnormalities. Obstet Gynecol 2007;109(1):217–27.

[23] American College of OG. ACOG Practice Bulletin No. 88, December 2007. Invasive prenatal testing for aneuploidy. Obstet Gynecol 2007;110(6):1459–67.

[24] Society for Maternal-Fetal Medicine, Electronic address pso, Dugoff L, Norton ME, Kuller JA. The use of chromosomal microarray for prenatal diagnosis. Am J Obstet Gynecol 2016;215(4):B2–9.

[25] Fan YS, Jayakar P, Zhu H, Barbouth D, Sacharow S, Morales A, et al. Detection of pathogenic gene copy number variations in patients with mental retardation by genomewide oligonucleotide array comparative genomic hybridization. Hum Mutat 2007;28(11):1124–32.

[26] Callaway JL, Shaffer LG, Chitty LS, Rosenfeld JA, Crolla JA. The clinical utility of microarray technologies applied to prenatal cytogenetics in the presence of a normal conventional karyotype: a review of the literature. Prenat Diagn 2013;33(12):1119–23.

[27] Scott F, Murphy K, Carey L, Greville W, Mansfield N, Barahona P, et al. Prenatal diagnosis using combined quantitative fluorescent polymerase chain reaction and array comparative genomic hybridization analysis as a first-line test: results from over 1000 consecutive cases. Ultrasound Obstet Gynecol 2013;41(5):500–7.

[28] Shaffer LG, Dabell MP, Fisher AJ, Coppinger J, Bandholz AM, Ellison JW, et al. Experience with microarray-based comparative genomic hybridization for prenatal diagnosis in over 5000 pregnancies. Prenat Diagn 2012;32(10):976–85.

[29] Van Opstal D, de Vries F, Govaerts L, Boter M, Lont D, van Veen S, et al. Benefits and burdens of using a SNP array in pregnancies at increased risk for the common aneuploidies. Hum Mutat 2015;36(3):319–26.

[30] Wapner RJ, Zachary J, Clifton R. Change in classification of prenatal microarray anaylsis copy number variants over time. Prenat Diagn 2015;35(Suppl. S1):Suppl. S1.

[31] Hillman SC, McMullan DJ, Hall G, Togneri FS, James N, Maher EJ, et al. Use of prenatal chromosomal microarray: prospective cohort study and systematic review and meta-analysis. Ultrasound Obstet Gynecol 2013; 41(6):610–20.

[32] McDonald-McGinn DM, Sullivan KE. Chromosome 22q11.2 deletion syndrome (DiGeorge syndrome/velocardiofacial syndrome). Medicine (Baltimore) 2011;90(1):1–18.

[33] Committee Opinion No.682. Microarrays and next-generation sequencing technology: the use of advanced genetic diagnostic tools in obstetrics and gynecology. Obstet Gynecol 2016;128(6):e262–8.

[34] Srebniak MI, Diderich KE, Joosten M, Govaerts LC, Knijnenburg J, de Vries FA, et al. Prenatal SNP array testing in 1000 fetuses with ultrasound anomalies: causative, unexpected and susceptibility CNVs. Eur J Hum Genet 2016;24(5):645–51.

[35] Faas BH, van der Burgt I, Kooper AJ, Pfundt R, Hehir-Kwa JY, Smits AP, et al. Identification of clinically significant, submicroscopic chromosome alterations and UPD in fetuses with ultrasound anomalies using genome-wide 250k SNP array analysis. J Med Genet 2010;47(9):586–94.

[36] Kleeman L, Bianchi DW, Shaffer LG, Rorem E, Cowan J, Craigo SD, et al. Use of array comparative genomic hybridization for prenatal diagnosis of fetuses with sonographic anomalies and normal metaphase karyotype. Prenat Diagn 2009;29(13):1213–7.

[37] Shaffer LG, Coppinger J, Alliman S, Torchia BA, Theisen A, Ballif BC, et al. Comparison of microarray-based detection rates for cytogenetic abnormalities in prenatal and neonatal specimens. Prenat Diagn 2008; 28(9):789–95.

[38] Shaffer LG, Rosenfeld JA, Dabell MP, Coppinger J, Bandholz AM, Ellison JW, et al. Detection rates of clinically significant genomic alterations by microarray analysis for specific anomalies detected by ultrasound. Prenat Diagn 2012;32(10):986–95.

[39] Donnelly JC, Platt LD, Rebarber A, Zachary J, Grobman WA, Wapner RJ. Association of copy number variants with specific ultrasonographically detected fetal anomalies. Obstet Gynecol 2014;124(1):83–90.

[40] South ST, Lee C, Lamb AN, Higgins AW, Kearney HM, Working Group for the American College of Medical G, et al. ACMG standards and guidelines for constitutional cytogenomic microarray analysis, including postnatal and prenatal applications: revision 2013. Genet Med 2013;15(11):901–9.

[41] Reddy UM, Page GP, Saade GR, Silver RM, Thorsten VR, Parker CB, et al. Karyotype versus microarray testing for genetic abnormalities after stillbirth. N Engl J Med 2012;367(23):2185–93.

[42] Aarabi M, Sniezek O, Jiang H, Saller DN, Bellissimo D, Yatsenko SA, et al. Importance of complete phenotyping in prenatal whole exome sequencing. Hum Genet 2018;137(2):175–81.

[43] Alamillo CL, Powis Z, Farwell K, Shahmirzadi L, Weltmer EC, Turocy J, et al. Exome sequencing positively identified relevant alterations in more than half of cases with an indication of prenatal ultrasound anomalies. Prenat Diagn 2015;35(11):1073–8.

[44] Carss KJ, Hillman SC, Parthiban V, McMullan DJ, Maher ER, Kilby MD, et al. Exome sequencing improves genetic diagnosis of structural fetal abnormalities revealed by ultrasound. Hum Mol Genet 2014;23(12):3269–77.

[45] Drury S, Williams H, Trump N, Boustred C, Gosgene, Lench N, et al. Exome sequencing for prenatal diagnosis of fetuses with sonographic abnormalities. Prenat Diagn 2015;35(10):1010–7.

[46] Fu F, Li R, Li Y, Nie ZQ, Lei TY, Wang D, et al. Whole exome sequencing as a diagnostic adjunct to clinical testing in a tertiary referral cohort of 3988 fetuses with structural abnormalities. Ultrasound Obstet Gynecol 2017.

[47] Lei TY, Fu F, Li R, Wang D, Wang RY, Jing XY, et al. Whole-exome sequencing for prenatal diagnosis of fetuses with congenital anomalies of the kidney and urinary tract. Nephrol Dial Transplant 2017;32(10):1665–75.

[48] Pangalos C, Hagnefelt B, Lilakos K, Konialis C. First applications of a targeted exome sequencing approach in fetuses with ultrasound abnormalities reveals an important fraction of cases with associated gene defects. PeerJ 2016;4:e1955.

[49] Vora NL, Powell B, Brandt A, Strande N, Hardisty E, Gilmore K, et al. Prenatal exome sequencing in anomalous fetuses: new opportunities and challenges. Genet Med 2017;19(11):1207–16.

[50] Yang Y, Muzny DM, Xia F, Niu Z, Person R, Ding Y, et al. Molecular findings among patients referred for clinical whole-exome sequencing. JAMA 2014;312(18):1870–9.

[51] Yates CL, Monaghan KG, Copenheaver D, Retterer K, Scuffins J, Kucera CR, et al. Whole-exome sequencing on deceased fetuses with ultrasound anomalies: expanding our knowledge of genetic disease during fetal development. Genet Med 2017;19(10):1171–8.

[52] Botstein D, Risch N. Discovering genotypes underlying human phenotypes: past successes for mendelian disease, future approaches for complex disease. Nat Genet 2003;33(Suppl):228–37.

[53] Rehm HL, Bale SJ, Bayrak-Toydemir P, Berg JS, Brown KK, Deignan JL, et al. ACMG clinical laboratory standards for next-generation sequencing. Genet Med 2013;15(9):733–47.

[54] Gahl WA, Markello TC, Toro C, Fajardo KF, Sincan M, Gill F, et al. The National Institutes of Health undiagnosed diseases program: insights into rare diseases. Genet Med 2012;14(1):51–9.

[55] Lazaridis KN, Schahl KA, Cousin MA, Babovic-Vuksanovic D, Riegert-Johnson DL, Gavrilova RH, et al. Outcome of whole exome sequencing for diagnostic odyssey cases of an individualized medicine clinic. The Mayo clinic experience. Mayo Clin Proc 2016;91(3):297–307.

[56] Henson M. Joint position statement from the International Society of Prenatal Diagnosis (ISPD), the Society of Maternal Fetal Medicine (SMFM) and the perinatal quality foundation (PQF) on the use of genome-wide sequencing for fetal diagnosis. Prenat Diagn 2018.

[57] Bernardo S, Marchionni E, Prudente S, De Liso P, Spalice A, Giancotti A, et al. Unusual association of SCN2A epileptic encephalopathy with severe cortical dysplasia detected by prenatal MRI. Eur J Paediatr Neurol 2017; 21(3):587–90.

[58] Trump N, McTague A, Brittain H, Papandreou A, Meyer E, Ngoh A, et al. Improving diagnosis and broadening the phenotypes in early-onset seizure and severe developmental delay disorders through gene panel analysis. J Med Genet 2016;53(5):310–7.

[59] Philippakis AA, Azzariti DR, Beltran S, Brookes AJ, Brownstein CA, Brudno M, et al. The matchmaker exchange: a platform for rare disease gene discovery. Hum Mutat 2015;36(10):915–21.

[60] Kalia SS, Adelman K, Bale SJ, Chung WK, Eng C, Evans JP, et al. Recommendations for reporting of secondary findings in clinical exome and genome sequencing, 2016 update (ACMG SF v2.0): a policy statement of the American College of Medical Genetics and Genomics. Genet Med 2017;19(2):249–55.

[61] Meng L, Pammi M, Saronwala A, Magoulas P, Ghazi AR, Vetrini F, et al. Use of exome sequencing for infants in intensive care units: ascertainment of severe single-gene disorders and effect on medical management. JAMA Pediatr 2017;171(12):e173438.

[62] Petrikin JE, Willig LK, Smith LD, Kingsmore SF. Rapid whole genome sequencing and precision neonatology. Semin Perinatol 2015;39(8):623–31.

[63] Hayward J, Chitty LS. Beyond screening for chromosomal abnormalities: advances in non-invasive diagnosis of single gene disorders and fetal exome sequencing. Semin Fetal Neonatal Med 2018.

[64] Horn R, Parker M. Opening Pandora's box? Ethical issues in prenatal whole genome and exome sequencing. Prenat Diagn 2017.

Noninvasive Prenatal Testing for Genetic Diseases

Jason C.H. Tsang*,†, *Y.M. Dennis Lo**,†

*Li Ka Shing Institute of Health Sciences, The Chinese University of Hong Kong, Shatin, Hong Kong, China †Department of Chemical Pathology, The Chinese University of Hong Kong, Shatin, Hong Kong, China

OUTLINE

Human Reproductive and Prenatal Genetics
https://doi.org/10.1016/B978-0-12-813570-9.00027-9

Abbreviations

ATAC-seq	assay for transposase-accessible chromatin using sequencing
CAH	congenital adrenal hyperplasia
ChIP-seq	chromatin immunoprecipitation sequencing
CVS	chorionic villous sampling
DHS-seq	DNaseI hypersensitivity site-sequencing
dsDNA	double-stranded DNA
EDTA	ethylenediaminetetraacetic acid
NIPT	noninvasive prenatal testing
PCR	polymerase chain reaction
RHDO	relative haplotype dosage analysis
RMD	relative mutation dosage analysis
SLE	systemic lupus erythematosis
SNP	single nucleotide polymorphism
ssDNA	single-stranded DNA

INTRODUCTION

Prenatal diagnosis aims to detect any functional or structural abnormalities of the developing fetus as early as possible during pregnancy so as to inform clinical decisions, such as the choice of delivery methods, the need for antenatal treatment, the risk to the mother, and the long-term disability of the child. Chromosomal aneuploidy is one of the most common causes of birth defects, with incidences in newborns at 0.3% [1]. As of 2017, there are more than 6,000 clinical phenotypes with a known genetic basis, according to the Online Mendelian Inheritance in Man (OMIM) database. The incidence of genetic birth defects is expected to increase with the growing use of sequencing technology to unravel many previously undiagnosed diseases. Moreover, as more and more families decide to have children later in life, the risk of birth defects accompanied with advanced parental age will become more prominent. All these factors contribute to the increasing demands for accurate and safe technologies for prenatal genetic screening.

INVASIVE AND NONINVASIVE APPROACHES OF PRENATAL TESTING OF GENETIC DISEASES

Fetal Tissue Sampling

Direct interrogation of the genetic constitution of the fetus via fetal tissue sampling is the most intuitive approach to achieve an accurate diagnosis of genetic disease. In pregnancy conceived through in vitro fertilization, cells can be directly biopsied from the embryos generated in vitro for preimplantation genetic diagnosis. Fetal tissue sampling is, however, not straightforward due to the protected in utero location of the fetus in a natural pregnancy. A number of invasive procedures, such as chorionic villous sampling (CVS), amniocentesis, and cordocentesis, have been developed in past decades for obtaining small amounts of chorionic villous tissue from the developing placenta as well as fetal epithelial cells found in the amniotic fluid and fetal blood, respectively. CVS is commonly performed in the first trimester (10–14 weeks) while amniocentesis and cordocentesis are performed in the second trimester (16–22 weeks and 19 weeks later, respectively). in vitro tissue culture can be performed to expand the cellular materials for testing. Fetal tissue sampling provides high-quality genetic materials that are amenable to many established genetic testing techniques such as karyotyping, genomic sequencing, haplotyping, or even functional testing such as enzymatic activity and biochemical profiling. In fetal aneuploidy testing, fetal cell karyotyping is currently the definitive diagnostic standard. Despite this, fetal tissue sampling is not suitable for population-wide prenatal screening because of the risk of its inherent invasiveness. CVS and amniocentesis can result in procedure-related miscarriages in 0.1%–1% of pregnancies, depending on operator experience [2, 3]. These procedures, therefore, are reserved for high-risk pregnancy requiring confirmatory testing.

Biochemical Markers and Ultrasonography

To circumvent the safety issues of direct fetal tissue sampling, researchers have developed indirect approaches by identifying the biochemical or structural epiphenomena associated with specific genetic diseases via maternal blood testing or fetal ultrasound scan. One of the most well-established areas is the use of maternal serum proteins/hormones and fetal ultrasonographic features in fetal aneuploidy screening. Increased accumulation of subcutaneous fluid in the dorsal part of the fetal neck, measured as increased nuchal translucency on an ultrasound scan, is associated with fetal aneuploidy [4]. Aberrant levels of maternal biochemical markers, such as alpha-fetoprotein, free human chorionic gonadotropin hormone, dimeric inhibin-A, unconjugated estriol, and pregnancy-associated plasma protein A, are also found in aneuploid pregnancies. Different screening strategies have been developed to combine the values of specific ultrasound features, maternal demographic risk factors, and biochemical markers to stratify high-risk women for definitive diagnosis by fetal tissue sampling [5]. Depending on the screening strategy and the combination of parameters, the detection rate of trisomy 21 varies from 64% to 96% [5]. However, specific ultrasound and maternal biochemical markers are not available in many other genetic conditions.

Circulating Fetal Cells in Maternal Circulation

The goal to develop a robust noninvasive method to directly access fetal genetic information appeared improbable in the conventional perception that the fetus, as a foreign allograft to the mother, must reside in a contained in utero environment regulated by the uteroplacental barrier to prevent rejection in pregnancy. Nonetheless, the unusual finding of nucleated fetal cells in the maternal system has inspired early researchers. A German pathologist, Georg Schmorl, first reported the finding of multinucleated trophoblast-like cells in the lungs of pregnant women who died of severe eclampsia in 1893 [6]. The phenomenon was initially interpreted as an exception occurring only in abnormal pregnancy. With the advancement of cell sorting and polymerase chain reaction (PCR) technology, later studies demonstrated that not only fetal trophoblasts but also fetal nucleated erythrocytes and fetal lymphocytes can be detected in the maternal circulation in early uncomplicated pregnancy [7–10]. These observations not only fueled the discussion of the biological consequences of fetomaternal cell trafficking, but also provided an ideal source of fetal cellular materials for noninvasive prenatal genetic testing [11–15]. Yet widespread clinical application of circulating nucleated fetal cells as a means for noninvasive genetic testing remains difficult even now because of the issues of cellular rarity and stability as well as the potential of cellular persistence. The concentration of circulating fetal cells in maternal blood has been estimated by fluorescent in situ hybridization to be 2–6 cells per milliliter in early second-trimester pregnancy [16]. The yield varies widely depending on the method for isolation, such as enrichment methods targeting fetal cell-specific antigens, male fetus-specific Y chromosomes, and cell size differences; it also often requires the analysis of 30 ml of maternal blood [17]. Despite earlier difficulty in fetal cell isolation [18], recent technological advances in single-cell whole genome amplification and the rare cell isolation method have significantly improved the recovery and genome-wide copy number study of circulating fetal cells from maternal blood [19, 20]. The report of the long-term persistence of fetal progenitor cells in the circulation of multiparous women is also a concern in its application [21]. In view of these, there is growing interest in obtaining fetal extravillous trophoblasts through an alternative transcervical route to reduce the background contamination from maternal blood cells. Initial results are promising with successful cell isolation at 5 weeks of gestation for aneuploidy testing [22].

Circulating Cell-Free Fetal Nucleic Acids in Maternal Circulation

An unexpected breakthrough occurred in the intense search for improvement of fetal cell-based technology in 1997 when Lo and colleagues discovered a significantly higher amount of male fetus-specific Y chromosome DNA in the acellular plasma and serum than in maternal whole blood [23]. With the rarity of circulating fetal cells in the maternal circulation, it therefore came as a surprise that fetal DNA can be readily detected in such cell-free fluids.

In fact, the recognition of the presence of cell-free nucleic acids in human serum can be traced to the early report by Mandel and Metais in 1948 [24]. Quantitative abnormalities of circulating cell-free DNA were later reported in a variety of pathological conditions, such as rheumatoid arthritis, systemic lupus erythematosus (SLE), pulmonary embolism, and cancer [25–32]. It is interesting that the fetus, in many aspects resembling a "controlled neoplastic growth" of the mother, would exhibit a similar property [23]. Since the first report on cell-free fetal DNA in 1997, it has emerged as a robust platform for noninvasive prenatal diagnosis of fetal chromosomal aneuploidy, monogenic diseases, and other complex gestational conditions. In the following sections, we will discuss our current understanding of the biological properties and clinical applications of cell-free fetal DNA.

THE BIOLOGICAL CHARACTERISTICS OF CELL-FREE FETAL DNA

Cell-Free Fetal DNA Can be Detected Early and Increases Throughout Pregnancy

Fetal-derived Y-chromosome DNA can be detected in the maternal plasma as early as 18 days after embryo transfer in assisted reproductive pregnancy; it becomes consistently present at 7 weeks of gestation [33, 34]. Longitudinal quantitative monitoring showed that the amount of fetal-derived DNA increased continuously throughout gestation and peaked before delivery [34–36]. Using duplex digital PCR technology targeting homologous ZFX and ZFY genes at the respective X and Y chromosomes, Lun et al. showed that the concentration of fetal DNA increases from 10% at the first trimester to 30% at the third trimester [35]. Importantly, cell-free fetal DNA is rapidly cleared from the maternal circulation after delivery, with an average half-life of 16 minutes [36] and complete clearance at 2 days postpartum [37]. Yu et al. showed that the clearance follows a biphasic kinetics, likely representing an early rapid clearance of the fetoplacental component and a late slow clearance of the residual placentodecidual and circulating fetal cellular component [37]. Rapid clearance ensures that the fetal genotype determined by cell-free fetal DNA analysis is devoid of interference from a previous pregnancy and the change of cell-free fetal DNA will be reflective of the changes at the fetal source in real time.

Nevertheless, the biological mechanism of cell-free fetal DNA clearance is not well understood. There are at least three potential contributing mechanisms: direct degradation by the extracellular nuclease system, passive filtration by the renal system, and active uptake by the reticuloendothelial system. Previous animal studies have showed that exogenous purified DNA is highly unstable in plasma and DNA endonuclease I (DNaseI) is abundant in plasma and serum [38–40]. Evidence from an incubation experiment of nonanticoagulated plasma, however, suggested that endogenous extracellular nuclease activity cannot fully account for the rapid clearance of cell-free fetal DNA [34]. There is also no significant degradative change in the size distribution of cell-free fetal DNA in postpartum monitoring [37]. Alternatively, cell-free fetal DNA molecules can be filtered at the glomeruli and cleared in urine. Indeed, cell-free fetal DNA is detectable in maternal urine, although it only accounted for a minority (0.2%–20%) of its plasma counterpart [37, 41]. The role of cell-mediated clearance in cell-free fetal DNA is less explored. However, it is well described in studies of innate immunity that tissue-resident macrophages and antigen-presenting cells are capable of sensing and taking up extracellular DNA derived from pathogens or apoptotic cells via DNA-binding toll-like receptors [40, 42].

Cell-Free Fetal DNA is Placental-Derived and Cell-Free Maternal DNA is Hematopoietic in Origin

The high concentration and rapid clearance of cell-free fetal DNA in maternal plasma suggested that it is continuously released in large amounts. One estimation suggested that 20,000 fetal genomes are released into maternal circulation per minute [34]. Candidate sources include direct release from circulating fetal cells such as fetal erythroblasts, transplacental trafficking from the fetal proper, or direct release from the placenta. Studies have found no correlations of circulating fetal erythrocytes, amniotic cell-free DNA, and maternal plasma cell-free fetal DNA level [43, 44]. Furthermore, the level of cell-free fetal DNA seems to be unaffected in rare anembryonic pregnancy, where the fetal proper fails to develop [45]. The placenta, which has a high surface area in direct contact with the maternal circulation, appears to be the most plausible origin of cell-free fetal DNA.

To ascertain the tissue origin of a cell-free DNA molecule, one can also take advantage of the genetic chimerism in organ transplant recipients, where the cell-free DNA released from the transplanted organ, if any, will be genetically marked by the donor-recipient polymorphisms. Lui et al. demonstrated "sex reversal" of the cell-free DNA profile of a female bone marrow recipient from the male donor, suggesting that the major source of background cell-free DNA in normal individuals is hematopoietic in origin [46]. Similarly, the liver, kidney, heart, and lung also contribute to plasma cell-free DNA in transplant recipients [47, 48]. One can also assess the tissue-specific epigenetic modifications on a DNA molecule, such as methylation of cytosine, and infer their tissue origin. Methylation of cytosine is an important epigenetic mechanism in the regulation of gene expression and commonly occurs in the genetic "island of CpG dinucleotides" [49]. The pattern of DNA methylation in the genome bears strong tissue specificity. Placenta-specific hypomethylated *SERPINB5* (coding for maspin) and hypermethylated *RASSF1A* DNA are present in maternal plasma [50, 51]. Erythroid tissue-specific hypomethylated DNA is also detectable in the plasma of normal individuals [52]. Extending this approach to the whole genome level, Sun et al. estimated that placental contribution to maternal plasma cell-free DNA rises from 10% at the first trimester to 30% at the third trimester, consistent with the early fetal DNA fraction estimated by Y-chromosome DNA sequences [53]. Current evidence therefore suggests that the majority of cell-free fetal DNA molecules are placenta-derived and maternal DNA is hematopoietic.

Cell-Free Fetal DNA is Nonrandomly Fragmented and Shorter Than Cell-Free Maternal DNA

Unlike cellular DNA existing in high molecular weight form, cell-free fetal DNA is highly fragmented. Evidence from variable amplicon-length PCR and pair-ended massive parallel sequencing showed that the majority of cell-free DNA is shorter than 180 base pairs [54, 55]. Interestingly, the size distribution is skewed, with a long tail at the small size range punctuated by periodic spikes at 10 base pair intervals and a peak at 166 base pairs [55]. This characteristic distribution corresponds to our understanding of DNA packaging, where the DNA molecule wraps 1.65 turns around the octameric nucleosome core particle spanning 146 base pairs, then completes the final turn on H1 histone proteins [56]. The periodic size ladder-like distribution resembles the classical DNA fragmentation pattern caused by caspase-dependent nucleases in cellular apoptosis [57].

Deeper dissection of this "nucleosomic" distribution revealed an intriguing shortening of cell-free fetal DNA [55, 58]. Lo and colleagues differentiated the fetomaternal origin of individual cell-free DNA molecules by detecting the presence of fetal-specific single nucleotide polymorphisms (SNPs), that is, SNPs that are fetal heterozygous and maternal homozygous. They found that the cell-free fetal DNA distribution peaks at 144 base pairs, 20 base pairs shorter than maternal DNA.

In spite of the fundamental significance of fetal DNA shortening, the biological mechanism underlying this observation is still elusive. Based on the observed 20-base pair linker DNA deficit and the usual DNA length spanning the H1 histone protein, one theory states that the shortening may be due to the loss of H1 histone protection in cell-free fetal DNA, thus increasing susceptibility to extracellular nuclease attack. Multiple H1 isoforms exists in humans, but their role in cell-free fetal DNA biology has not been studied [55]. An alternative theory states that the shortening of the length of the linker DNA in cell-free fetal DNA may be related to its "circulatory time" in the maternal circulation. Hematopoietic tissues are in constant contact with plasma while only a fixed volume of plasma is exposed to placenta-released cell-free fetal DNA per circulation cycle. One piece of supporting evidence includes the observation that similar shortening is also present in cell-free DNA derived from nonhematopoietic solid organ sources, such as donor liver-derived cell-free DNA in a transplant recipient and tumor-derived cell-free DNA in cancer patients [59, 60]. These two theories may focus on factors at the prerelease or postrelease stages of cell-free fetal DNA, and are not necessarily mutually exclusive.

The advance of pair-ended massive parallel sequencing technology not only allows accurate measurement of the fragment length of cell-free DNA, but also identifies the precise locations of the genomic origins of these DNA fragments. Molecular techniques, such as DNaseI hypersensitivity site-sequencing (DHS-seq) and assay for transposase-accessible chromatin using sequencing (ATAC-seq), have demonstrated that nucleosome positioning in the genome confers tissue-specific susceptibility to genomic fragmentation [61]. One important line of investigation would be to determine if cell-free fetal and maternal DNA fragments exhibit differential preference of fragmentation sites in the genome. Researchers have recently shown that the fragmentation patterns in cell-free DNA are not random and are associated with tissue-specific chromatin susceptibility and the expression levels of the genomic loci at the cellular level [62–66]. To remove noise from DNA amplification and provide robust statistical inference, Chan et al. performed ultradeep (200-fold genomic coverage) DNA sequencing on non-PCR amplified maternal plasma cell-free DNA and identified recurrent fetal and maternal preferred DNA fragmentation sites at single-base resolution [66].

Cell-Free Fetal DNA is Associated With Nucleosomes and Potentially Other Plasma Proteins

It is apparent that many of the fundamental properties of size and ends of cell-free fetal DNA are linked to nucleosome biology. Notably, cell-free DNA derived from mitochondria, cellular organelles with a circular genome devoid of nucleosome association, are significantly more fragmented than chromosome-derived cell-free DNA and do not exhibit the nucleosomic size distribution pattern [55, 59, 67]. A similar observation is also made in a DNA fragmentation study of cell-free urine DNA [68]. Considering the relatively protein-free environment in urine and the unexpected stability of cell-free DNA compared to purified DNA in plasma, it is postulated that certain DNA stabilizing agents exist in the plasma. Holdenrieder and colleagues demonstrated that the level of nucleosome-associated cell-free DNA correlates with cell-free DNA in the plasma [69]. Besides nucleosomes, cell-free DNA has also been reported to associate with a number of other circulating plasma proteins, such as lipoproteins and, most classically, the anti-dsDNA antibodies in SLE patients [70]. Chan et al. observed a bimodal size distribution pattern of plasma cell-free DNA in active SLE patients and suggested that the extra-short DNA peak is a result of impaired clearance of DNA-anti-dsDNA antibody complexes [71]. A systematic study of cell-free DNA-plasma protein interactome is lacking, especially in the context of pregnancy.

Another unexplored dimension of cell-free fetal DNA is the biochemical configuration of the molecules. Early studies have suggested the presence of ssDNA and dsDNA in serum using immunoassays [26]. There is a renewed interest in studying plasma cell-free DNA after ssDNA denaturation. A number of studies have reported substantial enrichment of short DNA fragments using ssDNA-based protocols [62, 67, 72–75]. However, the initial expectation of preferential enrichment of short DNA-predominant fetal DNA and cancer DNA was not realized, suggesting the potential existence of source-specific strandness preponderance in cell-free DNA [73, 74]. Because of the lack of a dsDNA or ssDNA cell-free DNA selection process in these studies, the relative abundance and biological significance of single-stranded and double-stranded cell-free DNA in plasma is still unknown.

Passive and Active Release of Cell-Free Fetal DNA

The description of the "nucleosomic-ladder" fragmentation pattern of cell-free DNA reinforced the idea that cell-free DNA is a product of natural cellular apoptosis. Based on evidence from animal studies, the nuclear DNA is fragmented by caspase-activated DNA nucleases (CAD) in apoptotic cells [57]. Apoptotic bodies containing the DNA fragments are released and phagocytosed by tissue-resident macrophages or other recruited phagocytes in a coordinated manner. Injection of apoptotic cells to macrophage-depleted mice suggested that macrophages are essential in the generation of the nucleosomic DNA ladder in the plasma [76]. In pathological conditions that involve heightened levels of cell death, the phagocytic activity at the local tissue environment may be overwhelmed. Elevated levels of cell-free DNA and its associated apoptotic content such as nucleosomes can act as inflammatory mediators to augment the recruitment of phagocytes to the site of pathology. Indeed, the role of circulating histones and nucleosomes as proinflammatory damage-associated molecular patterns is well described [77].

In addition to passive release due to cell death, there is also increasing understanding of active cell-free DNA release from immune cells as part of immune response [78]. In particular, activated neutrophils have been described to release their nuclear chromatin contents to create bactericidal histone-rich neutrophil extracellular traps against pathogens [79]. This model may play a role in the explanation of the elevated levels of plasma cell-free DNA and

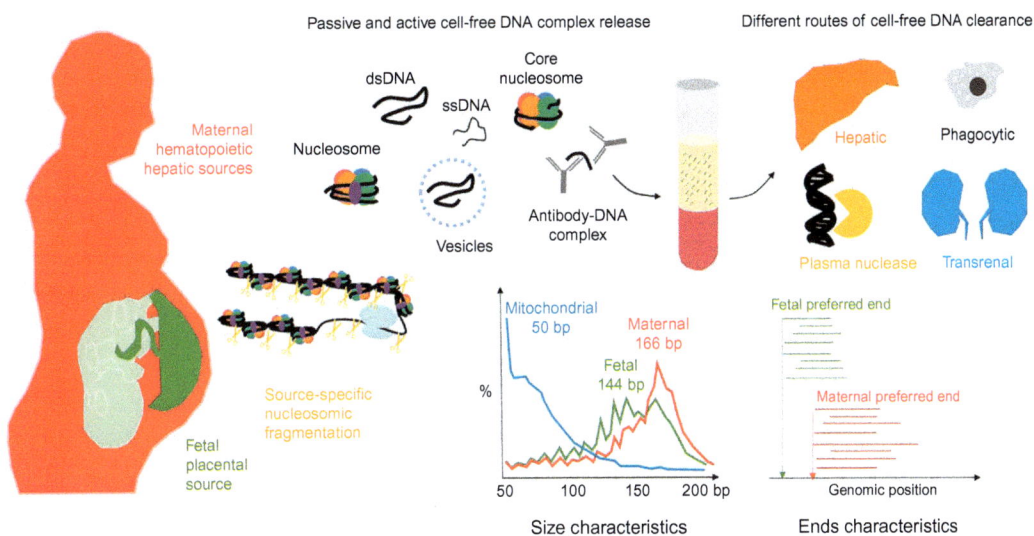

FIG. 27.1 Graphical abstract of cell-free fetal DNA biology.

nucleosomes in pathological conditions such as cancers, acute trauma, ischemic stroke, intense exercise, obesity, and certain gestational conditions that have been linked to inappropriate activation of maternal inflammatory responses such as the preterm onset of labor and preeclampsia (Fig. 27.1).

CLINICAL APPLICATION OF FETAL DNA

Preanalytical Consideration for Reliable Testing

It is important to identify and control the preanalytical factors in cell-free fetal DNA analysis to ensure robust and reliable testing results. In the original discovery, Lo et al. were able to detect the presence of Y-chromosome fetal DNA in heat-treated maternal plasma, serum, and whole blood with decreasing abundance [23]. Blood plasma is conventionally defined as the acellular aqueous portion of the blood after centrifugation and serum is defined as the acellular aqueous portion of blood after complete clotting. Unlike testing of most protein and hormone analytes, plasma is preferred to serum in cell-free fetal DNA analysis [34]. It is noted that although the total cell-free DNA amount is significantly higher in serum than plasma, the fractional concentration of fetal DNA is actually reduced [80]. The additional amount of cell-free DNA in serum likely comes from the activation and cell death of maternal leukocytes during clotting [80]. Moreover, serum possesses uninhibited nuclease activity compared to plasma pretreated with EDTA and thus may compromise long-term sample integrity [81]. Sample hemolysis, a surrogate marker of cell lysis, introduces maternal background contamination. Overnight fasting is not required, yet intense exercise has been linked to increased plasma cell-free DNA from blood cells [82]. Prior invasive procedures may increase cell-free fetal DNA release [83].

Citrate and EDTA have been shown to be superior to heparin as an anticoagulant for blood collection [84]. Residual heparin in heparinized plasma may bind to cell-free DNA and inhibit downstream detection [85]. EDTA has the additional advantage of being a more powerful bivalent ions chelator, which not only prevents clotting but also inhibits plasma nuclease activity. While cell-free fetal DNA in whole blood is stable up to 24h at room temperature, the average size and amount of total cell-free DNA increased significantly after 24h due to maternal cell lysis [86, 87]. Repeated freezing and thawing of plasma also induced fragmentation of cell-free DNA [87]. A number of commercial proprietary blood collection tubes have been developed to minimize cell lysis and DNA degradation during long-distance sample transportation. It is, however, advisable to separate the plasma from the cellular portion within 6h of collection and aliquot the plasma into a small volume to prevent a repeated freeze and thaw cycle.

Above all, it is critical to ensure that the plasma fraction analyzed is genuinely cell-free. A number of studies have reported the presence of circulating DNA packaged in different forms of microvesicles and exosomes. There are also reports of high molecular weight DNA associating with the cell surface of circulatory cells [88]. These DNA species may possess different biochemical properties from plasma cell-free DNA and may affect the interpretation of test results and cross-laboratory comparisons. High-speed centrifugation ($1600g$) followed by microcentrifugation ($16,000g$) or filtration ($0.2\,\mu m$) have been shown to be better than plasma separation by macromolecule density gradient in reducing the fluctuation of cell-free DNA quantification [89]. There are a variety of methods available for cell-free DNA extraction, but one has to note that the choice of DNA extraction methods should focus not only on the extraction yield, cost, and scalability, but also on the DNA size profiles of the extracted plasma cell-free DNA due to the short size preponderance of cell-free fetal DNA.

Applications in the Prenatal Screening of Chromosomal Aneuploidy

The human genome is diploid and organized into 23 pairs of chromosomes (22 pairs of autosomes and one pair of sex chromosomes). One copy of each chromosome pair is inherited through paternal and maternal origin after fertilization of haploid gametes. This configuration (karyotype) can be visualized under light microscope; a euploid male is designated as 46, XY and a euploid female is 46,XX. Aneuploidy is defined as the deviation from the normal diploid karyotype of 46 chromosomes. The most frequent aneuploidy occurs in autosomes, in particular trisomy 21 (Down syndrome), trisomy 18 (Edward syndrome), and trisomy 13 (Patau syndrome). Sex chromosome aneuploidy such as X monosomy (Turner syndrome) and XXY trisomy (Klinefelter syndrome) are also common. The risk of fetal aneuploidy increases dramatically with maternal age from about 1 in 525 at 20 to 1 in 62 at 40 [90].

Noninvasive fetal chromosomal aneuploidy testing is an important research direction in cell-free fetal DNA. This is, however, challenging due to the loss of fetal chromosomal integrity during cell-free fetal DNA release and mixing with maternal background cell-free DNA. An early study found that the total amount of maternal plasma cell-free fetal DNA is moderately elevated in trisomy 21 than euploid pregnancy by using Y chromosome-specific quantitative PCR assays [91].

Count-Based Approach for Fetal Chromosomal Aneuploidy Screening

Reasoning that the presence of trisomy will induce fetal allelic imbalance in the polymorphic regions in the affected chromosomes, Lo and colleagues were able to detect trisomy 18 by measuring the allelic imbalance of an SNP located at a placenta-specific hypomethylation region at chromosome 18 in the maternal plasma [92]. A similar scheme has been applied to detect trisomy 21 by detecting the allelic imbalance of an SNP present in a placenta-specific RNA transcript encoded at chromosome 21 using mass spectrometry and digital PCR technology [93]. Other polymorphism-based approaches have been developed to measure the allelic imbalance of multiple polymorphic sites on the targeted aneuploid and the reference chromosomes in maternal plasma to predict the risk of fetal aneuploidy [94–96].

In addition to an allelic imbalance at polymorphic sites, the total amount of DNA originating from the whole trisomic chromosome is also expected to be overrepresented in the maternal plasma compared to euploid pregnancy due to the presence of an extra fetal copy or unbalanced Robertsonian translocation. The overrepresentation can be quantified as a relative copy number ratio to an internal reference chromosome by counting targeted chromosome-specific DNA fragments using digital PCR technology [97, 98]. It can also be measured as a standardized Z score of the chromosomal representation of the target chromosome to an external reference euploid pregnancy population or to internal reference genomic regions by counting millions of maternal plasma cell-free DNA fragments using random whole genome [99–103] or chromosome-selective massively parallel sequencing technology [104]. In a similar light, monosomy aneuploidy can be detected by the underrepresentation of the monosomic chromosome in maternal plasma. This approach can be further refined to detect megabase long subchromosomal aberrations and microdeletion syndromes from the maternal plasma as a form of noninvasive fetal karyotyping [105] (Fig. 27.2).

Apart from count aberrations, the signal of an additional copy of a fetal chromosome also manifests as aberrations in the size distribution of the cell-free DNA from the aneuploid chromosome because of the change in short (<150 base pairs) fetal cell-free DNA contribution [58].

Screening Strategy and Discordant Testing Results

Noninvasive prenatal testing (NIPT) for fetal chromosomal aneuploidy by cell-free fetal DNA analysis has been validated in large-scale studies with robust clinical sensitivity and specificity [106–108]. In a metaanalysis on the general performance of all NIPT methods, Gil and colleagues calculated that the sensitivity of trisomy 21 detection is above 99.2% (trisomy 18: 96.3%, trisomy 13: 91%, and monosomy X: 90.3%) [109]. The lower detection sensitivity in trisomy 13 and 18 is related to the lower GC content in these two chromosomes and the GC bias of existing sequencing technology [110, 111]. The frequent "cryptic" mosaicism of monosomy X aneuploidy also contributed to the reduced performance [112]. The finding that fetal fraction is increased in trisomy 21 but reduced in trisomy 13, 18, and monosomy X may also explain the discrepancy [113].

Several professional bodies have recommended NIPT as a screening tool for common fetal chromosome aneuploidy to all singleton pregnancy women regardless of their risk of fetal aneuploidy [5, 114]. The strong (approaching 100%) negative predictive values of NIPT for

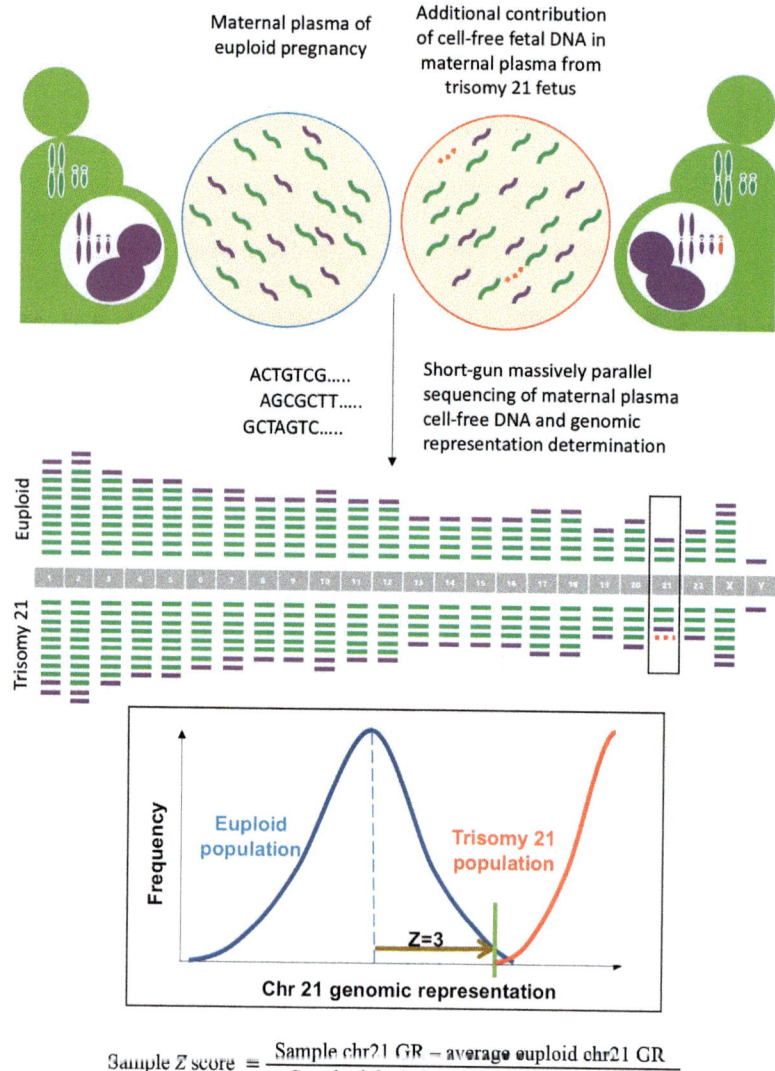

$$\text{Sample } Z \text{ score} = \frac{\text{Sample chr21 GR} - \text{average euploid chr21 GR}}{\text{Standard deviation of euploid chr21 GR}}$$

FIG. 27.2   Relative haplotype dosage analysis. Alpha-SNPs (C/C;D/C): maternal-specific SNPs in which the shared allele is linked to the maternal mutant haplotype; beta-SNPs (E/E;E/F): maternal-specific SNPs in which the shared allele is NOT linked to the maternal mutant haplotype. C1/D1, C2/D2, and C3/D3 denotes three different alpha-SNPs at three different genomic locations in the same mutation-linked haplotype block. The fetal inheritance of the maternal mutation (marked in purple triangle) is inferred indirectly by the fetal inheritance of the specific maternal haplotype. The fetal inheritance of the specific maternal haplotype is in turn detected by the cumulative allelic imbalance of alpha and beta-SNPs in the mutation-linked haplotype block.

trisomy 13, 18, and 21 have significantly reduced the number of invasive procedures by 50%–70% [108, 115], yet all positive results from NIPT should be confirmed by an invasive diagnostic test due to other factors that may lead to discordant results [108].

There is an ongoing discussion on the role of NIPT in fetal aneuploidy screening. Bianchi and colleagues reported that the positive predictive value of trisomy 21 by NIPT (45.5%) is far superior to biochemical screening (4.2%) [108]. Depending on the cost consideration and a clinical decision on target aneuploidy detection rate, NIPT can be implemented as a second-line screening test after abnormal first-trimester serum protein screening or as a first-line screening tool over serum protein screening [116, 117]. As a second-line screening test, the detection rate of fetal aneuploidy of NIPT will be limited by the performance of the first-line biochemical screening. As most of the laboratories only report results for a selected common aneuploidy, there is a concern that NIPT as first-line screening may miss other less common forms of fetal aneuploidy. There is evidence now that reporting of NIPT results using random maternal plasma DNA sequencing should be expanded to other rare forms of aneuploidy [118].

The widespread use of NIPT by cell-free fetal DNA analysis also identifies clinical scenarios where the test may provide misclassification. Misclassification can be caused by analytical and biological factors. Any preanalytical laboratory procedures that lead to loss of cell-free fetal DNA or contamination by maternal DNA can result in false-negative classification. Misclassification can also be intrinsic to the statistical threshold used in certain testing methods. It should be noted that with a Z-score cutoff of three covering 99.9% of the euploid population, 0.1% of the normal euploid fetuses will potentially be misclassified as aneuploid.

Biological factors for misclassification can be grouped into fetal and maternal sources. Because the placenta is the major source of cell-free fetal DNA in maternal plasma, the fetal genomic representation of cell-free fetal DNA is predominantly reflective of the genetic constitution of the placenta. Misclassification can occur if the karyotype is discordant between the placenta and the fetal proper, for example, confined placental mosaicism and fetal mosaicism [119]. The testing result will be dependent on the degree of mosaicism. Rare reports of twin resorption have also been reported to cause discordant results [120]. Maternal conditions, such as previous allogenic transplantation, SLE [71], undiagnosed maternal malignancy [121], and maternal aneuploidy/mosaicism [122, 123], can also lead to an abnormal maternal cell-free DNA background and discordant results. Low-dose molecular heparin [124] and intravenous immunoglobulins [125] have been associated with highly variable maternal background and low fetal fraction (Table 27.1).

Applications in the Prenatal Diagnosis of Monogenic Diseases

Monogenic diseases are caused by mutations of single genes and the inheritance of these diseases commonly follows Mendelian inheritance, with dominant or recessive phenotype manifestation. Conventional prenatal diagnosis of monogenic diseases with fetal tissues aims at directly defining the type of pathogenic mutations and determining whether both paternal and maternal inherited alleles are affected by the fetus, that is, heterozygous, compound heterozygous, or homozygous. However, this analytical principle cannot be directly translated to maternal plasma cell-free fetal DNA analysis. In particular, the presence of the maternal

TABLE 27.1   Factors Contributing to Discordant NIPT Results

Preanalytical

    Blood cell contamination

    Inappropriate anticoagulation

    Repeated freeze and thaw

    Laboratory error and misidentification

Analytical

    Intrinsic statistic distribution

Biological

    Fetal

        Confined placental mosaicism

        Fetal mosaicism

        Vanishing twins

        Low fetal fraction (Dating error, trisomy 13, 18, and monosomy X)

    Maternal

        Previous allogenic transplantation

        Recent blood transfusion

        Maternal malignancy

        Maternal aneuploidy

        Maternal obesity

        Autoimmune disease (e.g., SLE)

        Drug use (heparin, intravenous immunoglobulins)

mutation in the maternal plasma cell-free DNA cannot be concluded as evidence of fetal inheritance due to high maternal contribution to the plasma cell-free DNA pool. On the contrary, if the paternal mutation is present in maternal plasma cell-free DNA, it can be interpreted as evidence of fetal inheritance. Therefore, in cell-free fetal DNA analysis, it is critical to separate the analysis and interpret the result based on the parental origin of the mutation and the Mendelian inheritance of the disease.

Determining the Fetal Inheritance of Paternal Mutation by Direct Qualitative Detection

In autosomal dominant diseases or traits with paternal mutation, the presence of mutation in the maternal plasma confirms fetal inheritance while the absence of the mutation rejects fetal inheritance. Similar interpretation can be made on male-specific Y chromosome inheritance in prenatal fetal sexing. This principle has been successfully applied clinically in the noninvasive prenatal diagnosis of fetal rhesus D blood-group typing [126, 127].

In autosomal recessive diseases or traits with distinctive parental mutations, the absence of the paternal mutation excludes the possibility of fetal inheritance. However, the presence of paternal mutation does not confirm fetal inheritance of the disease, because the inheritance status of the maternal mutation is still undefined. Despite these issues, the result is still clinically useful in prenatal exclusion and selecting patients for further investigations and invasive confirmation such as in cystic fibrosis and beta-thalassemia [128, 129].

In autosomal recessive diseases or traits with identical paternal and maternal mutations or autosomal dominant diseases or traits with maternal mutation, the mutation will always be present in the maternal plasma and no conclusion on fetal inheritance can be made simply based on a qualitative detection result.

Determining the Fetal Inheritance of Maternal Mutation by Quantitative Analysis

A quantitative approach is needed to ascertain fetal inheritance of the maternal mutation [130, 131]. In autosomal dominant diseases or traits with maternal mutation or autosomal recessive diseases or traits with distinctive parental mutations, the ratio between the total copies of the mutant allele (M) and the wildtype allele (N) in the maternal plasma contributed by both the maternal and fetal cell-free DNA is expected to be balanced (M = N) if the genotype of the fetus (M/N) is identical to the mother (M/N). However, the allelic ratio will be imbalanced if the fetal genotype is different from the maternal genotype. If the fetus has not inherited the maternal mutant allele (N/N), there would be additional dosage of the wildtype allele in the maternal plasma contributed from the fetus, resulting in underrepresentation in the total copies of the mutant allele (M < N). The degree of expected allelic imbalance in maternal plasma depends on the fractional concentration of fetal DNA in the maternal plasma. This is based on the premise that both copies of the diploid maternal or fetal genome are released into the maternal plasma without allelic preference. A quantitative relative mutation dosage (RMD) approach has been developed to detect such mutant allelic imbalance [130, 131].

In autosomal recessive diseases or traits where the paternal and maternal mutation are identical, three fetal inheritance patterns are possible. If both paternal and maternal mutant alleles are inherited, the mutant allelic ratio in maternal plasma will be overrepresented (M > N). If no mutant allele is inherited, the wildtype allelic ratio will be overrepresented (M < N). If only one mutant allele is inherited, the allelic ratio will be balanced (M = N).

In X chromosome-linked diseases with maternal mutation, such as hemophilia A, the classification of the fetal inheritance pattern is easier. Fetal sex determination by Y-chromosome DNA detection in maternal plasma is usually performed to first exclude female babies, as they will not be affected by the disease even if they have inherited the maternal mutant allele. Further quantitative RMD analysis is performed for a male fetus. As a male fetus inherits only one copy of the X chromosome from the mother, the mutation-to-wildtype allelic ratio will always be imbalanced. If the male fetus has inherited the X-linked maternal mutation, the mutant allele will be overrepresented (M > N); otherwise, the wildtype allele will be overrepresented (M < N) (Table 27.2).

TABLE 27.2   Diagnostic Principle of Noninvasive Prenatal Testing of Monogenic Diseases by Cell-Free Fetal DNA Analysis

Noninvasive Prenatal Testing of Monogenic Disease		
Clinical Scenarios	Approach	Interpretation
Autosomal dominant (e.g., achondroplasia, neurofibromatosis, Rhesus blood group)		
Paternal mutation	Paternal mutation detection	Paternal mutation detected, fetus affected Paternal mutation not detected, fetus not affected
Maternal mutation	RMD analysis	M=N: Maternal mutation inherited, fetus affected N>M: Maternal mutation not inherited, fetus not affected
Autosomal recessive (e.g., cystic fibrosis, beta-thalassemia, sickle cell diseases, congenital adrenal hyperplasia)		
Identical parental mutations	RMD analysis	M>N: Both parental mutations inherited, fetus affected M=N: One parental mutation inherited, fetus not affected N>M: No parental mutations inherited, fetus not affected
Nonidentical parental mutations	Paternal mutation detection, followed by RMD analysis if paternal mutation is detected	Paternal mutation not detected, fetus not affected Paternal mutation detected, RMD interpretation similar to autosomal dominant maternal mutation
Sex chromosome-linked diseases (e.g., Hemophilia A)		
Female babies	Noninvasive prenatal sexing	No further testing
Male babies	Noninvasive prenatal sexing, followed by RMD analysis if baby is male	M>N: Maternal mutation inherited, fetus affected N>M: Maternal mutation not inherited, fetus not affected

It can be appreciated that reliable determination of the fetal inheritance of maternal mutation is more challenging than paternal mutation and demands highly sensitive and precise molecular tools for accurate measurement of the small imbalance of the mutation dosage. Statistical consideration is also needed to differentiate genuine imbalance from noise fluctuation due to inadequate allelic sampling. These have been successfully implemented in the noninvasive prenatal diagnosis of hemophilia and beta-thalassemia using digital PCR and target-capture sequencing technology [130–132]. Recently, Chan et al. demonstrated that the same relative allelic dosage principle can be expanded to the genome-wide scale to determine the inheritance of individual parental SNPs in the fetus with ultradeep maternal plasma DNA sequencing [66].

Universal Approach for Fetal Inheritance Determination by Relative Haplotype Dosage Analysis

Paternal mutation detection and relative mutation dosage analysis of maternal mutation offered a general analytical framework for noninvasive prenatal diagnosis of monogenic disease. Nevertheless, direct interrogation of the mutation can be difficult in certain genomic loci due to the presence of repetitive sequences, homologous pseudogenes, and undefined genomic rearrangement. Moreover, successful classification of allelic imbalance is statistically dependent on the available copies of mutant and wildtype alleles in the blood sample.

The classical diagnostic challenge includes the prenatal diagnosis of autosomal recessive congenital adrenal hyperplasia (CAH) due to 21-hydroxylase deficiency. Loss-of-function mutations in the *CYP21A2* gene abolish the function of 21-hydroxylase. This enzyme is important in the steroidogenesis of aldosterone and cortisol. The accumulation of steroid precursors due to enzyme deficiency will be diverted to androgen synthesis. The underproduction of aldosterone and cortisol and the overproduction of androgen can cause lethal salt wasting, hypoglycemia, and virilization of female babies. Prenatal steroid treatment before fetal sexual development at 9 weeks of gestation may suppress fetal androgen overproduction and prevent female virilization [133]. Early diagnosis of affected pregnancy is therefore clinically actionable. Additionally, the *CYP21A2* gene locus harbors a highly homologous (98% on exon level) nonfunctioning pseudogene *CYP21A2P* and is associated with high incidence of deletion and conversion mutations. The demand of early diagnosis at a low concentration of fetal DNA, the difficulty in assay design, and the fragmented nature of cell-free DNA impose significant technical challenges for RMD analysis.

These have inspired alternative solutions with indirect mutation inference by relative haplotype dosage analysis (RHDO) [55]. This approach capitalized on the high frequency of single nucleotide polymorphisms (SNP) between individuals (on average 1 SNP per 1000 base pairs) and the phenomenon of genetic linkage. SNPs or specific types of polymorphisms at different locations near a locus of interest are physically linked together in the DNA molecule. The combination of these polymorphisms forms a haplotype and each diploid individual inherits one haplotype from his parents. The closer the physical distance between the two polymorphisms, the more likely they are inherited together; the further away they are, the higher chance the linkage will be lost during meiotic recombination. As a result, the inheritance of a specific mutation can be indirectly inferred from the inheritance of closely linked SNPs. In CAH at-risk pregnancy with mother and father being heterozygous carriers of the *CYP21A2* mutation (N/M; N/M), the linkage of specific parental haplotype to the mutation is first determined in the parents by haplotype phasing. Fetal inheritance of the paternal mutation is then deduced by the detection of paternal-specific mutation-linked SNPs, that is, paternal heterozygous but maternal homozygous SNPs (A/B; A/A), that are linked to the paternal mutation (B) in maternal plasma. To deduce the fetal inheritance of the maternal mutation, SNPs that are maternal specific, that is, paternal homozygous but maternal heterozygous (C/C; D/C or E/E; E/F), are first identified. By convention, the maternal-paternal shared alleles that are linked to the maternal mutation (C) are labeled as α-SNPs while the shared alleles not linked to the maternal mutation (E) are β-SNP [55, 134]. Allelic counts of individual α-SNPs and β-SNPs from the affected loci can then be sequentially combined and statistically tested for imbalance, similar to RMD analysis [55, 134]. For instance, if the maternal haplotype block linked to the maternal mutation is inherited by the fetus, the allelic count of α-SNPs will be

Relative haplotype dosage analysis

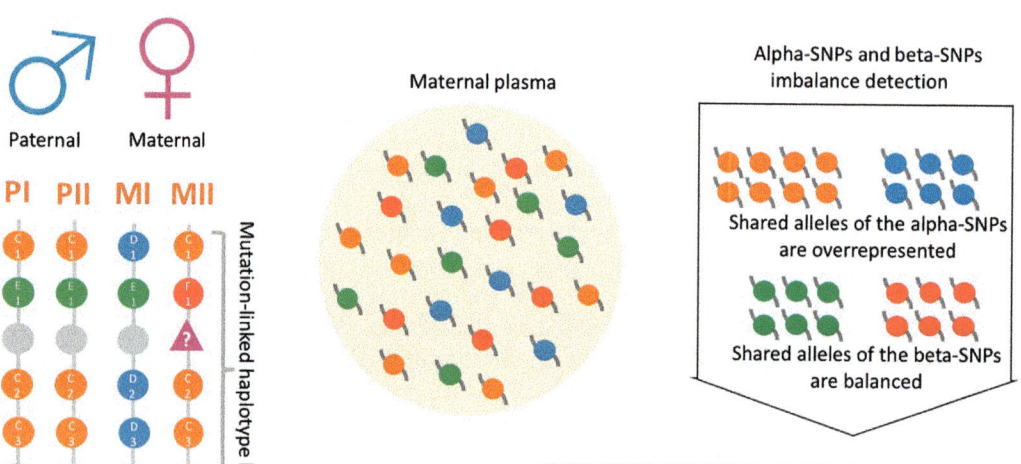

FIG. 27.3   Noninvasive prenatal screening of fetal trisomy 21 by cell-free fetal DNA analysis with shotgun massive parallel sequencing of maternal plasma DNA.

overrepresented (C > D) and the allelic count of β-SNPs will be balanced (E = F) in the maternal plasma. Contrarily, if the maternal haplotype block not linked to the maternal mutation is inherited by the fetus, the allelic count of α-SNPs will be balanced (C = D) and the allelic count of β-SNPs will be overrepresented (E > F) (Fig. 27.3).

The elegance of RHDO analysis is that multiple individual α-SNPs or β-SNPs can be combined and sequentially tested for statistical significance, therefore reducing the constraint on input cell-free fetal DNA fraction and information depth. Moreover, prior definition of the exact nature of the pathogenic mutation is not required as long as haplotype linkage can be established. Inheritance of multiple regions within the same haplotype block can also be inferred simultaneously. This approach has been successfully applied in the noninvasive genetic diagnosis of CAH [134], and the inference of SNP inheritance of the whole fetal genome [55]. With the advance of direct molecular haplotyping technology such as long-read DNA sequencing and short-read assembly methods, haplotype-based analysis of cell-free fetal DNA has the potential to become a universal approach for noninvasive prenatal diagnosis of monogenic diseases [135].

Noninvasive Detection of Fetal De Novo Mutations by Maternal Plasma DNA Sequencing

In most clinical settings, the decision of prenatal testing is initiated due to the presence of a proband in the family or a high population risk with sets of specific mutation targets. The development of an analytical approach for a genome-wide fetal genetic profile inference from maternal plasma cell-free DNA sequencing has ignited a discussion on the feasibility of non-invasive discovery of fetal de novo mutations in cell-free fetal DNA [55, 136, 137]. It has been estimated that every individual possesses about 74 de novo single nucleotide mutations that are absent in their parents [138], which may have arisen during gamete and embryo development. These mutations account for a significant portion of the sporadic monogenic diseases such as skeletal dysplasia, cancer syndrome, and neurodevelopmental disorders [138].

Theoretically, any mutation detected in the maternal plasma cell-free DNA that is absent in the parental genome is potentially a de novo mutation of the fetus. Empirically, this classification leads to significant overestimation because of artefactual mutations caused by DNA damage during DNA handling, polymerase error in sequencing library preparation, and base calling error during sequencing. These errors can be up to 0.3% and are expected to rise with higher sequencing depth and higher signal amplification with lower cell-free DNA input [55, 66]. Such high error rates are prohibitive in clinical applications. Recently, Chan et al. integrated nonamplification library preparation and a set of biologically informed computational filters accounting for the size preponderance of cell-free fetal DNA and fetal fraction in fetal de novo mutation detection [66]. They demonstrated significant improvement in the positive predicting values of fetal de novo mutations by 169-fold and the reduction of candidate mutations from more than 60,000 to 65 while maintaining detection sensitivity at 85%. They were able to detect the fetal de novo single nucleotide mutation in the *BRAF* gene in the maternal plasma of a pregnancy affected by cardiofaciocutaneous syndrome. These encouraging results have brought noninvasive whole fetal genome scanning one step closer to reality.

Applications in the Prenatal Screening of Complex Gestational Complications

Cell-free fetal DNA analysis has been successfully applied in the noninvasive prenatal diagnosis of monogenic diseases and fetal chromosomal aneuploidy. There are also efforts in exploiting the diagnostic values of cell-free DNA as a placenta tissue marker in gestational complications. The level of cell-free fetal DNA is elevated in preeclampsia [139–144], preterm labor [145–147], and intrauterine growth restriction [148–150]. These aberrations have been linked to the pathological cell turnover of placental trophoblasts and the systemic response of the mother [151]. Monitoring of cell-free DNA in maternal plasma may help predict the onset of preeclampsia [140, 143, 144] and the onset of labor [145, 146, 152]. However, the use of cell-free DNA in the prenatal screening of complex gestational diseases remains exploratory. The pathological mechanisms of these gestational complications are often heterogeneous. A better understanding of the etiologies of these complications is needed to refine the application of cell-free fetal DNA as a disease biomarker.

Diverse Methods for Fetal Fraction Estimation in Maternal Plasma Cell-Free DNA

Fetal fraction is the fractional concentration of cell-free fetal DNA in the maternal plasma. It is a fundamental quality-control safeguard and statistical parameter in cell-free fetal DNA analysis. In noninvasive monogenic disease testing by paternally inherited mutation detection, fetal fraction estimation ensures that negative results are not caused by insufficient cell-free fetal DNA in the sample [51]. The fetal fraction also determines the level of expected allelic or chromosomal imbalance in maternal plasma and thus informs the statistical decision in RMD, RHDO, and fetal aneuploidy analysis. Maternal plasma samples with a fetal fraction below 4% may reduce the reliability in NIPT [106]. Assessment of the fluctuation of fetal fraction across the genome in maternal plasma cell-free DNA can be used to deduce the zygosity of twin pregnancy [153]. The regional fetal fraction estimated by fetomaternal polymorphisms fluctuates widely in dizygotic twins due to the genetic dissimilarities between the twins while it remains stable in monozygotic twin pregnancy. It has been estimated that fetal fraction increases slowly at a rate of 0.1% per week in early gestation (10–20 weeks) and at 1% after 21 weeks of gestation [154]. The initial slow growth pattern of fetal fraction suggests that it may not be useful to repeat maternal blood sampling if the fetal fraction of the initial plasma sample is too low for reliable testing [114]. Maternal obesity has been consistently shown to be associated with reduced fetal fraction [154–156].

Fetal fraction can be estimated using different fetal-specific signals in maternal plasma cell-free DNA. In male-fetus pregnancy, the natural fetal markers will be the male-specific Y-chromosome DNA, such as *SRY*, *ZFY*, and *DYS14* [34, 35]. This approach is, however, not applicable to female-fetus pregnancy. Instead, the detection of fetal-specific SNPs (i.e., fetus: A/B and mother: A/A) generated by differentially homozygous polymorphisms of the parents (father: B/B and mother: A/A) will be informative [55]. The minor allele (B) will be solely contributed by the fetus and the fetal fraction can be calculated from total allelic counts of A and B in the maternal plasma as $2B/(A+B)$. Fetal fraction can also be estimated by the fraction of placenta-specific methylated cell-free DNA using methylation-specific quantitative PCR [50, 51]. These principles have been expanded to include thousands of polymorphic or epigenetic markers across the genome using massively parallel sequencing. Sun et al. identified 5000 tissue-specific methylation sites from public databases and estimated the fractional contribution of different tissues source to cell-free DNA in the maternal plasma [53]. The estimation of placental contribution correlates with that of fetal fraction estimation by fetomaternal polymorphisms. Alternative statistical models have also been developed to estimate fetal fraction using sequencing data with no parental or only maternal genotype information [157, 158]. A machine learning approach based on large-scale data training has been developed to predict fetal fraction directly from cell-free DNA sequencing profiles [159]. Other fetal-associated signals, including the short DNA fraction and recurrent fragmentation sites, have also been exploited to estimate fetal fraction [58, 63, 66].

Plasma Cell-Free Fetal RNA and Transcriptomics

Cell-free fetal RNA was discovered in maternal plasma shortly after the discovery of cell-free fetal DNA [160]. Male-specific transcripts from *ZFY* are detectable in the maternal plasma of male-fetus pregnancy by reverse-transcription PCR. The discovery is more unexpected than the presence of cell-free DNA in maternal plasma, considering the known fragility of RNA molecules in laboratory conditions and the nuclease-rich environment of the plasma. Later characterizations showed that placenta-specific cell-free mRNA is filterable by a 0.22-μm filter and is likely protected by microparticulate substances [161]. They are stable at 4°C for up to 24h [162]. Despite their stability, there is evidence that cell-free mRNAs are fragmented transcripts with a 5′ preponderance [163]. Moreover, cell-free placenta-specific small microRNA species can also be detected in maternal plasma [164]. Intriguingly, cell-free microRNA is not filterable and appears to be more stable than cell-free mRNA in a plasma incubation experiment [164].

Cell-free RNA provides an additional dimension of tissue specificity information to cell-free DNA analysis. A number of studies have shown that certain placenta-specific cell-free mRNA transcripts are elevated in preeclampsia and preterm labor [165–168]. The monitoring of cell-free RNA can predict the onset of the disease [169]. As discussed above, an RNA-SNP allelic imbalance of chromosome 21-encoded cell-free RNA can be utilized to detect fetal trisomy [93]. Transcriptome-wide profiling of cell-free RNA in maternal plasma by massively parallel sequencing has revealed the change of thousands of placenta-specific transcripts during pregnancy progression and identified hundreds of pregnancy-specific cell-free RNA markers [170–172]. The tissue specificity of the transcripts is often inferred from the expression profiles of whole tissues and may be skewed by the cellular heterogeneity of the tissue. In a proof-of-principle study, Tsang et al. dissected the cellular heterogeneity of term placentas using large-scale single-cell transcriptomic technology and identified cell-specific transcripts of different placental cell types [172]. Measuring the expression levels of these cell-specific gene sets allows the detection of placental cellular dynamics and the detection of trophoblastic dysfunction in early preterm preeclampsia. There is also a lot of interest in detecting other nonhost sources of cell-free RNA in maternal plasma, such as microbiota-derived cell-free bacterial RNA. This may provide a window to study antenatal infections during pregnancy [173].

CONCLUSION AND OUTLOOK

A safe and reliable approach for prenatal diagnosis of genetic diseases is essential in antenatal care. The discovery of cell-free fetal DNA in maternal plasma has revolutionized the approach to noninvasive prenatal testing over the past two decades. Cell-free fetal DNA is a uniquely versatile tissue marker that carries the genetic information of the genome and the epigenetic details reflective of the well-being of the cellular sources. Noninvasive fetal aneuploidy screening by cell-free DNA analysis has now been translated into routine antenatal care in many parts of the world. Noninvasive prenatal diagnosis of monogenic diseases is also feasible through paternal mutation detection and maternal mutation allelic imbalance analysis. Relative haplotype dosage analysis has the potential to become a universal noninvasive approach for monogenic disease testing. Massively parallel sequencing has enabled noninvasive profiling of the mutational, genetic, epigenomic, and transcriptomic profiles of the fetus. Many of these achievements have been made through our improved understanding of the biological properties of cell-free fetal DNA. We now know that cell-free fetal DNA molecules are relatively stable, primarily placenta-derived, and nonrandomly fragmented. They are shorter than maternal cell-free DNA and rapidly cleared from maternal circulation after delivery. Nonetheless, many important questions such as the cause of the size preponderance and mechanism of cell-free fetal DNA turnover remain elusive. Much effort is also needed to improve the detection of fetal de novo mutations by maternal cell-free DNA sequencing.

Cell-free fetal RNA transcriptomic and cell-free fetal DNA epigenomic analysis represent new research opportunities to expand the application of cell-free fetal nucleic acids to complex gestational complications. As the impact of cell-free fetal DNA research is transmitted from bench to bedside, it has also reignited many ethical discussions in prenatal diagnosis [174]. Breakthroughs in single-molecule sequencing have the potential to substantially reduce the cost and turnaround time for NIPT [175]. A cell-based approach for noninvasive prenatal testing is also revitalized by exciting developments in microfluidic and single-cell genomic technology. We can be optimistic that the continuous improvement in genomic and molecular technology will keep pushing the boundaries of noninvasive prenatal testing in the future.

References

[1] Hassold T, Hunt P. To err (meiotically) is human: the genesis of human aneuploidy. Nat Rev Genet 2001;2 (4):280–91.

[2] Tabor A, Alfirevic Z. Update on procedure-related risks for prenatal diagnosis techniques. Fetal Diagn Ther 2010;27(1):1–7.

[3] Akolekar R, Beta J, Picciarelli G, Ogilvie C, D'Antonio F. Procedure-related risk of miscarriage following amniocentesis and chorionic villus sampling: a systematic review and meta-analysis. Ultrasound Obstet Gynecol 2015;45(1):16–26.

[4] Nicolaides KH, Azar G, Byrne D, Mansur C, Marks K. Fetal nuchal translucency: ultrasound screening for chromosomal defects in first trimester of pregnancy. BMJ 1992;304(6831):867–9.

[5] Committee on Practice Bulletins-Obstetrics CoG, the Society for Maternal-Fetal M. Practice Bulletin No. 163. Screening for fetal aneuploidy. Obstet Gynecol 2016;127(5):e123–37.

[6] Lapaire O, Holzgreve W, Oosterwijk JC, Brinkhaus R, Bianchi DW. Georg Schmorl on trophoblasts in the maternal circulation. Placenta 2007;28(1):1–5.

[7] Herzenberg LA, Bianchi DW, Schroder J, Cann HM, Iverson GM. Fetal cells in the blood of pregnant women: detection and enrichment by fluorescence-activated cell sorting. Proc Natl Acad Sci U S A 1979;76(3):1453–5.

[8] Walknowska J, Conte FA, Grumbach MM. Practical and theoretical implications of fetal-maternal lymphocyte transfer. Lancet 1969;1(7606):1119–22.

[9] Douglas GW, Thomas L, Carr M, Cullen NM, Morris R. Trophoblast in the circulating blood during pregnancy. Am J Obstet Gynecol 1959;78:960–73.

[10] Bianchi DW, Flint AF, Pizzimenti MF, Knoll JH, Latt SA. Isolation of fetal DNA from nucleated erythrocytes in maternal blood. Proc Natl Acad Sci U S A 1990;87(9):3279–83.

[11] Lo YMD, Patel P, Wainscoat JS, Sampietro M, Gillmer MD, Fleming KA. Prenatal sex determination by DNA amplification from maternal peripheral blood. Lancet 1989;2(8676):1363–5.

[12] Price JO, Elias S, Wachtel SS, Klinger K, Dockter M, Tharapel A, et al. Prenatal diagnosis with fetal cells isolated from maternal blood by multiparameter flow cytometry. Am J Obstet Gynecol 1991;165(6):1731–7 Pt 1.

[13] Lo YMD, Bowell PJ, Selinger M, Mackenzie IZ, Chamberlain P, Gillmer MD, et al. Prenatal determination of fetal RhD status by analysis of peripheral blood of rhesus negative mothers. Lancet 1993;341(8853):1147–8.

[14] Bianchi DW, Mahr A, Zickwolf GK, Houseal TW, Flint AF, Klinger KW. Detection of fetal cells with 47,XY,+21 karyotype in maternal peripheral blood. Hum Genet 1992;90(4):368–70.

[15] Elias S, Price J, Dockter M, Wachtel S, Tharapel A, Simpson JL, et al. First trimester prenatal diagnosis of trisomy 21 in fetal cells from maternal blood. Lancet 1992;340(8826):1033.

[16] Krabchi K, Gros-Louis F, Yan J, Bronsard M, Masse J, Forest JC, et al. Quantification of all fetal nucleated cells in maternal blood between the 18th and 22nd weeks of pregnancy using molecular cytogenetic techniques. Clin Genet 2001;60(2):145–50.

[17] Jackson L. Fetal cells and DNA in maternal blood. Prenat Diagn 2003;23(10):837–46.

[18] Bianchi DW, Simpson JL, Jackson LG, Elias S, Holzgreve W, Evans MI, et al. Fetal gender and aneuploidy detection using fetal cells in maternal blood: analysis of NIFTY I data. Natl Inst Child Health Dev Fetal Cell Isolation Study Prenat Diagn 2002;22(7):609–15.

[19] Breman AM, Chow JC, U'Ren L, Normand EA, Qdaisat S, Zhao L, et al. Evidence for feasibility of fetal trophoblastic cell-based noninvasive prenatal testing. Prenat Diagn 2016;36(11):1009–19.

[20] Kolvraa S, Singh R, Normand EA, Qdaisat S, van den Veyver IB, Jackson L, et al. Genome-wide copy number analysis on DNA from fetal cells isolated from the blood of pregnant women. Prenat Diagn 2016;36(12):1127–34.

[21] Bianchi DW, Zickwolf GK, Weil GJ, Sylvester S, DeMaria MA. Male fetal progenitor cells persist in maternal blood for as long as 27 years postpartum. Proc Natl Acad Sci U S A 1996;93(2):705–8.

[22] Jain CV, Kadam L, van Dijk M, Kohan-Ghadr HR, Kilburn BA, Hartman C, et al. Fetal genome profiling at 5 weeks of gestation after noninvasive isolation of trophoblast cells from the endocervical canal. Sci Transl Med 2016;8(363):363re4.

[23] Lo YMD, Corbetta N, Chamberlain PF, Rai V, Sargent IL, Redman CW, et al. Presence of fetal DNA in maternal plasma and serum. Lancet 1997;350(9076):485–7.

[24] Mandel P, Metais P. Les acides nucléiques du plasma sanguin chez l'homme. C R Seances Soc Biol Fil 1948;142 (3-4):241–3.

[25] Sipes JN, Suratt PM, Teates CD, Barada FA, Davis JS. Tegtmeyer CJ. A prospective study of plasma DNA in the diagnosis of pulmonary embolism. Am Rev Respir Dis 1978;118(3):475–8.

[26] Koffler D, Agnello V, Winchester R, Kunkel HG. The occurrence of single-stranded DNA in the serum of patients with systemic lupus erythematosus and other diseases. J Clin Invest 1973;52(1):198–204.

[27] Leon SA, Ehrlich GE, Shapiro B, Labbate VA. Free DNA in the serum of rheumatoid arthritis patients. J Rheumatol 1977;4(2):139–43.

[28] Leon SA, Shapiro B, Sklaroff DM, Yaros MJ. Free DNA in the serum of cancer patients and the effect of therapy. Cancer Res 1977;37(3):646–50.

[29] Vasioukhin V, Anker P, Maurice P, Lyautey J, Lederrey C, Stroun M. Point mutations of the N-ras gene in the blood plasma DNA of patients with myelodysplastic syndrome or acute myelogenous leukaemia. Br J Haematol 1994;86(4):774–9.

[30] Chen X, Bonnefoi H, Diebold-Berger S, Lyautey J, Lederrey C, Faltin-Traub E, et al. Detecting tumor-related alterations in plasma or serum DNA of patients diagnosed with breast cancer. Clin Cancer Res 1999;5 (9):2297–303.

[31] Stroun M, Anker P, Maurice P, Lyautey J, Lederrey C, Beljanski M. Neoplastic characteristics of the DNA found in the plasma of cancer patients. Oncology 1989;46(5):318–22.

[32] Sorenson GD, Pribish DM, Valone FH, Memoli VA, Bzik DJ, Yao SL. Soluble normal and mutated DNA sequences from single-copy genes in human blood. Cancer Epidemiol Biomark Prev 1994;3(1):67–71.

[33] Guibert J, Benachi A, Grebille AG, Ernault P, Zorn JR, Costa JM. Kinetics of SRY gene appearance in maternal serum: detection by real time PCR in early pregnancy after assisted reproductive technique. Hum Reprod 2003;18(8):1733–6.

[34] Lo YMD, Tein MS, Lau TK, Haines CJ, Leung TN, Poon PM, et al. Quantitative analysis of fetal DNA in maternal plasma and serum: implications for noninvasive prenatal diagnosis. Am J Hum Genet 1998;62(4):768–75.

[35] Lun FMF, Chiu RWK, Chan KCA, Leung TY, Lau TK, Lo YMD. Microfluidics digital PCR reveals a higher than expected fraction of fetal DNA in maternal plasma. Clin Chem 2008;54(10):1664–72.

[36] Lo YMD, Zhang J, Leung TN, Lau TK, Chang AM, Hjelm NM. Rapid clearance of fetal DNA from maternal plasma. Am J Hum Genet 1999;64(1):218–24.

[37] Yu SC, Lee SW, Jiang P, Leung TY, Chan KCA, Chiu RWK, et al. High-resolution profiling of fetal DNA clearance from maternal plasma by massively parallel sequencing. Clin Chem 2013;59(8):1228–37.

[38] Savitsky JP. Plasma deoxyribonuclease and the quantitative effects of injected DNA. Proc Soc Exp Biol Med 1961;107:845–7.

[39] Kawabata K, Takakura Y, Hashida M. The fate of plasmid DNA after intravenous injection in mice: involvement of scavenger receptors in its hepatic uptake. Pharm Res 1995;12(6):825–30.

[40] Tsumita T, Iwanaga M. Fate of injected deoxyribonucleic acid in mice. Nature 1963;198:1088–9.

[41] Tsui NBY, Jiang P, Chow KC, Su X, Leung TY, Sun H, et al. High resolution size analysis of fetal DNA in the urine of pregnant women by paired-end massively parallel sequencing. PLoS ONE 2012;7(10)e48319.

[42] Jimenez-Dalmaroni MJ, Gerswhin ME, Adamopoulos IE. The critical role of toll-like receptors—from microbial recognition to autoimmunity: a comprehensive review. Autoimmun Rev 2016;15(1):1–8.

[43] Zhong XY, Holzgreve W, Hahn S. Cell-free fetal DNA in the maternal circulation does not stem from the transplacental passage of fetal erythroblasts. Mol Hum Reprod 2002;8(9):864–70.

[44] Zhong XY, Holzgreve W, Tercanli S, Wenzel F, Hahn S. Cell-free foetal DNA in maternal plasma does not appear to be derived from the rich pool of cell-free foetal DNA in amniotic fluid. Arch Gynecol Obstet 2006;273(4):221–6.

[45] Alberry M, Maddocks D, Jones M, Abdel Hadi M, Abdel-Fattah S, Avent N, et al. Free fetal DNA in maternal plasma in anembryonic pregnancies: confirmation that the origin is the trophoblast. Prenat Diagn 2007;27 (5):415–8.

[46] Lui YY, Chik KW, Chiu RWK, Ho CY, Lam CW, Lo YMD. Predominant hematopoietic origin of cell-free DNA in plasma and serum after sex-mismatched bone marrow transplantation. Clin Chem 2002;48(3):421–7.

[47] Lo YMD, Tein MS, Pang CC, Yeung CK, Tong KL, Hjelm NM. Presence of donor-specific DNA in plasma of kidney and liver-transplant recipients. Lancet 1998;351(9112):1329–30.

[48] De Vlaminck I, Valantine HA, Snyder TM, Strehl C, Cohen G, Luikart H, et al. Circulating cell-free DNA enables noninvasive diagnosis of heart transplant rejection. Sci Transl Med 2014;6(241):241ra77.

[49] Suzuki MM, Bird A. DNA methylation landscapes: provocative insights from epigenomics. Nat Rev Genet 2008;9(6):465–76.

[50] Chim SS, Tong YK, Chiu RWK, Lau TK, Leung TN, Chan LY, et al. Detection of the placental epigenetic signature of the maspin gene in maternal plasma. Proc Natl Acad Sci U S A 2005;102(41):14753–8.

[51] Chan KCA, Ding C, Gerovassili A, Yeung SW, Chiu RWK, Leung TN, et al. Hypermethylated RASSF1A in maternal plasma: a universal fetal DNA marker that improves the reliability of noninvasive prenatal diagnosis. Clin Chem 2006;52(12):2211–8.

[52] Lam WKJ, Gai W, Sun K, Wong RSM, Chan RWY, Jiang P, et al. DNA of erythroid origin is present in human plasma and informs the types of anemia. Clin Chem 2017;63(10):1614–23.

[53] Sun K, Jiang P, Chan KCA, Wong J, Cheng YK, Liang RH, et al. Plasma DNA tissue mapping by genome-wide methylation sequencing for noninvasive prenatal, cancer, and transplantation assessments. Proc Natl Acad Sci U S A 2015;112(40):E5503–12.

[54] Chan KCA, Zhang J, Hui AB, Wong N, Lau TK, Leung TN, et al. Size distributions of maternal and fetal DNA in maternal plasma. Clin Chem 2004;50(1):88–92.

[55] Lo YMD, Chan KCA, Sun H, Chen EZ, Jiang P, Lun FMF, et al. Maternal plasma DNA sequencing reveals the genome-wide genetic and mutational profile of the fetus. Sci Transl Med 2010;2(61):61ra91.

[56] Luger K, Mader AW, Richmond RK, Sargent DF, Richmond TJ. Crystal structure of the nucleosome core particle at 2.8 A resolution. Nature 1997;389(6648):251–60.

[57] Samejima K, Earnshaw WC. Trashing the genome: the role of nucleases during apoptosis. Nat Rev Mol Cell Biol 2005;6(9):677–88.

[58] Yu SC, Chan KCA, Zheng YW, Jiang P, Liao GJ, Sun H, et al. Size-based molecular diagnostics using plasma DNA for noninvasive prenatal testing. Proc Natl Acad Sci U S A 2014;111(23):8583–8.

[59] Jiang P, Chan CW, Chan KCA, Cheng SH, Wong J, Wong VW, et al. Lengthening and shortening of plasma DNA in hepatocellular carcinoma patients. Proc Natl Acad Sci U S A 2015;112(11):E1317–25.

[60] Zheng YW, Chan KCA, Sun H, Jiang P, Su X, Chen EZ, et al. Nonhematopoietically derived DNA is shorter than hematopoietically derived DNA in plasma: a transplantation model. Clin Chem 2012;58(3):549–58.

[61] Meyer CA, Liu XS. Identifying and mitigating bias in next-generation sequencing methods for chromatin biology. Nat Rev Genet 2014;15(11):709–21.

[62] Snyder MW, Kircher M, Hill AJ, Daza RM, Shendure J. Cell-free DNA comprises an in vivo nucleosome footprint that informs its tissues-of-origin. Cell 2016;164(1-2):57–68.

[63] Straver R, Oudejans CB, Sistermans EA, Reinders MJ. Calculating the fetal fraction for noninvasive prenatal testing based on genome-wide nucleosome profiles. Prenat Diagn 2016;36(7):614–21.

[64] Chandrananda D, Thorne NP, Bahlo M. High-resolution characterization of sequence signatures due to non-random cleavage of cell-free DNA. BMC Med Genet 2015;8:29.

[65] Ulz P, Thallinger GG, Auer M, Graf R, Kashofer K, Jahn SW, et al. Inferring expressed genes by whole-genome sequencing of plasma DNA. Nat Genet 2016;48(10):1273–8.

[66] Chan KCA, Jiang P, Sun K, Cheng YK, Tong YK, Cheng SH, et al. Second generation noninvasive fetal genome analysis reveals de novo mutations, single-base parental inheritance, and preferred DNA ends. Proc Natl Acad Sci U S A 2016;113(50):E8159–68.

[67] Burnham P, Kim MS, Agbor-Enoh S, Luikart H, Valantine HA, Khush KK, et al. Single-stranded DNA library preparation uncovers the origin and diversity of ultrashort cell-free DNA in plasma. Sci Rep 2016;6:27859.

[68] Cheng THT, Jiang P, Tam JCW, Sun X, Lee WS, Yu SCY, et al. Genomewide bisulfite sequencing reveals the origin and time-dependent fragmentation of urinary cfDNA. Clin Biochem 2017;50(9):496–501.

[69] Holdenrieder S, Stieber P, Chan LY, Geiger S, Kremer A, Nagel D, et al. Cell-free DNA in serum and plasma: comparison of ELISA and quantitative PCR. Clin Chem 2005;51(8):1544–6.

[70] Tan EM, Schur PH, Carr RI, Kunkel HG. Deoxybonucleic acid (DNA) and antibodies to DNA in the serum of patients with systemic lupus erythematosus. J Clin Invest 1966;45(11):1732–40.

[71] Chan RW, Jiang P, Peng X, Tam LS, Liao GJ, Li EK, et al. Plasma DNA aberrations in systemic lupus erythematosus revealed by genomic and methylomic sequencing. Proc Natl Acad Sci U S A 2014;111(49): E5302–11.

[72] Karlsson K, Sahlin E, Iwarsson E, Westgren M, Nordenskjold M, Linnarsson S. Amplification-free sequencing of cell-free DNA for prenatal non-invasive diagnosis of chromosomal aberrations. Genomics 2015;105(3):150–8.

[73] Vong JSL, Tsang JCH, Jiang P, Lee WS, Leung TY, Chan KCA, et al. Single-stranded DNA library preparation preferentially enriches short maternal DNA in maternal plasma. Clin Chem 2017;63(5):1031–7.

[74] Moser T, Ulz P, Zhou Q, Perakis S, Geigl JB, Speicher MR, et al. Single-stranded DNA library preparation does not preferentially enrich circulating tumor DNA. Clin Chem 2017;.

[75] Wu DC, Lambowitz AM. Facile single-stranded DNA sequencing of human plasma DNA via thermostable group II intron reverse transcriptase template switching. Sci Rep 2017;7(1):8421.

[76] Jiang N, Reich 3rd CF, Pisetsky DS. Role of macrophages in the generation of circulating blood nucleosomes from dead and dying cells. Blood 2003;102(6):2243–50.

[77] Kono H, Rock KL. How dying cells alert the immune system to danger. Nat Rev Immunol 2008;8(4):279–89.

[78] Thierry AR, El Messaoudi S, Gahan PB, Anker P, Stroun M. Origins, structures, and functions of circulating DNA in oncology. Cancer Metastasis Rev 2016;35(3):347–76.

[79] Brinkmann V, Reichard U, Goosmann C, Fauler B, Uhlemann Y, Weiss DS, et al. Neutrophil extracellular traps kill bacteria. Science 2004;303(5663):1532–5.

[80] Wong FCK, Sun K, Jiang P, Cheng YK, Chan KCA, Leung TY, et al. Cell-free DNA in maternal plasma and serum: a comparison of quantity, quality and tissue origin using genomic and epigenomic approaches. Clin Biochem 2016;49(18):1379–86.

[81] Barra GB, Santa Rita TH, de Almeida Vasques J, Chianca CF, Nery LF, Santana Soares Costa S. EDTA-mediated inhibition of DNases protects circulating cell-free DNA from ex vivo degradation in blood samples. Clin Biochem 2015;48(15):976–81.

[82] Tug S, Helmig S, Deichmann ER, Schmeier-Jurchott A, Wagner E, Zimmermann T, et al. Exercise-induced increases in cell free DNA in human plasma originate predominantly from cells of the haematopoietic lineage. Exerc Immunol Rev 2015;21:164–73.

[83] Samura O, Miharu N, Hyodo M, Honda H, Ohashi Y, Honda N, et al. Cell-free fetal DNA in maternal circulation after amniocentesis. Clin Chem 2003;49(7):1193–5.

[84] Lam NY, Rainer TH, Chiu RWK, Lo YMD. EDTA is a better anticoagulant than heparin or citrate for delayed blood processing for plasma DNA analysis. Clin Chem 2004;50(1):256–7.

[85] Sefrioui D, Beaussire L, Clatot F, Delacour J, Perdrix A, Frebourg T, et al. Heparinase enables reliable quantification of circulating tumor DNA from heparinized plasma samples by droplet digital PCR. Clin Chim Acta 2017;472:75–9.

[86] Angert RM, LeShane ES, Lo YMD, Chan LY, Delli-Bovi LC, Bianchi DW. Fetal cell-free plasma DNA concentrations in maternal blood are stable 24 hours after collection: analysis of first- and third-trimester samples. Clin Chem 2003;49(1):195–8.

[87] Chan KCA, Yeung SW, Lui WB, Rainer TH, Lo YMD. Effects of preanalytical factors on the molecular size of cell-free DNA in blood. Clin Chem 2005;51(4):781–4.

[88] Rykova EY, Morozkin ES, Ponomaryova AA, Loseva EM, Zaporozhchenko IA, Cherdyntseva NV, et al. Cell-free and cell-bound circulating nucleic acid complexes: mechanisms of generation, concentration and content. Expert Opin Biol Ther 2012;12(Suppl 1):S141–53.

[89] Chiu RWK, Poon LL, Lau TK, Leung TN, Wong EM, Lo YMD. Effects of blood-processing protocols on fetal and total DNA quantification in maternal plasma. Clin Chem 2001;47(9):1607–13.

[90] Hook EB. Rates of chromosome abnormalities at different maternal ages. Obstet Gynecol 1981;58(3):282–5.

[91] Lo YMD, Lau TK, Zhang J, Leung TN, Chang AM, Hjelm NM, et al. Increased fetal DNA concentrations in the plasma of pregnant women carrying fetuses with trisomy 21. Clin Chem 1999;45(10):1747–51.

[92] Tong YK, Ding C, Chiu RWK, Gerovassili A, Chim SS, Leung TY, et al. Noninvasive prenatal detection of fetal trisomy 18 by epigenetic allelic ratio analysis in maternal plasma: theoretical and empirical considerations. Clin Chem 2006;52(12):2194–202.

[93] Lo YMD, Tsui NBY, Chiu RWK, Lau TK, Leung TN, Heung MM, et al. Plasma placental RNA allelic ratio permits noninvasive prenatal chromosomal aneuploidy detection. Nat Med 2007;13(2):218–23.

[94] Liao GJ, Chan KCA, Jiang P, Sun H, Leung TY, Chiu RWK, et al. Noninvasive prenatal diagnosis of fetal trisomy 21 by allelic ratio analysis using targeted massively parallel sequencing of maternal plasma DNA. PLoS ONE 2012;7(5)e38154.

[95] Zimmermann B, Hill M, Gemelos G, Demko Z, Banjevic M, Baner J, et al. Noninvasive prenatal aneuploidy testing of chromosomes 13, 18, 21, X, and Y, using targeted sequencing of polymorphic loci. Prenat Diagn 2012;32(13):1233–41.

[96] Nicolaides KH, Syngelaki A, Gil M, Atanasova V, Markova D. Validation of targeted sequencing of single-nucleotide polymorphisms for non-invasive prenatal detection of aneuploidy of chromosomes 13, 18, 21, X, and Y. Prenat Diagn 2013;33(6):575–9.

[97] Lo YMD, Lun FMF, Chan KCA, Tsui NBY, Chong KC, Lau TK, et al. Digital PCR for the molecular detection of fetal chromosomal aneuploidy. Proc Natl Acad Sci U S A 2007;104(32):13116–21.

[98] Fan HC, Blumenfeld YJ, El-Sayed YY, Chueh J, Quake SR. Microfluidic digital PCR enables rapid prenatal diagnosis of fetal aneuploidy. Am J Obstet Gynecol 2009;200(5):543 e1–7.

[99] Fan HC, Blumenfeld YJ, Chitkara U, Hudgins L, Quake SR. Noninvasive diagnosis of fetal aneuploidy by shotgun sequencing DNA from maternal blood. Proc Natl Acad Sci U S A 2008;105(42):16266–71.

[100] Chiu RWK, Chan KCA, Gao Y, Lau VY, Zheng W, Leung TY, et al. Noninvasive prenatal diagnosis of fetal chromosomal aneuploidy by massively parallel genomic sequencing of DNA in maternal plasma. Proc Natl Acad Sci U S A 2008;105(51):20458–63.

[101] Chiu RWK, Sun H, Akolekar R, Clouser C, Lee C, McKernan K, et al. Maternal plasma DNA analysis with massively parallel sequencing by ligation for noninvasive prenatal diagnosis of trisomy 21. Clin Chem 2010;56(3):459–63.

[102] Straver R, Sistermans EA, Holstege H, Visser A, Oudejans CB, Reinders MJ. WISECONDOR: detection of fetal aberrations from shallow sequencing maternal plasma based on a within-sample comparison scheme. Nucleic Acids Res 2014;42(5):e31.

[103] Sun K, Chan KCA, Hudecova I, Chiu RWK, Lo YMD, Jiang P. COFFEE: control-free noninvasive fetal chromosomal examination using maternal plasma DNA. Prenat Diagn 2017;37(4):336–40.

[104] Sparks AB, Wang ET, Struble CA, Barrett W, Stokowski R, McBride C, et al. Selective analysis of cell-free DNA in maternal blood for evaluation of fetal trisomy. Prenat Diagn 2012;32(1):3–9.

[105] Yu SC, Jiang P, Choy KW, Chan KCA, Won HS, Leung WC, et al. Noninvasive prenatal molecular karyotyping from maternal plasma. PLoS ONE 2013;8(4)e60968.

[106] Chiu RWK, Akolekar R, Zheng YW, Leung TY, Sun H, Chan KCA, et al. Non-invasive prenatal assessment of trisomy 21 by multiplexed maternal plasma DNA sequencing: large scale validity study. BMJ 2011;342:c7401.

[107] Ehrich M, Deciu C, Zwiefelhofer T, Tynan JA, Cagasan L, Tim R, et al. Noninvasive detection of fetal trisomy 21 by sequencing of DNA in maternal blood: a study in a clinical setting. Am J Obstet Gynecol 2011;204(3):205 e1–11.

[108] Bianchi DW, Parker RL, Wentworth J, Madankumar R, Saffer C, Das AF, et al. DNA sequencing versus standard prenatal aneuploidy screening. N Engl J Med 2014;370(9):799–808.

[109] Gil MM, Quezada MS, Revello R, Akolekar R, Nicolaides KH. Analysis of cell-free DNA in maternal blood in screening for fetal aneuploidies: updated meta-analysis. Ultrasound Obstet Gynecol 2015;45(3):249–66.

[110] Chen EZ, Chiu RWK, Sun H, Akolekar R, Chan KCA, Leung TY, et al. Noninvasive prenatal diagnosis of fetal trisomy 18 and trisomy 13 by maternal plasma DNA sequencing. PLoS ONE 2011;6(7)e21791.

[111] Fan HC, Quake SR. Sensitivity of noninvasive prenatal detection of fetal aneuploidy from maternal plasma using shotgun sequencing is limited only by counting statistics. PLoS ONE 2010;5(5)e10439.

[112] Hook EB, Warburton D. Turner syndrome revisited: review of new data supports the hypothesis that all viable 45,X cases are cryptic mosaics with a rescue cell line, implying an origin by mitotic loss. Hum Genet 2014;133(4):417–24.

[113] Rava RP, Srinivasan A, Sehnert AJ, Bianchi DW. Circulating fetal cell-free DNA fractions differ in autosomal aneuploidies and monosomy X. Clin Chem 2014;60(1):243–50.

[114] Gregg AR, Skotko BG, Benkendorf JL, Monaghan KG, Bajaj K, Best RG, et al. Noninvasive prenatal screening for fetal aneuploidy, 2016 update: a position statement of the American College of Medical Genetics and Genomics. Genet Med 2016;18(10):1056–65.

[115] Larion S, Warsof SL, Romary L, Mlynarczyk M, Peleg D, Abuhamad AZ. Association of combined first-trimester screen and noninvasive prenatal testing on diagnostic procedures. Obstet Gynecol 2014;123(6):1303–10.

[116] Dondorp W, de Wert G, Bombard Y, Bianchi DW, Bergmann C, Borry P, et al. Non-invasive prenatal testing for aneuploidy and beyond: challenges of responsible innovation in prenatal screening. Eur J Hum Genet 2015;23 (11):1592.

[117] Cuckle H, Benn P, Pergament E. Cell-free DNA screening for fetal aneuploidy as a clinical service. Clin Biochem 2015;48(15):932–41.

[118] Pertile MD, Halks-Miller M, Flowers N, Barbacioru C, Kinnings SL, Vavrek D, et al. Rare autosomal trisomies, revealed by maternal plasma DNA sequencing, suggest increased risk of feto-placental disease. Sci Transl Med 2017;9(405).

[119] Choi H, Lau TK, Jiang FM, Chan MK, Zhang HY, Lo PS, et al. Fetal aneuploidy screening by maternal plasma DNA sequencing: 'false positive' due to confined placental mosaicism. Prenat Diagn 2013;33(2):198–200.

[120] Curnow KJ, Wilkins-Haug L, Ryan A, Kirkizlar E, Stosic M, Hall MP, et al. Detection of triploid, molar, and vanishing twin pregnancies by a single-nucleotide polymorphism-based noninvasive prenatal test. Am J Obstet Gynecol 2015;212(1):79 e1–9.

[121] Bianchi DW, Chudova D, Sehnert AJ, Bhatt S, Murray K, Prosen TL, et al. Noninvasive prenatal testing and incidental detection of occult maternal malignancies. JAMA 2015;314(2):162–9.

[122] Yao H, Zhang L, Zhang H, Jiang F, Hu H, Chen F, et al. Noninvasive prenatal genetic testing for fetal aneuploidy detects maternal trisomy X. Prenat Diagn 2012;32(11):1114–6.

[123] Wang Y, Chen Y, Tian F, Zhang J, Song Z, Wu Y, et al. Maternal mosaicism is a significant contributor to discordant sex chromosomal aneuploidies associated with noninvasive prenatal testing. Clin Chem 2014;60 (1):251–9.

[124] Gromminger S, Erkan S, Schock U, Stangier K, Bonnet J, Schloo R, et al. The influence of low molecular weight heparin medication on plasma DNA in pregnant women. Prenat Diagn 2015;35(11):1155–7.

[125] Hui L, Bethune M, Weeks A, Kelley J, Hayes L. Repeated failed non-invasive prenatal testing owing to low cell-free fetal DNA fraction and increased variance in a woman with severe autoimmune disease. Ultrasound Obstet Gynecol 2014;44(2):242–3.

[126] Lo YMD, Hjelm NM, Fidler C, Sargent IL, Murphy MF, Chamberlain PF, et al. Prenatal diagnosis of fetal RhD status by molecular analysis of maternal plasma. N Engl J Med 1998;339(24):1734–8.

[127] Scheffer PG, van der Schoot CE, Page-Christiaens GC, de Haas M. Noninvasive fetal blood group genotyping of rhesus D, c, E and of K in alloimmunised pregnant women: evaluation of a 7-year clinical experience. BJOG 2011;118(11):1340–8.

[128] Gonzalez-Gonzalez MC, Garcia-Hoyos M, Trujillo MJ, Rodriguez de Alba M, Lorda-Sanchez I, Diaz-Recasens J, et al. Prenatal detection of a cystic fibrosis mutation in fetal DNA from maternal plasma. Prenat Diagn 2002;22 (10):946–8.

[129] Chiu RWK, Lau TK, Leung TN, Chow KC, Chui DH, Lo YMD. Prenatal exclusion of beta thalassaemia major by examination of maternal plasma. Lancet 2002;360(9338):998–1000.

[130] Tsui NBY, Kadir RA, Chan KCA, Chi C, Mellars G, Tuddenham EG, et al. Noninvasive prenatal diagnosis of hemophilia by microfluidics digital PCR analysis of maternal plasma DNA. Blood 2011;117(13):3684–91.

[131] Lun FMF, Tsui NBY, Chan KCA, Leung TY, Lau TK, Charoenkwan P, et al. Noninvasive prenatal diagnosis of monogenic diseases by digital size selection and relative mutation dosage on DNA in maternal plasma. Proc Natl Acad Sci U S A 2008;105(50):19920–5.

[132] Hudecova I, Jiang P, Davies J, Lo YMD, Kadir RA, Chiu RWK. Noninvasive detection of F8 int22h-related inversions and sequence variants in maternal plasma of hemophilia carriers. Blood 2017;130(3):340–7.

[133] New MI, Abraham M, Yuen T, Lekarev O. An update on prenatal diagnosis and treatment of congenital adrenal hyperplasia. Semin Reprod Med 2012;30(5):396–9.

[134] New MI, Tong YK, Yuen T, Jiang P, Pina C, Chan KCA, et al. Noninvasive prenatal diagnosis of congenital adrenal hyperplasia using cell-free fetal DNA in maternal plasma. J Clin Endocrinol Metab 2014;99(6): E1022–30.

[135] Hui WW, Jiang P, Tong YK, Lee WS, Cheng YK, New MI, et al. Universal haplotype-based noninvasive prenatal testing for single gene diseases. Clin Chem 2017;63(2):513–24.

[136] Kitzman JO, Snyder MW, Ventura M, Lewis AP, Qiu R, Simmons LE, et al. Noninvasive whole-genome sequencing of a human fetus. Sci Transl Med 2012;4(137):137ra76.

[137] Fan HC, Gu W, Wang J, Blumenfeld YJ, El-Sayed YY, Quake SR. Non-invasive prenatal measurement of the fetal genome. Nature 2012;487(7407):320–4.

[138] Veltman JA, Brunner HG. De novo mutations in human genetic disease. Nat Rev Genet 2012;13(8):565–75.

[139] Lo YMD, Leung TN, Tein MS, Sargent IL, Zhang J, Lau TK, et al. Quantitative abnormalities of fetal DNA in maternal serum in preeclampsia. Clin Chem 1999;45(2):184–8.

[140] Leung TN, Zhang J, Lau TK, Chan LY, Lo YMD. Increased maternal plasma fetal DNA concentrations in women who eventually develop preeclampsia. Clin Chem 2001;47(1):137–9.

[141] Zhong XY, Holzgreve W, Hahn S. The levels of circulatory cell free fetal DNA in maternal plasma are elevated prior to the onset of preeclampsia. Hypertens Pregnancy 2002;21(1):77–83.

[142] Lau TW, Leung TN, Chan LY, Lau TK, Chan KCA, Tam WH, et al. Fetal DNA clearance from maternal plasma is impaired in preeclampsia. Clin Chem 2002;48(12):2141–6.

[143] Farina A, Sekizawa A, Rizzo N, Concu M, Banzola I, Carinci P, et al. Cell-free fetal DNA (SRY locus) concentration in maternal plasma is directly correlated to the time elapsed from the onset of preeclampsia to the collection of blood. Prenat Diagn 2004;24(4):293–7.

[144] Levine RJ, Qian C, Leshane ES, Yu KF, England LJ, Schisterman EF, et al. Two-stage elevation of cell-free fetal DNA in maternal sera before onset of preeclampsia. Am J Obstet Gynecol 2004;190(3):707–13.

[145] Leung TN, Zhang J, Lau TK, Hjelm NM, Lo YMD. Maternal plasma fetal DNA as a marker for preterm labour. Lancet 1998;352(9144):1904–5.

[146] Farina A, LeShane ES, Romero R, Gomez R, Chaiworapongsa T, Rizzo N, et al. High levels of fetal cell-free DNA in maternal serum: a risk factor for spontaneous preterm delivery. Am J Obstet Gynecol 2005;193(2):421–5.

[147] Dugoff L, Barberio A, Whittaker PG, Schwartz N, Sehdev H, Bastek JA. Cell-free DNA fetal fraction and pre-term birth. Am J Obstet Gynecol 2016;215(2):231 e1–7.

[148] Caramelli E, Rizzo N, Concu M, Simonazzi G, Carinci P, Bondavalli C, et al. Cell-free fetal DNA concentration in plasma of patients with abnormal uterine artery Doppler waveform and intrauterine growth restriction—a pilot study. Prenat Diagn 2003;23(5):367–71.

[149] Al Nakib M, Desbriere R, Bonello N, Bretelle F, Boubli L, Gabert J, et al. Total and fetal cell-free DNA analysis in maternal blood as markers of placental insufficiency in intrauterine growth restriction. Fetal Diagn Ther 2009;26(1):24–8.

[150] Alberry MS, Maddocks DG, Hadi MA, Metawi H, Hunt LP, Abdel-Fattah SA, et al. Quantification of cell free fetal DNA in maternal plasma in normal pregnancies and in pregnancies with placental dysfunction. Am J Obstet Gynecol 2009;200(1):98 e1–6.

[151] Chaiworapongsa T, Chaemsaithong P, Yeo L, Romero R. Pre-eclampsia part 1: current understanding of its pathophysiology. Nat Rev Nephrol 2014;10(8):466–80.

[152] Jakobsen TR, Clausen FB, Rode L, Dziegiel MH, Tabor A. High levels of fetal DNA are associated with increased risk of spontaneous preterm delivery. Prenat Diagn 2012;32(9):840–5.

[153] Qu JZ, Leung TY, Jiang P, Liao GJ, Cheng YK, Sun H, et al. Noninvasive prenatal determination of twin zygosity by maternal plasma DNA analysis. Clin Chem 2013;59(2):427–35.

[154] Wang E, Batey A, Struble C, Musci T, Song K, Oliphant A. Gestational age and maternal weight effects on fetal cell-free DNA in maternal plasma. Prenat Diagn 2013;33(7):662–6.

[155] Ashoor G, Poon L, Syngelaki A, Mosimann B, Nicolaides KH. Fetal fraction in maternal plasma cell-free DNA at 11-13 weeks' gestation: effect of maternal and fetal factors. Fetal Diagn Ther 2012;31(4):237–43.

[156] Ashoor G, Syngelaki A, Poon LC, Rezende JC, Nicolaides KH. Fetal fraction in maternal plasma cell-free DNA at 11–13 weeks' gestation: relation to maternal and fetal characteristics. Ultrasound Obstet Gynecol 2013;41(1):26 32.

[157] Jiang P, Chan KCA, Liao GJ, Zheng YW, Leung TY, Chiu RWK, et al. FetalQuant: deducing fractional fetal DNA concentration from massively parallel sequencing of DNA in maternal plasma. Bioinformatics 2012;28(22):2883–90.

[158] Jiang P, Peng X, Su X, Sun K, Yu SCY, Chu WI, et al. FetalQuantSD: accurate quantification of fetal DNA fraction by shallow-depth sequencing of maternal plasma DNA. npj Genom Med 2016;1(1).

[159] Kim SK, Hannum G, Geis J, Tynan J, Hogg G, Zhao C, et al. Determination of fetal DNA fraction from the plasma of pregnant women using sequence read counts. Prenat Diagn 2015;35(8):810–5.

[160] Poon LL, Leung TN, Lau TK, Lo YMD. Presence of fetal RNA in maternal plasma. Clin Chem 2000;46(11):1832–4.

[161] Ng EK, Tsui NBY, Lau TK, Leung TN, Chiu RWK, Panesar NS, et al. mRNA of placental origin is readily detectable in maternal plasma. Proc Natl Acad Sci U S A 2003;100(8):4748–53.

[162] Tsui NBY, Ng EK, Lo YMD. Stability of endogenous and added RNA in blood specimens, serum, and plasma. Clin Chem 2002;48(10):1647–53.

[163] Wong BC, Chiu RWK, Tsui NBY, Chan KCA, Chan LW, Lau TK, et al. Circulating placental RNA in maternal plasma is associated with a preponderance of 5′ mRNA fragments: implications for noninvasive prenatal diagnosis and monitoring. Clin Chem 2005;51(10):1786–95.

[164] Chim SS, Shing TK, Hung EC, Leung TY, Lau TK, Chiu RWK, et al. Detection and characterization of placental microRNAs in maternal plasma. Clin Chem 2008;54(3):482–90.

[165] Purwosunu Y, Sekizawa A, Farina A, Wibowo N, Okazaki S, Nakamura M, et al. Cell-free mRNA concentrations of CRH, PLAC1, and selectin-P are increased in the plasma of pregnant women with preeclampsia. Prenat Diagn 2007;27(8):772–7.

[166] Purwosunu Y, Sekizawa A, Koide K, Farina A, Wibowo N, Wiknjosastro GH, et al. Cell-free mRNA concentrations of plasminogen activator inhibitor-1 and tissue-type plasminogen activator are increased in the plasma of pregnant women with preeclampsia. Clin Chem 2007;53(3):399–404.

[167] Ng EK, Leung TN, Tsui NBY, Lau TK, Panesar NS, Chiu RWK, et al. The concentration of circulating corticotropin-releasing hormone mRNA in maternal plasma is increased in preeclampsia. Clin Chem 2003;49(5):727–31.

[168] Farina A, Chan CW, Chiu RWK, Tsui NBY, Carinci P, Concu M, et al. Circulating corticotropin-releasing hormone mRNA in maternal plasma: relationship with gestational age and severity of preeclampsia. Clin Chem 2004;50(10):1851–4.

[169] Purwosunu Y, Sekizawa A, Okazaki S, Farina A, Wibowo N, Nakamura M, et al. Prediction of preeclampsia by analysis of cell-free messenger RNA in maternal plasma. Am J Obstet Gynecol 2009;200(4):386 e1–7.

[170] Tsui NBY, Jiang P, Wong YF, Leung TY, Chan KCA, Chiu RWK, et al. Maternal plasma RNA sequencing for genome-wide transcriptomic profiling and identification of pregnancy-associated transcripts. Clin Chem 2014;60(7):954–62.

[171] Koh W, Pan W, Gawad C, Fan HC, Kerchner GA, Wyss-Coray T, et al. Noninvasive in vivo monitoring of tissue-specific global gene expression in humans. Proc Natl Acad Sci U S A 2014;111(20):7361–6.

[172] Tsang JCH, Vong JSL, Ji L, Poon LCY, Jiang P, Lui KO, et al. Integrative single-cell and cell-free plasma RNA transcriptomics elucidates placental cellular dynamics. Proc Natl Acad Sci U S A 2017;.

[173] Pan W, Ngo TTM, Camunas-Soler J, Song CX, Kowarsky M, Blumenfeld YJ, et al. Simultaneously monitoring immune response and microbial infections during pregnancy through plasma cfRNA sequencing. Clin Chem 2017;.

[174] Dondorp W, de Wert G, Bombard Y, Bianchi DW, Bergmann C, Borry P, et al. Non-invasive prenatal testing for aneuploidy and beyond: challenges of responsible innovation in prenatal screening. Summary and recommendations. Eur J Hum Genet 2015;.

[175] Cheng SH, Jiang P, Sun K, Cheng YK, Chan KCA, Leung TY, et al. Noninvasive prenatal testing by nanopore sequencing of maternal plasma DNA: feasibility assessment. Clin Chem 2015;61(10):1305–6.

28

Noninvasive Prenatal Testing by Cell-Free DNA: Technology, Biology, Clinical Utility, and Limitations

Francesca Romana Grati*, Komal Bajaj[†,‡], Giuseppe Simoni*, Federico Maggi*, Susan J Gross[§], Jose Carlos Pinto B. Ferreira[¶,‖,#,**]

*TOMA Advanced Biomedical Assays S.p.A., Busto Arsizio, Italy [†]New York City Health + Hospitals, New York, NY, United States [‡]Albert Einstein College of Medicine, New York, NY, United States [§]Department of Genetics and Genomic Sciences, Icahn School of Medicine at Mount Sinai, New York, NY, United States [¶]Genomed S.A., Warsaw, Poland [‖]Central Hospital of Maputo, Maputo, Mozambique [#]ICOR—Heart Institute, Maputo, Mozambique [**]Faculty of Medicine, Eduardo Mondlane University, Maputo, Mozambique

OUTLINE

Abbreviations

22q11.2DS	22q11.2 deletion syndrome
BMI	body mass index
cfDNA	cell-free DNA
CMA	chromosomal microarray
CPM	confined placental mosaicism
CVS	chorionic villus sampling
DANSR	digital analysis of selected regions
DR	detection rate
FF	fetal fraction
FORTE	fetal fraction optimized risk of trisomy evaluation
FPR	false positive rate
MPSS	massively parallel shotgun sequencing
NATUS	next-generation aneuploidy test using SNPs
NGS	next generation sequencing
NIPT/S	noninvasive prenatal testing/screening
NPV	negative predictive value
PPV	positive predictive value
SCA	sex chromosome aneuploidy
SNP	single nucleotide polymorphism
TFM	true fetal mosaicism
t-NGS	targeted NGS
WG	whole-genome

INTRODUCTION

Noninvasive testing based on quantification of the "fetal" cell-free circulating DNA in the maternal plasma (also known as cfDNA testing or noninvasive prenatal testing/screening, NIPT/S) aims to detect pregnancies with an increased risk for fetal trisomies 21, 18, and 13. The cfDNA fragments targeted by these tests are derived from both the mother and conceptus, specifically from the apoptosis of the cytotrophoblast, the external layer of the placenta [1–3]. The maternal cfDNA contribution greatly exceeds the placental contribution (also known as the "fetal fraction," FF). At 10 weeks of gestation, the average ratio of maternal-to-fetal contribution is approximately 9:1 [4, 5]. Current approaches to cfDNA cannot efficiently physically separate these two components for analysis. This limitation has two main implications. First, the assessment of trisomy presents a quantitative challenge of detecting the additional genetic information contributed by the third chromosome against an overwhelming background of maternal DNA. Therefore, cfDNA tests must investigate fetal trisomies using sophisticated algorithms that can detect minute increments of circulating chromosomal fragments.

The second consequence of dealing with minute amounts of circulating, placentally derived DNA fragments is that any chromosome imbalance affecting either the maternal or placental compartment may also be detected by cfDNA testing, thereby limiting the ability to discriminate between maternal and placental/fetal abnormalities. This biological phenomenon of a "needle in a haystack" has fundamental implications for the interpretation and management of positive/high risk cfDNA testing results and will be discussed below (see paragraph "biological reasons for cfDNA discordant results").

Several cfDNA testing strategies are available, and most of them are based on four testing technologies and their ramifications:

1. Massively parallel shotgun sequencing (MPSS).
2. Targeted NGS (t-NGS).
3. Digital analysis of selected regions (DANSR).
4. Single nucleotide polymorphisms (SNP)-based.

Initially, all four testing methodologies were based on next-generation sequencing (NGS) technology. Subsequently, DANSR technology shifted from NGS to a microarray platform. [6]

The analysis of the chromosome of interest is performed by two different approaches. For the t-NGS, DANSR, and SNP-based methods, the analysis is limited to a portion of the total circulated cfDNA fragments, based on the initial capture/amplification stage, using specific molecular probes that will only target circulating fragments of interest [7–9]. Using MPSS, the targeting of the region of interest is performed in the subsequent bioinformatic phase, but only after the sequencing, alignment, and counting of all circulating fragments derived from all chromosomes [10, 11].

DESCRIPTION OF DIFFERENT CFDNA TESTING TECHNOLOGIES

Massively Parallel Shotgun Sequencing (MPSS)

MPSS is a whole-genome (WG) technology based on NGS of ~25 million maternal and fetal cfDNA fragments belonging to all chromosomes. The main stages of the process are library preparation, sequencing of the cfDNA fragments, sequence alignment, fragment counting, statistical analysis, and reporting. With this technology, sample analysis is performed in a multiplex fashion so that several samples are sequenced in parallel in one NGS reaction [10, 11]. In MPSS, all cfDNA fragments from each individual sample are first molecularly barcoded to be recognized and assigned to the specific patient in the subsequent analytic stages. After barcoding, samples are "normalized" to obtain the same concentration of each DNA sample and mixed in a unique sequencing reaction (cell flow). As the total amount of sequences available in one reaction is fixed and predetermined by the size of the NGS reagent cartridge, the number of cfDNA fragments that can be sequenced for each sample (sequencing depth) depends on the number of samples and the ability to normalize DNA concentration across all samples. If the concentration of each sample is not normalized, a single sample can consume a disproportionate number of sequence reads in a flow cell, reducing reads available for reliably determining trisomy risk in the remaining samples. A fourfold variation in the median reads per sample might be observed for a 12-plex reaction [12]. These settings can determine the variability of the analytical conditions (sensitivity and specificity) depending on the NGS cartridge size, the number of sample inputs, and on a successful normalization procedure. The higher the sample input, the lower the sequencing depth and, consequently, the poorer the analytic performance [13]. Therefore, MPSS-based available tests may not offer the same analytical conditions and could vary from one laboratory to another even though they are all using the same technology.

After sequencing of the first nucleotides of each circulating fragment (usually from 25 to 36 base pairs, bp) the chromosome of origin of each individual read is determined by mapping the obtained sequence against a Human Reference Genome [10, 11]. Approximately 50% of these reads are usually mapped uniquely to the human genome covering 4% of the entire genome while the other reads remain unmapped. After alignment, the sequences belonging to each chromosome are counted and the counting data are analyzed to provide a fetal trisomy risk score. The counts of specific cfDNA fragments arising from a chromosome under investigation are compared to euploid and trisomic reference samples and usually presented as z-scores. The z-score refers to the number of standard deviations from the mean of a reference euploid data set. In disomic fetuses, the distribution of the z-score values is Gaussian with a mean of 0 and a standard deviation of 1. In trisomic fetuses, the mean z-score is higher than 0, as it is increased compared with the mean value of euploid reference pregnancies. The arbitrary z-score cut-off value usually utilized to call a sample as trisomic is 3. In addition, as discussed above, the z-score value in trisomic samples is directly proportional to the fetal fraction value [14, 15]. This implies that at low FF% (typically below 4%), the separation of disomic and trisomic samples is not as reliable as at higher FF levels. The separation between the distributions of z-scores from euploid and trisomic pregnancies at low FF% increases with deeper levels of sequencing. While deeper sequencing can increase performance at lower FF%, the validation process is more expensive and will, therefore, increase overall cost. Furthermore, the data on the precision of estimated fetal fractions is also currently limited.

Targeted NGS (t-NGS)

t-NGS is a variation of the MPSS in which only chromosomes of interest and a few selected reference chromosomes are analyzed for comparison. Specific complementary probes are used for the selection of the fragments of interest. Then, similarly to MPSS, the selected fragments are sequenced and then the resulting sequencing reads are aligned to the Human Genome Reference. The number belonging to each chromosome is counted and compared, after which a fetal risk score is obtained [9]. The selective sequencing of cfDNA fragments containing SNP loci allows the estimation of the FF% by SNP genotyping (see section "Fetal Fraction Estimation").

Digital Analysis of Selected Regions (DANSR)

This test uses two different sets of DANSR probes to capture the cfDNA fragments from chromosomes of interest. These two sets of probes provide different types of information: the trisomy risk score and the FF estimation. Risk score is provided by capturing and counting a selected number of fragments belonging to chromosomes 21, 18, and 13 and reference chromosomes (from 1 to 12). FF is determined by capturing and genotyping a selected number of cfDNA fragments belonging to reference chromosomes (from 1 to 12) and mapping SNP loci [6–8, 16]. The analysis of the DANSR products is performed with a microarray platform with a single-plex approach. Therefore, each sample is analyzed individually in each subarray, allowing for standardization of analytical conditions across laboratories performing this testing methodology [6]. Microarray output data are then analyzed with the fetal fraction optimized risk of trisomy evaluation (FORTE) algorithm, which estimates aneuploidy risk using an odds ratio comparing a model assuming a disomic fetal chromosome and a model assuming a trisomic fetal chromosome. FORTE incorporates different risk factors such as maternal and gestational age (the most relevant biological factors to determine a priori fetal trisomy risk) to provide a fetal trisomy risk score in the context of the specific FF% [8].

Single Nucleotide Polymorphisms (SNP)-Based

The SNP-based method targets specific chromosomes and differs from the other technologies as it does not require reference chromosomes for comparison. The test is based on targeted amplification and sequencing of cfDNA fragments belonging to chromosomes of interest. The test has gone through several iterations, with version 3 being introduced in 2016. In version 3, the current clinical test, the sequencing of maternal DNA from the buffy coat, the collection of paternal swab, and a portion of SNPs (from $\approx 20,000$ to $\approx 13,000$) required in version 1 and/or 2 were eliminated [9, 17–21]. In addition, reflexing cases with low FF to deeper sequencing (from 4.3 to 11 million reads) led to a decrease in the FF% cutoff (from 4% to 2.8%) [22].

Essentially, version 3 performs both a quantitative and a qualitative analysis of the obtained SNP profiles. With the quantitative analysis, it estimates the FF% using SNPs, like some of the other technologies. However, it is the qualitative analysis of the obtained sample's SNP profile that allows the formulation of the fetal trisomy risk score using NATUS (next-generation aneuploidy test using SNPs) algorithm. This algorithm uses Bayesian statistics and maximum likelihood estimation to first create hypotheses for various fetal genotypes. Subsequently, it tests these expected genotypes against actual observed results, thereby not requiring the use of a disomic reference chromosome. The NATUS algorithm considers parental genotypic information, crossover frequency data, and linkage disequilibrium to predict possible fetal euploid and aneuploid genotypes. The algorithm predicts what the sequencing data would be expected to look like for each of these hypothetical genotypes at various fetal cfDNA fractions. By comparing these hypothetical genotypes to the output of a specific test sample, the algorithm can call both fetal copy number and fetal fraction, thus generating a risk score [9, 23].

Fetal Fraction Estimation

The most used methods for fetal fraction estimation with MPSS are based on sex chromosome ratio or trisomy fraction [12, 24, 25] or on differences between maternal and fetal cfDNA characteristics such as methylation [26], distribution of maternal and placental cfDNA fragment size [27], indirect sequencing variables [28], DNA digestion of fetal and maternal cfDNA related to different nucleosome positioning [29], and preferred ending sites for maternal and fetal cfDNA [30].

With sex chromosome ratio or trisomy fraction, the FF is measured in male, monosomy X, or trisomic pregnancies from the count data of X and Y chromosomes or chromosome of interest, allowing the FF assessment of a subgroup (male and abnormal) pregnancies [12, 24, 25].

The remaining methods are based on differences in metabolic and enzymatic cell processes affecting maternal and trophoblastic chromatin remodeling whose physiology is essentially unknown. For example, the differential methylation method is based on five regions that are differentially methylated between maternal and placental DNA [26]. Using methylation-sensitive restriction enzymes to degrade the maternal (unmethylated) background fraction of the DNA sample, the remaining undigested fetal (placental) DNA fraction is then analyzed and quantified. This method resulted in a variable FF estimation due to the limited number of differentially methylated regions available for testing in addition to the fact that the digestion of nonmethylated cytosine might be incomplete, resulting in an overestimation of the FF [31].

Methods for FF estimation based on fragment size indirectly deduce this information from the overall plasma DNA size distribution for each maternal plasma DNA sample using a regression equation obtained from a training set of tested pregnancies, therefore without a specific molecular assay [27]. The principle behind this method is that there is a difference of average size of fetal and maternal fragments [32, 33]. Circulating DNA molecules show a predictable fragmentation pattern reminiscent of nuclease-cleaved nucleosomes, with the fetal DNA showing a relative enrichment of fragment distribution at 143 bp when compared with maternal DNA that showed an enrichment of fragment distribution at 166 bp. These methods were found to be less reliable than methods based on read counts on the Y chromosome [34]. This is likely due to the overlap of maternal and fetal fragment distributions at 143 bp. In addition, studies on the size distribution of circulating plasma fragments in unpregnant female patients with systemic lupus erythematosus revealed skewed molecular size-distribution profiles with a significantly increased proportion of short DNA fragments [35]. Little is known about possible concurrent metabolic processes skewing maternal cfDNA fragmentation during pregnancy (acute and chronic inflammation, exercise-induced overtraining, immune and autoimmune diseases, vascular diseases, cancer, etc.). Therefore, it was suggested to use this method to obtain only a basic indication of fetal fraction in female pregnancies [34].

Fetal fraction estimation by SNP genotyping is applied by DANSR and SNP-based cfDNA tests and by tNGS methods based on the selection of cfDNA fragments mapping SNP loci. This method is based on the quantification of SNP variants circulating in maternal plasma differing from those of the mother, thereby deriving from the paternal component released by the conceptus. SNP genotyping of a pure female sample DNA detects an SNP profile including the two homozygous genotypes, AA and BB, together with the heterozygous, AB (Fig. 28.1A).

The profile of the cfDNA of a pregnant woman, including a prevalent maternal component with a "contaminant" fetal component, shows additional genotypes (Fig. 28.1B) related to the paternal variants differing from that of the mother. This additional small component appears physically shifted from the maternal contribution. The quantity of this shift is proportional to the FF percentage and allows for the FF calculation. With DANSR technology, the fetal fraction is measured by testing 576 SNPs located on reference chromosomes 1–12 [6]. The

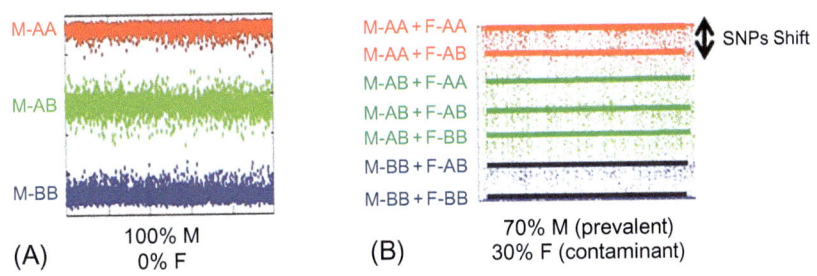

FIG. 28.1 General principle of fetal fraction estimation by SNP genotyping (targeted NGS; DANSR; SNP-counting); (A) SNP genotyping profile of a pure maternal sample (genomic maternal DNA extracted from the buffy coat); (B) SNP genotyping profile of a maternal plasma cfDNA sample with a fetal fraction of 30%. The entity of the shift of the maternal from fetal/paternal SNPs profiles is proportional to the amount of FF. AA and BB: homozygous SNP loci; AB: heterozygous SNP loci; M: mother; F: fetus.

validation of this method of FF quantification was done by SNP genotyping against Y fetal chromosome percent, conducted on 50,000 singleton male pregnancies. This validation showed a very high correlation coefficient ($r = 0.97$). The repetition of FF measurement on the second tube of the same sample draw conducted on 768 randomly selected samples showed a high reproducibility of the method with a correlation coefficient between the first and second measurement of 0.97 (oral communication, paper under revision). With SNP-based technology, the fetal fraction is measured against thousands of SNPs located on chromosomes of clinical interest and does not require reference chromosomes.

It is questionable whether all these methods, and hence the FFs reported in the different studies, are comparable. At present, no study concerning the reproducibility of FF measurement between laboratories has been found in the literature. It would be interesting to investigate the variation in results of FF analyzed in different laboratories and with different methods. Furthermore, it seems important to generate an international standardization of FF, ensuring comparability and consistency between laboratories.

A summary of the main characteristics of the different cfDNA testing technologies is presented in Table 28.1.

ANALYTIC VALIDITY, CLINICAL VALIDITY, AND CLINICAL UTILITY

Prior to the advent of cfDNA testing, aneuploidy screening was based on maternal serum markers alone in the second trimester and a combination of markers and ultrasound in the first trimester. With any new technology, it is critical to determine the performance characteristics of the new test, whether it works in the "real" world and, most importantly, whether the test truly impacts care. These analyses are called, respectively, analytic validity, clinical validity, and clinical utility. The CDC's Office of Public Health Genomics (OPHG) developed the ACCE Model Project from 2000 to 2004 to help guide how new genetic tests are evaluated (https://www.cdc.gov/genomics/gtesting/ACCE/), a model still used around the world.

TABLE 28.1 Summary of the Technical Characteristics of the Described Technologies

Technology	Fetal Trisomy Risk Determination			Fetal Fraction Determination		Technical Aspects				Examples of Tests
	Principle	Statistics	Triploids	Principle	DNA Contamination Detection—FF Estimation of Each Fetus in Dizygotic Twin Pregnancies	Sample Analysis	Counting Depth	Maximum Sample Throughput	Possibility to Add Further Clinically Validated Targets	
MPSS	Counting ratio	z-score/NCV	Not detectable	Multiple approaches other than SNP-genotyping	No as SNPs are not tested	Multiplex by NGS	Variable[a]	48 samples/run	Possible by unlocking bioinformatic analysis of the fragment counts belonging from other chromosomes	HiSeq/NextSeq (Illumina) or Ion Proton™ (Thermo Fisher Scientific) MPSS platform-based technologies which are offered under multiple local brands (e.g., VeriSeq, Verifi, Neobona, MaterniT, G-test, Tranquility, IONA, InformaSeq, Prelude, SafeTest, etc.)
Targeted-NGS	Counting ratio	z-score/NCV	Not detectable	SNP-genotyping or fragment size	Yes if SNPs are tested	Multiplex by NGS	Variable[a]	192 sample/run	Possible by adding new complementary molecular probes in the assay	Clarigo (Multiplicom, Agilent), Veracity (NIPD Genetics)
DANSR	Counting ratio	FORTE (incorporating FF%)—Odds ratio	Not detectable	SNP-genotyping	Yes	Singleplex by microarray	Standardized	384 samples/run	Possible by adding new complementary DANSR probes in the assay	Harmony Prenatal Test (Roche)
SNP-counting	SNP genotyping	NATUS (incorporating FF%)—Bayesian statistics with Maximum Likelihood Estimation	Detectable	SNP-genotyping	Yes	Multiplex by NGS	Variable[a]	48 samples/run	Possible if enough informative SNPs are present in the target chromosome region	Panorama (Natera)

[a] Depending on the size of NGS cartridge and sample input (see explanation in the text).

MPSS, massive parallel shotgun sequencing; FF, fetal fraction; NGS, next generation sequencing; DANSR, digital analysis of selected regions; SNP, single nucleotide polymorphism; NCV, normalized chromosome value; FORTE, fetal fraction optimized risk of trisomy evaluation; NATUS, next-generation aneuploidy test using SNPs.

Analytic Validity—Sensitivity and Specificity

There are numerous analyses involved in determining whether a test works in the laboratory. Many of these measures reflect upon laboratory quality. For example, are results reproducible or are quality assurance programs available? From a clinical perspective, the focus at the early stage of laboratory development revolves around two key performance characteristics: sensitivity and specificity.

Sensitivity (true positive rate, also frequently referred to as detection rate or DR) refers to the ability of a test to identify a genetic change when it is present. If a woman has an aneuploid fetus, how often will the new screening test be positive? The benchmark set by standard screening approaches were sensitivities of 50%–95% for fetal Down syndrome, depending on the screening strategy (http://www.acmg.net/docs/NIPS_AOP.pdf).

To assess the sensitivity of a new test such as cfDNA testing, the laboratory will typically use stored blood samples from mothers with the fetal aneuploidy of interest to determine how often the test identifies the samples correctly. It was already clear from 2011 onward that cfDNA testing was superior, with detection rates ranging from 98.6% to 100% for fetal Down syndrome [15, 19, 36–39]. While standard screening with biomarkers and/or ultrasound is primarily used to detect trisomy 21 and trisomy 18, cfDNA testing had the additional benefit of higher sensitivities for other aneuploidies as well, with some studies reporting sensitivities as high as 100% for trisomy 18 [38, 40–42] and 13 [40–43] and even many monosomy X studies reported 100% sensitivity [40–43].

Specificity (true negative rate; closely related to the false positive rate or FPR, which is equal to 1-specificity) refers to the ability of a test to report a negative screen result when in fact there is no aneuploidy while the FPR refers to the percentage of nonaneuploid samples that were initially reported as aneuploid.[1] The early cfDNA studies across all platforms demonstrated not only superior sensitivity, but also superior specificities for fetal Down syndrome ranging from 97% to 100% [19, 36–40]. Similarly, cfDNA testing also appeared to have very high specificities for other aneuploidies, between 93% and 100% for trisomy 18 [15, 38, 40–42], 84%–100% for trisomy 13 [15, 40–43], and 99%–100% for monosomy X [19, 38, 40, 42]. Compared to specificities of 95% for standard screening, cfDNA testing was a substantial improvement.

Systematic reviews have attempted to combine these various studies to best determine the performance characteristics of cfDNA testing. Iwarsson et al. [44] found that pooled sensitivity for Down syndrome was 0.993 and specificity was 0.999 for the general risk population, similar to performance characteristics in high risk groups. Sensitivity and specificity for trisomy 13 and 18 could not be calculated due to low numbers. Gil et al. [45] recently updated

---

[1] False positive rate (FPR), in this chapter, is the ratio between the number of euploid (normal) fetuses whose test results are interpreted as abnormal (false positives) and the total number of euploid fetuses (false positives plus true negatives); this equals one minus the specificity. This concept of FPR should not be confused with another way of interpreting this concept, in which it corresponds to the ratio between the number of false positives and the total number of cases reported as abnormal (false positives plus true positives), which would correspond to one minus the positive predictive value. Similarly, false negative rate (FNR) is the ratio between the number of abnormal aneuploid fetuses whose test results are interpreted as normal (false negatives) and the total number of aneuploid fetuses (false negatives plus true positives); this equal one minus sensitivity.

their previous metaanalysis on this subject. The respective weighted pooled detection rates and false-positive rates (FPRs) in singleton pregnancies were: 99.7% (95% CI 99.1%–99.9%) and 0.04% (95% CI 0.02%–0.08%) for T21; 98.2% (95% CI 95.5%–99.2%) and 0.05% (95% CI 0.03%–0.07%) for T18; 99.0% (95% CI 65.8%–100%) and 0.04% (95% CI 0.02%–0.07%) for T13; and 95.8% (95% CI 70.3%–99.5%) and 0.14% (95% CI 0.05%–0.38%) for monosomy X (MX). These data are the result of a pooled analysis of 35 relevant peer-review studies in which data on pregnancy outcome were provided for more than 85% of the study population to avoid reporting bias. Case-control studies, proof-of-principle articles, and studies in which the involved laboratory scientists were unblinded and thus aware of fetal karyotype or pregnancy outcome were excluded. These 35 studies are based on different molecular methodologies (MPSS, DANSR, and SNP-based).

Clinical Validity

Clinical validity speaks to the question of once there is evidence that a test works in a lab, how well will it perform in the clinical setting? A common error is the belief that lab results with excellent sensitivities and specificities will be reflected in the population at large. Sensitivity and specificity are derived from known, usually stored samples. However, women who undergo screening do not know if they have an affected fetus. In fact, in most circumstances it is likely that the fetus is normal. While sensitivity asks the question, "How often will a test be positive if we already know the status of the fetus?," what providers and patients really want to know is, "If the test is positive, is my baby really affected?," or what is known as "positive predictive value" (PPV). The opposite question is also important, "What is the chance my baby is really normal if the test says I am screen negative?," which describes the negative predictive value (NPV) of a test. To answer that question, the prevalence of a disorder becomes paramount. Even if a screening test has almost perfect performance characteristics, if the disorder of interest is rarely seen, a positive screen will, most times, be a false positive. Conversely, a test that has relatively poor screening performance characteristics may still perform reasonably well in a population where almost everyone has the disorder of interest, that is, a high-risk population. In the case of aneuploidy screening, Down syndrome is relatively common, especially in older maternal ages. However, trisomy 18 and 13 are relatively rare. In addition, there are biological complexities to cfDNA testing, such as placental mosaicism, that can affect the positive and negative predictive values by altering the sensitivities and specificities ascertained from known samples.

Due to the very high reported cfDNA testing sensitivities (i.e., many studies reporting 100%), there was the mistaken impression that cfDNA results were diagnostic. Following a few high-profile false positive cases [46], there was increasing recognition that studies were needed to accurately assess PPV and NPV, especially in the average risk population. In a prospective multicenter blinded study that included 18,995 women, the positive predictive value of cfDNA testing for Down syndrome was 80.9%, substantially better than standard screening that had a PPV of 3.4% [47]. However, this study demonstrated clearly that 100% sensitivity does not translate to 100% PPV, which is actually correlated more closely with specificity, or the symmetric FPR, rather than sensitivity.

For other aneuploidies, Dar et al. [21] found PPVs of 31.1% and 50.0% for trisomy 13 and monosomy X, respectively, in cytogenetically confirmed cases. The lower PPVs for chromosomes 13 and monosomy X may be related to placental mosaicism, as data indicates that these two chromosomes are more likely to exist in mosaic cell lines in the trophoblast [48, 49]. This higher rate of placental mosaicism in trisomy 13 and monosomy X may influence the choice of confirmatory diagnostic procedure. Amniocentesis, which, unlike CVS, interrogates fetal and not placental cells, may be more appropriate in these cases [50]. Because sensitivity and specificity do not adequately reflect the real-life performance of cfDNA testing and, due to the actual PPVs being far less than 100%, confirmatory testing is strongly recommended in the presence of a positive report.

Clinical Utility

While the above discussion clearly demonstrates that cfDNA testing for the major aneuploidies has met the bar for analytic and clinical validity, the outstanding question regarding clinical utility remains. Does the test impact care and outcomes? For example, an outstanding screening test with near perfect sensitivity would still not meet ACCE criteria if there was no management option for a screen-positive report. CfDNA testing, particularly for the trisomies, has prompted major practice shifts related to invasive testing. There are now several studies demonstrating a substantial decrease in amniocentesis and chorionic villus sampling (CVS) due to cfDNA testing. Johnson et al. [51] reported a national decline of 53.7% for diagnostic procedures between 2012 and 2015, although a lower drop at a major academic center due to increased testing for ultrasound anomalies generated by increased nuchal translucency measurements ≥3.5 mm. Larion et al. [52] reported similar declines following the introduction of cfDNA, with CVS dropping 68.6% and amniocentesis dropping 78.8%. This clear decline in procedures is used as a measure of clinical utility because it has significantly impacted care, so much so that there has been considerable concern regarding whether there will be an adequate number of CVS cases to train the next generation of fetal medicine experts [53] as well as cytogeneticists for karyotyping and regarding the missed chromosome abnormalities that are usually detected after an invasive procedure (see paragraph "Limitations of cfDNA testing").

Clinical Utility for Other Chromosomal Anomalies

There is no established clinical utility for trisomy 13 and 18, as they are never ordered alone and most will be picked up with sonography. Over time, other cfDNA testing became available but supporting data regarding clinical validity and utility are still limited.

Clinical Utility for Sex Chromosome Aneuploidy (SCA)

In the Gil et al. metaanalysis [45], for SCAs beyond monosomy X, the detection rate was 100% with a false positive rate of 0.004%. An ongoing issue is simply the small sample size of these studies, where even in this metaanalysis, there were only a combined total of 17 cases of SCA other than monosomy X. In some countries, screening for SCAs is not considered appropriate. Other professional organizations such as the American College of Obstetrics and Gynecology (ACOG) do not recommend SCA cfDNA screening as a front-line test, but it can be made available to patients upon request [54]. There is evidence that early treatment of Klinefelter syndrome early in childhood with testosterone can lead to better outcomes and studies may demonstrate the clinical utility of cfDNA testing for SCAs in the future [55].

Clinical Utility in Twins

Liao et al. [56] performed a metaanalysis in twins and reported a pooled sensitivity of 99% and a specificity of 100% for fetal Down syndrome, but the difference of detection rates between monozygotic and dichorionic twin pregnancies was not available as the number of monozygotic twins was limited. For trisomy 18 screening, the pooled sensitivity was 85% and specificity was 100%. There were only three cases of trisomy 13, but all were detected. In a recent metaanalysis, Gil et al. [45] similarly determined strong performance characteristics such that the detection rate for twins was 100% and specificity was 100%. Due to the limitation regarding numbers, a robust prospective clinical validation study will remain challenging.

Clinical Utility of 22q11.2DS

There are now a few studies describing success with cfDNA testing and the detection of deletions in the critical region of chromosome 22 that can lead to the relatively common 22q11.2 deletion syndrome (22q11.2DS), which occurs in approximately 1/1000 otherwise low-risk pregnancies. [57] This syndrome can have major consequences if not detected early, which is unfortunately most often the case. Validation studies thus far have been in high-risk populations where fetuses have been identified with cardiac anomalies. Detection rates in such settings have ranged from 44.2% to more than 80% [58, 59].

While there may be a case to be made for early detection, there are no studies available yet demonstrating performance parameters in an average risk population. Furthermore, ultrasound may be able to detect approximately 80% of fetal 22q11.2DS because of the association with significant cardiac abnormalities [60]. Therefore, the added benefit of cfDNA screening may be significantly reduced if detailed ultrasound can detect associated anomalies and provide patients the option of a full diagnostic microarray analysis. In addition, newborn screening may become a reality in the future, which would further help in providing early diagnosis and prevention of morbidity and mortality in affected children. Thus, further research is required before clinical validity in the average risk population and clinical utility data will be available to justify routine screening for all pregnant women.

Clinical Utility of Other Microdeletions

There are a few papers that have looked at cfDNA testing for non-22q deletions. In one such paper, using the SNP method, the authors had a 31.7% PPV, and the overall false positive rate for four other major deletion syndromes was 0.07% [58]. Using MPSS, 55 subchromosomal deletions were reported with an overall PPV ranging from 60% to 100% when follow-up was available with a false positive rate of 0.0017% [59]. However, like the previous discussion, there is minimal data on general risk populations and no prospective data as to whether there would be any tangible benefits with respect to outcome. A recent paper reported a PPV of 0 in the general population for some of these microdeletion syndromes [61]. Currently, these tests are available but professional societies generally caution that cfDNA testing for microdeletion syndromes is not currently recommended [62].

Clinical Utility in Genome-Wide Imbalances

Ehrich et al. [63] reported on the use of cfDNA testing for genome-wide screening. Using MPSS that interrogated the genome beyond the typical trisomies and SCAs, they could make positive calls on 554 patients out of 10,000. A total of 164 cases would have been missed using only the typical cfDNA testing panel. Of note, the authors do point out that the population was skewed toward higher risk. There are some important issues to consider beyond increased cost and the recurrent concern regarding generalizability beyond high-risk populations. Patients, especially those with fetal anomalies, should be offered invasive testing using microarray analysis, which will detect microdeletions that will go undetected with genome-wide cfDNA testing.

In addition, if anomalies are already present, the delivery team and neonatal unit will be sufficiently prepared regardless of whether a screening test is available, again limiting potential clinical utility.

A summary of the definitions of analytical performances, clinical performances, and clinical utility and the study designs requested to validate them is reported in Table 28.2.

LIMITATIONS OF cfDNA TESTING—ITS BIOLOGICAL REASONS

cfDNA Discordant Results

As outlined above, though cfDNA testing accuracy is high, it is not a diagnostic test. There is a small chance of a discordant (false-negative or false-positive) result when compared with the real fetal chromosomal constitution. There are different reasons for discordant results [64, 65]. In addition to technical issues (such as sample swap or lack of sufficient internal quality control metrics) [66], there are biological reasons for discordant results that can be of either fetal or maternal origin. The most relevant and frequent contributing fetal factors include insufficient/absent FF [28, 67] and fetoplacental mosaicism [64, 68–75].

The presence of a fetoplacental mosaicism in the conceptus, when the cytotrophoblast's chromosome constitution does not match the cytogenetic constitution of the fetus, is a major source for discordant cfDNA results [48, 68]. In particular, if a confined placental mosaicism (CPM) type I or III, that is, a combination of abnormal cytotrophoblast but normal fetal karyotype, is present, an FP cfDNA result may be detected by cfDNA testing. Conversely, in true fetal mosaicism (TFM) type V, a combination of a normal cytotrophoblast but an abnormal fetal karyotype, there is a possibility of an FN cfDNA result [65]. Increasingly, mosaicism percentage in the cytotrophoblast of CPM I/III conditions predicts progressively higher FP rates. Assuming that the cfDNA technologies can detect a 30% mosaic aneuploidy in the cytotrophoblast as "positive," the predicted chromosome-specific FPR consequent to the presence of a CPM I/III in the conceptus is 1/7372 for T21, 1/7472 for T18, and 1/3754 for T13 (see footnote 1). The estimated FNR due to the presence of TFM V is independent of the level of mosaicism because the cytotrophoblast is normal while the fetus is abnormal; it is 1/135 for T21, 1/64 for T18, and 1/136 for T13. For homogeneous sex chromosome aneuploidies, the combined FPR consequent to fetoplacental mosaicisms ranges between 0.04% and 0.07% while the FNR ranges from 0% to 5.7%, depending on the type of aneuploidy [48]. These data imply that series with less than 100 cases will not be reliable for assessing a "real world" detection rate.

TABLE 28.2   Summary of the Definitions of Analytical Performances, Clinical Performances and Clinical Utility and the Study Designs Requested to Validate Them

	Performance Measure	Relevant Questions	Study Design
Analytic validity: does this test work in the laboratory setting?	Sensitivity (also called "true positive rate")	How often is the test positive when a mutation is present?	Usually retrospective case control design: Assess proportion of positive results among affected (often stored) samples that is correctly called
	Specificity (also called "true negative rate")	How often is the test negative when a mutation is not present?	
Clinical validity: has this test been adequately validated on the populations to which it may be offered?	Prevalence	What percentage of the population in question is affected with the specific disorder? The NPV and PPV are both dependent on prevalence (see text)	Prospective blinded studies where patients receive both the new test and the previous "gold standard" test
	What is the positive predictive value (PPV)?	What is the chance that if the test is positive, the patient does have the disorder? PPV = number of true positives/(number of true positives + number of false positives)	
	What is the negative predictive value (NPV)	What is the chance that if the test is negative, the patient does not have the disorder? NPV = number of true negatives/(number of true negatives plus false negatives)	
Clinical utility: what is the impact of the test on patient care?	Are there effective interventions available in the case of a positive screen result?		Prospective, blinded, adequately powered randomized controlled trials to determine if there is any impact on clinical outcome (negative or positive)
	If applicable, are diagnostic tests available?		
	Is there general access to the intervention?		
	What are the financial costs and economic benefits associated with screening?		
	Are adequate educational materials in place?		
	What are the ethical, legal and social implications of this test?		

Modified from the ACCE Model List of 44 Targeted Questions Aimed at a Comprehensive Review of Genetic Testing (Centers for Disease Control and Prevention). https://www.cdc.gov/genomics/gtesting/acce/acce_proj.htm.

Less commonly, the presence of a vanishing twin can be the source for discordant results [76]. Cells derived from a vanished twin can still be present even after a fetal pole is no longer seen. Thus, if a vanishing twin was male with trisomy 21, Y DNA fragments and additional 21 sequences may be detected in the maternal circulation, even though the surviving twin is a normal female.

Maternally derived chromosomal abnormalities (including rare maternal occult malignancies) can also cause false-positive results (see paragraph below on "maternal incidental findings") [77–81]. Similarly, maternal organ/tissue transfer from a male donor can cause errors in fetal sex determination [82].

Another possible cause of discordant results is that of insufficient FF [31, 67], which can be influenced by several biological factors that impact its release into the maternal circulation [83–85]. FF is related to maternal weight and body mass index. Therefore, in obese women, there is an increased turnover of adipocytes that increases the relative amount of maternal cfDNA to fetal cfDNA, thereby decreasing the FF [86]. In addition, the total maternal circulating DNA increases with maternal weight, causing a decreasing FF because of the diluting effect. Therefore, maternal obesity might be a possible reason for unexplained false negative cfDNA testing results in laboratories not calculating FF [87, 88]. Another factor impacting FF is gestational age. Levels of fetal DNA increase by 0.1% per week up to 21 weeks of gestation. Beyond 21 weeks of gestation, it increases with a rate of 1% per week. The FF in maternal plasma also increases with serum pregnancy-associated plasma protein A (PAPP-A) and β-hCG; it is higher in women of Afro-Caribbean and East-Asian origin than in Caucasian or non-Hispanic origin, and decreases in IVF and twin pregnancies [89–91]. The presence of T13 and T18, but not T21, in the conceptus is reported to cause a decrease of the FF due to decreased placental mass [92]. Other chromosome abnormalities associated with smaller placenta such as triploidy are therefore also likely to cause a decrease of fetal fraction [19, 47].

No Results

Even considering the best possible performance characteristics, cfDNA tests can fail. There are different reasons for the absence of a result: (i) specimen handling issues (including administrative ones), (ii) insufficient (typically below 4%) FF, (iii) technical reasons (e.g., quality metrics of the extracted DNA, such as reduced amplification), and (iv) biological reasons. These situations apply, for example, in egg donor cases analyzed using SNP technologies in which there were wrong assumptions about the egg (e.g., a heterologous egg donation that was analyzed as a homologous donation).

The no result rate attributable to low FF ranges from 0.1% to 6.1%; the no result rate for reasons related to low quality assay ranges from 0.2% to 2.8% [45].

The reason for a no-call result is relevant information because it influences the overall test performance. A theoretical model predicted that choosing a cfDNA-based platform with effective FF metrics and the lowest no result rate reduces the number of pregnancies in which the test has lower performance due to insufficient FF while increasing PPV in the reported population [93].

PRACTICAL PROBLEMS RELATED TO THE LIMITATIONS OF cfDNA TESTING AND HOW TO MANAGE THEM

Loss of Detection of Additional Anomalies When Compared to Standard Screening

Since its inception, prenatal diagnosis of fetal chromosomic anomalies has been a two-tier approach. In the first tier, an assessment is made for the risk of trisomy 21, 18, or 13. In the second tier, a diagnostic technique is applied to screen positive cases. The diagnostic techniques can detect a much broader array of chromosomal anomalies, such as sex chromosome aneuploidies and structural anomalies. These additional defects would not have been diagnosed in the absence of a positive screening test, unless coincidently discoverable using prenatal ultrasound. However, this superior detection rate for this broad range of fetal chromosomal anomalies relies on a 5% false positive rate when using standard "pre-cfDNA" screening approaches for trisomy 21, 18, or 13.

The above reality has led to two very different ways of looking at prenatal diagnosis of fetal anomalies and the issue of the false positive rate (FPR). The first assumes that the goal of prenatal screening and diagnosis is to detect the maternal age associated anomalies, trisomy 21, 18, and 13. The second approach assumes that the goal of prenatal screening is to detect as many affected fetuses as possible, not only those with the three most common aneuploidies but also all fetuses carrying other detectable anomalies. Using this latter "serendipitous" approach, the false positive rate (FPR) is a benefit. Thus "cfDNA only," with its greatly lowered FPR, may miss some of these other nontrisomy 21/18/13 chromosomal anomalies when compared to standard screening [94]. To others, this decrease is considered beneficial because it not only lowers costs by reducing invasive diagnostic procedures, but it meets some patients' strong desire to avoid the risk of procedure-related fetal loss, even if that risk is relatively low.

Of note, the reasoning for the superiority of standard screening for the detection of other anomalies was based on a routine karyotype as the ultimate test for prenatal diagnosis of genomic defects. However, almost simultaneously with the development of cfDNA testing for screening for the common aneuploidies and SCAs, chromosomal microarrays (CMA) were being more widely adopted for pregnancies undergoing an invasive procedure. Because CMA can diagnose a much larger set of clinically relevant genomic defects, many of which are not currently screened for by most screening programs, it may strengthen the argument that favors detection of all genetic aberrations using the higher false positive rate associated with standard screening beyond the common aneuploidies.

Proponents of making invasive diagnostic testing available to all patients emphasize the risk of genetic anomalies other than the ones targeted by screening as well as the false negative rate for screening targets. Data from a study of the frequency of fetal karyotype abnormalities in women undergoing invasive testing, without ultrasound and other high-risk indications, provide the risks, per year of maternal age, for the anomalies not targeted by screening [95]. This so-called residual risk for clinically relevant anomalies other than trisomy 21, 18, and 13 in pregnancies between 15 and 20 weeks gestation ranged between 1/313 and 1/100 for women at 18 and 45 years old, respectively.

A dilemma raised by advocates of universally available prenatal diagnosis is whether the increase in detection rate of trisomy 21, 18, and 13, with or without the addition of the SCAs, heralded by cfDNA screening would be enough to compensate for the loss in the detection rate of the other anomalies caused by the decrease in the false positive rate. A study by Norton et al. partially addresses this question [96]. In this study, the results of invasive testing in a population subjected to serum/ultrasound screening-based tests (several modalities were included) were analyzed to assess the proportion of identified karyotype anomalies that were not targets of cfDNA-based testing. This study states that 16.9% of the karyotype anomalies diagnosed in women whose screening had been reported as high risk for aneuploidy, neural tube defects, Smith-Lemli-Opitz syndrome, congenital anomalies, or fetal demise and who have undergone fetal karyotype testing would not be detectable by cfDNA testing for the common trisomies and sex chromosome anomalies. Given the maternal age association with the targets of cfDNA screening, this proportion of karyotype anomalies was reported to vary between 20% and 25% for women under 25 years old to 8% in women aged 45 or older. However, there were multiple confounders with this study, including the inclusion of anomalies that may not be deleterious to the fetus and that only 37.8% of screen-positive women underwent diagnostic testing. A more recent study from the same group [97] reported a more accurate comparison between the detection rate of "sequential screening" and cfDNA testing for all chromosomal anomalies diagnosed pre- or postnatally. Their comparison was based on actual "sequential screening" data and on predicted performance of cfDNA testing. They reported a detection rate for sequential screening of 81.6% versus 70% or 72% for cfDNA testing. The two different results reported for cfDNA testing are related to whether fetal fraction was reported and the related number of inconclusive cases, assuming that these inconclusive cases would be considered as screen positive for Trisomy 13, 18, or triploidy. Still, these results are generalizable only to populations with similar maternal age distribution to that of the reported study. A per maternal age calculation would be more useful for counseling purposes.

As per the report of the study referred above of the frequency of fetal karyotype abnormalities in women undergoing invasive testing in the absence of scan anomalies [95], the addition of data from one of the largest studies of chromosomal microarrays in the prenatal setting [98] led to a prediction of the frequency of genomic defects detectable by chromosomal microarrays, after excluding the common trisomies currently incorporated into the common cfDNA panels. The frequencies of additional chromosomal anomalies beyond the common aneuploidies would vary between 1/51 (maternal age 18 years) and 1/38 (maternal age 45 years) at a gestational age greater than 15 weeks. Excluding SCAs, those risks would still be 1/55 and 1/47, respectively. These figures will surely decrease the proportion of diagnosable genomic defects that can be identified through any of the screening strategies. While cfDNA testing has expanded to include some submicroscopic deletions and duplications [99–101], cfDNA cannot at this time replace CMA. Studies addressing this possibility are based mainly on testing samples known to have such rare defects and not on samples of the general population.

A recent metaanalysis of studies addressing procedure-related fetal loss risk of amniocentesis and chorionic villous sampling has concluded that the risk associated with the procedures is likely to be much lower than previously reported [102, 103]. The studies selected for this analysis used more appropriate control groups than previous studies and are consistent with more contemporary technology. This is important because an increased risk for aneuploidy is associated with an increased risk for miscarriage and stillbirth [104] and remains a central argument to avoid invasive testing unless a woman is at high risk. These data add further strength to the proposal of offering diagnostic testing to everyone. While the safety issue may be addressed by this recent data, high cost and limited accessibility is a problem that remains unresolved.

In summary, there are two competing views regarding prenatal testing of fetal genomic defects. One values the diagnosis of the most common aneuploidies. For that view, cfDNA testing is the most efficient and accurate screening test and, as such, should be the screening test to use in everyone. The second view values the diagnosis of as many genomic anomalies as possible as per the most advanced diagnostic techniques. Presently, it is still not clear which of the two approaches—cfDNA based or serum/ultrasound based screening—has highest detection rate for genomic abnormalities.

Advances in cfDNA-based testing and next-generation sequencing of fetal DNA, plus the ethical dilemmas related to these new technologies, are likely to keep these debates unresolved for the immediate future.

The "No Results" Issue

Most laboratory tests have a failure rate and cfDNA testing is no exception. As briefly mentioned above, there are two main reasons for failure: technical-related failure and low FF. Technical-related failures may sometimes be easily explained. Obvious and easily identifiable reasons are related to collection, transport, and quality of the actual sampling and transportation method. This should be clear even prior to starting the assay. Other reasons include failure of DNA extraction, failure of sample amplification, or assay output that does not meet quality control metrics. Thus, if there are obvious identifiable errors at these levels, one reasonable way of dealing with the issue would be to repeat sampling.

The reason for FF failure is outlined in the detailed technical explanation above. Measuring FF is a best practice as it helps explain the reason for test failure and no result. It is also a good quality control metric in addition to the failure rate of the laboratory. There are other clinical reasons for knowing the FF and whether this was the reason for a potential failed test. The current consensus among the majority of cfDNA testing providers considers that samples with FF lower than a 4% result should not be reported due to the increased FNR. Furthermore, in cases in which there is no test failure despite lower FF, the results are likely to be less reliable, for obvious—albeit hypothetical—mathematical reasons [93]. Although there are studies that claim the same reliability despite lower FF samples [105, 106], such studies are actually reporting failure rates and do not report prospective outcomes such as the increase in false negative rates.

If a result is reported as failed or no result for low quality metrics, and if the FF is measured and is above 4%, a redraw may return a result, although the current literature is not explicit in distinguishing the success rate of a redraw specifically under these circumstances. By common sense, though, if the FF was above 4% but lower than 10%, it is likely that sampling later in the pregnancy will return a higher FF, making it more likely to provide an interpretable result. However, in obesity, this may not be the case. In obese patients, resampling for low quality metrics in combination with a FF level between 4% and 10% on the first sample is less likely to be successful [4, 107]. This is a limitation of this type of testing that should be raised as an issue during the pretest counseling. Obese patients (BMI > 30 kg/m^2) should be warned that their failure rate is likely to be higher, up to 24.3% [107], and potentially nonrecoverable from a redraw.

If a result is reported as failed and the FF is lower than 4% or not reported (thus assumed to be low), there is an increased risk for trisomy 18 and 13. Depending on the gestational age at the time of first draw and patient BMI, a choice must be made between retesting (favored in the case of early gestational age and low BMI), or an ultrasound assessment of the anatomy of the fetus looking for ultrasound markers of these chromosomal anomalies (favored if later gestational age and low BMI) or invasive-based diagnostic procedures (favored in patients with ultrasound anomalies, or in high BMI patients with a poor ultrasound assessment) [108].

It is therefore important when selecting the laboratory where to send their patients' samples that the physicians favor laboratories that report FF, that do not report results on samples with low FF, and that have the lowest failed results rate for technical reasons [93, 109].

The Problem of the "Maternal Incidental Findings"

An unintended consequence of cfDNA testing is the identification of maternal conditions. Some of these incidental findings are related to the presence of nondiagnosed cancer in the patient. This has been demonstrated mainly through case reports and, thus, is likely to be rare [110, 111]. Such a type of incidental finding should be suspected when a report returns a complex result suggesting the presence of several autosomal anomalies, as it would correspond to a mixture of DNA originating from normal and cancer cells, frequently rich in such multiple defects. A prospective study reports a risk of maternal occult malignancy of 19% for complex abnormalities detected by cfDNA testing [112]. However, there is not yet enough data to reliably determine the detection rate, false positive rate, and the PPV for occult cancer in women undergoing cfDNA testing for fetal aneuploidy screening.

Maternal mosaicism for autosomes has also been identified but is usually described as single cases or a series of case reports, also likely rare [113]. A normal fetal sonogram following a positive cfDNA test for trisomy 18 or 13, which should be readily detectable with prenatal ultrasound, could raise the possibility of maternal and not fetal mosaicism.

More frequent are the cases where the mother carries a mosaic cell line of 45,X or 46,XXX. Such cases are an important source of false positives for fetal SCAs in addition to confined placental mosaicism, as previously mentioned. The frequency of SCAs among abnormal cfDNA testing results has been reported to be around 40% [80]. The absolute frequency of SCAs has been found to be between 0.4% and 0.6% of patients undergoing cfDNA testing for prenatal screening [114–117]. Wang, Y et al. studied the X chromosome of mothers diagnosed with SCAs by cfDNA testing [80]. They found 8.7% (16/187) of those patients as being carriers of a maternal mosaic. Six had an abnormal chromosome X gain and 10 had an abnormal chromosome X loss.

The frequency of maternal mosaicism for the X chromosome loss is likely to increase with maternal age, as the somatic loss of the inactive X chromosome has been previously described as a relatively common occurrence in normal women. Mosaic 45,X was identified in 0.07% of girls younger than 16 and in 7.3% of women older than 65. [118]

Another less common source of false positives for fetal SCAs is the presence of chromosome X maternal copy number variation. Wang identified such a type of maternal genomic defect in two of 25 patients with high-risk results for cfDNA testing for SCAs and slight deviations of chromosomes X and Y from normal [119].

Thus, an abnormal cfDNA test related to the number of X chromosomes should probably be investigated in the mother before the fetus. However, it is important to note that a finding in the mother does not necessarily exclude the same abnormality in the fetus. Although the X chromosome related incidental findings are unlikely to significantly impact maternal health, these results may cause anxiety and thus it is important to warn patients undergoing cfDNA testing of this possibility.

CONCLUSION

The technical approach to cfDNA testing for fetal aneuploidy has evolved since its introduction into clinical practice in 2010. Similarly, our understanding of its potential applications and current limitations has challenged existing prenatal screening and diagnostic paradigms. Recent cfDNA testing success is narrowly focused on the high detection rate of the three most common aneuploidies with the much more dramatic decrease in the false positive rate. While a technically challenging test, cfDNA testing is easier to scale up than ultrasound-based screening. This makes it easier to standardize and, consequently, potentially decreases the error associated with variation in the provider quality.

However, the number of labs providing cfDNA testing keeps increasing, either by technology transfer from the primary developers or by development of their own "home-brew" technologies, usually by modifying the original assays. This increase in availability, competition, and consequent affordability expands the likelihood of error, especially because standardization and quality control programs are still in the process of being developed for this technology. Reports from a recent preliminary voluntary quality assessment survey to cfDNA testing reporting for MPSS-based technologies only, in preparation for the development of such a quality control program, have indeed uncovered suboptimal practices from some of the laboratories assessed [120]. This corroborates the need for caution and care that medical providers should take when selecting the laboratory whose services will be offered to their patients and ensuring appropriate pre- and posttest counseling is provided.

CONFLICT OF INTEREST

FRG is a full-time employee of TOMA laboratory without ownership shares. She is an advisory board member for Roche. GS and FM are owners of TOMA Advanced Biomedical Assays S.p.A. TOMA is a medical genetics laboratory performing prenatal diagnosis by cytogenetic analysis and chromosomal microarray on prenatal samples as well as cfDNA testing and serum screenings for fetal aneuploidy. SJG is an employee of Sema4 Genomics and a consultant for GENOOX. JCPF is a consultant for Genomed SA, which is a clinical genetics laboratory that provides molecular genetic tests, although not karyotypes or chromosome microarray analysis. KB reports no conflicts of interest.

References

[1] Tjoa ML, Cindrova-Davies T, Spasic-Boskovic O, Bianchi DW, Burton GJ. Trophoblastic oxidative stress and the release of cell-free feto-placental DNA. Am J Pathol 2006;169:400–4.

[2] Flori E, Doray B, Gautier E, Kohler M, Ernault P, Flori J, Costa JM. Circulating cell-free fetal DNA in maternal serum appears to originate from cyto- and syncytio-trophoblastic cells. Case report. Hum Reprod 2004;19:723–4.

[3] Faas BH, de Ligt J, Janssen I, Eggink AJ, Wijnberger LD, van Vugt JM, et al. Non-invasive prenatal diagnosis of fetal aneuploidies using massively parallel sequencing-by-ligation and evidence that cell-free fetal DNA in the maternal plasma originates from cytotrophoblastic cells. Expert Opin Biol Ther 2012;12(Suppl. 1): S19–26.

[4] Wang E, Batey A, Struble C, Musci T, Song K, Oliphant A. Gestational age and maternal weight effects on fetal cell-free DNA in maternal plasma. Prenat Diagn 2013;33(7):662–6.

[5] Ashoor G, Poon L, Syngelaki A, Mosimann B, Nicolaides KH. Fetal fraction in maternal plasma cell-free DNA at 11-13 weeks' gestation: effect of maternal and fetal factors. Fetal Diagn Ther 2012;31(4):237–43.

[6] Juneau K, Bogard PE, Huang S, Mohseni M, Wang ET, Ryvkin P, et al. Microarray-based cell-free DNA analysis improves noninvasive prenatal testing. Fetal Diagn Ther 2014;36(4):282–6.

[7] Stokowski R, Wang E, White K, Batey A, Jacobsson B, Brar H, et al. Clinical performance of non-invasive prenatal testing (NIPT) using targeted cell-free DNA analysis in maternal plasma with microarrays or next generation sequencing (NGS) is consistent across multiple controlled clinical studies. Prenat Diagn 2015; 35(12):1243–6.

[8] Sparks AB, Struble CA, Wang ET, Song K, Oliphant A. Noninvasive prenatal detection and selective analysis of cell-free DNA obtained from maternal blood: evaluation for trisomy 21 and trisomy 18. Am J Obstet Gynecol 2012;206(4):319.e1–9.

[9] Zimmermann B, Hill M, Gemelos G, Demko Z, Banjevic M, Baner J, et al. Noninvasive prenatal aneuploidy testing of chromosomes 13, 18, 21, X, and Y, using targeted sequencing of polymorphic loci. Prenat Diagn 2012;32(13):1233–41.

[10] Koumbaris G, Kypri E, Tsangaras K, Achilleos A, Mina P, Neofytou M, et al. Cell-free DNA analysis of targeted genomic regions in maternal plasma for non-invasive prenatal testing of trisomy 21, trisomy 18, trisomy 13, and fetal sex. Clin Chem 2016;62(6):848–55.

[11] Chiu RW, Chan KC, Gao Y, Lau VY, Zheng W, Leung TY, et al. Noninvasive prenatal diagnosis of fetal chromosomal aneuploidy by massively parallel genomic sequencing of DNA in maternal plasma. Proc Natl Acad Sci U S A 2008;105(51):20458–63.

[12] Fan HC, Blumenfeld YJ, Chitkara U, Hudgins L, Quake SR. Noninvasive diagnosis of fetal aneuploidy by shotgun sequencing DNA from maternal blood. Proc Natl Acad Sci U S A 2008 Oct 21;105(42):16266–71.

[13] Jensen TJ, Zwiefelhofer T, Tim RC, Džakula Ž, Kim SK, Mazloom AR, et al. High-throughput massively parallel sequencing for fetal aneuploidy detection from maternal plasma. PLoS ONE 2013;8:e57381.

[14] Chiu RW, Akolekar R, Zheng YW, Leung TY, Sun H, Chan KC, et al. Non-invasive prenatal assessment of trisomy 21 by multiplexed maternal plasma DNA sequencing: large scale validity study. BMJ 2011;c7401:342.

[15] Palomaki GE, Kloza EM, Lambert-Messerlian GM, Haddow JE, Neveux LM, Ehrich M, et al. DNA sequencing of maternal plasma to detect Down syndrome: an international clinical validation study. Genet Med 2011; 13(11):913–20.

[16] Sparks AB, Wang ET, Struble CA, Barrett W, Stokowski R, McBride C, et al. Selective analysis of cell-free DNA in maternal blood for evaluation of fetal trisomy. Prenat Diagn 2012;32(1):3–9.

[17] Nicolaides KH, Syngelaki A, Gil M, Atanasova V, Markova D. Validation of targeted sequencing of single-nucleotide polymorphisms for non-invasive prenatal detection of aneuploidy of chromosomes 13, 18, 21, X, and Y. Prenat Diagn 2013;33(6):575–9.

[18] Nicolaides KH, Syngelaki A, del Mar Gil M, Quezada MS, Zinevich Y. Prenatal detection of fetal triploidy from cell-free DNA testing in maternal blood. Fetal Diagn Ther 2014;35(3):212–7.

[19] Pergament E, Cuckle H, Zimmermann B, Banjevic M, Sigurjonsson S, Ryan A, et al. Single-nucleotide polymorphism-based noninvasive prenatal screening in a high-risk and low-risk cohort. Obstet Gynecol 2014;124(2 Pt 1):210–8.

[20] Samango-Sprouse C, Banjevic M, Ryan A, Sigurjonsson S, Zimmermann B, Hill M, et al. SNP-based non-invasive prenatal testing detects sex chromosome aneuploidies with high accuracy. Prenat Diagn 2013; 33(7):643–9.

[21] Dar P, Curnow KJ, Gross SJ, Hall MP, Stosic M, Demko Z, et al. Clinical experience and follow-up with large scale single-nucleotide polymorphism-based noninvasive prenatal aneuploidy testing. Am J Obstet Gynecol 2014;211(5):527.e1–527.e17.

[22] Ryan A, Hunkapiller N, Banjevic M, Vankayalapati N, Fong N, Jinnett KN, et al. Validation of an enhanced version of a single-nucleotide polymorphism-based noninvasive prenatal test for detection of fetal aneuploidies. Fetal Diagn Ther 2016;40(3):219–23.

[23] Hall MP, Hill M, Zimmermann B, Sigurjonsson S, Westemeyer M, Saucier J, et al. Non-invasive prenatal detection of trisomy 13 using a single nucleotide polymorphism- and informatics-based approach. PLoS ONE 2014;9(5):e96677.

[24] Fan HC, Quake SR. Sensitivity of noninvasive prenatal detection of fetal aneuploidy from maternal plasma using shotgun sequencing is limited only by counting statistics. PLoS ONE 2010;5(5):e10439.

[25] Rava RP, Srinivasan A, Sehnert AJ, Bianchi DW. Circulating fetal cell-free DNA fractions differ in autosomal aneuploidies and monosomy X. Clin Chem 2014;60(1):243–50.

[26] Nygren AO, Dean J, Jensen TJ, Kruse S, Kwong W, van den Boom D, Ehrich M. Quantification of fetal DNA by use of methylation-based DNA discrimination. Clin Chem 2010;56(10):1627–35.

[27] Yu SC, Chan KC, Zheng YW, Jiang P, Liao GJ, Sun H, et al. Size-based molecular diagnostics using plasma DNA for noninvasive prenatal testing. Proc Natl Acad Sci U S A 2014;111(23):8583–8.

[28] Kim SK, Hannum G, Geis J, Tynan J, Hogg G, Zhao C, et al. Determination of fetal DNA fraction from the plasma of pregnant women using sequence read counts. Prenat Diagn 2015;35:810–5.

[29] Straver R, Oudejans CB, Sistermans EA, Reinders MJT. Calculating the fetal fraction for noninvasive prenatal testing based on genome-wide nucleosome profiles. Prenat Diagn 2016;36:614–21.

[30] Chan KC, Jiang P, Sun K, Cheng YK, Tong YK, Cheng SH, et al. Second generation noninvasive fetal genome analysis reveals de novo mutations, single-base parental inheritance, and preferred DNA ends. Proc Natl Acad Sci U S A 2016;113:E8159–68.

[31] Takoudes T, Hamar B. Performance of non-invasive prenatal testing when fetal cell-free DNA is absent. Ultrasound Obstet Gynecol 2015;45(1):112.

[32] Chan KC, Zhang J, Hui AB, Wong N, Lau TK, Leung TN, et al. Size distributions of maternal and fetal DNA in maternal plasma. Clin Chem 2004;50(1):88–92.

[33] Lo YM, Chan KC, Sun H, Chen EZ, Jiang P, Lun FM, et al. Maternal plasma DNA sequencing reveals the genome-wide genetic and mutational profile of the fetus. Sci Transl Med 2010;2(61):61ra91.

[34] van Beek DM, Straver R, Weiss MM, Boon EMJ, Huijsdens-van Amsterdam K, Oudejans CBM, et al. Comparing methods for fetal fraction determination and quality control of NIPT samples. Prenat Diagn 2017;37(8):769–73.

[35] Chan RW, Jiang P, Peng X, Tam LS, Liao GJ, Li EK, et al. Plasma DNA aberrations in systemic lupus erythematosus revealed by genomic and methylomic sequencing. Proc Natl Acad Sci U S A 2014;111(49):E5302–11.

[36] Chiu RW, Akolekar R, Zheng YW, Leung TY, Sun H, Chan KC, et al. Non-invasive prenatal assessment of trisomy 21 by multiplexed maternal plasma DNA sequencing: large scale validity study. BMJ 2011;c7401:342.

[37] Palomaki GE, Deciu C, Kloza EM, Lambert-Messerlian GM, Haddow JE, Neveux LM, et al. DNA sequencing of maternal plasma reliably identifies trisomy 18 and trisomy 13 as well as Down syndrome: an international collaborative study. Genet Med 2012;14:296–305.

[38] Schnert AJ, Rhees B, Comstock D, de Feo E, Heilek G, Burke J, Rava RP. Optimal detection of fetal chromosomal abnormalities by massively parallel DNA sequencing of cell-free fetal DNA from maternal blood. Clin Chem 2011;57:1042–9.

[39] Ashoor G, Syngelaki A, Wagner M, Birdir C, Nicolaides KH. Chromosome-selective sequencing of maternal plasma cell-free DNA for first-trimester detection of trisomy 21 and trisomy 18. Am J Obstet Gynecol 2012;206:322.e1–5.

[40] Jiang F, Ren J, Chen F, Zhou Y, Xie J, Dan S, et al. Noninvasive fetal trisomy (NIFTY) test: an advanced noninvasive prenatal diagnosis methodology for fetal autosomal and sex chromosomal aneuploidies. BMC Med Genomics 2012;5:57.

[41] Chen EZ, Chiu RW, Sun H, Akolekar R, Chan KC, Leung TY, et al. Noninvasive prenatal diagnosis of fetal trisomy 18 and trisomy 13 by maternal plasma DNA sequencing. PLoS ONE 2011;6:e21791.

[42] Lau TK, Chen F, Pan X, Pooh RK, Jiang F, Li Y, et al. Noninvasive prenatal diagnosis of common fetal chromosomal aneuploidies by maternal plasma DNA sequencing. J Matern Fetal Neonatal Med 2012;25:1370–4.

[43] Guex N, Iseli C, Syngelaki A, Pescia G, Nicolaides KH, Xenarios I, Conrad B. A robust 2nd generation genome-wide test for fetal aneuploidy based on shotgun sequencing cell-free DNA in manternal blood. Prenat Diagn 2013;33:707–10.

[44] Iwarsson E, Jacobsson B, Dagerhamn J, Davidson T, Bernabé E, Heibert Arnlind M. Analysis of cell-free fetal DNA in maternal blood for detection of trisomy 21, 18 and 13 in a general pregnant population and in a high risk population—a systematic review and meta-analysis. Acta Obstet Gynecol Scand 2017;96(1):7–18.

[45] Gil MM, Accurti V, Santacruz B, Plana MN, Nicolaides KH. Analysis of cell-free DNA in maternal blood in screening for aneuploidies: updated meta-analysis. Ultrasound Obstet Gynecol 2017;50(3):302–14.

[46] Mennuti MT, Cherry AM, Morrissette JJ, Dugoff L. Is it time to sound an alarm about false-positive cell-free DNA testing for fetal aneuploidy? Am J Obstet Gynecol 2013;209(5):415–9.

[47] Norton ME, Jacobsson B, Swamy GK, Laurent LC, Ranzini AC, Brar H, et al. Cell-free DNA analysis for non-invasive examination of trisomy. N Engl J Med 2015;372(17):1589–97.

[48] Grati FR, Bajaj K, Zanatta V, Malvestiti F, Malvestiti B, Marcato L, et al. Implications of fetoplacental mosaicism on cell-free DNA testing for sex chromosome aneuploidies. Prenat Diagn 2017;37(10):1017–27. https://dx.doi.org/10.1002/pd.5138.

[49] Grati FR, Bajaj K, Malvestiti F, Agrati C, Grimi B, Malvestiti B, et al. The type of feto-placental aneuploidy detected by cfDNA testing may influence the choice of confirmatory diagnostic procedure. Prenat Diagn 2015;35(10):994–8.

[50] Malvestiti F, Agrati C, Grimi B, Pompilii E, Izzi C, Martinoni L, et al. Interpreting mosaicism in chorionic villi: results of a monocentric series of 1001 mosaics in chorionic villi with follow-up amniocentesis. Prenat Diagn 2015;35(11):1117–27.

[51] Johnson K, Kelley J, Saxton V, Walker SP, Hui L. Declining invasive prenatal diagnostic procedures: a comparison of tertiary hospital and national data from 2012 to 2015. Aust N Z J Obstet Gynaecol 2017;57(2):152–6.

[52] Larion S, Warsof SL, Romary L, Mlynarczyk M, Peleg D, Abuhamad AZ. Association of combined first-trimester screen and noninvasive prenatal testing on diagnostic procedures. Obstet Gynecol 2014;123(6):1303–10.

[53] Suskin BG, Sciscione AM, Teigen N, Jenkins TC, Wapner RJ, Gregg AR, Gross SJ, Bajaj K. Revisiting the challenges of training maternal fetal medicine fellows in chorionic villus sampling. Am J Obstet Gynecol 2016;215(6):777.e1–4.

[54] Committee Opinion No. 640. Cell-free DNA screening for fetal aneuploidy. Obstet Gynecol 2015 Sep;126(3):e31–7.

[55] Samango-Sprouse C, Stapleton EJ, Lawson P, Mitchell F, Sadeghin T, Powell S, Gropman AL. Positive effects of early androgen therapy on the behavioral phenotype of boys with 47,XXY. Am J Med Genet C Semin Med Genet 2015;169(2):150–7.

[56] Liao H, Liu S, Wang H. Performance of non-invasive prenatal screening for fetal aneuploidy in twin pregnancies: a meta-analysis. Prenat Diagn 2017;37(9):874–82.

[57] Grati FR, Molina Gomes D, Ferreira JC, Dupont C, Alesi V, Gouas L, et al. Prevalence of recurrent pathogenic microdeletions and microduplications in over 9500 pregnancies. Prenat Diagn 2015;35(8):801–9.

[58] Martin K, Iyengar S, Kalyan A, Lan C, Simon AL, Stosic M, et al. Clinical experience with a single-nucleotide polymorphism-based noninvasive prenatal test for five clinically significant microdeletions. Clin Genet 2017;93(2):293–300. https://dx.doi.org/10.1111/cge.13098.

[59] Helgeson J, Wardrop J, Boomer T, Almasri E, Paxton WB, Saldivar JS, et al. Clinical outcome of subchromosomal events detected by whole-genome noninvasive prenatal testing. Prenat Diagn 2015;35(10):999–1004.

[60] Besseau-Ayasse J, Violle-Poirsier C, Bazin A, Gruchy N, Moncla A, Girard F, et al. A French collaborative survey of 272 fetuses with 22q11.2 deletion: ultrasound findings, fetal autopsies and pregnancy outcomes. Prenat Diagn 2014;34(5):424–30.

[61] Petersen AK, Cheung SW, Smith JL, Bi W, Ward PA, Peacock S, Braxton A, Van Den Veyver IB, Breman AM. Positive predictive value estimates for cell-free noninvasive prenatal screening from data of a large referral genetic diagnostic laboratory. Am J Obstet Gynecol 2017; pii:S0002-9378(17)31187-0.

[62] Committee on Practice Bulletins—Obstetrics, Committee on Genetics, and the Society for Maternal-Fetal Medicine. Practice Bulletin No. 163. Screening for fetal aneuploidy. Obstet Gynecol 2016;127(5):e123–37.

[63] Ehrich M, Tynan J, Mazloom A, Almasri E, McCullough R, Boomer T, Grosu D, Chibuk J. Genome-wide cfDNA screening: clinical laboratory experience with the first 10,000 cases. Genet Med 2017;19(12):1332–7. https://dx.doi.org/10.1038/gim.2017.56.

[64] Hartwig TS, Ambye L, Sørensen S, Jørgensen FS. Discordant non-invasive prenatal testing (NIPT)—a systematic review. Prenat Diagn 2017;37(6):527–39.

[65] Grati FR. Implications of fetoplacental mosaicism on cell-free DNA testing: a review of a common biological phenomenon. Ultrasound Obstet Gynecol 2016;48(4):415–23.

[66] Jani J, Rego de Sousa MJ, Benachi A. Cell-free DNA testing: how to choose which laboratory to use? Ultrasound Obstet Gynecol 2015;46:515–7.

[67] Bevilacqua E, Guizani M, Cos Sanchez T, Jani JC. Concerns with performance of screening for aneuploidy by cell-free DNA analysis of maternal blood in twin pregnancy. Ultrasound Obstet Gynecol 2016;47:124–5.

[68] Grati FR, Malvestiti F, Ferreira JC, Bajaj K, Gaetani E, Agrati C, et al. Fetoplacental mosaicism: potential implications for false-positive and false-negative noninvasive prenatal screening results. Genet Med 2014;16:620–4.

[69] Hall AL, Drendel HM, Verbrugge JL, Reese AM, Schumacher KL, Griffith CB, Weaver DD, Abernathy MP, Litton CG, Vance GH. Positive cell-free fetal DNA testing for trisomy 13 reveals confined placental mosaicism. Genet Med 2013;15:729–32.

[70] Pan Q, Sun B, Huang X, Jing X, Liu H, Jiang F, et al. A prenatal case with discrepant findings between non-invasive prenatal testing and fetal genetic testings. Mol Cytogen 2014;7:48–53.

[71] Chen C, Cram DS, Xie F, Wang P, Xu X, Li H, et al. A pregnancy with discordant fetal and placental chromosome 18 aneuploidies revealed by invasive and noninvasive prenatal diagnosis. Reprod Biomed Online 2014;29:136–9.

[72] Mao J, Wang T, Wang BJ, Liu YH, Li H, Zhang J, Cram D, Chen Y. Confined placental origin of the circulating cell free fetal DNA revealed by a discordant non-invasive prenatal test result in a trisomy 18 pregnancy. Clin Chim Acta 2014;433:190–3.

[73] Zhang H, Gao Y, Jiang F, Fu M, Yuan Y, Guo Y, et al. Noninvasive prenatal testing for trisomy 21, 18 and 13: clinical experience from 146,958 pregnancies. Ultrasound Obstet Gynecol 2015;45:530–8.

[74] Choi H, Lau TK, Jiang FM, Chan MK, Zhang HY, Lo PS, et al. Fetal aneuploidy screening by maternal plasma DNA sequencing: "false positive" due to confined placental mosaicism. Prenat Diagn 2013;33:198–200.

[75] Pan M, Li FT, Li Y, Jiang FM, Li DZ, Lau TK, Liao C. Discordant results between fetal karyotyping and non-invasive prenatal testing by maternal plasma sequencing in a case of uniparental disomy 21 due to trisomic rescue. Prenat Diagn 2013;33:598–601.

[76] Gromminger S, Yagmur E, Erkan S, Nagy S, Schock U, Bonnet J, et al. Fetal aneuploidy detection by cell-free DNA sequencing for multiple pregnancies and quality issues with vanishing twins. Clin Med 2014;3:679–92.

[77] Snyder MW, Gammill HS, Shendure J. Copy-number variation and false positive results of prenatal screening. N Engl J Med 2015;373:2585.

[78] Kingsley C, Wang E, Oliphant A. Copy-number variation and false positive results of prenatal screening. N Engl J Med 2015;373:2583.

[79] van den Boom D, Ehrich M, Kim SK. Copy-number variation and false positive results of prenatal screening. N Engl J Med 2015;373:2584.

[80] Wang Y, Chen Y, Tian F, Zhang J, Song Z, Wu Y, et al. Maternal mosaicism is a significant contributor to discordant sex chromosomal aneuploidies associated with noninvasive prenatal testing. Clin Chem 2014;60:251–9.

[81] Bianchi DW, Chudova D, Sehnert AJ, Bhatt S, Murray K, Prosen TL, et al. Noninvasive prenatal testing and incidental detection of occult maternal malignancies. JAMA 2015;314:162–9.

[82] Bianchi DW, Parsa S, Bhatt S, Halks-Miller M, Kurtzman K, Sehnert AJ, Swanson A. Fetal sex chromosome testing by maternal plasma DNA sequencing: clinical laboratory experience and biology. Obstet Gynecol 2015;125:375–82.

[83] Canick JA, Palomaki GE, Kloza EM, Lambert-Messerlian GM, Haddow JE. The impact of maternal plasma DNA fetal fraction on next generation sequencing tests for common fetal aneuploidies. Prenat Diagn 2013;33(7):667–74.

[84] Wang E, Batey A, Struble C, Musci T, Song K, Oliphant A. Gestational age and maternal weight effects on fetal cell-free DNA in maternal plasma. Prenat Diagn 2013;33:662–6.

[85] Ashoor G, Syngelaki A, Poon LC, Rezende JC, Nicolaides KH. Fetal fraction in maternal plasma cell-free DNA at 11–13 weeks' gestation: relation to maternal and fetal characteristics. Ultrasound Obstet Gynecol 2013;41:26–32.

[86] Haghiac M, Vora NL, Basu S, Johnson KL, Presley L, Bianchi DW, Hauguel-de Mouzon S. Increased death of adipose cells, a path to release cell-free DNA into systemic circulation of obese women. Obesity (Silver Spring) 2012;20:2213–9.

[87] Lebo RV, Novak RW, Wolfe K, Michelson M, Robinson H, Mancuso MS. Discordant circulating fetal DNA and subsequent cytogenetics reveal false negative, placental mosaic, and fetal mosaic cfDNA genotypes. J Transl Med 2015;13:260–76.

[88] Zhang H, Gao Y, Jiang F, Fu M, Yuan Y, Guo Y, et al. Non-invasive prenatal testing for trisomies 21, 18 and 13: clinical experience from 146 958 pregnancies. Ultrasound Obstet Gynecol 2015;45:530–8.

[89] Brar H, Wang E, Struble C, Musci TJ, Norton ME. The fetal fraction of cell-free DNA in maternal plasma is not affected by a priori risk of fetal trisomy. J Matern Fetal Neonatal Med 2013;26(2):143–5.

[90] Poon LC, Musci T, Song K, Syngelaki A, Nicolaides KH. Maternal plasma cell-free fetal and maternal DNA at 11–13 weeks' gestation: relation to fetal and maternal characteristics and pregnancy outcomes. Fetal Diagn Ther 2013;33(4):215–23.

[91] Sarno L, Revello R, Hanson E, Akolekar R, Nicolaides KH. Prospective first-trimester screening for trisomies by cell-free DNA testing of maternal blood in twin pregnancy. Ultrasound Obstet Gynecol 2016;47(6):705–11.

[92] Revello R, Sarno L, Ispas A, Akolekar R, Nicolaides KH. Screening for trisomies by cell-free DNA testing of maternal blood: consequences of a failed result. Ultrasound Obstet Gynecol 2016;47(6):698–704.

[93] Grati FR, Kagan KO. Rate of no result in cell-free DNA testing and its influence on test performance metrics. Ultrasound Obstet Gynecol 2017;50(1):134–7.

[94] Norton ME, Baer RJ, Wapner RJ, Kuppermann M, Jelliffe-Pawlowski LL, Currier RJ. Cell-free DNA vs sequential screening for the detection of fetal chromosomal abnormalities. Am J Obstet Gynecol 2016;214(6). 727.e1–e6.

[95] Ferreira JC, Grati FR, Bajaj K, Malvestiti F, Grimi MB, Trotta A, et al. Frequency of fetal karyotype abnormalities in women undergoing invasive testing in the absence of ultrasound and other high-risk indications. Prenat Diagn 2016;36(12):1146–55.

[96] Norton ME, Jelliffe-Pawlowski LL, Currier RJ. Chromosome abnormalities detected by current prenatal screening and noninvasive prenatal testing. Obstet Gynecol 2014;124(5):979–86.

[97] Norton M, Wapner R, Kuppermann M, Jelliffe-Pawlowski L, Currier R. Cell free DNA analysis vs sequential screening as primary testing considering all fetal chromosomal abnormalities. Am J Obstet Gynecol 2015;212(1):S2.

[98] Wapner RJ, Martin CL, Levy B, Ballif BC, Eng CM, Zachary, et al. Chromosomal microarray versus karyotyping for prenatal diagnosis. N Engl J Med 2012;367(23):2175–84.

[99] Pescia G, Guex N, Iseli C, Brennan L, Osteras M, Xenarios I, Farinelli L, Conrad B. Cell-free DNA testing of an extended range of chromosomal anomalies: clinical experience with 6,388 consecutive cases. Genet Med 2017 Feb;19(2):169–75.

[100] Neofytou MC, Tsangaras K, Kypri E, Loizides C, Ioannides M, Achilleos A, et al. Targeted capture enrichment assay for noninvasive prenatal testing of large and small size sub-chromosomal deletions and duplications. PLoS ONE 2017;12(2):1–13.

[101] Benn P. Expanding non-invasive prenatal testing beyond chromosomes 21, 18, 13, X and Y. Clin Genet 2016;90 (6):477–85.

[102] Wulff CB, Gerds TA, Rode L, Ekelund CK, Petersen OB, Tabor A, Danish Fetal Medicine Study Group. Risk of fetal loss associated with invasive testing following combined first-trimester screening for Down syndrome: a national cohort of 147,987 singleton pregnancies. Ultrasound Obstet Gynecol 2016;47(1):38–44.

[103] Akolekar R, Beta J, Picciarelli G, Ogilvie C, D'Antonio F. Procedure-related risk of miscarriage following amniocentesis and chorionic villus sampling: a systematic review and meta-analysis. Ultrasound Obstet Gynecol 2015;45(1):16–26.

[104] Akolekar R, Bower S, Flack N, Bilardo CM, Nicolaides KH. Prediction of miscarriage and stillbirth at 11–13 weeks and the contribution of chorionic villus sampling. Prenat Diagn 2011;31(1):38–45.

[105] Fiorentino F, Bono S, Pizzuti F, Mariano M, Polverari A, Duca S, et al. The importance of determining the limit of detection of non-invasive prenatal testing methods. Prenat Diagn 2016;36(4):304–11.

[106] Artieri CG, Haverty C, Evans EA, Goldberg JD, Haque IS, Yaron Y, Muzzey D. Noninvasive prenatal screening at low fetal fraction: comparing whole-genome sequencing and single-nucleotide polymorphism methods. Prenat Diagn 2017;37(5):482–90.

[107] Yared E, Dinsmoor MJ, Endres LK, Vanden Berg MJ, Maier Hoell CJ, Lapin B, Plunkett BA. Obesity increases the risk of failure of noninvasive prenatal screening regardless of gestational age. Am J Obstet Gynecol 2016;215:370.e1–6.

[108] Aagaard-Tillery KM, Flint Porter T, Malone FD, Nyberg DA, Collins J, Comstock CH, et al. Influence of maternal BMI on genetic sonography in the FaSTER trial. Prenat Diagn 2009;26(10).

[109] Gregg AR, Skotko BG, Benkendorf JL, Monaghan KG, Bajaj K, Best RG, Klugman S, Watson MS. Noninvasive prenatal screening for fetal aneuploidy, 2016 update: a position statement of the American College of Medical Genetics and Genomics. Genet Med 2016;18(10):1056–65.

[110] Osborne CM, Hardisty E, Devers P, Kaiser-Rogers K, Hayden MA, Goodnight W, Vora NL. Discordant noninvasive prenatal testing results in a patient subsequently diagnosed with metastatic disease. Prenat Diagn 2013;33(6):609–11.

[111] Bianchi DW, Chudova D, Sehnert AJ, Bhatt S, Murray K, Prosen TL, et al. Noninvasive prenatal testing and incidental detection of occult maternal malignancies. JAMA 2015;314(2):162–9.

[112] Snyder HL, Curnow KJ, Bhatt S, Bianchi DW. Follow-up of multiple aneuploidies and single monosomies detected by noninvasive prenatal testing: implications for management and counseling. Prenat Diagn 2016;36(3):203–9.

[113] Song Y, Liu C, Qi H, Zhang Y, Bian X, Liu J. Noninvasive prenatal testing of fetal aneuploidies by massively parallel sequencing in a prospective Chinese population. Prenat Diagn 2013;33(7):700–6.

[114] Wang L, Meng Q, Tang X, Yin T, Zhang J, Yang S, et al. Maternal mosaicism of sex chromosome causes discordant sex chromosomal aneuploidies associated with noninvasive prenatal testing. Taiwan J Obstet Gynecol 2015;54(5):527–31.

[115] Lau TK, Cheung SW, Lo PSS, Pursley AN, Chan MK, Jiang F, et al. Non-invasive prenatal testing for fetal chromosomal abnormalities by low-coverage whole-genome sequencing of maternal plasma DNA: Review of 1982 consecutive cases in a single center. Ultrasound Obstet Gynecol 2014;43(3):254–64.

[116] Zhang B, Lu B-Y, Yu B, Zheng F-X, Zhou Q, Chen Y-P, Zhang X-Q. Noninvasive prenatal screening for fetal common sex chromosome aneuploidies from maternal blood. J Int Med Res 2017;45(2):621–30.

[117] Russell LM, Strike P, Browne CE, Jacobs PA. X chromosome loss and ageing. Cytogenet Genome Res 2007;116(3):181–5. https://dx.doi.org/10.1159/000098184.

[118] Reiss RE, Discenza M, Foster J, Dobson L, Wilkins-Haug L. Sex chromosome aneuploidy detection by noninvasive prenatal testing: helpful or hazardous? Prenat Diagn 2017;37(5):515–20.

[119] Wang S, Huang S, Ma L, Liang L, Zhang J, Zhang J, Cram DS. Maternal X chromosome copy number variations are associated with discordant fetal sex chromosome aneuploidies detected by noninvasive prenatal testing. Clin Chim Acta 2015;444:113–6.

[120] Deans Z, Khawaja F, Hastings R, Rack K, Patton S, Gutowska-Ding W, Allen S, Jenkins L, Chitty L, Sistermans E. Measuring the quality of NIPT for aneuploidies—results from the first pilot EQA. Prenat Diagn 2017;37(Suppl. 1):3–20.

Further Reading

Schmid M, Wang E, Bogard P, Zahn J, Hacker C, Wang S, et al. Prenatal screening for 22q11.2 deletions using a targeted microarray-based cell-free DNA (cfDNA) test. Prenat Diagn 2017;37(Suppl. 1):21–105 Poster P82.

Prenatal Diagnosis and Treatment of Genetic Steroid Disorders

Joe Leigh Simpson*, Svetlana Rechitsky[†], Ahmed Khattab[‡], Maria New[‡]

[*]Herbert Wertheim College of Medicine, Florida International University, Miami, FL, United States [†]Reproductive Genetic Innovations (RGI), Northbrook, IL, United States [‡]Icahn School of Medicine at Mount Sinai Hospital, New York, NY, United States

INTRODUCTION

Prenatal diagnosis serves as a complementary and increasingly essential tool in providing anticipatory guidance and genetic counseling for disease risk of inherited disorders, and is often requested by families with an affected proband. Fortunately, the use of prenatal diagnosis and fetal rescue therapy has become increasingly popular for disorders that may be lethal and associated with fetal malformations. Disorders that are amenable to therapy include, but are not limited to, pulmonary, cardiac, gastrointestinal, endocrine, renal, and immune disorders.

Although amniocentesis and chorionic villous sampling are widely used during invasive prenatal testing techniques, noninvasive prenatal testing has recently emerged and has become increasingly implemented. The genetic basis of multiple steroid disorders has been

Human Reproductive and Prenatal Genetics
https://doi.org/10.1016/B978-0-12-813570-9.00029-2

extensively described, many of which have been associated with significant morbidity and mortality. The advantages of prenatal diagnosis and management appear to be very attractive. The scope of this chapter will focus on prenatal diagnosis and management of congenital adrenal hyperplasia (CAH) as one of the most common genetic steroid disorders amenable to therapy.

PRENATAL DIAGNOSIS OF CONGENITAL ADRENAL HYPERPLASIA

CAH is a relatively frequent genetic steroid disorder [1] with an overall incidence of 1 in 15,000 live births, requiring mandatory newborn screening in all 50 states as well as in many other countries in the world. In more than 90% of cases, CAH is caused by autosomal recessive mutations in the CYP21A2 gene, which encodes the steroidogenic enzyme 21-hydroxylase. Three clinical CAH phenotypes of varying severity exist: salt-wasting, simple virilizing (both grouped as classical CAH), and nonclassical CAH. The latter has an even higher incidence of 1 in 27 live births in the Eastern European Jewish population [2]. A recent study demonstrated genotype-phenotype correlations in 1507 CAH patients [3]. In addition, computational studies modeled ~150 known mutations to predict the mechanism and extent of functional loss in 21-hydroxylase for each mutation [4].

The major clinical sign of classical CAH is genital ambiguity noted at birth in affected females (no genital ambiguity is noted in males), leading to psychological and psychosexual issues in adult life. The genital ambiguity in affected female fetuses is caused by excessive androgen production from fetal adrenal glands and is characterized by variable clitoral enlargement, labioscrotal fold fusion, and formation of a urogenital sinus. Surgical interventions in females to correct genital ambiguity have been successful only in part, and can result in urinary incontinence and sexual dysfunction [5–7]. For decades, the accepted protocol for prenatal treatment of CAH has been the administration of dexamethasone to the mother whose offspring are at risk for genital masculinization. Forest and Morel first showed that fetal hyperandrogenemia and genital ambiguity are preventable by treating high-risk pregnant mothers with low-dose oral dexamethasone (20 μg/kg/day of prepregnancy weight with a maximum daily dose of 1.5 mg/day) [8, 9]. Without binding to corticosteroid-binding globulin or being metabolized by placental 11β-hydroxysteroid dehydrogenase, dexamethasone crosses the placenta into the fetus to suppress androgen production and to prevent in utero virilization. To prevent genital ambiguity, prenatal treatment must be initiated before the ninth week of gestation.

It was not until the 1990s that prenatal medical management of high-risk expectant mothers for CAH with low-dose dexamethasone was initiated [10]. The idea was to suppress fetal androgen production during urogenital organogenesis starting at the ninth gestational week, thereby preventing or attenuating genital virilization. Despite successful outcomes in a cohort of more than 500 patients [10], two sets of concerns have arisen. First, not all high-risk fetuses currently being treated from the ninth gestational week benefit from treatment, namely all male and unaffected (normal or carrier) female fetuses as they are born with normal genitalia without treatment. The unaffected fetuses receive unnecessary dexamethasone therapy until the gender and genetic diagnosis of CAH become available about 2–3 weeks following chorionic villus sampling or amniocentesis (preformed at around 12 and 16 weeks of gestation, respectively). Thus, seven out of eight male and female fetuses are unnecessarily exposed to dexamethasone for 5–10 weeks. Second, there is an ongoing debate on whether dexamethasone, even when given at fivefold lower doses than that used in the third trimester to enhance lung maturity or prevent premature labor, results in reduced birth weight and mental impairment [10–14]. Animal studies using higher doses (100 µg/kg/day) of dexamethasone than those used for CAH (20 µg/kg/day) show low birth weights and impaired renal, pancreatic β-cell, and brain development in offspring [15]. Likewise, high-dose dexamethasone, when used later in pregnancy to delay preterm labor or promote fetal lung maturation, is associated with low birth weight as well as impaired brain development, hypothalamic-pituitary-adrenal axis, and fetal programming [16–18]. Additionally, prenatal low dose dexamethasone has been put on pause in Sweden [13], and, for the sake of safety, medical societies such as the US Endocrine Society have recommended that it be restricted to IRB-approved research settings [19]. These disagreements and the preference that the seven out of eight males and unaffected females not be exposed to prenatal dexamethasone compelled the development of a diagnostic protocol that would allow us to target dexamethasone therapy to the one out of eight mothers that carry affected females. With that said, a question remains: Why even treat? The alternative—genitoplasty—in childhood and often again in adolescence can cause urinary and sexual dysfunction in adults [5, 6]. Long-term studies of women who underwent genitoplasty in infancy or early childhood disclose impaired genital sensitivity, sexual difficulties, decreased intercourse frequency, and stress urinary leakage [7]. Other complications include strictures, fistulas, urinary infections, fibrosis, and scarring, all of which cause dissatisfaction with sexual life [20]. Early noninvasive diagnosis of CAH and subsequent prenatal dexamethasone can successfully avoid the psychological and psychosexual issues associated with genital ambiguity, allowing affected females to have more successful sexual lives.

CELL-FREE DNA ANALYSIS

A new era of noninvasive prenatal diagnosis and treatment began in 1997 when Dennis Lo reported the successful detection of cell-free fetal DNA in maternal plasma. This ushered in new possibilities for the noninvasive prenatal diagnosis of monogenic disorders, including CAH [21]. Before this discovery, noninvasive prenatal diagnosis was possible only by isolating nucleated fetal cells in the maternal circulation. However, these cells, known to remain in maternal circulation, can potentially confound prenatal genetic analysis in subsequent pregnancies [22, 23]. In contrast, cell-free fetal DNA disappears from the maternal circulation within 24 hours after birth. A pregnancy therefore does not capture cell-free fetal DNA from a former pregnancy [24]. Of note, a simple polymerase chain reaction (PCR) cannot be utilized on maternal plasma samples due to interference by vast amounts of maternal DNA. Further, in the case of CAH, long-range PCR cannot be performed in the highly fragmented fetal DNA to differentiate mutations in CYP21A2 from those in the homologous pseudogene CYP21A1P.

Dennis Lo developed the method through which inheritance of a CYP21A2 mutation can be established by targeted massively parallel sequencing (MPS) of cell-free fetal DNA in 3.6 mL plasma drawn from an expectant mother as early as six gestational weeks [25]. This technology has been applied with success to the prenatal diagnosis of β-thalasemia, achondroplasia, and cystic fibrosis, and is likely to be used in other monogenic disorders that may benefit from gene or stem cell therapy [26–30]. Again, this technology is not an assay for detecting specific CYP21A2 mutations. Instead, by performing targeted MPS of genomic DNA from the trio (both parents and the affected proband), it defines informative SNPs around the CYP21A2 locus so that haplotype blocks can be created to determine paternal and maternal allelic inheritance. Full diagnostic concordance between this noninvasive method and invasive diagnostic procedures or postnatal genetic testing was demonstrated in 14 families [25].

The diagnosis of CAH is made by targeted MPS of the genomic region flanking and including the CYP21A2 gene. For this, single nucleotide polymorphisms (SNPs) linked to the CYP21A2 gene in the parents and proband were mapped and then inspected for representation of the respective haplotype maps in the plasma of pregnant mothers. This innovative strategy of quantifying precisely by dosage analysis the amount of DNA inherited from the father and mother in the form of a "footprint" found in maternal plasma allowed elucidating the allelic composition of the fetus around the CYP21A2 locus. Equally innovative is the determination of fetal sex by identifying fetus-derived chromosome Y sequences in maternal plasma by MPS [25] that complements traditional PCR using a SRY probe [31], the results of which may be available within hours and can be used to eliminate unnecessary treatment of mothers carrying male fetuses.

PREIMPLANTATION GENETIC TESTING

Preimplantation genetic testing avoids the treatment of all unaffected embryos. One no longer must wait until the late first trimester to learn the diagnosis of CAH. Preimplantation genetic testing (PGT) is well established as an integral component of the prenatal genetic diagnosis armamentarium [32]. Assisted Reproductive Technologies (ART) readily offer preimplantation genetic testing (PGT) for both chromosomal abnormalities (PGT-A) and for monogenic disorders (PGT-M). PGT permits a couple wishing to avoid clinical termination of an affected pregnancy to do so. In a couple at risk for CAH, one can distinguish affected from unaffected embryos. Diagnosis prior to implantation requires a biopsy of all embryos, with only those known to be genetically normal transferred. One simply does not transfer affected embryos.

The preferred approach for PGT-M in 2018 is blastocyst biopsy. By 5–6 days, the embryo has greatly expanded in cell number (~120 cells), now existing in the form of a blastocyst. The inner cell mass (that would develop into the embryo per se) can be distinguished from the trophectoderm (placenta per se). Biopsy of the trophectoderm results in 5–10 cells, facilitating molecular accuracy compared to the single cell removed from biopsy of an eight-cell cleavage stage embryo.

PGT-M can be carried out for any single gene disorder whose chromosomal location is known. Even if the causative nucleotide mutation is not known, linkage analysis can provide a definitive result. Reproductive Genetics Innovation (RGI) has performed PGT for hundreds of different monogenic disorders, including CAH [33]. PGT-M for single gene disorders should be accompanied by PGT-aneuploidy (PGT-A) testing. This dyad increases the pregnancy rate by some 20% over PGT. Exhaustive details on PGT-M are provided elsewhere by one of the authors [33].

An embryo biopsy necessary for PGT logically must sometimes be associated with damage that may preclude survival. However, any damage seems to be either lethal or addressed by surviving potential cells. The anomaly rate in live births has not been shown to be increased [34].

CONCLUSION

This chapter describes the preimplantation diagnosis and prenatal diagnosis of the genetic adrenal disorder CAH. The development of preimplantation diagnosis of CAH provides information very early to pregnant women at risk. This technique provides a great advance in the counseling that can be given to those who wish to proceed with pregnancy. In contrast to the invasive prenatal diagnosis, the noninvasive method has several advantages. The noninvasive method requires only 3.6 mL of plasma from the pregnant mother, which could provide the diagnosis of CAH before the ninth week of gestation, as is necessary for effective and safe treatment. Further, only the affected female fetuses will be treated and all males and unaffected females will avoid unnecessary treatment. This strategy represents a generic approach for the noninvasive prenatal testing for an array of autosomal recessive disorders.

References

[1] Rare Disease Act of 2002. Public Law 107–280 2002.
[2] New MI. Extensive clinical experience: nonclassical 21-hydroxylase deficiency. J Clin Endocrinol Metab 2006;91:4205–14. Pub Med PMID: 16912124.
[3] New MI, Abraham M, Gonzalez B, Dumic M, Razzaghy-Azar M, Chitayat D, et al. Genotype-phenotype correlation in 1,507 families with congenital adrenal hyperplasia owing to 21-hydroxylase deficiency. Proc Natl Acad Sci U S A 2013;110:2611–6. PubMed PMID: 23359698; PubMed Central PMCID: PMC3574953.
[4] Haider S, Islam B, D'Atri V, Sgobba M, Poojari C, Sun L, et al. Structure-phenotype correlations of human CYP21A2 mutations in congenital adrenal hyperplasia. Proc Natl Acad Sci U S A 2013;110:2605–10. PubMed PMID: 23359706; PubMed Central PMCID: 3574933.
[5] Crouch NS, Minto CL, Laio LM, Woodhouse CR, Creighton SM. Genital sensation after feminizing genitoplasty for congenital adrenal hyperplasia: a pilot study. BJU Int 2004;93:135–8. PubMed PMID: 14678385.
[6] Nordenstrom A, Frisen L, Falhammar H, Filipsson H, Holmdahl G, Janson PO, et al. Sexual function and surgical outcome in women with congenital adrenal hyperplasia due to CYP21A2 deficiency: clinical perspective and the patients' perception. J Clin Endocrinol Metab 2010;95:3633–40. PubMed PMID: 20466782.
[7] Crouch NS, Liao LM, Woodhouse CR, Conway GS, Creighton SM. Sexual function and genital sensitivity following feminizing genitoplasty for congenital adrenal hyperplasia. J Urol 2008;179:634–8. PubMed PMID: 18082214.
[8] Forest MG, Betuel H, David M. Prenatal treatment in congenital adrenal hyperplasia due to 21-hydroxylase deficiency: up-date 88 of the French multicentric study. Endocr Res 1989;15:277–301. PubMed PMID: 2667968.
[9] Forest MG, David M, Morel Y. Prenatal diagnosis and treatment of 21-hydroxylase deficiency. J Steroid Biochem Mol Biol 1993;45:75–82. PubMed PMID: 8481354.
[10] New MI, Carlson A, Obeid J, Marshall I, Cabrera MS, Goseco A, et al. Prenatal diagnosis for congenital adrenal hyperplasia in 532 pregnancies. J Clin Endocrinol Metab 2001;86:5651–7. PubMed PMID: 11739415.
[11] Mercado AB, Wilson RC, Cheng KC, Wei JQ, New MI. Prenatal treatment and diagnosis of congenital adrenal hyperplasia owing to steroid 21-hydroxylase deficiency. J Clin Endocrinol Metab 1995;80:2014–20. PubMed PMID: 7608248.
[12] Hirvikoski T, Nordenstrom A, Lindholm T, Lindblad F, Ritzen EM, Lajic S. Long-term follow-up of prenatally treated children at risk for congenital adrenal hyperplasia: does dexamethasone cause behavioural problems? Eur J Endocrinol 2008;159:309–16. PubMed PMID: 18579553.

[13] Hirvikoski T, Nordenstrom A, Lindholm T, Lindblad F, Ritzen EM, Wedell A, et al. Cognitive functions in children at risk for congenital adrenal hyperplasia treated prenatally with dexamethasone. J Clin Endocrinol Metab 2007;92:542–8. PubMed PMID: 17148562.

[14] Hirvikoski T, Nordenstrom A, Wedell A, Ritzen M, Lajic S. Prenatal dexamethasone treatment of children at risk for congenital adrenal hyperplasia: the Swedish experience and standpoint. J Clin Endocrinol Metab 2012;97:1881–3. PubMed PMID: 22466333.

[15] Celsi G, Kistner A, Aizman R, Eklof AC, Ceccatelli S, de Santiago A, et al. Prenatal dexamethasone causes oligonephronia, sodium retention, and higher blood pressure in the offspring. Pediatr Res 1998;44:317–22. PubMed PMID: 9727707.

[16] Murphy KE, Hannah ME, Willan AR, Hewson SA, Ohlsson A, Kelly EN, et al. Multiple courses of antenatal corticosteroids for preterm birth (MACS): a randomised controlled trial. Lancet 2008;372:2143–51. PubMed PMID: 19101390.

[17] Reynolds RM, Seckl JR. Antenatal glucocorticoid treatment: are we doing harm to term babies? J Clin Endocrinol Metab 2012;97:3457–9. PubMed PMID: 23043197.

[18] Seckl JR. Prenatal glucocorticoids and long-term programming. Eur J Endocrinol 2004;151(Suppl. 3):U49–62. PubMed PMID: 15554887.

[19] Speiser PW, Azziz R, Baskin LS, Ghizzoni L, Hensle TW, Merke DP, et al. Congenital adrenal hyperplasia due to steroid 21-hydroxylase deficiency: an Endocrine Society clinical practice guideline. J Clin Endocrinol Metab 2010;95:4133–60. PubMed PMID: 20823466; PubMed Central PMCID: PMC2936060.

[20] Zainuddin AA, Grover SR, Shamsuddin K, Mahdy ZA. Research on quality of life in female patients with congenital adrenal hyperplasia and issues in developing nations. J Pediatr Adolesc Gynecol 2013;26:296–304. PubMed PMID: 23507003.

[21] Lo YM, Corbetta N, Chamberlain PF, Rai V, Sargent IL, Redman CW, et al. Presence of fetal DNA in maternal plasma and serum. Lancet 1997;350:485–7. PubMed PMID: 9274585.

[22] Lo YM. Fetal DNA in maternal plasma: biology and diagnostic applications. Clin Chem 2000;46:1903–6. PubMed PMID: 11106320.

[23] Bianchi DW, Zickwolf GK, Weil GJ, Sylvester S, DeMaria MA. Male fetal progenitor cells persist in maternal blood for as long as 27 years postpartum. Proc Natl Acad Sci U S A 1996;93:705–8. PubMed PMID: 8570620; PubMed Central PMCID: 40117.

[24] New MI, Abraham M, Yuen T, Lekarev O. An update on prenatal diagnosis and treatment of congenital adrenal hyperplasia. Semin Reprod Med 2012;30:396–9. PubMed PMID: 23044876.

[25] New MI, Tong YK, Yuen T, Jiang P, Pina C, Chan KC, et al. Noninvasive prenatal diagnosis of congenital adrenal hyperplasia using cell-free fetal DNA in maternal plasma. J Clin Endocrinol Metab 2014;99:E1022–30. PubMed PMID: 24606108; PubMed Central PMCID: PMC4037720.

[26] Chan J, O'Donoghue K, de la Fuente J, Roberts IA, Kumar S, Morgan JE, et al. Human fetal mesenchymal stem cells as vehicles for gene delivery. Stem Cells 2005;23:93–102. PubMed PMID: 15625126.

[27] Evans MI, Harrison MR, Flake AW, Johnson MP. Fetal therapy. Best Pract Res Clin Obstet Gynaecol 2002;16:671–83. PubMed PMID: 12475547.

[28] Mattar CN, Choolani M, Biswas A, Waddington SN, Chan JK. Fetal gene therapy: recent advances and current challenges. Expert Opin Biol Ther 2011;11:1257–71. PubMed PMID: 21623703.

[29] Rahim AA, Buckley SM, Chan JK, Peebles DM, Waddington SN. Perinatal gene delivery to the CNS. Ther Deliv 2011;2:483–91. PubMed PMID: 22826856.

[30] van Mieghem T, Baud D, Devlieger R, Lewi L, Ryan G, De Catte L, et al. Minimally invasive fetal therapy. Best Pract Res Clin Obstet Gynaecol 2012;26:711–25. PubMed PMID: 22682617.

[31] Tardy-Guidollet V, Menassa R, Costa JM, David M, Bouvattier-Morel C, Baumann C, et al. New management strategy of pregnancies at risk of congenital adrenal hyperplasia using fetal sex determination in maternal serum: French cohort of 258 cases (2002–2011). J Clin Endocrinol Metab 2014;99:1180–8. PubMed PMID: 24471566.

[32] Simpson JL. Preimplantation genetic diagnosis at 20 years. Prenatal Diag 2010;30:682–95.

[33] Kuliev A, Rechitsky S, Verlinsky O. Atlas of Preimplantation genetic diagnosis. 3rd ed. Boca Raton: CRC Press; 2014.

[34] Simpson JL. Children born after preimplantation genetic diagnosis show no increase in congenital anomalies. Hum Reprod 2010;25:6–8.

30

Application of Gene-Editing Technologies in Embryos and Their Potential for Gene Therapy

Cristina Eguizabal, Nuria Montserrat[†], Juan Carlos Izpisua Belmonte[‡]*

*Cell Therapy and Stem Cell Group, Basque Center for Blood Transfusion and Human Tissues, Galdakao, Spain [†]Pluripotent Stem Cells and Activation of Endogenous Tissue Programs for Organ Regeneration (PR Lab), Institute for Bioengineering of Catalonia (IBEC), Barcelona, Spain [‡]Gene Expression Laboratory, The Salk Institute for Biological Studies, La Jolla, CA, United States

Abbreviations

2N	diploid
3PN	tripronuclear
A1ATD	alpha1-antitrypsin deficiency
ART	assisted reproductive technology
CRISPR	clustered regularly interspaced short palindromic repeats
DMD	Duchenne muscular dystrophy
DSBs	DNA double strand breaks
EBV	Epstein-Barr Virus
ESCs	embryonic stem cells
G6PDH	glucose-6-phospahate dehydrogenase
GONAD	genome editing via oviductal nucleic acids delivery
HBB	hemoglobin subunit beta
HBV	hepatitis B virus
HDR	homology-directed repair
HIV	human immunodeficiency virus
HR	homologous recombination
HTI	hereditary tyrosinemia type I
iPSCs	induced pluripotent stem cells
IVF	in vitro fertilized
MII-SCC	metaphase II spindle-chromosome complex
MRT	mtDNA replacement therapy
mtDNA	mitochondrial DNA
NHEJ	Nonhomologous end joining
PAM	proto-spacer adjacent motif
PB1T	First Polar Body Transfer
PB1	first Polar Body
PB2T	second polar body transfer
PB2	second polar body
PGD	preimplantation genetic diagnosis
PNH	paroxysmal nocturnal hemoglobinuria
RDEB	recessive dystrophic epidermolysis
RVD	repeat variable di-residue
SCID	severe combined immmunodeficiency
SCNT	somatic cell nuclear transfer
sgRNA	single guide RNA
SSCs	spermatogonial stem cells
TALENs	transcription activator-like effector nucleases
XCGD	X-linked chronic granulomatous disease
ZFNs	zinc finger nucleases

INTRODUCTION

In recent years, several new genome-editing technologies have been developed. Zinc finger nucleases (ZFNs), transcription activator-like effector nucleases (TALENs), and the CRISPR/Cas9 RNA-guided endonuclease system are the most extensively described. Each of these technologies uses restriction enzymes to introduce a DNA double-stranded break at a targeted location, with homologous binding proteins or RNA guiding the nuclease to the target. Specific targeting is considered a significant improvement over current gene therapy methods, which lack such specificity. Several proof-of-concept studies have been performed to treat multiple disorders, including in vivo experiments in mammals and in vitro experiments in human embryos; these advancements are particularly relevant for future treatments in the field of reproductive medicine. At this stage, it is critical that more investigations into delivery strategies are conducted so that the therapeutic potential for gene editing will be achieved. In this review, we will briefly describe the mechanisms underlying each of these gene-editing technologies and evidence of therapeutic potential in the reproductive field. The topics and information reviewed here provide an outline of the groundbreaking research that is currently underway, but also highlights the potential for progress using these gene-editing technologies.

THE ERA OF GENE-EDITING TECHNOLOGY

DNA damage can occur in several situations: after exposure to ionizing radiation or chemotherapy, during the DNA replication process, or after experimental manipulation through the action of endonucleases. DNA double-strand breaks (DSBs) are often injurious to the cell, causing genome instability and mutations. Luckily, the DNA repair machinery helps to repair this damage, thus preventing DNA mutations that can cause disease. The ability of DNA damage to cause DNA responses forms the basis of the gene-editing concept. There are two main repair pathways for DSBs: (1) nonhomologous end joining (NHEJ) and (2) homologous recombination (HR) (Fig. 30.1) [1].

The NHEJ pathways are active during the cell cycle and are the most common mechanism for DSB repair. However, these pathways are error-prone and often result in genetic insertions or deletions (indels). HR is initiated only in the S or G2 phases of the cell cycle when a sister chromatid is available to provide a homologous donor template for the repair process. This process is referred to as homology-directed repair (HDR) when an exogenous DNA repair template is provided. In the 1980s, HDR was used to insert and repair genes in mammalian cells. These techniques were later applied to mouse embryos to generate transgenic mice, and the era of genetic engineering was born. These important early studies were performed by Drs. Oliver Smithies, Mario Capecchi, and Martin Evens, who together pioneered HR-mediated gene editing in mouse embryos. Later, they would share the 2007 Nobel Prize in Physiology or Medicine "for their discoveries of principles for introducing specific gene modifications in mice by the use of embryonic stem cells."

In the 1990s, gene targeting in mammalian cells using HR was further improved by Dr. Maria Jasin and colleagues when they used the yeast homing endonuclease I-*Sce*I [2]. This enzyme generates site-specific breaks in chromatin engineered to contain the 18-bp rare restriction site for this enzyme. Importantly, this site is not found in the mouse or human genomes, so the I-SceI cleavage event is highly specific. Cutting with this enzyme enhanced HR up to 500-fold at the target locus, foreshadowing the use of designer site-specific endonucleases to enhance gene editing. These next-generation endonucleases include the zinc finger nucleases (ZFNs) and the transcription activator-like effector nucleases (TALENs), which combined the DNA-binding specificity of zinc fingers or TALE transcription factors from plants with the DNA cutting activity of the *Fok*I endonuclease [3, 4] (Fig. 30.1).

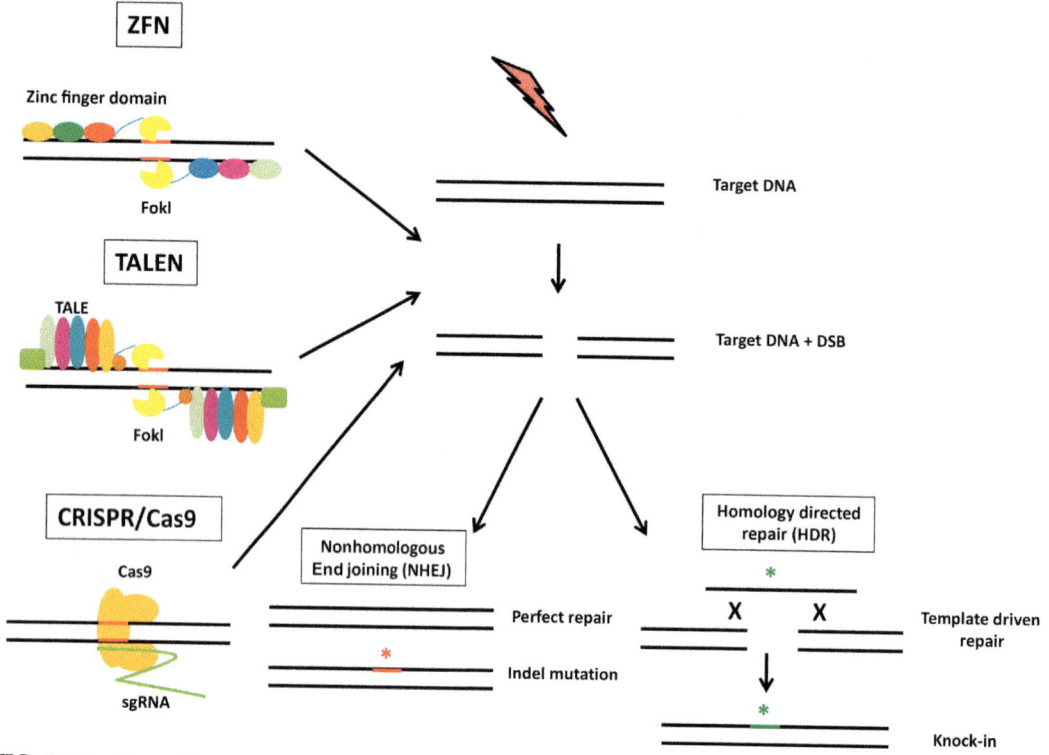

FIG. 30.1   Gene-editing tools to repair DNA DSBs (ZNFs, TALENs, and CRISPR/Cas9).

One of the major breakthroughs in the field of gene editing was achieved in the late 1980s, when a group of researchers discovered a strange topology at the 3′ end of the alkaline phosphatase gene in *Escherichia coli*. This consisted of 32 nucleotide sequences, flanked by short invariable palindromic repeats—the first known description of an array of clustered regularly interspaced short palindromic repeats (CRISPR). In 2005, several laboratories demonstrated that the unique spacer regions found in CRISPR arrays map to phage genomes, hinting that CRISPR may serve as an adaptive immune response to phage infection though an RNA-guided process [5–7]. The molecular mechanism of this immune response involves the transcription of CRISPR arrays into RNA, which is then cleaved and loaded into CRISPR-associated (Cas) proteins (Cas9). This RNA:protein complex then mediates RNA-guided dsDNA endonuclease activity. Furthermore, it was discovered that Cas9 can be reprogrammed to target novel sequences with an in vitro transcribed single guide RNA (sgRNA) (Fig. 30.1).

For many years, researchers had been searching for a tool for easily inducing mutations in a targeted fashion. While some progress had been made using engineered meganucleases, ZFNs, and TALENs, these systems had numerous limitations. They were either expensive, labor-intensive, or both as the targeting mechanisms were based on protein-nucleic acid interactions, and thereby required a custom-designed protein for each genetic locus of interest. The promise of RNA-guided nuclease activity afforded by CRISPR-based approaches led numerous groups to recognize immediately that this technology could represent a much more efficient way of inducing targeted DSBs in eukaryotes. DSBs produced either by previously available technologies or CRISPR-based systems are repaired by low-fidelity DNA repair pathways, leading to the production of indels—a class of mutations characterized by the random insertion or deletion of nucleotides at the site of the DSB.

THE VARIETY OF GENE EDITING TOOLS

As mentioned above, there are four basic nuclease technologies: meganucleases, ZFNs, TALENs, and CRISPR/Cas9 nucleases. The differences are described in detail below [8].

Engineered meganucleases are derived from a huge family of natural homing endonucleases [9]. A small number of these endonucleases have been designed to recognize natural target sites in the genome via a diversity of strategies, including structure-based design and yeast surface display [10, 11]. Historically, natural meganucleases have been the gold standard for specificity, but the challenge of engineering meganucleases for new target sites has limited their translational development. In addition, the specificity of engineered meganucleases has not been fully evaluated.

Zinc-finger nucleases (ZFNs) are artificial proteins in which a zinc-finger DNA-binding domain is fused to the nonspecific nuclease domain from the enzyme *Fok*I [12, 13]. To cut DNA efficiently, the nuclease domain must dimerize, a pair of ZFNs must be engineered for each target site, and these must be oriented correctly to allow dimerization. Zinc-finger-DNA binding domains can be engineered for novel target sites using a variety of strategies, including phage display, modular assembly, bacteria-based two-hybrid and one-hybrid systems, and combination approaches [14]. While ZFN design strategies are continuously being improved, engineering ZFNs that have high activity and specificity to endogenous target sites remain a challenge. The best-quality ZFNs created using a combination of phage display and modular display have entered into clinical trials, and T cells engineered using this system have been shown to be safe [15].

TAL effector nucleases (TALENs) are artificial proteins too. They share a similar structure to ZFNs in that an engineered DNA-binding domain is fused to the nuclease domain of the enzyme *Fok*I. In TALENs, the DNA-binding domain is engineered by assembling a series of TAL repeats [16], with each repeat mediating interaction with a single base through a two amino acid repeat variable di-residue (RVD), which can be described by a simple code [17]. Hence, creating a highly active TALEN is much simpler than creating a highly active ZFN, and simply involves using the code to assemble the correct TAL repeats needed to recognize a novel target sequence. In addition to TAL repeats that use natural RVDs, TAL repeats that use engineered RVDs are now being used to create TALENs. These engineered RVDs may have increased specificity over natural RVDs, although this requires further study. As with ZFNs, a pair of TALENs must be engineered to recognize each target site of interest. Even TALENs that use TAL repeats containing natural RVDs may have better specificity than ZFNs.

CRISPR/Cas9 nucleases (CRISPR is an abbreviation of "Clustered Regularly Interspaced Short Palindromic Repeats") are derived from a bacteria-based adaptive immune system [18]. In contrast to the other three tools, the CRISPR/Cas9 nuclease system does not derive specificity through protein-DNA interactions, but instead through RNA-DNA Watson-Crick base pairing. In the CRISPR/Cas9 system, a single-guide RNA (sgRNA) is designed such that the 20-bp recognition region of the sgRNA is identical to the desired target site. The target site must be nearby to a proto-spacer adjacent motif (PAM) sequence, which the Cas9 protein uses to identify target sites [19]. The multifunctional Cas9 protein, in complex with the sgRNA, is able to unwind double-stranded DNA, to interrogate whether the guide strand is sufficiently identical to the target site, and to then create a rounded DSB if there is sufficient identity. Consequently, CRISPR/Cas9 nucleases can be engineered very easily, and many of the designed nucleases are active at the desired target site.

THE RELEVANCE OF GENOME-EDITING TECHNOLOGIES FOR POTENTIAL THERAPEUTICS

Gene-editing technologies are powerful tools for basic research, but the ultimate goal is to translate these technologies into therapeutic applications. The potential use of gene editing in the clinic emerged from the idea that the best way to treat a monogenic disease would be to develop a method for correcting the disease-associated mutation, but that it would also be necessary to make more dramatic changes to the genome to cure diseases with multifactorial origins, which are more common. The use of genome editing to cure monogenic disease is conceptually simple, but the true power of genome editing tools is that they provide a mechanism for making more sophisticated genomic changes, which can be used to cure more common diseases or to modify their course. Indeed, there are currently several companies developing gene editing-based approaches to treat disease, including CRISPR Therapeutics, OvaScience, Cellectis, Editas Company, Intellia Therapeutics, Caribou Biosciences, Precision Biosciences, and Sangamo Therapeutics.

FORTHCOMING THERAPEUTIC APPLICATIONS OF NHEJ- AND HR-MEDIATED GENOME EDITING

The use of ZFNs for therapeutic purposes has primarily been focused on common inherited monogenic disorders, such as cystic fibrosis [20] and severe combined immunodeficiency (SCID) [21], as well as acquired infections, such as hepatitis B virus (HBV) [22] and human immunodeficiency virus (HIV) [15]. Studies have been conducted in a variety of animal models and human cell lines to test systems for NHEJ-mediated gene disruption and HDR-based gene correction. ZFNs are among the most mature gene-editing technologies, with the successful completion of early phase human clinical trials to treat HIV. A number of ZFN-based therapeutic applications have been investigated, including those to treat sickle cell anemia, hemophilia B, Parkinson's disease, Duchenne muscular dystrophy (DMD), Leigh's syndrome, X-linked chronic granulomatous disease (XCGD), ataxia, retinitis pigmentosa, paroxysmal nocturnal hemoglobinuria (PNH), and Down syndrome [23].

Comparable to ZFNs, the therapeutic potential of TALENs has been demonstrated in cellular and animal models for a variety of genetic and acquired disorders. Examples of both gene disruptions and gene corrections by NHEJ and HDR, respectively, have been reported. While TALEN technologies are not as mature as those utilizing ZFNs, and no clinical trials have been initiated, a number of studies are currently underway, including efforts to treat sickle cell anemia, alpha1-antitrypsin deficiency (A1ATD), DMD, recessive dystrophic epidermolysis bullosa (RDEB), polycythemia vera, SCID, mitochondrial disease, HIV-1 resistance, HBV replication, HPV infections, malaria, and various cancers [23].

Even though use of the *CRISPR/Cas9 system* as a gene-editing technique was first reported in January 2013, significant progress has been made in the last 4 years to demonstrate therapeutic potential for several diseases, including sickle cell anemia, cystic fibrosis, DMD, polycythemia vera, A1ATD, Barth's syndrome, cataracts, hereditary tyrosinemia type I (HTI), cardiovascular disease, HIV-1 resistance/infection/immunization [24], HBV, Epstein-Barr virus (EBV), HPV/cervical cancer, and osteosarcoma [23]. This is likely because the simplicity and relative affordability of the system makes it accessible to a wide array of researchers. Looking at the range of diseases targeted by CRISPR/Cas9-, ZFN-, and TALEN-based approaches, there is a clear effort to evaluate the utility of these three gene-editing technologies for treating inherited monogenic disorders and for antiviral therapy. For the CRISPR/Cas9 system, the first clinical trials may still be years away, but recent in vivo studies, including some performed in nonhuman primate embryos [25] and very recently in human embryos [26], demonstrate rapid progression toward that goal.

POTENTIAL APPLICATIONS OF GENE-EDITING TECHNOLOGY IN REPRODUCTIVE MEDICINE

There are numerous potential applications for the CRISPR/Cas9 technology in reproductive medicine, including those that target pluripotent stem cells, gametes, or embryos [27]. The first hypothetical choice for preventing a genetic disorder in a child is to use gene editing to correct the mutation in pluripotent stem cells (either induced pluripotent stem cells (iPSCs) or somatic cell nuclear transfer-human embryonic stem cells (SCNT-hESC)) that have been obtained from a patient with a specific disease. Pluripotent stem cells are ideal resources for studying gene-editing technologies because they are easy to grow in culture. Following gene correction, these pluripotent stem cells can be differentiated in vitro toward germ cells (oocytes or sperm) for potential use in assisted reproductive technology (ART). The ability to create stem cell-derived gametes (oocytes and sperm) in vitro was recently demonstrated in mice in two milestone papers by Hayashi et al. [28, 29]. Research is currently underway to achieve this result in humans with the eventual goal of overcoming certain types of infertility [30].

Efforts to generate human male gametes in vitro have encountered some difficulties, as it is presently very difficult to generate postmeiotic germ cells from spermatogonial stem cells (SSCs). Hence, it is likely that SSCs are the best cell type for treating male infertility (particularly if the patient cannot generate mature sperm). Recent studies in mouse SSCs have pushed the field closer toward using gene editing to treat male infertility [31, 32]. In contrast, female oocytes are amenable to genetic manipulation. However, the low number of oocytes typically collected from a patient, together with poor efficiency of the gene-editing process, represent challenges to applying this technology in the clinic. To date, it has been demonstrated in mice that gene-editing technologies (TALEN and CRISPR/Cas9) can be used to eliminate disease-associated mtDNA molecules from oocytes or embryos [33, 34]. Juan Carlos Izpisua Belmonte's research group has used mitochondria-targeting restriction enzymes and TALENs in mammalian oocytes for the first time. To determine whether these enzymes could be used to edit human mtDNA, they fused mouse oocytes with fibroblasts from patients with one of two mitochondrial disorders: (1) Leber's hereditary optic neuropathy and (2) neurogenic muscle weakness, ataxia, and retinitis pigmentosa. Following TALEN mRNA injection, mutant mtDNA was still detectable, although at lower levels. As the mutated mtDNA typically must be present in more than 60%–75% of a cell's mitochondria to cause disease symptoms, observed reductions were more than enough to ameliorate these phenotypes [33] (Fig. 30.2).

The correction of gene mutations in the germline could potentially enable patients to produce mutation-free gametes, and subsequently healthy embryos and progeny. In addition to correcting the mutation, this approach has the additional advantage of reducing the number of disease carriers, thereby reducing mutation frequency in the general population. However, there are only a few scenarios in which the application of CRISPR/Cas9 technology could benefit people at risk of having children with a monogenic disease. For example, in families with an autosomal recessive disease, two carrier parents have a 25% risk of their child being affected. In this case, preimplantation genetic diagnosis (PGD), which is a well-established technique that offers high-risk couples the possibility of having healthy children, or prenatal diagnosis could be used without adding risks associated with genetic manipulation. The same rationale can be applied to couples in which one partner is affected by an autosomal dominant disease, resulting in a 50% risk of disease transmission. However, for couples in which both are affected by the same monogenic disease, CRISPR/Cas9 could be used to ensure a healthy child via their own genetic material. Other potential candidates for germline editing are patients who are homozygous for an autosomal dominant disease, such as Huntington's disease, polycystic kidney disease, achondroplasia, Marfan syndrome, and Myotonic dystrophy type 1, among others [27].

CRISPR/Cas9 could also be used to correct chromosomal aberrations. A rare form of chromosomal rearrangement in humans, called Robertsonian translocations, can occur between acrocentric chromosome pairs, such as 13, 14, 15, 21, and 22. The Robertsonian translocation

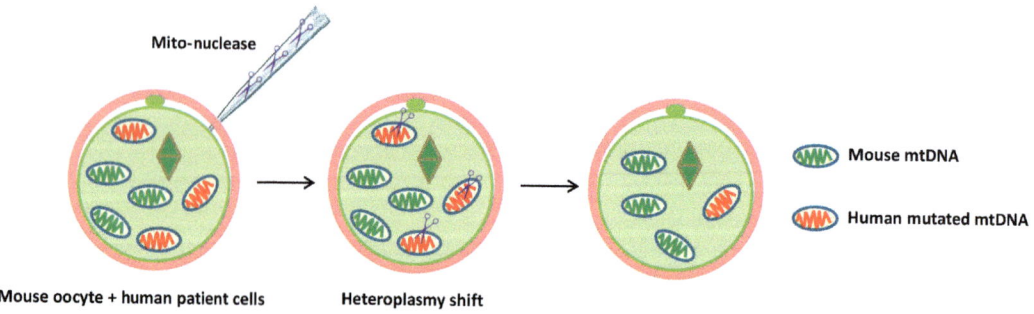

FIG. 30.2 Selective elimination of mitochondrial mutations in oocytes by mitochondria-targeted nucleases.

in chromosome 21 is, for example, a chromosomal aberration that exists in healthy carriers, but the progeny of this carrier may inherit an unbalanced trisomy 21, causing Down syndrome. However, the Robertsonian translocation between the two chromosomes 21 inevitably leads to trisomy 21. In the future, it may be possible to use CRISPR/Cas9 to separate the two chromosomes, and restore both the centromere and missing p arms. Other candidates for correction are genes related to infertility. To date, the most common and best-described genetic causes of infertility are chromosomal in nature: 45,X (Turner syndrome), 47,XXY (Klinefelter syndrome), and Y chromosome deletions (Oligozoospermia and Azoospermia). Finally, mutations in mitochondrial DNA (mtDNA) within the oocyte are good candidates for correction/elimination via CRISPR/Cas9, a technology we have already mentioned and will discuss later.

STATE OF THE ART IN GENE EDITING OF EMBRYOS

From 2015 until now, very few studies have been published documenting the specificity, efficiency, and fidelity of CRISPR/Cas9 editing in human embryos. Gene editing in embryos involves direct microinjection into the cytoplasm or pronuclei of zygotes or MII oocytes, and subsequent screening for embryos with a correctly edited genome that lack off-target modifications. In general, the efficiency of genomic editing in embryos is low, with the main problem being the generation of mosaic embryos as a result of inefficient nuclease cutting and/or inaccurate DNA repair before the embryo undergoes cleavage. Still, several studies in different animal models (from rats to pigs) have demonstrated the feasibility of gene editing in animals [35, 36]. The efficiency of genomic modifications into mammalian zygotes ranges from 0.5% to 40.9% per injected zygote when TALENs or Cas9 technologies are used [37].

In 2015, CRISPR/Cas9 editing was performed on human zygotes in China, and the efficiency, specificity, and fidelity of this system was analyzed. Liang et al. [38] injected 86 donated tripronuclear (3PN) zygotes with CRISPR/Cas9. Of these embryos, 82.6% survived the injection, 51.9% of the genome-edited zygotes were successfully spliced, and 5.6% exhibited the correct insertion of genetic material through homologous recombination. The gene-edited zygotes were mosaic, with results similar to findings in other model systems [39]. Furthermore, a large number of "off-target" mutations were identified. These likely were introduced by the CRISPR/Cas9 complex acting in other parts of the genome, or represented intrinsic abnormalities of the 3PN zygotes, or both.

One year later, another Chinese group injected 213 3PN zygotes, with the goal of introducing a mutation in the immune-cell gene, CCR5 [40]. This mutation confers HIV resistance as it prevents the virus from entering T cells. Genetic analysis showed that four of the 26 embryos were successfully modified (similar to the results reported by Liang et al. [38]). However, some embryos contained unmodified CCR5Δ32 mutation whereas others acquired a different mutation. The main advance of this paper is the precise introduction of a specific genetic modification in human zygotes using CRISPR/Cas9 and this future potential. Very recently, Tang et al. [41] published the first study demonstrating in diploid (2N) human embryos the successful correction of mutations within the hemoglobin subunit beta (HBB) gene and the glucose-6-phosphate dehydrogenase (G6PDH) gene [41]. Common defects in these genes cause different types of anemia. As mentioned previously, this had already been demonstrated in triploid human zygotes. The limitations associated with the Tang et al. study were the small number of embryos (10 embryos) compared with previous reports, and the generation of mosaic embryos.

A major milestone in the field of gene editing was recently reported, as Mitalipov's group improved the precision and safety of the CRISPR/Cas9 technique [42]. As mentioned above, CRISPR/Cas9-mediated editing of embryos and pluripotent stem cells frequently generates mosaics by repairing the mutation in only a fraction of the cells, and by introducing off-target mutations or extra mutations in the targeted gene (nonhomologous repair). However, Mitalipov's team improved the system by injecting CRISPR/Cas9 components together with the patient's sperm directly into normal MII oocytes. Until now, the Cas9 complex has typically been injected into the zygote. Remarkably, they found highly efficient homologous repair, no evidence of off-target genetic changes, and only a single mosaic embryo (in an experiment involving 58 human embryos) [26] (Fig. 30.3).

Finally, a novel technique, called genome editing via oviductal nucleic acids delivery (GONAD), has been effectively delivered to mouse preimplantation embryos within the intact mouse oviduct using a simple electroporation method. This resulted in the genetic modification of these embryos, but the edited embryos were mosaic (this has been seen in other model systems as well) and a substantial number of "off-target" mutations were identified [43].

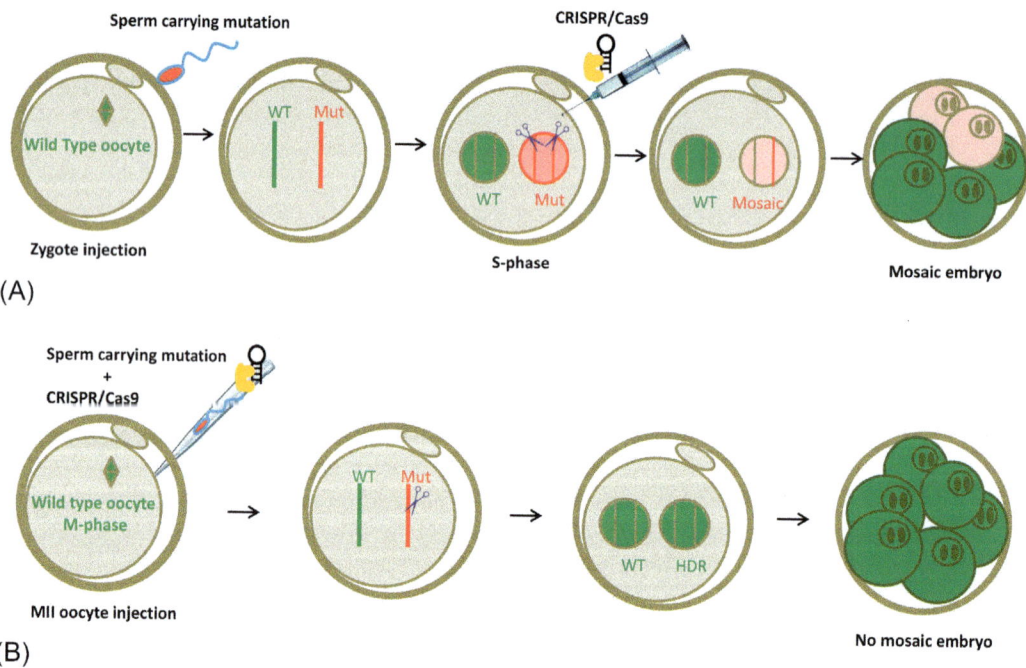

FIG. 30.3   Gene correction in human embryos. (A) Schematic of gene correction with CRISPR/Cas9 in zygotes, (B) CRISPR/Cas9 was coinjected with sperm into MII oocytes during ICSI. This allows genome editing to occur when a sperm contains a single mutant copy, thereby eliminating mosaicism.

Obviously, more research in animal models is needed to improve the safety and efficiency of the CRISPR/Cas9 system, particularly as the goal is to develop future treatments that involve the genetic editing of human embryos.

MITOCHONDRIAL DISEASES: GENE EDITING FOR THE TREATMENT OF GERMLINE GENETIC DISEASES

Mitochondrial diseases are among the most shocking inheritable diseases with the most severe forms causing death soon after birth. Mitochondrial diseases are caused by a mutation in a high proportion of the mitochondrial DNA (mtDNA) molecules present in the cells of the patient, called heteroplasmy. Mutations in mtDNA are transmitted exclusively via the oocyte, which can carry between 10,000 and 100,000 mtDNA copies. While at this time there are no cures for mitochondrial diseases, scientific advances in this field are providing new ways to prevent disease transmission. One strategy for preventing mitochondrial-disease transmission is preimplantation genetic diagnosis (PGD), a technique in which cells taken from in vitro fertilized (IVF) embryos prior to uterine implantation are screened to select healthy embryos. PGD significantly increases the probability of producing healthy offspring for heteroplasmic carriers of mtDNA mutations. However, PGD is not a valuable test for all patients. Though it provides information about the mtDNA, it offers no intervention when the mtDNA is found to be mutant, as is often the case with homoplasmic or heteroplasmic carriers close to the disease threshold. For this reason, the field of mitochondrial disease prevention is limited until technological interventions are ready for clinical use. In recent years, a novel method of mtDNA replacement therapy (MRT) has emerged as a promising approach for preventing mitochondrial disease transmission [44, 45]. There are currently four main techniques for mtDNA replacement: (1) germinal vesicular transfer, (2) metaphase II spindle-chromosome complex (MII-SCC) transfer, (3) pronuclear transfer, and (4) polar body transfer (Fig. 30.4).

(1) *Germinal vesicular transfer* involves transferring a healthy, immature oocyte nucleus from a patient carrying mtDNA mutations to a healthy donor egg. This potentially reduces the amount of inviable mtDNA, but more studies are required (Fig. 30.4A). Germinal vesicular transfer is performed during the arrested prophase of meiosis I, when two copies of each of the 23 pairs of chromosomes are present. Because of the large size of the nucleus at this stage, the harvest process is easier. The reconstructed oocyte is then matured in vitro prior to fertilization, or fertilized with sperm. However, during germinal vesicular harvest, small amounts of residual ooplasm may be transferred as well. Therefore, it is possible that some mutated mtDNA copies are present in the newly reconstructed oocyte.

(2) For *MII-SCC transfer*, the mature oocyte containing mutant mtDNA is progressed to metaphase II, when the chromosomal material is arranged along the metaphase plate. The nucleus is then harvested and implanted into a healthy, enucleated donor oocyte (Fig. 30.4B). This technique allows for the newly constructed oocyte to be fertilized by viable sperm after the transfer occurs, or by activation of the reconstructed oocyte. However, because of the imprecise nature of the spindle complex, this technique carries

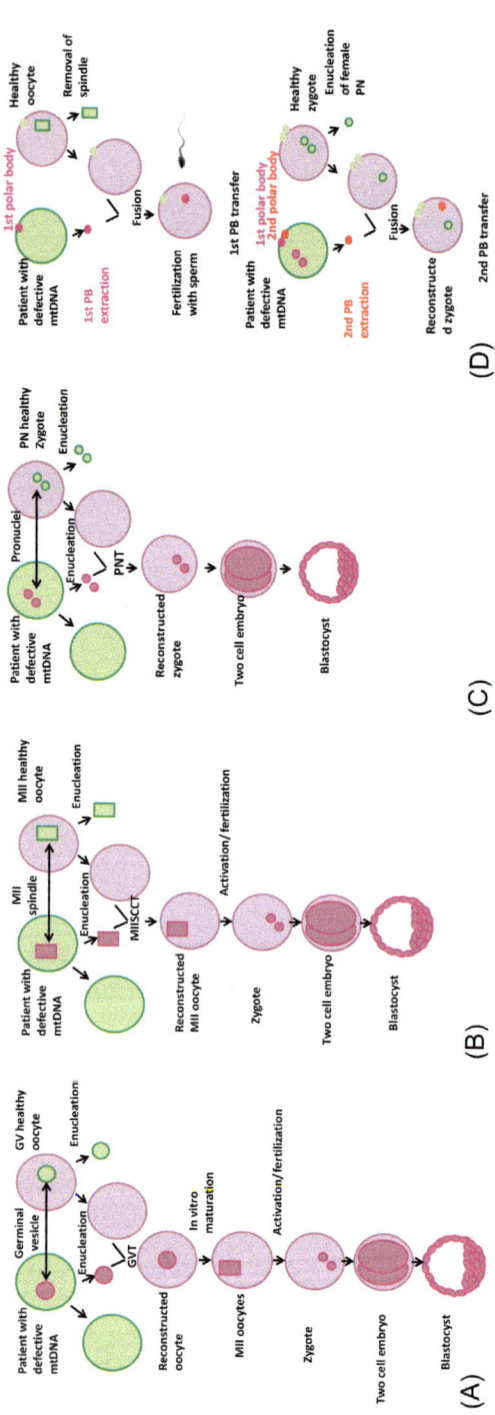

FIG. 30.4 mtDNA replacement techniques: (A) germinal vesicle transfer, (B) MII SSC transfer, (C) pronuclear transfer, and (D) first and second polar body transfer.

the risk of extracting more cytoplasm and increasing the amount of mutated mtDNA that is transferred. In 2009, this technique was performed in primates [42], resulting in four healthy monkeys in which mutant mtDNA was not detected. These were the first mammals born using MII-SCC transfer. In April 2016, the first child was born in Mexico from an asymptomatic mother that carried a mitochondrial mutation that caused Leigh syndrome. The child has 1% of its mother's mtDNA and is healthy at 3 months. Whether health problems arise later in life has yet to be determined [46].

(3) *Pronuclear transfer* is the process by which pronuclei from both the sperm and oocyte (before fusion) are removed from parental zygotes and placed in a donor zygote that was previously fertilized and enucleated (Fig. 30.4C). The advantage of this technique is that it allows for the extraction of two well-defined pronuclei after the sperm has been introduced into the oocyte, potentially reducing the amount of cytoplasm that is transferred with the pronuclei and therefore decreasing the carryover of mutated mtDNA. Pronuclear transfer, rather than MII-SCC transfer, may be the technique worth pursuing in the future to achieve better outcomes for patients with high levels of mutant mtDNA. Applying this technique in embryos with an abnormal number of pronuclei has successfully eliminated more that 98% of maternal mitochondria, which is enough to prevent disease manifestation. Pronuclear transfer also has been used in normal fertilized human embryos, resulting in 2%–5% of mtDNA carryover. In 2016, an infertile woman was pregnant with triplets via PNT, although none of the fetuses reached full term [47]. While it is difficult at this time to meaningfully compare MII-SCC and PN transfer techniques, both have advantages and show great promise.

(4) *Polar body transfer.* The first polar body (PB1) is formed during egg maturation. In this process, the DNA duplicates so that the egg contains four chromosome sets. Of these, two remain within the egg while the other two are packaged to form PB1. PBI is extruded and is not present in the resulting embryo. The second polar body (PB2) is formed during fertilization. One set of the remaining chromosomes is packaged, forming PB2, whereas the other set forms the nuclear DNA of the embryo together with the sperm DNA. Polar bodies contain very few mitochondria, which is an advantage for avoiding mitochondrial carryover. In polar body transfer, PB1 is transferred to an unfertilized enucleated donor egg (PB1T), or PB2 is transferred to a half-enucleated zygote (PB2T). Thus, the first strategy is strictly preventive whereas the second is not (Fig. 30.4D). In 2014, researchers compared PB1T with MII-SCCT and PB2T with pronuclear transfer and found that blastocyst-stage embryos were generated with the same efficiency between the PB1T and MII-SCCT groups (87% vs. 86%) whereas PB2T was less efficient than PNT (55% vs. 81%). In F1 and F2 offspring, they found undetectable levels of mtDNA carryover in MII-SCCT, PB1T, and PB2T. Levels of mtDNA carryover were higher in pronuclear transfer due to the performance of the technique or mtDNA amplification during zygote activation [48].

MITOCHONDRIAL GENE EDITING IN EMBRYOS USING TALENs

As described above, it is possible to use TALENS to specifically target mtDNA. These mito-TALENs can be used to cut mutated mtDNA, and have already been used to selectively

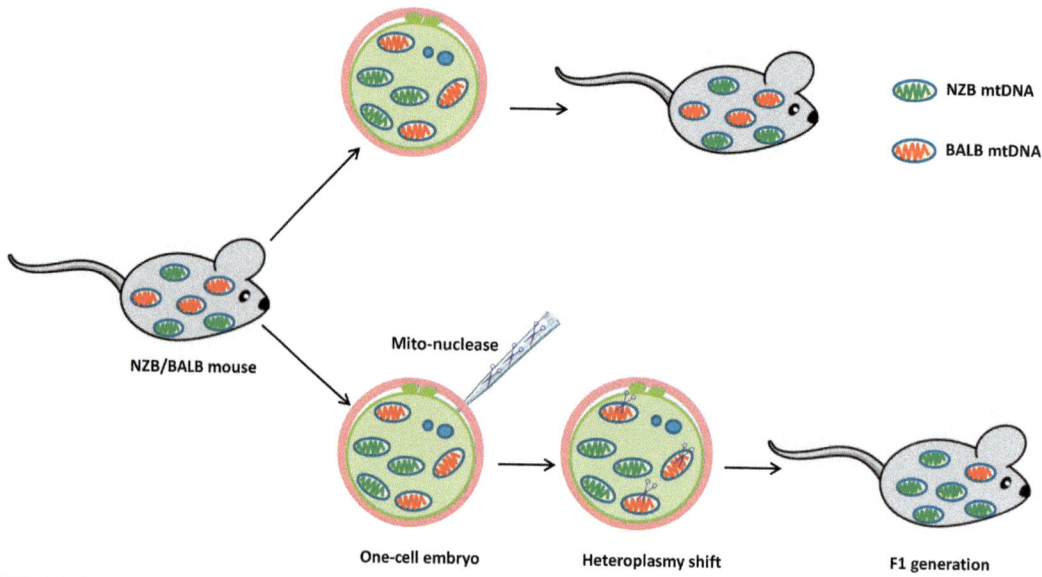

FIG. 30.5   Germline heteroplasmy shift prevents transmission of mtDNA haplotypes to offspring using mito-nucleases.

eliminate defective mtDNA in both unfertilized mouse eggs and murine zygotes. Furthermore, these genetically modified mice gave birth to two successive generations of healthy mice. When mRNA encoding mito-TALENs were injected into eggs from a heteroplasmic mouse that carried two mtDNA haplotypes (NZB and BALB), mtDNA heteroplasmy shift was achieved, and the edited embryos developed into normal mice. Importantly, the team did not observe any off-target effects [33] (Fig. 30.5).

MITOCHONDRIAL GENE EDITING IN EMBRYOS USING CRISPR/Cas9

In this review, we have enumerated the massive potential offered by the CRISPR/Cas9 technology, both in research and future clinical applications, particularly in the field of germline genetic modification (mitochondrial diseases). Successful applications of these technologies have made it possible to envision the curative potential of CRISPR/Cas9 for carriers of both heteroplasmic and homoplasmic mtDNA mutations. While the gene-editing machinery may not reach all mutant mtDNA, it may modify enough to bring the individual below the disease threshold, conferring therapeutic benefit. Although there is considerable hesitation within the scientific community in using this technique in the germline, studies using nonviable human embryos have already been performed [26, 38, 41, 48, 49]. The majority of these studies show that the technique works with low efficiency to generate on-target gene modification, but that off-target mutations are also generated and the resulting embryos are mosaic (with the exception of one recent study [26, 33]). CRISPR/Cas9 has already been successfully used to produce mitochondrial sequence-specific cleavage as proof of concept that this technique can target specific mitochondrial genes [33]. In the same study, researchers engineered a new version of the Cas9 enzyme, mitoCas9, whose localization is restricted to the mitochondrial matrix. This is an important step as it reduces the risk of off-target mutations in the embryo and prospective children.

CRISPR/Cas9 editing of embryonic mtDNA may represent a more socially acceptable alternative to "three-parent IVF." Instead of combining the genetics of three individuals, this technique may enable a couple to conceive without requiring donor genetic material. Another advantage of trying to treat mitochondrial disease before implantation is the small number of cells. This may make it easier to ensure a reduced mutation load if the embryo is in the eight- or 16-cell stage, rather than waiting to provide a therapeutic intervention after birth when there are exponentially more cells.

CONCLUSION

The development of new technologies based on sequence-specific nucleases has resulted in: (1) greatly improved efficiency in targeting specific genomic loci, (2) the avoidance of mosaicism in human embryos, (3) improved ease in constructing donor vectors, (4) the possibility of genome editing with oligonucleotides, (5) direct genome editing via the electroporation of embryos, and (6) the possibility of multiplexed genome editing. These technological breakthroughs dramatically increase the feasibility of editing the human germline, particularly for embryos that carry mitochondrial disorders. From our point of view, basic studies must be conducted in animal models to better understand targeted disease mechanisms before these technologies can successfully be used to edit DNA or mtDNA. In the case of mitochondrial diseases, for example, what are the mechanisms of nucleus/mitochondria communication, and how is the transmission of mtDNA to the offspring regulated? In particular, what controls variable transmission of heteroplasmy, and how does this affect symptoms of mitochondrial diseases? Because animal studies have limitations in predicting outcomes in humans, subsequent research in humans will be necessary. Similarly, before CRISPR/Cas9 technology can be translated to the clinic, outstanding problems must be resolved, primarily mosaicism and off-target effects, although these issues are rapidly being resolved. Nevertheless, even after these technical hurdles have been cleared, editing the human germline to prevent the birth of a child with a mitochondrial defect should only be considered when already established methods that do not entail genomic manipulation are not available.

Acknowledgments

Work in the laboratory of J.C.I.B. was supported by G. Harold and Leila Y. Mathers Charitable Foundation, the Leona M. and Harry B. Helmsley Charitable Trust, and the Moxie Foundation.

References

[1] Ceccaldi R, Rondinelli B, D'Andrea AD. Repair pathway choices and consequences at the double-strand break. Trends Cell Biol 2016;26(1):52–64.

[2] Rouet P, Smih F, Jasin M. Introduction of double-strand breaks into the genome of mouse cells by expression of a rare-cutting endonuclease. Mol Cell Biol 1994;14(12):8096–106.

[3] Miller JC, Tan S, Qiao G, et al. A TALE nuclease architecture for efficient genome editing. Nat Biotechnol 2011;29(2):143–8.

[4] Kim YG, Cha J, Chandrasegaran S. Hybrid restriction enzymes: zinc finger fusions to Fok I cleavage domain. Proc Natl Acad Sci U S A 1996;93(3):1156–60.

[5] Bolotin A, Quinquis B, Sorokin A, et al. Clustered regularly interspaced short palindrome repeats (CRISPRs) have spacers of extrachromosomal origin. Microbiology 2005;151:2551–61.

[6] Mojica FJ, Díez-Villaseñor C, García-Martínez J, et al. Intervening sequences of regularly spaced prokaryotic repeats derive from foreign genetic elements. J Mol Evol 2005;60(2):174–82.

[7] Pourcel C, Salvignol G, Vergnaud G. CRISPR elements in Yersinia pestis acquire new repeats by preferential uptake of bacteriophage DNA, and provide additional tools for evolutionary studies. Microbiology 2005;151(Pt 3):653–63.

[8] Porteus MH. Towards a new era in medicine: therapeutic genome editing. Genome Biol 2015;16:286.

[9] Chevalier BS, Stoddard BL. Homing endonucleases: structural and functional insight into the catalysts of intron/intein mobility. Nucleic Acids Res 2001;29:3757–74.

[10] Silva G, Poirot L, Galetto R, Smith J, Montoya G, Duchateau P, et al. Meganucleases and other tools for targeted genome engineering: perspectives and challenges for gene therapy. Curr Gene Ther 2011;11:11–27.

[11] Jarjour J, West-Foyle H, Certo MT, Hubert CG, Doyle L, Getz MM, et al. High-resolution profiling of homing endonuclease binding and catalytic specificity using yeast surface display. Nucleic Acids Res 2009;37:6871–80.

[12] Durai S, Mani M, Kandavelou K, Wu J, Porteus MH, Chandrasegaran S. Zinc finger nucleases: custom-designed molecular scissors for genome engineering of plant and mammalian cells. Nucleic Acids Res 2005;33:5978–90.

[13] Urnov FD, Rebar EJ, Holmes MC, Zhang HS, Gregory PD. Genome editing with engineered zinc finger nucleases. Nat Rev Genet 2010;11:636–46.

[14] Porteus MH, Carroll D. Gene targeting using zinc finger nucleases. Nat Biotechnol 2005;23:967–73.

[15] Tebas P, Stein D, Tang WW, Frank I, Wang SQ, Lee G, Spratt SK, Surosky RT, Giedlin MA, Nichol G, Holmes MC, Gregory PD, Ando DG, Kalos M, Collman RG, Binder-Scholl G, Plesa G, Hwang WT, Levine BL, June CH. Gene editing of CCR5 in autologous CD4 T cells of persons infected with HIV. N Engl J Med 2014;370:901–10.

[16] Bogdanove AJ, Voytas DF. TAL effectors: customizable proteins for DNA targeting. Science 2011;333:1843–6.

[17] Boch J, Scholze H, Schornack S, Landgraf A, Hahn S, Kay S, et al. Breaking the code of DNA binding specificity of TAL-type III effectors. Science 2009;326:1509–12.

[18] Doudna JA, Charpentier E. Genome editing. The new frontier of genome engineering with CRISPR-Cas9. Science 2014;346:1258096.

[19] Hsu PD, Lander ES, Zhang F. Development and applications of CRISPR-Cas9 for genome engineering. Cell 2014;157:1262–78.

[20] Lee CM, Flynn R, Hollywood JA, Scallan MF, Harrison PT. Correction of the DeltaF508 mutation in the cystic fibrosis transmembrane conductance regulator gene by zinc-finger nuclease homology-directed repair. BioResearch (open access) 2012;1:99–108.

[21] Lombardo A, Genovese P, Beausejour CM, Colleoni S, Lee YL, Kim KA, Ando D, Urnov FD, Galli C, Gregory PD, Holmes MC, Naldini L. Gene editing in human stem cells using zinc finger nucleases and integrase-defective lentiviral vector delivery. Nat Biotechnol 2007;25:1298–306.

[22] Cradick TJ, Keck K, Bradshaw S, Jamieson AC, McCaffrey AP. Zinc-finger nucleases as a novel therapeutic strategy for targeting hepatitis B virus DNAs. Mol Ther: J Am Soc Gene Ther 2010;18:947–54.

[23] LaFountaine JS, Fathe K, Smyth HD. Delivery and therapeutic applications of gene editing technologies ZFNs, TALENs, and CRISPR/Cas9. Int J Pharm 2015;494(1):180–94.

[24] Cradick TJ, Fine EJ, Antico CJ, Bao GCRISPR. Cas9 systems targeting betaglobin and CCR5 genes have substantial off-target activity. Nucleic Acids Res 2013;4:9584–92.

[25] Midic U, Hung PH, Vincent KA, Goheen B, Schupp PG, Chen DD, Bauer DE, VandeVoort CA, Latham KE. Quantitative assessment of timing, efficiency, specificity and genetic mosaicism of CRISPR/Cas9-mediated gene editing of hemoglobin beta gene in rhesus monkey embryos. Hum Mol Genet 2017;26(14):2678–89.

[26] Ma H, Marti-Gutierrez N, Park S-W, et al. Correction of a pathogenic gene mutation in human embryos. Nature 2017, https://dx.doi.org/10.1038/nature23305.

[27] Vassena R, Heindryckx B, Peco R, Pennings G, Raya A, Sermon K, et al. Genome engineering through CRISPR/Cas9 technology in the human germline and pluripotent stem cells. Hum Reprod Update 2016;22(4):411–9.

[28] Hayashi K, Ogushi S, Kurimoto K, Shimamoto S, Ohta H, Saitou M. Offspring from oocytes derived from in vitro primordial germ cell-like cells in mice. Science 2012;338(6109):971–5.

[29] Hayashi K, Ohta H, Kurimoto K, Aramaki S, Saitou M. Reconstitution of the mouse germ cell specification pathway in culture by pluripotent stem cells. Cell 2011;146(4):519–32.

[30] Vassena R, Eguizabal C, Heindryckx B, Sermon K, Simon C, van Pelt AM, et al. ESHRE special interest group stem cells. Stem cells in reproductive medicine: ready for the patient? Hum Reprod 2015;30(9):2014–21.

[31] Nickkholgh B, Mizrak SC, van Daalen SK, Korver CM, Sadri-Ardekani H, Repping S, van Pelt AM. Genetic and epigenetic stability of human spermatogonial stem cells during long-term culture. Fertil Steril 2014;102:1700–7.

[32] Wu Y, Zhou H, Fan X, Zhang Y, Zhang M, Wang Y, Xie Z, Bai M, Yin Q, Liang D, Tang W, Liao J, Zhou C, Liu W, Zhu P, Guo H, Pan H, Wu C, Shi H, Wu L, Tang F, Li J. Correction of a genetic disease by CRISPR-Cas9-mediated gene editing in mouse spermatogonial stem cells. Cell Res 2015 Jan;25(1):67–79.

[33] Reddy P, Ocampo A, Suzuki K, Luo J, Bacman SR, Williams SL, et al. Selective elimination of mitochondrial mutations in the germline by genome editing. Cell 2015;161(3):459–69.

[34] Wang S, Yi F, Qu J. Eliminate mitochondrial diseases by gene editing in germ-line cells and embryos. Protein Cell 2015;6(7):472–5.

[35] Shao Y, Guan Y, Wang L, Qiu Z, Liu M, Chen Y, WuL LY, Ma X, Liu M, et al. CRISPR/Cas-mediated genome editing in the rat via direct injection of one-cell embryos. Nat Protoc 2014;9:2493–512.

[36] Wells KD, Prather RS. Genome-editing technologies to improve research, reproduction, and production in pigs. Mol Reprod Dev 2017;84(9):1012–7. https://dx.doi.org/10.1002/mrd.22812.

[37] Araki M, Ishii T. International regulatory landscape and integration of corrective genome editing into in vitro fertilization. Reprod Biol Endocrinol 2014;12:108.

[38] Liang P, Xu Y, Zhang X, Ding C, Huang R, Zhang Z, Lv J, Xie X, Chen Y, Li Y, et al. CRISPR/Cas9-mediated gene editing in human tripronuclear zygotes. Protein Cell 2015;6:363–72.

[39] Yen ST, Zhang M, Deng JM, Usman SJ, Smith CN, Parker-Thornburg J, Swinton PG, Martin JF, Behringer RR. Somatic mosaicism and allele complexity induced by CRISPR/Cas9 RNA injections in mouse zygotes. Dev Biol 2014;393:3–9.

[40] Kang X, He W, Huang Y, Yu Q, Chen Y, Gao X, Sun X, Fan Y. Introducing precise genetic modifications into human 3PN embryos by CRISPR/Cas-mediated genome editing. J Assist Reprod Genet 2016;33(5):581–8. https://dx.doi.org/10.1007/s10815-016-0710-8.

[41] Tang L, Zeng Y, Du H, Gong M, Peng J, Zhang B, Lei M, Zhao F, Wang W, Li X, Liu J. CRISPR/Cas9-mediated gene editing in human zygotes using Cas9 protein. Mol Genet Genomics 2017;292(3):525–33.

[42] Tachibana M, Sparman M, Sritanaudomchai H, Ma H, Clepper L, Woodward J, Li Y, Ramsey C, Kolotushkina O, Mitalipov S. Mitochondrial gene replacement in primate offspring and embryonic stem cells. Nature 2009;461:367–72.

[43] Takahashi G, Gurumurthy CB, Wada K, Miura H, Sato M, Ohtsuka M. GONAD: genome-editing via oviductal nucleic acids delivery system: a novel microinjection independent genome engineering method in mice. Sci Rep 2015;5:11406.

[44] Fogleman S, Santana C, Bishop C, Miller A, CRISPR/ CDG. Cas9 and mitochondrial gene replacement therapy: promising techniques and ethical considerations. Am J Stem Cells 2016;5(2):39–52.

[45] Gómez-Tatay L, Hernández-Andreu JM, Aznar J. Mitochondrial modification techniques and ethical issues. J Clin Med 2017;24(6):3.

[46] Zhang J, Liu H, Luo S, Chavez-Badiola A, Liu Z, Yang M, Munne S, Konstantinidis M, Wells D, Huang T. First live birth using human oocytes reconstituted by spindle nuclear transfer for mitochondrial DNA mutation causing Leigh syndrome. Fertil Steril 2016;106:e375–6.

[47] Zhang J, Zhuang G, Zeng Y, Grifo J, Acosta C, Shu Y, Liu H. Pregnancy derived from human zygote pronuclear transfer in a patient who had arrested embryos after IVF. Reprod Biomed Online 2016;33:529–33.

[48] Wang T, Sha H, Ji D, Zhang HL, Chen D, Cao Y, Zhu J. Polar body genome transfer for preventing the transmission of inherited mitocondrial diseases. Cell 2014;157:1591–604.

[49] Kang X, He W, Huang Y, Yu Q, Chen Y, Gao X, Sun X, Fan Y. Introducing precise genetic modifications into human 3PN embryos by CRISPR/Cas-mediated genome editing. J Assist Reprod Genet 2016;33:581–8.

Glossary

DNA repair	A collection of processes by which a cell identifies and corrects damage to the DNA molecules that encode its genome.
Embryo	An early stage of development of a multicellular diploid eukaryotic organism.
Gene editing	A type of genetic engineering in which DNA is inserted, deleted, or replaced in the genome of a living organism using engineered nucleases.
Germline	In biology and genetics, the germline in a multicellular organism is the population of its bodily cells that are so differentiated or segregated that in the usual processes of reproduction, they may pass on their genetic material to the progeny.
Mitochondrial diseases	Those caused by mutations (acquired or inherited) in mitochondrial DNA (mtDNA), or in nuclear genes that code for mitochondrial components.
Monogenic disease	Disorders caused by the inheritance of a single defective gene, also called single-gene disorders.
Mosaicism	In genetics, a mosaic, or mosaicism, is the presence of two or more populations of cells with different genotypes in one individual who has developed from a single fertilized egg.
Nuclease	An enzyme capable of cleaving the phosphodiester bonds between monomers of nucleic acids.
Oocyte	A female germ cell involved in reproduction.
Pluripotent stem cells	Cell that can differentiate into other cell types; types of pluripotent stem cells include embryonic stem cells (ESCs) and induced pluripotent stem cells (iPSCs).
Sperm	A male germ cell involved in reproduction.

Index

Note: Page numbers followed by *f* indicate figures, *t* indicate tables, and *np* indicate footnotes.